AA003566

2nd Geoscience and Environmental Management Symposium (ICST 2021)

Held online due to COVID-19

E3S Web of Conferences Volume 325 (2021)

Yogyakarta, Indonesia
7 – 8 September 2021

Editors:

R. Che Oma
J.T. Sri Sumantyo
B. White
A. Cardenas Tristan
E. Haryono
D.R. Hizbaron
R.F. Putri

ISBN: 978-1-7138-3957-6

Printed from e-media with permission by:

Curran Associates, Inc.
57 Morehouse Lane
Red Hook, NY 12571

Some format issues inherent in the e-media version may also appear in this print version.

For additional information, please contact EDP Sciences – Web of Conferences
at the address below.

EDP Sciences – Web of Conferences
17, Avenue du Hoggar
Parc d'Activité de Courtabœuf
BP 112
F-91944 Les Ulis Cedex A
France

Phone: +33 (0) 1 69 18 75 75
Fax: +33 (0) 1 69 28 84 91

contact-edps@webofconferences.org

Additional copies of this publication are available from:

Curran Associates, Inc.
57 Morehouse Lane
Red Hook, NY 12571 USA
Phone: 845-758-0400
Fax: 845-758-2633
Email: curran@proceedings.com
Web: www.proceedings.com

TABLE OF CONTENTS

DISASTER RISK REDUCTION

LAND, WATER, AND NATURAL RESOURCES

ENVIRONMENTAL MANAGEMENT

ENVIRONMENTAL DEVELOPMENT TECHNOLOGY

ENVIRONMENTAL AND GEOLOGY ENGINEERING

HUMAN AND NATURAL RESOURCES

URBAN-RURAL RESOURCES AND LAND USE MANAGEMENT

GEO-RESOURCES AND GEO-HAZARD

https://doi.org/10.1051/e3sconf/202132500001

Preface

The 2nd Geoscience and Environmental Science Symposium was held as part of the 7th International Conference on Science and Technology (ICST 2021). The ICST 2021 is an annual meeting organized by Badan Penerbit dan Publikasi of Universitas Gadjah Mada (BPP UGM), Yogyakarta, Indonesia. This year conference was held on 7–8 September 2021. The scientific program consisted of plenary sessions and parallel sessions and included the following topics: Disaster Risk Reduction; Environmental and Geology Engineering; Environmental Development Technology; Environmental Management; Geo-resources and Geo-hazard; Human and Natural Resources; Land, Water, and Natural Resources; as well as Urban-rural Resources and Land Use Management.

We are grateful for the support of the Faculty of Geography and Faculty of Engineering, Universitas Gadjah Mada which scientifically support the symposium from the reviewers until the editing the papers published in the E3S Web of Conferences Proceeding. The ICST 2021 had about 786 attendees including the presenters and participants. Among those, the 2nd Geoscience and Environmental Science Symposium received 127 papers submitted to the symposium. All papers were subjected for at least 2–3 reviews by experts on the related topics. For this, only 106 papers were accepted, and 95 papers were selected for publications.

We believed that during the two days conference, we had enabled to facilitate plenty of networking opportunities and provide the participants with the opportunity to meet and interact with the leading scientists and researchers, friends as well as colleagues. We hope that this proceeding will be beneficial for the development of life science in general and the geoscience and environmental science in particular.

Welcoming Remarks from the Chairman

The honorable Rector of Universitas Gadjah Mada, Prof. Dr. Panut Mulyono.

Dear distinguished guests, our invited speakers, and participants of this conference.

On behalf of the Program Committee, I kindly welcome you to the 7th International Conference on Science and Technology (ICST 2021). It is an honor to be here with you today. The ICST 2021 is a part of the UGM Annual Scientific Conferences (UASC) series held by Badan Penerbit dan Publikasi (BPP), Universitas Gadjah Mada.

Our purpose is to provide an ideal academic platform for researchers to disseminate their research and the development of science and technology. This year we hold six symposia: the Computer and Information Technology Symposium; the Electronics, Power, Communication, Control, and Instrumentation Symposium; the Mechanical and Industrial Engineering Symposium; the Remote Sensing and Geomatics Symposium; the Geoscience and Environmental Management Symposium; and the Life Sciences, Materials, and Applied Chemistry Symposium.

This year, we have invited seven renowned speakers around the world: Prof. Tomohiro Yokozeki from the University of Tokyo, Japan; Prof. Gil Santos from De La Salle University, the Philippines; Prof. Shou-Yan Chou from National Taiwan University of Science and Technology, Taiwan; Dr. Ruchi Gupta from the University of Birmingham, United Kingdom; Prof. Nobuyuki Ichikuni from Chiba University, Japan; Dr. Vooi Voon Yap from Universiti Tenku Abdul Rahman, Malaysia; and Dr. Mohamad Shazwan Shah Jamil from Universiti Teknologi Malaysia. We thank the speakers for their expertise and knowledge that will bring great discussion during the conference.

This year, we received 290 submissions from 13 countries from Indonesia, Malaysia, the Philippines, India, Taiwan, Japan, Iran, France, United Kingdom, Russia, Saudi Arabia, the USA, and Mexico. All manuscripts went through a thorough review process. We will submit the accepted and presented papers to journals or proceedings such as AIP Conference Proceedings, E3S Web of Conferences, and Key Engineering Materials. All of them are indexed by Scopus.

This conference will not have become a reality without the support and assistance of many parties. On this occasion, I would like to sincerely thank the Rector of Universitas Gadjah Mada, BPP, DSSDI officers, all invited speakers, and the organizing committee for arranging all needed facilities for the success of this conference. We also would like to thank reviewers for their valuable comments and suggestions.

Although we cannot greet each other in our usual way, we still have another way to interact, discuss, share insights, and enhance our minds with other researchers and colleagues from various perspectives. We hope you find this event fruitful for your future career, for the sake of knowledge and humankind's benefit, and we hope you will enjoy this conference. Thank you.

Yogyakarta, September 7, 2021

Chairman of the Organizing Committee,
Dr.Eng. Titis Wijayanto, S.T., M.Des.

Welcoming Remarks
from the Rector of Universitas Gadjah Mada

Distinguished Keynote Speakers,
Distinguished Invited Speakers and Participants,
Conference's Committee,
Ladies and Gentlemen.

Good morning, good afternoon, good evening.
May peace be upon all of us.

On behalf of Universitas Gadjah Mada, it is my pleasure to welcome you all to the 7th International Conference on Science and Technology (ICST 2021) hosted by Universitas Gadjah Mada (UGM). First of all, I would like to extend my gratitude to distinguished keynote speakers. Please allow me to express my sincere appreciation for Prof. Tomohiro Yokozeki from University of Tokyo, Japan; Prof. Gil Nonato C. Santos from De La Salle University, Philippines; Prof. Shou-Yan Chou from National Taiwan University of Science and Technology, Taiwan; and Dr. Ruchi Gupta from University of Birmingham, United Kingdom.

History of UGM's education has earned a reputation as a pioneering university, pushing the boundaries between academics and professionals across the world, to serve and discover scientific progress as a valuable source of knowledge for the benefit of humankind. With its Annual Scientific Conference Series, Universitas Gadjah Mada builds upon this mission, holding annual gatherings for the brightest minds from Indonesia and abroad to share the latest findings in their respective fields. Through collaboration with our international partners, this series has seen enormous success, shaping the development of scientific networks, increasing Indonesian authors' prominence in publications with a global readership, and underscoring UGM's place as a standard-bearer of scientific development.

Today, the world is in a pandemic situation, which requires us to adapt to the current situation. The pandemic has changed many habits of human life, but we believe that this situation cannot diminish our intellectual ability to always contribute knowledge to the world. The wonderful thing about ICST 2021 is bridging the gap between disciplines to share their innovation, research, and ideas about our scientific issues today. UGM is proud to be leading the way in facilitating the interdisciplinary research dissemination of cutting-edge information. We are proud to conduct this conference, even though we conduct this conference virtually.

For the last 7 years, UGM hold ICST as a part of the Annual Scientific Conference to share the latest findings in their fields. It proves UGM's consistency to preserve the international academic's relation. We are honored and humbled to many experts who have attended this year's conference. We thank to the speakers for the expertise and knowledge that will bring to great discussion during the conference. Special thanks are also extended to the organizing committee members for their hard work, as well as the entire staff of UGM's Badan Penerbit dan Publikasi (BPP) for making and bringing the ICST 2021 to a reality. And finally, we would like to thank all the conference participants who will contribute to making this truly the most memorable ICST. For the unfinished business, this event should impact the fields of science, technology and also our humanity for the next many years. I am expecting this conference will also provide us better linkage and friendship.

I am sure you will have fruitful and rewarding exchanges. I hope you all enjoy this conference.

Thank you very much,
God bless us,

Rector of Universitas Gadjah Mada,
Prof. Ir. Panut Mulyono, M.Eng., D.Eng.

Statement of Peer review

In submitting conference proceedings to *Web of Conferences*, the editors of the proceedings certify to the Publisher that

1. They adhere to its **Policy on Publishing Integrity** in order to safeguard good scientific practice in publishing.
2. All articles have been subjected to peer review administered by the proceedings editors.
3. Reviews have been conducted by expert referees, who have been requested to provide unbiased and constructive comments aimed, whenever possible, at improving the work.
4. Proceedings editors have taken all reasonable steps to ensure the quality of the materials they publish and their decision to accept or reject a paper for publication has been based only on the merits of the work and the relevance to the journal.

Title, date and place of the conference

Conference title : ICST 2021 – The 2nd Geoscience and Environmental Management Symposium
Conference date : 7-8 September 2021
Conference place : Virtual conference, Yogyakarta, Indonesia

Proceedings editor(s):

Prof. Rohayu Che Omar (Universiti Tenaga Nasional, Malaysia)
(rohayu@uniten.edu.my)
Prof. Josaphat Tetuko Sri Sumantyo, Ph.D (Chiba University, Japan)
(jtetukoss@faculty.chiba-u.jp)
Prof. Ben White (Erasmus University, Netherlands)
(benwhite46@gmail.com)
Dr. Abraham Cardenas Tristan (Universidad Autónoma de San Luis Potosí, Mexico)
(abraham.cardenas@uaslp.mx)
Prof. Dr. Eko Haryono (Universitas Gadjah Mada, Indonesia)
(e.haryono@ugm.ac.id)
Dr. Dyah Rahmawati Hizbaron (Universitas Gadjah Mada, Indonesia)
(dyah.hizbaron@ugm.ac.id)
Ratih Fitria Putri, S.Si., M.Sc., Ph.D (Universitas Gadjah Mada, Indonesia)
(ratihfitria.putri@ugm.ac.id)

Date and editor's signature September, 18th 2021 Ratih Fitria Putri, S.Si., M.Sc., Ph.D.

17, avenue du Hoggar - PA de Courtabœuf – BP 112 - 91944 Les Ulis Cedex A (France)
Tél. : 33 (0)1 69 18 75 75 - Fax : 33(0)1 69 07 45 17 – www.edpsciences.org

A comparative study of supervised machine learning approaches for slope failure production

Ashanira Mat Deris[1], *Badariah* Solemon[2*], and *Rohayu* Che Omar[2]

[1]Fakulti Teknologi Kejuruteraaan Kelautan dan Informatik, Universiti Malaysia Terengganu 21030 Kuala Nerus, Terengganu
[2]Institute of Energy Infrastructure, Universiti Tenaga Nasional, Jln IKRAM-UNITEN, 43000 Kajang, Selangor, Malaysia

Abstract. Over the years, machine learning, which is a well-known method in artificial intelligent (AI) field has become a new trend and extensively applied in various applications to solve a realworld problem. This includes slope failure prediction. Slope failure is among the major geo-hazard phenomenon which gives the significant impact to the environment or human beings. The estimation of slope failure in slope stability analysis is a complex geotechnical engineering problem that involves many factors such as geology, topography, atmosphere, and land occupancy. Generally, slope failure can be estimated based on traditional methods such as limit equilibrium method (LEM) or finite equilibrium method (FEM). However, beside the methods are quite tedious and time consuming, LEM and FEM have their own limitations and do not guarantee the effectiveness when dealing against problem with various geometry or assumptions. Hence, the introduction of machine learning approaches provides the alternative tools for the prediction of slope failure. Current study applies two mostly used supervised machine learning approaches, support vector machine (SVM) and decision tree (DT) to predict the slope failure based on classification problem using historical cases. 148 of slope cases with six input parameters namely "unit weight, cohesion, internal friction angle, slope angle, slope height and pore pressure ratio and factor of safety (FOS) as an output parameter", was collected from multinational dataset that has been extracted from the literature. For development of the prediction model, the slope data was divided into 80% training data and 20% testing data. The prediction result from testing data was validated based on statistical analysis. The result shows that SVM model has outperformed DT model by giving the prediction accuracy of 97%. ith the advent of technology and the introduction of computational intelligent methods, the prediction of slope failure using the machine learning (ML) approach is rapidly growing for the past few decades. This study employs an "artificial neural network" (ANN) to predict the slope failures based on historical circular slope cases. Using the feed-forward back-propagation algorithm with a multilayer perceptron network, ANN is a powerful ML method capable of predicting the complex model of slope cases. However, the prediction result of ANN can be improved by integrating the statistical analysis method, namely grey relational analysis (GRA), to the ANN model. GRA is capable of identifying the influencing factors of the input data based on the correlation level of the reference sequence and comparability sequence of the dataset. This statistical machine learning model can analyze the slope data and eliminate the unnecessary data samples to improve the prediction performance. Grey relational analysis-artificial neural network (GRANN) prediction model was developed based on six slope factors: unit weight, friction angle, cohesion, pore pressure ratio, slope height, and slope angle, with the factor of safety (FOS) as the output factor. The prediction results were analyzed based on accuracy percentage and receiver operating characteristic (ROC) values. It shows that the GRANN model has outperformed the ANN model by giving 99% accuracy and 0.999 ROC value, compared with 91% and 0.929.

1 Introduction

Landslide is among the most destructive popular geological hazards worldwide. It poses a great danger to the environment as well as safety of human life and damaging the property and resources [1]. According to Korup and Stolle [2], landslide is generally defines as forms and processes that generated from the downward and outward movement of hillslide-forming components, such as dirt, rock or debris, caused by the gravity force, and usually which the existence of water. Landslides not only occurred in steep terrain, but also may triggers in gently sloping to almost flat terrain. The major causes of landslides are due to the slope failure. In order to mitigate the impacts and consequences of landslides, researchers are progressively investigating multivariate data analysis approaches in the fields of machine learning (ML) or data mining to estimate possible occurrences of slope failure from the historical distribution patterns. Traditionally, geotechnical engineers employ various approaches for analyzing the stability of slope, including limit equilibrium method (LEM), finite element method (FEM), upper bound limit analysis, maximum probability, genetic programming, etc [3,4]. However due to the some limitation of the traditional methods, the ML method was introduced to

* Corresponding author: badariah@uniten.edu.my

predict the slope failure in slope stability analysis. One of the limitations of the traditional methods is, it is difficult to mathematically describe the relationship between the significant factors of slope due to the complex mechanism of slope failure [5].

ML approaches such as artificial neural network (ANN), support vector machine (SVM) and decision tree (DT), has been successfully employed in this field to simulate the geotechnical problems [6]. This is due to the ability of these methods to establish the nonlinear equations between input and output set of data [7]. ANN is a sophisticated ML approach that mimicking the brain neurons to generate a solution for a problem. It has shows capability in prediction of complex model and has been widely used by previous researchers to analyze the slope stability [8, 9, 10]. However, ANN has been reported to have certain limitations such as low convergence speed and less generalization performance Samui and Kothari [7].

SVM is an effective ML approach based on the principal of "structural risk minimization" (SRM) to create the decision planes to define decision boundaries. SVM can guarantee greater accuracy for a prediction process in many practical applications in various field compared to other ML approaches [6, 11, 12, 13]. SVM has been successfully applied by previous researchers in various fields including financial [14], health [15]2019), agriculture [16], manufacturing [17] etc. In geo-engineering, SVM also is widely applied for slope stability analysis.

Samui [6] developed prediction models for slope stability using SVM and ANN. The factor of safety (FOS) value was predicted using regression problem while slope stability status was predicted using classification problem. The result shows that for both FOS and stability status, SVM gives better prediction compared with ANN by giving the accuracy of 85.71%. Qi and Tang [18] studied the performance of six different ML approaches to predict the stability of slope. The ML approaches used to develop prediction models were SVM, random forest (RF), gradient boosting machine (GBM), decision tree, logistic regression, and multilayer perceptron ANN (MPNN). The result found that SVM was efficient in terms of the accuracy and the true negative rate. With the advent of the new technology, SVM also evolved significantly to improve the performance of the prediction result. The prediction result is proven to be improved with the introduction of new variant of SVM. Kumar et al [19] applied three different SVM variants to predict the landslide prediction of Mandakini River Basin in India. The models were developed using basic SVM, proximal SVM (PSVM) and L2-SVM-modified finite Newton (L2-SVM-MFN). 2009 cases of landslide were divided into 50% training dataset and 50% testing dataset. The result showed that L2-SVM-MFN has outperformed PSVM and SVM with prediction value of 0.829, compared with 0.807 and 0.79, respectively. Lin et al [20] has developed prediction model of slope stability using ML approaches including SVM, random forest (RF), gravitational search algorithm (GSA) and NB. Six slope factors with 107 of domestic and worldwide slope cases were analyzed and measured to develop the prediction model. The result shows that GSA and RF have outperformed SVM and BN by giving the accuracy of 88.89%.

DT is a well-known ML approach and an efficient tool for prediction and classification. For the past decade, DT has been applied in many applications including geology [21, 18, 22] manufacturing [23], medical [24], agriculture [etc]. Bui et al [25] investigate the landslide susceptibility assessment in province of Huo Binh, Vietnam based on SVM, DT and Naive Bayes (NB). Ten factors were selected to computes the indexes of landslide susceptibility. The factors were soil type, slope angle, relief amplitude, rainfall slope aspect, lithology, distance to rivers, distance to roads, distance to faults, and land use. 118 landslides cases were divided into 70% training and 30% testing for all the models. The result shows that SVM prediction model has outperformed DT and NB. Park et al [26] analyzed 548 landslides in Gangneung-si, Korea to predict the landslides susceptibility using three different approaches of decision tree namely Chi-square automatic interaction detection (CHAID), exhaustive CHAID and Quick, Unbiased, and Efficient Statistical Tree (QUEST). 20 landslide factors were considered as input parameters with area under the curve (AUC) were used as output factor to develop the prediction models. The result shows that CHAID model has outperformed exhaustive CHAID and QUEST by giving the prediction result of 87.1% compared with 86.9% and 82.8% respectively. From the review of the previous literature, it shows that SVM and DT are able to give good prediction result for slope stability analysis including landslide prediction. In the current work, the SVM and DT are applied to predict the slope failure based on classification problem.

2 Methodology

2.1 Case study

To develop prediction models, historical cases of slope stability collected from a multinational dataset that has been extracted from the literature [27, 28, 29, 30]. The dataset consists of 148 slope cases with six input parameters namely "unit weight, internal friction angle, cohesion, slope angle, slope height and pore pressure ratio", and "factor of safety (FOS)" as output parameter. Two distinctive classes of FOS which are 1 and 0 refer to the "stable" and "unstable" slope, respectively. Table 1 shows the basic statistical of the slope cases.

Table 1. Basic statistical of the slope cases.

Slope parameters	Statistic		
	Minimum	Maximum	Standard Deviation
Unit weight (γ)	13.97	31.3	4.0193
Cohesion, (c)	4.95	300	47.1177
Internal friction angle (φ)	0	45	10.9554
Slope angle (β)	16	59	10.1382
Slope height (H)	3.6	511	138.2752

Pore water pressure ratio (ru)	0	45	3.6844

Two types of slope condition were identified from 148 slope cases: stable and unstable. Figure 1 shows the pie chart of the slope cases.

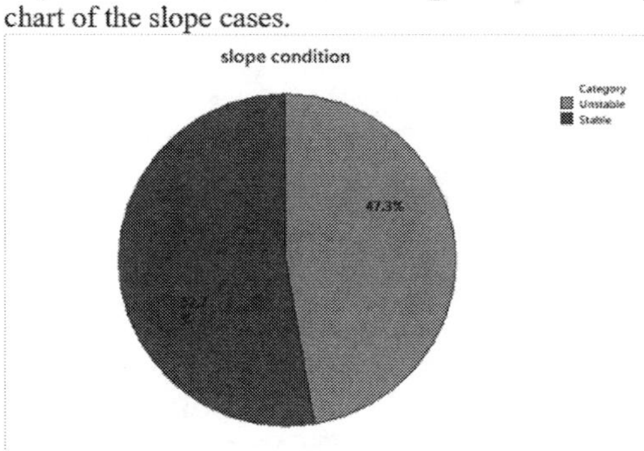

Fig.1. Pie chart of the slope cases.

From the figure, it can be seen that 52.7% of the slope are stable (78 cases) and 47.3% (70%) are unstable. Both types of slope condition are relatively balanced. To develop prediction models using ML approaches, the whole datasets need to be divided into two new subsets which are training and testing data. This is to ensure the generalization capability of the datasets. Basically, the training data is used to train the model and tuning the hyper-parameters while testing data is used for the prediction purpose to test its generalization capability. This study divides the datasets into 80:20 where 80% of the data is used for training and 20% is used for testing.

2.2 Support vector machine

SVM was founded by Vapnik [31], implementing the SRM principal is one of the most widely used ML approaches. The principal of SRM is an improvement of the Empirical Risk Minimisation (ERM) which is employed in neural network. The difference between SRM and ERM is where SRM minimise an upper bound of the expected risk while ERM minimise the error of the training data. This difference leads to the greater ability for SVM to generalize the goal of statistical learning to solve the problem. Generally SVM model consists of five major steps, which are illustrated in Figure 2.

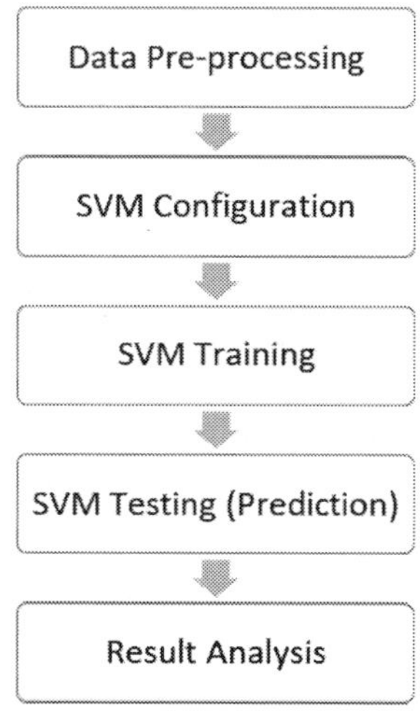

Fig. 2. Steps in SVM.

SVM problem can be solved based on classification or regression analysis [6]. For the classification problem, the objective for the algorithm is to obtain an optimal hyper plane with the maximum margin distance of two different classes in N-dimensional space that classifies the data points. Optimizing the gap from the margins offers some assurance so that potential data points can be identified with high levels of trust. Figure 3 shows the optimal hyper plane for two classifiers.

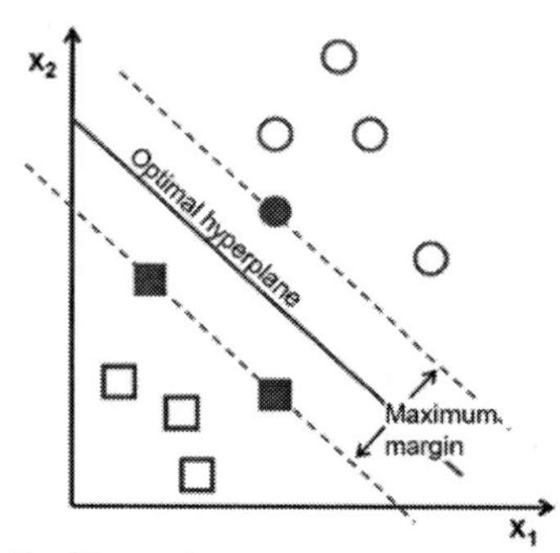

Fig. 3. Optimal hyperplane.

From Figure 3, it shows that the data point is linearly divided by the hyperplane with a feature class is defined by y= {1, -1}. SVM attempts to search for the largest margin between the two classes. Considers the training data as follows:

$$y_i\,(x_i w + b) - 1 \geq 0 \;\; \forall_i, \tag{1}$$

then the data points for the above equality hold lie on the hyperplanes $x_i w + b = 1$ and $x_i w + b = -1$. The margin can be expressed as follows:

$$margin = \frac{2}{||w||} \tag{2}$$

The maximum decision boundary margin can be calculated as in Eq (3).

$$max_{w,b} = \frac{2}{||w||} \tag{3}$$

Where x_i is the input parameters, y_i is the output parameter, w and b are the model parameters. The kernel function is applied in SVM to measure the correlation between two inputs when the classification problem is non-linearly separable. It can describe the inner product of pair wise samples to reduce the dimensions of the feature vector.

To develop SVM model, the dataset need to be divided into training and testing dataset. Among 148 cases, 80% or 118 slope case of the data was selected as training and 20% or 30 slope cases was selected as testing dataset. The selection of training and testing datasets also considers the 5-fold cross validation. The K-fold Cross Validation (KCV) is a widely used, efficient, simple and reliable re-sampling method [32]. KCV divides the dataset into k individual substances so that all cases are taken from each subset for training and testing. SVM performance generally depends on the choice of the kernel function that satisfies the Mercer's theorem. Kernel function enables the operation to be carried out in the input space instead of the potentially high dimensional future space. Therefore, the inner product does not need to be evaluated in the future space. There are four kernel functions that are normally used by researchers namely, radial basis function (RBF) also known as Gaussian kernel, linear kernel, sigmoid kernel and polynomial kernel [1].

i. RBF kernel
RBF is a universal kernel function and the most commonly used kernel function in SVM [1]. RBF kernel function is applied in the nonlinear mapping of SVM. RBF kernel function is given as follow:

$$K(x,y) = \exp(-||x-y||2/(2) \quad (4)$$

ii. Linear kernel
Linear kernel is generally described as:

$$K(x,y) = x\ y \quad (5)$$

iii. Polynomial kernel
The polynomial function is directional where the output is dependent on the direction of the two vectors in low dimensional space. It is because of the kernel dot product. Polynomial kernel function is given by:

$$K(x,y) = (x\ y + 1)p \quad (6)$$

iv. Sigmoid kernel
Sigmoid kernel function is also defined as Multi Layer Perception Kernel or Hyperbolic Tangent Kernel. Sigmoid kernel function is given as in Eq. (7).

$$K(x,y) = \tanh(kx\ y-) \quad (7)$$

2.3 Decision Tree (DT)

DT is a supervised ML approach that adapts the tree-like graph or model for a decision making. DT consists of two types of tree, classification and regression trees. The basic idea of DT is to develop a model based on learning process of several decision rules, which is called decision tree, from the overall data. The total population of the data is divided into two or more homogeneous sets depending on the most important input variables divider. DT process could save the overall processing time as there is no need for variable transformation due to the fact that the tree structure will remain the same either with or without the transformation [33]. Using DT, a complex modeling relationship between the variables can be easily interpreted by the decision makers.

The process of DT consists of two steps which are making the trees and pruning the trees [33]. To avoid the unnecessary nodes in the tree, pruning process may required in most cases. The generation of DT consists of a "root node", "intermediate nodes" and "leaf nodes" with the branches that connect all the nodes. The samples from the root nodes must comply according to the growing rules where all samples in the node are divided into the same group of subsets. This is to ensure that the samples features in the same subset are as homogeneous as possible and two distinct subsets samples are as heterogeneous as possible. The intermediate nodes also need to abide with this rule. Each of the leaf nodes will presents the samples class. The iteration of the nodes will be completed when the maximum depth is reached with the characteristic of all the nodes are in similar group [34]. Figure 4 shows the example of schematic view of a portion of decision tree.

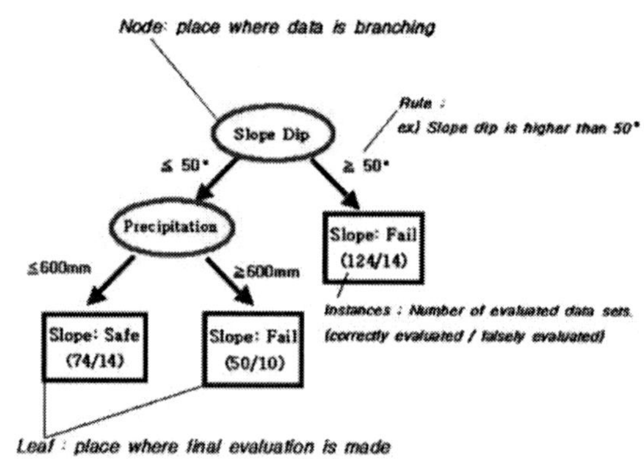

Fig. 4. Schematic view of decision tree [35].

3 Result and discussion

3.1 Confusion matrices

Confusion is a summary that associated with a classifier to explain the actual and predicted results. Confusion matrix is a basic tool to evaluate the confidence of classification result. The number of correct and incorrect prediction result are summarized for each class. Considers there are two classes for a classification problem, the confusion matrix as shown in Table 2.

Table 2. Confusion matrix.

		Predicted	
		Class 1	Class 2
	Class 1	TP	FN
Actual	Class 2	FP	TN

where
"TP = True Positive: The number of correct positive prediction

TN = True Negative: The number of correct negative prediction
FP = False Positive: The number of incorrect positive prediction
FN = False Negative: The number of incorrect negative prediction"

3.2 Prediction result

Table 3 shows the accuracy comparison of the prediction result for different kernel functions in SVM; Linear, RBF, Sigmoid and Polynomial. The accuracy for classification problems is calculated as in Eq. (8):

$$\text{Accuracy} = \frac{TP+TN}{TP+TN+FB} \quad \text{x } 100\% \qquad (8)$$

Table 3. Prediction results.

Kernel function	Accuracy
Linear	90.00
RBF	96.67
Sigmoid	63.34
Polynomial	93.34

From Table 3, it can be seen that RBF kernel function has outperformed Linear, Polynomial and Sigmoid function with the accuracy of 96.77%, compared with 90%, 93.33% and 63.33%, respectively. Table 4 shows the comparison prediction result of SVM-RBF and DT in terms of confusion matrices and Figure 5 shows the schematic view for the DT prediction.

Table 4. Confusion matrices of the prediction results.

		Actual		Accuracy
SVM-RBF		Positive	Negative	
Predictive	Positive	10	1	96.67
	Negative	0	19	
DT				
Predictive	Positive	15	0	83.34
	Negative	5	10	

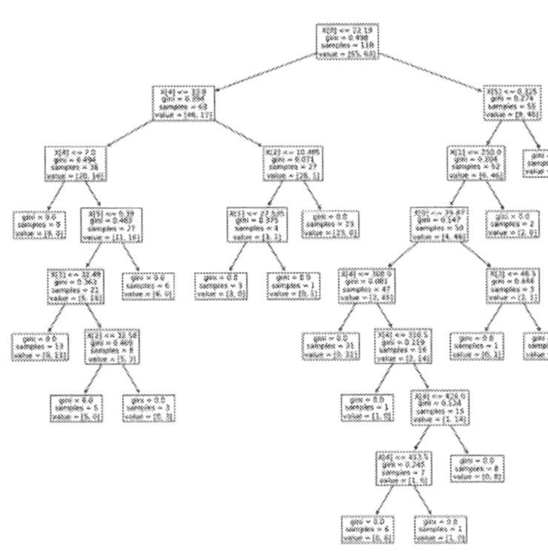

Fig. 5. Decision tree of slope prediction.

From the Table 4, the confusion matrix shows that for RBF-SVM, there are 11 cases of stable slope with 10 has correct prediction and 1 case is wrongly predicted while 19 cases of unstable slope with 19 cases are correctly predicted and zero case of inaccurately predicted. For DT, there are 15 cases of stable slope with all are correctly predicted while 15 cases of unstable slope with 10 cases are correctly predicted and 5 cases are wrongly predicted. As can be seen, the SVM model has higher accuracy compared to DT model which is 96.67%.

4 Conclusion

This study has implemented a novel approach for slope failure prediction using SVM and DT classification problem. Six slope parameters are identified ("Unit weight, cohesion internal friction angle, slope angle, slope height, pore pressure ratio, slope condition") from the 148 slope cases extracted from the historical worldwide slope data. The conclusions are as follows:

i. Among four kernel functions that are tested in SVM, RBF give better accuracy with 96.67%.
ii. The comparison between SVM-RBF and DT was found that SVM-RBF performed better than DT for slope failure prediction.

It should be noted that this research extends the body of knowledge in two aspects. First, the capability of SVM and DT as classifier to predict the slope stability and second, to verify that the performance of SVM model highly depends to the selection of kernel function.

This project was funded by UNITEN under Fundamental Research Grant Scheme FRGS/1/2018/STG08/UNITEN/02/1.

References

1. B. Feizizadeh, M.S. Roodposthi, T. Blaschke, J. Aryal, Arabian J. of Geosci.**10**, 122 (2017)
2. O. Korup, A. Stolle, Geology Today **30**, 1:26–33 (2014)
3. A. Mouyeaux, C. Carvajal, P. Bressolette, L. Peyras, P. Breul, C. Bacconnet, Compt. And Geotechnics **101**, 34–47 (2018)
4. L. He, Q. Tian, Z. Zhao, Q. Zhang, J. Zhao, Intr.J. Geomevhanics **186**, 04018044 (2018)
5. J. Zhou, E. Li, S.Yang, M. Wang, X. Shi, S. Yao, H. S. Mitri, Safety Sci. **118**, 505–518 (2019)
6. P. Samui, Env. Geology **56**, 2:55 (2008)
7. P. Samui, D. P. Kothari, Scientia Iranica **18**, 1:53–58 (2011)
8. M. Fakir, M. Ferentinou, *A Holistic Open Pit Mine Slope Stability Index Using Artificial Neural Networks*, in International Society for Rock Mechanics and Rock Engineering, January 2017 (2017)
9. M. J. Sulewska, Comp. Assisted Methods in Eng. and Sci. **18**, 4:231–241.
10. Z. G. Qian, A. J. Li, W. C. Chen, A. V. Lyamin, J. C. Jiang, Soils and Foundations **59**, 2:556–569 (2019)
11. A. M. Deris, A. M. Zain, R. Sallehuddin, Procedia Eng. **24**, 308–312 (2011)

12. V. H. Quej, J. Almorox, J. A. Arnaldo, L. Saito, J. of Atmospheric and Solar-Terrestial Physics **155**, 62–70 (2017)
13. S. Hosseini, B. M. H. Zade, Comp. Network, 107168 (2020)
14. L. Chao, J. Zhipeng, Z. Yuanjie, Expert Systems with Applications **123**, 283–298 (2019)
15. Y. Uchida, T. Funayama, Y. Kogure, *SVM classification of data obtained from a health conditon monitoring system using flexible force sensing resistors*. In Proceedings of the 5th Int. Conf. on Sensors Engineering and Electronics Instrumentation Advances (SEIA'1019), September 2019, Tenerife (Canary Islands), Spain, (2019)
16. J. Ruan, H. Jiang, X. Li, Y. Shi, F.T. Chan, W. Rao, IEEE Transactions on Industrial Informatics, **15**, 12:6510–6521 (2019)
17. J. Lu, X. Liao, S. Li, H. Ouyang, K. Chen, B. Huang, Complexity (2019).
18. C. Qi, X. Tang, Comp. & Indust. Eng. **118**, 112–122 (2018)
19. D. Kumar, M. Thakur, C. S. Dubey, D. P. Shukla, Geomorpholgy **295**, 115–125 (2017)
20. Y. Lin, K. Zhou, J. Li, IEE Access, 2169–3536 (2018)
21. X. N. Bui, H. Nguyen, Y. Choi, T. Nguyen-Thoi, J. Zhou, J. Dou, Scientific reports **10**, 1:1–17 (2020)
22. Y. Wu, Y. Ke, Z. Chen, S. Liang, H. Zhao, H. Hong, Catena **187**, 104396 (2020)
23. A. Kim, K. Oh, J. Y. Jung, B. Kim, Inter. J. of Computer Integrated Manufacturing **31**, 8:701–717 (2018)
24. S. J. Narayanan, R, Soundrapandiyan, B. Perumal, C. J. Baby, *Emphysema medical image classification using fuzzy decision tree with fuzzy particle swarm optimization clustering*, in Smart Intelligent Computing and Applications (pp. 305–313), 2019, Springer, Singapore (2019)
25. D. Tien Bui, B. Pradhan, O. Lofman, I. Revhaug, Mathematical problems in Engineering (2012)
26. S. J. Park, C. W. Lee, S. Lee, M. J. Lee, Remote Sensing **10**, 10:1545 (2018)
27. N. K. Sah, P. R. Sheorey, L.N. Upadhyaya, Intr. J. of Rock Mehanics and Mining Sci. & Geomechanics Abstracts **31**, 47–53 (1994)
28. K. P. Zhou, Z. Q. Chen, *Stability prediction of tailing dam slope based on neural network pattern recognition*, in 2009 Second International Conference on Environmental and Computer Science (pp. 380-383). December 2009, IEEE (2009)
29. J. Li, F. Wang, *Study on the forecasting models of slope stability under data mining*, in Earth and Space 2010: Engineering, Science, Construction, and Operations in Challenging Environments (pp. 765-776) (2010)
30. Y. Xiaoming, L. Xibing, *Bayes discriminant analysis method for predicting the stability of open pit slope*, in 2011 International Conference on Electric Technology and Civil Engineering (ICETCE) (pp. 147-150), April 2011, IEEE (2011)
31. V. Vapnik, The nature of statistical learning theory (Springer, New York, 1995)
32. D. Anguita, A. Ghio, S. Ridella, D. Sterpi, *K-Fold cross validation for error rate estimate in support vector machines*. In DMIN, July 2009, (pp. 291-297) (2009)
33. J. Dou, A. P. Yuus, D. T. Bui, A. Merghadi, M. Sahana, Z. Zhu, C. W. Chen, K. Khosravi, Y. Yang, B. T. Pham, Sci. of The Total Envi. **662**, 332–346 (2019)
34. C. Qi, X. Tang, Comp & Indust. Eng. **118**, 112–122 (2018)
35. S. Hwang, I. F. Guevarra, B. Yu, Eng. Geology, **104**, 1–2:126–134 (2009)
36. S. Rajeswari, K. Suthendran, Comp. and Electronics in Agriculture **156**, 530–539 (2019)
37. B. Choubin, E. Moradi, M. Golshan, J. Adamowski, F. Sajedi-Hosseini, A. Mosavi,Sci of The Total Env.**651**, 2087-2096 (2019)
38. Y. Radhika, M. Shashi, Inter. J. of Computing Theory Eng **1**, 1793–8201 (2009)

A Review Disaster Mitigation of Jakarta Land Subsidence Areas

Ratih Fitria Putri[1*], *Mya Dwi* Rostika[2], *Aji Wijaya* Abadi[1], and *Mega* Rakhmatika[1]

[1]Department of Environmental Geography, Faculty of Geography, Universitas Gadjah Mada, Indonesia
[2]Department of International Relations, Faculty of International Relations, Daito Bunka University, Japan

*Corresponding author: ratihfitria.putri@ugm.ac.id

Abstract. The Jakarta Capital Special Region of Jakarta is an area in the Java Island that has the highest level of vulnerability to land subsidence disasters. Land subsidence is caused by several factors such as excessive use of groundwater and natural compression of soil conditions. This study used a review analysis method based on a literature study on mitigation of land subsidence disasters in Jakarta city. The mitigation planning of land subsidence disasters in Jakarta will refer to government regulations on disaster management and Jakarta City spatial planning. Disaster mitigation is necessary to anticipate or reduce disaster risk. Disaster mitigation is a series of efforts to reduce disaster risk through physical development combined with awareness and capacity building in dealing with disaster threats. Education cannot be separated from the active role of the community and local government. High-densely populated areas will lead to more use of groundwater in Jakarta city. The excessive use of groundwater has resulted in more intensive land subsidence in Jakarta's urban area. The safe zone mapping of subsidence disasters is prominent to reduce and control land subsidence disasters. The government's efforts to lessen the impacts of land subsidence are by educating the population in disaster areas.

1 Introduction

Indonesia is an archipelago country that has a very high risk of natural disasters [1]. The disaster risks faced by Indonesia include natural disasters caused by geological factors (landslides, earthquakes, volcanic eruptions, and tsunamis), hydrometeorological factors (floods, landslides, droughts, hurricanes), disasters due to biological factors (human disease outbreaks, plant/livestock diseases) and technological failures (industrial accidents, transportation accidents, nuclear radiation, chemical pollution). Land subsidence is among natural disasters that take place annually and become a general concern to the big city in local government.

The Capital Special Region of Jakarta has the highest vulnerability to land subsidence disasters (Figure 1). Land subsidence is induced by various factors such as excessive groundwater withdrawal and natural compression of soil conditions. Land subsidence is highly potential in causing other problems such as changes in topographic gradients, damage to the soil surface, and reduction of groundwater-contained aquifer capacity. Land subsidence has an impact on the development of regional development conditions both locally and regionally. The phenomenon of land subsidence can increase the potential of flooding in an area [2]. In Jakarta, land subsidence is one of the significant causes of the January 2013 flood disaster [3].

The research area has 13 natural and artificial rivers for locals' drinking water and other domestic needs. Jakarta City has a tropical climate with annual rainfall reaches 1,755 mm or approximately 146 mm per month. Currently, Jakarta's development level has advanced rapidly following urbanization. According to the 2020 census data, the population density of Jakarta City is estimated at 15,964 people/Km^2 [4]. The Jakarta Capital Special Region map is showed in Figure 2.

As a result of losses caused by disasters generated by land subsidence, innovations in disaster mitigation planning to overcome and reduce disaster risk need to be carried out effectively and systematically to reduce inaccuracies in the disaster mitigation process and prevent overlaps in the implementation of disaster mitigation [5]. This study aims to analyze land subsidence mitigation in Indonesia as review process in local government disaster management.

Fig. 1. Land Subsidence Areas in the Jakarta City.

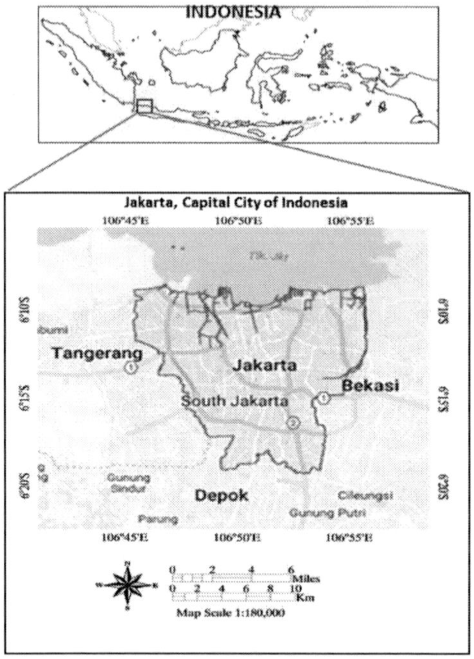

Fig. 2. Jakarta Special Capital Region Map.

2 Methods

This study applied a review analysis method according to a literature study on land subsidence disasters mitigation in Jakarta City. Data of land subsidence has been examined based on research using differential interferometry synthetic aperture radar [6]. The predetermined area for land subsidence analysis is in the northern part of the city of Jakarta. Disaster mitigation planning of land subsidence in Jakarta will refer to government regulations on disaster management and Jakarta City spatial planning [7].

3 Results and Discussions

3.1 Population Analysis

The burgeoning population density from year to year and the increasing number of settlements in Jakarta city are some causes of the land-use intensification in land subsidence-prone areas [5]. Land use in Jakarta City is a noticeable result of a long process of constant interaction, balance, and dynamics between the population activities on land and the environment limiting factors [6]. Land-use change can affect local ecological systems by water pollution, air pollution, climate change, and declining biodiversity. Population growth and progressing economic activity over time increase the necessity of resources, including space. Space remains fixed in a broad sense but, it will always change following changes in space utilization regarding physical, economic, and social composition. Changes in spatial use that do not consider the geo-biophysical balance will produce waste and result in natural disasters. Jakarta is one of the largest cities in Indonesia and population of 10,560,000 people in 2020. As the capital of Indonesia, Jakarta plays a vital role in the economy and the central government. The development of urban areas is a response to increasing population growth in Jakarta City. Therefore, it has become one of the triggers for various environmental problems in this decade.

This research will highlight three areas that have the highest level of subsidence. The areas to be analyzed are Mutiara Beach, Cengkareng, and Cakung. According to statistics of the population density of the Jakarta Province, the predetermined areas are densely populated. Mutiara Beach, Cengkareng, and Cakung have a population density of 8,635 people/ Km^2, 19,363 people/Km^2, and 11,919 people/Km^2 respectively. Jakarta's population tends to increase every year. The high population density occurred in Central Jakarta and West Jakarta at more than 18,000/Km^2. From the perspective of the population density, the denser the population, the higher the rate of land subsidence. Population density is directly proportional to the population. Therefore, the increasing number of residents from year to year considerably affects the increase in population density. One example of human activities that cause land subsidence is excessive groundwater extraction. The larger the population size in a certain area, the higher the demand for clean water to sustain their activities. Anointing is one way to fulfill the need for clean water in the community. Based on the 2020 statistics, the Cakung District consists of eight villages in which the Grinding Village has the largest population density value of 23,277 people/Km^2. West Cengkareng has the largest population density in Cengkareng District with a value of 20,217 people/Km^2. A population density of 29,868 people/Km^2 makes Penjaringan Village the most densely populated village in the Penjaringan District. Based on the value of the largest population density, it can be concluded that the highest potential for land subsidence is in the village in each of these sub-districts.

The problem of population and environmental damage are two things that pertain to each other. The occurrence of environmental damage which results in the imbalance of natural resources can affect human life at a macro level. Consequently, the imbalance between the population growth rate and the quality of natural resources can destroy all human life. Therefore, wise efforts are needed in the future to maintain environmental sustainability. This concept attempts to balance population with all its needs while sustaining environmental quality. Accordingly, the stability between population growth rate and environmental sustainability can be obtained.

3.2 Land Subsidence

Land subsidence is one of the impacts of the increasing development in Jakarta. Urbanization has resulted in a very high demand for residential land, making Jakarta city possessing various kinds of environmental problems. Jakarta is the largest city in the country that serves as the center of Indonesian politics and economic activity. The existence of these urban functions has resulted in the increasing potential for land subsidence in Jakarta. Three locations have experienced considerable land subsidence, namely Mutiara Beach, Cengkareng, and Cakung [5]. Mutiara Beach has land subsidence ranging from 8.5 - 10.9 cm/year. Land subsidence in Cengkareng ranges from 10 to 17.5 cm/year. Cakung experienced land subsidence of 7.5 – 9.5 cm/year.

Penjaringan District is in the northern part of Jakarta that has a coastal reclamation area. The reclamation process of the Jakarta coastal area is a government plan aimed at controlling flooding in the city of Jakarta and developing a rapidly growing business area. The reclamation was conducted by building a 60 km sea embankment in Jakarta Bay.

The land subsidence in Cengkareng is more impactful than in other areas in the city of Jakarta. Cengkareng is a district in West Jakarta with an area of 27.9 km2. The maximum land subsidence in the area is 17.5 cm/year. This subsidence event is triggered by human activities such as building construction and groundwater extraction. Cengkareng's land use consists of 1.10 residential areas and 97.25 ha industrial areas. The estimated rate of land subsidence in Cengkareng is in the range of 10.0-17.5 cm/year.

Cakung is one of the suburbs of Jakarta. Land subsidence in the Cakung area ranges from 7.5 to 9.5 cm/year. The majority of land in Cakung is used for residential areas and industrial zones. The residential areas make up 45.27% of the Cakung area while the industrial zones cover 24.33% of the rest. The district is addressed for industrial activities with the most frequent industrial activities operate in the Pulogadung area.

Land subsidence can trigger a disaster. Rob flood induced by land subsidence occurs in the Penjaringan District. The disaster risk is even more as Penjaringan District is highly developed and more vulnerable high-valued assets are exposed. The northern part of Penjaringan District has land use resulted from coastal reclamation. The north side of the district is an elite housing zone and exclusive apartments. This condition stimulated the increasingly extensive use of groundwater in the northern part of Penjaringan.

Cengkareng District's land has majorly been developed. Dense village settlements are the dominant land use. The load due to the existing land use can make the soil more compressed. Soil compression will cause land subsidence in Cengkareng District. Furthermore, over groundwater withdrawal from the settlements can accelerate land subsidence.

Cakung District is located on the easternmost part of East Jakarta, neighboring West Java Province. Its strategic location allows Cakung District has adequate transportation facilities. Warehouses and industries are the general land use found in the Cakung District. The presence of industries, warehouses, and houses in the Cakung District produces a large surface load. The high demand for water and the overuse of groundwater leads to groundwater depletion. When the volume of groundwater declined while the pressure from the surface increased, the phenomenon of land subsidence can occur more intensively.

3.3 Disaster Mitigation

Population adaptation to land subsidence has been a better strategy for managing the hazard. Some adaptations have been implemented by the community, such as creating traditional embankments and raising houses hence they will not be inundated by the flood [8]. Flooding is one of the disasters that resulted from land subsidence in Jakarta City. Disaster awareness of the land subsidence and its consequences has improved. The government attempts to reduce the risks of land subsidence by educating the population in disaster-prone areas so the capacity of the community increase. The education process can succeed if the community and the local government play active roles and support each other. Besides, spatial planning in Jakarta city must be appropriate so that there is no excessive groundwater extraction. The higher population density means higher settlement density as well. A densely populated area will drive more demand for groundwater in Jakarta City. Over groundwater extraction has stimulated more intense land subsidence in Jakarta's urban area. Therefore, safe zone mapping of subsidence is significant to reduce and control the disaster. The safe zone is determined by monitoring the land subsidence every year in Jakarta. The annual land subsidence measurements have to be carried out thoroughly to know the existing condition accurately.

4 Conclusions

Disaster mitigation is needed to anticipate or lessen the risk of natural disasters. Disaster mitigation is defined as a series of efforts to lower disaster risk that can be implemented through physical development, awareness enhancement, and capacity building in dealing with disaster threats. Disaster can occur at any time with or without warning. Therefore, we must be ready whenever and wherever we are. The most important thing is to keep following the directions of the authorized officers. Disaster mitigation can be applied in any form, such as introducing and monitoring disasters, developing a disaster awareness culture, and implementing cutting-edge technology to anticipate disaster risks. Those activities can help to improve spatial planning and development, reduce the impact of disasters, and upgrade public knowledge. Finally, the community has stronger resilience to cope with disasters that potentially occur in their surrounding environment.

References

1. R. F. Putri, A. W. Abadi, and U Kafafa. The correlation analysis between urbanization phenomena and landuse change in Jakarta Special Province. E3S Web of Conferences 200, 05003 (2020)
2. H. Z. Abidin, H. Andreas, I. Gumilar, M. Gamal, Y. Fukuda, Y. E. Pohan, and T. Deguchi, T. Land subsidence of Jakarta (Indonesia) and its relation with urban development, Nat Hazards, 59:1753–1771 (2011)
3. M. D. Anisuzzaman, Dicky, M. Arifuzzaman, and N. Shadrina. The Role of Clay Minerals n LandSubsidence: A Case Study in Jakarta, Indonesia Case Study in Jakarta, Indonesia". The 3rd International Conference on Ecological, Environmental and Biological Sciences Proceeding. 188-191 (2013)
4. BPS DKI Jakarta, *DKI Jakarta dalam Angka Tahun 2020*. Jakarta: Badan Pusat Statistik, 2020.
5. R. F. Putri, L. Bayuaji, J. T. S. Sumantyo, and H. Kuze, "Terrasar-X DInSAR for land deformation detection in Jakarta Urban area, Indonesia," *J. Urban Environ. Eng.*, vol. **7**, no. 2, pp. 195–205. (2013)
6. R. F. Putri, S. Wibirama, S. R. Giyarsih, A. Pradana, and Y. Kusmiati, "Landuse change monitoring and population density analysis of Penjaringan, Cengkareng, and Cakung Urban Area in Jakarta Province," *E3S Web Conf.*, vol. **76**. (2019)
7. Dinas Tata Ruang Provinsi DKI Jakarta. Rencana Tata Ruang Wilaya 2011-2030. Available in: http://webgis.tatakota-jakartaku.net/. Access on 3 June 2014.
8. G. Erkens and E. Stouthamer. The 6M approach to land subsidence. Proc. IAHS, 382, 733–740. (2020)

E3S Web of Conferences 325, 01003 (2021) https://doi.org/10.1051/e3sconf/202132501003
ICST 2021

Empirical analyses of excavation method and tunnel support system based on engineering geology of Tunnel No.8 at Jakarta – Bandung high-speed railway, West Bandung, Province of West Java, Indonesia

Ika Medika Ayu Oktarina[1], *Subagyo* Pramumijoyo[1], and *I Gde* Budi Indrawan[1*]

[1]Geological Engineering, Faculty of Engineering, Gadjah Mada University, 55281 Yogyakarta, Indonesia

Abstract. The Chinese Basic Quality (BQ) system was used to determine the rock mass classes and design tunnel support system of the Tunnel No.8 of the Jakarta-Bandung high-speed railway. As the BQ system is newly applied for tunnelling project in Indonesia and remains poorly understood, this research aims to determine rock mass classes at the tunnel construction site and propose the tunnel excavation method and support system alternative based on more widely used rock mass classifications of the Geological Strength Index (GSI) and Japan Society of Civil Engineers (JSCE) standard. Surface engineering geological mapping and evaluation of drill core samples were carried out to obtain accurate interpretation of subsurface engineering geological conditions, particularly the rock mass classes at the tunnel elevation. The results show that the tunnel construction area consists of very poor to fair classes of andesite breccia, tuff lapilli, and tuffaceous sandstone. The full-face method with auxiliary bench cut and multiple-bench cut method are to be applied at the tunnel sections consisting of rock mass classes CI and DII, respectively. The proposed tunnel support system is a combination of shotcrete, rock bolt, steel rib, and invert.

1 Introduction

The Jakarta – Bandung High-Speed Project, which was built to increase the potential for developing industry, trade, and tourism, has a 142.3 km line with plans to construct 13 tunnels, one of which is Tunnel No. 8. Tunnel No. 8 is located on DK 97+320 to DK 99+510 with a length of 2190m. The research location lay on Mandalasari Village, Cikalong Wetan District and Sumur Bandung Village, Cipatat District, West Bandung Regency, West Java Province. The research location can be seen in Figure 1.

The tunnel built to accommodate a double railway track is located on a hill with undulating topography and hills covered with vegetation. The inlet is located in a flat area, while the outlet is located in a hill area with dense traffic. A Tunnel is one of underground construction that requires many factors. Factors that need to be considered are the planning stage to the maintenance, and social impact after the tunnel is operational. That kind of factor makes the construction of a tunnel such complex construction. One of the crucial factors in planning and construction steps is the characterization of engineering geology conditions in a tunnel construction area and its surrounding.

The study area is dominated by Quaternary Formation of Older Volcanic Product (Qob) and some Miocene Cantayan Formation based on The Regional Geological Map of Cianjur Sheet[1]. The rock mass classification used in this project uses the BQ classification system adopted from the only national system suitable for most types of rock engineering in China. This study will identify rock mass classification using GSI.

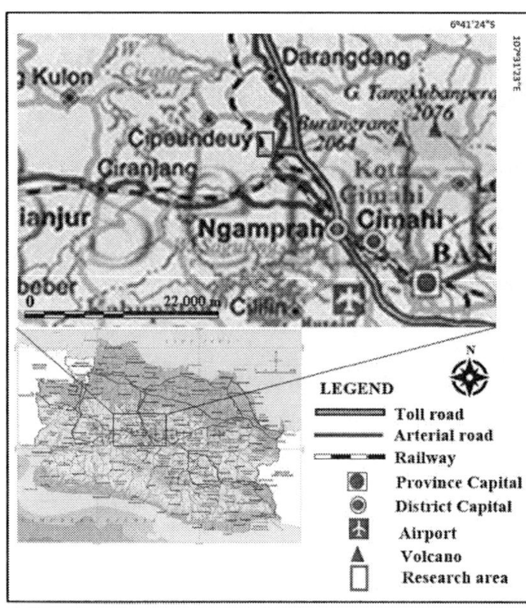

Fig. 1. Map of research location (Tunnel No 8).

* Corresponding author: igbindrawan@ugm.ac.id

Uniaxial Compressive Strength (UCS) values in the field and weathering conditions of surface rocks are estimated based on observations in the field and determined using empirical methods. These data are instrumental in determining the excavation method and support system at an early stage empirically.

2 Regional geology

According to van Bemmelen[2], the physiographic conditions of West Java are divided into 4 (four) zones: Jakarta Coastal Plains Zone, Bogor Zone, Bandung Zone, and the Southern Mountain Zone. Physio graphically, the research area is located in the Bandung Zone, where most of the zone forms younger volcanic deposits from the surrounding volcanoes.

The Regional Geological Map of Cianjur Sheet with a scale of 1:100,000 compile by Sudjatmiko[1] shows the regional stratigraphy of the Jatiluhur Formation (Mdm), Cantayan Formation (Mtts), and Older Volcanic Products (Qob). Marl and quartz sandstone member of Jatiluhur Formation the aged of Miocene consist of dark grey marl, marly claystone, and clayey shale with intercalations of quartz sandstone, quartzite, and marly limestone. Sandstone members of the Cantayan Formation's aged Miocene consist of well-bedded dirty sandstone, sandy shale, shaly clay, marine breccia, and conglomerate. Older volcanic products in the age of Pleistocene consist of breccia lahar lava, volcanic breccia, flow breccia, lahar deposits, and lava showing sheet and columnar jointing. The composition is between andesite and basalt. The Regional Geological Map shows that the research area is located in the Cantayan Formation (Mtts) and the Older Volcanic Products (Qob). Formation is shown in Figure 2.

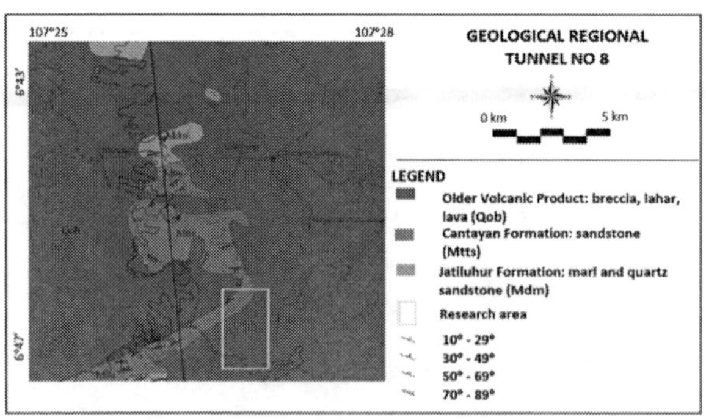

Fig. 2. The geological map of the research area is part of the Regional Geological Map of Cianjur sheet[1].

3 Methodology

Engineering geology conditions of the research area are compiled based on primary data collected during surface mappings, such as surface geological conditions and surface GSI value measurements. Surface engineering geological mapping was carried out to result from inaccurate interpretation of subsurface engineering geological conditions, including the rock mass classes at the tunnel elevation. The data were used as material in making Geological Maps, and surface rock mass class maps with a scale of 1:10,000. The subsurface database on core drill analysis for rock mass classes subsurface. The naming of pyroclastic rocks is based on the grain size of the constituent material based on Fisher Classification[3].

Determination of the degree of rock weathering in the field using a degree of rock weathering classification of rock mass according to International Society for Rock Mechanics (ISRM)[4]. The estimated value of Uniaxial Compressive Strength (UCS) in the field using an estimation table for UCS values in the field for intact rock according to Hoek[5].

Measurement of rock mass classes on the surface using basic chart of Geological Strength Index (GSI) system developed by Marinos and Hoek[6], while the subsurface measurement using quantitative chart of the Geological Strength Index (GSI) system which is regarding the factor of discontinuity conditions and value of 0.5 Rock Quality Designation (RQD) developed by Bieniawski and Hoek et al.[7,8].

The result of the determination of GSI value is correlated with the table correlation of the GSI classes value of rock mass by Sivakugan et al.[9](Table 1).

Table 1. Correlation of the GSI classes value of rock mass[9]

GSI value	Rock Mass Classes
<20	Very poor
21-40	Poor
41-55	Fair
56-75	Good
76-95	Very Good

Excavation methods and tunnel support systems are determined based on the rock/soil category listed in the Japan Society of Civil Engineers (JSCE)[10]. Apart from the rock/soil category, the tunnel excavation method is chosen by considering the advantages and disadvantages of each method and the appropriate media condition.

When excavating the tunnel, selecting the support and support system typically uses an empirical approach consisting of shotcrete, rock bolts, steel rib, and invert. The selection is also determined by the rock/soil category.

4 Result and discussion

4.1 Lithological condition

The lithology of the study area was determined based on observations of outcrops found around the tunnel site. Outcrops found along the tunnel body and its surroundings are dominated by settlements and plantations. The study area is dominated by volcanic rocks from the Qob Formation in the form of andesite breccia and tuff lapilli seen from the grain size of the constituent materials. Meanwhile, an outcrop of tuffaceous sandstone was found in an older formation, namely the Mtts formation (Figure 6).

4.1.1 Tuffaceous sandstone

Tuffaceous sandstone outcrops appear in Cantayan Formation member of sandstone (Mtts) based on the Regional Geological Map of Cianjur sheet. The distribution of this unit covers 6.5% of the research area with brownish-grey to blackish brown colour (Figure 3). This unit has a massive structure layered with fragments measuring 1-2 mm and a matrix of <0.01 mm in size. Based on observations in the field, this tuffaceous sandstone is in slightly weathered condition with discoloration of the rock and surface discontinuities.

Fig. 3. Slightly weathered tuffaceous sandstone.

4.1.2 Andesite Breccia

This unit is part of Older Volcanic Products (Qob). The geological map shows that andesite breccia units are spread almost 85% of the research area. Based on surface observation, andesite breccia units predominantly dark red-brow in the residual soil area and dark brown to dark yellow in highly to moderately weathered. Fragments 5-50 mm in size, even > 64 mm, were dominated by andesite fragment and matrix of < 2 mm in size.

Fig. 4. Moderately weathered andesite breccia.

4.1.3 Tuff Lapilli

Outcrops of tuff lapilli units showed on the top of the andesite breccia unit, which is also part of Older Volcanic Products (Qob). The distribution of tuff lapilli unit is around 8.5% of the total research area on the southwestern side. The size of the fragment is relatively uniform, which is about < 2 mm and matrix < 1 mm. The naming of this unit is based on Fisher's classification[3].

Fig. 5. Moderately weathered tuff lapilli.

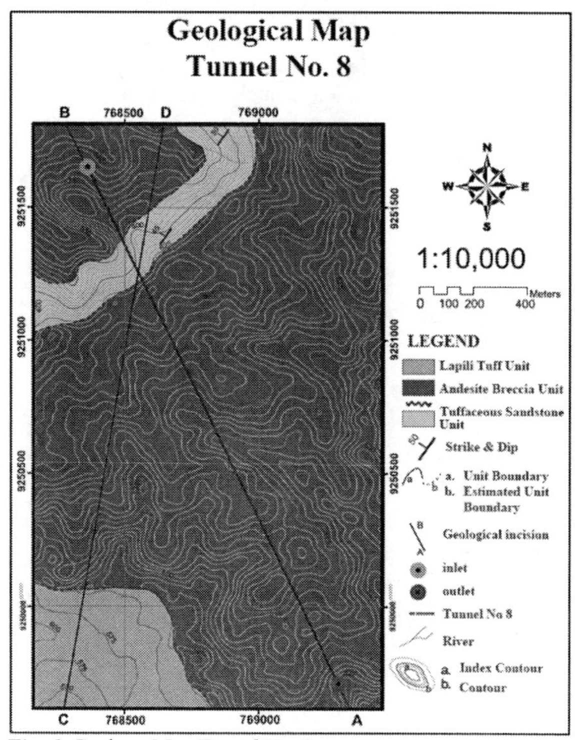

Fig. 6. Geology Map Tunnel No. 8.

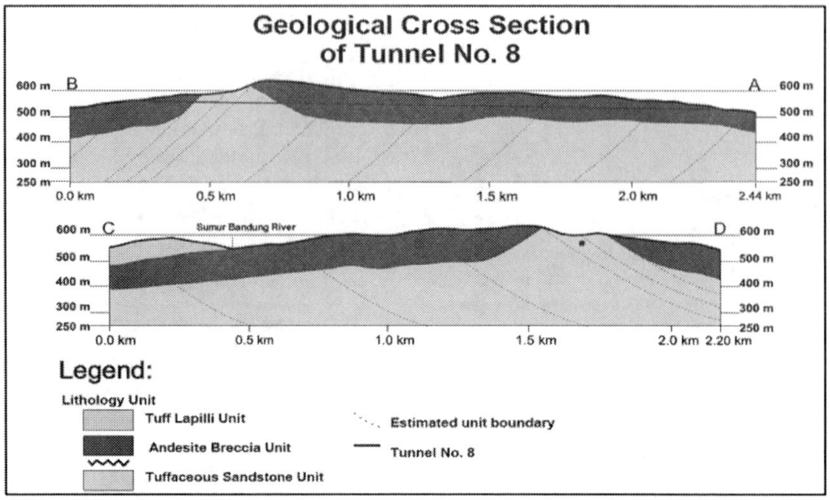

Fig. 7. Geology Crossed Section Tunnel No. 8.

4.2 Engineering classification of rock mass

The surface engineering geological map is prepared based on the degree of weathering and the quality of the GSI rock mass obtained from outcrop observations in the field. Qualitative assessment of rock weathering degree on surface outcrops at the Tunnel No. 8 construction based on ISRM[4] and an estimated value of UCS based on Hoek[5]. From these data, the geological engineering units found in this study area[4] can be divided into 5 (five) rock units: residual soil, moderately weathered tuff lapilli, highly weathered andesite breccia, moderately weathered andesite breccia, and slightly weathered tuffaceous sandstone (Figure 8).

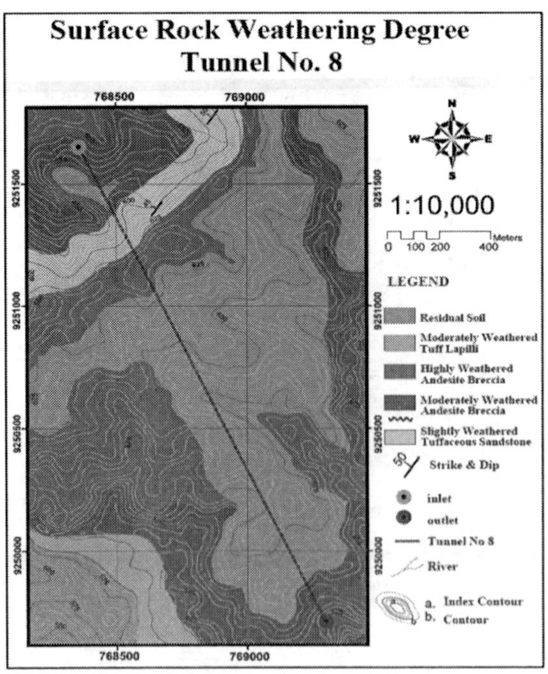

Fig. 8. Map of surface rock weathering degree in research area.

The value of Geological Strength Index (GSI) surface rock was obtained from field investigation on outcrop by referring to GSI basic chart according to Marinos and Hoek[6] (Figure 9). The GSI value on the basic chart is determined based on surface conditions and the discontinuous structure; then, these values are correlated with rock mass classes using the Sivakugan correlation table (Table 1). Based on surface observation, the value of GSI surface rock can be divided from very poor (GSI value of 0 – 20) to fair (GSI value of 41-55) (Figure 10).

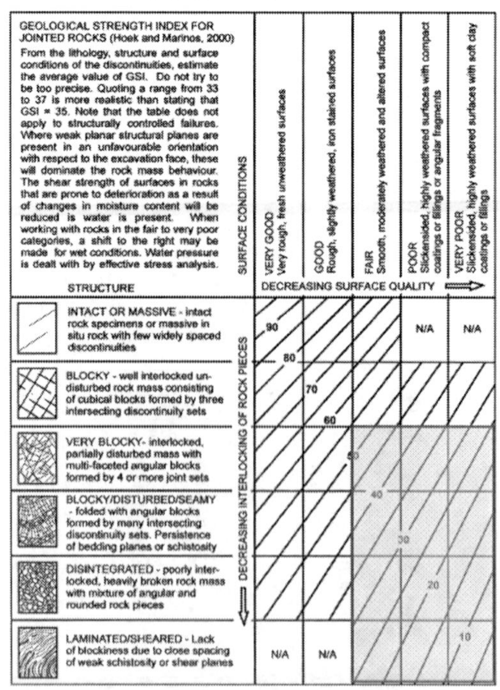

Fig. 9. GSI value of study area.

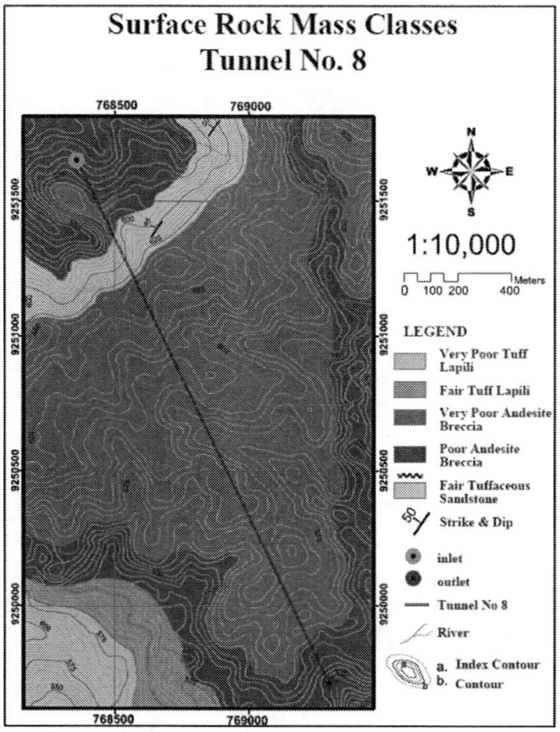

Fig. 10. Map of surface rock mass classes.

4.2.1 Residual soil unit

Most of the outcrops defined as residual soil are tea and cassava plantations. In this case, the Reddish-dark brown color due to the weathering process of older rocks is from andesite breccia. The thickness of this unit is around 5-15 m from the surface. Based on Hoek[5], residual soil can be state has UCS value <1 MPa.

4.2.2 Highly weathered andesite breccia unit

The colour of this unit is dark brown. The rock texture is poorly sortation, matrix-supported, with a fragment grain size of 0.01 m – 2m, a matrix grain size of < 2 mm with structural rocks of massive. Rock fragments composed of andesite and matrix are composed of sandy material. The value of intact rock strength is between 1-5 MPa, which is classified as R1 (very weak) with GSI value 10-20.

4.2.3 Moderately weathered andesite breccia unit

The color of this unit is brown to yellowish-brown with rock texture is poorly sortation, matrix-supported, the fragment grain size of 0.01 m – 2m, the matrix grain size of < 2 mm and with structural rocks of massive. Rock fragments composed of andesite and matrix are composed of sandy material. The value of intact rock strength is between 5-25 MPa with a GSI value 30-40.

4.2.4 Moderately weathered tuff lapilli unit

The color of this unit is light brown to yellowish-brown with rock texture is well sortation, fragment supported, composed by the pyroclastic material size of > 2 mm intact rock strength of moderately weathered tuff lapilli unit is 5-25 MPa with GSI value 45-50 (fair) and 15-20 (very poor).

4.2.5 Slightly weathered tuffaceous sandstone unit

The colour of this unit is yellowish-brown with rock texture is well sortation, fragment supported, composed by the pyroclastic material size of < 2 mm, grain size 0.05 – 0.1 mm from very fine sand to fine sand, intact rock strength of slightly weathered tuffaceous sandstone unit is 5-25 MPa with GSI value 40-50 (fair).

4.2.6 Subsurface condition

Subsurface conditions were estimated by the description of the core drill around the tunnel. Measurements were made on joint conditions and RQD measurements. GSI value of subsurface estimated by measurement using a quantitative chart of the GSI system based on Hoek et al. (2013) and Bieniawski (1989)[7,8], which is regarding the factor of joint conditions per 1 meter and value of 0.5 RQD. Observations of weathering conditions of core drills were also carried out, considering the surface conditions had a relatively high degree of weathering. Classification of subsurface rocks on core drill samples can be divided into 5 (five) units: highly weathered andesite breccia, moderately weathered andesite breccia, slightly weathered tuffaceous sandstone, and highly weathered claystone units.

The residual soil is predominantly reddish-dark brown, visible plant roots, in crushed condition (dry condition) or looks clayey (wet condition), RQD and GSI values are 0 (zero). Andesite breccia units were seen in the majority of core samples, from very poor to good quality. Moderately weathered andesite breccia can be found in core number 8 at depths of more than 50 meters from the surface with a predominantly blackish brown color, the RQD value is around 75-100 and combined with the joint condition values so that the GSI value is between 56 to 64 (good). Moderately weathered andesite breccia is also found in core numbers 7 and 8 at the depths of between 27 to over 50 meters, and combining the joint condition values with RQD, GSI values are obtained between 41-55 (fair). Highly weathered andesite breccia with greyish brown and brownish yellow color can be found in several core samples at a depth between 16-50 meters. The core conditions are mainly in a broken state with an RQD value of 10-65. This value is combined with the value of the joint condition, and then the GSI value is obtained between 21-40 (poor). Very poor breccia units were found in almost all core samples, with more than half of the rock is crushed or turned into the soil (highly

https://doi.org/10.1051/e3sconf/202132501003

weathered degree). They were located after the residual soil layer with a GSI value of less than 20.

Tuffaceous sandstone units are seen in core samples taken at The Mtts Formation site, particularly core number 3 with slight weathering conditions. Based on the value of RQD and joint conditions, the GSI values varied from very poor (0-20) to very good (76-95). Highly weathered claystone units were found in several core samples with GSI values ranging from very poor (0-20) to poor (21-40) (Figure 11).

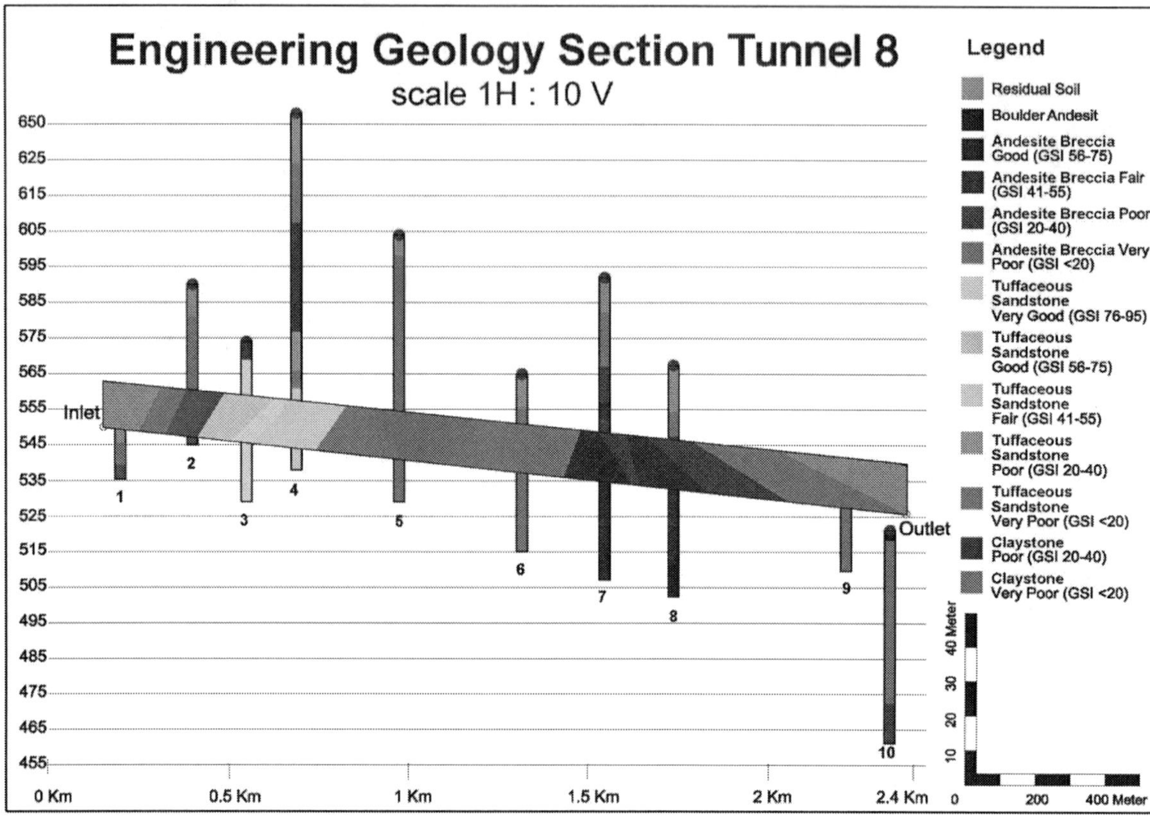

Fig. 11. Map of subsurface rock mass classes based on GSI value.

4.3 Excavation methods and tunnel support system

Rock/soil category is determined based on JSCE (2018)[10] by considering geological conditions and the condition of core drill in the tunnel elevation. It can be seen that both the tunnel elevation and the surface condition of the study area are dominated by andesite breccia and sandstone. The tunnel section of the Mtts Formation area has an RQD value of 40-100; the rock condition is relatively hard, although signs of weathering are visible, and there are no visible discontinuities caused by water. Based on the geological conditions and core drill conditions, the rock/soil category can be determined as class CI (represent tuffaceous sandstone). The tunnel section of the Qob Formation area has an RQD value of 10 or less, and the drill core is in the form of small pieces, but sometimes it is clay or sand mixed with rock fragments. Some of the rocks are fresh, but most of them are weathered or altered quite strongly. The rock/soil category can be defined as class DII (represent andesite breccia) based on these data.

The excavation method that can be used in rock class CI is the full-face method with auxiliary bench cut. Meanwhile, in class DII, the excavation method can be used multi-bench cut method. This excavation method considers the good media condition but interspersed with the bad media in the rock category CI class and considering poor media and requiring a small section of heading to stabilize the dig face in the rock category DII.

Based on JSCE (2018)[10], it can be decided that the suitable support type for rock class CI is shotcrete and rock bolt. The use of invert is also very effective, considering the condition of rocks that have many fractures. Whereas in rock class DII, in addition to using the support type as in-class CII, steel rib is also required to be installed immediately on the excavation face.

Scheme of support system based on JSCE (2018)[10], the length of the excavation rate is 1.5 m in rock category CI. The length of rock bolts is 4,0 m, with a distance at arc 1,2 m and longitudinal 1,5 m mounted on top heading and bench. Shotcrete thickness is 15 cm. The thickness of the lining is 40 cm on the arc and sidewall. The length of the excavation rate is 1,0 m or less in rock category DII. The length of rock bolts is 6,0 m, with a distance at arc 1,0 m and longitudinal 1,0 m or less mounted on top heading and bench. Shotcrete thickness is 25 cm. The lining thickness is 40 cm on the arc and the sidewall with 50 cm on the invert. Steel rib using H-200, both for top heading and bench, with a 1,0 m or less distance. The deformation permit on category DII is 10 cm.

5 Conclusion and recommendation

The lithology of the research area consists of 3 (three) units, namely tuff lapilli, andesite breccia, and tuffaceous sandstone. Weathering conditions of surface rocks vary from slightly weathered to residual soil. Based on the surface mapping, the rock mass classes are divided into very poor tuff lapilli, fair tuff lapilli, very poor andesite breccia, poor andesite breccia, and fair tuffaceous sandstone with GSI value from 0-20 (very poor) to 41-55 (fair). In comparison, subsurface rock mass classes vary from very poor (GSI value: 0-20) to very good (GSI value: 76-95). The body of the tunnel is dominated by andesite breccia and tuffaceous sandstone.

The excavation method used is the full-face method with an auxiliary bench cut (CI) and a multiple-bench cut method (DII). The support system used is shotcrete, rock bolts, and linings for rock category CI. While rock category DII added steel rib and invert. Deformation is only permitted in rock category DII. The inlet and outlet of the tunnel are the parts that need attention. Rock mass classes of inlet and outlet tunnels are very poor, so it is necessary further to analyses the slope stability of inlet and outlet tunnels.

Further study, particularly by numerical method, is required to compare the performance of the tunnel support system proposed in this study to the support system applied in the field, which is based on the China National Standard.

The authors would like to thank the owner of Jakarta – Bandung High-Speed Railway Project PT. Kereta Cepat Indonesia – China (PT. KCIC), High-Speed Railway Contractor Consortium (HSRCC), China Railway Engineering Corporation (CREC) and China Railway Design Corporation (CRDC) for research permit and data access permission. Financial support from The Ministry of Public Works and Housing of Indonesia is gratefully acknowledged.

References

1. Sudjatmiko, *Geological Map of The Cianjur Quadrangle, Jawa* (2003)
2. R.W. van Bemmelen, *The Geology of Indonesia Volume IA General Geology of Indonesia and Adjacent Archipelagoes* (The Hague, 1949)
3. R.V. Fisher, Earth-Sci. Rev. **1**, 287-298 (1966)
4. ISRM, Int. J. Rock Mech. Min. Sci. & Geomech. Abstr. **15**, 319-368 (1978)
5. E. Hoek, *Practical Rock Engineering* (RocScience, 2007)
6. P. Marinos, E. Hoek, *GSI: A geologically friendly tool for rock mass strength estimation*, in ISRM Int Symp (2000)
7. Z.T. Bieniawski, *Engineering rock mass classifications: a complete manual for engineers and geologists in mining, civil, and petroleum engineering* (John Wiley&Sons, New York, 1989)
8. E. Hoek, T.G. Carter, M.S. Diederichs. *Quantification of the Geological Strength Index Chart*, in US Rock Mech/Geomech Symp, **47** (2013)
9. N. Sivakugan, S.K. Shukla, B.M. Das, Rock Mechanics: an introduction (CRC Press, Florida, 2013)
10. JSCE. *Standard Specifications for Tunneling-2016: Mountain Tunnels*. Japan (2018)

https://doi.org/10.1051/e3sconf/202132501004

Engineering geological characterisation of rock masses for design of excavation method and support system of Tanju Tunnel, West Nusa Tenggara, Indonesia

Edden Umaga Dinata[1], *I Gde Budi* Indrawan[1*], and *Arifudin* Idrus[1]

[1]Department of Geological Engineering, Faculty of Engineering, Universitas Gadjah Mada

Abstract. This paper presents experiences with design and construction process of the Tanju Tunnel, West Nusa Tenggara. The objective of this research was to characterize the volcanic and intrusive rock masses for designs of the tunnel excavation method and support system. Engineering geological investigations were carried out, upon which the tunnel stand-up time was estimated and the tunnel excavation method and support system were determined based on the rock mass classifications of RMR (Bieniawski, 1989). The investigation results showed that the tunnel construction area consisted of colluvium, andesite, and tuff breccia. The rock masses were dominated by the andesite, which had UCS of intact rock ranging from 200 to 300 MPa classified as strong intact rock. Based on the RMR value, the andesite was classified as a good quality rock mass. The tunnel had unsupported roof stand-up time approximately 20.000 hours for a 3 m tunnel span. The recommended tunnel excavation method was full face, 1.0-1.5 m advance, and complete support 20 m from face, while the support systems were locally bolt in crown 3 m long and 2.5 m spacing, with occasional wire mesh, 50 mm thick shotcrete in crown where required.

1 Introduction

This research is carried out at the construction site of Tanju Tunnel, which is administratively, located in the village of Bara, Woja, Dompu District, West Nusa Tenggara (Fig. 1). The drainage tunnel is built to drain a 1.9 m^3/s water to fill Tanju Dam. The Tanju Dam has a very large reservoir, but its inflow potential is limited. In order to cover the lack of inflow into the reservoir, a 3 m diameter and 1.6 km long drainage tunnel is needed to convey water from the Rababaka River to Tanju Dam. The most effective and safe excavation method needs to be determined.

Several investigations were carried out in the tunnel construction design stage, including the rock mass characterizations for designs of the tunnel excavation method and support system. This paper presents results of engineering geological investigations to characterize the rock masses at the Tanju Tunnel construction site. Results of the rock mass characterization for designs of the tunnel excavation method and support system are described. The adopted excavation method and support system during the tunnel construction are presented. It is expected that this paper provides a better understanding of drainage tunnel design and construction in tropical regions, particularly in Indonesia.

2 Literature review

Based on the Regional Geological Map, the island of Sumbawa is composed of pyroclastic rocks and sedimentary rocks formed in the Miocene to Quaternary age [1].

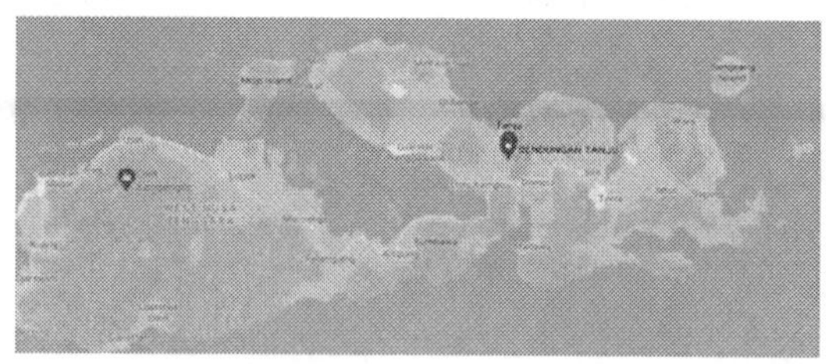

Fig. 1. Research location.

Part of the regional geological map of Sumbawa Island around the Tanju Tunnel construction site is shown in Fig. 2. Based on the regional geological map, the research location is composed of old volcanic rock products (Qvl) of Quaternary age and tuff breccia unit (Tmv) of Tertiary age. The Quaternary rock unit consists of volcanic sedimentary rocks, alluvial deposits and river sediments of the Recent age, while the Tertiary rock unit consists of alternating between breccias, lava and tuffs with a composition of andesite and basalt.

* Corresponding author: igbindrawan@ugm.ac.id

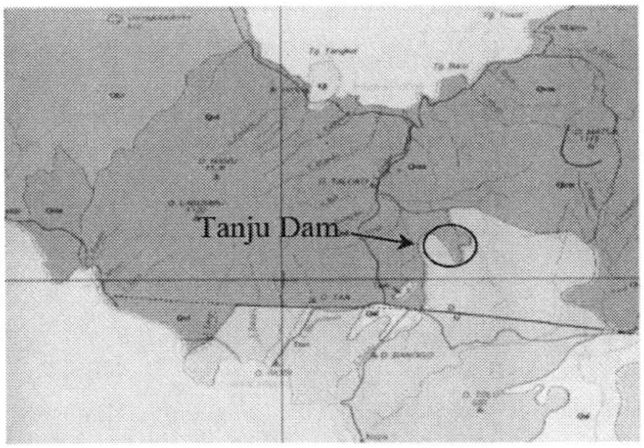

Fig. 2. Regional Geology Map of Sumbawa [1].

Rock masses are generally inhomogeneous, inelastic, discontinuous and anisotropic media [2]. Such properties make their characterization more complex and difficult [2]. Methods commonly used for rock mass classification include the Q index [3], Rock Mass Rating [4], and the Geological Strength Index [4]. These methods are very commonly used in projects, especially tunnels.

Table). This guideline can be applied with both conventional digging and blasting. A number of previous studies [6,7,8,9,10] used the RMR for designs of tunnel excavation method and support system.

3 Research methodology

The research was conducted through surface and subsurface geological mapping and rock mass classification mapping. The naming of volcanic rocks in the study area was carried out based on the classification of Fischer and Table .

The RMR [4] rock mass classification consists of 6 (six) parameters. RMR [4] system incorporates six parameters namely: 1) Uniaxial Compressive Strength (UCS) from intact rock (σc), 2) Rock Quality Designation (RQD), 3) Spacing of discontinuity, 4) Condition of discontinuity surface, 5) Ground water condition, 6) Orientation of discontinuity relative to structure. Value of those six RMR parameters afterward was determined based on the table by Bieniawski [4]. These parameters can be obtained by doing measurements of outcrops in the field or drilling cores. The assessment area must be divided into regions that have the same character, especially seen from the structural characteristics both quality and quantity. Based on the total RMR [4] value obtained from the calculation of the parameters, the rock masses are classified into 5 (five) classes of quality, as shown in Table . Bieniawski (1989) provides a recommendation for tunnel excavation methods and support systems based on the RMR value [4] (
Scmidt [11], while the naming of intrusive rocks was carried out based on the rock classification compiled by Travis [12]. Geological structures were analyzed based on analysis of satellite images and field observations. Determination of RMR parameters were carried out by a face mapping during tunnel construction [4]. The tunnel excavation method and support system were determined following recommendation as shown

Table 1. Rock Mass Quality Class based on RMR [4].

Rock Class	Very Poor	Poor	Fair	Good	Very Good
RMR Rating	< 20	21-40	41-60	61-80	81-100

Table 2. Guide for excavation and tunnel support systems of 10 meters span with the RMR system [4].

Rock Mass Class			**Support System**		
Class	**RMR Value**	**Excavation Method**	**Rock Bolts (d=20mm)**	**Shotcrete**	**Steel Sets**
Very Good	81-100	Full face 3 m advance	Generally, no support required except for occasional spot bolting		
Good	61-80	Full Face 1.0 - 1.5 m advance Complete support 20 m from face	Locally, bolts in crown 3 m long, spaced 2.5 m, with occasional wire mesh	50 mm in crown where required	None

Rock Mass Class			Support System		
Class	**RMR Value**	Excavation Method	**Rock Bolts (d=20mm)**	**Shotcrete**	**Steel Sets**
Fair	41-60	Top Heading and Bench 1.5 - 3 m advance in top heading Commence support after each blast.Complete support 10 m from face	Systematic bolts 4 m long, spaced 1.5-2 m in crown and walls with wire mesh in crown	50-100 mm in crown and 30 mm in sides	None
Poor	21-40	Top Heading and Bench 1.0 - 1.5 m advance in top heading Install support concurrently with excavation 10 m from face	Systematic bolts 4-5 m long, spaced 1-1.5 m in crown and wall with wire mesh	100-150 mm in crown and 100 mm in sides	Light to medium ribs spaced 1.5 m where required
Very Poor	<20	Multiple drifts 0.5 - 1.5 m advance in top heading. Install support concurrently with excavation. Shotcrete as soon as possible after blasting.	Systematic bolts 5-6 m long, spaced 1-1.5 m in crown and wall with wire mesh. Bolts invert	150-200 mm in crown, 150 mm in sides, and 50 mm on face	Medium to heavy ribs spaced 0.75 m with steel lagging and fore polling if required. Close invert

4 Results and discussion

The research area has hilly morphological conditions where the Tanju Tunnel cuts a ridge that extends in a relatively northeast-southwest direction. The ridge is about 2.2 km long and has a 320 m peak elevation. The slopes around the tunnel construction location are quite steep, ranging from 45° to 60°. The inlet of Tanju Tunnel is located at + 124.45 m elevation, while the outlet is located at +124.27 m elevation.

In general, the area of the Tanju Tunnel construction site consists of three rock units, namely colluvium, andesite, and tuff breccia. The andesite and tuff breccia near and at the ground surface are mostly weathered (Fig. 3 to Fig. 6). The colluvium in the form of a mixture of black clay and 2 - 10 cm andesite fragments is widely found on the soil surface, especially on the slopes of the hills downstream of the tunnel. The thickness varies from 2 to 6 m (Fig. 3). Moderately to higly weathered andesite [13] are often converted into soil mixed with 5 - 50 cm andesite fragments, which are abundant on the soil surface in the hilly area around the upstream tunnel.

Fig. 3. Colluvium consists of black clay and 2 – 10 cm andesite fragments.

The fresh to slightly weathered andesite is gray, hard and compact, which is the bedrock of the upstream tunnel. Most of the tunnels from upstream to middle will pass through this rock mass. Laboratory testing carried out on 2 (two) samples, namely the "drilling core" TR-1 at a depth between 61.65 - 62.00 m and 64.5 - 65.0 m, show that the andesite has 200 – 300 MPa uniaxial compressive strength (UCS) value and is classified as very strong to extremely strong rock based on the ISRM (1981) classification.

Fig. 4. Moderately to highly weathered andesite.

Fig. 5. Tuff breccia.

 https://doi.org/10.1051/e3sconf/202132501004

Fig. 6. Fresh to slightly weathered andesite.

Tuff breccia are brownish gray in color and have a relatively abundant of tuff matrix compared to the fragments, which are generally relatively small, about 2 to 5 cm in size. Tuff matrix is grayish brown to whitish, hard and massive. The fragments generally consist of andesite fragments, hard and relatively tapered. Weathered tuff breccia rocks in the form of a mixture of clay and 2 - 50 cm andesite fragments and tuff breccia are abundant in the downstream Tanju Tunnel. It is located below the colluvium to a depth of 21.50 m to more than 25 m below ground level. Due to the weak rock engineering properties, drainage section through this rock mass uses conduit, instead of tunnel, and will be backfilled later (Fig. 5). Engineering geological profile of the rock masses along the Tanju Tunnel is shown in Fig. 7.

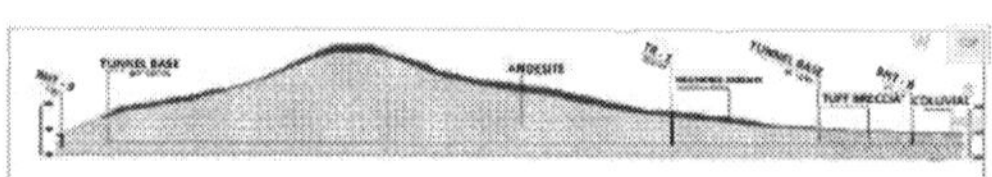

Fig. 7. Engineering geological profile along Tanju Tunnel.

At the tunnel elevation, the fresh to slightly weathered andesite has 90% RQD, less than 50 cm joint spacing, and damp groundwater condition. Based on the ratings of each parameter in Table 2, the andesite rock mass at the tunnel elevation has, consequently, a 65 total RMR value and is classified as good quality. Based on Bieniawski [4], the excavation method recommended for the good quality rock mass is full face, 1.0-1.5 m advance, and complete support 20 meters from face (Table 2).

As the andesite rock masses have 65 RMR value and the tunnel has 3 m roof span, Fig. 8 shows that the stand-up time for the unsupported tunnel during excavation will be 20.000 hours (27 months 13 days). Based on geological observations on the ground surface or through the results of existing drilling, there were no indications of weak zones, such as fracture, along the tunnel route. In addition, the andesite rock masses also have high UCS value of intact rock and good rock mass quality.

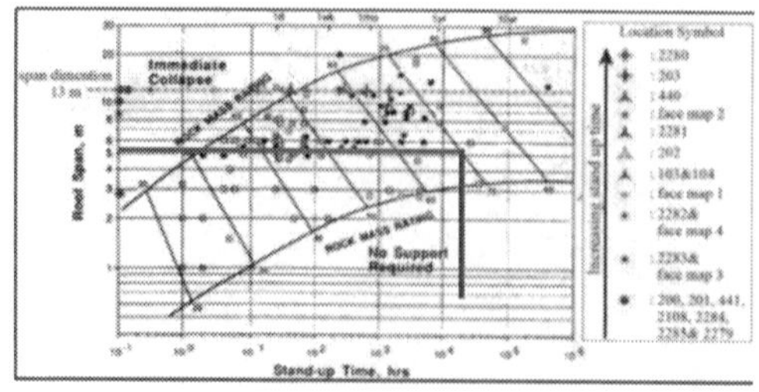

Fig. 8. Stand-up time of the excavated tunnel.

To facilitate a full face excavation method, as recommended by the RMR value, the rock masses were excavated by drilling and blasting method in the construction process (Fig. 9a, 9b, 9c). The typical rock mass condition after blasting is shown in Fig. 10. The long tunnel stand-up time predicted in Fig. 8 was confirmed during relative long drilling and blasting processes of the tunnel construction. As also recommended by the RMR value, the support system selected for the tunnel was wire mesh and shotcrete with locally bolt in crown (Fig. 11).

Fig. 9a. Tunnel excavation: Drilling.

Fig. 9b. Tunnel excavation: Charging.

Fig. 9c. Tunnel excavation: Blasting.

Fig. 10. Typical rock mass condition after blasting.

Fig. 11. Rock bolt, wire mesh, and shotcrete installations.

5 Conclusion and recommendation

The tunnel construction area consisted of colluvium, andesite, and tuff breccia. The rock masses surrounding the tunnel were dominated by the andesite, which had uniaxial compressive strength (UCS) of intact rock ranging from 200 to 300 MPa and, therefore, classified as strong to extremely strong intact rock. Based on the RMR value, the andesite was classified as good quality rock mass. Drilling and blasting is the most effective method to do the excavation based on the rock mass quality. The recommended tunnel excavation method based on the RMR value was full face, 1.0-1.5 m advance, and complete support 20 m from face, while the recommended tunnel support systems were locally bolt in crown 3 m long and 2.5 m spacing, with occasional wire mesh, 50 mm thick shotcrete in crown where required [4]. The tunnel had unsupported roof stand-up time approximately 20.000 hours for a 3 m tunnel span.

Further research is necessary to analyse stability of the Tanju Tunnel, such as by numerical method, to obtain better understanding the rock mass and support system interaction.

The authors would like to thank BWS Nusa Tenggara 1, Supervision Consultant, and PT. Nindya Karya for the permission to this study and to publish the results.

References

1. A. Sudrajat, S. Andi Mangga, N. Suwarna, Peta Geologi Lembar Sumbawa, Nusatenggara (1998)
2. F. Ceballos, C. Olalla, R. Jimenez, Relationship between RMRb and GSI based on in situ data (Eurock, Spain, 2014)
3. N. Barton, R. Lien, J. Lunde, Rock Mech. **6**, 4, 189-236 (1974)
4. Z. T. Bieniawski, Engineering rock mass classifications: a complete manual for engineers and geologists in mining, civil, and petroleum engineering (John Wiley & Sons, New York 1989)
5. E. Hoek, ISRM News Journal **2**, 4-16 (1994)
6. J. Ranasooriya, The reliability of rock mass classification systems as underground excavation support design tools (Doctoral dissertation, Curtin University, 2009)
7. A.R. Lowson and Z.T. Bieniawski, *Critical Assessment of RMR based tunnel design practices: a practical engineer's approach,* in Rapid Excavation & Tunneling Conference, Washington, DC, 23-26 (2013)
8. A. Rahmati, L.Faramarzi, M. Sanei, Frontiers of Structural and Civil Engineering, **8**, 4, 448-455 (2014).
9. V. Santos, P. F. Da Silva, M.G. Brito, Minerals **8**, 3, 78 (2018)
10. M.S. Akram, M. Zeeshan, M. Haroon, K. Mirza, Open Journal of Geology **9**, 11, 809-828 (2019)
11. J. McPhie, M. Doyle, R. Allen, Volcanic textures: a guide to the interpretation of textures in volcanic rocks (Tasmanian Government Printing Office, Tasmania, 1993)
12. R.B. Travis, *Classification of Rocks.* Colorado School of Mines, **50** (1955)

13. ISRM, *The complete ISRM suggested methods for rock characterization, Testing and Monitoring, Ankara*. ISRM Turkish National Group (1981)

Engineering geological investigations for evaluation of excavation method and stand-up time of the DK99-DK100 Jakarta – Bandung high-speed railway tunnel, Indonesia

Linda Ali[1,2], *I Gde* Budi Indrawan[1] , and *Hendarto* Hendarto[2]

[1]Gadjah Mada University, Geological Engineering Department, 55281 Yogyakarta, Indonesia
[2]Ministry of Public Works and Housing of Indonesia

Abstract. This paper presents the investigation of surface geology and subsurface engineering geology to analyze the excavation method and stand-up time of the DK99-DK100 Jakarta-Bandung high-speed railway Tunnel, Indonesia. Rock mass quality, tunnel excavation method, and stand-up time determined using Geological Strength Index (GSI), Basic Quality (BQ) systems, converted to Rock Mass Rating (RMR) and The Japan Society of Civil Engineering (JSCE) for comparison. The result shows that the study area consists of slightly to completely weathered andesite breccia and slightly weathered andesite lava. The rock masses at the tunnel elevation had very poor to poor quality and were associated with high weathering degrees. The recommended rock excavation method based on the GSI is digging. The recommended tunnel excavation method based on RMR is multiple drifts, top heading, and bench, while based on JSCE is bench cut method. The tunnel stand-up time is 30 minutes - 2 hours based on the RMR, while it is predicted to be unstable without support based on the BQ. The recommended design is expected to be applied effectively according to the geological conditions. It is expected to understand better the tunnel excavation method in poor rock masses, especially in Indonesia.

1 Introduction

The construction of the Jakarta-Bandung fast train is one of the national strategic projects to reduce congestion problems, air pollution, and travel time. The journey has taken a long time because of traffic jams and hilly areas shortened by using tunnel technology, and there are 12 tunnels, one of which is the DK 99-DK 100 tunnel with a tunnel length of 208 m (DK 99+820 to DK 100+050).

Design of the tunnel cut through mountainous areas consisting of various types of rocks layers and the influence of geological structure conditions. The condition of rock quality in each rock layer will affect what kind of tunnel excavation method is effectively applied. Therefore, it is necessary to conduct engineering geological investigations to obtain information on soil and rock mass characteristics, geological structures, and groundwater that may affect the construction site. The surface geological engineering mapping scale used is 1:4000 because, based on the level of accuracy and purpose, 1:5,000 to 1:25,000 for construction work, including excavation and construction materials [1].

An essential aspect in engineering geology is the condition of the rock mass at the site to be built. Engineering geological investigations of the rock mass are carried out on the surface and the subsurface and should be carried out at the beginning and during construction work. It is essential to ensure that field conditions are under the initial design and adjustments are made when necessary [2].

Rock mass quality analysis has been classified as surrounding rock class VI by the tunnel designer. However, based on the high uncertainty that may arise in rock masses classification, other methods' analysis needs to be compared. In this study, rock mass classification was carried out using Geological Strength Index (GSI) and Basic Quality (BQ) System, which is juxtaposed with the Rock Mass Rating (RMR) as mutually reinforcing indicators. The authors choose The GSI classification because it is easy to use in field observations and poor rock conditions due to prominent tectonic influences and high weathering rates suitable for use in research locations dominated by volcanic rock affected by tectonic processes. The excavation method is also determined by the empirical approach based on the Japanese classification Society of Civil Engineers (JSCE). The expected results of the excavation method and stand-up time in the DK 99 – DK 100 tunnel design are appropriate according to the geological conditions found in the field.

The research area is administratively located in Nyalindung Village, District Cipatat, West Bandung Regency, West Java Province, Indonesia; the location can be seen in Figure 1.

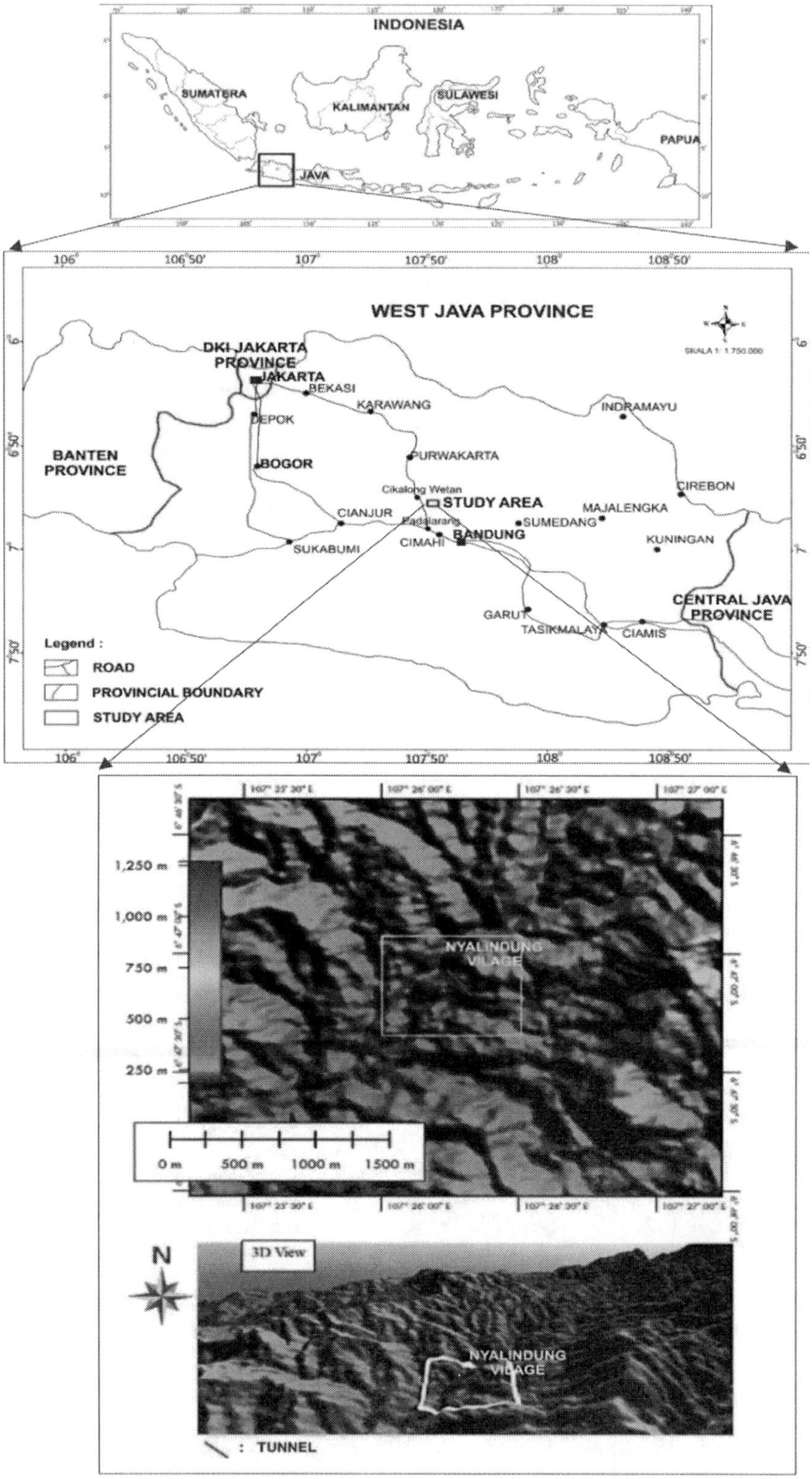

Fig. 1. Research area in West Java Province, Indonesia.

2 Methodology

The research includes surface geological mapping and subsurface engineering geological mapping. The surface geological mapping was carried out by megascopic observations of the rock outcrops in the field. The subsurface engineering geological mapping is carried out by tunnel face mapping and observing the drilling and correlating rock mass layers to develop a geological engineering profile. Rock mass weathering degree was determined based on the International Society for Rocks Mechanics, ISRM method [3].

Classifications of subsurface rock masses were carried out based on the GSI, BQ-System, and JSCE. The GSI determination was based on the rock description in the face mapping using GSI basic chart by Marinos & Hoek (2000), and core drill along the tunnel alignment with an assessment of the very poor rock masses using the GSI Index [4] by considering Broken Structural Domain (BSTR) and the Joint Condition Index (IJc) by the equation (1)

$$Ijc = (bs)+(W/2)+4 \quad (1)$$

Ijc is Joint Condition Index, bs is the value of Broken Structural Domain (BSTR), and W is the filling and weathering condition of the core material.

The GSI measurements at the tunnel elevation from face mapping and core drill analysis were converted into RMR by the equation (2) [4] about correlation GSI and RMR for poor and very poor rock mass [5].

$$RMR = 20In\ (GSI / 6) \quad (2)$$

The rock mass classifications based on BQ-System using two parameters: Rc and Kv, [6], [7], [8] and can be calculated based on the formula [8] as shown in equation (3) and (4) for BQ-Correction

$$BQ = 90 + 3Rc + 250Kv \quad (3)$$
$$[BQ] = BQ - 100\ (K_1 + K_2 + K_3) \quad (4)$$

Rc: Rock Strength (MPa) and Kv: Rock Mass Integrity (Vpm, km/s) from the elastic longitudinal wave velocity of rock. K_1: correction coefficient for the influence of underground water, K_2: the correction coefficient for the influence of the orientation of the main weak structural planes, and K_3: the correction coefficient for the influence of the initial stress state. The BQ system values are converted into RMR by equation (5) [6].

$$RMR = (BQ - 80,79)/(6,09) \quad (5)$$

Rock class/ category by the JSCE rock mass classification [9] requires a value of competence factor, determined by equation 6.

$$Competence\ Factor = qu / (\gamma.H) \quad (6)$$

qu: Unconfined compressive strength of the ground (kN/m^2), γ: unit weight of the ground (kN/m^3), and H: depth of cover (m).

The tunnel excavation method was determined [10] In addition to the RMR, the tunnel excavation method was also determined based on the Japan Society of Civil Engineering [9] for comparison. The rock excavation method is determined based on GSI value [11].

The determination of the stand-up time or the length of time the tunnel can last before the support system is installed uses two methods. The first method is based on the stand-up time chart developed [10] using the RMR correlation value from the GSI measurements. The second way is to use the stand-up time developed [8] based on the value of BQ.

3 Results and discussions

3.1 Surface geological mapping

Based on the regional geological map Sheet of Cianjur [12], the research area is influenced by the Older Volcanic Product Formation, consisting of breccia lahar, lava, volcanic breccia, flow breccia, and the composition is between andesite and basalt. The 1:4000 scale geological mapping showed the area formed by the andesite lava and andesite breccia rock unit. The breccia consists of andesite fragments from boulder to block size. The Geological map and profile are shown in Figure 2.

3.2 Subsurface engineering geological mapping

Analysis of core drilling and correlation rock mass layer shows the tunnel path consists of four geological engineering units, namely slightly to completely weathered andesite breccias, as shown in Figure. 3.

Based on the tunnel subsurface engineering geological profile in Figure. 3. The tunnel elevation is located in two engineering geological rock units: completely weathered andesite breccia and highly weathered andesite breccia rock unit

The completely weathered andesite breccia, indicating that all the material has been wrecked and has turned into the soil; however, the rock mass structure can still be seen. The highly weathered andesite breccia rock unit, indicating more than half of the rock is crushed or turned into soil. Engineering geological characteristics of rock mass form of crushed stone, breccia as powder and soil and the specific weight are 15 - 17 kN/m^3.

 https://doi.org/10.1051/e3sconf/202132501005

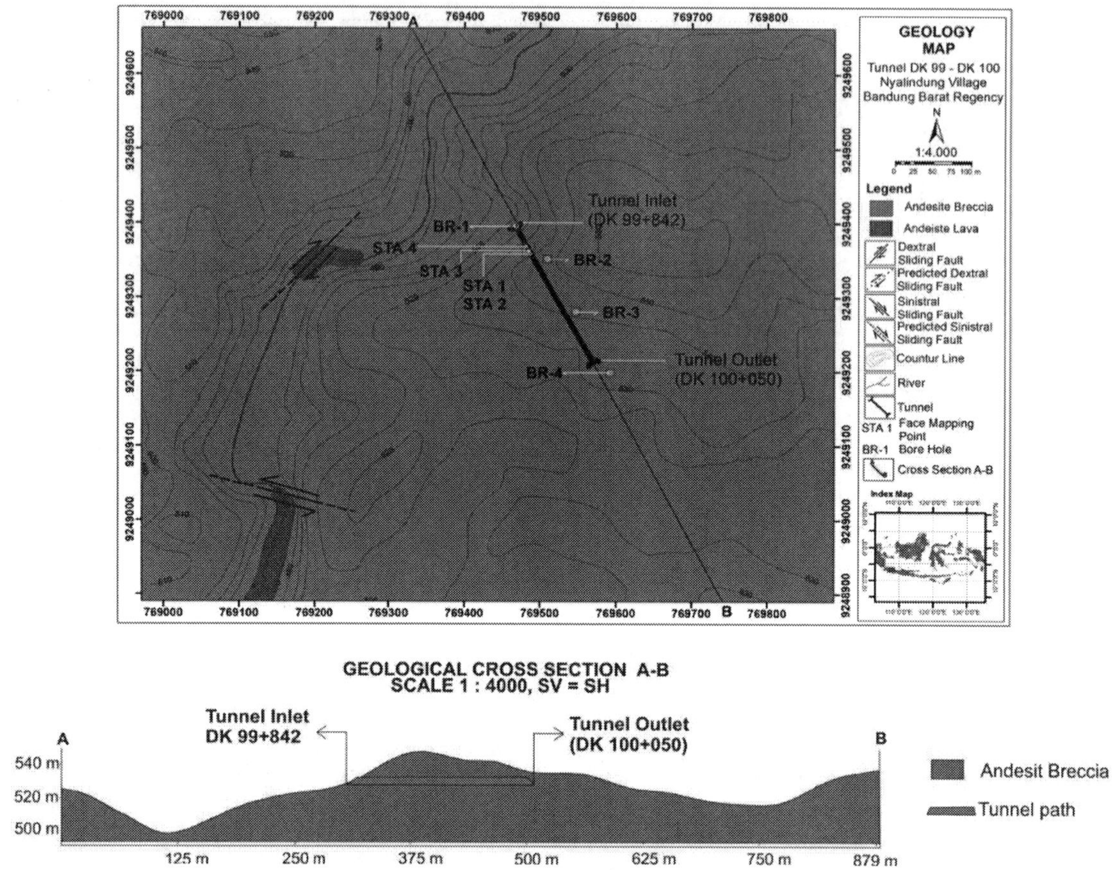

Fig. 2. Geological map and cross-section.

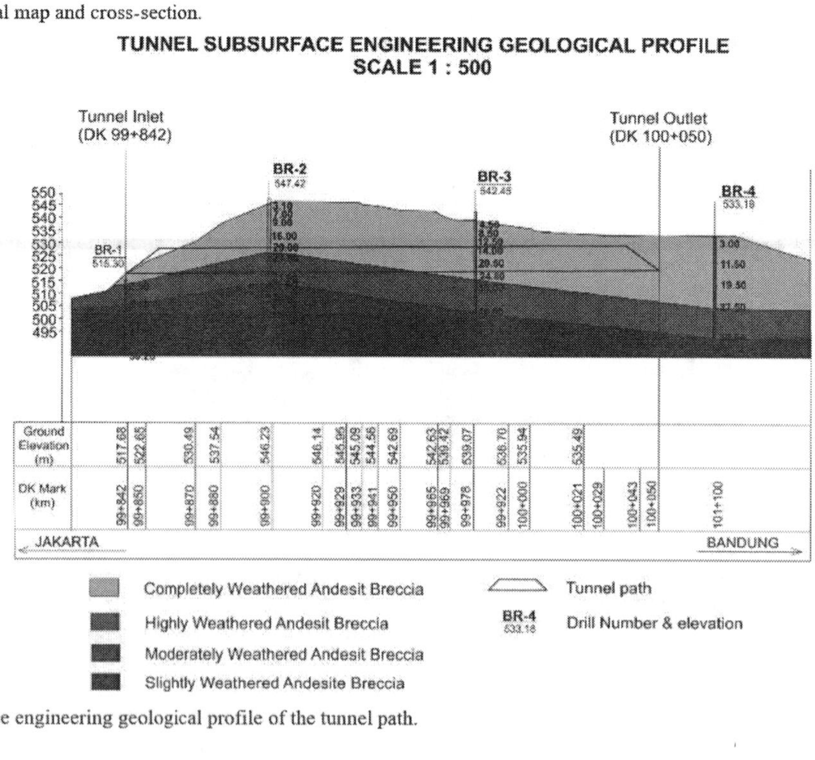

Fig. 3. Subsurface engineering geological profile of the tunnel path.

3.2.1 Geological strength index (GSI) classification

Rock mass characterization at the face of the tunnel by face mapping along the tunnel DK99+879, DK 99+878, 99+876, dan 99+869 as many as four observation points: STA 1 at DK 99+879, STA 2 at DK 99+878, STA 3 at DK 99+876, and STA 4 at DK 99+869. The GSI measurements are plotted on GSI basic chart [13] shown in Figure.4.

GSI values at the tunnel elevation were converted into RMR [4] about correlation GSI and RMR for poor and very poor rock mass are shown in Table 1 and Table 2.

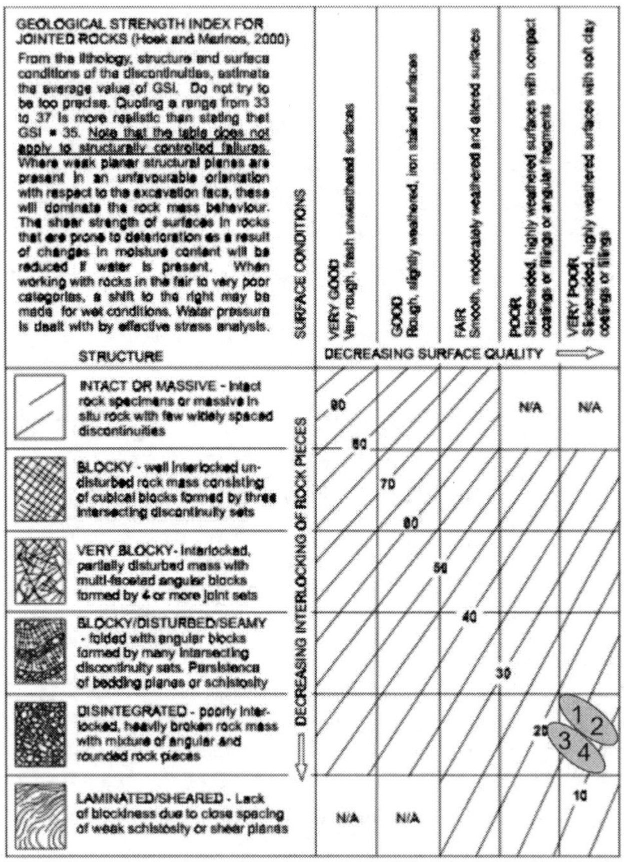

Fig. 4. GSI estimates from the geological face mapping observation on GSI chart [13].

Table 1. GSI and RMR result from face mapping at the tunnel elevation

No	STA Face Mapping	GSI	Rock Mass Class (Sivakaguan et.al 2013)	RMR	Rock Mass Class (Bieniawski, 1989)
1	STA 1	10-15	Very Poor	10.2-18.3	V (<20) Very Poor
2	STA 2	10-15	Very Poor	10.2-18.3	V (<20) Very Poor
3	STA 3	10-15	Very Poor	10.2-18.3	V (<20) Very Poor
4	STA 4	10-15	Very Poor	10.2-18.3	V (<20) Very Poor

Rock mass characterization of the core drill analysis consists of 4 points, namely BR 1 at DK 99+841.98, BR 2 at DK 99+898.23, BR 3 at DK 99+978.21, and BR 3 at DK 100+071.42.

The range of GSI measurements using the GSI Index for very poor and poor rock charts [4] on core analysis in BR-2 at tunnel elevation by the equation (2) shows the GSI range from 11 to 26 as in Figure. 5.

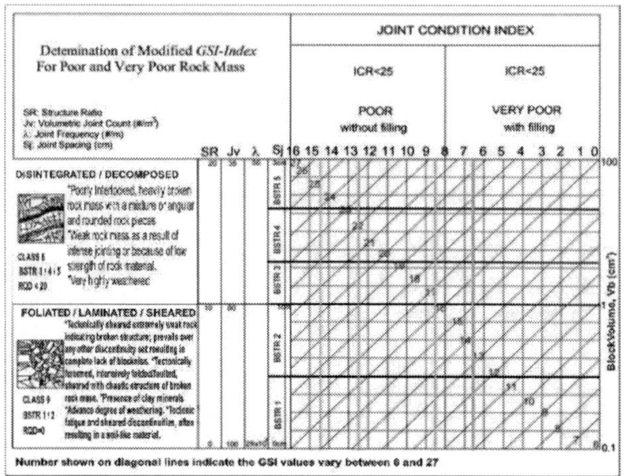

Fig. 5. GSI estimates from the core drill analysis by GSI Index for poor and very poor rock mass from [4].

Table 2. GSI and RMR result from core drill evaluation at the tunnel elevation

No	Borehole	Depth (m)	BSTR	Ijc	GSI	RMR	Rock Mass Class
1	BR-1	0-2.5	R0 Clayey Rock		0	0	V (<20) Very Poor
			BSTR 1	6.5	11-12	12.1- 13.9	
		20-25	BSTR 2	8.5	14-16	16.9- 19.6	
			BSTR 4	12.5	21-22	25.1- 26.0	
2	BR-2		BSTR 3	10.5	17-18	20.8- 22	IV (21 - 40) Poor
		26-30	BSTR 4	12.5	21-22	25.1- 26.0	
			BSTR 5	14.5	24-26	27.7- 29.3	
3	BR-3	13-24	R0 Clayey Rock		0	0	V (<20) Very Poor
4	BR-4	3-14	R0 Clayey Rock		0	0	

Analysis of core drilling and correlation rock mass quality shows the tunnel path consists of three rock mass qualities from very poor, poor, and fair, as shown in Figure 6.

Based on the tunnel subsurface rock mass quality profile in Figure 6. The tunnel elevation is located in Very poor and poor rock mass quality.

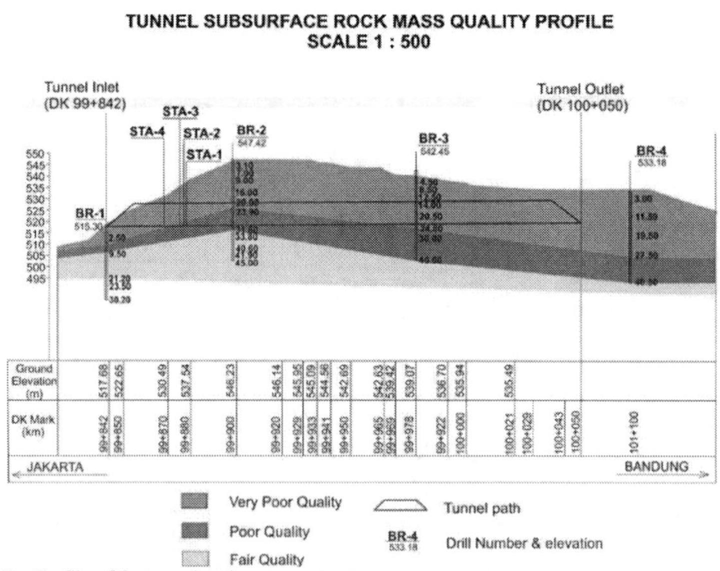

Fig. 6. Rock Mass Quality Profile of the tunnel path.

3.2.2 Basic Quality (BQ-System) classification

The BQ system values are converted into RMR and the results are shown in Table 3. In determining the rock class based on the BQ value, the author uses references from the Professional Standard by the Ministry of Railway of the People's Republic of China in the Code for Design on Tunnel of Railway, which has accommodated weak rock conditions.

Table 3. BQ and RMR result.

No	STA, Borehole & Depth (m)	BQ	Rock Class BQ Index (Feng & Hudson, 2011)	RMR	Rock Mass Class (Bieniawski, 1989)
1	STA 1	137-187	V (<250)	9.2-17.4	V (<20) Very Poor
2	STA 2	137-187	V (<250)	9.2-17.4	V (<20) Very Poor
3	STA 3	137-187	V (<250)	9.2-17.4	V (<20) Very Poor
4	STA 4	137-187	V (<250)	9.2-17.4	V (<20) Very Poor
5	BR-1 (0-2.5)	115-143	V (<250)	5.6-10.2	V (<20) Very Poor
6	BR-2 (20-30)	149-211	V (<250)	11.2-21.4	IV (21-40) Poor
7	BR-3 (13-24)	115-143	V (<250)	5.6-10.2	V (<20) Very Poor
8	BR-4 (3-14)	115-143	V (<250)	5.6-10.2	V (<20) Very Poor

3.2.3 Japan Society of Civil Engineering (JSCE, 2007) classification

Rock class/ category by the JSCE classification [3] requires a value of competence factor and the results are shown in Table 4.

In determining the rock class based on the JSCE, the author uses JSCE ground classification [9], which has been adopted as Guidelines for Excavation Planning Methods and Road Tunnel Strengthening System in Mixed Soil – Rock Media by the Ministry of Public Works and Public Housing of Indonesia [14]. Rock mass class E is classified with poor rock condition, and DII shows the rock with strong weathering, cores form of small pieces, sometimes form of clay, and with mixed rock fragment.

Table 4. JSCE Rock Class/Category result.

No	STA & Borehole	Depth (m)	Competence Factor	Rock Class/ Category
		8.5	0.65	E
1	STA 1	16	1.12	DII
		18.5	1.31	DII
		7.5	0.53	E
2	STA 2	16	1.27	DII
		17.5	1.48	DII
		6	0.53	E
3	STA 3	9	0.93	E
		16	1.59	DII
		2.5	0.73	E
4	STA 4	7	0.58	E
		12.5	0.73	E
5	BR-1		Under the tunnel elevation	
		18	0.98	E
6	BR-2	23.9	1.91	DII
		28	1.64	DII
		11.5	0.24	E
7	BR-3	20.5	0.84	E
		21.5	0.83	E
8	BR-4		Outside the tunnel elevation	

3.3 Excavation method

3.3.1 Rock excavation method

The rock excavation method based on the GSI value is shown in Figure. 7 and Table 5. GSI value and point load index are required to use this classification [11]. Based on the classification chart Is50 > 3 Mpa, and the Value of GSI 10 – 26, the result shows the rock excavation method from tunnel outlet until tunnel inlet is Digging.

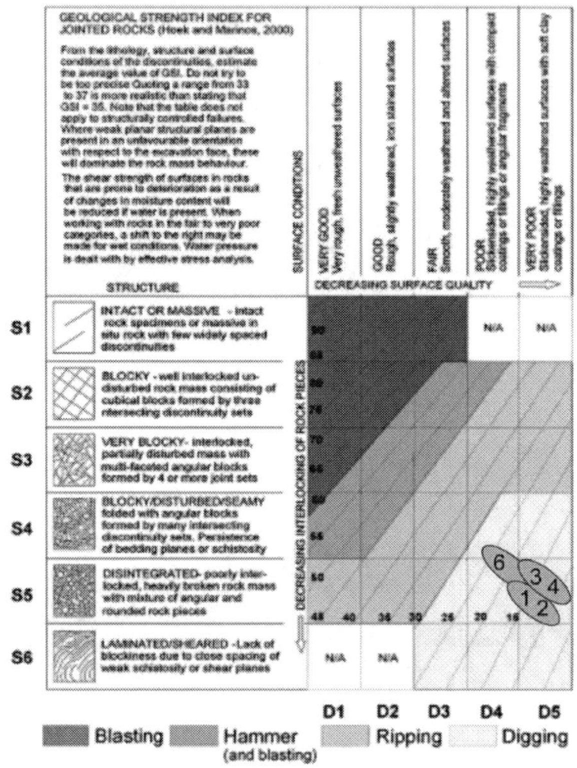

Fig. 7. Plotting GSI value in GSI rock excavation chart [11].

Table 5. Rock Excavation Method based on GSI value.

No	STA & Borehole (DK)	Rock Type	GSI	Rock Excavation Method (Tsiambos &Sroglou, 2009)
1	STA 1 (DK 99+879)		10-15	Digging
2	STA 2 (DK 99+878)		10-15	Digging
3	STA 3 (DK 99+876)		10-15	Digging
4	STA 4 (DK 99+869)	Andesite Breccia Completely Weathered	10-15	Digging
5	BR-1 (DK99+841.98)		Under the tunnel elevation	
6	BR-2 (DK99+898.23)		10-22 17-26	Digging Digging
7	BR-3 (DK99+878.21)		R0 (Clayey Rock)	Digging
8	BR-4 (DK100+071.42)		Outside the tunnel track	

3.3.2 Tunnel excavation method

Tunnel excavation methods [9,10] shown in Table 6. The recommended tunnel excavation method is divided into two sections based on rock mass quality at the tunnel path. Section one: DK 99+842 – DK 99+875 and DK 99+965 – DK 100+050 based on RMR is multiple drifts 0,5-1,5m advance in the top heading. Section two: DK 99+875 – DK 99+965 is top heading and bench, advancing 1,0-1,5m on top heading. Based on JSCE, multiple benches cut method from tunnel outlet (DK 100+050) until tunnel inlet (DK 99+842) as shown in Figure. 8. Multiple benches cut methods are shown in Figure. 9.

Table 6. Excavation method based on RMR and JSCE rock mass classifications.

No	STA & Borehole DK (km)	Excavation Method	
		JSCE	RMR
1	STA 1 (DK 99+879)	Bench Cut Method (Multiple Bench Cut Method)	Multiple drifts 0,5 - 1,5m advance in the top heading. Install support concurrently with excavation
2	STA 2 (DK 99+878)		
3	STA 3 (DK 99+876)		
4	STA 4 (DK 99+869)		
5	BR-1 (DK99+841.98)	Under the tunnel elevation	
6	BR-2 (DK99+898.23)	Bench Cut Method (Multiple Bench Cut Method)	Top heading and bench 1,0 – 1,5m advance in the top heading. Install support currently with excavation 10m from the face
7	BR-3 (DK99+878.21)	Bench Cut Method (Multiple Bench Cut Method)	Multiple drifts 0,5 - 1,5m advance in the top heading. Install support concurrently with excavation
8	BR-4 (DK100+071.42)	Outside the tunnel track	

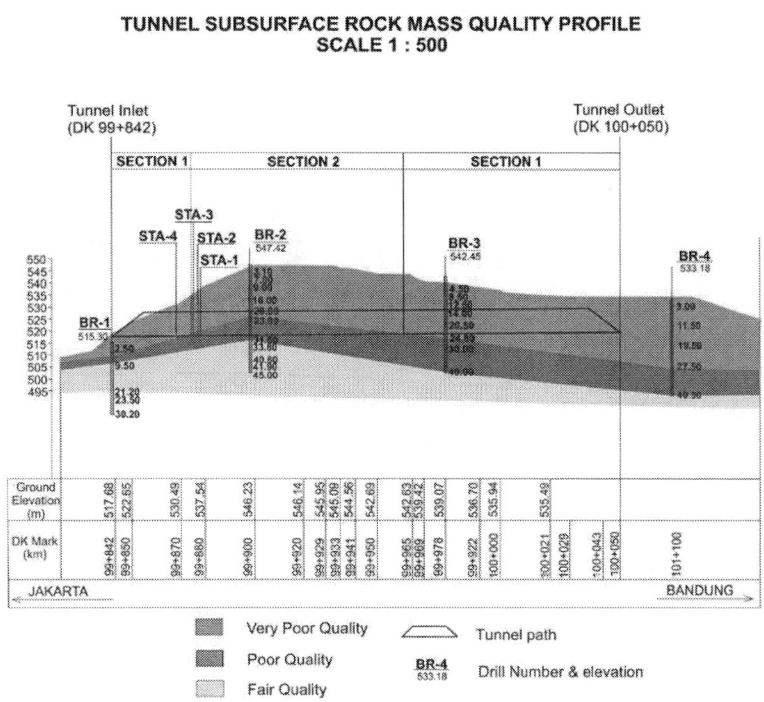

Fig. 8. The Section is divided based on rock mass quality at the tunnel elevation.

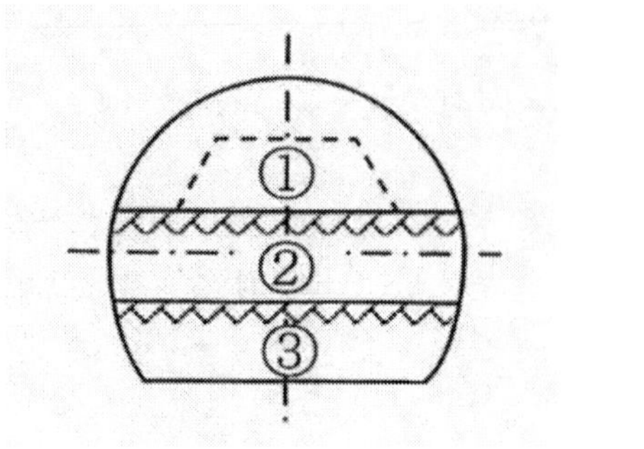

Fig. 9. Bench cut method (Multiple Bench Cut Method) [9] rock class/ category E and DII

The advantage of the Bench cut method / Multiple benches cut method (double bench) excavation method is that the face of the excavation area can be easy to stabilize.

3.4 Stand-up time

Stand-up time or time that a tunnel will stand unsupported after excavation was determined based on RMR value with unsupported roof span [10] as shown in Figure 10. Mark "A" show unsupported roof span is the width of the tunnel: 12,6m, while Mark "B" shows unsupported roof span is the distance from the face to the support by face mapping data: 1,2m and Stand-up time based on BQ [8] as shown in Table 7 and 8.

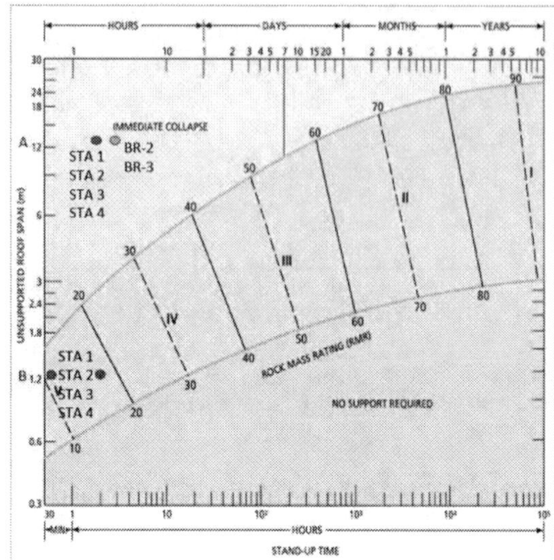

Fig. 10. Plotting RMR Value and Unsupported roof span [10]

Table 7. Tunnel stand-up time based on face mapping data.

No	STA Face Mapping	Stand-up Time		
		RMR with a span of the 12,6 meters	RMR with a span of the 1,2 meters from face mapping data	BQ System
1	STA 1 (DK 99+879)	Immediate Collapse	30 min - 2 hours	Class V (Not Stable without support or reinforcement)
2	STA 2 (DK 99+878)			
3	STA 3 (DK 99+876)			
4	STA 4 (DK 99+869)			

Table 8. Tunnel stand-up time based on core drill evaluation.

No	Borehole	Stand-up Time	
		RMR with a span of the12,6 meters	BQ System
1	BR-2 (DK99+898.23)	Immediate Collapse	Class V (Not Stable without support or reinforcement)
2	BR-3 (DK99+878.21)		

The controlling factor that influences the engineering geology conditions in the study area is the rock weathering degree. The geological control of the rock mass quality is the structure of the rock mass and the level of weathering, which plays an important role.

Very poor and poor rock mass quality in the tunnel path because the rock condition has undergone complete weathering. All rock material is crushed and turned into soil, but the rock mass structure can be seen; the rock mass structure shows disintegrated and poorly interlocked.

Very poor and poor rock mass quality in the research site located at a higher elevation or in hilly areas, at the same time, the decreasing depth indicates the rock mass quality is starting to improve with moderate quality structural conditions tend to be massive and slightly weathered, found in areas with elevation relatively low or flat and in areas along the river.

Another consideration is the research area regionally consisting of quaternary deposits and volcanic debris. The thickness of this quarterly deposit in regional formations [12] is estimated to be around 150 m. The tectonic setting of the West Java region can also be considered where there is an active fault to the south of the study area, causing the area to become a weak zone that affects the quality of rock mass.

The tunnel position is in poor to very poor rock mass quality, and it would be ideal if the tunnel is in a rock layer with very good quality. Therefore, it is necessary to pay attention to the calculation of tunnel elevation, design water discharge, overburden thickness, and load above the tunnel.

4 Conclusion

The study area consists of two rock units: andesite breccia and andesite lava rock unit. Rock weathering degree consists of slightly to completely weathered andesite breccia and slightly weathered andesite lava. The tunnel path is located at an andesite breccia rock unit with a completely weathered degree. The rock masses quality at the tunnel elevation consists of: very poor quality (GSI<20, RMR <20 Class V, BQ-Index<250, JSCE Class DII & E) and poor quality (GSI: 21-40, RMR: 21-40, BQ-Index<250, JSCE Class DII & E) controlled by the weathering degree of the rock. The tunnel position is in poor to very poor rock mass quality; under these conditions, the position of the tunnel can still be maintained by paying attention to the calculation of tunnel elevation, design water discharge, overburden thickness, and load above the tunnel.

The recommended rock excavation method based on GSI for DK99+842–DK100+050 is ripping. Recommended tunnel excavation method Based on RMR can be divided into two sections: section one on DK 99+842 – DK 99+875 and DK 99+965 – DK 100+050 is multiple drifts 0.5-1.5m in front of top heading, and section two on DK 99+875 – DK 99+965 is top heading and bench, advancing 1.0 – 1.5m on top heading, while based on JSCE it is multiple benches cut method from Tunnel Outlet (DK 100+050) – tunnel inlet (DK 99+842). Both excavation methods divide the excavated face into several benches to make it easy to carry out the necessary stabilization.

The tunnel is predicted to be unstable without support based on the BQ, while the tunnel stand-up time is 30 minutes – 2 hours based on the RMR. The RMR and BQ values in determining stand-up time show recommendations that are not much different. The stand-up time is short, and it is recommended to install the support system immediately after being excavated, so the correlation equation used is considered appropriate.

The authors would express gratitude to PT. Kereta Cepat Indoensia China (KCIC), PT. China Railway Engineering Corporation (CREC) and PT. China Railway Design Corporation (CRDC) for research permit and data access permission. The first author would also like to thank The Ministry of Public Works and Housing of Indonesia for the scholarship for the Master Study Program.

References

1. G. De Vallejo, & M. Ferrer, Geological Engineering (CRC Press, 2011).
2. Sugalang, Engineering geology guide (Ministry of Energy and Mineral Resources, Center for Groundwater Resources and Environmental Geology, Geological Agency, 2016)
3. ISRM, Int. J. Rock Mech, Sci.& Geomech. **15**(6), 319-368 (1978)
4. R. Osgoui and E. Unal, *Characterization of Weak Rock Masses Using GSI-Index and the Estimation of Support-Pressure*, in the - 40th US Rock Mechanics Symposium (2005)
5. Zhang Q Huang X, Zhu H, Li J 2019, Tunn. Undergr. Sp. Tech. **83,** 73-81 (2019)
6. M. Wang, Z. Wang, J. Tong, X. Zhang, Y. Dong, D. Liu, Jurnal of Central South University **28**(1), 247-263 (2021)
7. Ministry of Railways of the People's Republic of china, *Code for Design on Tunnel of Railway,* in Professional Standard of the people's republic of China, TB10003- p 100, (2005)
8. X. Feng, J. Hudson, Rock Engineering Design, (CRC Press, 2011)
9. Japan Society of Civil Engineering (JSCE), Standard Specification for Tunneling-2006: Mountain Tunnels. Tokyo, (2007)
10. Z. Bieniawski, Engineering rock mass classifications: A complete manual for engineers and geologist in mining, civil, and petroleum engineering (John Wiley and Sons, New York, 1989)
11. Tsiambaos, & Saroglau, Bull. Eng. Geol. Environ. **69(1)**, 13-27 (2009)
12. Sudjatmiko, Geological Map Of The Cianjur Quadrangle, Jawa (2003)
13. P. Marinos, E. Hoek, *GSI: a Geologically Friendly tool for rock mass strength estimation,* in Proceeding International Conference on Geotechnical & Geological Engineering, GeoEng2000, 1422-1442 (2000)
14. Ministry of Public Works and Public Housing, Guidelines for Excavation Planning Methods and Road Tunnel Strengthening Systems in Mixed Soil - Rock Media, 30/SE/M, (2015)
15. N. Sivakaguan, S. Shukla, B.M. Das, Rock Mechanics: An Introduction (CRC Press, 2013)

Engineering geological mapping for empirical design evaluation of Tunnel 10 of Jakarta - Bandung high-speed railway, Indonesia

Hanifah Hilda Herdiana[1, 2], *I Gde* Budi Indrawan[1*], and *Hendy* Setiawan[1]

[1]Department of Geological Engineering, Faculty of Engineering, Universitas Gadjah Mada, Yogyakarta, Indonesia
[2]Ministry of Public Works and Housing of Republic of Indonesia, Jakarta, Indonesia

Abstract. An engineering geological mapping was carried out at the construction site of the Tunnel 10 of Jakarta - Bandung High-Speed Railway to obtain data and information of the engineering geological conditions, particularly the rock masses. This research aims to determine the rock mass classes at the tunnel construction site and recommend the tunnel support system based on the Rock Mass Rating (RMR) and the Japan Society of Civil Engineers (JSCE) systems. This research is expected to better understand the rock mass classes, which were previously determined based on the newly applied Basic Quality (BQ) system for the tunnel support empirical design. The results showed that the research area consisted of young volcanic products, namely moderate to highly weathered tuff breccia and andesitic breccia. The uniaxial Compressive Strength (UCS) of rock mass varies between 1-25 MPa. The RMR value ranges from 21 to 40, indicating disintegrated and poor rock mass quality. The proposed tunnel support system is the combination of shotcrete, steel support for top heading and bench support, arch sidewall, and invert concrete.

1 Introduction

The Tunnel 10 of Jakarta – Bandung High-Speed Railway with 1230 meters of length and a maximum buried depth of 37 meters is located in Ngamprah District, West Bandung Regency, West Java Province of Indonesia. The tunnel is in the construction process and has three construction accesses from the entrance at DK 107+070, the center of the tunnel (inclined shaft) at DK 107+685, and the tunnel exit at DK 108+300. Tunnel excavation is carried out using the New Austrian Tunneling Method (NATM) with the three-bench excavation method and temporary transverse reinforcement during excavation [1].

According to the Map of Earthquake Hazard of Indonesia and Digital Elevation Model (DEM) generated from satellite imagery, Tunnel 10 is located between 2 (two) active structures, namely the Cimandiri fault and the Lembang fault, as well as the hilly topography. Cimandiri fault is left-lateral reverse fault with 0.4 – 1.0 mm/ year geodetic slip rate [2, 3] which located 2 kilometers southwest from the study area. Lembang fault is a strike-slip fault with dominant sinistral motion with an inferred fault within the study area. The slip rate of Lembang fault is 1.95 – 3.45 mm/ year [4]. The two active faults in the study area with different movements can form joints in the rock mass. The presence of joints in the rock mass affects the rock mass quality, which is one of the tunnel design parameters.

The Geologic Map of the Cianjur Quadrangle of Java [5] shows that the study area resulted from young volcanoes in the form of tuffs originating from Mount Tangkuban Prahu (Qyt). Engineering geological mapping shows that rock mass in the study area mostly in highly weathered condition. Based on the bore log data, the groundwater table along the tunnel route is relatively high. Therefore, the strength of the layers along the tunnel route and its vicinity are non-uniform.

According to the conditions above, it is crucial to evaluate the geological conditions to determine the rock mass quality and characterization and feasible tunnel support system design for Tunnel 10 by means of engineering geological mapping and bore log data. This study aims to present rock mass quality and characterization and propose a tunnel support system that fits the geological condition and complexity in the study area.

Rock mass quality and characterization in this study were determined with the RMR method while the tunnel support system was referred to the RMR method and the JSCE method. Rock mass quality calculation and tunnel support system determination used in this study have never been used in this study area and the Tunnel 10 construction project. The previous method used to empirically determine the Tunnel 10 support system is rock mass classification according to Basic Quality (BQ) system.

* Corresponding author: igbindrawan@ugm.ac.id

 https://doi.org/10.1051/e3sconf/202132501006

1.1 Study area

The study area is located about 31 km to the northwest of Bandung city (Fig. 1). Engineering geological mapping and surface rock sampling were carried out in the area of 1330 meters x 955 meters.

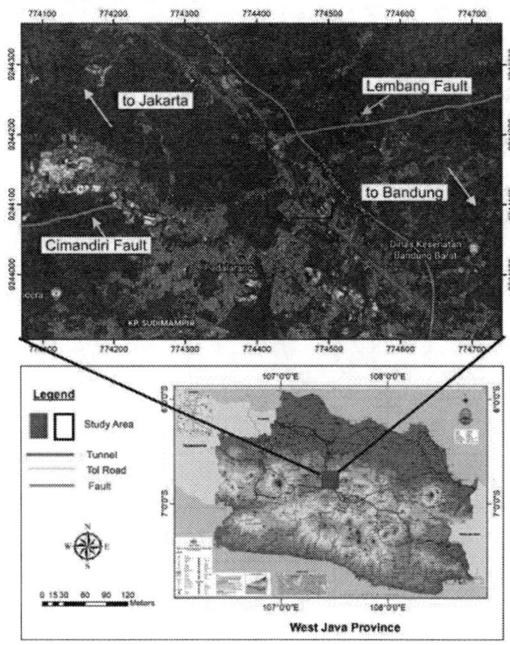

Fig. 1. Location of study area at Ngamprah District, West Bandung Regency, West Java Province of Indonesia [6].

2 Regional geology

2.1 Topography

The study area is covered by low hills with an undulating landscape (Fig. 2). Figure 2 shows that the lighter parts are hilly area and the darker parts are valley area. The tunnel route is located under rice fields, plantations, and settlements. According to the Design Drawing of No. 10 Tunnel (2018) of PT KCIC [1], the highest ground-level elevation in the tunnel alignment is 744 m, and the lowest is 684 m, with the slopes at the entrance and exit of the tunnel are 8° - 25° and 3° - 13°, respectively.

2.2 Lithology

Tunnel 10 is located in the Quaternary Pleistocene volcanic sediment layer (Qyt), consisting of clay, silt, silty soil, fine sand, medium sand, coarse sand, grit, fine angular gravel soil, coarse angular gravel soil, and black stone soil [12]. Engineering geological mapping shows that study area consists of two main lithology units namely the andesitic breccia and tuff breccia. The andesitic breccia composed of andesite fragment from breccia with silt to sand as its matrix, while the tuff breccia consist of tuff as a fragment with silt to sand as its matrix. In addition, the andesitic breccia and tuff breccia are distributed on the surface with moderate to highly weathering conditions.

2.3 Geological structures

Tunnel 10 is located between the main geological structures, namely the Cimandiri fault and the Lembang fault. The Map of Earthquake Hazard of Indonesia (2017) show that the Cimandiri Fault has an estimated slip rate geodetic ranging from 0.4 - 1 mm/ year [3], and the Lembang fault has an estimated slip rate of 3 - 14 mm/ year. The active segment of the Cimandiri fault is a dominant upward sinistral fault with a west-east and northeast-southwest orientations [7]. The location of the Cimandiri fault is in the southwest of the study area. The Lembang fault is a sinistral shear fault that cuts the Jakarta - Bandung High-Speed Railway line at DK 107+035 [4].

The alignment of the valleys in the study area shows the direction of northeast – southwest. The direction indicates that the structure pattern in the study area is influenced by the regional fault called the Maratus Pattern. The faults on the Java Island are caused by the collision of the Eurasian plate with the Indian-Australian plate. The collision resulted in the formation of fault patterns in Java, including the Maratus Pattern that is represented by the Cimandiri fault [8].

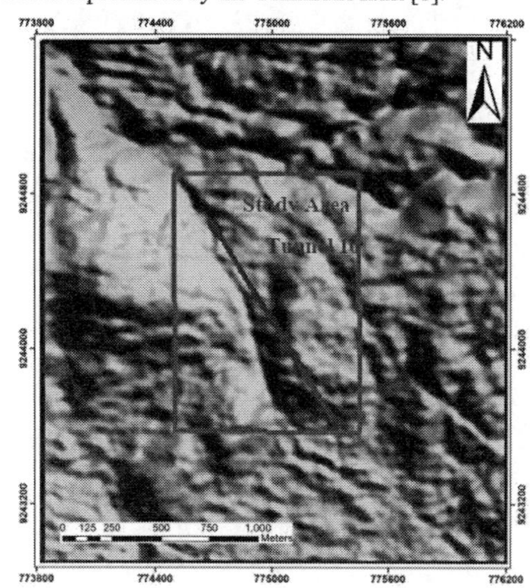

Fig. 2. Digital imagery of the study area and its vicinity.

3 Methodology

3.1 Engineering geological mapping

Engineering geological mapping was carried out to determine the morphological conditions, lithological distribution, and geological structure of the study area using a map scale of 1:5000. Surface rock samples were collected at various lithology and degree of weathering of rock mass. Combined with the bore log data, engineering geological mapping result was used to determine rock mass quality and to propose the tunnel support system based on morphological conditions, lithological distribution, and geological structure of the study area.

3.2 Rock mass classification and determination of tunnel support system using the RMR method

Classification of rock mass quality using the Rock Mass Rating (RMR) [9] method was carried out by determining the values of (1) strength of intact rock material, (2) the value of Rock Quality Designation (RQD), (3) spacing of discontinuities, (4) condition of discontinuities, (5) groundwater, and (6) discontinuity orientation. The strength of intact rock material in this study was determined by uniaxial compressive strength (UCS). The estimated field rock UCS values used in the RMR method obtained based on the ISRM (1981) [10]. Rock quality designation (RQD) values were obtained from the percentage of intact core pieces longer than 100 mm in the total length of core [9]. Rock discontinuity spacing values depend on the distance between each discontinuity. Meanwhile, discontinuity conditions were determined by the discontinuities surface and weathering condition. Groundwater conditions value was influenced by water inflow, water pressure, and its general condition. Discontinuity orientation values depend on strike and dip orientations, whether they are favorable or not for the structure planned in the study area. The RMR value range between 0 - 100.

The tunnel support system has a basic function to keep the tunnel and its opening stable [11]. Tunnel support system can be classified as temporary and permanent. The rock mass classification and the tunnel support system in accordance with the RMR method obtain based on Bieniawski (1989) [9].

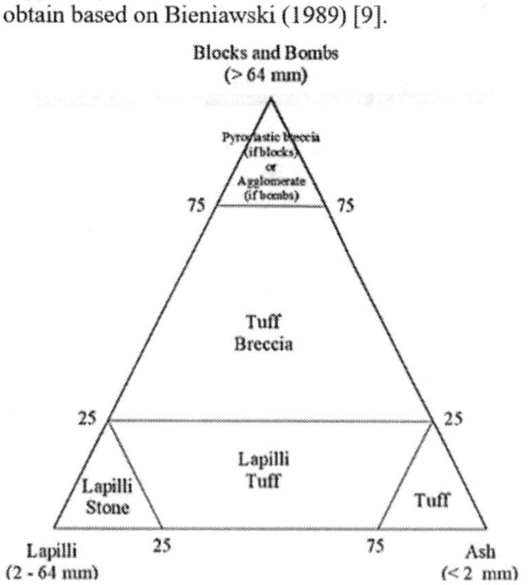

Fig. 3. Classification of pyroclastic rock [12].

The study area is a product of young volcanoes, thus the classification of pyroclastic rocks is based on Fisher's (1966) [12] classification as shown in Figure 2. The breccia rock that contains andesitic fragment classified by A. Streckeisen's (1976) [13] classification as shown in Figure 3. Geological structures were analyzed based on field observations and satellite imagery, while the degree of weathering of rock mass was determined based on ISRM (1981) [10].

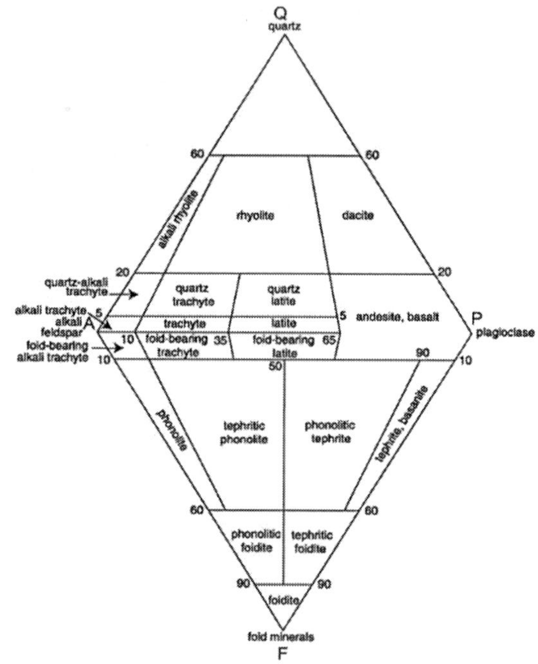

Fig. 4. Classification of plutonic rock [13].

3.3 Determination of tunnel support systems based on Japan Society of Civil Engineers (2018)

A safe tunnel support system based on the Japan Society of Civil Engineers [14] guidelines is carried out by determining the rock category/ class observed in the study area. Furthermore, it can be determined the types and specifications of the typical reinforcement required in the tunnel based on the rock category/ class.

4 Results and discussions

The study area is located in the hilly topography with a slope between 7 to 30 degrees. The lowest elevation is 649 meters, and the highest elevation is 760 meters above sea level in the study area. Meanwhile, the lowest ground-level elevation is 684 m, and the highest is 744 m along the tunnel route. According to the bore log data, the subsurface conditions indicate that the rock layers have been moderate to highly weathered. The rock layers show only a few traces of the original rock, such as andesite and tuff. The rock layers from top to bottom are generally clay, silt, and sand. Based on bore log data, the groundwater table around Tunnel 10 is measured as 0.13 m – 19.5 m beneath the surface, with water level variation considered as 2 – 3 meters.

The presence of the Cimandiri fault and the Lembang fault affect the stability of the rock mass conditions in the study area. The two faults affect the formation of joints in the rock mass and affect the direction of rock mass movement if there is a shift in the fault. The Cimandiri fault is a left-lateral reverse fault located 2 kilometers southwest of the study area with six active segments, namely Loji, Cidadap, Nyalindung, Cibeber, Padalarang, and Saguling, with a total length of 100 kilometers. The Cimandiri fault has a 0.4 – 1 mm/ year geodetic slip rate [2]. The Lembang Fault strike-slip fault with dominant sinistral motion movement of 1.95–3.45 mm/ year. The Lembang fault is divided into six active segments, namely Cimeta, Cipogor, Cihideung, Gunung Batu, Cikapundung and Batu Lonceng with a total length of 29 kilometers [4].

The observations of surface rock in the study area show that there are 2 (two) rock units from old to young, namely andesitic breccia and tuff breccia. The units of andesitic breccia and tuff breccia have an interlocking relationship. The stratigraphic relationship between andesitic breccia, tuff breccia, and the subordinate units is unknown. Based on observations, the geological structure developed at the study area was faults and joints, while the rock outcrop in the study area did not have a layering area.

In the north-western part of the study area, there are 2 (two) faults, the first inferred fault is a normal fault with a northwest-southeast direction (Fig. 7) with an N290°E / 70° direction, and the second is the approximately located - inferred fault with a northeast-southwest (ST 21 position in the study area shown in Figure 7) and an N30°E / 80° direction. The directions of dominant joint structures found in the study area are northeast (e.g., N77°E - N80°E and N63°E - N68°E) and southwest (e.g., N255°E - N250°E and N285°E - N288°E). Analysis data from the measurement of the fractional orientation of the rocks in the study area implies that the main force that controls the fracture has a northeast-southwest direction.

The spread of tuff breccia rock reached 58.48% of the total study area (shown in Figure 7 as a dark brown area). The tuff breccia rock observed on the surface has a moderate to a highly level of weathering so that the rock structure is not visible. The color of the rock is dark brown, composed of fragments measuring 0.06 - 2 mm with a fine-sized matrix - a very fine, poorly sorted with open packaging. Rock mass in Figure 5 indicates that ST 4 (Fig. 7) has a disintegrated structure with poorly interlocked, heavily broken rock mass with a mixture of angular and rounded rock pieces. The surface condition in ST 4 is poor with a slickensided, highly weathered surface with compact coatings or fillings or angular fragments. The surface condition of the rock mass is poor quality with a high degree of weathering.

The spread of andesitic breccia reached 41.52% of the total study area (shown in Figure 7 as a light brown area). Andesitic breccia has moderate to highly level of surface rock weathering so that the rock structure is not visible. The dominant rock color is light brown, composed of andesitic fragments from granule to boulder size with a matrix of granule size to very fine, poor sorting with open packaging. Figure 6 below shows that rock mass at ST 5 (Fig. 7) has a moderately weathered matrix with mainly andesite fragment size from coarse to boulder.

Fig. 5. Highly weathered tuff breccia in ST 4.

Fig. 6. Highly weathered andesitic breccia in ST 5.

The division of lithology, cross sectional of geological map, and geological engineering map in the study area is shown in Figure 7, Figure 8, Figure 9, and Figure 10.

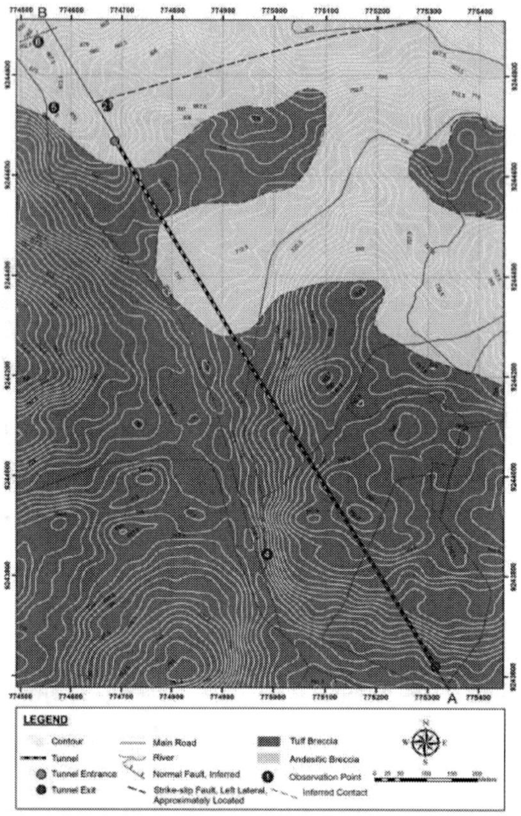

Fig. 7. Geological map of the study area.

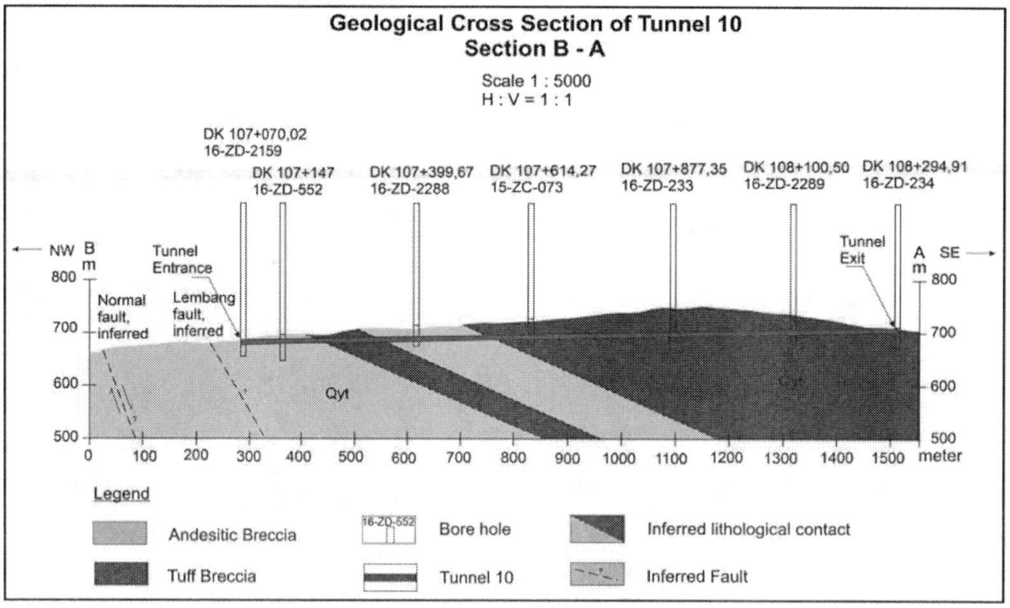

Fig. 8. Geological cross section of the study area.

https://doi.org/10.1051/e3sconf/202132501006

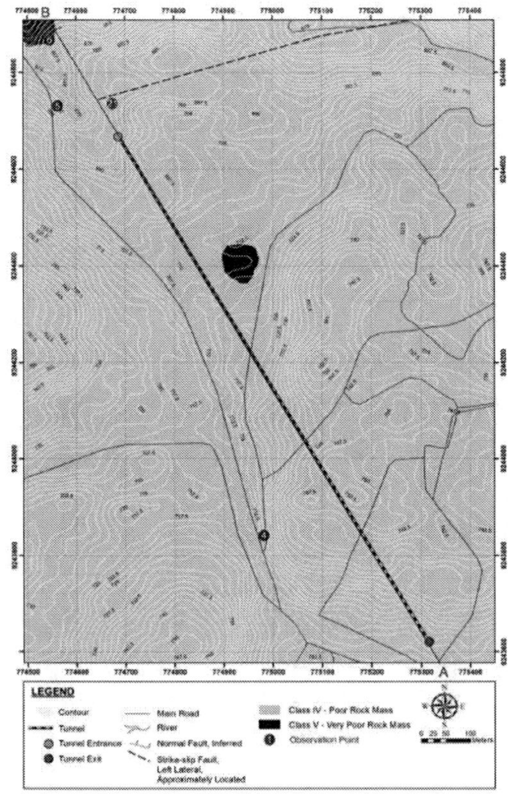

Fig. 9. Engineering geological map of the study area based on RMR.

The UCS value was determined in the field [10] as a rock mass quality parameter for the RMR method. The UCS values obtained ranged from 1 - 25 MPa. Table 1 shows an example of determining the rock mass quality value using the RMR method.

Table 1. The rock mass quality value at ST 4 and ST 5 determined by the RMR method

Parameters	ST 4	ST 5
	Rating	**Rating**
UCS (MPa)	1	4
RQD (%)	0	0
Spacing of Discontinuities	5	10
Condition of Discontinuities	20	20
Groundwater	0	0
Strike and dip orientations	0	0
RMR number	**26**	**34**

Based on the RMR calculation at ST 4 and ST 5, the total values of RMR are 26 and 34, respectively. This value indicates that the rock class belongs to class IV (RMR 21 - 40, poor rock). The rock mass in this category has a standing-up time of 10 hours for a 2.5 m span without support. According to Bieniawski (1989) [9], the support systems recommended for the tunnel in this rock mass category are rock bolts, shotcrete, and steel sets. The suitable excavation method is the top heading and bench with support installed while excavated.

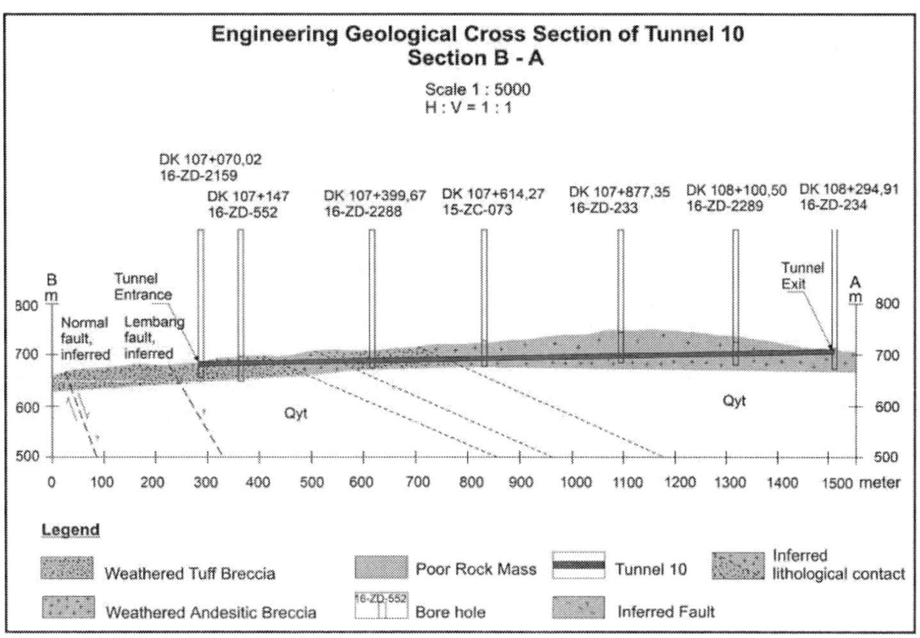

Fig. 10. Engineering geological cross section of the study area.

Along with the RMR method, the tunnel support system in this study is determined according to the characterization of the surface rock mass using the JSCE (2018) method [14]. Determination of rock mass category for tunnel support system by the JSCE obtained from the engineering geological mapping results and the RMR results. The engineering geological mapping data shows that the study area consists of two mainly lithological units, namely tuff breccia (58.48% of total study area) and andesitic breccia (41.52% from total of area) with poor and weathered condition. Thus, the rock mass classification in the tunnel route is categorized as tuff breccia in the weak class that belongs to the DII category in the JSCE method, with the L Massive rock type. Rock mass in this category is weak rocks that can experience large plastic and elastic deformations. The strength of the rock is considered minor with the very poor conditions of discontinuity. Thus, the strength of the rock bonds due to tunneling is reduced along the discontinuities, which can easily slip and increase movement. The excavation surface is considered unstable, and reinforcement such as shotcrete is required at the face of the excavation plane. The descriptions of the DII category in the JSCE method suit the rock mass condition in the study area.

The double bench excavation method is recommended for the DII category rock to stabilize the excavation surface area easily. The reinforcement system suggested by the JSCE method for the DII category rocks is shotcrete, rock bolts, steel supports, and work floors (invert), provided that the concrete work floor is immediately placed or closure of the excavation cross-section. A typical schematic of the support patterns and allowable deformations in DII category rocks is shown in Table 2.

The lithological unit in this study was obtained from correlation of engineering geological mapping, bore log data, and laboratory test results of surface and subsurface samples. The data show that most rock mass in the study is moderate to highly weathered, without a layering area in the rock outcrops. The subsurface condition was interpreted from bore log data and correlated with the engineering geological mapping information.

The geological structures that control the study area, namely the Cimandiri fault and the Lembang fault, have a different direction of the dominant sense of movement. The existence of the geological structures interpreted from satellite imagery (DEM) and the Map of Earthquake Hazard of Indonesia, validated by engineering geological mapping in the suspected location. This condition shows that the study area has complex geological conditions with accumulated volcanic products mostly in weathered states. Because of this consideration, the RMR method was used to determine the rock mass quality that mainly focused on quantifying the discontinuity conditions in the study area.

The rock mass quality assessment from observation points based on the RMR method is mainly shown as Poor Rock Mass (Class IV) with a hint of Very Poor Rock Mass (Class V) in two minor locations (Fig. 9). Thus, it can be concluded that the general rock mass condition in the study area is mainly in the poor category and highly weathered condition. Rock mass in this category has a small value of the UCS (< 25 MPa), cohesion (< 200 kPa), and friction angle (< 25 degrees). Poor rock quality can be easily deformed and destroyed by small energy, such as a knife or geological hammer.

Table 2. The excavation methods and tunnel support systems comparison in accordance with the RMR and JSCE methods.

Excavation Method and Tunnel Support Systems	The RMR method	The JSCE method
1. Excavation	Top heading and bench 1.0–1.5 m advance in top heading. Install support currently with excavation, 10 m from the face.	Double bench excavation method
2. Supports		
- Rock bolts	20 mm diameter, fully grouted	6.0 m of length
	Systematic bolts 4–5 m long, spaced 1–1.5 m in crown and wall with wire mesh	1.0 m spacing of circumferential direction, 1.0 m or less spacing of longitudinal direction in top heading and bench
- Shotcrete	100–150 mm in crown and 100 mm in sides	25 cm in crown and wall
- Steel sets	Light to medium ribs spaced 1.5 m where required.	H-200 beam in top heading and bench with 1.0 m or less spacing
- Lining	-	40 cm arch in sidewall and 50 cm in invert
3. Allowable deformation	-	10 cm

Rock mass condition based on the RMR classification and the JSCE category shows that the rock mass belongs to the poor category (Class IV and DII category). Figure 10 of the engineering geology cross-section of the study area shows that subsurface condition along the tunnel route is consists of weathered rock masses. This condition means that Tunnel 10 is located in a mainly poor rock mass with the possibility of numerous jointed rocks around the tunnel. The large plastic and elastic deformation can cause instability to tunnel excavation and structure. The rock mass already received generally strong weathering and alteration process with most of the materials turn into clay. Meanwhile, the soil materials mixed with pumice, tuff, and lapilli.

The comparison result between the RMR and the JSCE methods is shown in Table 2, including

 https://doi.org/10.1051/e3sconf/202132501006

comparison of the excavation method and the tunnel support system.

According to Table 2, typical tunnel support systems based on the RMR recommendation are rock bolts with 20 m diameter and fully grouted, shotcrete with 100 – 150 mm thickness in the crown and 100 mm thickness insides, and light to medium ribs spaced 1.5 m where required. Typical reinforcement schemes recommended for the tunnel according to JSCE (2018) are shotcrete with 25 cm thickness, steel support with H-200 beam for top heading and bench support, arch sidewall with 40 cm thickness, and invert concrete with 25 cm thickness. Based on the results, the JSCE provides higher safety factor, shown in the specification of its support system that is higher than the RMR recommendation. Besides, the RMR method does not set the allowable deformation limit to the excavation plane and does not add lining as a support system. Meanwhile the JSCE method sets the allowable deformation limit as 10 cm and recommends lining with a 40 cm arch in the sidewall and 50 cm in the invert.

5 Conclusion

This paper describes the lithology unit, distribution, and rock mass quality in the study area and proposes a suitable support system for Tunnel 10 based on the geological condition. The lithology of the study area consists of andesitic breccia and tuff breccia with a distribution of 58.48% and 41.52% of the total study area, respectively. The rock mass has a moderate to a highly degree of weathering. The UCS values ranged from 1 - 25 MPa, and the RMR values ranged from 21-40, which indicated the rock mass categorized as class IV (poor rock). The JSCE categorized the rock mass in the study area as DII category, shown as a poor rock mass.

According to the results, this study prefers to use the JSCE recommendation as a tunnel support system for Tunnel 10 with shotcrete with 25 cm thickness, steel support with H-200 beam for top heading and bench support, arch side wall with 40 cm thickness, and invert concrete with 25 cm thickness. However, numerical methods need to be done to determine the most effective and efficient tunnel support system for Tunnel 10. Detailed engineering geological mapping and research of faults and discontinuities around the tunnel are needed to determine their effects on tunnel construction and other infrastructure.

The first author wishes to thank the Ministry of Public Works and Housing of the Republic of Indonesia for the opportunity and financial support to conduct this study. The authors would like to thank PT Kereta Cepat Indonesia – China (KCIC), High Speed Railway Contractor Consortium (HSRCC), China Railway Group Limited (CREC), and China Railway Design Corporation (CRDC) for the research permission and data support.

References

1. PT Kereta Cepat Indonesia - Cina (KCIC), Design Drawing of No. 10 Tunnel (2018)
2. G.I. Marliyani, J. R. Arrowsmith, K. X. Whipple, J. Geophys. Res. Earth Surf. E **121**, 2287–2308 (2016)
3. Ministry of Public Works and Housing, Guidelines for Excavation Design Methods and Tunnel Reinforcement Systems in Mixed Soil-Rock Media (2015)
4. M.R. Daryono, D. H. Natawidjaja, B. Sapiie, P. Cummins, Tectonophysics. E **751**, 180 - 191 (2018)
5. Sudjatmiko, Geologic Map of the Cianjur Quadrangle of Java (1972)
6. Bakosurtanal, West Java Province Map (2003)
7. G.I. Marliyani et al, Neotectonics of Java, Indonesia: Crustal Deformation in the Overriding Plate of an Orthogonal Subduction System (2016)
8. F. Hilmi, I. Haryanto, Bull. Sci. Contrib.: Geol. **6** (2008)
9. Z.T. Bieniawski, Engineering Rock Mass Classifications (John. Wiley & Sons, New York, 1989)
10. ISRM, Rock Characterization Testing and Monitoring ISRM Suggested Method. E.T. Brown (Ed) (Pergamon Press, New York, 1981)
11. D.U. Deere, R. B. Peck, H. W. Parker, J. E. Monsees, B. Schmidt, Design of Tunnel Support Systems, (Whittles Publishing, 1970)
12. R.V. Fisher, Earth Sci. Rev. **1**(4), 287-298 (1966)
13. A. Streckeisen, Earth Sci. Rev. **12**(1), 1-33 (1976)
14. Japan Society of Civil Engineers (JSCE), Standard Specifications For Tunneling-2016: Mountain Tunnels (2018)
15. Ministry of Public Works and Housing, The Map of Indonesia Source and Earthquake Hazard (2017)

https://doi.org/10.1051/e3sconf/202132501007

Engineering geological studies for empirical design of tunnel 7 of Jakarta - Bandung high-speed railway, Indonesia

Riski Pramono[1,2], *I Gde Budi* Indrawan[1*], *Maulidya* Indah Junica[2]

[1]Department of Geological Engineering, Faculty of Engineering, Universitas Gadjah Mada, Yogyakarta, Indonesia
[2]The Ministry of Public Works and Housing, Jakarta, Indonesia

Abstract. The empirical design of tunnel 7 of the Jakarta-Bandung high-speed railway was based on the Basic Quality (BQ) system, which was the Chinese standard of rock mass classification. Due to the high uncertainty of a rock mass classification, comparative studies of rock mass quality and tunnel support systems based on other rock mass classifications are commonly required. The main objective of this research was to assess the rock mass quality and excavatability at the tunnel construction site and propose the tunnel support system based on three widely used methods, namely GSI, RMR, and Q-system. This study involved geological mapping, determination of surface and subsurface rock weathering, rock mass classification of drilling cores, and laboratory testing. The results show that the tunnel construction area consists of highly weathered volcanic and sedimentary rocks. The tunnel elevation consists of poor and extremely to very poor rock masses of sedimentary rocks. Hard digging is the optimum rock mass excavation method to use. Applications of rock bolt, wire mesh, and shotcrete are proposed for the tunnel support system.

1 Introduction

The Jakarta - Bandung high-speed railway Tunnel 7 Project is in West Bandung Regency, West Java Province (Fig. 1). The tunnel is a very environmentally friendly alternative compared to hill cutting. This tunnel is one of 13 tunnels traversed by the Jakarta - Bandung high-speed railway with a horseshoe-shaped and tunnel length of 1,225 km, horizontal diameter of 15 meters, and a vertical diameter of 13 meters. The field study conducted including geological mapping, engineering geological mapping, mapping of surface and subsurface rock weathering, observations of core drilling results, and sampling for laboratory testing.

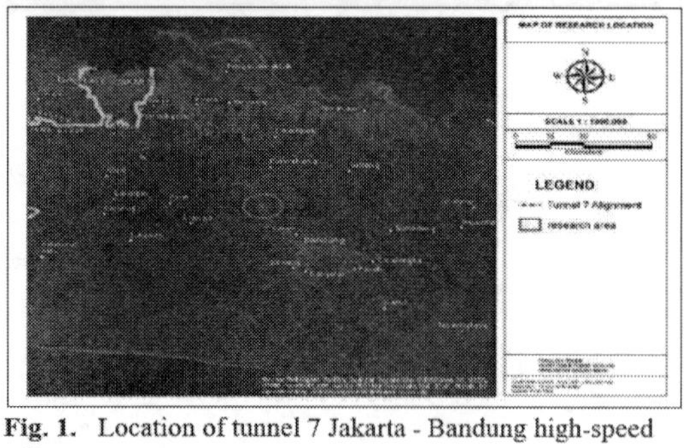

Fig. 1. Location of tunnel 7 Jakarta - Bandung high-speed railway

The planners designed this high-speed train tunnel based on the Professional Standard of The People's Republic of China "Code for Design on Tunnel of Railway." This guide uses the Basic Quality (BQ) rock mass classification issued by the Chinese government (Standard for Engineering Classification of Rock mass). Due to the high uncertainty in a rock mass classification, it is necessary to do a comparative analysis of rock mass quality using other widely used methods such as Rock Mass Rating (RMR), Q-system, and Geological Strength Index (GSI). The evaluation also needs to be done on the existing design as a comparison by using the empirical method of RMR and Q - System.

The empirical design method is based on rock mass classifications, which are widely utilized in rock engineering. Engineering rock mass classifications have improved their performance in recent years, and they are widely used in the design and construction of rock tunnels. The classification approach is sometimes the only practicable basis for designing complicated underground structures on many projects [1-2]. Rock mass classification systems are very useful tools for the preliminary design stage of a project when very little detailed information on rock mass is available. On the other hand, one or more rock mass classification methods may be used to create a picture of the composition and features of a rock mass to give initial support estimations and estimates of the rock mass strength and deformation properties [3].

Although the rock-mass rating (RMR) and tunneling quality index (Q) systems are used as empirical design tools in many rocks engineering

* Corresponding author: igbindrawan@ugm.ac.id

projects, tunnel design is internationally accepted. These systems were created and upgraded, particularly for this purpose [4]. In this study, the determination of rock quality and support system was carried out empirically using the Rock Mass Rating (RMR), Q system (Q), and Geological Strength Index (GSI). The characteristics of the surrounding rock, tunnel diameter, and geometry are the basic input parameters for a safe tunnel design. The results of this study are expected to contribute to tunneling on weak rock masses, provide input for stakeholders, and help promote tunnel technology in the transportation sector in Indonesia.

2 The research method

The behavior of the rock surrounding a tunnel is critical to its stability. For an appropriate design that can be depended on, the various forms of behavior necessitate distinct evaluations or rock engineering methods. Selecting appropriate design tools based on geological and geotechnical conditions is critical. It should be highlighted that ground behavior cannot be defined. As a result, proper design work cannot be carried out without a comprehensive understanding of geology, ground conditions, and site-specific characteristics. The selection of appropriate design tools is mostly determined by the real ground behavior, such as an accepted standard or any other need [5].

The engineering studies conducted in this research involved surface geological and engineering geological mapping, evaluation of rock cores, and laboratory tests for assessment of rock quality and analyses of excavation methods and tunnel support systems. The surface geological and engineering geological mapping was carried out to assist the interpretation of subsurface geological and engineering geological conditions. The research flow chart is shown in Figure 3.

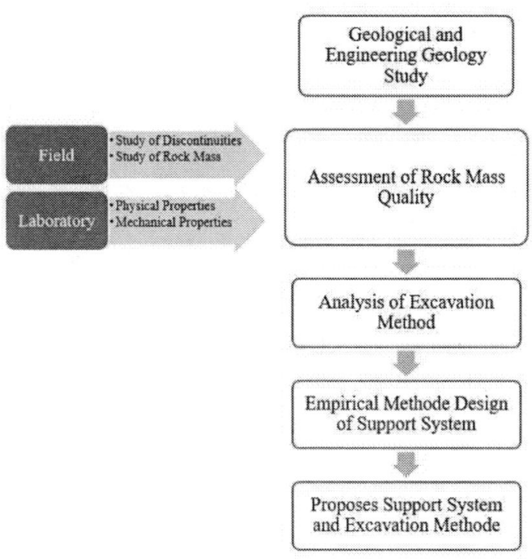

Fig. 2. Research flow diagram

3 Geological conditions

The Jakarta - Bandung high-speed railway Tunnel 7 is in the Bandung Zone based on the grouping that has been written by Van Bemmelen [6]. This zone has three lithological formations consisting of Old Volcanoes Results (Qob), Cantayan Formation (Mtts), and the Jatiluhur Formation (Mdm) based on the regional geological map sheet of Cianjur [7], as shown in Fig. 3.

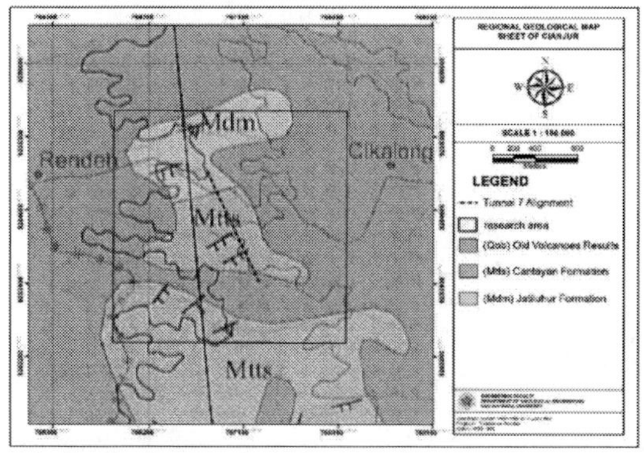

Fig. 3. Regional geological map of the research area

Results of surface geological mapping showed that the research area consists of volcanic breccia and laharic breccia of the Old Volcanoes (Qob) Formation, siltstone of the Cantayan Formation (Mtts), and sandstone and claystone bedding units of the Jatiluhur formation (Mdm), as shown in Fig. 4.

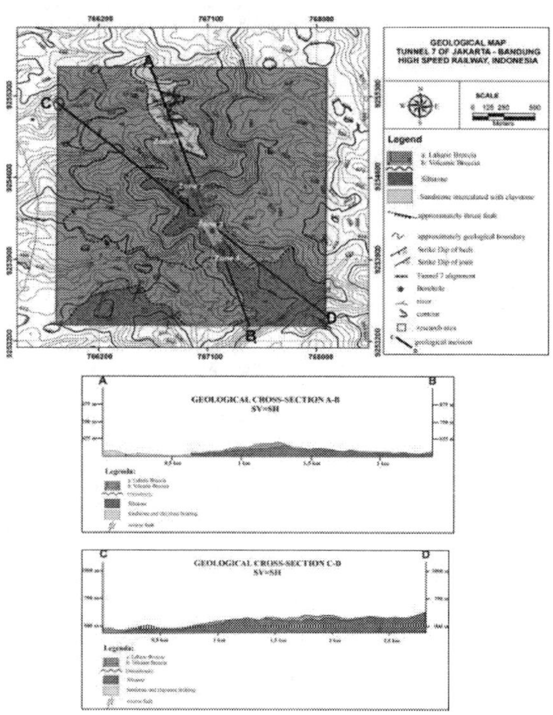

Fig. 4. Geological map of the research area

 https://doi.org/10.1051/e3sconf/202132501007

4 Engineering geological conditions

The geological engineering studies include both field and laboratory studies. The field studies consist of field observation, discontinuity surveys, and boreholes. Laboratory tests were conducted on samples collected from the field and the boreholes (Fig. 5.) The quantitative description of rock discontinuities, such as roughness, orientation, persistence, filling, and aperture, was determined in the field using the ISRM-recommended methods.

Fig. 5. Core drill zone 1 at 30-35 m depth (documentation of PT. KCIC)

The tectonic history of a rock mass has a significant impact on its quality. Radbruch-Hall (1975) evaluated the geology and tectonism (past and present) of each of California's physiographic provinces and their impact on slope stability and engineering practice. Folding and fracture were thought to make the granite mass more prone to weathering, erosion, and landslides. Current tectonic activity in California generates large earthquakes and triggers slope collapses [8]. Massironi et al. (2003) revealed that the tectonic history of the eastern Italian Alps resulted in ductile and brittle deformation structures that govern the various failure mechanisms functioning at three sites within a 10 km radius. The direction of the discontinuity and the position of the slope failure concerning the faults, folds, and orientation provided planes of weakness that favored lateral spreading, sackung development, or rockfall [9].

Engineering geological conditions describe the geological conditions, weathering rate, and rock quality. The geological surface engineering study site is divided into four units of engineering geology, namely highly weathered laharic breccias, highly weathered volcanic breccias, highly weathered siltstone, and highly weathered sandstone intercalated with claystone as shown in figure 6. Laboratory testing is carried out directly from drilling results according to the lithology and level of rock weathering. Tests were conducted on soil and rock index properties, shear strength, rock UCS, and triaxial rock strength. This testing is conducted to determine the parameters required in establishing the excavation method and tunnel stability. All laboratory tests were carried out in compliance with the ISRM recommended procedures.

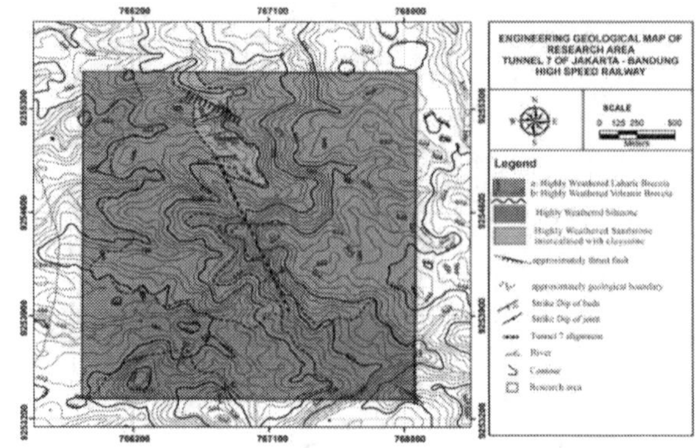

Fig. 6. Engineering geological map of the research area

5 Rock mass quality

5.1 Geological strength index (GSI)

According to Hoek and Marinos [10], the Geological Strength Index (GSI) can be determined directly from the outcrop covering two main parameters, namely the condition of the structure (structure) and the condition of its surface (surface condition). In this study, the GSI value was determined based on subsurface data from the results of drilling carried out at the tunnel location. Calculation of subsurface GSI, according to Hoek et al. [11], can be determined using the relationship between Joint Condition [12] and the value of Rock quality designation (RQD) [13]. The subsurface GSI value can be calculated using the following equation:

$$GSI = 1{,}5JCond_{89} + {}^{RQD}/_{2} \quad (1)$$

5.2 Rock mass rating (RMR)

According to Bieniawski [12], rock mass classification from the Rock Mass Rating (RMR) method uses six parameters obtained from field measurements or observations from drill data. The six parameters used in rock mass classification from the RMR) method are as follows: Uniaxial Compressive Strength (UCS) intact rock material, rock quality designation (RQD), discontinuity spacing, discontinuity conditions, groundwater condition, and discontinuity orientation.

The Rock Mass Rating (RMR) value can be found by calculating the values of all these parameters. The RMR classification can determine the excavation method and tunnel support systems according to the rock mass class. This paper calculates the RMR value to determine the support system. The RMR value is determined using the Hoek and Brown [14] correlation formula in equation 3.

5.3 Q-System

According to Barton [15], the Tunnelling Quality Index (Q) is a system that considers six parameters: RQD, amount of stock, roughness, change in stock, water

 https://doi.org/10.1051/e3sconf/202132501007

condition of the stock, and pressure factor. The Q-value can be calculated using the following equation [15]:

$$Q = \left(\frac{RQD}{Jn}\right)\left(\frac{Jr}{Ja}\right)\left(\frac{Jw}{SRF}\right) \quad (2)$$

This paper calculates the Q value to determine the support system based on the Q value. The Q value is determined using the Bieniawski [12] correlation formula in equation 4.

5.4 Correlation of RMR and GSI

The rock mass classification based on GSI is a development of RMR. Hoek and Brown [14] explain that the RMR classification based on Bieniawski [12] can estimate the GSI value as in the 1976 version. The rating of the groundwater under dry conditions is 15, and the orientation of the discontinuity is 0 in this case. The minimum value for the classification of the rock mass of RMR is based on Bieniawski [12] is 23. The final weight of this rock mass classification is called RMR_{89} Furthermore, it can be used to estimate the GSI value. If the value is GSI≥18 or RMR≥23, then:

$$RMR_{89} = GSI + 5 \quad (3)$$

5.5 Correlation of RMR and Q-System

Bieniawski [12] has provided various kinds of correlations between RMR and other parameters, but the most applicable correlation is the correlation between RMR and Q value, with the following formula:

$$Q = 10^{\frac{RMR-50}{15}} \quad (4)$$

5.6 Assessment of rock mass quality

Determination of the quality of the subsurface rock mass in this study was obtained from the borehole results. The quality of the rock mass is obtained from the 4-point borehole data, with the following results:

Table 1. Summary assessment of rock mass quality

Zone	Weathering rate	GSI	RMR	Rock Mass Quality	Q	Rock Mass Quality
1	Highly weathered	22	27	Poor	0.029	Extremely poor
2	Highly weathered	31	36	Poor	0.117	Very poor
3	Highly weathered	20	25	Poor	0.022	Extremely poor
4	Highly weathered	18	23	Poor	0.016	Extremely poor

6 Analysis of excavation method

The rock excavation method is determined by the type of rock material, especially its geotechnical properties. Discontinuity planes and intact rock are examples of this important property. Other factors, such as fractures, the contents of the fractures, and the strength of the discontinuity plane are other important.

Pettifer and Fookes [16] proposed a rock excavation method using blasting, ripping, and digging. The excavation method is recommended together with the equipment used. The excavation method recommendation is based on the value of the discontinuity plane spacing index (If) and the point load index (Is (50)). The value of Jv is obtained from the sum of the frequency of fractures or fractures that have the same direction pattern called a joint set. The calculation formula that can be used is the following equation:

$$If = \frac{3}{Jv} \quad (5)$$

$$Jv = \frac{1}{S1} + \frac{1}{S2} + \frac{1}{S3} \quad (6)$$

Table 2. If and Is50 load values to be plotted in the graph of Pettifer and Fookers [16]

Lithology	Rock Mass Classification	If	Is50 (Mpa)
Laharic breccia ☆	Poor rock	1.13	0.17
Volcanic breccia ☆	Poor rock	1.01	0.17
Claystone ☆	Poor rock	0.68	0.04
Siltstone ★	Poor rock	0.76	0.04
Sandstone ★	Poor rock	0.39	0.12

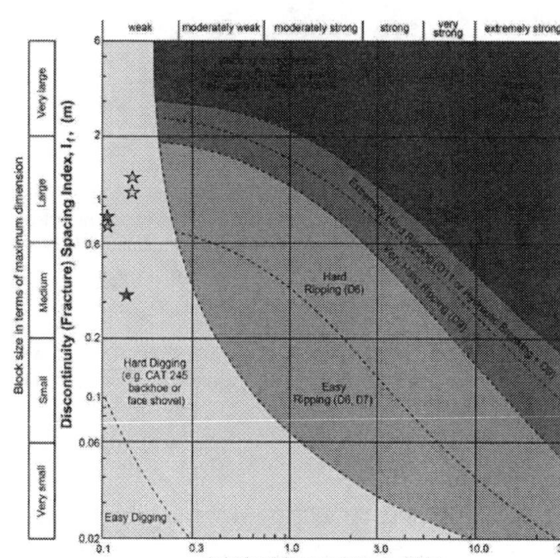

Fig. 7. The plot of rock excavation method.

Based on the plotting results in fig. 7, most of them indicate that the most suitable excavation method is hard digging.

https://doi.org/10.1051/e3sconf/202132501007

7 Empirical design of support system

Based on RMR classification systems, the necessary support systems for the zone along the tunnel route are presented in Table 3.

Table 3. The recommended supports are based on the RMR.

Zone	RMR	Rock Mass Quality	Support System Base on Q-System
Zone 1	27	Poor	Rockbolt: L = 5 m with spaced = 1.5 m in crown and wall
			Shotcretet: t = 150 mm (crown) t = 100 mm (wall)
			Wiremesh: in crown and wall
Zone 2	36	Poor	Rockbolt: L = 5 m with spaced = 1.5 m in crown and wall
			Shotcretet: t = 150 mm (crown) t = 100 mm (wall)
			Wiremesh: in crown and wall
Zone 3	25	Poor	Rockbolt: L = 5 m with spaced = 1.5 m in crown and wall
			Shotcretet: t = 150 mm (crown) t = 100 mm (wall)
			Wiremesh: in crown and wall
Zone 4	23	Poor	Rockbolt: L = 5 m with spaced = 1.5 m in crown and wall
			Shotcretet: t = 150 mm (crown) t = 100 mm (wall)
			Wiremesh: in crown and wall

The excavation and support system for all zone which has rock mass classification RMR is poor rock. So, the recommended excavation is gradual with a heading and benching system. The excavation and support systems for all zones based on the RMR rock mass classification are rock poor. Based on the quality of the rock mass, it is recommended that excavation be carried out in stages with a heading and benching system. Install completely grouted rock bolts with a length of 5 meters and spaced 1.5 meters at the crown and walls in a systematic way. Install wire mesh on the crown and wall, then shotcrete thickness of 150 mm on the crown and 100 mm on the wall. The installation of steel sets (steel ribs) with a 1.5 meter spacing was then combined.

Determination of the tunnel support system based on the Q system classification [15] is determined based on the Q value, the tunnel height or span, and the Excavation Support Ratio (ESR) value. The height of the tunnel used in this study is 13 meters, while the ESR value is 1 (railway tunnel) so that the ratio between the height and the ESR is 13. The support system at four zones in the study area is as shown in Fig. 8.

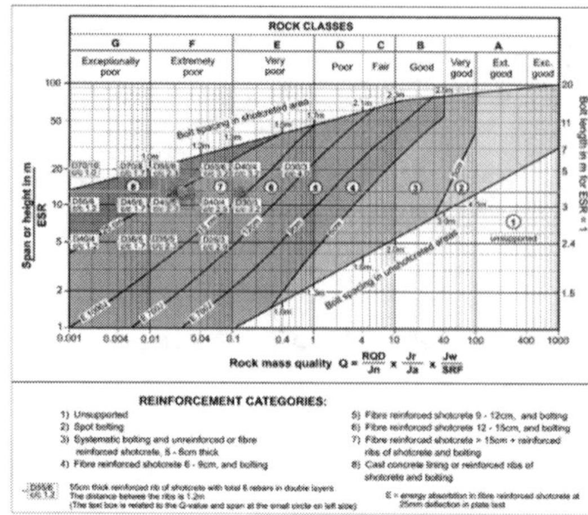

Fig. 8. The plot point for determining the support system is based on the Q system.

Based on Q value classification systems, the support systems for the zone along the tunnel route are presented in Table 4.

Table 4. The recommended supports are based on the Q System

Zone	Q Value	Rock Mass Quality	Support System Base on Q-System
Zone 1	0.03	Extremely Poor	Rockbolt: L = 3.8 m with spaced = 1.2 m in crown and wall
			Shotcretet: t = 240 mm (crown and wall)
Zone 2	0.12	Very Poor	Rockbolt: L = 3.8 m with spaced = 1.3 m in crown and wall
			Shotcretet: t = 230 mm (crown and wall)
Zone 3	0.02	Extremely Poor	Rockbolt: L = 3.8 m with spaced =1.2 m in crown and wall
			Shotcretet: t = 240 mm (crown and wall)
Zone 4	0.02	Extremely Poor	Rockbolt: L = 3.8 m with spaced =1.2 m in crown and wall
			Shotcretet: t = 240 mm (crown and wall)

8 Conclusions and discussion

This study involved geological mapping, determination of surface and subsurface rock weathering, rock mass classification of drilling cores, and laboratory testing. The study area consists of highly weathered laharic breccias, highly weathered volcanic breccias, highly weathered siltstone, and highly weathered sandstone intercalated with claystone. The tunnel elevation consists of poor rock masses based on the RMR and very poor to extremely poor rock masses according to Q-system.

The suggested rock mass excavation method is hard digging. The recommended supports are based on the RMR applied rock bolt length 5 meters with spaced 1.5 meters in crown and wall, wire mesh in crown and wall, and shotcrete thickness of 150 mm on the crown and 100 mm on the wall. The recommended supports are based on the Q-system applied rock bolt length 3.8 meters with spaced 1.2 – 1.3 meters in crown and wall, and shotcrete thickness of 230 - 240 mm on the crown and wall. Designing a tunnel with an empirical approach is very helpful in preliminary design. The further numerical study is required to determine the most optimum tunnel support design by comparing the tunnel support design determined by the Chinese standard of rock mass classification and that determined by the rock mass classifications used in this study.

The authors would like to thank PT. Kereta Cepat Indonesia China (KCIC), High-Speed Railway Consortium (HSRCC), China Railway Design Corporation (CRDC), and China Railway Group Limited (CREC) for the permission to conduct this research and supplying site investigation data. The first author would also like to thank The Ministry of Public Works and Housing of Indonesia for providing financial support.

References

1. R.K. Goel, B. Singh, Engineering Rock Mass Classification: Tunnelling, Foundations, and Landslides. Elsevier Science (2011)
2. FHWA. Technical Manual for Design and Construction of Road Tunnels Federal Highway Administration. US Department of Transportation: Washington DC (2009)
3. E. Hoek, Practical Rock Engineering: RocScience. (2007)
4. ISRM, Suggested Methods for The Quantitative Description of Discontinuities in Rock Masses: International Society for Rock Mechanics. International Journal of Rock Mechanics and Mining Science & Geomechanics Abstracts, **15**: 319-368 (1978)
5. A. Palmstrom, H. Stille, Ground behavior and rock engineering tools for underground excavations. Tunneling Underground Space Technol, **22**, 363–376 (2007)
6. V. Bemmelen, The Geology of Indonesia Vol. IA General Geology of Indonesia and Adjacent Archipelagoes, The Hague: Government Printing Office (1949)
7. Sudjatmiko, *Peta Geologi Lembar Cianjur, Jawa. Bandung*: Pusat Penelitian dan Pengembangan Geologi (2013)
8. D.H. Radbruch-Hall, Tectonic Activity in California, and its Relation to Engineering Geology. In: Geologica Carpathica XXVI, Bratislava, pp. 161-1 69 (1975)
9. M. Massironi, A. Bitacchi, G.V. Dal Piaz, B. Monopoli, A. Schiavo, Structural Control on Mass-Movement Evolution: A Case Study from the Vizze Valley, Italian Eastern Alps, Eclogae Geologicae Helvetiae, **96**: 85-98. (2003)
10. P. Marinos, E. Hoek, GSI: a geologically friendly tool for rock mass strength estimation, ISRM international symposium, International Society for Rock Mechanics and Rock Engineering (2000)
11. E. Hoek, T. G. Carter, M. S. Diederichs. Quantification of the geological strength index chart. 47th US rock mechanics/geomechanics symposium, American Rock Mechanics Association (2013)
12. Z. T. Bieniawski, Engineering rock mass classifications: a complete manual for engineers and geologists in mining, civil, and petroleum engineering, John Wiley & Sons (1989)
13. D.U. Deere, Technical description of rock cores for engineering purposes, Felsmechanik und Ingenieurgeologie (Rock Mechanics and Engineering Geology), **1**(1): 16-22, (1963)
14. E. Hoek, E.T. Brown, Practical estimates of rock mass strength, international journal of rock mechanics and mining sciences, **34**(8): 1165-1186 (1997)
15. N. Barton, R. Lien, J. Lunde, Engineering Classification of Rock Masses for the Design of Tunnel Support, Oslo: Norwegian Geotechnical Institute Publication **106** (1974)
16. G.S. Pettifer, P.G. Fookes, A Revision of The Graphical Method for Assessing the Excavatability of Rock, Quarterly Journal of Engineering Geology **27** (1994)

Integration of grey analysis with artificial neural network for classification of slope failure

Ashanira Mat Deris[1], *Badariah* Solemon[2*], and *Rohayu* Che Omar[2]

[1]Fakulti Teknologi Kejuruteraaan Kelautan dan Informatik, Universiti Malaysia Terengganu 21030 Kuala Nerus, Terengganu
[2]Institute of Energy Infrastructure, Universiti Tenaga Nasional, Jln IKRAM-UNITEN, 43000 Kajang, Selangor, Malaysia

Abstract. With the advent of technology and the introduction of computational intelligent methods, the prediction of slope failure using the machine learning (ML) approach is rapidly growing for the past few decades. This study employs an "artificial neural network" (ANN) to predict the slope failures based on historical circular slope cases. Using the feed-forward back-propagation algorithm with a multilayer perceptron network, ANN is a powerful ML method capable of predicting the complex model of slope cases. However, the prediction result of ANN can be improved by integrating the statistical analysis method, namely grey relational analysis (GRA), to the ANN model. GRA is capable of identifying the influencing factors of the input data based on the correlation level of the reference sequence and comparability sequence of the dataset. This statistical machine learning model can analyze the slope data and eliminate the unnecessary data samples to improve the prediction performance. Grey relational analysis-artificial neural network (GRANN) prediction model was developed based on six slope factors: unit weight, friction angle, cohesion, pore pressure ratio, slope height, and slope angle, with the factor of safety (FOS) as the output factor. The prediction results were analyzed based on accuracy percentage and receiver operating characteristic (ROC) values. It shows that the GRANN model has outperformed the ANN model by giving 99% accuracy and 0.999 ROC value, compared with 91% and 0.929.

1 Introduction

Landslide is a common geological hazard that occurs worldwide every year. Slope failure is the most contributed factor to the landslide, triggered by excessive rainfall, earthquake, or human intervention such as deforestation and construction. Slope failures can be described as a movement of soil or rock downward the earth's gravity [1]. Furthermore, this natural disaster can cause significant damage to the environment and properties and contribute to the human loss of life. To mitigate the impacts and damages caused by the landslides, the prediction of slope failure is a great consideration to which researchers have been investigated progressively for the past decades. Nevertheless, the prediction of slope failure remains a challenging task due to its complexity and uncertainties of geological factors and unbalance data samples. Traditional methods such as the limit equilibrium method (LEM), finite equilibrium method (FEM) are found to be computationally efficient, as reported by Ray [2]. However, these methods fail to have a comprehensive understanding of the slope's behavior due to the inherent drawbacks. Furthermore, the method often criticized its intensive computational power required and lacked computational efficiency for small probability levels [3]. Thus, with the introduction of the machine learning (ML) approach that is capable of modeling complex problems, the prediction of slope failure has been rapidly growing for the past few decades. ML approach is a powerful intelligent method capable of serving as a replacement for the traditional methods to predict slope failures.

ML approaches such as artificial neural network (ANN), decision tree (DT), Naïve Bayes (NB), support vector machine (SVM), Random Forest (RF), Logistic Regression (LR), etc. have been well studied and attracted interest among researchers [4-9]. ANN method is proposed based on the ML algorithm is capable of learning the relationship between the input factors and the FOS value from the historical slope cases. The previous ANN studies found that this intelligent computational tool can tackle complex problems in the geotechnical engineering field. Chakraborty and Goswami developed an ANN model to predict FOS value for slope stability analysis in Jorabat-Shillong Expressway, India [4]. Initially, 200 slope cases were analyzed based on four conventional methods: "Bishops, Morgenstern and Price, Janbu and Fellenius method," and the obtained FOS value was used to develop prediction models. The result found that ANN gives the correlation value of 99.63% compared to multilinear regression with 96.14%. Ray et al. have developed ANN prediction models to evaluate FOS of residual soil slope in Shiwalik, Himalaya. Eleven slope factors were investigated, and the factors that have a higher correlation coefficient were chosen as the input factors for the development of ANN prediction models

* Corresponding author: badariah@uniten.edu.my

[10]. The models were develop based on a multi-layer feed forward back-propagation network (FFBPN) with the optimal number of hidden neurons. The prediction results were evaluated using coefficient of determination (R2), root mean square error (RMSE), residual error and variance account. The author concluded that ANN could give a good performance and a straightforward, reliable, and valid ML approach for slope prediction. ANN also proven has outperformed other techniques. Bharati et al. applied ANN to predict the FOS value of dragline dump slope using 216 simulated dragline dump slope [11]. The prediction models were developed based on input parameters, including slope height, slope angle, and coal-rib height. The prediction performances were investigated based on the R and RMSE value. The prediction result of ANN was then compared with MLR, and it was found that ANN gives higher accuracy than the MLR model. It should be noted that although the previous study discussed above show their significance, there is still a problem that needs to be adequately conveyed, such as the significant relationship between each of the influencing parameters has not been extensively explored. Hence, this study attempts to integrate an ML method with the statistical method to improve the prediction performance.

A countless number of slope failure prediction models have been investigated and developed using physics and statistical-based modeling. Generally, physics-based models developed on creep theory depict soil and rock's fundamental relationship, while statistical models developed based on regression analysis of slope failure's historical occurrence [12]. However, physics-based models need further observations on-site and laboratory tests to identify mechanical and physical parameters and limit their broad applications. In contrast, the statistical model does not need the determination of physical parameters. It uses various statistical analysis methods to establish the response relationships between the slope failure occurrences and their associated contributing factors. Consequently, the statistical model has increased its popularity among researchers due to its simplicity and ease of implementation. Statistical analyses such as regression, grey system, and autoregressive-integrated moving average (ARIMA) have been widely applied in slope failure prediction. The integration of statistical analysis and the ML approach has been increasing in popularity in modeling landslide susceptibility. This approach attracts interest to address the challenge of landslide modeling for large regions, especially when the geotechnical data may not be adequate to implement the physics-based method [13].

The current study's objective is to develop an integration prediction model named GRANN to predict slope failure utilized a case study conducted by Sah et al. [14]. This model is the integration of machine learning, ANN, and statistical approach, GRA. The GRA is considered as a feature selection method, whereas the ANN is the classifier. The proposed model is expected to investigate the significant relationship between FOS and its influencing factors and eliminate the least influencing factor to improve the prediction performance. The result of GRANN was then compared with a single ANN model to observe its performance for slope failure prediction.

2 Methodology

2.1 Grey Relational Analysis

Professor Deng Julong has proposed an analysis method in Grey System theory which is called GRA [15]. In the grey theory, a level of information is between the black (unknown) and white (known), which means some of the information, is known, and some of the information is unknown. According to Lu and Rosenbaum, the possible examples of "grey" conditions in slope stability include ground quality characteristics that may refer to fuzzy values such as good, fair, poor, or slope fuzzy dampness values of damp, wet, and dry [16]. These slope's fuzzy conditions make it complex and challenging to measure the exact values of slope factors. Furthermore, the relationship between input factor and output factor may also be complicated to establish. Grey System provides various methods to deal with "grey" information or knowledge and thus can help predict the slope instability slope, considering its inherent subjectivity and ambiguity. GRA, one of the techniques mainly applied in Grey Systems, is used to illustrate how each of the input factors influences the output factor, FOS. This influence is denoted by grey relational grade (GRG) value. An input factor that gives a higher GRG value indicates a more significant relationship to the FOS. In comparison, a factor that offers a lower GRG value indicates a less significant relationship to the output factor. Figure 1 shows the simplified steps of GRA.

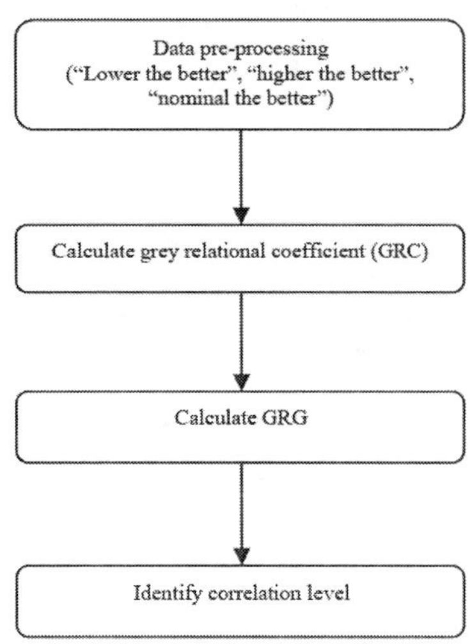

Fig.1. Simplified GRA steps.

Firstly, the data needs to be normalized in range [0,1] based on the three data characteristic namely "higher the better", "lower the better", and "nominal the better". This study applies "lower the better", as described in Eq. (1) [17].

$$x_i^*(k) = \frac{\max x_i^0(k) - x_i^*(k)}{\max x_i^0(k) - \min x_i^0(k)} \quad (1)$$

Where $x_i^*(k)$ is the current data to be normalized, $\max x_i^0(k)$ is the maximum value of the data, $\min x_i^0(k)$ is the minimum value of the data and x^0 is the desired target value. The relationship between the ideal and normalized experimental results is expressed GRC value, which is calculated as in Eq. (2):

$$\xi_i(k) = \frac{\Delta\min + \zeta.\Delta max}{\Delta_{0i}(k) + \zeta.\Delta max} \quad (2)$$

Where ζ is a distinguishing coefficient which is in range of [0.1], and generally, $\zeta = 0.5$ is used [18]. $\Delta_{0i}(k)$ is the deviation sequence which can be calculated as follows:

$$\Delta_{0i}(k) = \| x_0^*(k) - x_i^*(k) \| \quad (3)$$

$$\Delta\min = \min_{\forall j \in i} \min_{\forall k} \| x_0^*(k) - x_j^*(k) \| \quad (4)$$

$$\Delta\max = \max_{\forall j \in i} \max_{\forall k} \| x_0^*(k) - x_j^*(k) \| \quad (5)$$

Where $x_j^*(k)$ is the comparability sequence and $x_0^*(k)$ is the reference sequence and. After the GRC is determined, the value of GRG is calculated based on the formula in Eq. (6) :

$$GRG\ (y_i) = \frac{1}{n}\sum_{k=1}^{n} \xi_i(k) \quad (6)$$

GRG value shows the correlation level among the reference and the comparability sequence. The reference sequence refers to the output data, which is the value of FOS and the comparability sequence refers to the input factors of the slope. The value of GRG is in the range [0, 1] and is equal to 1 if the two sequences are identical. Typically, the input factor with a GRG value of less than 0.6 is considered a less influencing factor to the output. The higher GRG value shows that it influences the output value [19].

2.2 Artificial Neural Network (ANN)

ANN is a supervised ML method that imitates biological neurons in the human brain. ANN is a powerful tool to solve a complex problem, consists of artificial neurons connected for transmitting information in the neurons via the tendons. The most frequently used ANN architecture is multi-layer perceptron (MLP) consists of the input, hidden and output layer, connected to each other through weight and bias [20]. Each layer contains a number of neurons. The neuron in each input factors is multiplied with the corresponding weight to produce the product, as in Eq.(7):

$$net = \sum_{i=0}^{n} w_i x_i + \mathrm{b} \quad (7)$$

where n is the number of inputs, x_i is the input factor, w_i is the corresponding weight and b is bias. Then, the sum is processed using a nonlinear activation function. Note that if the sum is exceeds the threshold, then the neuron is activated. The output y_i as in Eq. (8):

$$y_i = f(net) \quad (8)$$

where f is the activation function. The optimal network architecture can be selected based on trial and error basis [21]. Number of hidden layer and the neurons for hidden layer is selected based on the problem itself. Figure 2 shows the architecture of MLP network.

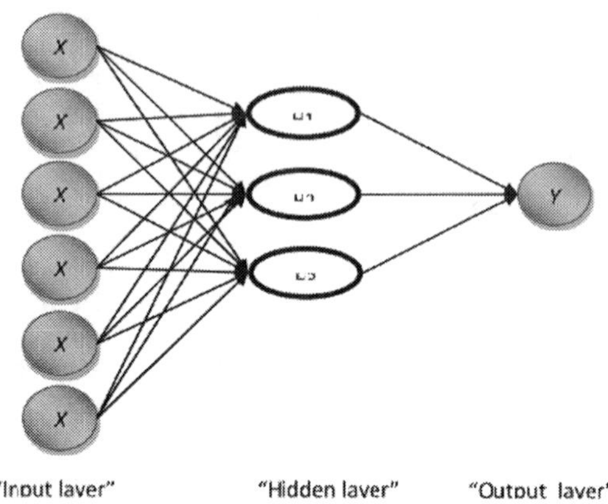

Fig. 2. MLP architecture.

This study applies the ANN prediction model to classify the FOS based on the slope input parameters using a FFBPN. Before the prediction process can be conducted, the slope cases data sets must be divided into training and testing datasets. The data division process is an essential aspect for any ML method where the selection of training and testing datasets will affect the prediction performance. The training dataset should represent the entire data set [22]. The model must be trained appropriately to determine the network's optimal weights and biases [21]. If the training dataset is limited, the network could not learn properly and if too large a training dataset is used, it is difficult to verify the generalization capability. Furthermore, it may lead to overfitting [10]. Hence, it is important to choose the proper training and testing datasets. The current study applies 70% slope cases for training and 30% slope cases for testing.

2.3 Case study

Historical cases of slope cases were extracted from Sah et al., (1994) to developed prediction models. The data consists of 46 cases of the circular slope with six input parameters, namely "unit weight, internal friction angle, cohesion, slope angle, slope height and pore pressure ratio" and FOS as the output factor. Two types of slopes were classififed from the 46 slope cases, which are "stable" and "fail". Figure 3 shows the bar chart of the stable and fail slope cases.

https://doi.org/10.1051/e3sconf/202132501008

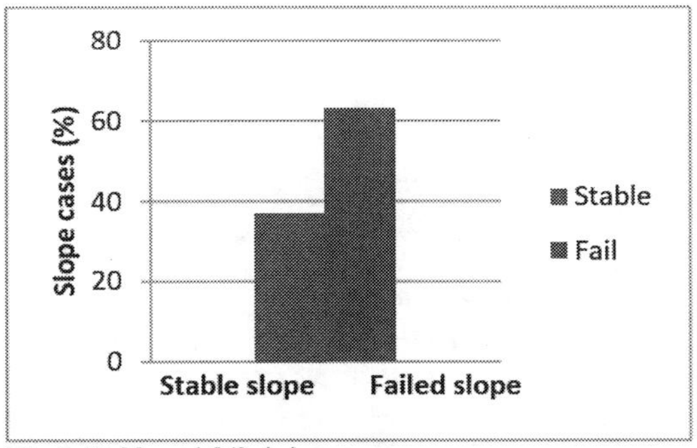

Fig. 3. Stable and failed slope cases.

3 Result and discussion

In GRANN model, the significant factors of six comparability sequences xi namely "unit weight, cohesion, internal friction angle, slope angle and pore pressure ratio" to the reference sequence x0, FOS value is determined in terms of GRG value. Table 1 shows the GRG value for all the input factors of circular slope.

Table 1. GRG values for the input parameters.

Input factors	GRA Analysis	
	GRG	Rank
Unit weight (γ)	0.6388	3
Cohesion, (c)	0.6392	2
Internal friction angle (φ)	0.6227	6
Slope angle (β)	0.6327	5
Slope height (H)	0.6481	1
Pore water pressure ratio (ru)	0.6375	4

From Table 1, it can be seen that slope height is the most influential factor to the FOS value with the GRG value is 0.6481 and internal friction angle is the least influence factor to the FOS value with the GRG value is 0.6227. Note that the GRG values of the input factors are in range [0.6, 0.7], which is considered as noticeable influenced [19, 23]. Based on the GRG value, the input factor that ranked last will be considered for the elimination from the input data. Elimination the input factor will reduce the input numbers and consequently will reduce the running time of the prediction model [19]. The classification accuracy is calculated as follows:

$$Accuracy = \frac{TP+TN}{TP+TN+FP+FN}\ x\ 100\% \qquad (9)$$

Where TP is "True Positive: the number of correct positive prediction, TN is True Negative: the number of correct negative prediction, FP: False Positive: the number of incorrect positive prediction and FN: False Negative: the number of incorrect negative prediction". Table 2 shows the confusion matrix of ANN and GRANN with the accuracy and ROC value while Figure 4 shows the graphs of ANN and GRANN prediction models.

Table 2. Confusion matrix of ANN and GRANN.

		Actual		Accuracy (%)	ROC
ANN		+	-		
Predictive	+	6	1	91.67	0.9285
	-	0	5		
GRANN				99.99	0.999
Predictive	+	8	0		
	-	0	4		

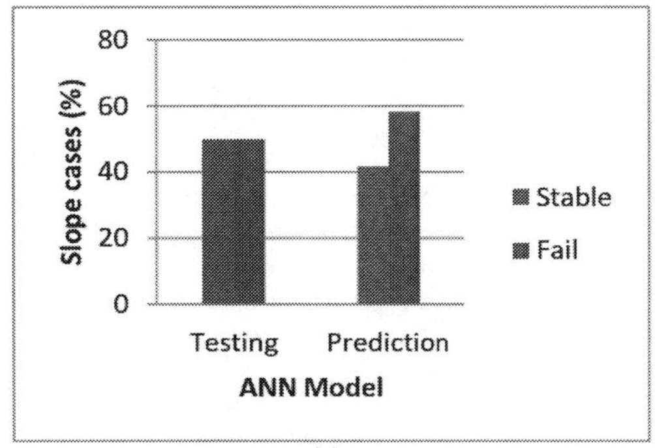

(a)

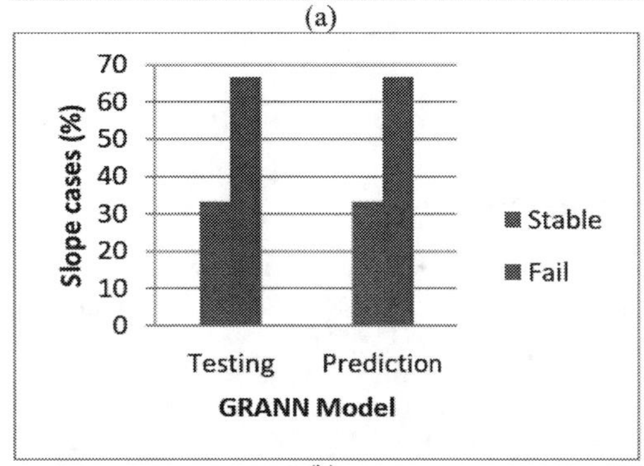

(b)

Fig. 4. Graphs of testing and prediction result for ANN (a) and GRANN (b) models.

Based on Table 2, the prediction result shows that for ANN, six stable slope cases are correctly predicted, and one slope case is wrongly predicted, while three failed slope cases are correctly predicted, and no stable slope case is incorrectly predicted. For GRANN, eight stable slope cases are correctly predicted and no stable slope case is incorrectly predicted, while four failed slope cases are correctly predicted, and no failed slope case is incorrectly predicted. ANN prediction model consists of six input parameters, while the GRANN prediction model consists of five input factors in which internal friction angle is eliminated from the input data. The integration of ANN was proven to improve the ANN model's performance by giving the accuracy and ROC value with 99% and 0.999, compared to ANN with 92% and 0.9444, respectively. Figure 4 shows that for the ANN model, the testing dataset gives 50% of stable slope and 50% of the failed slope, while the prediction result shows that 41.67% of stable slope and 58.33% of the failed slope. For the GRANN model, both the testing dataset and prediction result show similarity for stable and failed slopes, which are 33.33% and 66.67%, respectively.

4 Conclusion

This study attempts to develop an integration model of GRA and ANN to predict the slope failure of circular slopes. 46 slope cases were collected from previous study is used to develop prediction model based on ANN with six input factors; "cohesion, slope height, slope angle, internal friction angle, unit weight and pore pressure ratio". A three layer FFBPN with 6-3-1 architecture was selected for the prediction model. The dataset is divided into 70% for training and 30% of testing dataset. The prediction result of ANN gives 92% accuracy. To improve the prediction performance, GRA is integrated with ANN during the pre-processing phase where GRA analyze all the input factors. It was found that internal friction angle gives the least significant effect to the FOS value by giving the GRG value 0.6228, hence it will be eliminated from the input factors. The prediction result of GRANN gives 99% accuracy. GRANN show the capability of the model to analyze the slope data and eliminate the unnecessary data samples to improve the prediction performance.

This project was funded by UNITEN under Fundamental Research Grant Scheme FRGS/1/2018/STG08/UNITEN/02/1.

References

1. I. Chowdhuri, S. C. Pal, A. Arabameri, P. T. T. Ngo, R. Chakrabortty, S. Malik, P. Roy, Env. Earth Sci. **79**, 20:1–28 (2020)
2. A. Ray, V. Kumar, A. Kumar, R. Rai, M. Khandelwal, T.N. Singh, Natural Hazards **103**, 3: 3523–3540 (2020)
3. S. Y. Liu, L. T. Shao, H. J. Li, Comp. and Geotechnics **63**, 291–298 (2015)
4. A. Chakraborty, D. Goswami, Arabian J. of Geosciences **10**, 17, 1–11.
5. L. Y. Zhou, F. P. Shan, K. Shimizu, T. Imoto, H. Lateh, K.S. Peng, *A comparative study of slope failure prediction using logistic regression, support vector machine and least square support vector machine models*, in AIP Conference Proceedings (Vol. 1870, No. 1, p. 060012), August 2017, Publishing LLC (2017)
6. A. Aditian, T. Kubota, Y, Geomorphology **318**, 101–111 (2018)
7. X. Feng, S. Li, C. Yuan, P. Zeng, Y. Sun, KSCE J. of Civ. Eng. **22**, 3:941–950 (2018)
8. X. N. Bui, H. Nguyen, Y. Choi, T. Nguyen-Thoi, J. Zhou, J. Dou, Sci. reports **(10)**, 1:1–17 (2020)
9. M. Nouri, P. Sihag, F. Salmasi, J. Abraham, Geotechnical and Geological Eng., 1-14 (2021).
10. A. Ray, V. Kumar, A. Kumar, R. Rai, M. Khandelwal, Natural Hazards **103**, 3523–3540 (2020)
11. A. K. Bharati, A. Ray, M. Khandelwal, R. Rai, A. Jaiswal, (2021). Eng. with Comp, 1–9 (2021)
12. Y. Jiang, Q. Xu, Z. Lu, H. Luo, L. Liao, X. Dong, X. (2021). Natural Hazards and Risk **12**, 1:741–762 (2021)
13. J. N. Goetz, A. Brenning, H. Petschko, P. Leopold, P. (2015). Comp. & Geosci. **81**, 1-11 (2015)
14. N. K. Sah, P. R. Sheorey, L. N. Upadhyaya, Inter. J. of Rock Mechanics and Mining Sci. & Geomechanics Abstracts **31**, 47–53 (1994)
15. S. Liu, J. of Grey System **30**, 1:1–4 (2018)
16. P. Lu, M. S. Rosenbaum, Natural Hazards **30**, 3:383–398 (2003)
17. Y. Kuo, T. Yang, G. W. Huang, G. W. Comp. & indust. Eng. **55**, 1:80–93 (2008)
18. A. H. Bademlioglu, A. S. Canbolat, O. Kaynakli, Renewable and Sustainable Energy Reviews **117**, 109483 (2020)
19. R. Sallehuddin, S. M. Shamsuddin, S. Z. M. Hashim, *Forecasting small data set using hybrid cooperative feature selection*, in 2010 12th International Conference on Computer Modelling and Simulation (pp. 80-85), March 2010, IEEE.
20. Z. G. Qian, A. J. Li, W. C. Chen, A. V. Lyamin, J. C. Jiang, Soils & Foundations **59**, 2:556–569 (2019)
21. S. Rukhaiyar, M. Alam, N. Samadhiya, Int. J. Geotech. Eng. **12**, 556–566 (2018)
22. A. Trigila, C. Iadanza, C. Esposito, G. Scarascia-Mugnozza G, *Geomorphology* **249**, 119–136 (2015)
23. A. M. Deris, A. M., Zain, R. Sallehuddin, Meccanica **48**, 8:1937–1945 (2013)

Lithological modelling based on shear wave velocity using horizontal to vertical spectral ratio (HVSR) inversion of ellipticity curve method to mitigate landslide hazards at the main road Of Samigaluh District, Kulon Progo Regency, Yogyakarta

Skolastika Novita Widyadarsana[1], and *Eddy* Hartantyo[1*]

[1]Department of Physics, Universitas Gadjah Mada, Yogyakarta 55281, Indonesia

Abstract. Many landslides occur in Samigaluh District, Kulon Progo Regency, Yogyakarta, Indonesia. However, no research discusses landslides that often occur on the main road connecting the city of Yogyakarta and various tourist resorts in Samigaluh. This study aims at determining the soil vulnerability and lithology model at that main road as a contribution to landslide mitigation planning. This lithology model is based on shear wave velocity (Vs) and layer thickness derived by microtremor datasets. The data were processed by the inversion of the Horizontal to Vertical Spectral Ratio (HVSR) technique of the ellipticity curve method. The result of the study shows that the first layer is associated with topsoil which has Vs of 263 m/s, the second layer is clay which has Vs of 607 m/s, the third layer consists of clay, breccia, and pumice which has Vs of 1119 m/s, and the fourth layer is andesite bedrock which has Vs of 1721 m/s. Andesite is impermeable to water and can become a slip field for landslides. Clay, breccias, and pumice can absorb water so that their weight increases when it rains. When they are on an impermeable rock on a certain slope, a landslide occurs.

1 Introduction

Samigaluh District, Kulon Progo Regency, Yogyakarta, Indonesia is geologically a part of the Kulon Progo Mountains which is in the form of a dome with a flat top and steep surroundings nicknamed the "Oblong Dome". This district has two rock formations, namely the Kebo Butak Formation and the Jonggrangan Formation. The Kebo Butak Formation has a lithology of volcanic breccia with fragments of andesite, lapilli tuff, tuff, lapilli breccia, insertion of andesite lava flows, agglomerates, and volcanic sandstone. Meanwhile, the Jonggrangan Formation generally consists of conglomerate, tuffaceous marl, coral limestone, and calcareous sandstone with mollusc content and claystone with lignite insertion.

The topography of this area is dominated by hills with steep slopes make Samigaluh District has a high potential for landslide hazard. The steeper the slope and the higher the runoff rate, the value of soil stability will be smaller. A large number of residents and natural tourist resorts which are located around the main road of the Samigaluh District makes this main road a vital infrastructure that functions to connect Yogyakarta and tourist resorts in Samigaluh District. The slope of the Samigaluh District is dominated by a slope of 25-45% (steep) which is marked in yellow and a slope of >45% (very steep) which is marked in red [1]. The slope map of Samigaluh District is shown in Fig. 1.

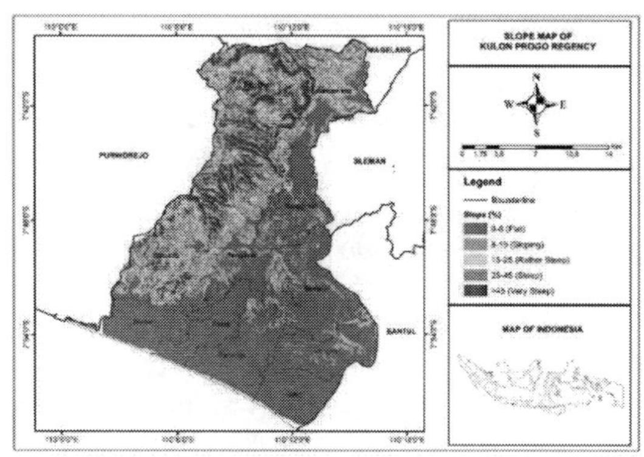

Fig. 1. Slope map of Kulon Progo Regency [2].

The type of landslide that occurs around the main road in Samigaluh District is the flow of debris because the landslide occurs due to high rainfall, which is then able to erode and move soil or soft rock on steep slopes [3].

According to [4], there were 26 landslides (54.17% of the total landslides in Kulon Progo Regency) occurred on the Samigaluh District main road. The recent landslide that occurred on the Samigaluh District main road are shown in Fig. 2(a) and (b).

*Corresponding author: hartantyo@ugm.ac.id

(a)

(b)

Fig. 2. The landslide occurred on the main road from the east of Samigaluh District: (a) on the right side and (b) on the left side.

Many landslides occur in Samigaluh District, but no research discusses landslides on the main road. There are many natural tourist resorts around the main road in Samigaluh District. The number of natural tourist resorts is one of the potential income in this district. Moreover, the Kulon Progo Regional Government has also designated Samigaluh District to be one of the Integrated Menoreh Agrotourism Rural Areas based on the Kulon Progo Regent's Decree Number 367/C/2016 [5]. So it is important to know the potential for landslides because the main road in Samigaluh District is a vital infrastructure to support transportation, economic development and ease of community mobility, goods and services. Research conducted by [6-7] shows the dominant frequency and amplification values in Samigaluh District. While research [8] shows the value of shear wave velocity and lithology located in Samigaluh District.

To mitigate the landslide hazard, we mapped the soil vulnerability and lithology based on shear wave velocity (Vs) and layer thickness derived by microtremor datasets. Microtremor in soil structure studies is used to determine subsurface conditions [8-9]. This method was chosen because the equipment used is relatively simple, low cost, does not require a lot of energy [10] and low cost [11]. Lithology plays an important role in the occurrence of landslides because it affects the strength and permeability properties of the bedrock associated with the slope [12].

2 Method

2.1 HVSR Inversion of Ellipticity Curve Method

The inversion process is a data matching process to find parameters that produce the best model with a small misfit value (minimum misfit) [13]. The ellipticity curve is one of the modelling methods that can be used to invert the HVSR curve. The principle of this method is to separate Rayleigh waves from other types of waves by combining the signal window from the microtremor data. Then calculate the energy of the sum of the vertical and horizontal signals from the signal window. This energy is used to estimate the ellipticity of Rayleigh waves. Rayleigh wave ellipticity as a function of frequency is related to the underground structure, that is the profile of shear wave velocity and sediment thickness [14]. Furthermore, the value of the shear wave velocity (Vs) will be used to estimate the subsurface lithology in the study area. Subsurface lithology modelling was obtained using Geopsy software.

2.2 Data acquisition

Several measurement requirements, as well as technical site selection and microtremor data collection techniques, are set by the SESAME European Research Project [15]. The research area is located in Samigaluh District, Kulon Progo Regency, Yogyakarta Special Region. The measurement area covers an area of 3000 x 4650 meters with a measurement distance of 450 meters and is carried out at 31 points. Data collection at each point was carried out for 40 to 50 minutes with a sampling frequency of 100 Hz. The selection of this study area is based on the distribution of tourism in the research area. The study area is shown in Fig. 3.

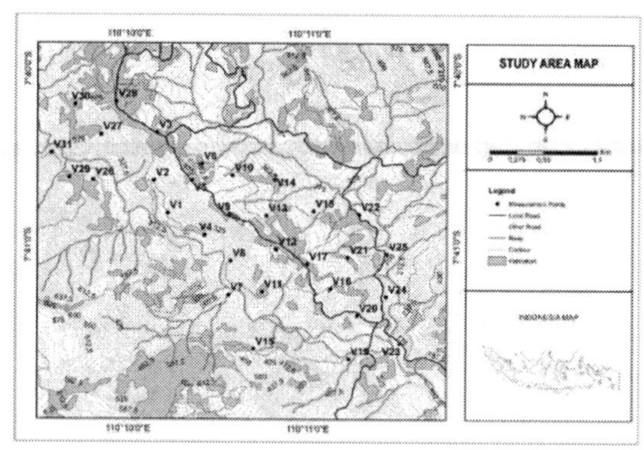

Fig. 3. Study area map [16].

2.3 Data analysis

The microtremor data is then processed with Horizontal to Vertical Spectral Ratio (HVSR) in Geopsy software which produces an HVSR curve. The HVSR curve as input data is inverted using the Dinver menu in the Geopsy software. Parameters in the form of P wave velocity (Vp), S wave velocity (Vs), and density are used as limits to obtain a model that approximates the actual subsurface structure. The parameters in this study refer to previous studies [8, 17-19]. Based on previous research, researchers made parameters in the research area which can be seen in Table 1.

Table 1. Parameters of study area.

Layer	Vp (m/s)	Vs (m/s)	ρ (kg/m³)
1	500-1500	100-1200	500-1500
2	800-2000	120-1200	1000-2000
3	1000-2500	150-1500	1200-2500
4	1200-3500	200-2000	1500-3500

3 Results and Discussion

The inversion of the HVSR curve produces ellipticity curves and ground profiles. The ellipticity curve shows the fit between the model created and the HVSR curve (misfit). The black colour on the ellipticity curve shows the HVSR curve, while the red colour shows the smallest displacement of the inversion result. Colours other than red and black on the ellipticity curve show the modelling curve with various misfit values. An example of an ellipticity curve display can be seen in Fig. 4.

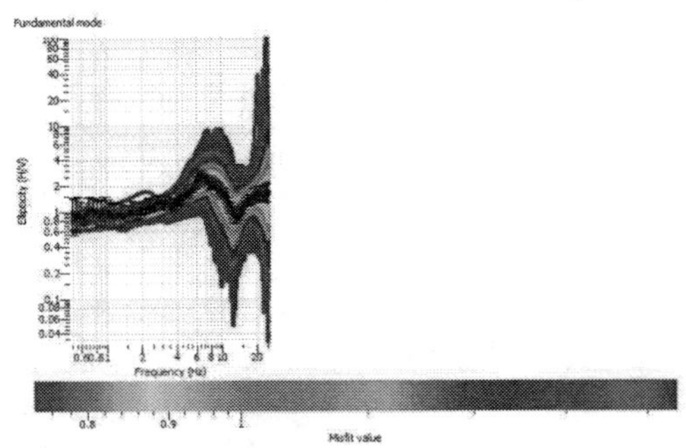

Fig. 4. Ellipticity curve of V4 site.

Ground profiles are modelling results that show the values of P wave velocity (Vp), S wave velocity (Vs), and density with varying depths. The black colour on the ground profiles display shows the smallest misfit of the inversion result. Colours other than black on the ground profiles show the modelling results with various misfit values. An example of the display of ground profiles can be seen in Fig. 5.

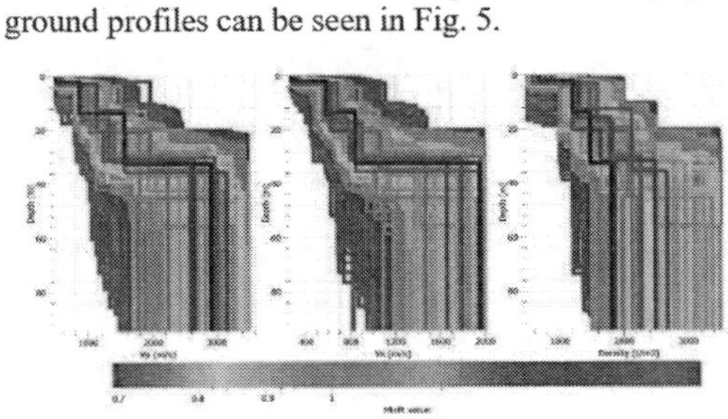

Fig. 5. Ground profiles of V4 site.

Based on the results of the inversion of the ellipticity curve, the average shear wave velocity value in the first layer is 263 m/s, the second layer is 607 m/s, the third layer is 1119 and the fourth layer is 1721 m/s. The modelling of the shear wave velocity can be seen in Fig. 6.

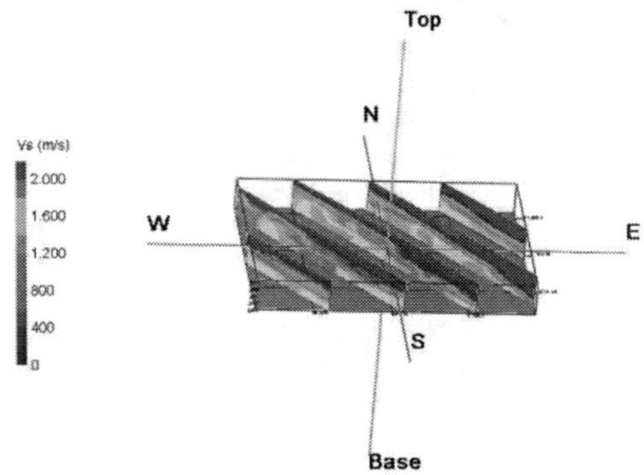

Fig. 6. Shear wave velocity values model in the study area.

The relationship between shear wave velocity and depth produces a value of Vs30. According to [21], the value of the shear wave velocity from the surface to a depth of 30 meters (Vs30) is used as a standard for amplification. The range of Vs30 values in the study area is 320-894 m/s. The resulting Vs30 value is then used for soil classification in the study area. Based on soil classification [22], the research area has 3 soil classifications, namely medium soil, very dense soil and soft rock and rock. The distribution of the shear wave velocity (Vs30) can be seen in Fig. 7.

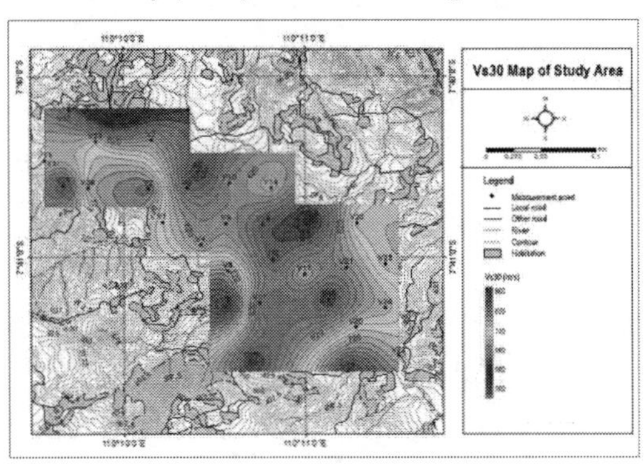

Fig 7. Map of the distribution of Vs30 values in the study area.

Every soft rock has a small shear wave velocity value compared to hard rock because the shear wave velocity value is directly proportional to the density. A velocity wave will enlarge as it passes through a dense medium, so rocks that have a high density will require a high shear wave velocity when passing through the rock.

The density value has an inverse relationship with the porosity value. Porosity is the degree of cavities between the rock particles. If the rock has many cavities, the density of the rock is getting smaller. So that the rock with high porosity has a low density.

Rocks with high porosity have a high risk of landslides because these rocks can absorb water easily. The small Vs30 value found on the main road in Samigaluh District indicates a high level of landslide susceptibility.

Based on the Vs value, the lithology in the study area consists of the first layer of topsoil, the second layer of clay, the third layer of soil, clay, breccia and pumice, and the fourth layer of bedrock in the form of andesite. To facilitate the visualization of lithology, an incision was made in the research area which was divided into 7 incisions based on the height and slope of the slopes and areas that have the potential for landslides. When it rains, areas that have thick sediment and are located at an altitude and steep slope will easily experience landslides. The trajectory map of the lithological incision of the study area can be seen in Fig. 8. and the lithological incision of the study area can be seen in Fig. 9.

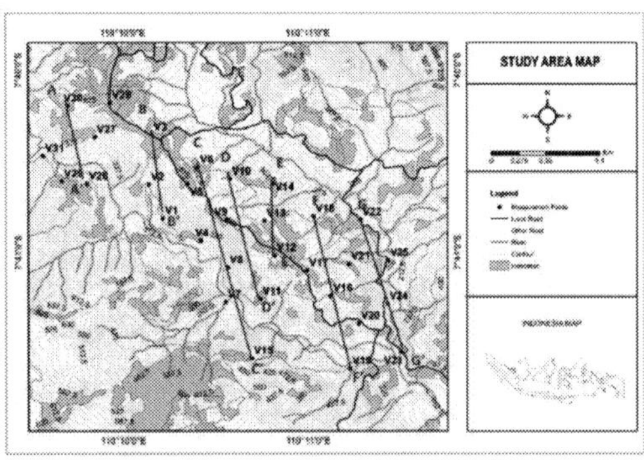

Fig. 8. Map of the trajectory lithological incision in the study area.

Many points in the research area such as V26, V2, V1, V6, V4, V8, V15, V10, V11, V14, V13, V16, V19, V24, V23 and several main roads are located on steep slopes that can occur landslides. In addition, there are several points located on steep slopes such as V30, V3, V9, and several main roads which of course can occur landslides on the edge of the main road.

The presence of andesite rock is a triggering factor for landslides because it is impermeable to water and can be used as a slip field for landslides. The type of soil that is clay, is the type of soil that easily passes water. In addition to clay, there are breccias and pumice which are capable of passing water in the subsurface layer of the study area.

With the nature of this type of soil, it makes the soil gain weight when it rains, so that if the soil is on an impermeable rock on a certain slope, the incoming water will be held back and the soil on a certain slope will have the potential to slide into a landslide.

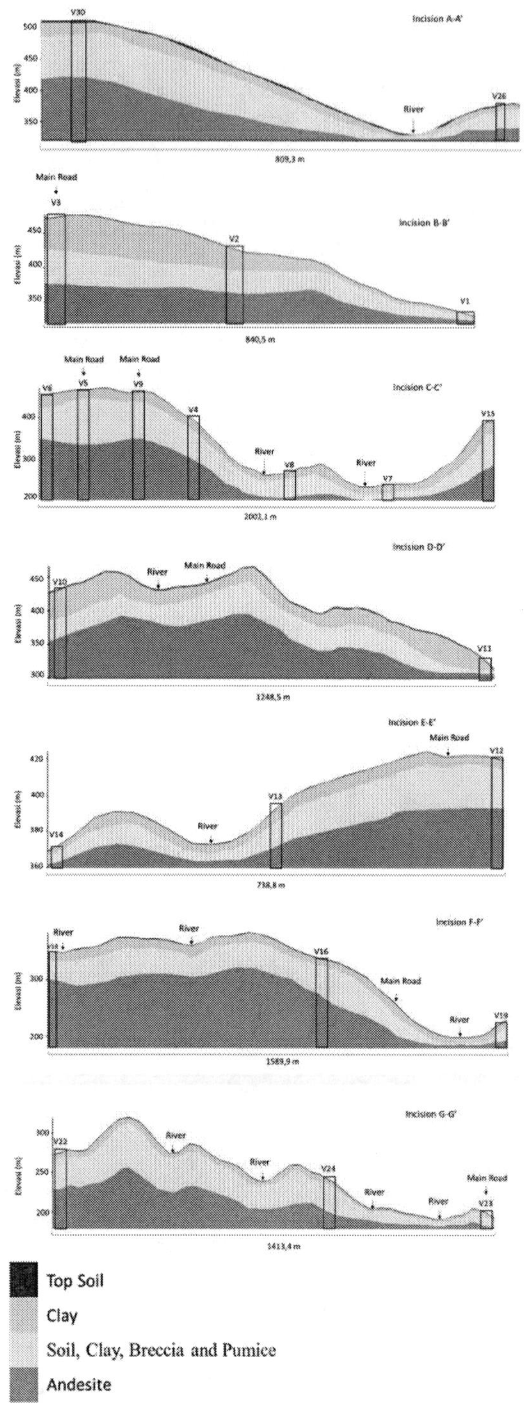

Fig. 9. Lithological incision in the study area.

The authors would like to thank BPBD Kulon Progo for the landslides data in this research area and the Geophysics Laboratory of FMIPA UGM for the research equipment.

References

1. D. M. Adfy *et al.*, J. Fis. Unand **10**, No.1, 8–14, (2021)
2. DEMNAS.URL:https.//tanahair.indonesia.go.id/demnas/#demnas
3. D. M. Cruden and D. J. Varnes, Spec. Rep. - Natl. Res. Counc. Transp. Res. Board **247**, 36–75, (1996)
4. E. S. Nugroho, D. S. Hadmoko, C. Van Westen, and N. Kingma, Forum Geogr **26**, No.2, 132–145, (2012)
5. Sutarmi, "Pemkab Kulon Progo Merintis Kawasan Pedesaan Agrowisata Menoreh Terpadu", (2020)
6. A. Syahputri and S. Sismanto, Jurnal Fisika Indonesia, **24**(2): 66 (2020)
7. B. Ruwanto, Y. Sumardi, and D, Darmawan, Jurnal Penelitian Saintek **21**(1): 66–74, (2016)
8. L. Katriani, D. Darmawan, B. Ruwanto, H. Lutfiana, and H. H. Prameswari, J. Phys. Conf. Ser. **1397**, No.1, (2019)
9. H. Sulistiawan, S. Supriyadi, and I. Yulianti, *Shear Wave Velocity Profiling Analysis for Site Classification Using Microtremor Single Station Method* in AIP Conference Proceedings (2014)
10. F. J. Sánchez-Sesma, Earth, Planets and Space **69**(1), (2017)
11. A. Gosar, Nat. Haz. Earth Syst. Sci **17**(6): 925–937, (2017)
12. F. E. S. Silalahi, Pamela, Y. Arifianti, and F. Hidayat, Geosci. Lett **6**, No. 1, (2019)
13. M. Hobiger, Polarization of Surface Waves: Characterization, Inversion and Application to Seismic Hazard Assessment, (2011)
14. M. Hobiger, P. Y. Bard, C. Cornou, and N. Le Bihan, Geophys. Res. Lett **36**, no. 14, pp. 1–5, (2009)
15. P. Y. Bard, K. Atakan, A. Anastasiadis, and R. M. Azzara, Guidelines for The Implementation of The H/V Spectral Ratio Technique on Ambient Vibrations-Measurements, Processing and Interpretations (SESAME Site Eff. Assess. using Ambient Excit, 2004)
16. Geospasial untuk Negeri, (2020)
17. P. Susilanto, D. Ngadmanto, B. Sunardi, and S. Rohadi, J. Lingkungan dan Bencana Geology **10**, No. 2, 41–50, (2019)
18. A. R. Maulana, F. Y. Arwa, N. B. Wibowo, Purwanta, and S. Kiswanti, Analisis Kg dan Pga Berdasarkan Model Vs30 USGS di Kabupaten Kulon Progo (2019)
19. N. Budi, I. Huda, Buletin Meteoroogi Klimatologi dan Geofisika **1**, No.2, 21–31 (2020)
20. A. I. Kanli, P. Tildy, Z. Prónay, A. Pinar, and L. Hermann, Geophys. J. Int **165**, No. 1, 223–235, (2006)
21. BSN, Tata Cara Perencanaan Ketahanan Gempa untuk Struktur Bangunan Gedung dan Nongedung (Badan Standardisasi Indonesia , Bandung, 2012)

Permutation entropy variation for 2007 effusive dome-forming eruption period of Kelud Volcano, Indonesia

Lusia Rita Nugraheni[1*], *Agung* Harijoko[1], *Wiwit* Suryanto[2], and *Hetty* Triastuty[3]

[1]Geological Engineering Department, Faculty of Engineering, Universitas Gadjah Mada, Indonesia
[2]Geophysics Sub-department, Physics Department, Faculty of Mathematical and Natural Sciences, Universitas Gadjah Mada, Indonesia
[3]Center for Volcanology and Geological Hazard Mitigation, Geological Agency, Indonesia

Abstract. The complexity of a system recorded in time series data can be measured statistically using permutation entropy (PE). The state of a system (e.g. regular, chaotic, random, etc.) that underlies the appearance of variations in time series can be determined with PE. Since volcanoes are considered as the complex dynamical system controlled by interactions of many processes. Permutation entropy can be applied to study the system mechanism of volcano. We utilized PE to study system mechanism of Kelud volcano in 2007 dome-forming eruption period, from 3 (KWH; KLD; UMBK) seismic stations with different distances from the crater lake. Then, we want to compare the results. The result of study shows that the PE pattern for each station is different. The unique PE pattern that can be used as an eruption precursor is only shown at KWH and KLD stations. This pattern began to appear 2.7 days before the eruption on 3 November 2007. Data from UMBK station doesn't show unique PE pattern. The factors such as sensor distance from magmatic activity center, size, and type of eruption probably influenced the final PE result. Using PE as the addition to volcano monitoring can maximize efforts in mitigation activities.

1 Introduction

One of the Quaternary active and hazardous volcanoes in Java Island, Indonesia is Kelud volcano (1731 m above sea level) [1]. The volcano has a crater lake that contains water at its summit. The tectonic setting of the volcano is the subduction zone between the Indo-Australia plate beneath the Eurasian plate [2]. Position of Kelud volcano is surrounded by Kawi and Butak volcanoes situated at the east of Kelud volcano, and also Arjuno-Welirang volcano that located at the northeast of Kelud volcano (Fig. 1).

Based on Center for Volcanology and Geological Hazard Mitigation (CVHGM) records, Kelud volcano has history of explosive and effusive eruptions. Explosive eruption of Kelud volcano usually was short-lived period and initiated by phreatomagmatic eruption, then followed by explosive magmatic eruptions [3]. According to eruptive records after 1000, explosive eruptions of the volcano have occurred in 1586; 1901; 1919; 1951; 1966; 1990; 2014 with explosive events of VEI between 3 to 5 [3]. Meanwhile for effusive eruptions of the volcano formed a lava dome. Effusive dome-forming eruption events have occurred in 1376; 1920; 2007 [3]. The VEI scale of effusive eruption was not bigger compared to explosive eruption.

Volcanoes are complex dynamical system [4]. The system is controlled by interaction of various processes, which are generally nonlinear and stochastic such as magma movements, gas and bubbles generation, etc.

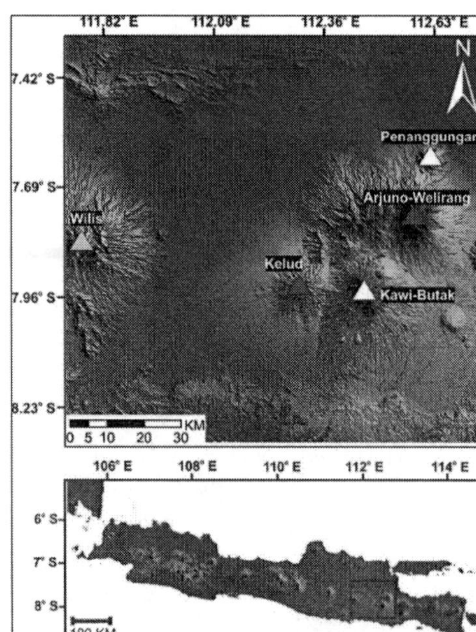

Fig. 1. Location map of Kelud volcano and its surroundings. Red – yellow – white triangles show volcano type A – B – C. Inset map shows volcanoes distribution at Java Island.

* Corresponding author: lusia.rita.n@mail.ugm.ac.id

Seismic waves like ambient noise seismic which propagates through the system of volcano can be considered as complex time series [5]. The randomness amount contained in complex time series can be measured statistically by permutation entropy (PE).

Glynn and Konstantinou (2016) introduced PE as one of methods to approach volcanic eruption forecasting. They applied PE to forecast the 1996 Gjálp fissure eruption at Iceland. Position of Gjálp fissure is between Grimsvotn volcano and Bardarbunga volcano with Mid Atlantic Ridge as its tectonic setting. The result of their study represents similar temporal pattern PE variation from 3 different seismic stations, namely 7 days before the eruption, PE reached minimum value [6]. This pattern can be considered as the precursor signal of the 1996 Gjálp eruption.

In this study, we applied PE on the 2007 dome-forming eruption event of Kelud volcano. We used data from 3 seismic stations with different distance from the crater lake. The aims of the study are to know the PE patterns of Kelud volcano especially for the 2007 effusive eruption event from seismic stations with different distance and to compare them. Furthermore, we want to know the effectiveness of PE for forecast effusive eruption at Kelud volcano.

2 Sequence of 2007 eruption event

The information on sequence of 2007 eruption event is summarized from CVHGM records in Hidayati, et.al. (2009); Hidayati, et.al. (2019); Nakamichi, et.al. (2019). According to CVHGM records, intense degassing at the lake floor in July 2007 was observed as the first sign of volcanic activity increased after 17 years in the rest period [3]. After that in August, the sign was followed by an increase of the crater lake water temperature and CO2 concentration. The colour change of crater lake water also recorded from typical green to yellowish in August [1, 3, 7]. Meanwhile from deformation monitoring showed that continuous uplift occurred at the crater rim from July until early September [7].

The increase of seismic activity was marked by first volcano-tectonic (VT) swarm that occurred on 9 September. CVHGM raised the level of Kelud activity from level I (normal) to level II (watch) [3]. Two days later on 11 September, VT earthquakes reached 13 events in 5 hours with hypocenters distribution at 2 – 4 km below the crater [1]. After that day, VT earthquake event occurred on average one each day. A remarkable VT earthquake occurred on 26 to 29 September. In this period, second VT swarm that contained larger amplitude occurred and amount of VT reached 61 events with no significant change of hypocenters distribution [1]. These events made CVHGM raised volcanic activity to level III (standby) [3]. The occurrence of VT earthquakes decreased after 29 September, with VT hypocenters distribution shifted to 0.5 – 1 km below the crater [1]. Low frequency (LF) earthquakes began to occur as VT earthquakes decreased. The occurrence of LF earthquake continued until 13 October. The first volcano tectonic type B (VB) swarm that contained 306 events occurred on 16 October for 7 hours [1]. This caused CVHGM raised volcanic activity again to level IV (warning) [1, 3]. The VB earthquake continued until 21 October with decreasing in number. The event was followed by seismicity quiescence that occurred from 22 to 23 October [1].

According to the record on 24 October, the occurrence of VT earthquakes was followed by occurrence of VB earthquakes [1]. Then on 31 October, VT swarm with larger amplitude reached 31 events occurred, and followed by 89 VB earthquake events. On the following day recorded 55 VT earthquake events were followed by 1437 VB events. Furthermore, on 1 November, there was also tremor earthquakes with 274 events (marked as line a in Fig. 3, Fig. 4, Fig 5) [1]. After that on 2 November, seismicity activity was signed by the occurrence of VT that reached 52 events and VB earthquakes that attained 690 events [1]. That event was followed by 8 hours of quiescent period [7]. The quiescence was ended by continuous tremor that began at 11:07 local time (marked as line b in Fig. 3, Fig. 4, Fig 5) [1]. The amplitude of continuous tremor was getting larger and at 14:43 continuous tremor was replaced by spasmodic tremor for 70 minutes (marked as line c in Fig. 3, Fig. 4, Fig 5) [3, 7]. The occurrences of tremor were interspersed by emission earthquakes. On 19:47 local time, VT earthquake was occurred again with the number and amplitude increasingly until 3 November 06:31 local time [3]. On the same day at 06:00, the temperature measurement of water crater lake showed the temperature has increased quite significantly [3].

On 3 November at 12:32 from seismogram showed continuous LF tremor began as rapid series of shallow LF and the event was followed by spasmodic LF tremor at 13:11 [3]. At 15:55 (marked as line d in Fig. 3, Fig. 4, Fig 5), the amplitude was getting larger and became saturated [1, 3, 7]. At the same time, there was sharp increase of temperature and sudden deflation of radial tilt. This moment was estimated as the time where extrusion lava occurred [3]. Unfortunately, the weather was heavy rain so it was too dangerous for field observation. A small fuming black dot in crater lake was observed from CCTV at 4 November [1, 3, 7].

3 Methodology

The activity of Kelud volcano is monitored by Kelud Volcano Observatory and operated by CVHGM. Since April 2007, the seismic activities are monitored by 5 seismic stations and deformation of the volcano is recorded by 2 tiltmeter stations. Each seismic station is equipped by L4-C vertical short-period seismometer [1]. The location of seismic stations is showed at Fig. 2.

This research used seismic records from monitoring of 3 seismic stations. There are Kawah (KWH) where located inside the crater, Kelud (KLD) where situated on the crater rim (about 600 m from active vent), and Umbuk (UMBK) where located about 4.5 km from the crater. The period of data is from 27 September until 5 November 2007 or 275 – 309 in Julian date.

The format data for input in PE calculation is MSEED. Because the original format data is WIN, so it must be converted to MSEED. After that, we cut the time series with window length of 10 minutes to generate several time series. These data were used as input in PE calculation.

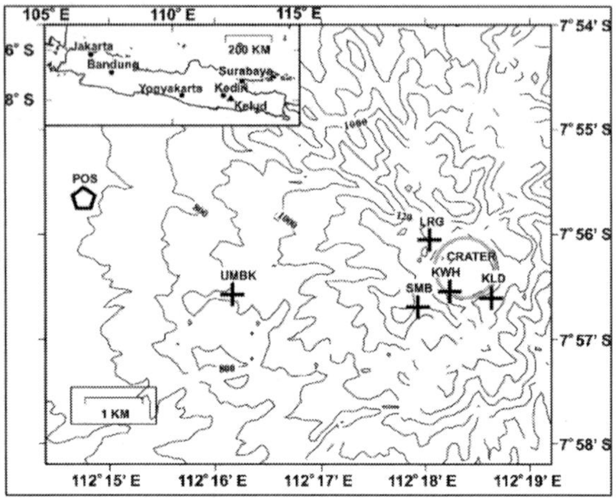

Fig. 2. Location map of Kelud volcano seismic stations and observation post. KWH = Kawah; KLD = Kelud; LRG = Lirang; SMB = Sumbing; UMBK = Umbuk; POS = observation post [7].

3.1 Calculation of permutation entropy

Permutation entropy (PE) was introduced by Bandt and Pompe on 2002 as statistical measurement method of complexity amount contained in the time series data [8]. This method can also be used to determine the state of system that underlies the appearance of variations in time series [9]. There are 2 important parameters in PE calculation, namely the embedding dimension (m) and the delay time (L). Embedding dimension describes the amount of information contained in each vector. Meanwhile, L represents delay time on successive of vectors.

PE transforms time series data into graph of permutation entropy value. A time series is arranged become several vectors based on value input in m and L parameters. Then the values contained in vector are mapped and substituted with unique permutation pattern π (0,1,..,m-1). Variation amount of the pattern are determined by m factorial ($m!$). After that, the probability distribution of each pattern is calculated as the input value in PE Equation (1) [9].

$$PE(m) = -\sum_{x=1}^{m!} \pi_x \, ln \, \pi_x \tag{1}$$

Normalizing PE(m) by $ln(m!)$ we obtain normalized quantity with interval value from 0 to 1. The value of PE normalization can describe how far the data from maximum complexity (PE = 1) or how close the data from minimum complexity (PE = 0). The zero value of PE indicates time series data are decreasing or increasing monotonously [6]. This condition shows the presences of regular system mechanism.

The values of PE depend on the combination of chosen parameter m and L [9]. For practical purposes, Bandt and Pompe recommend m = 3,...,7 [8]. Riedl, et.al. in 2013 summarized several studies of PE on various fields showing that most of them used m = 3-8, L from tens to hundreds for time series from measurements of dynamical processes being observed at a fixed rate per unit time, L = 1-5 for time series from measurements of processes that possess an inherent cycle represented by triggering events [10]. Previous studies such as [6] used m = 5 and L = 5 for monitor the 1996 Gjálp eruption; the same values for m and L were adapted by [11] to study the 2010 Merapi eruption; [5] applied m = 5,6 and L = 1,2,3,4,5 for monitor the January 2011 Shinmoedake eruptions sequence; [12] utilized m = 3,4,5 and L = 4,5,6 for study the 2014 Kelud eruption. Thus, we used variation of m = 3,5 and L = 2,3,4,5 to perform permutation entropy in this study.

4 Results and discussion

Based on the PE results of KWH, KLD, and UMBK stations show varying result. The PE graph of KLD station shows clear pattern at combination of parameter m = 5 and L = 2 (Fig. 3). PE value from 270 Julian date or 27 September to 304 Julian date end presents the state of volcano system on the maximum complexity during that period. It can be seen from the high PE value, which is close to 1. The differences PE pattern trend began to appear on 305 Julian date or 1 November, namely 2.7 days before eruption. It indicates that there has been a change in the system of volcano. PE value began to decrease, this trend corresponds with recorded tremor events (marked as a in Fig. 3). The properties of tremor which have monotonous wave pattern and lasting for long-duration indicated that the mechanism in volcano system began well-ordered. Then, the quiescent period that lasted for 8 hours caused PE value increased again (marked as pink highlight). This condition stated system on maximum complexity for a moment.

A sudden drop in the PE value (marked as b in Fig. 3) associated with the continuous tremor event that ended the quiescent period. Vertical line c in PE graph associated with amplitude of tremor became larger and the VT event occurred again. The trend of PE graph was increasing slowly but the maximum value was not high as before. Then, a sudden drop of PE value (marked as d in Fig. 3) coincided with the time that was interpreted as the period of first lava extrusion. At that time, the results of deformation and temperature monitoring also showed the precursor of eruption.

The PE graph of KWH station shows unclear pattern at all combination of parameter m and L. So, we tried to filter time series before it was used again in PE calculation. We filtered the data using LPF in 3 stages. Firstly, we tried to apply LPF 3 Hz. Then, the filtered data were processed using PE but the PE graph still showed unclear pattern. So, we tried to filter data using LPF 2 Hz and processed the filtered data again using PE.

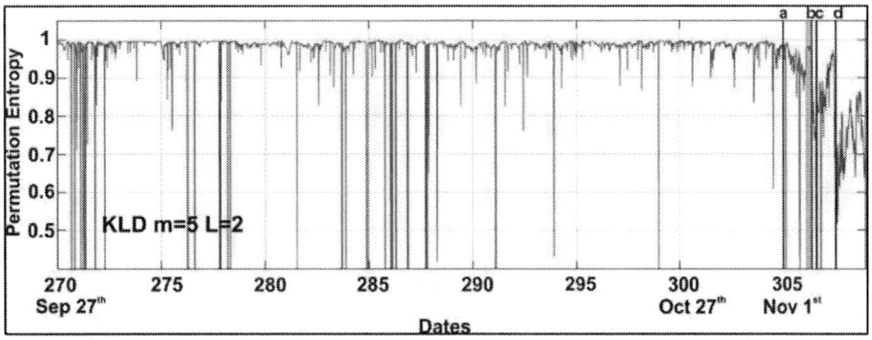

Fig. 3. Permutation entropy graph of KLD station where situated about 600 m from active crater.

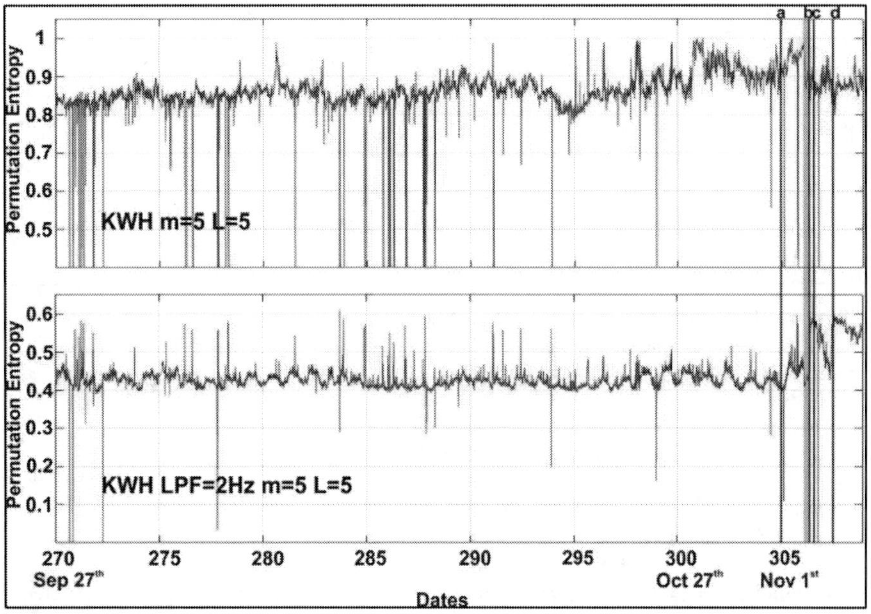

Fig. 4. PE graphs from KWH station that situated inside the crater. Top : PE graph from combination m=5 and L=5 without filtering step. Bottom : PE graph from combination $m = 5$ and $L = 5$ using filtered data LPF 2 Hz.

The PE graph of filtered data LPF 2 Hz shows clear pattern at combination $m = 5$ and $L = 5$ (Fig. 4). We also tried to filter data using LPF 1 Hz, but the final PE result doesn't show any trend.

The unique pattern of PE began appear at 305 Julian date (2.7 days before eruption) same as the result from KLD station (Fig 3 ; Fig. 4). But the both trend are not similar like the result of study [6]. The spectrogram of KWH showed high intensity at the dominant frequency <10 Hz from 270 – 304 Julian dates. However, KLD did not show same plot spectrogram at the same period. This may be due to small-scale activity in the Kelud center. Thus, signal from the activity had low energy. As a result, the signal can only be recorded by the nearest seismometer to crater lake (KWH). The volcano tectonic setting is interpreted to have influence on PE result. The tectonic setting of Gjálp fissure is mid ocean ridge and location of the fissure at neovolcanic zones. Neovolcanic zones are the active magmatic zones within divergent plate boundaries [13]. The magmatic system with mid ocean ridge tectonic setting is estimated to have larger scale than subduction tectonic setting. Based on [14], mid ocean ridges are the largest and most voluminous magmatic environment on Earth. So, it is possible that the installed seismometers at around neovolcanic zone will give the similar response.

The PE graphs from UMBK station that located 4.5 km from the crater presents unclear pattern before eruption (Fig. 5). This unclear pattern appear at all combination parameter, without and includes a filtering stage in processing. It is probably due to the location of sensor which is far from the center of magmatic activity (4.5 km). Thus, PE graph cannot describe the pattern that useful as one of eruption precursor. The type and size of eruption are interpreted will give influence because the amount of released seismic energy will different. Effusive eruption will release less amount of energy in comparison with explosive eruption.

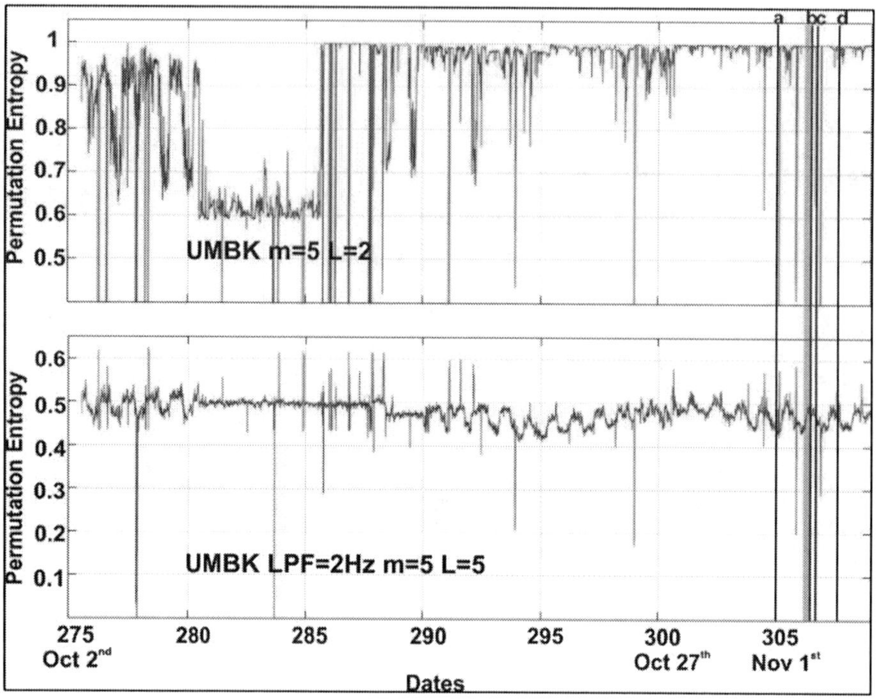

Fig. 5. PE graphs from UMBK station. Top : PE graph from combination $m = 5$ and $L = 2$ without filtering step. Bottom : PE graph from combination $m = 5$ and $L = 5$ using filtered data LPF 2 Hz.

Based on the result of this study, PE graph will show the pattern that useful as one of sign Kelud eruption precursor, especially dome-forming eruption, if the location of the sensor (short-period seismometer) is not too far, namely about 1 km from magmatic activity center. If the installed instrument is broadband seismometer probably data from sensor at UMBK can show PE pattern as dome-forming eruption precursor. Because the recorded frequency range is getting wider. From the result of this study, we know that using PE as an addition to existing monitoring method can maximize efforts to mitigate volcanic eruption.

5 Conclusion

PE patterns of dome-forming Kelud eruption period from 3 station are different. The unique PE pattern that appeared before eruption was only shown at Kawah (KWH) and Kelud (KLD) stations. According to this study, PE is effective to forecast the dome-forming eruption of Kelud with a note the sensor (short-period seismometer) distance from magmatic activity center is about 1 km. Using PE as addition volcano monitoring method can help mitigation activities.

We would like thank to Department of Geological Engineering, Engineering Faculty, Universitas Gadjah Mada for the "Hibah Departemen" funding. We want thank to Center for Volcanology and Geological Hazard Mitigation, Geological Agency, Indonesia for allow us to use Kelud seismogram data.

References

1. S. Hidayati, A. Basuki, Kristianto, I. Mulayana, IJOG **4**, 10 (2009)
2. R.V.V. Bemmelen, The Geology of Indonesia. Vol. IA: General Geology of Indonesia and Adjacent Archipelagoes (US Government Printing Office, 1949)
3. S. Hidayati, H. Triastuty, I. Mulyana, S. Adi, K. Ishihara, A. Basuki, H. Kuswandarto, B. Priyanto, A. Solikhin, J. Volcanol. Geotherm. Res. **382**, 18 (2019)
4. R.S.J. Sparks, Earth Planet. Sci. Lett. **210**, 15 (2003)
5. I. Nurfitriana, Permutation Entropy Variation of Seismic Noise prior to Eruptive Activity at Shinmoedake Volcano, Japan (Thesis, NCU, 2019)
6. C.C. Glynn, K.I. Konstantinou, Sci. Rep. **6:37733**, 9 (2016)
7. H. Nakamichi, M. Iguchi, H. Triastuty, M. Hendrasto, I. Mulyana, J. Volcanol. Geotherm. Res. **382**, 13 (2019)
8. C. Bandt, B. Pompe, Phys. Rev. Lett. **88**, 4 (2002)
9. M. Zanin, L. Zunino, O.A. Rosso, D. Papo, Entropy **14**, 26 (2012)
10. M. Riedl, A. Müller, N. Wessel, Eur. Phys. J. Special Topics **222**, 15 (2013)
11. A. Rakhman, Wahyudi, A.B. Santoso, H. Humaida, W. Suryanto, Int. J. Geophys. **2020**, 10 (2020)
12. D.M. Sudiyanto, Menuju Pemantauan Aktivitas Gunung api Secara Realtime Menggunakan Metode

Permutation Entropy: Studi Kasus Erupsi Gunung Kelud, 13 Februari 2014 (Unpublished Thesis, UGM, 2017) [in Bahasa Indonesia]

13. P.C. LaFemina, Plate Tectonics and Volcanism in The Encyclopedia of Volcanoes (Academic Press, 2015)
14. B. V. W. de Vries, M. V. W. de Vries, Tectonics and Volcanic and Igneous Plumbing Systems in Volcanic and Igneous Plumbing Systems (Elsevier, 2018)

Predicting remaining life of lattice steel transmission tower based on condition assessment and corrosion hazard data

Fathoni Usman[1]*and *Nor Hazwani* Nor Khalid[2]

[1]Institute of Energy Infrastructure, Universiti Tenaga Nasional, 43000 Kajang, Selangor, Malaysia
[2]Civil Engineering Department, College of Engineering, Universiti Tenaga Nasional, 43000 Kajang, Selangor, Malaysia

Abstract. This paper presents a novel approach for predicting the remaining life of lattice steel transmission towers in Malaysia. The transmission tower components assessment grouped into ground corrosive condition, foundation condition, lattice steel body of the tower, and conductor. Significant and scientific field investigation and visual assessment evaluate the state of the tower. The corrosion hazard, which is a product of climate and air pollution data, is used. Three separated lines and 84 towers were selected with different ages of the tower in service. The Weighing Scoring Method with Analytical Hierarchy Process to impose the weighed factor was used. The survival function analysis is used to predict service life, and the tower remaining life was determined. The paper shows how to use the model to predict the remaining life of a transmission tower and how this can be used to advise maintenance planning.

1 Introduction

Energy demand and consumption have a significant correlation with population growth. In the case of electricity, the demand is catapulted by the increase of household income which was contributed by 82% of domestic consumers out of 10.08 million consumers in Malaysia [1, 2]. The high voltage transmission towers are among the most precious asset in energy infrastructure spanning from the power plan crossing the urban, rural, and most remote areas to bring electricity to the transmission substations and then distribute to the end consumers. During its service life, the tower subjected to deterioration due to various factors. Environment with different corrosion hazard and the nature of its service and most of the tower located in remote area, the effort to ensure the good condition to provide less power disruption is the utmost. Appropriate maintenance strategies need to be advised and chosen to prevent any failure to the tower and transmission line [3]. The failure will not only cause financial loss but possible loss of life.

There are 24,607 Circuit-km of transmission line, 463 transmission substations, and with the circuit length, 123,290 GWh electricity has transmitted, and 18.808 MWh peak demand in Peninsular Malaysia reported in 2021 [2]. Environmental causes dominated power disruption in Peninsular Malaysia from 2000 to 2016 for 31%, 17% by structural failure, and 40% caused by conductor problems [1]. Environment load is still not fully understood and overlooked during design and along with the service life [4].

The total land area of Peninsular Malaysia is 132,078 km^2 which is 40% of the total land area of Malaysia. The climate dictated by the two monsoons, the Southwest Monsoon from May to September and Northeast Monsoon from October to March, comes with high received rainfall.

One of the most prevalent deteriorations caused by this complicated interplay of materials and nature is corrosion. The environmental load is defined as the degradation agent from the atmosphere (i.e., gaseous pollution, acid deposition, and rainwater acidity) and its local climate (i.e., the relative humidity, temperature, and amount of received precipitation) [5]. Table 1 shows the air pollution emission in Malaysia as published in the Malaysian Statistical Handbook 2019 [6]. These substances will hasten the process of degradation. Degradation can be caused by bad behavior or stormy conditions, but the terminology of deterioration focuses on the deteriorating or weakening of a specific thing. The emitted air pollution increased over the years.

Table 1. Air pollution emission 2015 – 2019 in thousand tonnes [6].

Type	2015	2016	2017	2018	2019
CO	1,759.2	1,941.0	2,123.3	2,210.6	2,235.1
NO_2	770.1	836.7	860.4	889.9	925.4
SO_2	190.1	221.5	218.7	257.5	270.3
PM	26.7	25.7	24.0	26.8	27.2

When deterioration begins, the service life of the structure and its components is reduced. With a relative humidity of more than 50% and an average temperature of 27°C, the average yearly time of wetness (TOW) in Peninsular Malaysia is 0.783 fractional hours [4]. The level of corrosion in Malaysia is categorized as class 5 by ISO 9223, where the layer of moisture film on the

* Corresponding author: fathoni@uniten.edu.my

surface of the metallic components can reside somewhat longer, and the deterioration process becomes more effective [7]. The most apparent effect of temperature is on the rate of the chemical reaction that results in degradation. Suppose their surface temperature goes below the dew point. In that case, the surface becomes moist and in the presence of corrosive pollutants, the concentrations of which are growing due to the stable influence of the temperature inversion, which is favorable to certain forms of material deterioration [8].

1.1 Tower and lines assessment

Energy infrastructure assessment is practically in place to ensure the reliability of the system to provide its service. Without an adequate understanding of deterioration processes, the effect of the defect to the other component of the tower, level or hazard of environmental exposure to the deterioration processes, a possible overlook on the existing condition of the tower will cause unexpected fault and failure.

Assessment outcome can be used to determine whether the component fits the assessment requirements. Each assessment method has a unique approach for collecting the results. Technical characteristic assessment is concerned with reporting the physical, mechanical, and qualities of a component that can be measured during scheduled patrolling activities using visual and assessment tools [9, 10]. The main steps of condition assessment include recording any damage and determining the causes of distress, assessing the extent of distress, and estimating the residual strengths of structural components and the system [7 – 9].

1.2 Health indices and predicted life

Health index is being used to describe the overall condition of the transmission tower and transmission line. The scoring or condition rating from assessing components of a tower and its sub-components is calculated [1]. Weighing Scoring Method (WSM) with Analytical Hierarchy Processes (AHP) was used in determining the contribution of each component of high voltage transmission tower in Thailand [10]. The AHP is a multi-criterion analysis as mentioned in [12] used to determine the suitable weighting factor for each assessed component. The process of AHP starts with preparing a pairwise comparison matrix and assigning the scale of relative importance for each assessed component with respect to the other component to determine the tower's overall condition. The next step is to normalize the matrix and measure the consistency of the criteria weight using principle Eigen λ_{max} as in Equation (1) and the consistency index, CI as in Equation (2).

$$Aw = \lambda_{max} \quad (1)$$

where A is an $n \times n$ non-negative matrix and w is a weight vector as in the eigenvector method [12].

$$CI = \frac{\lambda_{max}-1}{n-1} \quad (2)$$

where n is the number of the category.

The prediction of service life and remaining life of a lattice steel transmission tower is discussed in [13]. The mean estimate of service life is 63 years with 21 years standard deviation depending on the climatic, environment, maintenance practices, quality of galvanizing, and paint regimes. It was mentioned that the ground condition affects the life of the foundation as well. A study on metallic building components in Malaysia where the service load and environmental load have been taken into consideration gave the service life of steel material's component ranged from 31 years to 45 years in the industrial area [14]. The statistical and probabilistic approaches were used to determine the residual and remaining life of the transmission tower [3, 11, 15, 16]. In [3], the use of the Cox regression model of survival analysis model was used with respect to corrosion, which highly correlated deterioration due to climatic and atmospheric conditions.

2 Method and materials

In this study, the high voltage overhead transmission tower assessment was conducted on more than 80 towers in three different transmission lines. The non-destructive test and visual assessment were applied. The tower's condition was determined by eight parameters, i.e., corrosivity of the soil, chimney's condition representing the concrete part of the footing, galvanized steel member and the conductor, and the atmospheric corrosion level at the tower's location. Table 2 list the assessment method that was conducted on-site.

The level of deterioration is rated quantitatively in a range of numbers to measure the component's physical state. The ordinal numbers from 1 to 5 or 0 to 7 were used to reduce the likelihood of subjectivity in determining the level of performance due to too much detail distinguishing of specific components, with 1 being the worst condition and raising the number being the best condition. The description focuses on a quick write-up of the assessment, severity, and extent of any deterioration noticed. The description of the physical state would help in deciding on the rating and would also act as a crosscheck on the accuracy of the rating entry. A short description of the area where the tower is located was recorded as well. Photographs were recorded for validation purposes on the visual assessment data. Table 3 to Table 6 present the rating system that used in this study.

Table 2. Component of the tower and the assessment method.

Component	Assessment	Scoring
Soil	The resistivity of soil, based on ASTM G57	1 – 5
	Ph, moisture, and chemical content if resistivity reading below 3000 Ω-cm	
Concrete foundation	Carbonation based on CPC 18	1 – 5
	Rebound hammer based on ASTM C805	1 – 5

	UPV based on ASTM C597	1 – 5
	Visual assessment on any related concrete defect (i.e., crack, spall out, delamination, etc.) based on ACI 201, EN 1504	
Galvanized steel member	Visual assessment of corrosion severity.	1 – 5
	The thickness of the galvanize using handheld elcometer.	1 – 5
Conductor	The temperature at fitting measured using thermography camera.	1 – 5
Atmospheric corrosion level	Corrosion hazard map	1 – 5

Table 3. Evaluation criteria for soil.

Rating	Resistivity ohm - cm	Corrosivity
5	>20,000	Essentially non-corrosive
4	10,000 to 20,000	Mildly Corrosive
3	5,000 to 10,000	Moderately corrosive
2	3,000 to 5,000	Corrosive
1	1,000 to 3,000	Highly corrosive

Table 4. Evaluation criteria for concrete in rebound hammer test.

Rating	Average Rebound Number	Quality of Concrete
5	> 40	A very good hard layer
4	30 to 40	Good layer
3	20 to 30	Fair
2	< 20	Poor Concrete
1	0	Delaminated

Table 5. Evaluation criteria for concrete in UPV.

Rating	UPV (km/sec)	Concrete quality
5	> 4.0	Very good
4	3.5 to 4.0	Good, but maybe porous
3	3.0 to 3.5	Poor
2	2.5 to 3.0	Very poor
1	2.0 to 2.5	Very poor and low integrity

Table 6. Evaluation criteria for corrosion of lattice steel tower [4].

Rating	Description
7	New or Like New Condition. Galvanized layer clearly intact with no or little rust.
6	Good Condition. Galvanized layer intact and powdery rust evenly distributed.
5	Functioning as original design. The galvanized layer is still partially intact with few exposed spots on the steel member/connection with no localized corrosion.
4	Minimum Adequacy. Most of the galvanized layer is corroded and leaving the steel exposed with an even layer of rust.
3	Not Functioning as the original design. Loss of steel section due to rust (< 1 mm for member). Some localized corrosion takes place in partial depth.
2	Structurally Inadequate. Further loss of steel section due to rust (< 3 mm for member)
1	Potential Hazardous. Loss of steel section (5 mm for member). Steel corroded throughout the member/connection. Localized pitting is observed to full depth.
0	Dangerous. The steel section has corroded to the full extend. Many pit holes are found, and members can break off anytime.

The atmospheric corrosion level was published in a separate study. The corrosion hazard map was generated based on ten years' climate and air pollution data of wet deposition from 17 measurement stations in Peninsular Malaysia [4]. Figure 1. shows the location of 80 towers in the different transmission lines. In this study, a visual assessment on 8 members for each side of the tower was taken. The thickness of the galvanized Zink coating was measured at 32 points inside and outside the face of the lattice steel tower member.

Fig. 1. Location of transmission tower in this study.

3 Result and discussion

3.1 Corrosion

Corrosion was the common sign of deterioration observed during the assessment. The corrosion was on the galvanized coating with no significant further corrosion to the carbon steel member. The galvanized Zink coating observed has more than 100 microns of thickness. Figure 2 shows the distribution of corrosion on tower members. It shows that corrosion is observed more often on redundant members with less slanted or flat-oriented members. There is 25% soil with high and highly corrosivity conditions recorded and affecting the concrete condition. Figure 3 shows the Partial correlation graph of soil corrosivity and concrete condition.

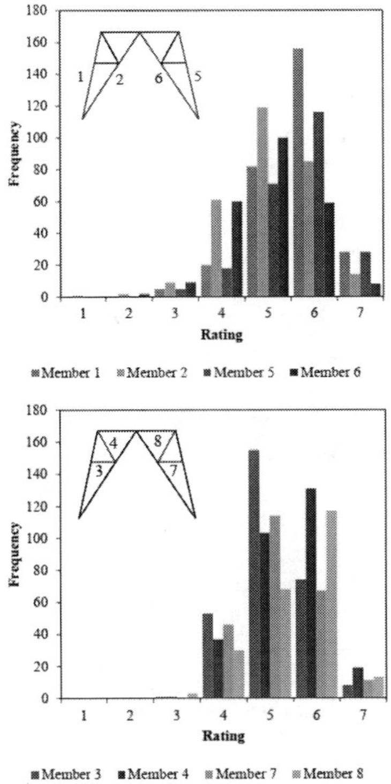

Fig. 2. Corrosion rating distribution for main member and bracing.

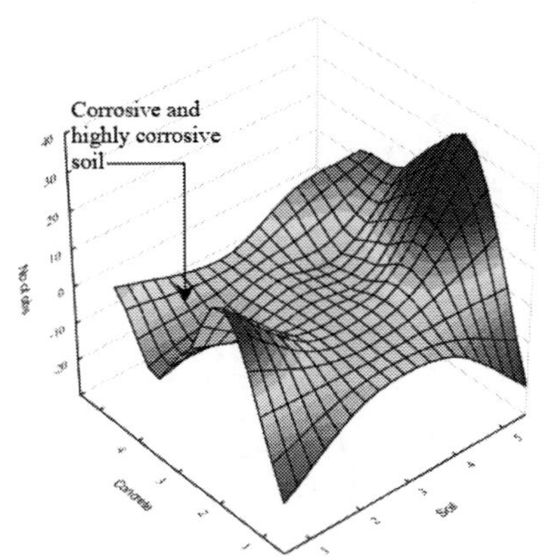

Fig. 3. Partial correlation graph of soil corrosivity and concrete condition.

3.2 Predicted life

The multi-criterion decision analysis with AHP produced weighing for each sub-component with 19.5%, 22.5%, 21.2%, 23.1%, 13.7% and the *CI* of 0.084 < 0.1. Figure 4 shows the linear fitting curve for the tower condition rating and age of the tower. It is found that the predicted age of a tower from this study can reach more than 70 years when it has a condition rating of 5 with an r-value of 0.5465. The predicted life of a tower will gradually reduce by about 20 years for each decrease of the tower condition rating.

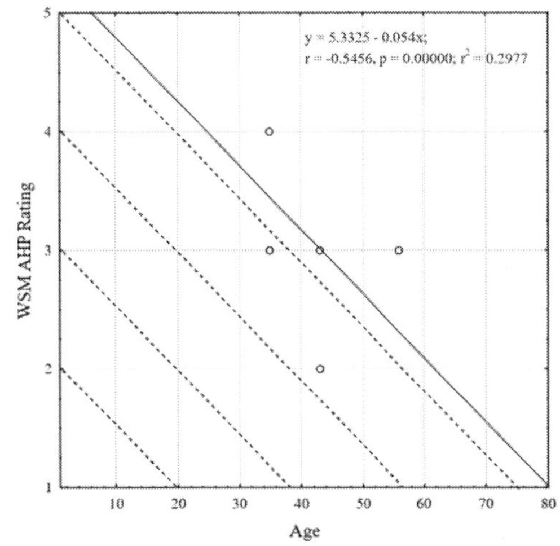

Fig. 4. Linear fitting of brushed data of condition rating and age of the tower.

The predictive remaining life of the lattice steel transmission tower in this study uses Kaplan-Meier survival analysis. The Kaplan-Meier model is based on estimating conditional probabilities at each time point when an event occurs to estimate the survival rate at each point in time. In the survival study, the tower condition rating and the age at which each rating is applied as the condition changes through time. Figure 5 shows the survival function plotted as a step function with the fitting curve of linear and quadratic mode. Visually, the range of predicted life of the tower from the survival function seems to be a good approximation of 60 to 78 years.

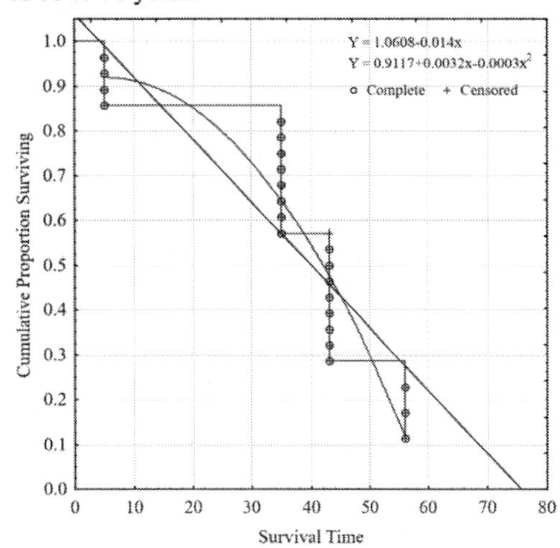

Fig. 5. The survival function plotted as a step function with fitting curve of linear and quadratic mode.

The Kaplan-Meier model is used when the expression of the baseline hazard is unknown, and it is just estimated at points in time where one or more reductions of the physical state of the tower have occurred. The tower's mean residual or remaining life (mrl) can be predicted using the fitting equation as in Equation 3.

 https://doi.org/10.1051/e3sconf/202132501011

$$mrl = \frac{\int_t^{\infty} F(x)dx}{F(x)} \quad (3)$$

where t is the age of the tower.

The *mrl* value from the survival function can be used by the maintenance unit of the asset management team to strategize and plan the frequency of inspection at a specific location and replacement of components in advance.

4 Conclusion

This paper presents assessments conducted on high voltage transmission towers at three different transmission lines. The Weighing Scoring Method (WSM) with multi-criteria decision analysis using Analytical Hierarchy Processes (AHP) was implemented to determine the tower condition rating. From the assessment, it was observed that higher corrosion occurred on more flattened members than slanted members. The soil corrosivity affects the concrete condition of the tower foundation. The Kaplan-Meier model of survival analysis was utilized in this study, and the tower's service life was estimated to be between 60 and 78 years. The remaining service life can be calculated using the mean residual life (*mrl*) that can be used to plan and strategize the maintenance and replacement works.

References

1. R. Hashim, F. Usman, and I. N. Z. Baharuddin, Resources **8**, 2 (2019)
2. Tenaga Nasional Berhad, *Integrated Annual Report 2020*, Kual Lumpur (2020)
3. M. Segovia *et al.*, *Predicting remaining life of transmission tower steelwork components*, in Proceeding of the 26th Eur. Saf. Reliab. Conference. ESREL (2017)
4. F. Usman, N. H. N. Khalid, R. C. Omar, and M. J. A. Jawas, *Atmospheric corrosion map for service life prediction of lattice steel tower using air pollution data and meteorological data*, in AIP Conference. Proceeding **2339**, 020189 (2021)
5. J. I. Silva Filho, L. A. M. C. Domingues, A. P. Ruffier, and A. A. Menezes, *Assessment of environmental statistics as an accessible breakthrough to improve OHTLS design*, in Proceeding of 41st Int. Conf. Large High Volt. Electr. Syst. 2006, CIGRE (2006)
6. Department of Statistics Malaysia, *Malaysia Statistical Handbook 2019*, October. Jabatan Perangkaan Malaysia (2019)
7. U. Fathoni, C. O. Rohayu, and C. M. Zakaria, *Environmental Zoning for Service Life Prediction of Building Components in Malaysia*, in Proceeding of IOP Conf. Ser. Earth Environ. Sci., **16**, 1 (2013)
8. J. J. Santana *et al.*, Metals (Basel) **9**, 10, (2019)
9. A. I. Wuller and A. Pharmatrisanti, *Transmission line assessment*, in Proceedings of 2012 IEEE International Conference on Condition Monitoring and Diagnosis, CMD 2012, September (2012)
10. Waraporn, Luejai, S. Thanapong, and S. Cattareeya, *Condition evaluation of high voltage transmission line in Thailand, in Lecture Notes,* in Electrical Engineering 598: Proccedings of the 21st International Symposium on High Voltage Engineering, Budapest (2017)
11. A. Van Der Wal, A. A. H. J. Ross, and T. D. C. Kema, *CIGRE Sess.* ***B2-204*** (2004)
12. J. A. Alonso and M. T. Lamata, Int. J. Uncertainty, Fuzziness Knowlege-Based Syst. **14**, 4 (2006)
13. CIGRE Working Group 37.27, *Ageing of the system - impact on planning*, in Session 2000 (2000)
14. F. Usman and Resdiansyah, S*ervice life prediction of building components*, in ICCBT — International Conference on Construction and Building Technology, Malaysia (2008)
15. A. A. H. J. Ross, B. Rhebergen, and V. A. N. O. J. P. M, *Probabilistic based condition assessment of overhead transmission lines*, CIGRE, France (2000).
16. J. Haque, U. Chandra, S. K. Agrawal, P. Kumar, and H. Vyas, *Use of statistical tools for assessment of residual life of transmission linesviz-a-viz steps taken to augment the reliability of transmission network of powergrid (India)*, in Proceeding of 41st International Conference on Large High Voltage Electric Systems 2006, CIGRE (2006)

Preliminary result for crustal properties derivation related to tectonics for hazard mitigation in Eastern Indonesia using Teleseismic P Coda

Anang Sahroni[1], *Leni* Sophia Heliani[2], *Cecep* Pratama[2], *Hidayat* Panuntun[3], and *Wiwit* Suryanto[1*]

[1]Seismology Research Group, Geophysics, Universitas Gadjah Mada
[2]Department of Geodetic Engineering, Faculty of Engineering, Universitas Gadjah Mada
[3]Geomatics Laboratory, Department of Geotechnology, Vocational College, Universitas Gadjah Mada

Abstract. Eastern Indonesia is tectonically complex, formed by different plates and microplates interactions from different origins. This complexity gives geoscientists a challenge to solve the 'jigsaw' of the complex interactions. The understanding of tectonic processes can lead to a breakthrough in both resource exploration and disaster risk reduction. We utilize teleseismic P wave coda for random coda from scattering and deterministic coda originated from the crust-mantle boundary (Moho) to derive the crustal properties, including thickness, Vp/Vs, and qualitative scattering characteristics. For the scattering properties, we apply Iterative Cross-Correlation and Stacking (ICCS) to align the waveform. At the same time, for the crust characteristic, we employ the Receiver Functions (RF) method alongside H-k stacking. The crustal thickness recovered from the RF and H-k stacking has a good correlation with the crustal origin, where the thickness in older and stable crust originated from Sundaland and Gondwana is thicker than a younger plate of the crust arc and subduction origin. The Vp/Vs is high in a region that is interpreted to be dominated by mafic lower crust originated from oceanic-oceanic subduction during Eocene, anisotropy, or by a magmatic anomaly. The P coda also correlated well with the subsurface magmatic anomaly by providing a unique pattern.

1 Introduction

Eastern Indonesia exhibits very complex geological and tectonic settings. This region is formed by the complex interaction between Sundaland as the eastern margin of Eurasian Plate, north moving convergence of Australian Plate, and west moving convergence of Pacific Plate [1]. This complexity is reflected by the forming of K-shaped Sulawesi and Halmahera Islands, drifting, and colliding of Sula, Buton, and other micro-continents westward, trapping and forming of the deep Banda Arc and generating major strike-slip structure of Sorong and Palu Koro [2, 3, 1] (Fig. 1A). The northern part features double subduction generating active volcano arc of Sangihe in the west and Halmahera in the east [1, 4].

The complexities bring a challenge to understand both the crustal physical characteristics, its origin, and its geodynamics. The understanding of the properties and origin can lead to a breakthrough in resource exploration (hydrocarbon exploration, mineral, geothermal) that are dependent highly on the tectonic framework; hazard can also be minimized by knowing geodynamics and possible process of magmatism beneath the crust. Several studies with different methods have been comprehensively described the tectonic evolution of this region [3,5]. Complex plate reconstruction using computer modelling can explain and model the interactions between Gondwana's microplates fragment and Sundaland from Jurassic-Cenozoic and correlate very well with the recent tectonic condition [3]. Seismic studies also provide a detailed view both in the spatial and temporal interaction of plates by using seismic tomography [5]. Seismic tomography being the bridge between mantle dynamics and crustal motions by evaluating each depth section and reconstruct by back projecting the plate motions and mantle structure variation. This method can also describe the velocity variation that is very related to the intrinsic physical characteristic of tectonic plates. Despite its powerful result, this method is very resource-extensive in the computational domain.

A simple seismic approach such as receiver function and teleseismic coda wave analysis can be the fastest and cheapest way in describing the crustal properties [6-9]. By using the phase conversion phenomenon in the velocity contrast between crust and upper mantle (Moho), we can derive and estimate the crustal thickness [8], which is important information in describing crustal and tectonic processes, including thickening, thinning, and its mantle implication. Teleseismic receiver function using vertical waveform to estimate the source waveform and remove it using deconvolution process [6, 8, 9]. Using the H-k stacking method, we further can

* Corresponding author: ws@ugm.ac.id

estimate the bulk Vp/Vs of the crust and describe the intrinsic properties. The coda wave also holds information on the scattering process in the subsurface; scattering is related to the distribution of small-scale heterogeneities such as magma packets as isolated strong velocity contrast [10].

Several studies have analyzed the Eastern Indonesia tectonic and crustal properties using the Receiver Function Method [11, 12], yet those previous methods have not provided any analysis about the random seismic coda and the heterogeneities relationship. The relationship of the derived crustal properties with the origin of the Plate or tectonic block also has not been discussed. This study will relate the derived crustal properties and coda pattern both for deterministic and random coda with the crustal origin and its heterogeneities characteristic.

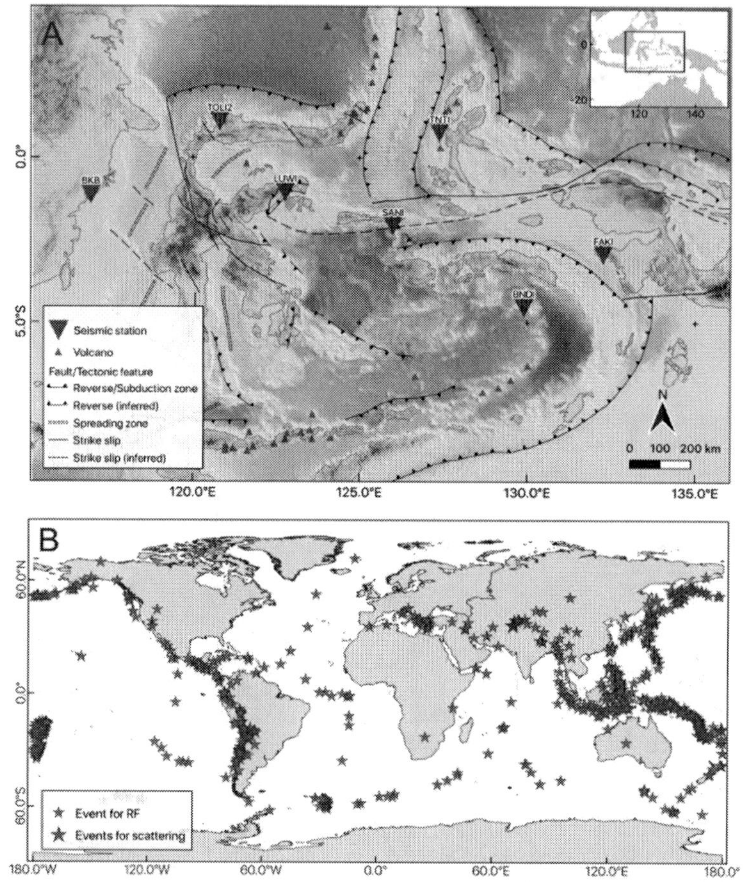

Fig. 1. Major active and past tectonic feature in Eastern Indonesia, after [1, 2] (A) and earthquake events for both scattering analysis (red star) and for Receiver Function analysis (green star).

2 Methodology

The P coda processing will be separated into two independent processes: scattering teleseismic P coda analysis and teleseismic P Coda Receiver Function. For the first process, we collect 7 Mw > 7.0 events from WebDC3 GFZ [13] (Fig. 1B) with epicentre to station distance between 50-80 degrees to obtain nearly vertical incident of P wave propagating into the lithosphere, to ensure that the coda does not contain any other deterministic body wave phases 25 s after P arrival, the depth is constrained to be greater than 110 [14]. The sampling rate is uniform in all of the data (40 Hz), but the recording instrument varies.

Before the alignment process, the instrument response is removed to get the true ground motion using Obspy [15]. The corrected data then normalized to each maximum value, and 0.1-8.0 Hz bandpass filtered [14]. We apply the Iterative Cross-correlation and Stacking (ICCS) algorithm [16] to remove the time delay between the station recording. This algorithm ensures consistency: multitrace cross-correlation by stacking all the traces and iteratively adjusting the time picks. The algorithm is started by predicting initial time picks (arrival time) of P according to the existing velocity model IASP91 [17]. The traces are then aligned and stacked using the initial time picks. The second step is cross-correlating and shifting each trace with the stacked trace from the previous step. The process continues until the iteration does not change the stack.

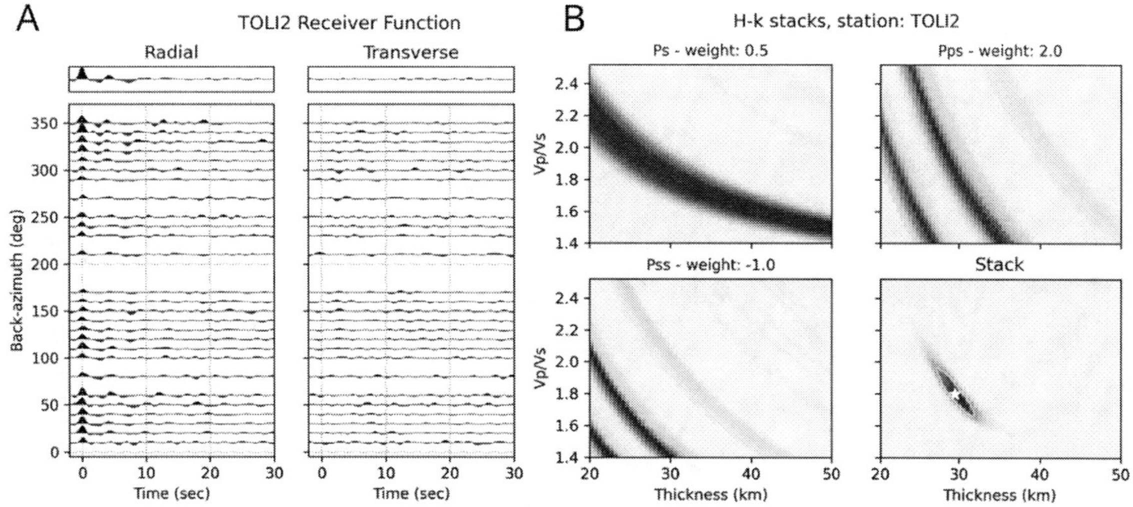

Fig. 2. Receiver Function for station TOLI2 plotted by back-azimuth, the upper part of the plot is stacked receiver functions (A), H-k stacking for TOLI2 showing Vp/Vs ratio of 1.80 and crustal thickness 29.5 km, stacked using Ps, Pps, and Pss phases with weight 0.5, 2.0, -1.0 respectively, uncertainty is denoted by the yellow line in the stack plot (B).

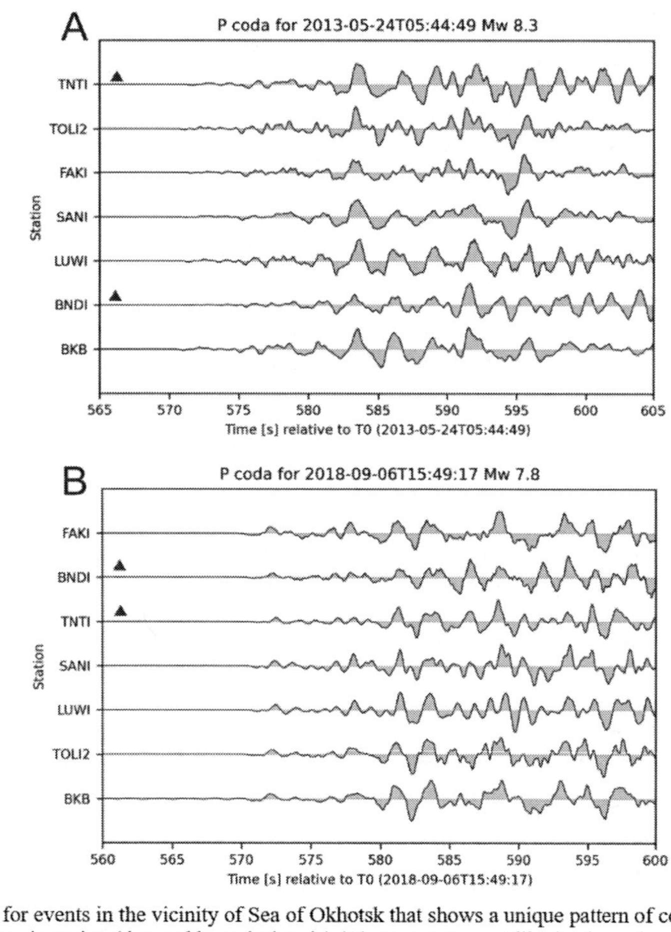

Fig. 3. Example of Coda for events in the vicinity of Sea of Okhotsk that shows a unique pattern of coda in the later part of the station located near a volcanic region (denoted by red triangle) (A), an event near Fiji Island, no clear, unique pattern as observed in A (B).

 https://doi.org/10.1051/e3sconf/202132501012

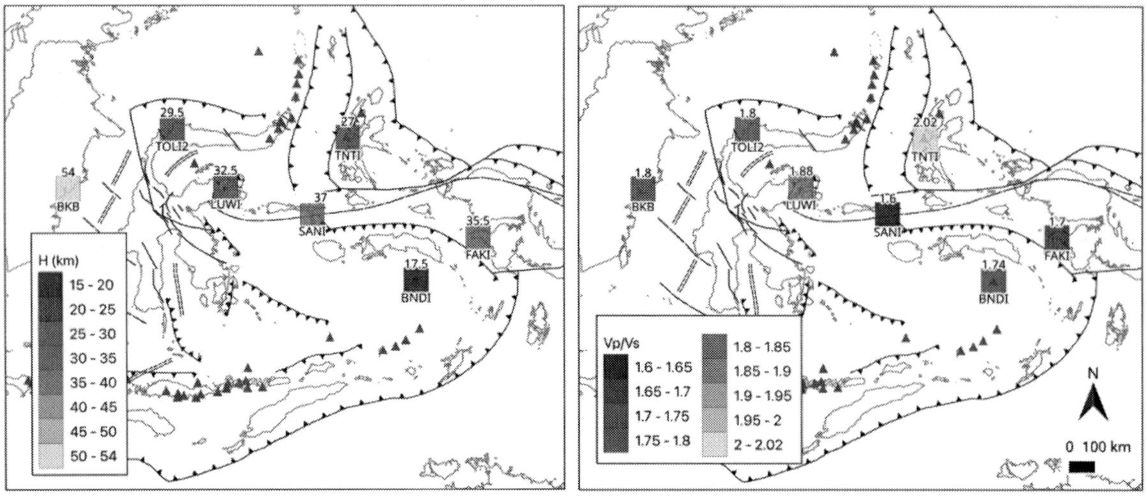

Fig. 4. Crustal thickness (A) and Vp/Vs (B) calculated using H-k stacking.

Table 1. Summary and comparison of crustal and P Coda characteristics.

Station	Crustal thickness (km)	*Vp/Vs*	P Coda notable feature (relative to the other stations)	Geological/tectonic characteristics
BKB	54.0	1.80	-	60 Ma collision zone [18, 19]
TOLI2	29.5	1.80	-	Volcanic arc, ophiolite, and accreted materials [1]
LUWI	32.5	1.88	Small fluctuation in 600-605 s, also observed in volcanic related seismic stations with stronger fluctuation : TNTI and BNDI	Past subduction zone [18, 1]
SANI	37.0	1.60	-	Continental basement [1]
TNTI	27.0	2.02 (high anisotropy?)	Prominent feature in 600-605 s : volcanic related ?	Volcanic in the western origin and ophiolite in the eastern origin [1]
BNDI	17.5	1.74	Prominent feature in 600-605 s : volcanic related ?	Volcanically active region
FAKI	35.5	1.70	-	Stable region, part of Papua, originated from Gondwana [3]

The P coda for receiver function was collected for the same station with different events criterion: distance between 30-90 degrees and magnitude greater than Mw 6.0; the data is collected from 2010-01-01 until 2021-05-28 [13] (Fig. 1B). Receiver function and H-k stack calculated using RfPy [9]. The seismogram first rotated to vertical, radial, and transversal (ZRT). The receiver function utilizes the vertical component seismogram to approximate the source waveform and will be removed from the horizontal waveform using deconvolution [7-9]. The general deconvolution process will include the seasonal ambient noise spectrum of microseism [20, 8]; this noise is reduced using the Wiener filtering approach [8]. The recovered radial and transversal waveforms stacked by back azimuth with spacing 10 degrees (Fig. 2) [6, 9] (Fig. 2A). The recovered radial waveform from the deconvolution process is convolved again with the vertical waveform and cross-correlated with the observed radial waveform; the correlation coefficient quantifies the process quality [9]. For the data quality, the 0.05-1 Hz bandpass filtered vertical waveform signal-to-noise ratio is calculated in the 30s before and after the IASP91 P arrival [17, 9].

Crustal thickness can be estimated directly by measuring the time difference between the first arrival P wave and its phase converted Ps. However, this simple method is very sensitive with Vp/Vs in the crust where error 0.1 s/km in Vp/Vs can lead to 4 km uncertainty in the estimated crustal thickness [19]. One way to reduce this dependency is by stacking multiple phases: Ps, PpPs (or Pps), and PsPs (or Pss) in the H and k domain by weighted sum [19]. [8] modified the stacking process by introducing the weighted product method in the amplitude corrected phases, where the negative amplitude is replaced by zero value. The maximum value obtained from stacking is the estimated Vp/Vs and crustal thickness (H).

The average Vp used is 6.0 km/s and stacked using 20 bins of slowness. The weighting combination is 0.5, 2.0, -1.0 for Ps, Pps, and Pps, respectively. The median value for every single-phase weighted stack is used to mitigate outlier that may distort the stacking process [9] (Fig. 2B). The k and H range is manually adjusted to get a reasonable value of k in crust [21] and H compared to the crustal thickness according to CRUST1.0: Global Model of Earth's Crust (http://igppweb.ucsd.edu/~gabi/crust1.html#reference).

3 Result and discussion

The crustal thickness of this study is in the range of 17.5-54 km, with BNDI (Banda Neira) being the area with the thinnest crust while BKB (Balikpapan) being the thickest (Fig. 4A) (Table 1). BKB is located northern of Meratus Suture zone that is resulted from the collision of East Java-West Sulawesi block to the South West Borneo block that has been amalgamated with Sundaland extended to northern in 60 Ma [18, 19]. The collision might result in crustal thickening, causing the area to have the thickest crust among the study area. This station shows strong scattering in the P Coda wave that we interpret to be related with the complex melange of the ophiolite (Fig. 3A and 3B), high Vp/Vs (1.8) (Fig. 4B and Table 1) is likely to be attributed to ~10 km sediments of Kutai Basin [22] or might be the contribution of small mafic part of the concealed ophiolite complex.

TOLI2 (Toli-Toli) located in the bending of the northern arm of Sulawesi Island; this arm is interpreted to be a volcanic arc, ophiolite, and accreted formed in plate margin [3] contain mostly volcanic, volcaniclastics, and granitic stocks [1] and was an Early Eocene subduction complex [3]. This young formerly ophiolite and the volcanic arc have 29.5 km crustal thickness, far from BKB, whose block originated from much older Gondwana's block [23, 3] and thickened by the accretion process. Though having a different value of crustal thickness, the Vp/Vs for this area is the same as BKB; we interpret that this high value of Vp/Vs is of mafic lower crustal contribution which forming island arc or accreted during the Early Eocene oceanic-oceanic subduction [3]. The P coda pattern is similar to BKB but showing higher fluctuation in the central part of the coda (window 585-595 s) in Fig. 3A and 3B. This difference might be attributed to more complex small-scale heterogeneities in TOLI2, either the granitic stocks or scattered magma pockets by the recent subduction zone. The surrounding LUWI also presents a past subduction zone and is the time marker of the early contact between the Australian Plate and Eurasian Plate [3] in 25 Ma. Sula Spur starts the contact as the fragment of the Australian Plate that collides with the volcanic arc of the northern Sulawesi arm. This contact formed subduction that accreted ophiolite and melange complex and formed the eastern arm [3, 1]. The crustal thickness is higher than its past generation; we suggest that this slight difference is mainly drifted by the difference of subducting plate origin and rate of subduction, wherein the Sula Spur acts this case. The Vp/Vs is also higher than TOLI2, which likely is caused by the anisotropy [24] in this massively deformed region.

SANI (Sanana) is part of Sula Spur that is composed mainly of granite and exposed as continental basement type [1], which might be the main contributor to its very low Vp/Vs (1.60) [21]. The crustal thickness is 37.0 km and is related to its Gondwana origin [3] and should be higher than TOLI2 and LUWI, which are much younger and are volcanic arc and ophiolite origin. The P coda shows no prominent and stable characteristic related to the observations in the other stations.

TNTI (Ternate), positioned in the Island of Maluku, has a very similar geological condition with Sulawesi, having volcanic in the western origin and ophiolite in the eastern origin [1]. TNTI lies near the currently active Gamalama Volcano; observation in this station obtains a very strange value of Vp/Vs 2.02; this anomaly can be interpreted as the existence of magma, based on its location. This high Vp/Vs can also be related to its origin that formed by oceanic-oceanic subduction and by the existence of the Molucca Sea Plate that subducts beneath this Plate. However, the latter could not be as the recovered crustal thickness is only 27 km. From the P coda pattern, magmatic activity likely to contribute to the prominent feature in the late time window (600-605 s) that is not observed in the non-volcanic related seismic stations (Fig. 3A). The late P coda prominent pattern is also observed in BNDI (Banda Neira), which is also located near a volcanically active region. Though its origin is still tried to be determined by a quantitative approach, the phenomenon is unique and spatially specific.

FAKI (Fak-Fak) being the typical characteristic of continental crust with 1.7 Vp/Vs and 35 km of crustal thickness. Tectonically, this area is part of Papua that is originated from Gondwana [3]. The coda shows a similar pattern with the other stations, but at certain time windows exhibit a weaker fluctuation.

4 Conclusion

The crustal thickness variation in the study area is very related to the origin of the block or plate where the observations are made. The crustal thickness for the older block originated from Gondwana (FAKI and SANI) is relatively thick, the BKB which should give the same result but thicken by the contact and amalgamation of Gondwana's microplates to the Sundaland, LUWI, TOLI2, and TNTI has thinner crust because of its young age and volcanic arc and ophiolite origin, while BNDI, the youngest arc having the thinnest crust. While the pattern Vp/Vs is harder to predict, we suggest that it is correlated with the crustal origin, anisotropy, or magmatic/fluid existence. The scattering properties observed by the P Coda can give qualitative insight into the subsurface heterogeneities; from the observation, the volcanic area relates to a unique pattern of the coda.

The crustal thickness and Vp/Vs value derived from receiver function analysis of deterministic the part of P

Coda correlates well with the crustal origin and estimated subsurface condition. The random part of the coda related to the subsurface's scattering properties shows prominent and unique features that are likely to be caused by magmatic origin. Crustal characteristics will provide important information on the past and ongoing geodynamic processes, the main actor that trigger earthquake and volcanic hazard.

The plot for P Coda analysis is generated using Matplotlib; the coastline base map is obtained from Data Basin, the volcano distribution is from LAPAN, bathimetry and topography map from GEBCO plotted using QGIS, graphic annotation and layouts done using Inkscape. We thank Wakhidatun Nisa for providing digitized tectonic features in the area. This research is funded by Penelitian Dasar Unggulan Perguruan Tinggi (PDUPT) 2021 granted to Wiwit Suryanto.

References

1. W. Hamilton, Tectonics of Indonesia Region (United States Government Printing, Washington, 1979)
2. I. M. Watkinson, Ductile Flow in the metamorphic rocks of central Sulawesi (Geological Society, London, n.d)
3. R. Hall, Tectonophysics, 1-41 (2012)
4. A. J. Katili, Mar. Pet. Geol. **8**, 70-83 (1991)
5. E.L. Advokaat, N.T. Marshall, S. Li, W. Spakman, W. Krijgsman, D.J van Hinsbergen, Tectonics **37**, 1 (2018)
6. A.P. Tarayoun, P. Audet, S. Mazzoti, A. Ashoori, J. Geophys. Res. Solid Earth **122**, 5268-5287 (2017)
7. A.W Frederiksen, C. Delaney, Geophys. J. Int. **201**, 1491-1506 (2015)
8. P. Audet, Seismol. Soc. Am., Bull. **100**, 1356-1362 (2010)
9. P. Audet, D.L. Schutt, A.J. Schaffer, C. Esteve, R.C. Aster, J.F. Cubley, Seismol. Res. Lett. **91**, 3076-3085
10. K. Aki, B. Chouet, J. Geophys. Res **80**, 3322-3341 (1975)
11. R. Harsoni, D. Ngadmanto, W. Suryanto, J. Phys. Conf. Ser **1825**
12. Syuhada, N.D Hananto, C.I. Abduallah, N.T. Puspito, T. Anggono, T. Yudhistira, Acta Geophys **64**, 2020-2050 (2016)
13. M. Bianchi, P.L. Evans, A. Heinloo, J. Quinteros, WebDC3 Web Interface (GFZ Data Services, Potsdam, 2015)
14. X. Ma, Z. Huang, Earth Planet and Space **72**, 13 (2020)
15. M. Beyreuther, R. Barsch, L. Krischer,T. Megies, , Y. Behr, J. Wassermann, Seismol. Res. Lett. **81**, 530-533 (2010)
16. X. Laou,S. van der Lee, S. Lloyd, , Seismol. Res. Lett. **84**, 85-93 (2013)
17. B.L.N. Kenneth, R. Engdahl Geophys. J. Int. **105**, 429-465 (1991)
18. B. Clements, R. Hall, J. Geol. Soc. London **168**, 1187-1200 (2011)
19. L. Zhu, H. Kanaomori, J. Geophys. Res. **105**, 2966-2980 (2000)
20. D.E. McNamara, R.P. Buland, Seismol. Soc. Am., Bull. **94**, 1517-1527 (2004)
21. N.I. Christensen, J. Geophys. Res. **101,** 3139-3156 (1995)
22. J.L. Chambers, I. I. R. Charte, J. Craig, S.J. Moss, D.W. Paterson, AAPG Memoir **82**, 614-634 (2004)
23. I. Metcalfe, Bull. - Geol. Soc. Malays. **63**, 27-60 (2017)
24. X.Q. Wang, A. Schubnel, J. Fortin, E.J. David, Y. Gueguen, H.K Ge, Seismol. Res. Lett. **39**, 1-6

Recent faulting along Gorontalo fault based on seismicity data analysis and lineament mapping

Hasan Arif Efendi [1,2*], *Gayatri* Indah Marliyani [2], and *Subagyo* Pramumijoyo[2]

[1]Badan Meteorologi Klimatologi dan Geofisika (BMKG), Stasiun Geofisika Gorontalo, Talumelito Gorontalo, Indonesia
[2]Universitas Gadjah Mada, Department of Geological Engineering, Jl. Grafika No. 2 Yogyakarta, Indonesia

Abstract. We focused our study to characterize the geometry and activity of Gorontalo fault. We analysed reviewed the ISC seismic catalogue and the BMKG relocated earthquake events available for the time period of 1960 to 2021, located along the expected location of this fault. In addition, we analysed continuous record from local seismic observatory available for the same period. Further, we mapped the lineaments using 8.3-m resolution DEMNAS data. Tens on shallow earthquakes occurred in the vicinity of this fault with a range magnitude of M 2 to 3. Our lineament analysis however does not reveal distinctive pattern that may indicate the fault manifestation at the surface. The NW-SE trending lineaments are coincidence with the mapped trace of Gorontalo Fault. The weak surface manifestation of the fault scarp may be related to the tropical climatic condition of the area which may obliterate the faulting topography. However, we observed alignment of the seismicity distribution with the mapped NW-SE lineament, indicating that the lineament is likely representing active fault and these earthquakes are associated with faulting along this fault. Our study provide indication that the Gorontalo Fault is active and further study is necessary to investigate subsurface geometry and mitigate its seismic hazards.

1 Introduction

1.1 Background

Gorontalo lies on the northern arm of Celebes Island known as Minahasa Peninsula. This region is located in an active tectonic region, as a result of convergence of Eurasian, Pacific and Australian plates. The triple plate junction in this area is characterized by the subduction of Celebes Sea plate in the north beneath the Minahasa Peninsula underlying Moluccas double subduction to the east [1] producing high seismicity. Based on historical data from the Indonesian Agency for Meteorology, Climatology, and Geophysics (BMKG), destructive earthquakes with magnitude higher than M 7 occurred in Gorontalo in 1990, 1991, 1997 and 2008 and was reported to occur along the Celebes subduction. There were also destructive events felt throughout Gorontalo area in 1939 and 1941, however the hypocentre location of these earthquakes was unknown and it is unclear whether they were originated from an inland fault or from the subduction.

There are several known active faults near the city of Gorontalo (716,603 inhabitants), exposing the community to earthquake hazards. One of the major fault is the NW-SE trending Gorontalo Fault but this fault is not well studied. Difficulties of characterizing the Gorontalo Fault is possibly because of the lack of detailed seismic monitoring data and diffused geomorphic expression of the fault. Nowadays, establishment of new seismic network and the availability of the 8.3-m resolution DEMNAS provide opportunity to better asses the fault zone. We focused our study to characterize the geometry and activity of this fault using these new available data. The aim of this study is to identify the correlation between lineaments and seismicity distribution. Results from our study will contribute toward better understanding of the Gorontalo fault zone and supporting the development of seismic hazard maps in the city of Gorontalo.

The purpose of this study is to introduce a new perspectives of Gorontalo Fault. The previous study data is not adequate enough to provide evidence of active deformation. A fault is categorized as active if it has moved within the past 10,000 years (Holocene) [1]. So, earthquake, lineament trace of remote sensing image, and visual deformation evidence are proof of an active fault as recent deformation.

1.2 Regional Geology of Gorontalo Fault

The Gorontalo Fault has been identified as a transcurrent fault, oriented to the NW-SE [2] and consist of several smaller fault segments [3–5]. The fault accommodates the different velocity between the east and west block of Minahasa Peninsula which is estimated to be around 11 mm/year based on geodetic modelling [6]. However, there were no major earthquakes recorded along

* Corresponding author: hasan.a.e@mail.ugm.ac.id

Gorontalo Fault [7] in historical record as well as during a temporary seismic network campaign in 1990 [6].

The activity of this fault is overlooked because of the absent of major earthquakes in historic years and the diffused expression of the fault in the topography [8,9]. The Fault consists of several branching segments and categorized as slow-slip faults [10]. There are no strong agreement about the fault is active or not, although recent movement of this fault was identified through GPS observation[11,12] and through the occurrence of small-medium earthquakes along this fault.

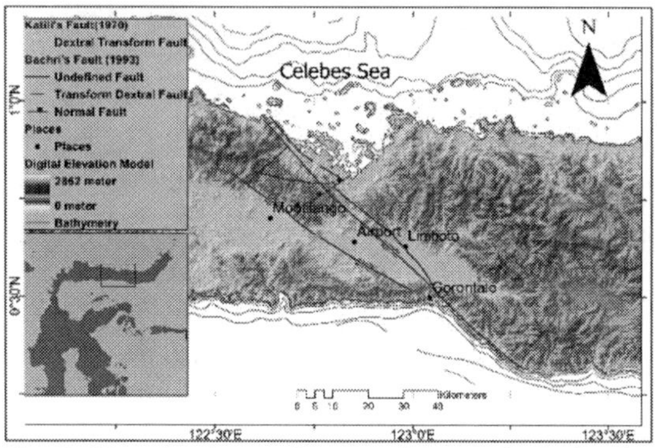

Fig. 1. Map showing study area of Gorontalo Fault where orange line represents Katili's active fault (1970) then blue line is geometry of Gorontalo Fault Zone were mapped by Bachri (1993).

2 Data and method

To characterize the geometry and activity of this fault we analyse the seismicity and topography along the expected trace of the fault zone. We retrieved 479 crustal shallow earthquake data (depth $\geq$ 30 km) from BMKG and International Seismological Centre (ISC) from 1960 to 2021. Based on the slab geometry inferred from ISC-EHB Bulletin[6] and SLAB2 model [13], the hypocentre below 30 km can be considered as intraplate events, and is likely caused by inland fault activity. The BMKG data include arrival time series recorded by the BMKG national network and data from the local observatory (BMKG Gorontalo geophysical station). Both of the central and local observatory network continuously monitor the seismicity, but the local observatory network, which was established in 2017, provide additional advantage through its capabilities to identify small magnitude earthquakes. We relocated the recorded events using Seiscomp3 software[14,15] to obtain better accuracy of the epicentre location. We also recalculate the hypocentre using HYP module of the SEISAN program[16] using CRUST 1.0 velocity model[17].

The fault expression on the Earth's surface can be identified by mapping linear feature in the topography. However, lineament may controlled by many different processes, including geological structure, lithological contacts, or human modification, therefore should be interpreted cautiously. We eliminate some of the lineaments that are not related to geological structure and focusing our observation to the suspected tectonic-related lineaments. The fault-related lineament usually can be identified through series of geomorphic offsets (river, hill, terraces, etc). The lineaments was manually delineated, some of the distinctive characteristics including: 1) the occurrence of fault scarp; expressed as prominent linear feature on topography, for example fluvial system creating linear valleys, series of aligned ridge and saddles 2) linear mountain fronts and linear coast 3) series of systematic drainage offset.

Table 1. Local crustal velocity model around Gorontalo Fault at 0.5N – 122.5E generated from CRUST 1.0[17].

Type	Vp (km/s)	Vp (km/s)	Depth (km)
Upper Sedimentary	1.75	0.34	0.48
Upper Crust	5.8	3.4	1.18
Middle Crust	6.3	3.62	10.98
Lower Crust	6.9	3.94	20.49
Mantle	7.84	4.37	30

For the lineament analysis, we use an 8.3 m resolutions digital elevation model (DEM) from DEMNAS (https://tanahair.indonesia.go.id/demnas/). We processed the DEM, creating various relief map using relief visualization (RVT) toolbox [18]. For the data integration and analysis, we use ArcGIS software. For the lineament orientation analysis, we only consider the long and firm lineaments. We conducted limited field check along the suspected tectonically-controlled lineaments to confirm their tectonic origin and to obtain any fault geometry indication.

We then overlay the suspected active fault map with seismicity distribution to see if the seismicity coincide with the lineaments and may represent recent faulting along the fault.

3 Result

We relocated 219 earthquakes from the BMKG data repository (2010 to 2021) and 111 events from the BMKG local observatory (2016-2017). In addition, we compiled it with 149 events from the ISC catalogue (1960 to 2009). We disregarded events recorded by only one station, for their lacking of arrival time data necessary for the relocation processes. We were able to reduce the average mean square error of the original catalogue from 0.6608 to 0.6286 and provide a better accuracy of their location. Some of the hypocentres were shifting to shallower depth. However, the distribution of the epicentre is scattered.

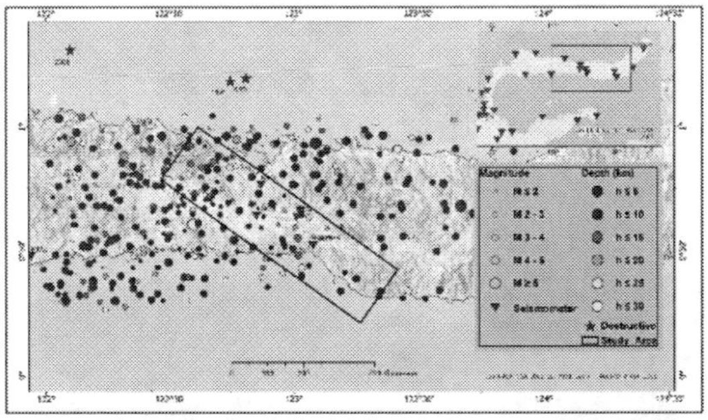

Fig. 2. Relocated seismicity of the BMKG repository and local BMKG station calculated using HYP module of SEISAN. Red stars indicate epicentre of destructive earthquake occurred along the North Celebes subduction (listed on ISC-EHB catalogue). Black box delineate our study area.

The distribution of lineament from our mapping indicate the lineaments are concentrated along the high relief topography oriented to the NW-SE, inferred as the Gorontalo fault zone. However, we were not able to locate the surface trace of the fault solely from the small-scale lineaments (Fig. 3). We also identify series of NE-SW lineaments that may represent another active fault zone.

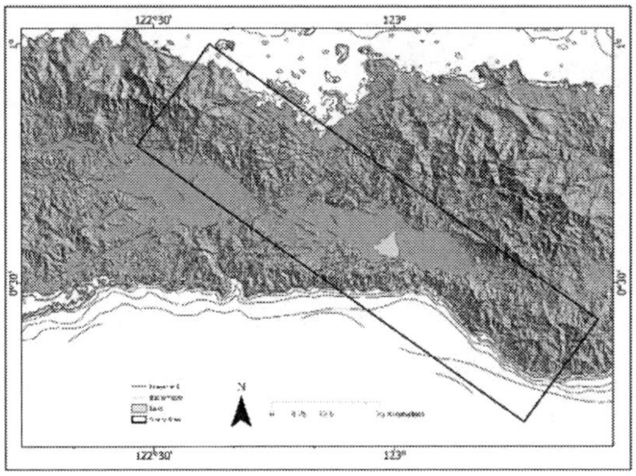

Fig. 3. Lineament map of the study area based on shaded relief DEMNAS data (tanahair.indonesia.go.id/demnas/). Red lines show the delineated lineaments that are likely tectonic-origin. Black box delineate our study area.

Following the lineament analysis, we conducted field checking along the suspected tectonic-origin lineaments. We documented recent deformation evidence of minor faults along the NE-SW trending lineaments in several area (Fig. 4). We also found two series of NW-SE trending triangular facets in Mootilango area (Fig.4). Near the Limboto Lake, we identify E-W oriented normal faults on the Quaternary limestone and loose Quaternary sediments. We also found several secondary faulting displacing several older sedimentary units with varying orientations.

4 Discussion

We interpreted the NW-SE trending major lineament representing the active trace of the Gorontalo fault zone represented by tens shallow earthquake occurred along our mapped Gorontalo fault zone. The clustered nature of the seismicity in this area may possibly related to the nature of sparse regional seismic network that may not able to provide accurate measurement of local events. The slab depth beneath our study area is at depth of 50 km [13], so the events that were originated above 30 km are likely sourced along active inland faults rather than subduction-related events. Based on this observation, we use 30 km as the maximum seismic depth for crustal events and use the distribution of events to characterize inland crustal faults. The spatial distribution of the earthquake indicate that the western side of Gorontalo is more active seismically in compare to its eastern side (Figure 3). The linear distribution of seismicity along the fault zone provide good indication that this fault is active.

Fig. 4. Photographs show indications of faulting along Gorontalo Fault. (a) linear topographic break showing triangular facets oriented to the NW at Mootilango area, camera face to the N. (b), (c) & (d) photographs of outcrops show faults dissecting Quaternary limestone at Gorontalo Bay and Limboto. (e) minor normal fault displacing loose Quaternary sediment near Limboto Lake.

The large scale Gorontalo Fault was originally identified based on the offset of the coastline toward its NW and SE ends [1]. However, the topographic expression of this suspected fault orientation is difficult to be observed with the mainland (Figure 4). The limitation of the surface manifestation of the fault scarp may be related to the tropical climatic condition of the area which may obliterate the faulting topography. The only observable geomorphic feature that may indicate this fault is a shutter ridge near the Gorontalo airport and few scarps and triangular facets observed in the Mootilango area (Figure 4). We estimate the geometry of the fault based on their field expression as well as from the depth distribution of shallow seismicity along the fault zone (Fig.5). Although most of the profile show diffused distribution, the seismic depth profile across one prominent lineament show indication of high angle NE-dipping fault (profile C-C', Figure 5). With the limitation of hypocentres distribution and the surface manifestation of the fault, characterization of the fault through subsurface investigation can be implemented for future study, for example using geophysical methods, to estimate the fault geometry.

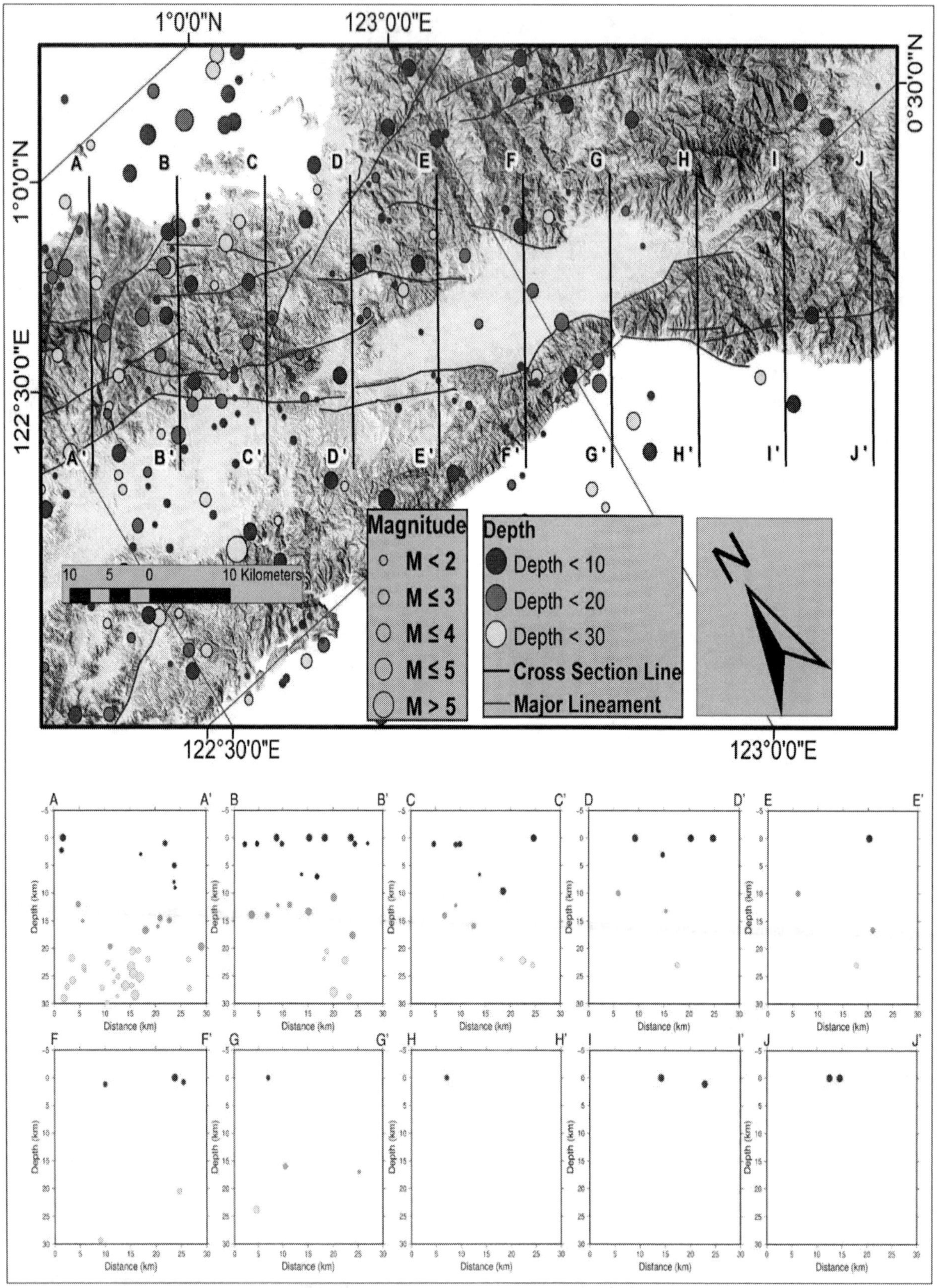

Fig. 5. Depth Distribution and cross section of shallow seismicity in the study area.

5 Conclusion

We presented the lineament and seismicity analysis of the Gorontalo fault zone. The alignment of the seismicity distribution with the mapped NW-SE lineament, indicating that the lineament is likely representing geometry of active fault and these earthquakes are associated with faulting along this fault. Several indications of recent faulting are also found in the field as the faults dissecting Quaternary age units. The activity of the fault also confirmed through the occurrence of shallow earthquakes that are clustered along the mapped lineament. All of these suggesting that the Gorontalo fault is active and is capable to produce earthquakes in the future. Further investigation to characterize the fault in detail is necessary to mitigate hazards associated with this fault. Future investigation of the subsurface data is necessary to obtain the fault geometry at depth and further asses their seismic potentials.

We thank BMKG for earthquake catalogue and BIG for DEMNAS data. This study was supported by Research Grant of Lembaga Pengelola Dana Pendidikan (LPDP) 2018.

References

1. E. A. Keller and N. Pinter, *Active Tectonics*, 2nd Edition (Prentice Hall, New Jersey, 2002)
2. J. A. Katili, Geol. Rundschau **59**, 581 (1970)
3. S. Bachri, Sukido, and N. Ratman, *Peta Geologi Lembar Tilamuta, Sulawesi, Skala 1:250.000* (Bandung, 1993)
4. T. Apandi and S. Bachri, *Peta Geologi Lembar Kotamoagu, Sulawesi, Skala 1:250.000* (Bandung, 1997)
5. J. Weston, E. R. Engdahl, J. Harris, D. Di Giacomo, and D. A. Storchak, Geophys. J. Int. **214**, 474 (2018)
6. S. S. Bawono, E. Partoyo, S. Wiryosujono, R. L. Situmorang, and R. J. . Supandjono, *Peta Geologi Lembar Limboto, Sulawesi, Skala 1:100.000* (Bandung, 1999)
7. P. C. Thenhaus, S. L. Hanson, I. Effendi, E. K. Kertapati, and S. T. Algermissen, Earthq. Spectra **9**, 97 (1993)
8. J. Surmont, C. Laj, C. Kissel, C. Rangin, H. Bellon, and B. Priadi, Earth Planet. Sci. Lett. **121**, 629 (1994)
9. P. Molnar and K. E. Dayem, Geosphere **6**, 444 (2010)
10. I. M. Watkinson and R. Hall, Geol. Soc. Spec. Publ. **441**, 71 (2017)
11. C. Rangin, X. Le Pichon, S. Mazzotti, M. Pubellier, N. Chamot-Rooke, M. Aurelio, A. Walpersdorf, and R. Quebral, Geophys. J. Int. **139**, 296 (1999)
12. A. Socquet, W. Simons, C. Vigny, R. McCaffrey, C. Subarya, D. Sarsito, B. Ambrosius, and W. Spakman, J. Geophys. Res. Solid Earth **111**, (2006)
13. G. P. Hayes, G. L. Moore, D. E. Portner, M. Hearne, H. Flamme, M. Furtney, and G. M. Smoczyk, Science (80). **362**, 58 (2018)
14. B. Weber, J. Becker, W. Hanka, A. Heinloo, M. Hoffman, T. Kraft, D. Pahlke, J. Reinhardt, J. Saul, and H. Thoms, in *EGU Gen. Assem. Conf. Abstr.* (2007)
15. GFZ German Research Centre for Geosciences and gempa GmbH, (2008)
16. J. Havskov, P. H. Voss, and L. Ottemöller, Seismol. Res. Lett. **91**, 1846 (2020)
17. G. Laske, G. Masters, Z. Ma, and M. Pasyanos, EGU Gen. Assem. 2013 **15**, 2658 (2013)
18. Ž. Kokalj and M. Somrak, Remote Sens. **11**, (2019)

Engineering geological characteristics based on rock mass rating (RMR) and geological strength index (GSI) in Tunnel Number 1 of the Sigli-Aceh toll road

Jabnes Satria[1,2], *I Gde* Budi Indrawan[1*], and *Nugroho* Imam Setiawan[1]

[1]Geological Engineering Departement, Engineering Faculty, Gadjah Mada University, Sleman, Yogyakarta, Indonesia
[2]PT. Hutama Karya (Persero), Jakarta, Indonesia

Abstract. This paper presents engineering geological investigation results in the form of rock mass characteristics for tunnel number 1 of the Sigli-Aceh toll road. The investigation was carried out through geological mapping, core drill evaluation, and laboratory tests. In this research, the rock mass rating (RMR) and Geological Strength Index (GSI) were applied for the rock mass classifications. The measurement of rock mass quality is then used to determine the excavation method and tunnel support system on the Sigli-Aceh toll road. The results showed that the research location consisted of calcareous sandstone with poor to good-quality (GSI (21.7 - 85.5), RMR (32.0 - 67.6)), and sandstone with good quality (GSI (86.3 - 86.9), RMR (64.0 - 65.0)). The poor quality rock masses were mainly caused by weathering effect. In addition, this research also analyzes the relationship between RMR and GSI based on the type and quality of rocks in the research location so that this correlation can be used in other areas with similar rock type and quality to this research location.

1 Introduction

Tunnel construction development is the most appropriate alternative to protect the environment and support the principles of sustainable development [1]. Tunnel construction development with sustainable development is currently being developed in Indonesia through the Ministry of Public Works and Housing (PUPR). Land problems have always been obstacles in development, such as roads, bridges, and other civil constructions. The problem of travel time, geological conditions, and various topographies of Indonesia, such as lowlands, hills, and mountains, is the determining factor for tunnel construction. One of the tunnels to be built is tunnel number one of the Sigli-Aceh toll road. The construction of this tunnel is located in Padang Tiji District, Pidie Regency, Aceh Province (Fig. 1).

Morphological units in the research location based on Brahmantyo et al. [2], which are homocline hills and homocline valleys in the Paguhulu River. Homocline hills have a slope of more than 35% with a steep category and the potential for landslides to occur on the slopes if tunnel excavation is carried out.

Several site investigations had carried out in the tunnel construction design stage. However, the engineering geological characteristics of rock mass at the research location have not been carried out in detail. The rock mass characteristics can be carried out through engineering geological investigations in the form of engineering geological mapping, core drill evaluation, and laboratory testing to determine the rock mass engineering classification.

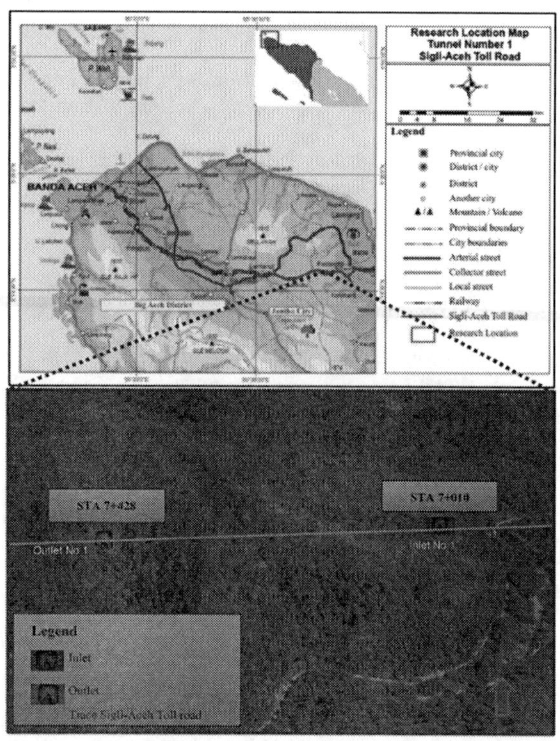

Fig. 1. Location of tunnel number 1 Sigli-Aceh toll road.

* Corresponding author: igbindrawan@ugm.ac.id

Rock mass engineering classification is an essential step to determining the excavation method and the stability of the tunnel. The rock mass rating (RMR) and Geological Strength Index (GSI) are the two most widely used rock mass classifications of the other rock mass classification systems. The rock mass rating (RMR) and Geological Strength Index (GSI) are very useful in determining different rock conditions. The empirical research to test the correlation between the rock mass of RMR and GSI was conducted previously [3, 16] by emphasizing its application for tunnel construction. Because the rock mass characteristics significantly affect the excavation method and the stability of the tunnel, it is necessary to carry out detailed research on the engineering geological characteristics based on the rock mass of RMR and GSI, that these rock mass classification has never been used before at this research location. In addition, the correlation between the rock mass RMR and GSI subsurface in Tunnel number 1 on the Sigli-Aceh toll road needs to be conducted so that the correlation results can be applied to other areas that having similar rock types and qualities with the research location.

2 Geological conditions

2.1 Lithology

The Banda Aceh Geological Map Sheet with a scale of 1: 250,000 was prepared by Bennet et al. [5] showed that the research location and its surroundings have a regional stratigraphy with an old to the young sequences consisting of the Meucampli Formation (Tlm), the Padangtiji Formation (Tuktp), the Kotobakti Formation (Tukt), the Seulimeum Formation (QTps), Lam Teuba Volcanic Rocks (QTvt), and Alluvium (Qh). The main formation types found at this research location are members of the Padangtiji Formation. At the tunnel location, there is one rock unit, namely calcareous sandstone with a sandstone interlayer. While for the lithology, it consists of calcareous sandstone and sandstone (Fig. 2).

(a) (b)

Fig. 2. Rock types from observations on outcrops at the research location: a) calcareous sandstone; b) sandstone.

2.2 Geological structure

The geological structure in Aceh Province based on the Hazard Map and Earthquake Sources of Indonesia [6] consists of two active faults: Nicobar Seulimeum-North and Aceh-North. Based on the interpretation of the Digital Elevation Model (DEM) map for the research location, it was found that there were dextral shear faults trending northeast-southwest. In addition, based on the measurement results of rock layers in the field, there are also two different directions of rock layers, namely the northeast and southeast. So that at the research location, there is an estimate of dextral shear faults.

3 Methodology

3.1 Sample collections and field parameters

The data used in this research consisted of engineering geological mapping, core drills, and laboratory results from subsurface samples. Analysis of rock mass quality at the research site used subsurface data from the results of drilling carried out as many as five boreholes, namely TB-11B with a depth of 40 m, TB-13B1 with a depth of 50 m, TB-15B with a depth of 30 m, TB-16B with a depth 60 m, and TB-17B with a depth of 50 m. Laboratory testing is carried out directly from drilling results according to the lithology and weathering degree of the rock [7]. Tests were conducted in the form of soil and rock index properties [8] and rock uniaxial compressive strength (UCS) [9]. This test is carried out to obtain the parameters used to determining engineering geological characteristics based on Rock Mass Rating (RMR) and Geological Strength Index (GSI).

3.2 Rock mass classification systems

3.2.1 Rock Mass Rating (RMR)

Rock Mass Rating (RMR) is an engineering classification of rock mass often used for practical purpose in the engineering works [10]. Rock mass rating (RMR) has six parameters used for classifying the rock mass, namely 1) uniaxial compressive strength (UCS); 2) designation of rock quality (RQD); 3) spacing of discontinuity; 4) condition of discontinuity; 5) conditions of groundwater; and 6) orientation of discontinuity. These parameter can be used to obtain the rock mass quality of RMR at the research location. Besides, the RMR classification can also determine the tunnel stand-up time, which is the safe time for the tunnel after excavation without any reinforcement. The RMR classification can also determine the excavation method and tunnel support systems according to the rock mass class.

3.2.2 Geological Strength Index (GSI)

The Geological Strength Index (GSI) is a suitable method for determining rock mass quality in bad conditions with a high weathering degree. According to Marinos and Hoek [11], the Geological Strength Index (GSI) can be determined directly from the outcrop

 https://doi.org/10.1051/e3sconf/202132501014

covering two main parameters, namely the condition of the structure (structure) and the condition of its surface (surface condition). In this research, the GSI value was determined based on subsurface data from the results of drilling carried out at the tunnel location. Calculation of subsurface GSI, according to Hoek et al. [12], can be determined using the relationship between Joint Condition [10] and the value of Rock Quality Designation (RQD) [3]. The subsurface GSI value can be calculated using the following equation:

$$GSI = 1.5JCond_{89} + RQD/2 \quad (1)$$

3.2.3 Correlation of RMR and GSI

The rock mass quality of RMR and GSI has a very close relationship because GSI parameters are generally based on RMR. The rock mass classification based on GSI is a development of RMR. Hoek and Brown [13] explain that the RMR classification based on Bieniawski [10] can estimate the GSI value as in the 1976 version. In this situation, the rating of the groundwater in dry conditions is equal to 15, and the orientation of the discontinuity is equal to 0. The minimum value for the classification of the rock mass of RMR is 23, according to Bieniawski [10]. The final weight of this rock mass classification is called RMR_{89} and can be used to estimate the GSI value. If the value is GSI≥18 or RMR≥23, then:

$$RMR_{89} = GSI + 5 \quad (2)$$

Besides, Zhang et al. [4] have also examined the rock mass quality relationship between RMR and GSI. This research was conducted to determine the correlation value between the rock mass of RMR and GSI through the latest RMR called RMR_{14} [14] by considering the uniaxial compressive strength (UCS) value and the properties of the intact rock. Based on the Monte Carlo simulation method, the quantitative correlation of RMR and GSI for UCS values ranging from 9.08 MPa < UCS < 74.87 MPa can be simplified into the following equation:

$$RMR_{14} = 0.642GSI + 0.494UCS^{0.65} + 0.179UCS + 11.985 \quad (3)$$

$$RMR_{89} = (RMR_{14} - 2)/1.1 = 0.584GSI + 0.449UCS^{0.65} + 0.163UCS + 9.077 \quad (4)$$

For the UCS value ≤ 9.08, then:

$$RMR_{14} = 0.642GSI + 0.494UCS^{0.65} + 13.365 \quad (5)$$

$$RMR_{89} = (RMR_{14} - 2)/1.1 = 0.584GSI + 0.449UCS^{0.65} + 12.150 \quad (6)$$

4 Results and discussion

The results of geological mapping in the form of observations on rock outcrops show that there is one rock unit, namely calcareous sandstone, with a sandstone interlayer. While for the lithology, it consists of calcareous sandstone and sandstone. For rock units and the rock layer directions at the research, location can be seen in Figure 3, while the geological section can be seen in Figure 4.

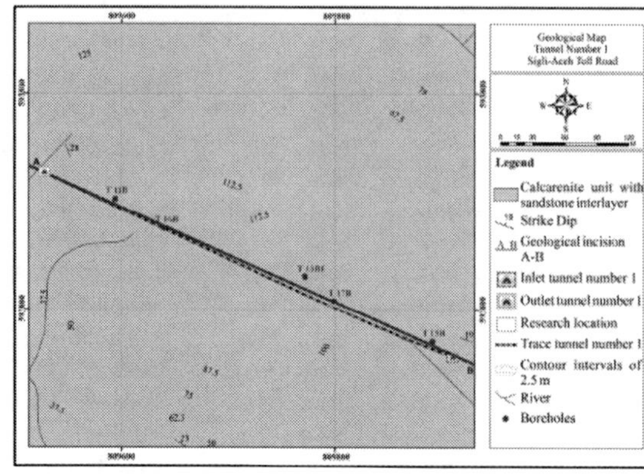

Fig. 3. Geological map of the research location.

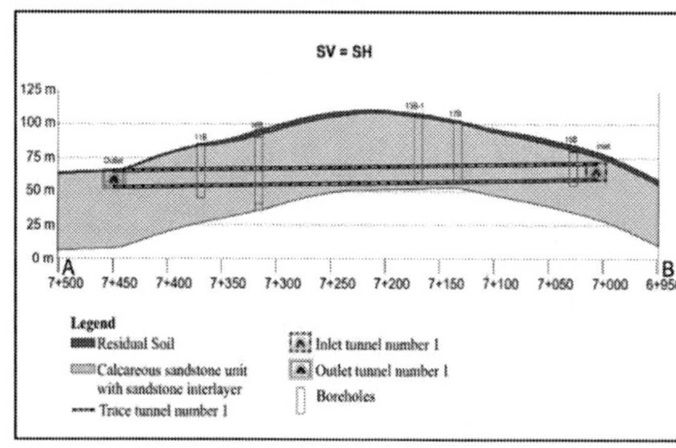

Fig. 4. Geological cross-section of A – B.

The rock mass classification used in determining the engineering geological characteristics is strongly influenced by rock discontinuities in the form of weathering degrees and joints. In this research, each borehole has rock discontinuities with high to slight weathering degrees. The rock discontinuity in the form of joints is directly proportional to the weathering degree. The higher the weathering degree at each borehole, the greater the number of joints. The weathering degree and joints have an important role in determining the quality of rock mass. The higher the weathering degree and joints in the rock, the worse the rock mass quality will be. Conversely, the lower the weathering degree and joints in the rock, the better the rock mass quality.

Boreholes of T-11B and T-15B are located at the inlet and outlet of the tunnel. The tunnel location at these boreholes has rock discontinuities with moderate weathering degrees. The boreholes of T-13B1, T-16B, and T-17B are located in along the tunnel alignment. The tunnel location at these boreholes has rock discontinuities with slight weathering degrees. The difference in the weathering degree at the borehole is due to the tunnel position from the topographic surface. For the tunnel position close to the topographic surface, so the rock discontinuities on that tunnel have a high weathering degree because these rocks are repeatedly exposed for a long time due to weather fluctuation and laboratory results at each borehole, the slake endurance test value is SDI <60%. Based on the Gamble classification [15], the SDI value <60% is low to very low durability, which indicates that the material will be very easily weathered into such as residual soil when exposed or when a flow of water enters the tunnel during excavation. If it is not reinforced immediately by shotcrete (to prevent direct contact with the atmosphere), the spalling or ravelling can occur at the walls and roof of the tunnel, which causing instability of the tunnel construction.

Based on rock discontinuity factors in the form of weathering degree and joints at the research location, the GSI rock mass quality from subsurface data (boreholes) has rock discontinuities with high to slight weathering degrees. The GSI rock mass quality [12] at the research location, based on its lithology, consists of calcareous sandstones with GSI values ranging from 21.7 to 85.5 and sandstones ranging from 86.3 to 86.9. Meanwhile, the RMR rock mass quality from the subsurface data (boreholes) has poor to good quality. The RMR rock mass [10], based on its lithology, consists of calcareous sandstone with RMR values ranging from 32.0 to 67.6 and sandstones ranging from 64.0 to 65.0.

The rock mass quality at the tunnel location (tunnel elevation) based on RMR and GSI has moderate to good quality. For GSI, rock mass values [12] ranged from 48.4 to 85.5. Meanwhile, the RMR rock mass [10] ranged from 46.9 to 67.6. In addition to rock discontinuities in the form of weathering degree and joints, the strength of intact rock material like the uniaxial compressive strength test also influences the rock mass quality. The uniaxial compressive strength value at the research location ranges from 0.352 to 9.32 MPa and is classified as weak to extremely weak. The rock mass quality and the weathering degree for each borehole can be seen in Figure 5, and the drill core material can be seen in Figure 6. While the engineering geological characteristics of rock mass at the research location based on lithology can be seen in Table 1.

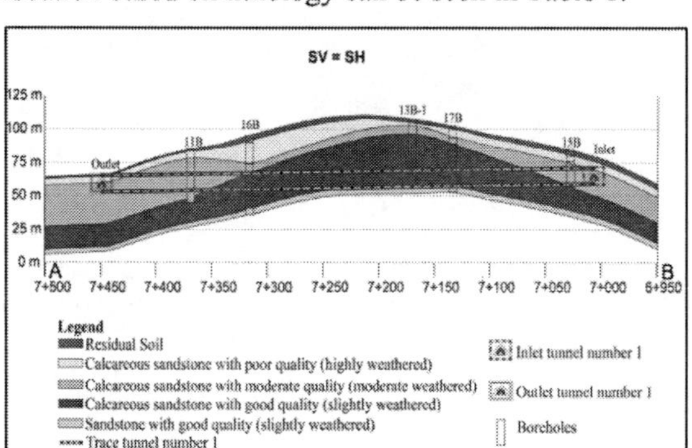

Fig. 5. The rock mass quality and the weathering degree for each borehole.

(a)

(b)

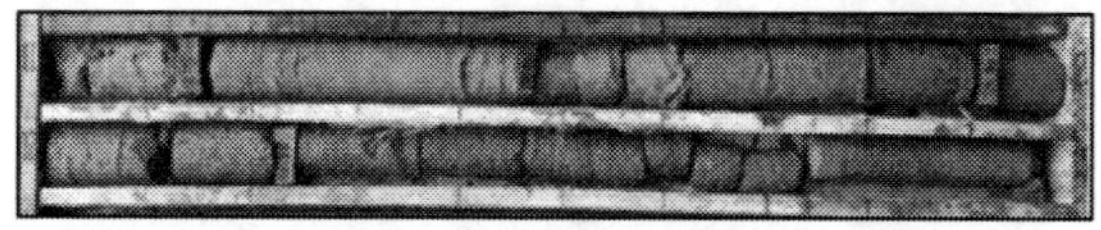

(c)

(d)

Fig. 6. Typical drill cores at the research location, material: a) residual soil; b) poor quality calcareous sandstone (highly weathered); c) fair quality calcareous sandstone (moderate weathered); d) good quality sandstone (slightly weathered).

The rock mass quality at the tunnel site is also very important to determine the most appropriate excavation method and tunnel support system. Using the RMR rock mass classification, we can determine an easy and safe excavation method for the tunnel. Tunnel excavation will be easier and safer if the tunnel axis is in the direction of the rock layer. Based on Figure 3, the research location generally has rock layers dipping to the northeast with an average angle of above 20^0. For the tunnel axis parallel to strike with a rock dipping angle of $20\text{-}45^0$, it will be fair favourable if tunnel excavation is carried out [10]. So, the tunnel construction will be quite easy and safe if carried out at this research location.

In addition, this research also analyzes the relationship between RMR and GSI. This correlation is carried out and can be used as reference in other areas that having similar rock types and quality with this research location. These correlations will be compared with Hoek and Brown [13] and Zhang et al. [4]. The correlation between the rock mass of RMR and GSI in the research location can be expressed by the following equation (Fig. 7).

$$RMR_{89} = 0.5207GSI + 20.753 \quad (7)$$

The comparison of rock mass values based on the classification of RMR [10], GSI [12], correlation of RMR and GSI according to [13] and [4] can be seen in Table 2. Based on equation 7, the calculation of the subsurface rock mass quality based on the correlation between the rock mass of RMR and GSI at the research location generally has similarities with Zhang et al. [4]. Meanwhile, when compared with the correlation according to Hoek and Brown [13], there are some differences in rock mass quality. Hence the correlation between the rock mass of RMR and GSI, according to Zhang et al. [4], can be used at this research location and in other areas with similar rock types and qualities.

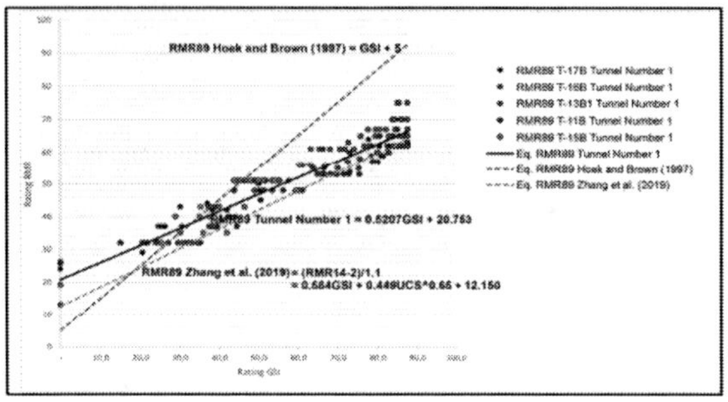

Fig. 7. The correlation equation RMR_{89} and GSI at the research location and comparison between correlation RMR_{89} and GSI at the research location with Hoek and Brown [13] and Zhang et al. [4]

5 Conclusions and recommendations

The tunnel number 1 of the Sigli-Aceh toll road consists of one rock unit, namely calcareous sandstone, with a sandstone interlayer. While for the lithology, it consists of calcareous sandstone and sandstone. The rock mass classification that will be used in determining the engineering geological characteristics is strongly influenced by rock discontinuities in the form of weathering degrees and joints. In addition, the uniaxial compressive strength also affects the rock mass quality. Rock mass based on the subsurface data (boreholes) has rock discontinuities with high to slight weathering degrees with uniaxial compressive strength (UCS) values ranging from 0.352 to 9.32 MPa classified as weak to extremely weak. The GSI rock mass [12] values based on lithology consist of calcareous sandstones with GSI values ranging from 21.7 to 85.5 and sandstones ranging from 86.3 to 86.9.

Meanwhile, the RMR rock mass quality from the subsurface data (boreholes) has poor to good quality. The RMR rock mass [10] values based on lithology consist of calcareous sandstone with RMR values ranging from 32.0 to 67.6 and sandstones ranging from 64.0 to 65.0. The rock mass quality at the tunnel site is also very influential in determining the excavation method and tunnel support system. The direction and angle of the rock slope are very influential on the construction of the tunnel. The tunnel construction at the research location based on RMR rock mass [10] will be fair favourable because the tunnel axis parallel to strike with a rock dipping angle is above 20^0.

The correlation of RMR and GSI rock mass at the research location ($RMR_{89} = 0.5207GSI + 20.753$) has a fairly good result because it has the same rock mass quality as the RMR and GSI correlation, according to Zhang et al. [4]. Meanwhile, compared with the correlation between the rock mass of RMR and GSI Hoek and Brown [13], there are several differences in rock mass quality because the correlation does not consider the value of uniaxial compressive strength (UCS) and intact rock properties. Therefore the correlation between the rock mass of RMR and GSI at the research location and according to Zhang et al. [4], can be used in other areas with similar rock types and quality to this research location.

Further research is needed, including determining the excavation method, and support system of tunnel number 1 on the Sigli-Aceh toll road based on the results of the rock mass characteristics in this research.

Table 1. Engineering geological characteristics of rock mass based on the lithology of the research location

Lithology	Characteristics
Calcareous sandstone with poor quality	The rock is whitish-grey color, grain size (fragment: 0.1-0.5 mm; matrix: <0.1 mm), rounded grain shape, closed packaging, fine sorting, sediment structure of bedding, composition: quartz, feldspar, biotite, lithic, calcareous material, and siliciclastic material. This lithology has a high level of weathering, GSI values 21.7-35.8, RMR_{89} 32.0-39.0, UCS <1 Mpa. Water content 13.06%, dry density 1.397 gr/cm^3, wet density 1.580 gr/cm^3.
Calcareous sandstone with moderate quality	The rock is whitish-grey color, grain size (fragment: 0.1-0.5 mm; matrix: <0.1 mm), rounded grain shape, closed packaging, fine sorting, sediment structure of bedding, composition: quartz, feldspar, biotite, lithic, calcareous material, and siliciclastic material. This lithology has a moderate level of weathering, GSI values 48.4-48.7, RMR_{89} 46.9-51.0, UCS ranges from 1-5 Mpa. Water content 14.24%, dry density 1.628 gr/cm^3, wet density 1.860 gr/cm^3.
Calcareous sandstone with good quality	The rock is whitish-grey color, grain size (fragment: 0.1-0.5 mm; matrix: <0.1 mm), rounded grain shape, closed packaging, fine sorting, sediment structure of bedding, composition: quartz, feldspar, biotite, lithic, calcareous material, and siliciclastic material. This lithology has a slight level of weathering, GSI values 65.7-74.4, RMR_{89} 61.9-67.6, UCS ranges from 5-25 Mpa. Water content 7.62%, dry density 1.6775 gr/cm^3, wet density 1.910 gr/cm^3.
Sandstone with good quality	The rock is Brownish color, grain size (fragment: 0.1-0.5 mm; matrix: <0.1 mm), rounded grain shape, closed packaging, good sorting, massive sedimentary structure, composition: quartz, feldspar, biotite, lithic, a siliciclastic material. This lithology has a slight level of weathering, GSI values 86.3-86.9, RMR_{89} 64.0-65.0, UCS ranges from 5-25 Mpa. Water content 9.56%, dry density 1.920 gr/cm^3, wet density 2.104 gr/cm^3.

Table 2. Rock mass quality based on RMR [10], GSI [12], correlation of RMR and GSI at the research location, correlation of RMR and GSI [13], and correlation of RMR and GSI [4]

Borehole	Lithology	Weathering Degree	Depth (m)		Rock Mass Quality								
					GSI	RMR	Rock Mass Quality	RMR	Rock Mass Quality	RMR	Rock Mass Quality	RMR	Rock Mass Quality
					Hoek et al. [12]	Bieniawski [10]		Research Location		Hoek & Brown [13]		Zhang et al. [4]	
T-17B	Residual soil		0	1.5									
T-17B	Calcareous sandstone	Highly weathered	1.5	4	23.4	34.3	Poor	32.9	Poor	28.4	Poor	26.2	Poor
T-17B	Calcareous sandstone	Highly weathered	4	9	33.4	39.0	Poor	38.1	Poor	38.4	Poor	32.1	Poor
T-17B	Calcareous sandstone	Moderate weathered	9	13	48.7	48.3	Fair	46.1	Fair	53.7	Fair	41.1	Fair
T-17B	Calcareous sandstone	Slightly weathered	13	45	84.5	63.3	Good	64.8	Good	89.5	Very Good	62.2	Good
T-17B	sandstone	Slightly weathered	45	50	86.9	65.0	Good	66.0	Good	91.9	Very Good	63.6	Good
T-16B	Residual soil		0	5.5									
T-16B	Calcareous sandstone	Highly weathered	5.5	22	35.8	34.5	Poor	39.4	Poor	40.8	Poor	33.5	Poor
T-16B	Calcareous sandstone	Moderate weathered	22	28	74.4	56.5	Fair	59.5	Good	79.4	Very Good	56.3	Good
T-16B	Calcareous sandstone	Slightly weathered	28	55	85.5	64.5	Good	65.3	Good	90.5	Very Good	62.8	Good
T-16B	sandstone	Slightly weathered	55	60	86.3	64.0	Good	65.7	Good	91.3	Very Good	63.3	Good
T-13B1	Residual soil		0	2.8									
T-13B1	Calcareous sandstone	Highly weathered	2.8	5	22.5	32.0	Poor	32.5	Poor	27.5	Poor	25.7	Poor
T-13B1	Calcareous sandstone	Moderate weathered	5	11	48.5	51.0	Fair	46.0	Fair	53.5	Fair	41.2	Fair
T-13B1	Calcareous sandstone	Slightly weathered	11	33	74.4	61.9	Good	59.5	Good	79.4	Good	56.3	Good
T-13B1	Calcareous sandstone	Slightly weathered	33	50	85.4	67.6	Good	65.2	Good	90.4	Very Good	62.7	Good
T-11B	Residual soil		0	1.5									
T-11B	Calcareous sandstone	Highly weathered	1.5	7	21.7	32.0	Poor	32.1	Poor	26.7	Poor	25.3	Poor
T-11B	Calcareous sandstone	Moderate weathered	7	18	48.4	47.8	Fair	46.0	Fair	53.4	Fair	41.1	Fair
T-11B	Calcareous sandstone	Slightly weathered	18	40	65.7	52.1	Fair	55.0	Fair	70.7	Good	51.2	Fair
T-15B	Residual soil		0	5									
T-15B	Calcareous sandstone	Highly weathered	5	9	31.7	38.0	Poor	37.3	Poor	36.7	Poor	31.1	Poor
T-15B	Calcareous sandstone	Moderate weathered	9	30	48.4	46.9	Fair	46.0	Fair	53.4	Fair	41.1	Fair

We acknowledge PT. Hutama Karya (Persero) and PT. Adhi Karya (Persero) Tbk to give us permission in conducting research at the Sigli-Aceh toll road tunnel 1. Thank you to Dr. Hendy Setiawan for guiding and providing valuable inputs in refining the manuscript. This research is financially supported by PT. Hutama Karya (Persero) in the scheme of a super-specialist program by the Ministry of Public Works and Housing year 2020.

References

1. Parker, Harvey W, *Good for sustainable development, and vice versa*, in International Tunnelling Association (ITA), Singapore (2004)
2. B. Brahmantyo and B. Salim, Jurnal Geoaplika **1**,2:071-078 (2006)
3. D.U. Deere and D. W. Deere, Rock quality designation (RQD) after twenty years, 11-19 (US Army Corps of Engineers, Washington,1989)
4. Q. Zhang, X. Huang, H. Zhu, J. Li, Tunn. Undergr. Sp. Tech, **83**:73-81 (2019)
5. J.D. Bennet, D.McC. Bridge, N.R. Cameron, A. Djunuddin, S.A. Ghazali, D.H. Jeffery, W. Kartawa, W. Keats, N.M.S. Rock, S.J. Thomson, R. Whandoyo, *Peta geologi lembar Banda Aceh, Sumatra* (1981)
6. PUSGEN, *Indonesia earthquake source and hazard map (Peta sumber dan bahaya gempabumi Indonesia)*(2017)
7. ISRM, Int. J. Rock Mech. Min. Sci. Geomech. Abstr. **15**:319-368 (1978)

8. SNI 03-3637-1994, Method of testing the unit weight of fine-grained soil with the test object mold, 1-4 (Public Works Research and Housing, Indonesia, 1994)
9. SNI 2825, The instruction of uniaxial compressive strength test for rock, 1-8 (Association of National Standardization, Indonesia, 2008)
10. Z. T. Bieniawski, Engineering rock mass classifications: a complete manual for engineers and geologists in mining, civil, and petroleum engineering (Wiley, New York 1989)
11. P. Marinos and E. Hoek, *GSI: a geologically friendly tool for rock mass strength estimation,* in Proceeding GeoEng2000 Conference, Melbourne, 1422-1442 (2000)
12. E. Hoek, T. G. Carter, M. S. Diederichs, *Quantification of the geological strength index chart,* in US Rock Mechanics/Geomechanics Symposium, San Francisco (2013)
13. E. Hoek and E. T. Brown, Int. J. Rock Mech. Min. Sci. **34**,8:1165-1186 (1997)
14. B. Celada, I. Tardáguila, P. Varona, A. Rodrígueza, Z. T. Bieniawski, *Innovating tunnel design by an improved experience-based RMR system* in Proceedings of the World Tunnel Congress, 1-9 (2014)
15. J.C. Gamble, *Durability-plasticity classification of shales and other argillaceous rocks.* University of Illinois, Urbana-Champaign (1971)

Slope stability analyses by circular failure chart and limit equilibrium methods: the inlet and outlet of diversion tunnel of Bolango Ulu Dam, Indonesia

Alesandro Sejo Luden[1,2], *I Gde* Budi Indrawan[1*], and *Dwikorita* Karnawati[1]

[1]Departement of Geological Engineering, Universitas Gadjah Mada, Yogyakarta, Indonesia
[2]Ministry of Public Works and Housing of Indonesia, South Jakarta 12110, Indonesia

Abstract. The objective of this research is to evaluate the stability of the natural slopes at the inlet and outlet portals of the Bolangu Ulu diversion tunnel, Gorontalo. The natural slopes were considered stable, and therefore slope stability analyses were not carried out previously in the tunnel portal design. The slope stability analyses were carried out using the Circular Failure Chart (CFC) and Limit Equilibrium Methods (LEM). Input data for the slope stability analyses were obtained from field mapping and laboratory testing of soil and rock samples. The results show that the portal slopes consist of diorite and residual soil. Both stability analysis methods yield nearly the same results. The slope at the outlet section had the factor of safety (FOS) values of 1.29 and 1.30 based on the CFC method and LEM, respectively, indicating the slope in a stable condition. However, the slope at the inlet section had the FOS values of 1.01 and 1.07 based on the CFC method and LEM, respectively, indicating the slope in a critical condition. The results suggest that stabilization of the portal slopes, particularly the portal slope at the inlet section, is required to prevent slope failures under static and earthquake loads.

1 Introduction

Slope failures are a serious geo-environmental hazard that endangers human populations, transportation routes such as highways and rail lines, and civil structures such as dams, buildings, and other structures [1]. Predicting and analyzing the slope's stability is a serious challenge for geotechnical engineers [2]. It plays an essential role for a designer to plan engineering structures, namely dams, tunnels, roads, etc. In tunnel construction, slope stability analysis is required to ensure the portal is stable during the construction and operation [3]. It also helps the designer to estimate the construction cost that will be involved. Many researchers focused on assessing and developed various methodologies for slope stability.

Slope stability analysis is performed to assess the safety factor of potential and probable slope failures [4]. The safety factor is comparing shear strength to maintain slope stability and the shear strength that causes. From the comparisons, it is obtained the value of slope stability expressed in numbers. In the condition that Factor of Safety (FOS) > 1, the slope will be stable. However, if the FOS < 1, the slope becomes unstable, and slope failures occur [5]. The slope is considered stable when the FOS is greater than one, and the driving shear stress is lower than the resistive shear stress [1]. When this ratio approaches 1, shear strength and shear stress are approximately equal, and the slope is on the verge of failure or critical condition. If the slopes become unstable, remedial and corrective can be proposed to develop a construction area, namely rock slope stabilization programs [1].

This paper presents the results of stability analyses of natural slopes at the inlet and outlet portals of the Bolango Ulu Dam's diversion tunnel using the Circular Failure Chart and Limit Equilibrium Method. It is expected that the results of this research contribute to the safe designs of the tunnel portal slopes.

1.1 Study area

Bolango Ulu Dam is located in Bone Bolango Regency, Gorontalo Province, Indonesia (Figure 1). The inlet and outlet section of the diversion tunnel is considered for the slope stability study.

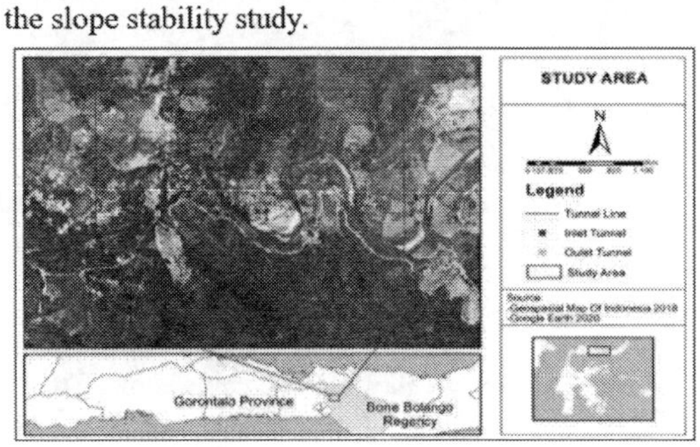

Fig. 1. Location map

* Corresponding author: igbindrawan@ugm.ac.id

 https://doi.org/10.1051/e3sconf/202132501015

2 Methodology

2.1 Sample collections and field parameters

The data used in this study include the geological structure and the slope, geological mapping, core drill, and engineering properties from laboratory test results. Two types of samples were collected for laboratory analysis, namely surface samples and core samples. Surface samples were collected along every section taken up for stability analysis and collected from the subsurface for different sections chosen for analysis. Rock and soil samples were obtained from various areas of the slope profile at each slope segment. The soil samples were taken from a depth of 1.0 m after removing the overburdened soil, while the rock sample was collected from the drill core.

The collected samples were put through a direct shear test. These representative samples were put through a direct shear test with five different normal loads to obtain their shear strength values. The measurements were plotted in normal stress (x) and shear stress (y) to get representative shear strength parameters. All samples were subjected to direct shear strength testing to get the value of the angle of internal friction (φ) and determine cohesion (c), as per the procedure mentioned in ASTM D–2850 directive [6]. The core cutter analysis was used to obtain the bulk density (γ) of the soil samples. Core samples were used to obtain the unit weight and density of soil materials and the grain size of soil samples. The slope depth of overburden material and face angle were analyzed and measured through site work.

2.2 Circular Failure Chart (CFC)

The circular failure charts are used to find an essential combination of slide surface and tension crack for a wide range of slope geometries and groundwater conditions [5]. The circular Failure Chart method is the easiest method to analyze and find out the Factor of Safety. This method is a semi-empirical approach method and required minimum laboratory testing. Its required parameters are density, shear strength, height, and steepness of the cut slope. These parameters will give the safety factor of the particular slope after plotting the corresponding circular failure chart. Circular failure happens when the small particles in soil or completely exposed rock mass are small compared to the slope's size and are not interlocked due to their shape [7].

In the Circular Failure Chart (CFC), several groundwater flow schemes are given to adjust the pore water pressure in the subsurface and stresses due to water present in tension fractures [1]. For that purpose, Figure 2 depicts a set of possible field conditions that have been chosen and shown in a combined manner [1].

To obtain the FOS, follow the techniques explained below and depicted in Figures 3 and 4. The first step is to choose rock strength parameters that apply to the slope's material. Then determine the groundwater conditions on the slope and select the chart closest to these conditions using Figure 2. The next step is to compute the dimensionless ratio with the formula below:

$$c/(\gamma \text{ H } \tan \varphi) \quad (1)$$

Furthermore, find this value on the chart's outer circular scale. The value found to the junction with the curve corresponds to the slope angle and follows the radial line. The last step is to find the suitable value of c/(H FS) or tan/FS, whichever is more convenient and calculate the FOS.

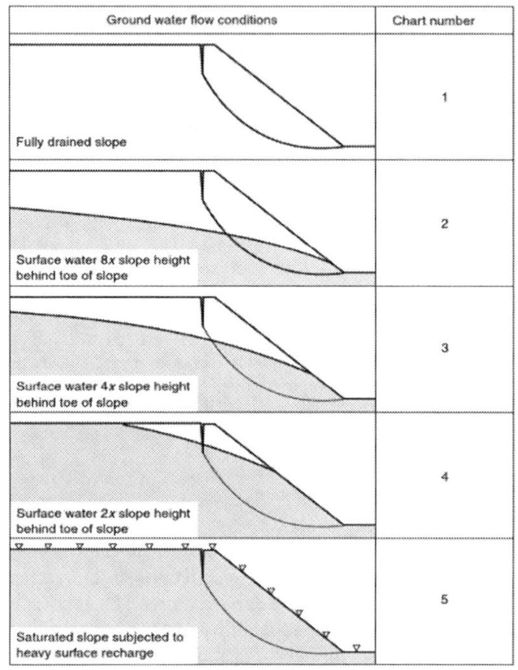

Fig. 2. Groundwater conditions chart [5].

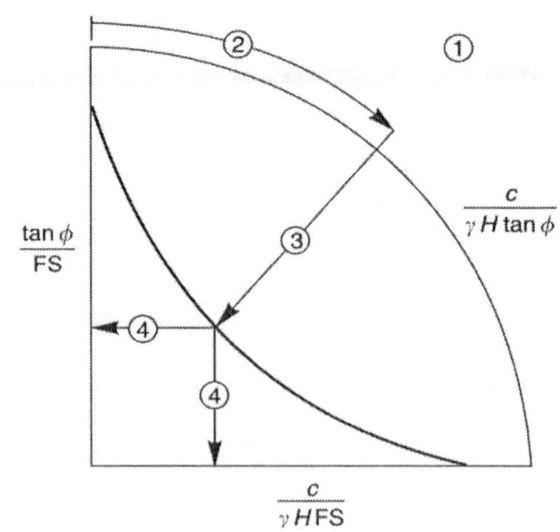

Fig. 3. Calculation of FOS from CFC chart [5].

 https://doi.org/10.1051/e3sconf/202132501015

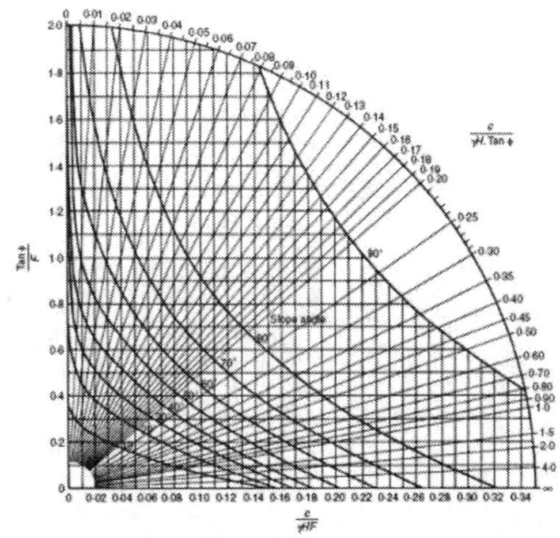

Fig. 4. Model CFC chart [5].

2.3 Limit Equilibrium Method (LEM)

LEM is a method that uses the principle of force equilibrium [8]. This analysis method assumes that the area of failure that possibly occurs in circular failure or noncircular failure. The calculation method is divided the landslide plane into slices. Figure 5 shows the forces that are considered on every slice.

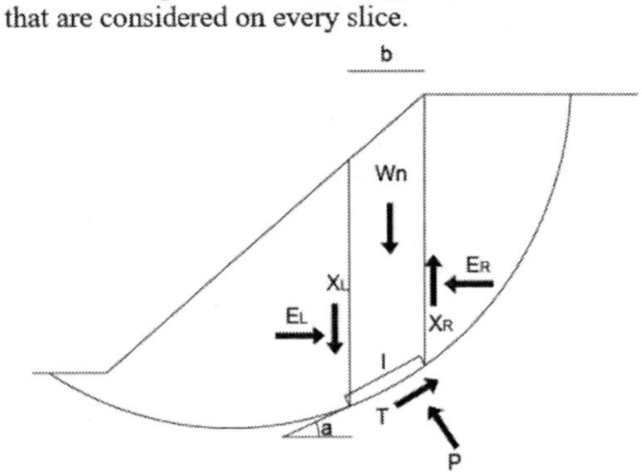

Fig. 5. The forces considered.

LEM analysis was performed using Slide software from Rocscience. Analysis slope stability is carried out using a method developed by Fellenius (the standard method of the slice), Bishop, Janbu, Morgenstern-Price, Spencer, and Sarma [9, 10, 11]. Fellenius introduced the first method in which the forces between the slices are ignored and only considered the moment equilibrium. Bishop [9] then develops more complex ways by incorporating a working force around the wedge plane but calculates with moment equilibrium. Bishop [9] also came up with the Simplified Bishop Method where the force is normal between the slices is considered, but the inter shear force of the slices is ignored. Janbu [10] developed the method which is similar to Bishop's simple method. The difference is that the Janbu [10] method was derived from the horizontal force equilibrium.

3 Result and discussion

3.1 Geological condition

3.1.1 Lithology

The main rock formation consisting of the study area is the diorite and alluvial deposits. Diorite occupies about 85% of the study area. This unit is exposed along the slopes on the left and right of the Bolango river. This rock unit is grey-brown to grey-white, phaneritic, quartz, and the minerals biotite, andesine, and pyrite. The weathering level is light to strong, and the hardness level is weak to very strong. The soil resulting from weathering of diorite is in the form of clay sand, greyish-brown in colour, medium to coarse-grained, strongly weathered to perfect, relative density decomposed to dense. This diorite unit is equivalent to the Bone Diorite Unit on the Kotamobagu geological map sheet, estimated to be of the late Miocene age [12].

Alluvial unit is found on the left and right from upstream to downstream of the Bulango river and becomes floodplains as water discharge rises. The characteristic is fine to lumps, with a percentage of coarse to fine sand is 30%, gravel about 50%, 10% lumps, and 10% mud. This unit is characterized by the condition of loose and rounding blocks.

3.1.2 Geological structure

The geological structures found in the study area are normal faults and shear joints. Normal faults were found by the presence of slickensides or fault mirrors shaped like smooth planes and striations that seemed to be polished (Figure 6). The normal fault is in the form of a left shear fault (sinistral) with a direction of N 35°E. The direction is unidirectional of the river's bend, where the river's bend can indicate a geological structure. The fault at the study site cuts the dam axle continuously downstream of the inlet of the diversion tunnel. The existence of this geological structure is not expected to have a significant effect on the stability of the tunnel because the existing faults are no longer active.

The joint structure found in the study area generally develops in diorite units, namely shear joints. These shear joints have the characteristics of straight joint planes; sometimes, there is a slight shift, the fracture plane is tight, and the angle of the joint pair that intersects each other is 60° or more [13].

Fig. 6. The presence of slickensides or fault mirrors.

Based on the lithology and geological structure found during the field work, the geological map of the study area is prepared and shown in Figure 7.

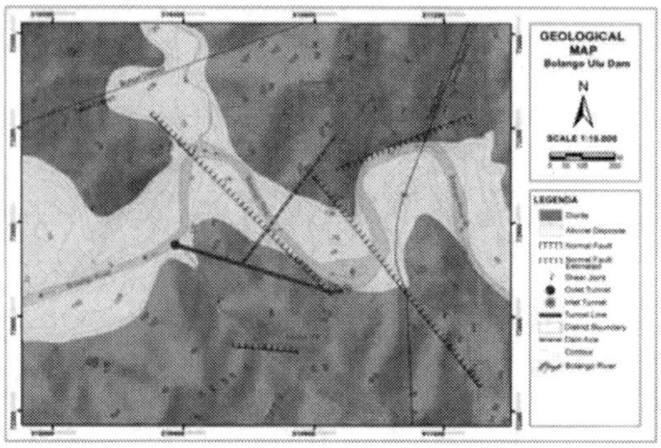

Fig. 7. The geological map of the study area.

3.2 Estimation of FOS using CFC method

Four samples in two sections (inlet and outlet tunnel) were chosen for comprehensive stability investigations and FOS evaluation utilizing the CFC method to estimate the factor of safety [7]. The most suitable chart for groundwater conditions and slope geometry in the study area is chart number 2. Table 1 shown the input parameters used to determine the dimensionless ratio *($c/(\gamma H \tan \varphi)$)*. The charts for the circular failure method are shown in Figures 9-12. The safety factor at the inlet section is very close to 1. This condition indicated that these slope sections are at risk of collapsing and failing during heavy rain or when the excavation tunnel begins. This result confirms the presence of slope Failures in the outlet section during the field investigation (Figure 8). The outlet section FOS critical value is 1.18 indicating the slope in favourably stable condition. The critical FOS values of all slope sections at various saturation conditions are given in Table 2. The details of each slope section are discussed below.

3.2.1 IL - 1 (Residual soil)

The slope section IL-1 is located at the inlet section of the diversion tunnel. This soil sample was collected from the surface. The height of the slope is about 12.5 m. According to the given observation, the critical factor of safety is 1.01. This number of safety factors indicates that the slope is in critical condition.

3.2.2 IL - 2 (Diorite)

The slope section I-2 is the second sample point in the inlet portal tunnel selected for analysis. This sample is a rock sample collected from the drill core. The inclination of the general slope is about 50^0. In all charts, the FOS value is bigger than 1.0 indicates that the slope is stable (Table 2).

Fig. 8. The Existing slope failure in the study area.

3.2.3 OL - 1 (Residual soil)

This slope section is located on the outlet side of the diversion tunnel. It is about 10 m by height and 45^0 in the angle of the slope. The value of cohesion (c) and angle of internal friction (φ) are given in table 1. The factor of safety is calculated in chart number 3 condition. The FOS value is 1.29 indicates the slope is stable.

3.2.4 OL - 2 (Diorite)

The last slope section that was carried out for this study is section Outlet 2. This sample is a rock sample collected from the drill core—this section same as section IL-1 located in the outlet section of the diversion tunnel. The parameters are given in table 1. The result shows that in all charts, the FOS>1 indicates the slope is stable. The factor of safety details is given in table 2.

Table 1. Input parameters for slope stability analysis.

Slope Section	Layer	Slope height (H)	Slope face angle	Unit weight (γ)	Cohesion (c)	Internal friction (φ)	Dimensionless ratio (c/(γ H tan φ)
		m	*Degree*	*kN/m3*	*kN/m2*	*Degree*	
IL- 1	Residual soil	12.5	50.0	17.71	20.20	20.93	0.24
IL- 2	Diorite	12.5	50.0	27.49	201.00	59.49	0.34
OL-1	Residual Soil	10.0	45.0	19.03	22.16	22.45	0.28
OL-2	Diorite	10.0	45.0	28.05	226.00	62.86	0.41

Table 2. FOS values for chart number 2.

Slope Section		Layer	FOS	Critical FOS
Inlet	IL-1	Residual soil	1.01	1.01
	IL-2	Diorite	5.39	
Outlet	OL-1	Residual Soil	1.29	1.29
	OL-2	Diorite	7.36	

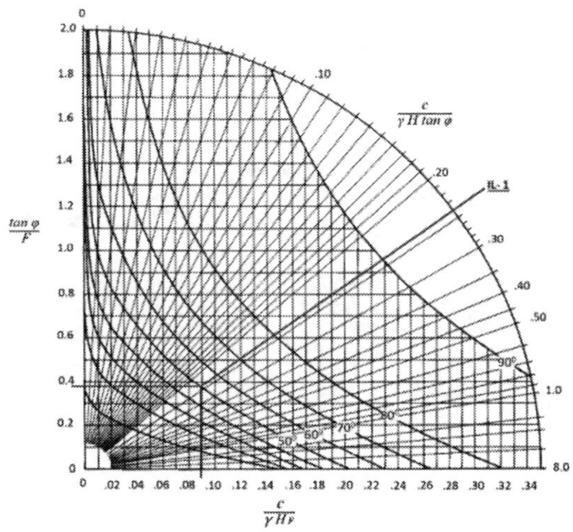

Fig. 9. CFC chart number 2 for slope IL-1.

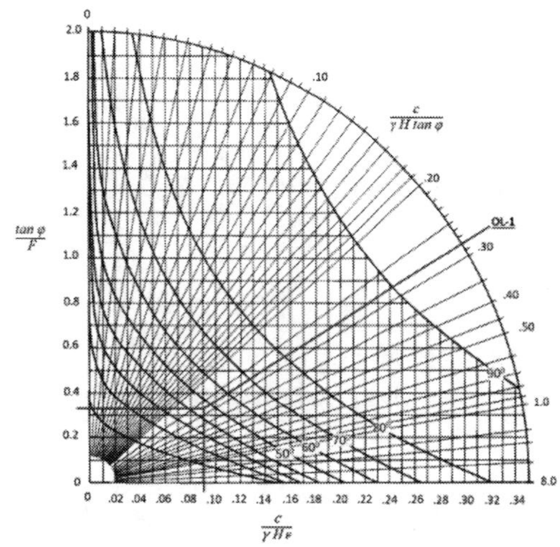

Fig. 11. CFC chart number 2 for slope OL-1.

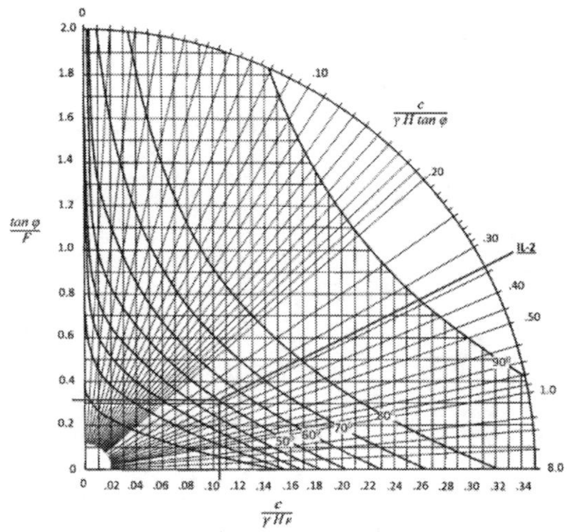

Fig. 10. CFC chart number 2 for slope IL-2.

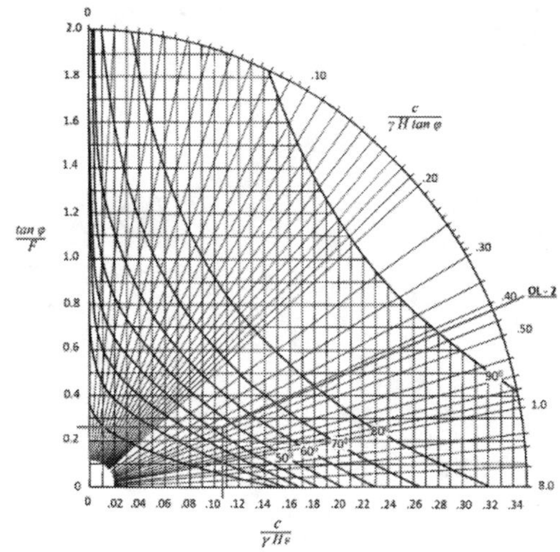

Fig. 12. CFC chart number 2 for slope OL-2.

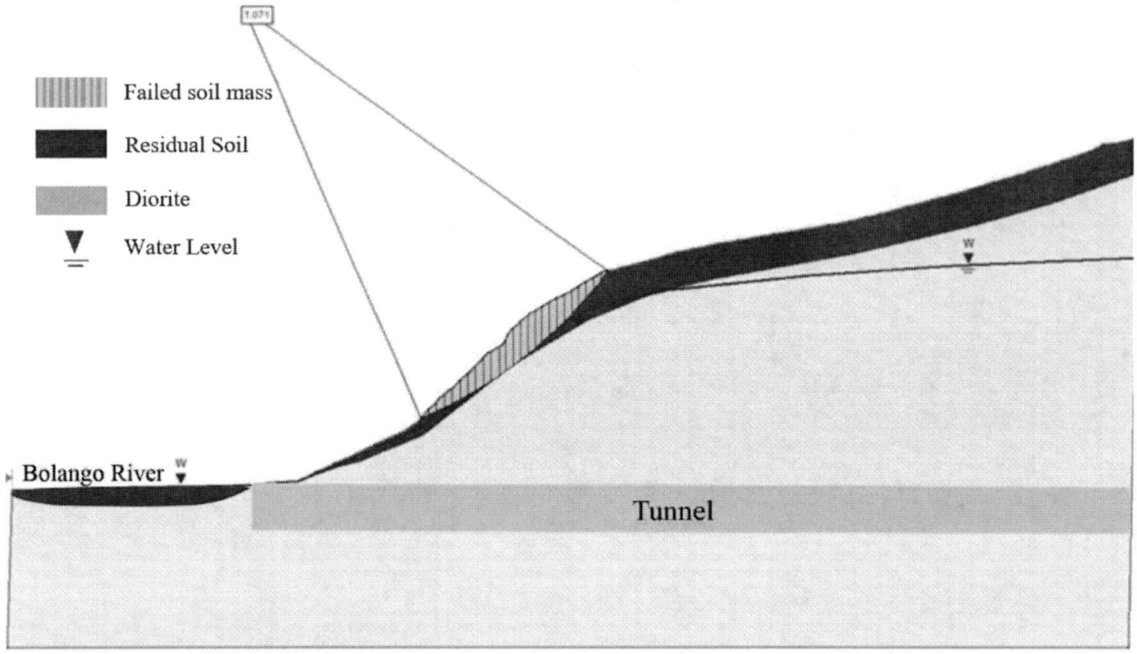

Fig. 13. Slope stability analysis result of inlet section based on LEM.

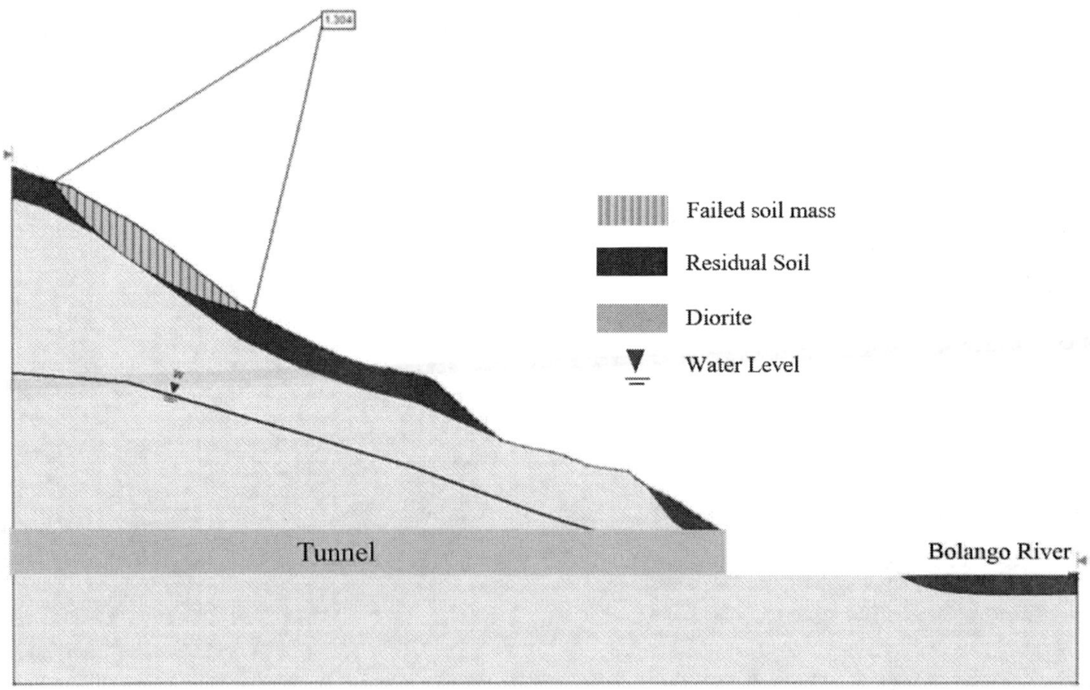

Fig. 14. Slope stability analysis result of outlet section based on LEM.

3.3 Estimation of FOS using LEM

The LEM approach was also used to determine the FOS for each slope section. In each part, both samples (soil and rock) are calculated. Rock and soil properties such as cohesiveness, unit weight, and friction angle of slope forming material were considered. The Factor of Safety values of every slope section using the LEM is shown in Table 3 and Figures 13-14.

Table 3. FOS values for LE method.

Slope Section	FOS values
Inlet	1.07
Outlet	1.30

The results show that the slope at the outlet section is stable (FOS is more than 1). However, the slope at the inlet section is in critical condition (FOS is close to 1). This result is similar to the Circular Failure Chart Method.

4 Conclusion and recommendation

The geological conditions of the research area consist of diorite and alluvial deposits. The diorite occupied 85% of the research area. There are normal faults and shear joints discovered in the study area. The presence of this geological structure is not expected to have a substantial effect on the stability of the tunnel because the existing faults are no longer active. The unit weight value of the inlet section is 17.71 kN/m3 for residual soil, and for diorite is 27.49 kN/m3. The cohesion value of residual soil is 20.2 kN/m2 and 22.16 kN/m2 for the inlet and outlet section, while for diorite is 201 kN/m2 and 206 kN/m2. The internal friction of residual soil is 20.93° and 22.45° for the inlet and outlet section, while for diorite is 59.49° and 62.86°.

The Circular Failure Chart and Limit Equilibrium Method produce a similar result. The inlet section of Bolango Ulu's diversion tunnel is nearly unstable because FOS is very close to 1, while the outlet section is favourably stable. The inlet section requires stabilization before the tunnel excavation to prevent immediate or long-term slope failure. The stabilization measures include decreasing the degree of slope, installing soil nailing and retaining walls, or applying shotcrete. However, further study is required to verify the stabilization measures.

The first author would like to thank the Ministry of Public Works and Housing Indonesia for the research funding and the River Basin Organization of Sulawesi II for allowing this research.

References

1. M. Kannan, R.R. Rao, S.M. Ahmed, Int. J. Eng. Res. Tech. **7**(1) (2019)
2. A. J. Li, R.S. Merifield, A. V. Lyamin, Int. J. Rock Mech. Min. Sci. **45** (2008)
3. R. Yang, P. Guan, K. Cai, W. Xiao, X. Si, Z. Zhaofeng, J. Eng. Sci. Technol. **8**, 14–20 (2015)
4. R.R. Djunaedi. Jurnal Student Teknik Sipil **1**, 2:55–64 (2020)
5. D.C. Wyllie, C.W. Mah, *Rock slope engineering: Civil and mining, 4th edition* (Spon Press, New York, 2017)
6. ASTM D2850 - 95, ASTM Int. **15**, 1–6 (2015)
7. S. Anbazhagan, V. Ramesh, S.E. Saranaathan, Nat. Hazards **86**, 3, 1081–104 (2017)
8. D. Juven, G. Herman, *Analisa Stabilitas Lereng Dengan Limit Equilibrium dan Finite Element Method* (2016)
9. A.W Bishop, Géotechnique **5**, 1, 7–17 (1955)
10. N. Janbu, Int. J. Rock Mech. Min. Sci. Geomech. Abstr. **12,** (1975)
11. E. Spencer, Géotechnique **17**, 1, 11–26 (1967)
12. T. Apandi , S. Bahri, *Peta geologi skala 1:250.000 lembar Kotamobagu* (1997)
13. K. McClay, The Mapping of Geological Structures (John Wiley & Sons, New York, 1988)

 https://doi.org/10.1051/e3sconf/202132501016

Slope stability analysis of the drainage tunnel portal in Tanju Dam, Dompu District, West Nusa Tenggara

Edden Umaga Dinata[1], *I Gde Budi* Indrawan[1*], and *Arifudin* Idrus[1]

[1]Department of Geological Engineering, Faculty of Engineering, Universitas Gadjah Mada

Abstract. This paper presents design results of the tunnel portal slopes at the Tanju Dam, Dompu, West Nusa Tenggara. The objective of this research was to analyse the stability of the tunnel portal slopes using circular failure chart (CFC) method, limit equilibrium method (LEM), and finite element method (FEM). Input parameters were obtained from drill core evaluations and laboratory tests. By considering the rock mass rating (RMR) values of rock masses, which are categorized as class II, at the two slopes, adjustments for the cohesion and inner friction angle values are made. The inlet slope (IL) have cohesion values of 350 kPa and 40° inner friction angle and the outlet slope (OL) have cohesion values of 400 kPa and 45° inner friction angle. The CFC method shows that the IL and OL have safety factor (FS) values of 3.5 and 3.44, respectively. The LEM shows that the IL and OL have the FS values of 3.69 and 3.65, respectively. Meanwhile, the FEM shows that the IL and OL have FS values of 4.78 and 4.79, respectively. The stability analysis results indicate that designed slopes are stable.

1 Introduction

The study location is carried out at the drainage tunnel construction site of tanju dam. In Administrative, the tunnel is located in the village of Bara, Woja, Dompu District, West Nusa Tenggara Province (Fig. 1). Drainage is built to drain the drain 1.9 m3/sec water to fill Tanju reservoir. This research is purposed to determine the slope stability of the designed slope of the drainage tunnel in Tanju dam.

The research area is in morphology with hilly conditions around the tunnel which is a ridge that extends in a relatively northeast-southwest direction, with a ridge length of about 2.2 km and a peak elevation of 320 m [1]. The morphology around the location is quite steep with an angle of 45° – 60°.

Prediction and analysis of slope stability is a matter of concern for geotechnical engineers. This is due to the importance of slope stability analysis for planners to plan the slope of a slope in constructions such as dams, road tunnels and others.

Fig. 1. Research location.

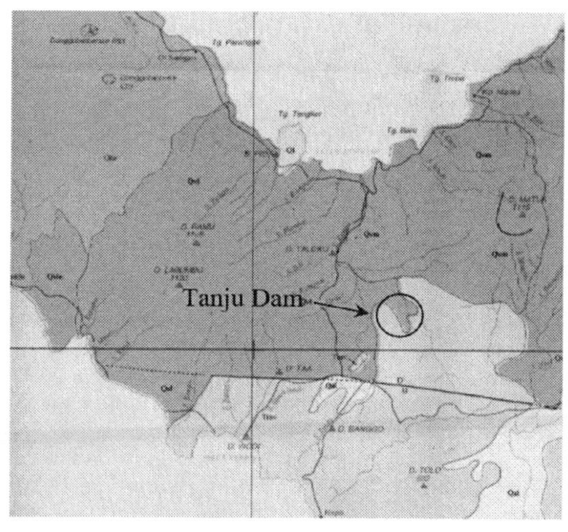

Fig. 2. Regional Geology Map of Sumbawa [2].

2 Geology condition

2.1 Regional geology

Based on the Regional Map (Fig. 2.), it shows that the research location is composed of old volcanic rock products (QTv, Qv) of Quaternary age and alluvial deposits [3]. This group consists of breccias lava and tuffs with a composition of andesite and basalt. The research location is mainly at the location of the planned tunnel which is composed of rocks that are even older

* Corresponding author: igbindrawan@ugm.ac.id.

than Tertiary and Quaternary. Tertiary rocks are predominantly composed of clastic volcanic sedimentary rocks, volcanic rocks and intrusions. Meanwhile, the quaternary rocks consist of volcanic sedimentary rocks, alluvial deposits and river sediments of the Resen age.

2.2 Geology structure

The hilly condition around the tunnel is a ridge that extends in a relatively northeast-southwest direction, with a ridge length of about 2.2 km and a peak elevation of 320 m. The morphology around the location is quite steep with an angle of 45° – 60° [1]

In general, the area around the Tanju Tunnel plan consists [4] of 5 (five) rock units, namely (Fig. 3.):

a. Coluvial
b. Weathered Andesite
c. Andesite
d. Weathered tuff breccia
e. Tuff Breccia

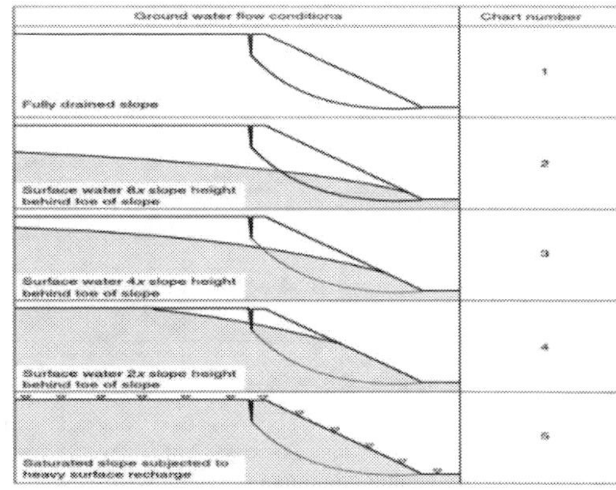

Fig. 3. Groundwater condition graphic (Duncan C. Wyllie & Christoper W. Mah, 2004) [4].

2.3 Rock mass rating

The RMR value [5] in this research area is 65 which can be classified as good rock as in the RMR [5] and GSI [5] classification tables (Table 1.), entered into the Type II Tunneling Support System. RMR value used only to determine the rock mass quality.

Table 1. RMR and the parameters [2].

A. CLASSIFICATION PARAMETERS AND THEIR RATINGS

1	Strenght of Intact Rock of Intact Rock	Point load strenght index	>10 Mpa	4-10 Mpa	2-4 Mpa	1-2 Mpa	For this Range - Uniaxial Compresive is prefered		
		Uniaxial Comp. Strenght	>250 Mpa	100-250 Mpa	50-100 Mpa	25-50 Mpa	5-25 Mpa	1-5 Mpa	< 1 Mpa
	Rating		15	12	7	4	2	1	0
2	RQD			90-100%	75-90%	50-75%	25-50%	<25%	
	Rating			20	17	13	8	3	
3	Spacing of Disc			>2m	0.5-2 m	20 - 60 cm	6 - 20 cm	< 6 cm	
	Rating			20	15	10	8	5	
4	Condition of Discontinuities		Very rough surface not continous no separation unweathered wall rock	Slightly rough surface separation < 1mm Slightly weathered walls	Slightly rough surface separation < 1mm Highly weathered walls	Slickersided surface or Gauge < 5mm thick or separation 1-5mm continous		Soft gauge >5mm thick or separation >5mm continous	
	Rating		30	25	20	10		0	
5	Groundwater			Completely dry	Damp	Wet	Dripping	Flowing	
				15	10	7	4	0	

B. RATINGS ADJUSMENT FOR DISCONTINUITY ORIENTATIONS

Strike and Dip Orientation		Very favourable	Favourable	Fair	Unfavourabl e	Very Favourable
Rating	Tunnels & Mines	0	-2	-5	-10	-12
	Foundations	0	-2	-7	-15	-25
	Slopes	0	-5	-25	-50	

C. ROCK MASS CLASSES DETERMINED FROM TOTAL RATINGS

Rating	100-82	80-61	60-41	40-21	<21
Class Number	I	II	III	IV	V
Discription	Very Good Rock	Good Rock	Fair Rock	Poor Rock	Very Poor Rock

D. MEANING OF ROCK CLASSES

Class Number	I	II	III	IV	V
Average Stand-up time	20 years for 15m span	1 year for 10m span	1 week for 5m span	10hrs for 2.5 m span	30min for 1m span
Cohesion of rock mass (kpa)	>400	300-400	200-300	100-200	<100
Friction angle of rock mass (deg)	>45	35-45	25-35	15-25	<15

E. GUIDELINES FOR CLASSIFICATION OF DISCONTINUITY CONDITION

Disc length	<1m	1-3m	3-10m	10-20m	>20m
Rating	6	4	2	1	0
Separation (Apeture)	none	<0.1mm	0.1-1mm	1-5mm	>5mm
Rating	6	5	4	1	0
Roughness	very rough	rough	slightly rough	smooth	slikensided
Rating	6	4	2	1	0
Infilling	none	Hard Filling <5mm	Hard filling >5mm	soft filling <5mm	soft filling >5mm
Rating	6	4	2	2	0
Weathering	Unweathered	Slightly Weathered	mod weathered	highly weat	Decomposed
Rating	6	5	3	1	0

F. EFFECT OF DISCONTINUITY STRIKE AND DIP ORIENTATION TUNNELING

<table>
<tr><td rowspan="8">Effect of disc to the tunnel</td><td colspan="5">Strike perpendicular to tunnel axis</td></tr>
<tr><td>drive with dip (dip 45-90)</td><td>Drive w/ dip (dip 20-45)</td><td colspan="2">Drive against dip (45-90)</td><td>Drive against dip 20-45</td></tr>
<tr><td>very favourable</td><td>Favourable</td><td colspan="2">Fair</td><td>Unfavourable</td></tr>
<tr><td>0</td><td>-2</td><td colspan="2">-5</td><td>-10</td></tr>
<tr><td colspan="5">Strike paralel to tunnel axis</td></tr>
<tr><td>Dip 20-45</td><td colspan="2">Dip 0-20</td><td colspan="2">Dip 45-90</td></tr>
<tr><td>Fair</td><td colspan="2">Fair</td><td colspan="2">Unfavourable</td></tr>
<tr><td>-5</td><td colspan="2">-5</td><td colspan="2">-10</td></tr>
</table>

3 Research method

3.1 Sample and parameters

The data used in this study include the geological structure on the slopes, geological mapping, core drill and laboratory test results. The samples that have been taken were carried out by laboratory tests [1] in the form of property index and engineering properties. The engineering properties parameter is obtained from the triaxial test so that the cohesion value and the inner shear angle are obtained.

3.2 Circular failure chart (CFC)

CFC charts are used to find the critical point of longitudinal and fracture surface tensions of various slope geometries and groundwater conditions [4]. This method is the easiest method to find the slope safety factor. This method is a semi-empirical method that requires minimal laboratory data such as material density, shear angle, cohesion, slope height and slope.

The CFC method is highly dependent on groundwater conditions. So that the analysis in this method is very dependent on determining field conditions. This can be seen in (Fig. 3). To determine the safety factor of the CFC method, you can follow the steps below (Fig. 4).

- The first step: determine the rock strength parameters that compose the slope.
- Second step: Use (Fig. 3) to determine groundwater conditions according to field conditions.
- The third step: Calculate the ratio value with the formula $c / (\gamma H \tan \varphi)$ and plot the calculation results in the CFC diagram that we use as shown in Fig. 5.
- Step four: Draw a straight line from the value in step three to the curved line that describes the amount of the angle of the slope.
- Step five: Use the following formula φ / FS or $c / (\gamma H FS)$ to find a convincing safety factor.

https://doi.org/10.1051/e3sconf/202132501016

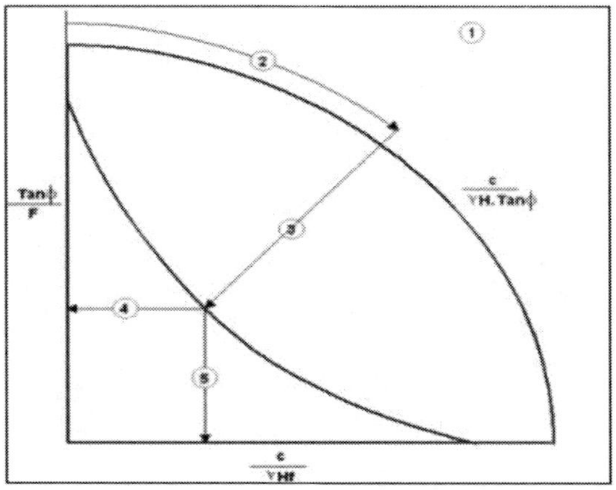

Fig. 4. Slope safety to CFC [5].

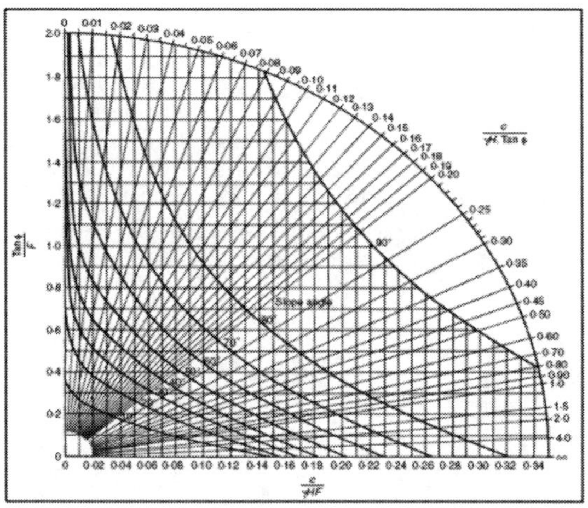

Fig. 5. CFC Model Graphic [5].

3.3 Finite element method (FEM)

The method of slope stability analysis used in this study is the finite element shear strength reduction technique (SSR-FEM). In this method, the available soil shear strength parameters are automatically reduced until failure occurs. So that the safe factor (SF) slope stability becomes:

$$\sum M_{sf} = \tan \phi_{input} / \tan \phi_{production} \quad (1)$$

$$= C_{input} / C_{production} \quad (2)$$

$$SF = \frac{current\ shear\ strength}{landslide\ shear\ strength} \quad (3)$$

$$= \text{Value of } \sum M_{sf} \text{ at time of landslide} \quad (4)$$

Note:
C_{input} = Soil Cohesion
ϕ_{input} = Ground shear angle (°)
$C_{reduction}$ = Reduction of Soil Cohesion
$\phi_{reduction}$ = Reduction of Ground shear angle (°)
(Duncan C. Wyllie & Christoper W. Mah,) [4]

3.4 Limited equlibrium method (LEM)

LEM is a method that uses the principle of force equilibrium. This method of analysis first assumes the area of failure that can occur. There are two assumptions of the landslide area [4], namely: the landslide plane is circular and the sliding plane is assumed to be non-circular (it can also be planar).

The calculation is carried out by dividing the land that is in the landslide plane into sections, therefore this method is also known as the method of slice. The various existing solutions for this wedge method have been developed over the years, from Fellenius, Taylor, Bishop, Morgenstern-Price to Sarma and others.

In this LEM the safety factor, SF, is in principle calculated from the ratio between the shear strength of the soil, τf with the thrust τ or the ratio between the moment of resistance, RM, to the moment of thrust, DM, as shown by the formula below.

$$SF = \frac{\tau f}{\tau} \ or \ SF = \frac{RM}{DM} \quad (5)$$

(Duncan C. Wyllie & Christoper W. Mah,) [4].

4 Results and discussion

The slopes at the inlet and outlet of the Tanju Tunnel conduit were analyzed using the CFC method. For each location analyzed, one sample was taken for laboratory analysis. Sampling was done using the rotary drilling method. Samples were taken at the Inlet (IL) and Outlet (OL) locations at a depth of 8 m. From the results of the field description, the naming of rocks is andesite. To find the safety value, the triaxialysis test was carried out to produce the cohesion value (c) and the inner shear angle (φ) according to ASTM D-2264. The groundwater condition in the field is also considered, as well as the condition of the rock mass.

4.1 Determine the safety factor of slope using cfc, lem, and fem method

By following the steps described in the previous sub-chapter, the authors used dry groundwater conditions (Fig. 3). So that the graph used is graph number 1 (Fig.5). This determination is adjusted to the conditions in the field where the slopes at the two research locations show dry conditions. *Rocscience slide V 6.0* used to calculate LEM and *Rocscience slide V 8.0* used to calculate FEM.

4.1.1. Inlet slope

The inlet slope condition has a geometry of 12 meters high and a slope of 45 degree. From the results of the drilling investigation, the RMR value of rock which is included in the good rock class II class is obtained. With laboratory data can be seen in (able 2). In determining the safety factor, the rock engineering parameters used are adjusted to the rock mass condition or RMR.

So the authors use a cohesion value of 350 kPa and an inner shear angle of 40°. refers to (Table 1).
From the calculation using the formula c / (γ H tan φ), the result is 1.4. So that a line or plot is drawn on the CFC graph, it can be found tan φ / FS = 0.28
Then: tan φ / FS = 0.28
tan 40 / 0.28 = FS
FS = 3.5

From these calculations, it is obtained SF of 3.5 which is indicated that the slope is stable.

From the calculation of stability using the LEM (Bishop) method with the help of the Rocscience slide V 6.0 program and the Mohr-Colomb collapse parameter, it was found that an SF of 3.69 was included in the safe category, which can be seen in (Fig. 6).

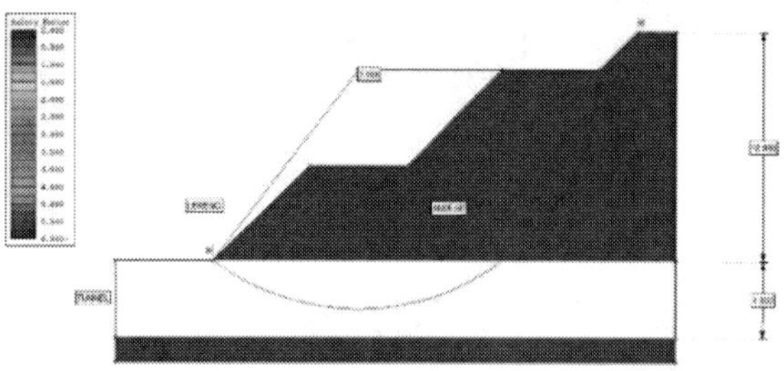

Fig. 6. Inlet slope Safety Factor as the result of LEM Calculation using Bishop Method

From the calculation of stability using the FEM method with the help of the Rocscience Phase V 8.0 program and the Mohr-Colomb collapse parameter, it is found that an SF of 4.78 is included in the safe category, which can be seen in (Fig. 7).

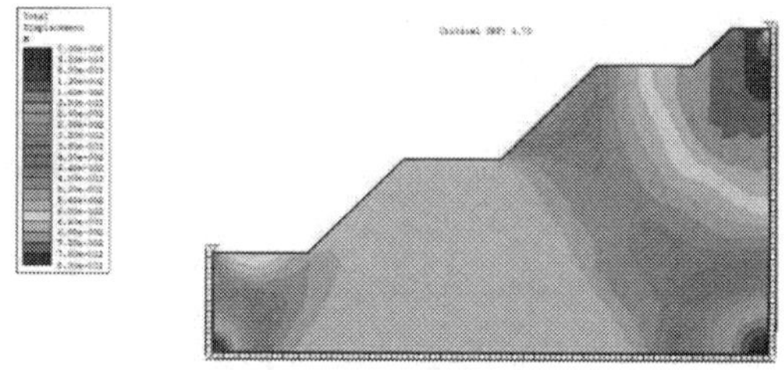

Fig. 7. Inlet slope Safety Factor as the result of FEM Calculation.

4.1.2. Outlet slope

The condition of the outlet slope has a geometry of 12 meters high and a slope of 45. From the results of the drilling investigation, the RMR value of rock is included in the good rock class II. With laboratory data can be seen in (Table 2). In determining the safety factor, the rock engineering parameters used are adjusted to the rock mass condition or RMR.

Table 2. Laboratorium results [1].

SUMMARY OF ROCK TEST					
PROJECT				D/D TEROWONGAN N TANJU	
SAMPLE NO				INLET	OUTLET
DEPTH			M	8-8,7	7-7,7
I	SPECIFIC GRAVITY & ABSORPTION				
	Bulk Dry Basis			2.562	2.607
	Saturated Surface Dry Condition			2.595	2.625
	Apparent			2.650	2.656
	Absorption		(%)	1.28	0.72
II	SPECIFIC GRAVITY OF SOIL			-	-
III	NATURAL DENSITY		gr/cm³	2.553	2.614
IV	WATER CONTENT		%	1.04	0.30
V	DRY DENSITY		gr/cm³	2.526	2.606
VI	VOID RATIO			0.049	0.019
VII	SATURA TED DENSITY			2.573	2.625
VIII	SUBMERGED DENSITY			1.573	1.625
IX	DEGREE OF SATURATION			56.39	41.53
X	UNIAXIAL COMPRESSIVE STRENGTH				
	Comp. Strenght qu		kg/cm²	357.34	610.20
	Modulus Elasticity	Axial Ea	kg/cm²	4,216E + 04	7,540E +04
		Diametral Ed	kg/cm²	1,044E + 05	1.735E + 05
	Poisson's Rasio μ		-	0.404	0.435
	Axial Strain		%	0.87	0.84

So the authors use a cohesion value of 400 kPa and an inner shear angle of 45°. refers to (Table 1).
From the calculation using the formula c / (γ H tan φ), the result is 1.4. So that a line or plot is drawn on the CFC graph, the obtained tan φ / FS = 0.29
Then: tan φ / FS = 0.29
tan 45 / 0.29 = FS
FS = 3.44

From these calculations, it is found that SF is 3.44 which is included in the safe category.

From the calculation of stability using the LEM (Bishop) method with the help of the Rocscience slide V 6.0 program and the Mohr-Colomb collapse parameter, the SF of 3.65 can be seen (Fig. 8).

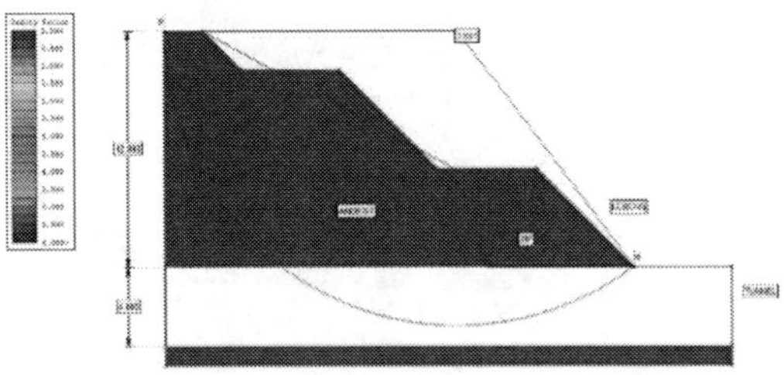

Fig. 8. Outlet slope Safety Factor as the result of LEM Calculation using bishop method.

From the calculation of stability using the FEM method with the help of the Rocscience Phase V 8.0 program and the Mohr-Colomb collapse parameter, it was found that an SF of 4.79 was included in the safe category, which can be seen in (Fig. 9).

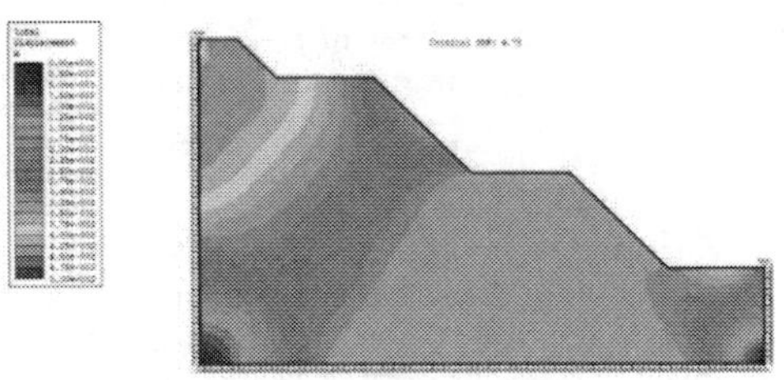

Fig. 9. Outlet slope Safety Factor as the result of FEM Calculation.

5 Conclusion and recommendations

Based on the field conditions, the slopes in the research location are categorized as dry, so the number 1 graph model is used in the CFC method. On the inlet slope (IL), laboratory results show the density value of 2.526 g/ cm³ or 24.77 kN / m³, cohesion of 25.4 kg / cm³ or 2451.66 kPa and the inner angle of shear of 46°. On the outlet slope (OL), from laboratory results, the density value is 2.606 g/ cm³ or 25.60 kN / m³, cohesion is 37.846 kg / cm³ or 3711.42 kPa and the inner shear angle is 54°. By considering the RMR conditions of the two slopes that are included in the class 2 category, adjustments for the cohesion value and inner shear angle are made according to (Table 1).

From the calculation of the CFC value, the safety factor value is 3.5 on the inlet slope (IL) and 3.44 on the outlet slope (OL), both of which are included in safe conditions. From the calculation of the LEM value, the value of the safety factor is 3.69 on the inlet slope (IL) and 3.65 on the outlet slope (OL), both of which are in a safe condition. From the calculation of the FEM value, the safety factor value is 4.78 on the inlet slope (IL) and 4.79 on the outlet slope (OL), both of which are in a safe condition. From the calculation of CFC, LEM, and FEM can be value the slope stability of the designed drainage tunnel is in safe condition

Research can be strengthened by various methods such as the slope mass rating (SMR) to determine the type of reinforcement and protection.

The authors would like to thank BWS Nusa Tenggara 1, Supervision Consultant and PT. Nindya Karya as the Contractor for the permission to conduct the research.

References

1. Anonym. *Laporan Akhir Detail Engineering Design (DED) Rababaka Komplek Kabupaten Dompu Nusa Tenggara Barat.* Mataram. PT. Indra Karya (2018)
2. A. Sudradjat, N. Suwarna, Mangga, S. Andi, Peta geologi lembar Sumbawa, *Nusa Tenggara Geological map of the Sumbawa Quadrangle, Nusa Tenggara* (1998)
3. C. I. Abdullah, N. A. Magetsari, H. S. Purwanto, J. Math. Fundam. Sci. ITB, 35(2), doi:10.5614 (2003)
4. D. C. Wyllie & C. W. Mah, *Rock slope engineering : civil and mining* (New York 2004)
5. Z. T. Bieniawski, Engineering Rock Mass Classification A Complete Manual for Engineers and Geologist in Mining, Civil and Petroleum Engineering (John. Wiley & Sons, NewYork 1989)

Study of meteorological drought and its impact on rainfed paddy productivity in Gunungkidul Regency

Qooi Insanu Putra, *and Emilya* Nurjani*

Department of Environmental Geography, Faculty of Geography, Universitas Gadjah Mada, Indonesia

Abstract. Gunungkidul Regency is known as an area that often experiences drought. On the other hand, Gunungkidul Regency is also the regency with the highest amount of rice production in the Special Region of Yogyakarta Province. Rainfed paddy farming is the most widely developed type of paddy in Gunungkidul Regency where irrigation needs are determined by rainfall. Decreased rainfall that triggers meteorological drought can disrupt rainfed-based agriculture. This study aims to analyze the distribution of meteorological drought and analyze the impact of meteorological drought on rainfed paddy productivity in Gunungkidul Regency during the period 2001 – 2019. Meteorological drought identification was carried out using the Standardized Precipitation Index (SPI). Results of the SPI classification was mapped using Spline-Tension interpolation for spatial analysis of the distribution of meteorological drought. Spatial analysis and comparison graphs were used to analyze the relationship between drought and rainfed paddy productivity. The widest meteorological drought in Gunungkidul Regency occurred in November 2006. The highest frequency of drought events occurred in Paliyan Sub-district for 50 months while the lowest occurred in Ponjong Sub-district for 30 months. Most sub-districts in Gunungkidul Regency have a positive relationship between meteorological drought. Most of rainfed paddy productivity decreased when El Nino occurred.

1 Introduction

Drought is a condition of lack of water resources in an area caused by rainfall conditions that are below the normal average at a certain time scale continuously [1]. The climate anomaly caused by the El Nino phenomenon can exacerbate the occurrence of droughts because it will accelerate the onset of the dry season and the next rainy season will come later than normal conditions [2,3]. In relation to drought, El Nino is a phenomenon that is one of the biggest factors causing drought in Indonesia because El Nino has an impact on reducing the amount of rainfall from its normal condition [4]. On the other hand, La Nina has an impact on increasing rainfall so that from a drought perspective, La Nina events will minimize the occurrence of drought events.

Several studies have shown a strong correlation between the ENSO phenomenon and the accompanying climate anomalies with food problems in the world, which mainly occur in developing countries with the majority being agricultural countries that rely on the agricultural sector [5,6]. One type of drought that can be an early indicator of drought is meteorological drought [7].

Global climate disturbances caused by the ENSO phenomenon have had an impact on the life sector in Indonesia, especially related to food issues [2]. Agricultural sector is one of the sectors affected because it has a direct relationship with various weather and climate factors, making it quite vulnerable to weather and climate disturbances [4]. The effects of climate change on the agricultural sector are multidimensional, ranging from resources, agricultural infrastructure, agricultural production systems, to aspects of food security and self-sufficiency, as well as the welfare of farmers and society in general [8].

For farmers, the occurrence of rain becomes a reference for planting which will be used to increase the water content or soil moisture [4]. Changes in the pattern and distribution of rainfall can result in a shift in the beginning of the season and the planting period [9]. According to [10], a decrease in rainfall can reduce the potential for one period of rice planting. These results provide information that climate disturbances can have a real impact on the agricultural sector because they can reduce agricultural production and productivity. It is feared that the decline in paddy production and productivity in a sustainable manner will disrupt food security in the future.

One of the impacts of weather and climate disturbances on agriculture is the expansion of crop failure land in Gunungkidul Regency which is strongly suspected to be caused by weather conditions that cannot be predicted with certainty. Gunungkidul Regency experienced a significant expansion of crop failure land over the past three years due to the early dry season.

According to [11], Gunungkidul Regency is one of the areas known as a barren area and prone to drought

* Corresponding author: emilya.nurjani@ugm.ac.id

caused by the low intensity of rainfall in the dry season. On the other hand, Gunungkidul Regency is also one of the regencies with the highest rice production in the Special Region of Yogyakarta (DIY) with the total production in 2019 reaching 201,654 tons [12]. In percentage terms, Gunungkidul Regency supplies around 37.79% of the total rice production in the DIY Province [12].

Weather and climatic conditions in Indonesia which are influenced by various global climate dynamics as described above are expected to cause various impacts including the intensity of meteorological drought events and disturbances to the productivity of rainfed paddy in the research location. The importance of the agricultural sector in Gunungkidul Regency as the fulfilment of the largest food needs in DIY Province makes it important to study the meteorological drought events and also its causes because it is thought to have a direct impact on various vulnerable sectors, mainly related to rainfed paddy productivity.

Based on these conditions, this study will identify the distribution of meteorological drought events in Gunungkidul Regency and see its impact on rainfed paddy productivity which is the largest food source in Gunungkidul Regency.

2 Methods

2.1 Research location

The research was conducted by taking Gunungkidul Regency as the research location. Gunungkidul Regency is one of the regencies in the Special Region of Yogyakarta Province which is located in the UTM position zone 49M 425415mT – 493513mT to 9091415mU – 9139903mU. Gunungkidul Regency has an area of 1,485.36 km² which makes it the largest regency in the Special Region of Yogyakarta Province with a percentage of 46% of the total area of the Special Region of Yogyakarta Province. Gunungkidul Regency consists of 18 sub-districts including Panggang, Purwosari, Paliyan, Saptosari, Tepus, Tanjungsari, Rongkop, Girisubo, Semanu, Ponjong, Karangmojo, Wonosari, Playen, Patuk, Gedangsari, Nglipar, Ngawen and Semin. The Administrative area of Gunungkidul Regency is shown in **Figure 1.**

Gunungkidul Regency was chosen as the research location by considering the characteristics of the dry area so that it is prone to drought [13]. This condition makes the agricultural land that is widely developed in Gunungkidul Regency is rainfed or dryland-based agricultural [14].

In addition, Gunungkidul Regency is a regency that has the largest total harvested area and rice production in the Special Region of Yogyakarta Province. Food and agriculture data from the Central Bureau of Statistics (CBS) shows that during 2001 – 2019, Gunungkidul Regency became the regency with the largest paddy harvest area in the Special Region of Yogyakarta Province. According to [12], agricultural data during that period also shows that around 75% of the harvested area for paddy farming in Gunungkidul Regency is rainfed paddy farming. This is different from the condition of other regencies and cities in Special Region of Yogyakarta Province which on average only has a percentage of rainfed paddy farming area of 1% of the total area.

Rainfed paddy farming is dependent on rainfall to meet its irrigation needs so that it is prone to drought. The existence of an anomaly of rainfall that causes drought, is indicated to be the main factor causing farmers to fail to harvest in Gunungkidul Regency at several harvest season.

2.2 Research method

This research is classified as a quantitative descriptive research, that is, the analysis is described from the results of processing numerical data which results in a pattern of interrelationships between one another and compared with the other research and literatures. The final result of this research is a description of the study of meteorological drought and its impact on rainfed paddy agriculture in Gunungkidul Regency. To realize these results, the stages of research that will be carried out are: (1) Data Collection, (2) Data Processing, and (3) Data Analysis.

2.3 Research data

Research analysis was carried out during 2001 - 2019 considering the condition of regional expansion in Gunungkidul Regency which was divided into 18 sub-districts since regional autonomy was enacted in 2001. This is related to the availability of data in each sub-district which will become the unit of analysis for research analysis. In addition, the year selection is also based on the frequency of ENSO events where during 2001 – 2019 there have been several El Nino and La Nina events in weak, moderate, and strong classifications so that they may have an impact on meteorological drought in Gunungkidul Regency.

The research was conducted by processing various basic data that became the research material. This study uses secondary data obtained from various agency publications and research literature studies, including: monthly rainfall, rainfed paddy productivity, shapefile (.shp), and Southern Oscillation Index (SOI) data.

Monthly rainfall data during 2001 – 2019 was obtained from the publication of the Agriculture and Food Office of Gunungkidul Regency at rain stations spread over 18 sub-districts in Gunungkidul Regency. Data on rainfed paddy productivity during 2001 – 2019 was obtained from the Agriculture and Food Office of Gunungkidul Regency in the CBS Publications of 2001 – 2019. Shapefile data was collected from National Mapping Agency of Indonesia that used as mapping data. Monthly SOI data during 2001 – 2019 is sourced from the official website of the Australian Bureau of Meteorology. In this study, SOI data are used as additional data used for identification and analysis of the causes of meteorological drought especially on El Nino phase of ENSO [4].

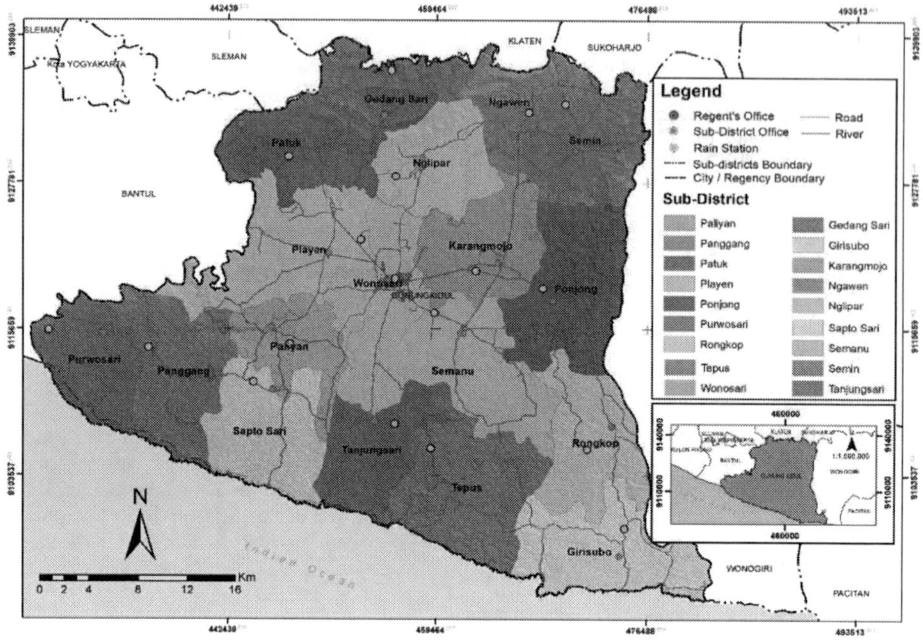

Fig.1. Administration map of Gunungkidul Regency

2.4 Data processing

First step in the research is to fill in the empty rainfall data. Rainfall data has an important role because it is the main indicator of meteorological drought analysis. One method that can be used is the Inverse Square Distance method. The Inverse Square Distance method is a method of filling in rainfall data by utilizing rain data from adjacent stations [15]. Then, Standardized Precipitation Index (SPI) method is used to determine the classification of meteorological drought at each rain station. According to [16], the classification of SPI values is presented in **Table I.**

Table 1. Classification of SPI

Num	Classification	SPI Value
1	Extremely Wet	≥ 2
2	Very Wet	1.5 – 1.99
3	Moderately Wet	1.0 – 1.49
5	Normal	(-0.99) – 0.99
6	Moderately Dry	(-1.0) – (-1.49)
7	Severely Dry	(-1.5) – (-1.99)
8	Extremely Dry	≤ (-2)

This study uses a six-month SPI (SPI-6) scenario and four-month paddy season (SPI-4) scenario in answering the objectives. The SPI-6 scenario was used to mapping the distribution of drought in various sub-districts in Gunungkidul Regency. The SPI-6 scenario describes a short to medium term meteorological drought that is commonly used to identify droughts affecting agriculture, irrigation, and surface water resources and is ideal for comparing droughts in areas with two different periods [7,16,17].

SPI-4 scenario in the paddy farming season to be adapted to rainfed paddy planting conditions in Gunungkidul Regency which has a paddy growing season of approximately 110-120 days from October to January [14,18]. Therefore, SPI-4 in January describes the conditions of the previous 4 months during farming season. The adjustment to the SPI scenario, especially SPI-4 in answering the third objective, is due to the characteristics of the SPI which has the flexibility to be adjusted according to the research objectives [19,20].

Drought mapping was carried out using ArcGIS 10.3. This study uses the Tension Spline interpolation method. Tension Spline was chosen because it has a lower RMSE value than other interpolation methods. Tension Spline interpolation method is considered to have smoother results and lower RMSE levels than other methods [21-24].

Drought mapping focuses on the worst drought in every month throughout 2001 – 2019. The worst drought every month from January to December are known through the results of SPI-6 data processing where the month of the year with the highest number of droughts is identified as the most severe drought month.

2.5 Research analysis

Data analysis was carried out through various techniques to examine each unit of research analysis. The unit of analysis used for all purposes emphasizes the administrative unit in the form of a sub-district. The analytical technique used in this research is spatial and associative analysis.

Spatial analysis was used to explain the distribution of meteorological drought in Gunungkidul Regency. Spatial analysis was carried out on maps that shown the maps of the most severe of meteorological droughts in each month during period 2001 – 2019. The mapping is only carried out in the month of the worst drought from the results of the SPI matrix so that it will shows of the worst drought for 12 months from January to December with the year of occurrence that is not always the same.

Associative analysis was used to identify the graph of the relationship between meteorological drought and rainfed paddy productivity. Analysis of the impact on rainfed paddy productivity was also carried out in a spatio-temporal manner to explain the average paddy productivity, especially during certain ENSO phases such as El Nino, La Nina and Normal. This is done to describe descriptively and spatially whether every El Nino occurrence will have an impact on paddy productivity through an intermediate variable in the form of a decrease in rainfall so as to increase the incidence of meteorological drought which is ultimately thought to have an impact on the productivity of rainfed paddy.

3 Result and discussion

3.1 Frequency of meteorological drought events in each sub-district of Gunungkidul Regency during 2001 – 2019

Drought frequency states the number of drought events during a certain time [25]. Drought events in Gunungkidul Regency were identified through the results of the SPI classification at 18 observation rain stations showing moderately dry, severely dry and extremely dry classes during the period 2001 – 2019. The SPI processing used in the identification of meteorological drought is the 6-month SPI scenario (SPI- 6). The processing results show that each rain station has a varying frequency of meteorological drought events as shown in **Table 2.**

Table 2. Frequency of meteorological drought events in Gunungkidul Regency 2001 - 2019

Num	Name of Sub-districts	Meteorological Drought Events (Month)
1	Paliyan	50
2	Purwosari	46
3	Tepus	43
5	Karangmojo	43
6	Semin	43
7	Saptosari	42
8	Wonosari	41
9	Patuk	40
10	Semanu	38
11	Gedangsari	38
12	Nglipar	38
13	Ngawen	37
14	Rongkop	36
15	Playen	35
16	Tanjungsari	33
17	Panggang	32
18	Girisubo	31

During 2001 - 2019, Paliyan Sub-district was the sub-district that experienced the most meteorological drought events, reaching 50 drought events, followed by Purwosari Sub-District with 46 drought events. Tepus, Karangmojo and Semin sub-districts are the third sub-districts that have the highest number of drought events. The three sub-districts have the same number of drought events, namely as many as 43 drought events. The districts with the least number of drought events are Girisubo and Ponjong Sub-districts with 31 and 30 drought events, respectively.

Frequency of drought events in each of these sub-districts when viewed spatially does not show a pattern because the frequency of occurrence is spread throughout all sub-districts in Gunungkidul Regency. Paliyan and Purwosari Sub-districts as the sub-districts with the highest number of meteorological droughts are located on the south side of Gunungkidul Regency. However, Purwosari Sub-district is directly adjacent to the east side with Panggang Sub-district, which is the district with the third lowest number of droughts in Gunungkidul Regency. Likewise, the Sub-districts of Tepus, Karangmojo and Semin which are the third sub-districts with the highest number of droughts have different spatial positions. Tepus Sub-district is located in the southern part, Karangmojo Sub-district is located in the middle part, while Semin Sub-district is located in the northern part of Gunungkidul Regency.

The absence of a spatial pattern in the variation in the frequency of meteorological drought events is the same as the results of [21] which examines meteorological drought in Bangkalan Regency where meteorological drought events spread randomly in every part of the region so that they do not show a certain spatial pattern. This is more because the parameters used in the meteorological drought analysis only come from rainfall. According to [26], the other possible factors that influence the development of drought are differences in conditions of wind speed and direction, evapotranspiration, temperature, land cover, land use, water retention and regional landforms. These factors can affect the decrease in rainfall so that there are differences in the frequency of the number of meteorological droughts in various areas in Gunungkidul Regency.

3.2 Distribution of meteorological drought every month with the worst meteorological drought in Gunungkidul Regency during 2001 – 2019

Spatial analysis of the distribution of meteorological drought in Gunungkidul Regency during the years of 2001 – 2019 emphasized the occurrence of the worst drought in each month. The worst drought events in each month were identified from the matrix table of the results of the SPI data processing in the SPI-6 scenario every month during the years 2001 – 2019. The worst drought for 12 months from January to December was not always the same year because every year it produced months with different frequency, duration and intensity of drought. To simplify the analysis, the analysis is divided into two sub-subsections from January to June and July to December.

3.2.1 Analysis of the worst meteorological drought every month during January - June

The worst meteorological drought in January during 2001 – 2019 occurred in January 2007. **Figure 2** shows the distribution of each drought class from moderately dry, severely dry and extremely dry which has a varying area which is sequentially reaching an area of 205.12 km^2, 463.73 km^2, and 542.88 km^2. The total area of drought events that occurred in January 2007 reached 1211.73 km^2 or about 82.1% of the total area of Gunungkidul Regency. Some other areas were observed to have a normal class with an area of 246.6 km^2. In addition, there is a small area that has a wet class with an area of 18.27 km^2 or about 1.2% of the total area of Gunungkidul Regency.

Drought in January 2007 continued, resulting in February 2007 being the worst drought that occurred in Gunungkidul Regency in February during the period 2001 - 2019. The drought in February 2007 began to reduce the area of drought from the previous month where the total drought area of which originally reached 82% to 79.4% of the total area of Gunungkidul Regency or around 1171.70 km^2. Extremely dry drought class experienced a significant decrease from around 36% in January 2007 to 25.7% in February 2007.

The worst meteorological drought that occurred in January and February 2007 is thought to have been due to a decrease in rain due to the El Nino phenomenon. According to the result of [27] in Klaten Regency which is geographically located directly adjacent to Gunungkidul Regency also showed a similar situation where there was a significant decrease in rainfall in January 2007 due to the influence of the El Nino phase during late 2006 to early 2007.

The drought in March 2010 was the worst drought that occurred in March during 2001 – 2019. The total area of drought that occurred in March 2010 reached 1238.2 km^2 or about 83.9% of the total area of Gunungkidul Regency. The percentage of the area of drought occurrence is almost similar to the area that occurred in January 2007 but both have spatially different area patterns as shown in **Figure 2**.

The worst drought in the following month occurred in April 2010 which was a continuation of the drought in March 2010. Compared to the previous month, the total area of drought in April 2010 decreased from 1238.19 km^2 to 1081.05 km^2 or around 73, 2% of the total area of Gunungkidul Regency. This indicates a reduction in drought intensity from the previous month.

The next worst drought occurred in May 2005. Different from the previous worst months, drought in May 2005 had a smaller total area of only 692.81 km^2 or 46.9% of the total area of Gunungkidul Regency. **Figure 2** shows the extent of drought during the month of May 2005. It can be seen that the areas experiencing drought are spread only in a small part of the north, middle and south of Gunungkidul Regency, including the Sub-districts of Semin, Nglipar, Karangmojo, Wonosari, Playen, Paliyan and Saptosari.

June 2005 became June with the worst drought throughout 2001-2019 and was a continuation of the drought in May 2005. The drought in June 2005 experienced an increase in drought events from the previous month with a total area of 1164.14 km^2 or around 78,9% of the total area of Gunungkidul Regency. The expansion of the drought indicates that the intensity of drought in the Gunungkidul Regency has increased.

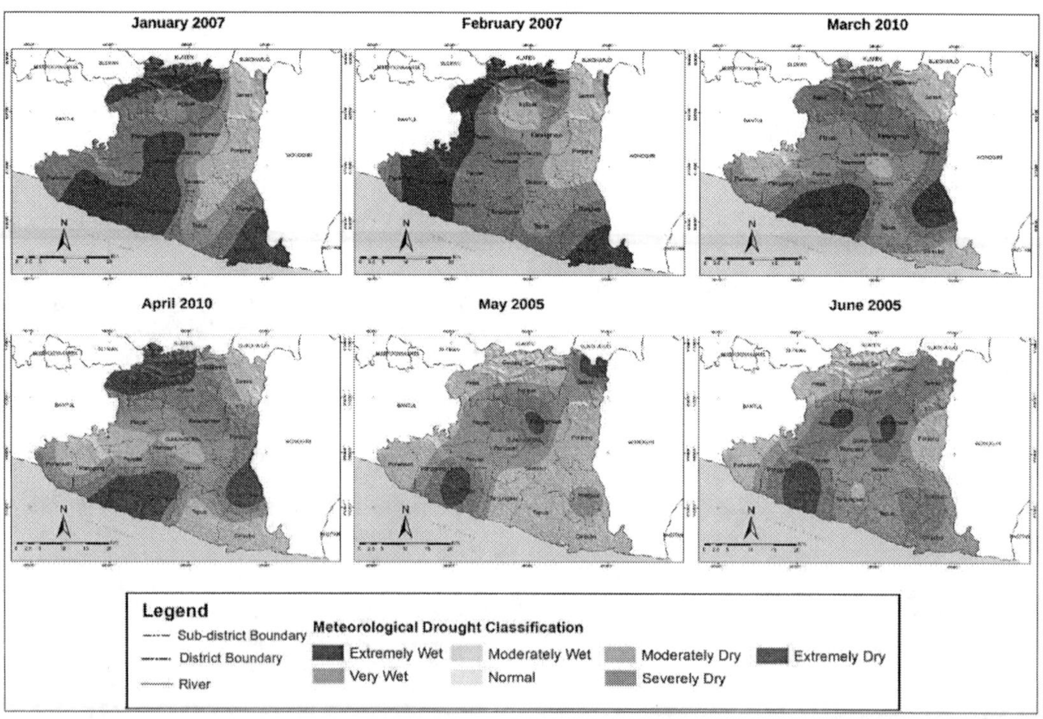

Fig. 2. The worst meteorological drought in January - June in Gunungkidul Regency during 2001 – 2019

3.2.2 Analysis of the worst meteorological drought every month during July - December

The drought in May and June 2005 turned out to still continue until July 2005 so that it became the month of July with the worst drought than any other July during 2001 – 2019. Compared to the previous month, the drought in July 2005 began to experience a decrease in drought intensity. This can be seen from the total area of drought which began to decline to 906.23 km^2 or 61% of the area of Gunungkidul Regency. The spatial pattern of drought events shown in **Figure 3** is not much different from the spatial pattern of drought events in the previous month. However, in the southern part of Gunungkidul Regency, which includes Tepus Sub-istrict and parts of Semanu and Rongkop Sub-districts, the extent of drought events has narrowed. Meanwhile, the area of drought in other parts tends to be stagnant and there is not much visible change compared to the previous month.

The drought in August 2008 became the next worst drought. In that month, an area of 48.7% of the area of Gunungkidul Regency or an area of 718.68 km^2 experienced drought. The distribution of drought events in August 2008 is shown in Figure 4.8. The thing that distinguishes the drought distribution pattern in August 2008 compared to the worst months before is the condition of the drought area which tends to strengthen on the east side of Gunungkidul Regency including Ponjong Sub-district and Rongkop Sub-district with a extremely dry drought class. This condition is certainly different from the period of the worst months previously discussed where the eastern side of Gunungkidul Regency rarely experiences drought events.

The drought conditions in August 2008 were strengthened by [28] which examined the 3-Month SPI in Gunungkidul Regency where it was said that August 2008 was one of the months with the worst drought index in the short term throughout 2007-2016. According to [28], the results of the SPI-3 in August 2008 were extremely concentrated in several areas such as Patuk, Ngawen, Gedangsari, Ponjong, Semanu, Semin, Karangmojo, Rongkop and Tepus. This pattern is slightly different from the SPI-6 produced in this study. However, this can be an indication that August 2008 was the month of August with the worst drought in both the short- and medium-term drought scenarios. One of the causes of the decrease in rain in August 2008 was due to the positive Indian Ocean Dipole [29,30].

The drought that occurred in September 2019 resulted in a drought area of 1181.46 km^2 or about 80% of Gunungkidul Regency. Classes of drought that occurred included normal, moderately dry and severely dry with the percentage of area to the area of Gunungkidul Regency respectively reaching 18.5%, 44.1%, 35.9%. In that month, no extremely dry class was observed and there were a small number of areas with wet class reaching 1.13%.

The drought in October 2019 experienced an expansion of drought events compared to the previous month in the same year. The increase in the area of drought occurred quite significantly where the area of drought reached 1445.58 km^2 or about 97.9% of the area of Gunungkidul Regency experienced drought. It can be seen in **Figure 3** that drought almost occur in all areas in Gunungkidul Regency.

An increase in the intensity of drought in October 2019 occurred in all of Gunungkidul Regency where all areas that were observed to be normal in the previous month experienced drought events from moderately dry to severely dry. The normal observed area only occurs on the northeast side of Gunungkidul Regency which in the previous month was observed in the wet classification. An increase in drought intensity also occurred on the western side of the southern part of Gunungkidul Regency, precisely in parts of Purwosari and Panggang Sub-districts. The area has increased from being severely dry to extremely dry. The widespread drought that occurred in 2019 coincided with the weakening of the SOI value which is an indication of a stronger El Nino. This results in a longer dry season. In years with a normal period, September should have increased rainfall. However, in 2019 from September to October there was a meteorological drought which indicated a decrease in rainfall and a longer dry season.

The worst drought that occurred in the following month occurred in November 2006 which was also the month with the worst drought among all months during 2001 – 2019. The entire area of Gunungkidul Regency experienced drought events and all of them were classified as moderately to extremely dry. The area classified as severely dry into the widest drought class reaches 1081.92 km^2 which is then followed by moderately dry and extremely dry drought class with an each area of 358.95 km^2 and 35.29 km^2.

Drought conditions in November 2006 in Gunungkidul Regency which became the most severe of all months during 2001-2019 were in line with several studies on meteorological droughts conducted in the Java and Bali regions. Result of [21] on meteorological drought in Bangkalan Regency during 2001 – 2015 also resulted in the widest drought in November 2006 where the entire Bangkalan Regency experienced drought events. Likewise, with [23] stated that the meteorological drought in Central Java during 1980 - 2010 where it was said that November 2006 was one of the months with the most severe drought because it almost occurred in all regencies in Central Java. Also, the result of [31] on the island of Bali also indicated the same thing where in November 2006 most of Bali experienced drought in all SPI scenarios.

The worst drought in December still occurred in the same year as the previous month. The difference is that the area of drought began to decrease with an area of 1125.85 km^2 or equivalent to 76.3% of the total area of Gunungkidul Regency. It can be seen in **Figure 3** that a number of areas that initially experienced drought began to return to normal as happened in the eastern and north-eastern parts of Gunungkidul Regency including Ponjong, Semin and parts of Semanu and Rongkop Sub-districts. Weakening conditions of drought intensity were also observed in the northern part where areas that were previously classified as severely dry began to show changes to moderately dry classification, such as in the Sub-districts of Gedangsari, Ngawen, Nglipar and parts of Patuk Sub-district. The same thing also happened in the southern part covering the Sub-districts of Paliyan, Purwosari, Tepus and Saptosari. As for the southeast, precisely in Girisubo Sub-district, there is an increase in drought intensity.

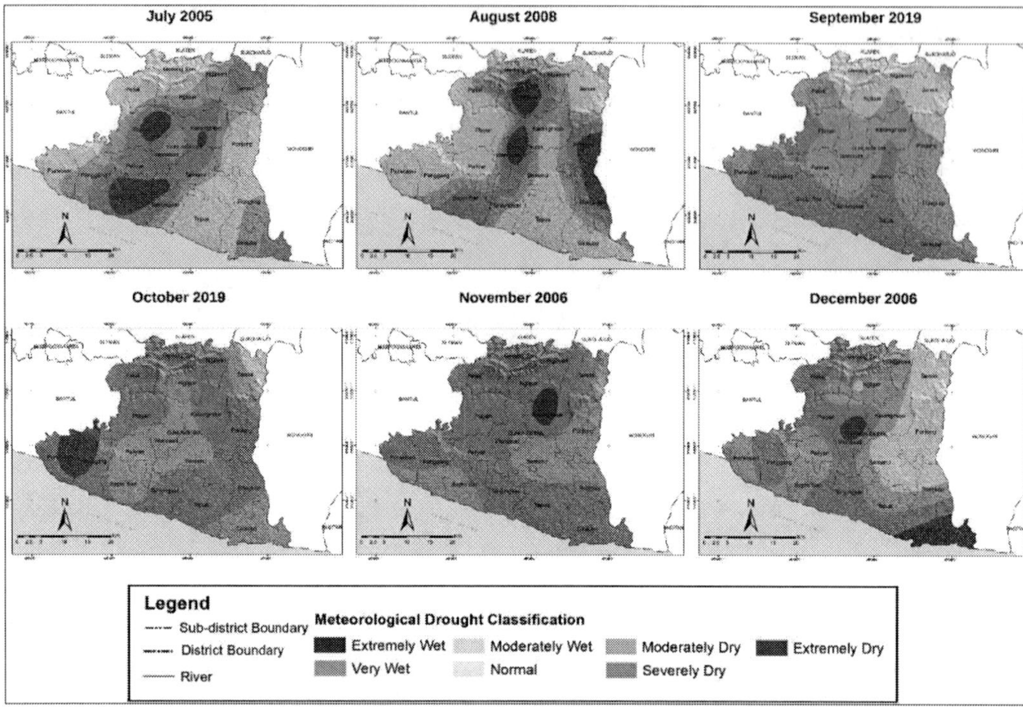

Fig. 3. The worst meteorological drought in July – December in Gunungkidul Regency during 2001 – 2019

3.3 Extreme meteorological drought in Gunungkidul Regency during 2001 - 2019

Extreme drought events are drought conditions that are classified as *extremely dry* droughts. The extreme drought that occurs in the medium term from the SPI-6 scenario indicates an indication of reduced surface water resources, reduced irrigation discharge and agricultural impacts and is an early warning before the occurrence of further agricultural and hydrological droughts [7]. The results of the analysis of the drought in Gunungkidul Regency during 2001 – 2019 showed that the extent of extreme drought varied. Not all years produce extreme drought areas as discussed in the previous discussion points where drought areas in the worst months often indicate large areas of drought without any areas with extremely dry droughts.

Table 3 shows the area of the largest extreme drought that occurred during 2001 – 2019. The extreme drought occurred in January 2007 with extreme drought reaching 542.88 km². The area of extreme drought that year was the largest compared to other extreme droughts with a significant difference in area. This can be seen from the area of extreme drought which then only has an area of 378.79 km², 343.51 km², 336.08 km² and 228.08 km². The spatial analysis of each of these areas has been discussed in the previous discussion points along with meteorological drought events in other drought classifications.

Extreme drought can be related to the geomorphology of landforms and land use in analyzing the potential impacts that occur [23]. All extreme drought events in 2007 and 2010 as shown in **Table 3** were concentrated in the southern and northern parts of Gunungkidul Regency. If these results are analyzed from the perspective of geomorphological conditions in Gunungkidul Regency, the southern part is a karst area while the northern part is a Baturagung hill area. Extreme drought in the southern region can threaten the existence of seasonal lakes to meet the residents' domestic water needs, while drought in the northern region can result in a decrease in springs and river flow. In relation to land use, extreme drought in all areas in Gunungkidul Regency can threaten dryland agriculture because the area of agricultural land spreads over almost the entire area of Gunungkidul Regency.

Temporarily, meteorological drought in Gunungkidul Regency is more common during the wet month period than the dry period. The area of Gunungkidul Regency generally has a C3 and D3 climate type which has a wet period that occurs around November - April. The widespread meteorological drought conditions classified as extreme drought during the wet period indicated a significant decrease in rainfall from normal conditions. This can threaten agriculture because the wet period is identical to the planting period in Gunungkidul Regency, resulting in an increase in the impact of drought. One of the factors that cause the decrease in rainfall is the El Nino phenomenon.

Table 3 also shows that the incidence of extreme drought in Gunungkidul Regency coincided with the year of El Nino. However, the extreme drought is more dominant when El Nino is weak to moderate. When a strong El Nino is observed no significant drought occurs. In fact, no months with the worst drought events were observed during a strong El Nino of 2015. This is an indication that the decrease in rain in Gunungkidul Regency is not only influenced by El Nino or it could also indicate that the increase in meteorological drought depends on the time and duration of El Nino events.

Table 3. Extreme meteorological drought in Gunungkidul Regency 2001 - 2019

Num	Month (Result of SPI-6 Scenario)	Year	Area (km^2)	ENSO Phase
1	August - **January**	2007	542.878	Moderate El Nino
2	September - **February**	2007	378.79	Weak El Nino
3	October - **March**	2007	343.51	Weak El Nino
4	November - **April**	2010	336.08	Moderate El Nino
5	December - **May**	2010	228.06	Weak El Nino

3.4 Meteorological drought relationship with rainfed paddy productivity in Gunungkidul Regency

The analysis of rainfed paddy productivity in Gunungkidul Regency is illustrated through an annual graph during 2001 – 2019 in **Figure 4 – Figure 6** . This is to see fluctuations in rainfed paddy productivity, especially when various strength fluctuations occur during the ENSO phases. The year of occurrence of ENSO is represented by the annual SOI value while the SPI-4 graph value is used to represent the meteorological drought that occurs. The graph of the value of rainfed paddy productivity is represented by the several sub-districts which represents sub-districts with varying locations from the southern, central and northern sides of Gunungkidul Regency covering Panggang, Patuk, Karangmojo, Paliyan and Ngawen sub-districts.

Through the graph in **Figure 4 – Figure 6** it can be see when there is a strengthening of El Nino as the annual SOI value decreases, it is followed by a decrease in the value of SPI-4 which then has an impact on the decrease in rainfed paddy productivity. This condition was clearly seen, especially in 2004, 2005, 2007, 2017 and 2019. Most of these years were the year when the El Nino period was moderate.

Different things were shown when there was a strong El Nino in 2015. In 2015, it was observed that paddy productivity did not experience a significant change in line with the condition of the SPI-4 value in that year nor was it too different from the previous year. This is related to the previous discussion points where the 2015 El Nino did not have much effect on the decrease in monthly rainfall in Gunungkidul Regency because El Nino occurred during the dry month. The stagnation of rainfed paddy productivity in 2015 is also in accordance with [27] that the conditions of rainfed paddy productivity in Klaten Regency which also shows no significant change from previous years due to rainfall conditions as the main irrigation source for lowland paddy. dry season did not experience a significant decrease when El Nino occurred in 2015.

Related to paddy varieties, Ciherang and Situbagendit are evenly distributed throughout the region [32]. One of the adaptations in Gunungkidul Regency was the provision of water resources for horticultural irrigation. Drilling wells are provided especially in the central and northern parts of Gunungkidul Regency which have shallow groundwater [33,34]. Therefore, variations in the relationship between drought and paddy productivity may vary in each region due to different type of paddy varieties and the development of adaptation in each sub-district.

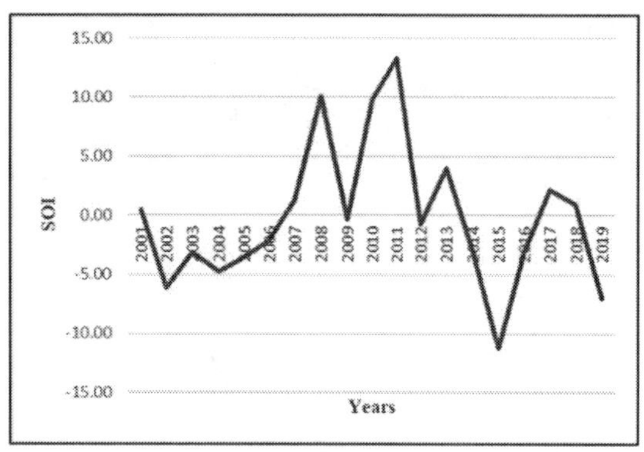

Fig. 4. Southern oscillation index as an indicator of the ENSO phase during 2001 – 2019.

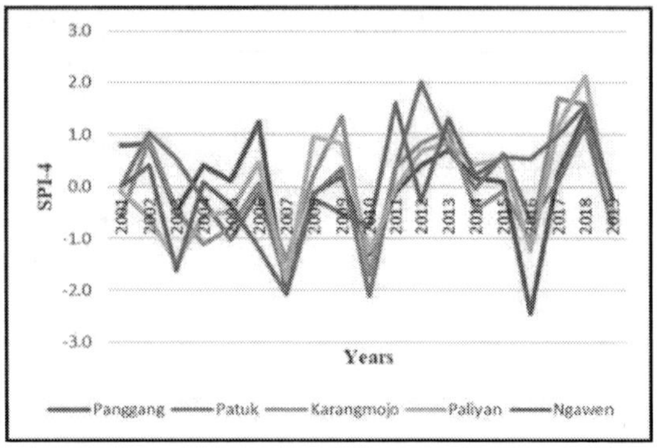

Fig. 5. SPI-4 in January as an indicator of drought during agricultural season in Gunungkidul Regency during 2001 – 2019

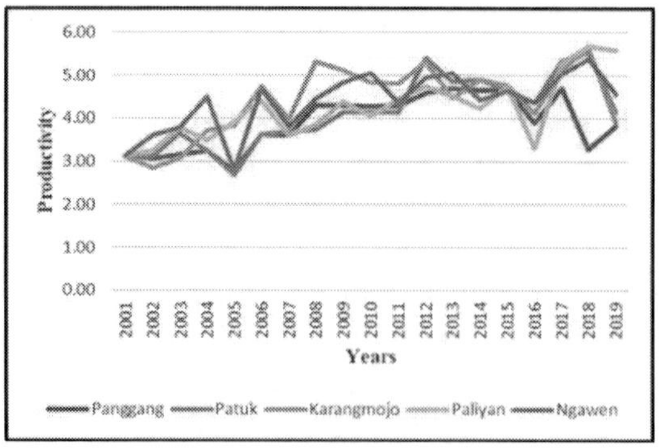

Fig.6. Rainfed paddy productivity of several sub-district of annual first growing season (October – January) in Gunungkidul Regency during 2001 – 2019.

3.5 Rainfed paddy productivity during Normal, El Nino and La Nina years in Gunungkidul Regency

Analysis of rainfed paddy productivity in Gunungkidul Regency was also carried out spatially through the distribution of the average rainfed paddy productivity during Normal, El Nino and La Nina periods so that they could be compared temporally and spatially. The results of the average productivity values during Normal, El Nino and La Nina periods are shown in **Figure 7.** The average productivity in each period is then divided into five classes from very low, low, medium, high and very high to facilitate the representation into map. The intervals of productivity values (n) in tons/ha for each class sequentially are: Very Low ($n < 3.83$), Low ($3.83 < n < 4.14$), Medium ($4.14 < n < 4.46$), High ($4.46 < n < 4.77$) and Very High ($n > 4.77$).

Figure 7 shows the spatial conditions of the average value of rainfed paddy productivity in Gunungkidul Regency during the normal, El Nino and La Nina phases. During El Nino, most areas in Gunungkidul Regency experienced a decrease in productivity from normal years where many sub-districts were observed which were classified as low and very low productivity values, especially in sub-districts located in the southern part of Gunungkidul Regency. This can happen because the southern region is a karst region that has deep groundwater, making it difficult to build irrigation wells for agricultural needs, so that the decrease in rain due to El Nino can directly impact irrigation.

On the other hand, during the La Nina, there was an increase in the value of paddy productivity in all areas of Gunungkidul Regency. It was observed that there were no sub-districts that were classified as very low as happened during the Normal and El Nino periods. There is a significant difference in productivity during the El Nino and La Nina phases in line with [35] that many agricultural sectors in developing countries have a high dependence on climate so that extreme climate, especially from the ENSO - El Nino, can reduce its productivity.

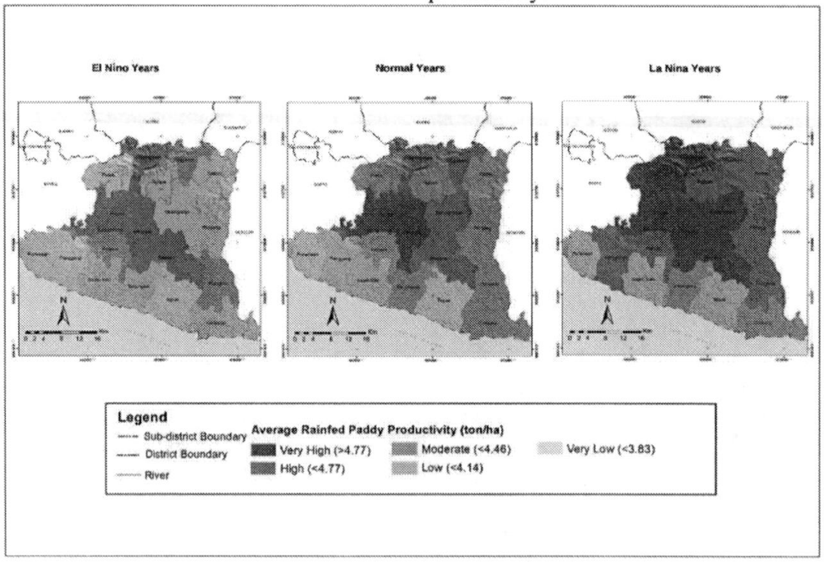

Fig. 7. The worst meteorological drought in July – December in Gunungkidul Regency during 2001 – 2019

3.6 Resource management recommendations based on research results

The condition of the majority of people who have a high dependence on agricultural products in Gunungkidul Regency is recommended to be able to carry out an intercropping system, especially in areas that have a significant relationship between ENSO, drought and rainfed paddy productivity. Intercropping systems can minimize crop failures compared to monoculture systems because through intercropping the failure of one commodity can be replaced with other commodities that are planted together [36]. In relation to intercropping, one way that can minimize the impact of subsequent drought is through a paddy agroforestry system where paddy is planted intercropping with other seasonal crops which are then planted with border trees such as Teak, Silk Tree and Mahogany [34]. The agroforestry system can be a nature-based solution in minimizing drought where limiting trees planted with seasonal agricultural commodities can maintain the temperature of the surrounding microclimate and can increase the ability of the soil to capture and hold water and reduce evaporation so as to maintain soil moisture. better for the growth of annual crops such as paddy [37,38].

One of the most important things in overcoming various impacts of climate anomalies, including ENSO that causes meteorological drought, especially in the agricultural sector, is the adaptability of each farmer. However, the adaptation of farmers will be hampered in their development without support from the government [39]. Therefore, the strengthening of community adaptation that can be carried out through structural forms of mitigation is through the provision of adequate irrigation infrastructure for the community. Increasing water availability can be done by storing rainwater, preserving land cover, and other building engineering to be able to control surface runoff [40]. In addition, it is important for the government to provide access to nutrients and better paddy varieties to anticipate stronger drought intensity in the future where traditional methods may no longer be able to cope. Often farmers still maintain traditional adaptation procedures because agricultural inputs such as nutrients, fertilizers and seeds are expensive with insufficient availability making it difficult for farm households to reach [36].

4 Conclusion

The widest meteorological drought during 2001 – 2019 occurred in November 2006 where the entire area of Gunungkidul Regency experienced a drought. Meteorological drought tends to occur more frequently during the month during the wet period than during the dry period. The highest frequency of drought events occurred in Paliyan Sub-district with the number of months of drought reaching 50 months while the lowest frequency of drought events occurred in Ponjong Sub-district with the number of dry months reaching 30 months.

Rainfed paddy productivity in Gunungkidul Regency with meteorological drought has a positive relationship. During El Nino years, paddy productivity decreases and vice versa when La Nina occurs, productivity will increase. However, not all sub-districts have the same relationship between paddy productivity and meteorological drought. The difference in the paddy varieties and development of different adaptations in each sub-district may causes variations in the relationship in each region.

Further research related to drought in Gunungkidul Regency can examine the condition of the development of community resilience to the various impacts of drought that have often occurred for years in Gunungkidul Regency. The difference in the development of adaptation to drought in various regions may be one of the differences in the variation in the impact of drought, especially on the agricultural sector.

This research is part of the 2021 UGM RTA grant on behalf of Dr. Emilya Nurjani M.Si. Thank you for those who have helped with this research as well as all parties from the Agriculture and Food Office of Gunungkidul Regency, the Central Statistics Agency of Indonesia, National Mapping Agency of Indonesia and Australian Bureau of Meteorology who have published the data so that it can be used for the purposes of this research.

References

1. L.Maheswara, Jurnal Manajemen Bencana Kekeringan, **01,** 1–13 (2016)
2. R. L. Naylor, D. S. Battisti, D. J. Vimont, W. P. Falcon, M. B. Burke, **104** (19), 7752-7757 (2007)
3. S.M. Vicente-Serrano, E. Aguilar, R. Martínez, *et al,* Clim Dyn **48,** 405–427 (2017)
4. Sudibyakto, *Manajemen Bencana di Indonesia Ke Mana?* (2011)
5. W. Cai, S. Borlace, M. Lengaigne, P. V. Rensch, M. Collins, G. Vecchi, A. Timmermann, A. Santoso, M. J. McPhaden, L. Wu, M. H. England, G. Wang, E. Guilyardi, F.-F. Jin, Nature Climate Change, **4**(2), 111–116 (2014)
6. F. Kogan, *Monitoring Drought from Space and Food Security*, In book: Remote Sensing for Food Security, 75-113) (2019)
7. K. Jenkins, R. Warren, Theoretical and Applied Climatology, **120**(1–2), 41–54 (2015)
8. Badan Penelitian dan Pengembangan Pertanian, Pedoman Umum Adaptasi Perubahan Iklim Sektor Pertanian (2011)
9. Sugihardjo, *Model Adaptasi Ekologi Petani Sebagai Strategi Pengelolaan Usahatani Akibat Perubahan Iklim (Kasus Di Daerah Aliran Sungai Cemoro, Jawa Tengah),* in Doctoral Dissertation, Universitas Sebelas Maret, Surakarta (2016)
10. E. Runtunuwu, H. Syahbuddin, Jurnal Tanah Dan Iklim, **26,** 1–12 (2007)

 https://doi.org/10.1051/e3sconf/202132501017

11. F. Apriani, Y. D. Setianingsih, U. M. P. Arum, K. A. Susanti, S. I. Wicaksono, A. Faruk, Khazanah, **6**(2), 13–22 (2007)
12. Badan Pusat Statistik, *Kabupaten Gunungkidul Dalam Angka 2019* (2019)
13. I. Wredaningrum, Jurnal Bumi Indonesia, **3**(4), 1–10, (2014)
14. A. Cahyadi, H. Nugraha, F. Nucifera, In E. Sudarmadji, Haryono, T.N. Adji, M. Widyastuti, R. Harini, E. Nurjani, A. Cahyadi, H. Nugraha (Eds.), *Ekologi Lingkungan Kawasan Karst Indonesia: Menjaga Asa Kelestarian Kawasan Karst Indonesia* (2013)
15. F. Prawaka, A. Zakaria, S. Tugiono, Jurnal Rekayasa Sipil Dan Desain, **4**(3), 397–406 (2016)
16. World Meteorological Organization, *Standardized Precipitation Index User Guide* (2012)
17. Z. Pei, S. Fang, L. Wang, W. Yang, Water, **12**(7) (2020)
18. B. Sutaryo, S. Widodo, *Kajian Keragaan Beberapa Varietas Unggul Baru Padi Gogo di Lahan Sub-Optimal Gunungkidul*, In Prosiding Seminar Nasional Lahan Suboptimal 2018, Palembang 18-19 Oktober 2018 "Tantangan Dan Solusi Pengembangan PAJALE Dan Kelapa Sawit Generasi Kedua (Replanting) Di Lahan Suboptimal", **8**, 978–979 (2019)
19. P. Dahal, N. S. Shrestha, M. L. Shrestha, N. Y. Krakauer, J. Panthi, S. M. Pradhanang, T. Lakhankar, Natural Hazards, **80**(3), 1913–1932 (2016)
20. R. Zhao, H. Wang, C. Zhan, S. Hu, M. Ma, Y. Dong, Theoretical and Applied Climatology, **139**(3–4), 1363–1377 (2020)
21. K. A. Cahyono, *Kajian Kekeringan Meteorologis di Kabupaten Bangkalan Jawa Timur*, In Bachelor Thesis, Universitas Gadjah Mada, Yogyakarta (2018)
22. D. Manatsa, G. Mukwada, E. Siziba, T. Chinyanganya, Theoretical and Applied Climatology, **102**(3), 287–305 (2010)
23. H. Nurrohmah, *Analisis Kekeringan Meteorologis Menggunakan Standardized Precipitation Index (SPI) Di Provinsi Jawa Tengah*, In Bachelor Thesis, Universitas Gadjah Mada, Yogyakarta (2017)
24. A. H. Payab, U. Türker, Environmental Earth Sciences, **78**(10), 1–19 (2019),
25. J. Spinoni, G. Naumann, H. Carrao, P. Barbosa, J. Vogt, International Journal of Climatology, **34**(8), 2792–2804 (2014)
26. U. Šebenik, M. Brilly, M. Šraj, Acta Geographica Slovenica, **57**(1), 31–49 (2017)
27. I. I. Putri, E. Nurjani, Jurnal Bumi Indonesia, 1–12 (2018)
28. S. Sulastri, *Analisis Indeks Curah Hujan untuk Identifikasi Daerah Rawan Kekeringan di Kabupaten Gunungkidul*, In Bachelor Thesis, Universitas Gadjah Mada, Yogyakarta (2017)
29. A. Fadholi, Jurnal Ilmu Lingkungan, **12**(2), 43 (2013)
30. Y. Ikawati, *Kekeringan Sampai November Akibat IOD Positif* (2008)
31. M. A. Yosilia, *Analisis Hubungan El Nino dengan Kekeringan Meteorologis Menggunakan SPI (Standardized Precipitation Index) di Pulau Bali*, In Bachelor Thesis, Universitas Gadjah Mada, Yogyakarta (2014)
32. Riyandoko, E. Mulyoutami, Kiprah Agroforestri **24**, 9–10 (2016)
33. A. Anshori, D. Riyanto, Suradal, Journal of Agri-Food, Nutrition and Public Health, **1**(2), 55–61 (2020)
34. J. Nugroho, M. Zid, M. Miarsyah, Journal of Environmental Sustainability Management, 438-447 (2020)
35. A. Limsakul, Environment and Natural Resources Journal, **17**(4), 30–42 (2019)
36. M. Maryana, S. Priyanto, *Peran Pertanian Tumpangsari Untuk Mendukung Kedaulatan Pangan*, In Prosiding Seminar Nasional Kedaulatan Pangan, Fak, Pertanian Universitas Pembangunan Nasional Veteran Yogyakarta (2012)
37. J. Cook, *3 Steps to Scaling Up Nature-Based Solutions for Climate Adaptation* (2020)
38. M. Kumar, H. Singh, In: S. Dhyan, A. Gupta, M. Karki, Disaster Resilience and Green Growth (2020)
39. A.B. Sekaranom, E. Nurjani, F. Nucifera, Sustainability, **13**(13), 7069 2021
40. I. Narulita, Jurnal Tanah Dan Iklim, **41**(1), 45–60 (2017)

Surface and subsurface engineering geological mapping for empirical designs of excavation method and support system of Tunnel 6 of Jakarta-Bandung high speed railway, Indonesia

Octavika Malda[1], *I Gde* Budi Indrawan[1*], and *Akmaluddin* Akmaluddin[1]

[1] Gadjah Mada University, Department of Geological Engineering, Jl. Grafika Bulaksumur No.2, Yogyakarta, Indonesia

Abstract. Empirical design of support system at the Tunnel 6 of the Jakarta – Bandung high-speed railway was based on the Basic Quality (BQ) system, which had not been adopted in Indonesia. This research was carried out to better understand the rock mass quality at the tunnel construction site by comparing rock mass quality determined by the BQ system to that determined by two more popular rock mass classifications, namely the Geological Strength Index (GSI) and Rock Mass Rating (RMR). Surface and subsurface engineering geological mapping were carried out and tunnel excavation method and support system were proposed. The engineering geological model of the BQ, GSI, and RMR systems showed that the sedimentary rock masses of the Miocene Jatiluhur Formation generally had poor to very poor quality, while those of the Quaternary Volcanic Formation had very poor to good quality. Based on the RMR, the stand-up time values of the sedimentary rock masses were predicted to be relatively low as compared with those of the Quaternary Volcanic Formation, implying requirement of a relatively quick support system installation after excavation. In general, a combination of systematic bolt, shotcrete and steel ribs is the recommended support system for this tunnel.

1 Introduction

Tunnel 6 is the longest of the 13 tunnels of the Jakarta-Bandung high-speed railway. Tunnel 6 stretches for 4.4 km and crosses the administrative area of West Java Province: Purwakarta Regency on the inlet side and West Bandung Regency on the outlet side (Fig. 1). Tunnel 6 was built using the NATM method with a horseshoe geometry having dimensions of 13 meters high, 14 meters wide and ± 25% trace grade from the inlet to the outlet. The morphology at the tunnel location is dominated by undulating hills with the ridge direction generally east-west and partly northeast-southwest.

The investigation data used in the construction of this tunnel is in the form of surface geological mapping activities, core samples of 16 drilling locations and 4 face mapping data. The rock composition at the tunnel construction site is volcanic rock from the Quartenary Formation of Older Volcanic Product (Qob) which is overlay unconformity with the fine-grained clastic sedimentary rocks of the inserts of the Miocene Jatiluhur Formation Sandstones (Mdm)[1]. The angular unconformity relationship between the two formations at the study area shows the complexity of the geological conditions so that accuracy is needed to develop geological and engineering geological models in this work. This condition becomes more complicated where the lateral distribution of the rock on the surface is volcanic from Qob Formation, while at the elevation of the tunnel It is a fine-grained clastic sedimentary rock from the Jatiluhur Formation which is very limited exposed on the surface.

The existing investigative data is used as one of the variables in determining the quality condition of the rock mass. The rock mass quality classification system is a very useful instrument for the initial design stage of a project when the available information is limited[2]. The rock mass quality classification system used in this project is the BQ classification system, which is relatively new in Indonesia. In order to build a model regarding the condition of rock mass quality it is important to use more than one classification system in order to describe the composition and characteristics of the rock mass from various perspectives to provide a proper initial estimate of the support system of the tunnel[2]. The rock mass classification used in this study is the RMR89 and GSI. The model built from a rock mass classification system is basically an empirical model, however it is very useful in the early stages of a tunnel construction work[2]. Several rock parameters were obtained from direct field test results with reference to the field direct method test from ISRM 1981[3]. It is expected that the results of this research improve understanding the use of BQ system for determination of rock mass quality, particularly in tunnelling projects in Indonesia.

* Corresponding author: igbindrawan@ugm.ac.id

https://doi.org/10.1051/e3sconf/202132501018

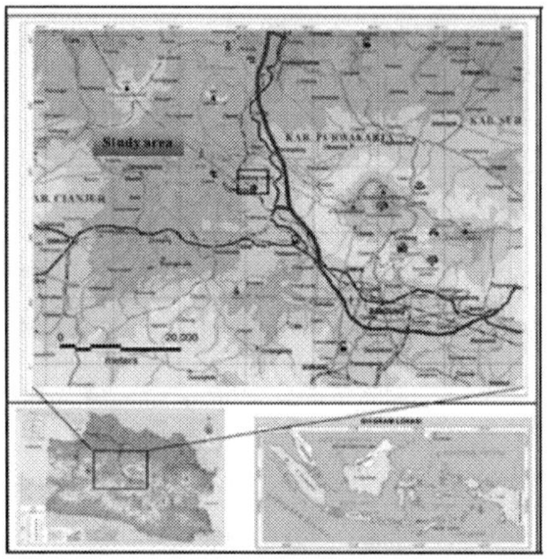

Fig. 1. Research area[4]

2 Methodology

In this study, the type of data is divided into 2 data models, surface data and subsurface data, both of which serve as variables in forming geological models, classification of rock mass quality, engineering geological models and empirical references to determine the initial design of the tunnel support system, standup time and determining the excavation method that will be used in the construction of the tunnel 6.

Surface data from the results of surface mapping are in the form of outcrop data, lithological types, discontinuity measurement data and classification of surface rock mass quality. The subsurface data is in the form of drilling data as many as 16 drill holes, 4 face mapping locations. Weathering rate parameter data of surface and subsurface rocks refers to ISRM[3]. The rock strength parameter refers to the direct field test method Hoek., et al[5]. The flow of this study can be seen in Figure 2.

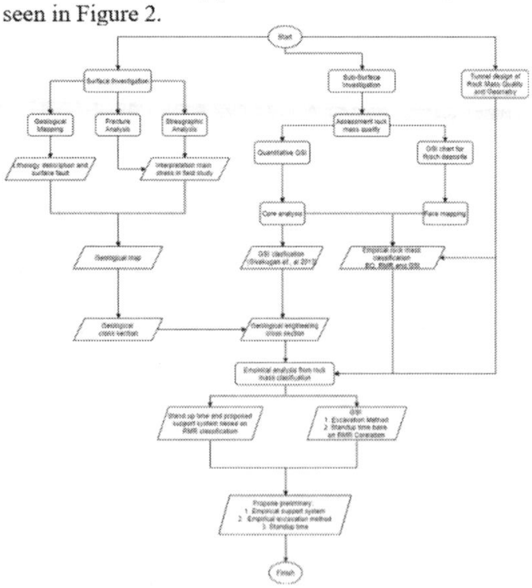

Fig. 2. Research flow diagram

3 Classification of rock mass quality

In this paper, there are 3 classifications that will be used, it is the BQ system classification[6], RMR89[7] and GSI [8–11].

3.1 Basic Quality (BQ) system

Rock mass classification Basic Quality (BQ) system is a rock mass classification system that is mandatory in China[12]. Referring to the code for design of railway tunnel: TB 10003-2016[6], the BQ value is determined by two parameters: rock strength (Rc) and rock mass integrity (Kv), as shown in Eq. (1):

$$BQ = 100 + 3Rc + 250Kv \tag{1}$$

Where Rc is the value of saturated UCS and Kv is the value of elastic longitudinal wave velocity of rock mass and of intact rock. To provide more actual rock mass quality data the basic BQ value (eq. 1) needs to be adjusted as follows: three coefficients, groundwater condition K1, orientation of weakness zone related to the excavation K2 and in-situ stress condition K3, are considered to corrected BQ value as[6]:

$$[BQ] = BQ-100(K1+K2+K3) \tag{2}$$

3.2 Rock Mass Ratting (RMR) system

RMR has changed historically at least 5 times so it is important to include the year or version of the RMR that is used as a reference in a study[13]. This study uses the 1989 version of the RMR classification[7]. RMR89 has 6 parameters and each parameter has a different value from the sum of these values, the rating is obtained for a rock mass, the six parameters are the compressive strength value of intact rock (UCS/ Point Load), RQD, discontinuous spacing, discontinuous conditions, groundwater conditions at the measurement location and the orientation of the discontinuous plane to the measurement location.

3.3 Geological Strength Index (GSI)

The essence of the GSI is a careful description of engineering geology of the rock mass, which is essentially qualitative, because it is believed that a amount of fracture (joint) largely meant for weak rock mass conditions and complex[14]. The determination of the GSI value in the field is based on the basic GSI chart proposed by using two variable, structural parameters on the surface and the quality of the surface of the outcrop[7]. In this study also used the GSI classification table developed to assess a rock mass from tubidite or flysch deposits[11]. As for determining the value of GSI of the drilling results based on the method of quantification of GSI based on two factors: joint condition from RMR89 and value of 0.5RQD[9].

4 Geology study area

The geological conditions of the research location on the surface are dominated by volcanic rocks from the Qob Formation in the form of volcanic breccias, laharic breccias and several locations where andesite lava is present. The Qob Formation overlaps the Mdm Formation which is generally fine-grained sedimentary rocks such as marl, shale, claystone with sandstone intercalated which are deep sea turbidite deposits or flysch[15].

Based on the results of surface geological mapping, the position of the rocks that make up the Mdm Formation has a dip direction to the south with a varying angle between 30°-55° which shows the relationship of angular unconformity with the Qob formation at the top. The reconstruction of the surface geological model based on drilling data, face mapping and surface mapping shows that the tunnel elevation is generally located in the sedimentary rock units of the Mdm formation and partly in the Qob formation (Fig. 3 and 9).

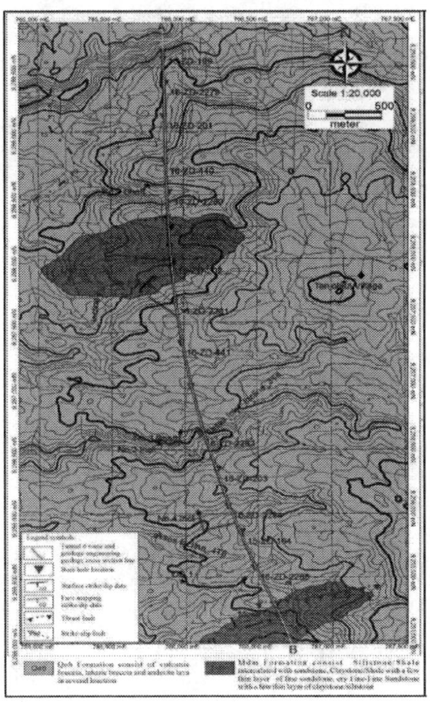

Fig. 3. Geological map of study area

4.1 Lithology

The results of surface geological mapping show the Qob Formation unit consisted of:

1. Volcanic breccias with fragment supported fragment compositions in the form of igneous andesite rock and a small matrix of coarse sand (Fig. 4a).
2. The fundamental difference between laharic breccias and volcanic breccias is the difference in the amount of fragments where in the laharic breccias the fragments are still floating andesite rocks on a medium-coarse sand matrix, the size of the fragments is smaller than volcanic breccias (Fig. 4b).
3. In this formation there are several lava andesite (Fig.4c).

The Mdm formation at the study site consists of claystone, marl, calcareous shale with intercalated sandstones. This formation shows the characteristics of deep sea deposits or flysch with the appearance of graded bedding sedimentary structures and pelagic claystone (Fig.4d)[16].

Fig. 4. Example of outcrops in study area figure a-c lithology from Qob formation and sandstone from Mdm formation; a. volcanic breccia, b. laharic breccia, c. lava andesite and d. sandstone with graded bedding structure

4.2 Geological structure

From the appearance of aerial imagery or the National Digital Elevation Models (DEMNas), the research locations generally have a general direction of relative West-East alignment (Fig. 5). The direction of this ridge reflects the direction of the "Java pattern" structure which reflects the direction of the main stress influenced by the subduction of the Indian plate which is in the south of Java[17]. This is also reflected in the position of the rock in a direction to the south, thrust faults with a plane position N065 ° E / 68 ° (Fig. 5) and the stereographic analysis of the fractures in the sandstone layer of the Mdm formation which shows the direction of the main stresses N329E ° and N147 ° E or relative north-south (Fig. 6).

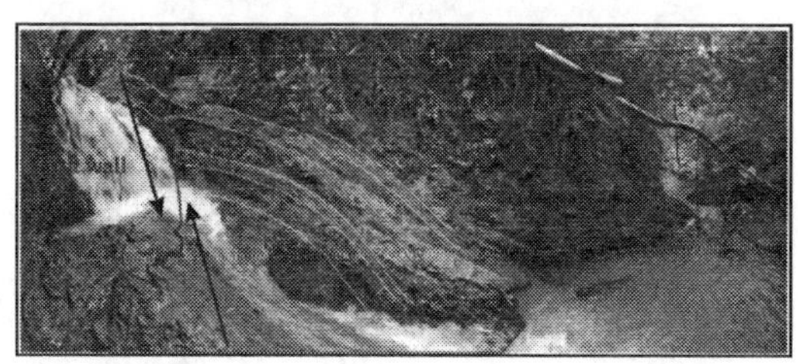

Fig. 5. Dander thrust fault one of the supporting data in forming a geological and engineering geology model

https://doi.org/10.1051/e3sconf/202132501018

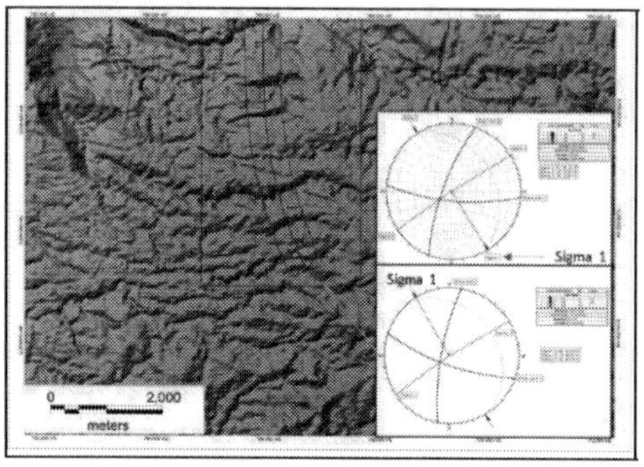

Fig. 6. Lineament and stereographic analyses indicate the main tectonic stress has a relative north to south direction

5 Engineering geology of study area

With geological conditions that are quite complex and the data on outcrops are limited, it is necessary to correlate and calibrate the surface and subsurface conditions.

5.1 Surface condition

The quality condition of the surface rock mass uses the GSI basic chart and the GSI chart for heterogeneous rock masses such as flysch deposits. The use of the GSI classification in this study is because this classification is easier to use, considers geological conditions and provides the required geological information [7].

There are several outcrops that are considered to still reflect the quality of the rock mass in each formation at the research location. The Qob formation is as follows:

1. Volcanic breccias (Fig.4a), generally have moderate weathered to highly weathered weathering rates refer to ISRM, 1981[3]. Surface structure is disintegrated, GSI value is based on basic chart 25-30.
2. Laharic breccias (Fig.4b), weathering rates are generally highly weathered according to ISRM, 1981[3]. The surface structure is disintegrated, the GSI value is based on the basic chart 20-25.
3. Andesite lava (Fig.4c), surface conditions are generally in the form of rock blocks surrounded by red weathered sandy material. High weathered weathering conditions refer to ISRM, 1981[3]. Very blocky surface structure with join sets of 4 to more than 4, GSI values range from 33-38.

The Mdm Formation is dominated by claystone outcrops with intercalated of sand, siltstone and shale with rock mass qualities as follows:

1. Shale with thin inserts of sandstones (Fig. 7) weathering conditions of highly weathered rocks with poor discontinuous surface conditions. The structure and composition of the outcrop entered into type IV of the GSI chart for heterogeneous rocks or flysch deposits, GSI values 25-30.
2. The outcrop is an thrust fault zone with a composition dominated by claystone with several layers of fine-very fine sandstone based on plotting on chart IV of the GSI chart for heterogeneous rocks or flysch deposits, GSI values 25-30 (Fig. 6).

Fig. 7. Outcrops of claystone/shale with a few thin layer of fine sandstone indicated flysch deposite product.

5.2 Sub-surface condition

The minimal amount of surface outcrops, weathering conditions on the outcrops due to climate and tunnel elevations that are deep or far from the surface, so it is necessary to interpret subsurface conditions using data from core drilling results. The ongoing development process allows additional data collection of face mapping activities from within the tunnel location as additional information that is essential for calibrating surface and subsurface models.

From the 16 existing bore holes, the classification of rock mass quality has been calculated using the BQ system, RMR and GSI classifications. For the assessment of rock conditions from the results of face mapping in this paper using the RMR and GSI classifications.

Observations from 16 drilling points show the following data:

1. The volcanic breccias of the Qob Formation with relatively in broken conditions, this may be due to the level of weathering and the composition of the grain supported which is pebble to boulder, but fragments of boulder from andesite rocks are still visible.

2. The laharic breccias of the Qob Formation generally have better core conditions with the composition generally in the form of a supported matrix with the size of the fragments in the form of cobble-pebble so that they are still recorded intact in the drilling data.
3. At one of the drillings location, it was identified the presence of lava in the form of igneous andesite rocks between the layers of volcanic breccia, this can be seen from the condition of the core samples which are generally massive with a thickness of more than 1 meter at the tunnel elevation.

Observations from the 4 face mapping locations only show the constituent rocks of the Mdm Formation which show lithology in the form of massive shale without sandstone intercalated with very clear fissure plane, then there are sandstones with high brittle intensity (Fig. 8).

Calculation of rock mass quality data from 16 bore holes at tunnel elevation and from the results of face mapping shows that the tunnels are in a very poor dominant rock mass condition in Mdm or Qob formations, but in some locations there are still good to fair rock conditions at the Qob Formation is generally in the form of laharic breccias and andesite lava inserts, while the Mdm formation is in the form of sandstones with a few siltstone intercalation (Table 1). The position of the rock based on the results of face mapping is relatively similar position with the surface outcrops. The strike direction relative to the west-east and the dip angel is between 30°-60°, this is used to make geological models and subsurface geology (Fig. 8). Existing data, the model of subsurface geological and engineering geological conditions can be seen in Figure 9.

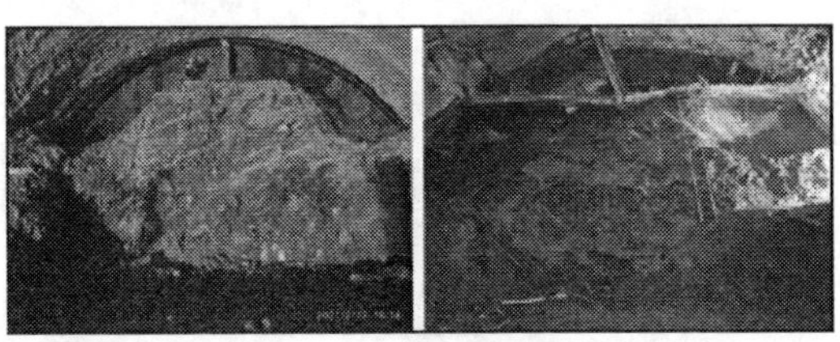

Fig. 8. Excavation face condition from tunnel 6, location measured the rock mass classification from GSI and Q system, left location face mapping 1 and right loaction face mapping 4

6 Data evaluation

Based on surface and subsurface data such as constituent lihology, geological structure and an assessment of the quality of the existing rock mass, analysis can be carried out to evaluate the possibility of empirical models of the support system, excavation methods and standup time or the ability of the tunnel opening to be stable during the installation of the support system.

6.1 Stand-up time

In determining the stand-up time of 16 drilling data at tunnel elevation using GSI values which are correlated using 3 equations, this is done to obtain the average value of the equation to validate so that the value is not too high or too low in the design. E. Hoek and Brown[18] stated, that the correlation between RMR89 and GSI can be done provided that the value of groundwater conditions is in a dry position, the value for the orientation of the discontinuous plane is considered 0 and this equation is suitable for GSI values> 25, the equation used is:

$$GSI = RMR - 5 \quad (3)$$

Because generally the rock conditions in the study area have a low GSI value or GSI <25, this paper adopts the equation proposed by Osgoui and Ünal[19]. This equation is considered suitable for low GSI or GSI <25 values and this equation is obtained from sedimentary rocks such as siltstone, clay to shale, the equation is as follows:

$$GSI = 6e^{0.05RMR} \quad (4)$$

The next equation from Ceballos., et al[20], where this equation has a value of R2 = 0.89 or close to 1. The value of R2 or the coefficient of determination which simply reflects the value that is getting closer to 1 is a good value reflecting that the x-axis coefficient affects directly with the y axis, the equation is as follows:

$$GSI = 1.13\ RMR - 11.63 \quad (5)$$

Whereas the RMR89 value from face mapping is not correlated and obtained from direct measurement results, the RMR89 correlation value and the RMR89 value from the face mapping measurement results can be seen in the Table. 1.

After obtaining the average RMR89 value from 3 correlation equations or direct measurements, the method used to estimate stand-up time is to use the rock mass classification curve from Bieniawski[7] with the technical aspects used, namely: RMR89 value, span dimension and based on data The design span dimension is 13 meters.

Based on the modeling results at stand-up time using plotting on the rock mass classification curve from Bieniawski[7], drill point 2280 has an estimated stand-up time of up to 63 days, drill point 203 is estimated to be 27 days, drill point 440 is estimated to be 13 days while 12 drill points another showed a stand-up time of less than 2 days until collapsing immediately after the excavation began (Fig. 10). The result of face mapping data showed the stand-up time value is low and and can collapse immediately when excavation is carried out (Table. 2).

https://doi.org/10.1051/e3sconf/202132501018

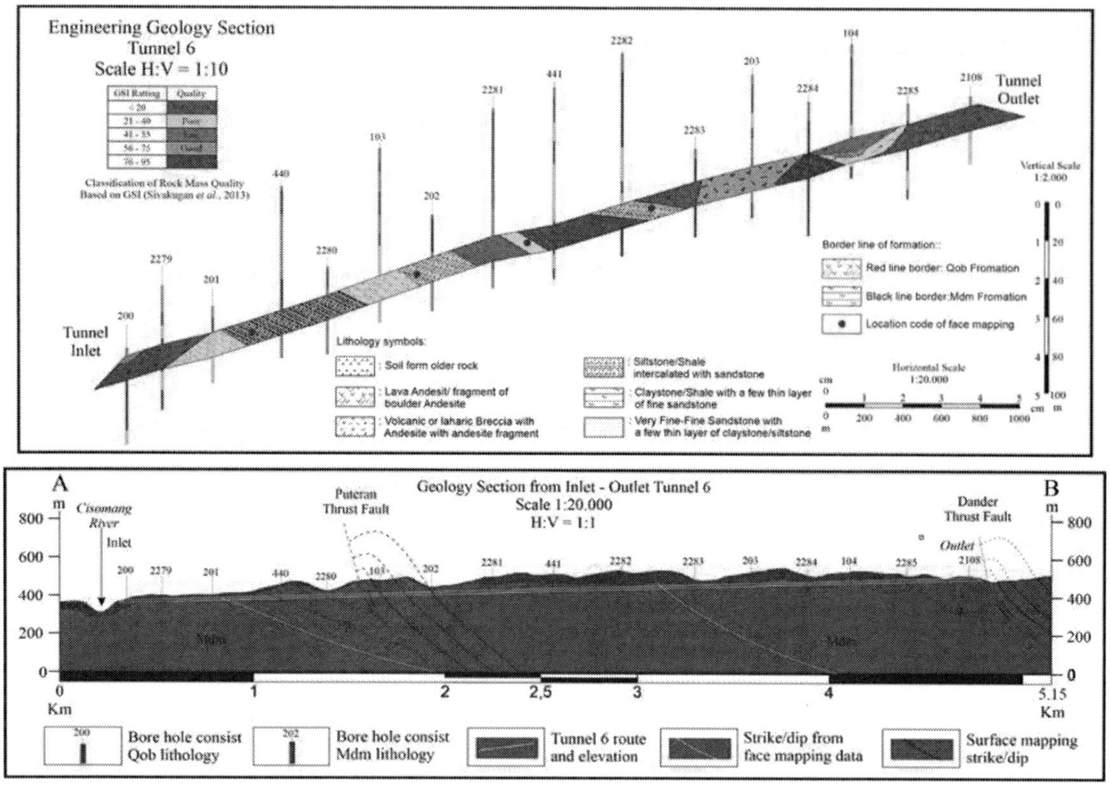

Fig. 9. Cross section model of engineering geology (above) and geological cross section model (below).

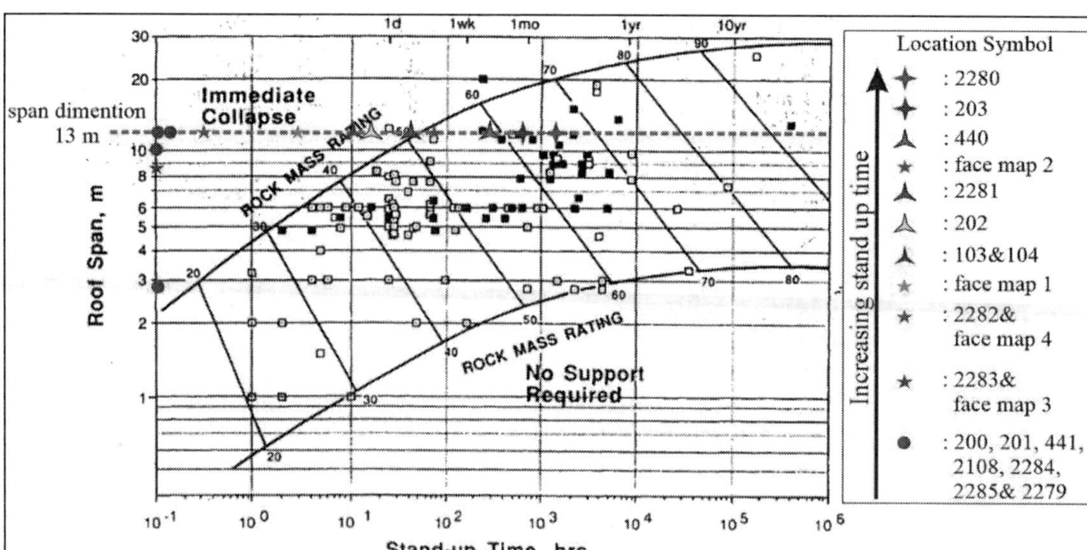

Fig. 10. The results of plotting the RMR value on the Bieniawski stand-up time chart[7].

Table 1. The results of the rock mass quality assessment using several classifications on the elevation of the tunnel.

Formation	Bore Hole Number/ Face Map. Loc	Tunnel Km	Depht (m)		GSI		Class GSI	Corelation GSI to RMR89			Avg. RMR		RMR face mapping		BQ System	
			From	To	Quantitative	Flysch deposite	Sivakugan et al., [10]	* Eq (1)	** Eq (2)	*** Eq (3)	Value	Class	Ratting	Class	Class	Description
Qob	200	88.915	4	17	8		Very Poor	13	4	17	11	Very Poor			V	Very Poor
	2279	89.100	31	44	22		Poor	27	26	30	27	Poor			V	Very Poor
Mdm	201	89.361	14	27	6		Very Poor	11	1	16	9	Very Poor			V	Very Poor
	Face Map. 1	89.428	17	30		40	Poor						39	Poor	V	Very Poor
	440	89.716	65	78	62		Good	67	47	65	59	Fair			IV	Poor
	2280	89.960	15	28	64		Good	69	47	67	61	Good			IV	Poor
	103	90.233	67	80	41		Fair	46	39	47	44	Fair			IV	Poor
	Face Map. 2	90.426	60	73		32	Poor						53	Fair	IV	Poor
	202	90.500	22	35	44		Fair	49	40	49	46	Fair			V	Very Poor
	2281	90.819	67	80	51		Fair	56	43	55	51	Fair			IV	Poor
	Face Map. 3	91.085	70	83		20	Very Poor						21	Poor	V	Very Poor
	441	91.145	72	85	19		Very Poor	24	23	27	25	Poor			V	Very Poor
	2282	91.499	79	92	24		Poor	29	28	32	29	Poor			IV	Poor
	Face Map. 4	91.630	80	93		25	Poor						29	Poor	IV	Poor
Qob	2283	91.869	17	30	15		Very Poor	20	18	24	21	Poor			V	Very Poor
	203	92.166	48	61	65		Good	70	48	68	62	Good			IV	Poor
Mdm	2284	92.459	27	40	17		Very Poor	22	21	25	23	Poor			V	Very Poor
Qob	104	92.681	50	63	41		Fair	46	39	47	44	Fair			V	Very Poor
	2285	92.974	11	24	19		Very Poor	24	23	27	25	Poor			V	Very Poor
	2108	93.300	5	18	8		Very Poor	13	5	17	12	Very Poor			V	Very Poor

* E. Hoek dan Brown[18], ** Osgoui dan Ünal[19], *** Ceballos., et al [20].

Table 2. Summary of stand-up time based on RMR[7] and excavatabilitybased on modifed GSI chart[21]

Bore Hole Number/ Face Mapping Location	Rock Formation	RMR			GSI		Estimated Point Load**	Excavability#
		Ratting	Class	Estimated standup time	Value	Quality*		
200	Qob	11	Very Poor	Immediate collapse	8	Very Poor	R2	Digging
2279		27	Poor		20		R3	
201	Mdm	9	Very Poor		6		R2	
Face Map. 1		39	Poor	1 hrs	40	Poor	R3	Ripping
440		59	Fair	± 13 days	62	Good		Hydrolic Breaker
2280		61	Good	± 80 days	64			
103		44	Fair	± 10 hrs	40	Fair		Ripping
Face Map. 2		53		± 3 days	32	Poor	R2	Digging
202		46		± 20 hrs	44	Fair		Ripping
2281		51		± 2 days	51			
Face Map. 3		21	Poor	Immediate collapse	20	Very Poor	R1	Digging
441		25			19			
2282		29			24	Poor	R2	
Face Map. 4		29			25			
2283	Qob	21			15	V. Poor	R3	
203		62	Good	± 27 days	65	Good	R4	Hydrolic Breaker
2284	Mdm	23	Poor	Immediate collapse	17	Very Poor	R1	Digging
104	Qob	44	Fair	± 10 hrs	41	Fair	R2	Ripping
2285		25	Poor	Immediate collapse	19	Very Poor	R1	Digging
2108		12	Very Poor		8		R1	

*: Sivakugan et al., [10], **: Hoek., et al[5], #: Tsiambaos and Saroglou[21].

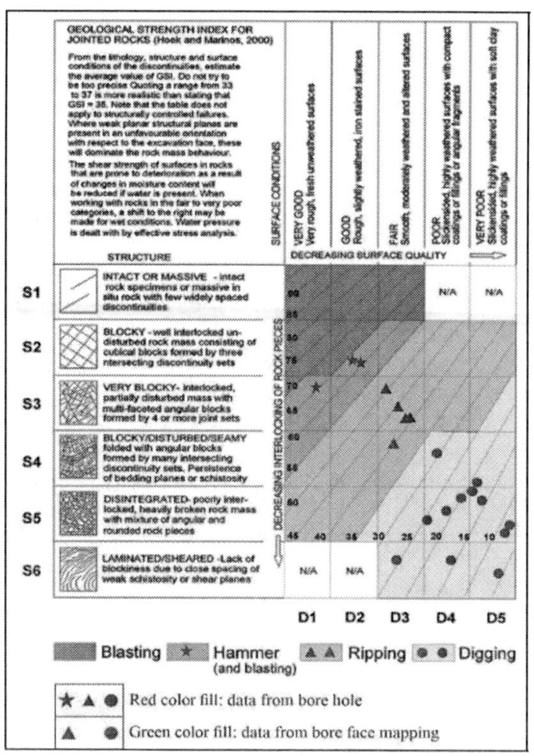

Fig. 11. The excavation method based on Tsiambous and Saroglou graphic chart for Is <3MPa[21]

https://doi.org/10.1051/e3sconf/202132501018

Table 3. Empirical support system based on RMR classification[7].

Bore Hole Number/ Face Mapping Location	Rock Formation	Rock Mass Ratting (RMR) Classification														
		Quality	Rock Bolt			Rock Bolt Postion			Shotcrete Thicknes (cm)			Wire Mesh		Steel Sets		
			Configuration	Length (m)	Spacing (m)	Crown	Wall	Invert	Crown	Wall	Face	Crown	Wall	Type	Spacing (m)	additional support
200	Qob	Very Poor	Systematic	5-6	1-1.5	√	√	√	15-20	15	5	√	√	med-heavy	0.75	forepoling
2279		Poor		4-5	1-1.5	√	√	-	10-15	10	-	√	√	light-med	1.5	-
201	Mdm	Very Poor		5-6	1-1.5	√	√	√	15-20	15	5	√	√	med-heavy	0.75	forepoling
Face Map. 1		Poor		4-5	1-1.5	√	√	-	10-15	10	-	√	√	light-med	1.5	-
440		Fair		4	1.5-2	√	√	-	5-10	30	-	√	-	No required		
2280		Good	Locally	3	2.5	√	-	-	5	-	-	√	-	No required		
103		Fair	Systematic	4	1.5-2	√	√	-	5-10	30	-	√	-	No required		
Face Map. 2																
202																
2281																
Face Map. 3		Poor		4-5	1-1.5	√	√	-	10-15	10	-	√	√	light-med	1.5	-
441																
2282																
Face Map. 4																
2283	Qob															
203		Good	Locally	3	2.5	√	-	-	5	-	-	√	-	No required		
2284	Mdm	Poor	Systematic	4-5	1-1.5	√	√	-	10-15	10	-	√	√	light-med	1.5	-
104	Qob	Fair		4	1.5-2	√	√	-	5-10	30	-	√	-	No required		
2285																
2108		Very Poor		5-6	1-1.5	√	√	√	15-20	15	5	√	√	med-heavy	0.75	forepoling

6.2 Excavatability

In evaluating the selection of an effective excavation type it is important to consider the physical, mechanical and behaviour characteristics of the geo-material to be excavated[21]. In this study, the evaluation of the selection of the type of excavation method refers to the GSI value of the rock mass at the tunnel elevation and the rock strength value from the direct test results using the classification of Hoek., et al[5].

GSI values and rock strength values from the direct test results will be plotted on the GSI chart of the excavatability assessment of a rock mass from G. Tsiambaos & H. Saroglou[21]. The plotting results can be seen in Figure 19. The selection of the type of excavation method based on the plot results is dominated by the digging method, ripping at 4 drilling locations and 1 face mapping location, while the hammer breaker method can be carried out at three drilling point locations (Fig. 11).

6.3 Support system

To determine the type and dimensions of the support system in this study, it refers to the average RMR correlation value from the analysis of 16 core samples and the RMR value from the direct measurement results at 4 face mapping locations. The results of the analysis showed 12 locations with very poor to poor quality, 6 locations with fair quality and 2 locations with good quality based on the RMR classification. Empirical support system design based on RMR for very poor-poor rock mass quality is a combination of rockbolt, shotcrete and steel sets, while for rock mass conditions fair-good steel sets are not required (Table 3). The approach to determining the empirical support system based on the RMR value was chosen because this classification provides details of the dimensions, spacing and configuration of the support system for tunnel construction.

7 Discussion

To build subsurface models, the bedding plane of surface outcrops is very useful in calibrating the model that was formed. The interpretation of the subsurface geological structure uses calibration from the position of the surface area assisted by the position of the bedding plane on face mapping measurements such as the reconstruction of the Dander thrust fault and the Puteran thrust fault, this is also supported by drill data at drill location 103 which shows the relative poor rock mass quality this is indicated due to the influence of the position of the drill location 103 is on the hanging wall of the Puteran thrust fault.

The surface outcrops conditions at several locations are quite helpful in providing an interpretive model of the rock mass conditions below the surface, especially in the Qob formation in the form of volcanic breccia and lava. The surface outcrops are also useful in calibrating rock mass quality assessments, especially in terms of calculating the value of subsurface GSI classification.

The BQ classification system used in the construction of this tunnel shows a more conservative rock mass quality compared to the other two classifications. On the other hand, the BQ rock mass classification system is a relatively new classification used in Indonesia. Therefore, in this study, other classifications such as GSI and RMR are used to obtain a more comprehensive assessment of the rock mass quality in this tunnel. Each rock mass classification has its own limitations and advantages so that the use of more than one classification is very useful to complement each other and provide additional information in tunnel construction. There are three aspects that are evaluated in this paper that is the stand-up time value, the type of excavation method and the empirical support system, which are the basic things needed in the early stages of tunnel construction.

8 Conclusion

Based on the engineering geological mapping, evaluation of rock coring, and face mapping, the research conclusions are:

1. The tunnel construction area consisted of volcanic breccia, laharic breccia, and andesite lava of the Qob formation, and claystone, marl, and calcareous shale with intercalated sandstones of the Mdm formation.
2. The sedimentary rock masses of the Mdm formation along the tunnel route had relatively poor to very-poor quality.
3. The stand-up time of the sedimentary rock masses is relatively low, implying reinforcement is necessary before excavation. The stand-up time of the andesite lava is higher than that of the rock masses of the Mdm formation, while the volcanic breccia and laharic breccias can collapse immediately during excavation.
4. The excavation method based on evaluation using GSI values and compressive strength from the direct field test results show that for sedimentary rocks the conventional excavation method can still be applied, while the Qob formation in andesite lava and breccia with good quality is more suitable to use a ripper and hydraulic breaker.
5. Empirical support system design based on RMR for very poor-poor rock mass quality is a combination of rockbolt, shotcrete and steel sets, while for rock mass fair-good conditions steel sets are not required.
6. The excavation methods and supports systems proposed in this study can change during tunnel construction process because the empirical models proposed in this study are developed with limited data.

The authors wish to express deep gratitude to PT. Kereta Cepat Indonesia China (PT. KCIC), High Speed Railway Consortium Company (HSRCC), China Railway Engineering Corporation (CERC), and China Railway Design Corporation (CRDC) for the permission to do this research and publish the site investigation data. Financial support from The Ministry of Public Works and Housing of Indonesia is gratefully acknowledged.

References

1. Sudjatmiko. *Geological Map of The Cianjur Quadrangle* (2003)
2. M. Genis, H. Basarir, A. Ozarslan, E. Bilir, E. Balaban, Eng. Geol. **92**(1), 14-26 (2007)
3. ISRM. ISRM suggested methods for Rock characterisation, testing and monitoring (Pergamon Press, New York, 1981)
4. Bakosurtanal. *West Java Province Map* (2003)
5. E. Hoek, P.K. Kaiser, W.F. Bawden. Support of Underground Excavations in Hard Rock (CRC Press, 2000)
6. State Railway Administrator. TB 10003 - 2016/J 449 - 2016: *Code for Design of Rai1way Tunnel. Beijing* (2016)
7. Z.T. Bieniawski. Engineering rock mass classifications: a complete manual for engineers and geologists in mining, civil, and petroleum engineering. 1st ed. (John. Wiley and Sons, New York, 1989)
8. P. Marinos, E. Hoek. *GSI: A geologically friendly tool for rock mass strength estimation,* in ISRM Int Symp (2000)
9. E Hoek, T.G. Carter, M.S. Diederichs. *Quantification of the geological strength index chart,* in 47th US Rock Mech/Geomech Symp (2013)
10. N. Sivakugan, S.K. Shukla, B.M. Das. Rock Mechanics An Introduction. (Taylor & Francis Group , New York, 2013)
11. V. Marinos, T.G. Carter, Eng. Geol. **239**, 282-297 (2018)
12. S feng. Guo, S wen. Qi, C. Saroglou. Journal of Central South University **27**, 3090-3102 (2020)
13. B. Singh, R.K. Goel. Engineering Rock Mass Classification Tunneling, Foundations, and Landslides Bhawani (Elsevier, 2011)
14. P. Marinos, V. Marinos, E. Hoek. *Geological Strength Index (GSI). A characterization tool for assessing engineering properties for rock masses,* in Proceeding Underground Work under Special Condition (2007)
15. Abdurrokhim. Bull. Sci. Contrib. Geol, **15** No.2 (2017)
16. A. Bouma. Development in Sedimentology **3**, 247-256 (1964)
17. A.H. Satyana, M.E.M. Purwaningsih. *Structural Curvature of Central Java: A Horizontal Fault Segmentation,* in Proceeding of IAGI (Ikatan Ahli Geologi Indonesia), 1-14 (2002)
18. E. Hoek, E.T. Brown. Int. J. Rock Mech. Min. Sci. **34**, 1165-1186 (1997)
19. R. Osgoui, E. Ünal. *Rock reinforcement design for unstable tunnels originally excavated in very poor rock mass,* in Proceeding **31st** ITA-AITES World Tunnel Congress (2005)
20. F. Ceballos, C. Olalla, R. Jiménez. *Relationship between RMRb and GSI based on in situ data,* in ISRM Eur Reg Symp, Vigo, Madrid (2014)
21. G. Tsiambaos, H. Saroglou, Bull. Eng. Geol. Environ. **69**(1), 13-27 (2010)

The b-value of the Kelud Volcano in the Last Three Decades

Sintia Windhi Niasari[1*], *Lusia* Rita Nugraheni[2], and *Puspita* Dian Maghfira[2]

[1]Physics Department, Faculty of Mathematics and Natural Sciences, UGM, 55281 Yogyakarta, Indonesia
[2]Geological Engineering Department, Faculty of Engineering, UGM, 55281 Yogyakarta, Indonesia

Abstract. Kelud volcano is located in the Kediri sub-district, East Java Province, Indonesia. This volcano is still active, with total population, in the radius of 10 km, is around 10 thousand people. Kelud volcano is a popular tourist destination. On the weekend, total visitor can reach 5,000 people per-day. These people are at high risk when the Kelud volcano erupts. The last eruption of the Kelud volcano occurred in 2014 and was explosive eruption. Previously, there was an effusive eruption in 2007. These two types of eruption have its own geo hazard risk. Thus, predict the eruption type could help hazard mitigation. In this study, two data sets of earthquakes, 1990-2007 and 2008-2020, were analysed to determine the b-value and its relationship to the eruption type of the Kelud volcano. The calculation of the b-value uses the Gutenberg-Richter relationship. Calculation of the b-value in 2007, when there was an effusive eruption, showed a value of 2.27, while in 2014 (when there was an explosive eruption) was 1.85. After 2009, the curve of the b-value against time shows decrease. As a long term precursor of the Kelud activity, this b-value curve should be analysed continuously, besides volcano tectonic seismicity monitoring.

1 Introduction

Kelud volcano is an active volcano which has an altitude of 1731 m.a.s.l. This active volcano is located in Kediri, East Java, Indonesia (Figure 1). Total population, in the radius of 10 km, is around 10 thousand people. Additionally, Kelud volcano is a popular destination in East Java Province, where up to ten thousand people visit this tourist destination on the weekend.

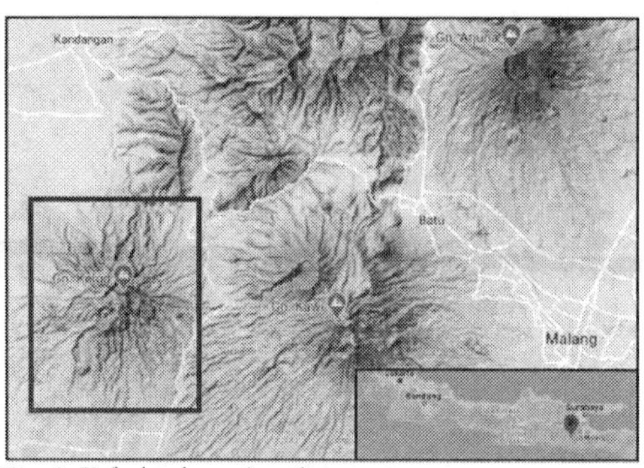

Fig. 1. Kelud volcano location.

The Kelud volcano is an active and dangerous volcano in Indonesia with eight eruptions of VEI 4 (Volcano Eruption Index scale of 5) [1]. This volcano is a result of Australian plate subduction beneath Eurasian plate. The subduction produces magma, from partial melting, that moves upward to the surface through weak zone, such as fault. Before the volcano erupts, magma migrates through conduit from magma chamber. If the magma contains high dissolve gasses, then effusive eruption occur [2]. In the contrary, when the magma is more viscous, pressure may build up and resulted in explosive eruption.

Kelud volcano eruption type consists of both explosive and effusive [1]. Centre of Volcanology and Geological Hazard Mitigation reported 33 eruption of the Kelud volcano [1]. The first eruption was on 1000, while the last eruption was on February 13, 2014.

An explosive eruption was occurred on 10 February 1990 [1]. An effusive eruption on 3 November 2007 was forming a lava dome in the crater [1]. And then, a large explosion took place in the evening of 13 February 2014, destroying the 2007 lava dome [3].

Both eruptions in 2007 and 2014 have similar precursory seismicity, but the depth range and the magnitude are different [1]. The depth range of VT for 2007 eruption was 1–2 km, while on the 2014 eruption was about 0–4 km. The magnitudes of VTs in 2007 are generally smaller than 2014.

The b-value is the seismotectonic parameter of an area and shows the relative number of small and large earthquake [4].The b-value has been studied for long- to short-term predictions of future earthquakes [5]. The b-value decreases before large earthquakes [6].

The USGS earthquake data catalog around the Kelud volcano, with the event period from 1973 until 1990 and 1990-2007, were used to identify any correlation between b value variations of the Gutenberg-Richter law with the eruption type [7]. The first data set (1973 – 1990), which connecting to the explosive erupion in 1990, showed low b-value of 0.9. The other data set (1990 – 2007), related to effusive eruption in 2007, resulted in b-value of 1.0. The authors conlcuded that

* Corresponding author: sintia_windhi@ugm.ac.id

low b-value prior to the explosive eruption of 1990 is described as a symptom of stress build-up accompanied by a symptom of increased speed due to the emergence and development of cracks.

The b value variations can be a good indicator of the state of the volcano and an efficient long term precursor [8]. Thus, in this study, two data sets of earthquakes, 1990-2007 and 2008-2020, were analysed to determine the b-value and its relationship to the eruption type of the Kelud volcano. We proposed that the b-value of explosive eruption in 2014 is lower than the b-value of effusive eruption in 2007.

2 Methods

The data used in this study were from the USGS earthquake catalog (https://earthquake.usgs.gov/earthquakes/search/). The study area are limited to latitude of -3° until -17° and longitude of 111.8° until 112.8°.

After the data were downloaded, magnitudes of the earthquakes are converted to the same magnitude, i.e. moment magnitude (Mw) using empirical equation SNI 1726:2012 [9]. And then, the data were de-clustering to separate the main earthquake from the initial earthquake and aftershocks. The calculation stage of the magnitude of completeness (Mc) is carried out to check the completeness of the tectonic data in general, before calculating the b-value. The Mc value is calculated using the Maximum-Curvature method by finding the maximum value of the first decrease in the frequency-magnitude distribution curve. Two earthquakes datasets (1990-2007 and 2008-2020) are then analyzed.

The b-value is obtained from the equation of the relationship between the magnitude and number of earthquakes formulated by [10], using linear equation:

$$\text{Log } N = a - bM \quad (1),$$

with N is number of earthquake, M is earthquake magnitude, a and b are constants depending of number of earthquake event. High b-value in an area indicates brittle rock. Additionally, decreasing b-value indicates increasing critical failure stress [11].

3 Results and discussion

Figure 2 displays similar hypocentre distribution pattern between two earthquake datasets, i.e. deeper hypocentre to the north. Earthquakes with a hypocentre between 0 - 100 km are predominantly in the ocean. In the contrary, earthquakes with a hypocentre of more than 100 km are predominantly on land.

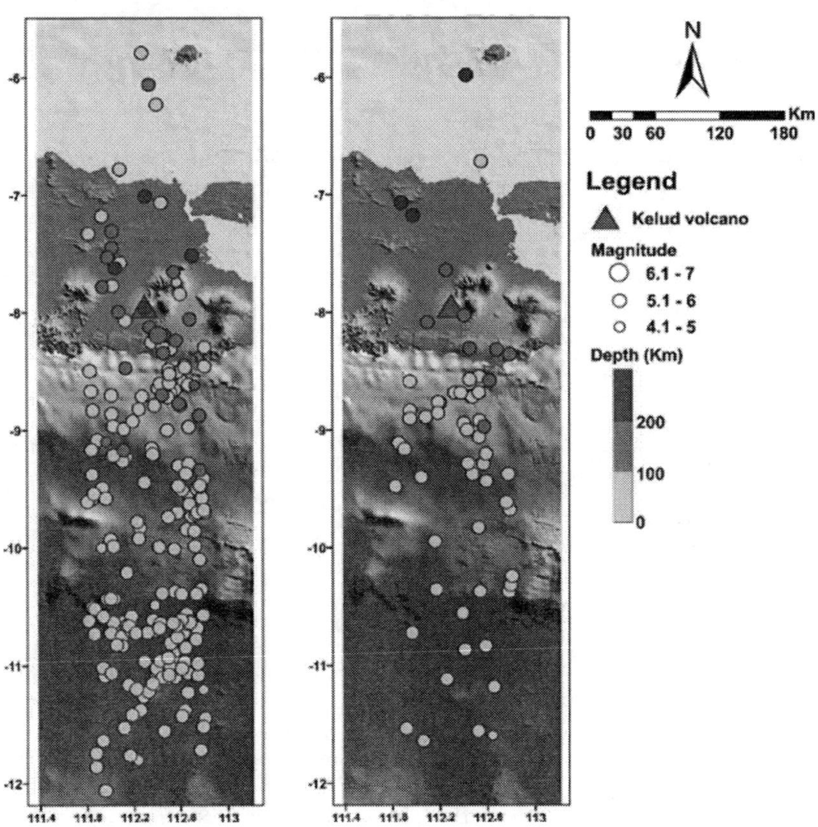

Fig. 2. Seismic events (circle) of the Kelud volcano (red triangle) and to the south until the Java trench. Left: from 1990-2007 earthquake dataset. Right: 2008-2020 earthquake dataset.

Figure 2 also shows the epicentre of two earthquakes locating close to the Kelud volcano. The first earthquake was occurred in 2006 with a depth of 232 km and a magnitude of 5.2. The second earthquake was happened in 2011 with a hypocentre depth of 165 km and a magnitude of 5.4.

For datasets 1990-2007 and 2008-2020; the b-value is 2.27 and 1.85, respectively. The b-value resulted in this study is higher than the b-values as referred by [12] for many areas of the world. The b = 1.01 as computed by [7] is smaller than our findings for dataset 1990-2007. For 2014-2019 data, the b-value 1.85 shows a decreasing (Figure 3).

Following [7], decreasing b-value around a volcano indicates accumulation of stress accompanied by instability of crack growth. From the temporal variation graph, it can be seen that at 2 different types of eruptions of the Kelud volcano, there are differences in the b-value of tectonic earthquakes. The b-value in the period of effusive eruption was greater than the period during the explosive eruption. This is probably related to the amount of energy that will be released. The same pattern of b-values also corresponds to the results of the study by [7], where the b-value before the explosive eruption in 1990 was smaller than the b-value before the effusive eruption in 2007.

The analysis of the b value in time does not change immediately before an eruption [3, 7, 8]. Therefore, the b value variations seem to be a long term precursor of the volcanic activity rather than a short precursor.

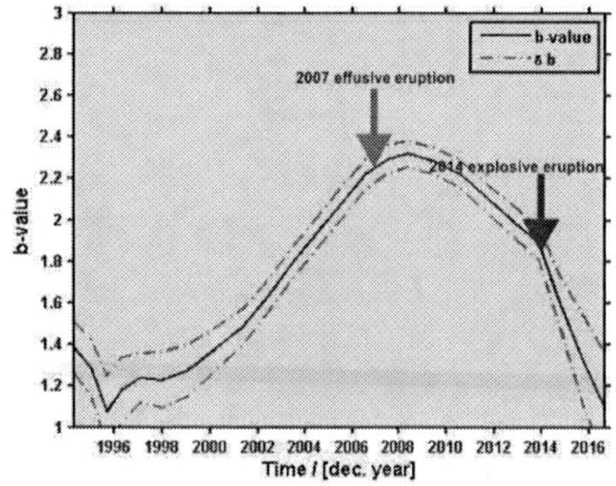

Fig. 3. B-value from 1990 until 2019 around Kelud volcano. Light blue arrow indicates when effusive eruption happened on 2007, while dark blue arrow points to explosive eruption on 2014. The explosive eruption on 2014 displays lower b-value compare to the effusive eruption on 2007.

A decrease in b-value is observed since 2009. It is possible that the decreases in b-value in this region result from small cracks due to change in the stress field near the magma chamber. In this way b-value calculations can give us more information about the physical process below the volcano. Given these findings we suggest that systematic evaluation of b-values become a more regular part of monitoring efforts. We also suggest analysing b-value using earthquakes data from local seismometer to double check and compare with b-value from earthquake data catalogue of USGS.

4 Conclusion

The b-value has been studied for earthquakes that occurred in the Kelud volcano region from 1990 to 2020. The b-value has been estimated that follows Gutenberg–Richter relationship. The b-values in 2007, when there was an effusive eruption, showed a value of 2.27, while in 2014 (when there was an explosive eruption) was 1.85. This result is in good agreement with previous study that found the b-value before the explosive eruption in 1990 was smaller than the b-value before the effusive eruption in 2007.

Decreasing b-value after 2009 indicates there is accumulation stress below the volcano. This decrease likely reflects a combination of increased crack length due to the growth of small faults and cracks related to the 1990 and 2014 eruptions. As a long term precursor, b-value variation is needed to calculate through time, as addition to volcano tectonic seismicity monitoring that gives short precursor.

The authors would like to thank to USGS for the data. We also want to thank to Faculty of Mathematics and Natural Sciences for the funding for this research.

References

1. S. Hidayati, H. Triastuty, I. Mulyana, S. Adi, K. Ishihara, A. Basuki, H. Kuswandarto, B. Priyanto, A. Solikhin, JVGR **382**, 50–67 (2019)
2. R. Cas, J. Wright, Geological Magazine **125**, 189–200 (1988)
3. F. Maeno, S. Nakada, M. Yoshimoto, T. Shimano, N. Hokanishi, A. Zaenudin, M. Iguchi, JVGR **382**, 24–41 (2019)
4. K. Aki, Bull Earthq Res Inst Tokyo Uni **43**, 237–239 (1965)
5. A. Kumar, S. Rai, A. Joshi, H. Mittal, R. Sachdeva, R. Kumar, V., Ghangas, Earthq Sci. **26**(2), 99–105 (2013)
6. T. Hirata, T. Satoh, K. Ito, K, Geophys J R Astron Soc. **90**, 369–374 (1987)
7. K. Brotopuspito, Wahyudi, Jurnal Berkala MIPA, **17**, 3 (2007)
8. A. Mantovani, S. D'Amico, M. Barbano, *The b value as a long term precursor of volcanic activity: a case study at Mt. Etna*, in Proceedings of the Convegno Nazionale GNGTS, November 2012, Potenza, Italy (2012)
9. M. Irsyam, M. Asrurifak, B. Hendriyawan, W. Triyoso, A. Firmanti, A., Geomech Geoengin Int J 5(1), 35–47 (2010)
10. B. Gutenberg F. Richter F., Seismicity of the Earth and its associated phenomenon (Princeton University Press. Princeton, 1954)

11. S. Rao, K. Lakshmi, K., Current Science **89**(9), 1582 (2005)
12. W. Lee, S. Stewart, Principles and application of micro-earthquake networks (Academic Press, New York, p 216, 1981)

 https://doi.org/10.1051/e3sconf/202132501020

The impacts of Seroja Tropical Cyclone towards extreme weather in East Nusa Tenggara

Andung Bayu Sekaranom[1,2,*], *Narastravika Henardi* Putri[2], and *Fatih Cinderaswari* Puspaningrani[1]

[1]Research Center for Disaster, Universitas Gadjah Mada, Yogyakarta-Indonesia
[2]Department of Environmental Geography, Faculty of Geography, Universitas Gadjah Mada, Yogyakarta-Indonesia

Abstract. This paper aims to discuss the Seroja Tropical Cyclone and its impact on extreme weather. Seroja tropical cyclones occur from April 4 to 5th 2021, in the East Nusa Tenggara (NTT) region. Based on data from the Meteorology, Climatology and Geophysics Agency (BMKG), the initial position of the Seroja tropical cyclone was in the Savu Sea, southwest of Timor Island. Since April 1, 2021, the NTT region has become the center of low pressure that triggers the formation of this cyclone. When a tropical storm occurs, the intensity of rainfall which initially reaches less than 60 mm/day, increases rapidly to more than 100 mm/day on April 4 to 5, 2021. This is the impact of the low-pressure center that triggers the formation of tropical cyclones in the region.

1 Introduction

Weather conditions in Indonesia are influenced by various factors, including tropical cyclones [1]. Tropical cyclones have a direct impact on the areas as well as the surrounding [2]. Tropical cyclones also cause flooding due to increased rainfall. Tropical cyclones have different effects on rainfall depending on the characteristics of the cyclones and environmental conditions [3]. Tropical cyclones are one of the atmospheric phenomena that occur due to transferring heat from the equator to the higher latitudes, characterized by low air pressure that triggers strong winds. Tropical cyclones can impact increased rainfall, flood disasters, and damage to infrastructure over coastal and land areas [4]. Tropical cyclones cause destructive winds to areas with a distance of 250 miles or 155 km from the cyclone center, with wind speeds reaching 150 miles/hour or 93 km/hour [5]. Tropical cyclones form over large oceans with a sea surface temperature of at least 26.5°C [6]. Tropical cyclones have varying durations ranging from a few hours to 14 days with an average of 6 days. The process starts from the tropical cyclone formation stage to the death stage when moving towards land or turning into subtropical areas [7].

The formation of tropical cyclones is supported by several atmospheric conditions, such as warm sea surface temperature, atmospheric instability, and high humidity in the lower to intermediate layers of the atmosphere. The Coriolis force also can maintain low pressure at its center, combined with pre-existing atmospheric disturbances and vertically low shear winds. Tropical cyclones develop near the equator and then move further from the equator. Tropical cyclones that form in the northern hemisphere move west or northwest, while tropical cyclones that form in the southern hemisphere move west or southwest. Tropical cyclones usually form over warm tropical waters [8].

Previous researches have identified some phenomena where tropical cyclone affect areas closer to the equator [1,3]. This is extremely rare condition since tropical cyclone often develop further, particularly at 10° latitude or higher. Therefore, there is a gap of reseach question why this tropical cyclone develop closer to the equator. This information is crucial in prediction the future condition whether tropical cyclone tends to develop close to the equator. This paper aims to identify the characteristics of Seroja Tropical Cyclones occurring from April 4 to 5, 2021, in the East Nusa Tenggara (NTT) region. Based on data from the Meteorology, Climatology and Geophysics Agency (BMKG), the position of Seroja Tropical Cyclones is in the Savu Sea, southwest of Timor Island. BMKG stated that this tropical cyclone caused an increase in rainfall around the cyclone and extreme weather in NTT. The increase in rainfall with moderate to high intensity causes flash floods in the region. Therefore, this study is important to develop disaster mitigation and adaptation plans in the future.

2 Data and method

The analysis presented in this paper is based on near-real-time data immediately after the Seroja Tropical Cyclone. The data used is mainly data from satellite-based rainfall estimation products. The use of satellite estimation rainfall data is selected because it can produce fast information immediately after the event occurs [9]. Although there are differences in rainfall

* Corresponding author: andungnbayu@geo.ugm.ac.id

based on estimates of weather satellites and meteorological stations due to differences in measurement principles, the results of these satellite rainfall estimates are not much different from meteorological station data [10,11]. This is supported by the characteristics of the study area, which are affected by the Southeast Asian Monsoon, where the rainy season occurs between November-May and the dry season occurs between June-October. In general, satellite-based rainfall estimates in Indonesia will be higher in areas affected by monsoons than in areas near the equator with two rain peaks and areas affected by local circulation [12-14].

The stages carried out in the analysis consisted of two steps, namely 1) Analysis of extreme rainfall and 2) Estimating the impact caused by the sacred tropical cyclone. Satellite rainfall estimation data from the GSMaP product (https://sharaku.eorc.jaxa.jp/GSMaP/) were used to analyze the impact of Seroja Tropical Cyclones on extreme rainfall. The product used is daily data derived from near-real-time products (GSMaP NRT). The analysis time used is from March 29, 2021, to April 11, 2021. The average rainfall calculation is based on landfalling precipitation in a bounding box between 118.5 to 125.5 degrees longitude and 7.5 to 12.5 degrees latitude. The estimation of the impact caused by the Seroja tropical cyclone was carried out based on the publication of data from the National Disaster Management Agency of Indonesia (BNPB) and releases from the mass media shortly after the cyclone occurred. Data processing at both stages is carried out using geographic information system (GIS) software.

3 Extreme precipitation due to the Seroja Tropical Cyclone

During the last seven days before Seroja Tropical Cyclone was formed, the East Nusa Tenggara (NTT) region tended to have a warmer sea surface temperature than the surrounding areas. This condition indicates the amount of convective activity that triggers the formation of rain clouds. Warm sea surface temperatures also support atmospheric conditions when tropical cyclones occur. Since April 1, 2021, the NTT region has become the center of low pressure. This is based on the air pressure in the NTT region ranging from 1006-1008 mb and continues to decline until it reaches the Tropical Depression of 994 mb on April 4, 2021. Seroja tropical cyclones, which have air pressure reaching 994 mb according to the Saffir-Simpson scale, are included in the minimal category. The maximum wind speed is estimated to be between 33-42 m / s. Seroja tropical cyclones fall into the 1st category where the wind is classified as dangerous and can cause damage to the exterior of a well-constructed house, uproot small trees, break tree branches, and cause damage to the power grid.

As the wind moves from the center of high pressure to the center of low pressure in the south part of the study area, the center of low-pressure depression then turns into a tropical cyclone. The lower air pressure makes air masses tend to gather, resulting in heavy rains and the potential to cause flooding in the NTT region. BMKG stated that this sacred tropical cyclone caused an increase in rainfall around the cyclone and extreme weather in NTT. The increase in rainfall with moderate to high intensity causes flash floods in NTT. This is also indicated by the high amount of rainfall on April 4 to 5, 2021, based on the Global Satellite Mapping of Precipitation (GSMaP) rainfall data, as shown in Figure 1.

Figure 1 shows that the rainfall in NTT is high during the Seroja Tropical Cyclone. It can be seen that the rainfall has continued to increase during the last few days before the occurrence of a tropical cyclone, but this increase is not significant and volatile. When a tropical cyclone occurs, the intensity of rainfall which initially reaches less than 60 mm/day, increases rapidly to reach more than 100 mm/day on April 4 to 5, 2021. This happens as an impact of the low-pressure center that triggers the formation of cyclones. Tropical cyclones then move away from the study area and reduce their impact on rainfall starting April 6, 2021. From April 6, 2021, rainfall seems to have decreased even closer to 0 mm/day (no rain). This shows that the cyclone has moved away from the region where rainfall has returned to a significant decline. Spatially, the distribution of rainfall can be seen in Figure 2 and Figure 3.

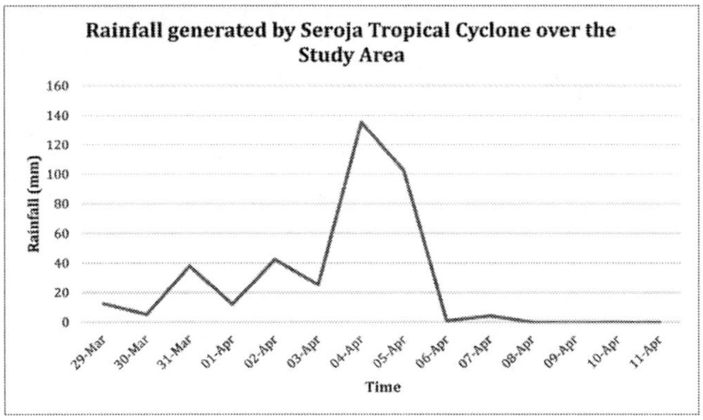

Fig. 1. Graph of rainfall in East Nusa Tenggara (source: results of GSMaP rainfall estimation data processing in 2021)

 https://doi.org/10.1051/e3sconf/202132501020

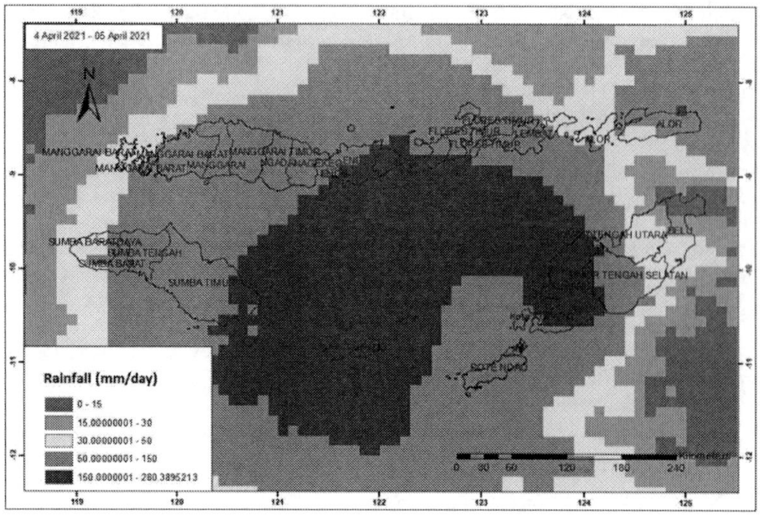

Fig. 2. Spatial distribution of rainfall in East Nusa Tenggara, April 4-5, 2021 (source: results of GSMaP rainfall estimation data processing for 2021)

From Figure 2, it can be seen that on April 4 to 5, 2021, high to very high concentrations of rainfall occurred in districts/cities adjacent to the Savu Sea as the center for the formation of the tropical cyclone. Districts with very high rainfall are Sabu Raijua, Kupang, most of Sikka and Ende, and a small part of East Sumba and South Central Timor. Districts / cities with high rainfall are Rote Ndao, Kupang, East Flores, Ngada, East Manggarai, Manggarai, Central Sumba, West Sumba, most of West Manggarai, Lembata, South Central Timor, East Sumba, a small part of Ende and North Central Timor District.

The Seroja Tropical Cyclone that begins to move away to the southwest as associated with decreasing rainfall, as shown in Figures 3, impacts rainfall in the area around the center of the cyclone formation. In contrast to Figure 2, the rainfall on April 6 to 7, 2021, is classified as low and even very low with a rainfall intensity of less than 15 mm/day. During that time, the cyclone had moved to the southwest, away from the NTT region. This is because tropical cyclones in the southern hemisphere tend to move west or southwest and away from the equator. Rainfall on April 6 to 7, 2021, can be seen spatially in Figure 3.

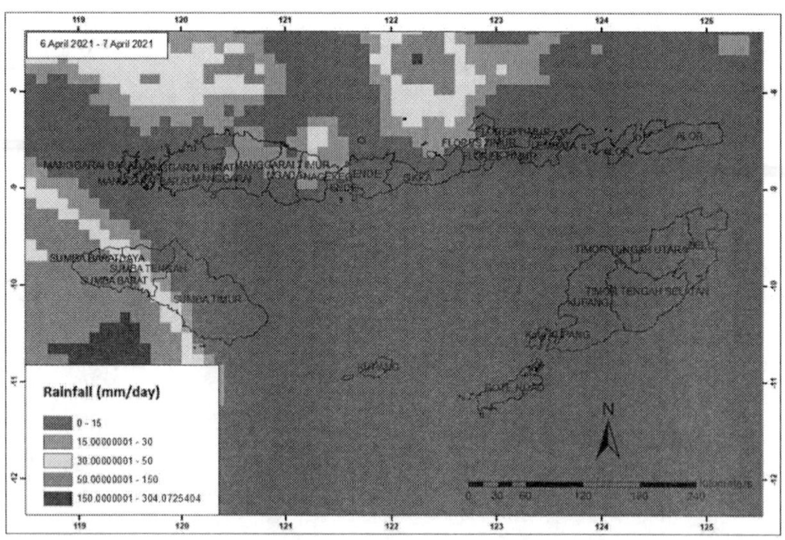

Fig. 3. Spatial distribution of rainfall in East Nusa Tenggara, April 6-7, 2021 (source: results of GSMaP rainfall estimation data processing for 2021)

Based on Figure 3, most rainfall in the NTT region shows a very low rainfall intensity (0-15 mm / day). In some areas, including Sumba Tengah, Manggarai Tengah, Manggarai Timur, and Ngada, they have low to moderate rainfall. High rainfall with an intensity of 50-150 mm / day is only found in parts of Southwest Sumba and West Sumba. This indicates that the Seroja Tropical Cyclone has a considerable impact on weather conditions in NTT. Rainfall after a cyclone is classified as low and has a much different intensity than at the cyclone time. As seen in Figure 5, from April 6 to 7, 2021, high to very high rainfall is in the southwestern region of NTT. It then moves following the movement to the southwest.

4 Damage estimation

Apart from heavy rains, cyclones can also cause storm surges and winds that lead to disasters [15]. Tropical cyclones cause enormous impacts such as strong winds and heavy rain, which can cause flooding, high waves, and storm surges or storm surges in the areas they pass through. According to BMKG, Seroja tropical cyclones have the opportunity to cause sea waves to be more than six meters high in the southern part of the study area. Districts/cities affected by the tropical cyclone of Seroja according to BNPB include the City of Kupang, West Sumba, East Sumba, Kupang, South Central Timor, Belu, Ngada, Alor, Lembata, East Flores, Sikka, Ende, Rote Ndao, Sabu Raijua, and Malacca Regency. More details on the areas affected by the Seroja tropical cyclone can be seen in Figure 4.

Seroja tropical cyclones cause rain to occur continuously and result in flash floods in East Nusa Tenggara (NTT). The disaster caused significant casualties, damage to buildings, and economic losses. Due to this disaster, the number of victims has continued to increase for about a week since the disaster. Based on data compiled from publications by the official BNPB and several other mass media sources, as of April 18, 2021, 181 people died, 225 injured victims, and 48 missing victims. Meanwhile, 13050 people were displaced, with 428,986 affected residents. The number of dead, missing, and displaced victims in detail can be seen in Table 1.

Seroja tropical cyclones, apart from causing casualties, also caused damage to buildings in various locations. Based on data collected from BNPB as well as information from online mass media, up to April 18, 2021, it is estimated that the number of damaged houses in the NTT region will reach more than 66,000 housing units. 17,124 units of houses were severely damaged, 13,652 units moderately damaged, and 35,733 units slightly damaged. The composition of severely damaged houses, moderately damaged, and slightly damaged were 26%, 21%, and 54% of the total houses affected by the disaster, respectively, or can be seen in Table 2.

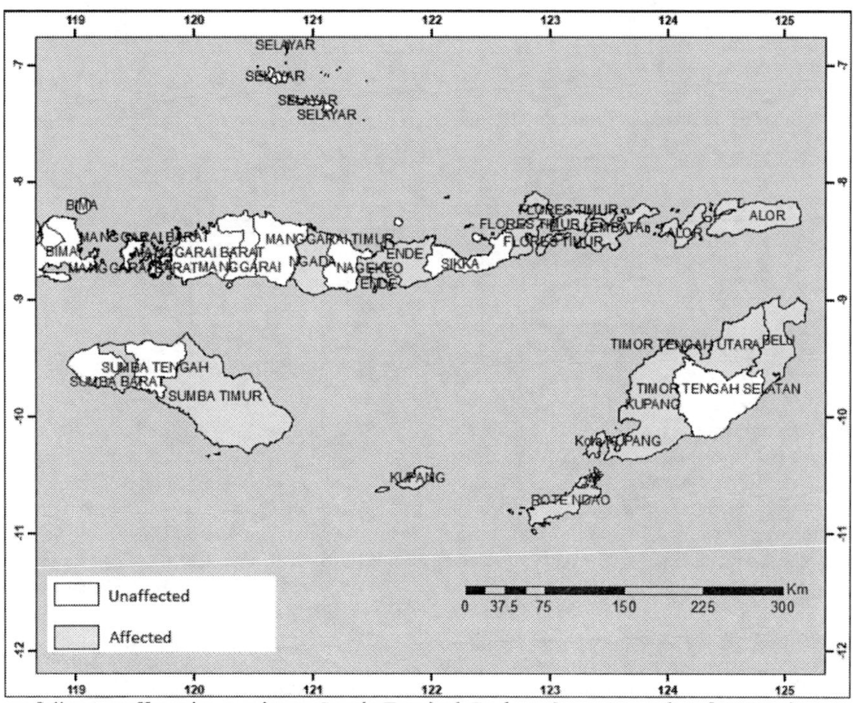

Fig. 4. Map of disaster affected areas due to Seroja Tropical Cyclone (source: results of BNPB data processing in 2021)

 https://doi.org/10.1051/e3sconf/202132501020

Table 1. Number of affected person due to Seroja Tropical Cyclone (source: data processing of BNPB and mass media reports)

Districts/Cities	Person affected			
	Death	Injured	Losses	Evacuated
Kota Kupang	6	8	1	265
Kabupaten Sumba Barat	-	-	-	284
Kabupaten Sumba Timur	-	-	-	570
Kabupaten Kupang	12	26	3	1698
Kabupaten Timor Tengah Selatan	-	-	-	690
Kabupaten Belu	-	-	-	644
Kabupaten Ngada	-	3	-	-
Kabupaten Alor	28	25	13	-
Kabupaten Lembata	46	53	22	1146
Kabupaten Flores Timur	72	76	2	2118
Kabupaten Sikka	1	-	-	-
Kabupaten Ende	1	-	-	20
Kabupaten Rote Ndao	1	34	-	5556
Kabupaten Sabu Raijua	3	-	7	59
Kabupaten Malaka	11	-	-	-
Total	**181**	**225**	**48**	**13050**

Tabel 2. Number of damaged building due to Seroja Tropical Cyclone (source: data processing of BNPB and mass media reports)

level of damages	Unit	Percentage (%)
High	17124	26
Medium	13652	21
Low	35733	54
Total	**66509**	**100**

The damage to buildings caused by the cyclone is limited to houses belonging to residents and facilities and infrastructure for residential areas, including damage to roads, bridges, electricity networks, telephone and internet networks. In addition, there was damage to other buildings, both private and government-owned. However, the condition of the houses and other buildings is not yet known in detail the extent of the damage so that the number of material losses from building damage cannot be ascertained. In Kota Kupang, the loss was estimated at 47 billion rupiahs. In Lembata Regency, at least 70 families lost their agricultural land, lodgings in plantation locations, and dozens of livestock that were swept away by the floods so that the losses were estimated at hundreds of millions of rupiah. In Kupang Regency, the fisheries and tourism sectors suffered considerable losses, including failed seaweed harvests reaching 7 billion rupiahs, damage to fishing boats reaching 780 million rupiahs, and damage to tourism facilities reaching 100 million rupiahs. The losses caused by the tropical cyclone of Seroja in the Rote Ndao Regency have primarily occurred in the productive economic sector, including agriculture and fisheries. In the agricultural sector, the damage to rice fields covering an area of 1800 hectares is estimated at 71 billion rupiahs. In the fisheries sector, damage to docks and boats and lost boats reached 141 billion rupiahs.

The data above shows an example of the significant impact of a catastrophic event caused by a tropical cyclone. Although the center of a tropical cyclone is not on land, the tail of the sacred tropical cyclone brings extreme rainfall. The analysis results show that during the formation of Seroja tropical cyclones, rainfall reached extreme levels above 100 mm/day for two consecutive days. Compared with the average monthly rainfall during the rainy season, which is generally only 150 to 300 mm/month, the total rain received during these two days almost matches the monthly rainfall. This, of course, has an impact on the surface hydrology system, which very rarely receives rainfall of this magnitude. In this case, the characteristics of the watershed also contributed to the occurrence of flash floods, which resulted in the death toll of more than 200 people and economic losses of up to billions of rupiahs.

5 Conclusions

Disaster events induced by tropical cyclones in Indonesia still have the potential to occur in the future. Although tropical cyclones tend to form in areas about 10° to 15 ° degrees from the equator and move away, their tails can carry enough moisture to trigger extreme rain, likewise for the case of Seroja Tropical Cyclone, which formed around the East of Rote Island about 10°S in latitude. Several tropical cyclones formed at a latitude of about 10°S have also impacted the occurrence of flash floods and landslides in southern Indonesia, for example, Cempaka and Dahlia Cyclones in 2017.

Tropical cyclones can form due to an increase in sea surface temperature. As a result, the air temperature in the lower atmosphere is getting warmer, followed by an increase in humidity which triggers a low pressure center followed by high rainfall. Climate change in the form of a global temperature increase will theoretically result in a lower frequency of tropical cyclones due to a reduced temperature gradient between the air near the surface and the layer above it. However, the strength of cyclones (in the form of wind speed and rainfall) will increase over time because warmer waters contain a more significant energy content in the form of heat. The potential for the formation of tropical cyclones is also predicted to shift closer to the equator as the earth's temperature increases due to climate change [12].

The impact of potential disasters due to tropical cyclones needs to be minimized. An example is by carrying out disaster risk mapping and educating people in disaster-prone areas. The first for coastal areas is the danger of storm surge due to tides accompanied by increased wind speed. The high wave early warning system is beneficial so that local populations/fishermen stay away from activities around the coast. The existence of coastal protection plants is also beneficial for reducing increased abrasion due to storm surges. For areas further inland, extreme rainfall is a significant concern. Extreme rainfall can cause flooding and trigger landslides, such as in the 2017 Cempaka tropical cyclone in Pacitan - Central Java. Mitigation efforts, in this case, need to pay attention to watershed ecosystems. They were mainly related to forest conservation which has the function of water regulation and preventing landslides.

This publication is supported by grants of "Hibah Desentralisasi Penelitian Dasar Perguruan Tinggi (PDUPT)" by Directorate of Higher Education, Indonesian Ministry of Education. The grant entitled "Optimization of Rainfall Estimation Products based on Characterization of Precipitation Organization" (contract number 1632/UN1/DITLIT/DIT-LIT/PT/2021) with Dr. Andung Bayu Sekaranom as principal investigator.

References

1. S. H. J. Tongkukut, W. L. Kawuwung, S. Sutikno, Jurnal MIPA Universitas Sam Ratulangi, **8**, 21-27 (2019)
2. N. S. Haryani, Zubaidah, Jurnal Ilmiah Widya, **29**, 54-58 (2012)
3. R. Prasetya, A. As'ari, W. Dayantolis, Jurnal Fisika Dan Aplikasinya, **10**, 1-9 (2014)
4. D. Surinati, D.A. Kusuma, Oseano, **43**, 1 – 12 (2018)
5. A. Suryantoro, Majalah Sains dan Teknologi Dirgantara, **3**, 21-32 (2008)
6. M.D. Syaifullah, Jurnal Sains dan Teknologi Modifikasi Cuaca, **16**, 61 – 71 (2015)
7. N. Dyahwathi, S. Effendy, E. Adiningsih, Agromet, **21**, 61-72 (2007)
8. A. M. Dewi, A. Kristianto, Jurnal Meteorologi Klimatologi dan Geofisika, **4**, 8-16 (2017)
9. A. B. Sekaranom, H. Masunaga, J. Appl. Meteorol. Climatol., **56**, 1867–1881 (2017)
10. A. B. Sekaranom, H. Masunaga, J. Appl. Meteorol. Climatol., **58**, 37–54 (2019)
11. A. B. Sekaranom, E. Nurjani, M. P. Hadi, M. A. Marfai, Quaest. Geogr., **37**, 97–114 (2018)
12. A. B. Sekaranom, E. Nurjani, R. Harini, A. S. Muttaqin, Indones. J. Geogr., **52**, 143–153 (2020)
13. A. B. Sekaranom, E. Nurjani, *The development of Articulated Weather Generator model and its application in simulating future climate variability, in proceeding of* IOP Conference Series: Earth and Environmental Science, **256** (2019)
14. A. B. Sekaranom, E. Nurjani, I. Pujiastuti, *Cloud structure evolution of heavy rain events from the East-West Pacific Ocean: A combined global observation analysis*, in proceeding of IOP Conference Series: Earth and Environmental Science, **148** (2018)
15. H. Xu, X. Zhang, X. Xu, Advances in Meteorology, **13**, 1-13 (2013)

Understanding Community Collective Behaviour Through Social Media Responses: Case of Sunda Strait Tsunami, 2018, Indonesia

Farah Pasha Salsabilla[1], *Dyah* Rahmawati Hizbaron[1*]

[1]Department of Environmental Geography, Faculty of Geography, Universitas Gadjah Mada, 55281, Bulaksumur, Yogyakarta Indonesia

Abstract. The Sunda Strait Tsunami which occurred on 22nd December 2018 is one among too many examples of a rapid on-set disaster that attracted public attention through Twitter. This sudden event had a massive impact on parts of the west coast of Banten Province, Indonesia. Therefore, this research aimed to evaluate the collective response reflected on Twitter due to the 2018 Sunda Strait Tsunami. Previous studies shows the utilization of crowd sourcing data from social media for community capacity and quick assessment of disaster impacts. Therefore, the characteristics of people's responses on social media based on spatio-temporal attributes needs to be understood first to build better understanding about the information that can be used for emergency response strategies consideration. This research method involved a spatial statistics approach, while data collection and descriptive analysis were carried out based on Twitter word cloud data. This analysis showed that temporally, the highest number of tweets was generated at the beginning of the disaster period with downward trend into the end of phase. As for spatially, people in directly affected areas by the disaster tend to give negative sentiments as their expression of sadness and fears towards the disaster. The content of the tweets involved asking for help, reporting on the current situation, and confirming the news on accounts belonging to government agencies. Furthermore, people in areas that were not directly affected produced tweets with more positive sentiment with expressions of condolences, sympathy, gratitude and invitations for volunteers and social actions.

1 Introduction

On December 22, 2018, at 21.03 Local Time or 14.03 UCT, a tsunami disaster with waves reaching 13.5 meters struck the coastal areas of South Lampung, Banten, and the surrounding islands of Anak Krakatau Volcano. The Tsunami post-survey at Banten showed 5.8 meters run-up with an inundation distance up to 330 meters away[1]. The hardest-hit area was the west coast of Banten Province and the southern coast of Lampung Province[2]. However, the Tsunami also caused an impact to the surrounding islands, such as Sertung, Rakata, and Panaitan Islands located around the Anak Krakatau Volcano. This tsunami wave was eventually identified to have been triggered by the volcanic flank failure at the Anak Krakatau Volcano[2][3] which created sub-aerial flanks[4]. Banten was likely less inundated due to the existing trace of coastal defense along the seaside.

The western coastal area of Banten was hit by tsunami waves without any significant warning, causing many casualties and damage. Figure 1 indicated some areas damaged due to the disaster occurrence, especially at Pandeglaged and Serang District, Banten, West Java, Indonesia (Figure 1). It caused a 437 death toll, 1,143 people injured, and 5,361 displaced. Furthermore, physical damage to 558 houses, 8 hotels, 60 stalls and 350 boats occurred[5]. The infrastructure was also disconnected which caused damage to 248 power stations and 42 electricity poles[6]. Sumur Subdistrict in Pandeglang District, was the hardest hit area, which was indicated by a significant change in land use from settlements and vegetation cover to open land[7]. Additionally, the impact of the tsunami also caused the cutting off of several road accesses and isolated areas to Pandeglang and Serang.

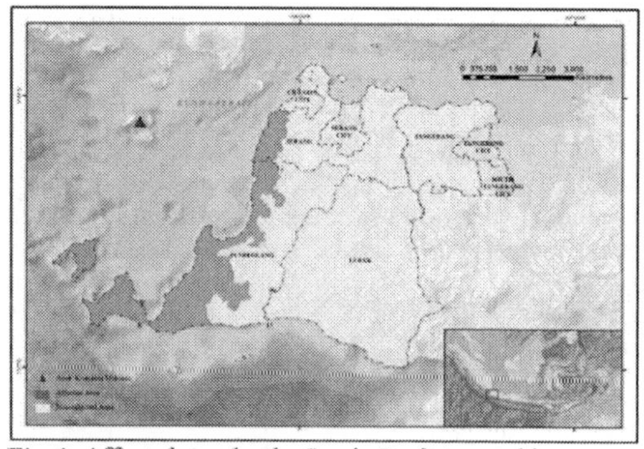

Fig. 1. Affected area by the Sunda Strait tsunami in Banten Province, Indonesia

* Corresponding author: dyah.hizbaron@ugm.ac.d

 https://doi.org/10.1051/e3sconf/202132501021

To reduce the impact of disasters, information sourced from the community at the scene of the post-disaster situation is very important in disaster response and recovery[8]. Therefore, as a product of technological developments in the field of communication, social media has advantages in terms of real-time data availability and unlimited audience reach. Its popularity and ability in information dissemination renders it a potential channel in disaster management communication involving a wide audience, and the provision of valuable sources of information for rapid assessment. Social media for emergency response management is widely used for distributing the latest information on disasters, building public awareness, and raising moral and material support[9].

The emergency response shown by a community is an important social dimension of disaster management. The collective response pattern formed will add references for policymakers in considering a more effective emergency response strategy by involving community dynamics. This study aims to evaluate the response characteristics of the 2018 Sunda Strait Tsunami, specifically in the area of Banten, Indonesia. Furthermore, the research objective was based on utilizing crowdsourcing data from Twitter, processed and analyzed in a spatio-temporal perspective. It applied social media data to capture the local capacity during an emergency. As the scientific documents on the subject of disaster capacity due to Sunda Strait Tsunami were not sufficient for capture, this research takes a contributing initiative.

2 Research Framework

As communication technology interventions are increasingly attached to the life of modern society, people tend to be more open in sharing their opinions and activities on social media. This underlies the research idea of spatial-temporal based crowdsourcing which sees social media as a potential data source to be utilized, including for the implementation of disaster emergency response. Previous research was conducted to evaluate the preparedness of school students to the Tsunami at Banten Province, and ascertained a low disaster response capacity[10]. Furthermore, in times of disaster emergency response there is always a certain marginalized group which cannot be reached, as in the case of Japan Tsunami dated back at 2011[11]. How does social media enable us to capture on-site and off-site responses which supported the local capacity to deal with disaster? Previous research relevant to the subject in the case of urban response to tropical cyclone using Twitter indicated that the response varied according to the emergency phases[12]. Therefore, this research highlights the importance of spatial temporal data from crowd sourcing that reflects the inclusivity of responses in dealing with Tsunami.

Information on the tsunami disaster quickly spread on various communication platforms such as Twitter, which triggered various responses from users in various regions. When a disaster occurs, people tend to be more active in using social media to exchange information[13]. Nowadays, it has significantly transformed from its original needs for personal communication[14] to become part of people's daily lives[15]. It has therefore been considered sufficient to describe social dynamics[16]. Social media enables the communication of various responses uploaded by users during the disaster period. The collection of reactions of each individual on social media illustrates the collective response pattern of the community towards disasters, both in the affected and non-affected areas. Furthermore, understanding the community's emergency response was carried out at a collective level, which aims to understand the general response patterns of community groups, especially in Banten Province which are actively using Twitter. The determination of collective response of the community when facing a disaster needs to be done as an element of social dimension of the community to facilitate its use as a reference in the formulation of an effective disaster emergency response strategy.

Twitter is one of the most frequently used platforms for monitoring natural disasters[17]. As a micro-blogging social media, Twitter has the advantage of presenting concise and actual information. It is also equipped with a post topic sorting feature based on hashtag (#), interaction via retweets, replies and likes, attachments of photo and video media, time information and a geo-tag feature that shows the location of the tweet being uploaded, which are important keys in exploring the public's response to data. The community utilizes this social media by sharing information such as notifications of disaster events, increasing awareness and sensitivity, reporting on actual conditions at the disaster site, building emotional support, forming a collective movement to help the needs of the affected community, and information on the mitigation and evacuation process[18][9]. In addition, community proactivity at the scene of the incident is considered a sensor that shares actual information continuously, becoming the basis for the use of crowd-sourcing data in disaster management. Information shared from the community in the field provides a big picture of the actual conditions due to disasters. This significantly influences decisions on emergency response by elements in authority[19].

The response formed immediately after a disaster significantly influences the success of disaster management[20]. In this case of Sunda Strait tsunami which had rapid on set characteristics without early warning announcements, the capability of the community and qualified policy makers to respond and act quickly when the disaster occurred affected the effectiveness of disaster management compared to only depending on existing structural mitigation[21]. The development and intervention of information technology is inevitable, therefore the development of methods by integrating technology has the potential to bring progress in the application of disaster emergency responses that require time efficiency.

Destructive events such as disasters disrupt normal stability which then becomes a stimulant for the formation of collective response in a community during the emergency phase. In this phase, emergency response is emphasized in efforts to prevent the worsening impact of a disaster. A series of emergency response actions include activities to rescue and evacuate victims, property, fulfilment of basic needs, protection, handling of refugees, and the restoration of vital facilities and infrastructure. Furthermore, emergency response is influenced by the knowledge, capabilities and competence of each element involved in disaster management efforts, especially in crisis management when a disaster occurs[22]. The successful implementation of preparedness strategies in emergency response is mainly influenced by information, communication, and the availability of resources[23]. Understanding the characteristics of a community, including collective responses when facing disasters and their interactions with the environment, is an important aspect of disaster risk reduction[24][25]. The collective action underlies the importance of local capacity to face disaster[26].

The collective response of the affected community is manifested from a series of activities and roles, resources, and authority that the community[27]. The roles and activity aspect involves a series of actions taken immediately after a disaster occurs, and emphasizes community efforts to minimize the impact of a disaster. The second aspect relates to resource management, explaining how a community manages the available resources to deal with crises during the emergency response period, including resource allocation and distribution. The third aspect, specifically the role of the authority explains the form of the contribution of the competent agency in overcoming disasters, which is shown by the deployment and involvement of disaster management personnel, provision of basic living necessities, and activities that support the acceleration of disaster impact recovery.

Collective response determines the collective behavior of the community. The grassroot movements build-up to pursue a collective goal during crises, and involve the actions carried out to survive through the disaster[28]. This pattern shown by a community when facing disasters demonstrates their capability in reducing, absorbing, adapting and accelerating efforts to recover from the impact of disasters[29], or exacerbating the impact and hindering the post-disaster recovery process[30]. Meanwhile, disaster management is much more effective when the community organizes an emergency response quickly and accurately compared to only relying on existing physical mitigation structures[21]. An understanding of the collective behavior of a community is used by policymakers to determine the efficiency of preparedness strategies and emergency response, evaluate and allow for policy reformulation to increase community resilience to disasters in the future[31][32].

Alongside technological developments, collective response patterns can be reflected by a community on social media, especially when there are events that have a massive impact such as disasters[33]. Social media has currently transformed from what was originally intended for personal communication needs to become part of people's daily lives[15]. Therefore, the information in it represents the activities and social dynamics of society[34]. Previous research showed that there are differences in the collective responses of people in affected areas and the ones not directly affected by the disaster. Social media users which are closer to the disaster scene tend to provide valuable information[35][36]. In affected communities, volunteers from cyberspace communities use social media to connect and contribute as portals and information passers[37]. This is the concept of a spatially based Voluntary Geographical Information (VGI) data source by utilizing information from public uploads as field sensors.

Twitter is one of the most frequently accessed social media not only widely used during emergency response situations but also changes the way people create, identify, and re-share disaster information[38]. This is a social media with micro-blogging principles that sufficiently and concisely presents data. The social media also offers a hashtag (#) feature to facilitate sorting topics, photo, and video attachments, to the source attribute where the tweet was uploaded via geo-tagging which is crucial in VGI data for spatial analysis of disasters that can be activated by users. The features offered by Twitter social media can be used for disaster analysis, especially during the response or emergency phase. Furthermore, information in form of responses or public opinions represented on social media were generally categorized into three main aspects, namely spatial, temporal and content[39]. The spatio-temporal analysis is related to the time attribute and geo-tag attached to the tweet. The tweet content can also be analyzed through text-based data processing, which is manifested in sentiment analysis and wordcloud form.

3 Method

3.1 Data Collection

Data crawling techniques were used to obtain raw data in form of tweets related to the topic of the Sunda Strait tsunami. Some restrictions were applied to obtain the results of the raw tweet data following the research objectives, namely through the language used, keywords, and time frames. Three keywords in Indonesian were used to find suitable topics and for narrowing the search range specifically for Indonesian-language tweets. The keywords used were "Tsunami Selat Sunda", "Tsunami Banten", and "Krakatau". This selection was based on keywords that were often used by government agencies and the mass media in reporting the progress of the 2018 Sunda Strait tsunami disaster[12]. The data crawling process was carried out during the Sunda Strait tsunami emergency response period, namely 22 December 2018 - 4 January 2019. The data crawling process was carried out using the Python programming language with the Twint module, which generated attributes including username, date and time, tweet content, tweet location, interaction counts, media attachments and hashtags.

3.2 Data Pre-processing

The pre-processing stage is important in text data processing techniques and is based on Natural Language Processing. Raw data tweets are unstructured data that require adjustments to facilitate the process of understanding the tweet contents. Tweets with geo-tag attributes originating from Banten Province were separated for further processing through the reduction and stemming stages. The reduction stage was carried out to duplicate tweets, unnecessary punctuation and link removal. The stemming process was performed for uniformity in lowercase and spelling improvements without changing the core meaning of the tweet. Meanwhile, the reduction and stemming processes were carried out using the Python programming language.

3.3 Data Processing

The identification of the characteristics of tweet sentiment that have gone through pre-processing was carried out through sentiment analysis with the Naïve Bayes algorithm. This classification works by calculating the odds predictive, based on probabilistic statistical calculations that refer to previous data reading experiences. The Naïve Bayes algorithm separates each attribute in the data and assumes all of them to be independent variables. Furthermore, this study applies supervised learning to the Naïve Bayes algorithm in classifying sentiment categories labeled positive, neutral and negative for each tweet. The classification process consisted of making an algorithm model using the python programming language at the data training stage and testing the model through data testing. A total of 90% or 327 tweets were used for training data, while 10% or 36 tweets for testing the Naïve Bayes classification. Each tweet was quantified by its polarity weight indicating positive (1), negative (-1) or neutral (0) emotion.

The sentiment characteristics were reviewed from a spatial and temporal perspective to determine the dynamics of community response. From the temporal perspective, the emergency response period was divided into 3 phases, namely phase 1 on December 22-24 2018, phase 2 on December 25-30 2018 and phase 3 on December 31 2018 - January 4 2019. This distribution was based on fluctuations in the number of daily tweets that indicated a shift in the locus of public attention on Twitter in response to the Sunda Strait tsunami. The geo-tag attribute embedded in each tweet was able to refer to the location up to the sub-district level. The location information contained in the geo-tagged attribute signified the location from which the tweet was uploaded, and was analyzed spatially through a Volunteered Geographical Information (VGI) based map via ArcGIS. The spatial analysis emphasized the tweet characteristics of the affected area compared to the non-affected area, and the distributions were determined through the data released by Regional Disaster Management Agency (BPBD) in Banten Province.

The response analysis was based on the tweet contents presented in wordcloud. The words that appear in wordcloud represent the most frequently used words by the community in responding to the Sunda Strait tsunami on Twitter. A larger size of a word presented in wordcloud signified a more frequent use of that word by Twitter users. Understanding the information presented in the worcloud was performed by linking the interrelationships between unstructured words that often appear, as an indication that represents the focus of people's attention at that time.

4 Results and Discussion

The tsunami disaster that occurred in December 2018 became one of Indonesia's Trending Topics at the time through the hashtags #PrayforBanten and #PrayforAnyer. These trending topics showed that the Sunda Strait tsunami received great attention from Twitter users with many related topic interactions. Some users were observed to use Trending Topics to raise irrelevant topics and embed hashtags or keywords related to the Sunda Strait tsunami. Therefore, the process of sorting tweets through reduction and stemming was highly necessary to obtain data following the research objectives. The data crawling process and data pre-processing on 3 keywords "Tsunami Sunda Strait", "Banten Tsunami" and "Krakatau" on Twitter from 22 December 2018 - 4 January 2019 produced 842 Indonesian-language tweets that were known to have geo-tagged attributes. Furthermore, about 43% or 363 tweets were known to come from Banten Province.

4.1 Temporal Characteristics of Tweet Sentiment

The sentiment dynamics of the Banten Province community's tweets towards the disaster are shown in graphical form (Figure 2). Tweets with positive, negative, and neutral sentiment generally showed a pattern of significant increase at the beginning of the period (Phase 1) which then declined over time. The decrease in the number of tweets represents a decrease in public attention over time compared to the initial phase of the emergency response period.

https://doi.org/10.1051/e3sconf/202132501021

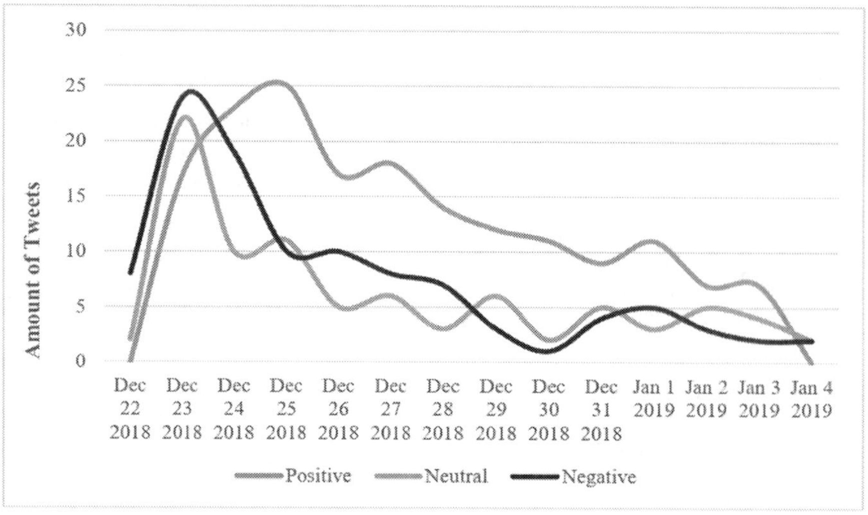

Fig. 2. Temporal distribution of tweet sentiments during the Sunda Strait tsunami emergency phase

Phase 1 (22 - 24 December 2018) showed the concentration of the number of tweets with the highest positive, neutral and negative sentiments compared to other periods. Tweets with negative and neutral sentiments reached their peak on 23 December 2018. Tweets with negative sentiments dominated this period, signifying that the tsunami disaster provided stimulation which was reflected through negative perceptions by the public. The negative sentiment was shown through tweet responses with negative contents such as panic, fear, and sadness over the occurrence of the tsunami.

Tweets with positive sentiment were dominant in phase 2, specifically December 25-30 2018, while tweets with negative and neutral sentiments began to decline. This condition may indicate the response of the community that attempts to gather support and recover from the impact of the disaster. The increase in positive sentiment tweets became a stimulant for community integration through social media which in turn influences actions on the ground. This was demonstrated through the formation of many solidarity actions to help communities in affected areas, both in form of invitations to raise donations, charity events, and volunteering.

Phase 3 started on December 31, 2018 - January 4, 2019, showed fluctuations in the number of tweets that were positive, neutral or negative, but generally had a downward trend. Negative and neutral sentiment had slightly increased on 31 December 2018, alongside the momentum ahead of the New Year. The New Year's Day became a locus of new topics that attracted the attention of the public, associated with reflective attitudes and optimistic efforts regarding the occurrence of the Sunda Strait tsunami disaster. The number of tweets with positive sentiment were still generally dominant in this phase, which showed that public support on social media for the efforts to recover from the impact of the Sunda Strait tsunami was still continuous.

4.2 Spatial Characteristics of Tweet Sentiment

The spatial perspective based on the location attributes contained in each tweet showed that as many as 143 out of 363 tweets were uploaded from areas directly affected, covering 13 sub-districts from 2 regencies, namely Pandeglang and Serang (Figure 3). The post-disaster conditions were not yet conducive to sufficiently grab the attention of people in the affected areas to be more focused on efforts to evacuate themselves. The areas with rural characteristics were also known to have lower internet penetration rates [40]. This was due to various factors ranging from the capabilities of human resources to communication technology, capital that supports network involvement such as devices, to the availability of existing communication network infrastructure in general. The communities in the directly affected areas which were assumed to have predominantly rural characteristics were generally characterized by closer ties and social interactions in the real world[41]. Furthermore, the use of social media in their daily lives is not as intensive as people with urban characteristics.

Meanwhile, the areas that were indirectly affected by the tsunami generated more tweets during the emergency response period, namely 220 tweets spread across Tangerang City, South Tangerang City, Cilegon City, Serang City, Lebak Regency, and Tangerang Regency. This spatial pattern generally showed that areas with urban characteristics tend to be more reactive on social media in response to an issue[12]. This is supported by technological literacy capabilities and more adequate communication infrastructure capital.

https://doi.org/10.1051/e3sconf/202132501021

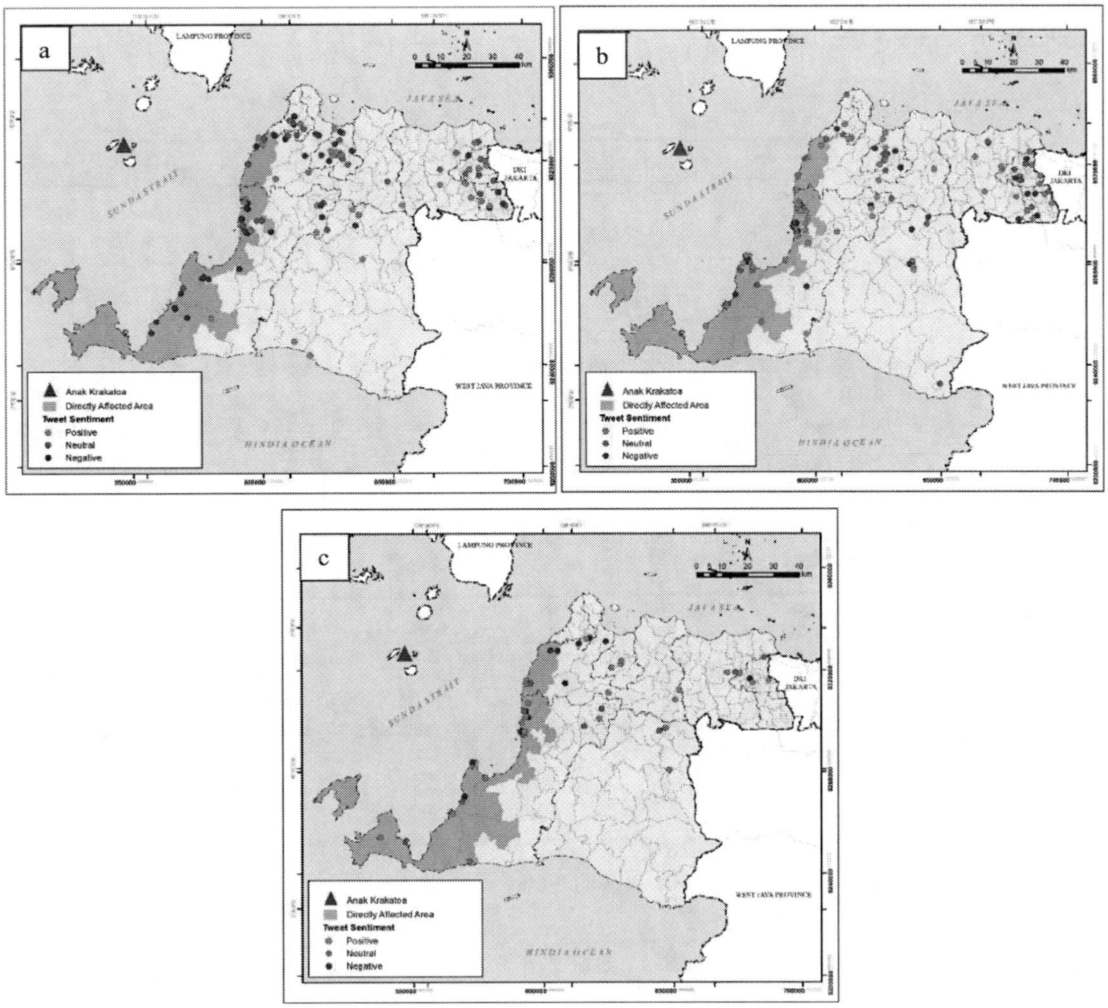

Fig. 3. Spatial distribution of tweet sentiment regarding the Sunda Strait tsunami during Phase 1 (a), Phase 2 (b) and Phase 3 (c)

The tsunami-affected areas showed the dominance of tweets with negative sentiment in Phase 1, namely 22 - 24 December 2018 (Figure 3(a)). This negative sentiment was a manifestation of the response to the tsunami impact. People also tend to be more reactive on social media based on their proximity to the location of the disaster[34]. Tweets with positive sentiment were known to form concentration patterns in urban areas, such as Cilegon City, Serang City, Tangerang City, and surrounding areas. This positive sentiment was manifested in form of expressions of gratitude for safe conditions, sympathy through condolences, and moral support for communities in disaster-affected areas.

Meanwhile, tweets with neutral sentiment did not show any concentration in certain areas in phases 1, 2, or 3. In phase 2 there was a shift in the concentration of tweets with positive sentiment dominating in disaster-affected areas (Figure 3(b)). This condition showed a shift in the perception of the community that was beginning to accelerate the recovery from the disaster. At the end of the disaster emergency response period, tweets with positive, neutral, or negative sentiments did not show a tendency to concentrate in a particular area in phase 3 (Figure 3(c)). Tweets with positive sentiment were found scattered both in the affected and non-affected areas, which signified that support for the recovery of the affected areas was still ongoing. Furthermore, several tweets show a distinct sporadic distribution of trend patterns across different phases. The public sentiment during a disaster period does not always reflect the actual impact of disasters they experience because social media users have their own preferences to respond to a disaster topic in a certain sentiment[42].

4.3 Community Emergency Response Towards Sunda Strait Tsunami on Twitter

4.3.1 Community Response in Directly Affected Areas

Fig. 4. Wordcloud tweets from people in directly affected by the tsunami areas

Figure 4 shows the wordcloud containing the responses of Twitter users which came from the impact during the Sunda Strait tsunami emergency response period. People living in disaster-affected areas tend to post tweets related to emergencies around them and efforts to save themselves[35]. This sparked public reactions through the words 'ask', 'info', and 'hoax' which represents their confusion and curiosity. This further indicated that people in affected areas were able to use social media to confirm the truth of news from relevant sources. Furthermore, the appearance of the words 'help' and 'please', represented the community's efforts to ask for help. These words also involved reports on the current condition of their surroundings with attached photos or videos and embedding the words 'condition' and 'portrait' alongside the names of the areas being referred to such as 'banyuasih', 'sumur', 'cigeulis' and 'carita'. Several words containing sentimental expressions such as 'panic' and 'hope' represented an anxiety reaction as well as asking for prayer and support when faced with the tsunami disaster.

a. Community Activities and Roles in Directly Affected Areas

The role-sharing agreement process was not explicitly implied via community tweets. However, based on the actual reportage uploaded by the public shortly after the news of the tsunami disaster, it was observed that there was independent mobility of residents to places that were considered safer. This self-help evacuation by the community was carried out in response to avoiding the danger of a tsunami. Hills, fields to places of worship were some of the locations chosen by the community to save themselves based on experiences taught between generations when faced with the tsunami disaster. This collective mobility process was generally coordinated and carried out independently of a person appointed and trusted by the local community. The existence of independent evacuation efforts until the formation of evacuation point clusters signified the collective response of the community which was influenced by the urge to save themselves from the danger of a tsunami. No further information was obtained regarding the process of determining the gathering point for independent evacuation initiated by the community, whether it was a collective agreement or an official gathering point directed by the government. However, the community demonstrated the ability and role of local wisdom in identifying the environment and dangers around them. The concentration of the community at easily identifiable points facilitated the process of handling the impact of further disasters by the authorities such as distribution of aid, and further evacuation.

b. Resources Management in Directly Affected Areas

News of the tsunami disaster in the Sunda Strait attracted a lot of attention from people in various regions. Various forms of assistance and support came from individuals and groups/organizations both in form of volunteers, provision of basic living needs, and donations. The existence of social media was also enabled people in affected areas to report the latest needs required by victims in evacuation. Regarding the fulfillment of basic living needs, the community generally relied on assistance from the government or the results of donations channeled from the nearest post. The post was known to have been established by both government and non-government organizations. Therefore, the role of aid collection and distribution in the affected areas was mostly carried out by volunteers and representatives of government and non-government organizations. Through social media, conversations and tweets (posts on Twitter) reported the most recent conditions and emergency needs at evacuation points. Much of this information was conveyed by official accounts of non-governmental organizations to personal accounts in areas directly affected by the tsunami. The information exchange related to logistical needs at certain points quickly spread through social media and helped the effectiveness of aid distribution, thereby making it more targeted. Sumur Subdistrict in Pandeglang Regency is an example of an isolated area that quickly received assistance after many reports from Twitter users became a trending topic. Assistance received from various sources was managed jointly at government posts and by non-government volunteers, as well as public kitchens scattered in the areas directly affected by the tsunami to meet the needs of refugees.

c. Authoritative Government Role in Directly Affected Areas

The government through its various functional agencies and institutions carried out a synergistic multi-sectoral disaster management function. Social media accounts belonging to government agencies are one of the sources of information that were relevant and actual in understanding more deeply the emergency response of communities in affected areas. Through these accounts, active emergency response management was carried out by various multi-sectoral lines of government agencies. The national forces collaborating with the rescue team contributed to the evacuation and recovery of vital objects. Furthermore, the Red Cross Association of Cilegon City through @PMI_Cilegon showed their contribution in providing health facilities and basic living needs for refugees. The process of restoration and disaster recovery was mostly informed by the Regional Disaster Management Agency in Serang Regency through @BPBD_Kab_Serang account. This agency showed the process of coordination through meetings to evaluate the acceleration of recovery from the impact of the Sunda Strait tsunami disaster. It was observed that on several occasions the local community participated and worked together in clearing roads from the debris of the tsunami. However, it did not describe in detail the influence of the government in shaping the community's collective behavior significantly.

4.3.2 Community Response in Indirectly Affected Areas

Based on Figure 5 below, the responses from unaffected areas were more homogeneous, generally manifested by expressions of gratitude for their safe and secure conditions[35]. These responses showed moral support and invitations to participate in social solidarity actions, as denoted by 'safe' ,' condolence ',' hope ', and' prayer. Communities in areas not directly affected by the tsunami were observed to be predominantly urban in nature. Among the tweets posted from areas not affected by the tsunami, some were discovered to originate from the accounts of experts. The head of BMKG, Dwikorita Karnawati, through her account @dwiko_rita, often posted tweets related to developments on the role of BMKG in handling the impact of the Sunda Strait tsunami. Meanwhile, the involvement of experts in providing information through social media also influenced the community's perspective of the disaster. Credible sources of information from experts suppressed the development of fake news which may trigger public panic, thereby ensuring the smoothness of the disaster impact management process.

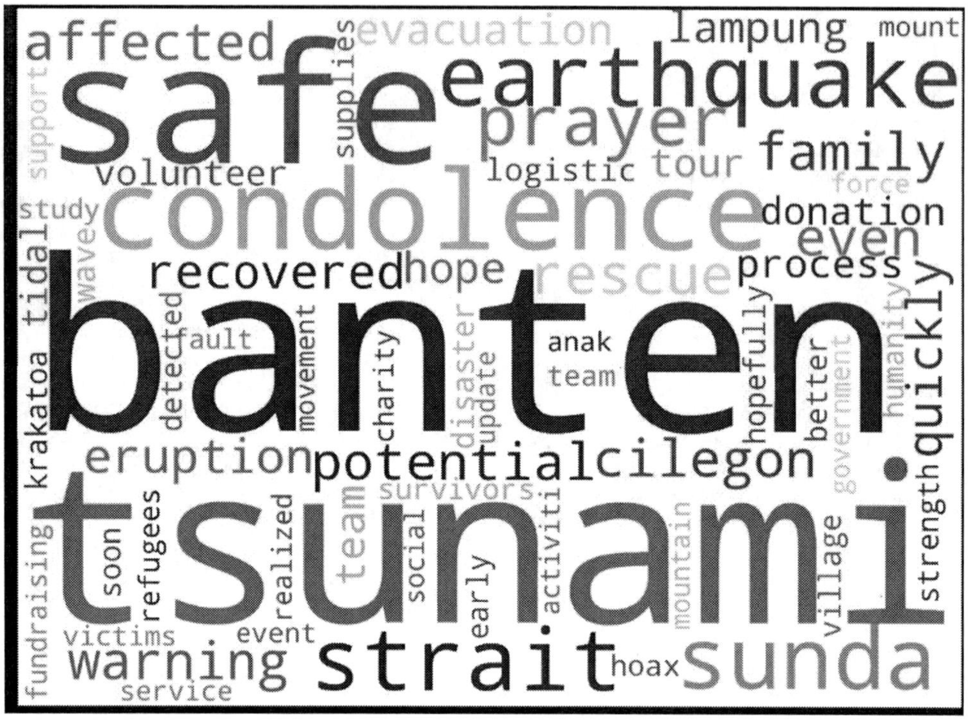

Fig. 5. Wordcloud tweets from people in indirectly affected by the tsunami areas

a. Community Activities and Roles in Directly Affected Areas

The community activities performed in areas not directly affected were dominated by social actions such as fundraising to charity events, both initiated by individuals and representatives of certain organizations. This showed a mutual response to requests for assistance from people in affected areas which were connected via Twitter. The outpouring of support and prayers conveyed through social media at the same time expanded the dissemination of information on the Sunda Strait tsunami disaster, which helped to increase awareness and generated other similar collective responses. Various fundraising actions to charity events sprung up in various regions which emphasized the growing sense of sympathy for helping victims of the tsunami disaster. Furthermore, support and assistance from communities in areas not directly affected were realized through the donation of resources, both in form of logistics and personal staff as volunteers. This signified the role of social media in gathering and expanding support from various regions for tsunami victims and helping to accelerate the handling of the tsunami impact in the directly affected areas.

b. Resources Management in Indirectly Affected Areas

The communities in areas not directly affected by the Sunda Strait tsunami distributed aid through collection points initiated by various government and non-government organizations. The collected aid was distributed and delivered by group representatives directly to the affected areas and channeled through the local official posts. Communities from affected areas that contributed to aid distribution generally uploaded some pictures of the aid delivery as a symbol of the task completion report and persuaded the community to also donate to help victims of the disaster on their Twitter accounts. The community also used social media to monitor the conditions and needs of victims in the affected areas to enable efficient, equitable and targeted aid distribution. This signified the existence of mutual interactions carried out by the community in the context of disaster management involving the use of Twitter.

c. Authoritative Government Role in Indirectly Affected Areas

Involvement of the authorities did not appear to be significant in areas not directly affected by the disaster. The government did not play an intense role in influencing the pattern of community responses in unaffected areas. The role played by the government in areas that were not directly affected generally led to efforts to increase awareness by disseminating chronological information. This enhanced the evacuation process by informing on the actual situation of the affected areas through their representatives in the field. Several agencies, for example, the Red Cross Association of Serang City, were also known to receive and manage donations deposited by the community to be distributed to affected areas. Furthermore, through their social media accounts, government agencies supported social movements initiated by the community, including by opening and spreading charity information to help people in affected areas.

Social media connects each user without being limited by space and time. Through it, interactions were formed between people in affected and unaffected areas. The public was also able to easily convey and receive information, which significantly influenced communication efficiency during the disaster. The large exchange of information flows during a disaster period has both advantages and the need for vigilance. Furthermore, information shared by the community in affected areas is important to consider in the formulation of a more effective emergency response strategy. Social media through this phenomenon also shows its role in accommodating social integration and solidarity among people which are physically disconnected in the real world. On the other hand, the government authorities are required to take part in preventing the development of fake news (hoax) which may harm the emergency response process. The existence of official accounts owned by the government as well as contributive experts that conveyed valid and relevant information can help the community in increasing disaster literacy as well as sorting information wisely.

5 Conclusion

The Sunda Strait Tsunami was a disaster that massively destructive and suddenly encouraged the formation of collective response in society. The communities in the directly and indirectly affected areas gave different responses. Mutual interaction was formed through social media which connected individuals and groups in the areas directly and indirectly affected. The communities outside the affected area utilized social media for monitoring the current situation and determining locations that still needed assistance from the actual uploads shared by the community in the affected area. The collective community response was manifested in form of social integration and solidarity, which was facilitated by social media. Massive responses and exchange of information on social media during disasters may influence real-world action. However, the limited number of tweets with the geo-tag attribute limited the information that can be explored to further understand the collective response of people on social media, and how it affects their actions.

The author expresses their gratitude to Universitas Gadjah Mada for facilitating the compilation, as well as to D.R. Hizbaron as the supervisor for supporting this research. The topic raised is part of the undergraduate thesis compiled by the author by discussing similar topics in more detail. The author is also grateful to readers for the feedback and constructive suggestions given on this article.

References

1. A. Muhari, M. Heidarzadeh, H. Susmoro, H. D. Nugroho, E. Kriswati, Supartoyo, T. Arikawa, Pure Appl. Geophys. **176**, 5219-5233 (2019) https://doi.org/10.1007/s00024-019-02358-2
2. Syamsidik, Benazir, M. Luthfi, A. Suppasri, L.K. Comfort, Nat. Hazards Earth Syst. Sci., **20**, 549-565 (2020) https://doi.org/10.5194/nhess-20-549-2020
3. P.S. Putra, Aswan, K.A. Maryunani, E. Yulianto, S.H. Nugroho, V. Setiawan, Pure Appl. Geophys. (2020) https://doi.org/10.1007/s00024-020-02446-8
4. M. Heidarzadeh, T. Ishibe, O. Sandanbata, A. Muhari, A.B. Wijanarto, Ocean Eng., **195** (2020) https://doi.org/10.1016/j.oceaneng.2019.106733
5. National Disaster Management Agency (BNPB), Sunda Srait Tsunami. Retrived 25 May 2020, from https://bnpb.go.id/publikasi/infografis/tsunami-selat-sunda.html
6. Humanity Road, Tsunami in Banten Province, Retrived May 29 2020, from https://www.humanityroad.org/situation-reports/indonesia/indonesia-sunda-strait-tsunami
7. W.L. Farahdita, S.R. Siagian, IOP Conf. Ser: Earth Environ. Sci. **429** 012052 (2020) https://doi.org/10.1088/1755-1315/429/1/012052
8. J. Fang, J. Hu, X. Shi, L. Zhao, Int. J. Disaster Risk Reduct. **34**, 275–282 (2018) https://doi.org/10.1016/j.ijdrr.2018.11.027
9. M. Jamali, A. Nejat, S. Ghosh, F. Jin, G. Cao, Int. J. Inf. Manage. **44**, 25–37 (2019)
10. Muzani, C. Setiawan, M. Zid, F.M. Akmal, IOP Conf. Ser: Earth and Environ. Sci., **412**(1) (2020) https://doi.org/10.1088/1755-1315/412/1/012014
11. K. Pongponrat, K. Ishii, Int. J. Disaster Risk Reduct., **27**, 133–141 (2018) https://doi.org/10.1016/j.ijdrr.2017.09.047
12. U. Widyanarko, D.R. Hizbaron, IOP Conf. Ser: Earth and Environ. Sci., **451** (2020) https://doi.org/10.1088/1755-1315/451/1/012098
13. S. Coetzee, S. Steiniger, B. Köbben, A. Iwaniak, I. Kaczmarek, P. Rapant, H. Moellering, H., *Advances in Cartography and GIS Science*. (M. P. Peterson, Ed.), Lecture Notes in Geoinformation and Cartography. Cham: Springer International Publishing (2017)
14. Y.A. Ahmed, M.N. Ahmad, N. Ahmad, N.H. Zakaria, Telemat. Inform., **37**, 72–112 (2019)
15. W. Shiau, Y.K. Dwivedi, H. Lai, Int. J. Inf. Manage., **43**, 52–63 (2018)

16. S. Krishna, T. Bhavaraju, C. Beyney, C. Nicholson, Int. J. Disaster Risk Reduct., **39**, 101251. (2019) https://doi.org/10.1016/j.ijdrr.2019.101251
17. C. Reuter, G. Backfried, M. Kaufhold, F. Spahr, *ISCRAM turns 15: a trend analysis of social media papers 2004-2017*, Proceedings of the 15th ISCRAM Conference (2018)
18. Q. Deng, Y. Liu, H. Zhang, H., X. Deng, Y. Ma, Nat. Hazards, **84**(2), 1241–1256 (2016) https://doi.org/10.1007/s11069-016-2484-9
19. T.B. Quillinan, F. Brazier, H. Aldewereld, F. Dignum, V. Dignum, L. Pen serini, N. Wijngaards, *Developing Agent-based Organizational Models for Crisis Management*, Int. Conf. AAMAS, Budapest, Hungary, 45-52 (2009)
20. E. Benjamin, A.M. Bassily-Marcus, E. Babu, L. Silver, M.L. Martin, Mt Sinai J. Med., **78**, 306-318 (2011)
21. K. Liao, (2012). Ecol. Soc., **17**(4), 48 (2012) http://dx.doi.org/10.5751/ES-05231-170448
22. National Research Council, *Facing hazard and disasters: understanding human dimensions,* Washington D.C., The National Academies Press (2006)
23. W.N. Carter, *Disaster management: a disaster manager's handbook*, Manila, Philipines, Asian Development Bank (2008)
24. J. Ingleton, *Natural disaster management: a pre-sentation to commemorate the international decade for natural disaster reduction (IDNDR) 1990-2000.* Leicester, England: Tudor Rose Insight (1999)
25. B. Wisner, P. Blaikie, T. Cannon, Davis, *At risk: natural hazards, people's vulnerability and disasters (sec. edi),* Routledge, London (2003)
26. D.R. Hizbaron, N. Ismayani, F.N. Ernawan, N.N.A. Puspitasari, Y. Yulianda, R. Jati, Environ. Dev. Sustain. (2021) https://doi.org/10.1007/s10668-021-01249-8
27. G.A. Kreps, Ann. Rev. Sociol., **10**, 309–330 (1984)
28. R.A. Stallings, E.L. Quarantelli, Public Adm., **45** (1985)
29. C. Chen, L. Xu, D. Zhao, T. Xu, P. Lei, Saf. Sci, **128** (2020)
30. J. Drury, D. Novelli, C. Stott, Resilience, **1**(1), 18–37 (2013)
31. C.K. May, Ocean Coast Manage., **169**, 86–95 (2019)
32. S.L. Cutter, L. Barnes, M. Berry, C. Burton, E. Evans, E. Tate, J. Webb, Glob. Environ. Change, **18**, 598–606 (2008)
33. R.V. Shari, T. Buehner, M.J. Palenchar, J. Contingencies Crisis Manage., **19**(2) 110–122 (2011)
34. S. Krishna, T. Bhavaraju, C. Beyney, C. Nicholson, Int. J. Disaster Risk Reduct., **39**, 101251 (2019) https://doi.org/10.1016/j.ijdrr.2019.101251
35. J.P. De Albuquerque, B. Herfort, A. Brenning, Int. J. Geogr. Inf. Sci., (April) (2019) https://doi.org/10.1080/13658816.2014.996567
36. A. Acar, Y. Muraki, Y. Int. J. Web Based Communities, 7(3), 392–402 (2011)
37. M. Latonero, I. Shklovski, I., Int. J. Inf. Syst. Crisis Response Manage., **3**(4), 67–86 (2011)
38. L. Li, Q. Zhang, J. Tian, H. Wang, Int. J. Inf. Manage., 38, 34–41 , (2018)
39. Y. Hu, Geogr. Compass, **12**(11) (2018)
40. S. Kemp, Digital 2019: Indonesia - DATAREPORTAL – global Digital insights. Retrieved April 06, 2021, from https://datareportal.com/reports/digital-2019-indonesia?rq=digital+2019+indonesia (2019)
41. T. Rockenbauch, P. Sakdapolrak, Ecol. Soc., **22**(1) (2017)
42. F. Yuan, M. Li, R. Liu, Int. J. Disaster Risk Reduct., **51**(July), 101798 (2020) https://doi.org/10.1016/j.ijdrr.2020.101798

Vulnerability analysis toward river flooding in the Comal Watershed, Indonesia

Yunus Aris Wibowo[1*], *Muh Aris* Marfai[2], *Muhammad Pramono* Hadi[3], *Lintang* Ronggowulan[4], and *Puspita Indra* Wardhani[1]

[1]Geography Education Department, Universitas Muhammadiyah Surakarta, Indonesia
[2]Geospatial Information Agency (BIG), Indonesia
[3]Geography Faculty, Universitas Gadjah Mada, Indonesia
[4]Geography Education Department, Universitas Sebelas Maret Surakarta, Indonesia

Abstract. River flooding is a disaster that almost every year hits the downstream Comal watershed. This incident resulted in material, psychological, and even casualties. Therefore, a vulnerability analysis is needed to reduce the impact of flooding. This study aims to analyze the vulnerability to river flooding in the downstream Comal Watershed area, Indonesia. Vulnerability variables used are social, economic, physical, and ecological (environmental) vulnerability following the Indonesian National Disaster Management Agency (BNPB) Regulation 2012. Data were obtained from government agencies and interpretation of SPOT 5 satellite imagery. Data analysis were carried out using scoring and weighting. Moreover, the data classification were using the mean and standard deviation equations. The analysis showed that the vulnerability to river flooding in the Comal watershed area was dominated by a moderate vulnerable class of 10,832.15 hectares (49.69%) and distributed in 26 villages. Furthermore, the high level of vulnerability also covered a large area by 8,773.11 hectares (40.24%), and distributed across 29 villages. Therefore, proper river flooding mitigation planning with the characteristics of the area is needed.

1 Introduction

Flood is a phenomenon that annually hits most cities in the world. It can be triggered either by natural processes or by human activities. Natural processes include hydrometeorological hazards such as tropical cyclones, storms, and high rainfall, while human activities include dam failure and land conversion [1,2]. The global impact of flooding is a disruption of life and the economy in exposed areas [3,4]. Furthermore, it is influenced by the intensity and frequency of hazards/threats and vulnerabilities in flood-prone areas.

The basic concept of vulnerability is a broad and appropriate framework for considering the underlying and indirect causes of disaster events. Vulnerability is simply a condition of a society that makes them potentially impacted by adverse hazards. Vulnerability is seen as a result of danger and is determined by the variables of vulnerability, sensitivity, and consequences of potential hazards [5–8]. The vulnerability can be seen from various variables that can increase the level of vulnerability to a hazard. [9–13] mentions four vulnerability variables, namely social, economic, physical, and ecological/environmental vulnerability.

Vulnerability variables always have different risk components depending on the types and characteristics of the vulnerability variables [6]. Risk components are all objects, individuals, animals, activities, and processes that can be negatively impacted by harmful natural phenomena, either directly or indirectly [13]. The frequency of exposure to hazards/threats tends to increase the community's vulnerability because there will gradually be a weakening of the capacity either physically, socially, or economically. In the long term, this can decrease coping capacity and lead to community resilience to disasters.

The Comal River flood disaster is a natural phenomenon that annually occurs during the rainy season downstream of Comal Watershed. It can be categorized as a severe disaster because it has caused many casualties and losses. One of the worst flood disasters in the last 20 years occurred from January to February 2014. Based on data recorded by the Regional Disaster Management Agency and the Social, Department of Social, Labor, and Transmigration Service of Pemalang Regency, floods in 2014 resulted in 66,810 residents displaced, tens of hectares of agricultural land were submerged, 16,259 houses were submerged, 53 of which were severely damaged with an estimated loss of Rp 35,000,000.00 for each house and three bridges in Longkeyang Village were broken. Therefore, flood disaster management measures are urgently needed in the Comal watershed area. One of the first steps in overcoming the flood disaster is to conduct vulnerability analysis of the downstream Comal watershed.

In line with the development of science and technology used in disaster analysis, several views have

* Corresponding author: yaw222@ums.ac.id

https://doi.org/10.1051/e3sconf/202132501022

emerged regarding the appropriate method for vulnerability analysis [14]. At least, there are two most popular approaches, namely the techno-scientific approach and the social approach [15–18]. The techno-scientific approach focuses on physical variables and tends to neglect the indirect effects of disasters. Conversely, the social approach prioritizes human variables through a social process. This approach places human variables as the main focus to be saved in disaster management.

This study focused on analyzing the vulnerability of the area to river flooding in the Comal watershed, Indonesia. It is conducted by emphasizing social, economic, physical, and ecological (environmental) variables. The vulnerability analysis is carried out quantitatively with a spatial approach. These variables are selected with consideration to obtain comprehensive and reliable analysis results. Vulnerability analysis not only focuses on specific variables and direct impacts but also analyzes all variables of vulnerability that affect a region. Furthermore, the analysis is also conducted based on variables that may cause long-term or indirect impacts. Comprehensive vulnerability analysis can produce an accurate risk assessment and mitigation planning for flood disasters [19,20].

2 Research method

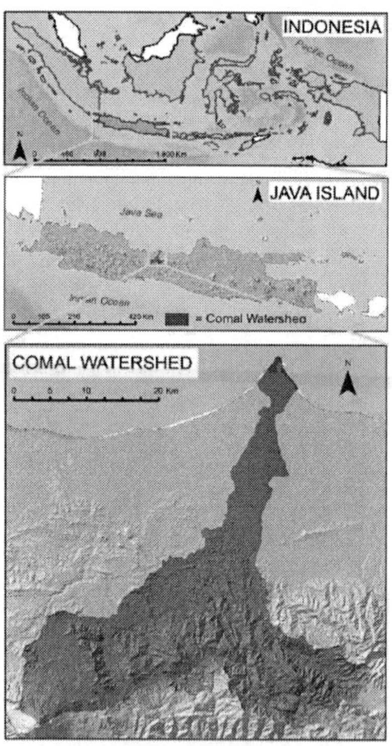

Figure 1. Study location

The downstream Comal watershed is located in Pemalang Regency, Central Java Province, Indonesia (Fig. 1). This watershed is one of the watersheds with a priority for handling in Indonesia. The priority includes land and forest rehabilitation, including reforestation, soil, and water conservation, whether carried out vegetatively, agronomically, structurally, and in management [21]. The Comal River flooding vulnerability level in the downstream Comal watershed is analyzed based on four physical, social, economic, and (ecology) environmental variables. Each vulnerability variable has a different weight depending on the level of fatality. Social variables weight 40%, economic 25%, physical 25%, and ecology (environmental) 10% based on modification of [9]. Each variable consists of exposure parameters that reflect the vulnerability characteristics of each variable in the research location. Moreover, each variable consists of several parameters (element at-risk) that have a weight determined based on fatality level when exposed to hazards/threats. Element at-risk data were obtained from various official government sources and primary data analysis.

The unit of analysis in this research is the village area. This is done to obtain detailed results and to describe the vulnerability of the region. Furthermore, this research covers 59 villages from five sub-districts downstream of the Comal watershed. The analysis process is carried out by scoring based on the scoring guidelines from [9].

Vulnerability analysis begins with grouping the vulnerability parameters (element at-risk) of each vulnerability variable based on its class. Each parameter of vulnerability (element at-risk) is classified into five categories. The level is obtained from calculations using the mean and standard deviation equations.

Mean deviation : $\bar{x} = \frac{1}{n}\sum_{i=1}^{n} x_i$ (1)

Standard deviation : $S = \sqrt{\frac{1}{n-1}\sum_{i=1}^{n}(x_i - x)^2}$ (2)

The next stage is scoring on each parameter of vulnerability (element at-risk) of each variable. The score given is very high = 5, high = 4, medium = 3, low = 2 and very low = 1 (Table 1). The score is multiplied by the weight of each risk component for each vulnerability variable. The multiplication results are added according to each vulnerability variable (social, economic, physical, and environmental). The total weight (score * weight of the risk component) on each variable of vulnerability (social, economic, physical, and environmental) is summed, and the results are used as the basis for determining the class of total vulnerability. The total vulnerability class is calculated by using the mean and standard deviation equations, such as determining the class of risk components/parameters for each vulnerability variable.

Table 1. The vulnerability levels, scores, and weights

Classes	Score	Weights
Very high	5	4.2 – 5
High	4	3.4 – 4.2
Moderate	3	2.6 – 3.4
Low	2	1.8 – 2.6
Very low	1	< 1,8

Source: Data analysis

Vulnerability = (0.4 × social vulnerability score) + (0.25 × economic vulnerability score) + (0.25 × physical vulnerability score) + (0.1 × environmental vulnerability score) (3)

3 Result dan Discussion

3.1 Result

3.1.1 Social vulnerability

The social vulnerability in this study consists of population density with a weight of 60%, the gender ratio of 10%, age group ratio of 10%, disability ratio of 10%, and the low-income ratio of 10%. Population density has the highest weight because the population is the subject most affected by the disaster. Data for each parameter was obtained from the Central Bureau of Statistics Pemalang Regency and the Indonesian Geospatial Information Agency.

The population density was calculated based on data obtained from the Central Bureau of Statistics Pemalang Regency and village administrative area data obtained from the Indonesian Geospatial Information Agency. The gender ratio is the ratio of the female population to the total population in an administrative area. The data were obtained from the Central Bureau of Statistics Pemalang Regency. The age group ratio in this study was the age group that is prone to hazards or disasters, namely the population aged 0-4-year-old and the population aged ≥ 65-year-old. The proportion of persons with disabilities in this study was categorized into several types, namely disabilities, deaf and speech impaired, blind, multiple disabilities, mental disabilities, and disabilities from chronic disease.

Furthermore, the ratio of poor people used in this study was a standard according to the Indonesian National Family Planning Coordinating Board. According to the the National Family Planning Coordinating Board, the criteria for low-income families are Pre-Prosperous Families and Prosperous Families 1. The analysis results using the average deviation, standard deviation, scoring, and weighting indicated that the downstream Comal watershed area was dominated by a very high level of vulnerability to river floods (Table 3).

Table 2. The social vulnerability levels

Vulnerability level	Number of Villages	Large	
		Hectare	Percent (%)
Very high	37	8,527.66	39.12
High	9	4,063.70	18.64
Moderate	10	6,768.17	31.04
Low	3	2,442.22	11.20
Total	59	21,801.75	100.00

Source: Data analysis

3.1.2 Economic vulnerability

Economic vulnerability in this study included the productive land area with a weight of 60% and the number of Economic Facilities 40%. The productive land in the downstream Comal watershed was used as gardens/plantations, shrimp ponds, fishponds, wetland agricultural land (paddy fields), and dryland agriculture (moor/fields) with the primary commodities of rice, long beans, cassava, jasmine, maize, bitter melon, and chilies. The area of productive land was obtained from the interpretation of SPOT 5 High-Resolution Satellite Imagery, whose accuracy has been tested by field survey. Furthermore, the second parameter (element at risk) was the number of economic facilities. The economic facilities in this study include the number of markets, industries, hotels, shops, or stalls. These data were obtained from the Central Bureau of Statistics Pemalang Regency.

The results of data analysis showed that the low vulnerable class dominated the level of economic vulnerability to river flooding in the downstream Comal watershed with 27 villages (Table 4). It was influenced by the productive land area and the low number of economic facilities. Furthermore, economic vulnerability with high and very high vulnerable levels was not very dominant because of the influence of the weight of the productive land area parameter, which was more prevalent than the number of economic facilities.

Table 3. The economic vulnerability levels

Vulnerability level	Number of Villages	Large	
		Hectare	Percent (%)
Very high	4	3,650.79	16.75
High	14	10,809.40	49.58
Moderate	5	1,762.43	8.08
Low	28	5,579.13	25.59
Total	59	21,801.75	100.00

Source: Data analysis

3.1.3 Physical vulnerability

The Physical vulnerabilities in this study consist of settlement density with a weight of 40%, number of critical and public facilities by 30%, and road length by 30%. Settlement density is obtained from the number of houses in one village (unit) divided by village area (hectares). The number of settlements, critical facilities, and public facilities were obtained from the Central Bureau of Statistics Pemalang Regency. At the same time, the road length data was obtained from the visual interpretation of topographic maps and field surveys.

The density of settlements includes permanent, semi-permanent, and non-permanent houses [22]. At the same time, the number of public and critical facilities contained places of worship, schools, and health service infrastructures. Furthermore, the length of the road was the length of all the streets of the various road classes in the research location. The analysis results indicated that the level of the physical vulnerability to river flooding in the downstream Comal Watershed consisted of four vulnerability levels which was dominated by very high vulnerable levels (Table 5).

Table 4. The physical vulnerability levels

Vulnerability level	Number of Villages	Large	
		Hectare	Percent (%)
Very high	27	8,553.55	39.24
High	23	9,429.73	43.25
Moderate	8	3,292.37	15.10
Low	1	526.10	2.41
Total	59	21,801.75	100.00

Source: Data analysis

3.1.4 Ecological (environmental) vulnerability

In this study, environmental vulnerability included parameters (element at risk) of forest area with a weight of 70%, bush/shrubs by 10%, and swamp area by 10%. The forests in the research location consisted of community forests, state forests, and mangrove forests. The community forest is state-owned land managed by the local community, while the state forest is a forest that is owned and operated by the state. Mangrove forest is a forest with typical plants in tidal areas. These data were obtained from the SPOT 5 satellite imagery interpretation results that had been tested for accuracy through a field survey. Moreover, the level of environmental vulnerability to the Comal River flood was dominated by the very low vulnerable level. It was due to not all the parameters of the environmental vulnerabilities that exist in all villages (Table 6).

Table 5. The ecological (environment) vulnerability levels

Vulnerability level	Number of Villages	Large	
		Hectare	Percent (%)
High	5	3,817.19	17.51
Moderate	1	427.86	1.96
Low	1	1,051.07	4.82
Very low	52	16,505.63	75.71
Total	59	21,801.75	100.00

Source: Data analysis

3.1.5 The level of vulnerability toward the Comal River flood

The level of vulnerability to the Comal River flooding was obtained from the sum of the weights of each vulnerability variable, namely the social, economic, physical, and ecology (environmental) variables (Fig. 2). The social variables have a weight of 40%, economic variables 25%, physical variables 25%, and environmental variables 10%. The weight of each of these variables is based on the Head of the Indonesian National Disaster Management Agency Regulation 2012 [9]. The results showed the level of vulnerability toward the Comal River flooding consisted of four vulnerability levels and was dominated by high vulnerable levels (Table 7).

Table 6. The level of vulnerability toward the Comal river flood

Vulnerability level	Number of Villages	Large	
		Hectare	Percent (%)
Very high	3	1,670.39	7.66
High	29	8,773.11	40.24
Moderate	26	10,832.15	49.69
Low	1	526.10	2.41
Total	59	21,801.75	100.00

Source: Data analysis

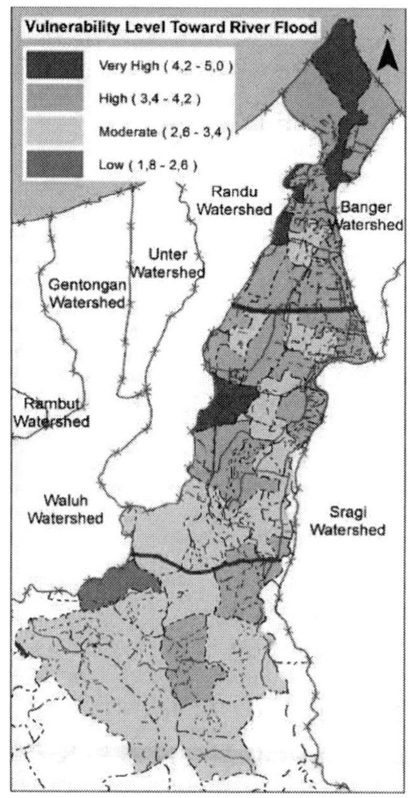

Figure 2. The vulnerability toward Comal river flooding

3.2 Discussion

The Research conducted by [3] used landuse as a parameter of vulnerability. [1] used the poor, gender, and education level as the parameters, while [23][24] formulated social vulnerability parameters including Poverty, Gender disparities and low educational attainment, access to drinking water and health hazards, rapid population growth as well as transboundary insecurity. Moreover, the differences in data use will affect the outcome of social vulnerability. The more variations in population data used, the more accurate the results will be in describing conditions in the field [7].

Therefore, this study used the population parameters by [1,9,25] because it was considered to represent all aspects related to the population.

The most dominant social vulnerability level in the downstream Comal watershed was the very high and high vulnerable level. The high social vulnerability to Comal River flooding describes the lack of capacity of the population in dealing with floods. Therefore, socialization and simulations of river flooding disasters were needed to increase people's understanding of flood disasters. Furthermore, the role of the local government was very much needed because the increased knowledge of the population about flood disasters will increase their capacity in dealing with flood disasters [13,26–29].

The parameter of economic vulnerability in this study significantly affects the condition of the population. The assumption used was that the income of people who work as farmers, fishpond farmers, traders, service entrepreneurs will be disrupted. It happened when the floods hit, and they cannot run the kind of economic activity under normal circumstances. This means that no income was earned, or it can even be assumed that losses were due to damaged rice fields, damaged ponds, and difficulty selling merchandise and services due to flooding. It was very different from residents who work as state civil servants, bank employees, police, Indonesian national army, or other professions with regular income each month. Furthermore, residents with this profession will continue to earn income as in normal circumstances without a disaster even though the settlement or place of work of residents of that profession is affected by flooding.

The parameters in physical vulnerability were the facilities needed and used by the population in their daily life. Therefore, the higher the frequency and intensity of floods that hit these parameters, the activity of the population will also be disrupted and even completely paralyzed. One of the most dominant parameters was the density of houses/settlements. The higher the population density, the denser the settlement [30]. The high density of population and settlements causes new needs to emerge. These needs include housing, economic, social, and educational facilities. The facilities and infrastructure to meet these needs appear along with the development of these areas. Very high and high vulnerable levels dominated the level of physical vulnerability to flooding in the downstream Comal watershed. Very high and high vulnerable levels dominate the sum of weights for all parameters of physical vulnerability to the Comal River Flooding. It was very reasonable when viewed from the parameters that make up physical vulnerability because almost all villages in the downstream Comal watershed have developed into dense settlements.

The parameters in ecological (environmental) vulnerability were determined based on their natural function. If these parameters were affected by flooding, the natural function and ecosystem in the area will be disturbed or even damaged. Environmental variables included natural assets that can reduce the negative impact of disasters. If environmental variables are disturbed or damaged, it will take a very long time to rehabilitate and restore them to their original condition. Therefore, the deeper the vulnerability assessment, the wider the parameters of environmental vulnerability in an area. This will affect the high level of vulnerability of the region.

Protected forests, mangrove forests, and swamps are habitats for a wide variety of flora and fauna, while production forests have an economic function for their managers. Protected forests, production forests, mangrove forests, shrubs also have other natural parts to slow down the flow or velocity of water flow when a flood occurs and hold the inherent material washed away when the surge occurs. Therefore, if a flood hits and causes mangrove forest damage, it will cause huge losses, not only from an economic perspective but also from an ecological perspective. This study used the vulnerability variable following the Head of the Indonesian National Disaster Management Agency Regulation 2012 [9].

Very low vulnerable classes dominated the level of environmental vulnerability to flooding in the downstream Comal watershed. This was because not all ecological vulnerability parameters were found in all villages. Therefore, the environmental vulnerability classes in the downstream Comal watershed were at the same level. Only a few villages had these parameters, and this was due to the conversion of land functions and different characteristics of the area. Swamps and mangrove forests were only found around the mouth of the Comal River, namely in the villages of Mojo and Kendaldoyong. The forest was located in the village in the Bodeh Subdistrict area, one of the areas with the highest elevation in the downstream Comal watershed.

Very high and medium vulnerable classes dominated regional vulnerability to flooding in the downstream Comal watershed. This was because the social and physical vulnerability had a dominant influence over economic and environmental vulnerability. It was because not all parameters of economic and environmental vulnerability exist in all villages. Areas with a very high vulnerability class consisted of Mojo and Kendaldoyong Villages. In Mojo and Kendaldoyong Village, there were all the parameters of vulnerability, especially environmental vulnerability, such as forests and swamps that were not found in other villages. Meanwhile, in Kendalsari Village, although it was a very high vulnerable class, the dominant variables were social, economic, and physical vulnerability.

4 Conclusion

The vulnerability to Comal River flooding were classified into four classes of vulnerability, namely very high, high, medium, and low. The variables used in determining total vulnerability are social, economic, physical, and ecological (environmental) vulnerability. Each variable consisted of various parameters adjusted to the type of vulnerability variable and the characteristics of the research location. The total vulnerability to Comal river flooding in the downstream Comal watershed was dominated by the moderate vulnerable class of 10,832.15 hectares (49.69%) and distributed in 26 villages. This was because the area was an area with fertile and flat topography, so that it has developed into a center of government, settlement, economic center, and education center. Therefore, it needs efforts to increase community capacity and disaster mitigation planning to reduce the impact of disasters. This was because the high level of vulnerability also covered a large area, namely 8,773.11 hectares (40.24%), and was distributed across 29 villages. Furthermore, although the vulnerability variables were determined based on the Head of the Indonesian National Disaster Management Agency Regulation 2012, the determination of the parameters of each variable has not been carried out by a more in-depth method. It is hoped that further research can carry out a more in-depth analysis so that the vulnerability analysis results are more accurate. Financial support for this study was provided by the Indonesian ministry of education and culture.

References

1. E. Mavhura, A. Collins, and P. P. Bongo, Disaster Prev. Manag. **26**, 41 (2017)
2. S. Mahato, S. Pal, S. Talukdar, T. Kanti Saha, and P. Mandal, Geosci. Front. **12**, (2020)
3. W. Chen, X. Wang, S. Deng, C. Liu, H. Xie, and Y. Zhu, J. Hydrol. **575**, 454 (2019)
4. S. Thapa, A. Shrestha, S. Lamichhane, R. Adhikari, and D. Gautam, J. Hydrol. Reg. Stud. **30**, 100704 (2020)
5. H. Bang, L. Miles, and R. Gordon, Foresight **21**, 266 (2019)
6. A. Dewan, *Flood in a Megacity: Geospatial Techniques in Assessing Hazards, Risk and Vulnerability* (2013)
7. H. Chang, A. Pallathadka, J. Sauer, N. Grimm, R. Zimmerman, C. Cheng, D. Iwaniec, Y. Kim, R. Lloyd, T. McPhearson, B. Rosenzweig, T. Troxler, C. Welty, R. Brenner, and P. Herreros, Sustain. Cities Soc. **68**, 102786 (2021)
8. R. Jha and H. Gundimeda, Int. J. Disaster Risk Reduct. **35**, 101074 (2019)
9. the Indonesian National Disaster Management Agency, 62 (2012)
10. UNISDR, (2009)
11. C. J. Van Westen, D. Alkema, M. C. J. Damen, N. Kerle, and N. C. Kingma, *Risk City Exercise Book* (United Nations University – ITC School on Disaster Geoinformation Management, Twente, 2011)
12. S. Kumpulainen, Spec. Pap. Geol. Surv. Finl. 65 (2006)
13. ADPC, *Urban Governance and Community Resilience Guides: Risk Assessment in Cities* (2010)
14. M. Imran, K. Sumra, S. A. Mahmood, and F. Sajjad, Int. J. Disaster Risk Reduct. **41**, 101288 (2019)
15. H. B. Abbas and J. K. Routray, Disaster Prev. Manag. **23**, 395 (2014)
16. A. El-Zein, T. Ahmed, and F. Tonmoy, Ecol. Indic. **121**, 106988 (2021)
17. E. A. Adams and H. Nyantakyi-Frimpong, Heal. Place **67**, 102500 (2021)
18. T. Sarmah, S. Das, A. Narendr, and D. B. Aithal, Int. J. Disaster Risk Reduct. **50**, 101659 (2020)
19. V. S. A. Hendrawan and D. Komori, Int. J. Disaster Risk Reduct. **54**, 102058 (2021)
20. M. Musiyam, Int. J. GEOMATE **19**, (2020)
21. The Republic of Indonesia Ministry of Forestry, *Decree of The Minister of Forestry of The Republic of Indonesia Number 328 Menhut-II 2009 Concerning Determination of Priority Watersheds in the Framework of the Medium Term Development Plan (RPJM) 2010-2014* (Indonesia, 2009), p. 8
22. H. Desalegn and A. Mulu, Heliyon **6**, (2021)
23. M. A. A. Hoque, S. Phinn, C. Roelfsema, and I. Childs, Appl. Geogr. **98**, 22 (2018)
24. H. Bang, (2009)
25. A. Karunarathne and G. Lee, Int. J. Disaster Risk Reduct. **49**, 101679 (2020)
26. M. A. Marfai, A. Cahyadi, H. Fatchurohman, F. S. C. Rosaji, and Y. A. Wibowo, IOP Conf. Ser. Earth Environ. Sci. **256**, (2019)
27. M. A. Marfai, Sunarto, N. Khakim, H. Fatchurohman, A. Cahyadi, Y. A. Wibowo, and F. S. C. Rosaji, in *E3S Web Conf.* (2019)
28. F. Ullah, S. Azmat, S. A. Ali Shah, S. E. Saqib, M. Yaseen, S. Muhammad, and Haider, Int. J. Disaster Risk Reduct. **52**, 101967 (2021)
29. N. Pricope, J. Halls, and L. Rosul, J. Environ. Manage. **237**, 387 (2019)
30. A. K. Christian, B. D. Dovie, W. Akpalu, and S. N. A. Codjoe, Urban Clim. **35**, 100759 (2021)

Coastal erosion risk assessment in the coastal area of Muara Asam Asam, Tanah Laut Regency

Rizki Adriadi Ghiffari[1]*and *Haryo* Dwito Armono[2]

[1]Department of Development Geography, Faculty of Geography, Universitas Gadjah Mada, Yogyakarta, Indonesia
[2]Marine Engineering Department, Faculty of Marine Technology, Institut Teknologi Sepuluh Nopember, Surabaya, Indonesia

Abstract. Coastal erosion is the process of land erosion in coastal areas due to waves and ocean currents which reduce land and can adversely affect socio-economic activities in coastal areas. Coastal erosion risk assessment in the estuary is a development of previous research, because the erosion and accretion processes do not only occur along the shoreline, but also on the side facing river water bodies, due to the confluence of fluvial and marine processes. The landward shift of the shoreline in the Muara Asam Asam coastal area has reached 90 meters in the last 28 years, and has been detrimental because the area is a densely populated area with major economic activities in the fisheries and agriculture sectors. This study aims to determine the coastal erosion risk level zones based on the relationship among vulnerability and consequence parameters, through distance decay weighted based method. The results indicate that Muara Asam Asam has high risk zones of coastal erosion, especially in densely populated residential areas and dry land agriculture on the west side of the estuary, due to the lack of implementation of preventive measures through hard coastal structure and coastal zoning policy to protect socio-economic activities and coastal ecological environments.

1 Introduction

The coastal area is a very important environment for the community, because it supports a large number of activities related to fishing, navigation, trade and tourism. Therefore most of the world's population lives in coastal areas, including most important cities, also in coastal areas [1].

The presence of anthropogenic processes in coastal areas raises concerns related to hazards affecting coastal areas such as tidal waves, storms, tidal flooding, coastal erosion, and sea level rise. To address the potential consequences of these hazards, several methods for assessing the level of vulnerability and risk of coastal erosion have been developed. Since 1991, when the IPPC general method was presented [2], vulnerability and risk assessments have continued to be developed, such as videotape-assisted based vulnerability analysis [3], coastal vulnerability index [4], the smart line approach [5], the coastal area simulator [6], and the latest is the coastal area risk assessment framework [7].

In its basic definition, risk must be understood as an estimate of the consequences of a particular event [8]. On the other hand, in the risk assessment of coastal areas, the occurrence of coastal erosion is defined as a hazard, which is a hazardous phenomenon that can cause loss of life, injury or other health impacts, property damage, loss of livelihood and public service facilities, social and economic disruption, or environmental damage [9].

Therefore, the assessment of the risk of coastal erosion depends on the likelihood and level of hazard density and the hazard consequences of a coastal erosion event [10]. For coastal areas, the negative impacts of hazards are expressed in terms of human, economic and environmental impacts, as well as social / political impacts. Vulnerability reveals the characteristics and conditions of the community, system or asset that make it vulnerable to damage caused by coastal erosion hazards [9]. Coelho (2005) developed a methodology for assessing the risk of coastal erosion. This method was chosen in this study because there is no other complex model which in its formulation allows for rapid application at the research site by adjusting the number and detail of parameters based on the availability of field data.

The coastal erosion phenomena in Muara Asam-Asam Village has entered an increasingly severe stage, due to the destruction of mangrove forests which have been used as coastal protection plants, as well as the opening and expansion of ponds and other agricultural land [11]. Coastal erosion in this area is also influenced by sediment transport activity generated by currents and waves due to seasons and winds. The occurrence of longshore currents that distribute sediment, so that in some locations there is coastal erosion, and in other locations there is accretion / sedimentation.

The incidence of extreme coastal erosion began in early 1983 (Results of Interview with Village Officials). Moreover, in the time period until this study was carried out, the impact of this disaster had significantly affected people's lives and livelihoods.

The aim of this study is to identify the coastal erosion risk zones according to the characteristics of

* Corresponding author: rizki.adriadi.g@ugm.ac.id

vulnerability and consequences. This classification aims to support decision-making in accordance with the characteristics of the area, and the identification of areas that need more attention (priority zones) in the face of coastal erosion hazards.

2 Method

This study uses the methodology proposed by Coelho (2005) to assess vulnerability, consequences and risks associated with coastal erosion. This method is focuses specifically on coastal erosion, rather than most of other methods that generally concerned on sea level rise [12]. Apart from being easy to modify, such as changing the weight of each parameter or changing its parameters, this method also can be implemented in various geographic information system software, making it easy to access in spatial analyzes related to coastal area management.

Coastal erosion risk assessment divides into two parts. The first is a vulnerability assessment that estimates, qualitatively and quantitatively, the level of vulnerability of a coastal area to coastal erosion [13]. This assessment focuses on the physical characteristics of the coastal area and the potential threats to erosive agents. The second part is an assessment of the consequences of the hazard event, by looking at the social, environmental, cultural and economic aspects in the study area [13].

The data collection stage consists of the acquisition of a digital elevation map that is stereopair extracted from aerial photographs acquired via an unmanned aerial vehicle (UAV), which are then used for the preparation of topographic maps. In addition, measurements of wave height and tidal range along the shoreline and river estuaries are also carried out as part of the vulnerability parameters.

Some of the parameters are obtained from secondary data extraction such as geological parameters derived from the interpretation of the geological map. The spatial pattern plan and land use map from the Regional Spatial Plan of Tanah Laut Regency are used to be extracted into ecological parameters, historical sites and land cover. Meanwhile, there is the use of Landsat 5,7,8 satellite imagery to obtain the value of the annual average coastal erosion / accretion rate parameter. Specifically for this parameter, the data is processed through The Digital Shoreline Analysis System (DSAS) software, so that it can analyze spatial statistics, any changes to the shoreline every year from 1989 to 2021.

Other data such as geomorphology, anthropogenic action, population density and economic level are interpreted based on the appearance of the UAV photo image, which is compared with the statistical data of Jorong District in Figures of 2021, as well as the results of field observations when taking the tie point of aerial photo images.

3 Result and discussion

3.1 Study area

The research area is a coastal area of Asam Asam River Estuary, Muara Asam Asam Village which is located in Jorong District, Tanah Laut Regency, South Kalimantan Province, shown in figure 1. Asam Asam Village is one of the oldest settlements in Tanah Laut where the majority of the population is the Banjar Tribe. The name of this village is already in the Hikayat Banjar which was written in 1663 which is referred to as Hasam Hasam [14].

Based on its physical characteristics, this village is located at the estuary of the Asam Asam river, precisely on land formed by sediment from the confluence of fluvial energy from the river and the energy of ocean currents [15]. With these unique characteristics, the coastal area of Muara Asam Asam Village changes significantly every year, both by the incidence of coastal erosion and accretion [16]. This is important to study considering there are thousands of villagers whose homes in densely populated area that are in risk of coastal erosion.

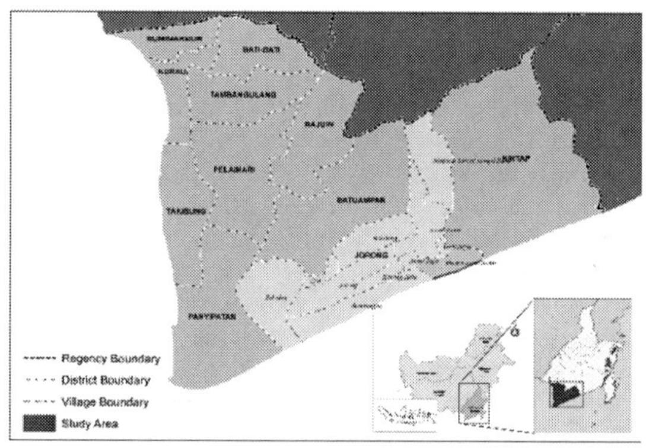

Fig. 1. Study area orientation map

3.2 Vulnerability assessment

Data from each parameter in the level of vulnerability analysis is classified first based on the classification rules presented in Table 1. After that, all parameters are inputted in the CERA application to perform raster operations using the Map Algebra method according to the weight calculation algorithm presented in Table 2. Distance decay principle is used to determine the different weight on each value of each parameter based on their pythagorean distance to the shoreline.

Table 1. Classification of vulnerability parameters

No	Parameter	Level of Vulnerability				
		Very Low	Low	Moderate	High	Very High
1	Distance to shoreline (m)	> 1000	200 - 1000	50 - 200	20 - 50	< 20
2	Topography (m)	> 30	20 - 30	10 - 20	5 - 10	< 5
3	Geology	Igneous rocks	Metamorphic rocks	Sedimentary rocks	Coarse textured sediments	Smooth textured sediments
5	Geomorphology	Mountains	Rocky cliffs	Eroded cliffs that extend along the shoreline	Beaches, fluvio marin plains	sand dunes, estuaries
6	Land cover	Forests	Agricultural crops	Bare lands	Rural settlements	Urban settlements, industrial areas
7	Anthropogenic actions	Intervention on shoreline stabilization	Intervention without reduction of sediment sources	Intervention with reduction of sediment sources	Without intervention or reduction of sediment sources	Without intervention, but still any reduction in sediment sources
8	Maximum significant wave height (m)	< 3	3 - 5	5 - 6	6 – 6,9	> 6,9
9	Maximum tide distance (m)	< 1	1 - 2	2 - 4	4 - 6	> 6
10	Annual average on coastal erosion/accretion rate (m/year)	> 0 (Accretion)	(-1) – 0 (Erosion)	(-3) – (-1) (Erosion)	(-5) – (-3) (Erosion)	< (-5) (Erosion)

Source: [17]

Table 2. The weight of each parameter of the level of vulnerability, based on its distance to the shoreline

No	Parameter	Distance to Shoreline (m)		
		< 100	100 - 5000	> 5000
1	Distance to Shoreline (m)	0.214	$\frac{6}{28}+\frac{0.665(d-100)}{4900}$	0.879
2	Topography (m)	0.214	$\frac{6}{28}+\frac{0.665(d-100)}{4900}\frac{6}{22}$	0.033
3	Geology	0.214	$\frac{6}{28}+\frac{0.665(d-100)}{4900}\frac{6}{22}$	0.033
5	Geomorphology	0.071	$\frac{2}{28}+\frac{0.665(d-100)}{4900}\frac{2}{22}$	0.011
6	Land cover	0.071	$\frac{2}{28}+\frac{0.665(d-100)}{4900}\frac{2}{22}$	0.011

https://doi.org/10.1051/e3sconf/202132501023

No	Parameter	Distance to Shoreline (m)		
		< 100	100 - 5000	> 5000
7	Anthropogenic actions	0.036	$\frac{1}{28} + \frac{0.665(d-100)}{4900}\frac{1}{22}$	0.005
8	Maximum significant wave height (m)	0.071	$\frac{2}{28} + \frac{0.665(d-100)}{4900}\frac{2}{22}$	0.011
9	Maximum tide distance (m)	0.071	$\frac{2}{28} + \frac{0.665(d-100)}{4900}\frac{2}{22}$	0.011
10	Annual average on coastal erosion/accretion rate (m/year)	0.036	$\frac{1}{28} + \frac{0.665(d-100)}{4900}\frac{1}{22}$	0.005

Source: [17]

Based on the results of field measurements, the highest tide distance of sea water on the coast of Muara Asam-Asam Village is a maximum of 2 meters from the coastline, this includes low classification in the level of vulnerability to coastal erosion. Parts of the river close to the estuary are also affected by tides, but only at a maximum level of 1 m or classified as very low in terms of coastal erosion vulnerability

The maximum wave height on the coast around Muara Asam-Asam Village is in the medium category, because it ranges from 5-6 m in bad weather conditions. This of course will greatly affect the level of vulnerability of community activity zones along the coastline, because the movement of wave energy is also sometimes followed by mass movements of ocean currents that can erode the shoreline or carry sediment in the form of sediment from the seabed.

Basically, the coastal area of Muara Asam Asam Village has similar topographic characteristics because it has a land elevation ranging from 1-4 m above sea level and with a gentle slope to flat (0-2%). This is because almost all of the land in this area is formed by sediment deposits from the sea and rivers, not through tectonic or denudational processes.

Based on the results of the shoreline change analysis through the DSAS software, it was found that along the shoreline of Muara Asam Asam Village, there were varied levels of erosion and accretion when viewed from the trend of shoreline changes between 1989-2017. The part with the highest level of erosion is around the center of the settlement of Muara Asam Asam Village, which is located some distance from the estuary, as well as the eastern part of the river estuary. This is an indication of the movement of the dominant ocean currents from the southwest to the northeast, thus diverting the river flow and forming an estuary on the east side. In fact, based on the identified trends, there is a tendency for river estuaries to move as far as 90 m to the east in a period of 28 years, which means that if this trend continues, new land will emerge on the eastern side of the settlement center of Muara Asam Asam Village, and vice versa, the east side. the estuary which is an agricultural area and open land will be eroded to become part of the river estuary.

In general, the coastal area of Muara Asam Asam Village is dominated by densely populated residential areas on the west side of the river, and open land and shrubs on the east side of the river. This makes the western region more vulnerable and at risk of coastal erosion than the eastern region.

Anthropogenic action consists of efforts to reduce erosion rates and coastal accretion. Based on the results of field observations, there is a coastal construction that is parallel to the coastline and functions in reducing the kinetic energy of waves and ocean currents, in the form of a wave breaker. This construction performance is less than optimal due to its proper placement in the middle of the tidal area, so that the erosion of the pan still occurs because the waves break before meeting the breakwater, even in some locations, the breakwater that has been installed is then buried in the sand due to tidal energy and sea waves. However, for the analysis the level of vulnerability for the presence of breakwaters is classified as low, because it is better than other locations in the study area without any protection against coastal erosion.

The results of the level of vulnerability analysis produced a map of the level of vulnerability to coastal erosion in Muara Asam Asam Village which can be seen in Figure 2.

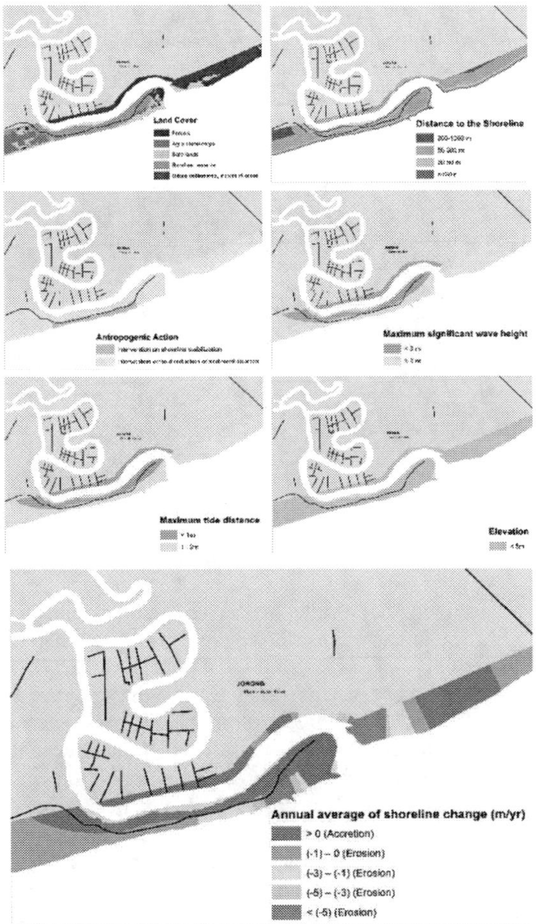

Fig. 2. Coastal erosion vulnerability assessment parameters

Based on the results of the analysis of the level of vulnerability, it was found that the zones with a high level of vulnerability were almost evenly spread across the coast of Muara Asam Asam Village with different distances to the mainland (Figure 3). The zone with the highest very high vulnerability level is located in the center of the settlement facing the sea and near the river estuary, in contrast to the eastern side near the industrial area which has a smaller area of very high vulnerability. Meanwhile, the other sections are classified into high and medium levels of vulnerability.

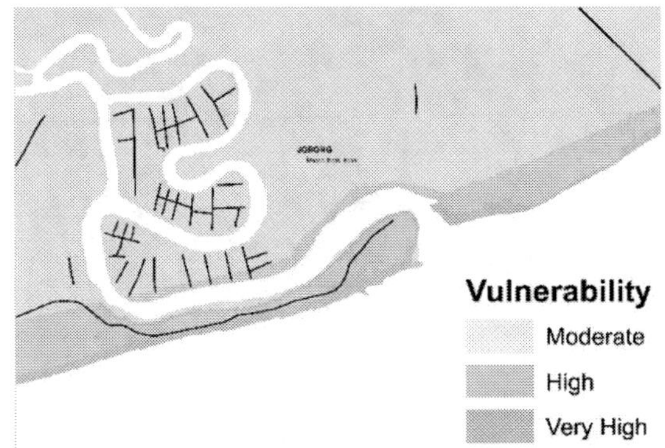

Fig. 3. Coastal erosion vulnerability map

3.3 Consequence assessment

The level of consequence is based on the estimation of the adverse impacts that may arise from the erosion hazard that occurs in Muara Asam Asam Village. The more important a location is, which is characterized by population density, distribution of ecological zones, the presence of historical sites, and the presence of massive economic activity, then the location will have major consequences if coastal erosion continues, because it will affect the economic, social and environmental structure. at that location.

The four parameters selected in the consequence level analysis are based on previous studies conducted in Aveiro, Portugal and in Macaneta Spit, Mozambique [18]. Parameters of population density, economic level, distribution of ecological areas, and distribution of historical buildings. Each is classified into 5 different values, according to the classification shown in Table 3.

Table 3. Classification of consequence parameters

No	Parameter	Level of Consequence				
		Very Low	Low	Moderate	High	Very High
1	Population density (Jiwa/Km^2)	500	500 - 1000	1000 - 2000	2000 - 4000	> 4000
2	Number of workers	0	0 – 10	10 - 30	30 - 50	> 50
3	Ecological protection zones	Not an ecological protection zone	Strategic agricultur al zones	Ecological conservati on zones	Ecological protected zones	National parks
4	Historical sites	No historical site	There are several historical construct ions	There are historical constructio ns and community activities in it	Regional historic construction	National Monument

Source: [18]

The following is an empirical assessment of the four parameters used as input in assessing the level of consequences of coastal erosion in Muara Asam Asam Village (Figure 4).

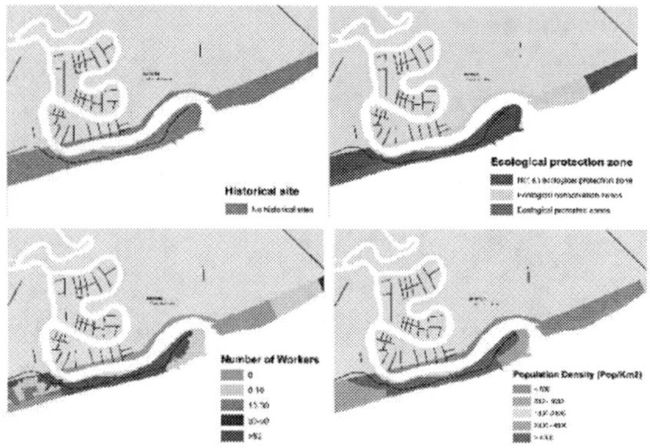

Fig. 4. Compiling parameters of consequence assessment

The economic activity is measured based on the estimated number of workers in each land unit indicated as the location of economic activity, such as agricultural land, settlement areas, ports, and industrial areas. Economic activities in Muara Asam Asam Village are dominated by the agricultural and fisheries sectors and are concentrated around the center of settlements on the west side of the river and towards the mainland. Therefore, in its coastal area, the dominant sector is fisheries, which is marked by the number of public ports and fishing boats as well as a main mode of river transportation in Tanah Laut Regency.

The level of population density is generally classified as high to very high in the settlement center of Muara Asam Asam Village, and low in other parts, because there are almost no buildings in the agricultural area or bare land on the east side of the river and the westernmost side of the study area. There are no historical sites around the coast of Muara Asam Asam Village, so the entire study area is classified as low on this parameter.

The ecological protected area is extracted based on the spatial plan map of the Tanah Laut Regency, so that the classification is very low (not an ecological protection area), low (agricultural area), medium (ecological conservation area) scattered in mangrove forest on the north side of the river, and highly classified in the coastal protection area located at the western part of the study area. The consequence assessment to coastal erosion produces an output in the form of a consequence level map shown in Figure 5.

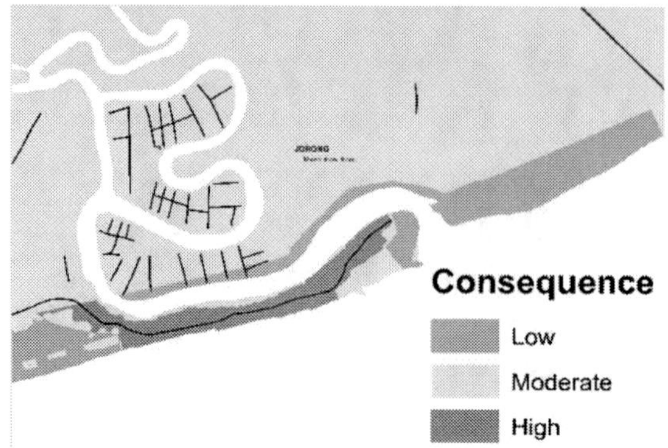

Fig. 5. Coastal erosion consequence map

Based on the results of the assessment shown in Figure 5, it is known that the consequence level map has a different value distribution from the level of vulnerability, because in this analysis it does not use the proximity factor to the coastline as a parameter or weight compiler. The weights for each parameter in the consequence analysis are the same, resulting in a proportional mix. The high consequence level is centered on the residential area in Muara Asam Asam Village, because from various parameters, this area is the biggest consequence if coastal erosion continues, because all economic and social activities are linked and centered in the settlement area of Muara Asam Asam Village.

3.4 Risk assessment

The level of risk is the relationship between the level of vulnerability and the level of consequences for the catastrophic erosion that occurred in Muara Asam Asam Village. This relationship is arranged in the risk level matrix shown in Table 4.

Table 4. Risk matrix

		Consequence Index				
		Very low	Low	Moderate	High	Very high
Vulnerability Index	Very low	Very low risk	Very low risk	Very low risk	Low risk	Moderate risk
	Low	Very low risk	Very low risk	Low risk	Moderate risk	High risk
	Moderate	Very low risk	Low risk	Moderate risk	High risk	Very high risk
	High	Low risk	Moderate risk	High risk	Very high risk	Very high risk
	Very high	Moderate risk	High risk	Very high risk	Very high risk	Very high risk

Source: [18]

From the results of this analysis is the final result of the risk assessment of coastal erosion in Muara Asam Asam Village and it is known that the distribution of zones that have a high level of risk requires special attention by making them a priority area for handling at the recommendation formulation stage. The results of the risk level analysis are shown in Figure 6.

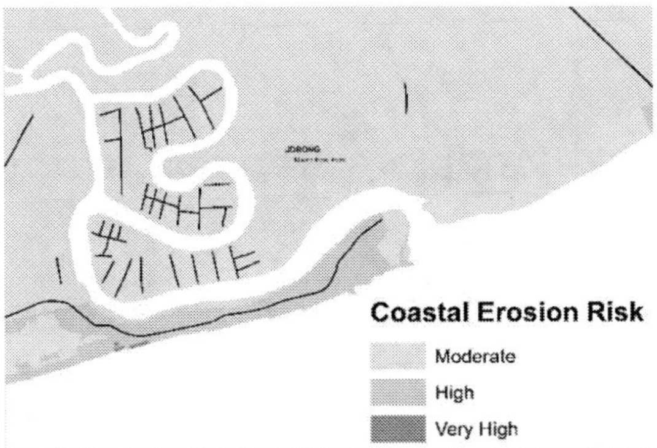

Fig. 6. Coastal erosion risk map

Based on the spatial description presented in Figure 14, the high-risk zones is concentrated on the western side of the settlement area, which in this part has not been maximally protected by erosion-retaining coastal construction. Meanwhile, the high-risk classification is evenly distributed in the residential area of Muara Asam Asam Village, which is located on the west side of the river and part of the industrial area on the easternmost side of the study area. Meanwhile, other areas, namely agricultural areas, open land and mangrove forests, are classified as zones with a medium risk level. So it can be concluded that the control of coastal erosion should be centered on the side of settlements classified as high and moderate risk, as well as in industrial areas.

4 Conclusion

The results of the study can be concluded that the coastal area of Muara Asam Asam has a high risk of coastal erosion, especially in the coastal settlement area on the west side of the study area, due to the lack of implementation of coastal hard structures such as groynes, detached breakwaters and artificial submerged reefs that can be built to significantly reduce coastal erosion. This hazard also threatens industrial activities on the eastern part of Asam-Asam Village, due to high annual erosion rate. It is necessary to combine the development of erosion-preventing infrastructure with a coastal zoning policy to protect the socio-economic activities of the community as well as the ecological environmental resources in the coastal area of Muara Asam Asam as part of sustainable coastal development.

The author would like to thank the Local Government of Tanah Laut Regency, Especially the Office of Public Works, Spatial Planning and Land, who have provided us with the opportunity to research the dynamics of the coastal area in Muara Asam-Asam.

References

1. C. Brown, E. Corcoran, P. Herkenrath, and J. Thonell, *Marine and Coastal Ecosystems and Human Well-Being* (UNEP, Nairobi, 2006)
2. IPPC, IPCC First Assess. Rep. **48**, 1334 (1992)
3. S. P. Leatherman, R. J. Nicholls, K. C. Dennis, and K. C. Dennis2, Source J. Coast. Res. J. Coast. Res. SI 15 (1995)
4. E. R. Thieler and E. S. Hammar-Klose, Open-File Rep. 00-179 1 (1999)
5. C. Sharples, Wollongong Pap. Marit. Policy **16**, 235 (2006)
6. M. Mokrech, S. Hanson, R. J. Nicholls, J. Wolf, C. M. Fontaine, S. Nicholson-cole, S. R. Jude, J. Leake, A. R. Watkinson, M. D. A. Rounsevell, J. A. Lowe, J. W. Hall, M. Mokrech, S. Hanson, R. J. Nicholls, J. Wolf, M. Walkden, C. M. Fontaine, S. N. Simon, R. J. James, P. Stansby, A. R. Watkinson, M. D. A. Rounsevell, J. A. Lowe, and J. W. Hall, J. Coast Conserv. **15**, (2011)
7. O. Ferreira, C. Viavattene, J. Jiménez, A. Bole, T. Plomaritis, S. Costas, and S. Smets, E3S Web Conf. **7**, 1 (2016)
8. M. H. Faber, *Statistics and Probability Theory In Pursuit of Engineering Decision Support*, 18th ed. (Springer, Norfolk, Virginia, U.S.A., 2012)
9. UNISDR, *2009 UNISDR Terminology on Disaster Risk Reduction* (United Nations International Strategy for Disaster Reductiom (UNISDR), Geneva, 2009)
10. ISO, *Risk Management — Vocabulary* (Swiss, 2009)
11. E. Iriadenta, Fish Sci. **3**, (2013)
12. J. Hinkel and R. J. T. Klein, Glob. Environ. Chang. **19**, 384 (2009)
13. C. D. B. Coelho, Riscos de Exposição de Frentes Urbanas Para Diferentes Intervenções de Defesa Costeira, Universidade de Aveiro, 2005
14. J. J. Ras, *Hikayat Banjar Translated by Siti Hawa Salleh* (Percetakan Dewan Bahasa dan Pustaka, Selangor Darul Ehsan, Malaysia, 1990)
15. E. Iriadenta, Fish Sci. **4**, 64 (2013)
16. Darmiati, I. W. Nurjaya, and A. S. Atmadipoera, **12**, 211 (2020)
17. C. Coelho, M.-J. Granjo, and C. Segurado-Silva, Recur. Hídricos **28**, (2007)
18. P. Narra, C. Coelho, F. Sancho, and J. Palalane, Ocean Coast. Manag. **142**, 1 (2017)

A rapid flood inundation model in the floodplain area due to an extreme rainfall event

Indah Salsabiela[1], *Kuswantoro* Marko[1*], *Mangapul* P. Tambunan[1], and *Faris* Zulkarnain[1]

[1]Department of Geography, University of Indonesia, Depok 16424, West Java, Indonesia

Abstract. Extreme rainfall in East Jakarta on February 19, 2021 caused flooding in a number of subdistricts. The research was conducted in the central part of Kali Sunter, which flows through three subdistricts, namely Cipinang Melayu, Cipinang Muara, and Pondok Bambu. The purpose of the study was to do flood hazard modeling and analyze the characteristics of flood-affected areas based on land use and topography. Inundation and flood hazard maps is done by: calculating the flood discharge using the SCS-CN method, flood inundation modelling using HEC-RAS, and analyzing the characteristics of the inundated area. This combination is effective for rapid modeling during extreme rainfall events. Based on the research, the distribution of the highest flood hazard area is in RW 004 Cipinang Melayu, with the widest inundation affecting small and medium-sized houses. The characteristics of the affected area are that there is green and empty land which reduces the potential for water to inundate buildings or other land uses. Buildings located in low-hazard housing areas tend to be more organized and relatively medium to large in size. While the types of housing in the Cipinang Melayu with a high level of danger tend to be dense and small to medium in size, but the majority have two floors as a form of flood adaptation.

1 Introduction

Flooding is one of the major natural disasters that could create severe catastrophic impacts for population, infrastructure, economy, and environment [1-4]. Developing countries are likely to be more prone to flooding considering lower incomes, more housing facilities below standard, and lack of warning systems and preparedness [5]. Extreme and heavy rainfalls happened in Jakarta because of various environmental and climatic conditions [4, 7]. Causes of the floods in Jakarta are mainly because of heavy rainfall, strongly supported by climate change [7]. Climate change plays a role in changing the cycle of hydrology by affecting precipitation, soil moisture, surface runoff, and the recharge rates of groundwater [6]. In developing countries such as Indonesia, hydrological balance is also disrupted by other factors such as rapid growth of settlements and urbanization which makes the increasing amount of population in cities [8]. Jakarta is also facing another major challenge which is land subsidence. Flooding usually occurs in a relatively high rate of land subsidence areas, and could somehow affect the flooding in both inland and coastal areas [9, 10]. Subsidence within inland areas usually occurs near the [1]river, and along with river canal changes and malfunction of the flow system, it will cause riverine flooding aggravation [9].

The study was held in the middle part of Sunter River, which flows through three sub-districts in East Jakarta, namely Cipinang Melayu of Makasar District, Cipinang Muara of Jatinegara District, and Pondok Bambu in Duren Sawit District. As of 19 February 2021, some parts of East Jakarta, was heavily flooded because of extreme rainfall event. Land use/land cover change (LUCC) in the city is causing the reduction of reservoir areas, which will escalate flood hazard in distinctive places [11]. Rapid development consequently changing the former land use into built-up areas for industrial, residential, and commercial purposes. This expansion and intensification caused by urbanization, added by the topographic condition where most of the areas are floodplains, will result in an infrastructure deficiency which includes flood control. The most affected people are the ones in lowland, from human settlement to livelihood areas [4]. Therefore, it is urgently needed to do flood hazard analysis rapidly in order to help rescue the affected areas through spatial modelling [1]. Flood risk mitigation plans should consider both land use change and land subsidence [12]. One of the products of flood modelling is the inundation map, which is an important tool that can be used by many stakeholders and policymakers to make emergency decisions and plans.

In this study, HEC-RAS is a hydraulic modelling used to produce micro-scaled flood inundation maps effectively. The integration between GIS and HEC-RAS Model makes it easier to produce the maps and count the flood discharge. This study is aimed to find out the potential extension and depth of impacted areas by

* Corresponding author: kuswantoro@sci.ui.ac.id

modeling the distribution of flood hazard areas and analyze the characteristics of flood-affected areas based on land use and topography. Another purpose of this study is to help produce flood hazard maps within floodplain areas to measure the hazard and potential risk in the future.

2 Methods

2.1 Study Area

This research is conducted in the middle part of Sunter River segmentation, which includes a river length of 5.56 km with an area of 5 km^2. The topography of Sunter River as a whole has a height difference of 108.5 m, with maximum height of 120 m and minimum height of 11.5 m. Sunter River is one of thirteen rivers that flows through Jakarta. Located in 6° 10' – 6° 22' LS dan 106° 51' – 106° 55' BT with 36,6 km long and the area of 76 km^2, the upper stream of Sunter River is located in Tapos, Depok, West Java, and ends in Koja, North Jakarta and run through sub-districts, such as Cipinang Melayu, Pondok Bambu, and Cipinang Muara. The surface elevation ranges from 0-34.58 m with the slope of 0.3%. Sunter River generally has 8 tributaries with the type of bird's-feather watershed, shown in Figure 1. The width of Sunter River ranges from 3-10 m, and is considered as a small river.

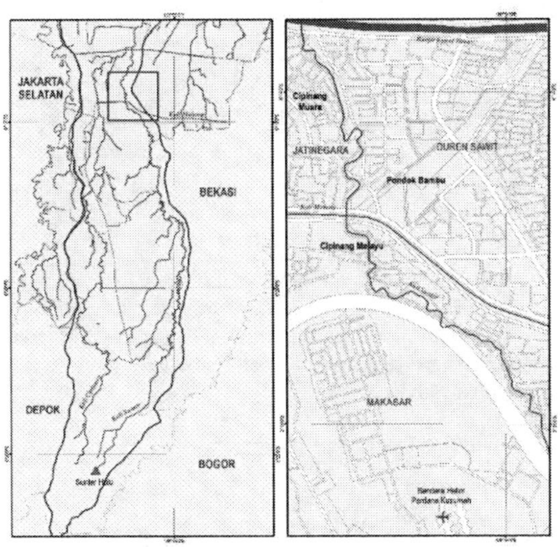

Fig. 1. Study Area and Upper-stream.

This river went through normalization in 2011, meaning it has been designed for the betterment of the channel condition through fixing the river bank and the dimension [13]. Normalization was meant to provide enough river flow capacity to keep the water flowing from the upstream, especially during extreme weather and intense rainfall capacity. Other benefits are to prevent the banks or cliffs from erosion, as well as to return back to the initial river profile to help mitigate the flood.

Generally, the climate of DKI Jakarta is influenced by monsoons with dry season period from June to September and rainy season period from November to March. In this study area, rainfall with high intensity tends to occur from January to March, with the peak intensity in February, as seen on Fig. 2. This happens because February is in the Equator thermal line or the intertropical convergence area. Therefore, the evaporation rate raises as temperature increased, which then causes intense and heavy rains. Observations of climate elements are carried out by the Meteorology, Climatology and Geophysics Agency (BMKG) which include observations of air temperature, humidity, wind speed, air pressure, amount of rainfall and number of rainy days. DKI Jakarta Province has two climate monitoring stations, including: the Kemayoran Meteorological Station in Central Jakarta and the Tanjung Priok Meteorological Station in North Jakarta. According to the BMKG in the 2020 Jakarta Statistics book, DKI Jakarta Province is one of the cities in Indonesia with high temperatures, which generally reaches 35.6°C at the maximum value and 24°C at the minimum value with an average humidity of 77.7% at Kemayoran Station and 75.5% at Tanjung Priok Station.

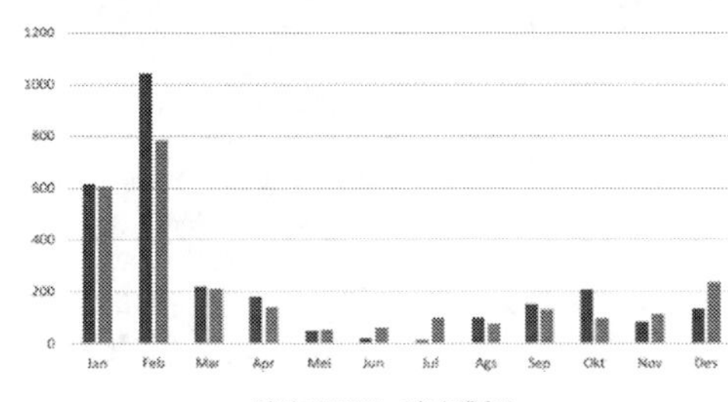

Fig. 2. Rainfall data in Jakarta, 2020.

History of floods in Jakarta has been going on for a long time, but there are three flood phenomena that are among the worst floods happened in the last two decades, which are in 1996, 2002, and 2007 [14]. The flood was classified as severe, both on a regional and national scale, due to its impacts to the wellbeing and mortality. Rainfall data used in this study is rain data when a flood occurs on 19 February 2021, aiming to calculate efficiently. Rain data were obtained from the BMKG Station and Jakarta Flood Monitor. Data verification is done by observing the nearest rain station that represents the research area and with good data conditions.

According to the UN World Population Prospects in Macrotrends, population growth rate in Jakarta has increased over the past decade [15]. In 2011, the population rate was 1.11% and in 2021, it is now 1.35%. Land use change keeps increasing because of the development in the city, such as build-up areas, specifically residential areas, industrial areas, and commercial areas. This condition could potentially harm the people who are living near the floodplain, or the ones who reside within the lower stream area.

During the calculation of major floods, land use is a very influential factor on surface water runoff. Fluvial landforms have the potential as an area that is prone to flooding. These vulnerable areas are not necessarily affected by flooding, depending on several other triggers such as land use change. The majority of land use within the study area is built-up area, while the area of wet land or water bodies relatively less or decreased.

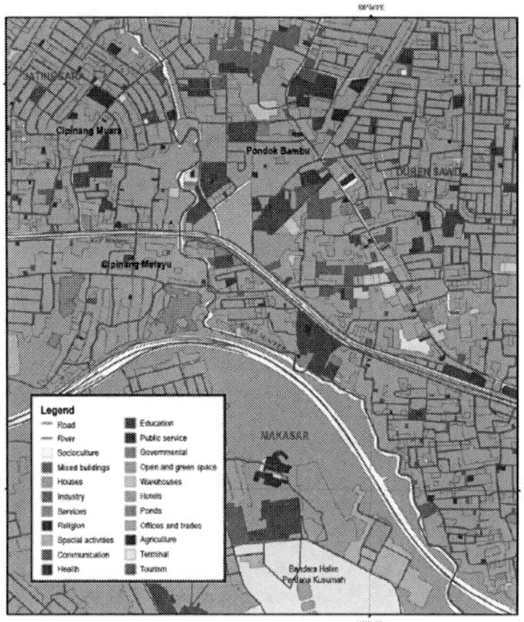

Fig. 3. Land use of the study area.

There are 21 types of land use for the research area on a scale of 1:13,500, as seen on Table 1.

Table 1. Land use type per subdistricts.

Types of Land Use	Sub-district		
	Cipinang Muara	Cipinang Melayu	Pondok Bambu
Socio-culture	4	-	1
Mixed buildings	-	7	20
Houses	235	369	497
Industry	1	2	19
Services	14	32	45
Religion	25	37	40
Special activities	-	7	-
Others	-	5	3
Health	11	6	21
Education	24	31	39
Public services	8	2	13
Government	1	2	5
Vegetation and open space	91	146	197
Warehouses	8	7	25
Hotels	-	1	1
Ponds	-	5	1
Offices and trades	39	47	147
Agriculture	-	-	1
Terminal	6	7	4
Tourism	17	48	46

The upper-stream part of the Sunter River has 7 types of land use on a scale of 1:125,000, as shown in Figure 4. The land uses include: lakes/situ, jungle forests, domestic airports, plantations/gardens, settlements, rice fields, and fields/fields. In this map scale, it is also found that the dominance of the land use is housing and settlements. Hulu Kali Sunter is a field or golf course with surrounding settlements. Hulu Kali Sunter is located on a higher ground, namely in Tapos District, Depok City, West Java, along the border between East Jakarta City and Bekasi City.

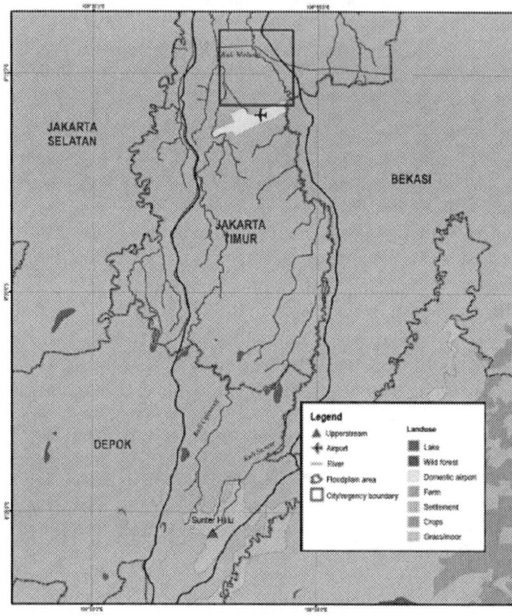

Fig. 4. Land use of the Upper-stream area.

Along the upstream to the study area, the type of soil around Kali Sunter is dominated by volcanic plains. In general, the rock structure in DKI Jakarta is clay and is a quaternary volcanic deposit. Soil texture is used to consider HSG.

Table 2. Soil hydrological classification based on USDA.

Group	Discharge potential when wet	Infiltration	Soil Type	Soil Texture
A	Low	High	Sandy soil, gravel	Sand, clay sand, sandy loam
B	Semi-low	Medium	Medium grain soil	Silt, clay
C	Semi-high	Slow	Medium-fine grained soil	Sandy clay
D	High	Very slow	Clay with high swellability, soil with high permanent water table, clay soil	Clay (assorted)

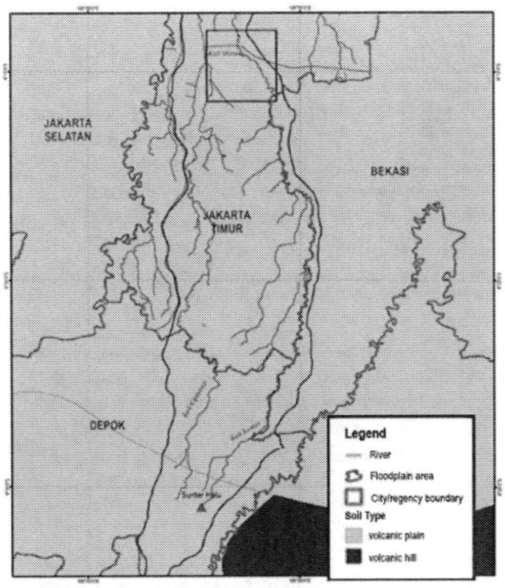

Fig. 5. Soil type of the Upper-stream area.

2.2 Hydraulic and Flood Modelling

In this study, hydraulic modelling is implemented to help evaluate the importance of surface flow, such as for forecasting and producing inundation maps. The software that is used is HEC-RAS version 5.0.1, a product of the US Army Corps. of Engineers Hydrologic Engineering Center for River Analysis System, as a tool to model and perform one-dimensional hydraulic calculations in a natural and constructed channel network, for 1D steady and 1D or 2D unsteady flow. In this study, 1D steady flow simulation is used to model the system that is intended to calculate water surface profiles which can handle a single river reach. This software was chosen due to its effectiveness and capability of modelling mixed flow regime water surface profiles.

HEC-RAS is an integrated and interactive software with graphical user interface (GUI) to separate analysis, data storage and management, graphics, and other facilities. It can be used for multi-tasking and multi-user network environments. There are four hydraulic analysis components in the system, including: steady-flow water surface profile; 1D and 2D unsteady-flow simulations; sediment transport computations; and water temperature transport modelling [15]. Steady flow simulation is used in this study, with special features and parameters for steady flow components, such as: obstructions, bridges, culverts, weirs, spillways, levees, and other structures within the floodplain can be considered in the computations. The steady flow simulation is designed for further evaluation in the study of floodplain management and flood control.

HEC-GeoRAS is an integrated extension tool to support using HEC-RAS in ArcGIS, targeted for geographers and planners because it combines engineering and GIS. This study uses HEC-GeoRAS version 10.5 suitable for ArcMap version 10.5.1. Some benefits of using HEC-GeoRAS are: having spatial links when inputting data, doing hydraulic modelling, and mapping floodplain or inundation; utilizing detailed topographic information; able to identify basin areas and the interactions effectively. Combining HEC-RAS and GIS improves modelling efficiency and produces a better and more accurate visual in the results.

2.3 Flood Discharge Calculation

Flood discharge is calculated by rainfall-runoff depth and curve number with the method known as Soil Conservation Service Curve Number (SCS-CN) by the U.S. Department of Agriculture. This method is one of the most popular methods to calculate the volume of surface runoff in urban watersheds with a certain amount of rainfall because it is easily applicable, accessible, stable, and useful for ungauged watersheds [16]. The parameters for the curve number are: soil-vegetation-land use (SVL) complex; hydrologic condition; antecedent moisture condition; and climate of the watershed. The total of rainfall precipitation (*P*) with the potential of soil retention (*S*) are divided into runoff discharge (*Q*) and initial abstraction (*Ia*) with the equation as follows.

$$Q = \frac{(P - I_a)^2}{(P - I_a) + S} \qquad (1)$$

To calculate the depth of excess rainfall, initial abstraction correlation value is obtained with Equation 2 and to calculate antecedent moisture condition, S refers to Equation 3 with the conversion from inch to mm.

$$I_a = 0.2\,S \qquad (2)$$

$$S = 25.4\left(\frac{1000}{\mathrm{CN}} - 10\right) \qquad (3)$$

In this study, the soil type belongs to Group D, according to the Hydrologic Soil Group. Group D soils have a low infiltration rate when thoroughly wetted, from 0.05-0.15 inch/hour. The soils mainly contain a constraint layer that limits the downward movement of water. The textures are usually clay loams, shallow sandy loam, and soils low in organic content, which indicate a slow rate of water transmission. Urban lands tend to have low permeability and the characteristics of the watersheds area and its surroundings are residential areas, commercial areas, and roads. Hydrologic condition depends on how much vegetation covers the land. A good hydrologic condition allows more infiltration than a poor hydrologic condition because it helps protect the watershed from erosion for the purpose of soil conservation. The initial abstraction consists of interception, surface detention, evaporation, and infiltration [16]. The relation between rainfall intensity and time duration is that the more intense the rainfall, the lesser will be the time duration, and it applies alternatively.

2.4 Flood Inundation Mapping

It requires three main steps to do the inundation mapping, which include: pre-processing using HEC-GeoRAS, HEC-RAS to process the exported data from ArcMap, and post-processing to map flood inundation areas [17].

The data required are: digital elevation model (DEM) with 10x10 m spatial resolution to create triangular irregular network (TIN) and a satellite image of the study area sourced from Google Earth and taken through SAS Planet. The required parameters to process hydraulic models in HEC-RAS are: stream centerline, main channel banks, cross-section lines, which are called channel geometry. A total of 19 cross sections were taken over the single reach modelled as seen in Figure 6.

Pre-processing requires DEM and a satellite image to facilitate the river geometry digitization process using the HEC-GeoRAS extension in ArcMap. The layers of geometry required are: river stream centerlines, riverbank lines, cross-sectional cut lines, flow path centerlines. Other geometry layers are considered optional, such as: bridges/culverts, ineffective flow areas, blocked obstructions, land use, levee, inline structures, and lateral structures layer. The collected geometry files were exported to the HEC-RAS model to edit the geometry per cross-sections, determine the discharge flow, and generate the steady flow simulation and water surface profiles. The files are exported back to ArcMap to visualize flood inundation maps and later to analyze and determine the affected area and area at risk.

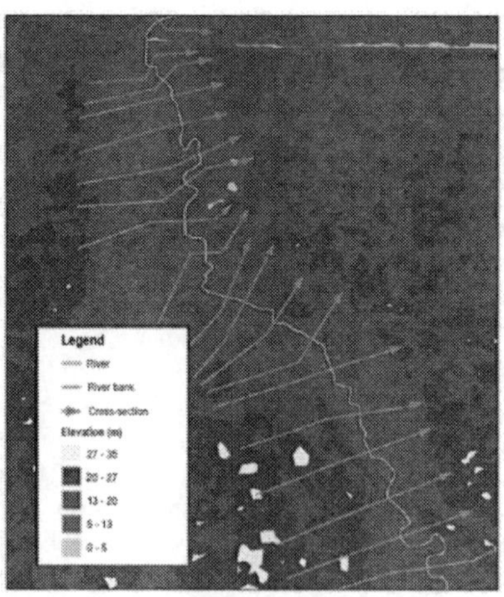

Fig. 6. Cross-section of study area.

3 Results

Calculation of flood discharge is a calculation with hydrological factors such as rainfall and the characteristics of the river segment. Characteristics to determine the curve number (CN) in the calculation of flood discharge using SCS-CN method in the form of topography, land use, and soil type. The output from the calculation of flood discharge is used to delineate the flood inundation. Rainfall data based on BMKG from data from the Halim Perdanakusumah Rainfall Station on February 19, 2021 at 15:00 WIB is 160 mm/day.

CN value is obtained based on land use and soil hydrological groups. The type of land use in the upstream area is a built-up urban area with a subtype of open space, namely golf courses with grass cover exceeding 75%. Soil characteristics in the upper stream and study area determine the class of soil hydrology. Soil belongs to group D which tends to have saturated hydraulic conductivity from the least transmissive layer, the depth of the impermeable layer is in the range of 50-100 m. The hydrological condition of class D soil has a high potential for runoff when wet conditions with a slow infiltration rate, therefore making it an area that is easily flooded. Class D soils have between 40% clay and <50% sand and have textures of loam, silt loam, sandy loam, loamy loam, and silty loam. By looking at the curve table, the CN used is 80, seen on Table 3.

Soil retention potential or *S* is the capacity or ability of a river to sustain extreme rainfall. The runoff will remain zero until the accumulated rainfall exceeds the initial abstraction (*Ia*). From the manual calculations, the maximum flood discharge results for some time when the flood is in progress is 102.98 m^3/s.

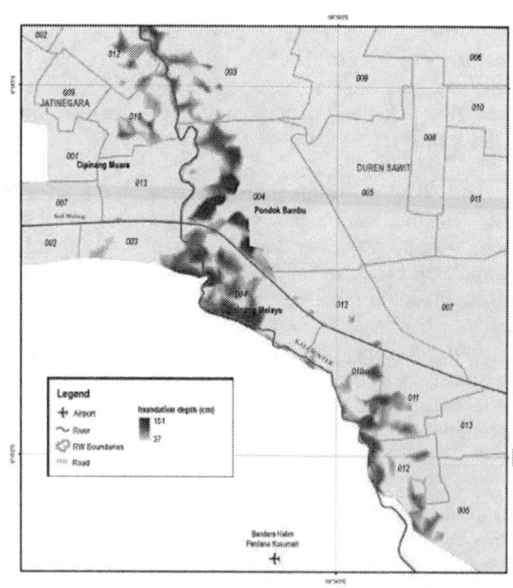

Fig. 7. Flood Inundation Map.

Table 3. Curve Number on Technical Release (TR-55) USDA.

Description	Curve Numbers for HSG			
Land cover type and hydrologic condition	**A**	**B**	**C**	**D**
Fully-developed urban areas				
Open spaces (lawns, parks, golf courses, cemeteries, others):				
Poor conditions (grass cover <50%)	68	79	86	89
Moderate conditions (grass cover 50% - 75%)	49	69	79	84
Good condition (grass cover >75%)	39	61	74	80

From the HEC-RAS modeling, it is known that the total flow area based on the cross section is 143.63 m^2 on the left side of the river body (overbank flow) and 12.64 m^2 on the right. The total flow rate is 102.93 m^3/s with the flow in the puddle on the left is 98.75 m^3/s and 4.18 m^3/s on the right. The surface width of the inundated cross-section on average is 164.20 m. The height of the water surface calculated from the energy equation is 14.96 m and the critical water level is 14.02 m. The average total flow velocity based on the cross section is 0.66 m/s. The transport discharge is 4,570 m^3/s with the main flow length of 428.56 m.

The dominant flood inundation is on the left side of the river body which includes Cipinang Melayu and Cipinang Muara Sub-districts. After normalization happened in 2011, the river segment that separates the Cipinang Muara from Pondok Bambu is higher than the housing in the eastern part or the settlements around the Pondok Bambu. The maximum depth is 1.51 m in the main stream, while the maximum inundation depth is 1.11 m and the minimum depth is 0.37 m.

Characteristics of flood-affected areas are determined based on land use. The large number of residential types of land use indicates the high population of a village and the number of buildings in the coverage area of the study. With a total of 497 residential buildings, Pondok Bambu is the sub-district with the most types of residential buildings when compared to the other two. In detail, the types of land use affected by flooding in the scale of Rukun Warga (RW) or local community will be discussed.

The level of flood hazard is determined by the depth of water inundation at a location or land use with a high population. The same water level in residential areas with open land has a different level of danger. Settlements are more vulnerable to flooding than open land because the safety and welfare of the community is threatened. Although Kalimalang River near Cipinang Melayu area and BKT was not affiliated with the cause of the flood that happened. The reason is because the canal and Kalimalang flow condition was stable and there was no discharge. BKT acted as an outlet for the flood near Cipinang Indah Residence.

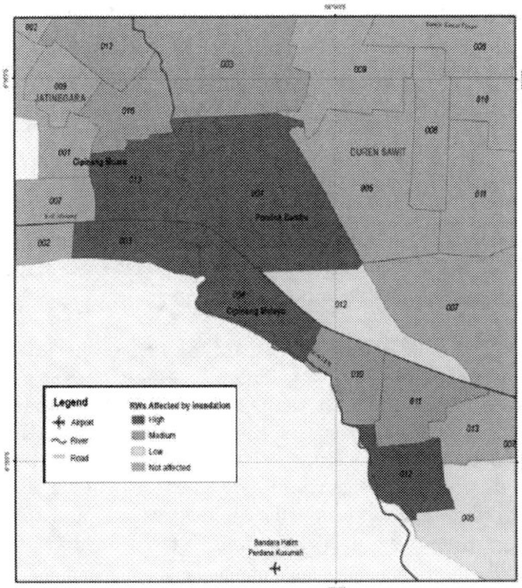

Fig. 8. Flood Inundation Map per RW.

The total of 10 RWs are passed by or directly adjacent to the Sunter River, namely RW 012, 013, 016 from Cipinang Muara, RW 003, 004 from Pondok Bambu, and RW 003, 004, 005, 010, 012 from Cipinang Melayu. From the result of overlaying the administrative map with the flood inundation map, we get a map of the RW affected by flooding with levels, as shown in Figure 8.

Pondok Bambu Village was hit by floods in 2 RWs, namely RW 003 and RW 004. The height of inundation in both RWs varied, at RW 003 it was 31-70 cm high and in RW 004 it was 151 cm. Types of land use affected in RW 003 are houses of very small type or size and medium houses, shop houses, green space and open space, and city parks. The type of house that dominates in RW 003 is a medium-sized house located in Cipinang Indah Residence, and RW 003 has green land and large vacant land, both on the banks of the river and around the housing estate. In contrast to RW 004, the flood inundation has an impact on public facilities compared to houses or housing. There are several types of housing in RW 004 with small types. The types of land use affected in RW 004 are as follows.

Access from the Kali Malang Channel Inspection Road to Jalan Cipinang Indah Raya II goes downhill. On the border between RW 003 and RW 004, there is green land and vacant land which is quite large and is located lower than buildings and roads, so that puddles are not trapped on concrete land and are more easily absorbed into the ground. There is an upper secondary education service building located on the edge of the river, but the school is not affected because the location of the

https://doi.org/10.1051/e3sconf/202132501024

building is higher than the road, so that runoff water does not inundate the school, but goes to a green open land located on Jalan Kutilang. Small-type houses located on the banks of the river or the border between Kelurahan Pondok Bambu and Kelurahan Cipinang Muara are higher than road level, so they have bridge access to it.

RW 012 and RW 016 are dominated by housing of Villa Cipinang Indah which has various types of housing, ranging from small, medium, to large types. Meanwhile, in RW 013, most of the houses affected are small in size, namely houses on the banks of the river, bordering the Pondok Bambu Village.

The most affected areas are the one in the lowest slopes, such as RW 004, Cipinang Melayu, areas near Mall Cipinang Indah, and Cipinang Indah Residence near East Flood Canal or Banjir Kanal Timur (BKT). According to rescuer team, it reached 90-150 cm depth at its peak and the maximum depth was calculated to be 151 cm. The flood mostly affected houses and other infrastructures, such as commercial industries and public facilities.

Fig 9. Flood condition in RW. 04, Cipinang Melayu.

4 Conclusions

The distribution of the highest flood hazard area is in RW 004, Cipinang Melayu Village. Among the three sub-districts, Cipinang Melayu is also the one with the largest inundation. In this RW, there are several types of land use, such as residential, green land, vacant land, houses of worship, and higher education. The dominance of land use is small and medium-sized houses. In RW 012, 013, and 016 Kelurahan Cipinang Muara, the inundated areas are categorized as low, with the affected buildings in the form of houses and green land. Meanwhile, RW 003 and RW 004 in Pondok Bambu Village have a moderate to high level of danger.

Characteristically, the western part, such as Cipinang Muara and part of the Cipinang Melayu Village, is higher than the eastern part, such as Pondok Bambu. Green land is useful in water infiltration, so the wider the green land, the more water will be absorbed into the soil, so it does not cause inundation. A large vacant land is located in RW 004 Pondok Bambu Village, so water will only inundate the land and reduce the potential for water to inundate buildings or other land uses. The types of buildings and dwellings in the Cipinang Muara Village housing estates tend to be more organized and relatively medium to large in size and have one or two floors. While the types of housing in the Cipinang Melayu Village tend to be compact and small to medium in size, but the majority have two floors.

Extreme rainfall which causes rivers to overflow is the main cause of flooding in the Cipinang Melayu Village and surrounding areas. HEC-RAS is useful as an effective hydraulic modeling tool so it can be used for real-time modeling. HEC-GeoRAS is an ArcGIS integrated extension for hydraulic modeling with available georeferenced data. Flood inundation mapping is important for mitigation purposes and the results can be informed quickly.

With limited data and time, this study only reviews the physical aspects, starting from flood hazard modeling to analyzing the characteristics of the affected area. The results of this study can be used as a reference in the development of research on the social characteristics and resilience of the affected community if there are adequate data related to the loss of flood events.

This research is one of the requirements for graduation in the Department of Geography, Faculty of Mathematic and Natural Science, University of Indonesia. The authors would like to thank the Directorate of Research and Community Engagement (DPRM) University Indonesia for the research funding program Publikasi Terindeks Internasional (PUTI) Saintekes in 2020 No. NKB-4876/UN2.RST/HKP.05.00/2020

References

1. A. Cook, V. Merwade, J. Hydro. **377(1-2)**, 131–142 (2009)
2. F. Hajibayov, B. Ozkul, F. Terzi, Geographical Information Science Research- UK (GISRUK) April 2017, University of Manchester, UK (2017)
3. A. Bannari, G. Kadhem, *IGARSS 2019,* in IEEE International Geoscience and Remote Sensing Symposium, 28 July-2 Aug. 2019, Yokohama, Japan (2019)
4. A. Pathan, P. Agnihotri, Intl. J Recent Tech. Eng. **8 (1)**, 1410–1417, (2019)
5. I. Susanti, *Improving Lesson Learnt in Disaster Management,* in MATEC Web of Conferences 229, International Conference on Disaster Management, ICDM, 2-4 May 2018, Padang, Indonesia (2018)
6. A. El-Naqa, M. Jaber, J Civ. Env. Eng. **8 (5)**, (2018)

7. B. Takhur, R. Parajuli, A. Kalra, S. Ahmad, R. Gupta, World Environmental and Water Resources Congress, 240–251, California, USA (2017)
8. Y. İcaga, E. Tas, M. Kilit, Acta Geobalcanica **2(2)**, 111–118 (2016)
9. H.Z. Abidin, H. Andreas, I. Gumilar, I.R. Wibowo, *Changes in Flood Risk and Perception in Catchments and Cities* in Proceeding IAHS, **370**, 15–20, IUGG, 22 June-2 July 2015, Prague, Czech Republic (2015)
10. H. Takagi, M. Esteban, T. Mikami, D Fujii, U. Climate **17**, 135–145 (2016)
11. Z. Liu, L. Cuo, Q. Li, X. Liu, X. Ma, L. Liang, J. Ding, Water **12 (4)**, 1198 (2020)
12. I.R. Moe, S. Kure, N.F. Januriyadi, M. Farid, K. Udo, S. Kazama, S. Koshimura, Hydro. Rsch. Letters **11 (2)**, 99–105 (2017)
13. E.K. Ovilia, E. Primadianti, D Yatmadi, Const. Mat. J. **2 (3)**, 219–228 (2020)
14. M. Tambunan. *Strengthening Sustainbility Development*, in MATEC Web of Conferences 229, in International Conference on Disaster Management, ICDM, 2-4 May 2018, Padang, Indonesia (2018)
15. UN World Population Prospects, Jakarta, Indonesia Metro Area Population 1950-2021. www.macrotrends.net. Retrieved 2021-04-15.
16. C.T. Ackerman, GIS Tools for Support of HEC-RAS using ArcGIS®, HEC-GeoRAS User's Manual (US Army Engineers-HEC, California, 2011)
17. S.K. Mishra, V. Singh, Soil Conservation Service Curve Number (SCS-CN) Methodology (2003)
18. E. Kusratmoko, K. Marko, M. Elfeki, J. Geo. Env. Earth Sci. Intl. **5 (3)**, 1–10 (2016)

Modified RMR to determine rock mass quality, study case of diversion tunnel in Meninting Dam, West Nusa Tenggara, Indonesia

Brilyan Jati Wijaya[1,2], *I Wayan* Warmada[1]*, and *I Gde* Budi Indrawan[1]

[1] Departement of Geological Engineering, Gadjah Mada University, Yogyakarta, Indonesia
[2] Regional Infrastructure Development Agency, Ministry of Public Works and Housing, Indonesia

Abstract. The Meninting Dam under construction on Lombok Island, West Nusa Tenggara, Indonesia, requires a good planning to build a diversion tunnel to support its development and mobilization. The diversion tunnel is planned to be built through rocks with medium to poor rock mass quality. The planning stage involves various parameters, i.e., the rock mass classification, using either the RMR or GSI method. Converting values from one method to another makes planning work easier. However, the constraints found were the limitations of the observational data, such as discontinuity conditions. The objective of this article is to discuss the alternative depiction of discontinuity conditions in rock mass using RMR method. An alternative equation was developed to obtain a prediction model for determining the RMR value, based on GSI data. The evaluation showed that the mathematical models developed in this research had a small gap of error compared to other values. The models then can be used to predict RMR value based on GSI data and vice versa, with a higher degree of accuracy and precision according to the actual rock surface conditions, especially in the construction site of diversion tunnel at Meninting Dam.

1 Introduction

The construction of a dam is intended to optimize water resource utilization and to reduce flood disasters in rural areas. When building a dam, there is a stage to divert river or water flow. The deflection of that water flow is aimed to create dry conditions for constructing the main dam and support the mobilization of heavy equipment and building materials.

In conducting diversion of the river, BWS Nusa Tenggara I, as the stakeholder, utilizes a diversion tunnel. The diversion tunnel at the Meninting Dam is being built through rock units of pyroclastic breccias and lapilli tuff as shown on the geological regional map by Mangga et al [1]. Wiyasri [2] in 2020 stated that rock mass quality in that site had fair to poor rock mass quality as shown in Fig.2. The rock mass quality at the construction site had been determined using RMR method. Comparable values need to be calculated to describe and to ensure the rock mass quality, since RMR could not provide some results for complex computation i.e., failure criterion [3]. One of the methods which can be used to determine rock mass quality is GSI method. GSI offers simplicity in determining rock mass quality, as it based on qualitative observation. But problems were found when correlating RMR and GSI, some sub parameters which used in RMR were reduced in GSI, and they made the degree of accuracy and precision declined. Thus, a model needs to be developed to close the gap and to make the correlation model more reliable.

*Corresponding author: warmada@ugm.ac.id

1.1 RMR

Rock Mass Rating (RMR) is a method used to classify the quality of rock mass. RMR_{89} was first developed by Bieniawski [4] in 1973-1989 and continues to be developed until the recent years. RMR is a method to determine rock mass quality by combining qualitative and quantitative observations. In RMR_{89}, there are 5 parameters used to determine the quality of rock mass, as shown on Table 1, they are the number of unconfined compressive strength, the amount of Rock Quality Designation (RQD), spacing of discontinuities, discontinuities condition, and the presence of groundwater, each of them then categorized into R_1, R_2, R_3, R_4, and R_5.

The condition of the discontinuities (then called R_4) consists of several sub-parameters, including the length of the discontinuities (persistence), the aperture of the discontinuities, the condition of the gap filling material, the degree of roughness, and weathering. Then RMR_{89} value is summarized from each rating of R_1 to R_5. In its development, RMR_{89} also adopts R_6, where this parameter describes the effect of discontinuities orientation. The weight of the parameter R_6 is from to below zero, considering that the discontinuities orientation tends to disserve the rock mass quality.

https://doi.org/10.1051/e3sconf/202132502001

Table 1. A set of parameters of RMR_{89}.

No	Parameter		Range Values						
1	Strength of intact rock material	Point load strength index (MPa)	> 10	4 - 10	2 - 4	1 - 2	For this low range, uniaxial compressive test is preferred		
		UCS (Mpa)	> 250	100 - 250	50 - 100	25 - 50	5-25	1-5	<1
	Rating		15	12	7	4	2	1	0
2	RQD (%)		90 - 100	75 - 90	50 - 75	25 - 50	< 25		
	Rating		20	17	13	8	3		
3	Spacing of discontinuities		> 2 m	0.6 - 2 m	200 - 600 mm	<60 mm			
	Rating		20	15	10	8	5		
4	Condition of Discontinuities	Persistence	< 1 m	1 - 3 m	3 - 10 m	10 - 20 m	> 20 m		
		Rating	6	4	2	1	0		
		Aperture	None	< 0.1 mm	0.1 - 1 mm	1 - 5 mm	> 5 mm		
		Rating	6	5	4	1	0		
		Degree of Roughness	Very rough	Rough	Slightly rough	Smooth	Very smooth		
		Rating	6	5	3	2	0		
		Infilling	None	Hard < 5 mm	Hard > 5 mm	Soft < 5 mm	Soft > 5 mm		
		Rating	6	4	2	2	0		
		Degree of Weathering	Fresh	Slightly weathered	Moderately weathered	Highly weathered	Completely weathered		
		Rating	6	5	3	1	0		
5	Groundwater	Inflow per 10 m tunnel length	None	< 10	10 - 25	25 - 125	> 125		
		Ratio: joint water pressure/major principal stress	0	< 0.1	0.1 - 0.2	0.2 - 0.5	> 0.5		
		General conditions	Completely dry	Damp	Wet	Dripping	Flowing		
		Rating	15	10	7	4	0		

1.2 GSI

In the rock mechanics, Hoek-Brown failure criterion often utilized to determine rock mass deformation. The Hoek-Brown [3] failure criterion uses several input parameters which some can be obtained from RMR. However, the values of those parameters derived from RMR observations did not work well on rock with very poor quality, since then Hoek [5] stated that minimum value of RMR_{89} is 23.Thus, in the following years Hoek and Brown [6] developed a rock mass quality classification method called GSI. GSI (Geological Strength Index) is a method to determine rock mass quality based on visual observations and it is qualitative, as stated by Sonmez and Ulusay [7]. Fig. 1 shows the chart used to interpret rock mass quality. In principle, the determination of rock mass quality using GSI method is carried out by observing the surface and the structural conditions on the rock surface, as stated by Sonmez and Ulusay [8].

Before correlating GSI and RMR_{89}, a conceptual comparison of the use of both methods is carried out as follows.

1.3 Parameters

Ceballos et al [9] compared the parameters used in classifying rock mass quality, both in RMR_{89} and GSI as shown in the Table 2 below. In RMR_{89}, there are 5 parameters in minimal that used to determine the rock mass quality (excluding orientation of discontinuities). Each parameter in the RMR has its own weight, with a total value of 100. Meanwhile in GSI, the determination of rock mass quality is worked qualitatively by observing the condition of the very rock mass structure and the condition of the rock surface, where those two parameters can be considered to have the same weight.

Table 2. Parameter used in GSI and RMR_{89}

Parameter	Weight	
	RMR_{89}	GSI
UCS	15	0
Structure of Rock Mass	40	50
Condition of Discontinuities	30	50
Presence of Water	15	0

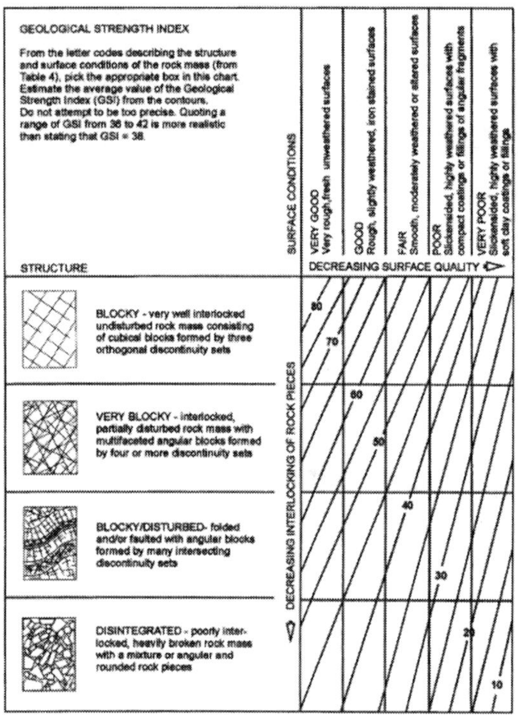

Fig. 1. Chart of GSI that used for determining rock mass quality.

Apart from the parameters, the two methods above are also distinguished by their intended use. RMR was created and developed to determine the quality of the rock mass as a matter for tunnel planning. Some of tunnel planning using this method for deciding the excavation method, supporting, standing time analysis, and tunnel lining calculation.

Meanwhile, GSI was developed to determine the value of rock mass quality in more simpler ways. Then value of that process is usually used to determine geotechnical calculations, such as deformation analysis and shear strength, as stated by Ceballos et al [9].

1.4 Objectives

Based on GSI and RMR values that had been obtained from direct observation, a model can be developed to correlate them both. Some correlation models had been developed on the previous research and can be used to show the adjacency between each other's. Then, to convert the values from a method to another is highly possible to do. But in the other hand, an issue found while describing the parameter of discontinuities condition on RMR. It is quite hard to interpret the condition of discontinuities when observable data are limited, for example when the available data is only from the core box. Thus, the objective of this research is to develop an alternative way in describing discontinuities condition, and then correlate the RMR and GSI vales and form it into model which solves the issue. The model formed shall consider the degree of accuracy and precision.

2 Materials and methods

2.1 Geological condition

Rock mass quality determination in the diversion tunnel of Meninting Dam was carried out through quantitative and qualitative observations on the rock samples acquired from drilling at 4 boreholes, as shown in the Fig. 2 below. The dataset from 4 boreholes consists of 205 subsurface rock samples that stored per meter in the core box.

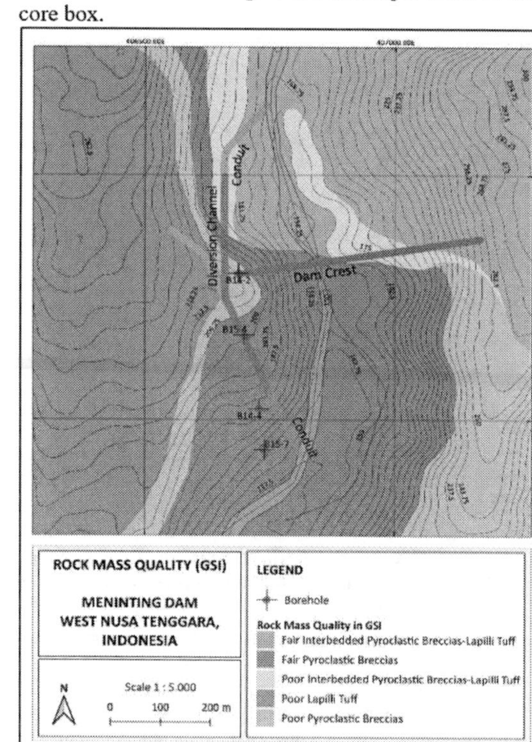

Fig. 2. Geological map and its rock mass quality using GSI method.

Determination of rock mass quality using GSI was carried out by observing the rock structure in the core box and each surface condition. The rating of each parameter was given per meter. Measurements were also conducted to the rock samples which were more than 10 cm in length in a meter. The accounted data then used to summarize the percentage of RQD. Equation (4) below is used to calculate the final value of GSI, where J_{cond} is observation rating of joint condition.

$$GSI = 1{,}5J_{cond} + RQD/2 \quad (4)$$

Some values of the rock mass quality also determined by RMR_{89}. Some samples are having whole parameters, especially when the boreholes position intersects with the face of excavation, in that term the facemapping can be conducted.

2.2 Existing correlations

By interpreting rock mass quality directly using RMR_{89} and GSI in the site with a comprehensive data available, the mathematical model can be drawn. That model describes the correlation between the value of both methods and can be used to predict the conversion between them in the future. However, limited data for determining the RMR_{89} value often causes the results to be less accurate and precise.

Meanwhile, GSI value that is obtained through visual observation considered to have higher degree of accuracy, it is because GSI can describe the actual conditions of a rock mass. Hence, many researchers had compiled many mathematical models which can describes the correlation of RMR_{89} and GSI. There are several mathematical models that have been developed to predict the value of RMR_{89} based on the value of GSI observed, some of them summarized below. Sequently from equation (1) to (3) is developed by Hoek Brown [6], Cosar [10], Singh and Tamrakar [11].

$$RMR_{89} = GSI + 5 \quad (1)$$
$$RMR_{89} = 2.38GSI - 54.93 \quad (2)$$
$$RMR_{89} = 1.36GSI + 5.90 \quad (3)$$

In the other hand, rock mass quality determination using RMR_{89} considers the intact rock strength (R_1), rock mass structure (R_2 and R_3), discontinuities condition (R_4), and the presence of groundwater (R_5) according to Table 1. The orientation of discontinuities (R_6) was not considered because this parameter was not represented conceptually in GSI equation.

R_1 was obtained through unconfined compressive testing, R_2 and R_3 could be obtained through the RQD and space of discontinuities, as well as the number of joints per meter as stated in RMR_{14} developed by Celada et al [12]. R_4 could be obtained through surface interpretation on rock samples, while R_5 is assumed to have a value of 15, namely for dry rock mass.

2.3 Calculation of Discontinuities Condition

The problem which found while determining rock mass quality using RMR_{89} is the lack of precision in the observable data that related to discontinuities condition (R_4). In determining the weight for R_4, there are several parameters that tend to have less precision when they were just observed through drilling rock samples, i.e., the length of each discontinuity (persistence) and their apertures. Obtaining those parameters just by observing the drilling rock samples did not represent its actual condition. Question about the precision of R_4 became a dilemma then.

To answer the question above, it is necessary to make a correlation between the facemapping data observed from tunnel excavation and the interpretation of drilling rock samples. It becomes significant when the borehole intersects with excavation surface. The correlation between them both will produce more precise information. Frankly, the lack of boreholes makes this difficult to do. An alternative is needed to increase the degree of precision and the degree of confidence in the value of R_4.

Meanwhile, GSI is obtained by visually observing the condition of rock structure that displayed on its surface, then it is namely Surface Condition Ratings (SCR). This parameter is closely related to the interpretation which carried out to produce R_4 in the RMR method. Zhang et al [13] compared the weights that used to determine R_4 and SCR, as shown in the Table 3.

Table 3. Comparison of sub parameters between RMR_{89} and GSI to determine discontinuities condition.

Sub parameter	R4 (RMR_{89})	Rate	SCR (GSI)	Rate
Persistence	< 1 m	6	-	-
	1-3 m	4	-	-
	3 - 10 m	2	-	-
	10 - 20 m	1	-	-
	> 20 m	0	-	-
Aperture	None	6	-	-
	< 0.1 mm	5	-	-
	0.1 - 1 mm	4	-	-
	1 - 5 mm	1	-	-
	> 5 mm	0	-	-
Roughness	Very rough	6	Very rough	6
	Rough	5	Rough	5
	Slightly rough	3	Slightly rough	3
	Smooth	2	Smooth	1
	Very smooth	0	Very smooth	0
Infilling	None	6	None	6
	Hard < 5 mm	4	Hard < 5 mm	4
	Hard > 5 mm	2	Hard > 5 mm	2
	Soft < 5 mm	2	Soft < 5 mm	2

Sub parameter	R4 (RMR_{89})	Rate	SCR (GSI)	Rate
	Soft > 5 mm	0	Soft > 5 mm	0
Weathering	Unweathered	6	Unweathered	6
	Slightly weathered	5	Slightly weathered	5
	Moderately weathered	3	Moderately weathered	3
	Highly weathered	1	Highly weathered	1
	Completely weathered	0	Completely weathered	0

Based on the comparison in the Table 3, a mathematical model shall be generated. The first method in this research is to find an equation that able to predict the value of R_4 using SCR value. In this work, there are several data that already have R_4 value, and they are considered having a good precision. That dataset then being correlated with SCR value at the very similar locations. Correlating those data using mathematical method is adopted to generate the trendline and the equation. The result of the correlation is shown in the next chapter.

2.4 Calculation of modified RMR and RMR_{89}

The second method in this research is finding the value of RMR. In this paper, RMR is divided into two, they are RMR_{89}, which is based on observation data, and modified RMR, (then called RMR_m). The RMR_m is the RMR that calculated differently using modified R_4 that derived from the SCR according to the previous chapter. From that modification, the equation which going to be used to summarize RMR_m is changed into the one below.

$$RMR_m = R1 + R2 + R3 + \text{modified } R4 + R5 \quad (5)$$

3 Results and Discussion

This chapter shows some results, i.e., the value of discontinuities conditions, the value of RMR_m and RMR_{89}, and discusses about the degree of accuracy and its precision.

3.1 The Generated Models

Based on Fig. 3 below, the trendline is formed from those scattered points. The points are the data of R_4 and SCR that were obtained from observation. A mathematical model can be generated as in the equation (6) below. Then this equation can be used to describe and calculate the value of discontinuities condition, when all the data that we have is Surface Condition Ratings (SCR) that was observed using GSI method. Then, the equation for RMR_m is substituted into equation (7).

$$R_4 = 1.0951SCR + 10.328 \quad (6)$$

$$RMR_m = R1 + R2 + R3 + 1.0952SCR + 10.328 + R5 \quad (7)$$

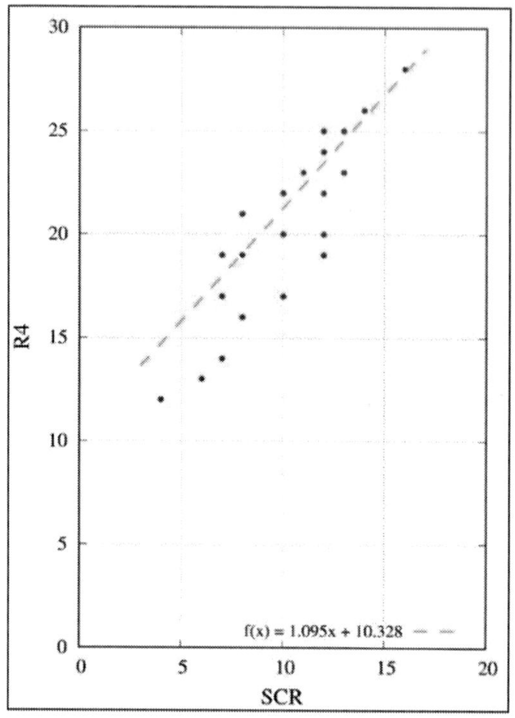

Fig. 3. Correlation between R4 (on RMR89) and SCR (on GSI).

After conducting some tests on that mathematical model, the data generated by the equation is presented in Fig. 4. The figure below displays some trendlines that drawn from the scattered of observation data using RMR_{89} and RMR_m. Both lines then visually compared to several trendlines from the equations that have been mentioned by other researchers, as shown in equation (1), (2), and (3).

In accordance with the distribution of data that observed using RMR_{89} and RMR_m, it shows that there are variances ranging from 0-5 points. The differences come bolder on data with comprehensive information, especially mention to their persistence and aperture. Values of the observed data using RMR_{89} with comprehensive information available give a significantly different position on chart, while the observed data without comprehensive information available tend to be close to the value of RMR_m.

From the information above, the relative position of observation data from RMR_{89}, RMR_m, and their trendlines mean that they have a correlation. Mathematical models then be derived as shown in the equation (8) and (9) below.

$$RMR_{89} = 0.334GSI + 38.78 \quad (8)$$

$$RMR_m = 0.3497GSI + 37.971 \quad (9)$$

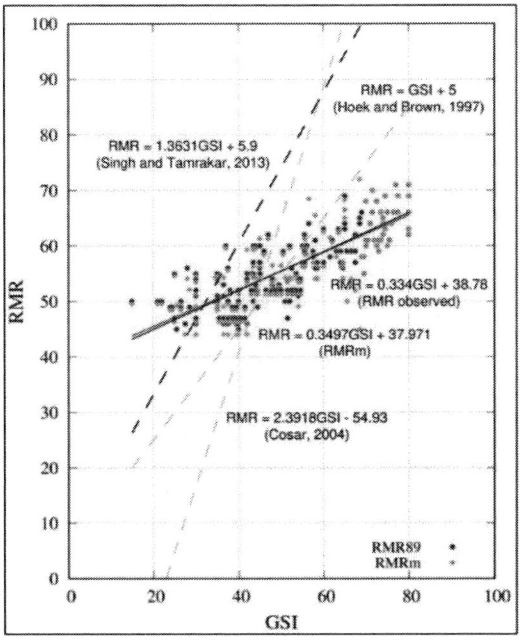

Fig. 4. Scattered data from observation using RMR_{89} and RMR_m

3.2 Evaluation of the Precision and Accuracy

Adjacency between RMR_{89} based on observation, RMR_m based on observation, and the prediction generated from both equation model (8) and (9) are presented in Fig. 5. There is a significant difference in values between RMR_{89} based on observation and RMR_m based on observation, while the predicted data have small differences in value.

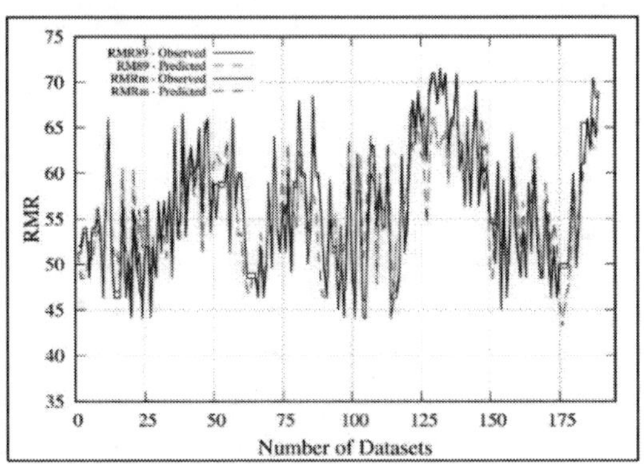

Fig. 5. Precision of observation data and predicted data, both using RMR_{89} and RMR_m

To ensure the degree of accuracy of both mathematical model (8) and (9), the MSE and RMSE ought to be calculated. The calculation is done by determining the relative comparison between the predicted values (both generated from RMR_{89} and RMR_m) and the RMR_{89} data based on observation. RMR_{89} dataset that were obtained from observation are assumed to be the correct ones and being absolute.

Table 4. MSE and RMSE.

Eq.	MSE	RMSE
Hoek-Brown (1997)	126,867	11,263
Cosar (2004)	1515,903	38,934
Singh and Tamrakar (2013)	1223,221	34,974
RMR_{89}	14,164	3,763
RMR_m	14,232	3,772

Table 4 above shows the degree of accuracy between the equation (8) and (9). Both mathematical models of RMR_{89} and RMR_m are having the good accuracy, which is indicated by a small value in the RMSE. RMSE from those mathematical models, is also close to equation that developed by Hoek and Brown [3]. Fig. 6 below also shows the relationship between the error of the values generated by the RMR_{89} and RMR_m mathematical models, the narrow gap between them means that both equations have similar accuracy.

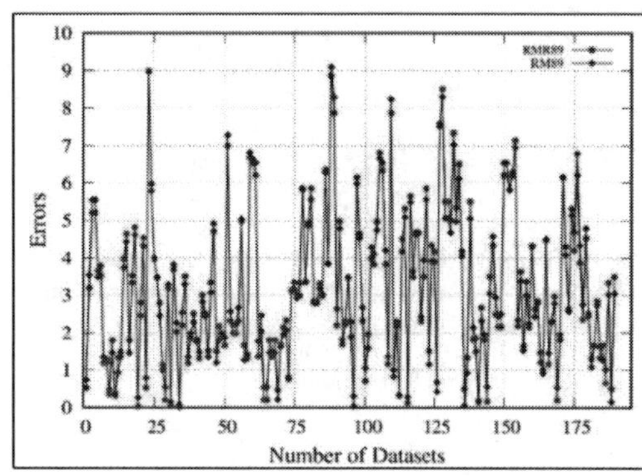

Fig. 6. Gap of error between RMR_{89} and RMR_m.

4 Conclusions

This study explains the correlation among some methods to determine rock mass quality. The methods used in this study are GSI, RMR_{89}, and RMR_m, where RMR_m is the RMR_{89} developed by Bienawski [4] with changes in R_4 parameter. The modified R_4 is calculated according to the weighting score of SCR (on the GSI). Through a series of experiments, the following conclusions are drawn below.

1. Based on 205 datasets of drilling rock samples and facemapping data, a series of observations can be conducted to determine rock quality based on the GSI and RMR_{89} method.
2. Due to limited information for compiling R_4 according to the rules developed by Bienawski [4], it can be solved by using an alternative equation to describe the discontinuities condition. R4 = 1.0951SCR + 10.328 can be applied, where SCR is the condition of discontinuities observed using the GSI's term. So that RMR_m = R1 + R2 + R3 +

1.0951SCR + 10.328 + R5 is stated, and the values can be computed.
3. Based on the scatter chart of data that been observed using RMR_{89} and RMR_m, trendlines and mathematical models are derived. Both equations are $RMR_{89} = 0.334GSI + 38.78$ and $RMR_m = 0.3497GSI + 37.971$, and they can be implemented to predict RMR value based on GSI data.
4. Two mathematical models above, can be applied to imply the correlation between RMR and GSI based on rock mass in the construction site of diversion tunnel of Meninting Dam. The RMR_{89} mathematical model will work effectively if comprehensive information about persistence, aperture, roughness, infilling, and weathering are available. Meanwhile, the mathematical model of RMR_m can be an alternative equation when information about discontinuities condition is not thoroughly available.
5. Based on the evaluation of precision and accuracy, it can be stated that both mathematical models are having small gaps of error to each other as well as adjacency to RMR_{89} based on data observation and equation of RMR_{89} developed by Hoek and Brown [6].
6. In the near future, conversion from GSI to RMR, or vice versa, can be conducted with higher degree of accuracy. Those data then can be used for determining the excavation method, supporting, and the study of rock mass deformation precisely, especially in Meninting Dam.

The first author would like to thank to Ministry of Public Works and Housing for the full support of research funding through 2020 Super Specialist Scholarship Program, and the support from BWS Nusa Tenggara I for the permit to conduct this research are gratefully acknowledged.

References

1. S.A. Mangga, et al. Peta Geologi Regional Lembar Pulau Lombok, Nusa Tenggara Barat Skala 1:250.000 (2010)
2. Y. Wiyasri, Evaluasi Kondisi Geologi Teknik untuk Perencanaan Terowongan Saluran Penngelak Bendungan Meninting Lombok Barat. Master thesis. Gadjah Mada University, Indonesia (2020)
3. E. Hoek., E.T. Brown, J. Heuristics **16** (2), 167–188. (1988)
4. Z.T. Bieniawski, Engineering Rock Mass Classifications: A Complete Manual for Engineers and Geologists in Mining, Civil and Petroleum Engineering (John. Wiley & Sons, New York 1989)
5. E. Hoek, *Strength of rock and rock masses*. ISRM News J. 2 (2), 4–16 (1994)
6. E. Hoek, E.T. Brown, Int. J. Rock Mech. Min. Sci. Geomech. Abstr. **34**, 1165–1186 (1997)
7. H. Sonmez, R. Ulusay, Int. J. Rock Mech. Min. Sci. **36** (6), 743–760 (1999)
8. H. Sonmez, R. Ulusay, Yerbilimleri **26**, 77–99 (2002)
9. F. Ceballos et al, *Relationship between RMRb and GSI based on in situ data*, in ISRM European Rock Mechanics Symposium,Vigo (2014)
10. S. Coşar, Application of Rock Mass Classification Systems for Future Support Design of the Dim Tunnel Near Alanya. PhD thesis. Middle East Technical University, Turkey (2004)
11. J.L. Singh, N.K. Tamrakar, Bull. Dept. Geol. **16**, 29–42 (2013)
12. B. Celada, I. Tardáguila, P. Varona, A. Rodríguez, Z.T. Bieniawski, *Innovating tunnel design by an improved experience based RMR system*, in World Tunnel Congress. Proceedings, Foz do Iguaçu, Brazil, 1–9 (2014)
13. Q. Zhang et al, Tunn. Undergr. Sp. Tech. **83**, 73-81 (2019)

Comparison of data mining algorithms for pressure prediction of crude oil pipeline to identify congeal

Agus Santoso[1*], *F. Danang* Wijaya[1], *Noor* Akhmad Setiawan[1] , and *Joko* Waluyo[2]

[1]Department of Electrical Engineering and Information Engineering, Faculty of Engineering, Universitas Gadjah Mada, Jl. Grafika 2, Yogyakarta, Indonesia
[2]Department of Mechanical and Industrial Engineering, Faculty of Engineering, Universitas Gadjah Mada, Jl. Grafika 2, Yogyakarta, Indonesia

Abstract. Data mining is applied in many areas. In oil and gas industries, data mining may be implemented to support the decision making in their operation to prevent a massive loss. One of serious problems in the petroleum industry is congeal phenomenon, since it leads to block crude oil flow during transport in a pipeline system. In the crude oil pipeline system, pressure online monitoring in the pipeline is usually implemented to control the congeal phenomenon. However, this system is not able to predict the pipeline pressure on the next several days. This research is purposed to compare the pressure prediction of the crude oil pipeline using data mining algorithms based on the real historical data from the petroleum field. To find the best algorithms, it was compared 4 data mining algorithms, i.e. Random Forest, Multilayer Perceptron (MLP), Decision Tree, and Linear Regression. As a result, the Linear Regression shows the best performance among the 4 algorithms with $R^2 = 0.55$ and RMSE = 28.34. This research confirmed that data mining algorithm is a good method to be implemented in petroleum industry to predict the pressure of the crude oil pipeline, even the accuracy of the prediction values should be improved. To have better accuracy, it is necessary to collect more data and find better performance of the data mining algorithm

1 Introduction

Applications of data science and data mining in many fields become very popular recently, one of which is implementation in the petroleum industries [1]. The data mining is able to support the decision making to prevent a massive loss in the operation of the petroleum industries. One of the most common problems in the operation of the crude oil pipeline system is flow congestion, which is caused by the crude oil flow from liquid to pasta phase. In extreme condition, the crude oil flow will lead to the solid phase. This is called congeal flow in the petroleum industry. Congeal is a serious global problem in the petroleum industry, since it leads to significant effects. Congeal phenomenon wastes the operation cost of crude oil exploration in the world and needs to spend millions of USD to solve this problem [2, 3].

To prevent the congeal occurrences, petroleum industries usually implement online monitoring system which watch out and record the pressure of the crude oil pipeline and other parameters [4]. However, such system is designed to record the occurred events and not able to predict the future events as well as the pipeline pressure on the next several days.

The existing studies on congeal prediction mostly focused on the prediction of wax deposition by using static data resulting from controlled experiments [5-8]. Several data mining algorithms indicated relatively good accuracy in predicting congeal flow due to wax deposition [7, 8]. However, the experimental results cannot be directly applied in the operation field of the crude oil pipeline system, because of the dynamic system of the pipeline. Objective of this research is to carry out analytics of the real historical petroleum field data using several algorithms of data mining and to compare the results for finding the best one to predict the pressure of the crude oil pipeline in future.

2 Methodology

The pressure of crude oil pipeline may be predicted using several algorithms of data mining. In this research, 4 algorithms of data mining were adopted. The first algorithm is Random Forest, which is an ensemble method for regression using a large number of decision trees [9]. The second one, Multilayer Perceptron (MLP), is a type of artificial neural network (ANN) with an individual node named as perceptron organized in a series of layers. Each layer is categorized into three

* Corresponding author: agussantosokul@gmail.com

layers of nodes: an input layer, a hidden layer and an output layer [9].

The third algorithm is Decision Tree, that is a tree-like model which can support a decision making based on the conditions [9,10]. The last algorithm, Linear Regression, can be used to make a prediction of dependent attributes based on the several independent variables [10].

Data set employed in this research were a series real historical data acquired from the field of crude oil pipeline of a petroleum company in Indonesia using an online monitoring system. The data set contains the daily actual pressure measured at a certain point in the crude oil pipeline system in Psi. The series length of the daily data is 200 days, as shown in Fig. 1 (a). In this pipeline system, the critical pressure is assumed at 200 Psi. In addition, the daily weather forecast conditions, including ambient temperature and precipitation forecast, were also used as the input data. Figure 1 (b) shows the daily weather forecast conditions in conjunction with the daily actual pressure of the crude oil pipeline system. Example of the data set is shown in Table 1.

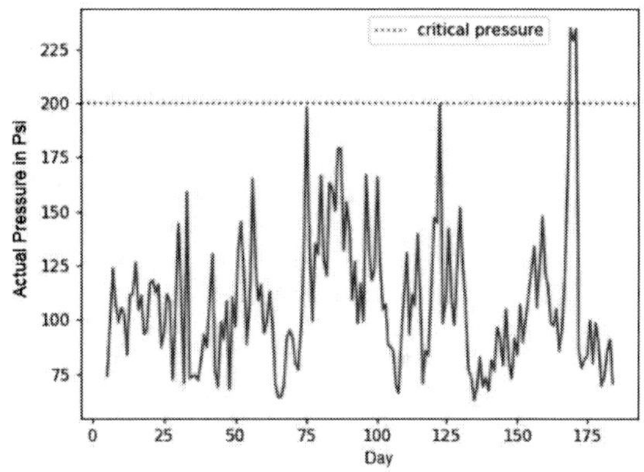

(a)

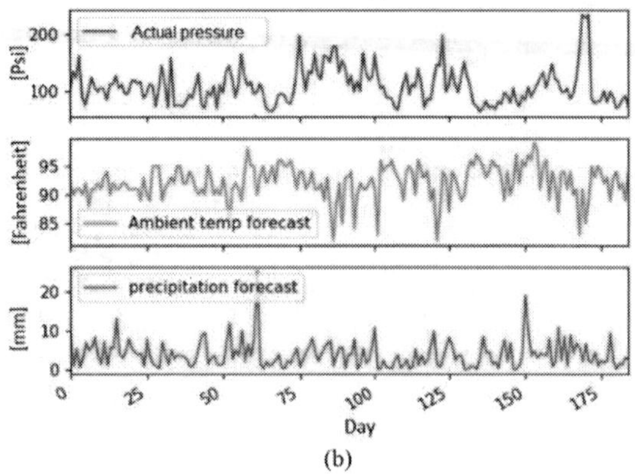

(b)

Fig. 1. (a) The daily actual pressure measured at a certain point in the crude oil pipeline system, (b) The daily weather forecast conditions including ambient temperature and precipitation forecast

Table 1. Example values of the data set.

Day	Act. Press	Amb Temp.	Precipitation
1	97,67	92	8,7
2	134,2	90	0,9
3	122,51	91	5,3
4	162,7	91	0,5
5	87,45	90	2,7
6	74,38	92	6,9
7	96,87	88	5
8	123,75	91	5,8
9	107,04	91	8,2
...	...	...	...

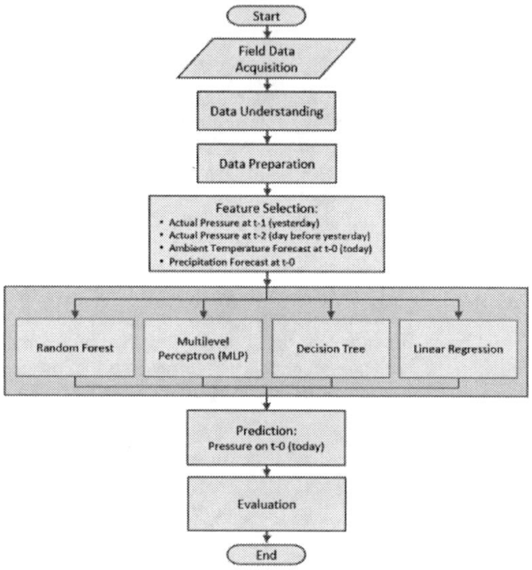

Fig. 2. Method for pressure prediction of crude oil pipeline using data mining algorithms.

Method employed for the pressure prediction using the data mining algorithms in this research is shown in Fig. 2. The pipeline actual pressure at a day (for example today, t-0) was predicted by the 4 data mining algorithms using 4 input parameters, i.e. the actual historical pressure of 1 day before (yesterday, t-1) in Psi, the actual historical pressure of 2 days before (day before yesterday, t-2) in Psi, the ambient temperature forecast at t-0 in Fahrenheit, and the precipitation forecast at t-0 in mm. Output is the predicted pressure of today (t-0).

The data set was separated in 2 groups. The first group, the series data from day 1 until day 120, was used for the training of the data mining algorithms, while the second group containing the rest series data (from day 121 until day 155) was used for pressure prediction and the blind test evaluation.

Performances of each data mining algorithms employed in this research were measured using the coefficient of determination or R squared (R^2) method, and the root-mean-square deviation (RMSD) or root

https://doi.org/10.1051/e3sconf/202132502002

mean square error (RMSE) method. The coefficient of determination method (R^2) shows whether the predicted pressure is fit with the actual pressure. Values of the coefficient of determination normally ranges from 0 to 1. The value near 1 shows better fitness between the predicted and the actual pressure values. While the RMSE method shows the differences between the predicted and the actual pressure values. The smaller value of the RMSE shows better fitness.

3 Result and discussion

The first algorithm used in the pressure prediction of crude oil pipeline was Random Forest. The algorithm delivers the pressure prediction value for the next day based on the historical data in the data set. Figure 3 (a) shows the predicted pressure of the crude oil pipeline using Random Forest algorithm as a function of time (brown plot) from day 121 until day 155. For comparison, the actual pressure (blue plot) is also shown in the figure. It is confirmed that the values of predicted pressures are going up and down with almost similar trend with those of the actual pressure. Even though there are some error values in the predicted pressure, it can be revealed that data mining algorithm can be used in petroleum industry to predict the crude oil pipeline pressure.

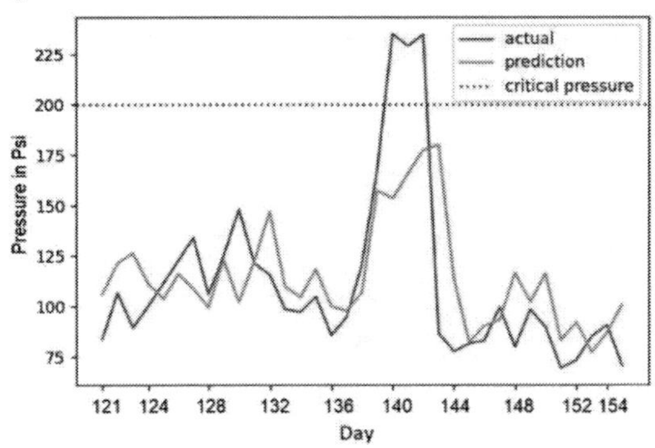

(a)

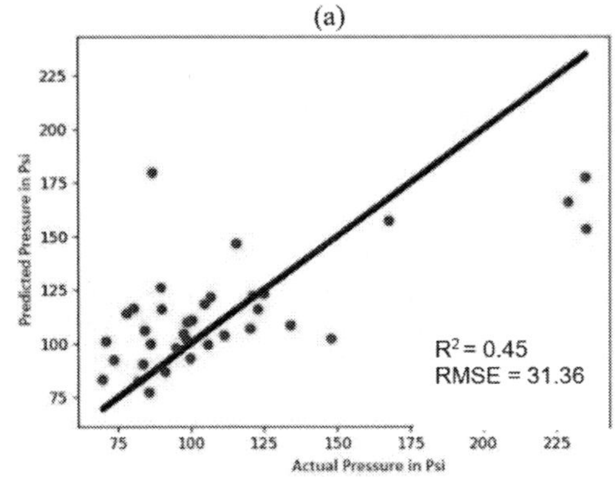

(b)

Fig. 3. (a) The predicted pressure using Random Forest algorithm and the actual pressure of the crude oil pipeline from day 121 until day 155, (b) Blind test pressure prediction results using Random Forest algorithm.

Figure 3 (b) shows the relationship between the blind test pressure prediction results using Random Forest algorithm and the actual pressure. According to the figure, the predicted values seem accurate at the pressure range below 130 Psi, but rather large deviation for pressure larger than 130 Psi. For blind test evaluation of the accuracy of the pressure prediction, R squared (R^2) and root mean square error (RMSE) tests were adopted in this research. It is attained that the R^2 and RMSE values for the pressure prediction using Random Forest algorithm are 0.45 and 31.36 respectively. The values show that the prediction results still do not accurate enough, therefore the method should be improved.

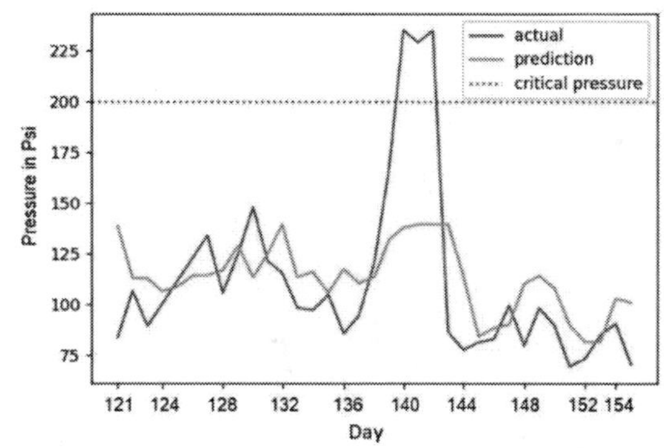

(a)

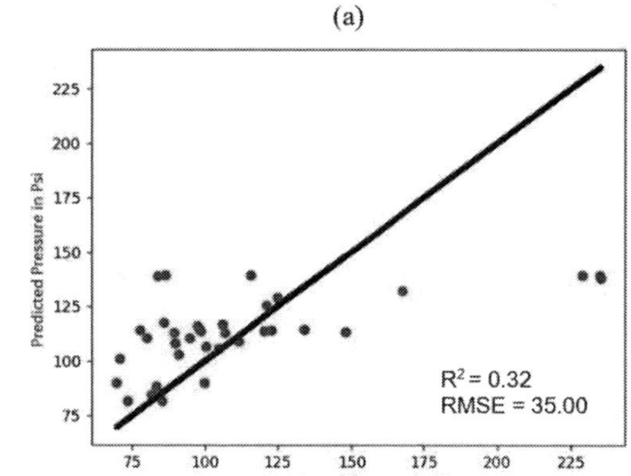

(b)

Fig. 4. (a) The predicted pressure using Multilayer Perceptron (MLP) algorithm and the actual pressure of the crude oil pipeline from day 121 until day 155, (b) Blind test pressure prediction results using MLP algorithm.

The next algorithm to be used in this research is Multilayer Perceptron (MLP). The brown plot of Fig. 4 (a) shows the pressure prediction values of the crude oil pipeline using MLP algorithm from day 121 until day 155. As the previous figure, the blue plot in the figure shows the actual pressure. The figure shows that the pressure prediction using MLP algorithm seems less accurate than that of using Random Forest algorithm. This conclusion is confirmed by the relationship between the blind test pressure prediction results and the actual pressure as shown in Fig. 4 (b). From this figure, it is known that correlation coefficient R^2 is less than that of using Random Forest algorithm, i.e. 0.32 and the RMSE values is larger than that of using Random Forest algorithm, i.e. 35.00. The larger RMSE means less accurate due to the larger error.

To find the best algorithm to predict the pressure of the crude oil pipeline, this research also employed the Decision Tree and Linier Regression algorithms. The results are shown in the Fig. 5 and Fig. 6. Table 2 shows the pressure prediction result summary of the four algorithms used in this research. Accordingly, the pressure prediction using Decision Tree algorithm indicates better performance than using Random Forest and MLP algorithms, showing by the larger R^2 and the smaller RMSA values. However, the pressure prediction using Linier Regression algorithm shows the best performance comparing to the other 3 algorithms. This tendency agrees with other researches [11].

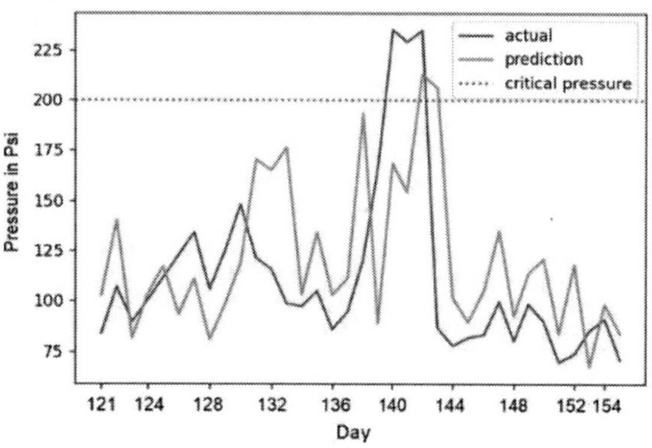

(a)

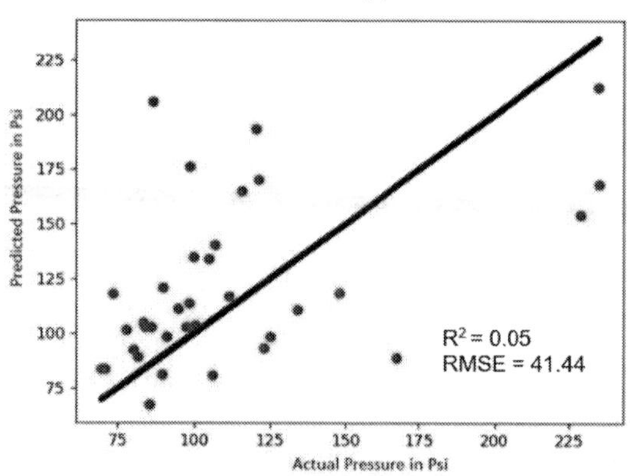

(b)

Fig. 5. (a) The predicted pressure using Decision Tree algorithm and the actual pressure of the crude oil pipeline from day 121 until day 155, (b) Blind test pressure prediction results using Decision Tree algorithm.

Table 2. Summary of the pressure prediction results using data mining with Random Forest, Multilayer Perceptron (MLP), Decision Tree, and Linear Regression prediction models.

Metrics	Algorithm			
	Random Forest	MLP	Decision Tree	Linier Regression
R2	0.45	0.32	0.05	0.55
RMSE	31.36	35	41.44	28.34

The 4 data mining algorithms successfully predicted the next day pressure of the crude oil pipeline. However, according to the blind test results, the accuracy should be improved to obtained more confidence predictions. Improvement of the prediction accuracy may be carried out by several methods, such as enrichment of the data transformation, adding the new and more comprehensive data set, and finding the better data mining algorithm and tuning up the parameters.

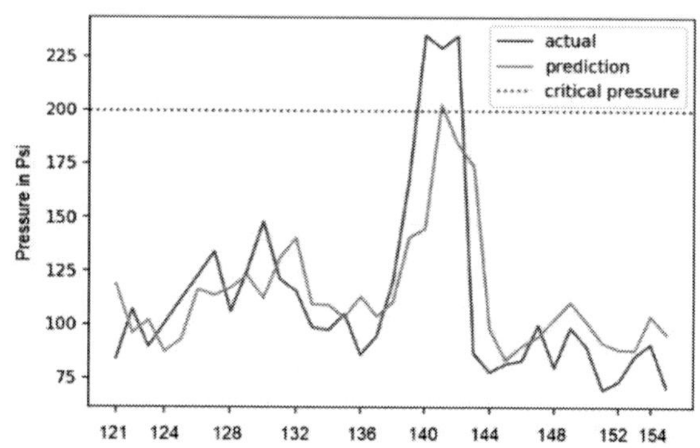

(a)

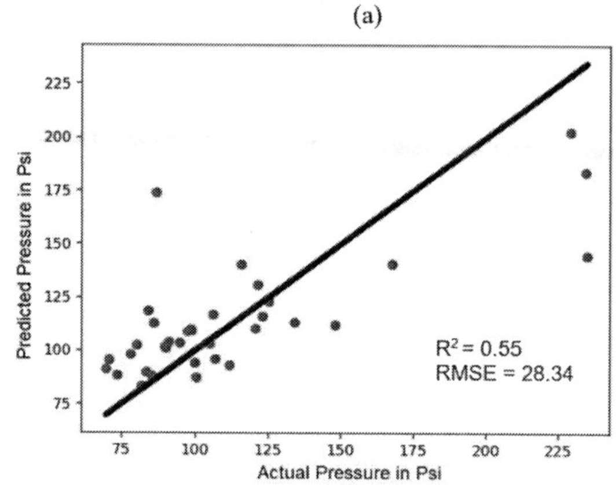

(b)

Fig. 6. (a) The predicted pressure using Linear Regression algorithm and the actual pressure of the crude oil pipeline from day 121 until day 155, (b) Blind test pressure prediction results using Linear Regression algorithm.

4 Conclusion

The pressure predictions of crude oil pipeline have been analysed using 4 algorithms of data mining: Random Forest, Multilayer Perceptron (MLP), Decision Tree, and Linear Regression algorithms. The predicted pressures using the 4 algorithms are agree and having similar trend with the actual pressures, even the accuracies which represented by the R^2 are not high, about 0.55 or less. Comparison of the results confirmed that the Linear Regression with $R^2 = 0.55$ and RMSE = 28.34 is the best algorithm among the 4 data mining algorithms adopted in this research for this particular data set. Therefore, it can be concluded that data mining algorithms are useful to predict the crude oil pipeline in the petroleum industry. However, there are still many rooms for continuing this research to improve the results. In future, further recommended works may be focused on the improving the accuracy and reducing the error by data transformation and data set improvement, also finding the better data mining algorithm and tuning up the parameters.

References

1. P. Bangert, Machine Learning and Data Science in the Oil and Gas Industry (Gulf Professional Publishing, 2021)
2. A. K. Manshad, H. Rostami, H. Rezaei, S. M. Hosseini, H. Niknafs and A. H. Mohammadi, Heavy Oil, 161-173 (Nova Science Publishers, New York (2017)
3. Z. Q. Chu, J. Sasanipour, M. H. Saeedi, A. Baghban, and H. Mansoori, J. Petrol. Sci. Tech **35**, 1974-1981 (2017)
4. E.B. Priyanka, C. Maheswari, S. Thangavel, Flow Meas. Instrum **62**, 144-151 (2018)
5. A. Kamari, A. K. Manshad, F. Garagheizi, A. H. Mohammadi, "A Robust Model for Determination of Wax Deposition in Oil Systems", ACS, I&EC (2013)
6. G. Zhang, G. Liu, J. Petrol. Sci. Eng **70**, 1-9 (2010)
7. T. J. Behbahani, A. A. M. Beigi, Z. Taheri, B. Ghanbari, J. Petroleum **1**, 223-230 (2015)
8. E. O. Obanijesu and E. O. Omidiora, J. Petrol. Sci. Tech **26**:16, 1977–1991 (2008)
9. J. H. Moon,,Y. S. Kim, M. J. Son, and E. J. Hwang, Energies **11**(12), 3283 (2018)
10. Y. S. Kim, Expert Syst. Appl **34**, 1227-1234 (2008)
11. Z. Hu, M. Wu, K. Hu dan J. Liu, J. Petrol. Sci. Tech **33**, 1499-1507 (2015)

The potency of water availability and its use by the community around the Mount Merbabu National Park

Lugas Prasetyo[1] and *Ambar* Kusumandari[2*]

[1]Alumny of the Faculty of Forestry UGM, FRCDepartment, Yogyakarta, Indonesia
[2]Lecturer of the Faculty of Forestry, UGM, FRC Department, Yogyakarta, Indonesia

Abstract. Water is one of the basic needs for every living thing, including humans. This study aims to determine the availability of water in the two springs, the utilization of water by the community around MMNP, the vegetation characteristics of the spring catchment areas, and the Water Use Index (WUI). Data of water availability was taken using the volumetric method. Water utilization by the community is collected by doing the interviews with the respondents. The vegetation characteristics is analyzed based on the vegetation density by using the NDVI method. Meanwhile, the Water Use Index (WUI) is obtained from a comparison between water demand and water supply. The results showed that the water debit available in the Tuk Simuncar and Tuk sipendok was 23.22 liters/second and 31.20 liters/second which were classified as class IV. The area of Tuk simuncar is dominated by high density vegetation (75 Ha), while Tuk Sipendok is dominated by moderate density vegetation (102.28 Ha). The community around MMNP uses 211 liters of water for household needs, 59 liters for livestock, and 10 liters for agriculture. The Water Use Index (WUA) of the two springs has a value of 0.3 which is included in the low level.

1 Introduction

Water is an important element for the welfare of society. Currently there is increasing distance between population and water sources due to population growth, which is causing rapid expansion of some water networks. At the same time, global water consumption has quadrupled in the last 50 years and it is estimated that this value will continue to increase [1]. The water needed by all living things can naturally be provided by the forest.

Forest land use is usually associated with protecting water resources from contamination and reducing the cost of drinking water supply [2]. Stabilizing water supplies is one of the important functions of forests. Although forests consume large amounts of water as they grow, it increases the water storage capacity of the groundwater. Thus, functionally it becomes the largest water source in terrestrial ecosystems. In view of global water security issues and their relation to economic development, the importance of forests in maintaining water supplies in the current era is highly recognized [3]. Forests have a role as controller of the water cycle which has the concept of guaranteeing water yield. The amount of water that can be harvested is closely related to the availability of water in an area. Water availability is the volume of water contained in the hydrological cycle in an area, which is a combination of rainwater, surface water, and ground water [4].

The forest area in the Mount Merbabu National Park (MMNP) is one of the water catchment areas or water sources for the community at the foot of Mount Merbabu. This is in accordance with the Decree of the Minister of Forestry No. SK.135/Menhut-II/2004, which states that the basis for determining the forest area on the slopes of Mount Merbabu to become a National Park area is a source of water for the surrounding community, protected flora and fauna habitat, and has attractive natural and cultural potential, so it is necessary to protected and preserved [5]. Thus, many springs in the Mount Merbabu area are used to support the lives of the surrounding community. Among them there are 2 springs that are used by the community in Boyolali Regency to meet their needs, namely Tuk Simuncar and Tuk Sipendok.

According to [6], communities from 13 villages bordering the Gunung Merbabu National Park in Boyolali Regency are dependent on forests for water extraction to meet their needs. Of the total people who use water in MMNP 89% of them use water directly from springs, either their own, neighbors or public wells whose water availability depends on water supply from Mount Merbabu. This makes the community around MMNP very dependent on water sources from the MMNP area. The disconnection of the clean water pipe network originating from Mount Merbabu, such as what occurred during a forest fire in the Mount Merbabu National Park area, made people have to take water from other hamlets [7]. This incident shows that the community around MMNP is very dependent on water sources on Mount Merbabu, including Tuk Simuncar and Tuk Sipendok.

Tuk Simuncar and Tuk Sipendok are springs located in the MMNP area, the two springs are used by 8 villages

* Corresponding author: ambar_kusumandari@ugm.ac.id

in Boyolali Regency, namely Jeruk Village, Candisari Village, Kembang Village, Kembangkuning Village, Ngagrong Village, Ngargoloka Village, Sampetan Village, and Seboto Village. The eight villages rely heavily on water from the two springs to meet their needs. The high dependence of the community on Tuk Simuncar and Tuk Sipendok if not accompanied by adequate water availability can potentially disrupt community activities. Therefore, this research is needed that discusses the potential availability of Tuk Simuncar and Tuk Sipendok water in meeting the needs of the surrounding community. The purpose of this study was to determine the availability of water, vegetation characteristics of the spring catchment area, utilization, and the Water Use Index (IPA) in Tuk Simuncar and Tuk Sipendok Mount Merbabu National Parks.

2 Material and methods

2.1 Location

This research was conducted in August – December 2020 with locations, namely Tuk Simuncar and Tuk Sipendok water sources, as well as people who use both water sources, including: Jeruk Village, Kembangkuning Village, Ngagrong Village, Kembang Village, Seboto Village, Candisari Village , Ngargoloka Village, and Sampetan Village.

2.2 Materials

The materials used in this study are:

1. Two springs that are Tuk Simuncar and Tuk Sipendok discharge.
2. Land coverage of Mount Merbabu National Park.
3. Community who live surrounding the two springs.

The equipment used in this study consists of volumetric tools for discharge measurement, tally sheet, plastic, ruler, analytical scales, clinometer, roll meter, hagameter, tape meter, and camera.

2.3 Methodes

2.3.1 *Discharge*

Discharge data is taken using the volumetric method. This method is used to measure the flow rate of the two springs. In the volumetric method the entire flow in the study plot is collected in a container over a certain period of time. The discharge is expressed by Q (m3/s) which calculates the volume V (m3) at a certain time t (s), the discharge is read directly from the container or calculated indirectly [8].

The data that has been obtained then continues to process by using the following equations and classifications.

$$Q = \frac{V}{t} \quad (1)$$

Q = Discharge (m3/s)
V = Storage volume (m3)
t = Observation time (s)

Table 1. Classification of Spring Discharge According to Meinzer (1923) in [9].

Class	Discharge
I	$\geq 10\ m^3/s$
II	$1 - 10\ m^3s$
III	$0.1 - 1\ m^3/s$
IV	10 – 100 l/s
V	1 – 10 l/s
VI	0.1 – 1 l/s
VII	10 – 100 l/s
VIII	< 10 l/s

Data on vegetation characteristics of the two springs catchment areas was obtained through remote sensing approach using the NDVI (Normalized Difference Vegetation Index) method which describes vegetation density and supervised classification to obtain a land cover map in the Gunung Merbabu National Park area. NDVI is a dimensionless index that describes the difference between visible and near infrared reflections from vegetation cover and can be used to estimate green density in an area of land [10]. Supervised classification is a classification of pixel values based on an example of a training area with known object types and spectral values [11]. The NDVI vegetation index that has been obtained is then classified based on previous research by [12] Dasuka et al. (2016) as shown in the following table.

Table 2 Value of NDVI

Number	Density class	NDVI
1	Low	0.296 - 0.612
2	Medium	0.612 – 0.72
3	High	> 0.72

Data regarding the use of springs by the community was collected by interview method. The interview method used is a structured interview using an interview guide. Structured interviews are interviews conducted by using by asking questions to respondents using questions that have been prepared or guide questions [13]. Based on data from Resort Ampel TNGMb water users Tuk Simuncar and Tuk Sipendok as many as 4.061. so that the number of samples obtained is 98 respondents. The Slovin formula used is as follows [14].

$$n = \frac{N}{1 + Ne^2} \quad (2)$$

n = number of samples
N = total population
E = estimated error rate

The amount of water used by the community is obtained through the following formula.

Water needs/capita/day = (water needs per family per day)/(number of family members)

The total water requirement of the population per day = water demand/capita/day×the total population

The population's water needs are also carried out by projecting population growth in 8 villages that use the Tuk Simuncar and Sipendok springs for the next 10 years. namely 2021 - 2031. Population growth projections are carried out using the arithmetic method. this method assumes that the population in the future will increase by the same amount every year [15]. The formula used in the arithmetic method is:

$$P_t = P_0(1 + rt) \quad (3)$$

and $$r = \frac{1}{t}\left(\frac{P_t}{P_0} - 1\right) \quad (4)$$

Information:
Pt = total population in year t
P0 = total population in base year
r = population growth rate
t = time period between base year and year t (in years)

2.3.2 Water Use Index (IPA)

The Water Use Index (IPA) is a comparison between water demand and the existing water supply in a place [16]. IPA is used to determine the balance between water supply and community needs. The data used to obtain the IPA value comes from flow rate data as a water supply variable and data on the use of Tuk Simuncar and sipendok water by the community as a variable water demand. The value of IPA can be obtained using the following formula.

$$\text{IPA} = \frac{\text{Demand}}{\text{Supply}} \quad (5)$$

The IPA value that has been obtained is then classified based on the Minister of Forestry Regulation No. P.61/Menhut-II/2014. as follows.

Table 3. IPA Classification

Number	IPA	Class
1	IPA ≤ 0.25	Very low
2	0.25 < IPA ≤ 0.5	Low
3	0.5 < IPA ≤ 0.75	Medium
4	0.75 < IPA ≤ 1.00	High
5	IPA > 1.00	Very high

3 Results and Discussion

3.1 . Water Availability at Tuk Simuncar and Tuk Sipendok

Tuk Simuncar and Tuk Sipendok are two water sources located in the Mount Merbabu National Park area (Figure 1). The two water sources are included in the utilization zone of Mount Merbabu National Park. this is in accordance with the Minister of Forestry Regulation No. P.56/Menhut-II/2006. that the locations of water sources. especially those that have the potential to be utilized for the benefit of conservation. the community. and water service providers are indeed recommended to become zones for the use of water resources.

The existence of Tuk Simuncar and Tuk Sipendok at the Mount Merbabu TN area is in accordance with the statement of [17]. that the availability of water on Mount Merbabu is indeed abundant even in the long dry season. This is because Mount Merbabu is formed from pyroclastic rock (which is loose) and this landform can store water through former lava flows. However. Mount Merbabu is a rain shadow area so that in the dry season the water discharge is threatened to decrease. The amount of water discharge from Tuk Simuncar and Tuk Sipendok can be seen in the following table.

Table 4. Discharge of Tuk Simuncar and Tuk Sipendok

Num ber	Location	Discharge (l/s)	Class
1	Tuk Simuncar	23.22	IV
2	Tuk Sipendok	31.20	IV

* Classification of spring discharge according to Meinzer (1923) in [9] .

The table above shows that the available water discharge in Tuk Sipendok is greater than Tuk Simuncar. namely 31.20 lt/s and 23.22 lt/s. Based on the classification of Meinzer (1923) in [9]. both water sources belong to the same class. namely class IV. The difference in flow rate in a spring is influenced by three factors. namely the permeability of the aquifer (ground water level). the recharge area that fills the aquifer. and the amount of ground water recharge (Davis & de Wist. 1966 in [18]. From this statement. it is not surprising that the water discharge of Tuk Sipendok is greater than Tuk Simuncar because of its larger catchment area. which is 192.9 Ha compared to 137.6 Ha.The magnitude of the water discharge issued by Tuk Sipendok compared to the water discharge of Tuk Simuncar is in line with the large potential possessed by the two springs. This is in accordance with the statement of [7]. that the magnitude of a flow rate can be used to determine the potential of water to be utilized. so that the greater the discharge of a water flow. the greater its potential to support human activities. One of the reasons for choosing Tuk Simuncar and Tuk Sipendok as water sources for the community is because the two springs are included in the type of water flow that flows throughout the year or commonly called chronic springs. According to [19]. parennial springs are springs that release water throughout the year and are not affected by rainfall. so that people can take advantage of the flow of water throughout the year without worrying about losing water sources.

The existence of Tuk Simuncar and Tuk Sipendok needs to be preserved so that water supplies from both water sources remain sustainable. Preservation of water sources in the national park area. according to [20]. it can be done by providing clarity of regulations and policies by the manager so that there is a policy or agreement made between the manager and the water user. Efforts to maintain the sustainability of Tuk Simuncar and Tuk Sipendok can be realized through appropriate management strategies from the Gunung Merbabu National Park Office. According to the Long-Term Management Plan (RPJP) of Gunung Merbabu National Park for the 2019-2023 period. maintaining the availability of water resources is the first management goal. which includes Tuk Simuncar and Tuk Sipendok. The strategies owned by the Gunung Merbabu National Park Office regarding these objectives are periodic inventory and monitoring of water resources. facilitation of permits for the use of water resources. provision of supporting infrastructure. and provision of land cover maps.

According to the Minister of Forestry Regulation No. P.64/Menhut-II/2013. the volume of water that can be utilized from water sources within the national park area for non-commercial purposes is a maximum of 50% of the water volume that has been determined. so that according to the regulation. the flow data of Tuk Simuncar and Tuk Sipendok water volume that can be utilized is 11.61 liters/second and 15.6 liters/second. Determination of the volume of water utilization is carried out by taking into account the use of existing water. the carrying capacity of water resources. the number and distribution of the population around the area as well as the projected growth. calculation and projection of water resource needs. as well as objects and natural tourist attractions Utilization of water can only be carried out after obtaining a permit. in which a water utilization permit (IPA) can be submitted by government agencies. community groups. or social institutions. With this regulation. it is hoped that water user groups can apply for water utilization permits (IPA) from Tuk Simuncar and Tuk Sipendok so that the community has legality in using both water sources. Then. after the community groups apply for a water utilization permit (IPA). the Gunung Merbabu National Park Office can follow up by issuing water utilization permits (IPA) for Tuk Simuncar and Tuk Sipendok in accordance with applicable regulations.

3.2. Vegetation characteristics

In accordance with the management mandate of Mount Merbabu National Park. the existence of forests in Mount Merbabu National Park is an area that acts as a water catchment area. In the hydrological cycle. forest is a subsystem that has a function to be able to maintain the continuity of flow and the quality of the water that comes out (water yield). This is a role given by the density. composition and structure of forest vegetation and litter on the forest floor which makes it easier for water to enter the soil. thereby increasing the capacity of groundwater storage [21].

In accordance with the management mandate of Mount Merbabu National Park. the existence of forests in Mount Merbabu National Park is an area that acts as a water catchment area. In the hydrological cycle. forest is a subsystem that has a function to maintain the continuity of the flow and the quality of the water that comes out (water yield). This is a role given by the density. composition and structure of forest vegetation and litter on the forest floor which makes it easier for water to enter the soil. thereby increasing the capacity of groundwater storage [21].

It is important to know the vegetation density in the catchment area of the springs in order to ensure the availability of vegetation that acts as a water catcher and reservoir. In this study. vegetation density was obtained using the Sentinel 2 satellite imagery which was processed using the Normalized Difference Vegetation Index (NDVI) method. Sentinel 2 satellite imagery was chosen because it is based on spatial and temporal resolution according to research needs. According to ESA (2018) in [22]. Sentinel 2 imagery has a spatial resolution ranging from 10 m and a temporal resolution of 10 days with one satellite and 5 days with 2 satellites. NDVI is an index that is often used to determine vegetation density through level of greenery or the photosynthetic activity of vegetation. The NDVI vegetation index is based on the association of red and infrared waves so that it can display the greenness index of vegetation.

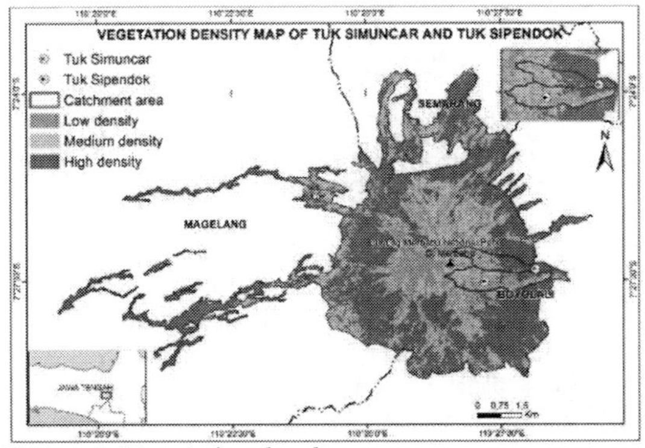

Figure 1. Vegetation density map

Table 5. Number of NDVI Pixels of the Springs Area

Num ber	Vegetati on density	NDVI	Number of pixels		Area (Ha)	
			Tuk Sim unca r	Tuk Sipen dok	Tuk Simu ncar	Tuk Sipend ok
1	Low	0.296 – 0.612	312	1051	12.48	42.04
2	medium	0.612 – 0..2	1298	2557	51.92	102.28
3	high	> 0.72	1875	1291	75	51.64

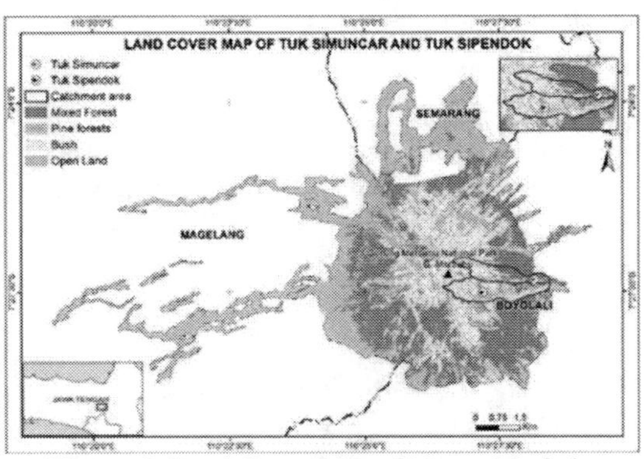

Figure 2. Land Cover Map of Mount Merbabu National Park

The picture above shows that the vegetation that grows in the Tuk Simuncar and Tuk Sipendok catchments is mixed forest. pine forest. and shrubs. Based on Table 6. the composition of land cover in the Tuk Simuncar catchment consists of mixed forest covering an area of 53.95 Ha with a percentage of 39.6%. pine forest covering an area of 2.14 Ha with a percentage of 1.6%. shrubs covering an area of 69.24 Ha with a percentage of 50. .8%. and 10.86 hectares of open land with a percentage of 8%. Meanwhile. the land cover in the Tuk Sipendok DTA consists of mixed forest covering an area of 30.27 Ha with a percentage of 15.7%. pine forest covering an area of 7.28 Ha with a percentage of 3.8%. shrubs covering an area of 117.52 Ha with a percentage of 60.9%. and open land area of

37.87 Ha with a percentage of 19.6%. Mixed forest has a relationship with water management. namely when it rains on mixed forest land cover. various crown and root stratifications are able to hold and absorb water well so that the surface runoff formed is low [23].

Pine forest dominates the land cover in Mount Merbabu National Park because historically the Gunung Merbabu National Park area was formerly a Perhutani area which has the main commodity of pine stands. This pine forest. which has been widely developed on the island of Java since the Dutch colonial era. has controversy regarding its role as a regulator of water management. Pines are considered to cause drought in the dry season because they consume a lot of water. but on the other hand. pines have deep roots. playing a role in the process of interception and high evapotranspiration [24]. Shrubs as the land cover that dominates the Tuk Simuncar and Tuk Sipendok catchments are shrubs. Related to water management. shrubs are a type of land cover that has a low value in terms of soil moisture capacity to store water reserves or water holding capacity [25].

Table 6. Land Cover Composition in Tuk Simuncar and Tuk Sipendok catchments

Land coverage	Tuk Simuncar		Tuk Sipendok	
	Area (Ha)	%	Area (Ha)	%
Pine forest	2.14	1.6	7.28	3.8
Mixed forest	53.95	39.6	30.27	15.7
Bush	69.24	50.8	117.52	60.9
Bareland	10.86	8.0	37.87	19.6

Based on Table 5 and Table 6. it can be seen that the Tuk Sipendok catchment area is dominated by vegetation with medium density and bushland cover. This condition was caused by forest fires that occurred in 2019. Fires that burned vegetated land caused the land to become open which was then overgrown by bushes. This situation can threaten the existence of Tuk Sipendok because it is prone to landslides. Therefore. the Gunung Merbabu National Park Office is expected to be able to educate the public on the importance of preventing forest fires. Another effort that can be made by Balai TN is to conduct stricter supervision so that people do not carry out activities in zones that are not permitted. such as jungle zones and core zones. The efforts of the Gunung Merbabu National Park Office must also be accompanied by awareness by the community not to carry out activities or carry items that can trigger forest fires such as burning grass and bringing matches.

3.3 Water Utilization by Communities Around MMNP

Communities around Mount Merbabu National Park are very dependent on springs in the Mount Merbbau National Park area. including Tuk Simuncar and Tuk Sipendok. This is because on the slopes of Mount Merbabu it is very difficult to make dug wells and drilled wells so that people only rely on water from the Mount Merbabu NP area. including Tuk Simuncar and Tuk Sipendok. Tuk Simuncar is used by 10.493 people living in Candisari Village. Kembang Village. Ngargoloka Village. and Sampetan Village. while Tuk Sipendok is used by 13.551 people living in Jeruk Village. Kembang Village. Kembangkuning Village. Ngagrong Village. and Seboto Village.

Six of the eight villages that use water from Tuk Simuncar and Tuk Sipendok have used the PAMSIMAS (Community-Based Drinking Water and Sanitation Provision) management system. PAMSIMAS is one of the government programs (central and regional) with the support of the World Bank. to improve the supply of drinking water. sanitation. and increase the value and behavior of healthy living by building/providing sustainable community-based drinking water and sanitation facilities and infrastructure by the community. The PAMSIMAS program is implemented by the central government through the provision of assistance directly to the village level [26].

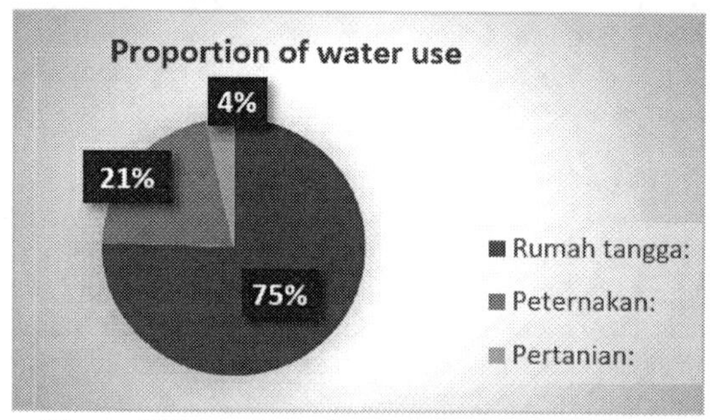

Figure 3 Proportion of water use for Tuk Simuncar and Tuk Sipendok

Based on the figure above. it can be seen that Tuk Simuncar and Tuk Sipendok are used by the community to meet household needs (75%). livestock (21%). and agriculture (4%). Household needs dominate the community's water needs. Household needs require the most water because it is the basic need of every family in carrying out their activities.

Domestic water needs us a water used in private residences to meet daily needs. household water needs are used for drinking. cooking. bathing. washing toilets (MCK). washing cars/motorcycles. watering plants. and others [27]. Research by [28]. showed that household size. typology of residence. equipment efficiency. presence of children under 12 years of age. presence of a swimming pool and dishwasher are statistically significant variables affecting household water use. The need for water for households dominates the use of water by the community. This is in accordance with Law no. 17 of 2019 concerning Water Resources that the fulfillment of the right to water is prioritized for daily basic needs. smallholder agriculture. and the use of water resources for business needs to meet daily basic needs through a drinking water supply system. The use of water for household needs is important because it is a basic need for every human being to be able to carry out activities properly.

According to the data that has been obtained. at least every citizen who uses Tuk Simuncar and Tuk Sipendok needs 58 liters of water every day. The total water need of 58 liters per person is close to the water needs according to Table 1. which ranges from 60 - 80 liters / day. this adjustment is because the people who use water from Tuk Simuncar and Tuk Sipendok live in rural areas with a population of < 20.000 people. However. according to Soedjoko (2011) in [29]. the availability of clean water which is the right of every human being is 50 liters/capita/day. so that when compared with this figure. the results of the study are appropriate.

In agriculture and animal husbandry. the Tuk Simuncar and Tuk Sipendok water users on average use 59 liters of water/day for agriculture and 10 liters/day for agriculture. The average community owns 2 cows. 5 goats. and 42 chickens. In the research area. livestock ownership is also influenced by the harvest season. as is the case when data collection is the tobacco harvest season. During the tobacco harvest season. many people sell their livestock because they are focused on taking care of the tobacco being harvested because it takes a lot of time. so the need for water for livestock can increase when people are not taking care of the harvest. While livestock water needs are influenced by plant age. based on interviews with farmers. water needs when plants are young is more than when plants are old. this is because when plants are young they need a lot of water to accelerate growth.

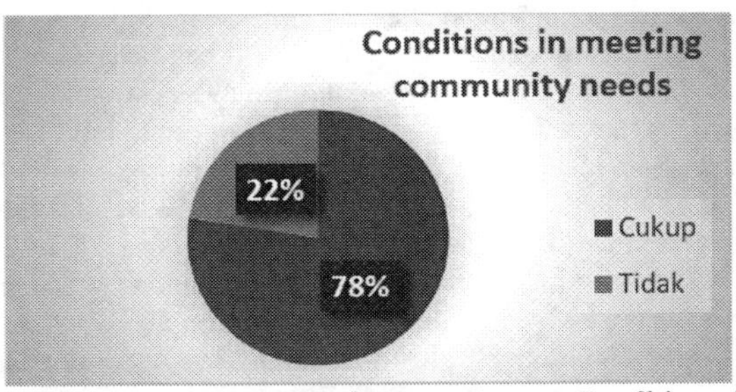

Figure 4. Percentage of Water Source Conditions in Meeting Community Needs

Through the picture above. it can be seen that as many as 78% of the community think that Tuk Simuncar and Tuk Sipendok are sufficient in meeting the needs of the community. while the other 22% think otherwise. Whether or not the water is sufficient or not is influenced by the village where they live. because the village they live in is related to the distance from the water source. The long distance from the water source can cause the water to not get to the residents' houses smoothly. Water that does not reach people's homes is caused by the height of the village compared to the water source. flow rate. and pipe leaks. According to [30]. the flow rate in a pipe is influenced by high pressure. pipe dimensions. pipe length. inlet dimensions. and outlet dimensions. Pipe leaks can be caused by water pressure that is too high and intentional by humans. this is done intentionally by making holes in the water pipe to be taken for the perpetrator's needs. such as watering the fields. Therefore. the placement of water pipes should be in locations that are often accessed by the public to minimize the possibility of pipe holes in the context of theft of water.

The high demand of the community for Tuk Simuncar and Tuk Sipendok water flows must be accompanied by good management as well. The availability of water for Tuk Simuncar and Tuk Sipendok can be maintained if the management is appropriate. including community management in making water networks. Although six of the eight villages that use water from Tuk Simuncar and Tuk Sipendok have been formed in the PAMSIMAS system. it needs to be accompanied by good communication between water user groups so that relations between groups run well and can exchange ideas in terms of supporting sustainable water use.

One way to maintain the sustainability of the Tuk Simuncar and Tuk Sipendok springs is to plan for proper water use. According to [31]. planning for the use of springs must be carried out by considering:

1. Long-term needs for springs. based on existing conditions of use and plans for developing springs in the future. so that they can be utilized sustainably.
2. The natural spring discharge is taken with the correct technique.
3. Benefit for the community. with the understanding that in addition to financial benefits. the use of springs must also provide social benefits. especially for people who have benefited before the springs were developed.
4. Conservation of catchment areas. to ensure the sustainability of the use of springs.

3.4 Water Use Index (IPA)

Water from Tuk Simuncar and Tuk Sipendok has been used by communities around Mount Merbabu National Park with a total of 10.493 users for Tuk Simuncar and 13.551 people for Tuk Sipendok. The number of water users needs to be adjusted to the availability of water in both water sources so as not to exceed their carrying capacity. One way to determine the suitability between the number of users and the available water is to monitor water use. According to the Minister of Forestry Regulation No. P.61/Menhut-II/2014. monitoring of water use is carried out to find out the description of the amount of water demand compared to the quantity of water available in a water flow.

Monitoring of water use can be done by calculating the value of the Water Use Index (IPA). The IPA value of a water flow is said to be good if the amount of water used in a water flow is still less than its potential so that the water flow still produces water that can come out to its downstream area. on the contrary it is said to be bad if the amount of water used is greater than its potential so that the volume of water used is produced from water sources for the downstream area are few or none [16]. The value of IPA in Tuk Simuncar and Tuk Sipendok can be seen in the calculation below:

$$\text{IPA Tuk Simuncar} = \frac{\text{Demand}}{\text{Supply}} = \frac{\text{number of user x demand of each person}}{\text{discharge} \times 86400} = \frac{10493 \times 58}{23.22 \times 86400} = 0.3 \quad (6)$$

$$\text{IPA Tuk Sipendok} = \frac{\text{Demand}}{\text{Supply}}$$

$$= \frac{\text{number of user x demand of each person}}{\text{discharge} \times 86400}$$
$$= \frac{13551 \times 58}{31.20 \times 86400}$$
$$= 0.3$$

3.5 Potential Water Availability in 2020 – 2029

The current water debit of Tuk Simuncar and Tuk Sipendok is still able to meet the needs of the surrounding community. but it is also important to know how the potential is for the future so that the management of water sources is in accordance with their designation. Over time. the community's water needs have the potential to change. which changes are related to the increasing number of people in the future. For this reason. it is necessary to do population projections to determine changes in the number of water users. The population projection in question is a population forecast based on a scientific calculation based on components that affect population growth in the future [15]. Population projections are carried out based on the number of residents in each village using Tuk Simuncar and Tuk Sipendok water for the last three years. The increasing number of water users must be adjusted to the available water debit in Tuk Simuncar and Tuk Sipendok.

According to Davis & de Wist (1966) in [18]. the amount of flow discharge in a spring is determined by the amount of ground water filling. According to [32]. rain is the main source of groundwater. therefore in this study a spring discharge forecast was carried out based on rainfall data in the Tuk Simuncar and Tuk Sipendok catchments for the last 10 years obtained from the Chirps satellite with the help of Google Earth Engine.

Table 7. IPA of Tuk Simuncar Projections for 2020 – 2029

Year	User	Water demand (l/sec)	Water supply (l/sec)	IPA	Class
2020	9.784	6.57	23.22	0.3	Low
2021	10.056	6.75	22.58	0.3	Low
2022	10.328	6.93	21.94	0.3	Low
2023	10.600	7.12	21.29	0.3	Low
2024	10.872	7.30	20.65	0.4	low
2025	11.143	7.48	20.01	0.4	Low
2026	11.415	7.66	19.37	0.4	Low
2027	11.687	7.85	18.73	0.4	Low
2028	11.959	8.03	18.09	0.4	Low
2029	12.231	8.21	17.44	0.5	Medium

Table 7. shows that the number of Tuk Simuncar water users is projected to continue to increase until 2029 with a total of 12.231 users. of which it is predicted that 8.21 liters/second of water will be required to meet daily needs. To meet the water demand. it is predicted that in 2029 Tuk Simuncar's water discharge will be 17.44 liters/second. From the value of water demand and supply. it can be seen that the Tuk Simuncar Water Use Index (IPA) value in 2029 is 0.5 which is in the medium class. Medium class means that the amount of water taken from Tuk Simuncar has a proportion of 50% of the total discharge. so that there is still water flowing downstream.

Table 8. IPA of of Tuk Sipendok Projection for 2020 – 2029

Year	User	Water demand (l/sec)	Water supply (l/sec)	IPA	Class
2020	12.534	8.41	31.20	0.3	Low
2021	12.779	8.58	30.34	0.3	Low
2022	13.024	8.74	29.48	0.3	Low
2023	13.268	8.91	28.61	0.3	Low
2024	13.513	9.07	27.75	0.3	Low
2025	13.757	9.24	26.89	0.3	Low
2026	14.002	9.40	26.03	0.4	Low
2027	14.247	9.56	25.16	0.4	Low
2028	14.491	9.73	24.30	0.4	Low
2029	14.736	9.89	23.44	0.4	Low

Table 8 shows that the number of Tuk Sipendok water users is projected in 2029 to be 14.736 people with the amount of water needed at 9.89 liters/second and it is estimated that the available water in Tuk Sipendok is 23.44 liters/second. from this figure it can be seen that the value of IPA Tuk Sipendok in 2029 is estimated to be 0.4 or included in the low class. Low grade indicates that water taken from water sources has a proportion of less than 50% so that there is still a lot of water that can flow to the downstream area.

The need for water from both water sources tends to increase every year which is influenced by the dynamics of the population. While water supply is closely related to changes in rain patterns. [33]. changes in rain patterns and evapotranspiration in many tropical islands will cause changes in water resources. runoff and seasonal distribution patterns. The meeting between demand and supply of water indicates that the IPA value is equal to 1. or the water taken is equal to the supply. Based on Figure 20. it can be seen that IPA is estimated to be worth 1 in 2040 for Tuk Simuncar and 2042 for Tuk Sipendok.

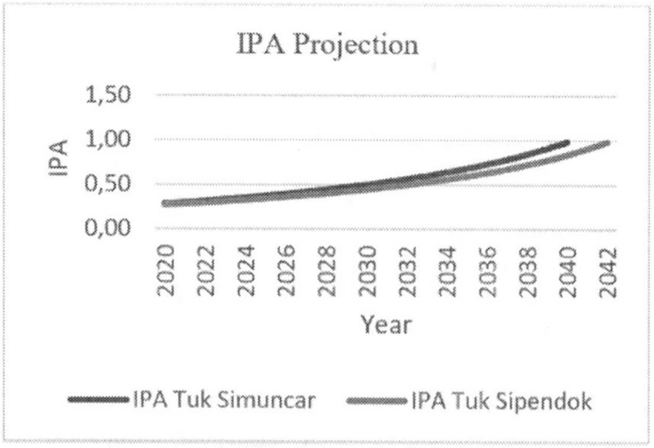

Figure 5. Graph of Projected IPA Values

The availability of water for Tuk Simuncar and Tuk Sipendok in the future. apart from being determined by natural factors. also needs to be balanced with the efforts of the management of Mount Merbabu National Park (MMNP) and the community. related to the application of water conservation techniques. Water conservation through effective management and efficient use is an activity that is urgently needed and urgent. Water management based on its existence as a natural resource is part of a water conservation program that fully maintains. rehabilitates. maintains and utilizes existing water sources effectively and efficiently in order to play a role in improving people's welfare [34]. Water conservation is one of the main keys in ensuring water availability and increasing water supply in line with increasing demands for water [35].

The study of the available flow rate. the condition of the vegetation in the catchment area. the use of community water. and the balance between water demand and water supply can refer to the potential availability of water from Tuk Simuncar and Tuk Sipendok. From the flow rate data and community water utilization. it can be seen the IPA value which shows the comparison between demand and water supply. The known IPA value can indicate an opportunity for further utilization. if the IPA value is still low. it means that there is still an opportunity to utilize the available water more. Meanwhile. data on vegetation characteristics of water catchment areas is used to support the potential for water availability that can be owned by Tuk Simuncar and Tuk Sipendok. Good vegetation conditions such as having high density with mixed forest cover can increase soil aggregate and infiltration capacity. and reduce the threat of landslides.

4 Conclusions

Based on the description in this study. it can be concluded that:

1. The available water debits in Tuk Simuncar and Tuk Sipendok are 23.22 liters/second and 31.20 liters/second. The water discharge for the future may change according to natural conditions. based on the results of the projected water discharge at the two water sources. it is estimated that in 2029 the water discharge of Tuk Simuncar and Tuk Sipendok will decrease so that it has a water discharge of 17.44 liters/second and 23.44 liters/second. To maintain the sustainability of water sources. it is necessary to make regulations and policies related to the use of water by taking into account the applicable provisions.
2. The water catchment area of Tuk Simuncar is dominated by vegetation with a high density of 75 Ha. while at Tuk Sipendok it is dominated by vegetation with a medium density of 102.28 Ha. When viewed from the dominance of vegetation cover. both Tuk Simuncar and Tuk Sipendok are dominated by shrubs. Related to water management. shrubs are a type of land cover that has a low value in terms of soil moisture capacity to store water reserves or water holding capacity. With this condition. it is hoped that the Gunung Merbabu National Park Center and the community can prevent land fires that can damage vegetation in the water catchment area.
3. Communities around Gunung Merbabu National Park use Tuk Simuncar and Tuk Sipendok water to meet household water needs (75%). livestock (21%). and agriculture (4%). Of the total use of water. every day requires as much as 58 liters of water / person to meet their needs. Household needs are important because they are basic needs for every human being to be able to carry out activities properly. The use of water by the community around Mount Merbabu TN should be accompanied by permits that are in accordance with applicable regulations so that people get legality in using water from the Mount Merbabu TN area.
4. The Water Usage Index (IPA) of Tuk Simuncar and Tuk Sipendok is currently 0.3. which is in the low class. This figure means that the amount of water taken by the two water sources is still relatively low. so it still leaves a lot of water to flow to the area below. The current IPA value is different from the projected IPA value in 2029. which is 0.5 (middle) for Tuk Simuncar and 0.3 (low) for Tuk Sipendok which indicates the increasing number of users but decreasing water supply.

The authors would like to thank for the support funding from RTA UGM year 2021 in facilitating the field data collection and presentation of this research. Special thanks to Mount Merbabu National Park Agency for giving the research permission at the study area.

References

1. B. Coelho, A. Andrade-Campos, Efficiency achievement in water supply systems—A review. Renewable and Sustainable Energy Reviews **30**, 59 – 84 (2014)
2. J. Fiquepron, S. Garcia, A. Stenger, Journal of Environmental Management **126**, 113 – 121 (2013)
3. C. Lim, C. Song, Y. Choi, S. Woo Jeon, W. Lee, Journal of Environmental Management **248** (2019)
4. S. Sosrodarsono, K. Takeda, Hidrologi untuk Pengairan (Pradna Paramita, Jakarta, 2003).
5. Aditya, D.N. Gilang, M. Faddel, Sunarto, Pros Sem Nas Masy Biodiv **5**(1), 362-368 (2019)
6. H. Gunawan, M. Bismark, K. Haruni, Jurnal Penelitian Hutan dan Konservasi Alam **10**(2), 103-119 (2013)
7. A.N. Nugroho, M. Z. Faza, P.A. Winarso, *Analisis Kondisi Meteorologi Terkait Kejadian Kebakaran Hutan di Lereng Gunung Merbabu*, In Prosiding SNFA (Seminar Nasional Fisika dan Aplikasinya), 24 – 34 (2018)
8. M. Chang, Forest Hidrology Third Edition (CRC Press, London, 2013)
9. D.K. Todd, L. W. Mays, Groundwater Hydrology Third Edition (John Wiley & Sons Inc., New Jersey, 2005)
10. L.H. Schinasi, T. Benmarhnia, A.J. De Roos, Environmental Research **161**, 168 – 180 (2018)
11. N.H. Nawangwulan, S. Bambang, S. Bandi, Jurnal Geodesi Undip **2**(2), 127-140 (2013)

12. Y.P. Dasuka, S. Bandi, Hani'ah, Jurnal Geodesi Undip **5**(2), 1-8 (2016)
13. U. Silalahi, Metode Penelitian Sosial (Rafika Aditama, Bandung, 2009)
14. N. Duli, Metodologi Penelitian Kuantitatif: Beberapa Konsep Dasar Untuk Penulisan Skripsi & Analisis Data dengan SPSS (Budi Utama, Yogyakarta, 2019)
15. D. Handiyatmo, S. Idha, R. Hasnani, Pedoman Penghitungan Proyeksi Penduduk Dan Angkatan Kerja (Badan Pusat Statistik, Jakarta, 2010)
16. G. Oktavia, Y.L. Handayani, Siswanto, Jom FTEKNIK **5**(2), 1-6 (2018)
17. Purwaningsih, Berk. Penel. Hayati Edisi Khusus **4A**, 23-28 (2010)
18. S. Sofiah, P. F. Abban, Berk. Penel. Hayati Edisi Khusus **4A**, 1-3 (2010)
19. S. Purnama, Hidrologi Air Tanah (Penerbit Kanisius, Yogyakarta, 2010)
20. A. Yuniarsih, D. Marsono, S. Pudyatmoko, R. Sadono, Jurnal Penelitian Hutan dan Konservasi Alam **11**(3), 239-259 (2014)
21. L.P. Harahap, U.H. Angelia, Nilai Ekonomi Kelembagaan Menjaga Hutan dan Air untuk Irigasi (Puspantara, Medan, 2018)
22. D. Ariani, Y. Prasetyo, B. Sasmito, Jurnal Geodesi UNDIP **9**(1), 207 – 216 (2020)
23. D.L. Setyowati, Forum Geografi **24**(1), 39-56 (2010)
24. Y. Indrajaya, W. Handayani, Info Hutan **5**(3), 231-240 (208)
25. N.B. Laksono, S. Suprayogi, Jurnal Bumi Indonesia, **8**(3) (2010)
26. M.T. Astuti, M. Rahdriawan, Jurnal Teknik PWK, **2**(4) (2013)
27. J.D. Putuhena, Pengelolaan Daerah Aliran Sungai (DAS) Dimensi Pengeloaan Berkelanjutan di Kota Ambon (Pascasarjana Manajemen Hutan Unpatti, Ambon, 2019)
28. K. Rathnayaka, S. Maheepala, B. Nawarathna, B. George, H. Malano, M. Arora, P. Roberts, Resources, Conservation and Recycling **92**, 85–94 (2014)
29. Sepsiaji, D. Nurhadi, Awang S. A. dan W. Andayani, Kelangkaan Air: Mitos Sosial. Kiat. dan Ekonomi Rakyat (DEBUT Press, Yogyakarta, 2005)
30. E. Sulistyanto, K. Suharno, S. Widodo, Wahana Ilmuwan **2**(2), 13-22 (2016)
31. S. Arsyad, E. Rustiadi, Penyelamatan Tanah. Air. dan Lingkungan (Yayasan Obor Indonesia, Jakarta, 2008)
32. S.D. Devy, H. Heru, P.E.P. Dony, *Pemodelan Airtanah Daerah Penambangan Batubara Pit Terbuka Di Muara Lawa. Kabupaten Kutai Barat. Kalimantan Timur*, In Prosiding Seminar Nasional Kebumian Ke-7 Jurusan Teknik Geologi, Fakultas Teknik Universitas Gadjah Mada, 30 – 31 Oktober (2014)
33. R.M. Delinom, D. Marganingrum, Sumber Daya Air dan Lingkungan: Potensi, Degradasi, dan Masa Depan (LIPI Press, Jakarta, 2007)
34. M.K. Sallata, Info Teknis Eboni **12**(1), 75-86 (2015)
35. R.J. Kodoatie, S. Roestam, Tata Ruang Air (Andi, Yogyakarta, 2010)

Geological mapping and stand-up time estimation based on core drill evaluation using RMR89 on Tunnel 3 Sigli Banda Aceh Toll Road

Hanif Khoirul Latif [1,2], *Salahuddin* Husein[1*], and *I Gde Budi* Indrawan[1]

[1]Gadjah Mada University, Yogyakarta, Indonesia
[2]Regional Infrastructure Development Agency, Ministry of Public Works and Housing, Indonesia

Abstract. Tunnel 3 of the Sigli Banda Aceh Toll Road has been designed based on several site investigations. The site investigations have not included surface geological investigations. Subsurface geological investigations have been carried out but with less detailed drill core evaluation. This will cause less accuracy on the assessed tunnel stand-up time. This research aims to evaluate the engineering geological conditions at the tunnel construction site by conducting geological mapping with 1:5,000 scale and re-evaluating drill core using RMR_{89} method more thoroughly so that it can produce more accurate assessment of the tunnel stand-up time. The geological mapping result shows that the research area has three rock units, namely lapili, tuffaceous sandstones, and interbedded sandstone-siltstones, with moderate to fresh weathering degree. The drill core evaluation shows that research area has poor to good rock mass classes (34 – 67). The drill core evaluation results also indicate possible geological structures that may influence subsurface rock mass strength. The tunnel stand-up time obtained from the lowest RMR_{89} value for each drill core shows a value of 5 - 90 hours, implying immediate application of support system after tunnel excavation.

1 Introduction

The Sigli - Banda Aceh Toll Road consists of six sections. Section one is section Seulimeum - Padang Tiji. The Seulimeum - Padang Tiji section is planned for 26 km. The construction of the Sigli – Banda Aceh Toll Road section one, 3 tunnels included, is carried out by PT. Hutama Karya with PT. Adhi Karya as contractor and PT. Indra Karya as a planning consultant. This research is conducted in Tunnel 3 which is located at station 15 + 400 to station 15 + 750 with a length of 325 m.

In the construction stage of Tunnel 3, a series of geology and geological engineering investigations have been carried out by PT. Indra Karya. The investigation consisted of topographic survey, drill core taking, geoelectrical investigation, and rock mass quality classification. The results of the investigation stated that Tunnel 3 has a type of sandstone rock with a weak compressive strength value. Meanwhile, the classification of subsurface rock mass quality on the drill core using the rock mass rating (RMR_{89}) [1] method illustrates that the rock mass quality is in the fair to poor class. The stand-up time obtained from the results of the rock mass quality class has a range from 3 hours to 100 hours [2, 3].

The various investigations that have been carried out by the planning consultant have only been carried out specifically at the location of the tunnel alignment. The investigation also has not included elements of surface geological investigations. Only subsurface investigations have been carried out. In fact, to produce sufficient knowledge regarding geological conditions and engineering geology in detail, it will not be sufficient to rely solely on subsurface aspects. Surface aspects are also needed to support subsurface findings. Investigations on the quality of rock mass using the RMR_{89} classification that have been carried out by the planning consultant have also not been carried out in detail. The investigation was only carried out by assessing RMR_{89} from the average drill core condition obtained. The RMR_{89} value obtained only from the average drill core condition is feared not precise enough in describing the quality conditions of the rock mass in the field so that it can cause misleading in determining the estimated stand up time of the tunnel.

Based on the lack of detailed geological and engineering geology investigations in the construction of Tunnel 3, this research seeks to provide comparative data by conducting surface geological investigations through large-scale geological mapping at the tunnel construction site. Investigation of subsurface geological conditions will also be carried out again by evaluating the drill core in detail using the RMR_{89} method to determine the quality of the subsurface rock mass at the research area. The calculation of the stand-up time will be carried out based on the results of the classification of rock mass quality using the RMR_{89} method.

*Corresponding author: shddin@ugm.ac.id

 https://doi.org/10.1051/e3sconf/202132502004

2 Methodology

2.1 Geological Condition

The research area is part of Geologic Map of Banda Aceh Quadrangle by Bennet, et al, [4] with 1 : 25.000 map scale. The rock formation on research area is part of Padang Tiji Member and around the research location there are the Alluvium Formation, Lam Teuba Volcano Formation, Indrapuri Formation, Seulimeum Formation, Kotabakti Formation and Padang Tiji Members (See Fig.1).

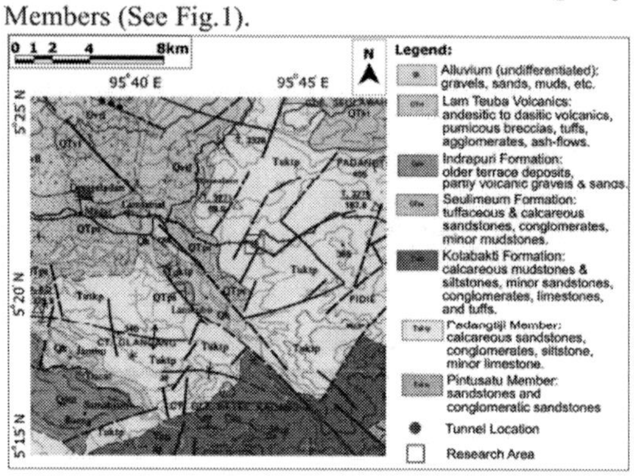

Fig. 1. Regional geological map of research area [4]

In Fig. 1, it can also be observed that there is a geological structure in the form of a fault in the research area. The fault is a branch of the Seulimeum Fault. The Seulimeum fault is an active fault that cuts through Plio - Pleistocene and volcanic sedimentary rocks from Mount Seulawah Agam [5].

2.2 Geological Mapping

Geological mapping is a highly interpretive scientific process by which a geological mapper seeks to understand the composition and structure of the surface and subsurface geology of the earth.

In conducting geological mapping, a geological mapper will record in a field book about geological features such as strikes and dips, rock types, weathering levels, sketches of rock outcrops and so on. The more narratives obtained in the field will make it easier for a geological mapper to compile a geological map.

In this research, geological mapping was carried out with a mapping area of 1 x 1 km with a mapping time of 30 days. The geological trajectory map is made based on the contours obtained from the digital elevation model (DEM) data. The geological trajectory map covers the entire research area with 37 field observation stations.

2.3 Rock Mass Rating

The RMR classification (RMR_{89}) was first proposed by Bieniawski [1] using six parameters to determine the rock mass rating. The six parameters are uniaxial compressive strength (UCS) for intact rock, rock quality designation (RQD), discontinuity spacing, discontinuity conditions, groundwater conditions and discontinuity orientation.

Each of the six parameters has the following weightings: 0 to 15 for the strength of the intact rock material, 3 to 20 for RQD, 5 to 20 for the spacing of discontinuity, 0 to 30 for discontinuity conditions, 0 to 15 for groundwater conditions, and 0 to (-12) for discontinuity effects. The maximum value of the weighting of the RMR parameter is 100. Classes in RMR classification are divided into 5 classes from very poor to very good. In this research, RMR will be used in determining subsurface rock mass classification based on four drill core points (T35A, T35B, T36A, T36B).

The strength parameters of intact rock material based on UCS used in this study were obtained from the value of UCS test in the laboratory for four drill core points at a certain depth produced by the planning consultant (see Table 1). The depth point is selected based on the selection of samples that represent the condition of the level of rock weathering and rock strength from the drill core taken. RMR_{89} rating for the strength parameters of intact rock materials can be seen in Table 2.

Table 1. UCS value of drill cores

No	Core Number	Depth (m)	UCS value (Mpa)
1	T35A	6.50 – 7.00	0.248
		21.10 – 22.00	10.741
		29.25 – 30.00	1.903
		35.35 – 36.00	10.027
		39.20 – 40.00	7.877
		43.00 – 43.65	1.494
2	T35B	6.10 – 6.60	2.930
		22.40 – 23.00	1.894
		30.30 – 30.90	5.680
		34.40 – 35.00	3.265
		37.20 – 37.80	4.262
		42.40 – 43.00	4.784
3	T36A	1.80 – 2.50	4.403
		19.10 – 19.70	4.045
		26.10 – 26.90	8.589
		33.40 – 34.00	5.571
		38.40 – 38.90	6.347
		45.00 – 45.60	6.985
4	T36B	6.40 – 6.60	3.016
		9.70 – 9.90	2.736
		12.30 – 12.80	3.772
		14.20 – 14.60	6.671
		17.60 – 18.00	3.398
		22.60 – 22.80	5.517

The RQD parameter used in this study is the RQD value per one meter on each drill core. The calculation of the RQD value was carried out using the RQD measurement guide in Fig. 2 [6]. Meanwhile, the formula used to calculate the RQD value can be seen in formula (1) [6]. RMR_{89} rating for RQD parameter can be seen in Table 3.

Table 2. UCS rating used at RMR_{89}[1]

UCS (MPa)	Rating
>250	15
100 – 250	12
50 – 100	7
25 – 50	4
5 – 25	2
1 – 5	1
<1	0

$$RQD = \frac{\Sigma Length\ of\ core\ pieces > 10cm\ length}{Total\ length\ of\ core\ run} x100\% \quad (1)$$

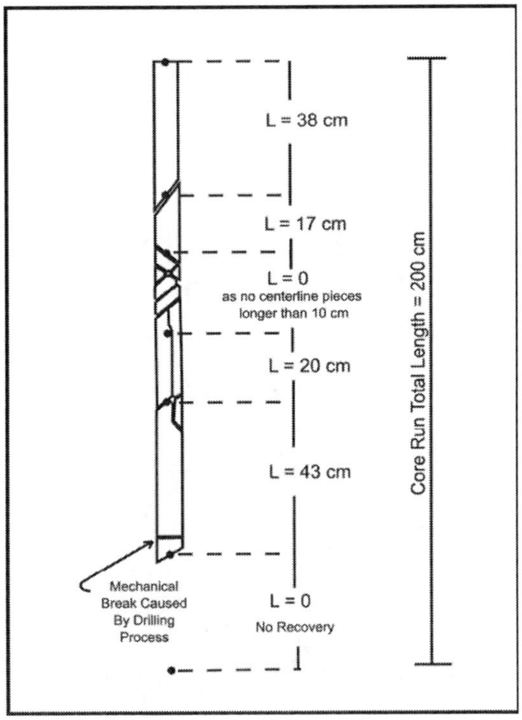

Fig. 2. RQD measurement guidance [6]

Table 3. RQD rating at RMR89 [1]

Drill core quality RQD (%)	Rating
90 – 100	20
75 – 90	17
50 – 75	13
25 – 50	8
<25	3

The discontinuity spacing parameter used in this study was obtained from the perpendicular distance between two successive discontinuity planes (joints, faults, folds, bedding planes, etc.) along the measurement line on the drill core. Discontinuity spacing rating in RMR_{89} is classified based on the range of distances between discontinuities. The wider the distance between the discontinuities, the higher the rating given. This is because discontinuities with a close distance will weaken the strength of the rock mass. Discontinuity spacing rating on RMR_{89} can be seen in Table 4.

Table 4. Rating of discontinuity spacing used at RMR_{89} [1]

Discontinuity spacing (m)	Rating
> 2	20
0.6 – 2	15
0.2 – 0.6	10
0.006 – 0.2	8
<0.006	5

The discontinuity condition parameter used in this study has five sub-parameters, namely persistence, separation, roughness, infilling and the level of weathering of the discontinuity plane. Persistence is the length of the discontinuity plane. Separation is the distance between the discontinuity plane faces. Roughness is the roughness of the discontinuity face. The rougher the discontinuity planes, the more likely they are to interlock. Infilling is a material that fills the discontinuity plane. The material can be hard or soft material. The hardness or softness of the filler material will have a different value in the RMR_{89} rating. Weathering level is the degree of weathering conditions of discontinuities in rock. Ratings for each sub-parameter in the discontinuity condition parameters can be seen in Table 5.

Table 5. RMR_{89} rating of sub-parameters of discontinuity condition [1]

Parameter	Ratings				
Presistence (m)	<1	1-3	3-10	10-20	>20
	6	4	2	1	0
Separation (mm)	None	<0.1	0.1 – 1.0	1 – 5	>5
	6	5	4	1	0
Roughness	Very	Rough	Slight	Smooth	Slicken sided
	6	5	3	1	0
Infilling (mm)		Hard Filling		Soft filling	
	None	<5	>5	<5	>5
	6	4	2	2	0
Weathering	None	Slight	Moderate	High	Decom posed
	6	5	3	1	0

The groundwater condition parameters used in this study were obtained from rubbing the surface of the drill core with the palm of the hand. In addition, the assessment of this parameter is also carried out by taking into account the assessment of the planning consultant on the parameters of the groundwater condition of the drill core. There are several ground water conditions, namely completely dry, damp, wet, dripping, and flowing. The rating of groundwater conditions at RMR_{89} can be seen in Table 6.

Table 6. Rating of groundwater condition at RMR_{89} [1]

Groundwater (general condition)	
Condition	Rating
Complitely dry	15
Damp	10
Wet	7
Dripping	4
Flowing	0

The discontinuity plane orientation parameter is represented by strike and dip. The orientation of the discontinuity plane is categorized in terms of favourable and unfavourable (see Table 7). Ratings for discontinuity orientation can be seen in Table 8.

Table 7. Category of discontinuity plane orientation [1]

Strike Perpendicular to Tunnel Axis			
Drive with Dip		Drive against Dip	
Dip 45° – 90°	Dip 20° – 45°	Dip 45° – 90°	Dip 20° – 45°
Very favorable	Favorable	Fair	Unfavorab le
Strike Parallel to Tubbel Axis		Irrespective of Strike	
Dip 20° – 45°	Dip 45° – 90°	Dip 0° – 20°	
Fair	Very unfavorable	Fair	

Table 8. RMR_{89} ratings for discontinuity plane orientation [1]

Ratings of Discontinuity Plane Orientation for Tunnels	
Descriptions	Ratings
Very favorable	0
Favorable	-2
Fair	-5
Unfavorable	-10
Very unfavorable	-12

2.4 Stand Up Time

The term stand-up time was first used by Lauffer [7] in Lauffer's rock mass classification and then further developed by Lauffer [8] for excavation using tunnel boring machines purpose. Stand-up time is defined as the length of time the tunnel can support itself without additional supporting structures [9]. The stand-up time estimation is very important because it will affect the excavation cycle, the required support structure, and the tunnel excavation method.

The stand-up time developed by Lauffer is then correlated with rock mass classification and the most frequently applied in tunnel construction work is the RMR and stand-up time correlation by Bieniawski [1]. The correlation by Bieniawski can be seen in Fig. 3.

In this research, the correlation between stand up time and RMR by Bieniawski is used to determine the stand-up time estimation. The RMR value used to correlate with stand-up time is the lowest RMR value of each drill core point at the tunnel elevation (30 m).

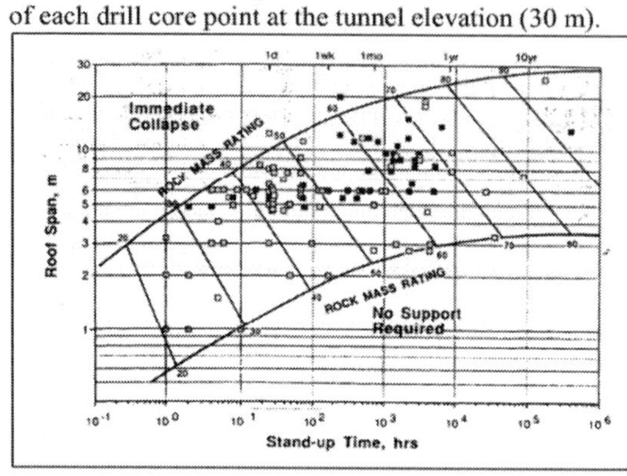

Fig. 3. Stand-up time and RMR value relation [1]

3 Result and Discussion

Based on the geological mapping carried out in the research area, it was found that the research area has three rock units, namely lapili, tuffaceous sandstones, and interbedded sandstone-siltstones (see Fig 4, 5 and 6).

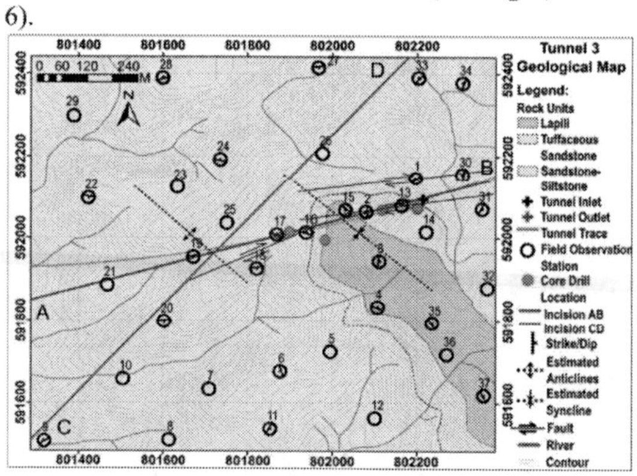

Fig. 4. Geological map of Tunnel 3

Lapili have a moderate level of weathering, tuffaceous sandstones have a moderate to slight level of weathering, while sandstone-siltstones have a slight to fresh level of weathering. The rock type and weathering level have a good correlation with the observations on the drill core. The results of observations of rock types in the drill core state that the research area consists of two rock units, namely tuffaceous sandstone and interbedded sandstone-siltstone. Lapilli were not found in the drill core because the drill core extraction areas were not in the lapilli zone as shown in Fig. 4. From observations on the drill core it was also found that the rocks in the study area experienced moderate to fresh weathering degrees (see Table 9).

Table 9. Lithology and degree of weathering based on drill core evaluation

No	Core	Depth (m)	Rock Unit	Weathering Level
1	T36B	0 - 15	Interbedded Sandstone-Siltstone	Slightly Weathered
		15 - 30	Interbedded Sandstone-Siltstone	Fresh
2	T35A	0 - 10	Tuffaceous Sandstone	Moderately weathered
		10 - 29	Tuffaceous Sandstone	Slightly Weathered
		29 - 40	Interbedded Sandstone-Siltstone	Slightly Weathered
		40 - 51	Interbedded Sandstone-Siltstone	Fresh
3	T 35 B	0 - 31	Tuffaceous Sandstone	Moderately weathered
		31 - 46	Interbedded Sandstone-Siltstone	Slightly Weathered
		46 - 49	Interbedded Sandstone-Siltstone	Fresh
4	T 36 A	0 - 22	Tuffaceous Sandstone	Moderately weathered
		22 -34	Interbedded Sandstone-Siltstone	Slightly Weathered
		34 - 56	Interbedded Sandstone-Siltstone	Fresh

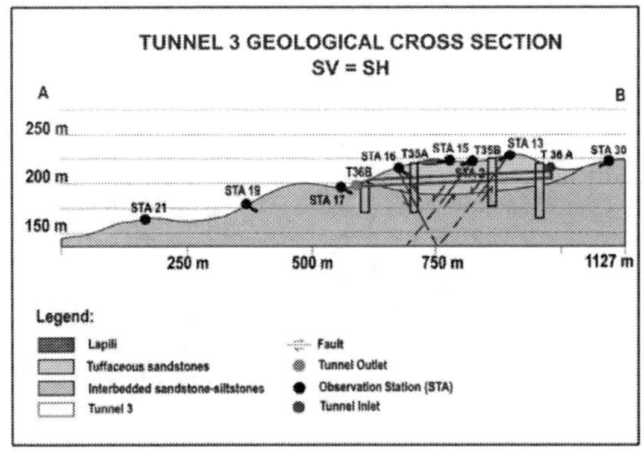

Fig. 5. Geological cross section AB

From the geological mapping carried out, it was also found that there were several geological structures such as folds and faults. These faults are faults found in STA 2, STA 13, and STA 16. These faults intersect the tunnel path as shown in Fig. 5. In Fig. 5, it can be seen that the direction of the fault dip is not only cutting the tunnel line but also cutting the drill core bits T35A and T35B.

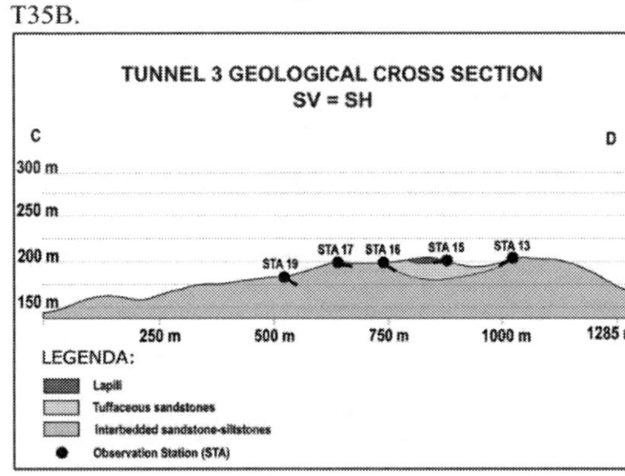

Fig. 6. Geological cross section CD

The results obtained from the subsurface rock mass classification using RMR_{89} as shown in the table 10 indicate that the research area has a rock mass between 34 to 67. The rock mass value indicates that the quality of the subsurface rock mass in the study area consists of tuffaceous sandstone with poor to good quality, and sandstones from poor to good quality. The variation in the quality of the rock mass is possible because of the geological structure in the form of faults as discussed in the previous paragraph. In the T35A drill core, there was a decrease in rock mass quality at a depth of 40m - 51m where there was an intersection with the fault. The same thing also happened to the T35B drill core which experienced a decrease in rock mass quality at a depth of 7m – 13m.

The results of subsurface rock mass rating using the RMR method are then used as the basis for calculating the stand-up time by taking the lowest RMR rating of each drill cores at the tunnel elevation (30m). The lowest rating is correlated using the stand up time chart developed by Bieniawski [1]. The roof span required

https://doi.org/10.1051/e3sconf/202132502004

for correlation is obtained from the tunnel planning roof span of 10 m. The correlation results illustrate that Tunnel 3 of Sigli Banda Aceh Toll Road according to the drill core T35A has a stand-up time of 90 hours, according to drill core T35B 10 hours, according to drill core T36A 90 hours, and according to drill core T36B 5 hours. More detail regarding the stand-up time for Tunnel 3 of Sigli Banda Aceh Toll Road can be seen in Fig. 7, Fig. 8, Fig. 9, and Fig. 10.

Table 10. Rock mass quality on research area

Core	Depth (m)	Lithology	RMR89 Rating	RMR89 Quality
T36B	0 - 9	Interbedded Sandstone-Siltstone	34 - 44	Poor to Fair
	9 - 14	Interbedded Sandstone-Siltstone	44 - 51	Fair
	14 - 30	Interbedded Sandstone-Siltstone	36 - 45	Fair
T35A	0 - 15	Tuffaceous Sandstone	42 - 58	Fair
	15 - 29	Tuffaceous Sandstone	55 - 67	Fair to Good
	29 - 40	Interbedded Sandstone-Siltstone	55 - 66	Fair to Good
	40 - 51	Interbedded Sandstone-Siltstone	45 - 55	Fair
T35B	0 - 7	Tuffaceous Sandstone	42 - 54	Fair
	7 - 13	Tuffaceous Sandstone	37 - 42	Poor to Fair
	13 - 17	Tuffaceous Sandstone	42 - 51	Fair
	17 - 25	Tuffaceous Sandstone	37 - 47	Poor to Fair
	25 - 31	Tuffaceous Sandstone	42 - 57	Fair
	31 - 45	Interbedded Sandstone-Siltstone	44 - 61	Fair to Good
	45 - 49	Interbedded Sandstone-Siltstone	39 - 54	Poor to Fair
T36A	0 - 22	Tuffaceous Sandstone	42 - 59	Fair
	22 - 28	Interbedded Sandstone-Siltstone	50 - 55	Fair
	28 - 35	Interbedded Sandstone-Siltstone	55 - 62	Fair to Good
	35 - 50	Interbedded Sandstone-Siltstone	63	Good
	50 - 56	Interbedded Sandstone-Siltstone	51 - 63	Fair to Good

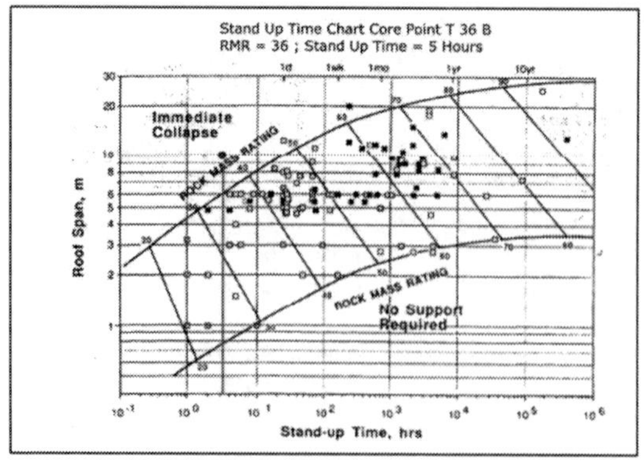

Fig. 7. Stand-up time based on lowest RMR on T36B

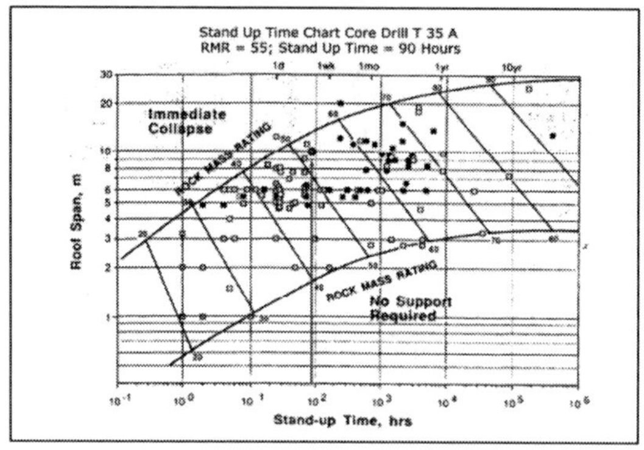

Fig. 8. Stand-up time based on lowest RMR on T35A

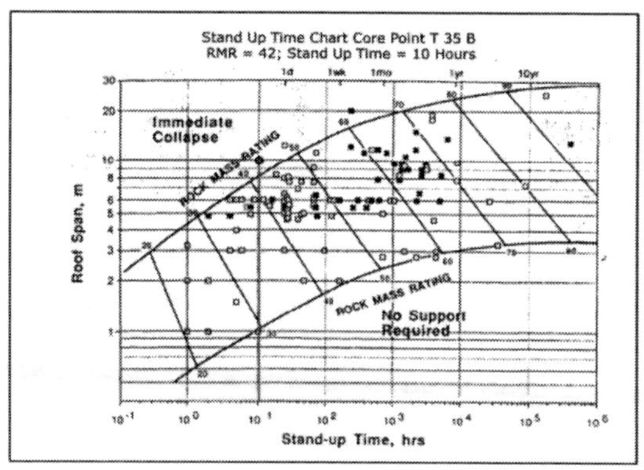

Fig. 9. Stand-up time based on lowest RMR on T35B

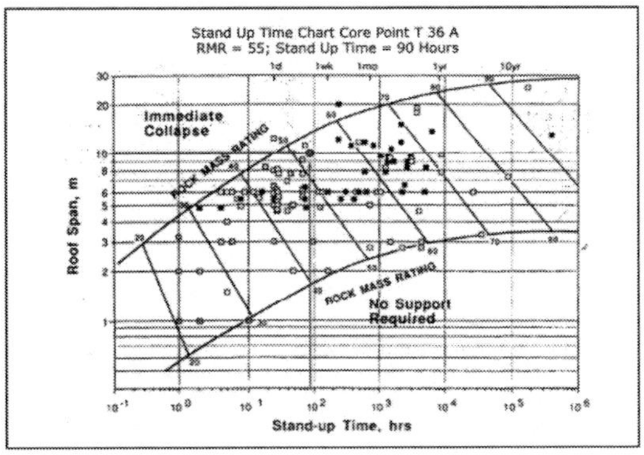

Fig. 10. Stand-up time based on lowest RMR on T36A

4 Conclusion

Based on geological mapping, the research area has three rock units, namely lapilli, tuffaceous sandstone, and interbedded sandstone-siltstone. Lapilli has a moderate level of weathering, tuffaceous sandstone has a moderate to slight weathering level, while the interbedded sandstone-siltstone has a slight to fresh weathering level. The results of the geological mapping have a correlation with the results of drill core evaluations on the type of rock unit and the level of weathering. From the type of rock unit and the level of weathering, it can be seen that there is no rock with a high level of weathering. This can be considered in the planning of tunnel excavation method.

Evaluation of subsurface rock mass quality using RMR_{89} method shows the level of strength from poor to good with a rating range of 34 - 67. The results of the subsurface rock mass quality classification indicate the possibility of geological structure influence on the quality of rock mass. It is indicated with a decrease in quality at a certain depth that intersects with a geological structure in the form of faults. Evaluation of the quality of the subsurface rock mass supported by geological mapping is expected to be more accurate data in determining tunnel stand-up time.

The lowest RMR_{89} value in the study area for each drill core used to determine stand up time shows Tunnel 3 on the Sigli – Banda Aceh Toll Road can survive without reinforcement for 5 to 90 hours. According to the result, it is necessary to immediately strengthen the tunnel during the excavation process to avoid collapse and construction failure.

Further research is needed including determining the excavation method and tunnel stability analysis based on the RMR_{89} value and stand up time data that have been obtained.

Acknowledgments

The authors greatfully acknowledge to the Ministry of Public Works and Public Housing for the full support of research funding through the 2020 Superspecialist scholarship program at the Geological Engineering Studies of Tunnel and Underground Space, Master Program of Geological Engineering, Gadjah Mada University, and to PT. Hutama Karya (Persero) and also PT. Adhi Karya (Persero) for the availability of data and convenience during data collection.

References

1. Z. T. Bieniawski, *Engineering Rock Mass Classifications: A Complete Manual for Engineers and Geologists in Mining, Civil and Petroleum Engineering*. New York (1989)
2. PT. Adhi Karya (Persero) Tbk. and PT. Indra Karya (Persero), *Final Report Pekerjaan Penyelidikan Geologi Jalur Terowongan Seksi 1 Proyek Pembangunan Jalan Tol Trans Sumatera Ruas Sigli - Banda Aceh*. Unpublished (2020)
3. PT. Indra Karya (Persero), *Hasil Studi Kelayakan Terowongan Jalan Tol Ruas Sigli-Banda Aceh – Proyek Pembangunan Jalan Tol Trans Sumatera Ruas Sigli-Banda Aceh Seksi 1*. Unpublished (2020)
4. J.D. Bennet, D.McC. Bridge, N.R. Cameron, A. Djunuddin, S.A. Ghazali, D.H. Jeffery, W. Kartawa, W. Keats, N.M.S. Rock, S.J. Thomson, R. Whandoyo, *Peta Geologi Lembar Banda Aeh, Sumatra* (1981)
5. A.J. Barber, M.J. Crow, J.S. Milsom, *Sumatera: Geology, Resources and Tectonic Evolution*. London (2005)
6. D. U. Deere and D. W. Deere, The rock quality. des., J. Rock Class. Syst. for Eng. Purposes, 91–101 (1988)
7. H. Lauffer, Gebirgsklassifizierung fur, J. Geologic und Bauwesan, **24**, 46–51 (1958).
8. H. Lauffer, Zur gebirgs., J. Felsbau, **6**, 137–149 (1988)
9. V. M. Nguyen and Q. P. Nguyen, Analytical sol., J. *Tunn. Undergr. Sp. Technol.*, **47**, 10–15 (2015)

THE EFFECT OF LAND CHARACTERISTICS ON AGRICULTURAL PRODUCTIVITY USING MULTIPLE LINEAR REGRESSION METHOD IN MADIUN REGENCY, EAST JAVA

Elya Pratiwi Istifaroh[1] , *Rika* Harini[1]
[1]Department of Environmental Geography, Faculty of Geography, Universitas Gadjah Mada, Indonesia

Abstract. Agriculture is an important sector to support national food availability. Soil and climatic conditions in Indonesia are very suitable for agriculture. However, the land characteristics in each region are different, Especially in Madiun Regency, where this study is conducted. This study aims to determine what commodities are suitable for the characteristics of the land based on land suitability and how land characteristics influence agricultural productivity. The data used in this research is secondary data. Determination of agricultural commodities is carried out by cross-checking between land characteristics and guidelines for agricultural commodity land requirements. There are six types of agricultural commodities that are suitable for land characteristics in Madiun Regency. The effect of land characteristics consisting of several parameters on agricultural productivity was sought using multiple linear regression methods. The processing results are known that 8 variables of land characteristics significantly affect agricultural productivity by 93%. The independent variables were tested using the Simultaneous F-Test and were known to have a simultaneous (together) effect on agricultural productivity. In addition, a Partial T-Test was conducted, and it was found that there were only two variables that had a partial effect, namely the soil CEC characteristics and the sand characteristics.

1 Introduction

The Indonesian territory, which is located in the equatorial area and has two seasons, greatly supports soil fertility in various parts of Indonesia. Vast land areas are in provision for forests, agricultural land, and agricultural land. Natural resources and abundant soil fertility is an advantage for the people of Indonesia. Indonesia is known as an agricultural country because most of its people work as farmers. The agricultural sector is the mainstay for Indonesia in order to fulfill its national food supply.

Since ancient times, the international community has ogled the agricultural sector in Indonesia, so it needs more advanced development to get maximum results from the agricultural sector. Indonesia's dense population also encourages greater food demand, which requires more agricultural products (productivity). One of the factors that affects agricultural productivity is fertility [1]. Therefore, knowledge about land fertility or land characteristics needs to be deepened to implement strategies in farming.

Characteristics of land in Indonesia's territory are very varied. It can be clearly seen from the types of agricultural commodities in Indonesia which are also diverse. One of the studied areas in this research is Madiun Regency. According to [2], it has a remarkable carrying capacity of agricultural land to provide food self-sufficiency for its residents. In addition, the need for agricultural land in Madiun Regency is very well met by production land, which is still very large with the support of locations in the lowlands. Although the characteristics of the land in Madiun Regency tend to be the same, there are differences because the parameters of the land characteristics are quite large, such as rainfall, humidity, temperature, soil pH, CEC, nutrients, etc.

Based on land characteristics data, it can be known what agricultural commodities are suitable for the land, so that we can plant agricultural commodities according to the criteria for the land to get maximum productivity. In addition, according to [3],the determination of land suitability needs to be carried out as a guide for agricultural development that is following the principles of land conservation. Land characteristics consisting of several parameters can affect the results of agricultural productivity. However, we do not know how much influence it has and what characteristic parameters affect agricultural productivity. This study aims to determine how land characteristics influence agricultural productivity.

2 Methodology

This research has the following research stages:

2.1 Research Time and Place

This research was conducted in Madiun Regency using units of analysis per subdistrict. The time used in this study is one year, namely 2018 because this study uses secondary data that considers the completeness of the data. The research location in Madiun Regency can be seen in the map image below:

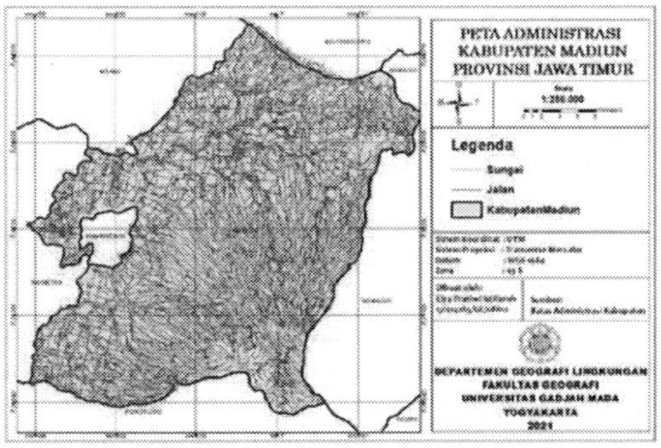

Fig. 1. Madiun Regency research location map.

2.2 Population and Sample

The population in this study is all land in Madiun Regency, while the research sample uses data on land characteristics of each sub-district in Madiun Regency, totaling 15 sub-districts.

2.3 Data

There are two variables for the data used in this study namely the influencing variable and the affected variable. The influence variable uses land characteristic data, which consists of several parameters, namely as follows:

Table 1. Data Types and Source.

Parameters of Land Characteristics	Source
Rainfall	BMKG *online*
Temperature	BMKG *online*
Humidity	BMKG *online*
Soil pH	Study of literature[4]
Soil CEC	Study of literature [4]
c-organic	Study of literature [4]
Soil texture (sand, dust, clay)	Study of literature [4]
Soil nutrient content (N, P, K, Na, Ca, Mg)	Study of literature [4]

In addition to land characteristics data, there is also an affected variable in this study: agricultural productivity. This study uses seven types of commodity limits which are also the leading commodities in Madiun Regency, including types of rice, corn, soybeans, green beans, peanuts, sweet potatoes, and cassava. The agricultural productivity data was obtained through the BPS of Madiun Regency in 2019[5].

2.4 Processing and data Analysis

The purpose of this research is to find out what commodities are suitable for land characteristics in Madiun Regency using the cross-check technique concerning land requirements in the "Technical Guidelines for Land Evaluation for Agricultural Commodities" or Technical Guidelines published by the Center for Research and Development of Agricultural Land Resources, Ministry of Agriculture in 2011. Then, from several suitable commodities per sub-district, the most suitable commodities for the parameters of the characteristics of the land can be concluded. The most suitable commodities per sub-district were analyzed by using descriptive qualitative approach.

The effect of land characteristics variables on agricultural productivity variables were tested using SPSS software. The first stage is carried out on the data, namely the normality test of the data using the Shapiro Wilk method. This method is in accordance with this study with data of less than 50 samples [6]. The data is considered normally distributed if the significance value is > 0.05, and vice versa. Then the land characteristics data, which consists of several parameters and agricultural productivity data which are normally distributed, are carried out with multiple linear regression to determine the effect. This numerous linear regression method aims to determine the effect of two or more independent variables on the dependent variable [7]. The magnitude of the effect can be seen through the value of R2, where if R2 is close to 1, then the influence of the independent variable is considered perfect. On the contrary, the closer the value of R2 is to 0, the weaker the effect will be. In addition, the effect of independent variables can also be known through the significance level in the Model Summary table, where if the value of Sig. F. Change < 0.05, then there is a significant effect, and otherwise.

After knowing the effect of independent variables in general, it can be continued with hypothesis testing, namely the Simultaneous F Test and Partial T-Test. A simultaneous F-test is used to determine the effect of independent variables simultaneously (together). In contrast,a Partial T-test is used to test the impact of independent variables (alone) partially. The two tests have the same method, namely by comparing F count and F table or T count and T table values.

3 Result and Discussion

3.1 Commodities that match the characteristics of the land in Madiun Regency

Commodities that match the characteristics of the land in the Madiun Regency per sub-district analysis unit produce several commodities in accordance with each sub-district's land characteristics. The results of this study can be seen in the table below:

Table 2. Suitability of Agricultural Land.

Districts	Commodity Directive					
	Cassava	Rice	Sweet potato	Soy	Green beans	Corns
Balerejo		√				
Dagangan	√					
Dolopo	√		√			
Geger	√		√			
Gemarang	√		√			
Jiwan		√				
Kare	√					
Kebonsari	√		√			
Madiun		√				
Mejayan		√				
Pilang kenceng		√		√	√	
Saradan						√
Sawahan				√		
Wonosari	√	√	√			
Wungu	√	√	√			

Based on the results of a cross-check between several parameters of land characteristics concerning land requirements, it produces commodity directions in accordance with the characteristics of the land in Madiun Regency as shown in the table above. Some sub-districts are suitable for one type of commodity, and some other sub-districts are suitable for three commodities at once. 8 sub-districts are suitable for only one type of commodity, while those that are suitable for two commodities are 4 sub-districts, and those that are suitable for three commodities are 3 sub-districts.

The table above shows six suitable commodities in the Madiun Regency, where previously seven commodities were selected to suit the characteristics of the land. One commodity that is not in accordance with the characteristics of the land in all sub-districts in the Madiun Regency is peanuts because this commodity has several land requirements that are not in accordance with the characteristics of the land.

The most suitable commodity for the most part in the Madiun Regency area is cassava which is suitable for 8 sub-districts out of 15 sub-districts. Furthermore, rice commodities are suitable for 7 sub-districts, sweet potato commodities are suitable for 6 sub-districts, soybean commodities are suitable for 2 sub-districts, and green beans and corn are suitable for 1 sub-district. The sub-districts that are suitable for three types of commodities are Pilangkenceng, Wonosari, and Wungu sub-districts.

Directions for the use of agricultural land for various types of agricultural commodities can be used as a reference for farmers to choose what commodities are in accordance with the characteristics of the land. However, farmers still have to apply cultivation patterns as in the study [8] regarding farmers in Saradan District who always replace commodities according to cultivation pattern to replace missing nutrients.

3.2 The Relationship between the Suitability of Agricultural Land and Agricultural Productivity

The type of commodity that is suggested for agricultural land use can be considered as a strategy to maximize the result of its produce. However, nowadays many farmers choose types of agricultural commodities without knowing in detail how the characteristics of the area are, so that sometimes there are differences between the results of the land suitability analysis and the reality on the ground. Below is the comparison between land suitability and agricultural productivity yields of various commodities.

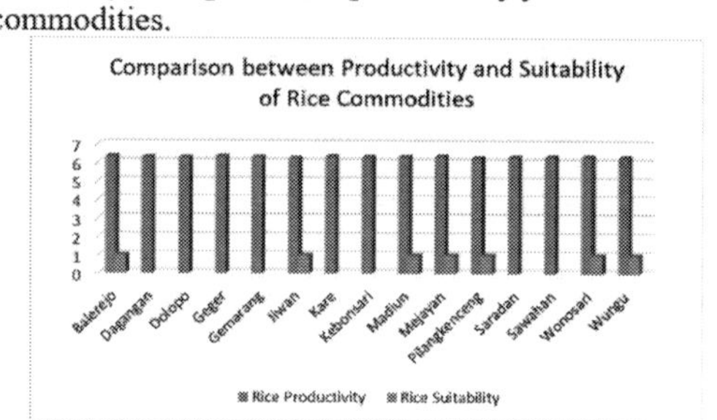

Fig. 2. Comparison between Productivity and Suitability of Rice Commodities.

The diagram above shows the comparison between the productivity and the corresponding sub-districts of rice commodities. The blue bars represent the rice productivity in each district, while the orange bars represent the suitability of the land areas for rice produce in each district. It can be seen in the graph that rice productivity tends to be equal and evenly distributed in all districts. However, according to the results of land suitability, only a few sub-districts are suitable for rice commodities. In addition, some sub-districts which serve for the use of rice land but have low rice productivity among other sub-districts, namely Jiwan District and Pilangkenceng District. This can be due to many factors, such as the lack of farmers' strategies in rice cultivation.

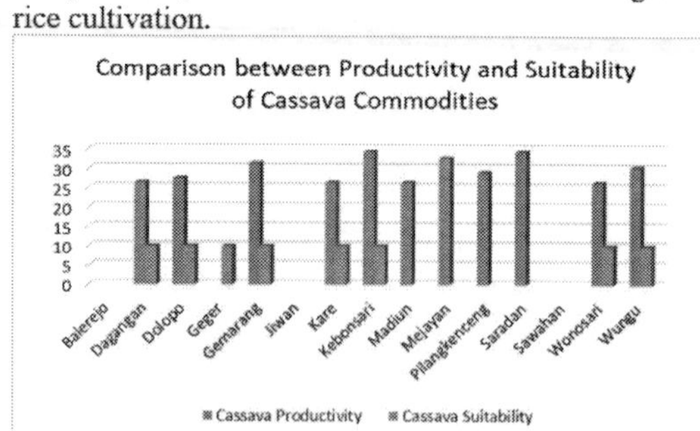

Fig. 3. Comparison between Productivity and Suitability of Cassava Commodities.

Another commodity that is the direction of land use in Madiun Regency is cassava. These commodities are often found in the Madiun Regency area as staple crops of the ancients. Cassava productivity in 2017 based on data from BPS was produced by several sub-districts except Geger, Jiwan, and Sawahan sub-districts. However, based on the direction of land use in Geger District, it is in fact suitable for cassava commodities. This is not in accordance with actual circumstances where there is no cassava productivity at all. This can happen due to several factors, such as farmers' desire to choose cassava to be planted on their agricultural land because the selling price is low.

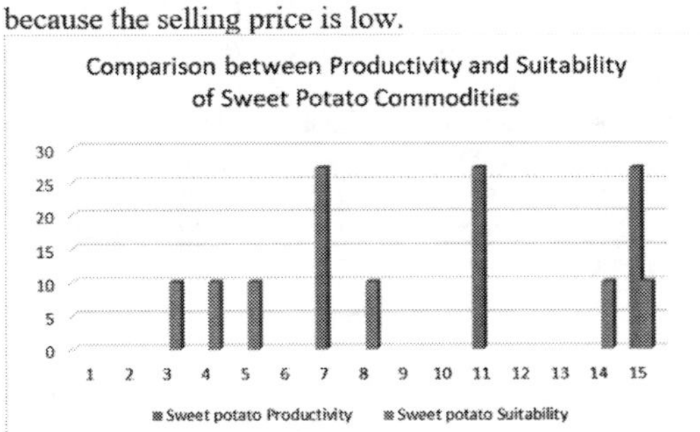

Fig. 4. Comparison between Productivity and Suitability of Sweet Potato Commodities.

Sweet potato farming in Madiun Regency is only spread over three sub-districts that have high productivity. Based on the results of land suitability analysis, there should be six sub-districts suitable for sweet potato commodities. This can happen like a cassava commodity, that sweet potato is also not a staple commodity that is consumed daily today. This causes farmers not to choose sweet potato or cassava commodities, even though the characteristics of the land area suitable for these commodities.

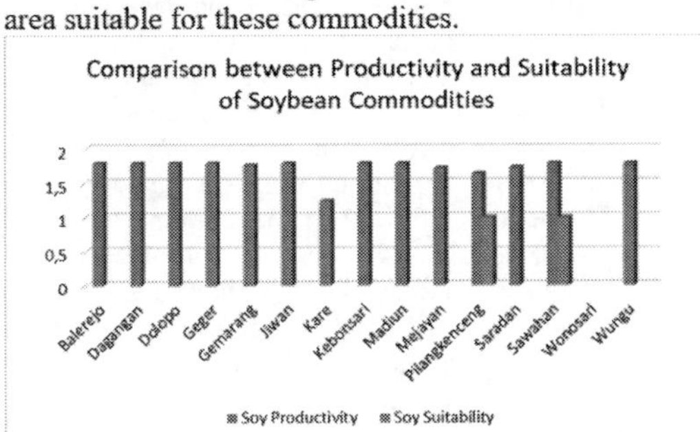

Fig. 5. Comparison between Productivity and Suitability of Soybean Commodities.

Soybean agricultural commodity is a commodity that is often consumed by the community because it contains high protein. Almost all sub-districts in Madiun Regency have soybean productivity except for Wonosari District. According to the results of land suitability, there are only two sub-districts that are most suitable for soybean commodities, namely Pilangkenceng District and Sawahan District. Both sub-districts have the same soybean productivity as other sub-districts, although other areas have less suitable land characteristics for soybeans. In detail, Sawahan Sub-district is suitable for soybean commodity according to its productivity results, classified as higher than other sub-districts. For Pilangkenceng Sub-district however, despite being in accordance with the the characteristics of the land, it has the second-lowest productivity after Kare District.

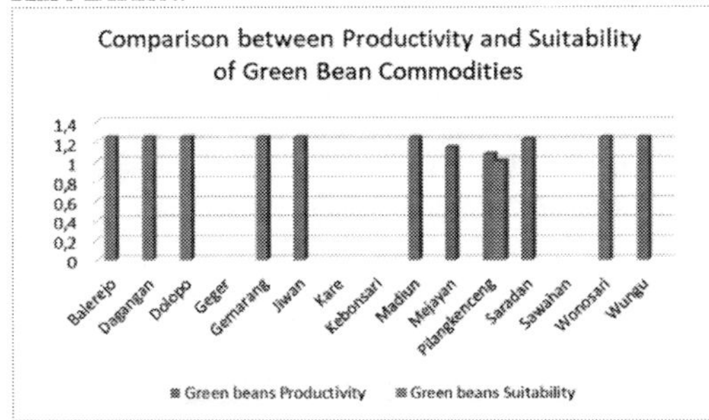

Fig. 6. Comparison between Productivity and Suitability of Green Bean Commodities.

The chart above is a comparison between land suitability for mung bean and its productivity in Madiun Regency. The green bean commodity in Madiun Regency is only in accordance with the characteristics of the land in the Pilangkenceng District. However, based on productivity data, Pilangkenceng District has the lowest green bean productivity in Madiun Regency. The comparison is inversely proportional which can be influenced by many factors such as plant-attacking pests, extreme weather that causes farmers to fail to harvest, etc.

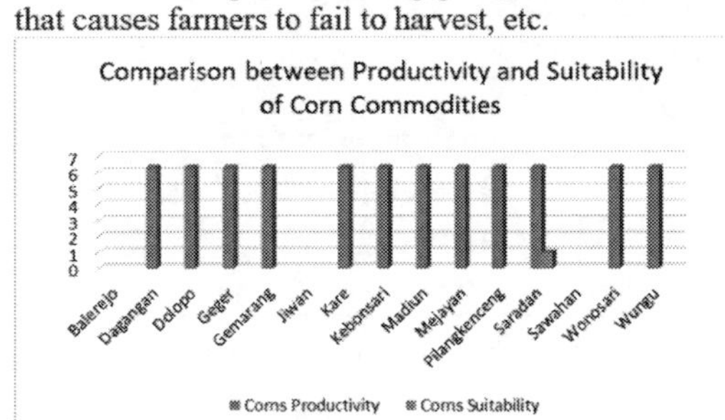

Fig. 7. Comparison between Productivity and Suitability of Corn Commodities.

The last agricultural commodity that matches the characteristics of the land in Madiun Regency is corn. The picture above shows that there is only one area suitable for corn commodities, which is Saradan District. However, it can be seen that corn productivity is primarily produced in almost all sub-districts in Madiun Regency. Although other areas based on land characteristics are not suitable for corn farming, they still have fairly high corn productivity. In addition to having unsuitable land characteristics, this great productivity can be supported by farmers' strategies in cultivating corn, such as by cultivating the land or regulating irrigation to produce high productivity.

3.3 The Effect of Land Characteristics on Agricultural Productivity

The relationship between land characteristics and agricultural productivity can occur because of a causal relationship between the two variables. The correlation allows the influence of one variable that affects other variables. The influencing variable in this study is land

 https://doi.org/10.1051/e3sconf/202132502005

characteristics, and the affected variable is agricultural productivity.

Before multiple linear regressions, the data normality test was conducted so that the data produced valid results. Data normality test was carried out on all data including characteristics data and agricultural productivity data. Normality conducted through SPSS software can be seen in the table below:

Table 3. Normality Test.

	Shapiro-Wilk		
	Statistic	df	Sig.
Ln_Temperatur	.859	15	.024
Ln_Kelembaban	.815	15	.006
Ln_Curahhujan	.967	15	.812
Ln_pH	.842	15	.013
Ln_KTK	.961	15	.714
Ln_corganik	.938	15	.353
Ln_N	.943	15	.417
Ln_P	.941	15	.390
Ln_K	.603	15	.000
Ln_Pasir	.970	15	.862
Ln_Debu	.951	15	.533
Ln_Liat	.901	15	.099
Ln_Padi	.901	15	.098

*. This is a lower bound of the true significance.

a. Lilliefors Significance Correction

The results of the data normality test showed that there were 8 data on land characteristics that became the influencing variable, and one data on agricultural productivity, which is rice productivity, became the affected variable. After getting the data that is normally distributed, multiple linear regressions can be done as shown in the table below:

Table 4. Multiple Linear Regression Test.

Model Summary[b]

Model		1
R		.932[a]
R Square		.868
Adjusted R Square		.693
Std. Error of the Estimate		.00223
Change Statistics	R Square Change	.868
	F Change	4.943
	df1	8
	df2	6
	Sig. F Change	.034

a. Predictors: (Constant), Ln_Rainfal, Ln_Sand, Ln_P, Ln c-organic, Ln N, Ln Dust, Ln Clay, Ln CEC

b. Dependent Variable: Ln_Rice Productivity

The Model Summary table above results from multiple linear regression of 8 land characteristics variables and rice agricultural productivity variables. The relationship between these variables shows an influence of land characteristics variables as indicated by an R-value of 0.932a, which means that 93% of land characteristics variables affect rice productivity and other factors influence the rest. The significance of land characteristics influence can be identified through the value of Sig. F Change above. A significance value of less than 0.05 indicates that the independent variable has a significant effect, and the results of the analysis above show the value of Sig. F Change is 0.034 which means that the variable of land characteristics significantly affects rice productivity. The effect of the independent variable was tested again using 2 tests to determine the impact simultaneously and partially. Below is a table of the results of the Simultaneous F Test of the variable land characteristics.

Table 5. F-Simultaneous Test.

ANOVA[a]

Model		Sum of Squares	df	Mean Square	F	Sig.
1	Regression	.000	8	.000	4.943	.034[b]
	Residual	.000	6	.000		
	Total	.000	14			

a. Dependent Variable: Ln_Rice Productivity

b. Predictors: (Constant), Ln_Rainfal, Ln Sand, Ln_P, Ln_c-organic, Ln_N, Ln_Dust, Ln_Clay, Ln_CEC

The ANOVA table above is a table of Simultaneous F test results from 8 land characteristic variables. This test is proven by the calculated F value and F table. The calculated F value can be seen in the table above, which is 4.943, while the F table value is obtained from the distribution of the r-table value at a significance of 5% or 0.05 is 3.73. The condition for the simultaneous F test is that if F count > F table. The results of the F test above show that the calculated F value is 4.934 > F table 3.73, so it can be seen that the variables of land characteristics have a simultaneous (together) effect on agricultural productivity. In addition, testing the effect of the independent variable partially using the T-test as shown in the table 6.

Variables of land characteristics, totaling 8 variables, may have their respective effects on agricultural productivity or partial (alone) effects. The partial effect can be seen in the t value in table 5. It is the same as the Simultaneous F test, namely by comparing the calculated T values and T tables. The calculated T value can be seen in the table 6, while the T table value is sought from the distribution of the t table value with a probability of 0.025, and in this study, the T table value is 2.447. The calculated T value is greater than the T table, and then the variable has a partial effect. This test shows that 2 variables have T count more than the T table, namely the CEC variable and the sand variable.

Cation exchange capacity (CEC) is one of the characteristics of the land that is important in increasing nutrient sources by binding soil nutrients. The level of CEC in the soil always affects soil fertility and its impact

Table 6. Partial-T Test
Coefficients[a]

Model		Unstandardized Coefficients		Standardized Coefficients	t	Sig.	Correlations			Collinearity Statistics	
		B	Std. Error	Beta			Zero-order	Partial	Part	Tolerance	VIF
1	(Constant)	1.846	.068		27.015	.000					
	Ln_CEC	.031	.006	3.089	5.286	.002	.187	.907	.783	.064	15.559
	Ln_c-organic	-.001	.003	-.058	-.274	.793	.111	-.111	-.041	.494	2.024
	Ln_N	-.005	.004	-.258	-1.203	.274	.195	-.441	-.178	.477	2.097
	Ln_P	-.002	.001	-.233	-1.244	.260	-.292	-.453	-.184	.624	1.604
	Ln_Sans	.013	.003	2.773	5.166	.002	.093	.904	.766	.076	13.118
	Ln_Dust	.001	.005	.034	.127	.903	-.191	.052	.019	.298	3.357
	Ln_Clay	-.002	.005	-.253	-.514	.625	.102	-.205	-.076	.091	10.987
	Ln_Rainfal	-.019	.008	-.473	-2.455	.049	-.116	-.708	-.364	.592	1.688

a. Dependent Variable: Ln_Rice Productivity

on crop productivity. This study is in line with research conducted by [9] which found that the increase in soil CEC due to the application of granular solid organic fertilizer with NPK fertilizer caused an increase in rice yields by 54% without the treatment.

Characteristics of soil texture, especially sand, showed a partial effect on rice productivity. This result is also in line with research [10] which stated that sandy soil texture influences rice production. The positive impact of sandy soil is that the rice harvest index is significantly higher than clay soil where pest disease attacks on clay soil is higher.

4 Conclusion

There are 6 referred commodities in Madiun Regency based on land suitability analysis, namely rice, corn, soybeans, peanuts, sweet potatoes, and cassava. Rice commodity is suitable for 7 sub-districts, namely Balerejo, Jiwan, Madiun, Mejayan, Pilangkenceng, Wonosari and Wungu sub-districts. Sweet potato commodity is suitable for 6 sub-districts, namely Dolopo, Geger, Gemarang, Kebonsari, Wonosari, and Wungu sub-districts. Cassava commodity is suitable in 8 sub-districts, namely Dagangan, Dolopo, Geger, Gemarang, Kare, Kebonsari, Wonosari, and Wungu sub-districts.Soybean commodity is only suitable in 2 sub-districts, namely Pilangkenceng and Sawahan Districts. Corn commodity is only suitable for 1 sub-district, which is in Saradan District. The green bean commodity is only suitable in 1 sub-district, namely Pilangkenceng District which is in Saradan District. The green bean commodity is only suitable in 1 sub-district, namely Pilangkenceng District.

Land characteristics significantly affect rice productivity by 93%. The influence of 8 variables of land characteristics simultaneously affects agricultural productivity. Still, there are only 2 variables of land characteristics that have partial impact, namely the characteristics of CEC and sand characteristic.

References

1. S. Hadi, 1 (2011)
2. R. D. S. Kuncoro, Pros. Semin. Nas. Geogr. UMS **8**, 370 (2017)
3. S. W. Tjokrokusumo, J. Teknol. Lingkung. **3**, 136 (2002)
4. M. Muhammad and U. Wasit, Semin. Nas. Dalam Rangka Dies Natalis UNS Ke 43 Tahun 2019 **3**, 42 (2019)
5. Badan Pus. Stat. (2019)
6. M. A. Oktaviani and H. B. Notobroto, J. Biometrika Dan Kependud. **3**, 127 (2014)
7. D. N. A. Janie, *Statistik Deskriptif Dan Regresi Linier Berganda Dengan SPSS* (Semarang University Publisher, 2012)
8. S. L. Wibowo, *Analisis Efesiensi Alokatif Faktor-Faktor Produksi Dan Pendapatan Usahatani Padi (Oryza Sativa L.) (Studi Kasus Di Desa Sambirejo, Kecamatan Saraden, Kabupaten Madiun* (Fakultas Pertanian, Universitas Brawijaya, 2012)
9. E. S. Puspita, 2013 (2013)
10. L. Hadiawati and A. Suriadi, 311 (2013)

Study of Groundwater Vulnerability to Contamination in Part of the Klaten District Area, Central Java-Indonesia

Yohana *Mariska Puteri* Adhesti[1] and *Margaretha* Widyastuti[1*]

[1]Department of Environmental Geography, Faculty of Geography, Universitas Gadjah Mada, 55281 the Special Region of Yogyakarta, Indonesia

Abstract. Water demands and risks of groundwater pollution in the Klaten District consistently increase as the human population grows over the years. This study aimed to examine the intrinsic and specific groundwater vulnerability levels (low, moderate, high) and analyze the spatial distribution and degree of contribution of the parameters to the vulnerability in some parts of the district area. The data included depth to the water table (measured directly at selected wells), rainfall records in 2009–2019, land-use maps (RBI), 30m DEM (SRTM), geological maps and geoelectric data. Groundwater vulnerability was analyzed with the Susceptibility Index by overlaying five parameters, which according to the results varied across the district: shallow to deep groundwater, aquifers that were composed of sandstone, sand and gravels and rock materials in between, topography of <2%and 2–6% and diverse land-use types from agricultural fields to built-up lands. Most of the research area had moderate intrinsic and specific vulnerability (43.93% and 50.53%), followed by high (38.24% and 46.43%) and low (17.83% and 3.04%). Also, depth to the water table and land use are the most influencing parameters. The results of this study are expected to provide a reference for the government to consider factoring in priority areas based on groundwater vulnerability levels in formulating the district's spatial plan.

1 Introduction

Water is one of the earth's components essential to support and sustain the life of living organisms. To meet basic daily needs humans extract water from different sources. In some regions groundwater is particularly preferred as it has relatively better quality than other water sources and is easy to reach [1]. However, as population multiplies water demands are expected to increase and more anthropogenic activities are projected to generate waste and introduce pollutants into the groundwater, potentially reducing its quality.

Groundwater vulnerability measures the ease with which pollutants introduced onto the soil surface can reach groundwater tables [2]. It differentiates into two types, namely specific groundwater vulnerability and intrinsic groundwater vulnerability [3]. The former is attributable to physical conditions, such as hydrogeological characteristics that protect groundwater from being polluted, while the latter deals with the absence or presence of a source of pollutants [4].

Klaten District is flanked by two big cities: Surakarta and Yogyakarta that are accelerating urbanisation and development in the area [5]. This has increased and diversified human activities in occupying and utilising land and has created potential sources of environmental pollutants that affect specific groundwater vulnerability.

*Corresponding author : mwiwik@ugm.ac.id

Agriculture makes one of the activities that dominantly occupy the district and thus contribute to the largest share of groundwater extraction. As stated in the Klaten District Regulation Number 9 of 2012 Article 7, groundwater withdrawal is prioritised for agricultural purposes. However, not a few farming practices in the district use chemical fertilisers excessively [6], causing the groundwater to be more prone to pollution. Furthermore, densely occupied settlements in urban areas and several industrial activities can contaminate this water source. If left untreated, the water quality will deteriorate and not fulfil the need for clean water.

Therefore, it is necessary to assess groundwater vulnerability in the study location. The purpose of this research was to examine the spatial distribution of groundwater vulnerability levels and their parameter sand analyse which parameters contributed the most to the groundwater vulnerability in part of the Klaten District area. The results and findings are expected to assist in groundwater management and treatment to protect and preserve water quantity and quality, sustaining the intended use of groundwater by the community.

2 Research location

The research was conducted in part of the Klaten District, Central Java, covering a 182.67-km^2 area. It consists of 16 sub-districts: Klaten Utara, Klaten Tengah, Klaten Selatan, Kalikotes, Jatinom, Ngawen, Karanganom, Polanharjo,Ceper, Trucuk, Delanggu, Pedan, Cawas, Karangdowo, Juwiring and Wonosari (Fig. 1). Astronomically it is located at 7°38'16.41"–7°43'55.89" S and 110°35'26.58"–110°44'55.94" E. Klaten is included in the Karanganyar-Boyololali Groundwater Basin and is one of the districts in the province that is flanked by Merapi Volcano and Seribu Mountains [7]. It is composed of three geological formations: Mandalika Formation (Tomm), Undivided Volcanic Rocks (Qvm) and Alluvium (Qa) [8].

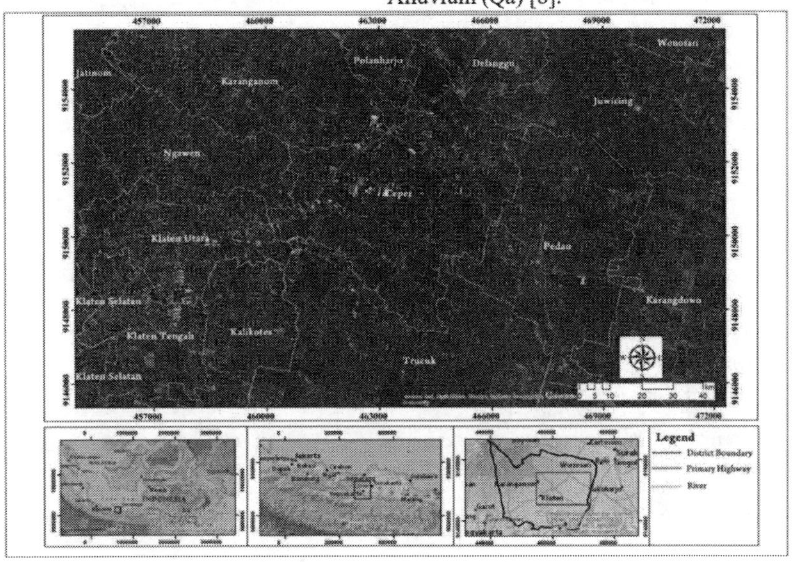

Fig. 1. Map of the study area.

3 Methods

3.1. *Susceptibility Index*

The research used the Susceptibility Index (SI), which is a function of the five parameters mentioned below:

1. Depth to the water table (D)
2. Groundwater recharge (R)
3. Aquifer media (A)
4. Topography (T)
5. Land use (LU)

SI is one of the developments of the DRASTIC method and was introduced by Ribeiro [9]. It adds a new parameter, land use (LU), to the equation and thus enables its users to incorporate anthropogenic factors and influences in the calculation [9]. In other words, SI produces specific instead of intrinsic groundwater vulnerability. Furthermore, with the addition of a new weighting system the measured vulnerability is expected to be more precise in characterising the hydrological environment in any case study [10].

These parameters had different weight values and thus different effects. In detail, the weight value of each parameter is presented in Table 1 below [9].

Table 1. The weight values of the Susceptibility Index (SI) parameters.

Parameters	Symbols	Weight values
Depth to the water table	D	0.186
Groundwater recharge	R	0.212
Aquifer media	A	0.259
Topography	T	0.121
Land use	LU	0.222

The above parameters were overlaid in ArcGIS to produce groundwater vulnerability. For the intrinsic vulnerability, the first four parameters were overlaid: depth to the water table, groundwater recharge, aquifer media and slope. As for the specific vulnerability, it combined them with an additional parameter: land use. This process used Eq. 1 to produce an index, which basically sums the multiplication of parameter scores by their respective weight values:

$$SI = 0.186Dr + 0.212Rr + 0.259Ar + 0.121Tr + 0.222LU \quad (1)$$

The SI values were statistically grouped into low, moderate or high groundwater vulnerability based on the range and the class interval calculated using Eq. 2 and 3 below:

$$Range = X_{max} - X_{min} \quad (2)$$

$$Class\ interval = Range : Number\ of\ classes \quad (3)$$

3.2. Depth to the Water Table

Depth to the water table was measured directly in the field (primary data) and then processed in ArcGIS software. The measurement used systematic and grid sampling, which evenly divided the study area into 1x1 km^2 grids, resulting in a sample size of 198 wells. These data were divided into seven classes, each has a range of values and a score, as presented in detail in Table 2 [9].

Table 2. Classification and scores of depths to water table (D).

No.	Range of values (m)	Scores
1.	<1.5	100
2.	1.5–4.6	90
3.	4.6–9.1	70
4.	9.1–15.2	50
5.	15.2–22.9	30
6.	22.9–30.5	20
7.	>30.5	10

3.3. Groundwater recharge (R)

For groundwater recharge, the research processed rainfall records (secondary data) to obtain the groundwater recharge by factoring in the local geological properties. The groundwater recharge coefficients for different lithological units are presented in Table 3 [11]. Meanwhile, the criteria for rainfall classification are summarised in Table 4 [9].

Table 3. Groundwater recharge coefficients of several geological characteristics.

No.	Lithological Units	Groundwater Recharge Coefficients (%)
1.	Impervious rocks	10–20
2.	Alluvial and pyroclastic deposits	60
3.	Lava	80
4.	Sandstones – clay – clay deposits and limestones	20
5.	Limestones	90
6.	Pyroclastics	40

Table 4. Classification and scores of rainfall data as a proxy of groundwater recharge (R).

No.	Range of values (mm)	Scores
1.	<51	10
2.	51–102	30
3.	102–178	60
4.	178–254	80
5.	>254	90

3.4. Aquifer media (A)

Details on the aquifer media were obtained by combining geoelectric measurements results and secondary data. The five measurement points were selected by purposive sampling that took into account the landforms at the study site. The geoelectric measurement results were then compared with the secondary data. Table 5 shows several types of aquifer media and their respective scores for the SI calculation [9].

Table 5. Types and scores of various aquifer media (A).

No.	Types of aquifer media	Scores
1.	Massive shales	20
2.	Metamorphic/igneous rocks	30
3.	Metamorphic/weathered igneous rocks	40
4.	Glacial deposits	50
5.	Sandstone layers, limestones, shale layers	60
6.	Massive sandstones	60
7.	Massive limestones	80
8.	Sand and gravels	80
9.	Basalt	90

3.5. Topography (T)

Topographyswere extracted from 30m DEM data (SRTM, secondary data). The criteria for slope classification in the SI calculation are shown in Table 6 [9].

Table 6. Topography (T) classification.

No.	Classes (%)	Scores
1.	<2	100
2.	2–6	90
3.	6–12	50
4.	12–18	30
5.	>18	10

3.6. Land use (LU)

Land use represents human activities taking place in a defined area. Land-use types and distributions were obtained from the 1:25,000 Indonesian Land-Use Map (RBI,secondary data). As shown in Table 7, each land-use type has a different score in the SI calculation [9].

Table 7. Land use (LU) classification.

Types	Scores
Agricultural areas	
Annual crops, rice fields	90
Permanent crops	70
Heterogenous agricultural areas	50
Pastures and agro-forestry areas	50
Artificial areas	
Waste-generating industries, landfills	100
Mines, shipyards, open-pit mines	80
Urban areas, airports, ports, railway stations, industrial and commercial complexes, green open spaces	75
Natural areas	
Aquatic ecosystems (swamps, lagoons, tidal zones)	50
Forests and semi-natural zones	0
Bodies of water	0

3.7. Validation to Nitrate

This step correlated intrinsic and specific groundwater vulnerability with nitrate concentrations (NO_3)—a groundwater quality parameter that had previously been processed and analysed from 26 water samples in the laboratory. These samples were selected

https://doi.org/10.1051/e3sconf/202132502006

by purposive sampling and examined for their qualities.

4 Results and discussion

4.1. Depth to the water table

Depth to the water table determines the length of contact between pollutants and groundwater [12]. The deeper the water table from the surface, the longer it takes for pollutants to reach the groundwater [8].The measurement results showed that the depth to the water table in the research location varied from <1.5 m to 15.2 m. Further, the groundwater was shallow in alluvial plains in the east and became deeper northward (up to 15.2 m), as identified in Karanganom and Polanharjo. Both sub-districts had deep groundwater because they are located close to the foot of the volcano and many springs.

4.2. Groundwater recharge

Groundwater recharge is the amount of rainwater seeping into the ground [8]. The greater the recharge, the higher the possibility of the groundwater to be contaminated [11]. Based on the analysis, the groundwater recharge in the study area was>254 mm/yearor the highest in the classification for the SI calculation. Thus, there is a high likelihood for the groundwater to be contaminated.

4.3. Aquifer media

Aquifer media consists of layers of materials whose hydrogeological properties shape their respective permeability values, which determine the infiltration rate of pollutants [12]. Based on the geoelectric measurements at five points and some secondary data, it was found that the aquifer media of the study site was a dominant layer of sandstone and sand and gravel.

4.4. Slope

Based on the analysis results, the area had two slope classes: <2% and 2–6%. Areas with low Topography retain runoff water longer, allowing higher infiltration and greater potential for pollution [13].

4.5. Land use

Based on the data analysis, more than half of the research location was mainly used for irrigated rice fields (55.6% of the total area). This type of land use was distributed evenly, except in the City of Klaten where settlements and industrial buildings were ubiquitous. Settlements were the land use with the second-highest areal percentage. Like the irrigated rice fields, they were scattered in each sub-district, though appearing in a much higher density in urban regions like the City of Klaten. The rest of the research location was used for dry cultivated fields, plantations and industrial and office buildings that form cluster patterns along the main Yogyakarta-Solo Road.

4.6. Intrinsic groundwater vulnerability

Based on the parameter overlay results, there are three classes of intrinsic vulnerability: low, moderate and high (Fig. 2). About 17.83% of the total research area had low vulnerabilities spread in the northern and southern sub-districts:part of Klaten Tengah, Kalikotes, Ngawen, Karanganom, Polanharjo, Ceper and Delanggu. Low vulnerability indicates that pollutants cannot easily contaminate groundwater [14]. Here, it is the combined result of small Topographys (2-6%), aquifer media in the form of sandstones, groundwater recharge of >254 mm/year and shallow to very deep groundwater (depths to water tables = 1.5–15.2 m). The deeper the water table from the ground's surface, the longer it takes for pollutants to reach the groundwater [8].

The moderate vulnerability level was identified in 43.93% of the total research area, distributed in the northern and western sub-districts:a part of Jatinom, Ngawen, Klaten Utara, Klaten Selatan, Klaten Tengah, Kalikotes, Trucuk, Karanganom, Ceper, Pedan, Delanggu, Juwiring and Wonosari. Compared with the low vulnerability, moderate here means that pollutants introduced on the surface can easily reach and contaminate groundwater. It is the combined result of flat slope(<2%), aquifer media in the form of sandstones, shallow to medium-deep groundwater (depths to water tables = 1.5–9.1 m) and groundwater recharge of >254 mm/year. Smaller Topographys and shallower water tables than what characterise low vulnerability have made the regions fall into the category of moderate vulnerability.

High vulnerability covered up to 38.24% of the research area, particularly in the eastern sub-districts: part of Klaten Selatan, Klaten Utara, Kalikotes, Trucuk, Ceper, Pedan, Cawas, Juwiringand Karangdowo. High vulnerability results from a combination of environmental properties that allow pollutants to seep into the soil and reach groundwater easily. Here, these properties are flat terrains (Topography of <2%) and sand-and-gravel aquifer media. Sand and gravels have good porosity and are excellent media for storing and releasing water [15], facilitating pollutants to enter the aquifer zone much quickly.

4.7. Specific groundwater vulnerability

Specific vulnerability adds intrinsic vulnerability parameters with land use to represent waste-generating activities that potentially introduce pollutants to groundwater. There are wide variations between the intrinsic and specific vulnerabilities. The specific groundwater vulnerability map of the research location is shown in Fig. 3.

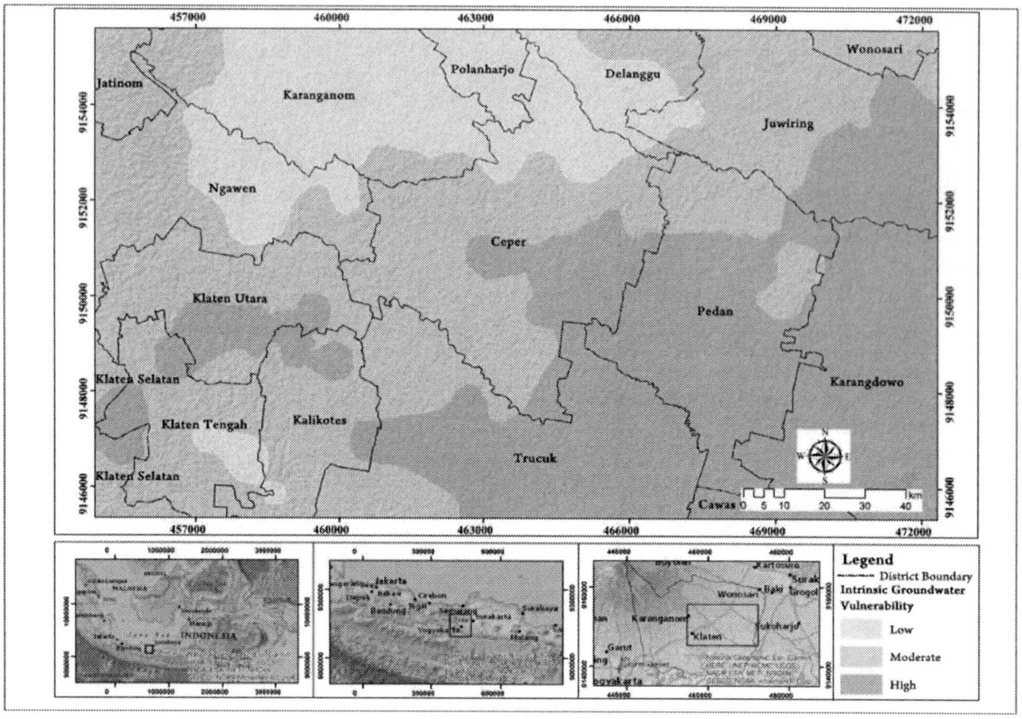

Fig. 2. Map of the intrinsic groundwater vulnerability in part of the Klaten District area.

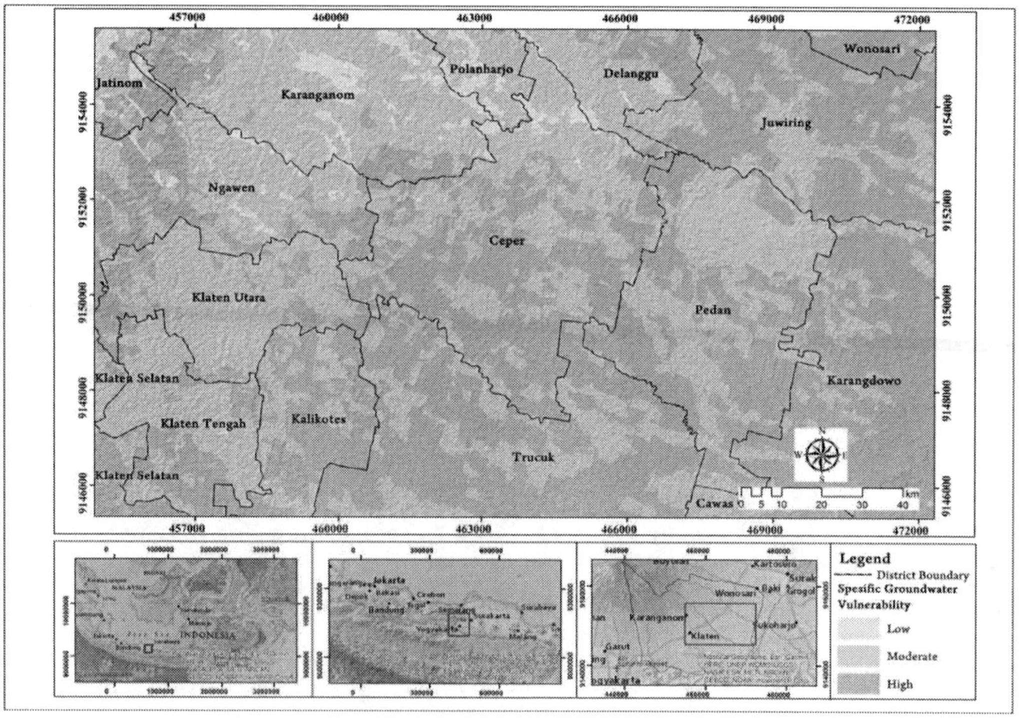

Fig.3. Map of the specific groundwater vulnerability in part of the Klaten District area.

 https://doi.org/10.1051/e3sconf/202132502006

Sub-districts with moderate vulnerability were located in the west and north, covering 50.53% of the total research area, including the City of Klaten (Klaten Utara, Klaten Tengah and Klaten Selatan). They are characterised by vast irrigated rice fields, shallow groundwater and an aquifer composed of sandstones. The moderate vulnerability also indicates that the contaminants generated on the surface can likely reach and pollute groundwater. However, over time and with extensive regional development, the sub-districts may develop characteristics that worsen their moderate groundwater vulnerability into high.

The high specific vulnerability covered 46.43% of the total research area, especially in the eastern sub-districts: Trucuk, Ceper, Juwiring, Wonosari, Pedan and Karangdowo. In these regions irrigated rice fields and settlements were distributed widely and interspersed with industrial, commercial and office buildings, thus potentially generating more waste by-products than other land-use types. When coupled with level terrains, shallow groundwater and sand-and-gravel aquifer, these land-use types create high vulnerability. In this state the groundwater is at higher risk of being polluted compared with moderate and low vulnerability levels.

In contrast to the current study, the SI-based research in the Daule aquifer, Ecuador, presents six vulnerability classes on the resultant map: extremely low, very low, low, moderate, high and very high [9]. Through extensive analysis, it has been found that regions with high vulnerability levels are mainly used for irrigated rice fields and have high groundwater recharge on the alluvial plain. This finding is similar to the current study results in that the high specific groundwater vulnerability in Klaten is attributable to the vast irrigated rice fields and high groundwater recharge. Meanwhile, the moderate and low vulnerabilities in the Daule aquifer coincide with three other land-use types: forests, bodies of water and semi-natural areas.

Another study using the SI method has been conducted in the Takelsa aquifer, Northeast Tunisia,in which the groundwater vulnerabilities are presented in four classes: low, moderate, high and very high [16]. Like the current study, most of the Takelsa aquifer (90%) is moderately and very highly vulnerable to pollution; only a small part of the recharge area (10%) has low vulnerabilities. The high vulnerability is the combined result of level terrains, extensive agricultural land use, high groundwater recharge, sandy soils and shallow depths to water tables. In the current study these factors are also responsible for the high specific groundwater vulnerability in Klaten.

There has also been another study conducted in the alluvial plains of Semarang (Central Java, Indonesia), where the groundwater vulnerability is classified into low, moderate and high [17]. The analysis results have shown that nearly half of the area (41.3%) is highly vulnerable to pollution, particularly in the north where ponds, settlements, industries and rice fields are extensive. The high vulnerability results from level slopes, aquifer media composed of sand and shallow groundwater. Meanwhile, more than half of the alluvial plains (50.3%) are moderately vulnerable. They are used mainly for settlements and rice fields and have gently sloping terrains, shallow groundwater and aquifers composed of sand. Only a small part of the area (8.4%), including the southern regions, has low vulnerabilities because of a mixed land-use type (some residential areas and forest covers), aquifer media in the form of igneous rocks, steep slopes and deep groundwater.

4.8. The most influencing factors of groundwater vulnerability

The most influencing factors of groundwater vulnerability can widely vary across regions. For instance, groundwater recharge and slope are two factors that most significantly affect the groundwater vulnerability in the Daule aquifer [9]. Meanwhile, in the Takelsa aquifer, factors with such influence are groundwater recharge, aquifer media and depths to water tables [17]. The current research has found land-use types to be the most influencing factor aside from depths to water tables. As presented in Table 8, the analysis results of the groundwater vulnerability factors showed that these two parameters had high weight values and high total index values. The total index was adjusted to the classification types of each parameter in the research location.

Table 8. The Susceptibility Index parameters in part of the Klaten District area

NO.	Parameters	Rating	Weight values	Total index
1.	Depths to water table	50–100	0.186	57.66
2.	Groundwater recharge	90	0.212	19.08
3.	Aquifer media	60–80	0.259	36.26
4.	Topography	90-100	0.121	22.99
5.	Land use	50–90	0.222	81.03

4.9. Validation to Nitrate

The vulnerability maps modelled in this study (see Fig. 1 and 2) were validated with the nitrate test results (see Fig.4). Nitrates are associated with the by-products of human activities such as domestic waste disposal, leaching from landfills and excessive use of fertilisers [18]. The intrinsic and specific groundwater vulnerabilities were 19.23% and 11.54% accurate based on the validation results. These are considered low because the minimum thresholds for intrinsic and specific vulnerability accuracies are 63% and 71%, respectively [19].

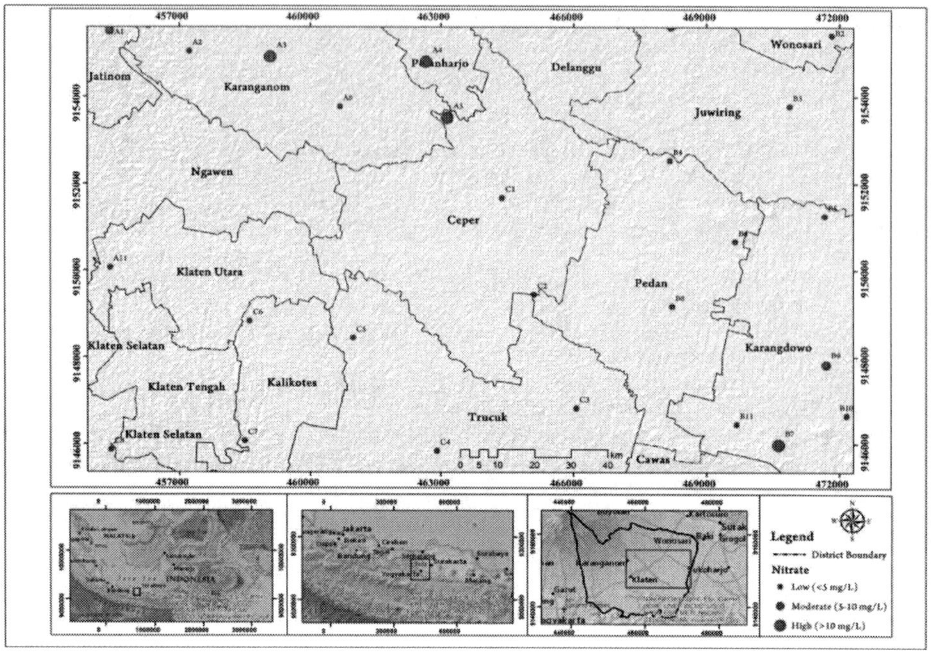

Fig.4. Map of the nitrate test result distribution in part of the Klaten District area.

Groundwater with low nitrate contents (<5 mg/L) dominated the study area and was found in sub-districts with high vulnerability. This finding does not match with the modelled groundwater vulnerability, which indicates low accuracy. Low nitrate levels can be attributed to the dynamic movement of groundwater and the complex mechanisms with which nitrates enter the groundwater. The amount of nitrate that enters the aquifer zone is also influenced by several factors, including fertiliser application, nitrification, denitrification and transportation by rainwater [20].

The low accuracyis also because some of the parameters are not entirely representative of the research area. This is similar to a previous study in Semarang City, where the groundwater recharge is categorised into >254 mm/year [21], although the tropical climate in the city produces a much higher rainfall of up to >2000 mm/year. Different climates can cause large differences in rainfall and groundwater recharge; therefore, in this context, the groundwater vulnerability modelled with the SI method has low accuracy.

5 Conclusion

The Susceptibility Index measures groundwater vulnerability using five parameters: depth to the water table, groundwater recharge, aquifer media, slope and land use. In part of the Klaten District area, these parameters have varied spatial distribution. The shallowest groundwater is distributed in the middle to the eastern sub-districts, while the deepest is in the north. The groundwater recharge of the entire research area is >254 mm/year. The aquifer media are composed of sandstone (in the western sub-districts) and sand and gravel (from the middle to eastern).The Topographys are <2% and 2–6%, and some parts of the district are mainly used for irrigated rice fields (distributed in several sub-districts) and dense settlements (in the western and eastern sub-districts). Based on the Susceptibility Index (SI) calculation method, the intrinsic and specific groundwater vulnerabilities are divided into three classes. The areal percentage of the intrinsic groundwater vulnerability is as follows: low (17.83%, covering the northern and a few sub-districts in the south), moderate (43.93%, in the north and west) and high (38.24%, in the east). Meanwhile, the areal percentage of the specific counterparts is as follows: low (3.04%, covering the northern sub-districts), moderate (50.53%, the west and north) and high (46.43%, in the middle to the east). The most influencing factors of groundwater vulnerability in the study area are depths to the water table and land use.

Due to the high groundwater vulnerability in the study area, it is necessary to take protective measures, for example, by:

1. Developing priority areas based on groundwater vulnerability.
2. Formulating waste-related regulations in sub-districts with high groundwater vulnerability
3. Establishing regulations that control groundwater protection

Acknowledgements

The authors would like to thank Universitas Gadjah Mada for funding the research through the 2021 Final Project Recognition Grant (*Rekognisi Tugas Akhir*, RTA). Gratitude is also expressed to the Faculty of Geography,Universitas Gadjah Mada, for supporting and providing the opportunity to carry out this research.

References

1. Sudarmadji, Perembetan Pencemaran dalam Airtanah pada Akuifer Tak Tertekan di Daerah Lereng Gunungapi Merapi, *Research Report,* Inter-university Centre for Engineering Sceince (PAU Ilmu Teknik) UGM, Yogyakarta, (1990).
2. Kusuma, K.I, Studi Kerentanan Air Tanah Menggunakan Metode DRASTIC di Urban Area Kota Semarang, *Bachelor Thesis,* Faculty of Engineering, Diponegoro University, Semarang, (2009).
3. National Research Council, *Groundwater Vulnerability Assessment, Contaminant Potential Under Conditions of Uncertainty*, National Academy Press, Washington DC, (1993).
4. Riyanto, I. A., & Widyastuti, M, Kerentanan Intrinsik dan Spesifik Airtanah terhadap Pencemaran di Kecamatan Banjarnegara dan Sekitarnya, *Jurnal Bumi Indonesia*, *5*(4), 1-17, (2016).
5. Prakoso, F. B., Analisis Tingkat Perkembangan Wilayah Kecamatan di Kabupaten Klaten Tahun2005-2015, *Bachelor Thesis*, Faculty of Geography, Muhammadiyah University of Surakarta, (2018).
6. Prakoso, T. S. (2018). *Sawah di Klaten Tak Lagi Subur.* Accessed on 1 September 2020, at http://www.koransolo.co/2018/08/28/sawah-di-klaten-tak-lagi-subur/
7. Pemerintah Kabupaten Klaten, *RTRW Kabupaten Klaten tahun 2011-2031, Perda 11/201,* Pemkab Klaten, Klaten, (2011).
8. Putranto, T. T., Ali, R. K., & Putro, A. B., Studi Kerentanan Airtanah terhadap Pencemaran dengan Menggunakan Metode DRASTIC pada Cekungan Airtanah (CAT) Karanganyar-Boyolali, Provinsi Jawa Tengah, *Jurnal Ilmu Lingkungan*, *17*(1), 158171, (2019).
9. Ribeiro, L., Pindo, J. C., & Dominguez-Granda, L, Assessment of Groundwater Vulnerability in The Daule Aquifer, Ecuador, Using The Susceptibility Index Method, *Science of The Total Environment*, *574*, 1674-1683, (2017).
10. Hsu, C., Sandford, B, The Delphi Technique: Making Sense of Consensus, *Practical Assessment, Research & Evaluation* 12(10), ISSN 1531–7714, (2007).
11. Abdillah, A. dan Adji, T.N., Kajian Kerentanan Airtanah Terhadap Pencemar di Daerah Aliran Sungai Serang, *Jurnal Bumi Indonesia*, 7(2), 28-41, (2018).
12. Jang, W.S., Engel, B., Harbor, J., Theller, L., Aquifer Vulnerability Assessment for Sustainable Groundwater Management using DRASTIC., *Water*, 9 (792), (2017).
13. Amit, G., Ashwani, K.T., Santosh, D., A GIS based DRASTIC Model for Assessing Groundwater Vulnerability of Katri Watershed, Dhanbad, India. Model, *Earth Syst. Environ.* 1 (11), 1–14, (2015).
14. Sugianti, K., Mulyadi, D., & Maria, R, Analisis Kerentanan Pencemaran Air Tanah dengan Pendekatan Metode DRASTIC di Bandung Selatan, *Jurnal Lingkungan dan Bencana Geologi*, *7*(1), 19-33, (2017).
15. Sukadana, I. G., & Indrastomo, F. D, Kombinasi Pengukuran Radioaktivitas Batuan dan Geofisika dalam Menentukan Akuifer Airtanah Potensial di Desa Sumbermanjing Kulon, Pagak, Malang, Jawa Timur, *EKSPLORIUM*, *32*(2), 125-138, (2011).
16. Ghouili, N., Jarraya-Horriche, F., Hamzaoui-Azaza, F., Zaghrarni, M. F., Ribeiro, L., & Zammouri, M., Groundwater Vulnerability Mapping using the Susceptibility Index (SI) Method: Case Study of Takelsa Aquifer, Northeastern Tunisia, *Journal of African Earth Sciences*, *173*, 104035, (2021).
17. Marjuanto, A. A., Putranto, T. T., &Sugianto, D. N., Mapping of Groundwater Vulnerability Index in the Alluvial Plain of Semarang City Using the Susceptibility Index Method, In *E3S Web of Conferences* (Vol. 125, p. 01010), EDP Sciences, (2019).
18. Sudaryanto, S., & Suherman, D., Degradasi Kualitas Airtanah Berdasarkan Kandungan Nitrat di Cekungan Airtanah Jakarta, *Riset Geologi dan Pertambangan*, 18(2), 61-68, (2008).
19. Putranto, T. T., & Marjuanto, A. A., Assessment of Contamination Vulnerability of Groundwater using Susceptibility Index Method: Semarang City as Case Study, *Journal of Engineering Science and Technology*, *16*(1), 438-454, (2021).
20. Zhang, D., Li, G., Yang, S., Zhang, X., and Guo, H., Bio-geological Processes of Nitrogen Transport and Transformation in The Aeration Zone and Aquifer, *Hydrological Sciences Journal.* 54: 316-326, (2009).
21. Abdel madjid, B., & Omar, S., Assessment of Groundwater Pollution by Nitrates using Intrinsic Vulnerability Methods: A Case Study of The Nil Valley Groundwater (Jijel, North-East Algeria), *African Journal of Environmental Science and Technology*, *7*(10), 949-960, (2013)

The study of soil infiltration in west part of Semarang City, Indonesia

Muhammad Nursalam Octariko[1], and *Wawan* Budianta[1*]

[1]Department of Geological Engineering, Gadjah Mada University, Yogyakarta, Indonesia

Abstract. This research examined the types and characteristics of soil, including the porosity, the water content, the void ratio, and the degree of saturation to the value of the infiltration rate in the soil located in the western part of Semarang City, Indonesia. The infiltration rate test was carried out at 49 measurement points, based on a 2×2 km^2 grid division in the research area. The infiltration rate measurement in the field was carried out using a turf-tech infiltrometer. The data from the infiltration rate measurement results in the field were calculated using the infiltration rate estimation model from Kostiakov, which was further classified by the Konhke classification. The results of the infiltration rate measurement showed that the research area had infiltration rate values that varied, from very slow to very fast, with the values ranged from 0 mm/hour to 1601.33 mm/hour. The analysis results showed that the type of soil, the water content, the degree of saturation, and the porosity related to the infiltration rate in the research area. The soil type parameter had the smallest relationship to the infiltration rate with the R^2 value of 0.2923. The soil porosity had the largest relationship to the infiltration rate in the research area, with an R^2 value of 0.4854.

1 Introduction

Infiltration is one of the main components in the hydrological cycle that determines the amount of rainwater that enters the soil and becomes surface runoff; also, infiltration plays an essential role in the availability of groundwater [1, 2]. The western part of Semarang City is an ideal location to study the effect of land characteristics on the infiltration rate since it often experiences floods [3, 4]. Population growth that continues to rise, followed by changes in land use related to urbanization, influences the produced land characteristics [5]. The geological condition of the western part of Semarang is mainly composed of volcanic rocks in the north and alluvium in the south [6]. Each type of rock and soil has different physical properties related to infiltrating water [7]. Soil with sand-grain-sized composition will have greater infiltration capacity than clay due to the larger void size of the sand [8].

2 Materials and methods

The study area was located in western part of Semarang city, Indonesia as shown in Figure 1. The infiltration rate was measured using a Turf-tec infiltrometer. The measurement was made on the Horizon-A of the weathered soil, then every value generated from the infiltrometer at each measurement point was recorded. At each point, measurement was carried out for 45 minutes. The measurement point was determined by dividing the work area into a grid measuring 2×2 km^2 with 49 grid points—each grid required at least one infiltration rate measurement data (Figure 2). The turf-tech was plugged into the soil to a depth of 5 to 10 cm. Then the outer and inner cylinders were filled with water to a height of 5 – 20 cm when infiltration occurred, water was added until it reached a quasi-steady flow, or the decrease in water inside the cylinders became constant. To estimate the value of the actual infiltration rate, the infiltration rate was estimated using infiltration model equation. One of the equations of the infiltration model that is often used is the Kostiakov empirical model. The Kostiakov infiltration equation model is an empirical model based on field measurements [7]. The Kostiakov equation was chosen in this research due to its simple form compared to other methods. The Kostiakov equation relates infiltration to time as power function:

$$F_p = at^b \rightarrow y = ax^b \tag{1}$$

where cumulative infiltration (Fp in mm), time after infiltration begins (t in hours), a and b are the parameters obtained through fitting curves that should be evaluated from the measured infiltration data, since those parameters have no physical meaning [9]. Microsoft Excel was used to calculate the infiltration data and making the graphs in this research. Soil properties including grain size, water content, void ratio and degree of saturation was tested based on soil standard test refer to Liu and Evett [10].

* Corresponding author: wbudianta@ugm.ac.id

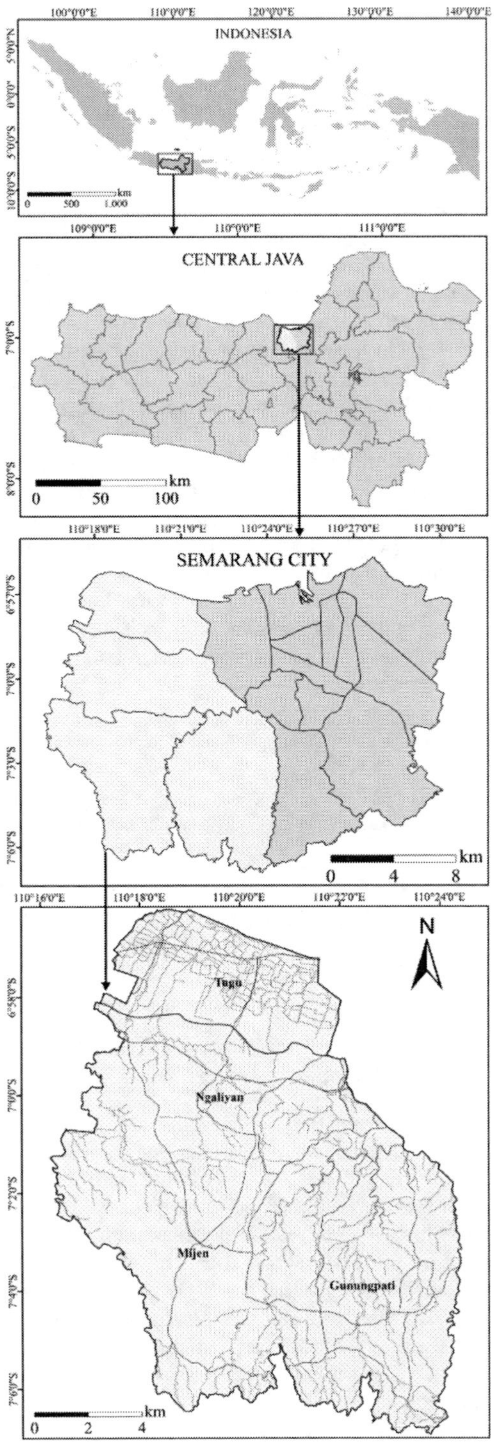

Fig. 1. The study area.

3 Result

3.1. Infiltration rate

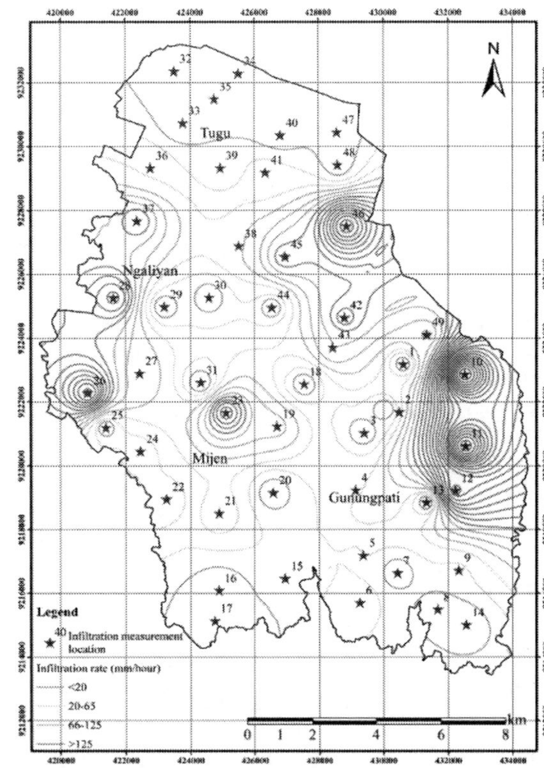

Fig. 2. Infiltration value map and measurement point location.

Figure 2 is a contour map of the infiltration rate in the research area, divided into four classes of infiltration rate contours; the classification of infiltration rate contour classes is based on the simplified classification of Konhke (1968)[11]. The contour of the infiltration rate <20 mm/hour (orange) is included in the classification of Konhke (1968)[11] into the very slow to moderately slow infiltration rate class; the contour of the infiltration rate of 20 to 65 mm/hour (yellow) is included in the medium infiltration rate class; the infiltration rate contour >66 to 125 mm/hour (light blue) belongs to the quite fast infiltration rate class, and the infiltration rate contour >125 mm/hour belongs to the fast to very fast infiltration rate class..

3.2 The influence of soil characteristic

3.2.1. The effect soil texture

Figure 3 show that the exponential regression model had the highest coefficient of determination, namely $R^2 = 0.2866$, which means that 28.66% of the variation of the infiltration rate variable could be explained by the variable composition of the fine fraction; while the remaining 71.34% was influenced by other variables. It can be seen from the negative trend, indicating that the lower the infiltration rate, the finer the resulting soil type. These results supported the existing theory and previous study, which shows that the higher the fine material in the soil, the infiltration rate will decrease due to the smaller void size [12]. The soil texture parameter had the smallest effect on the infiltration rate, and this is

 https://doi.org/10.1051/e3sconf/202132502007

because, in the field conditions, other parameters also affected the infiltration rate. When rainwater hit the soil surface, the infiltration was affected by the texture of soil, but all components that made up the soil simultaneously affected the infiltration rate. It can be seen in the graph that some plotting points did not follow the established trend between infiltration rate and soil type (the grey circle in Figure 3), points with coarse-grained soil types produced very slow infiltration rate values due to the water content and the degree of saturation in the high soil. Moreover, the results of plotting the fine-grained soil, which showed several points with grained soil types, actually provided a fast infiltration rate due to the low water content and degree of saturation in the soil and very high water content porosity in the soil.

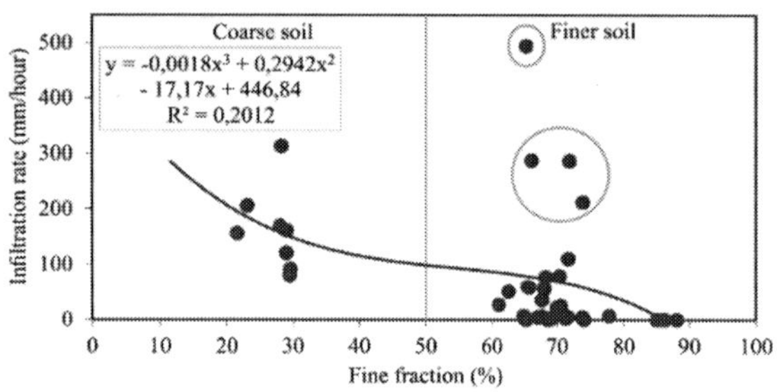

Fig. 3. Correlation between infiltration rate and soil texture.

3.2.2 The effect of soil porosity

Figure 4 show that the polynomial regression model had the highest coefficient of determination, namely $R^2 = 0.4854$, which means that the porosity variable could explain 48.54% of the variation of the infiltration rate variable in the soil. In comparison, other variables influenced the remaining 51.46%. It can be seen from the positive trend, indicating that the higher the infiltration rate, the greater the porosity in the soil. These results supported the previous study, which show that the value of soil porosity was directly proportional to the value of the infiltration rate [13]. At some points, the plotting results did not follow the trend between the infiltration rate and the porosity in the soil (the grey circle in Figure 4), such as at the points with low soil porosity, which resulted in a high infiltration rate due to the influence the water content and the saturation in the low soil, and located on the coarse-grained soil type, as well as on the results of plotting the porosity in high soils which gave a low infiltration rate due to the high water content and the degree of saturation in the soil, as well as being in the fine-grained soil.

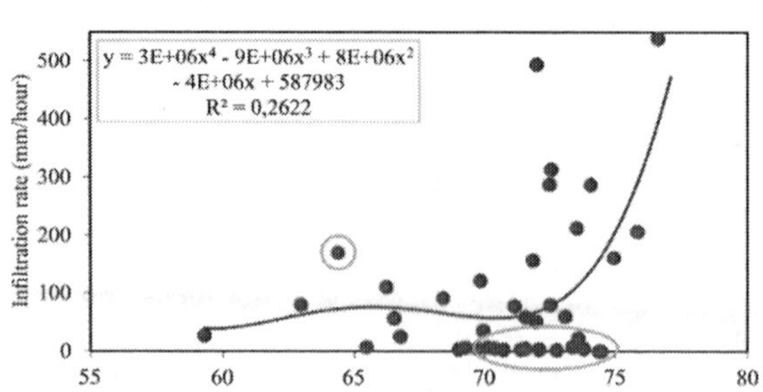

Fig. 4. Correlation between infiltration rate and soil porosity.

3.2.3 The effect of soil water content

Figure 5 show that the exponential regression model had the highest coefficient of determination, namely $R^2 = 0.4560$, which means that 45.60% of the variation of the infiltration rate variable could be explained by the variable water content in the soil. In comparison, the remaining 54.40 % was influenced by other variables. It can be seen from the negative trend, indicating that the lower the infiltration rate, the greater the water content in the soil. This relationship confirmed by another study show that the higher the water content in the soil, the lower the infiltration rate [14]. At some points, the plotting results did not follow the trend between the infiltration rate and the water content in the soil (the grey circle in Figure 5), such as at the points with low water content in the soil, which resulted in a low infiltration rate due to the influence of porosity in the low soil, as well as the results of plotting the high water content in the soil which gave high infiltration rate due to the very high porosity in the soil.

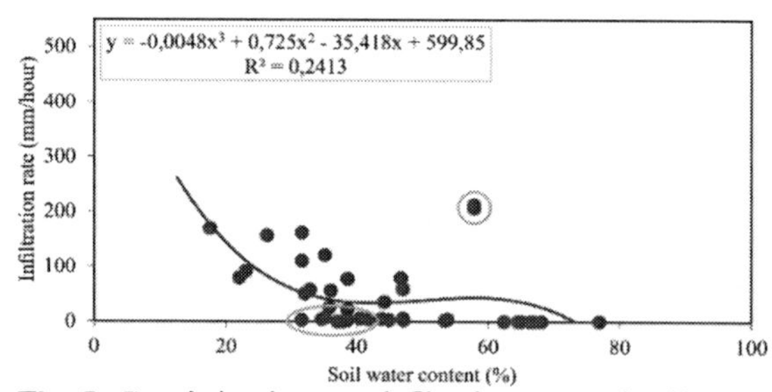

Fig. 5. Correlation between infiltration rate and soil water content.

3.2.4 The effect of soil void ratio

Figure 6 show that the polynomial regression model has the highest coefficient of determination, namely $R^2 = 0.3438$, which means that 34.38% of the variation of the infiltration rate variable can be explained by the void ratio variable in the soil, while the remaining 65.62 % influenced by other variables. It can be seen from the positive trend indicating that the higher the infiltration rate, the greater the void ratio in the soil. These results supported the previous research, which explains that the value of the soil is directly proportional to the value of the infiltration rate [14]. At some points, the plotting results do not follow the trend between the infiltration rate and the void ratio in the soil (grey circle in Figure 6), as at points with a low void ratio in the soil, and it results in a high infiltration rate due to the influence of water content and saturation in low soils, and being in coarse-grained soil types, as well as the results of plotting a high void ratio in the soil gives a low infiltration rate due to the high water content and degree of saturation in the soil, as well as being in fine-grained soils.

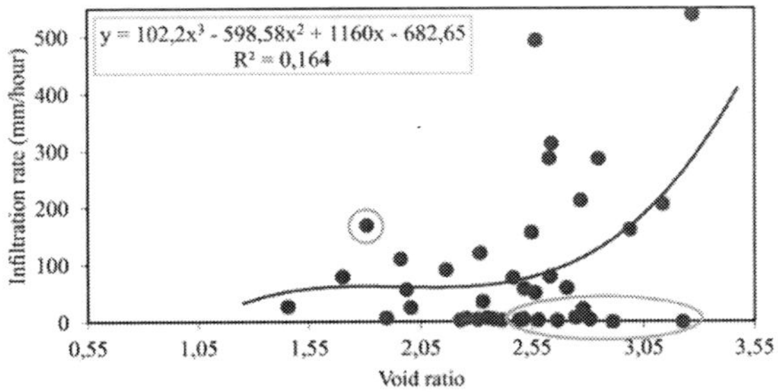

Fig. 6. Correlation between infiltration rate and void ratio.

3.2.5 The effect of degree of saturation

Figure 7 show that the exponential regression model had the highest coefficient of determination, namely R^2 = 0.3956, which means that 39.56% of the variation of the infiltration rate variable could be explained by the degree of saturation in the soil, while other variables influenced the remaining 60.44%. It can be seen from the negative trend, indicating that the lower the infiltration rate, the greater the degree of saturation in the soil. This supported the previous finding that the more saturated the initial conditions of soil, the lower the infiltration rate [15]. At some points, the plotting results did not follow the trend between the infiltration rate and the degree of saturation in the soil (the grey circle in Figure 7), as at the points with a low degree of saturation in the soil, which resulted in low infiltration rate due to the influence of porosity in the low soil, as well as the results of plotting the degree of saturation in the high soil which gave high infiltration rate due to the very high porosity in the soil.

4 Discussion

The test of the simple regression model above aimed to examine the extent of effects of each parameter from the five parameters which had been determined in this research on the infiltration rate. The reality was that there were still many other parameters remain that affected the infiltration rate, such as soil density, organic element content, temperature, structure, land slope, land use, and others [16-19].

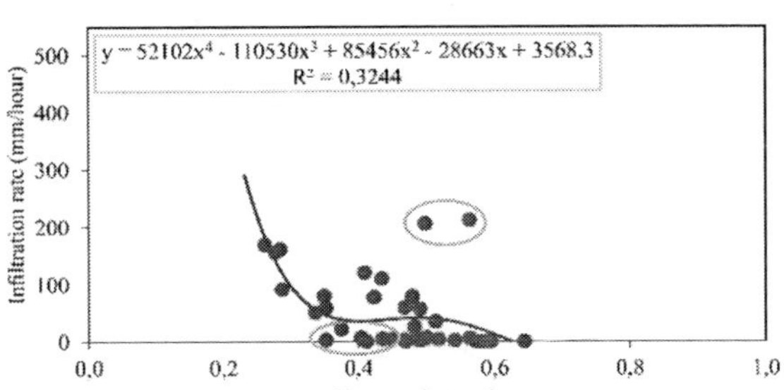

Fig. 7. Correlation between infiltration rate and degree of saturation.

The coefficient of determination (R^2) is an index to determine how much variability in the dependent variable can be explained by the regression model. In The result of all parameter investigated in this study show that the R^2 value has a relatively low value (<0.50), with the lowest R^2 value in the soil texture parameter, which was 0.2923. In previous research conducted by Fauzan and Har [20] regarding soil physical properties, slope, and lithology on the value of the infiltration rate in Padang City, also gave low R^2 results. Low correlation results were also produced in research by Fitzgerald et al. [21] and concluded that there might have been other parameters with better correlation, either the infiltration measurement method using the infiltrometer was not reliable, caused by plugging the infiltrometer into the soil, which disturbed the physical properties of the soil, therefore the soil inside the ring did not represent the type of soil around it. Another factor is due to the effect of size scale of the used ring infiltrometer, in which according to Li et al. [22], the double-ring infiltrometer is having a size-scale impact where the larger the ring size in the used infiltrometer, the scale effect decreases, and the scale effect does not occur when using a 40-cm diameter inner ring.

5 Conclusion

The results of the infiltration rate measurement show that, in the western part of Semarang City, the infiltration rate varies from very slow to very fast, with the infiltration rate of 0 to 1601.33 mm/hour. The values of the very slow to moderate infiltration rate class are spread in the Tugu District and partly in the Mijen District. In contrast, the values of the fast to very fast infiltration rate class are distributed in the Ngaliyan District and partly in the Gunungpati District. Meanwhile, the correlation results graphically show that the soil type, water content, degree of saturation, and porosity correlate to the infiltration rate in the western part of Semarang City. The parameter of soil texture has the smallest relationship to the infiltration rate with an R^2 value of 0.2923, and the porosity in the soil has the largest relationship to the infiltration rate in the research area with an R^2 value of 0.4854.

Authors acknowledge thank to Universitas Gadjah Mada for the supporting research funding.

References

1. W. Brutsaert, Hydrology: An introduction (Cambridge University Press, Cambridge, 2005).
2. S.K. Gupta, Modern hydrology and sustainable water development (Wiley-Blackwell, Oxford, 2011).
3. S. Wilonoyudho, Indones. J. Geogr **42**, 2 (2010).
4. A.W. Sejati, I. Buchori, I. Rudiarto IOP Conference Series: Earth and Environmental Science **123**, 1 (2018)

5. J.S. Deng, K Wang, Y. Hong, J.G. Qi, Landsc. urban plan **92**, 3–4 (2009)
6. R. Thanden, H. Sumadirdja, P. Sutisna *Geological map of Magelang and Semarang, sheet, Java, in scale 1:100.000 P3G, Bandung, Indonesia* (1996)
7. R. E. Smith, K. R. Smettem, P. Broadbridge, D. A. Woolhiser, Infiltration theory for hydrologic applications (American Geophysical Union, Washington DC, 2002).
8. M. K. Shukla, Soil physics: An introduction (CRC Press, Boca Ranton, 2014).
9. S. Xue, F. Tong, S. Hao, J. Wang, Adv. Sci. Technol. Water Resour, **36** (2016)
10. C. Liu, J. B. Evett, Soil properties: testing, measurement and evaluation (Prentice Hall, New Jersey, 1984)
11. H. Kohnke, Soil physics (McGraw-Hill, New York, 1968)
12. A. Marquart, L. Goldbach, N. Blaum, Ecohydrology **13**, 8 (2020)
13. D. Harisuseno, E. N. Cahya, J. Water Land Dev. **77**, 88 (2020)
14. F. Pittman, A. Mohammed, E. Cey, Hydrol. Process **34**, 3 (2020)
15. E. Damiano, R. Greco, A. Guida, L. Olivares, L. Picarelli, J. Eng. Geol, **220** (2017)
16. F. Haghnazari, H. Shahgholi, M. Feizi, IJAAR **6**, 5 (2015)
17. L. Gang, W. Xiangyun, Chinese Agr. Sci. Bull., **7** (2008)
18. M. R. Pahlavan-Rad, K. Dahmardeh, M. Hadizadeh, G. Keykha, N. Mohammadnia, M. Gangali, C. Brungard, Catena, **194** (2020)
19. J. Díaz-Sanz, S. Robert, C. Keller, Geoderma, **376** (2020)
20. A. Fauzan, R. Har, Jurnal Bina Tambang **3**, 4 (2018)
21. P.D. Fiztgerald, G.G. Cossens, D.S. Rickard, *Hydrology* **10**, 2 (1971)
22. M. Li, T. Liu, L. Duan, Y. Luo, L. Ma, J. Zhang, Y. Zhou, Z. Chen, Water **11**, 1457 (2019)

Analysis of groundwater sufficienty for domestic and livestock requirements as an anticipation to sea water intrusion in Kebumen coast

Ignasius Loyola Setyawan Purnama[1*], *Salsabiila* Bayu Nugrahaeni[1], *La Ode* Saleh Isa[1] and *Vincentia* Anindha Primacintya[2]

[1]Faculty of Geography, Universitas Gadjah Mada, Indonesia
[2]Accounting Programme, Master of Science and Doctoral, Faculty of Economics and Business, Universitas Gadjah Mada, Indonesia

Abstract. This study has three objectives, i.e calculating the availability of groundwater in the study area, calculating water demand for domestic and livestock and evaluating the safe yield of groundwater extraction. To achieve these three objectives, groundwater availability, safe yield and actual utilization are calculated. Groundwater availability is calculated based on the multiplication of the area, aquifer thickness and the specific yield. Water demand for domestic is determined base on the number of population and the amount of water demand of each person in each day. Water demand for livestock is calculated based on the number of livestock and water consumption for each livestock in each day, while the safe yield is calculated based on the multiplication of area, annual groundwater level fluctuation and specific yield. The results of research showed that the availability of groundwater in the study area reached 440,517,770 m^3/year, while the safe yield for its extraction was 76,641,600 m^3/year. Because the water demand for domestic and livestock is 6,495,100 m^3/year, so that the availability of groundwater in the study area is still sufficient to support water demand for these sectors. The water demand of the two sectors still have not exceeded the safe yield water extraction, so it will not cause decreasing of groundwater pressure which can cause to sea water intrusion.

Keywords: groundwater, sea water intrusion, Kebumen coast

1 Introduction

Even though water is only composed of two atoms, namely H and O, however water is the main source of life needed for all human activities such as domestic, agricultural, livestock and industrial requirement [1]. It can be said that water is a blessing because its existence affects the quality of life of each human being [2]. According to [3], the increasing requirement for water related to population growth has led to awareness from the community to manage the existence of limited water resources.

One source of water that is very important in supporting everyday human life is groundwater. Until now groundwater is still the main source of water in various regions, especially for non-agricultural activities such as for domestic, industrial and livestock needs [4-5].

The high exploitation of groundwater, if it does not cause land subsidence, can also cause sea water intrusion. This second impact is more frequent, so it has become a serious problem in cities located in coastal areas [6-12]. Therefore, groundwater extraction must not exceed the safe yield. According to [11,13], it is incorrect to assume that the safe yield is equivalent to the amount of groundwater recharge because the safe yield is only part of the increase in groundwater. Some of the other water will be lost from aquifers in various ways.

An interesting phenomenon is that in fact the existence of interfaces is not only detected in densely populated urban areas, but also in rural areas where the population is scarce, such as in the coastal area of Kebumen. The results of the geoelectric sounding carried out by [14] in the area, show that there are interfaces in three geoelectric sounding route, i.e. (1) the route Salak-Petanahan-Munggu-Gadung-Petanahan coast, (2) the route of Tambakrejo-Jeblok-Prajuritan-Adikarto-Kebumen coast and (3) the route of Sinungrejo-Sidoluhur-Bener-Kaibon-Ambal and Kutowinangun coast. Base on data from geoelectric sounding, the detection of interfaces in the coastal area of Kebumen is because groundwater sources in this area only come from rainwater that falls in the beach ridge, so the fresh water is not strong enough to push sea water. In addition, the back swamp area in this area also often enters sea water at high tide.

The research of [15] did not detect any interfaces again in these three geoelectric sounding route. It was said that the undetected interfaces in the study area were caused by the deeper position of the interfaces due to the increasingly strong pressure of fresh groundwater. The increasing strength of this fresh water pressure is due to the increasing number due to the intensive accumulation of rainwater infiltration into groundwater and the intense flushing of sea water by fresh water. In addition, the back swamp area in this area may no longer enter sea water at high tide

According to [16], the Kebumen coast is part of the Southern Mountains of Java, which has been submerged and then covered by alluvium material. The alluvial plains which are the research locations are mostly composed of sand deposits, silty sand and clay

*Corresponding author : Setyapurna@geo.ugm.ac.id

https://doi.org/10.1051/e3sconf/202132502008

sand. Generally, these alluvial plains can be divided into sand dune, beach ridges and ex backswamp route.

The sand dune is a route directly adjacent to the coastline composed of unconsolidated sand. The width of this route is not always the same but depending on the availability of sand and the strength of the wind. The beach ridge consists of sand material which has begun to develop into soil. This route can be either single or multiple beach ridge. For the coast to the east of the Karangbolong Mountains to Parangtritis Beach, generally there is only one beach ridge route, so that in the study area there is only one beach ridge route.

The lagoon is generally an area that lies lower than the surrounding area. At present, this area is either a rice field area or is still a permanently inundated swamp. In this lagoon route, it often experiences problems related to inundation of the area during the rainy season and intrusion of sea water from the river estuary at high tide. In addition, because it is an ex swamp, the quality of groundwater is also not good because it is often found water with a strong odor.

Based on this background, this study aims to calculate groundwater availability, to calculate water demand for domestic and livestock and to analyze the safe yield of groundwater extraction in the study area so that seawater intrusion does not occur.

2 Methodology

2.1 Calculation of water availability

Groundwater availability is calculated using a static approach using the formula:

$$Vat = Sy \times Vak$$

where Vat is the volume of groundwater that can be extracted from the aquifer, Sy is the specific yield or percentage of water that can be released from the aquifer and Vak is the volume of the aquifer, which is the area of the district multiplied by the thickness of the aquifer. The value of Sy is determined using the Sy table based on the type of aquifer material from the results of the geoelectric sounding carried out by [15]. Likewise, the thickness of the aquifer is also determined based on the geoelectric sounding, whose location is shown in Figure 1.

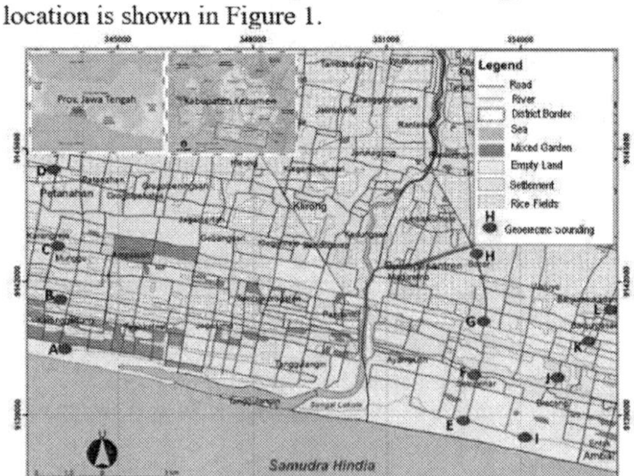

Fig. 1. The route of geoelectric sounding [15]

2.2 Calculation of safe yield

In unconfined aquifers, the safe yield of groundwater can be indicated by the fluctuation of groundwater, the area of the aquifer and the specific yield, and is calculated by the equation:

$$\text{Safe yield} = F.\ A.\ Sy$$

where F is the groundwater level fluctuation obtained through interviews with residents, and A is the area of aquifer obtained from administrative area data.

2.3 Determination of water demand

In this study, the water demand that is calculated is the requirement of water that sourced from groundwater i.e domestic and livestock. The water demand for domestic is determined according to the number of the population and the amount of water demand of each person/day. According to Mangku Sitepoe [in 17], water demand in big cities is generally > 150 liters/person/day, in medium cities 80-150 liters/person/day, districts 60-80 liters/person/days and villages ranging from 30-60 liters/person/day. Based on these criteria, because the research area is a district with a fairly dense population, the water requirement of 100 liters/person/day is determined.

Water demand for livestock is calculated based on the number of livestock and water consumption of each livestock/day, where the types of livestock that are taken into account for water demand are cattle-buffalo-horses (big livestock), sheep (small livestock) and poultry. The standard of water requirement for big livestock is 40 liters/livestock/day, small livestock is 5 liters/livestock/day and poultry is 0.6 liter/livestock/day [18].

3 Result and discussion

3.1 Location, boundary and area

Administratively, the research location is located in coastal area of Kebumen Regency, covering Petanahan, Klirong and Bulupesantren Districts. Judging from its boundaries, in the west it is bordered by Puring District, in the east by Ambal District, in the south by the Indian Ocean and in the north by Sruweng, Pejagoan and Kebumen Districts. The research area has an area of 136.86 km^2

3.2 Climate

The coastal area of Kebumen has an annual rainfall amount of 2,500 to 5,000 mm/year, with the average of 2.647 mm/year. Base on its number, the rainfall of Kebumen coastal area is classified as high rainfall. For monthly rainfall, the highest average monthly rainfall occurs in November at 426 mm, and the lowest rainfall occurs in August at 18 mm.

Regarding climatic conditions, the climate on the coast of Kebumen cannot be separated from the overall climatic conditions of Java Island which are influenced by monsoon. During the months of October to April when the sun is located in the southern hemisphere, mainland Australia is a center of low pressure so that the winds blow from mainland Asia which is the center of high pressure. When passing through the equator, this wind is deflected from the east of the ocean to the west of the ocean. When passing through Java, this wet monsoon winds into the west wind so it is often called the wet west monsoon. During this period, Java Island in general and the Kebumen coast in particular experienced the rainy season. On the other hand, from May to October, mainland Asia became a center of low pressure and Australia became a center of high pressure. The monsoon winds blow from Australia and do not carry water vapor, so that parts of Java Island experience the dry season.

Koppen divides climate types based on the amount of rainfall and air temperature into type A or tropical rain climate, type B or dry climate, type C or warm climate, type D or cold snow forest climate and type E or polar climate. To show the characteristics of climate types in more detail, Koppen subdivides the main climate types by adding lowercase letters f, s, w, m and a. The letter f indicates constant humidity or rain that occurs every month is sufficient, the letter s shows the dry period in summer, the letter w shows the dry period in winter, the letter m shows the monsoon rains with one dry season and the letter a shows the average the temperature in the hot month is greater than 22°C.

Specifically for Climate Type A, it can be divided in more detail by calculating the amount of rain in the dry month in a year and the amount of annual rainfall. A dry month is a month with less than 60 mm of rainfall. It is called a humid month when the rainfall ranges from 60-100 mm, and it is called a wet month when the rainfall is more than 100 mm.

As a result of this division, the types Af, Am and Aw were known. Af is a lower tropical climate or tropical rainforest climate. Climate Af has no dry season, because all months have rainfall greater than 60 mm. Type Am is a monsoon tropical rainforest climate. This type of climate has one or more dry months, however the rain that occurs can still compensate for the dry season, so that the existing vegetation is not so affected by this short dry period. Type Aw is a tropical, savanna rainy climate. This type of climate is characterized by the presence of dry months for 4-8 months, so that the rain that occurs in the wet period cannot compensate for the dry period.

Based on the division of climate types from Koppen, the climate type on the coast of Kebumen is Am. As previously explained, in the Am type, the amount of rainfall that falls in the wet month can compensate for the short dry season. So even though the Am climate has one or two dry months, the rainfall in other months is quite high, so that the vegetation is not so affected by the short dry period.

3.3 Geology and geohydrology

The main rock material constituent of the Kebumen coast is alluvium. In the western part, the existence of this material is cut off by the limestone karst area in the Karangbolong area. As is well known, Karangbolong is part of the southern mountains that does not experience drowning. The lower part of this karst area consists of old andesite breccias that are inconsistent with a layer of limestone. In the northern part of this area, limestone is obtained from the Sentolo Formation and old andesite breccias with tuff layers inserted.

Base on its geohydrological conditions, the coast of Kebumen is included in the Kebumen-Purworejo groundwater basin. In the eastern part, this groundwater basin is bordered by Wates groundwater basin, in the west by the Banyumudal groundwater basin, in the south by the Indian Ocean and in the north by non-aquifer areas. The Kebumen-Purworejo groundwater basin has only one aquifer layer in the form of a unconfined aquifer with a discharge of around 130 million m^3/year [19]. Determining the existence of an area in groundwater basin is the initial stage in groundwater management. By assuming a groundwater basin as a natural underground reservoir, the utilization of groundwater on the coast of Kebumen will affect the overall condition of groundwater in this Kebumen-Purworejo groundwater basin.

3.4 Groundwater availability

Groundwater availability in the study area is obtained based on the analysis of aquifer units. Based on this analysis, the availability of groundwater in each district is determined, by multiplying of the aquifer thickness, the specific yield and the area of each district. The safe yied of groundwater extraction is the result of the multiplication of the area, the specific yield and the fluctuation of groundwater. The calculation results of these parameters are shown in Table 1.

Based on Table 1, it can be seen that the highest groundwater availability is in Petanahan District, followed by Klirong and the lowest is in Bulupesantren. Petanahan District has groundwater availability of 502,208,000 m^3/year, Klirong of 363,300,000 m^3/year and Bulupesantren of 341,390,000 m^3/year. The difference in groundwater availability is due to the different area and thickness of the aquifer.

Table 1. Groundwater availability of each district in the study area

No.	District	Area (m^2)	Sy (%)	Aquifer Thickness (m)	Groundwater Fluctuation (m)	Groundwater Availability (m^3/year)	Safe Yield (m^3/year)
1.	Petanahan	44,840,000	28	40	2	502,208,000	25,110,400
2.	Klirong	43,250,000	28	30	2	363,300,000	24,220,000

3.	Bulupesantren	48,770,000	28	25	2	341,390,000	27,311,200

Source : Petanahan, Klirong, Bulupesantern District in Number 2020 and calculation result

3.5 Utilization of groundwater resources

Observing Table 2, the amount of water demand for domestic in the study area is 16,308,200 liters/day. In terms of each district, Klirong District has the largest domestic water demand, namely 5,512,400 liters/day, Petanahan District 5,451,000 liters/day and Bulupesantren District has the smallest domestic water demand, namely 5,344,800 liters/day.

In the study area there are many livestock businesses, both large livestock, small livestock and poultry. The types of big livestock cultivated by the population include dairy cows, beef cattle, buffalo and horses, while small livestock types are cultivated by goats, sheep and rabbits. For poultry, several types of poultry that are cultivated are purebred chickens, broilers, ducks, manila ducks, quails and geese. For big livestock, the water requirement is 711,920 liters/day, small livestock 349,235 liters/day and poultry 425,430 liters/day..

Table 2. The water demand of domestic and livestock in the study area in 2019

No.	District	Domestic		Livestock			
		Population (person)	Water Demand (liters/day)	Big	Small	Poultry	Water Demand (liters/day)
1.	Petanahan	54,510	5,451,000	3,804	36,568	327,852	531,711
2.	Klirong	55,124	5,512,400	5,703	5,481	172,877	359,251
3.	Bulupesantren	53,448	5,344,800	8,291	27,798	208,336	595,632

Source : Petanahan, Klirong, Bulupesantern District in Number 2020 dan calculation result

3.6 Evaluation of groundwater availability and safe yield

The results of the calculation show that the amount of water demand in the study area is 6,495,100 m^3/year, while the safe yield of groundwater extraction is 76,641,600 m^3/year (Table 3). Petanahan District has the highest water demand of 2,183,689 m^3/year, followed by Bulupesantren District of 2,168,258 m^3/year and the lowest water demand is in Klirong District of 2,143,153 m^3/year. Based on this data, it can be seen that the water demand in the three sub-districts are not much different.

Table 3. Water demand and safe yield of groundwater extraction

No.	District	Water Demand (m^3/year)	Safe Yield (m^3/year)
1.	Petanahan	2,183,689	25,110,400
2.	Klirong	2,143,153	24,220,000
3.	Bulupesantren	2,168,258	27,311,200
	Total	6,495,100	76,641,600

For safe yield, the highest safe yield was in Bulupesantren District at 76,641,600 m^3/year, followed by Petanahan District at 25,110,400 m^3/year and the lowest was in Klirong District at 24,220,000 m^3/year. The difference in safe yield values in the three sub-districts is more influenced by the different areas.

Based on this data, it can be said that the groundwater extraction that has been carried out by residents has not endangered the availability of groundwater, so there is no decrease in groundwater pressure which can cause sea water intrusion.

However, because the strong pressure of fresh groundwater into sea water is not always due to its small extraction, it must always be maintained so that rainwater as input is always stored in the beach ridge and lagoon. In addition, the back swamp in this area must also be maintained so that sea water does not enter at high tide.

4 Conclusion

1) Availability of groundwater in the study area 440,517,770 m^3/year, with the safe yield of groundwater extraction is 76,641,600 m^3/year
2) The groundwater demand in the study area was 6,495,100 m^3/year
3) The extraction of groundwater that has been carried out by residents has not endangered the availability of groundwater, so there is no decrease in groundwater pressure which can lead to seawater intrusion.

Acknowledgement

This article is the result of a study financed by the Final Assignment Recognition (Rekognisi Tugas Akhir/RTA) Programme Universitas Gadjah Mada Fiscal Year 2021.

Reference

1. A. A. Fenta, A. Kifle, Hydrogeology Journal **23**, 195-206 (2014)
2. M. Fontana, F. Grassa, G. Cuisimano, R. Favara, Environ Geol **57**, 885-898 (2008).
3. A. F. Moench, P. M. Barlow, Journal of Hydrology **230**, 192-2010 (2000).
4. I. L. S. Purnama, N. Rahardjo, B. Sulaswono, Jurnal Rekayasa Lingkungan **4**, 131-140 (2008)
5. I. L. S. Purnama, Jurnal Rekayasa Lingkungan **5**, 61-67 (2009).
6. D. K. Saha, K, Choudhury, *J. Ind. Geophys. Union* **9**, 241-247 (2005)
7. D. K. Todd, L. W. Mays. 2005. *Groundwater Hydrology*, New York : John Wiley & Sons (2005)
8. I. L. S. Purnama, Indonesian Journal of Geography, **5**, 206–216 (2019)
9. I. L. S. Purnama, A. Cahyadi A, Taiwan Water Conservancy **67**, 57-67 (2019).
10. I. L. S. Purnama, B. Sulaswono, Majalah Geografi Indonesia, **20**, 52-66 (2006)
11. C. W. Fetter, *Applied Hydrogeology*, Upper Saddle River New Jersey : Prentice-Hall, Inc. (2001)
12. K. R. Rushton, *Groundwater Hydrology : Conceptual and Computational Models,* Chichester, West Sussex : John Wiley & Sons Ltd, (2003).
13. I. L. S. Purnama, M. A. Marfai, Journal of Natural Resources and Development. **2012**, 25-32 (2012).
14. S. Simoen, Suyono, I. L. S. Purnama, Lembaga Penelitian UGM, Yogyakarta (1993).
15. I. L. S. Purnama. Majalah Geografi Indonesia **31**, 1-7 (2017).
16. R. W. V. Bemmelen, *The Geology of Indonesia : General Geology of Indonesia and Adjacent Archipelagoes*, The Hague : Martinus Nijhoff (1970).
17. Y. Priyana, D. Safriningsih, Forum Geografi **19**, 81-87 (2005).
18. Departemen Pekerjaan Umum. Proyek Pembinaan Pengelolaan Sumberdaya Air, Direktorat Pendayagunaan Sumberdaya Air, Direktorat Jenderal Pengairan, Departemen Pekerjaan Umum, Jakarta (1997).
19. H. Setiadi, Direktorat Tata Lingkungan Geologi dan Kawasan Pertambangan, Bandung (2003).

Analysis of Support Systems and Excavation Methods Based on Rock Mass Quality in the Intake Tunnel of the Jlantah Dam

Doni Apriadi Putera[1*], *Heru* Hendrayana[1], and *Hendy* Setiawan[1]

[1]Geological Engineering Departement, Engineering Faculty, Gadjah Mada University, Sleman, Yogyakarta, Indonesia

Abstract. Research on the classification of rock mass quality in the intake tunnel Jlantah dam has not been carried out in detail because the research focuses on the location of the main dam so that empirical excavation methods and support systems have not been carried out. The rock mass quality will be used as a parameter in determining the excavation method and tunnel support system that will be used in the Jlantah Dam intake tunnel. The investigation was carried out through engineering geological mapping, core drill evaluation, and supported by laboratory test data based on the Rock Mass Rating (RMR) and Q-system rock mass classification. The rock mass at the research location based on the RMR classification is in class IV (poor rock). Based on the Q-system method, a very poor rock class is obtained. Based on the analysis of the RMR and Q-system methods, the suitable support system for engineering geological conditions such as the intake tunnel of the Jlantah Dam is shotcrete 10 cm thick, steel set with a distance of 1.5 m and rockbolt length of 1.6 m with a distance of 1.5 m. The proper excavation method for the tunnel intake is top heading and bench.

1 Introduction

The research location is at the Jlantah Dam construction site. Administratively, this dam is located in Tlobo Village, Jatiyoso District, Karanganyar Regency, Central Java Province (fig 1). The Jlantah Dam will produce raw water of 150 liters / second with a total storage capacity of 10.97 million m3, and this dam is also used as a Micro Hydro Power Plant (PLTMH) of 0.625 MW [6].

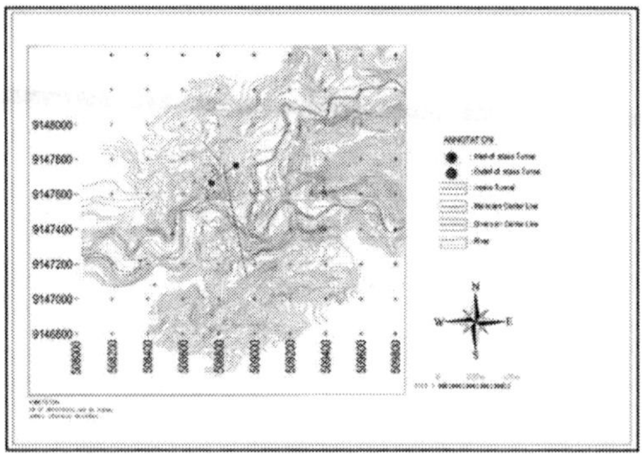

Fig 1. Location of Jlantah dam intake tunnel

Several site investigations have been carried out in the design stage of the dam construction. Still, the study of rock mass characterization in the intake tunnel was not carried out in detail. Details of the intake tunnel design have been made, however, the empirical determination of the tunnel excavation method and the support system has not been carried out [6]. It is necessary to select a safe and stable excavation method and tunnel support system based on the classification of rock mass quality. Several analytical methods can be performed to identify the rock mass characteristics of a tunnel. The determination of rock mass classification empirically used in this study uses the RMR system [3] and the Q-System [1].

2 Geological Conditions

Based on the Geological Map of the Ponorogo Sheet compiled by Samodra & Sampurno [7], the rock formations at the tunnel location and its surroundings consist of Lava Lawu (Ella) Jobolarang Lava (Qvjl), and Lava Sidoramping (Qvsl) formations. Observations of rock outcrops indicate that the research area consists of 4 rock units from old to young, namely: Volcanic Breccia Unit, Slightly and Moderately weathered Lapilli Tuff Unit, and Silt Sand Soil Unit. The map of the distribution of rock units and the geological cross-section of the research location can be seen in Figures 2 and 3. The rock units in the research area are included in the Lawu Volcano sediment with a Holocene age. The path of the intake tunnel is located in the Tuff Lapilli Unit. The rock mass classification in the study area has rock discontinuity with high to slightly weathering rates.

*Corresponding author: doni.a.p@mail.ugm.ac.id

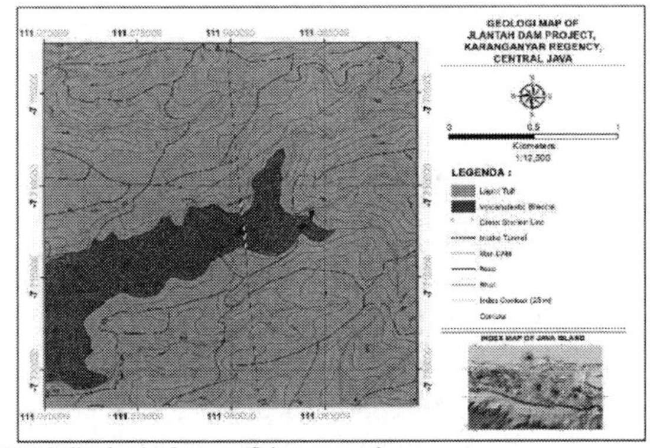

Fig. 2 Geological map of the research area

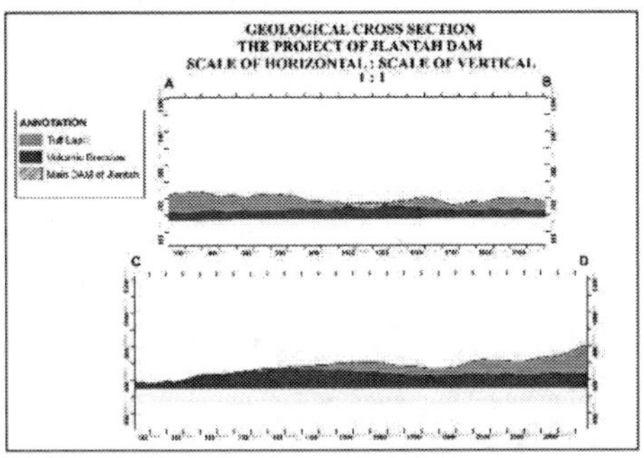

Fig. 3. Geological cross section of Jlantah dam

The Jlantah Dam intake channel has a total design length of 185 meters. The intake channel is a combination of a 172-meter long tunnel and a 13-meter long conduit with dimensions of 4.2 m [6].

3 Methodology

The main reason for using RMR is its ease and flexibility in various practical purposes in engineering (Bieniawski, 1989). According to Bieniawski (1989), the rock mass classification of the RMR method uses 6 (six) parameters with measurements that can be carried out in the field and interpretation of drill hole data. These parameters include 1) Uniaxial Compressive Strength (UCS) of intact rock material, 2) Rock Quality Designation (RQD), 3) discontinuity spacing, 4) discontinuity conditions, 5) groundwater conditions, and 6) discontinuity orientation. The rock mass quality parameters are used to determine the RMR value at the research site based on Bieniawski, (1989) [3]. In addition, Bieniawski also determined the support system in the form of type, diameter, and length of rock bolt, steel set, and shotcrete based on the RMR value as described in Bieniawski [2].

The initial Q-System, is a system that takes into account six parameters: RQD, amount of fracture, the roughness of the fracture, change in stock, water conditions on the stock, and the pressure factor [1]. The basic geotechnical parameters according to Barton [1] are block size, minimum shear strength between blocks, and active pressure. The basic geotechnical parameters are represented by the following ratios [5]: 1) Relative block size = RQD / Jn, 2) Relative strength of friction = Jr / Ja, and 3) Active pressure = Jw / SRF. The determination of rock mass quality is obtained from the following equation [1]:

$$Q = \frac{RQD}{Jn} \; x \; \frac{Jr}{Ja} \; x \; \frac{Jw}{SRF} \qquad (1)$$

Description: RQD => 10; Jn = number of stockings; Jr = roughness value; Ja = value of change in stock; Jw = water value and reduction factor; SRF = reduction of stress factors in fracture, strength / stress ratio in hard boulders and swelling / squizing rocks.

The range of Q values varies between 0.001 and 1000. The correlation of the Tunneling Quality Index (Q) with the behavior and support requirements of underground excavation called Equivalent Dimensions (De) is defined as follows:

$$De = \frac{Height\ of\ tunnel}{Excavation\ Support\ Ratio} \qquad (2)$$

The ESR value for the water tunnel is 1.6 as recommended by Barton [1].

4 Results and Discussion

The morphology in the study area is mainly influenced by the volcanic activity of the old and young Lawu Mountains, with the characterization of steep to very steep slopes. The lowest elevation is the slat where the Jlantah River passes. From the results of the surface geological mapping that has been carried out along the Jlantah River and the Puru River, starting from the plan of the Jlantah Dam to the inundation plan area, it is known that the regional geological structure pattern covering the investigation area has a west-east straight line and is not a fault. The local geological structures found are the stocky structure of the breccias and poor bedding on the breccias sandstone insert.

Rock Mass Rating is obtained from the sum of all the rock mass parameter weights obtained from the research location. The RMR value obtained from the rock test results, field observations then carried out the rock mass classification. RMR and Q-system data were taken from face mapping of tunnel excavations at 6 locations and from observations of core drill results as much as 2 points, namely BS7 and BS11. Face mapping is carried out after the excavation is complete and before the support system is installed. Observation of core drill results is carried out by directly assessing core drilled rock and comparing it with the photo documentation at the beginning and supported by laboratory testing to obtain index and mechanical properties.

Based on observations on the results of the core drill (figure 6 & 7) and the face of the tunnel (figure 4 & 5), the

https://doi.org/10.1051/e3sconf/202132502009

RMR value of the tunnel varies between 18 - 35 so that it can be classified into 3 rock mass classes, namely 1) lapilli tuff with very poor quality (RMR value 18 - 20), 2) poor quality lapilli tuff (RMR 21 - 27) and 3) medium quality lapilli tuff (RMR 28 - 35).

Fig. 4. Appearance of lapilli tuff on the inlet tunnel face

Fig. 5. Appearance of lapilli tuff on the outlet tunnel face

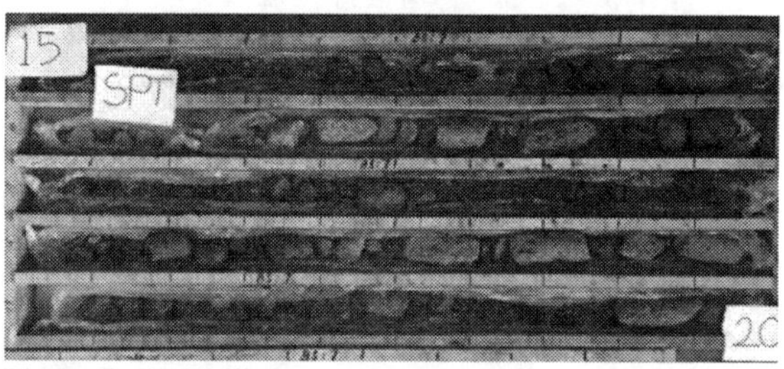

Fig. 6. Typical drill cores at the research location, material lapilli tuff poor quality

Fig. 7. Typical drill cores at the research location, material lapilli tuff with medium quality

The tunnel support system recommendation can determine how long a tunnel is safe without a support system. Based on the RMR value and recommendations from Bieniawski [2], the recommended method of excavation is top heading and bench. The recommended support system is a steel set with a distance of 1.5 m, shotcrete with a thickness of 10 - 15 cm, and a rock bolt with a length of 4 - 5 m with a distance of 1 - 1.5 m.

The tunnel stability analysis criteria can be expressed in the form of a graph of the relationship between the RMR and the unsupported span to determine the stand-up time value (Figure 8). The recommended standup time value for tunnels with very poor quality lapilli tuff rock mass class is 0.8 m long without reinforcement for 1 hour. The standup time value for the poor quality lapilli tuff rock mass class is 1 m long without reinforcement for 1.5 hours. The standup time value for the poor quality lapilli tuff rock mass class is 1.5 m without reinforcement for 3 hours.

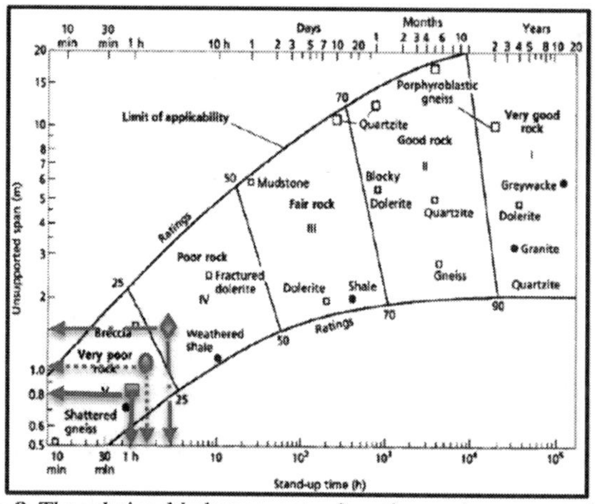

Fig. 8. The relationship between stand-up time and unsupported span for rock mass classes [3]

The Q-system value is obtained from the sum of all rock mass parameter weights obtained from the research location. The Q-system value obtained from the rock test results, field observations then carried out the rock mass classification.

Based on observations on the results of the core drill and the face of the tunnel excavation, the Q-system value of the tunnel varies between 0.23 - 0.58 so that it can be classified into 3 rock mass classes, namely 1) lapilli tuff with very poor quality (Q value 0.23 - 0.35), 2) poor quality lapilli tuff (Q value 0.36 - 48) and 3) medium quality lapilli tuff (Q value 0.49 - 58).

The tunnel diameter is 4.2 m and the ESR value is 1.6 so that the Equivalent Dimension (De) value is 2.6. The Q value and De value will be input on the Q-system graph to get a buffer system recommendation based on the Q-system (Figure 9). The support system for very poor quality lapilli tuff is fiber-reinforced shotcrete 5 - 9 cm thick and rock bolt 1.6 m long with a distance of 1.4 m. The support system for poor quality lapilli tuff is fiber-reinforced shotcrete 5 - 9 cm thick and 1.6 m long rock bolt with a distance of 1.5 m. The support system for medium quality lapilli tuff is fiber-reinforced shotcrete 5 -

9 cm thick and 1.6 m long rock bolt with a distance of 1.6m.

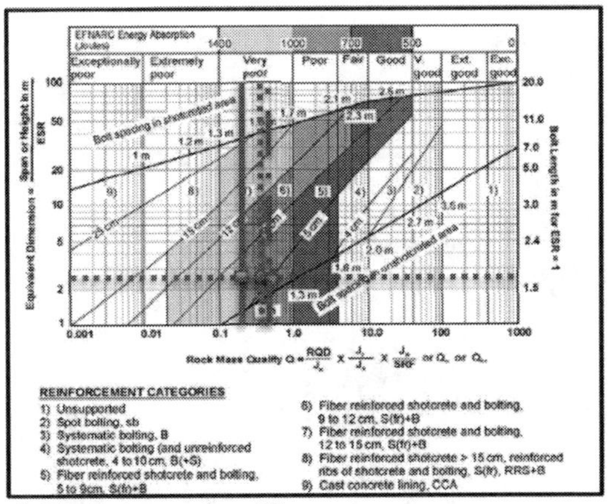

Fig. 9. The Q support chart [5]

Technical geological parameters such as discontinuity conditions, water conditions and the RQD value of a rock will determine the value of rock mass class so that the geological condition of a rock will determine the appropriate excavation method and support system for the rock. For example, geological processes affect the occurrence of rock discontinuity areas, causing the level of rock weathering and rock joint conditions. Another example is the direction and angle of rock layers (dipping) based on RMR which affects the value of rock mass quality. In the case of the tunnel intake, the Jlantah Dam has a rock mass quality value of poor to very poor, so the recommendation for the excavation method used is top heading and bench.

5 Conclusions and Recommendations

Based on the RMR method, the rock mass at the research location is in class IV (poor rock) and class V (very poor rock) with an RMR value ranging from 18 - 35. Based on the Q-system method, it is obtained rock mass with very poor rock class and value The Q-system ranges from 0.23 - 0.58.

Based on the classification of the RMR value, the recommended method of excavation is top heading and bench. The recommended support system is a steel set with a distance of 1.5 m, shotcrete with a thickness of 10 - 15 cm, and a rock bolt with a length of 4 - 5 m with a distance of 1 - 1.5 m. The standup time value is 0.8 - 1.5 m without reinforcement for 1 - 3 hours. The recommendation for the support system according to the Q-System is fiber-reinforced shotcrete 5 - 9 cm thick and 1.6 m long rock bolt with a distance of 1.4 - 1.6 m.

Based on the analysis of the RMR and Q-system methods, the suitable support system for engineering geological conditions such as the intake tunnel of the Jlantah Dam is shotcrete 10 cm thick, steel set with a distance of 1.5 m and rockbolt length of 1.6 m with a distance of 1.5 m. The proper excavation method for the tunnel intake of the Jlantah Dam is top heading and bench. In practice, it is suggested to use these two systems according to needs, sometimes one another complements or corrects each other to get the right decision.

Thank you very much to BBWS Bengawan Solo for research and Mr. I Gde Budi as the supervising lecturers of Masters in Geological Engineering, Gadjah Mada University, so that this journal can be completed. We acknowledge PT. Waskita Karya (Persero) for the financial support for education and research through the super-specialist program held by PUPR in the year 2020.

References

1. N.R Barton, R. Lien, J. Lunde, *Engineering classification of rock a masses for the design of tunnel support*, New York. (1974)
2. Z.T. Bieniawski, Rock mechanics design in mining and tunneling (p. 272). Rotterdam (1984).
3. Z.T. Bieniawski, *Engineering Rock Mass Classification*, NewYork: John Wiley & Sons (1989)
4. D.U. Deere and D.W. Deere, *RQD Index*, U.S. Corps of Engineers, Washington DC (1967)
5. E. Grimstad and N. Barton, *Updating of the Q system for NMT*, in Proceedings of the International Symposium on Sprayed Concrete- Modern Use of Wet Mix Sprayed Concrete for Underground Support, Fagernes: Oslo, page. 46-66 (1993)
6. PT. Aditya, *Studi Investigasi Tambahan Untuk DD Waduk Jlantah*, Kartasura (2017)
7. Sampurno and H. Samodra, *Peta Geologi Lembar Ponorogo*, Bandung (1997)

Analysis of underground river network connectivity in Barat Cave, Karst Karangbolong, Central Java, using the Artificial Tracer Test Method

Alpine Prima Priambada[1,2*], *Tjahyo Nugroho* Adji[1,2], *Eko* Haryono[1,2], *Romza Fauzan* Agniy[1], *Ahmad* Cahyadi[1,2], *Muhammad Qodri* Al-Ghozali[1,2], *Aulia Ika* Rahmawati[1,2], *Danung Shodikh* Makhrizal[1,2], *Rakhmat Dwi* Putra[3], *Danang Riza* Fauzi[1,2], *Elisabeth Supi* Astuti[1,2], *Andy* Setyawan[1,2], *Gangsar Edi* Laksono[1], *Muslih* Biladi[3], *Faizal* Musthofa[3]

[1]Karst Research Group, Faculty of Geography, Universitas Gadjah Mada, Indonesia
[2]Department of Environmental Geography, Faculty of Geography, Gadjah Mada University, Indonesia
[3]Department of Geographic Information Science, Faculty of Geography, Universitas Gadjah Mada

Abstract. The connectivity of the Barat Cave underground river system needs to be known to support the implementation of proper environmental management so that water resources can be maintained sustainably. However, the mapping of underground river paths is often hindered by conditions of narrow cave passages, a barrier blocking the flow of water (siphon), deep underground lakes, underground waterfalls, as well as paths filled with water. This research was conducted in Barat Cave, Karangbolong Karst Area. The purpose of this study is to determine the upstream-downstream connectivity system in this underground river and define the characteristics of the passageway based on quantitative analysis of the transport parameters from the tracer test results. This underground river network analysis needs to be done because previous research has never analyzed this underground river network. The research method used in this study is divided into three stages, namely the pre-field stage, the field stage, and the post-field stage. The pre-field step includes determining the location of the study, collecting secondary data, and studying the literature. The field stage consists of a hydrogeological survey to find information on the presence of caves, springs, sinking stream ponors, or luweng in the study area, instantaneous discharge measurements, and tracer tests. The post-field stage includes data processing and analysis. The results showed that the Barat underground river system originated from the Kalimas sinking stream, Mblabak Cave, Pendok Cave, and Pagilangan sinking streams, then merged into a single tunnel without a flow breaker to the Barat Cave, Pengantin Cave, and appeared in the Kalikarak springs to become a surface river, with a tunnel pattern in the form of curvilinear branchwork. The transport parameters for the underground system tracing of the Barat cave have an advection value of 86.528 m/hour, a dispersion of 0.092 m^2/second, a dispersivity of 3.38 meters, and a recovery of 63%. The transport value of the tracing test parameter is influenced by the characteristics of the passageway and underground river flow conditions.

1 Introduction

Karst is a landscape that has hydrological and geomorphological systems with its characteristics. Karst is a term used to describe a unique landscape with caves and underground water systems that develop and are formed as a result of the dissolving process in soluble rocks [1], such as limestone, dolomite, gypsum, halite, and conglomerates [2]. Water and its involvement in the dissolution process are the most important factors in forming karst landscapes [3].

One of the unique characteristics of hydrological conditions in karst areas is the scarcity or absence of surface drainage/rivers [4]. This occurs because the surface flow is lost through the sinkhole, which then becomes an underground flow that flows through the conduit voids that are formed in the karst rock. Karst surface water that is swallowed and enters through the ponor in the valley/doline includes a conduit passageway and can develop as open channels or vadose pipes [4]. Conduit tunnels that flow with water will comprise an underground river. This condition causes the karst aquifer to have heterogeneous-anisotropic properties; that is, its hydraulic conductivity varies widely, and its magnitude is controlled by the direction or size of secondary porosity or developing voids [5].

The formation of underground rivers is inseparable from the appearance of karst caves [6]. A cave is a naturally occurring underground cavity large enough for humans to enter, in which it can be fully or partially filled with sediment or water [7]. A cave is a naturally occurring cavity in rock that acts as a channel for water flow between the entry point (river flow) and the outlet point (spring or seepage) [8]. Caves connected will form a river system, and when the caves are flowed by water, an underground river is formed [9].*

* Corresponding author:
alpine.prima.priambada@mail.ugm.ac.id

Underground river channel formed due to the dissolution process. Water that passes through underground rivers will come out as karst springs, but this water can also go directly into the sea. The search for underground rivers is often hindered by several conditions such as narrow passageways, a reduction in the dimensions of the passageway accompanied by a barrier obstructing the flow of water (siphon), deep underground lakes, underground waterfalls, passages full of water. To overcome this problem, artificial tracer tests can be used to map underground river connectivity.

Tracer test is a method used to define underground river connectivity [6, 10, 11, 12 13]. The analysis of the connectivity of the cave passage can be modelled by interpreting the breakthrough curve (BTC) of the tracer recorded by the fluorometer sensor, which is used to capture the tracer concentration.

This research was conducted in the underground river of Barat Cave, Karangbolong Karst Area, Kebumen Regency (Fig. 1). Geologically, this cave is located in the Kalipucang Formation, which is composed of shale, limestone, and limestone sandstones, with polymic breccias as inserts. The upper part of this formation consists of coral limestone with carbonate content reaching 95.5% to 99% to be classified as pure limestone [14-16]. This formation is Middle Miocene aged and deposited in an open shallow marine environment.

The underground river Barat Cave has an important value for the surrounding community in meeting water needs for bathing, washing and consumption, and supporting economic activities in the tourism sector. The water that flows in this underground river is used as a source of daily water. Therefore, this underground river system needs to be studied further so that proper environmental management can be applied to maintain its water resources sustainably. This research aims to define the underground river system of Barat Cave and to know the characteristics of the passageway based on the quantitative analysis of the transport parameters from the tracer test results. By knowing underground rivers and the attributes of their passages, proper environmental management can be implemented so that water resources can be preserved.

2 Research Method

The implementation of this research is divided into three stages, namely the pre-field stage, the field stage and the post-field stage. The pre-field step includes literature studies, determining the location of the study, and collecting secondary data. The determination of the study location takes into account the importance of underground rivers, the presence of annual (perennial) flows with large discharge, and accessibility to the site. These considerations underlie the choice of the Barat underground river as the object of research. Secondary data collected includes morphological data in the form of DEM (Digital Elevation Model) from Alos Palsar and hydrogeological data, which provides for mapping data from caves, springs, ponors, luweng, or existing connectivity routes between caves.

The field stage is more emphasized on obtaining primary data. Primary data collection is divided into field surveys to determine the presence of caves, springs, sinking streams, luweng and ponor in the study area, discharge measurements at monitoring points, and tracing tests on underground river flows. Discharge data, target peak concentration, and horizontal distance between the injection point and the point of observation are used in calculating the quantity of tracer poured, according to the formula [17]:

$$M = 1{,}9.10^{-5}(LQC)^{0{,}95} \tag{1}$$

where:
M = quantity of tracer substance required (kg);
L = the distance from the injection point to the observation point (km);
Q = discharge (L/s);
C = peak concentration target (μg/L)

Tracing tests were carried out using fluorescent dyes, namely uranine and tinopal, while the instrument used to record the tracer concentration was a fluorometer. The use of fluorescent dyes as tracing agents is based on their conservative nature, safety, low cost, and high detectability [17]. The properties of the tracer used are presented in Table 1.

Table 1. Properties of tracer uranine and tinopal

No	Tracer	Limit of detection	Safety	Weaknesses
1	Uranine	10^{-3} (μg/L)	Safe	absorbs at low pH
2	Tinopal	10^{-3} (μg/L)	Safe	easily absorbed

After conducting a survey of caves, springs, sinking streams, luweng, and ponor around the study area, the injection point is determined. The injection point is determined by considering the direction of water flow and the slope of the rock formation. Another factor that is taken into consideration is the direction and pattern of valley alignment at the study location. The injection points that are used as locations for tracer substances pouring include the Pagilangan sinking stream (IP1), the Kalimas sinking stream (IP2), and the Mblabak Cave (IP3). Tracer injection was performed at three different times. The I1 injection was carried out on October 10 2020, using uranine, the I2 injection was carried out on October 13 2020, using the tinopal substance, and the I3 injection was carried out on October 16 2020, using the tinopal substance. FL30 fluorometer was installed in the Barat Cave (O.P.) from October 13 2020 to October 16 2020, to record tracer substances poured in IP1 and IP2. Meanwhile, tracing agents from IP3 were captured using active charcoal.

 https://doi.org/10.1051/e3sconf/202132502010

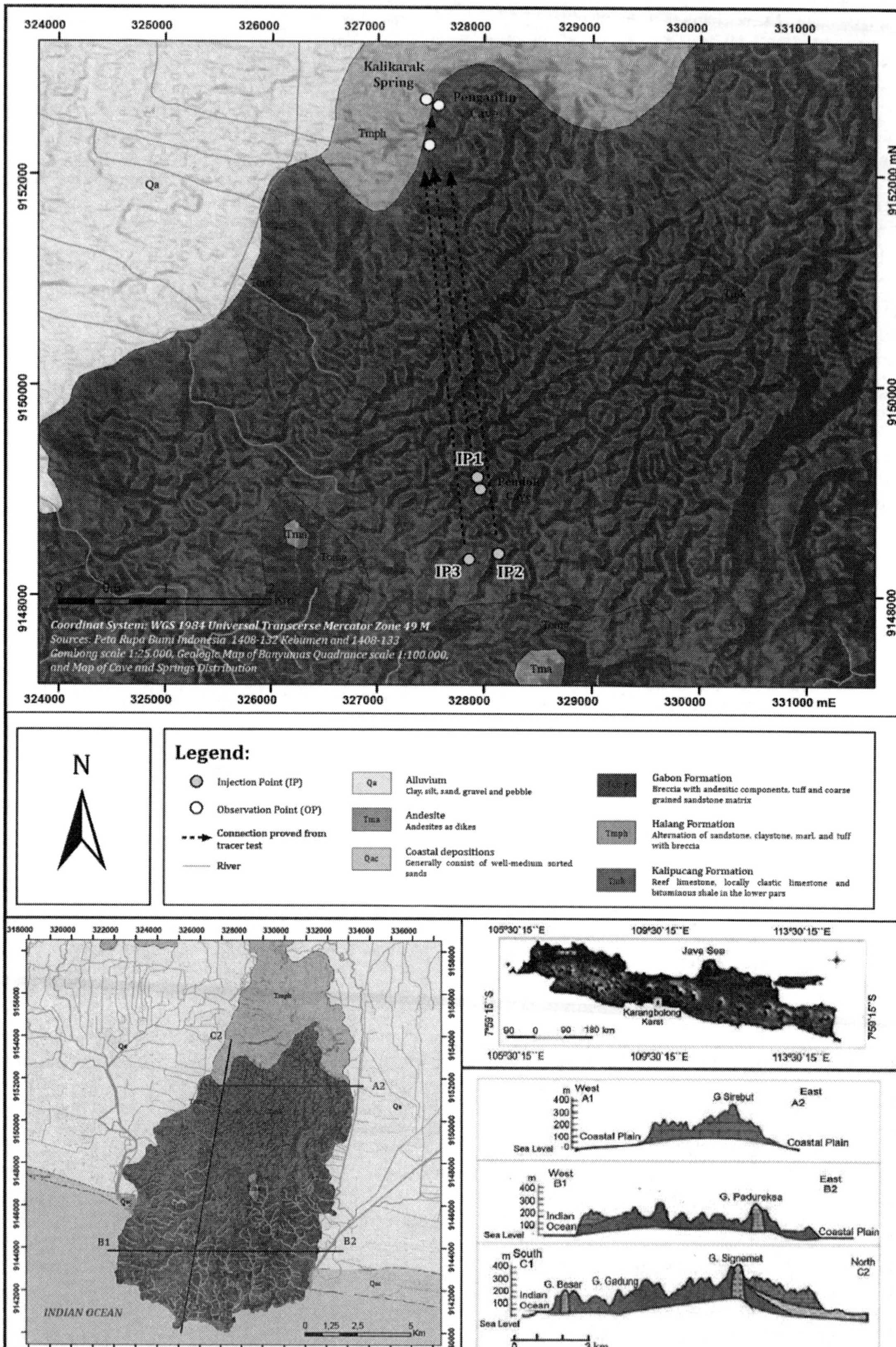

Fig. 1. Location of injection points, monitoring points, prediction of flow direction, and geological conditions of the study area. The geological condition of the study area shows the slope of the Kalipucang Formation to the north in the cross-section of C1-C2 [21]

The post-field stage includes the processing and analysis of data obtained from field surveys and measurement results. Tracing test results are processed using Qtracer2 software. Qtracer2 is a software that is easy to use for tracing analysis in karst areas or other areas [18]. This Qtracer2 software processes tracing test data using the Advection-Dispersion Model to find the value of the transport parameter from the tracing test results.

Tracing test results analysis uses quantitative analysis, which emphasizes the interpretation of the transport parameters, including advection, dispersion, dispersivity, and recovery. This quantitative analysis is based on a detailed study of the breakthrough curve (BTC) resulting from the tracer recorded [19]. Meanwhile, there are parameters obtained from the analysis on the BTC, namely the time it was detected for the first time, the peak time, the maximum concentration, and the peak velocity.

The general equation regarding advection-dispersion is shown in the following equation [17]:

$$\frac{\partial c}{\partial t} = D_l \frac{\partial^2 c}{\partial x^2} - v \frac{\partial c}{\partial x} \qquad (2)$$

where :

c = concentration

t = time

D_l = dispersion longitudinal coefficient

v = flow velocity

x = longitudinal distance

The equation for calculating recovery is shown in the following equation [20]:

$$R_m = \frac{M\ out}{M\ in} = \frac{\int_{t=0}^{tn} (C(t).Q(t))dt}{M\ in}\ x\ 100\% \qquad (3)$$

where :

M_{in} = amount of the trace substance in IP

M_{out} = amount of the trace substance in IP

t_0 = initial detection time

t_n = the end time of detection

Q(t) = discharge in time-t

C(t) = concentration of the tracer in time-t

3 Results and Discussion

The underground river connectivity of Barat Cave is defined by tracing tests. The discharge measurement was carried out at the OP before carrying out the tracing test, namely on October 9, 2020. The measured discharge of 75 l / second was then used to determine the quantity of the tracer to be poured. Tracer injection was carried out at IP1, IP2, and IP3. IP1 and IP2 are sinking streams located in the transitional area between the Gabon Formation and the Kalipucang Formation, while IP3 is a cave that acts as a flow entry during the rainy season. The three injection points show the direction of flow north to the West Cave (OP) and the slope of the Kalipucang Formation, which slopes to the north (Fig. 1).

Meanwhile, the parallel valleys in the northwestern part of the Karangbolong Karst Area, which is around the study site, have a northwestward direction [21]. The direction of the parallel valley leads to the Barat Cave. Tracing test results prove that the three injection points have connectivity with the Barat Cave. Table 2 shows the characteristics of the tracing test in the Barat Cave underground river system.

The tracing test results curve analysis is then carried out by looking at the breakthrough curve (BTC) formed from the IP1 and IP2 tracing tests. The IP3 tracing test could not be analyzed because the tracer was recorded using charcoal. The shape of the breakthrough curve (BTC) resulting from the tracing test is influenced by the tracer character, flow conditions, and the structure of the aquifer [22]. That way, the void branching analysis can be conducted by looking at the peak of the breakthrough curve (BTC) tracing results. The tracing test results produce a breakthrough curve (BTC) with one peak, which means that the injection point and the observation point are connected through a single conduit void without any flow breakers. Breakthrough curve (BTC) tracing test results in IP1 and IP2 are shown in Fig. 2.

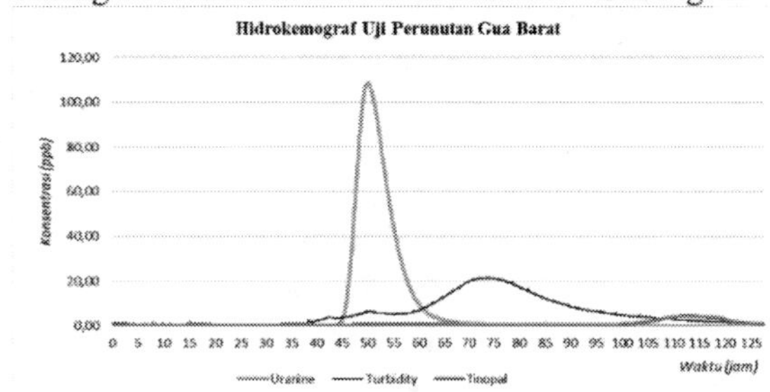

Fig. 2. Breakthrough curve (BTC) tracing test results in IP1 and IP2, which are monitored at the OP.

The underground river flow that passes through the Barat Cave has a direction towards Pengantin then exits through a single outlet at Kalikarak Springs, which then becomes a surface river. The appearance of uranine in Kalikarak Spring is presented in Fig. 3. Thus, the underground river connectivity system in the Barat Cave has a curvilinear branchwork void pattern. The curvilinear void pattern consists of several voids as tributaries that merge into a single, larger tunnel downstream [23]. In general, each primary water source such as doline or ponor contributes to one conduit channel, although more than one water input can supply one channel so that the curvilinear passageway resembles a dendritic surface river flow pattern [23,24].

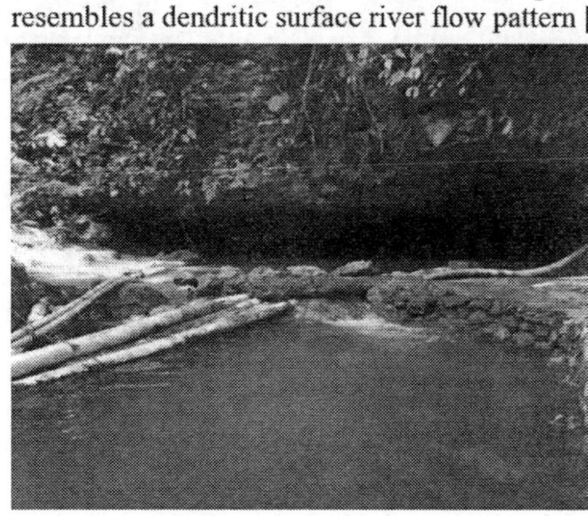

Fig. 3. The appearance of uranine in Kalikarak Springs

Table 2. Parameters of tracing test results in the Barat Cave underground river system

Time of injection		10/10/2020 14:35:00 WIB	13/10/2020 14:00:00 WIB	16/10/2020 15:45:00 WIB
Injection points		IP1	IP2	IP3
Observation points		OP		
Quantity (gr)		300 gram (uranine)	110 gram (tinopal)	100 gram (tinopal)
Fluorometer type		FL 30	FL 30	Charcoal
Peak target of concentration		110 ppb	35 ppb	
Parameter	**Unit**	**Value**		
First time of detection	min	2840		
Time to peak	min	3145	2595	
Maximum concentration	ppb	108	4.51	
Distance	meter	3150	4030	3845
Peak velocity	m/min	1.00		
ADM				
Dispersion	m^2/sec	0.091988		
Dispersivity	meter	3.8272		
Recovery	%	63		
Advection	m/hour	87		
R^2	-	0.9834		

The breakthrough curve (BTC) characteristic shows a significant increase in tracer concentration to reach the peak concentration. The time needed to achieve the peak concentration from the first time it was detected 305 minutes. Meanwhile, the characteristics of the recession show a significant decline but experience a late slowdown or retardation. Retardation can be characterized by a breakthrough curve (BTC) that experiences tailings. This condition is indicated by the length of time it takes to reach a concentration of 0 ppb from the peak concentration, which is 1210 minutes, but it is driven by rain events which are indicated by an increase in turbidity.

Retardation indicates that some tracer particles have a longer residence time in the flow, probably due to the flow through the underground lake causing flow retardation [25]. This condition causes some of the tracer dissolved substances to separate from the others due to the duality of flow, namely turbulent flow due to water hitting the passage walls or obstructions so that a slowed flow is formed and the main flow has a higher speed. Besides, retardation is also caused by the process of entering the water in IP1 into the underground river through two different ways, namely infiltration and entering through small cracks along with the flow.

The results of the tracing test on IP1 were processed using the Advection-Dispersion Model (ADM) using Qtracer2 software. The processing results show that the underground river Barat Cave has an advection value of 86.53 m/hour. The advection value reflects the horizontal displacement movement so that the higher the advection value, the faster the flow of flow from the injection point to the monitoring point and vice versa. Conduit development conditions influence advection value. Flow-through large conduits or underground lakes can cause a decrease in flow velocity as a result of an increase in a cross-sectional area [26]. This fact is in accordance with the flow rate equation ($Q = V \times A$); when the cross-sectional area is larger, the flow velocity is getting smaller. Apart from that, the hydraulic gradient also affects the advection value [26]. The underground river Barat Cave has a hydraulic gradient value of 0.08 when calculated from IP1 to OP. The higher the hydraulic gradient value, the faster the advection value tends to be. In addition, this condition is also influenced by the conditions of the underground river passage.

The results of Qtracer2 software processing show a dispersion value of 0.091988 m^2/second or equivalent to 331.16 m^2/hour and a dispersivity value of 3.8272 m. The dispersion value represents the tracer's spread over the flow per unit time, while the dispersivity represents the tracer's deviation from the waterway carrying it. The dispersion and dispersivity values are influenced by the characteristics of the conduit passageway and the flow conditions so that these two values can provide a general description of the flow tunnel conditions.

The dispersion and dispersivity values will increase more if there is considerable heterogeneity [25]. Heterogeneity is the variation in conduit geometry, including the presence of underground ponds or lakes, obstructions, underground waterfalls, or whirlpools that cause variations in flow velocity [27]. More specifically, the geometric irregularity/shape of the conduit is the variation in the size of the aisle, the winding passageway, and the hall with a pipe or siphon-like form [25]. The siphon/sump is part of the cave's passage, which was flooded [28]. A cave passage is called a siphon when there is a reduction in the hallway's dimensions accompanied by a barrier blocking water flow.

The occurrence of a conduit that has an irregular geometry will increase the dispersion value, which is triggered due to turbulence along the channel. Meanwhile, the dispersivity is more influenced by a large number of velocity variations in the cross-section of the underground river flow. The presence of a moving and immobile flow will increase the dispersivity value due to the drift/deflection of the solute tracer [29]. This condition is often found in the upper reaches where the bottom of the cave channel is not fully filled and in underground lakes where there is a moving and immobile flow that causes tracer substances to deviate from the flow of water carrying it.

The recovery value shows the number of tracer substances recorded at the monitoring point. The result of the calculation shows that the recovery value is 63%. Of the 300 grams of uranine poured in IP1, 189.36 grams of uranine were recorded in the OP. Determination of the location of the fluorometer placement is the reason for the relatively low recovery value. This reduction is due to the existence of a PAMSIMNAS pipe installed at the top of the fluorometer placement location. With this pipe, part of the tracer substance is not recorded because it enters the pipe, thereby reducing the recovery value. This fact is reinforced by the charcoal installed in other springs around the Barat Cave which does not detect tracer substances.

4 Conclusion

The underground river of Barat Cave has a curvilinear branchwork pattern that starts from the Kalimas sinking stream, the Mblabak Cave, the Pendok Cave, and the Pagilangan sinking stream, then merging into a single tunnel without breaking the flow to the Barat Cave, the Pengantin Cave, and exit at Kalikarak Springs as a surface river. The transport parameter value is obtained from the Advection-Dispersion Model processing using Qtracer2 software. The transport parameter for the underground system tracing test of the Barat Cave has an advection value of 86.528 m/hour, a dispersion of 0.092 m^2/second, a dispersivity of 3.38 meters, and a recovery of 63%. The transport value seems to be influenced by the characteristics of the passageway and flow conditions along the underground river. The transpor parameters value indicates slow water flow with relatively homogeneous geometric variations and the presence of an underground pool.

References

1. D.C. Ford, P.W. Williams, *Karst Geomorhology and Hydrology.* London: Chapman and Hall. (2007)
2. H. Karimi, *Hydrogeology of Karstic Area,* (2012) dalam G. A. Kazemi, *Hydrogeology A Global Perspective.* Croatia: In Tech (2012)
3. P. E. van Beynen, *Karst management.* Springer Science & Business Media (2011)
4. E. Haryono, T. N. Adji, *Geomorfologi dan Hidrologi Karst,* Kelompok Studi Karst, Fakultas Geografi, Universitas Gadjah Mada (2004)
5. R. F. Agniy, A. Cahyadi, *Analisis Evolusi Hidrogeokimia Airtanah di Sebagian Mataair Karst Kabupaten Rembang Bagian Selatan,* in Prosiding Seminar Nasional Innovation in Environmental Management. Semarang: Program Pascasarjana Universitas Diponegoro (2015)
6. T. N. Adji, R.F. Agniy, A. Cahyadi, E. Haryono, A. Nurkholis, *Karakteristik sistem aliran sungai bawah tanah di kawasan Karst Jonggrangan dengan tracer test,* INA-Rxiv, 18 Sept. 2019. doi:10.31227/osf.io/9r5kh (2019)
7. A. Bogli, *Karst Hydrology and Physical Speleology.* Berlin: Springer-Verlag (1980)
8. Gillieson, *Caves: Processes, Development, and Management.* Oxford: Blackwell Publishers (1996)
9. S. B. Kusumayudha, *Hidrogeologi Karst dan Geometri Fraktal Daerah Gunungsewu.* Yogyakarta: Adicita Karya Nusa (2005)
10. A. Cahyadi, R. F. Agniy, *Analisis Breakthrough Curve untuk Karakteristerisasi Pelorongan Di Sistem Sungai Bawah Tanah Pindul Kabupaten Gunungkidul,* INA-Rxiv. August 29. doi:10.31227/osf.io/3bs2v (2017)
11. I. A. Riyanto, A. Cahyadi, T. N. Adji, E. Haryono, M. Widyastuti, R. F. Agniy, W. A. Fathon, *Analisis konektivitas dan karakterisasi pelorongan dengan uji perunutan pada mataair epikarst sub-sistem panggang, kawasan karst Gunungsewu,* INA-Rxiv, 2 Dec. 2018. doi:10.31227/osf.io/9d3s4 (2018)
12. A. Cahyadi, E. Haryono, T. N. Adji, M. Widyastuti, M. Naufal, F. Ramadhan, I. A. Riyanto, *Analisis Konektivitas dan Karakteristik Lorong pada Sistem Hidrogeologi Mataair Beton, Kawasan Karst Gunungsewu, Kabupaten Gunungkidul dengan Uji Perunutan.* Jurnal Geografi, **12**(2), 105-114 (2020)
13. R. F. Agniy, A. Cahyadi, A. Nurkholis, *Analisis Karakteristik Akuifer Karst dengan Uji Perunutan dan Pemetaan Gua,* OSF Preprints, 2 Oct 2017 doi:10.31219/osf.io/dfxjh (2017)
14. S. Asikin, *Geologi Struktur Indonesia,* Bandung: ITB Press (1992)
15. Ruswanto, I. Badri, A. Anwar, *Inventarisasi Geologi Lingkungan Kawasan Karst Gombong, Kabupaten Kebumen, Jawa Tengah,* Laporan Proyek, Bandung: Direktorat Tata

Lingkungan Geologi dan Kawasan Pertambanagan (2002)

16. B. D. Saputra, *Morfometri Dolina di Wilayah Karst Gombong Selatan,* Skripsi, Depok: Universitas Indonesia (2008)
17. R. Benischke, N. Goldscheider, C. Smart, *Tracer Techniques* (2007) dalam N. Goldscheider, D. Drew, *Methods in Karst Hydrogeology*, London: Taylor & Francis Group (2007)
18. N. Goldscheider, J. Meiman, M. Pronk, C. Smart, *Tracer tests in karst hydrogeology and speleology,* International Journal of speleology, **37**(1), 27-40 (2008)
19. M. S. Field, *The QTRACER2 program for tracer-breakthrough curve analysis for tracer tests in karstic aquifers and other hydrologic systems*, National Center for Environmental Assessment-Washington Office, Office of Research and Development, US Environmental Protection Agency (2012)
20. V. T. Tam, N. D. Trung, D. T. Ke, O. Batelaan, *Interpretation of a cave system based on tracer experiment, geostructure and cave development analysis,* in Trans-Karst 2004, International Transdiciplinary Conference on Development and Conservation of Karst Regions (pp. 202-206). Research Institute of Geology and Mineral Resources (RIGMR); Hanoi, Vietnam (2004)
21. E. Haryono, T. P. Sutanto, Suratman, Sutikno, *Polygonal Karst Morphology of Karangbolong Area, Java-Indonesia,* Acta Carsologica, **46**(1) (2017)
22. C. Smart, *Artificial tracer techniques for the determination of the structure of conduit aquifers*, Groundwater, **26**(4), 445-453 (1988)
23. A. N. Palmer, *Origin and Morphology of Limestone Caves*, Geological Society of America Bulletin **103**, 1-21 (1991)
24. J. Jouves, S. Viseur, B. Arfib, C. Baudement, H. Camus, P. Collon, Y. Guglielmi, *Speleogenesis, geometry, and topology of caves: A quantitative study of 3D karst conduits,* Geomorphology, **298**, 86-106 (2017)
25. M. Hauns, *Modeling tracer and particle transport under turbulent flow conditions in karst conduit structures,* Thesis, Université de Neuchatel: Freiburger Schriften Zur Hydrologie (2000)
26. U. Lauber, W. Ufrecht, N. Goldscheider, *Spatially resolved information on karst conduit flow from in-cave dye-tracing*, Hydrology & Earth System Sciences Discussions, **10**(9) (2014)
27. M. Hauns, P. Y. Jeannin, O. Atteia,. *Dispersion, retardation and scale effect in tracer breakthrough curves in karst conduits,* Journal of hydrology, **241**(3-4), 177-193 (2001)
28. M. S. Field, *A Lexicon of Cave and Karst Terminology with Special Reference to Environmental Karst Hydrology*. United States Environmental Protection Agency, Washington, 214 p (2002)
29. A. Ender, N. Goeppert, N. Goldscheider, *Spatial resolution of transport parameters in a subtropical karst conduit system during dry and wet seasons,* Hydrogeology Journal, **26**(7), 2241-2255 (2018)

Laundry wastewater characteristics and their relationship with river water quality as an indicator of water pollution. Case study: Code Watershed, Yogyakarta

Galih Dwi Jayanto[1], *Margaretha* Widyastuti[2*], *M. Pramono* Hadi[2]

[1]Master Program on Planning and Management of Coastal Areas and Watersheds, Faculty of Geography, Universitas Gadjah Mada, Special Region of Yogyakarta 55281, Indonesia

[2]Department of Environmental Geography, Faculty of Geography, Universitas Gadjah Mada, Special Region of Yogyakarta 55281, Indonesia

Abstract. Laundry services are rapidly becoming ubiquitous in rural and urban areas, generating high domestic waste volumes. Their growth has raised concerns about the quality of natural waters, particularly the rivers into which most laundry businesses discharge their wastewater. The research set out to study laundry wastewater characteristics and their relationship with the river water quality as water pollution indicators in Code Watershed (Indonesia). It lies in three administrative units where many human activities take place in the watershed's upper to the lower reaches. BOD and COD of 25 sampled laundry businesses were analyzed descriptively and compared to the laundry wastewater standards stipulated in Regulation No. 7 of 2016 and the class II water quality standards issued in Governor Regulation No. 20 of 2008. The results showed that the BOD of four laundry businesses and the COD of 18 laundry businesses exceeded the predetermined standards. Overall, rivers in Code Watershed meet the class II river water criteria for BOD at three of the seven sampling points and COD at all points.

1. Introduction

In the Special Region of Yogyakarta (SRY), Indonesia, increasing community activities along the Code River, mostly those of domestic and small-scale industrial sectors, have modified the water quality. Monitoring in 2013 confirms ammonia and COD are low at the inlet but high at the outlet, up to the point where the quality standards are exceeded. These findings indicate that the influence of waste disposal from, for instance, settlements, lodging, small-scale industries, and health services, becomes more significant downstream [1]. Meanwhile, another monitoring by the SRY Environment and Forestry Service reveals an increasing trend in detergent contents from 2017 through 2019 and 2020, above the class II water quality standards. A significant increase also appears in BOD. Accordingly, the source of pollutants in the Code River that passes through urban areas comes from domestic waste, e.g., greywater, soap, food wastes and scraps, and fecal wastes [2].

These contaminants can adversely affect river water and its surroundings, that the constantly decreasing water quality alters processes and mechanisms in abiotic, biotic, and social environments in Code River [3]. High nitrate and manganese concentrations indicate the abiotic environment's damaged functions, disrupting the other environmental constituents: reduced biota diversity and elevated community vulnerability to diseases caused by deteriorating river water quality and poor sanitation.

Laundry business constitutes a small-scale industry with a persistently growing number and economic significance in several countries, such as Indonesia, India, and China. Nevertheless, this industry's development raises various kinds of problems as it generates about 3.881 million liters of wastes per day, comprising 40% of the total industrial wastes, or 20,000–25,000 liters/person/year [4]. Meanwhile, in Turin, Italy, data shows that the laundry industry produces an average of 400 m^3 of wastewater per day for every 15 liters of water used. Because the wastewater consists of organic substances (soap, detergents, chlorinated solvents, aromatics, biological substances, fats, and grease) and inorganic substances (heavy metals, sand, silt, clay, metal ions, and particles), it makes the refining process intricate and highly time-consuming [5].

With the increasing use of chemicals in the industry, river water pollution has quickly become an alarming issue [6]. The characteristics of laundry wastewater are divided into three types, namely domestic laundry, industrial laundry and hospital laundry [7] (Table 1). Based on the table, it is known that the distribution of wastewater quality levels from the three activities. Laundry wastewater from laundry services (industry) has a higher level than other activities.

In this study, only selected parameters were used for analysis, only parameters that had a high content in laundry wastewater were selected. Wastewater from laundry activities still contains high concentrations of

* Corresponding author: m.widyastuti@geo.ugm.ac.id

dissolved materials (TSS), BOD, COD, and also LAS (linear alkylbenzene sulphonate) [8,9]. The COD and MBAS content after leaving the washing machine were 186 - 2418 mg/L and 25.0 - 33.9 mg/L decreased slightly to 122 - 14488 mg/L for COD and 6.50-10.3 mg/L for MBAS in water bodies/rivers [10].

Table 1. Characterisation of domestic,industrial and hospital laundries wastewater

Parameters	Domestic laundry	Industrial laundry	Hospital laundry
pH	9.3 - 10	9.0 - 11	11.4 - 11.6
EC (μs/cm)	190-1400	640-3000	808-2000
TDS (mg/L)	400-6000	420	456-800
TSS (mg/L)	200-987	400-1000	66-71
TH (mg/L CaCO3)	-	44	53-68
TA (mg/L CaCO3)	83-200	128	302-375
TOG (mg/L)	8.0-35	71.5-11790	25-26
Phosphate (mg/L)	4-27.6	3.43	10.8-167
BOD5 (mg/L)	48-1200	218-9810	44-50
COD (mg/L)	375-4155	80-212000	477-876
Turbidity (NTU)	14-400	40-150	87.9

Source: (Lade & Zainab, 2018)

Based on the background, the purpose of this study was to determine the characteristics of the laundry business wastewater and its relationship to the water quality of the Code river. Where the current development of the laundry business is very large and has the potential to pollute the river environment.

2. Research Area Description

Based on the inventory of laundry business locations in SRY, Code Watershed was selected as the research location because it is located in the middle of the region, dividing the City of Yogyakarta into two (Fig 1). It stretches as far as 45.58 km, traversing three administrative units: from Sleman Regency (precisely, Merapi Volcano National Park in Hargobinangun, Pakem) to the City of Yogyakarta then Bantul Regency (i.e., Jl. Imogiri Timur, Trimulyo, Jetis). Two main rivers flow in the watershed, namely Boyong and Code. Based on public knowledge, Boyong starts from Merapi Volcano to Jl. Ring Road Utara then continues downstream as Code River.

3. Methodology

The research began by taking 25 samples of laundry business wastewater and Code River water at ten observation points simultaneously. These samples were determined using purposive sampling, i.e., based on the distance between the laundry business and the river, while the observation points were selected based on the laundry sample density. The research tools included 500 ml water sample bottles to safely store and transport the collected water to the laboratory and a Global Positioning System (GPS) to determine the location. The laboratory analysis results were analyzed descriptively and compared with the wastewater standards and the class II river water quality standards based on the predetermined stream order. The parameters observed were Biological Oxygen Demand (BOD) and Chemical Oxygen Demand (COD).

3.1 Laundry Wastewater Characterization

The wastewater quality test results were analyzed and observed side-by-side with the wastewater standards issued in Regulation No. 7 of 2016 [11], where the highest allowable BOD and COD levels are 75 mg/L and 159 mg/L, respectively (Table 2).

Table 2. Wastewater Parameter Standards for Laundry Industries

Parameters	Maximum Allowable Presence (mg/L)	Highest Pollution Loads (kg/ton)
BOD_5	75	1.5
COD	150	3
TSS	100	2
TDS	2,000	40
Detergent	5	0.1
Temperature	± 3°C of air temperature	
pH	6.0–9.0	
Largest wastewater discharge (L/kg)	20	

Source: SRY Regulation No. 7 of 2016

3.2. Code River Water Quality Analysis

The river water quality characteristics were compared with the SRY Government's Regulation No. 20 of 2008 [12] (Table 3), which classifies the quality based on a predefined stream order. Based on the SRY Government's Regulation No. 22 of 2007 [13], the Boyong River (upper reach of the watershed) is a first-order stream, while the Code River (middle and lower reaches) is second-order.

Table 3. BOD and COD Standards by Stream Orders

Parameters	Units	First	Second	Third	Fourth
BOD	mg/L	2	3	6	12
COD	mg/L	10	25	50	100

Source: SRY Government's Regulation No. 20 of 2008

3.3. Comparative Analysis of Laundry Wastewater and River Water Quality

The relationship between laundry wastewater and Code River water quality was analyzed descriptively using tables and graphics. Code Watershed was divided into ten segments; each had a water quality observation point. These segments were also identified as experiencing point source pollution from the wastewater generated by laundry businesses.

 https://doi.org/10.1051/e3sconf/202132502011

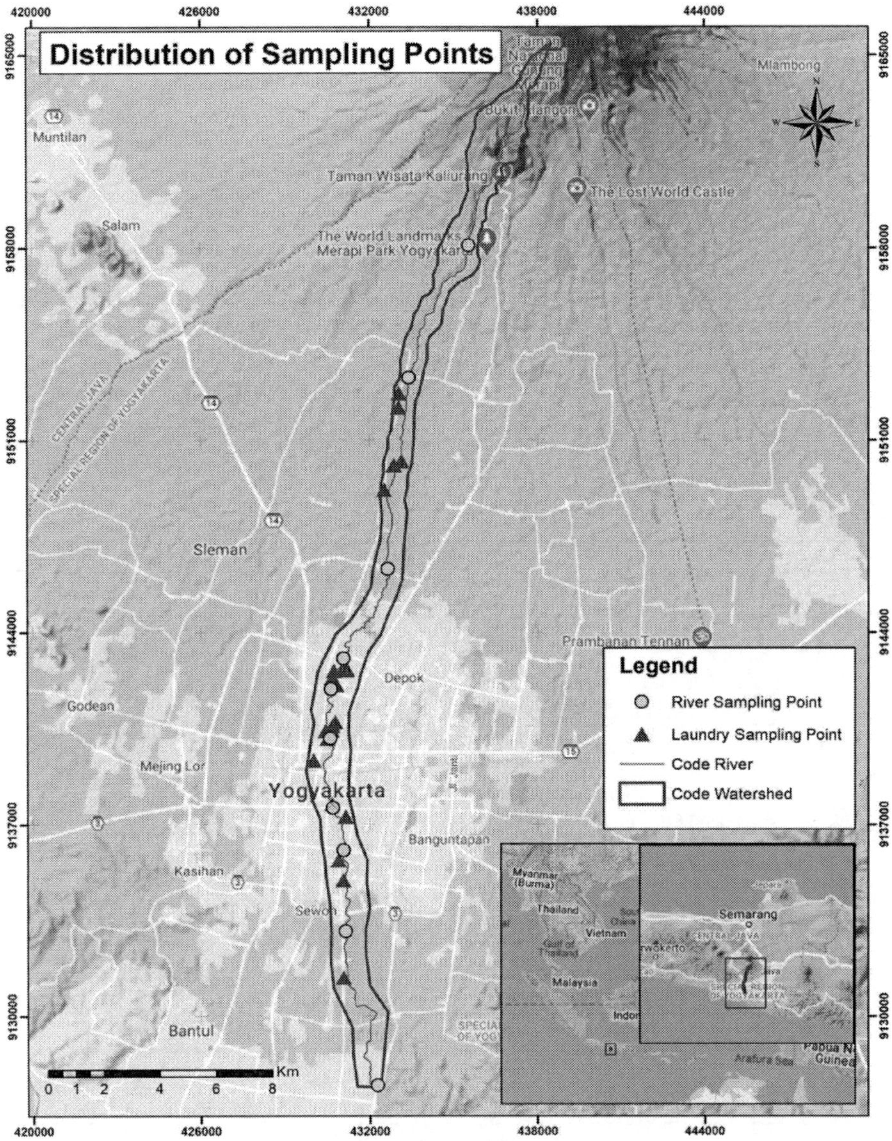

Fig 1. Water Sampling Points in the Code River

4. Results and Discussion

4.1. Laundry Wastewater Characteristics

Table 4 shows the laboratory test results of 25 laundry wastewater samples. Compared with the wastewater quality standards set in Regulation No. 7 of 2016, five samples contained high BOD levels exceeding the upper threshold, 75 mg/L. These five samples were that of Rumah Cuci (165 mg/L), Malika (93.3 mg/L), Ceria (141 mg/L), Hayu (125 mg/L), and Icigo (264 mg/L). As indicated by the bold figures in Table 4, 18 samples had COD in the range of 136–366 mg/L, exceeding the laundry wastewater quality standard, 150 mg/L.

High BOD and COD levels are in line with previous research in India that found high levels of BOD (230–626 mg/L), COD (376–910 mg/L), and suspended solids (445–1,550 mg/L) in laundry wastewater [9], and these conditions are believed to be due to the use of various detergents [14]. The interview with laundry owners revealed that cloth washing requires different types of detergents, increasing the presence of surfactants (complex organic compounds) in the liquid waste and the amount of oxygen required by microorganisms to break down organic matters in the water.

4.2. Code River Water Quality

Table 5 shows the BOD and COD levels of the Code River water samples. Sampling points S1-S3 and S4-S10 were, respectively, compared with class I and class II water quality standards issued in the SRY Government's Regulation No. 20 of 2008. Five samples contained high BOD levels exceeding the predetermined standard: S2 (Pulowatu Bridge, 2.14 mg/L), S4 (Ring Road Utara Bridge, 6.99 mg/L), S7 (Sayidan Bridge, 4.76 mg/L), S8 (Lowanu Bridge, 5.63 mg/L), and S10 (Kembangsong Bridge, 3.33 mg/L). Meanwhile, the COD levels varied from 6.32 to 11.4 mg/L, meaning that none of the COD in all ten river segments is above the class I and class II water quality standards.

S2 (Pulowatu Bridge), where the BOD exceeded its allowable level in class I water, is located in the upper reach of the Code River. Similarly, S4 (Ring Road Utara Bridge) is in the middle segment, i.e., the center of the economic growth in the Sleman Regency. Meanwhile, the other three sampling points (S7, S8, and S10) are located in the lower reach, and the excessively high BOD levels at these points are potentially caused by the accumulation of pollutants upstream. A previous study found that the Code River segment in Kotabaru (City of Yogyakarta) had a high BOD level, i.e., 100.1 mg/L [15], and it is located close to the lower reach of the river.

Table 4. BOD and COD Levels of the Sampled Laundry Business Wastewater in the Code River

No	Laundry Business Names	Coordinates		Wastewater Quality (mg/L)	
		X	Y	BOD	COD
1	Gading Laundry	433056.53	9152706.17	10.2	**272**
2	Nabila Laundry	432870.02	9150095.37	13.4	**158**
3	Amanah Laundry	433136.02	9150236.97	14.3	**253**
4	Djamil Laundry	432537.89	9149194.23	14.3	**136**
5	Rumah Cuci Laundry	433040.64	9152194.03	**165**	**354**
6	Purple Bubble Laundry	431077.00	9142700.00	19.4	107
7	Medio Laundry	431167.00	9142670.00	5.64	**190**
8	Omah Degan Laundry	430711.05	9142590.74	27.4	78.8
9	Spongebob Laundry	431179.00	9142616.00	51.6	**392**
10	Express Laundry	431073.95	9142652.43	1.56	**183**
11	Kawankoe Laundry	430759.00	9142175.00	6.54	50.6
12	Mom's Laundry	430653.00	9142337.00	20.6	**259**
13	B & K Laundry	430750.00	9142136.00	14.4	**253**
14	Icigo Laundry	430803.00	9142081.00	**264**	**334**
15	Pelangi Laundry	430764.00	9140728.00	19.8	69.5
16	Nana Laundry	430631.00	9140432.00	23.5	44.2
17	Malika Laundry	430405.00	9140384.00	**93.3**	101
18	Laundry 31 C	430669.85	9140591.29	20.1	75.8
19	Laundry Kiloan 34	430492.00	9140124.00	23.8	**171**
20	Ceria Laundry	429996.00	9139322.00	**141**	**348**
21	Hayu Laundry	430650.00	9137731.00	**125**	**366**
22	Gangsar Laundry	431137.00	9137274.00	41.1	**341**
23	Barokah Laundry	430884.00	9135680.00	12.0	**221**
24	Youshan Laundry	431050.00	9134953.00	3.98	**240**
25	Uno Fresh Laundry	431053.00	9131401.00	70.4	**240**

Source: Laboratory test results, 2020

Table 5. BOD and COD Levels of the Sampled Water in the Code River

Codes	Sampling Points	Coordinates		Measured Parameter Levels (mg/L)		Water Quality Standard (mg/L)	
		X	Y	BOD	COD	BOD	COD
S1	Gantung Boyong Bridge	435443.24	9158022.7	0.19	5.69	2	10
S2	Pulowatu Bridge	433451.18	9153253.9	**2.14**	3.16	2	10
S3	Kamdanen Bridge	432545.7	9146251.5	1.33	9.48	2	10
S4	Ring Road Utara Bridge	431096.92	9143052.1	**6.99**	8.85	3	25
S5	New UGM Bridge	430553.64	9141844.8	0.71	7.58	3	25
S6	Sardjito Bridge	430674.37	9140154.6	0.71	10.7	3	25
S7	Sayidan Bridge	430614	9137619.2	**4.76**	7.58	3	25
S8	Lowanu Bridge	431036.56	9136049.7	**5.63**	11.4	3	25
S9	Pandeyan Bridge	431157.29	9132850.3	2.14	6.32	3	25
S10	Kembangsongo Bridge	432424.97	9127538.2	**3.33**	9.48	3	25

Source: Laboratory test results, 2020

Table 6. BOD and COD Level Comparison of the Laundry Wastewater and River Water Samples in the Code River

No	Laundry Business Names	Laundry Wastewater Quality (mg/L)		Stream Section with Suspected Point Source Pollution	River Water Quality Downstream of the Laundry Wastewater Outlet (mg/L)	
		BOD	COD		BOD	COD
1	Gading Laundry	10.2	**272**	Kamdanen Bridge	1.33	9.48
2	Nabila Laundry	13.4	**158**			
3	Amanah Laundry	14.3	**253**			
4	Djamil Laundry	14.3	**136**			
5	Rumah Cuci Laundry	**165**	**354**			
6	Purple Bubble Laundry	19.4	107	New UGM Bridge	0.71	7.58
7	Medio Laundry	5.64	**190**			
8	Omah Degan Laundry	27.4	78.8			
9	Spongebob Laundry	51.6	**392**			
10	Express Laundry	1.56	**183**			
11	Kawankoe Laundry	6.54	50.6			
12	Mom's Laundry	20.6	**259**			
13	B & K Laundry	14.4	**253**			
14	Icigo Laundry	264	**334**			
15	Pelangi Laundry	19.8	69.5	Sardjito Bridge	0.71	10.7
16	Nana Laundry	23.5	44.2			
17	Malika Laundry	**93.3**	101			
18	Laundry 31 C	20.1	75.8			
19	Laundry Kiloan 34	23.8	**171**	Sayidan Bridge	**4.76**	7.58
20	Ceria Laundry	**141**	**348**			
21	Hayu Laundry	**125**	**366**			

No	Laundry Business Names	Laundry Wastewater Quality (mg/L)		Stream Section with Suspected Point Source Pollution	River Water Quality Downstream of the Laundry Wastewater Outlet (mg/L)	
		BOD	COD		BOD	COD
22	Gangsar Laundry	41.1	**341**	Lowanu Bridge	**5.63**	11.4
23	Barokah Laundry	12.0	**221**	Pandeyan Bridge	2.14	6.32
24	Youshan Laundry	3.98	**240**			
25	Uno Fresh Laundry	70.4	**240**	Kembangsongo Bridge	**3.33**	9.48

Notes: bold figures indicate exceeded quality standards

4.3. Comparison between Laundry Wastewater and River Water Quality

Table 6 shows the spatial distribution of the wastewater and river water quality samples at seven sampling points: S3 (Kamdanen Bridge), S5 (New UGM Bridge), S6 (Sarjito Bridge), S7 (Sayidan Bridge), S8 (Lowanu Bridge), S9 (Pandeyan Bridge), and S10 (Kembangsongo Bridge). Fig. 3 and Fig. 4 illustrate the distribution of BOD and COD levels of the laundry wastewater and Code River water quality. Nearly all of the wastewater samples showed that BOD and COD decreased downstream when Boyong River reaches Code River. Spatially, in the Kamdanen Bridge segment, there was a waste input from five laundry businesses, with BOD levels varying from 10.2 to 165 mg/L and COD in the range of 136–354 mg/L. At the point where Boyong River stops and continues as Code River, the BOD and COD levels lowered to 1.33 mg/L and 9.48 mg/L, below the class 2 river water quality standards.

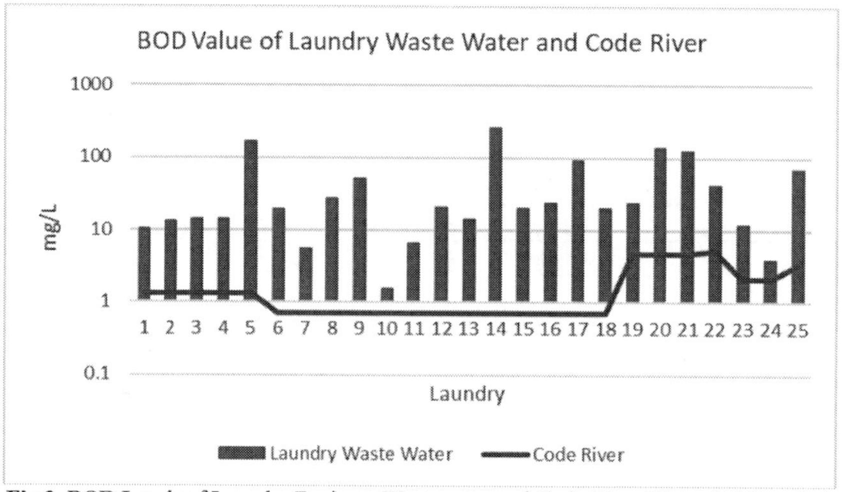

Fig 2. BOD Levels of Laundry Business Wastewater and Code River Water (25 Samples)

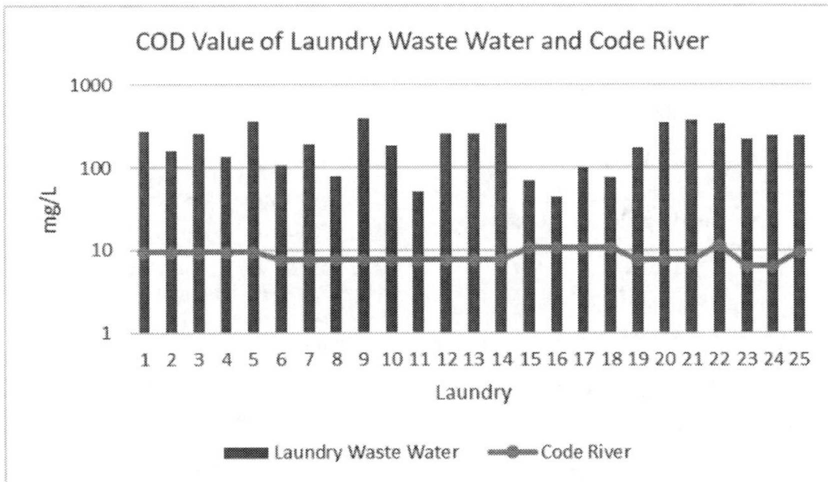

Fig 3. COD Levels of Laundry Business Wastewater and Code River Water (25 Samples)

S5 (New UGM Bridge segment) received waste input from nine laundry businesses and, therefore, had BOD in the range of 1.56–264 mg/L and COD between 50.6 and 392 mg/L. At the point where Boyong continues as Code River, the BOD and COD lowered to 0.71 mg/L and 7.58 mg/L, respectively, below the class II river water quality standard. The same case applies to S6 (Sardjito Bridge segment), which received wastewater from four laundry businesses; BOD and COD levels decreased from 19.8–93.3 mg/L and 44.2–75.8 mg/L to 0.71 mg/L and 10.7 mg/L, below the quality standards. Three laundry businesses disposed of their wastewater in S7 (Sayidan Bridge segment), leading to high BOD (23.8–141 mg/L) and COD (171–366 mg/L) that later decreased to 4.76 mg/L and 7.58 mg/L, respectively, where the Code River channel begins. At this point, the BOD is still above the specified river water quality standard, while the COD is below it.

S8 (Lowanu Bridge segment) received waste input from one laundry business, and like the previous three sampling points, its BOD and COD lowered from 41.1 mg/L and 341 mg/L to 5.63 mg/L and 11.4 mg/L, respectively, where the Code River channel starts. These BOD levels are above the predetermined standard. The same case applies to S9 (Pandeyan Bridge segment) that received wastewater from two laundry businesses; BOD and COD levels decreased from 3.98–12.0 mg/L and 221–240 mg/L to 2.14 mg/L and 6.32 mg/L. One laundry business discharged its wastewater into S10 (Kembangsongo Bridge segment), leading to high BOD (70.4 mg/L) and COD (240 mg/L) that later decreased to 3.3 mg/L and 9.48 mg/L, respectively, where Boyong continues as Code River. The BOD levels in this segment have exceeded the predefined class II river water quality standard.

5. Conclusion

The wastewater disposed of laundry businesses around the Code River has high levels of BOD and COD at several sampling points because the various detergents used in cloth washing elevate the amount of oxygen needed by microorganisms to break down organic matter. Based on the Code River water quality test results, the BOD at some sampling points has exceeded its allowable level in river water, but the COD is below the predetermined quality standard. The point at which Boyong River stops and Code River channel starts is marked by lowered BOD and COD levels because of many factors. Further research is required to explain this relationship in more detail.

Acknowledgment

The authors would like to thank Universitas Gadjah Mada for financially assisting this research through the Thesis Recognition Grant (Hibah Rekognisi Tugas Akhir) in the fiscal year of 2020.

References

1. Widodo, B., Kasam, Ribut L, and Ike A. Strategi penurunan pencemaran limbah domestik di Sungai Code DIY. Jurnal Sains dan Teknologi Lingkungan, vol **5** (1): 36–47 (2013)
2. Dinas Lingkungan Hidup Kota Yogyakarta (DLHK). Kajian Beban Pencemaran Sungai Code Tahun 2018. 92 (2018)
3. Puspitasari, Dinarjati Eka. Dampak pencemaran air terhadap kesehatan lingkungan dalam perspektif hukum lingkungan (studi kasus Sungi Code di Kelurahan Wirogunan, Kecamatan Mergangsan dan Kelurahan Prawirodirjan Kecamatan Gondomanan Yogyakarta). Mimbar Hukum, Vol **21** (1): 23-34 (2009)
4. Wendland, C., & Albold, A. Sustainable and cost-effective wastewater systems. **19** (2010)
5. Ciabattia, I., Cesaro, F., Faralli, L., Fatarella, E., & Tognotti, F. Demonstration of a treatment system for purification and reuse of Laundry wastewater.

Desalination, **245**(1–3), 451–459. https://doi.org/10.1016/j.desal.2009.02.008 (2009)

6. Agarwal, A. Water pollution problems posed by small industries: A case study of India and China. Water Science and Technology, **45**(8), 47–52. https://doi.org/10.2166/wst.2002.0142 (2002)
7. Lade, O., & Zainab, G. Potential of recycling laundry wastewater for domestic use. Journal of Civil Engineering and Sustainable Water Supply, **4**, 56–60 (2018)
8. Braga, J. K., & Varesche, M. B. A. Commercial Laundry Water Characterisation. American Journal of Analytical Chemistry, **05**(01), 8–16. https://doi.org/10.4236/ajac.2014.51002 (2014)
9. Sheth, K. N., Patel Mittal., & Desai, M. D. A study on characterization & treatment of Laundry effluent. International Journal for Innovative Research in Science & Technology Vol **4** (1): 50–55 (2017)
10. Ardiyanto, Putra., and Yuantari, Maria G.C. Analisis limbah Laundry informal dengan tingkat pencemaran lingkungan di Kelurahan Muktiharjo Kidul, Kecamatan Pedurungan Semarang. Jukung Jurnal Teknik Lingkungan vol **2** (1): 1–12 (2016)
11. Regulation of the Special Region of Yogyakarta No. 7 of 2006 on Wastewater Quality Standards (2006)
12. Regulation of the Governor of the Special Region of Yogyakarta No. 20 of 2008 on Water Quality Standards (2008)
13. Regulation of the Governor of the Special Region of Yogyakarta No. 22 of 2007 on Stream Order Determination of the Special Region of Yogyakarta (2007)
14. Watiniasih, N. L., Purnama I. G. H., Padmanabha, G., Merdana, I.M., Antara I. N. G. Managing Laundry wastewater. IOP Conference Series: Earth and Environmental Science 248 012084. doi:10.1088/1755-1315/248/1/012084 (2018)
15. Widyasari, Titiek. Beban pencemaran sumber limbah di Sungai Code. Jurnal Teknik Sipil Vol **5** (2): 93–169 (2009)

Assessing the spatial distribution of urban soil lead contamination in Yogyakarta City, Indonesia

Andria Tri Sekarningsih[1], *Wawan* Budianta[1,*] *I Wayan* Warmada[1], *Hirofumi* Hinode[2]

[1]Department of Geological Engineering, Universitas Gadjah Mada, Yogyakarta, Indonesia

[2]Tokyo Institute of Technology, Japan

Abstract. Yogyakarta is currently a developing city in Indonesia, and urbanization is a very relevant issue. One of the impact of urbanization was soil pollution which becomes an important and interesting environmental concern. This research aimed investigate lead (Pb) pollution level distribution using spatial analysis on lead (Pb) distribution in the soil at Yogyakarta city and its surroundings. The determination of the soil sampling location point was carried out using the gridding method, and the soil sampling was obtained at a depth of 10 cm. The concentration of lead (Pb) in the soil samples was measured using ICP-AES. The spatial analysis was carried out by making a level of distribution map of the lead (Pb) and then overlaying it with the data of residential, traffic, and industrial densities. These research results indicated that anthropogenic factors influenced the lead (Pb) concentration in the soil's soil more than natural factors. The anthropogenic factors of settlement and traffic densities were thought to contribute most to the lead concentration in the soil at the research site. The industrial presence factor was less likely to contribute to the high Pb concentration in the soil at the study area.

1 Introduction

Heavy metals in soil, especially lead (Pb), can potentially have severe impacts on the environment and human health [1, 2]. Until today, Pb has still been considered a pollutant that can cause soil and environmental pollution [3]. Yogyakarta is one of the currently developing cities because it is one of the world's tourist destinations; thus, urbanization is a very relevant issue [4]. An implication of urbanization in Yogyakarta is air pollution by lead (Pb), which is 0.3656 – 75.7 mg/m^3, which has passed the allowable threshold of 2 mg/m^3 [5]. Another more important thing is the decreasing in the surface soil quality in Yogyakarta City and its surroundings. Research conducted by Budianta (2012) [6] showed that the Pb content on the soil surface in urban and suburban sites of Yogyakarta varied from 25 to 95.50 mg/kg, with an average value of 64 mg/kg. The Pb concentration in the city center has about 62 mg/kg, five times that in the suburbs (14 mg/kg). Based on this description, research studying the surface soil contaminated by lead (Pb) in Yogyakarta City for 2021 was then required, as well as to determine the contamination level development by conducting a more comprehensive spatial analysis of the lead (Pb) distribution in the soil at the research site. The study area was located in Yogyakarta city, Indonesia and its surroundings area as shown in Figure 1.

2 Materials and methods

The soil sampling location was determined using the gridding method, where the research site was on a grid with each outer grid of 1.5 km x 1.5 km (Figure 2). The sampling point was attempted to be in the middle of the grid, representing each grid site. The soil sampling was carried out by hand drill at a depth of 10 cm, assuming that it meant the topsoil layer. The determination of the sampling location point was done using GPS. Also, the soil sampling was carried out by considering avoiding locations close to pollutant sources, such as trash bins or incineration sites, to prevent anomalies in the lead (Pb) content that was too high; therefore, a more general overview of lead (Pb) content in the site was obtained. For the lead (Pb) analysis, the soil sample taken in the field was put into a sample plastic and was prepared for laboratory analysis. All soil samples were then dried at room temperature for approximately one week. Then they were smoothed or homogenized using an agate mortar and pestle to obtain a fine sample (<2 mm) referring to Argyraki and Kelepertzis (2014) [1]. The lead (Pb) analysis in the soil sample was carried out using ICP-AES (Inductively Coupled Plasma – Atomic Emission Spectrometry) at the Tokyo Institute of Technology, Japan.

* Corresponding author: wbudianta@ugm.ac.id

https://doi.org/10.1051/e3sconf/202132503001

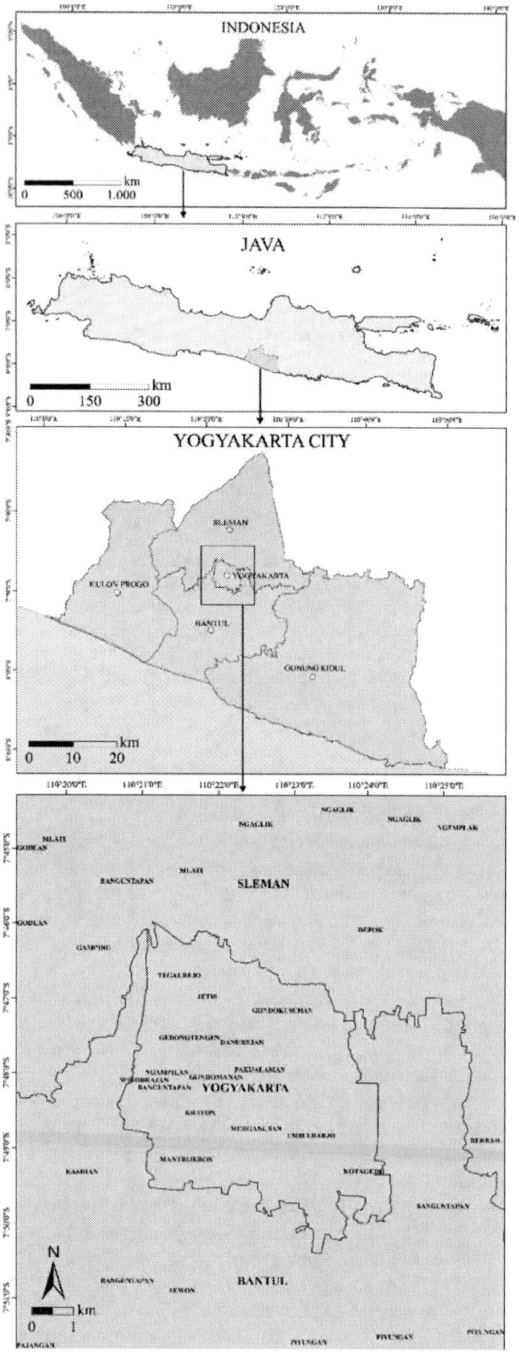

Fig. 1. The study area

The lead (Pb) concentration was measured after the soil sample was dissolved in strong acid (HF-HCl and HNO_3), followed by the dissolution using microwave as descibed by Bettinelli et al, 2000 [7]. The Pb distribution analysis in the soil sample was carried out spatially. The lateral distribution was carried out based on the lead (Pb) content in the sample from all sampling points by making a level of distribution map of the lead (Pb). The results of the lateral distribution analysis associated with several parameters –mainly anthropogenic factors such as settlement, traffic, and industry densities— were expected to provide information about the factors that affected the lead (Pb) distribution in the soil at the research site.

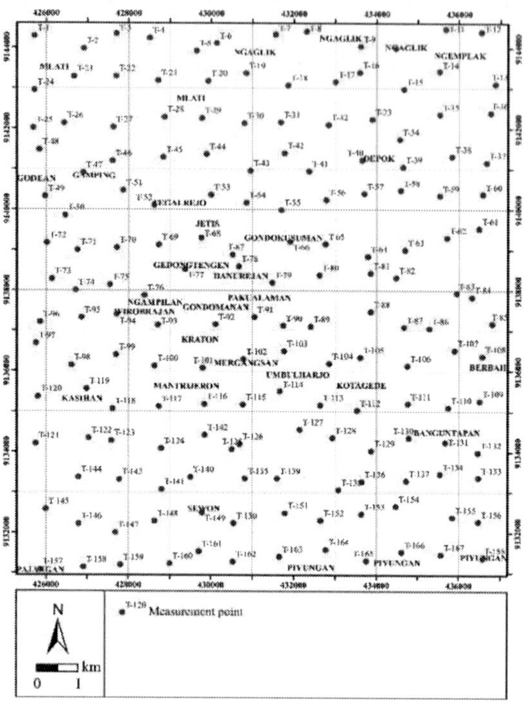

Fig. 2. The sampling point location

3 Result and Discussion

The geological conditions of the Yogyakarta urban area are mostly composed of volcanic materials from young Merapi volcano [8]. These materials were from the sedimentation of the lava flow, pyroclastic flow, and clastic volcanic sedimentation during the Quaternary (Pleistocene-Holocene) period [8]. Physically, the volcanic material that covers the urban area of Yogyakarta (Yogyakarta City and parts of Sleman and Bantul regencies) has not been consolidated and is loose [9]. The lithology of this young Merapi volcanic deposit is composed of fine-medium and gravel sand, where the fine-medium sand is composed of alternating fine and medium sand, and there are some insertions of silt and clay [9]. The outcrop of volcanic material can be well seen on the river slopes that pass through Yogyakarta. The rock-forming materials in Yogyakarta City are derived from volcanic material from Merapi volcano, which is limestone-alkaline, basaltic, andesite with high K-content with a SiO_2 composition ranging from 52 – 57%. The mineralogy of the Merapi volcano eruption material is plagioclase, clinopyroxene (augite-salite), brown hornblende, olivine, and hypersthenes [10]. In addition to these volcanic materials, volcanic ash also plays a role in

https://doi.org/10.1051/e3sconf/202132503001

forming soil and rock constituent materials at the research site, containing silica minerals. The most common elements are sulfate, chloride, sodium, calcium, and magnesium fluoride [10]. Most of the volcanic material has undergone weathering, so that several places in the urban area of Yogyakarta have formed quite thick soil above the urban periphery of Yogyakarta. Spatially, the distribution pattern of lead (Pb) in the surface soil at the study area can be seen in Figure 3.

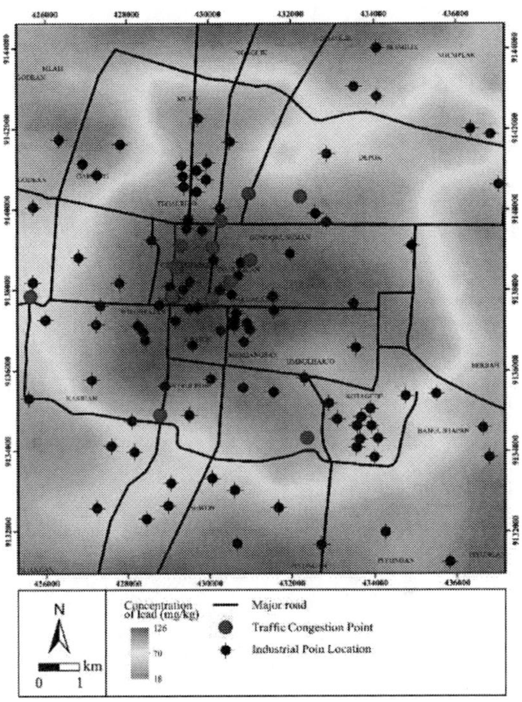

Fig. 3. Spatial distribution of lead (Pb)

The Pb concentration at the center of the research site, the city center, commonly had a higher Pb concentration than the surrounding area. Figure 3 is an overlay map of the distribution of lead (Pb) concentration in the soil at the research site and the map of roads, residential densities, and industrial locations. The spatial analysis was carried out by observing the lead (Pb) distribution pattern and then associated with several factors considered influential, both natural and anthropogenic. In detail, it will be discussed one by one in the following description. Natural or geogenic factors in the presence of lead (Pb) in the soil can occur due to the enrichment process of the original rock that has undergone weathering at the research site. As explained in the previous description, most of the study area consist of volcanic material from Merapi volcano, which was not consolidated and a loose material. The rock constituent materials in Yogyakarta City come from volcanic materials of Merapi volcano, which are limestone-alkaline, basaltic, and andesite with high K-content. Elements in volcanic soils generally are 1.8-5.9% Al, 1-2.4% Mg, 2.6-2.8% Si, and 1.4-9.3% Fe [11]. From this description, it can be interpreted that the lead (Pb) content in volcanic rocks naturally has a low value [12]. Generally, the increase in the lead (Pb) concentration in the soil at the study area can be concluded to be influenced by human activities. The second factor influencing the distribution and increase of lead (Pb) concentration in the soil at the study area was the residential and population factors. One of the materials that humans often use as a source of the lead (Pb) is the use of paint. Things that need to be considered for residential factors are house paint, the pattern of settlement density distribution, age of settlement, condition, type of the house, and housing distance to traffic density. The population density of the study area revealed that the settlement density was generally concentrated in the center of the map, or the center of Yogyakarta City, which generally had an older age of settlement compared to their surroundings. The next factor was vehicle emissions that might increase the lead (Pb) concentration in the soil at the study area. Figure 3 shows that the main roads and congestion traffic points were more dominant in the city center, where vehicle emission was more prevalent in this area. This vehicle emission was interpreted to have contributed significantly to the high concentration of lead (Pb) at this area. The distance from the main road, which was the source of vehicle emissions, and the high level of settlement in this area made two factors that simultaneously contributed significantly to the high concentration of Pb in the soil at study area. Several study shows that the vehicle emission has significant contribution on the lead (Pb) contaminated soil in urban area [13, 14]. Another factor that was interpreted to be contributing was other human activities, such as industry and waste disposal. The map of Figure 3 shows that the city center dominated the presence of the industry at the study area despite there were some industrial locations outside the city at the study area. However, Figure 3 also shows several points of the industrial presence outside the city that less contributed to the high concentration of Pb in the soil at that location.

4 Conclusion

The spatial analysis results of overlaying several maps can be concluded that the anthropogenic factors influenced the lead (Pb) concentration in the soil at the study area more than the natural factors. The geological and soil conditions at the study area, which were composed of young volcanic material, would contribute to low concentrations of lead (Pb) in the soil due to the low lead (Pb) content in host rocks at the study area. The anthropogenic factors such as settlement and traffic densities were interpreted to contribute most to the lead (Pb) concentration in the soil at the study area. The presence of the industrial factor did not seems have contribution to the high concentration of lead (Pb) in the soil at the study area.

Authors thank to Universitas Gadjah Mada for the supporting research funding.

References

1. A. Argyraki and E. Kelepertzis, Sci. Total. Environ. **482** (2014)
2. M. Gaberšek and M. Gosar, J. Geochem. Explor. **187** (2018)
3. M. Andersson, R.T. Ottesen and M. Langedal, Geoderma **156**, 3-4 (2010)
4. W. Pradoto, Intern. J. Arts. Sci. **4**, 27 (2011)
5. Z. Abidin and S. Sunardi, Indones. J. Chem **9**, 3 (2010)
6. W. Budianta, J. South East Asia Appl. Geology. **4**, 2 (2012)
7. M. Bettinelli, G.M. Beone, S. Spezia and C. Baffi, Anal. Chim. Acta. **424**, 2 (2020).
8. W. Rahardjo, H.M.D Rosidi and Sukandarrumidi (1995)
9. S.M MacDonald, Greater Yogyakarta Groundwater Resources Study (1984)
10. G. Camus, A. Gourgaud, P.C. Mossand-Berthommier and P.M. Vincent, J. Volcanol. Geotherm. Res. **100** (2000)
11. Sudaryo and Sucipto, *Identifikasi dan penentuan logam pada tanah vulkanik di daerah Cangkringan Kabupaten Sleman dengan metode analisis aktivasi neutron cepat*, in Prosiding Seminar Nasional V SDM Teknologi Nuklir, 5 November (2009)
12. H. Meuser, Contaminated urban soils (Springer, London, 2010)
13. I.D. Yesilonis, R.V. Pouyat, and N.K. Neerchal, Environ. Pollution, **156**, 3 (2008)
14. D.S. Manta, M. Angelone, A. Bellanca, R. Neri, and M. Sprovieri, Sci.Total. Environ. **300**, 1-3 (2002)

Characterization of carbonate rocks using seismic wave for tunnel design stability

Hairin Taha[1], *Nurul* Atiqah Dzulkifli[2], *Rohayu* Che Omar[1*], *Ratih* Fitria Putri[3*] *Rasyikin* Roslan[1] and *Intan* Nor Zuliana Baharuddin[1]

[1]Institute of Energy Infrastructure, Universiti Tenaga Nasional, 43000 Kajang, Malaysia
[2]College of Graduate Studies, Universiti Tenaga Nasional, Kajang, 43000, Malaysia
[3]Department of Environmental Geography, Faculty of Geography, Gadjah Mada University, Yogyakarta, 55281, Indonesia

Abstract. The design and construction of a tunnel depends on the mechanical properties of the rock mass around the tunnel. Seismic method can be used to characterize the dynamic properties of rocks. The technique is mostly conducted in geophysical surveys and geotechnical investigations. The method utilizes reflected sound waves that can be used to describe the dynamic properties of rocks. Physical properties of carbonate rocks such as water content, density, hardness, permeability, porosity, wave velocity and abrasivity can be assessed and estimated using P-wave velocity. One of the important characteristics in rock is its ability to remain stable. In this research, seismic refraction survey was applied to measure the strength of carbonate rocks for tunnel stability design. The findings revealed that the regression between the primary velocity and the uniaxial compressive strength R^2 was 0.8592, indicating that the rock was firm and solid. Observation by physical visual test showed that the rock samples with yellowish-grey and light grey colours were categorized in the weathering grade II and III, respectively. The results have concluded that the rocks in the proposed area met the full requirements for tunnelling construction.

1 Introduction

The materials composed in an intact rock can be quantified and characterized through physical test or laboratory test. In general, rock materials can be categorized into two classification, physical and mechanical properties. The properties of rock materials are almost the same to the physical content of rock that form minerals and variety of mineral bonding. The physical properties of rock materials include water content, density, hardness, permeability, porosity, wave velocity and abrasivity [1, 2] whereas the strength of rock is classified under mechanical properties. Before initiating any construction, a preliminary ground investigation has to be conducted to get information on the soil profile and condition in the proposed area so that safety measures and precautions can be exercised during the construction period.

In this case, various ground investigations were conducted along the MRT lines during the early phase of construction that consisted of 400 boreholes and geophysical surveys. During the preliminary soil sampling, it was noted that the construction area was formed from the merging of Kenny Hills sedimentary rock and KL Limestone formation which was composed of highly endured karst [3]. Nevertheless, the rocks at the KL Limestone region consist of uncertain formation of karstic features such as steeply inclined bedrock, cavities, pinnacles and floater which are more challenging that needed extra inspection and procedures [4].

This zone has an outrageous karstic feature containing marbles and cavities which needs to be considered and evaluated by the specialists or engineers in term of ground condition and the rock properties before doing underground construction. In this case, several tests were required to analyse the rock strength before embarking on any construction to determine if the rock condition and its physical properties has any effect on the stability of the tunnel that might create any negative impact or outcome.

This paper describes an *in situ* study and laboratory testing of rock strength for tunnel stability design by characterizing the properties of rock samples in the Kuala Lumpur Limestone formation. The parameters of the rock mass were studied and analysed to determine if the rock mass will directly affect the wall stability of tunnel during the construction. The data was obtained by *in-situ* testing and performing laboratory test to determine the strength of intact rock. Geophysical investigation was carried in the field while the borehole samples were tested in the laboratory. Several testing including destructive and undestructive tests were conducted to test the rock strength.

* Corresponding author: rohayu@uniten.edu.my, ratihfitria.putri@ugm.ac.id

2 Methodology

2.1 Experimental site

In this work, the experimental site was at Bukit Bintang station which was located between two rock formations, KL Limestone Formation and Kenny Hills formation that largely consists of karstic limestone [5]. It was observed that the limestone mass was covered with subsurface cavities and underground chambers that can be potentially dangerous for tunnelling excavation and construction. To ensure the safety of the tunnelling process, safety measures and extra precaution need to be considered to avoid unfortunate incidents. Figure 1 and Figure 2 show the study area along the Kenny Hills and KL Limestone formation where the proposed underground tunnel would be constructed.

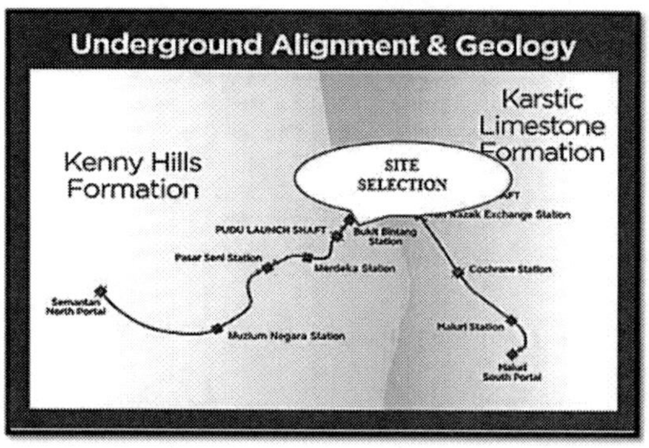

Fig. 1. Study location

2.1.1 Geophysical assessment

Geophysical survey is commonly used to investigate the underground structures of tunnelling prior to any construction works to study the underground subsurface and the rock properties. Seismic refraction and resistivity were carried out for cross-hole tomography analysis to investigate the rock mass [6]. The seismic refraction survey is a widely used method in geophysical technique to explore the subsurface condition and characteristics by utilizing the refraction of seismic waves whereas the electrical resistivity technique is used to characterize the sub-surface materials based on their electrical properties. In this work, geophysical survey was conducted along Line 1 to collect information on the physical properties of the rocks. Figure 3 shows the flowchart of the research process.

Fig. 2. Geophysical investigation on Line 1.

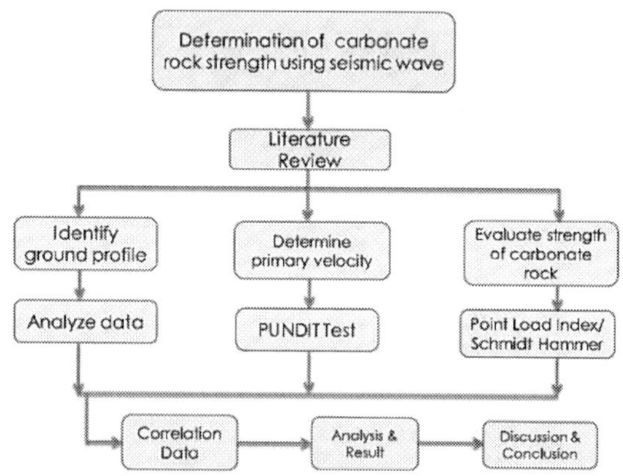

Fig. 3. Flow chart of the study.

2.1.2 Sample preparation

Figure 4 shows the samples of rock coring (limestone) that belong to two different boreholes which were CUB 040 and CUB 073 that were acquired from the site. Each of the rock samples were cut approximately 150 mm following ASTM standard using a diamond cutter, as the rock coring diameter was 55mm. There were two intact rocks in box CUB 040 and other 13 samples from box CUB073. Most of the rocks in box CUB 040 had cracks and defects.

https://doi.org/10.1051/e3sconf/202132503002

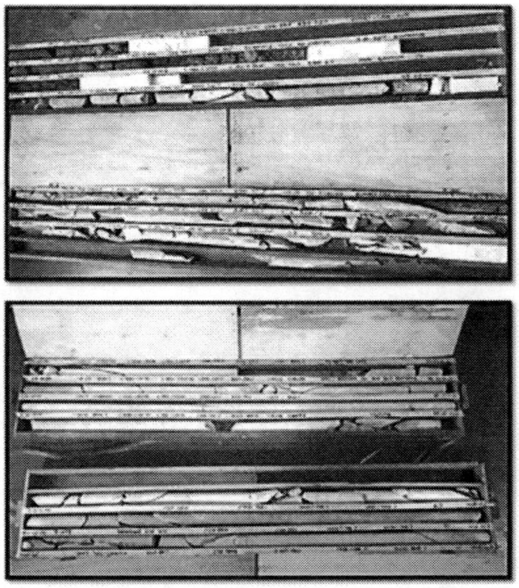

Fig. 4. Rock coring samples from borehole CUB 040 and CUB 073.

2.1.3 Ultrasonic Wave Velocity

Ultrasonic pulse velocity (UPV) is a non-destructive method that measure the primary velocity of an electronic pulse that pass through the rock (Figure 5). It is also known as PUNDIT test. The test was run by putting the connector between the transducer and the receiver on contact to achieve zero reading. When it was stabilized, the aluminium cylinder was switched with the real specimen samples. Both transducer - receiver was positioned at the outer surface of the rock samples. Data was recorded when the reading in the meter shows a constant value. The primary velocity (Vp) was obtained by dividing the length of samples with average transit time.

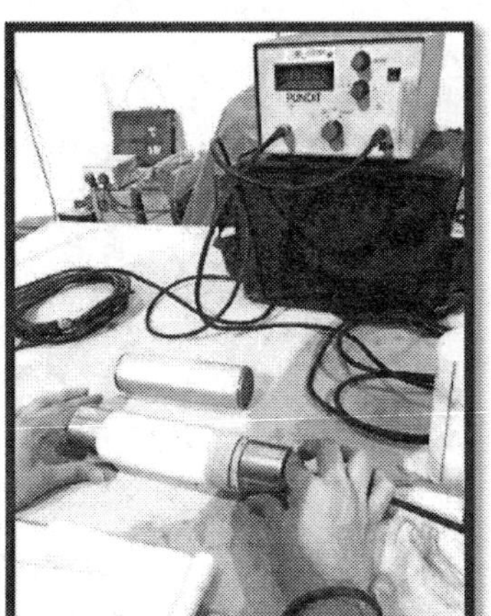

Fig. 5 UPV test.

2.1.4 Schmidt hammer Test

The Schmidth hammer test or rebound hammer was used to measure the uniaxial compressive strength (UCS) of the rock samples (Figure 6). The test was conducted according to ASTM standard instead of ISRM standard because the core diameter of the specimen was less than 55 mm [8]. In this test, the hammer was firmly held perpendicular to the length of the specimens, at about 90° normal. The surface contact is represented by the length of sample. The data was calculated by basic mathematical statistic due to reading variabilities of each sample. The average reading of rebound hammer, was taken approximately 10 times around the surface contact of each sample.

Fig. 6. Schmidt hammer test

2.1.5 Point Load test

The uniaxial compressive strength (UCS) of the rock samples was measured by Point Load test which is a destructive test. Basically, the specimen will experience tension before it breaks when the maximum limit is reached. In this work, a diametrical shape representing the tunnel boring machine (TBM) was selected. The rock samples was placed in perpendicular position to the tip of point load as shown in Figure 7, to portray the tension received from both direction of the tunnel . For a measurement of diametrical test, De is equivalent to D that represent the equivalent of core diameter. The corrected size of the point load index Is (50) is defined as Is if the D_e is 50mm [8]. Point Load Strength Index (Is) was measured by dividing the point load test value with the diameter core of the samples [9]. The UCS value is calculated by multiplying with the conversion factor as shown below;

$$Is(50) = \frac{P}{De^2} \quad (1)$$

Where ;
I_s = Corrected point load strength index (MPa or psi)
P = Load failure, in MN or lbf (max pressure x jack piston area)

D_e = Equivalent core diameter (meters or inches)
(D_e = D for diametrical test).

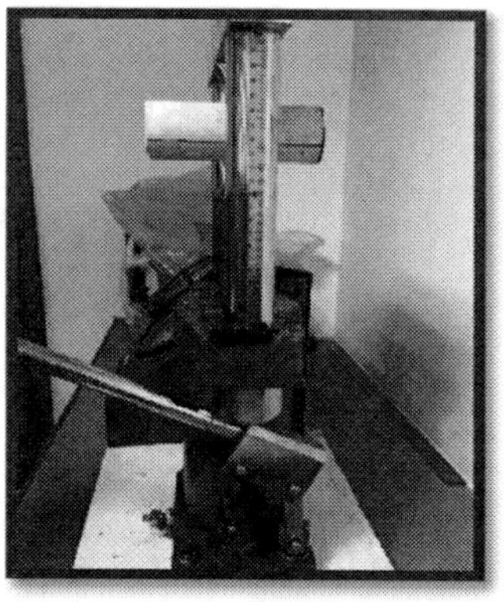

Fig. 7. The point load index test from the side view.

3 Results and discussion

3.1 Weathering grade

There are two types of rock weathering of rocks which are mechanical weathering and chemical weathering [10]. The weathering characteristics of the rock samples were assessed based on the physical appearances and colours. It was observed that most of the rock samples were yellowish- grey. According to the standard reference, the weathering index for the 14 rock samples were graded between Grade II and III. Figure 8 shows the two rock samples with different colours from box CUB 073, which indicated different grade of weathering. Sample F2 was light grey in colour indicating weathering grade II while B5 was yellowish grey in colour, indicating grade III of weathering.

Fig. 8. Two different rock coring samples.

3.1.1 Schmidt hammer test

Table 1 showed the minimum UCS value which was 25.49 KPa and the maximum value of UCS was 42.84 KPa. According to the standard reference [11], Schmidt hammer index for limestone ranged around 35- 51. But the UCS readings for some samples were slightly low due to the weathering factors.

Table 1. Results of Schmidt Hammer Test

Sample	Average value of Schmidt Hammer, N	UCS = 4.24 e ^ 0.059 N	Mode	Mean	Median	Min UCS	Max UCS
1	32.4	28.68					
2	35.3	34.03					
3	31.4	27.04					
4	35.3	34.03					
5	33.1	29.88					
6	33.8	31.15					
7	39.2	42.84					
8	39.2	42.84	39.2	33.8	34.6	25.49	42.84
9	38.1	40.14					
10	32.6	29.02					
11	30.9	26.25					
12	30.3	25.34					
13	39.2	42.84					
14	30.4	25.49					
15	37.9	39.67					

3.1.2 Point load test

The point load test showed that the rock sample, CUB 073- B5 had the highest value force at failure which was 6.69 kN that gave the highest result of UCS value. The condition of the rock samples CUB 073-B5 after ruptured is shown in Figure 9.

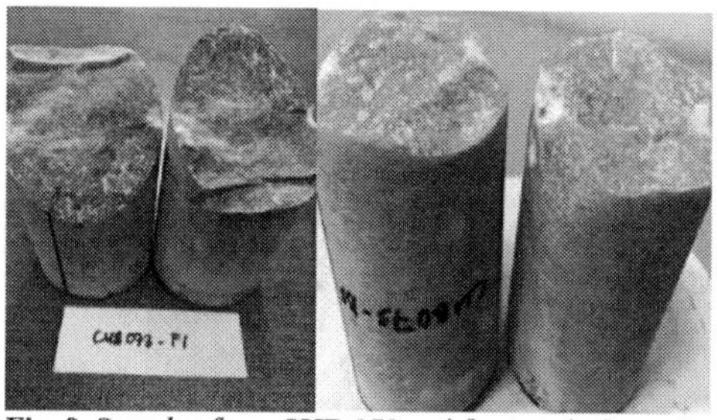

Fig. 9. Samples from CUB 073 and fractured after point load test.

3.1.3 Geological survey to rock testing correlation

For the analysis, geophysical surveys and borehole data were correlated with the laboratory results of rock testing. Seismic refraction and electrical resistivity tomography analysis were conducted along Line 1. The data obtained from the experimental site showed that the surface wave velocity, Vs below 400 m/s was inferred as low density medium whereas for Vs greater than 400 m/s is considered high density medium. Therefore, the Vs with the approximate value 400 m/s was interpreted as the interface between soil and rock layers. The result of MASW was further reconfirmed using the EI survey that was carried out at the same line. Figure 10 shows the comparison of MASW and EI colour range.

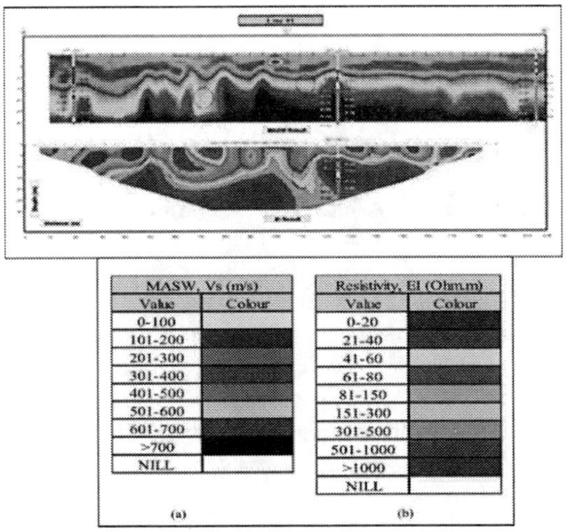

Fig.10. Seismic wave analysis.

3.1.4 Borehole analysis

The rock was identified at depth of 13.50m at borehole BPU- BH1. The data of geophysical survey data showed that the carbonate rock was slightly weathered from 30 m to 36 m depth which only can be penetrated by surface wave velocity at this level. Meanwhile, analysis of surface wave velocity and electrical resistivity and were done at borehole BPU – BH2. The results indicated that the rock was a little weathered at 17 m depth to 36 m depth. The electrical resistivity showed that the carbonate rock possessed the highest wave propagation at 9 m depth to 13.50 m depth. It was concluded that the rock sample at this borehole was intact without any fracture. As for borehole BPU-BH4 with rock from 10 m depth, the velocities of the seismic refraction was measured from 300 m/s up to 700 m/s down to borehole depth. From this analysis, it can be deduced that, higher seismic value as it reach deeper into hard soil stratum.

3.1.5 Graph Correlation between tests

The correlation from the experimental results was obtained based on all collected data from the testing is tabulated in Table 2. The graph in Figure 11 shows that the regression between destructive and undestructive test was about 0.7648 meanwhile in Figure 12, the graph demonstrated higher velocity reading of rock as the density increased. Figure 14 shows the graph between the density of rocks and Schmidth hammer test. It shows that the regression between two is R^2 = 0.8468 and Figure 15 show that the regression between the primary velocity and the uniaxial compressive strength gives the value of R2 =0.8592, indicating that the rock is firm and intact with higher velocity.

Table 2. Summary of overall results

	Vp (km/s)	PLT (MPa)	SH (kN)	Density (kg/m³)	UCS from PLT (MPa)
1	3.505	0.970	30.3	2.710	15.5
2	4.319	0.970	30.4	2.753	15.5
3	4.337	3.685	30.9	2.773	59.0
4	4.557	3.685	31.4	2.781	59.0
5	4.594	3.685	32.4	2.787	59.0
6	4.62	4.170	32.6	2.789	66.7
7	4.858	4.267	33.1	2.804	68.3
8	5.088	4.461	33.8	2.815	71.40
9	5.201	4.849	35.3	2.822	77.40
10	5.302	5.237	35.3	2.828	83.80
11	5.432	5.334	37.9	2.83	85.30
12	5.596	5.528	38.1	2.842	88.40
13	5.771	5.819	39.2	2.845	93.10
14	5.845	6.750	39.2	2.857	108.00
15	6.62		39.2	2.862	

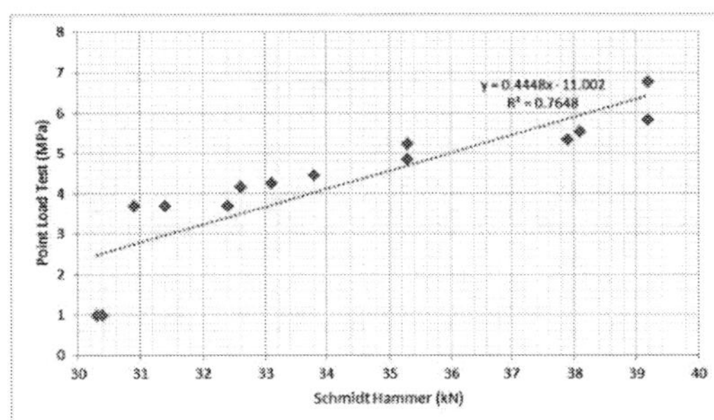

Fig. 11. Graph of Schmidt hammer vs point load index test.

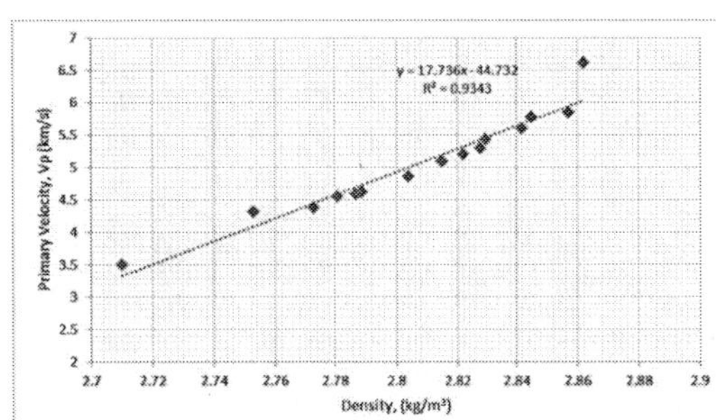

Fig. 12. Graph of density vs primary velocity.

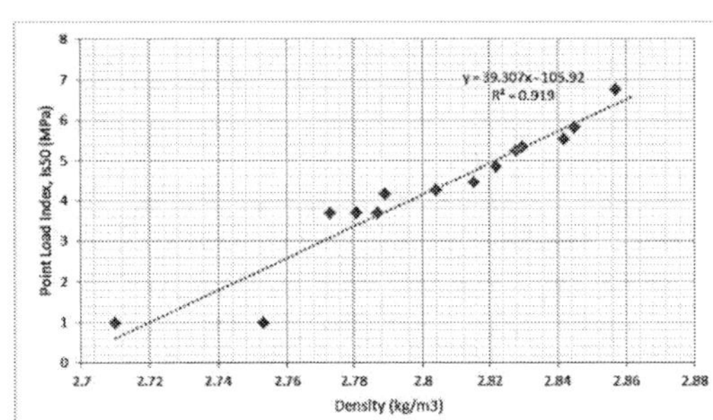

Fig. 13. Graph of density vs point load index test.

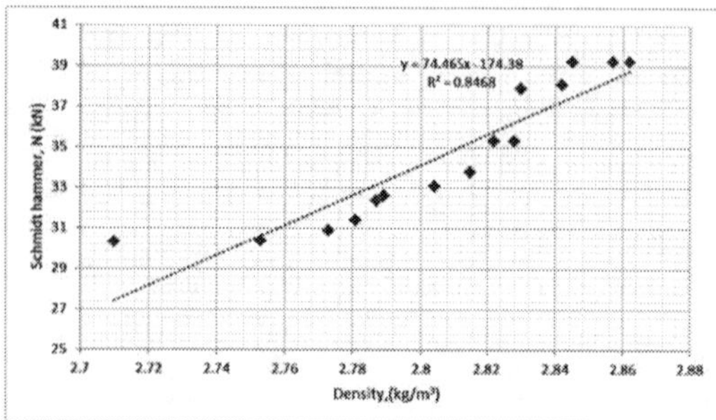

Fig. 14. Graph of density vs Schmidt hammer.

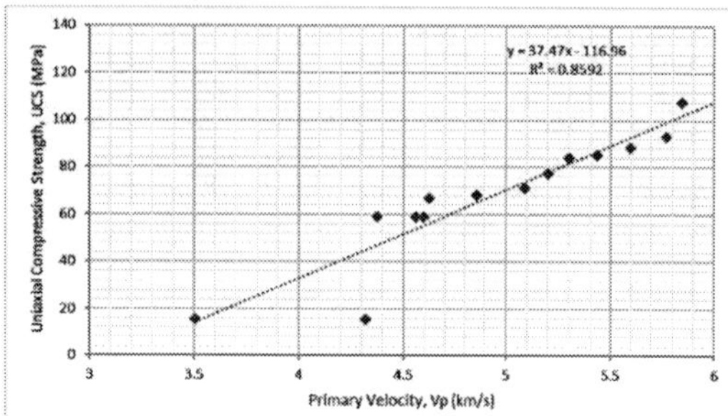

Fig. 15. Primary velocity, Vp vs UCS of Point Load Index

4 Conclusion

This work investigated the subsurface tomography of a proposed site for tunnel construction by collecting information on borehole data and seismic refraction survey. The results showed that most of the subsurface rocks were made up of limestone. Based on the physical observation and visual test, most of the rock samples with yellowish-grey colour were classified in the weathering grade II and the rocks with light grey colour rocks were classified in class III. In conclusion, the data collected from the UCS of the rock samples can be used as a reference for the design of rock tunnelling project and construction.

The author would like to thank UNITEN's Innovation & Research Management Centre (iRMC) for funding this study through research grant 20200102CRGJ. The author would like to thank the Department of Environmental Geography, Faculty of Geography, for their assistance in interpretation, as well as Gamuda Bhd for providing samples for this study.

References

1. A. A. El Aal, Egyptian Journal of Petroleum **26**, 3:565–578 (2017)
2. L. Zhang, Engineering properties of rocks, (The American Institute of Architects (AIA), Butterworth-Heinemann, 2016)
3. H. Zabidi, N. L. Khalil, N. E. Q. M. Anua, S. Suri, Journal of Physics: Conference Series **1242**, 1:012044 (2019)
4. H. Zabidi, M. H. De Freitas, Engineering Geology **117**, 3–4:159–169 (2011)
5. G. T. Lai, A. S. Serasa, A. G. Rafaek, N. Simon, A. Hussin, T. R. T. Mohamed, L.K. Ern, Electronic Journal of Geothechnical Engineering, **22**:1681–1692 (2017)
6. D. Stead, A. Wolter, Journal of Structural Geology, 7: 1–23 (2015)
7. T. Yilmaz, B. Ercikdi, Nondestructive Testing and Evaluation **31**, 3:247–266 (2016)
8. A. Aydin, International Journal of Rock Mechaics and Mining Series **46**, 3:627–634 (2009)
9. M. Akram, M. A. Bakar, Pakistan Journal of Engineering and Applied Sciences (2016)
10. G. Vlastelica, P. Miščević, N. Pavić, Gradevinar, **68**, 12:955–965 (2016)
11. E. Hoek, E. T. Brown, J Geotech. Engng Div, 16–28 (1980)

Effects of Liquid Wastes from Tofu Industries on Water Quality in Parangan River, Magelang District, Central Java-Indonesia

Wiwik Widyaningrum[1] and *Margaretha* Widyastuti[1*]

[1]Department of Environmental Geography, Faculty of Geography, Universitas Gadjah Mada, 55281 Yogyakarta Special Region Indonesia

Abstract. Continued entries of liquid wastes generated by tofu industries into nearby rivers might deteriorate the water quality if disposed of without prior treatments. The research objectives were to determine the liquid waste quality from tofu productions near Parangan River (Magelang, Indonesia), assess the river water quality and investigate the effects of the liquid waste on pollution load and water quality status. With this information, the research can help determine pollutant load allocation in the river and provide the basis for formulating water quality management policies if it is not contaminated or pollution controls if otherwise. This research employed purposive sampling, direct measurements in the field, laboratory tests, descriptive-comparative analysis, mass balance and pollution index method. The results showed that the pH, temperature, COD, BOD, TSS and ammonia levels of the liquid wastes had surpassed their maximum permissible levels. Similarly, the river's COD, BOD and ammonia contents did not meet the water quality standards. With BOD, a pollution load indicator, exceeding the total maximum daily load, the water quality statuses at monitoring points S2, S3 and S4 were categorised into slightly polluted. It means that the liquid waste generated by the tofu industries has detrimental effects on the water quality of the Parangan River.

1 Introduction

Water is a natural resource that is utilised vastly to meet the basic needs of living organisms worldwide. Without good management, damages to sources of extractable water like decreased quality are inevitable. Water quality measures the suitability of a water body for a particular designated use based on physical, chemical and biological parameters [1]. In this context, pollution occurs when parameter values or classes do not meet the applicable water quality standards. Quality, among other factors, determines if and to what extent a source can sustain its function and intended use.

Industrial activity is among the crucial components in meeting the economic needs of a community [2]. On the other hand, it is also one of the leading causes of water pollution due to the waste produced. In Magelang District, Indonesia[1], tofu industries are small-scale enterprises that have yet to incorporate liquid waste treatment prior to disposal into production costs. This liquid waste contains high pollutants and suspended and dissolved solids that experience physical, chemical and biological changes [3]. With these characteristics, it will substantially reduce the water quality if discarded directly into the river and lead to decreased productivity, environmental capacity and carrying capacity of the river water [23].

Candimulyo and Tegalrejo Sub-districts are the centres of tofu industries in the Magelang District. The factories have formed a linear pattern following the Parangan River channel, which somehow facilitates direct waste disposal to the water body. Each has a pipeline or an outlet that drains into the nearby channel. As a result, the local community no longer uses or extracts water from the river because the tofu industry has made it the dumpsite for the liquid waste by-products.

This area was selected because the tofu industry did not pre-treat the liquid waste before disposal of to the river. If left unchecked, this practice can reduce the water quality. Further, there has not been a study focusing on this issue on the Parangan River. Therefore, this research was designed to assess the liquid waste quality generated by the tofu industry based on Provincial Regulation of Central Java Number 5 of 2012, analyse the river water quality per the class II water standards in the Government Regulation Number 82 of 2001 and investigate the pollution load and water quality status of the Parangan River.

2 Research Method

The tools and materials used in this study were Avenza maps, sample bottles, water checkers, ranging poles, cooler boxes, measuring tapes, Abney level, camera, research checklist, plastic strings, labels, stationery and distilled water.

The sampling points for liquid waste were selected by purposive sampling that considered the scale of tofu production. The amount of soybean used was divided using standard deviation classification into three scales of production: low, medium and high. The river water sampling time was also determined by purposive sampling that factored in the time of day when the liquid waste was discarded. Here, the water collection referred to the surface water sampling methods issued in SNI 6989-57-2008.

The river water for the quality analysis was sampled at four points: S1, S2, S3 and S4. S1 represents the river segment upstream of the liquid waste outlets. S2 marks the river segment where the liquid waste starts to mix with the river water, while S3 receives a significant input of the liquid waste. S4 is located at the outlet of the

* Corresponding author: mwiwik@ugm.ac.id

Parangan River. The river water sampling locations in this study are shown in a map in Fig. 1.

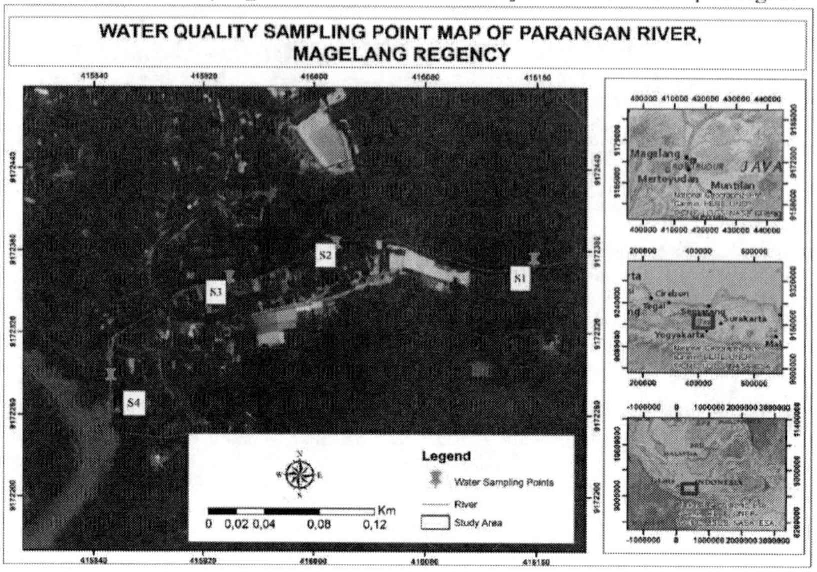

Fig. 1. Map of the four water sampling points in the Parangan River

The water samples were collected twice at different times by taking into account the water travel time in river channels. According to [4,5], water travel time, T (in minutes), is a function of river length, L (m), and flow velocity, V (m/s), or as presented in Eq. 1 below.

$$L=T/60V \tag{1}$$

The discharge was measured directly in the field using the slope area method because the river had rocky beds. The general discharge formula, Q (m^3/s), is flow velocity, V (m/s), times the wet cross-sectional area, A (m^2), or as presented in Eq. 2 below.

$$Q = A \times V \tag{2}$$

Data on the liquid waste and river water qualities, including temperature, colour, odour, pH and electrical conductivity (EC), were measured directly in the field using a water checker. Meanwhile, other parameters like total suspended solids (TSS), chemical oxygen demand (COD), biological oxygen demand (BOD), ammonia (NH_3-N) and dissolved oxygen (DO) were tested in the laboratory (Table 1). Before the laboratory analyses, the samples were preserved by cooling to suppress physical, chemical and biological changes.

Table 1. The water sample analysis methods in the laboratory

Parameters	Test Methods
COD	SNI 6989.2-2019
BOD	SNI 6989.72-2009
TSS	In-house method
NH_3-N	SNI 06-6989.30-2005
DO	APHA 2017, section 4500-OG

The parameter values were analysed descriptively and compared with the quality standards, namely Provincial Regulation Number 5 of 2012 for the liquid waste [6] and Government Regulation Number 82 of 2001 for the river water designated for class II purposes, e.g., water recreation infrastructure/facilities, freshwater aquaculture, livestock farming and agricultural irrigation [7]. The pollution load was calculated with the mass balance method using the formula in Eq. 3. The mean constituent concentration of the combined flow, CR, is a function of the constituent concentration in the i-th flow, Ci, and the discharge of the i-th flow, Qi. CR is also the constituent mass in the i-th flow, Mi, divided by Qi

$$CR = \frac{\sum Ci\ Qi}{\sum Qi} = \frac{\sum Mi}{\sum Qi} \tag{3}$$

The river water quality status was determined based on the Pollution Index (PI) according to the Decree of Minister of Environment Number 115 of 2003 using Eq. 4 below

$$Pij = \sqrt{\frac{\left(Ci/_{Lij}\right)^2 M + \left(Ci/_{Lij}\right)^2 R}{2}} \tag{4}$$

Pij is the pollution index for the water designation j, Ci is the concentration or value of the water quality parameter i, Lij is the concentration or value of the water quality parameter listed in the quality standard for the water designation j, $(Ci/Lij)M$ is the maximum value of Ci/Lij and $(Ci/Lij)R$ is the average value of Ci/Lij.

3 Results and Discussion

3.1. Centre of Tofu Industry Activities

There are 42 home-based small-scale tofu producers on the north and south sides of the Parangan River (Fig. 2). Each operates 29 days a month and processes, on average, 400–500 kg of soybeans per day. This industrial activity has been running for decades.

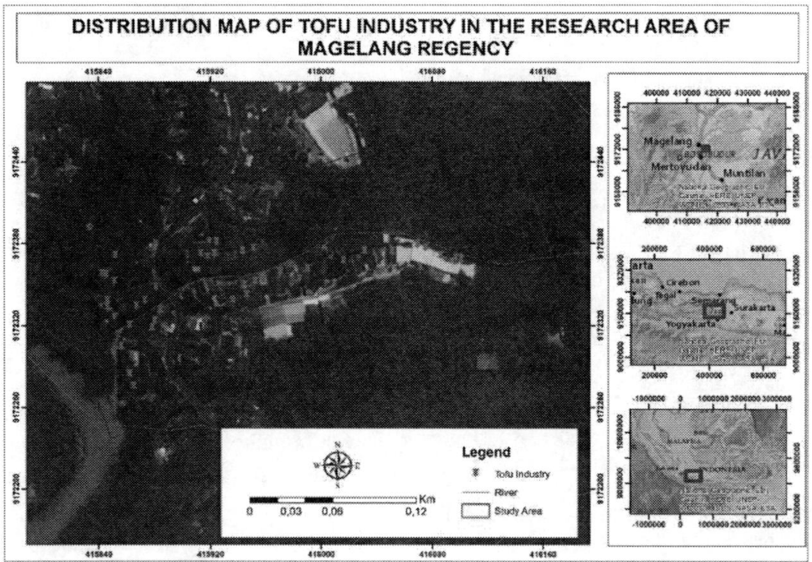

Fig. 2. Distribution map of the tofu producers on the south and north sides of the Parangan River, Magelang District

Only four tofu producers drained their waste by-products into fish ponds as treatment facilities before the treated waste flowed into the Parangan River through the pond outlets. Most others disposed of the liquid waste directly to the river. Based on the interview results, the latter was dominant because there was no space left to build the wastewater treatment plant (WWTP) and the presumption that the liquid waste would not contaminate the river and that after all this water body would continuously flow.

3.2. Liquid Waste Quality

In this research, the liquid waste quality parameters were measured in the field and from the samples collected during three different scales of production, namely low, medium and high. The measured temperatures varied between 43.2 and 47.1°C (Fig. 3) or were above the upper limit value in the quality standard, 38°C [6]. The entry of waste with high temperatures into the water will interfere with the growth of aquatic biota [8].

The pH values were in the range of 3.53–4.67 (Fig. 4) and did not meet the quality standard, namely 6–9 [6]. This acidic waste is believed to have originated in the vinegar used in tofu making, especially the clumping and soaking process [8].

The liquid waste samples were turbid, yellow fluid with white suspension and a pungent smell (Table 2). The strong odour is caused by volatile substances released from the fluid [9].

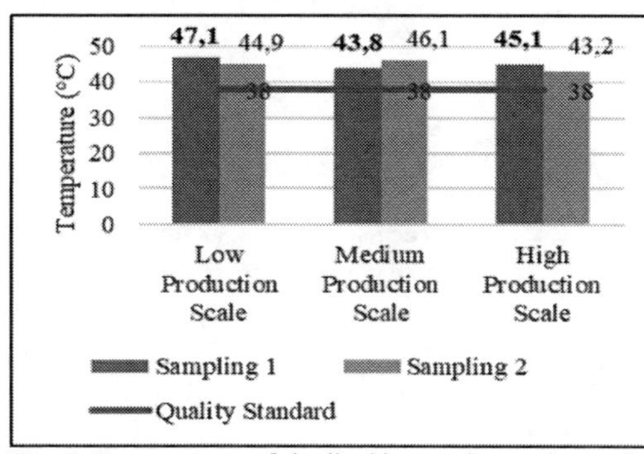

Fig. 3. Temperatures of the liquid waste by-products of the tofu industry during low, medium and high production scales

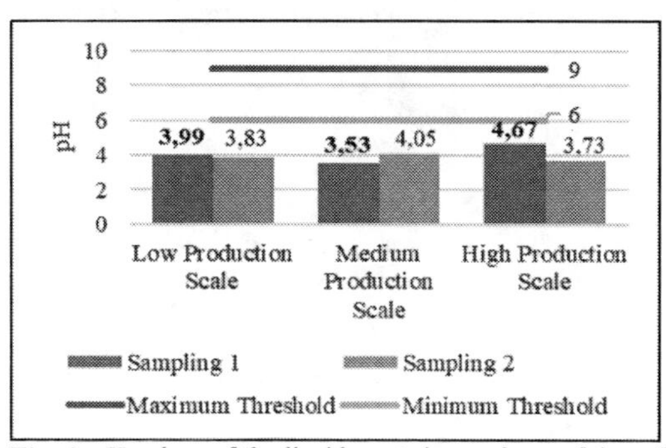

Fig. 4. pH values of the liquid waste by-products of the tofu industry during low, medium and high production scales

https://doi.org/10.1051/e3sconf/202132503003

Table 2. The colours and odours of the liquid waste by-products of the tofu industry during low, medium and high production scales

Production Scales	Sample 1		Sample 2	
	Colours	**Odours**	**Colours**	**Odours**
Low	Turbid, yellow	Pungent	Turbid, yellow	Pungent
Medium	Turbid, yellow	Pungent	Turbid, yellow	Pungent
High	Turbid, yellow	Pungent	Turbid, yellow	Pungent

Electrical conductivity (EC) describes the total concentrations of the dissolved ions in water. The EC values of the samples varied from 1508 to 1726 µs/cm (Fig. 5). The highest EC, 1.726 µs/cm, was detected during high production. High EC occurs when large amounts of elements are dissolved in the water.

In general, the samples had very high Chemical Oxygen Demand (COD), i.e., 7150–8850 mg/L (Fig. 6). This range far exceeded the maximum permissible level of COD in the quality standard, 275 mg/L [6]. In this case high COD is attributable to the presence of high organic compounds that are barely decomposed by microorganisms [10].

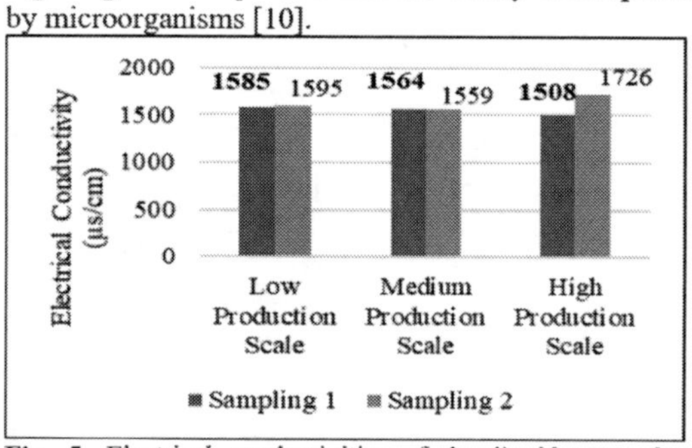

Fig. 5. Electrical conductivities of the liquid waste by-products of the tofu industry during low, medium and high production scales

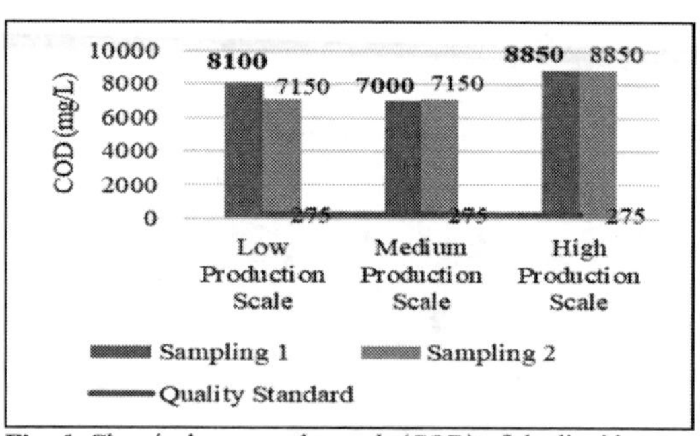

Fig. 6. Chemical oxygen demands (COD) of the liquid waste by-products of the tofu industry during low, medium and high production scales

The samples also had varying levels of Biochemical Oxygen Demand (BOD), i.e., 4880–6810 mg/L (Fig. 7), that did not meet the quality standard, 150 mg/L [6]. The high BOD is caused by the high protein contained in the liquid waste, which is 40–60 % of the raw material used in tofu making [11].

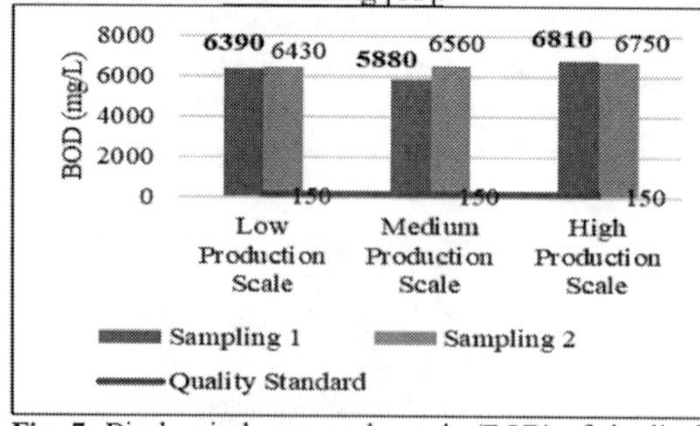

Fig. 7. Biochemical oxygen demands (BOD) of the liquid waste by-products of the tofu industry during low, medium and high production scales

Total Suspended Solid (TSS) in all samples varied between 1100 and 3280 mg/L (Fig. 8), exceeding its upper limit value, 100 mg/L [6]. The high TSS is believed to be the result of solid residues (soybeans) suspended in the liquid waste. TSS closely correlates with turbidity: the higher the TSS, the more turbid the water [1].

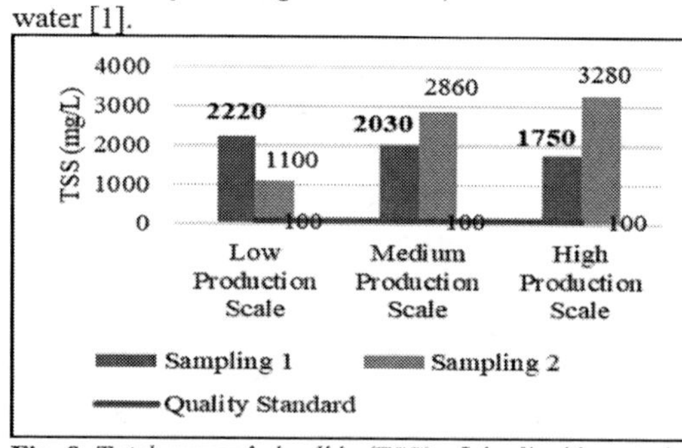

Fig. 8. Total suspended solids (TSS) of the liquid waste by-products of the tofu industry during low, medium and high production scales

The liquid waste generated by the tofu industry also contains ammonia, which comes from the anaerobic digestion of nitrogen [1]. The samples showed 0.1390 to 5.9987 mg/L of NH_3-N (Fig. 9), with the highest NH_3-N found in the first sample of the medium production scale. The high NH_3-N level is responsible for the foul smell of the liquid waste [12].

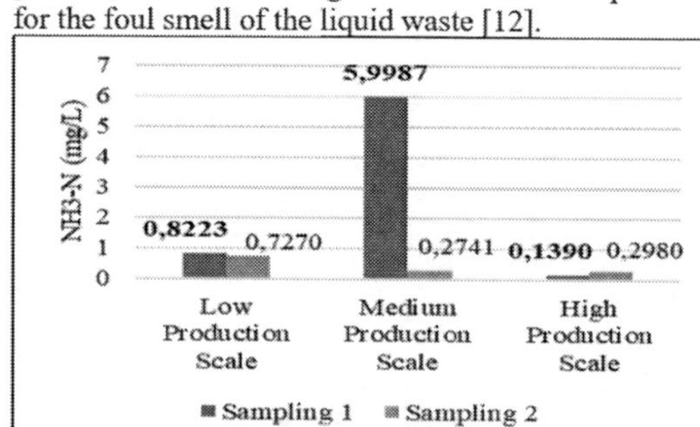

Fig. 9. Total ammonia contents (NH_3-N) of the liquid waste by-products of the tofu industry during low, medium and high production scales

 https://doi.org/10.1051/e3sconf/202132503003

Overall, the temperature, pH, COD, BOD, TSS and NH_3-N of the samples measured and collected during low, medium and high production scales have exceeded their upper limit values in the quality standards [6]. Also, it was found that each parameter varies in levels and values, but this does not depend on the tofu production scale. In other words, the production scale does not affect the quality of the liquid waste by-products. However, suppose any liquid waste with parameter values exceeding their maximum permissible levels is discarded without proper treatments. Then, many environmental problems may arise from increased water turbidity, high organic matter, low DO, pathogenic bacterial growth and unpleasant odour [19].

3.3 River Water Quality

The temperatures were measured directly in the field, and the readings showed a variation between 26.3 and 27.2°C (Fig. 10). Aquatic organisms grow and metabolise optimally at 25–30°C [17]. This means that the river water temperature still meets the class II water quality standards, namely a deviation of 3°C from nature's temperature [7]. The two samples collected at S4 showed the highest temperature.

A water body normally has pH 6–9 and most aquatic biotas prefer pH 7–8.5 [1]. Measurements at four sampling points showed that the river's pH ranged from 6.52 to 7.22 (Fig. 11), which is still within the acceptable pH for class II water and meets the requirements for the growth of fish and aquatic biota [2, 7]. Low pH can reduce dissolved oxygen levels and cause the death of aquatic biota [18].

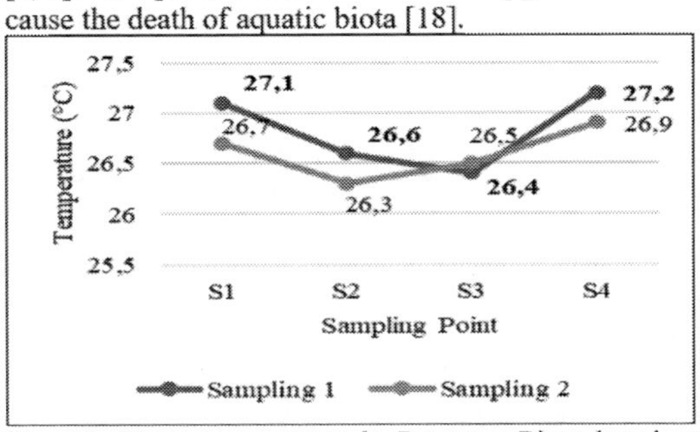

Fig. 10. Water temperatures in Parangan River based on direct measurements at four sampling points (in duplicate)

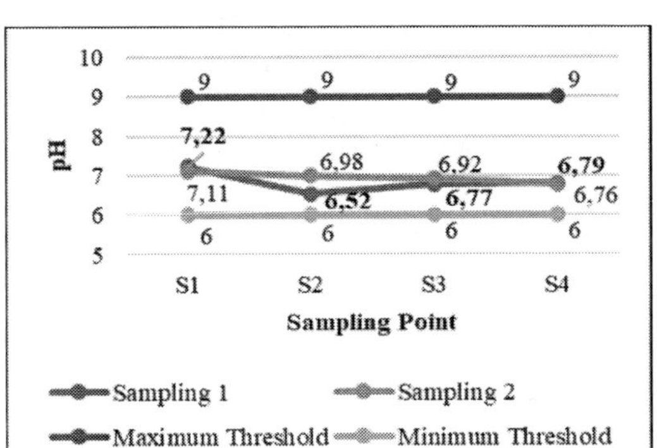

Fig. 11. Water pH in Parangan River based on direct measurements at four sampling points (in duplicate)

Colours and odours are two physical parameters observed directly in the field. At all observation/sampling points, the water in Parangan River had a slightly turbid, grey colour and non-pungent smell (Table 3). The slightly turbid colour is caused by the entry of the liquid waste.

Table 3. Water colours and odours in Parangan River based on direct measurements at four sampling points (in duplicate)

Sampling Points	Sample 1		Sample 2	
	Colours	**Odours**	**Colours**	**Odours**
S1	Clear	Not pungent	Clear	Not pungent
S2	Light brown, slightly turbid	Not pungent	Light brown, slightly turbid	Not pungent
S3	Grey, slightly turbid	Not pungent	Grey, slightly turbid	Not pungent
S4	Grey, slightly turbid	Not pungent	Grey, slightly turbid	Not pungent

In natural waters, the electrical conductivities (EC) are normally in the range of 20–1500 µs/cm [13]. The measurement results showed that the EC values of the river did not vary widely, i.e., 200–233 µs/cm (Fig. 12), and were relatively low because the measurement was conducted in the rainy season. Water bodies in areas with high rainfall intensity are known to have low EC [14]. Government Regulation Number 82 of 2001 does not include the lower or upper limit values for EC in rivers designated for class II purposes [7].

Chemical oxygen demand (COD) describes the amount of oxygen required to oxidise organic matter chemically. In waters organic matters can occur naturally or come from household and industrial activities [1]. The laboratory analysis results showed that the COD levels of the samples were in the range of 14.3–33 mg/L (Fig. 13). Based on class II water quality standards [7], the maximum allowable COD is 25 mg/L. This limit was exceeded at two sampling points, namely S2 and S4, with COD of 33 mg/L and 25.4 mg/L, respectively. Higher COD indicates that the waste contains greater concentrations of organic matter and the river experiences severer pollution [11, 18].

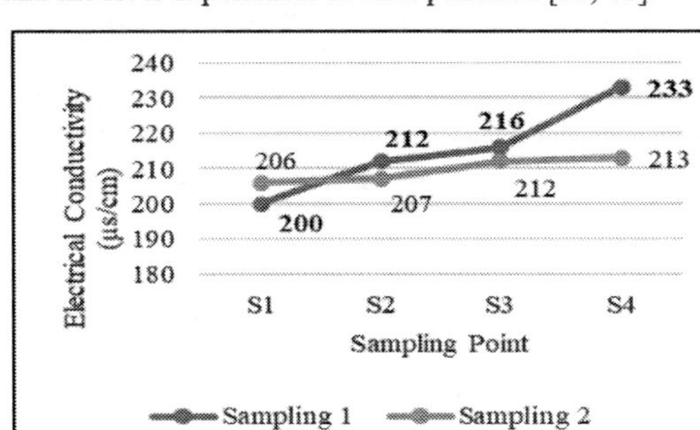

Fig. 12. Electrical conductivities in Parangan River based on direct measurements at four sampling points (in duplicate)

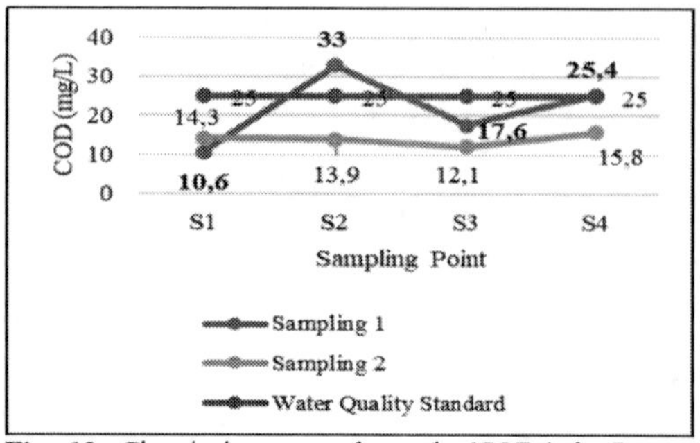

Fig. 13. Chemical oxygen demands (CODs) in Parangan River based on the laboratory analysis of samples collected in duplicate at four sampling points

Biochemical oxygen demands (BODs) describe the amount of oxygen that aerobic microbes require to oxidise organic matter [1]. In Parangan River, the BODs ranged from 1 to 7.8 mg/L (Fig.14). Based on the quality standards [7], all samples have higher BOD than the maximum permissible level, 3 mg/L, except for S1 or upstream of the liquid waste outlet. The high BOD is attributable to the entry of liquid waste from the tofu industry that contains a substantial amount of organic matter. Higher BOD indicates a higher presence of organic matter in a water body [15].

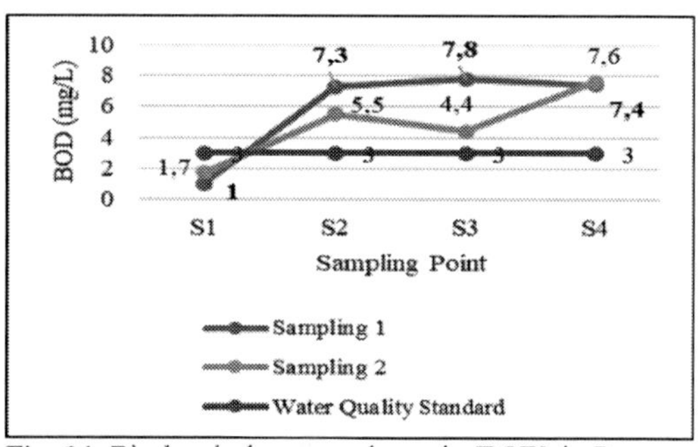

Fig. 14. Biochemical oxygen demands (BOD) in Parangan River based on the laboratory analysis of samples collected in duplicate at four sampling points

Total suspended solids (TSS) measure particles that are larger than >1 µm in diameter, floating or drifting in the water, water-insoluble and are retained if passed through a 0.45µm filter paper [1]. The laboratory analysis results showed TSS in the range of 1–5 mg/L (Fig. 15). Because the permissible maximum TSS is 50 mg/L [7], the water in the Parangan River can be used for class II purposes. Furthermore, the low TSS indicates that the liquid waste does not contain solids that affect water quality [18].

Another parameter of water contamination by the liquid waste from the tofu industry is ammonia. The laboratory analysis results showed that the ammonia levels of the samples were <0.0062–0.4998 mg/L (Fig. 16) or high because of the liquid waste outlets at the sampling site. Government Regulation Number 82 of 2001 does not include the upper limit values for ammonia presence in rivers used for class II purposes. However, because of its inert toxicity, the total ammonia found in waters should not exceed 0.2 mg/L [16]. High ammonia presence can interfere with the growth of aquatic biota [18].

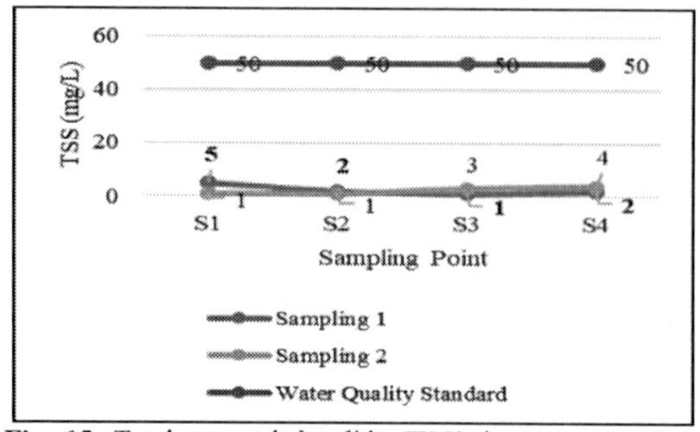

Fig. 15. Total suspended solids (TSS) in Parangan River based on the laboratory analysis of samples collected in duplicate at four sampling points

Dissolved oxygen (DO) represents the amount of oxygen produced in photosynthesis (by aquatic biota and microorganisms) and the entry of air into water [19]. Higher DO indicates better water quality. The DO of the samples varied from 4.4 to 8.2 mg/L (Fig. 17), which meet the stipulated lower limit value of at least 4 mg/L [7]. However, DO levels were found to decrease from point S1 to S4, meaning that the water quality worsens further downstream. This is because the liquid waste disposed of without pre-treatment contained a significant amount of organic matter [24].

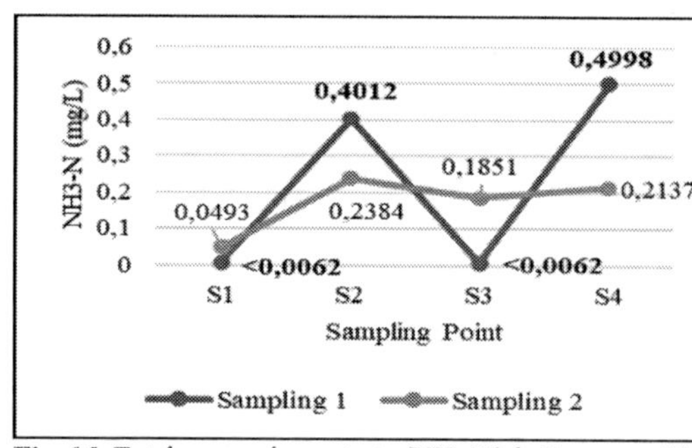

Fig. 16. Total ammonia contents (NH_3-N) in Parangan River based on the laboratory analysis of samples collected in duplicate at four sampling points

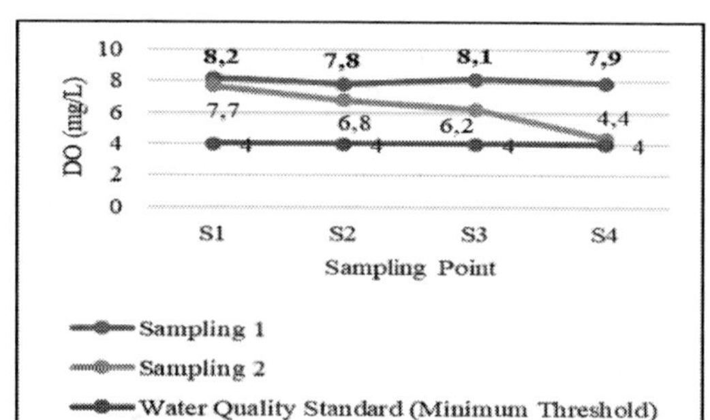

Fig. 17. Dissolved oxygen (DO) levels in Parangan River based on the laboratory analysis of samples collected in duplicate at four sampling points

The water quality in the Parangan River showed a fluctuating pattern at each sampling point. However, based on the analysis results of the duplicate samples, there was a decreasing trend in quality from the first to the last points downstream. Some influencing factors included the liquid waste disposed of to the river and the river discharge. If left uncontrolled, this water contamination can only become worse. The same case has been identified at Muara Murong River in Jombang (East Java, Indonesia) and Tikala River in Manado (North Sulawesi, Indonesia), whose water qualities deteriorate because of the continuous disposal of untreated waste to these rivers [2, 18].

3.4 River Water Quality Status

Here, status defines and categorises water quality conditions into good-quality to polluted within a particular length of time by comparing the parameters with the specified water quality standards [7]. Based on the direct measurements in the field and the laboratory analysis results, the parameter values or classes at S1 meet the standards, meaning that the water at this sampling point has good quality. On the contrary, some parameters at S2, S3 and S4 were outside their accepted range, indicating slightly polluted water. These three points represent river segments that continuously receive liquid waste from the tofu-making process; hence, poorer water quality conditions than S1 (upstream of the waste outlet).

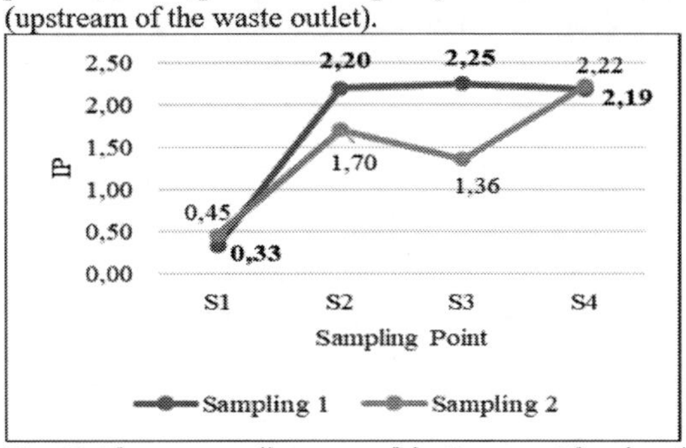

Fig. 18. The water quality status of the Parangan River based on the Pollution Index

Similarly, the Pollution Index (PI) values of S2, S3 and S4 (higher than 1.70) were lower than S1 (0.33–0.43) (Fig. 18). This finding confirms that the untreated liquid waste discharged at point S2 decreases river water quality. The PI values can change if the observed water quality parameters increase [20]. However, PI varies not only according to the waste input but also flow velocity. Severe contamination occurs in a river segment with fast flows, whereas slight pollution is found in a river segment with low velocity. In this context, low stream velocity allows an effective contact between the substrate and microorganisms, while high stream velocity leads to shorter contact and less optimal self-purification [21].

3.5 Total Maximum Daily Load

Pollution load is the amount of a polluting element that a water body, e.g., a river, carries at a given time [7]. In this research, the pollution load calculation produced the total maximum daily load (TMDL) at S4 (Table 4). The results showed that BOD, as an indicator of pollution loads, was the only parameter at S4 that exceeded the specified upper limit in class II water quality [7]. Other parameters like pH, temperature, COD, TSS, NH_3-N and DO were still below the TMDL. The first sample at S4 (S4-1) showed that the COD level was below but close to the specified threshold. For this reason, it is imperative that treatments and preventive strategies be enforced to lower any parameters that nearly exceed their respective upper limit values.

Table 4. The pollution loads in the Parangan River

Sampling Points	Discharge (m^3/s)	pH	Temp. (°C)	COD (mg/L)	BOD (mg/L)	TSS (mg/L)	NH_3-N (mg/L)	DO (mg/L)
S1-1*	0.98	7.22	27.1	10.6	1	5	0.0062	8.2
S2-1	1.44	6.52	26.6	33	7.3	2	0.4012	7.8
S3-1	1.88	6.77	26.4	17.6	7.8	1	0.0062	8.1
S4-1	4.30	6.79	26.63	21.17	6.08	2.25	0.1387	8.02
S1-2*	0.98	7.11	26.7	14.3	1.7	1	0.0493	7.7
S2-2	1.44	6.98	26.3	13.9	5.5	1	0.2384	6.8
S3-2	1.88	6.92	26.5	12.1	4.4	3	0.1851	6.2
S4-2	4.30	6.98	26.48	13.20	4.15	1.87	0.1720	6.74
Water Quality Standards	-	**6–9**	**3 (deviation)**	**25**	**3**	**50**	-	**4**

Notes: *-1, -2 = the *i*-th sample collected at the observation points

Pollution load is inseparable from natural conditions and human activities in the vicinity. High BOD is associated with the liquid waste generated by the tofu industry that contains high organic matter. The level of BOD as an indicator of pollution load depends on the amount of the liquid waste disposed of to the river.

The TMDL calculation produces the maximum permissible amount of waste or pollutants present in the river. This way, the water body can self-purify optimally, and relevant authorities can perform proper river management and immediate restoration of impaired waters where this maximum amount has been exceeded [22]. Overall, the TMDL in the Parangan River is within the safety buffer because the number of parameters that exceeded the maximum pollution load is fewer than those that did not.

4 Conclusion

The liquid waste quality generated by the tofu industry has exceeded the specified limits for temperature, pH, COD, BOD, TSS and ammonia contents described in Provincial Regulation No. 5/2012. The water in the Parangan River has also exceeded the upper limit values of BOD and COD in class II water quality standards (Government Regulation Number 82 of 2001). Compared with the river segment upstream of the waste outlet (S1), the segments downstream (S2–4) have higher TSS and ammonia content and lower DO, indicating contamination by the liquid waste.

In terms of status, S1 has good-quality water, while S2, S3 and S4 are slightly polluted. Parangan River can accommodate pollution loads (as evident from pH, temperature, COD, TSS, NH_3-N and DO) while still meeting the specified water quality standards, except for those indicated by high BOD levels.

Acknowledgement

The authors would like to thank Universitas Gadjah Mada for funding this research through the Final Project Recognition Grant (*Hibah Rekognisi Tugas Akhir*) scheme for the fiscal year 2021. The authors are also grateful to the Faculty of Geography, Universitas Gadjah Mada, for providing the opportunity to conduct this research.

References

1. Effendi, H., *Telaah Kualitas Air Bagi Pengelolaan Sumber Daya dan Lingkungan Perairan*. Kanisius, Yogyakarta (2003)
2. R. D. Sandi., Analisis Kualitas Air dan Distribusi Limbah Cair Industri Tahu Di Sungai Murong Kecamatan Jogoroto Kabupaten Jombang, *Swara Bhumi.*, **5**(9), 59–66 (2019)
3. H. Husni, and M. Esmiralda., Uji Toksisitas Akut Limbah Cair Industri Tahu Terhadao Ikan Mas (Cyprinus carpio Lin) (Studi Kasus: Limbah Cair Industri Tahu "SUPER", Padang), *Bachelor Thesis*. Faculty of Engineering, University of Andalas, Padang (2011)
4. C. Asdak., *Hidrologi dan Pengelolaan Daerah Aliran Sungai*. Gadjah Mada University Press, Yogyakarta (2002)
5. L. R. Mubarok., Kajian Karakteristik Pencemar Bagi Hulu Sungai Belik, Daerah Istimewa Yogyakarta, *Bachelor Thesis*. Faculty of Geography, Universitas Gadjah Mada, Yogyakarta (2018)
6. Anonymous, *Peraturan Daerah Jawa Tengah Nomor 5 Tahun 2012 tentang Baku Mutu Air Limbah,* Sekretariat Negara, Jawa Tengah, (2012)
7. Anonymous, *Peraturan Pemerintah Nomor 82 Tahun 2001 tentang Pengelolaan Kualitas Air dan Pengendalian Pencemaran Air*. Sekretariat Negara, Jakarta, (2001)
8. Nurhasan, and B. B. Pramudyanto., *Pengolahan Air Buangan Tahu*. Yayasan Bina Karta Lestari dan Wahana Lingkungan Hidup Indonesia, Semarang (1991)
9. BPPT., Teknologi Pengolahan Limbah Tahu-Tempe Dengan Proses Biofilter Anaerob Aerob. *Laporan Kegiatan*. Kelompok Teknologi Pengolahan Air Bersih dan Limbah Cair, Jakarta (1997)
10. D. D. Kesuma., Pengaruh Limbah Industri Tahu Terhadap Kualitas Air Sungai di Kabupaten Klaten. *Bachelor Thesis*. Faculty of Geography, Universitas Gadjah Mada, Yogyakarta (2012)
11. R. P. Dewa., and S. Idrus., Identifikasi Cemaran Air Limbah Industri Tahu Di Kota Ambon. *Majalah Biam*, **13**(2), 11–15, (2017)
12. E. Wiryani., *Analisis Kandungan Limbah Cair Pabrik Tempe*. Lab. Ekologi dan Biosistematik Jurusan Biologi FMIPA Universitas Diponegoro, Semarang, (2007)
13. C. E. Boyd., *Water Quality in Warmwater Fish Ponds*. Fourth Printing. Auburn University Agricultural Experiment Station Alabama, USA, (1988)
14. C. E. Boyd., *Water Quality: An Introduction*, 3rd ed. Springer Nature Switzerland, Cham, (2020)
15. E. K. Sari., and O.E. Wijaya., Penentuan Status Mutu Air dengan Metode Indeks Pencemaran dan Strategi Pengendalian Pencemaran Sungai Ogan Kabupaten Ogan Komering Ulu. *J. Ilmu Lingkungan*, **13**(3), 486-491, (2019)
16. M. Widyastuti, and E. Haryono, Water Quality Characteristics of Jonge Telaga (Doline Pond) as Water Resources for the People of Semanu District Gunungkidul Regency. *Indonesian Journal of Geography*, **48**(2), 157–167, (2016)
17. D. Mardhia., and V. Abdullah., Studi Analisis Kualitas Air Sungai Brangbiji Sumbawa Besar. *J. Biologi Tropis*, **18**(2), 182–189, (2018)
18. Sepriani, J. Abidjulu., and H. S. J. Kolengan., Pengaruh Limbah Cair Industri Tahu Terhadap Kualitas Air Sungai Paal 4 Kecamatan Tikala Kota Manado. *Chem. Prog*., **9**(1), 29–33, (2016)
19. H. Effendi, R. Seroja, and S. Hariyadi., Response Surface Method Application In Tofu Production Liquid Waste Treatment. *Indones. J. Chem*., **19**(2), 298 – 304, (2019)
20. Y. Martinus, W. Astono, and D. Hendrawan., IOP Conf. Ser.: *Earth Environ. Sci.* **106** 012022 (2018)
21. A. Suriadikusumah, O. Mulyani, R. Sudirja, E. T. Sofyan, M. H. R. Maulana, and A. Mulyono., Analysis Of The Water Quality At Cipeusing River, Indonesia Using The Pollution Index Method, *Acta Ecologica Sinica*, **41**(3), 177–182, (2021)
22. Y. Rahayu, I. Juwana, and D. Marganingrum., Kajian Perhitungan Beban Pencemaran Air Sungai Di Daerah Aliran Sungai (DAS) Cikapundung Dari Sektor Domestik. *Jurnal Rekayasa Hijau*, **2**(1), 61 – 71, (2018)
23. S. Y. Saragih., Evaluasi Kualitas Air Sungai Bukit Tujuh di Desa Bukit Tujuh Torgamba Kabupaten Labuhanbatu Selatan, *Skripsi*. Fakultas Ilmu Sosial Universitas Negeri Medan, Medan (2015)
24. A. Azwar., Soemarno, and M. Purnomo., Kajian Kualitas Air dan Status Mutu Air Sungai Metro di Kecamatan Sukun Kota Malang. *Jurnal Bumi Lestari*, **13**(2), 265 – 274, (2013)

NPP: Protecting the environment

Haryono Santosa[1*], *Muhammad* Hadi[2], and *Yudi* Imardjoko[3]
[1] Universitas Gadjah Mada, The Graduate School, Jl. Teknika Utara, Yogyakarta, 55284, Indonesia
[2] Universitas Gadjah Mada, Faculty of Geography, Sekip Utara, Bulaksumur, Yogyakarta 55281, Indonesia
[3] Universitas Gadjah Mada, Nuclear Engineering and Engineering Physic Department, Jl. Grafika 2, Yogyakarta 55281, Indonesia

Abstract. Nuclear energy is a good option for the environment, but the potential for NPP to harm the environment is still evident. Safety is used by NPP to protect its environment. Low environmental risk was demonstrated, but the last two accidents resulted in disaster. Environmental protection parties need to participate in reducing environmental hazards of NPP by knowing why NPP is harmful and how to compensate. The perspective of protecting the environment is needed. An exploratory study of NPP's environmental protection was conducted to obtain answers. Publications from institutions engaged in NPP and expert opinions are sources of information. The results obtained indicate that there is optimism on the success of confining fission products, but very little control has been done on the quantity. No environmental protection is carried out when accidents occur, so environmental goods and services are sacrificed. Human protection is the focus. References to environmental radiological protection are still limited to animal and plant species and have not touched environmental goods and services. This allows for an imbalance between the potential radiological load of the NPP and the environmental capacity. It is necessary to develop the idea that NPP is not significantly dangerous in an accident.

1 Introduction

Nuclear energy is the type of energy that is needed today because it is low-carbon and conservative in natural resources. But this superior characteristic does not improve its contribution to the energy mix of Indonesia and the world. Many factors determine the acceptance of nuclear energy, one of which is the ability of a nuclear power plant (NPP) to protect the environment. The adverse environmental impacts caused by the accident of the Fukushima Daiichi NPP in Japan (2011) and the Chernobyl NPP in Ukraine (1986) [1, 2] are technological failure events that should not be forgotten so that it does not happen again [3]. The concept and technological implementation of NPPs in protecting the environment is important in this regard. Improvements are needed to make nuclear power plants capable to protect the environment in the future, so that the use of nuclear energy can increase.

The environmental perspective needs to take into account the environmental protection methodology applied to NPP technology. This makes environmental protection parties need to understand the methodology applied so far. Thinking about how NPP technology protects the environment is still lacking. This can be seen from the research articles produced between January 1955 to June 2021. The topics of nuclear power plants are 33,430 articles, nuclear power plants and accidents 9,151 articles, nuclear power plants and accidents and environmental protection (not waste) 1,224 articles, nuclear power plants and accidents and environmental protection and safety principles 79 articles, NPP and accidents and environmental protection and safety and emergency principles 23 articles [4]. The last topic relates to how nuclear power plants protect their environment in the event of an accident, but mostly relates to human protection and the relationship between NPPs and earthquake events.

An exploratory study of the publications of institutions engaged in nuclear power plants and expert opinion was carried out in order to understand NPP technology in protecting the environment. Environmental protection opinion is expected to be the result of this study.

2 Nuclear power plant (NPP)

NPP works like steam power plants. NPP fuel is nuclear fissile uranium 235 (^{235}U) which is called nuclear fuel, and fission is its combustion reaction. Some nuclear fuel containing ^{235}U is fed into the reactor vessel and forms the reactor core. Each ^{235}U fission reaction generates 200 MeV of heat ($3.2.10^{-1}1$ watt-second per fission) [5], so to produce 1 MWe of electricity, $9.3.10^{16}$ fission/second is required (33.3% thermal-electric efficiency). Nuclear fuel is designed to be able to provide a certain amount of heat energy called burn up. The values generally range from 7 to 50 gigawatt days per tonne of nuclear fuel (GWd/tHM). This makes nuclear fuel in the reactor for a long period of time (≥ 1 year) before being removed from the reactor as spent fuel [6]. This characteristic indicates the presence of a

* Corresponding author: hbsantosa@gmail.com

very large excess of ^{235}U nuclei in the reactor at all times.

The heat balance between fission heat generation and extraction is formed in the nuclear power plant reactor. A heat removal system is provided to take the fission heat generated in the nuclear fuel. The heat taken is used to generate steam for turning the electric turbine. This system is also called a nuclear fuel cooling system because it will lower or maintain the temperature of the nuclear fuel. The heat balance is designed to be balanced because the imbalance can heat the nuclear fuel excessively [5].

The fission reaction of ^{235}U produces fission fragments in addition to generating fission heat. The fission fragments are unstable, therefore they decay to form other atomic nuclei by emitting beta and gamma radiation which ionize biological cells. The combination of fission fragment radionuclides and their decay products is called a fission product. The fission products are in nuclear fuel. The fission product radionuclides can be grouped into inert (noble gases: Xe, Kr), volatile (halogens, alkali metals, tellurium groups) and non-volatile (alkali earths, noble metals, rare earths and refractory oxides) groups [5]. Another group that can be added is radionuclides that emit neutrons in their decay (neutron precursors) [7]. All nuclear fuel in the reactor forms a core, then all of its fission products form the radioactive activity of the reactor core or core inventory. The inventory value of large NPP cores (> 500 MWe) can reach hundreds of EBq (10^{18} Bq) [5]. The greater the NPP's power generation capacity, the greater the inventory.

All fission products are contained in nuclear fuel. This confinement is a standard function provided to restrain its movement. The existence of this function makes NPP have a basic function of operation and safety, namely controlling the generation of fission heat, taking heat (cooling biofuels) and confining fission products [8].

Commercial NPP uses certain types of nuclear reactors that can be distinguished from the energy of the neutrons to trigger fission reactions, moderators, heat removal (nuclear fuel coolant) and steam generation methods [9]. Fission reactions in nuclear power reactors can occur with thermal neutrons (low energy) or fast neutrons (high energy), so there are known thermal reactors and fast reactors. Thermal reactors require a moderator to lower the energy of the neutrons. The material used is H_2O, D_2O or graphite. Heat can be taken up by water or gas (CO_2). Steam generation can take place inside the reactor by boiling, but it can take place in a steam generating device outside the reactor. The second method of taking heat requires that the heat removing fluid does not experience boiling, therefore high pressure is applied. The combination of these parameters produces different nuclear reactors for NPP. A summary of the types of commercial nuclear reactors is shown in Figure 1.

The description of NPP technology shows three important characteristics. The NPP has a ^{235}U nucleus which is larger than its instantaneous requirement. This excess is the potential for heat generation to be faster and larger than it should be (nuclear excursions). Second, nuclear power plants have fission products with high radioactive activity. Third, NPP has a heat balance that must be maintained in balance in order to maintain the integrity of nuclear fuel as a containment for fission products. Nuclear power plants have the basic safety function of controlling fission heat generation, removing heat (cooling nuclear fuel) and confining fission products.

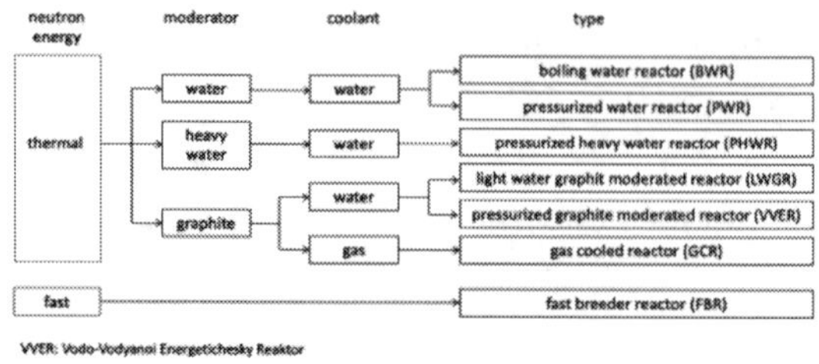

Fig.1. Types of NPP reactors (modification of [9]).

3 Hazards of NPP

NPP technology was developed to obtain abundant clean energy [10], but it is accompanied by danger to life. Long before fission reactions were discovered, the dangers of ionizing radiation were known [11]. Furthermore, the experience of the atomic bomb explosion destroying life through the extremely fast fission reaction of ^{235}U and ^{239}Pu, is seen as a danger from the use of nuclear energy. This view is evident in the statement Alvin Weinberg, a nuclear physicist, told the Special Committee on Atomic Energy, United States Congress in late 1945, "atomic power can heal and kill, can fertilize and enrich an area and destroy it, can broaden human horizons and forced him back into the cave [12]. The experience of operating a research power reactor has strengthened awareness of the dangers of NPP. In 1952 there was a nuclear excursion accident at the NRX research reactor in Canada. The excess of ^{235}U nucleus in the reactor underwent an improper fission reaction (nuclear excursion) due to failure to control the fission reaction. This failure resulted in a heat balance disturbance which increased the nuclear fuel temperature significantly. Excessive heat accumulation results in the melting of nuclear fuel and the release of fission products [13]. This incident raises awareness that accidents at nuclear reactors cannot be ruled out so that more confinement is needed than nuclear fuel alone. Strong and tight structure confining reactor vessel and cooling systems are recommended [8].

Two views on how to start NPP evolved during the development of the technology. The first view directly realizes the NPP with the belief that there is always a solution to the problems that arise. The second view chooses to solve problems first before realizing the NPP. This view found a solution to the problem of nuclear excursions. This settlement managed to tame the wildness of the nuclear reactor. [14]. The first view gave birth to the world's first NPP in 1954 in Obninsk, Russia. The second view gave birth to NPPs a little slower. The two views are united on the economic aspect. Both argue that the larger the capacity of a nuclear power plant, the lower the cost of production, so that it can compete with fossil energy prices [14].

Events that occurred in the pre-NPP period show that NPP is dangerous because it contains a large number of fission products whose radiation is harmful to humans, has the potential to damage nuclear fuel so that it releases fission products into the environment (accidents), and tends to prioritize energy production rather than limiting its fission product content.

4 Safety of NPP

The operation of a NPP is considered to be in an operational state or an accident based on the control, hazards and potential accidents. Operational conditions include normal operation and deviations. This deviation is calculated not to cause significant equipment damage to the safety equipment. Accident circumstances include design base accidents and design extension conditions. A design basis accident is a postulated event for which anticipation and compensation have been provided. The escape of fission products that occur is considered to be within acceptable limits. as the basis for the design of NPPs. Design extension conditions are postulated events that do not form the basis of the design and are still considered in the design process and the release of the fission product is kept within acceptable limits. This condition can be without significant nuclear fuel damage or with melting of the reactor core [15]. The summary of NPP status is shown in Figure 2.

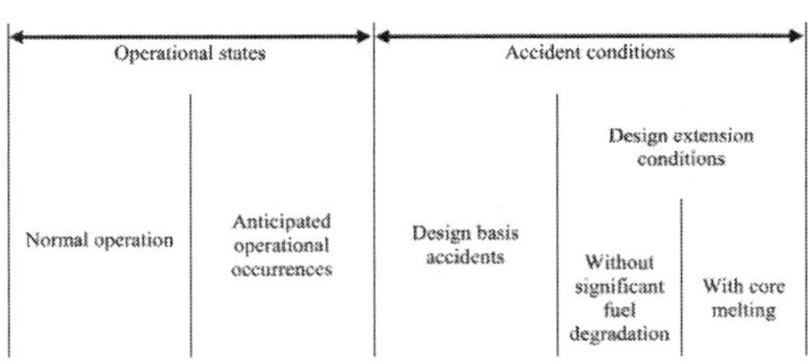

Fig. 2. States of NPP [15].

NPP technology was developed to anticipate the dangers and potential accidents. The existence of danger in the NPP gave birth to international safety standards. Its fundamental safety goal is "to protect people and the environment from the harmful effects of ionizing radiation." [16]. Fundamental safety objectives guide the attainment of the highest safety possible through the prevention of accidents with hazardous consequences caused by loss of control over the reactor core (heat of fission and decay), ensuring that the radiological consequences of all accidents are below the relevant limits that are reasonably achievable, and ensuring that the probability of an accident with serious radiological consequences is very low [17].

The main way to prevent accidents and reduce their consequences when they occur is developed through the principle of defense in depth. Five layers of defense that are independent of each other make up this principle. The first defense is aimed at preventing deviations from normal operation and the failure of items critical to safety. The second defense aims to detect and control deviations from normal operating conditions to prevent anticipated operational events at the plant from escalating into accidental conditions. The third defense aims to keep the state of development within the design base accident so that compensation is available. The fourth defense aims to reduce the consequences of accidents resulting from the failure of the third level of defense. The fifth level of defense aims to reduce the radiological consequences of potentially accidental radioactive releases. On-site and off-site emergency response plans, procedures and facilities are used for this purpose. This concept is applied to all activities related to safety at the design, organizational, supervisory and behavioral levels [17].

The implementation of the defense in depth principle is always assessed in a deterministic and probabilistic manner [8]. Deterministic assessment produces safety criteria as a result of causality. Probabilistic assessment is used to deal with the random nature of accidents with both internal and external causes. The principle of defense in depth is the standard approach to NPP safety [17]. This principle has grown and developed since 1956 [18]. A number of changes have occurred after the NPP accident [8, 18]. The NPP safety framework is within the scope of international standards consisting of fundamental safety objectives and principles, general and specific requirements and safety guidelines [16].

5 Protection and safety

The protection of the public and the environment from the harmful effects of ionizing radiation is largely influenced by the safety characteristics and recommendations of the International Commission on Radiological Protection (ICRP).

5.1 Public protection

NPP hazard management to protect the public and the environment is carried out through the concept of "protection and safety". This concept is defined as "the protection of people against exposure to ionizing radiation or exposure due to radioactive materials and the safety of the source, including the means to achieve this, and the means to prevent accidents and to reduce the consequences of accidents if they occur" [15]. Protection is defined as protection of persons from exposure to ionizing radiation, whereas safety is defined as control of radiation sources (fission products) [15]. If protection and safety is seen as a way, then this method includes ways to achieve it, prevent accidents and reduce the consequences of accidents if they occur. The word accident in the description is defined as "any undesired event, including operating errors, equipment failure and other accidents, the consequences or potential consequences of which cannot be ignored from a protection and safety point of view" [15]. If the meaning of control is used, then protection (radiation) means controlling radiation exposure and its effects, while safety means maintaining control over the source (radiation). The word source in this case includes nuclear installations in addition to functions for other radioactive materials, where nuclear reactors are included in nuclear installations [15].

Safety is primarily concerned with maintaining control over radiation sources (fission products), whereas protection (radiation) is primarily concerned with controlling radiation exposure and its effects. Obviously the two are closely related: radiation protection (or radiological protection) is much simpler when the source in question is under control, so safety must contribute to protection [15].

5.2 Environmental protection

The environment is seen as an entity that does not include humans. Environment as "the conditions in which humans, animals and plants live or develop and which sustain all life and development; especially conditions such as those affected by human activities." [15]. Environmental protection has its own definition, namely "protection and conservation: non-human species, both animal and plant, and their biological diversity; environmental goods and services such as food and feed production; resources used in agriculture, forestry, fisheries and tourism; facilities used in spiritual, cultural and recreational activities; media such as soil, water and air; and natural processes such as the carbon, nitrogen and water cycles." [15].

Protection of the environment is now contained in the ICRP recommendations recently. The purpose of environmental protection is to prevent or reduce the frequency of harmful radiation effects on biota, species conservation; or the health and status of natural habitats, communities, and ecosystems. Environmental protection in the sense of preventing pollution of environmental media (soil, water, sediment, and air) which are environmental resources of value to humans or protecting nature has not yet become the center of attention. Environmental media is seen as a pathway for radiation exposure in humans and biota [19].

6 NPP accident and environmental impact

NPP's nearly seven decades of operation suffered three serious accidents that released fission products into the environment. The last accident occurred at NPP Fukushima Daiichi units 1 (460 MWe), 2 (784 MWe) and 3 (784 MWe) in 2011. The accident occurred when the three NPP units were turned off. The shutdown of nuclear reactors automatically takes place immediately after an earthquake on the east coast of Japan [1]. This accident resulted in radioactive contamination of the terrestrial, aquatic and oceanic environments. The total contaminated area reaches thousands of square kilometers, but for the oceans it is difficult to determine. No fatalities occurred, but the evacuation of large numbers of residents and the partial relocation of some did occur [20]. The collapse of environmental goods and services occurs in contaminated areas. This indicates the escape of radioactive fission products that occur exceeds the acceptable limit. Huge costs arise for the sake of population emergencies and environmental restoration, but it still leaves an area of land that cannot be re-used [21]. Natural events in the form of tsunami waves are designated as the cause of external accidents [1].

Questions about the ability of NPP technology to protect the environment emerged and intensified after the accident. The Japan Nuclear Safety Institute invites all parties to think deeply about safety practices in the form of compliance that have been implemented so far [22]. The NPP utility recognizes weaknesses in the applied safety culture, which weakens the application of the defense in depth principle [23]. The findings of the Fukushima NPP weaknesses and suggestions for improvement are all in the realm of layered defense [24, 25]. Improvements to international safety standards and global safety regimes took place after the accident under the coordination of the IAEA [26].

More severe environmental impacts occurred after the NPP Chernobyl unit 4 in Ukraine had an accident in 1986. The accident occurred during a test run of the emergency power supply system [27, 28]. Fission products with 14 EBq activity released to the environment of a number of countries. Land and air contamination of more than 200,000 km^2 occurred. The evacuation of hundreds of thousands of residents in the contaminated area is underway. Some of them had to relocate their residences [29]. Environmental restoration is not easy to do because it requires a very large cost. The large land area is calculated not to be taken advantage of until hundreds of years into the future [30]. International assistance is needed to restore the environment [31]. The results of the investigation showed that there were weaknesses in all aspects related to safety, including the application of the principle of defense in depth [28].

The Chernobyl accident gave birth to a global nuclear safety regime. Previous safety practices were national. Each country that develops nuclear power technology uses their own safety standards [32].

The NPP accident that released radioactive fission products happened before Chernobyl. The accident occurred at Three Mile Island NPP unit 2 in the United States in 1979. The escape of fission products resulted in mild environmental radioactive contamination. Imperfections in nuclear power plant design and lack of operator knowledge were identified as the causes of accidents [33, 34]. Light environmental impact was assessed as the effectiveness of the layered defense principle [35].

All three NPP accidents show the release of radioactive fission products of more than acceptable activity limits can occur. The public and the environment are victims of these accidents [36]. The cause of the accident is evolving. Its development has involved unpredictable natural events. The application of the principle of defense in depth to deal with the causes of accidents that are not easily predictable reduces its effectiveness. The content of fission products in nuclear power plants is not a safety parameter.

7 Discussion

Fission products are a hazard to NPP for the public and the environment. The greater the NPP power generation capacity, the greater the content of fission products. The high content of radioactive fission products as a public and environmental hazard tends not to be a controlled safety variable, even though the greater the activity, the greater the decay heat. The nuclear fuel damage in the Three Mile Island unit 2 and Fukushima Daiichi accidents was caused by the decay heat that radiates when cooling fails.

The release of fission products into the environment from nuclear power plants is determined by the effectiveness of their containment. The containment layer of fission products in the form of nuclear fuel, reactor vessel and cooling system, as well as the containment structure of the reactor vessel and cooling system was not effective enough in the Fukushima Daiichi nuclear power plant accident. The heat and high pressure generated when some of the nuclear fuel melts is able to penetrate the enclosure provided. This shows that the NPP has a self-destructive power that exceeds the provided compensation power.

Accident prevention through defenses 1, 2 and 3 is potentially ineffective when the trigger or cause of the accident is not well known. This can be seen in the Fukushima Daiichi accident. Insufficient compensation for tsunami waves provided. The TMI and Chernobyl accidents occurred due to internal causes, while the Fukushina Daiichi accidents were caused or triggered by external factors (natural events). How capable NPP managers are in recognizing internal and external factors that could lead to accidents is a never ending question.

NPP technology is designed, built and operated with the intention of protecting people and the environment through accident prevention, reducing the consequences of fission product released from accidents that occur, and reducing the radiological consequences of fission product escape in the environment. Two things can be seen from the protection procedures. Objects that are considered important to be protected are humans and non-human living objects (biota, animals, plants). Environmental media such as soil, water and air are not objects of protection, but are seen as a pathway for exposure to living things. Contamination that occurs in soil, water and air as a result of the Fukushima Daiichi and Chernobyl accidents can clearly disrupt people's lives and destroy environmental services in the long term. If viewed from the perspective of the path of radiation exposure to living things, its existence can last a very long time and become a long-term radiological latent hazard as well. An environmental protection focus that does not include the protection of environmental goods and services may mean allowing environmental violence to occur.

8 Conclusion

- NPP technology has recognized the radiological hazards to humans and non-human species as well as environmental goods and services.
- Efforts to protect hazard recipients have been carried out through accident prevention, mitigating the escape of fission products and mitigating the radiological consequences of escapes, but still requires the participation of external parties to do so.
- Nuclear power plant technology has the potential to be called a technology that commits environmental violence because it does not make efforts to save the environment from the collapse of the value of its goods and services for life.
- Efforts to protect by regulating the content of fission products or power generation capacity in such a way as to limit the release of fission products due to accidents need to be carried out. The idea of designing a nuclear power plant technology based on an acceptable fission product released needs to be ensured for its implementation.

References

1. IAEA, The Fukushima Daiichi Accident, Technical Volume 1: Description And Context Of The Accident, Wina, Austria (2015)
2. IAEA, Environmental Consequences of The Chernobyl Accident and Their Remediation: Twenty Years of Experience, Report of the Chernobyl Forum Expert Group 'Environment', Wina, Austria (2015)
3. G. Santayana, *Reason in Common Sense*, p. **284**, volume 1 of *The Life of* Reason
4. http://ugm.summon.serialssolutions.com/
5. E. Lewis, Nuclear Power Reactor Safety (John Wiley and Sons, New York, 1977)
6. H. Feiveson, Z. Mian, M.V. Ramana and F. von Hippel (Eds.), Managing Spent Fuel from Nuclear Power Reactors: Experience and Lessons from Around the World, International Panel on Fissile Materials (2011)
7. M.C. Brady, Evaluation and Application of Delayed Neutron Precursor Data, Los Alamos, N.M.: Los Alamos National Lab (1989)
8. IAEA, The Safety of Nuclear Power, A report by the International Nuclear Safety Advisory Group, INSAG-5, Wina, Austria (1992)
9. B. Wealer, S. Bauer, N. Landry, H. Seiß, C.R. von Hirschhausen, Nuclear power reactors worldwide: Technology developments, diffusion patterns, and country-by-country analysis of implementation (1951-2017), DIW Data Documentation, No. 93, Berlin: Deutsches Institut für Wirtschaftsforschung (DIW), (2018)
10. N.L. Char and B.J. Csik, IAEA Bulletin: The IAEA at 30, 3/1987, 19-25 (1987)
11. The British Journal of Radiology 7, New Series, No. 83, November, (1934)

12. J.S. Walker and T.R. Wellock, A Short History of Nuclear Regulation, 1946–2009, NUREG/BR-0175, Rev. 2, U.S. Nuclear Regulatory Commission (2010)
13. W.B. Lewis, The Accident to the NRX Reactor on December 12, 1952, Chalk River, Ontario:Atomic Energy of Canada Limited (1953)
14. A.H. Snell, Geneva Diary, August 8–20, 1955, American Scientist, Vol. 43, No. 4 (October 1955), pp. **505-520** Published by: Sigma Xi, The Scientific Research Honor Society, Stable URL: https://www.jstor.org/stable/27826696 Accessed: 12-06-2021 06:09 UTC
15. IAEA, IAEA Safety Glossary, Terminology Used in Nuclear Safety and Radiation Protection 2018 Edition, Wina, Austria (2019)
16. IAEA, Fundamental Safety Principles , Safety Fundamentals, IAEA Safety Standards Series No. SF-1, Wina, Austria (2006)
17. IAEA, Safety of Nuclear Power Plants: Design, Specific Safety Requirements, IAEA Safety Standards Series No. SSR-2/1 (Rev. 1)
18. M. Drouin, B.Wagner, J. Lehner, V.Mubayi, Historical Review and Observations of Defense-in-Depth, NUREG/KM-0009, Washington D.C: U.S. Nuclear Regulatory Commission (2016)
19. ICRP, Protection of the environment under different exposure situations. ICRP Publication 124, Ann. ICRP **43**(1) (2014).
20. IAEA, The Fukushima Daiichi Accident, Technical Volume 4: Radiological Consequences, Wina, Austria (2015).
21. IAEA, The Fukushima Daiichi Accident, Technical Volume 5: Post Accident Recovery, Wina, Austria (2015)
22. S. Matsuura, New Paradigm for Nuclear Safety, Japan Nuclear Safety Institute (2013)
23. The Development of and Lessons from the Fukushima Daiichi Nuclear Accident, 1st edition, Tokyo Electric Power Company, Inc (2013)
24. J. Buongiorno, R. Ballinger, M. Driscoll, B. Forget, C. Forsberg, M. Golay, M. Kazimi, N. Todreas, J. Yanch, Technical Lessons Learned from the Fukushima-Daichii Accident and Possible Corrective Actions for the Nuclear Industry: An Initial Evaluation, Cambridge, M.A.: Massachusetts Institute of Technology (2011)
25. A. Omoto, The accident at TEPCO's Fukushima-Daiichi Nuclear Power Station: What went wrong and what lessons are universal? Nuclear Instruments and Methods in Physics Research A, **731**, (2013) 3–7
26. Progress in the Implementation of the IAEA Action Plan on Nuclear Safety, Report by the Director General, GOV/INF/2015/13-GC(59)/INF/5 (31 July 2015)
27. Department of Energy; Electric Power Research Institute; Environmental Protection Agency; Federal Emergency Management Agency; Institute of Nuclear Power Operations; Nuclear Regulatory Commission, Report on the Accident at the Chernobyl Nuclear Power Station, Washington, D.C.: U.S. Nuclear Regulatory Commission (1987)
28. IAEA, The Chernobyl Accident: Updating of INSAG-1, INSAG-7, A report by the International Nuclear Safety Advisory Group, Wina, Austria (1992)
29. IAEA, Environmental Consequences of The Chernobyl Accident and Their Remediation: Twenty Years of Experience, Report of the Chernobyl Forum Expert Group 'Environment'. Wina, Austria (2006). ISBN 92–0–114705–8
30. V. Kashparov, Chernobyl: 30 Years of Radioactive Contamination Legacy, Ukrainian Institute of Agricultural Radiology (UIAR) of National University of Life and Environmental Sciences of Ukraine (NUBiP of Ukraine), Kiev, Ukraine (2016)
31. CHERNOBYL disaster: Why are the consequences still observed? and Why is the international assistance still critical? Belarus Foreign Ministry (April 2009), retrieved from www.mfa.gov.by
32. IAEA, Strengthening The Global Nuclear Safety Regime, INSAG-2, A report by the International Nuclear Safety Group, Wina, Austria (2006)
33. Three Mile Island Accident, Backgrounder, Office of Public Affair (2018), www.nrc.gov
34. 1979 Annual Report, Washington D.C.: U.S. Nuclear Regulatory Commissionn (1979)
35. L. Battist, J. Buchanan, F. Congel and H. Peterson,Population Dose And Health Impactof The Accident at The Three Mile Island Nuclear Station, Preliminary Estimates Prepared by Ad Hoc Interagency Dose Assessment Group, NUREG-0558, Washington, D.C.: U.S. Nuclear Regulatory Commission (1979)
36. E. Molodstova, Pace Envtl. L. Rev. **185** (1994)

https://doi.org/10.1051/e3sconf/202132503005

Post-Consumer plastic packaging waste management in Indonesia: a producer responsibility approach

Siti Kardian Pramiati[1]*, *Tri* Edhi Budhi Soesilo[1], and *Haruki* Agustina[1]

[1]School of Environmental Sciences, Universitas Indonesia, Central Jakarta 10430, Indonesia

Abstract. In a country with a large and diverse population such as Indonesia, the waste problem is difficult to solve, especially in urban areas. The growth of the population in urban areas has been accompanied by their rapid development of infrastructure and industrial growth. The dynamic lifestyle of urban communities changes their consumption behavior to products that are more practical in packaging. There are fundamental things that arise in the waste management system in Indonesia, including limited waste management capacity in the regions, inadequate infrastructure, implementation of regulations, and a lack of public concern, including the manufacturing industry, on waste management issues. Through the approach of the concept of extended producer responsibility (EPR). However, EPR concept implementation in developing countries is still lacking, where the collection and recycling system has not been adequately addressed. On the other hand, the industrial sector does not yet have enough attention to expand its product responsibilities for the post-consumption stage. This article presents an overview of current practices of post-consumer packaging waste management through an extended producer responsibility mechanism in Indonesia. More specifically, this study focused on consumer goods manufacturing industries. Based on the literature study, various industrial efforts have been shown to adopt the EPR concept in handling its product packaging waste.

1 Introduction

Solid waste management has become a major concern for many urban managers around the world, including in Indonesia, a country with a high population density. Solid waste management in Indonesia is primarily caused by the high volume of waste products, the limited capacity of final disposal sites, waste management technology cost, and lack of public awareness that can bring new problems such as water pollution and flood disaster.

Based on the 2018 Environmental Indifference Index, the most indifference of Indonesians to the environment is in the dimension of waste management, compared to indifference to the energy management, water-saving, or public transportation dimensions. The high level of indifference to waste management is shown by the behavior of households in Indonesia in terms of waste handling. More than half of households in Indonesia use environmentally unfriendly methods when managing waste. The survey results show that as many as 53% of respondents burn trash, 5% throw garbage into rivers and 2.7% throw garbage everywhere. Public ignorance is also reflected in behavior towards plastic waste, as many as 81.4% of people do not care about plastic waste [11].

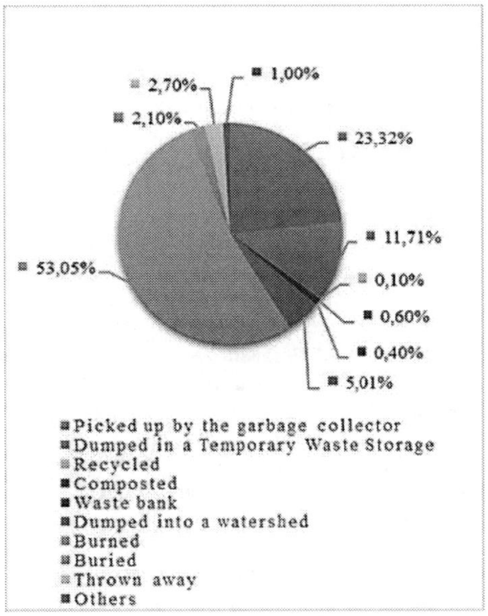

Fig 1. Percentage of Indonesian households based on frequent waste handling (Central Bureau for Statistics, 2018)

Only about 60% of waste in big cities in Indonesia can be transported to final processing sites (TPA), whose

* Corresponding author: dhiankui@gmail.com

main operation is landfilling. The landfilling system is the mainstay of a city in solving its waste problem [11]. According to the National Waste Management Information System Data (2018), the amount of waste in DKI Jakarta that is dumped in the TPA reaches 11 thousand tons per day and 273 tons is still not manageable. Changes in the lifestyle of urban communities, including changes in consumption patterns, can affect the generation of waste generated every day, including plastic packaging waste. The generation of plastic waste reaches 16.69% of the composition of Indonesia's national waste [11], even the DKI Jakarta Provincial Environmental Agency stated that DKI Jakarta produces 357 tons of plastic waste every year.

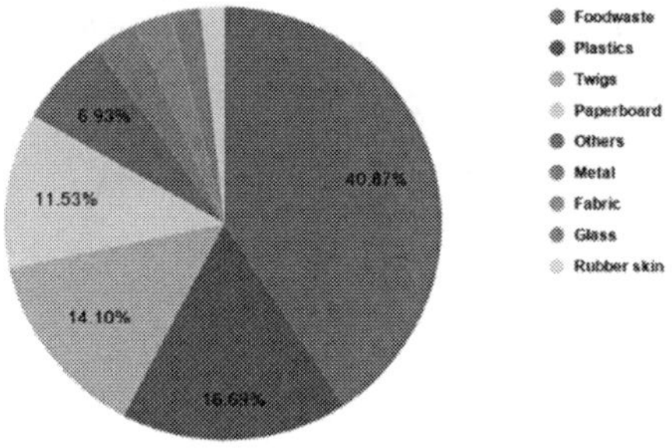

Fig.2. Solid waste compotition by type (MoEF,2020)

Almost all food packaging and goods use plastic and plastic bags. The use of plastics has been discovered in the 20th century. And growing tremendously from just a few hundred tonnes/year in the 1930s, to 150 million tonnes/year in the 1990s and 220 million tonnes/year in 2005. Plastic production on an industrial scale began in the 1940s, then estimated to grow at an average of 10 percent per year. The increase in global production occurred from 1.3 million tonnes in 1995 to around 300 million tonnes in 2014 [17].

A total of 400 million tons of plastic are produced worldwide every year for various purposes, including as a wrapping material because it is lightweight and functional. About 150 million tons of plastic waste are in world waters. Its growth is no less great, reaching 8 million tons per year [13]. Plastics have become popular in the industrial sector because of their special characteristics, flexibility to shape according to needs; lightweight and easy to transport; durability; safe from chemical contamination and its impacts; sealability; weather and temperature resistance; and more importantly, it is cheap.

The use of plastic materials in Western European countries reaches 60kg/person/ year, in the United States, it reaches 80kg/person/year, while in India only 2kg/person/year. Indonesia has stated as the second largest country contributed to plastic pollution in a marine environment, reaching 187.2 million tons after China, which reached 262.9 million tons. Meanwhile, in third place was the Philippines, which contributed almost 83.4 million tonnes of plastic waste, followed by Vietnam with 55.9 million tonnes, and Sri Lanka with 14.6 million tonnes per year [11]. This phenomenon can occur because every year the Indonesian sea is estimated to receive 70-80 percent of plastic waste sent from land which is used for human consumption. The amount is between 480 thousand and 1.29 million tons of plastic waste from a total of 3.22 million tons of garbage that enters marine and coastal areas [13]. Plastic pollution in Indonesia is predicted to increase in line with the development of the consumer goods industries in Indonesia, considering that this is one of the fastest-growing sectors. In the first quarter of 2019, the growth of the beverage processing industry reached 24.2% on an annual basis. Some materials are considered to have the same characteristics as plastic. Quite a lot of ingredients have better quality if compared to plastic. Unfortunately, this material is more expensive than plastic so that industry is reluctant to replace raw materials because it will increase production costs and reduce profit margins [5].

2 Material and method

A systematic literature review was carried out to record how packaging plastic waste management schemes were organized in Indonesia, as well as their main results. The scientific literature on plastic waste is extensive and most of them describe plastic management technology; however, no comprehensive review was carried out to assess the role of producer responsibility in managing their packaging waste, which was the main focus of this article. This article also covered gray literature and some case studies of producer initiatives in several consumer goods industries. Data presented in this article came from previous researches that could be more updated in the next researches data with the same keyword focuses.

3 Result and discussion

Since 2005 the Government of Indonesia has shown concern for environmental problems, particularly waste management, following the issuance of Law Number 18 of 2008 regarding Solid Waste Management. The main mandate of waste management in this law is to change the paradigm of waste management from collect-transport-waste to reduce at source and recycle resources. To replace the end of pipe approach, the government implemented the 3R principle approach (reduce, reuse, recycle), extended producer responsibility (EPR), processing and utilizing waste into resources, both as raw material and a source of renewable energy,

The policy on EPR has been mandated in Law Number 18 of 2008 regarding Solid Waste Management, particularly in Article 1 which reads: "Producers are required to manage packaging and/or the goods they produce which cannot or are difficult to decompose by natural processes". Although this article does not textually mention EPR, the presence of this law is a legal basis for demanding the roles and responsibilities of producers in efforts to reduce and handle waste because producers, through the products and packaging of the products they produce, are one of the sources of producing waste.

In general, EPR is described as a pollution prevention policy that focuses on the production system rather than the production facility. Thus the responsibility for products extends beyond the emissions and wastes generated by extraction or manufacturing processes to include product management of the product after disposal. The concept of EPR or sustainable producer responsibility is the principle of environmental protection policies to reduce the environmental impact that comes from the product life cycle by extending the responsibility of the manufacturer for the product life cycle by recalling and final disposal of the product after the sale [11][11].

To meet the needs of dynamic urban communities, the consumer goods industry sector is growing rapidly producing goods with practical and economical packaging. Most consumer goods manufacturers choose plastic to package their products. This plastic packaging has become a big problem for the environment in Indonesia. The characteristics of plastic are made durable, flexible, lightweight and price, sealable, puncture and water resist [9][17]. Several types of packaging waste cannot be recycled. With such characteristics, packaging waste becomes a multidimensional problem. This is important to resolve and find solutions together, not only from the government. It takes a strong commitment from various parties to collaborate to solve these problems, including the commitment of the manufacturing industry as a producer.

The Ministry of Environment and Forestry of the Republic of Indonesia designs targets for producers to reduce 30 percent of waste from products and/or product packaging by 2029. The target of reducing plastic packaging waste in Indonesia has become part of the Roadmap for Reducing Plastic Waste by Producers [17]. Several commitments to reduce plastic waste have been demonstrated by the consumer goods manufacturing industry in Indonesia, in various ways.

3.1 Take-back Post-Consumer Packaging Waste

Some popular brand owner entities have taken initiatives to run a take-back program. A collaboration between several consumer goods manufacturing industries has been carried out to help reduce the amount of packaging waste through the Dropbox Packaging Trash program. This collaborative action begins with the placement of five Packaging Waste Dropbox in a number of supermarket outlets in the Jakarta, Bogor, Depok, Tangerang and Bekasi areas. The type of waste collected at Dropbox is valuable inorganic waste which consists of three categories, namely the paper packaging waste category, the plastic packaging waste category, and the glass packaging waste category. All of the collected waste will be transported by the pick-up partner, namely the Waste Bank, to then be channeled and processed by recycling centers into goods that are more useful and have economic value. Multinational bottled beverage product companies have shown their commitment by collecting post-consumer packaging and trying to recycle them equivalently to the number of packages sold. This program established a partnership with informal sectors to help collecting packaging waste [11]. In fact, the informal sector plays an important role in Indonesia's recycling sector [9][5].

3.2 Redesign the packaging

In addition to the take-back packaging waste program, some entities initiate redesigning of packaging into sustainable packaging. A multinational bottled water company targets 100 percent of its packaging to be recycled by 2025 [17]..The results of this innovative plastic waste recycled packaging are claimed to be safe for consumption. The packaging has met the standards set by BPOM, Halal, SNI and FSSC 22000.Thus, the quality of the recycled packaging is the same as bottles virgin PET or new PET, so the bottled water products are of the same quality.

3.3 Develop package free shopping services

As a form of commitment to reduce plastic waste, a fast moving consumer goods company presents a refill station as an example of applying the circular economy concept. This concept emphasizes the importance of elements of reuse and recycling, and reduces the use of plastics. The concept then developed into a consumer goods store without packaging that accustoms consumers to shopping with their own containers. This method is quite effective and efficient to reduce the use of plastic packaging for daily necessities, while still paying attention to safety and quality control from the factory.

3.4 Build recycling facilities

A food company giant association shows its commitment by building a waste management technology infrastructure for an investment of 90 million US dollars to improve waste management and technology. Some brand owners had conducted voluntary business and non-business activities to reduce the impact of their product packaging and improve the capabilities of the existing recycling sector. The challenge faced by the existing initiatives is the poor quality of the collected materials. The quantity collected is also small and the scale of the activity is limited. Consumer lifestyle can also contribute to the quality of the materials. Another challenge mentioned in the previous research was the high transportation costs associated with the take-back program considering the geographical challenges [11].

4 Conclusion

The industry contributions in implementing sustainable systems in consumer lifestyle can also contribute to the quality of the materials. The high transportation costs associated with the take-back program considering the geographical challenges packaging plastic waste management plays an important role in plastic waste management in Indonesia. The target of reducing plastic waste can be achieved if industrial sector as producers continue to innovate and educate consumers to adopt an eco-lifestyle in using plastic and manage household waste. Producers in Indonesia need to work together with related stakeholders to implement a sustainable system to help reducing the amount of packaging plastic waste in Indonesia effectively. Each producer may consider different systems in handling the plastic packaging waste in post consumer level, however developing a collective awareness in managing plastic waste properly is the main idea of plastic waste management in Indonesia.

This article was supported by The Grant of Indexed International Publication for Post Graduate Studies (PUTI Q4) Universitas Indonesia 2020, with contract number: NKB-2563/UN2.RST/HKP.05.00/2020.

References

1. A.P.N. Sulami, T. Murayama, S. Nishikizawa, Environment and Natural Resources Journal, **16**(1), 70–81 (2018)
2. B. Raj, Plastics and Their Role in Food Packaging, Food Packaging Technology Department Central Food Technological Research Institute, 020, 17–46 (2018)
3. Central Bureau of Statistics, *Laporan Indeks Perilaku Ketidakpedulian Lingkungan Hidup Indonesia 2018* (2018)
4. E. Damanhuri, T. Padmi, *Pengelolaan Sampah. Program Studi Teknik Lingkungan ITB* (2010)
5. F. Zahra., T. P. Damanhuri, Studi Teknik Lingkungan, Jurnal Teknik Lingkungan, **17**(April), 59 (2011)
6. G. Hickle, Business Strategy and The Environment, **26**(1) (2015)
7. I. Muise, M. Adams, R.Côté, & G.W. Price, Conservation and Recycling, **109**, 137–145 (2016)
8. J.R. Jambeck, Q. Ji, Y.G. Zhang, D. Liu, D.M. Grossnickle, Z.X.Luo, Science, **347**(6223), 764–768 (2015)
9. M. Chaerul, A.R. Fahruroji, A. R., & T. Fujiwara, Journal of Material Cycles and Waste Management, **16**(3), 509–518 (2014)
10. Ministry of Environment and Forestry Regulation No. P.75/MENLHK/SETJEN/KUM.1/10/2019, Ministry of Environment and Forestry (2019) https://doi.org/10.1017/CBO9781107415324.004
11. Ministry of Environment and Forestry, Sistem Informasi Sampah Nasional
12. Qonaah, Jurnal Komunikasi, 48-55 (2019)
13. T.Tasaki, N.Tojo, T. Lindhqvist, International Survey on Stakeholders ' Perception of the Concept of Extended Producer Responsibility and Product Stewardship (Issue May) (2015)
14. World Economic Forum, Mengurangi polusi plastik secara radikal di Indonesia: Rencana aksi multipemangku kepentingan. Laporan Mendalam, April, 47 (2020)
15. World Econonic Forum, White paper Circular Economy in Cities report 2018 (2018).
16. Y.A. Hidayat, S. Kiranamahsa, M.A. Zamal, AIMS Energy, **7**(3), 350–370 (2019)
17. H. Metropolia, Zuzana Sedlacekova Food Packaging Materials Comparison of Materials Used for Packaging Purposes (2017)

https://doi.org/10.1051/e3sconf/202132503006

Proportion of exported non-oil and gas commodities for manufacturing industries, case study in the City of Cilegon

Julia Thio, *Fahma* Husnalawati, *Fatma* Kusuma Porodani, *Afifudin*, *Anggun* Nurzahwa Haris, *Jaa'ana* Nurur Riza, *Ridho* Dayan Bagus, and *Ratih* Fitria Putri*

Department of Enviornmental Geography, Faculty of Geography, Universitas Gadjah Mada, Indonesia

Abstract. Export activities are important to the economic growth of the country, especially for developing countries. The city of Cilegon in Banten Province is the main city for the manufacturing industry in Indonesia, mainly for its iron and steel products. This study aims to determine the proportion of exported goods for the manufacturing industry in the City of Cilegon and make suggestions for further development. Data used are BPS publications 2016-2020, LQ is calculated to know the basis of the economic sector in the city. According to the calculation, the manufacturing industry in the City of Cilegon shares 59.70%, the highest among other regions. Commodities that dominated the export trade are footwear, plastic products, iron and steel products, while the US, China, and Japan were the countries that had the highest FOB. To escalate export performance and intensity, it is necessary to consider market and competitors' networks, improvement of education and transport infrastructure, along with environmental and social-cultural impacts.

1 Introduction

Each country has different kinds of resources (either its natural and human resources). Therefore trade between countries cannot be avoided. Apart from making a profit, trade between countries (international trade) is meant to fulfil the citizens' needs. The difference in the quantity and quality of products from a country leads them to trade with other countries [1]. Globalization and open trade have helped the world economy to grow, especially for the origin country. According to Kalafsky and Gravers [2], exports have long been considered an important factor to enhance economic development by policymakers (particularly export-oriented manufacturing). Also, Brightnell [3] said, "By distributing commodities in surplus to places where they are in deficit, a new form of global equality is imagined here through transoceanic shipping and borderless flow of luxury commodities".

Export activities are meaningful for developing countries, because of their contribution through positive externalities, gain benefits from foreign buyers and creating a competitive market [1]. The higher the revenue from the export-base sector, the higher the profit and its economic growth. Therefore, it can be said that exports are determining the rate of regional economic growth. Exporting commodities requires knowledge about foreign markets and fixed costs for business [4]. Previous research shows [4] that exporting firms would locate near others and form agglomeration to share transport infrastructure, knowledge of foreign markets, marketing, exhibition, banking, and exporting credit insurance. In the context of Indonesia, it can be seen in the City of Cilegon.

The city of Cilegon in Banten Province is the main city for the manufacturing industry in Indonesia. The manufacturing industry, especially the chemical and steel sector, in Cilegon gives a big portion for its GRPD (Gross Regional Domestic Product). Data from BPS (Badan Pusat Statistik/Central Bureau of Statistics Republic of Indonesia) showed the manufacturing industries had 57% of the GRDP proportion in the City of Cilegon [5] [6], which is the highest compared to other sectors and regions in Banten Province. According to district boundary by government and high demand places to invest, there are several industrial districts: JIIPE (Gresik, East Java), Sentul (Bogor, West Java), Tugu Wijaya (Semarang, Central Java), Kertajati International Industrial Estate Majalengka (West Java), Cilegon (City of Cilegon, Banten), etc. [7].

The Ministry of Industry [8] has been supporting the manufacturing industry in the City of Cilegon, which is infrastructure development to promote national and international investments. Furthermore, the increasing infrastructure to support the manufacturing activities could intensify the quality of the goods and could compete with national and international markets.

Even though Cilegon has many big manufacturing industries, the productions are unable to meet all the community needs. Indonesia, especially Banten Province, has high rates of imported goods such as live animals, foodstuffs, and oil [9] [10] [11] [12] [13]. Export and import activities are very reasonable to do, in addition to meeting the needs of a country that has

* Corresponding author: ratihfitria.putri@ugm.ac.id

 https://doi.org/10.1051/e3sconf/202132503006

problems with high demand but less supply and to increase the country's foreign exchange. It is also carried out to establish trade cooperation with other countries so that the local trade market can be known by many countries.

Therefore, this study aims to determine the proportion of exported goods for the manufacturing industry. We are also going to give suggestions for further development. Before that, we are going to calculate the LQ (Location Quotient) to prove the basis of the economic sector in the study area.

The city of Cilegon is used as a study area because this city is known as an Industrial City or the City of Steel, where the industrial area has been integrated and its infrastructure constructed (Fig. 1). The City of Cilegon itself is used by the government as an example of an industrial district outside Java [8].

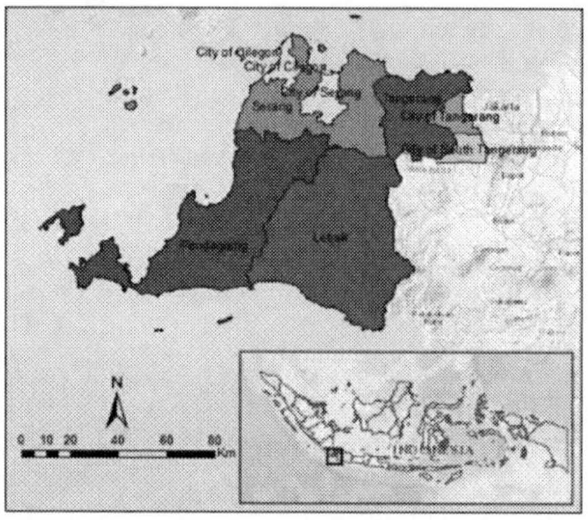

Fig. 1. Banten Province of administration boundary
Source: Data processing (2021)

The city of Cilegon has various kinds of infrastructure supporting industry activities in Banten and it is close to trading distribution roads, including Soekarno-Hatta International Airport, Merak Port, Jakarta-Merak Highway, Jakarta-Rangkasbitung-Merak Railway Network, and the Port of Bojonegara [8]. Those points add up the value of the city, which give proof of the strategic location for domestic and foreign investment. One of the industrial successes in the City of Cilegon is a Joint Venture investment between PT. Krakatau Steel with PT. Pohang Steel Corporation from Korea. This success triggered other companies to participate in expanding their business such as PT. Nippon Shokubai Indonesia, PT. Candra Asri, and PT. Asahimas Petrochemical Indonesia.

This research is limited to the discussion of exported commodities by considering the market price of goods and country of destination. The discussion will not consider the natural and social environmental impacts of manufacturing activities.

There are four sections in this paper. The first section will explain the research background and the aims. In section two, the data and methods we use to analyze are explained, while in section three there are tables from data collection and calculation. The last section (4) is the conclusion and suggestions.

2 Data and method

2.1 Data

Secondary data for this research is obtained from BPS and other relevant literature. There are two types of data to analyze the economy of the manufacturing industry: Gross Regional Domestic Product (GRDP) per business field and GRDP from the City of Cilegon, also the publication of export goods from the year 2015—2019. GRDP is needed to calculate LQ (Location Quotient) to determine the base of the economic sector in the City of Cilegon. To analyze exported commodities, the data used are export commodities according to its raw material through Harmonized System Code or HS 2 digits 2012 version (simpler compared to SITC system), and the countries which have the trade. Furthermore, the data is processed using descriptive techniques in the form of visual graphs and tables, followed by an analysis of each topic based on related literature.

2.2 Method

GRDP in 2015—2019 based on constant market prices at city and province-level are used to know the Location Quotient (LQ) number. LQ is used to determine the level of specialization of the basic or the prime sector by comparing it with other industrial sectors on the regional economic scale [14]. In general, LQ is formulated as follows [15]: variable *vi* resembles sector *i* GRDP in a region (district), *vt*—total GRDP in the region (district), *Vi*—GRDP from sector *i* in higher administration level (province), and *Vt*—total GRDP in higher administration level (province). If the LQ <1, it indicates that sector *i* is not the region basis (non-base), while LQ > 1, sector *i* is the base economy for the region.

$$LQ = \frac{vi/vt}{Vi/Vt} \; or \; LQ = \frac{vi/Vi}{vt/Vt} \qquad (1)$$

3 Results and discussion

The manufacturing industry is the basis of the economic sector in the City of Cilegon (Table 1). The processing industry in Banten contributes the highest GDP compared to other sectors. With the basis of the processing industry in the City of Cilegon and decent ports, it supports the existence of the City of Cilegon as an investment destination. The City of Cilegon has an economic base on three sectors, there are manufacturing; electricity and gas; water supply, sewerage, waste management, and remediation activities, scored 1.76; 3.12; and 2.66 respectively (Table 1). Electricity and gas sector along with water supply, sewerage, waste management and remediation activities sector indeed have had high LQ value, as all the industries need energy and water.

Table 2 describes the percentage of the GRDP manufacturing share in total compared with the province GRDP. Manufacturing in the City of Cilegon contributes 59.70% among other regions. The existence of PT. Krakatau Industrial Estate Cilegon (KIEC) which processes steel is one of the factors that the manufacturing industry is the base economy in the City of Cilegon. The Indonesia government itself hopes that Banten, especially the CIty of Cilegon, contribution to international trade could improve the local economy growth and its population welfare, as well as other regions in Indonesia too [12].

Export rates are dependent on global economic conditions. The slowdown in the economy can lead to an economic slowdown which affects the export development of Banten and also Indonesia in general. Vice versa, if the global economy experiences faster growth, it will drive world demand and then affect the increase in demand for Banten's export goods. Exports in Banten in 2015-2019 experienced a fairly good condition, although in 2019 the FOB value decreased due to the global crisis [12].

Table 1. LQ calculation the City of Cilegon in 2019

Sector	LQ	Category
City of Cilegon		
A. Agriculture, Forestry and Fishery	0.04	Non-Basis
B. Mining and Quarrying	0.07	Non-Basis
C. Manufacturing	1.76	Basis
D. Electricity and Gas	3.12	Basis
E. Water Supply, Sewerage, Waste Management, and Remediation Activities	2.66	Basis
F. Information and Communication	0.19	Non-Basis

Source: BPS (2020)

Table 2. GRDP sector manufacturing in 2019

Region	GRDP per sector (Rp)	Total GRDP (Rp)	Percentage (%)
City of Cilegon	44,411,694.70	74,385,223.50	59.70
Pandeglang Regency	1,293,388.7	19,895,912	6.5
Lebak Regency	2,099,454.91	20,908,728.37	10.04
Tangernag Regency	35,904,811.20	97,142,198.47	36.96
Serang Regency	26,890.41	54,404.47	49.43
City of Tangerang	39,599,102.87	110,592,776.31	35.80%
City of Serang	1,065,416.90	22,813,096.36	4.67
City of South Tangerang	5,021,653.81	60,145,115.32	8.33
Banten Province	**154,849,855.98**	**456,740,827.85**	**33.90**

Source: BPS (2020)

Non-oil and gas processed commodities have been dominating the city trade. Over the last 5 years since 2015 (Fig. 2), there were 10 top commodities that dominated the trade. The first rank was for footwear, followed by plastic products, and iron and steel. From Fig. 2, it is clear that for 5 years, the FOB had increased and decreased. In 2015 until 2018, there was positive economic growth in the Province of Banten. Nevertheless, in 2019 the revenue was slightly decreased because of the global crisis [12]. Fig. 3 shows the detailed proportion of exported commodities in the Province of Banten in 2019.

The City of Cilegon is known for its steel company and production, but its proportion was not much compared to other commodities. This anomaly could be due to the source of data. The exported and imported data are only available at the province level. Iron and steel commodities are considered as top 10 commodities since 2014 (US$ 505,77 million). In 2013 and the previous years, iron and steel commodities had FOB value less than US$ 100 million (in 2013, the value was US$ 85,45 million) [17]. The spike from 2013 to 2014 leads the commodities to reach the top 5 for the next few years. It happened as PT Krakatau Steel started to collaborate with POSCO Korea in December 2013 and was named as PT Krakatau POSCO. The new collaboration used the best technologies, with their capacity to produce 3 million tons of iron and steel plates per year [18]. It technically escalated productivity.

Based on Fig. 4, the United States, China and Japan have become non-oil and gas export destinations with FOB (Free on Board) value of more than US$ 700 million since 2015. These three countries have become the main destinations for non-oil and gas exports compared with other countries (less than US$ 500 million). The value of Banten Province's non-oil exports have always increased since 2015 - 2018, but there is a slight downward in 2019. The highest non-oil export is the United States with a FOB value exceeding US$ 1000 million.

The United States, China, and Japan are Indonesia's trading partner countries, which are considered to have had a strong economic cooperation relationship and have been the destination of Indonesia's export market for a long time [16]. It is also said that Japan has been the main export partner to Indonesia since the late 1990s and around the 2010s, China had started becoming one of the main export destinations [19]. The other countries are mainly from Asia: Philippines, South Korea, Thailand, Vietnam, Malaysia, India, Singapore; also other countries Germany and Australia. Asias' countries have dominated the destination of free trade from Indonesia since the distance is close and have political-economic engagement.

The United States, China, and Japan are also part of international agreements with Indonesia. AFTA (ASEAN Free Trade Area) had economic partnerships with China, Australia, Japan, India, South Korea, and the United States. The two agreements are: the China-ASEAN Free Trade Agreement (CAFTA) and the General System of Preferences (GSP). The agreements allow easier procedure, regulation, access to markets, and transfer of information between countries. It helps the exporter country to reduce the cost for the trade.

There are many possibilities that the same category of products is being exported and imported (two-way intra-industry exchange), but with different value-added. In Indonesia, the value-added for the products is less than 20%, which is lower than China, Thailand, and Malaysia [20] due to a lack of advanced technology. Most of the export manufacturing in Indonesia is resource-based and low technology-based products.

Looking back at the crisis in 2019 and how the investment in manufacturing has weakened are likely to put Indonesia's economy at risk [21]. It could lead to decreasing productivity and diminishing the amount of labour. Employment in the manufacturing sector (and other services) has accelerated poverty alleviation [20]. And unfortunately, as the fourth populous country, Indonesia contributed only 0.84% of the world exports [22].

Improving the export performance and intensity is needed to keep the development growth. However, the manufacturing impacts on environmental and socio-cultural cannot be neglected (even though this paper does not mention it). The other things that need to be considered to enhance the export performance and intensity are: industrial location, clients and competitors networks, institutional networks, and marketing [23]. Rodgriguez-Pose et. al. [19] also mentioned that in Indonesia, the conditions of the firm's province with the neighbouring provinces, agglomeration, education, and transport infrastructure are influencing Indonesians' export performance. Other than increasing export performance and intensity, escalating the product's quality is also needed.

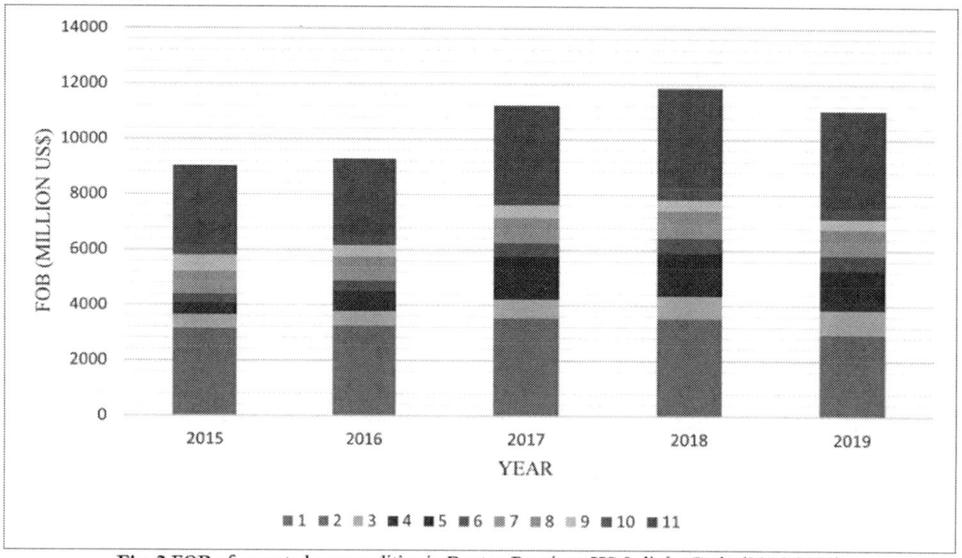

Fig. 2 FOB of exported commodities in Banten Province HS 2 digits Code (2015-2019)
Source: BPS (2016-2020)

Notes: (1) Footwear, (2) Plastic products, (3) Iron and steel, (4) Organic chemicals, (5) Various Processed foods, (6) Machinery/aircrasft mechanics, (7) Rubber products, (8) Electrical machinery and equipment, (9) Copper, (10) Paper and cardboard, (11) others

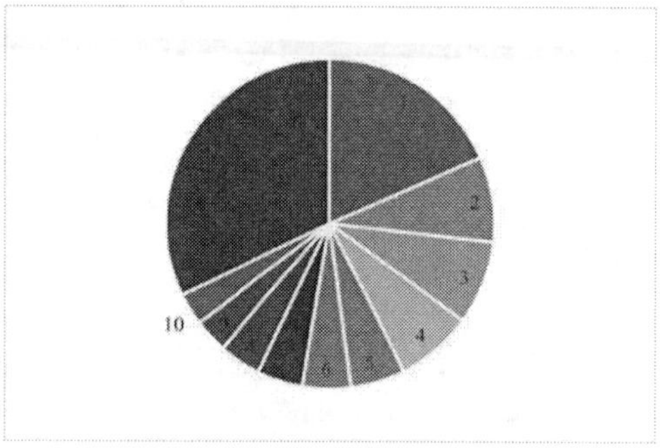

Fig. 3 Exported commodities in Banten Province HS 2 digits code (2019)
Source: BPS (2020)

Notes: (1) Footwear, (2) Plastic products, (3) Iron and steel, (4) Organic chemicals, (5) Various Processed foods, (6) Machinery/aircrasft mechanics, (7) Rubber products, (8) Electrical machinery and equipment, (9) Copper, (10) Paper and cardboard, (11) others

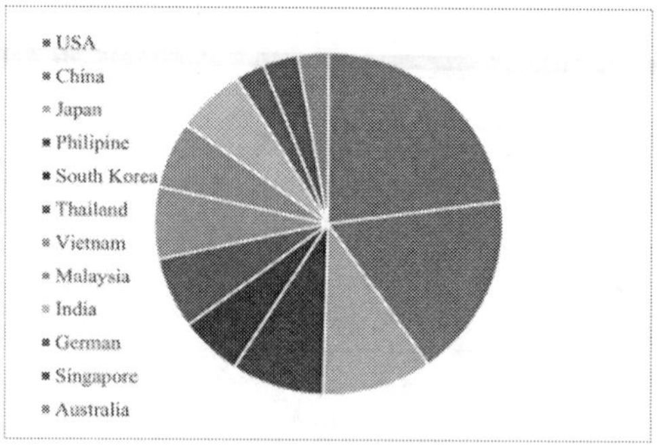

Fig. 4 Countries destination to export the commodities in Banten Province (2019)
Source: BPS (2020)

4 Conclusion

Export activities in the Province of Banten are dominated by non-oil and gas commodities for industrial products, no exception for the City of Cilegon. The commodities which gave the biggest share of FOB in 2019 were footwear, plastic products, and iron and steel products. Iron and steel products have been included in the top 10 exported commodities since 2014 due to the company escalating productivity. The United States, China, and Japan are the top 3 countries that had high FOB. The high FOB for those three countries was influenced by bilateral or regional policies in international trade that facilitate and allow easier procedure, regulation, access to markets, and transfer of information between countries.

Further research regarding the improvement of footwear, organic chemical, and plastic products are needed, especially for its environmental impacts. Also, the Province of Banten itself has footwear, textile commodities which share high FOB for more than 10 years. Study and other development should be conducted to ensure the province's economic growth for its potential commodities.

Studies regarding the export macroeconomy in Indonesia is also crucial. There are limited updated studies for export commodities in terms of macroeconomy and geography. Because such studies are necessary for export planning, policy, and the development of exported products themselves.

References

1. A. H. Yahya, N. Mohd, S. Syahnur, Aceh Int. J. Soc. Sci, **4**: 56–67 (2015)
2. R. V Kalafsky, W. Graves, R. V Kalafsky, W. Graves, Southeast Geogr., **56**, 2 (2016)
3. T. Birtchnell, Area, **45**, 4:436–442(2013)
4. C. He, Q. Guo, X. Ye, GeoJournal, **81**, 6: 947–964 (2016)
5. BPS, *Kota Cilegon Dalam Angka 2016* (2017)
6. BPS, *Kota Cilegon Dalam Angka 2020* (2021)
7. Kementerian Perindustrian, *Links Kawasan Industri* (2016)
8. Kementerian Perindustrian, *Kawasan Industri Banten Jadi Proyek Percontohan di Luar Jawa* (2016)
9. BPS, *Ekspor dan Impor Provinsi Banten 2016* (2017)
10. BPS, *Ekspor dan Impor Provinsi Banten 2017* (2018)
11. BPS, *Ekspor dan Impor Provinsi Banten 2018* (2019)
12. BPS, *Ekspor dan Impor Provinsi Banten 2019* (2020)
13. BPS, *Ekspor dan Impor Provinsi Banten 2020* (2021)
14. Emilia, Imelia, *Modul Ekonomi Regional*. Jambi: Fakultas Ekonomi Univ. Jambi (2006)
15. Kadiriah, Ekonomi Perencanaan. Jakarta: LPFE Universitas Indonesia (1985)
16. Sabaruddin, J. Ilm. Hub. Int., **12**, 2: 205-219 (2017). DOI: 10.26593/jihi.v12i2.2654.205-219.
17. BPS, *Ekspor dan Impor Provinsi Banten 2014* (2015)
18. KrakatauPOSCO, www.krakatauposco.co.id/ (2019)
19. A. Rodríguez-Pose, V. Tselios, D. Winkler, T. Farole, World Dev., **44**, 269868: 225–240 (2013). DOI: 10.1016/j.worlddev.2012.12.002.
20. F. Kimura, L. Chen, Bulletin Indonesian Economic Study, **54**, 2: 165–192 (2018). DOI: 10.1080/00074918.2018.1505412.
21. P. J. Burke, M. D. Siyaranamual, Bulletin Indonesian Economic Study, **55**, 3: 269–293 (2019). DOI: 10.1080/00074918.2019.1690410.
22. T. Y. Wicaksono and C. Mangunsong, Bulletin Indonesian Economic Study, **56**, 4: 1–41 (2021). DOI: 10.1080/00074918.2020.1866745.
23. J. A. Belso-Martínez, Eur. Plan. Stud., **14**, 6: 791–810 (2006). DOI: 10.1080/09654310500496115.

 https://doi.org/10.1051/e3sconf/202132503007

Relationship of the Significance of Freshwater Use to Health Protocols During the Covid-19 Pandemic in Bantul Village, Yogyakarta

Fahima Ulumia[1], *Moh* Syahrul Irfan Fahmi[2], *Relinda* Dewi Astabella[1], *Tjahyo* Nugroho Adji[1], *Ratih* Fitria Putri[2*], *Heein* Yang[3]

[1] Enviromental Geography, Department of Environmental Geography, Faculty of Geography, Universitas Gadjah Mada, Indonesia
[2] Regional Development, Department of Development Geography, Faculty of Geography, Universitas Gadjah Mada, Indonesia
[3]Departement of Electrical and Computer Engineering, Ajou University, Korea

Abstract. The Covid-19 pandemic has an impact on the use of freshwater. Freshwater has an essential role in preventing the spread of Covid-19 related to the implementation of health protocols also Clean and Healthy Living Behavior. The important factors that influence water use are socio-economic conditions, including income level, type of work, and education level. The spread of Covid-19 significantly occurs in urban areas. Bantul village is the center of government, economy and one of the planned areas of urban settlements in Bantul Regency. Bantul village is also a hinterland of the Yogyakarta agglomeration area. This study was conducted to determine the influence of socio-economic conditions on the significance of changes in water use during the Covid-19 pandemic to implement health protocols in Bantul village. The data was obtained through interviews with 99 respondents based on random sampling techniques. This research was conducted with quantitative methods and descriptive analysis techniques. The result shows an increase in freshwater use during the pandemic, particularly washing hands and taking a shower. The highest increase in freshwater use is in Kurahan Hamlet. While in Grujugan Hamlet, there has been no increase. The increase in freshwater use indicates the suitable implementation of health protocols.

1 Introduction

The Covid-19 pandemic has caused profound impacts on human life and the environment, including freshwater use globally [1]. The primary benefit of freshwater for residents is the need for domestic water used daily, such as drinking, cooking, bathing, washing, and cleaning equipment. The use of freshwater for daily activities helps people avoid various diseases. Freshwater supports the sustainability of all lives, especially during the Covid-19 pandemic. During the Covid-19 pandemic, frequent handwashing with soap and water is recommended as one of the most effective measures to reduce the spread of infection. Regular handwashing with soap and running water is a preventive measure of Covid-19 transmission recommended by WHO and the Ministry of Health [2]. The current Covid-19 pandemic has caused the use of freshwater to increase, especially in central areas of activity and high population density.

The amount of domestic water use in each region is different. The amount of water use is influenced by population and type of activity [3]. The high population and activity lead to higher water use. According to the research by [4], domestic water use is influenced by income levels and technology. The urban center of China has more per capita domestic water use than its suburbs, while American urban centers have lower per capita water use than their suburbs. Residents of American urban centers have high incomes and water-saving technology so that the water use per capita of the population is less.

Bantul sub-district is the capital of Bantul Regency. Bantul sub-district consists of 5 villages, namely Palbapang, Ringinharjo, Bantul, Trirenggo, and Sabdodadi. Among the five villages, Bantul Village has the highest population density. Bantul Village is in the center of the Bantul Sub-district and becomes the center of activity in the Bantul Regency. Population density is not the leading cause and has no significant effect on the spread of Covid-19 [5]. However, high population

* Corresponding author: ratihfitria.putri@ugm.ac.id

density can lead to faster disease transmission with a more compact and complex spread chain. Bantul Village as a center of activity is also an additional trigger in the spread of Covid-19. Thus, preventing the spread of the Covid-19 virus by carrying out health protocols, especially the use of clean water, is very important. Bantul Village residents have socio-economic characteristics, water needs, and diverse freshwater usage patterns. Based on this description, this study aims to (1) identify the socio-economic characteristics of the population and its effect on the level of water needs; (2) identify the socio-economic characteristics of the population towards the application of health protocols; and (3) analyze the significance of the increased water needs during the pandemic to the application of health protocols.

Research for correlations of the community's socio-economic conditions on the significance of changes in water use during the Covid-19 pandemic to implement health protocols has never been carried out by other researchers. Previous research was limited to observing the socio-economic influence of the community on water use patterns. The existence of this research can deepen understanding related to water use patterns in the community so that it is known the factors that influence and the extent to which people use it for implementing health protocols.

2 Methods

This research is included in the descriptive quantitative type to look for correlations of the community's socio-economic conditions on the significance of changes in water use during the Covid-19 pandemic to implement health protocols in Bantul Villages. In this case, water use is defined as the daily water requirement in the household for daily activities such as bathing, washing, and watering. Water use is based on the average daily clean water use and the number of family members or by using the following formula [6]:

$$\text{Domestic water needs: } \frac{[Water\ Needs/Families/Days]}{[Number\ of\ people\ per\ Family\ (person)]} \quad (1)$$

The population used in this study was the Head of Family in Bantul Villages which has 16,535 families [7]. In this case, the number of samples is determined using the Slovin formula This formula allows a researcher to sample the population with a certain level of accuracy [8].

$$\text{n} = \frac{N}{N\,(d)^2+1} \quad (2)$$

N = Population size
n = Size of sample required
d = Default error used (0,1)

The formula is known as the number of 99 respondents with the level of error used by 10%. This is because the population is known to be quite large. There are limitations in terms of the source of funds, time, and energy available. Then by using the proportional random sampling technique, the distribution of sample proportions in 12 hamlets in Bantul Villages (figure 1).

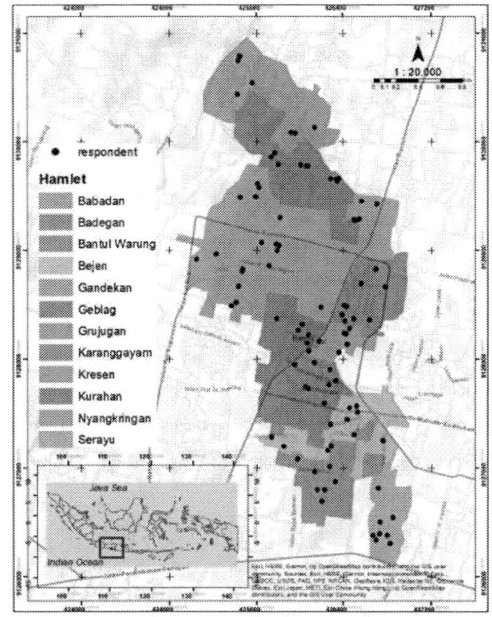

Fig. 1. Distribution of research respondents in Bantul Village

Sampling data is done by sharing questionnaires by creating a list of questions related to the focus of the research. Population water usage data is processed using SPSS (Statistical Package for Social Sciences) application. Analysis of water usage data before and during the Covid-19 Pandemic was conducted with significance tests using the Wilcoxon test. This test is used for normal non-redistributed data characters. Statistical test results are analyzed using descriptive analysis. This is the easiest to interpret and perfectly suited for collating and summarizing quantitative data [9]. However, the analysis was carried out without intending to make conclusions that apply to the public or generalization [10].

3 Results and Discussions

3.1 The Amount of Domestic Water Use

The average amount of domestic water use per capita in urban areas is 171 liters/day/capita and that in rural area is 114 liters/day/capita [11]. Based on the survey, the amount of domestic water use per capita of Bantul Villages varies greatly, ranging from 54 liters/day to 1044 liters/day. The amount of water use per capita is classified into three classes: low (66-325.5 liters/day/capita), medium (325.6-651 liters/day/capita), and high (651.1-1,042.5 liters/day/capita). Based on 99 respondents as representatives of the population in Kalurahan Bantul, 91 respondents with moderate per

 https://doi.org/10.1051/e3sconf/202132503007

capita water use (91.9%), one respondent with low per capita water use (1%), and seven respondents with high per capita water use (7.1%). This proves that it is infrequent for the population to have too little or too much water uses than the National Standard that has been set. Based on the analysis results, variations in domestic water use are not influenced by socio-economic conditions. This is evidenced by the absence of a pattern formed when domestic water use data is crosstab with income levels, types of jobs, and education levels (table 1).

Table 1. Crosstab results (a) the level of education and water use. (b) the type of work and water use. (c) the income level and water use.

Education \ Classification of normal water use	High	Medium	Low	Total
No education	1	0	5	6
Primary School	1	0	10	11
Junior High School	2	0	7	9
High School	2	1	47	50
Bachelor	0	0	1	1
Master	7	1	91	99
Total	1	0	5	6

(a)

Profession \ Classification of normal water use	High	Medium	Low	Total
Unemployment	0	0	2	2
Breeder	0	0	1	1
College student	0	0	3	3
Entrepreneur	0	0	3	3
Farmer	1	0	10	11
Government employees	0	0	5	5
Housewife	1	1	27	29
Laborer	1	0	3	4
Private	2	0	17	19
Retired	2	0	9	11
Teacher	0	0	3	3
Trader	0	0	8	8
Total	7	1	91	99

(b)

Income Classification \ Classification of normal water use	High	Medium	Low	Total
Low (<1500000)	3	0	25	28
Medium (1500000-2500000)	2	0	19	21
High (2500000-3500000)	2	1	10	13
Very high (>3500000)	0	0	37	37
Total	7	1	91	99

(c)

Source: Data processing, 2021

3.2 Pandemic and Significance of Water Use

Pandemics have a substantial impact on water use in areas with larger populations. Overall, the effect is in the form of increased domestic water use in urban areas. The results follow the study by [12], which stated an increase in water use in the settlement category. The increase in the use of clean water in Bantul Village for bathing needs occurred by 4.6%, while the increase in the use of clean water for hand washing purposes occurred very significantly by 101.4%. This figure increased dramatically compared to the pre-pandemic situation of Covid-19. Table 2 describes in more detail the increasing use of clean water for various domestic needs from pre-pandemic conditions and during pandemics.

Considering the variation of average water consumption before and during the pandemic, there was an increase in water use of 13.5 liters/day/capita. Figure 2 shows a map showing the distribution of the significance of total water consumption, both for bathing and handwashing in each Hamlets. The consumption of clean water began to increase even before the quarantine period began. Nevertheless, all categories appear to be affected by measures to limit the spread of pandemics, especially the implementation of health protocols and clean and healthy living behaviors.

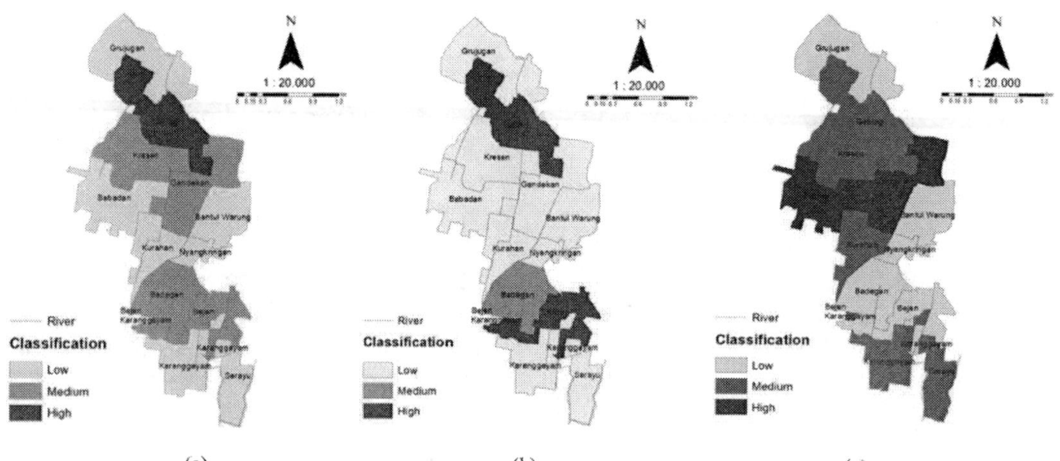

(a) (b) (c)

Fig. 2. Distribution of significance of clean water use during the Covid-19 pandemic. (a) The total significance of water use. (b) Significance of water used for bathing purposes. (c) The significance of the use of water for handwashing.

Source: Data processing, 2021

Table 2. Significance of Clean Water Use During the Covid-19 Pandemic

Freshwater Uses	Before the Covid-19 pandemic (liter/day/capita)	During the Covid-19 pandemic (liter/day/capita)	Increase in freshwater use during the pandemic (liter/capita/day)	The signification of freshwater use during the pandemic (%)
Take a shower	178.48	186.71	8.22	4.60
Wash clothes	31.52	31.61	0.09	0.70
Drinking water	1.86	1.86	0.00	0.00
Watering plants	7.08	7.08	0.00	0.00
Cook	5.24	5.34	0.10	0.75
Wudu'	10.70	10.79	0.09	0.65
Washing vehicles	0.89	0.89	0.00	0.00
Hand washing	4.89	9.85	4.96	101.4
Other	0.42	0.45	0.03	0.22

Source: Data processing, 2021

Figure 2a shows the total significance of high-category clean water use occurring in Geblag Hamlet, while in the category of significance is occurring in Kresen, Gandekan, Badegan, and Bejen Hamlets. The remaining seven hamlets have a low significance category for water use during the Covid-19 pandemic. The significance of the use of clean water is inseparable from the application of health protocols by the people of Bantul Village, one of which is hand washing. During the Covid-19 pandemic, frequent handwashing with soap and water is recommended as one of the most effective measures to reduce the spread of infection. Figure 2c shows almost all communities in each of the Hamlets in Bantul Village experiencing the significance of using water for handwashing.

3.3 Implementation of Health Protocols

The increase in clean water using during the pandemic is caused by several factors, including the use of water for bathing and hand washing which is included in the application of health protocols. The amount of water reportedly used for handwashing that has allowed reduction of fecal contamination ranges from 0.5-2 liters per person. In addition, WHO and UNICEF also recommend washing hands in running water with soap and rubbing for at least 20 seconds [2]. Water use is significant for handwashing per capita during the Covid-19 and pre-Covid-19 pandemics of 4.9 liters and 9.9 liters (figure 3), respectively.

The study found that, based on 99 respondents, there are 98.9% had implemented health protocols, while 1.1% did not implement health protocols during the Covid-19 pandemic (figure 4). This indicates that people in some districts in Bantul Village have implemented suitable health protocols following the government's recommendations. The increase in the use of water for handwashing during Covid-19 is also coherent with other studies. According to research by [13], people begin to wash their hands during activities outside routinely and after returning from outside during the pandemic situation.

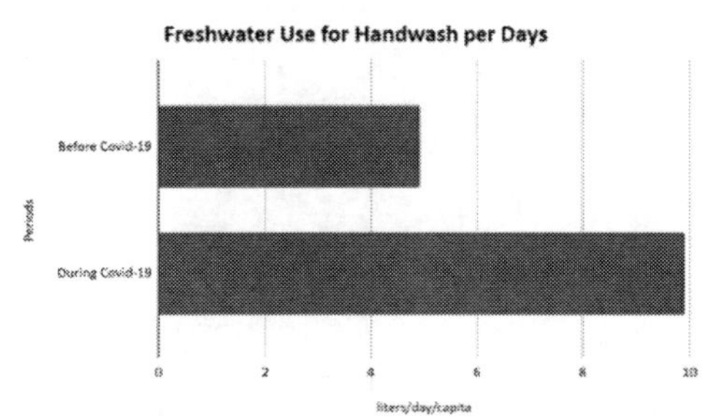

Fig. 3. Use of clean water for hand washing purposes per day during the Covid-19 and pre-Covid-19 pandemics.

Source: Data processing, 2021

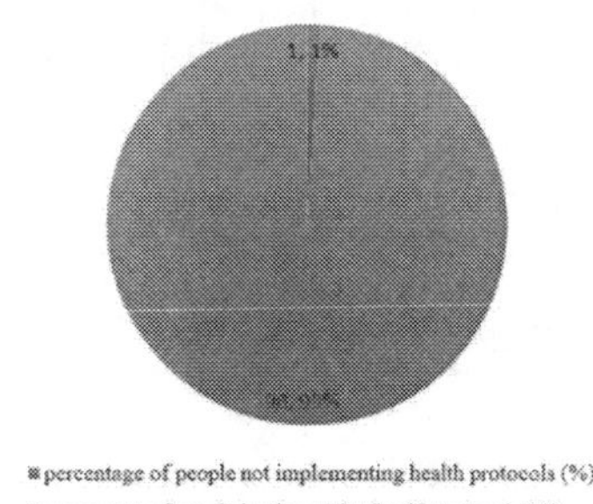

Fig. 4. Percentage of health protocol and implementation in Bantul Village.

Source: Data processing, 2021

3.4 Socio-economic and Significance of Water Use

The significance of Bantul Villages domestic water use is mostly at low levels. The survey results showed that out of a total of 99 respondents, 96 respondents had the significance of domestic water use at a low level (0-77.2 liters/day/capita), while two respondents at the moderate level (77.3-154.4 liters/day/capita) and one respondent at the high level (154.5-231.7 liters/day/capita).

Table 3. shows crosstab results between education level, income level, and type of work with the significance of water use. Crosstab results showed no pattern between socio-economic characteristics and the significance of household water use. Low or high levels of education, employment, and incomes have low domestic water use significance. This indicates that socio-economic characteristics do not affect the significance of water use.

Table 3. Crosstab results (a) the level of education and significance of water use. (b) the income level and significance of water use. (c) the type of work and significance of water use.

Education \ Classification of water use significance	High	Medium	Low	Total
No education	0	0	6	6
Primary School	0	0	11	11
Junior High School	0	0	9	9
High School	1	0	49	50
Bachelor	0	2	20	22
Master	0	0	1	1
Total	1	2	96	99

(a)

Profession \ Classification of water use significance	High	Medium	Low	Total
Unemployment	0	0	2	2
Breeder	0	0	1	1
College student	0	0	3	3
Entrepreneur	0	0	3	3
Farmer	0	0	11	11
Government employees	0	0	5	5
Housewife	0	0	29	29
Laborer	0	0	4	4
Private	0	0	19	19
Retired	0	1	10	11
Teacher	1	1	1	3
Trader	0	0	8	8
Total	1	96	2	99

(b)

Income Classification \ Classification of water use significance	High	Medium	Low	Total
Low (<1500000)	0	0	28	28
Medium (1500000-2500000)	0	0	21	21
High (2500000-3500000)	0	0	13	13
Very high (>3500000)	1	2	34	37
Total	1	2	96	99

(c)

Source: Data processing, 2021

3.5 Socio-economic and Health Protocols

The socio-economic characteristics of Bantul villagers are classified based on the level of education, the level of income, and the type of work. Tables 4, 5, and 6 show crosstab results between education level, income level, and job type with the type of implementation of health protocol. Although, in general, all respondents apply hand washing or handwashing and bathing as a health protocol that is often applied. It can be seen the type of health protocol based on the most dominant respondents.

Table 4. Crosstab results between the income level and the implementation of health protocols.

Implementation of health protocols \ Income Classification	I1	I2	I3	I4	Total
Hand washing	12	3	1	12	28
Hand washing, hand sanitizer	0	0	0	2	2
Hand washing, mask	1	1	2	1	5
Hand washing, mask, hand sanitizer	0	0	0	1	1
Hand washing, mask, physical distancing	0	1	1	1	3
Hand washing, physical distancing	1	0	0	0	1
Hand washing, take a shower	10	12	8	6	36
Hand washing, take a shower, disinfectant	0	0	0	1	1
Hand washing, take a shower, mask	2	3	1	10	16
Hand washing, take a shower, mask, hand sanitizer	0	0	0	2	2
Hand washing, take a shower, mask, physical distancing	1	1	0	0	2
Hand sanitizer	0	0	0	1	1
Not implement	1	0	0	0	1
Total	28	21	13	37	99

I1. Low (<Rp 1.500.000); I2. Medium (Rp 1.500.000-Rp 2.500.000); I3. High (Rp 2.500.000- Rp 3.500.000); I4. Very High (>Rp 3.500.000)

Source: Data processing, 2021

Table 4 shows the crosstab results between income levels and the implementation of health protocols. Most of the respondents have a high-income level (>Rp 3,500,000). However, most respondents applied slightly more handwashing than other health protocols. In addition to hand washing and bathing, respondents have been able to apply health protocols in the form of the use of masks. It is seen in every crosstab result that the application of health protocols in the form of hand washing, bathing, and wearing masks has often been applied by respondents in all socio-economic characteristics.

Table 5. Crosstab results between the type of work and the implementation of health protocols.

Implementation of health protocols \ Profession	P1	P2	P3	P4	P5	P6	P7	P8	P9	P10	P11	P12	Total
Hand washing	2	1	1	1	3	1	7	1	4	2	1	4	28
Hand washing, hand sanitizer	0	0	0	0	0	0	0	0	0	2	0	0	2
Hand washing, mask	0	0	0	0	0	0	3	1	0	1	0	0	5
Hand washing, mask, hand sanitizer	0	0	0	0	0	0	0	0	0	1	0	0	1
Hand washing, mask, physical distancing	0	0	0	1	0	0	1	0	0	1	0	0	3
Hand washing, physical distancing	0	0	0	0	1	0	0	0	0	0	0	0	1
Hand washing, take a shower	0	0	2	1	5	2	12	2	7	3	0	2	36
Hand washing, take a shower, disinfectant	0	0	0	0	0	1	0	0	0	0	0	0	1
Hand washing, take a shower, mask	0	0	0	0	2	1	4	0	6	0	2	1	16
Hand washing, take a shower, mask, hand sanitizer	0	0	0	0	0	1	0	0	0	1	0	0	2
Hand washing, take a shower, mask, physical distancing	0	0	0	0	0	0	1	0	0	0	0	1	2
Hand sanitizer	0	0	0	0	0	1	0	0	0	0	0	0	1
Not implement	0	0	0	0	0	0	1	0	0	0	0	0	1
Total	2	1	3	3	11	7	29	4	17	11	3	8	99

P1. Unemployment; P2. Breeder; P3. College Student; P4. Entrepreneur; P5. Farmer; P6. Government employee; P7. A housewife; P8. Laborer; P9. Private; P10. Retired; P11. Teacher; P123. Trader.

Source: Data processing, 2021

Table 6. Crosstab results between the level of education and the implementation of health protocols.

Implementation of health protocols \ Education	E1	E2	E3	E4	E5	E6	Total
Hand washing	0	2	2	18	6	0	28
Hand washing, hand sanitizer	0	0	0	1	1	0	2
Hand washing, mask	0	1	0	3	1	0	5
Hand washing, mask, hand sanitizer	0	0	0	0	1	0	1
Hand washing, mask, physical distancing	0	0	0	1	2	0	3
Hand washing, physical distancing	0	0	0	1	0	0	1
Hand washing, take a shower	4	6	4	18	4	0	36
Hand washing, take a shower, disinfectant	0	0	0	0	0	1	1
Hand washing, take a shower, mask	1	1	3	6	5	0	16
Hand washing, take a shower, mask, hand sanitizer	0	0	0	1	1	0	2
Hand washing, take a shower, mask, physical distancing	1	0	0	1	0	0	2
Hand sanitizer	0	0	0	0	1	0	1
Not implement	0	1	0	0	0	0	1
Total	6	11	9	50	22	1	99

E1. No education; E2. Primary School; E3. Junior High School; E4. High School; E5. Bachelor E6. Master

Source: Data processing, 2021

Table 5 shows the crosstab results between job types and the implementation of health protocols. Of the 99 participants, 29 were housewives. Most housewives apply more hand washing and bathing. Table 6 shows the crosstab results between education levels and health protocols. Of the 99 participants, 50 respondents have the last level of high school education. Most respondents applied hand washing and hand washing and bathing with the same proportion.

The crosstab results do not show any pattern between socio-economic characteristics and health protocols applied by the residents of Bantul Village. This indicates that socio-economic characteristics do not affect the type of health protocols applied by Bantul Village residents. Finally, the type of health protocol carried out by the residents of Bantul Village generally is not influenced by socio-economic characteristics. All respondents with different socio-economic characteristics (education level, income level, and type of occupation) applied handwashing or handwashing and bathing as a form of frequently applied health protocol measures.

4 Conclusion

The Covid-19 pandemic had an impact on freshwater use in larger areas. Overall, the real impact is the increasing use of freshwater in the domestic sector by residents in urban areas. Bantul Village residents have socio-economic characteristics, water needs, and diverse clean water usage patterns. Low and high levels of education, employment, and income have low household water use significance. This indicates that socio-economic characteristics do not affect the significance of water use. The amount of household freshwater use per capita of Bantul Village is very high, ranging from 54 liters/day to 1,044 liters/day. In addition, socio-economic characteristics do not affect the type of health protocols applied by Bantul Village residents. However, the variations of the average water consumption before and during the pandemic occurred, there was a water usage significance of 13.5 liters/day/capita.

The significance of the use of freshwater during the pandemic is most for bathing and washing hands. The increase of freshwater use is very high in Kurahan Hamlet, while in Grujugan Hamlet, there is no increase. The increase in freshwater use indicates the suitable implementation of health protocols. Based on 99 respondents, there are 98.9% had implemented health protocols, while 1.1% did not implement health protocols during the Covid-19 pandemic. So further studies are required not only in Bantul Village but also worldwide. New findings of this study will benefit future research to comprehensively assess the socio-economic impact associated with the significance of freshwater use in developed and developing countries due to new health protocols also clean and healthy living behavior during the Covid-19 outbreak.

References

1. R. Pant, K.Bishwakarma, R. F. U. Qaiser, L. Pathak, G. Jayaswal, B. Sapkota, K. B. Pal, L. B. Thapa, Koirala, M., Rijal, K., R. Maskey, J. Environ. Manage, **289** (2021)
2. WHO and UNICEF, Water, sanitation, hygiene, and waste management for the COVID-19 virus [website], (2020) (https://apps.who.int/iris/bitstream/handle/10665/331846/WHO-2019-nCoV-IPC_WASH-2020.3-eng.pdf?ua=1, accessed 9 June 2021)
3. World Health Organization, Water Sanitation and Health [website], (2020) (https://www.who.int/teams/environment-climate-change-and-health/water-sanitation-and-health/environmental-health-in-emergencies/humanitarian-emergencies, accessed 9 June 2021)
4. Y. Z. Wang, Y. Franz, K. Zhang, X. Ding, K. Jia, G. Yuan., Resour. Conserv. Recycl., **169** (2021)
5. D. Wahyuni, J. Geo. Edu. dan Ling., **5**, 1 (2021)
6. P. N. Cahyo, P. Hadi, T. N. Adji, Maj. Geogr. Indones., **30**, 2 (2016)
7. Badan Pusat Statistik, Bantul Subdistrict in Figures, in Katalog BPS (2020)
8. E. Stephanie, Slovin's Formula Sampling Techniques (Houghton-Mifflin, New York, 2003).
9. G. Marshall, L. Jonker, The College of Radiography, **16** (2010)
10. Sugiyono. Metode Penelitian Kuantitatif, Kualitatif, dan R&D, (Alfabeta, Bandung, 2017)
11. Asian Development Bank, Indonesia: Country Water Assessment (2016)
12. K. Andreza, E. Henning, M. P. Brikalski, F. V. de Luca, A. C. Konrath, Resour. Conserv. Recycl., **163** (2020)
13. A. Sayeed, M. H. Rahman, J. Bundschuh, I. Herath, F. Ahmed, P. Bhattacharya, Groundw. Sustain. Dev., **13** (2021)

Study of the Economic Valuation of Gemah Beach Tourism During the Covid-19 Pandemic in Tulungagung Regency, East Java

Vivi Tiara Mandela*, *Sudrajat* Sudrajat *Rika* Harini

Geography Study Program, Faculty of Geography, Gadjah Mada University, Yogyakarta, Indonesia

Abstract. The engulfing COVID-19 pandemic in various regions affected various sectors, one of which was the Indonesian tourism sector. The policy of social restrictions and travel restrictions led to a decrease in tourists at almost all tourist attractions. The purpose of this study is to identify the type of economic value generated by Gemah Beach tourism objects during the Covid-19 pandemic and evaluate the total economic value generated by Gemah Beach tourism objects during the Covid-19 pandemic. This research method uses a survey method by taking a sample of a number of tourists, business people, managers of Gemah Beach tourism objects. The type of data collected consists of primary data and secondary data. The economic valuation method is used to calculate the total economic value of the Gemah Beach tourism object obtained from direct use value, indirect use value, preferred use value, existence value, and bequest value. Data analysis used descriptive quantitative. The results of the study still found various types of economic activities carried out by business actors at the Gemah Beach tourism object during the COVID-19 pandemic, despite the government's social and economic restrictions. Meanwhile, based on the results of the economic valuation of the Gemah Beach tourism object during the COVID-19 pandemic, a total economic value of IDR 28,131,949,529 was obtained. The total economic value, according to tourism managers, is lower than before the COVID-19 pandemic due to activity restrictions and a decrease in the number of visitors. Optimizing the economic and natural potential of the Gemah Beach tourist attraction with the fulfillment of facilities and infrastructure in accordance with the COVID-19 health protocol can be used as a strategy in dealing with the COVID-19 pandemic so that the tourism industry can continue to grow.

1. Introduction

Tulungagung Regency is geographically a regency that is directly facing the Indian Ocean and has the potential of the region and nature that supports it to be developed. Tulungagung Regency has a coastline of 54 km, extending from Besuki District to Pucanglaban District, The magnitude of the potential possessed by Tulungagung Regency is a driving force for increasing tourist visits every year. This is evidenced by the increasing number of foreign and domestic tourist arrivals in Tulungagung Regency in 2014-2018, Based on data from the Culture and Tourism Office of Tulungagung Regency in 2019, the number of visitors to Gemah Beach reached 114,875 tourists. However, since the COVID-19 pandemic emerged in early 2020, the tourism industry in Tulungagung Regency has experienced a decrease in the number of visitors [1]. This is due to the policy of travel restrictions and restrictions. Even with the existence of quarantine policies to suppress the spread of the virus, it has become an extreme vulnerability and challenge for the tourism industry itself [2].

Limited mobility during the pandemic caused a decrease in tourist visits to Gemah Beach in Tulungagung Regency. The economic value of utilizing Gemah Beach tourism as a tourist attraction is often neglected; thus, data on quantitative economic value is still lacking or even non-existent. In addition, the ignorance of some parties to the value of natural resources and the environment, especially at the Gemah Beach tourist attraction, will lead to disproportionate policy decisions. Systems theory and crisis management argue that crises need to be tackled collectively. Therefore the government needs policies and strategies for tourism destinations as tourism in the future [3]. Especially in the COVID-19 pandemic situation where restrictions on social and economic activities are government policies that must be obeyed. In fact, during the COVID-19 pandemic, economic activity in tourist areas had stopped when the government imposed a ban on activities outside the home. As a result, tourism activities have stopped because of the closure of tourist attractions and prohibiting tourist visits.

The COVID-19 pandemic situation where restrictions on social and economic activities are government policies must be adhered to. In fact, during the COVID-19 pandemic, economic activity in tourist areas had stopped when the government imposed a ban on activities outside the home. As a result, tourism activities have stopped due to the closure of tourist attractions and prohibiting tourist visits. The decrease in the number of visitors economically will have an impact on economic activity and a decrease in the resulting economic value. Therefore, to see the extent of changes in economic value that occurred due to the Covid-19 pandemic, economic valuation is very necessary. An economic valuation can be an important instrument in

*Corresponding author:vivi.tiara.mandela@mail.ugm.ac.id

https://doi.org/10.1051/e3sconf/202132503008

increasing public appreciation and awareness of goods and services produced by natural resources and the environment [4]. This means, economic valuation is also an effort to provide a quantitative value for goods and services produced by natural resources and the environment at market value and non-market value [5]. Economic valuation is essential [6] because environmental resources play an important role in human well-being as well as to control and prevent damage at tourist sites. Based on this, this research was conducted with the aim of identifying the types of

$$Y = a + \beta_1 X_1 + \beta_2 X_2 + \beta_3 X_3 + \beta_4 X_4 + \beta_5 X_5 + ei \quad (1)$$

economic value sharing generated by Gemah Beach tourism objects and evaluating the total economic value generated by Gemah Beach tourism objects during the Covid-19 pandemic.

2. Methodology

This research was conducted at the Gemah Beach tourist attraction, Tulung Agung Regency. The population in this study were tourists, business actors and tourist managers in the Gemah Beach area. The number of samples for tourists were 93 tourists who were taken by accidental sampling and for business actors were taken by census, while for managers were taken by purposive sampling. There are two types of data used in this study including primary data and secondary data. Primary data is data obtained from observations, interviews with respondents and indept interviews. Secondary data is data obtained by researchers indirectly through intermediary media. Secondary data in this study were obtained from books, journals, previous research, online news and official documents from the government or tourism object managers.

The economic valuation method used in this study is to calculate the total economic value of the tourism object including direct use value, indirect use value, optional use value, existence value, and bequest value of Gemah Beach tourism object in Tulungagung Regency [7]. The total economic value (TEV) of the Gemah Beach tourist attraction is the total sum of all the benefits from utilization at that location, with the following formula :

$$TEV = (DUV + IUV + OV) + (XV + BV) \quad (1)$$

Details :

DUV = Direct Use Value
IUV = Indirect Use Value
OV = Option Value
XV = Existence Value
BV = Bequest Value

Furthermore, to determine the value of each benefit value, the following analytical methods are used :

2.1 Travel Cost Method (TCM)

The travel cost method is a function of tourism demand based on the frequency of visits as the dependent variable described in a linear equation and the independent variable as the explanatory variable. Factors that influence tourism demand from Gemah Beach attractions are identified using the Individual Travel Cost Method (ITCM) which is part of the Travel Cost Method (TCM). Multiple linear regression analysis [8] was used in this study to determine the factors that influence the demand for visits to the Gemah Beach tourist attraction, statistically formulated as follows :

Details :

Y = Tourism visit request
X_1 = Total cost of travel to Gemah Beach tourist attraction (Rupiah/person)
X_2 = Distance from home to tourist attraction
X_3 = Total income of respondents
X_4 = Respondents's Education Level
X_5 = Respondents Age
$\beta_1 - \beta_5$ = Regression coefficient for factor X_1- X_3
ei = Error term

Furthermore, to calculate the economic value of the Gemah Beach tourist attraction uses the lower limit formulation, i.e. the lowest cost incurred by tourists, and the upper limit is the highest cost that tourists spend when travelling at the Gemah Beach tourist attraction. Therefore the integral equation for consumer surplus is as follows:

$$CS = \int_{p0}^{p1} f\,(Px)\,dP \quad (2)$$

2.2 Contingent Valuation Method (CVM)

The CVM method uses WTP (Willingnes to Pay) as a parameter for calculating the total benefit. Similar to the consumer surplus estimation approach, in calculating the WTP of tourists use the contingency assessment method [9] it can be done by using the average tourist willingness to pay for environmental services and multiplied by the total tourists in a certain year, the formula for calculating the WTP of tourists is as follows :

$$TB = WTPi\ x\ Pt \quad (4)$$

TB is the total benefit; WTPi is the individual WTP value; Pt is the total population/number of tourists in the t year relevant to the analysis or the number of tourists in the one year of the covid-19 pandemic

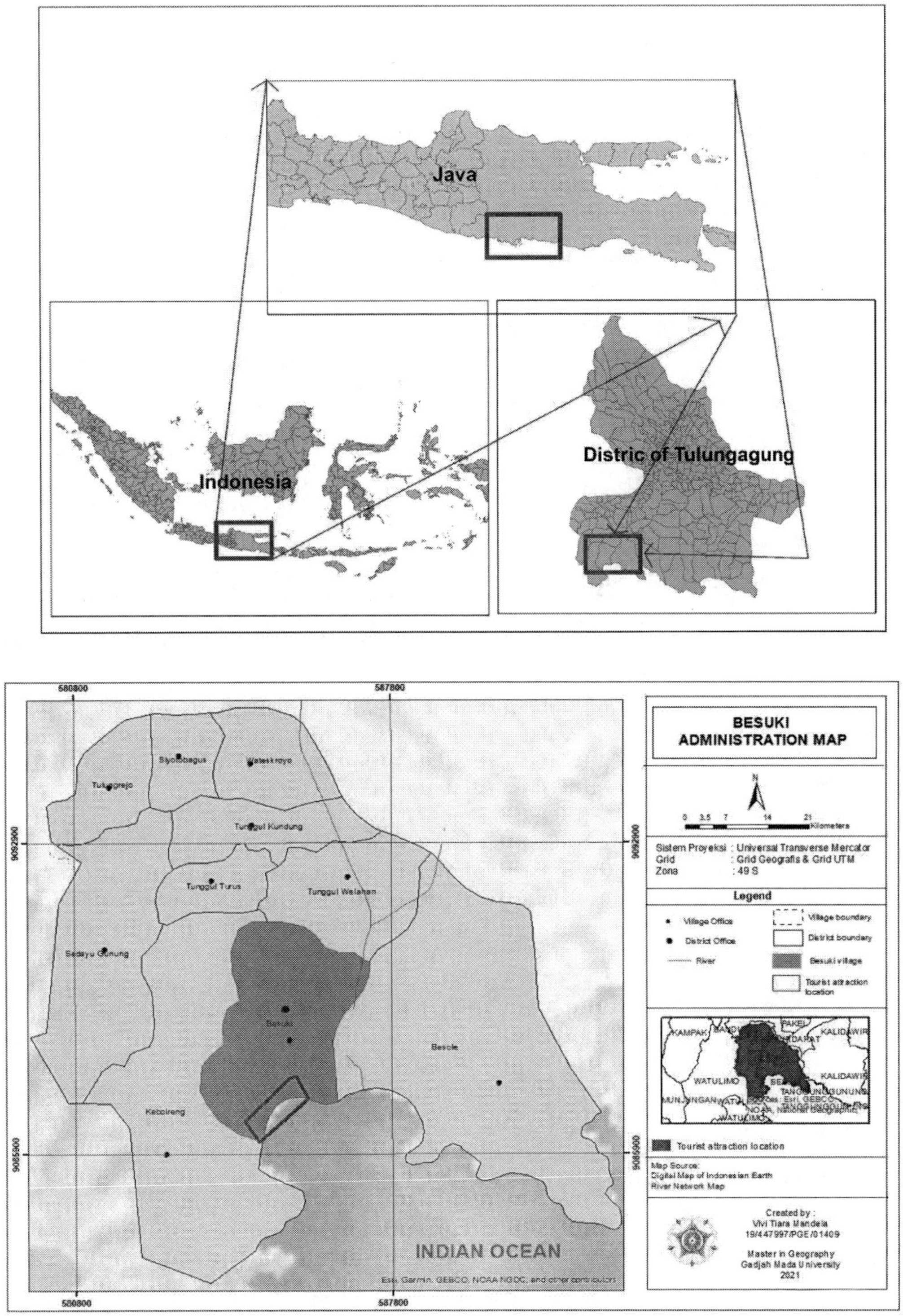

Fig. 1. The research location at Gemah Beach Tourism Object Tulungagung Regency, East Java

https://doi.org/10.1051/e3sconf/202132503008

3. Result and Discussion

3.1. Types of Economic Benefit Value of Gemah Beach Tourism Object

Gemah Beach tourism object is one of the mainstay tourist attractions in Tulungagung Regency, the restoration of this tourist attraction during the Pandemic period is a priority for the current Tulungagung Regency government (Tulungagung Regency Tourism Office, 2021). The existence of the Gemah Beach tourist attraction plays an important role in improving the local community's economy. Gemah Beach attractions are socially, economically and environmentally beneficial. The economic benefits of these attractions can be calculated through the economic valuation method. The economic valuation method requires a process of identifying the economic values that exist in the tourist attraction. Identification of the economic value of the Gemah Beach tourism object refers to the use value and non-use value of its use. The economic value identified by Gemah Beach tourism object can be seen in Table 1. Below.

Table .1 The economic value identified by Gemah Beach tourism object

Type of Economic Value		Benefits
Use value	Direct use value	Tourist attractions
	Indirect use value	The economic benefits of the surrounding community include traders, employees or managers, retribution and parking.
	Option use value	Means of environmental preservation and conservation
Non-use value	Existence value	The existence of Gemah Beach tourist attraction as a promotional medium for tourism objects in Tulungagung Regency
	Bequest value	The magnificence and cleanliness of the Gemah Beach tourist attraction costs money so that its sustainability can be enjoyed by future generations

Source: Data Processing Results, 2021

The total economic value of Gemah Beach tourism object is calculated from direct use value, indirect use value, optional use value, existence value, and bequest value. The direct use value of the Gemah Beach tourist attraction is the value of the direct benefits received by tourists. The direct use value of tourism objects according to Fauzi (2014) is one of the economic values related to tourism activities to fulfill recreational needs. In Figure 2 it appears that the number of tourist visits from year to year is relatively increasing, but from 2020 to 2021 the number has decreased due to the covid-19 pandemic.

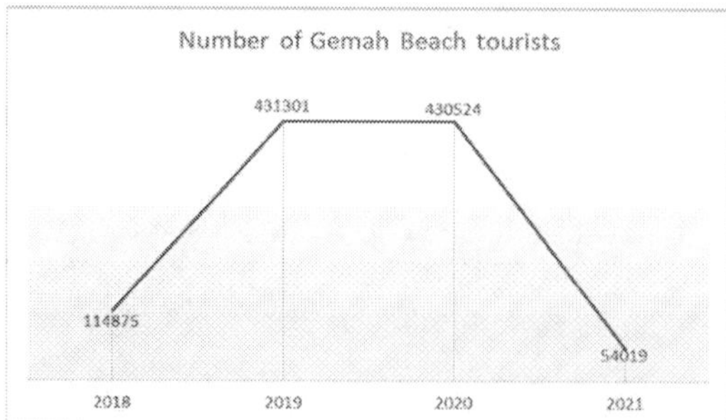

Fig. 2. Graph of the Number of Tourists at Gemah Beach Tourism Object

The indirect use value of the Gemah Beach tourism object is an economic benefit that is received and felt by the surrounding community and other parties. The results of the research are some of the economic benefits felt by the community in activities in the field of trade and services. In addition, the contribution of the community around the tourism object as a tourism object manager, as a parking guard, ticket counter guard and cleanliness is an economic benefit from the indirect use value of Gemah Beach tourism object. Another economic benefit which is an indirect benefit value is retribution income from tickets and tourist parking.

The option use value of the Gemah Beach tourist attraction is the value of economic benefits related to the preservation of the natural environment and conservation. Furthermore, the Gemah Beach tourist attraction is also used as a place to carry out health activities such as joint exercises on Sunday mornings with fresh air. Meanwhile, the value of the existence of the Gemah Beach tourist attraction is the value of the economic benefits associated with the existence of a tourist attraction in that location. Then, the value of the economic benefits of the existence of the Gemah Beach tourist attraction is as a promotional media for the surrounding tourist objectsthen the benefits of the existence of the Gemah Beach tourist attraction can be a promotional media for others such as Bayem Beach tourism, Klatak Beach, Niama Beach, Popoh Beach and JLS (South Cross Road).

The bequest value of the Gemah Beach tourist attraction is the value of the benefits of preserving the magnificence and cleanliness that is enjoyed by the next generation. The bequest value is calculated to find out how much value will be given for the sake of preserving the magnificence and cleanliness of the Gemah Beach tourist attraction.

3.2. Valuation of Economic Value of Gemah Beach Tourism Object

In this study, the calculation of the direct use value of the Gemah Beach tourism object used the travel cost method. The travel costs method is calculated by the amount of costs incurred by each tourist to visit the Gemah Beach tourist attraction. These costs consist of transportation costs, accommodation, consumption, souvenirs, entrance tickets, parking, and rides issued by tourists while at the Gemah Beach tourist attraction.

Table 2. Coefficients

Coefficients[a]					
Model	Unstandardized Coefficients		Standardized Coefficients	t	Sig.
	B	Std. Error	Beta		
(Constant)	4.649	.587		7.916	.000
total cost of travel	2,64E-03	.000	.140	2.162	.033
distance	-.026	.002	-.753	-10.849	.000
income	-1,97E-04	.000	-.126	-1.351	.180
education	.184	.136	.141	1.356	.178
age	-.014	.012	-.094	-1.173	.244

a. Dependent Variable: The frequency of the number of visits

Based on the results of the calculation of travel costs with regression analysis, the regression equation with a constant value of 4.649 is as follows :

$$Y = 4,649 + 2,643E - 6\,X_1 - 0,26\,X_2 - 1,968E - 7 + 0,184X_4 - 0,012\,X_5 \quad (5)$$

Among the five independent variables tested (x_1) total cost of travel, (x_2) distance, (x_3) income, (x_4) education and (x_5) age, only variable (x1) total travel cost biaya dan (x_2) distance which significantly affects the frequency of the number of visits. The influence of these two variables has a negative effect on the frequency of the number of visits. This means that the higher the travel costs incurred by visitors and the farther the distance to a tourist attraction, the less the frequency of the number of tourist visits. The results of the study are in line with what was stated [10] ; [11] that tourist visits will be influenced by travel costs incurred by tourists and the distance to tourist sites as a tourist choice. Furthermore, to calculate consumer surplus, the results of the regression analysis are then entered into the integral equation as follows :

$$CS = \int_{31.000}^{531.000} 4,649 - 0,000002643\,X_1 \quad (6)$$

Consumer surplus is calculated based on the average value of the number of tourist visits during the covid 19 pandemic from March 2020 – March 2021. The result of the calculation is that the consumer surplus value is Rp. 878,551,-/individual/visit.In the period from March 2020 - March 2021, during the COVID-19 pandemic with a 4-month lockdown, the number of visitors to the Gemah beach tourism object was only 26,009 people. Based on the number of visitors, the economic potential of the Gemah Beach tourist attraction as a direct use value during the covid 19 pandemic from March 2020 - March 2021 is IDR 22,850,237,669,-. This value is relatively lower than the economic value obtained before the COVID-19 pandemic. This happened because the number of visitors in that period was only 26.009 people, while in 2019 – 2020 before the covid pandemic the number of visits reached 431.301 people.

The next approach used to calculate the value of economic benefits from the Gemah Beach tourism object uses an indirect use value approach. The indirect value approach is calculated using the total benefit method of the Gemah Beach tourist attraction. The calculation of the total benefit method is based on the average annual profit of the parties who indirectly feel the economic benefits of the Gemah Beach tourist attraction. The economic activity actors are calculated because they get economic benefits from the existence of Gemah beach tourism objects, namely traders, employees/managers, parking revenues and retribution. Based on the average value of the profits obtained from the perpetrators of these economic activities at the Gemah Beach tourist attraction during the covid-19 pandemic from March 2020 - March 2021, an economic value of IDR 4,809,397,500 was obtained. The percentage of the economic value of economic activity actors in the Gemah Beach tourism object can be seen in Figure 3 below.

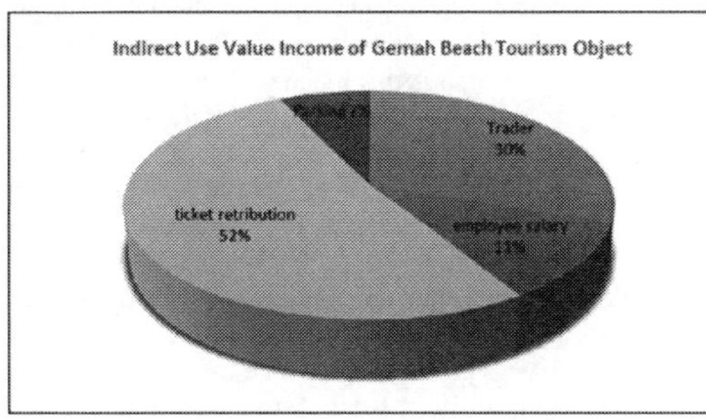

Fig. 3. Percentage of Indirect Use Value Contribution

Based on the figure 3, the income from the entrance fee has the greatest economic value benefit compared to income in other fields with a percentage of 51.75%, while the second-largest income is economic income from traders who trade in the Gemah Beach tourist area with a percentage of 30.27% of the total indirect economic value. Meanwhile, for other parties, only 10.83%, i.e. employee salaries, and parking revenues, only 7.15% of the total value of the indirect economic value of the Gemah Beach tourist attraction during the covid-19 pandemic. The contribution of economic value from the ticket retribution means that during the COVID-19 pandemic, visitors who came to Gemah Beach did not spend a lot of money, especially related to buying food. This is in line with what was stated by [12] ; [13] ; [14] The income of the UMKM sector, including the culinary sector, during the COVID-19 pandemic, experienced a decline in turnover because buyers were afraid. Therefore, to increase sales turnover, there must be a special strategy so that consumers feel comfortable and safe shopping or consuming the food they buy.

To calculate the economic benefits of the use value of the choice of Gemah Beach tourism object, the contingent valuation method (CVM) is used. The option use value is calculated based on the willingness to pay (WTP) of tourists for services provided by natural resources and the environment in order to maintain their sustainability [15]. The results showed that the willingness to pay of tourists for the sustainability of the Gemah Beach tourism object was an average of IDR 6,120 / tourist. Based on this, if the number of visitors is 26,0009 people during the period March 2020-March 2021, the use value of the choice of the Gemah Beach tourist attraction is Rp. 159,172,020,-. This phenomenon means that Gemah Beach tourists have a concern for the sustainability of the existence of tourism objects. This is in line with what was stated by [16] that tourists have a concern for the preservation of the tourism environment as a form to maintain the beauty and sustainability of the tourism object. However, the role of managers and stakeholders in maintaining the environmental sustainability of a tourist area is also very important.

Meanwhile, the value of economic benefits of the bequest value of the Gemah Beach tourism object, the willing to pay (WTP) method is used. The bequets value is calculated from the amount of costs used to maintain the magnificence and cleanliness of the Gemah Beach area given by tourists, with the aim that tourism objects can be enjoyed by future generations. The results of the study obtained the value of bequest with an average WTP of Rp. 5,920/tourist. If the average WTP value is IDR 5,920 and the number of tourist visits during the COVID-19 pandemic is 26,009 people, then the bequest value for the preservation and cleanliness of the Gemah Beach tourist attraction is IDR 153,970,320. This phenomenon also provides an image that tourists have a sense of caring for the magnificence and cleanliness of the Gemah Beach area. This is in line with what was stated by [17] that currently tourists are starting to have a high awareness of the cleanliness and magnificence of the tourist area. This can be seen from the results of research in Bukit Matang Kaladan Banjar, where the average WTP value given by tourists is Rp. 11,260/tourist. Thus, it can be said that tourists in general already have a high enough concern for the cleanliness and beauty of the current tourist area and to be enjoyed by future generations.

Meanwhile, to see the total economic value (TEV) of various economic values during the COVID-19 pandemic, all the use values obtained from the Gemah Beach tourism object are added up in value. In more detail, the total economic value of Gemah Beach tourism object can be seen in Table 2 below.

Table 3. Total Enomic Value

Use Value Type	Total/Rupiah	Percentage (%)
Direct Use Value (DUV)	Rp 22.850.237.669	81,22
Indirect Use Value (IUV)	Rp 4.809.397.500	17,09
Option Use Value (OV)	Rp 159.172.020	0,57
Existence Value (XV)	Rp 159.172.020	0,57
Bequest Value (BV)	Rp 153.970.320	0,55
Total Enomic Value (TEV)	Rp 28.131.949.529	100

Source: Data Processing Results, 2021

The calculation results in Table 1 show that the total economic value (TEV) of the Gemah Beach tourist attraction during the Covid-19 pandemic from March 2020-March 2021 is IDR 28,131,949,529. The contribution of each value to the total value varies; the direct use value contributed the most, which was 81.22% of the total economic value, then the indirect use value contributed 17.09% of the total economic value. Meanwhile, the other value, i.e. the option use-value, the existence value, and the bequest value, contributed less than 1% and ranged from 0.5% of the total economic value of the Gemah Beach tourism object.The results of these calculations provide an image that the COVID-19 pandemic has affected the total economic value (TEV) of Gemah Beach tourism due to a decrease in the number of visitors. Thus, it is quite logical that the number of visitors will affect the direct use value and indirect use value of the Gemah Beach tourism object.

The economic value generated by Gemah Beach tourism during the covid-19 pandemic can then be used as a reference in developing tourist objects in the face of the covid-19 pandemic. The strategy that can be done in dealing with the pandemic is to optimize economic potential by completing facilities and infrastructure in accordance with the Covid-19 health protocol and continue to urge tourists and managers to continue to comply with health protocols while traveling at the Gemah Beach tourist attraction in order to reduce the spread of the virus. the virus and the tourism industry continues to operate as before.

4. Conclusion

During the COVID-19 pandemic from March 2020-March 2021, the types of economic activities at Gemah Beach tourism objects are still quite diverse, even though there are social restrictions and lockdown policies for 4 months. Meanwhile, the results of the economic valuation obtained the total economic value (TEV) of the Gemah Beach tourism object during the COVID-19 pandemic from March 2020-March 2021, a total economic value of IDR 28,131,949,529 was obtained. The largest contribution to the total economic value (TEV) is provided by direct use value, which is 81.22% and indirect use value, which is 17.09%, while option use value, existence value and inheritance value only contribute less than 1%. Optimizing the economic and natural potential of the Gemah Beach tourist attraction by fulfilling facilities and infrastructure that are in accordance with the COVID-19 health protocol can be used as a strategy in dealing with the COVID-19 pandemic so that the tourism industry can continue to run and the spread of the virus can be controlled.

Acknowledgement

On this occasion the authors would like to thank the UGM Publisher and Publication Board (PBB) for providing RTA 2021 Grant funds, so that the authors can attend the ICST 2021 seminar and complete the thesis.

References

[1] W. Fan *et al.*, "Holmes Edward C., Zhang Yong-Zhen. A new coronavirus associated with human respiratory disease in China," *Nature*, vol. 579, no. 7798, pp. 265–269, 2020.
[2] H. Zhang, T. Cho, and H. Wang, "The Impact of a Terminal High Altitude Area Defense Incident on Tourism Risk Perception and Attitude Change of Chinese Tourists Traveling to South Korea," *Sustainability*, vol. 12, no. 1, p. 7, 2020.
[3] M. Sigala, "Tourism and COVID-19: Impacts and implications for advancing and resetting industry and research," *J. Bus. Res.*, vol. 117, no. June, pp. 312–321, 2020, doi: 10.1016/j.jbusres.2020.06.015.
[4] G. Garrod and K. G. Willis, "Economic valuation of the environment," *Books*, 1999.
[5] E. B. Barbier, M. Acreman, and D. Knowler, "Economic valuation of wetlands: a guide for policy makers and planners," 1997.
[6] S. Zandi, S. M. Limaei, and N. Amiri, "An economic evaluation of a forest park using the individual travel cost method (a case study of Ghaleh Rudkhan forest park in northern Iran)," *Environ. Socio-Economic Stud.*, vol. 6, no. 2, pp. 48–55, 2018, doi: 10.2478/environ-2018-0014.
[7] M. Suparmoko, D. Sudirman, Y. Setyarko, and H. S. Wibowo, "Valuasi Ekonomi Sumberdaya Alam dan Lingkungan(Cetakan Pe)," *Yogyakarta: BPFE-Yogyakarta*, 2014.
[8] A. Fauzi, "Valuasi ekonomi dan penilaian kerusakan sumber daya alam dan lingkungan," *IPB Pr, Bogor*, 2014.
[9] W. Sumatera, T. Kusumastanto, and A. Fahrudin, "Penilaian Ekonomi Wisata Pesisir Kawasan Carocok Painan , Kabupaten Pesisir Selatan , Sumatera Barat Economic Assesment of Coastal Tourism in Carocok Painan , Pesisir Selatan Pendahuluan Tinjauan Literatur," vol. 18, no. 1, pp. 1–21, 2018.
[10] S. Indayani and B. Hartono, "Analisis Pengangguran dan Pertumbuhan Ekonomi sebagai Akibat Pandemi Covid-19," *J. Perspekt.*, 2020.
[11] F. Zulpikar, P. D. E., T. V. Shelvatis, K. K. Komara, and M. Pramudawardhani, "Valuasi Ekonomi Objek Wisata Berbasis Jasa Lingkungan Menggunakan Metode Biaya Perjalanan di Pantai Batu Karas Kabupaten Pangandaran," *J. Reg. Rural Dev. Plan.*, vol. 1, no. 1, p. 53, 2017, doi: 10.29244/jp2wd.2017.1.1.53-63.
[12] L. D. Septiningrum, "Manajemen Strategi Untuk Meningkatkan Penjualan Food And Beverage Di Era Pandemi Covid 19," *JMBI UNSRAT (Jurnal Ilm. Manaj. Bisnis dan Inov. Univ. Sam Ratulangi).*, vol. 8, no. 1, 2021.
[13] W. laura Hardilawati, "Strategi Bertahan UMKM di Tengah Pandemi Covid-19," *J. Akunt. dan Ekon.*, vol. 10, no. 1, pp. 89–98, 2020, doi: 10.37859/jae.v10i1.1934.
[14] I. Fitriyani, N. Sudiyarti, and M. N. Fietroh, "Strategi Manajemen Bisnis Pasca Pandemi Covid-19," *Indones. J. Soc. Sci. Humanit.*, vol. 1, no. 2, pp. 87–95, 2020.
[15] A. Bharali and R. Mazumder, "Application of travel cost method to assess the pricing policy of public parks: the case of Kaziranga National Park," *J. Reg. Dev. Plan.*, vol. 1, no. 1, pp. 44–52, 2012.
[16] G. J. Ghobadi and M. S. Verdian, "The environmental effects of tourism development in Noushahr," *Open J. Ecol.*, vol. 6, no. 9, pp. 529–536, 2016.
[17] M. R. Fikri and N. Rahmini, "Analisis Willingness To Pay Pada Wisata Bukit Matang Kaladan Desa Tiwingan Lama Kecamatan Aranio Kabupaten Banjar Kalimantan Selatan," *JIEP J. Ilmu Ekon. dan Pembang.*, vol. 3, no. 1, pp. 158–170, 2020.

The energy charts for occupants' behavioural improvement through BIM

Rasyikin Roslan[1], *Rohayu* Che Omar[2*], *Hairin* Taha[1], *Intan* Nor Zuliana Baharuddin[1], and *Ratih* Fitria Putri[3]

[1]Institute of Energy Infrastructure, Universiti Tenaga Nasional, 43000 Kajang, Malaysia
[2]Department of Mechanical Engineering, Universiti Tenaga Nasional, 43000 Kajang, Malaysia
[3]Department of Environmental Geography, Faculty of Geography, Universitas Gadjah Mada, Yogyakarta, 55281, Indonesia

Abstract. BIM is an easy method to design sustainable buildings for professionals like architects and engineers. By using the real measurable environment, software tools can shape building concepts such as geometry, spatial relations, geographical details and the design of structural composites and buildings, and the usage of energy annual like energy costs and life-cycle costs can be estimated and calculated. This study revealed the findings of the energy consumption within one year can provide valuable insights into the projected resources in the UNITEN Administrative Building. The application of these data would simplify energy cost calculation, an essential component of the costs of operating a property. With the purpose of maximizing energy efficiency in green buildings, the outcomes can be used as guidance to advice occupant management in optimizing occupant actions.

1 Introduction

Crucial decisions on the construction of sustainable buildings are initially achieved conceptually during the development process. In general, energy modelling analysis is performed when the design and documents relating to the buildings have been created [1].

The American Architects Institute (AIA) has acknowledged the application of Building Energy Modelling (BEM) in projects involving energy-efficient building for its stakeholders. BEM enables designers plan and develop energy-efficient and high-performance buildings. Performance-based modelling provides housing owners with lower project lifecycle costs such as change orders, lower initial costs, maintenance and operating costs and at the same time, building occupants can achieve comfort and healthy indoor environment [2].

Systems adopting Green building certification can be applied as standard guidance for building comparisons, record progress and documentation of information and strategies used in buildings [3]. There are various systems of green building certifications and methodologies that are applicable for classification purposes such as the Building Research Establishment Environmental Assessment Method (BREEAM), the Building and Environmental Performance Assessment Criteria (BEPAC, Canada), the Leadership in Energy and Environmental Design (LEED,US), and the Green Star (GBCA, Australia) [3] which are widely used to determine the level achievement of environmental targets and to guide the process of planning. Besides, comprehensive environmental assessment guides such as the World Green Building Council and the Building Design Guide were also used to evaluate the building performance for green index building [1] and [3].

The experimental method is proposed in the design using different efficiency criteria specific to the construction of buildings such as building materials for roofs and walls and the ratios of window-to-wall [4], [5], [6]. The results suggest that data on climate plays an important role in selecting construction considerations that are ideally suited for the productive use of energy in buildings. Specifically, in contrast to the main climate group, sub-type climate classifications have a limited effect [4], [6] and [7].

As example for window-to-wall ratio, the architectural factor improvement results in an increment of 15% energy use in buildings. The insights gleaned from this study include the impact of space, material thickness and exterior openings, and the types of building envelopes in all climate classifications, which are included to design low-energy buildings [4], [5], [7] and [8].

Building efficiency analyses are essential aspects of sustainable building design. Performance analysis generally helps to determine how buildings work as a result of daylighting in terms of their built material and the light environment. The built material use shows a progressive reduction of environmental effects during the building's lifecycle, such as energy usage. Building information modelling (BIM) has described as an essential method for building performance analyses from the design to the maintenance process [9], [10] and [11].

The Revit Architecture 2017 and Low Carbon Cities Framework (LCCF) and the KeTHHA 2015 assessment system are used to determine the best energy

* Corresponding author: rohayu@uniten.edu.my

performance and low carbon city assessment using alternative material configurations [12].The Revit modelling highlights the different building materials and design principles by defining alternatives using innovative approaches to significantly minimize the building's operating energy usage during its annual life cycle and quality requirements for cities with low carbon dioxide. By the conclusion of the research study, an energy usage and energy performance chart help to educate staff and strategies for enhancing energy efficiency that impact occupants' behavioural improvement is crucial.

2 Methodology

2.1 Experimental site

UNITEN Administration Building (2°58'39.2"N, 101°43'53.1"E) use as a sample of case study to evaluate the performance of the building with respect to energy analysis as shown in Figure 1 and Figure 2. Autodesk Revit and Green Building studio were used to simulate the UNITEN administration building to generate the energy model for energy performance and green building index. The location and analysis of the site are the main elements that can affect the position of the buildings. These contributing factors could be used to determine the spatial orientation and facade of the buildings and to develop the construction of the surrounding landscape.

The main research tool used by architects and engineers in the visualization of 3D models is the Building Information Modelling (BIM) system. The physical and technical features of a visual building through BIM energy analysis use to calculate the energy efficiency and thermal comfort of a building [13] and [14]. The data processed include the geometry and orientation of the building and their thermal properties, the types and categories of materials used, the lighting, cooling, heating, and ventilation systems including the recent utility rates and weather data and observations.

Fig.1. UNITEN administration building.

Fig.2. Front part of UNITEN administration building.

3 Result and discussion

3.1 Autodesk Revit Software and GBS analysis

Primary and secondary data, including the as-built administrative UNITEN building drawings to be analysed, were collected. The primary data collected using laser scanning mapping and the secondary data from hardcopy transformed to digitalization to suit the standard used in building information modelling (BIM). The drawing is registered in Autodesk Revit software and transforms into a 3D model as shown in Figure 3, featuring building components like windows, walls and floors that are created for energy simulation to suit various designs and building elements.

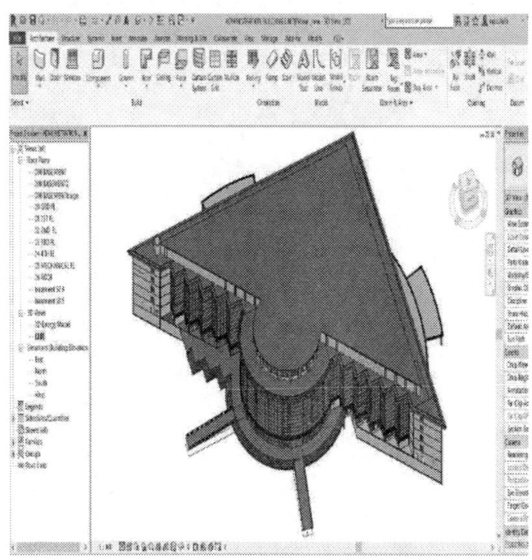

Fig.3. A 3D model.

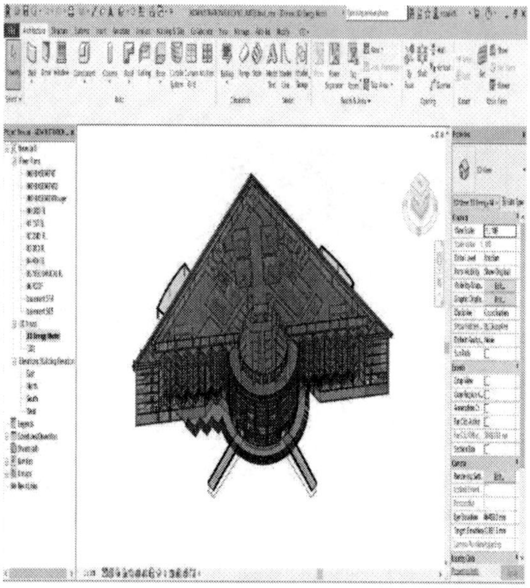

Fig. 4. Energy model in 3D.

3.2 Green Building Studio software analysis

The energy modelling and analysis of the building were processed using the Green Building Studio (GBS) which is a cloud-based simulation tool as illustrated in Figure 3 and Figure 4. The extensible green construct Language gbXML file format was used to connect BIM software to external simulation tools.

The results provided a simulation of building performance that allows users to improve energy consumption and efficiency by reducing carbon emissions to ensure sustainable buildings. The GBS results show that in this building occupied up to 1,208 people, with the use of average lighting power density 10.67 W/m2 lower than typical value, average equipment power density up to 10.61W/m2, specific fan flow 3.9 LPerSec/m2, total fan flow 22,214 LperSec, specific cooling and specific heating lower than typical value, i.e. 0m2/kW and total cooling and heating capacity less and not more than 12, 484701kW, as shown in Figure 5.

The summary of the results were then analysed for the amount of electrical energy consumption, water usage and annual energy costs and fuel emissions report for carbon dioxide (CO2) as demonstrated in Figure 5, Figure 6, Figure 7 and Figure 8. The annual energy cost indicates as RM1, 598,722.00, with a life cycle cost RM566, 574,590.00. Once the energy analysis and conceptual design are completed, the building is then evaluated and analysed using the life cycle assessment (LCA) module based on sustainability requirements. The calculation of the main operating energy including embodied energy such as the refine and deliver energy, extracted energy and emissions of air, water and land over the life cycle of the UNITEN administration building was carried out by using the Impact Estimator (IE) in this software as shown in Figure 5 and Figure 6. Apart of that, other environmental effects of the UNITEN administration building also displayed the inherent trade-offs which are associated to the increase in insulation that decrease operational energy consumption (Figure 6 and Figure 8).

The energy chart and analysis in Figure 8 showed the significant factors that affect the energy consumption of the analyzed model in the Administrative UNITEN building. The results showed that most annual electric use derived from lighting systems (19.1%), air condition system (58.70%) and others (22.3%). It showed that usage of energy that contribute to the annual electric end-use was projected by the behavior of electric consumers in the building. The cause was due to the lack of preparation of building occupants in the utilization of electrical equipments, unpredictable actions of occupants, preference of comfort criterion for occupants, lack of the goal of energy consumption by the occupants, the behavioural difference between occupants, lack of understanding and the lack of coordination on the particular required behaviors.

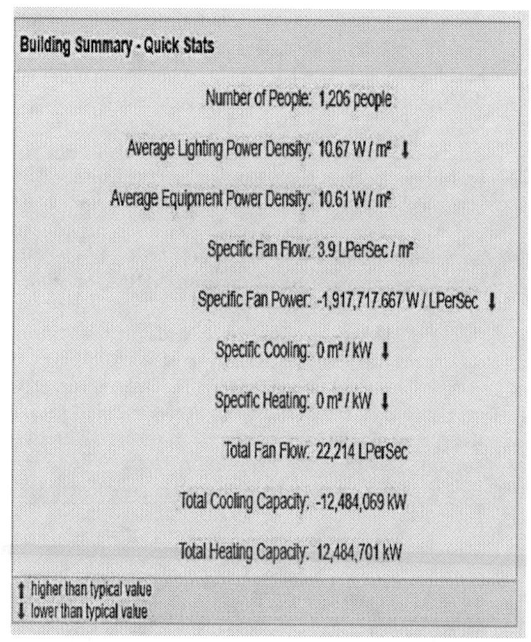

Fig. 5. The GBS building summary.

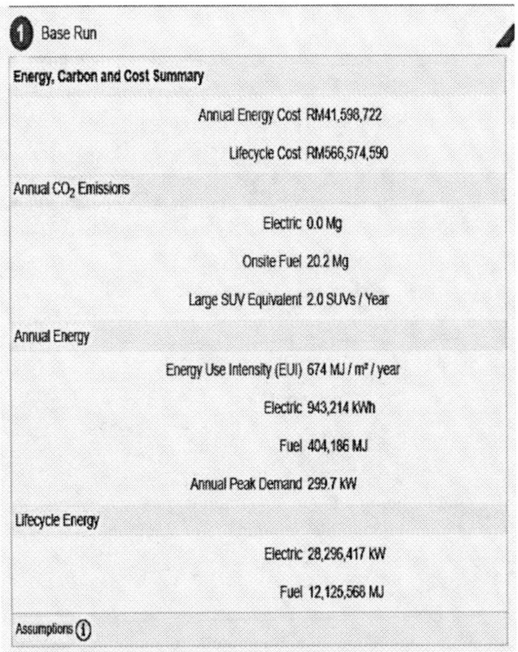

Fig. 6. The energy, carbon and cost summary.

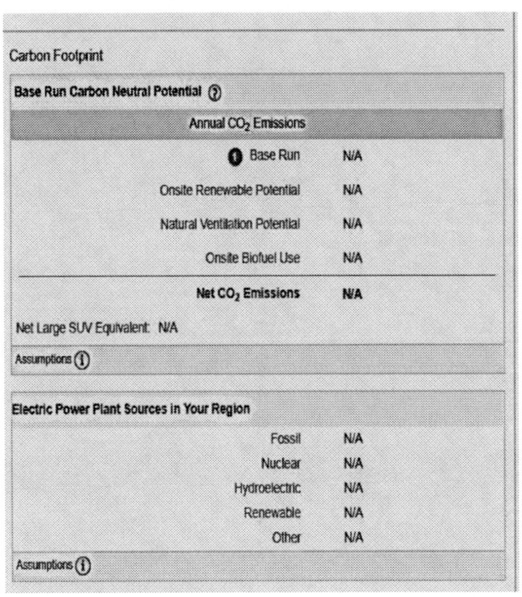

Fig. 7. The carbon footprint analysis.

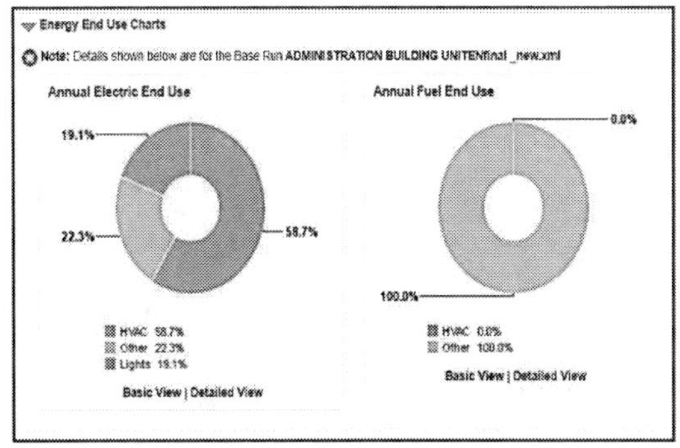

Fig. 8. The energy chart.

4 Conclusion

The findings of the energy analysis offer a valuable perspective on the expected energy to be obtained or lost in the UNITEN administrative building within one year. The application of these data would simplify the estimation of energy cost, which contributed largely to the running costs of any building. Life Cycle Cost consolidates environmental and energy impacts, which had a significant effect on the future building portion, as against traditional building estimates. In addition to obvious environmental advantages, Green Building Certification is less costly to maintain and better suited for industrial and residential occupants. Therefore, the results used to educate the occupant's behaviour and strategy that effect occupant behavioural change is vital in order to enhance the energy performance towards green building initiative.

This study was funded by UNITEN internal grant (J 510050715). The authors thank Mr Aiman for his assistance in data processing and compilation. Thanks, Universiti Gadjah Mada, Indonesia for geographical, behavioural analysis study and partnering as co-authorship.

References

1. F. Jalaei, A. Jrade, ITCon **19**,211–220 (2014)
2. The American Institute of Architects (AIA), Integrating energy modelling in the design process, (The American Institute of Architects: Washington, DC, USA, 2012)
3. N. Wang, K. M. Fowler, R.S. Sullivan, Green building certification system review, (Pacific Northwest National Laboratory, Battelle for USDOE and US GSA; Whole Building Design Guide (WBDG), 2012)
4. J. Hong, G. Shen, Y. Feng, W. S. T. Lau, C. Mao, Cleaner Production **103**, 249–259 (2015)
5. A. Allouhi, Y. El Fouih, T. Kousksou, A. Jamil, Y. Zeraoulli, Y. Mourad, Cleaner Production **109**, 118–130 (2015)
6. M. K. Najjar, V. W. Y. Tam, L. T. D. Gregorio, A. C. J. Evangelista, A.W.A. Hammad, A. Haddad, Energies **12**, 1515 (2019)

7. F. H. Abanda, L. Byers, Energy **97**, 249–259 (2016)
8. A. Alwisy, S. BuHamdan, M. Gul, Energy and Buildings **178**, 347–259 (2018)
9. H. Khan, M. Asif, Sustainabilty **9**, 4:640 (2017)
10. M. V. Shoubi, M. V. Shoubi, A. Bagchi, A. S. Barough, Ain Shams Engineering Journal **6**, 1:41–55 (2015)
11. J. B. Kim, W. Jeong, M. J. Clayton, J. S. Haberl, W. Yan, Automation in Construction **50**, 16–28 (2015)
12. F. Jalaei, A. Jrade, Sustainable Cities and Society **18**, 95–107
13. C. C. Ohueri, W. I. Enegbuma, K. K. Kuok, N. M. Wong, L. T. Ng, R. Kenley, *Preliminary evaluation of synergizing BIM and Malaysian carbon reduction and environmental sustainability tool,* in International Conference on Sustainability in Energy and Buildings, June, Springer, Cham (2018)
14. H. J. Kuo, S. H. Hsieh, R. C. Guo, C. C. Chan, Energy and Buildings **130**, 95–107 (2016)

The relevancy of law enforcement toward water quality in Kali Buntung's downstream area

Lisa Elizabeth Gunawan* and *Irwan Bagyo* Santoso

Department of Environmental Engineering, Faculty of Civil, Environment and Geo Engineering, Institut Teknologi Sepuluh Nopember, Indonesia

Abstract - *Kali Buntung* actually is a man-made *afvoer* river. However, due to city development, the location now become a strategic asset for Sidoarjo Region. Due to that reason, this river experience external forces such as rapid land conversion, social, law enforcement, upstream quality, rainfall and surrounding businesses that make the rivers' burden much heavier. This paper aims to investigate the external forces matters and the calculating the impact of those external forces towards the water quality in *Kali Buntung*'s downstream area. Using Structural Equation Model (SEM) as the basis to produce such intended result, the indicator data will be processed using SMART PLS statistic software. The result of this research was: (1) the upstream water quality is very significant with the water quality in downstream. (2) rainfall have a significant effect toward upstream water's revitalization process along the way the water flows toward the downstream. (3) social and business aspects have significant influence to the water quality in downstream. (4) law enforcement has positive impact on downstream's water quality, even surpass either the rainfall or business factors. (5) the law enforcement to businesses may reward greater impact than to the common inhabitant surrounding. (6) since land conversion is inevitable, law enforcement might be the only way to keep enjoying the favorable impact of business capability and agriculture field to reduce the pollutant in *Kali Buntung*'s downstream.

1 Introduction

Basically, water bodies, to be specific in this paper, rivers, has their own mechanism to improve their own conditions. Therefore, when it comes to water bodies condition's improvement, in general will be 2 (two) major aspects play the role which are the internal and the external. In internal case, water bodies have so-called "self-purification" mechanism to heal themselves[1]. In virgin area such as untouch forest, its actually enough for take care of themselves and their surroundings. However, not many rivers have this exclusivity. Some other rivers, which located in crowded cities, may not be able to do so since they are heavily impacted by the external forces. Those external forces make the rivers' burden much heavier so that it disables the rivers to collect themselves just the mother nature does. This paper aims to investigate the external forces matters and the calculating the impact of those external forces towards the water improvement conditions of "*Kali Buntung*" in Sidoarjo region in east java, Indonesia.

Kali Buntung actually is a man-made *afvoer* river. However, due to city development, the location now become a strategic asset for Sidoarjo Region. Nowadays, it serves as sewer, irrigation, fishpond and also residential area. It's an industrial sewer in upstream but also used for food production in the downstream. It's has a very crucial role as the bio-accumulator for the community surrounding. To be able to serve for those contradictive needs, it's supposed that *Kali Buntung* has a better quality at least, second grade class in Indonesian law. Hence, its necessary to handle this condition by managing the external forces correctly.

In this paper, the data will used both secondary data obtained from the government offices and primary data from on-field observation. The data includes both the quality and the quantity at *Kali Buntung* in Balungbendo as the beginning point and (6) Ketegan as the end point. Those will be compared to the impact of (1) rainfall, (2) technology aspect, (3) law aspect, (4) social aspect, and (5) land conversion aspect of surrounding in *Kali Buntung* area. Those will be process using Structural Equation Modelling Partial Least Square to analyse which aspect matters the most for improving *kali-Buntung* water condition's.

Structural Equation Model (SEM) is a cause and effect analysis used to describe link between observation indicator link between observation variable (indicator variable) and variable that cannot be measured directly (latent variable)[2]. It application's is broad and possible to be further explored. In Putranto and Santoso (2018) it used for choose among factors for drainage system solution. While, in this research is, SEM is used to analyze the effect and the significancy of those aforementioned 6 aspects. By having the analysis, the writer may come out with effective suggestions for the

* Corresponding author: lisaegunawan@gmail.com, irwan080565@gmail.com

best way to improve *kali-Buntung* downstream water condition's. Both researches are looking for the solution for non-direct measured variable but different in the object. Within the social discussion, the latent variables which rarely being seen physically are important factors in actual. Sometimes, those even more than the countable physical one. In the other hand, environment in urban area sit in social shade, thus, any insight from this side of perspective may bring new horizon. This research offered a solution for water quality in Kali Buntung area, in future, it also possible to be a valuable study for other rivers with the same case in Indonesia.

2 Materials and methods

2.1 Independent variables

A set of observed variables[3] that may be valued, each act as indicators for a specific aspect mentioned above. Those variables derived from the existing condition in research location, furthermore, categorized in accordance to the applicable set of theories.

2.2 Smart PLS studies

Structural Equation Model (SEM) used for analyse the effect and the significant value[4] of independent variables used in the model. Those calculation including the interaction between variable, the number of sample, and various considerations as well. Since the number or the samples are limited, not every SEM software can run this properly. SMART PLS software allows the SEM calculation even with sample as few as 30 samples only, compare with others' which require at least 200 samples to run properly.

3 Result and discussion

3.1 Location

This research is taking place in area surrounding Kali Buntung in Sidoarjo. The river is stretch less than 25 kilometre long. However, as every river, it may impact the surrounding area. In this case, the impact area is indicated by the red line in picture below.

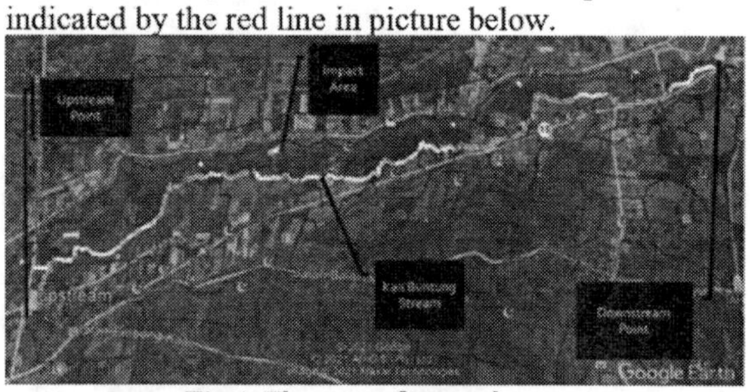

Fig.1. The map of research area

The upstream and the downstream point taken from local government's usual surveillance points as those points already have the data needed in the research.

3.2 Model Concept

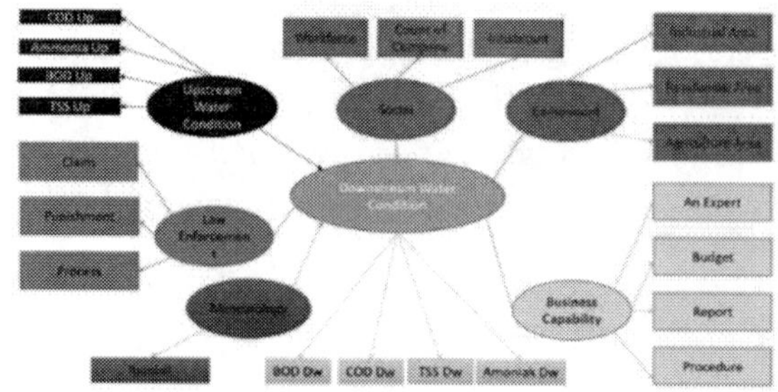

Fig. 2. The model concept mindmap

The model was derived from applicable theories collected by the writer. Later will be tested using the outer model step in SEM such as validity test, reliability test and multicollinearity test. Once passed, it will be eligible for further process in inner SEM model.

4 Output of structural equation modelling analysis

Even though the aimed end result (dependent variable) is only 1, which are the water quality of the *Kali Buntung* Downstream, the model produces 2 types of output in 7 steps. In usual case, this relation assumed as linear model sampling, but to do so it is required large amount of sampling. As mentioned in previous section, the sampling in this research is mainly divided by historical timeline, using SMART PLS, it allows both relatively small amount of data and non-linear sampling because it will eventually further process using bootstrapping method.

Bootstrapping process allows resampling process that may solve the data abnormality. Refer to 22 observed variables in total, and the aimed accuracy level of 80%, the value of T-Table is 1.316.

4.1 Handed Model

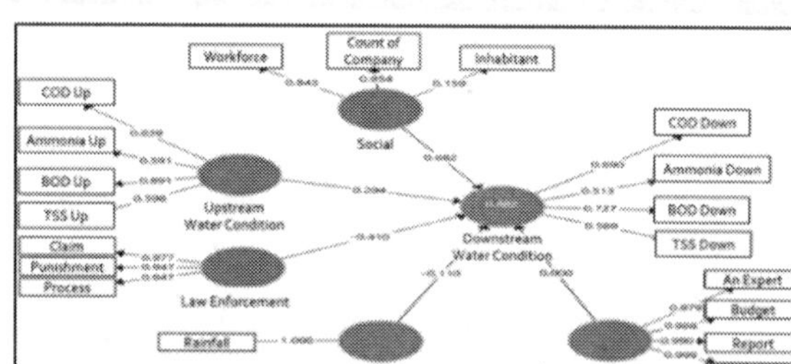

Fig.3. The Outer Model: Upstream-Downstream-Rainfall – Social – Business – Law Enforcement

As the business and rainfall, law enforcement also has a favorable impact toward pollutant reduction in downstream area. Even more, with the presence of law enforcement, it able to shift the business capability and rainfall place as the most impactful factor among the others which shown by the number of law enforcement

(-0.410) compare with rainfall (-0.110) and plunge business capability into zero impact. Not to mention, reduce the unfavorable impact of social and upstream water in adding the downstream pollutant.

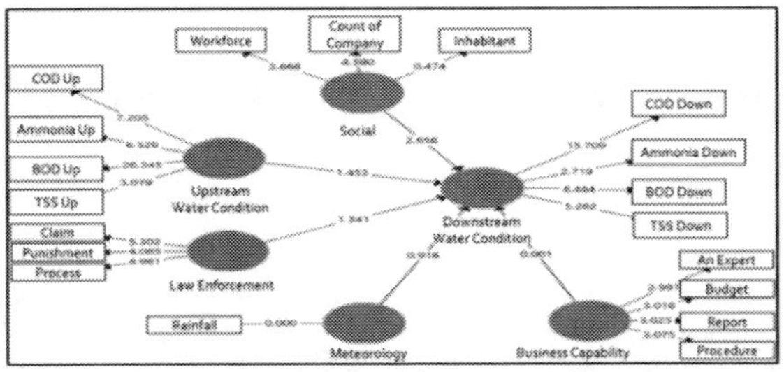

Fig.4. The Bootsrapping Model: Upstream-Downstream-Rainfall – Social – Business – Law Enforcement

In addition, law enforcement even with the competition with the other factor, still can counted as a significant factor (1.341>1.316) along with the Social (2.658>1.316) and Upstream (1.453>1.316) afterward. It means that, in this research, this factor does a powerful tools for revitalizing *Kali Buntung* downstream water quality. Only one out of those variables within those three significant factors proves insignificant, which is the civil population surrounding. It means that actually, business has more significancy in affecting the downstream water quality.

4.2 The End Game: 5-Handed Model

From the entire aspects observed, only one aspect proves insignificant, which is the land conversion aspect, eventhough in actual, green area is very good for the water quality. Moreover, since law enforcement, at the end, has the highest contribution and also proved significant, it even able to surpass business and rainfall as the most impacting factor. Suppose that this aspect is favorable control tools for water quality in downstream.

In addition, law enforcement actually a very interesting factor to explore further. It supposes can also direct the land and business aspect through government permit and regulation. Business shall bound to the environment permit as well as land to the land permit.

5 Suggestion

Even though some aspects in inner model do not look promising, but once it being breakdown again, they actually did it in their own way. Instead, they do impact the upstream water quality significantly, prove by their substantial amount of outer value. When it talks about the downstream water quality, it is really closely related to the upstream water quality. In wholistic river principle, garbage in garbage out really does matter. It is impossible to have good downstream water quality without good upstream quality either.

Since the land conversion and city development is inevitable factor in most civilization case, there is no way to keep the amount of agriculture field in promising area even though it is the best way to keep the water quality. It is true that meteorology has a significant effect too, but still, in accordance to the figure above, it has the lesser effect compare with law enforcement. Moreover, rainfall is complicated factor to be played since it also can trigger the flood and draught disaster. Therefore, the least optional effort is by controlling business institution, social population and land usage throughout the enforcement of the law. And it can only be done through law enforcement. Furthermore, the law enforcement to businesses may be an agile tool to embrace the other factors too. it can also direct the land and business aspect through government permit and regulation. As the business shall bound to the environment permit as well as land to the land permit.

6 Conclusion

This research aims to determine the effect and the significant indicators from land conversion, social, law enforcement, business and meteorology as well as give suggestion about what to do to keep the water quality of prominent *Kali Buntung* downstream area. The analysis resulted, that the 5 handed model works best.

The model suggests that social, law enforcement and upstream water quality is very significant for the downstream water quality. But ever since, wholistically river flows from the upstream to the downstream, the control in upstream will lead to the better downstream. Especially when the upstream is heavily impacted by the social and business which runs by the population, controlling the population through law enforcement may be the only way to save the water quality condition. It not also applicable in *Kali Buntung* case, but suppose the same conduct applies to other numerous amounts of similar case in Indonesia.

Reference

1. S. Tian, Z. Wang, & H. Shang, *Study on the Self-purification of Juma River* (2011)
2. T.A. Brown, *Confirmatory factor analysis for applied research*. The Guilford Press (2006)
3. J.C. Putranto. & I.B Santoso, *Water Inundation Control Modeling for Greges Drainage System in Surabaya, Indonesia* (2018)
4. D.F. Wicaksono, B. Rahadi, L.D. Susanawati, *The Evaluation of Waste Water Pollution of Paper Factory in Sungai Klinter Kabupaten Nganjuk*(2013)
5. A.K. Novitasari, *Analysis on Pollution Sources at Kali Surabaya*(2015)
6. M.A.A. Latif, M.S Pallu, J. Patanduk, *The Study of Water Quantity and Quality in Sungai Tallo for Water Source* (2012)

Assessment of Social Vulnerability to Kiematubu Volcano in Tidore Island, North Maluku

Widiyana Riasasi[1], *Muh Aris* Marfai[2,3], *Bachtiar W* Mutaqin[2,*] *Danang Sri* Hadmoko[2], *Franck* Lavigne[4], *Audrey* Faral[4], and *Helvetia* Wijayanti[2]

[1]Geography Department, Universitas Amikom Yogyakarta, 55281 Sleman, Indonesia
[2]Coastal and Watershed Research Group, Faculty of Geography, Universitas Gadjah Mada, 55281 Bulaksumur, Indonesia
[3]Geospatial Information Agency, 16911 Cibinong, Indonesia
[4]Université Paris 1 Panthéon Sorbonne, Laboratoire de Géographie Physique UMR 8591, 92190 Meudon, France

Abstract. Kiematubu volcano is often considered as non-volcanic, even though it consists of basalt material, since it has never erupted before. In fact, that small volcanic islands have a high risk due to their restriction on means and resources. The study aims to assess the social vulnerability of the community in Tidore Island that may be exposed to the eruption of the Kiematubu volcano. There has not been previous research in Tidore Island regarding volcanic vulnerability yet. The social vulnerability is an initial assessment of disaster management, which will affect in optimizing community's capacity then minimizing the disaster impacts. Social parameter of demographic condition, health facilities, and education facilities was weighted to assess social vulnerability. The result shows that the social vulnerability class of Tidore Island is dominantly low, approximately 80%, the rest is middle and high, with a percentage of 13% and 7%, respectively. Mostly, the low vulnerable villages are due to less population density. However, the highly vulnerable villages, Gamtufkange and Indonesiana, consist of very high and high population density. The southeast part of Tidore Island, where both villages are situated, is the center of human activities, such as governmental, trades, and education.

1 Introduction

Indonesia's location lying on the ring of fire causes volcanic range formation (Figure 1). Moreover, Indonesia is also known as an archipelagic country, which consists of more than 17,000 islands. Aside from flood, earthquake, land subsidence, or landslide, a volcanic eruption is a natural disaster that frequently occurred in Indonesia because the active volcanoes are scattered in major (Sumatera, Java, Sulawesi) and small islands near Maluku, Nusa Tenggara, Halmahera) [1,2].

Maluku Kepulauan is situated in the eastern part of Indonesia with many small islands and has active volcanoes due to the location precisely on the subduction area of Eurasia and Indo-Australia tectonic plates. Gamalama volcano in Ternate, Gamkonora and Ibu volcanoes in West Halmahera, and Dukono volcano in North Halmahera had advisory (*waspada*) stage in 2017 [3]. Gamalama volcano is the most active one that had occurred 77 times eruption since 1510 [4].

Ternate is a small island with a population density of more than 200,000 in an area of 111.4 km². It makes the population has a high vulnerability to eruption hazard. At the southeast part of Ternate, lay a small island called Tidore with a 1,550 km² area. The island is part of Tidore Kepulauan municipality, with Kiematubu volcano situated in the center of the Tidore region. Kiematubu volcano has the highest summit of 1,730 meters in Maluku.

Fig. 1. Volcanic range formation in Indonesia. Different color means different status for each volcano from green (normal) to red (warning) (Courtesy: MAGMA Indonesia, 2021).

* Corresponding author: mutaqin@ugm.ac.id

Despite having various beneficiaries of natural resources, such as agriculture, livestock, and tourism, the volcano also has a potential hazard, such as eruption and its post-disaster hazard, i.e., famine, epidemic disease, or lahar [5,6]. Inhabitant living in small volcanic islands has a higher risk of being exposed to eruption disaster than those living in major islands. Kiematubu volcano has never erupted in history, nevertheless according to the first ascent expedition by C.G.C. Reinwardth in 1821, found that the material of the volcano was basalt [7]. The only tectonic disaster in history based on previous historical records by the Dutch scientists that had been occurred earthquake with violent shocks in 1855 in Ternate and Tidore. It caused 25 houses to collapse and 24 fatalities due to blocks of rock sliding from a hill in Tidore [7].

Even though the Kiematubu volcano is considered at a rest-phase stage, eruption's potential still exists. Inhabitant that lives in potential hazard area is vulnerable, mostly the poor, gender minorities, the disabled, the elderly, and those who are unable to access safety resource [8,9]. Restriction to access resources, i.e., transportation, communication, and limited natural resources, often becomes an obstacle in regional development in a small island [4,10], including in terms of disaster management.

Disaster risk reduction requires improvement in community capacity. To improve the capacity, it is necessary to measure the level of vulnerability in the potential hazard area. There are several types of vulnerabilities to assess the disaster risk, such as physical, economic, and social elements. Social vulnerability tends to be a multidimensional concept because there is no agreement about social vulnerability within the social science community [11]. Approaches were implemented to assess the social vulnerability using indicators to identify the traits of a community on responding hazard. This research aims to assess the social vulnerability of the community in Tidore Island that may be exposed to the eruption of the Kiematubu volcano.

2 Methods

Administratively, Tidore Kepulauan municipality is part of North Maluku province. Its area scatters in three different islands, namely Tidore Island, Mare Island, and Halmahera Island. However, the study particularly assessed Tidore Island, where the Kiematubu volcano is situated (Figure 2). There are 31 villages' lies on Tidore Island, which is included in four sub districts.

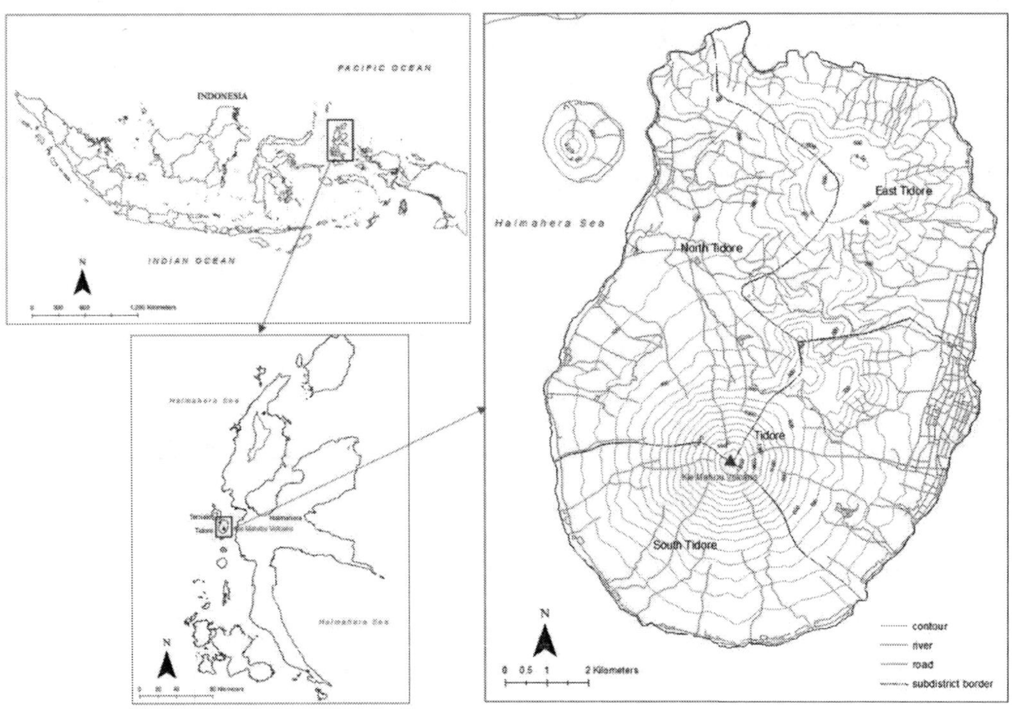

Fig. 2. Administrative map of Tidore Island.

Tidore Kepulauan municipality data in 2020 was employed along with spatial data of topographical maps and high-resolution satellite imagery data. Some parameters to access social vulnerability consist of demographic condition, education and knowledge, health, institution, and social insurance [2,11]. Detailed variables of each parameter are shown in Table 1. Assessment of social vulnerability used the village as the unit of analysis.

Table 1. Parameters of social vulnerability assessment, modified from Ref. [2,11].

Parameter	Variable
Demographic Condition	Population density
Education and Knowledge	Number of the formal institution (elementary)
Health	Number of health facilities (hospital)

Variable of education and knowledge indicated the number of an existing elementary school in a village. Elementary school is a stage of education with basic knowledge, such as writing and reading. The least capacity of the human to coordinate in emergency or crisis situations is understood and capable of reading and writing. The health parameter used the number of health facilities, particularly hospitals. The existence of a hospital is important, especially in the response stage when a disaster occurred.

The variable was weighted based on its level of impact affected by the hazard. The existence of education institutions will increase community capacity; therefore, the village has a school was weighted 1. If the school does not exist, weighted 0. A similar process was employed to the existence of a hospital in the village. According to the National Standard of Indonesia, it divided the category into four in the case of population density [12], as presented in Table 2.

Table 2. Category of population density, modified from the National Standard of Indonesia.

Category	Population density (population/hectare)	Weight
Low	< 150	3
Middle	151 - 200	2
High	201 - 400	1
Very high	> 400	0

Statistic calculation was performed to determine social vulnerability class—furthermore, the highest total value of all variables minus the lowest value equal to the class range. Moreover, the range was divided into three social vulnerability classes: low, middle, and high. The bigger number of weights, the higher class of vulnerability. Weights of social vulnerability are shown in Table 3.

Table 3. Social vulnerability class.

Total Weights	Vulnerability Class
0 - 1	Low
2 - 3	Middle
4 - 5	High

3 Results and discussion

Vulnerability defines a characteristic or circumstance of a community, structure, assets, or system that is susceptible to be damaged or disrupted by a hazard, or simply means as the potential for loss [11, 13,14]. The measurement of vulnerability cannot be precise because various perspectives determine the subject/object, which is multidimensional and differential, scale-dependent, and dynamic [15,16]. However, the most recognized approach to assess vulnerability consists of five dimensions: physical/functional, economic, social, environmental, and political/institutional [17].

The study mainly focused on assessing social vulnerability due to potential hazards of Kiematubu volcano, using secondary data of Tidore Kepulauan municipality in 2020, and employing Geographic Information System (GIS) as the tools. The result shows that the social vulnerability class of Tidore Island is dominantly low, approximately 80%, the rest is middle and high, with a percentage of 13% and 7%, respectively. Detail social vulnerability class of the village is shown in Table 4.

Table 4. Social vulnerability of Tidore Island.

Sub district	Village	Class
North Tidore	Afa Afa	low
South Tidore	Dokiri	low
East Tidore	Dowora	low
Tidore	Folarora	low
Tidore	Goto	low
North Tidore	Gubu Kusuma	low
Tidore	Gura-Bunga	low
North Tidore	Jaya	low
East Tidore	Kalaodi	low
East Tidore	Mafututu	low
North Tidore	Mareku	low
North Tidore	Ome	low
North Tidore	Rum	low
North Tidore	Rum Balibunga	low
North Tidore	Sirongo Folaraha	low
Tidore	Soa Sio	low
Tidore	Soadara	low
South Tidore	Toloa	low
Tidore	Tomagoba	low
South Tidore	Tomalou	low
South Tidore	Tongowai	low
Tidore	Topo	low
Tidore	Topo Tiga	low
East Tidore	Tosa	low
South Tidore	Tuguiha	low
North Tidore	Bobo	middle
North Tidore	Fobaharu	middle
South Tidore	Gurabati	middle
Tidore	Seli	middle
Tidore	Gamtufkange	high
Tidore	Indonesiana	high

Mostly, the low vulnerable villages are due to less population density. However, the highly vulnerable villages, Gamtufkange and Indonesiana, consist of very high and high population density. The southeast part of Tidore Island is the center of human activities, such as governmental, trades, and education, where both villages are situated. There are two universities located in Gamtufkange. Notwithstanding, health and education facilities do not exist in those villages. The map of the vulnerability class can be seen in Figure 3.

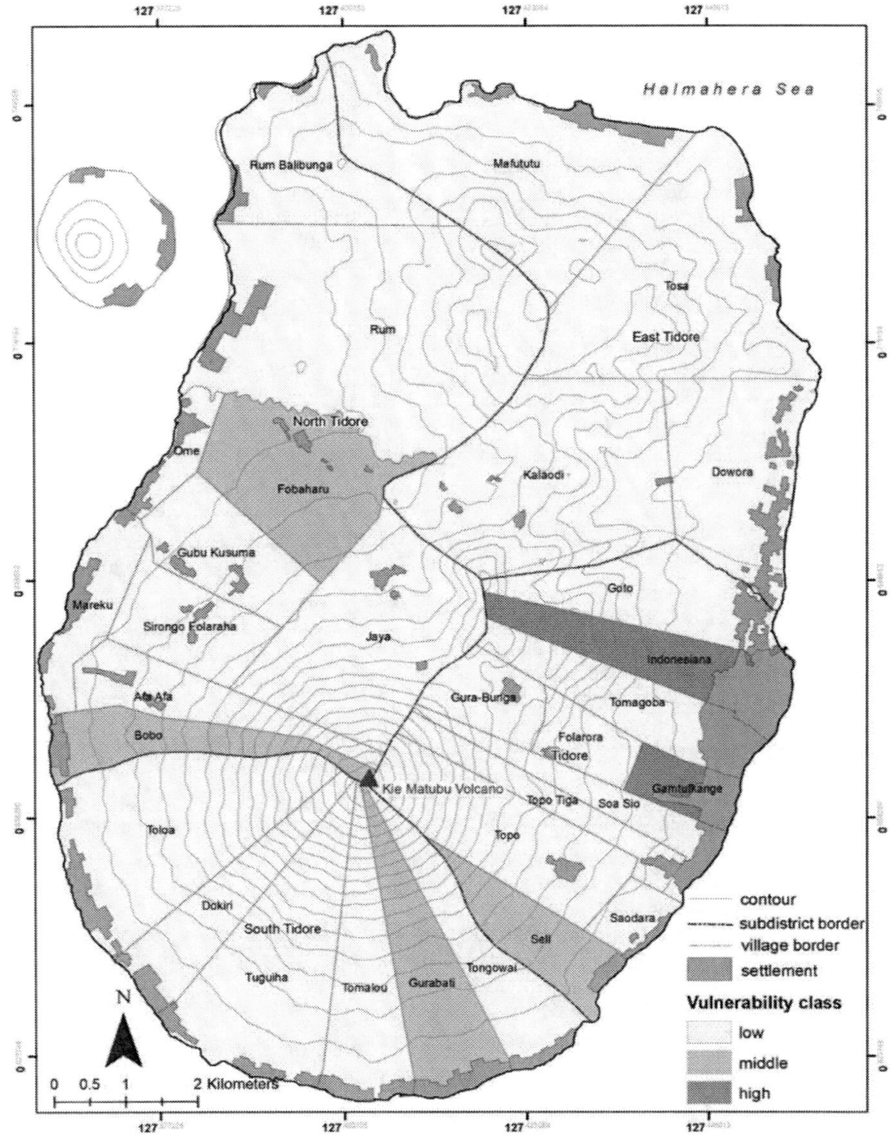

Fig. 3. Social vulnerability map of Tidore Island.

Nevertheless, disaster preparedness in Tidore Island has been quite good because each subdistrict has a health facility (hospital). The South Tidore subdistrict has a hospital in Tomalou, the North Tidore's hospital is in Ome, the East Tidore's is in Tosa, and the Tidore subdistrict's hospital is in Tomagoba village. Meanwhile, education facility of elementary school exists in most villages of the island.

To reduce the risk of disaster, it is necessary to minimize the vulnerability and optimize the community's capacity that is potentially affected by the disaster. Regardless, the Kiematubu volcano has never erupted in history, and disaster preparedness is still necessary to minimize the impacts of a future eruption. As in Sendai Framework for disaster risk reduction, enhance disaster preparedness for effective response is stated as one of the priorities for action [18].

Social vulnerability assessment is more focused on measuring the social facilities' aspects against potential disaster. Hence, the supporting aspects must be advanced to reduce vulnerability, such as improving health facilities service and effectively developing education facilities. Assessment of social vulnerability occupied the second rank based on the research in Gamalama volcano conducted by Ref. [2], after economic vulnerability, then followed by the physical vulnerability.

Collaboration with its stakeholders is an integral part of vulnerability assessment because it aims to reduce vulnerability by enhancing the lack and fixing the existing [19]. As a small island, Tidore faces different challenges in disaster management compared to major islands, such as the evacuation scenario or access to resources. On the disaster management, the government of Indonesia has established the National Agency for Disaster Management (BNPB) at the national level and the Local Agency for Disaster Management (BPBD) at the province and regency levels. Therefore, all stakeholders, including the community and scientists, need to be integrated into disaster risk reduction action [20].

4 Conclusions

Based on the assessment of the social vulnerability to the Kiematubu volcano in Tidore Island, most villages are categorized as low vulnerable. Indonesiana and Gamtufkange village are classified as having a high social vulnerability, that both are the center of human activities in Tidore Island. Kiematubu volcano has never erupted before; nonetheless, disaster preparedness is necessary to minimize disaster risk.

The study was initial research for the social vulnerability of the Kiematubu volcano. Further social vulnerability with a more comprehensive parameter should be assessed to get more valid information that will be useful for the community and related stakeholders to reduce the social vulnerability. Furthermore, companies with other vulnerabilities, i.e., economy and physical vulnerability, assessment of vulnerability in all aspects will complete the assessment to formulate disaster risk reduction.

This research was funded by the grant of National Basic Research from the Indonesian Ministry of Research, Technology and Higher Education - GASTROPODE Project. We thank anonymous reviewers for their constructive comments on the manuscript. The authors also acknowledge the Regional Disaster Management Agency of North Maluku.

References

1. M.A. Marfai, L. King, L.P. Singh, D. Mardiatno, J. Sartohadi, D.S. Hadmoko, A. Dewi. Environmental Geology **56**, 2 (2008)
2. E.T.W. Mei, I.M. Sari, A. Fajarwati, D. Safitri. *Assessing the Social Economic and Physical Vulnerabilities to Gamalama Volcano*, in Proceedings of the 1st International Conference on Geography and Education, 29 October 2016, Malang, Indonesia (2016)
3. B. Nurgianto. *Empat Gunung Berapi di Maluku Utara Berstatus Waspada*, https://nasional.tempo.co/read/881559/empat-gunung-berapi-di-maluku-utara-berstatus-waspada (accessed on 20 May 2021)
4. A. Hidayat, M.A. Marfai, D.S. Hadmoko. International Journal of Geomate **18**, 66 (2020)
5. E.T.W. Mei, A. Harijoko, N.I. Setiawan, H. Wijayanti, W. Riasasi. IOP Conference Series: Earth and Environmental Science **412** 012012 (2020)
6. C.J. Tanguy, C. Ribière, A. Scarth, W. Tjetjep. Bull Volcanol. **60** (1998)
7. A. Wichmann. KNAW Proceedings **21** (1919)
8. B. Balgos, J.C. Gaillard, K. Sanz. Gender and Development **20**, 2 (2012)
9. E.T.W. Mei, F. Lavigne. Geological Society Special Publication **361**, 1 (2012)
10. E. Wilkinson, E. Lovell, B. Carby, J. Barclay, R.E.A. Robertson. Resources **5**, 21 (2016)
11. S.L. Cutter, B.J. Boruff, L.W. Shirley. Social Science Quarterly **84**, 2 (2003)
12. National Standard of Indonesia. Tata cara perencanaan lingkungan perumahan di perkotaan. SNI 03-1733-2004 (2004)
13. H. Khan, L.G. Vasilescu, A. Khan. Management and Marketing Journal **6**, 1 (2008)
14. UNISDR. UNISDR Terminology on disaster risk reduction (2009)
15. C. Vogel, K. O'Brien. Bulletin on Global Environmental Change and Human Security **13** (2004)
16. J. Birkmann. *Indicators and criteria for measuring vulnerability: theoretical basis and requirements*, in Measuring vulnerability to natural hazards towards disaster resilient societies (2006)
17. R. Ciurean, D. Schröter, T. Glade. *Conceptual Frameworks of Vulnerability Assessments for Natural Disasters Reduction*, in Approaches to Disaster Management - Examining the Implications of Hazards, Emergencies, and Disasters (2013)
18. UNISDR. Sendai Framework for Disaster Risk Reduction 2015-2030 (2015)
19. D. Schröter. *Vulnerability to changes in ecosystem services*, in Assessing Vulnerability to Global Environmental Change – Making Research Useful for Adaptation Decision Making and Policy (2005)
20. F. Lavigne, V. Jean, Suwardji, Kusnadi, Hiden, Edong, Syamsuddin, M. Reghezza-Zitt, B.W. Mutaqin, M.N. Malawani. Earthquakes and tsunamis in Lombok, NTB : from hazard assessment to crisis management **1** (2020)

Geomorphology of the small island of Tidore and Hiri (North Maluku, Indonesia)

Bachtiar W Mutaqin[1*], *Muh Aris* Marfai[1,2], *Danang Sri* Hadmoko[1], *Franck* Lavigne[3], *Audrey* Faral[3], *Helvetia* Wijayanti[1] and *Widiyana* Riasasi[4]

[1]Coastal and Watershed Research Group, Faculty of Geography, Universitas Gadjah Mada, 55281 Bulaksumur, Indonesia
[2]Geospatial Information Agency, 16911 Cibinong, Indonesia
[3]Université Paris 1 Panthéon Sorbonne, Laboratoire de Géographie Physique UMR 8591, 92190 Meudon, France
[4]Geography Department, Universitas Amikom Yogyakarta, 55281 Sleman, Indonesia

Abstract. Tidore and Hiri Islands in North Maluku Province is the result of a complex tectonic setting. In contrast with Ternate Island and its well-known volcano, Gamalama volcano, there is still a lack of research about volcanic information or volcanic landform in Tidore and Hiri Islands. Even though the two islands also have volcanoes, i.e., Hiri and Tidore/Kiematubu volcano. This study aims to provide geomorphological information, especially in Tidore and Hiri Islands, since this information is hard to find whereas it is very important to disaster mitigation and landuse planning. We used remote sensing images, digital elevation models (DEM), and geological maps to classify geomorphological information of the small island of Tidore and Hiri in the North Maluku based on geomorphological aspects, i.e., morphology, morphogenesis, morpho-chronology, and morpho-arrangement. As a volcanic island, the slope in Tidore and Hiri Island is dominated by a slightly steep (8º-16º) and a steep slope (16º-35º), while the relief is dominated by hills and mountainous. Volcanic landforms on Tidore and Hiri Islands are characterized by relatively symmetrical cone-shaped volcanic cones, which are accumulations of falling pyroclastic material and lava ejected from magma vents. Volcanic islands include the case of Tidore and Hiri Island have radial centrifugal river flow patterns. The material on Tidore and Hiri Islands is dominated by Holocene volcanic rocks. In Tidore, there is alluvial material on the west and east coasts of the island. This study result also can be developed into more detailed geomorphological maps, or landscape evolution in a volcanic island, or spatial planning.

1 Introduction

North Maluku in the eastern part of Indonesia is a tectonically active region since it is situated in the convergence of Australian, Eurasian, and Philippine Sea plates [1]. As a result, there are several volcanoes and small islands in the North Maluku province, including Tidore and Hiri Islands. These two islands are located between the main island of Halmahera and the Halmahera arc and stretch along a volcanic arc from Morotai in the northern part to the Bacan Island in the southern part of the North Maluku province (Figure 1).

Compared with other volcanoes in this region, the Gamalama volcano in Ternate Island is categorized as one of the most active volcanoes in the North Maluku Province during the Holocene era [2]. That is why there are numerous researches in Ternate Island related to Gamalama volcano, for example, evacuation route planning during the Gamalama eruption, landslide-induce tsunami due to Gamalama instability, or disaster mitigation in Ternate [3,4,5].

In contrast with Ternate Island and its well-known volcano, Gamalama volcano, there is still a lack of research about volcanic information or volcanic landform in Tidore and Hiri Islands. Even though the two islands also have volcanoes, i.e., Hiri and Tidore/Kiematubu volcano [1,6,7]. Furthermore, Tidore Island also has a caldera form with a 2-km diameter in the northern part of Tidore Island. Tidore Island is made up of two quite distinct volcanic structures. Kiematubu is a stratovolcano with a stunningly conical and circular peak on Tidore Island's southern part [8,9,10].

With a height of 1,730m, Kiematubu volcano is the highest volcano compared with other volcanoes in the North Maluku province. The vast and lower landform in the northern part of Tidore Island can be identified as a caldera from Holocene era [7,10].

Meanwhile, Hiri Island, which located 9-km in the north of Gamalama volcano, is a small island with 3-km of diameter and dominated by forest. Although it was identified as Holocene stratovolcano, nothing is known about this volcano [7,10]. With a height of 630-m, Hiri volcano is known as the northernmost volcano along the sequence of small volcanic islands which is located separate from the main island of Halmahera in the North Maluku province (Figure 2). This study aims to provide geomorphological information, especially in Tidore and Hiri Islands, since this information is hard to find whereas it is very important to disaster mitigation and landuse planning [11].

* Corresponding author: mutaqin@ugm.ac.id

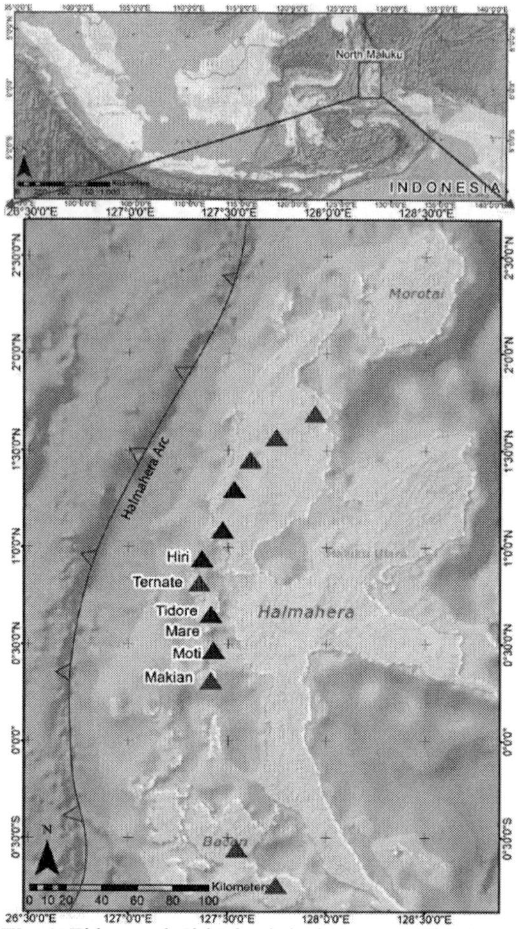

Fig. 1. Tidore and Hiri Islands in the North Maluku.

Fig. 2. View of Hiri Island from Sulamadaha Beach in the northern part of Ternate Island.

2 Material and methods

Spatial data as the main input in this research include remote sensing images, digital elevation models (DEM) from the Indonesian Geospatial Agency, and secondary data on geological maps from the Indonesian Geological Agency. Terrain features and landforms can be identified through satellite imagery, DEM data, and GIS-based mapping [12,13]. The advantage of multispectral satellite imagery is that it can be used to identify features in a large area and has been used in various studies related to mapping of terrain features and landforms [6,11,13,14]. DEM data is also used in river flow extraction and recognizes flow patterns that indicate a landform [14].

The combination of analysis of satellite imagery, DEM, aerial photography, and geological maps can also be used to identify the geomorphic classification of surface materials and landform types [15]. We classify geomorphological information of the small island of Tidore and Hiri in the North Maluku based on geomorphological aspects, i.e., morphology, morphogenesis, morpho-chronology, and morpho-arrangement.

3 Results and discussion

Morphological information in this study uses relief and slope analysis. There are two data in determining the relief, namely the slope and elevation. The elevation is directly obtained from the DEM data, while the slope needs to proceed from the DEM data [14]. Relief is classified into 4, namely plain, hillock, hill, and mountainous. As a volcanic island, the slope in Tidore and Hiri Island is dominated by a slightly steep (8º-16º) and a steep slope (16º-35º) [16]. While the relief is dominated by hills and mountainous (Figure 3).

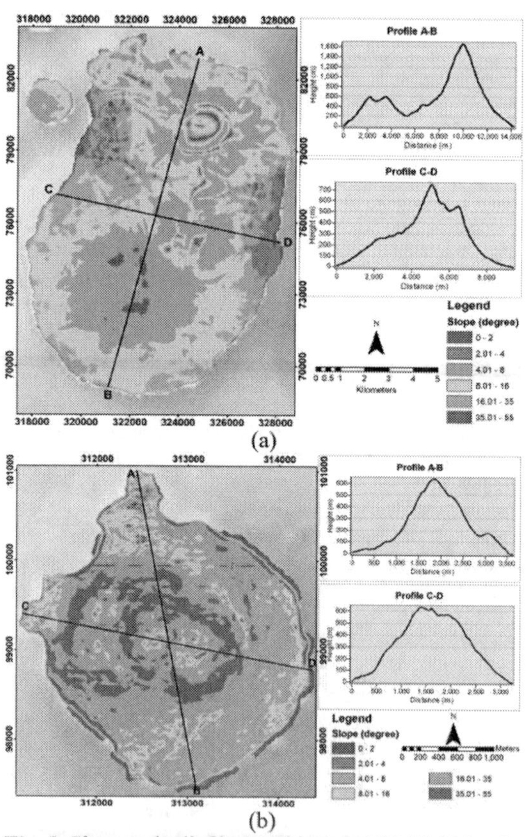

Fig. 3. Slope and relief in (a) Tidore and (b) Hiri Islands.

 https://doi.org/10.1051/e3sconf/202132503012

The middle and upstream areas in the form of mountains and hills are used as forests and plantations, while the plains relief is used as agricultural land and settlements. The morphology of the plains has a limited area around the coast, both in Tidore and Hiri Islands. Hence, settlements and public infrastructure are concentrated along the coast (Figure 4).

(a)

(b)

Fig. 4. (a) Settlements and public infrastructure are concentrated along the coast of Tidore which has plains relief; and (b) Mountains and hills are used as forests and plantations in Hiri Island.

Volcanic landforms on Tidore and Hiri Islands are characterized by relatively symmetrical cone-shaped volcanic cones, which are accumulations of falling pyroclastic material and lava ejected from magma vents [12,13]. The caldera on the north side of Tidore Island has a diameter of 2 km and was formed by the explosive eruption of the main magma vent. In addition to a very large eruption, a caldera can also be formed due to the collapse of the walls and bottom of a volcanic crater. Occasionally, craters or calderas have been filled with lava flows, lava domes, or water, and often have been eroded in older volcanic cones.

Flow pattern information requires river flow data on the identified island. The flow can be obtained by extracting DEM data into flow accumulation (Figure 5). In the flow accumulation data, pixels with high flow accumulation values are locations of concentrated flows and can be used to identify river flows, both permanent (perennial) and periodic (intermittent) rivers. If the river flow on the island can be detected, the flow pattern can also be interpreted. The results of this river flow detection are checked for correctness through field activities.

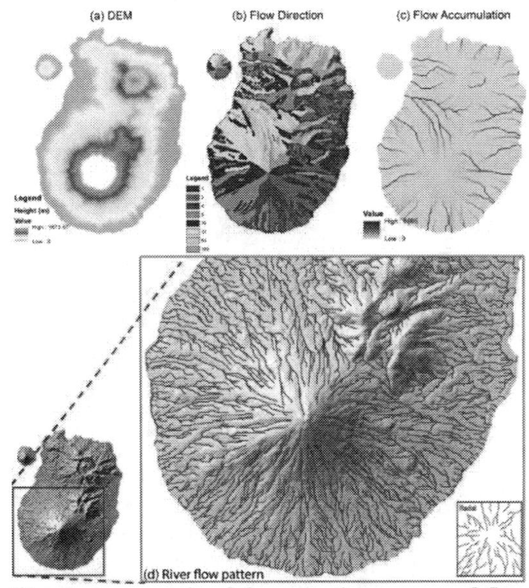

Fig. 5. An example of the extraction process of flow accumulation from DEM data for interpretation of characteristics of river flow patterns in Tidore Island: (a) DEM, (b) Flow direction, (c) Flow accumulation, and (d) River flow pattern.

At the locations of the field survey, river flows were found in accordance with the detection made through remote sensing data. Thus, the results of flow extraction through remote sensing data produce data that is in accordance with the conditions in the field and can be used as a method of extracting flow patterns (Figure 6).

Fig. 6. An example of field survey in point number 30 for river flow pattern in Tidore Island.

Volcanic islands have a characteristic river flow pattern in the form of radial centrifugal, while non-volcanic islands have different river flow patterns, such as dendritic [17]. This also includes the case of Tidore and Hiri Island where both islands have radial centrifugal river flow patterns.

Material identification in Tidore and Hiri Island is carried out through geological map analysis and validation during field surveys. Tidore and Hiri islands were formed by volcanism processes, which are characterized by their relatively spherical shape with lava, pyroclastic, or sub-aerial depositional rock material [7,12,18,19]. The material on Tidore and Hiri Islands is dominated by Holocene volcanic rocks. In Tidore, there is alluvial material on the west and east coasts of the island [6,7].

4 Conclusions

Here we present the geomorphological information of Tidore and Hiri Island in the eastern part of Indonesia. This study result is important as a basis data for disaster mitigation and planning, considering the difference of terrain and landform features in these islands. Tidore and Hiri Island were classified as volcanic islands due to their morphology, morphogenesis, morpho-chronology, and morpho-arrangement.

As a volcanic island, the slope in Tidore and Hiri Island is dominated by a slightly steep (8^0-16^0) and a steep slope (16^0-35^0), while the relief is dominated by hills and mountainous. Volcanic landforms on Tidore and Hiri Islands are characterized by relatively symmetrical cone-shaped volcanic cones, which are accumulations of falling pyroclastic material and lava ejected from magma vents. Volcanic islands include the case of Tidore and Hiri Island have radial centrifugal river flow patterns. The material on Tidore and Hiri Islands is dominated by Holocene volcanic rocks. In Tidore, there is alluvial material on the west and east coasts of the island. This study result also can be developed into more detailed geomorphological maps, or landscape evolution in a volcanic island, or spatial planning.

This research was funded by the grant of National Basic Research from the Indonesian Ministry of Research, Technology and Higher Education - GASTROPODE Project. We thank anonymous reviewers for their constructive comments on the manuscript. The authors also acknowledge the Regional Disaster Management Agency of North Maluku.

References

1. E. Gunawan, F. Ghozalba, Syauqi, Y. Widiastomo, I. Meilano, N.R. Hanifa, Daryono, S. Hidayati, Geotech. Geol. Eng. **35**, 425 (2017)
2. A. Hidayat, M.A. Marfai, D.S. Hadmoko, GeoJournal, **In Press**, (2020)
3. S.F. Syiko, A.R. Tuminingtyas, A. Yudono. Procedia Environ. Sci. **17**, 344 (2013)
4. Saaduddin, J. Neuberg, M.E. Thomas, J. Hill. *The Mt Gamalama Instability Level in Generating Landslide-Induced Tsunami in Ternate Island, Indonesia*, in Understanding and Reducing Landslide Disaster Risk, WLF 2020, Springer (2021)
5. A. Hidayat, M.A. Marfai, D.S. Hadmoko. Int. J. GEOMATE. **18**, 66 (2020)
6. B.W. Mutaqin, W. Handayani, F.S.C. Rosaji, D. Wahyuningtyas, M.A. Marfai. Jurnal Geografi. **13**, 2 (2021)
7. T. Apandi, D. Sudana. Geologic map of Ternate quadrangle, North Maluku, scale 1:250.000 (Geological Research and Development Center, Bandung, 1980)
8. K. Cosburn, M. Roy. J Volcanol Geotherm Res. **407**, 107123 (2020)
9. D. Karátson, M. Favalli, S. Tarquini, A. Fornaciai, G. Wörner. J Volcanol Geotherm Res. **193,** 3-4 (2010)
10. Global Volcanism Program. Volcanoes of the World, v. 4.10.0 (14 May 2021) (Smithsonian Institution, 2021)
11. A.Quesada-Román, P.M. Pérez-Briceño. J. Maps. **15**, 2 (2019)
12. M. Kervyn, F. Kervyn, R. Goossens, S.K. Rowland, G.G.J. Ernst. Geol. Soc. Spec. Publ. **283**, 5 (2007)
13. J. Napieralski, I. Barr, U. Kamp, M. Kervyn. *Remote sensing and GIScience in geomorphological mapping*, in Treatise on Geomorphology, Elsevier (2013)
14. B.W. Mutaqin, F. Lavigne, Y. Sudrajat, L. Handayani, P. Lahitte, C. Virmoux, Hiden, D.S. Hadmoko, J.C. Komorowski, N.D. Hananto, P. Wassmer, Hartono, K. Boillot-Airaksinen, Geomorphology **327**, 338-350 (2019)
15. J.-C. Thouret. Earth-Sci. Rev. **47**, 1-2 (1999)
16. B.W. Mutaqin, M.A. Marfai, D.S. Hadmoko, H. Wijayanti, F. Lavigne, A. Faral, J. Hunan Univ. Nat. Sci. **48**, 6 (2021)
17. J.C. Santamarta, R.J. Lario-Bascones, J. Rodríguez-Martín, L.E. Hernández-Gutiérrez, R. Poncela. IERI Procedia. **9**, 135 (2014)
18. H.P. Chan, C.P. Chang, T.H. Lin, M. Blackett, H. Kuo-Chen, A.T.S. Lin. Int. J. Remote Sens. **41**, 4 (2020)
19. J.J. Coello-Bravo, A. Márquez, R. Herrera, M.J. Huertas, E. Ancochea. J Volcanol Geotherm Res. **401**, 106980 (2020)

An overview of Persian Gulf environmental pollutions

Majid Khazali[1]* and *Lobat* Taghavi[2]

[1]Department of Energy Engineering and Economy, Faculty of Natural Resources and Environment, Science and Research Branch, Islamic Azad University, Tehran, Iran
[2]Departmet of Environmental Sciences, Faculty of Natural Resources and Environment, Science and Research Branch, Islamic Azad University, Tehran, Iran

Abstract. Aquatic environments are among the most vulnerable habitats that affect all Organisms. The Persian Gulf is home to numerous plant and animal species, and the biological safety is at risk due to its numerous human activities. Traffic of numerous oil and commercial vessels in the area, waste water, desalination, proximity of numerous industrial, refinery and petrochemical units to shores and rivers, oil rigs, fishing, and tourism activities are potential hazards to the environment and should lead to severe chemicals and physicals pollution into the Persian Gulf ecosystem. All this while the Persian Gulf is highly vulnerable due to various natural in the reason of such as low water circulation due to isolation, high evaporation, and low water depth. In this article, previous studies on the status of pollutants in the Persian Gulf have been reviewed. Much of the research on Persian Gulf pollution has focused on hydrocarbons and heavy metals. However, according to recent research, the amounts of these pollutants have been lower than the allowed levels. Also, further research into the impacts of plastics and microplastics on the Persian Gulf ecosystem should be undertaken.

1 Introduction

The relationship between humans and the environment is very tense, today due to population growth and increasing energy consumption, the extent of damage to nature goes far beyond self-healing and self-purification. The mentioned damages often could not be repaired and refined even in a small amount by their own. Aquatic environments are among the most vulnerable habitats that affect all Organisms. The Persian Gulf is located in the Middle East and is joined to the sea of Oman through a narrow strait called Hormuz. The existence of such a disjunction in geography limits the rate of species exchange between the regions [1]. The complete replacement of water in this region lasts 3–5 years. As a result, it takes a long time for the Persian Gulf to exterminate pollutants [2]. The Persian Gulf, which is located between the Arabian Peninsula and the Islamic Republic of Iran, and to the Asian Southwest, is an Indian Ocean extension [3]. Seven countries constitute the Arabian Peninsula: Kuwait, Bahrain, Qatar, Saudi Arabia, Yemen, Oman, and United Arab Emirates (UAE). Islamic Republic of Iran, Iraq and six of the above-mentioned countries produce nearly 25 percent of the total oil consumed by the world. With 35 meters average depth (hardly exceeding 100 meters), the Persian Gulf is considered as a rather shallow water body. The Persian Gulf is a semi-enclosed water body, and the Strait of Hormuz connects it to the global open waters [4–6]. The Gulf hosts some rather susceptible ecosystems related to a naturally highly stressful environment with poor flushing properties, elevated evaporation rates, UV exposure, salinity, and temperatures. Thus, it is more likely that the contaminant materials become subject to slower dispersion, and constrained dilution, and thereby, remain in the region for longer periods of time [7,8]. The integrity of the ecosystem is highly influenced by the activities of humans.

In the Persian Gulf countries, the pollution and contamination of air, soil, and water by inorganic and organic pollutant materials is a key issue. One could attribute the local pollution of the Gulf to extraction of oil from platforms, rapid growth of industrialization, and the traffic of transportation by heavy ships as well as oil spills, although the offshore generated pollution is the second general pollution origin in the same regions followed by discharge of contaminant materials varying from basic or acidic toxins to mercury and the other compounds with organic origin, including Polycyclic Aromatic Hydrocarbons (PAHs) and n-alkanes [2]. At all times, one of the main environmental hazards for human well-being has been marine pollution.

In the Persian Gulf, numerous species are active while they are subject to physiological constraints [4]. Therefore, the extra stresses in the reason of various pollutants to the aquatic environment are of potential extreme outcomes [7,9,10], For example, tanker spills and accidents begin by wars, terminals, land-based urban and industrial sources, offshore oil exploration, and agriculture and recreational development. It is evident that the Persian Gulf seafood such as shrimps and fishes are of high value for earning revenues from exports as well as for being consumed locally. Thus, due

* Corresponding author: majid.khazali@srbiau.ac.ir

to various socio-economic motivations, keeping the good quality of the aqueous environment is of utmost importance [9,11]. Additionally, since many countries neighboring the Persian Gulf rely on seawater desalination as a drinkable water source for industrial and domestic use, the quality of seawater is extreme importance. Every day, in the Persian Gulf region, a water volume of 14,000,000 m^3 is desalinated [12], and after processing, the hot saline wastes are released into the coastal regions which shallow depths.The late modeling attempts have demonstrated that the anticipated desalination capacity growth and the resultant increased discharge of saltwater into the Persian Gulf are of potential significant region-wide effects on the already high salinity and temperatures, leading to increased temperatures of sea ranging from 0.5°C to 1.4°C, and elevated salinity by 10 to 18 Practical Salinity Unit (PSU) at scales pertaining to the whole basin up to 2050 [12].

During the past 30 years, three wars have occurred at the Persian Gulf: the Islamic Republic of Iran– Iraq war (1980–1988), the first and the second Persian Gulf Wars (1991 and 2003, respectively). Thus, a large oil spill occurred in the Persian Gulf in 1991, and as a result, 6 to 8 million Kuwaiti crude oil barrels were discharged into the Persian Gulf in addition to numerous spills due to typical oil operations and those associated with tankers [7,9,13,14]. The oil spills occurred due to the first Persian Gulf War is deemed as the worst oil spill throughout history. Thus, a large volume of research was concentrated on the same spill that approved the oil spill impact radius of 400 km from the point in which the spill occurred to the coastlines of Saudi Arabia and the rapid degradation of the major contaminants [15]. In this paper, previous studies on the types of pollutants in the Persian Gulf have been reviewed, with the aim of identifying major pollutants in the region, the risks and strategies for their prevention and removal.

2 Methods

In this review, relevant papers published in past decades regarding Heavy Metals, Polycyclic aromatic hydrocarbons (PAHs) and plastic and micro-plastics pollution in sediments in estuary regions of the Persian Gulf have been discussed. In addition, papers on sedimentary Heavy Metals, PAHs and micro-plastic in core sediments have been collected and analyzed to reveal metal pollution in different historical periods. The review of Heavy Metals, PAHs and plastic pollution in sediments from the Persian Gulf sea will contribute to the comprehensive understanding of the distribution, sources, temporal variability, and ecological risks of pollution in major estuaries and bays of the Persian Gulf sea. The results will help identify potential factors that result in pollution in the aquatic environment and provide references for decision-making on protecting marine ecosystems and public health.

2.1 Polycyclic Aromatic Hydrocarbon

PAHs are classified as chemical compounds with semi-volatile nature composed of a minimum of two fused rings of benzene in different arrangements, primarily originated from biogenic, pyrogenic, and petro-genic sources [2]. ΣLMWs is the sum of the entire 2-3 ring parent-PAHs; however, ΣHMWs defines the sum of the entire 4-6 ring parent-PAHs. According to the International Agency of Cancer Research, of the sixteen PAHs categorized as primary high-risk pollutants, some are potential carcinogenic threats for human [16].

In coastal sediments and biota from Qatar, Bahrain, Oman, and UAE, the PAHs and Total Petroleum Hydrocarbons (TPH) were studied [17]. With the exception of the sediments sampled close to the Bahraini oil refinery of the BAPCO, TPH and PAHs were comparatively lower in the studied biota and sediments compared to those of the global reports, including those presented by the United Nations Environment Programme (UNEP), and the National Status and Trends (NS&T) sites in Alaska.

Additionally, [16] presented a thorough evaluation of the PAHs compositional patterns and distribution and derivatives of them found in tissues of three edible fish species (Scomberomorusguttatus, Lutjanusargentimaculatus, and Lethrinusmicrodon) sampled from the coral reef ecosystems of the Persian Gulf in Iran. The heterogeneous background pollution to PAHs was approved. Also, in fish tissues, the dominance of LMW-PAHs with respect to HMW-PAH compounds was evident. The fish liver showed higher concentrations of PAHs and derivatives of them compared to their muscles, with the lowest and the highest concentrations of PAH found in S. guttatus, and L. argentimaculatus, respectively. In fish tissues, lipid contents and PAHs and also theoctanol-water partition coefficients (Kow) of PAHs congeners showed a significant positive correlation. In addition, there was a high considerable positive correlation between the PAHs levels in the tissues of the fish and their total weight and lengths. Generally, the current findings present valuable basic data for future studies on the PAHs eco-toxicity in aquatic organisms. In fact, in order to find explicit associations between the fish health and their exposure to PAHs, and the resultant implications for the human well-being, it is necessary to conduct more detailed research on both cellular and tissue levels [16].

2.2 Heavy metals

Due to their persistence, potential of bioaccumulation and toxicity in numerous species, metals are of greater concern amongst the countless organic and inorganic compounds released into the aqueous environments [18]. Metals could be re-mineralized or absorbed by particles under different biogeochemical cycles if released into marine environments, leading to sub-lethal or even lethal impacts on ecosystems. Considering the diverse physio-chemical processes, such as chelation, precipitation, and adsorption, sediments play the role of scavenger agents, and thus, they act as the secondary contamination source in marine regions with the potential of releasing metals back to the column of water in which they become accessible for organisms through desorption, sediment resuspension, and redox reactions. Due to their mobility, toxicity, longer biological half-lives, and bioaccumulation via the food chains, the dissolved metals are of a high threatening potential for aquatic, and also, human life. Thus, it is necessary to estimate their levels in marine environments precisely. The metal contamination has had adverse impacts on the health of coral reefs of the Persian Gulf, Iran, due to the coral islands, known as metals sinks, are located adjacent to the sites of petroleum industry [19].

Seafood is considered as a good source of various vitamins, essential fatty acids, minerals, and proteins decreasing heart attack risk and diseases. However, the higher concentration of organic and inorganic contaminants (metalloids and toxic metals) found in seafood has led to concerns on human well-being when consuming them. The seafood contaminated by metals has turned into a critical subject, which is due to their longtime persistence, stability, and non-biodegradability in food chains. The factors affecting metals bioaccumulation in fish include their ecological requirements, nutritional habits, duration of fish life, metals concentration in sediments and water, physicochemical characteristics of water, and seasonal variations. Amongst heavy metals, arsenic (As), cadmium (Cd), copper (Cu), lead (Pb), and mercury (Hg) are of the greatest potential of being accumulated in fish tissues to concentrations more than the allowed limits. Acute and chronic exposure to such heavy metals mayjeopardize human life. For instance, chronic exposure to lead may lead to gastrointestinal colitis and cerebral diseases, the senses of sight, hearing, and touch may be impaired by mercury, cadmium may lead to cardiovascular diseases and bone pains [20].

The scholars in [20] studied the concentrations for the four metals of arsenic, lead, cadmium and mercury in gill and fillet of Scarusghobban (parrotfish) of the Persian Gulf (Iran). In both gill and fillet, the Pb concentrations were higher, although the mercury concentrations had the minimum concentrations. The metal concentrations analyzed were much lower than ($p < 0.05$) the limits recommended by FAO and allowed by national standards. Compared to fillets, the recorded metals accumulation in gills was higher. There was no significant correlation between metals in both gill and fillet ($p > 0.05$). For both child and adult parrotfish, the values of the Total Target Hazard Quotient (TTHQ) and Target Hazard Quotient (THQ) for fillets were lower than 1, showing no significant non-carcinogenic risks.

Contrary to expectations, the values of Incremental Lifetime Cancer Risk (ILCR) for children and adults showed that they were of threshold risk of being carcinogenic. Also, the research [20] showed that in Hendijan, Iran, the children were at higher carcinogenic and non-carcinogenic risks than the adults. Thus, in order to protect seafood consumers against health risks, the disposal oflethal toxic metals to aqueous environments must be consistently monitored and controlled [20].

2.3 Plastics and microplastics

Over the recent decades, the global plastics production has increased immensely from 1,500,000 tons annually in the 1950s to 322,000,000 tons annually in 2015.Also, the plastics are among the most plentiful aquatic wastes found at beaches. Due to the intense and ongoing disposal of plastic-based materials into the environment, their particles are accumulated in aquatic environments and coastlines. Despite the efforts made to cleanse the wastes from the aqueous environments and constraints on disposal along the shorelines and at the seas, the quantity of plastic debris disposal has been increasing so far. This process seems to beinevitable, since most plastics are not degradable biologically, and their amounts found in aqueous environments will increase as the time passes. Ocean currents may also scatter the plastic debris from industrialized and highly populated areas to even the unpopulated and the most distant coastal regions [21].

Dobaradaran et al. [21] researches result demonstrated that given the number of items made of plastic found at the investigated coastlines, the plastic debris found in Bushehr beaches along the Persian Gulf, particularly given the categories of the smaller size (5 mm), was a great challenge. In spite of the fact that presenting a time schedule to cleanse the Bushehr beach may decrease the amount of plastic debris on the coastline, this process is only useful for the particles of the sizes greater than 5 cm. Due to the dominant inefficient disposals in the Persian Gulf region, the urban managers of the same area are required to resort to more fundamental methods including public training, particularly for people of younger ages through the school capabilities and utilizing media. Given the volume of plastics disposed every year, the same report shows that considerable volumes of metals such as chromium, cadmium, lead, nickel, iron, aluminum, copper and manganese may enter into the aqueous environments annually as well that may create potential acute risks for aqueous organisms, and enter the food chains and ultimately the human bodies. Further studies on transportation, production, the chemical and physical properties, fate, and impacts of Microplastics (MPs) on the Persian Gulf region are highly recommended. One could use the estimation of baseline depicted in this paper as a starting point to reach the necessary data for managing the Persian Gulf region. Ultimately, it is noteworthy to mention that effective removing of microplastic particles from sediments and sand in seawater and coastal areas is not feasible. Thus, the only approach taken to decrease their amounts is reducing the release of plastics into aquatic environments [21].

Additionally, the initial report on the presence of microplastics in shellfishes of the Persian Gulf [22] showed that in the Iranian coastal waters, molluscan shellfish, including deposit feeders, carnivores, and feeders accumulate and ingest MPs in high concentrations among the predatory species, reflecting the MPs trophic transfer within the food web. Most of the discovered plastics were plastics fragments and microfibres that all frequently found everywhere within the aqueous environment as well. Therefore, through

 https://doi.org/10.1051/e3sconf/202132503013

trophic transfer, shellfish consumers, whether the humans or other animals found in wildlife, could be vulnerable to MPs. The findings of Naji et al. [22] demonstrate that more studies should be conducted on the effect of microplastics on shellfish within the aqueous environment of the Persian Gulf, particularly on those caught adjacent to the highly populated regions to be consumed as a seafood. Given the fact that the environmental circumstances could be changed with the passage of time, establishing long-term and spatial trends in microplastic contamination seems to be essential to determine the vulnerability to the ingestion of microplastics and the high-risk regions for mollusks and other aqueous biotas.

3 Result and discussion

The Persian Gulf is known for its unequaled ecosystems, which are vulnerable to the sharing of coasts between various countries, interferences in the reason of human activities and thus, the increasing challenges acting as barriers for conservation attempts and management of the environment. Another feature of the Persian Gulf is the constrained exchange of water with the Indian Ocean, leading to increased potential of catchment and coastal habitat destruction, environmental damages made by pollution, unsustainable utilization of resources, and the loss of biodiversity.

Given the economic activities development, including agriculture, mining, tourism and shrimp aquaculture, population increase and growing desalination, the issues related to Persian Gulf pollution may become worse and lead to considerable public health and ecological risks within the next 20 years. The pollution has already led to changes discovered in mollusks, algal communities, aqueous mammals, and fishes at some sites. The effects of pollution on invaluable coastal habitats (wetlands, lagoons, and estuaries) and the fisheries of Persian Gulf should be particularly investigated due to compared to marginal seas, they have been poorly investigated. The effect of mercury, arsenic, POPs, and contaminants of great concern, including Polybrominated Diphenyl Ethers (PBDEs) and the compounds derived from plastics should be particularly investigated.

The same review has evidently referred to the fact that the pollution to petroleum hydrocarbons, trace elements, and the heavy metals in the marine and coastal regions of Persian Gulf constitute the most research in the area compared to those related to air and soil

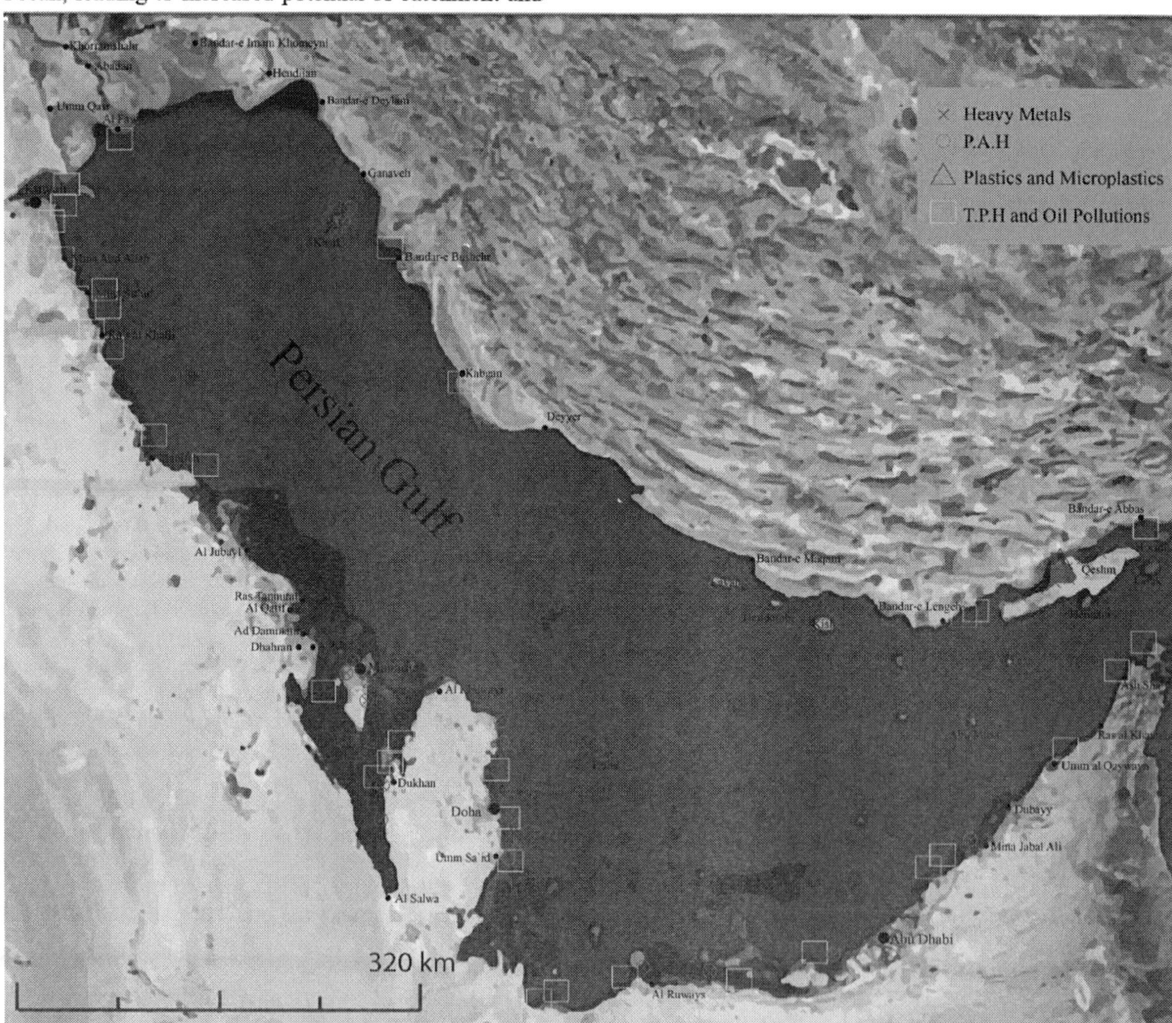

Fig. 1. Locations of sampling of various pollutants in the Persian Gulf.

pollutions. In addition, less attention has been paid to thermal pollution, increased salinity and other dangers of Persian Gulf water desalination. Contrary to the fact that the contamination to hydrocarbons and metals has imposed significant environmental stresses to the Persian Gulf, and it has experienced three wars over the late 30 years, most of the research considered in this paper showed comparatively low levels contaminations to petroleum residues and heavy metals in various organisms and habitats, and that significant improvements have been made in the region regarding the petroleum hydrocarbon concentrations that have reached their baseline levels before the oil spills in 1991. Generally, there are few research that have indicated higher concentrations of some pollutions such as comparatively low to moderate chronic contaminations in the reason of metals and oil in some regions, although the chronically contaminated regions were constrained to some hot spots, including the BAPCO site within Bahrain. However, the protective attempts and efficient sustainable management of marine, coastal, air, and soil environments within the Persian Gulf region are very important and crucial for efficient management of resources and deciding on environmental issues.

Figure 1 shows the location of Researches samples studied in this paper. As it turns out, more research has been done on heavy metal contamination and PAHs on the north coast of the Persian Gulf, while on the south coast, most research has focused on Total petroleum hydrocarbon (T.P.H) and oil pollutants. As mentioned, most research in the field of oil pollutants is related to previous years, and in recent years, most research has focused on heavy metals, PAHs and microplastics.

The restoration expenditures are dramatic, and often exceeding the capacities of developing countries. Thus, the initial stage and the crucial objective of effective and efficient management and conservation of the Persian Gulf are protecting it against pollution. The strategies used for the protection of the region will be dependent on regional curative and preventive conventions, strategic action plans for sustainable conservation and use of the resources of the region by complying with the regulations of international entities including the United Nation Development Program (UNDP), and the UNEP, and also the donor organizations such as the World Bank, and the Islamic Development Bank. Currently, the only Sea Forum within the Persian Gulf area is the Regional Organization for Protection of Marine Environment (ROPME), which has made considerable attempts towards the protection and management of the Persian Gulf region. The ROPME has eight member states: Saudi Arabia, Qatar, Bahrain, UAE, Iraq, Oman, Kuwait, and the Islamic Republic of Iran; at the same time the ROPME SEA Area (RSA) is consisted of the Sea of Oman, the Persian Gulf, and the Omanis south eastern coasts located at the vicinity of the Arabian Sea. Thus, ROPME could be considered as one of the most appropriate alternatives for later corporations within the region. The key procedures for determining the first priority regional actions with considerable long- and short-term outcomes for sustainable conservation and use, are regional review of pollution topics pertaining to the Persian Gulf region, such as the root reasons, immediate reasons, impacts and symptoms, analysis of the environmental issues of the region, threats evaluation and severity and scale on the basis of a series of national reports.

4 Conclusion

In the present study, the Persian Gulf sea marginal systems from different locations along Iran were selected for Heavy Metals, PAHs and plastic source identification and risk evaluation. The levels of Heavy Metals, PAHS and plastics in the Persian Gulf sea sediments are much higher than their background values, which suggests the involvement of anthropogenic emissions. The concentrations of pollutants in different regions are consistent with local key industries and industrialization levels.

The result suggested that the pollutants in the most study areas were not contaminated-weakly contaminated, but was contaminated-heavily contaminated in some pollution hot spots. It requires local governments to formulate more effective policies and regulations to protect the marine environment. Gouvernements also should set 'zero tolerance for this issue and compel the industries to use biodegradable materials such as starch instead of nondegradable material. This biodegradable material will then be decomposed by microorganisms and ultimately reducing the lifetime of these plastics in the surroundings. In industries, the process of recycling or upgrading of plastic litter should be encouraged.

References

1. N. Niamaimandi, Z. Bahmyari, N. Sheykhsagha, E. Kouhgardi, and R. G. Vaghei, Reg. Stud. Mar. Sci. **15**, 26 (2017)
2. A. Ranjbar Jafarabadi, A. Riyahi Bakhtiari, L. Hedouin, A. Shadmehri Toosi, and T. Cappello, Ecotoxicol. Environ. Saf. **163**, 307 (2018)
3. U. NATIONS, G. O. E. O. G. NAMES, and Twenty-third, *Historical, Geographical and Legal Validity of the Name: PERSIAN GULF* (Vienna, 2006)
4. C. R. C. Sheppard, Mar. Pollut. Bull. **27**, 3 (1993)
5. M. S. Massoud, F. Al-Abdali, A. N. Al-Ghadban, and M. Al-Sarawi, Environ. Pollut. **93**, 271 (1996)
6. R. Michael Reynolds, Mar. Pollut. Bull. **27**, 35 (1993)
7. I. M. Madany, A. A. A. Wahab, and Z. Al-Alawi, Water. Air. Soil Pollut. **91**, 233 (1996)
8. S. De Mora, S. W. Fowler, E. Wyse, and S. Azemard, Mar. Pollut. Bull. **49**, 410 (2004)
9. C. Sheppard, M. Al-Husiani, F. Al-Jamali, F. Al-Yamani, R. Baldwin, J. Bishop, F. Benzoni, E. Dutrieux, N. K. Dulvy, S. R. V. Durvasula, D. A. Jones, R. Loughland, D. Medio, M.

Nithyanandan, G. M. Pilling, I. Polikarpov, A. R. G. Price, S. Purkis, B. Riegl, M. Saburova, K. S. Namin, O. Taylor, S. Wilson, and K. Zainal, Mar. Pollut. Bull. **60**, 13 (2010)

10. I. Al-Saleh, I. Al-Doush, and A. Echeverria-Quevedo, Bull. Environ. Contam. Toxicol. **63**, 451 (1999)
11. A. R. G. Price, C. R. C. Sheppard, and C. M. Roberts, Mar. Pollut. Bull. **27**, 9 (1993)
12. AGEDI, (2016)
13. T. W. Kureishy, Mar. Pollut. Bull. **27**, 183 (1993)
14. P. Literathy, Mar. Pollut. Bull. **27**, 349 (1993)
15. S. de Mora, I. Tolosa, S. W. Fowler, J. P. Villeneuve, R. Cassi, and C. Cattini, Mar. Pollut. Bull. **60**, 2323 (2010)
16. A. Ranjbar Jafarabadi, A. R. Bakhtiari, Z. Yaghoobi, C. K. Yap, M. Maisano, and T. Cappello, Chemosphere (2018)
17. I. Tolosa, S. J. De Mora, S. W. Fowler, J. P. Villeneuve, J. Bartocci, and C. Cattini, Mar. Pollut. Bull. **50**, 1619 (2005)
18. A. Ranjbar Jafarabadi, A. Riyahi Bakhtiari, M. Maisano, P. Pereira, and T. Cappello, Sci. Total Environ. **640–641**, 1500 (2018)
19. A. Ranjbar Jafarabadi, A. Riyahi Bakhtiari, N. Spanò, and T. Cappello, Mar. Pollut. Bull. **137**, 185 (2018)
20. Y. Fakhri, N. Saha, A. Miri, M. Baghaei, L. Roomiani, M. Ghaderpoori, M. Taghavi, H. Keramati, Z. Bahmani, B. Moradi, A. Bay, and R. H. Pouya, Food Chem. Toxicol. **118**, 348 (2018)
21. S. Dobaradaran, T. C. Schmidt, I. Nabipour, N. Khajeahmadi, S. Tajbakhsh, R. Saeedi, M. Javad Mohammadi, M. Keshtkar, M. Khorsand, and F. Faraji Ghasemi, Waste Manag. **78**, 649 (2018)
22. A. Naji, M. Nuri, and A. D. Vethaak, Environ. Pollut. **235**, 113 (2018)

Numerical modelling of geocell reinforced slope by use of plastic bottle waste

Afif Agri Diansyah, *Fikri* Faris* , *Ahmad* Rifa'i, and *Roberth* Evander Meidudga

Department of Civil and Environmental Engineering, Universitas Gadjah Mada, 55281, Yogyakarta, Indonesia

Abstract. Nowadays, waste management should be one of the focuses in controlling a healthy and sustainable environment. It is undeniable that plastic bottle waste has covered the earth's surface. One of the solutions to reduce the polluting impact is to use it as a geocell in the geotechnical field. The previous researchers more study about slope reinforcement using fabricated geocell. In this study, geocell is made by utilizing plastic bottle waste. Slope stabilization using geocell is a kind of internal stabilization. Other kinds of slope stabilization, such as using sheet piles and gravity walls, are external stabilization. The purpose of this study is to obtain an overview of the effect of the geocell reinforced slope. From experimental studies using plate load test, results obtained that geocell reinforce soil can increase the bearing capacity of the soil. Numerical modelling uses PLAXIS 2D modelled geocell reinforce slope as a slab that can spread the load widely over the reinforced area, also known as beam model. Young's modulus of geocell-soil composite is obtained from experimental studies, empirical equation and validated using axisymmetry in PLAXIS 2D. The results of this study, generally geocell reinforce slope can restrain the slope from failure deformation by redistributing the loads in a large area. The significant effect is shown in sandy soil with a high improvement factor (*IF*), different to the reinforcement in clayey soil with a low improvement factor (*IF*). It implies that plastic bottle waste can be used as a geocell which can increase its economic value. Since it is applicable, it can reduce the pollution caused by plastic bottle waste.

1 Introduction

Plastic waste became a severe problem faced by developing countries such as Indonesia. One type of plastic waste found in daily life is plastic bottles in several forms of its function, primarily as a beverage package. The most material commonly used is *polyethylene terephthalate* (*PET*). Hence it is solid waste, it takes a very long time to decompose. The needed time to naturally decompose is between 100-400 years [1]. The potential pollution to the environment and natural disasters caused by this kind of waste is hazardous. It has contributed to global warming, floods, water pollution and many others. Therefore, this problem requires many solutions to minimize the potency and impact of the natural disasters caused by plastic bottle waste. On the other side, some failures in construction occur due to neglecting geotechnical aspects. Such failures are subsidence of foundation, railroad damage, failure of embankments, slope and others. Based on the environmental and geotechnical aspects above, this research utilizes *PET* into geocell used for soil reinforcement applied in various types of reinforcement. Geocell is a term of geosynthetics type. It has a different shape from other geosynthetics such as geogrid, geoweb, geomesh, etc. Geocell has a 3-D shape and resembles a honeycomb. The Confinement system provided by each cell to the soil inside forms a composite that produces a higher strength than ordinary soil (without geocell) [2]. Geocell can consist of various types and materials. It, which come from fabricating, have a standardized technical characteristic. Hand-made geocell can also use fabricated material such as geogrids which are then shaped into geocell by some manual process. Although plastic bottles are factory-made, the tensile strength of these plastic bottles can vary. Geosynthetics soil reinforcement became one of the solutions to increase soil strength which generally the problem was the bearing capacity or stability [3].

Geocell were first developed as soil reinforcement by the U.S army engineering corps to stabilize sand quickly [4]. Soil reinforcement is often considered a common problem, but the failure impact can take a high cost to rehabilitate or reconstruct. Soil reinforcement certainly requires additional costs and time, but the benefit obtained can extend life service and reduce maintenance costs. Currently, geocell has been widely used in geotechnical fields. It also can be applied to various soil types for embankment, retaining walls, railroad foundation, slope reinforcement from slide and erosion. The advantages of using geocell are fast processing time, low costs and being environmentally friendly.

* Corresponding author: fikri.faris@ugm.ac.id

Previous researchers have carried out some experimental studies to study the effect of using geocell as soil reinforcement. However, in practice, it does not rely solely on experimental studies to design and analyze the geocell reinforcement, let alone on complex geometries. Hence experimental studies take quite a long time, and a numerical approach is required to simplify design and analysis in a fairly short time [5]. Numerical modelling of geocell as the original form indeed a quite difficult way cause the honeycomb shape has a complex geometry and the geocell thin sheet becomes an obstacle to modelled the structure. Another approach to model geocell numerically is with equivalent composites, where the geocell-soil layer can be modelled as a slab with stiffness parameters [6-9,5]. A numerical model of geocell reinforcement can also be modelled in 2-D, which representing the 3-D conditions. [10] used a 2-D model to design embankment with geocell based on laboratory studies. According to the literature study above, geocell reinforcement is able to increase the bearing capacity and reduce subsidence. In terms of using plastic bottle waste as geocell reinforcement, very few related reports are published, specifically on its application as slope reinforcement. [11] reported that geocell-soil reinforcement can behave as a retaining wall and strengthen the wall layers.

Slope stability analysis has been studied extensively, but the behaviour of slope reinforced by geocell is still interesting to explore. This paper aimed to investigate the geocell-reinforced slope made from plastic bottle waste using the 2-D model. The study varies the slope angle, geocell dimension and number of installed layers. This study modelled geocell as a slab that produces flexural and membrane stress. Numerical analyses were also conducted to verify the experimental test results.

2 Experimental studies

An experimental test was conducted to obtain the performance of geocell reinforcement against fine sand. Two geocell sizes were used as 5 cm and 7.5 cm in height. These sizes were chosen to optimize the use of the plastic bottle. **Fig. 1.** showed the geocell from intact bottle cut into pieces with two dimensions as 5 cm height, and 7.5 cm height because the adequate size of bottle to be used is 15 cm. For 5 cm height, one bottle can turn into three parts and 7.5 cm height can turn into two parts. The same in height pieces were then jointed with rivet nails. A schematic view of the plate load test setup is shown in **Fig. 2.** The external box size is 1100 mm in length, 1100 mm in width, 1000 mm in height and made from iron. Hydraulic jack attached to loading frame and manually operated. The load is distributed uniformly by the steel plate with 120 mm length, 120 mm width, 10 mm thick. In this test, two dial gages are installed to measure soil settlement. The load is applied gradually in constant increments. The experimental results showed that the geocell increase the bearing capacity of the soil. **Fig. 3.** shows the load increasing from unreinforced to reinforced soil. A 7.5 cm height of geocell indicates the highest bearing capacity. The sand used in the experimental test have the parameters such as coefficient of uniformity (C_u) 2.207, coefficient of concavity (C_c) 1.09, medium diameter (D_{50}) 0.58 mm, maximum diameter 2 mm, specific gravity (G_s) 2.83, internal friction (φ) 32, USCS classification as sand poor graded (SP).

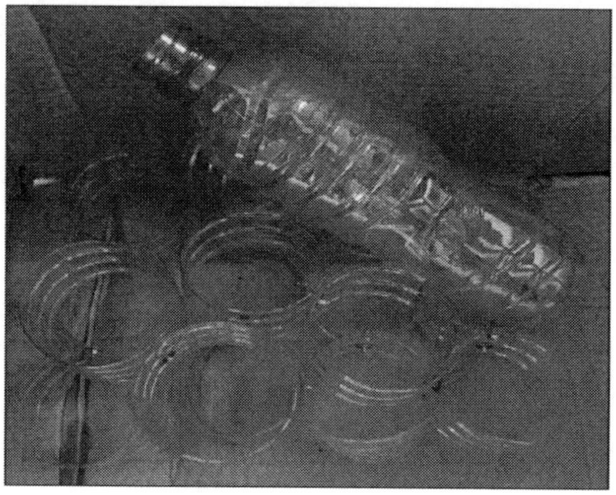

Fig. 1. Geocell materials from the intact bottle and cut into pieces.

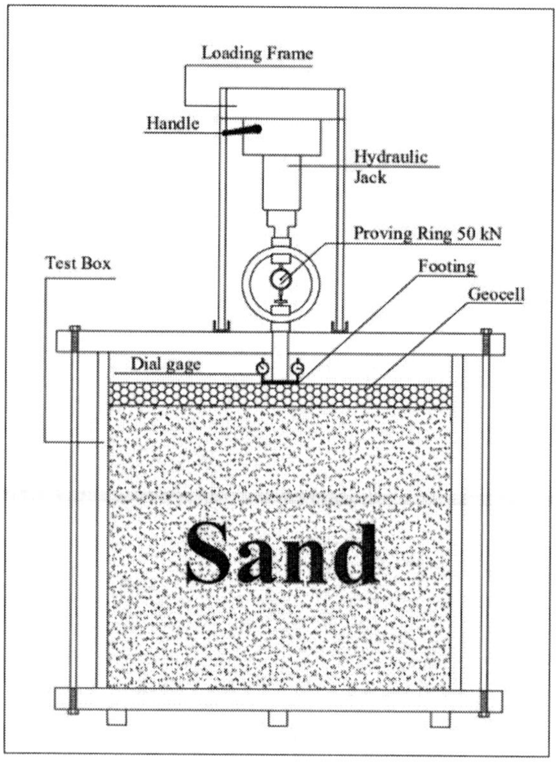

Fig. 2. Schematic view of the test setup.

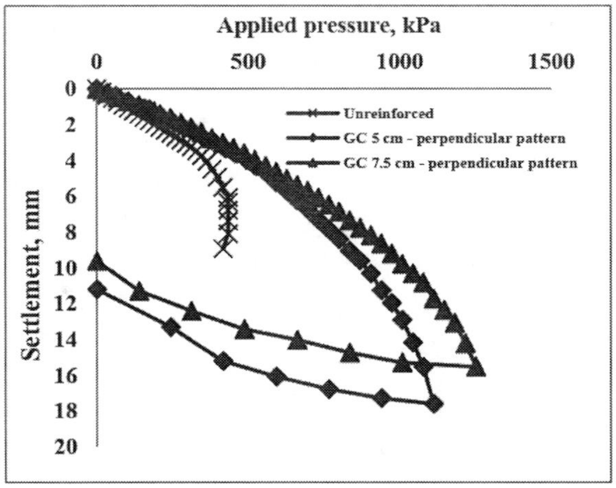

Fig. 3. Applied pressure-settlement.

Schematic view of geocell layer reinforced slope or embankment shown in **Fig. 4.** By a load of slide mass driven by active earth pressure, the geocell layer gives a tension reaction (*T*) to hold the slope deformation from its active pressure. This scheme is common to use in slope analysis with geosynthetic reinforcement. The notations (*T*), (*M*) and Q, respectively, are the geocell reinforcement tension, bending and shear force of geocell reinforcement. Other notations are load *q*(y) and *P*(*y*) act on super-face and sub-face of geocell. Interface resistance is represented as $\tau_t(x)$ and $\tau_b(x)$ respectively act at the top and bottom of the geocell layer [9].

3 Numerical modelling

2-D modelling using Plaxis 2D V8.6 to perform geocell reinforce slope behaviour. The slope is modelled as plain strain. Plaxis 2D can model geocell reinforcement as a slab that produces stiffness and can spread the load widely. The fact soil eternally behaves non-linear under load. Several levels of sophistication can model the non-linear stress-strain. Though the soil performs a non-linear behave, it is still able to simplify. Mhor-Coulomb model is well-known to be considered as a simple model. M-C model proposed linear elastic perfectly-plastic with five basic input parameters Young's modulus (E), Poisson's ratio (v), cohesion (c), friction angle (φ) angle of dilatancy (ψ). **Fig. 5.** shows the simplification of non-linear to elastic perfectly-plastic behaviour from standard drained triaxial tests. In reality, on the ground, the soil always behaves non-linearly, resembling the standard drained triaxial tests (**Fig. 5** a).

Plaxis2D can perform the non-linear behaviour of soil, but more soil parameters are required. However, the results are more reliable than the simplified model (elasto-plastic). The variables that must also evaluate become more. Besides, to simplify analysis with simpler parameters than non-linear. The Mohr-Coulomb model also can be used to perform the elasto-plastic behaviour (**Fig. 5** b). The primary difference between the non-linear and the elastic perfectly plastic was the stress-strain curve from different constitutive equations and the number of parameters. In this study, the material model used for soil as an M-C model to perform slope reinforcement using geocell in the numerical model. Other sophisticated models can be used in Plaxis, such as Duncan-Chang hyperbolic, Hardening soil, Advanced Mhor-Coulomb model etc.

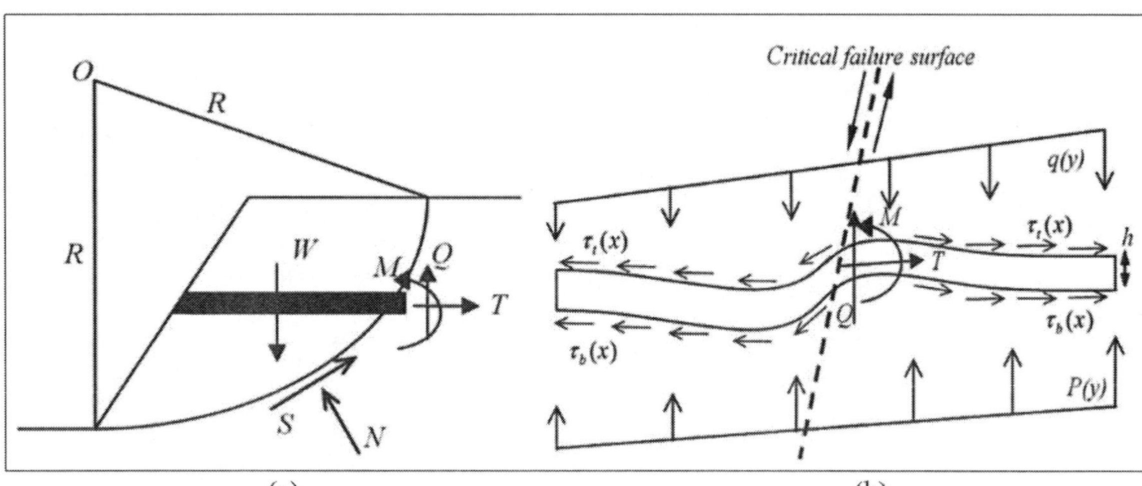

Fig. 4. Schematic view of geocell reinforced slope; (a) failure mechanism (b) forces act along the geocell reinforcement. [9]

https://doi.org/10.1051/e3sconf/202132504001

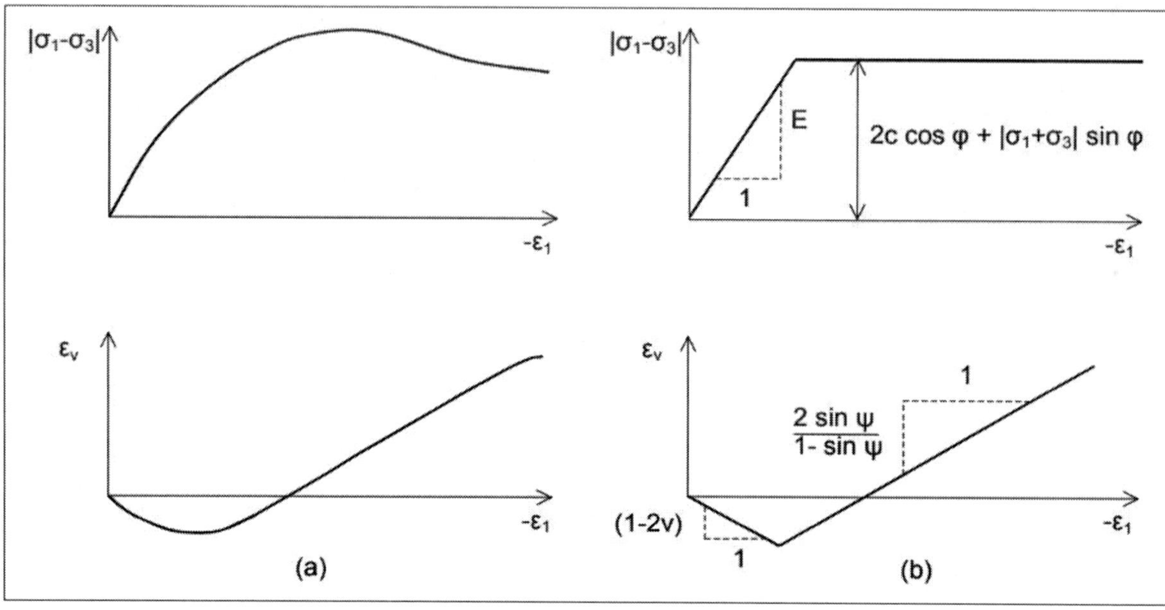

Fig. 5. (a) Results from standard drained triaxial tests (b) elasto-plastic model. [12]

Where (σ_1) is axial stress, (σ_3) is constant confining pressure, (ε_1) is axial strain, (ε_3) volumetric strain. Confining pressure can be calculated at-rest earth pressure or K_0 condition by multiply K_0 and (σ_1).

$$\sigma_3 = K_0 \times \sigma_1 \tag{1}$$

where

$$K_0 = 1 - \sin\varphi \tag{2}$$

3.1 Estimation of geocell stiffness parameter

The stiffness parameter of the geocell can be obtained by validating the experimental test result using axisymmetry in Plaxis2D. Axisymmetry modelled half of the box width in **Fig. 2.** with shape adjustment from square to circle because axisymmetry assumed the box as half-circle. The axisymmetry model was used to estimate the geocell stiffness by several trials.

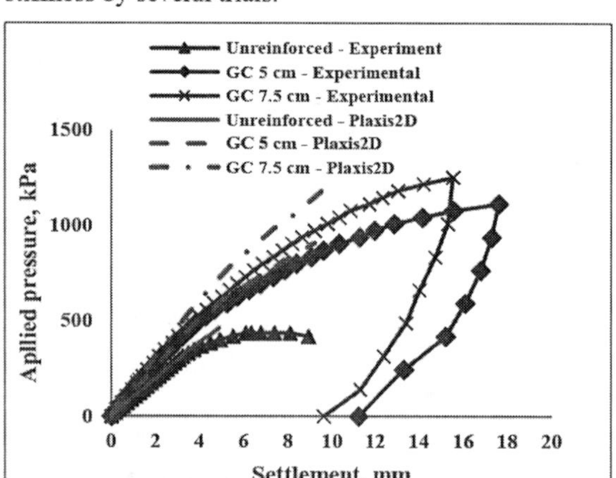

Fig. 6. Comparison between numerical and experimental results.

The geocell is modelled as a plate or slab with *EI* and *EA* as the main parameter, bending and axial stiffness, respectively. *E* is the Young's modulus of geocell, *I* is the second moment of inertia, and *A* is the area of geocell. **Fig. 6.** shows the result between experimental and numerical pressure-settlement in this study.

3.2 Geometry of slope and reinforcement

The modelled slope height is 10 meters, with slope variations (α) 30°, 45° and 60°. Geocell were installed below the slope surface. The top width of slope soil is 15 meters, and the bottom width of slope soil follows the slope. The width of bed soil is 50 meters. **Fig. 7.** shows the slope geometry used in this model; the typical geometry was proposed by [9]. The detailed geometry of geocell installation are shown in **Table 1**. (u) is the distance from the top surface to the first layer of geocell, and (d) is the distance between geocell layers.

Table 1. Reinforcement geometry (α) 30°, 45°, 60°.

State	Number of layer	*u* (m)	*d* (m)
Unreinforced	-	-	-
Reinforced	1	5	-
	2	2.5	5
	3	2	3
	4	2	2

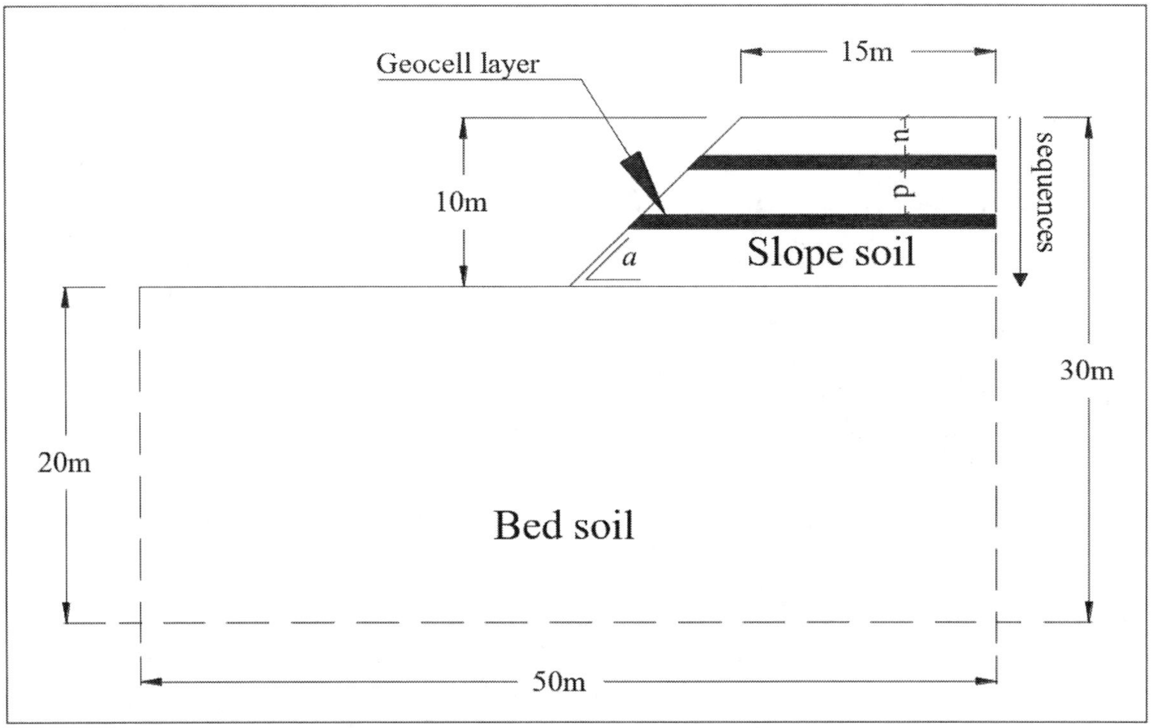

Fig. 7. Slope geometry.

3.3 Properties material model

The material model uses the Mohr-Coulomb failure criterion, where the soil material is considered to have an elasto-plastic behaviour or linear path failure. Hence it uses the M-C criterion, as mentioned in point 3.1, the geocell stiffness obtained from validating the experimental using axisymmetry in Plaxis2D by several trials. From the validating use of axisymmetry, the Young's modulus of geocell was obtained as in **Table 2**. The linear function extracted from axisymmetry Plaxis2D is shown in **Fig. 8**. Soil and geocell properties used are shown in **Table 2.** The soil interface is input manually as a coefficient in soil properties where geocell are placed. The range of coefficient between 0 – 1 is that the value is one, there is no effect of disturbance by the presence of geocell. Besides, if the value is under one, the soil parameter cohesion (c) and friction angle (φ) interact along the geocell surfaces are reduced by the coefficient. Groundwater is not used in this study to obtain the geocell effect clearly.

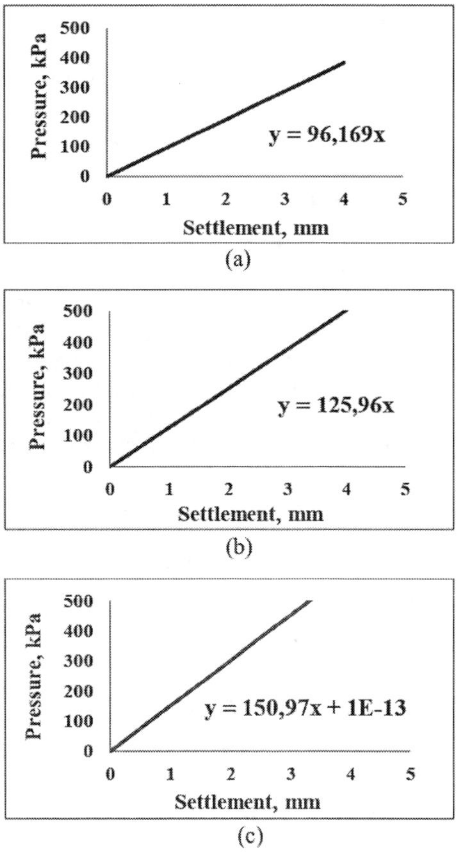

Fig. 8. pressure-settlement in term elastic condition; (a) unreinforced (b) GC 5 cm (c) GC 7.5 cm.

To estimate the initial value of Young's modulus of geocell-soil can be obtained from the equivalent stiffness equation proposed by [13]. This initial value is used to found the closest deal with the trial and error method until the numerical value in **Fig. 6** is obtained.

$$E_g = 4(\sigma_3)^{0.7}(K_e + 200M^{0.16}) \qquad (3)$$

The modulus parameter above is the horizontal stress at the middle of the geocell layer (σ_3), Young's modulus parameter (K_e), secant modulus of geocell material (M). The equation parameter above corresponds to the modulus number of the hyperbolic model proposed by [14]. It should note that the determination of each parameter in this study was also carried out through a series of literature studies.

Table 2. Material properties used in the numerical model.

Soil properties	**Bed soil**	**Slope soil**
Unit weight (kN/m³)	17.6	17.6
Young's modulus (kPa)	10000	7500
Poisson's ratio	0.3	0.3
Sandy soil		
Internal friction (°)	31	30.5
Cohesion (kPa)	1	1
Clayey soil		
Internal friction (°)	5	5
Cohesion (kPa)	25	25
Geocell layer properties		
Height (m)	0.05	0.075
Diameter of cell (m)	0.064	0.064
Young's modulus (kPa)	107250	206855
Second moment of area (m⁴)	1.042×10^{-5}	3.52×10^{-5}

4 Results and discussion

Based on the framework in section 3, this study was conducted to determine the geocell that reinforce a slope, how influential this reinforcement exerts, and the effect of increasing the safety factor. Highly likely that slope angle variations, the two cell dimensions, and the number of layers determines the reinforcement behaviour. **Fig. 9.** shows both unreinforced and reinforced slope deformation in sandy soil. The unreinforced slope experienced a partial failure at half of the top width slope from the sloping side; the collapsed masses are concentrated only a half body. In contrast with unreinforced, the one layer geocell reinforcement shows a uniform deformation tendency along the sloping plane. This condition represents that geocell can be transferring load in all directions, which strengthens. **Fig 10.** shows the slope deformation in clayey soil; generally, its behaviour is not much different from sandy soil. In clayey soil, after reinforcement, the total displacement is concentrated to the right side away from the slip surface because clay is very expansive soil. Clay with low cohesion is easy to deform and very compressible. **Fig. 11.** and **Fig. 12.** shows the shear strain of the unreinforced and one layer reinforced in sandy soil and clayey soil. From the shear strain, the case of sandy soil shows the deformation is closer to the translation slide, it is represent the nature of sand with flakes collapse or crumble. The clayey soil shows the deformation is more relative to the rotation slide, it is represent the nature of clay with block collapse.

As shown in **Fig. 11.** and **Fig. 12.** generally, geocell reinforcement reduces the internal shear strain of the slope. In sandy soil, the maximum shear strain of the unreinforced slope is 24 percent, and the one layer reinforce 17 percent. In clayey soil, the maximum shear strain of the unreinforced slope is 170 percent, and the one layer reinforces 18 percent, which the maximum value occurs at slope foot. The shear strain of unreinforced slope is continuously from top to the slope feet follows the slip surface and as seen over the half of top width are collapses. Different behaviour by the one layer reinforce, the shear strain grade cut off by geocell layer and start again under geocell. As seen, the presence of geocell divided the critical surface and reduced strain. It is represented that geocell absorb the load by collapses mass and transferred along the line, which is more stable. Hence the deformation can be uniformly driven by geocell.

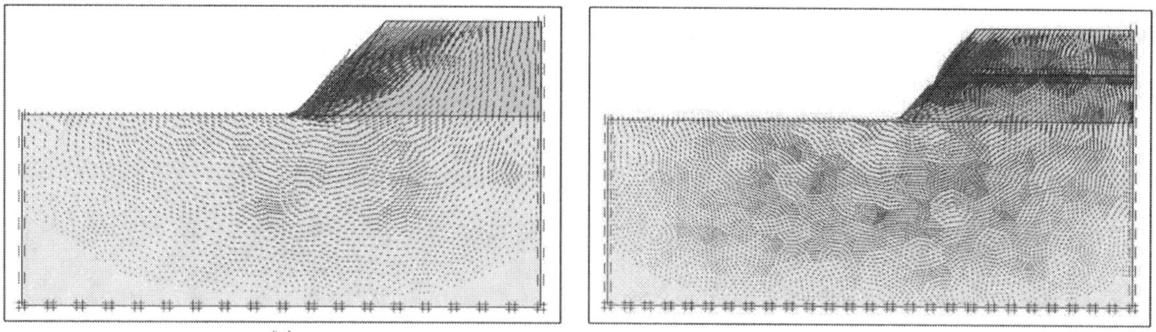

(a) (b)

Fig. 9. Displacement vector of sandy soil at (α) = 60°; (a) unreinforced (b) one layer reinforced.

https://doi.org/10.1051/e3sconf/202132504001

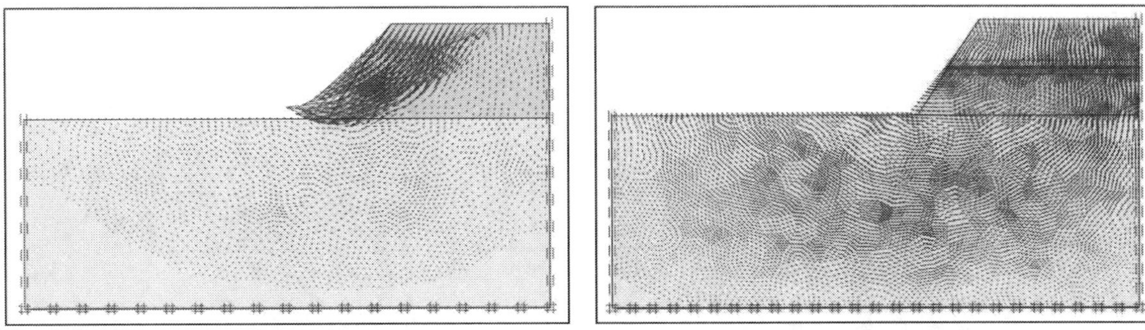

(a) (b)

Fig. 10. Displacement vector of clayey soil at (α) = 60°; (a) unreinforced (b) one layer reinforced.

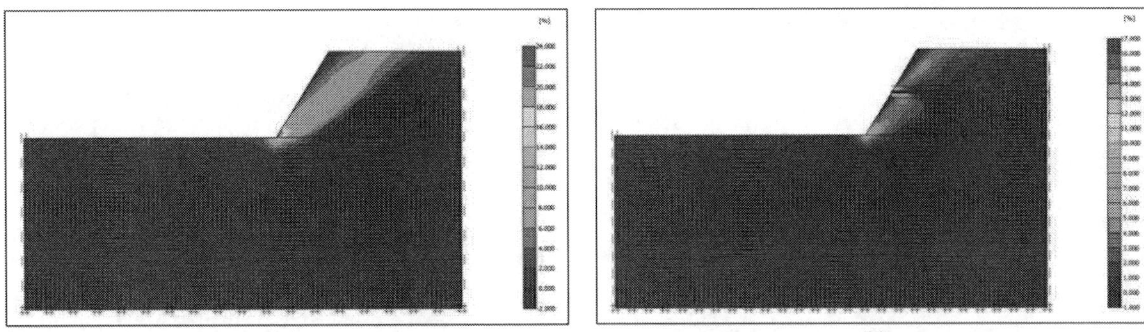

(a) (b)

Fig. 11. Total shear strain of sandy soil at (α) = 60°; (a) unreinforced (b) one layer reinforced.

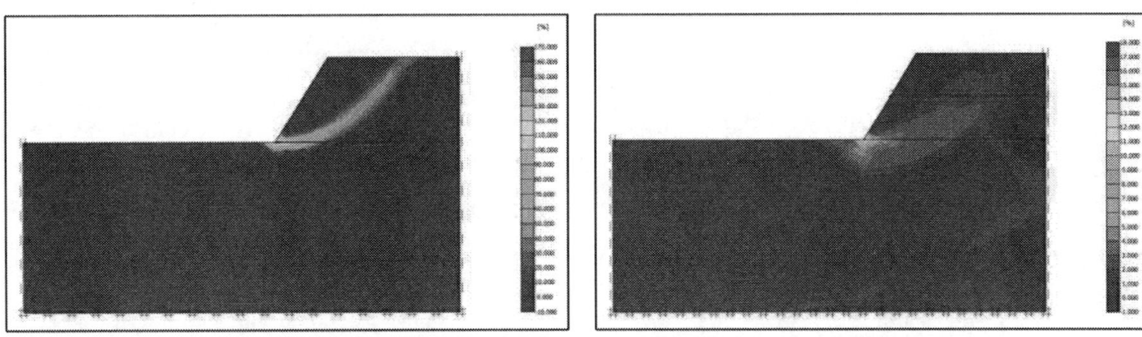

(a) (b)

Fig. 12. Total shear strain of clayey soil at (α) = 60°; (a) unreinforced (b) one layer reinforced.

The reinforcement behaviour can be evaluated by a dimensionless number as an improvement factor (*IF*), which is the ratio of the safety factor of reinforced to unreinforced.

$$IF = \frac{FOS_{(reinforced)}}{FOS_{(unreinforced)}} \quad (4)$$

Where $FOS_{(reinforced)}$ is the safety factor of reinforced slope and $FOS_{(unreinforced)}$ is the safety factor of unreinforced. **Fig. 13.** and **Fig. 14.** shows improvement factor of each variation modelled both in sandy soil and clayey soil. Reinforcement in sandy soil has a significant effect of increasing the safety factor. Unlike clayey soil, the reinforcement effect is not so substantial because clay is compressible, and the vertical deformation of clay with low strength is significant. The more layers installed, the higher (*IF*) are obtained based on the number of layers. By the height of the cell, the 7.5 cm gave a higher safety factor. The significant differences in the improvement factor are indicated by slope angle variation. The 60° produces the highest value of (*IF*). It is because the geocell layer intersects the critical slip surface, where the active soil mass is highest on the 60° slope. The improvement factor (*IF*) of one layer GC 5 cm reinforcement in sandy soil with slopes of 30°, 45° and 60° are respectively 1.05, 1.07 and 1.35, while the GC 7.5 cm with slopes of 30°, 45° and 60° respectively 1.11, 1.16 and 1.34. The improvement factor (*IF*) of one layer GC 5 cm reinforcement in clayey soil with slopes of 30°, 45° and 60° are respectively 1.08, 1.04 and 1.02, while the GC 7.5 cm with slopes of 30°, 45° and 60° respectively 1.03, 1.01 and 1.01.

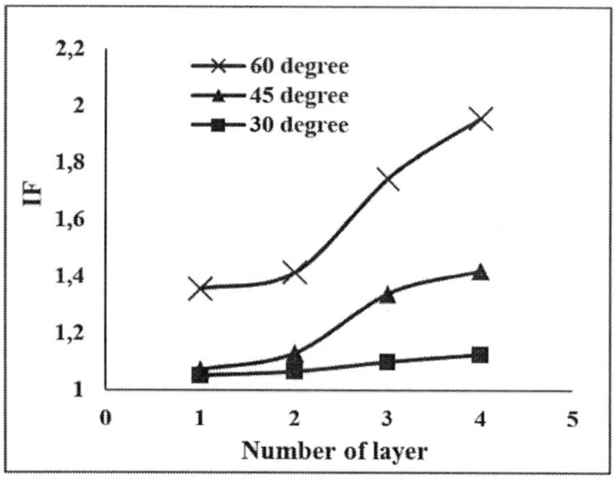

(a)

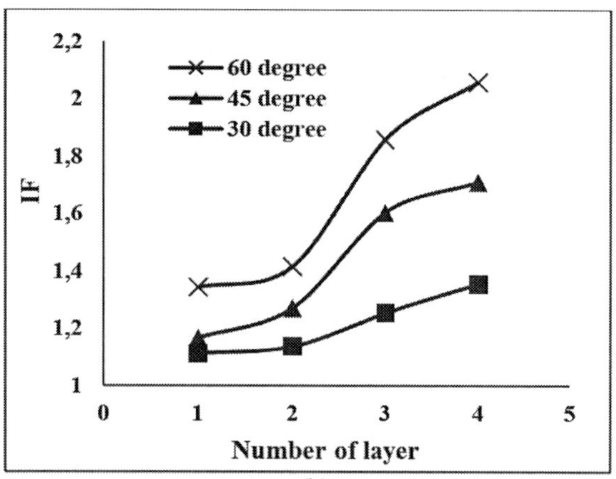

(b)

Fig. 13. Improvement factor of the geocell reinforcement in sandy soil with different heights of; (a) GC 5 cm (b) GC 7.5 cm.

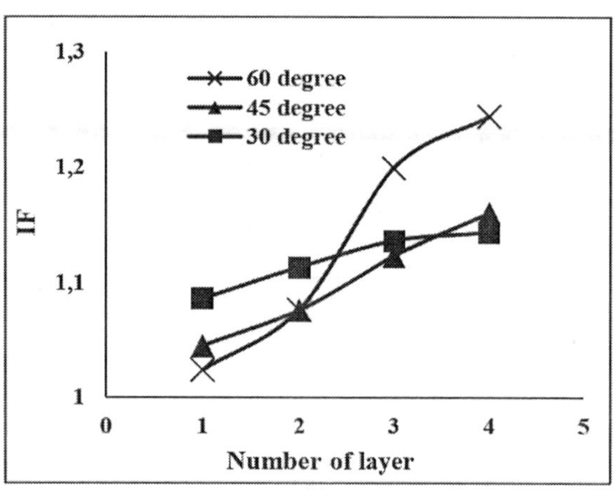

(a)

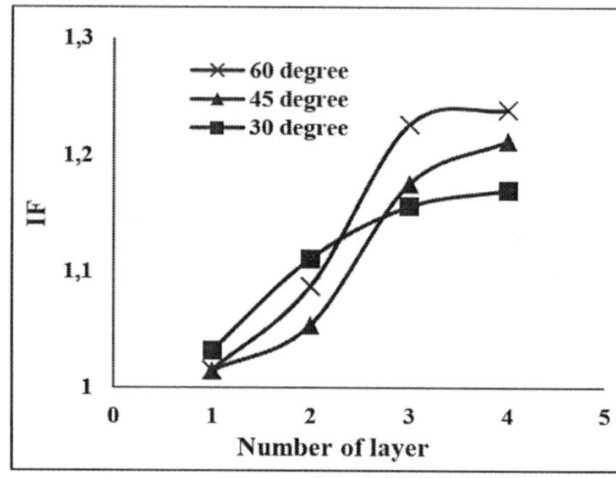

(b)

Fig. 14. Improvement factor of the geocell reinforcement in clayey soil with different heights of; (a) GC 5 cm (b) 7.5 cm.

The numerical analysis also obtained the axial force acting on each geocell layer. The numerical results show that the maximum force is different in each typical geometry and the number of layers used. The maximum axial forces of one layer GC 5 cm in sandy soil with slopes variations of 30°, 45° and 60° are respectively, 13.91 kN/m, 50.31 kN/m and 82.4 kN/m. Then the 7.5 cm with slopes variations of 30°, 45° and 60° respectively, 34.65 kN/m, 61.47 kN/m and 103.69 kN/m. The maximum axial forces in clayey soil with one layer 5 cm reinforcement with slopes of 30°, 45° and 60° respectively, 27.12 kN/m, 43.12 kN/m and 33.20 kN/m. Then the 7.5 cm with slopes of 30°, 45° and 60° respectively, 67.17 kN/m, 81.75 kN/m and 126.79 kN/m.

As shown in **Fig. 15.** the 5 cm geocell in sandy soil has a smaller axial force than 7.5 cm because of the different thicknesses. The 7.5 cm height produce a higher Axial stiffness than 5 cm height, so it also absorbs a higher axial force than 5 cm height. Hence the axial forces are smaller, almost the same with the improvement factor even though the differences are slight in the 60° slope but significant enough at 30° and 45° slope. Unlike the clayey soil, there are almost no axial forces differences by the variated slope angle, **Fig. 16**. shows the axial forces are overlaps between the variation in angle. The very contrasting difference of axial force between the 5 cm and 7.5 cm height because the 7.5 cm height produce a higher axial stiffness so it can absorb a higher force than 5 cm height. Almost all the 7.5 cm height geocell are in the bottom layer in clayey soil have the highest axial forces above one hundred kilonewtons. It denotes that the clayey soil with the above parameters in **Table 2.** has a large enough deformation and poor safety factor. Actually, the internal force in geocell like axial, shear and moment depend on many factors. The type of geocell, material, the geometry of slope and placement, how many layers are placed, and the soil strength parameters. In this study, the shear force and bending moment act on geocell in every number of layers, and variated angles are very small. So, the shear force and bending moment can be ignored.

https://doi.org/10.1051/e3sconf/202132504001

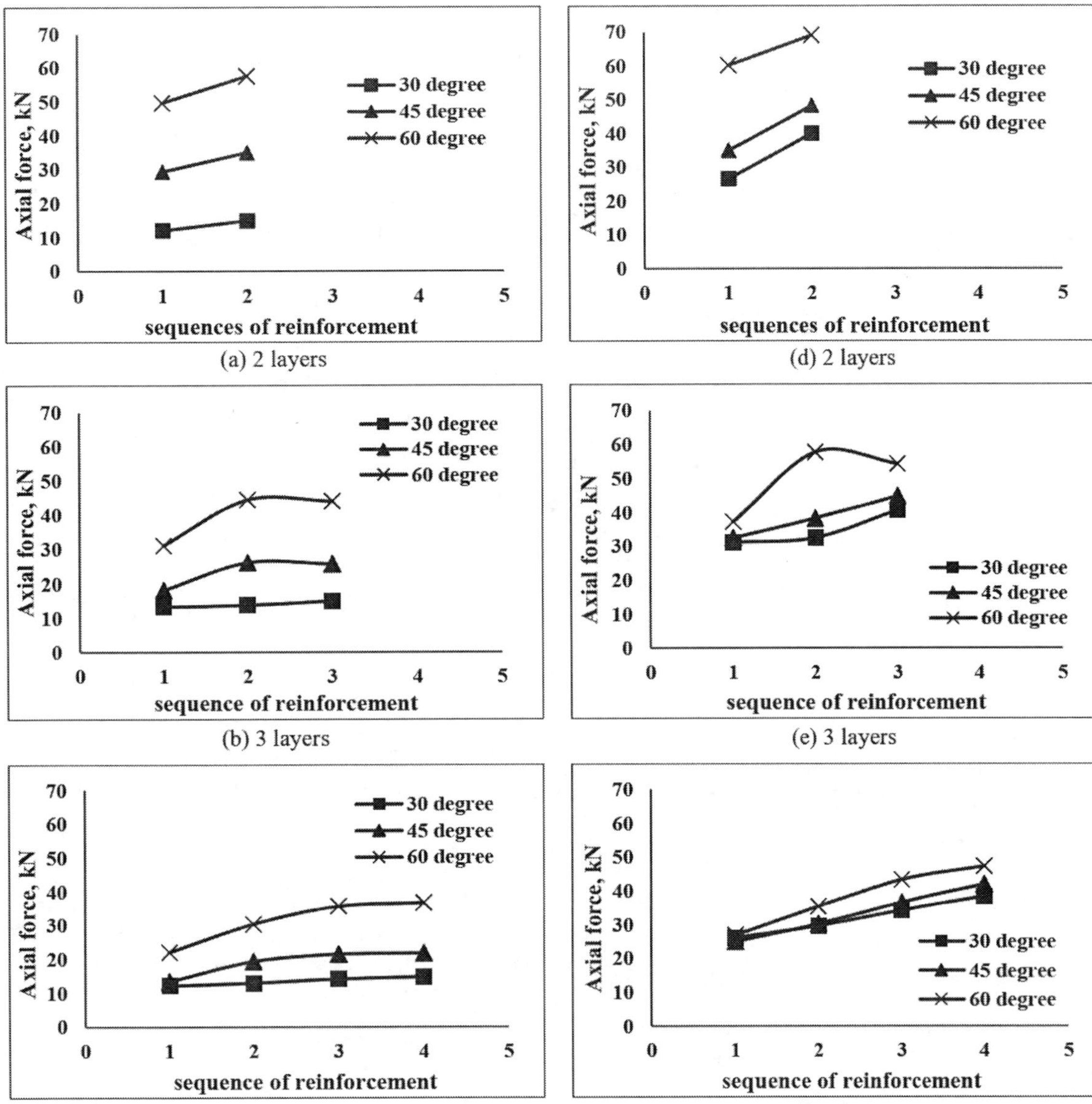

(a) 2 layers (d) 2 layers

(b) 3 layers (e) 3 layers

(c) 4 layers (f) 4 layers

Fig. 15. Maximum axial forces of the geocell in sandy soil with; (a b c) 5 cm height (d e f) 7.5 cm height.

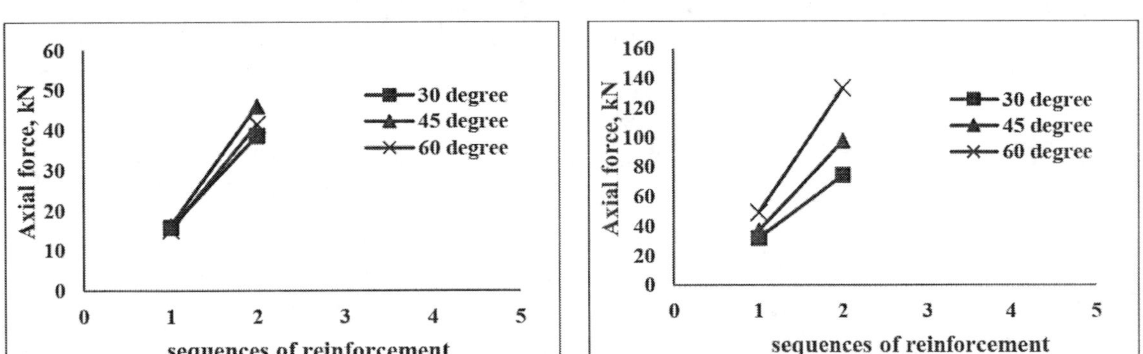

(a) 2 layers (d) 2 layers

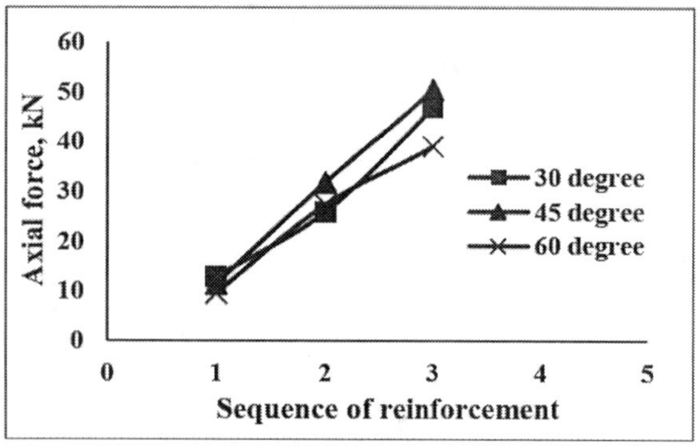

(b) 3 layers

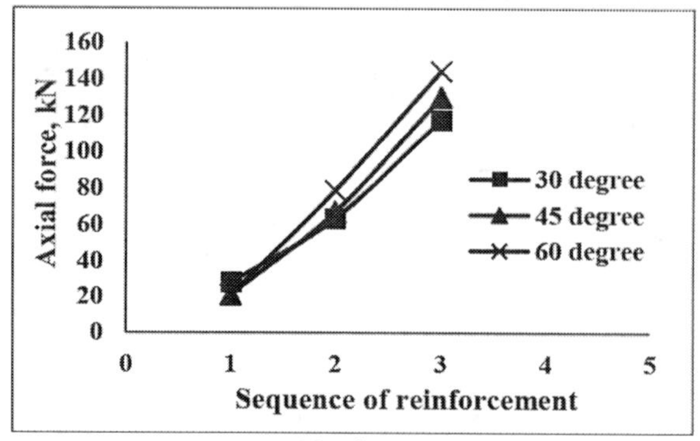

(e) 3 layers

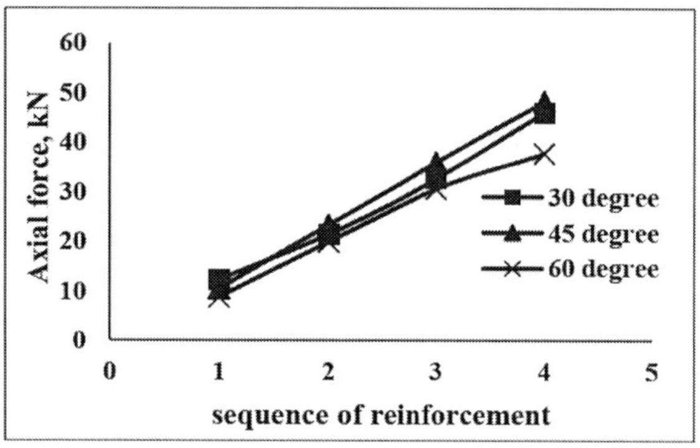

(c) 4 layers

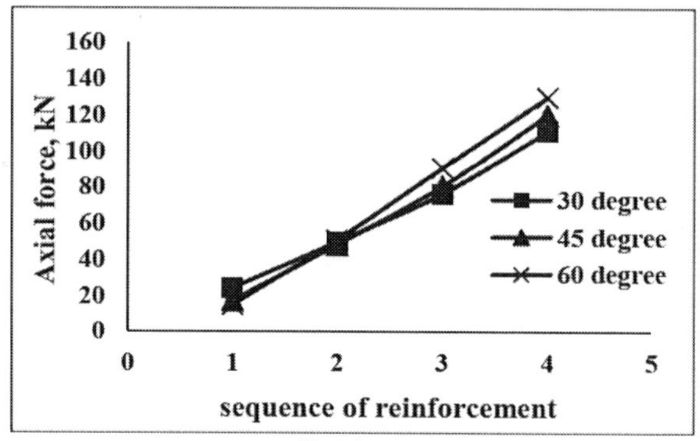

(f) 4 layers

Fig. 16. Maximum axial forces of the geocell in clayey soil with; (a b c) 5 cm height (d e f) 7.5 cm height.

5 Conclusion

This paper presents numerical modelling of geocell reinforced slope modelled as a slab. The purpose of this paper is to find out how effectively waste plastic bottles can be used as geocells in slope reinforcement. The experimental test showed bottled waste geocells increase soil stiffness by providing additional lateral confining, Increase bearing capacity. Generally, the geocell reinforce slope can improve the safety factor.

In sandy soil, the result shows that geocell has a significant effect on the improvement factor. The number of layers installed affects the improvement factor (*IF*) value; the more layers established; the higher improvement factor is obtained. A significant effect is seen on the 60° slope as seen with the highest improvement factor. The effect of geocell reinforcement on the 30° slope is not significant because the 30° is relatively stable by the angle of internal friction of 30.5°.

Unlike the sandy soil, in the clayey soil case, the effect of geocell reinforcement is not very significant. The improvement factor shows that geocell gives a small improvement to safety factor; although the axial force act on geocell is very high, the total deformation of the slope is still high even with the presence of the geocell. It means the clayey soil need a more stiff material or a larger dimension of geocell to restrain deformation and can equalizing the improvement factor with the sandy soil.

The shear strain of unreinforced slope in clayey soil is very high than sandy soil. The two results above between sandy soil and clayey soil show that geocell is more effective to use in sandy soil than in clayey soil. Various results are also obtained by varying the slope, the number of reinforcement layers and the geocell height. Those variations above determine the improvement factor (*IF*).

List of symbols

A area of geocell (m^2)
C_u coefficient of uniformity
C_c coefficient of concavity
c cohesion of soil (kPa)
D_{50} medium diameter (mm)
d distance between geocell layers (m)
E Young's modulus (kPa)
E_g equivalent stiffness of geocell reinforced soil (kPa)
$FOS_{(reinforced)}$ safety factor of reinforced soil
$FOS_{(unreinforced)}$ safety factor of unreinforced soil
G_s specific gravity
I second moment of inertia (m^4)
IF improvement factor
K_0 at-rest lateral earth pressure coefficient
K_e Young's modulus parameter of the unreinforced soil (kPa)
M secant modulus of geocell material (kN/m)
$P(y)$ subgrade reaction at bottom of geocell

Q shear force of geocell reinforcement
$q(y)$ uniform load act on super-face of geocell
T tension force of geocell reinforcement
u distance from the top ground surface to the first layer of geocell (m)
α slope angle (°)
φ friction angle of the soil (°)
σ_1 axial stress (kPa)
σ_3 confining pressure (kPa)
$\tau_b(x)$ interface resistance at bottom of geocell
$\tau_t(x)$ interface resistance at top of geocell
ε_1 axial strain
ε_3 volumetric strain
v Poisson's ratio of soil
ψ angle of dilatancy (°)

References

1. United States Environmental Agency, Environmental factoids. Available from: www.thebalancesmb.com., 2021. The decomposition of waste on landfills. https://www.thebalancesmb.com/how-long-does-it-take-garbage-to-decompose-2878033. [accessed 19 February 2021]
2. T.S. Ingold, The geotextiles and geomembranes manual (Elsevier Science Publishers LTD, 1994)
3. W.F. Van Impe, Soil Improvement Techniques and Their Evolution (Taylor & Francis, 1989).
4. X.M. Yang, J. Han, R.L. Parsons, D. Leshchinsky, Springer-Verlag, Front. Archit. Civ. Eng. China, **4** (2), 233-240 (2010)
5. A. Hedge, T.G. Sitharam, Geotextiles and Geomembranes, **43**, 171-181 (2015)
6. R. J. Bathrust, M.A. Knight, Comput. Geotech. 22 (3/4), 205-219 (1993)
7. G.M. Latha, A. Somwanshi, Geotext. Geomembr. **27**, 409-422 (2009)
8. A. Hedge, T.G. Sitharam, Int. J. Geotech. Eng. **7** (4), 347-354 (2013)
9. I. Mehdipour, M. Ghazavi, R.Z. Moayed, Geotext. Geomembr. **37,** 23-24 (2013)
10. G.M. Latha, K. Rajagopal, Canadian Geotechnical Journal **44** (8), 917-927 (2007)
11. H.I. Ling, D. Leshchinsky, J.P Wang, Y. Mohri, A. Rosen, Journal of Geotechnical and Geoenvironmental Engineering, ASCE **135** (4), 515-524 (2009)
12. Plaxis, Plaxis 2D V8 manual reference (2002)
13. K. Rajagopal, N.R. Krishnaswamy, G.M. Latha, Computer Methods and Advances in Geomechanics, 1251-1254 , Balkema (2001)
14. J.M Duncan, C.Y Chang, *Non-linear analysis of stress and strain in soils*, Journal of the Soil Mechanics and Foundations Division, ASCE **96**, 1629-1653 (1970)

Recommendations of excavation and support systems of Pamukkulu Dam diversion tunnel based on correlation of rock mass classification between RMR and GSI

Dico Nasrulloh[1,2], *Agung* Setianto[1*], and *I Gde Budi* Indrawan [1]

[1]Department of Geological Engineering, Engineering Faculty, Gadjah Mada University, Yogyakarta, Indonesia.

[2] Ministry of Public Works and Housing, Jakarta, Indonesia.

Abstract. This paper presents the results of geological engineering research conducted to determine the character of rock masses, recommendations of tunnel excavation method and support system based on stand- up time estimates in unsuported conditions. The investigation was conducted by observing rock mass quality based on the newest bore log sample test results in 2019 using Rock Mass Rating (RMR) and Geological Strength Index (GSI) rock mass classification. The results showed that area consist of lithology in the form of porphyryc lava basalt and pyroclastic volcanic breccia. Rock mass has a slightly weathering alteration rates. Intact rocks have Uniaxial Compressive Strength (UCS) values ranging from 100-250 Mpa to >250 Mpa and are a category of strong rocks. Rock mass has fair to good rock quality class III-II based on RMR values between 53-69, GSI values between 48-64. The roof span required is obtained from the tunnel planning roof span of 10 meter, with a stand-up time of 70 hours without support system and immediate collapse for 5 days. The recommended excavation methods are excavation by drill and blast on top heading and bench: 1,5-3 meter advance in top heading tunnel face, and then can be recommended support system using rock bolts (20 mm diameter, fullly bonded): systematic bolts 4 meter long, spaced 1,5-2 meter in crown and bench with wiremesh in crown then shotcrete: 50-100 mm in crown, and 30 mm in sides, without steel ribs support.

1 Introduction

Geological investigation of subsurface engineering at the site of the diversion tunnel plan using the newest bore log test is carried out to know in detail the type of rock/soil, structure, the strength of rocks, permeability and thickness of the soil layer that will be used as a guideline to determine the design and geometry trace of the tunnel.

Therefore, adequate geological data is needed to support the right methods and stages to be able to reduce the risk of construction failure.

The purpose of the classification of rock mass is to group the types of rock mass based on his behavior, as the basis for understanding the character of each class, provide quantitative data to determine excavations methods and tunnel's support system.

This research aims to observe and analyze bore log test result to determine the quality of rock mass using Rock Mass Rating (Bienawski,1989) and Geological Strength Index (Hoek&Brown, 2002) in the Pamukkulu's Dam diversion tunnel.

*Corresponding Author: agung_setianto@ugm.ac.id

And then determine the stand-up time of tunnel without support system to provide recommendations on excavation methods and tunnel support systems on tunnel's rocks that are effectively based on quality profile of rock mass.

Previous studies have been conducted to investigate the quality of rock mass with criepi's method. High uncertainty in that rock mass classification method requires more analysis of rock quality with other methods namely RMR and GSI, using the latest bore log test data investigation in 2019.

1.1 Location

Pamukkulu Dam is located in Kale Ko'Mara Village, North Polombangkeng District, Takalar Regency, South Sulawesi Province, Indonesia. Diversion tunnel using horseshoe shape with dimensions, diameter 7 meters, 370 meter long section.

 https://doi.org/10.1051/e3sconf/202132504002

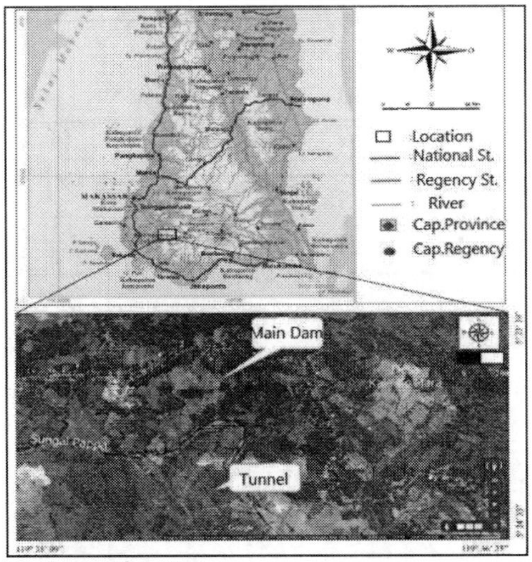

Fig. 1. Location Map

2 Geological condition

2.1 Lithology

The research area is located in the southern part of Sulawesi, based on a regional geological map of Lembar Ujungpandang, Bantaeng and Sinjai, Sulawesi (Sukamto and Supriatna, 1982), it appears that the research area was influenced by several lithological formations consisting: Camba Formation (Tmc), Tonasa Formation (Temt), Coastal Alluvium Deposits (Qac) and Baturape Volcano Rocks–Cindako (Tpbv), is shown in Figure 2 below.

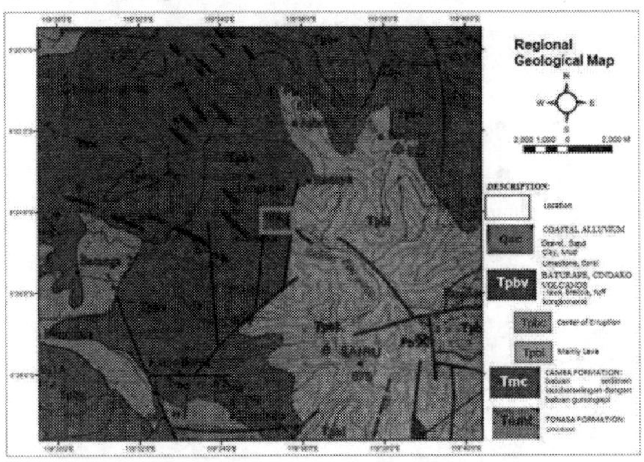

Fig. 2. Regional Geological Map sheet of Ujungpandang, Bantaeng and Sinjai, Sulawesi (Sukamto and Supriatna, 1982)

Research area dominated by residual soil, slightly weathered basalt lava group, fresh lava-breccia group and tuff lithology group, highly weathered breccia.

2.2 Geological structure

The geological structure in the research area is a fault down with N - S and W-E orientation. Fault is located on the northern part of the main dam, and extends along the coverage area of the dam.

Concerning active faults, the location is approximately 100 km west of the Walanae active fault northwest (NW) – southeast (SE).

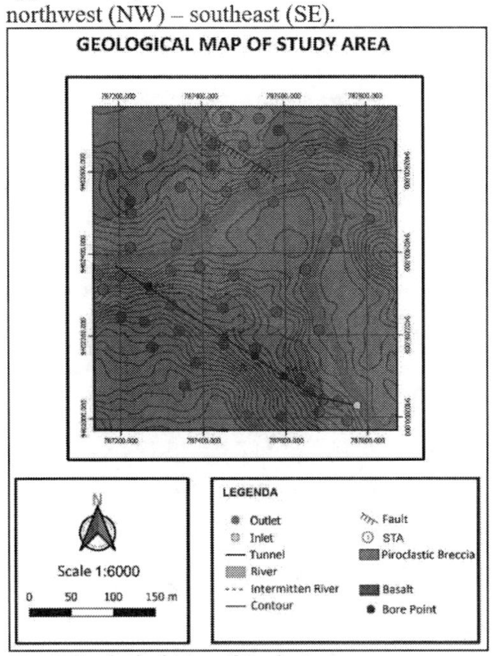

Fig. 3. Geological Map of Study Area

3 Methodology

3.1 Rock mass rating (RMR)

The main reason for using the Rock Mass Rating (RMR) classification method its ease and flexibility in a variety of practical purposes in engineering (Bieniawski, 1989). Rock Mass Rating is defined and calibrated based on observations and experiences of use

in the excavation civil construction and tunnels design.

This engineering classification system, Bieniawski developed in 1973-1989 utilizes the following six rock mass paramaters, including:

1. Uniaxial Compressive Strength of intact rock material
2. Rock Quality Designation (RQD)
3. Condition of discontinuities, given as:
   - Persistance
   - Separation
   - Roughness
   - Infilling
   - Alteration/Weathering
4. Groundwater conditions
5. Orientation of discontinuities

All of these are measurable in the field and can also be obtained from borehole data. The rating of each of these parameters is summarised to give a value of Rock Mass Rating (RMR). All parameters are measurable in the field and some may also obtain of borehole data.

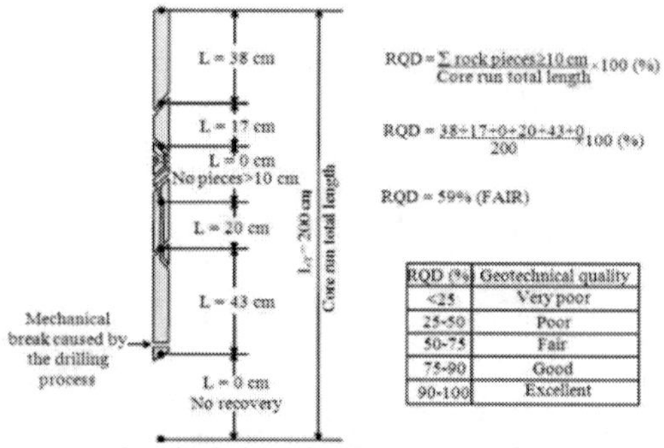

Fig. 4. Example calculation of RQD (Deere,et.al,1988)

Table 1. RMR classification parameters and ratings (Bieniawski, 1989)

PARAMETER			Range of values // ratings						
1	Strength of intact rock material	Point-load strength index	> 10 MPa	4 - 10 MPa	2 - 4 MPa	1 - 2 MPa	For this low range uniaxial compr. strength is preferred		
		Uniaxial compressive strength	> 250 MPa	100 - 250 MPa	50 - 100 MPa	25 - 50 MPa	5 - 25 MPa	1 - 5 MPa	< 1 MPa
	RATING		15	12	7	4	2	1	0
2	Drill core quality RQD		90 - 100%	75 - 90%	50 - 75%	25 - 50%	< 25%		
	RATING		20	17	13	8	5		
3	Spacing of discontinuities		> 2 m	0.6 - 2 m	200 - 600 mm	60 - 200 mm	< 60 mm		
	RATING		20	15	10	8	5		
4	Condition of discontinuities	Length, persistence	< 1 m	1 - 3 m	3 - 10 m	10 - 20 m	> 20 m		
		Rating	6	4	2	1	0		
		Separation	none	< 0.1 mm	0.1 - 1 mm	1 - 5 mm	> 5 mm		
		Rating	6	5	4	1	0		
		Roughness	very rough	rough	slightly rough	smooth	slickensided		
		Rating	6	5	3	1	0		
		Infilling (gouge)	none	Hard filling		Soft filling			
			-	< 5 mm	> 5 mm	< 5 mm	> 5 mm		
		Rating	6	4	2	2	0		
		Weathering	unweathered	slightly w.	moderately w.	highly w.	decomposed		
		Rating	6	5	3	1	0		
5	Ground water	Inflow per 10 m tunnel length	none	< 10 litres/min	10 - 25 litres/min	25 - 125 litres/min	> 125 litres /min		
		$p_w / \sigma 1$	0	0 - 0.1	0.1 - 0.2	0.2 - 0.5	> 0.5		
		General conditions	completely dry	damp	wet	dripping	flowing		
	RATING		15	10	7	4	0		

p_w = joint water pressure; $\sigma 1$ = major principal stress

Table 2. Rating adjustment for discontinuity orientations (Bieniawski, 1989)

		Very favourable	Favourable	Fair	Unfavourable	Very unfavourable
RATINGS	Tunnels	0	-2	-5	-10	-12
	Foundations	0	-2	-7	-15	-25
	Slopes	0	-5	-25	-50	-60

Table 3. Rock mass classes are determined from total ratings (Bieniawski, 1989)

Rating	100 - 81	80 - 61	60 - 41	40 - 21	< 20
Class No.	I	II	III	IV	V
Description	VERY GOOD	GOOD	FAIR	POOR	VERY POOR

Table 4. Meaning of rock mass classes (Bieniawski, 1989)

Class No.	I	II	III	IV	V
Average stand-up time	10 years for 15 m span	6 months for 8 m span	1 week for 5 m span	10 hours for 2.5 m span	30 minutes for 1 m span
Cohesion of the rock mass	> 400 kPa	300 - 400 kPa	200 - 300 kPa	100 - 200 kPa	< 100 kPa
Friction angle of the rock mass	< 45°	35 - 45°	25 - 35°	15 - 25°	< 15°

3.2 Geological strength index (GSI)

GSI was introduced by Hoek (1994) and Hoek, Kaiser and Bawden (1995), can be used to estimating rock mass strength for different geology conditions, but it is less effective for poor or highly weathered rocks.

GSI can estimate the strength and deformation characteristics of rock masses that are required for underground excavation analysis. (Hoek and Brown,1980).

One of the most important factors in estimating rock mass through GSI is to get Uniaxial Compressive Strength (UCS) values that describe level of weathering rocks that can known through laboratory test or using field estimates of UCS for intact rock pieces (Hoek et.al, 1998) [Table 3].

The next development of GSI is combined into 2 parameters the structure of the surface conditions in the form of roughness, weathering/alteration and fillers and block properties is described in Figure 5. (Hoek and Marinos, 2000).

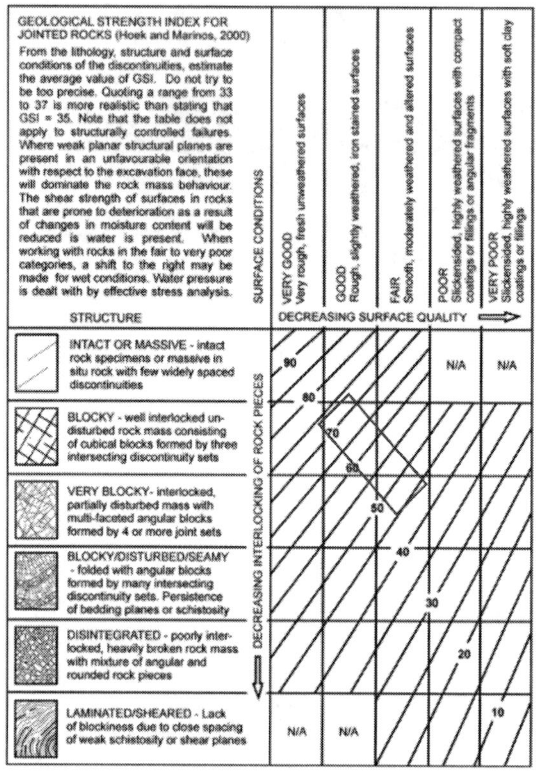

Fig. 5. Estimation value of GSI (Hoek and Marinos, 2000).

One of the most decisive factors in estimating rock mass through GSI is to obtain a Uniaxial Compressive Strength of intact rock material value that describes the level of weathering of rocks. The approximate value of UCS in the field according to Hoek et al. (1998) is seen in Table 5 below.

Table 5. Field estimates of UCS intack rock pieces (Hoek et.al, 1998)

Grade*	Term	Uniaxial compressive strength (MPa)	Point load index (MPa)	Field estimate of strength	Examples
R6	Extremely strong	>250	>10	Specimen can only be chipped with a geological hammer	Fresh basalt, chert, diabase, gneiss, granite, quartzite
R5	Very strong	100–250	4–10	Specimen requires many blows of a geological hammer to fracture it	Amphibolite, sandstone, basalt, gabbro, gneiss, granodiorite, limestone, marble, rhyolite, tuff
R4	Strong	50–100	2–4	Specimen requires more than one blow of a geological hammer to fracture it	Limestone, marble, phyllite, sandstone, schist shale
R3	Medium strong	25–50	1–2	Cannot be scraped or peeled with a pocket knife, specimen can be fractured with a single blow from a geological hammer	Claystone, coal, concrete, schist, shale, siltstone
R2	Weak	5–25	b	Can be peeled with a pocket knife with difficulty, shallow indentation made by firm blow with point of a geological hammer	Chalk, rocksalt, potash
R1	Very weak	1–5	b	Crumbles under firm blows with point of a geological hammer, can be peeled by a pocket knife	Highly weathered or altered rock
R0	Extremely weak	0.25–1	b	Indented by thumbnail	Stiff fault gouge

[a] Grade according to Brown (1981)
[b] Point load tests on rocks with a unaxial compressive strength below 25 MPa are likely to yield ambiguous results

3.3 Correlation between RMR and GSI

Classification of rock mass with the GSI system is a development of the rock mass classification of the RMR system. Hoek and Karzulovic (2000) explained that the mass of rocks with GSI>25 and RMR>23, GSI values can be calculated by the following Equation:

$$GSI=RMR_{89}-5 \quad (1)$$

The RMR_{89} value is the base of RMR value (Bieniawski, 1989) by providing a value of groundwater with a value weight of 15 (dry conditions) and a joint orientation value with weight of 0 (zero). The equation above should not be used for rock mass with very poor quality or with a GSI value<25.

Table 6. Quality class of rock mass based on a linear correlation between RMR and GSI rating (Hoek and Karzulovic, 2000)

Rock Mass Clasification	Rock Mass Quality				
	Very Poor	Poor	Fair	Good	Very Good
RMR	<21	21-40	41-60	61-80	81-100
GSI	<12	12-30	31-49	50-68	69-87

3.4 Stand-up time

The term stand-up time was first used by Lauffer in Lauffer's rock mass classification and then further developed by Lauffer for excavation using tunnel boring machines purposes. Stand-up time is defined as the length of time the tunnel can support itself without additional supporting structures.

The stand-up time estimation is very important because it will affect the excavation cycle, the required support structure, and the tunnel excavation method. The stand-up time developed by Lauffer is then correlated with rock mass classification and the most frequently applied in tunnel construction work is the RMR and stand-up time correlation by Bieniawski. The correlation by Bieniawski can be seen in Fig. 6.

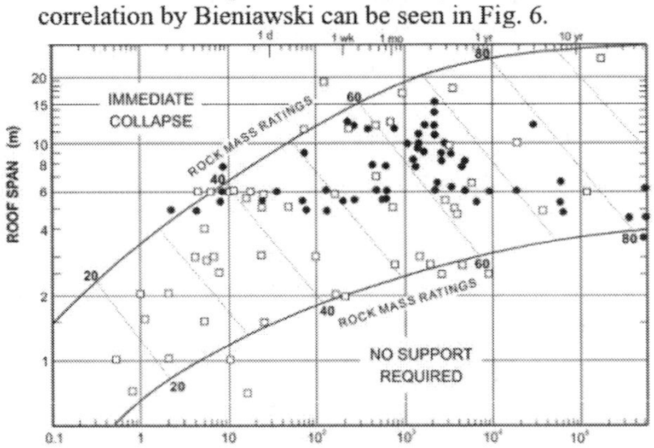

Fig.6. RMR classification of rock masses. (Countour lines indicate limits of applicability) (Bieniawski, 1989).

This study used the correlation between stand-up time and RMR to determine how long the roof span tunnel can withstand collapse without the tunnel's support system. The RMR value used to correlate with stand-up time is the lowest RMR value of each borehole point due to the correlation between GSI and RMR on the site.

3.5 Excavation methods and support system

Based on the lowest RMR value of each borehole point on diversion tunnel use to determine guide for excavation and support system in rock tunnels, provide that the tunnel is horseshoe shaped, roof span width 10 meters, vertical stress below 25 Mpa, using drill and blast methods.

Table 7. RMR classification guide for excavation and support in rock tunnels (Bieniawski, 1989).

Rock mass class	Excavation	Support		
		Rock bolts (20 mm diam., fully bonded)	*Shotcrete*	*Steel sets*
1. Very good rock RMR: 81-100	Full face: 3 m advance	Generally no support required except for occasional spot bolting		
2. Good rock RMR: 61-80	Full face: 1.0-1.5 m advance; Complete support 20 m from face	Locally bolts in crown, 3 m long, spaced 2.5 m with occasional wire mesh	50 mm in crown where required	None
3. Fair rock RMR: 41-60	Top heading and bench: 1.5-3 m advance in top heading; Commence support after each blast; Commence support 10 m from face	Systematic bolts 4 m long, spaced 1.5-2 m in crown and walls with wire mesh in crown	50-100 mm in crown, and 30 mm in sides	None
4. Poor rock RMR: 21-40	Top heading and bench: 1.0-1.5 m advance in top heading; Install support concurrently with excavation - 10 m from face	Systematic bolts 4-5 m long, spaced 1-1.5 m in crown and walls with wire mesh	100-150 mm in crown and 100 mm in sides	Light ribs spaced 1.5 m where required
5. Very poor rock RMR < 21	Multiple drifts: 0.5-1.5 m advance in top heading; Install support concurrently with excavation; shotcrete as soon as possible after blasting	Systematic bolts 5-6 m long, spaced 1-1.5 m in crown and walls with wire mesh. Bolt invert	150-200 mm in crown, 150 mm in sides, and 50 mm on face	Medium to heavy ribs spaced 0.75 m with steel lagging and forepoling if required. Close invert

4 Result and discussion

In this research has been investigated soil and rocks below the surface of the tunnel site using the following stages of work below:

1. Geological mapping and determination of bore log point location;
2. Conduct core bor tests at predetermined points
3. Description of rocks, determination of rock mass class and value, using correlation of RMR and GSI methods;
4. Determine how long the roof span tunnel can withstand collapse without the tunnel's support system [Fig.6];
5. Determine excavation methods and support system in rock tunnels[Table 7].

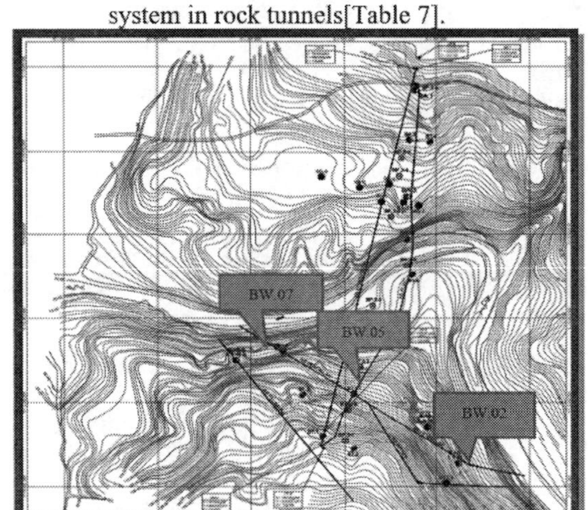

Fig. 7. Bor hole point test location in tunnel area (Soil&Rock Investigation, 2020)

Table 8. Bore log test distribution. (Soil&Rock Investigation, 2020)

No.	Bore Hole	Depth(m)	Elevation (mdpl)	Located
1	BW.02	30	100.032	Inlet
2	BW.05	40	110.512	Plinth&Tunnel
3	BW.07	40	105.918	Outlet

Referring to rock mass classification (RMR and GSI) conducted observations on samples of core bor test results with the following results:

Table 9. Lithology description and rock mass rating on BW-02 (Inlet Tunnel)

No	Depth (m) From	Depth (m) To	Lithology	RMR (Bieniawski, 1989) Classification Parameters: UCS	RQD(%)	Joint Spacing	Cond.of discontinuities: Disc.	Sep.	Roughs	Infilling	Weather	G.Water	AJ	RMR Total	GSI (Hoek Brown, 2002)
1	0	8	Soil Clay												
	Rating														
2	9	15	Basalt Porfiri	100-250	65,8	0,2-0,6	3-10	0.1-1	Slight.R	>5	Slight.W	Drip	Favour	53	48
	Rating			12	13	10	2	4	3	2	5	4	-2		
3	16	30	Basalt Porfiri	100-250	75,2	0,2-0,6	3-10	0.1-1	Slight.R	>5	Slight.W	Wet	Favour	60	55
	Rating			12	17	10	2	4	3	2	5	7	-2		

Source: Calculation

It was found that this area consists of lithology in the form of residual soil (clay and highly weathered sand) and porphyry lava basalt. Generally rock mass has a slightly weathering rate, and RMR value between 53-60 present category of class No.III describe as fair rock [Table 3].

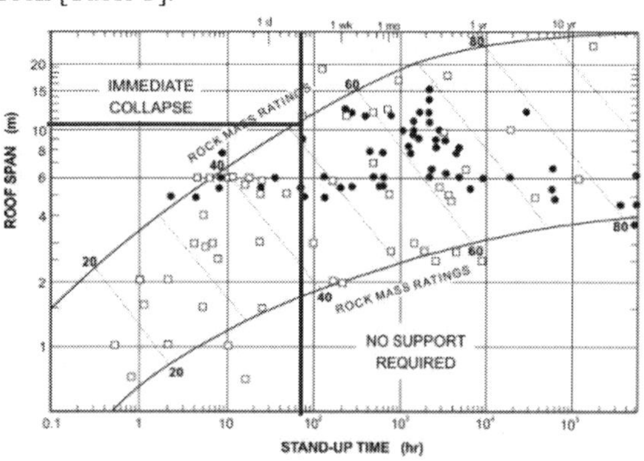

Fig. 8.Stand-up time based on BW.02 point (Inlet Tunnel)

Based on [Fig.8], the lowest RMR value at BW.02 borehole point is correlated using stand-up time chart by Bieniawski. The roof span required is obtained from the tunnel planning roof span of 10 m, which has a stand-up time of 70 hours without rock support system and immediate collapse for 5 days.

Table 10. Lithology description and rock mass rating on BW-05 (Plinth&Tunnel)

No	Depth (m) From	Depth (m) To	Lithology	RMR (Bieniawski, 1989) Classification Parameters: UCS	RQD(%)	Joint Spacing	Cond.of discontinuities: Disc.	Sep.	Roughs	Infilling	Weather	G.Water	AJ	RMR Total	GSI (Hoek Brown, 2002)
1	0	8	Soil Clay												
	Rating														
2	9	15	Basalt Porfiri	100-250	65.8	0.2-0.6	3-10	0.1-1	Slight.R	>5	Slight.W	Drip	Favour	53	48
	Rating			12	13	10	2	4	3	2	5	4	-2		
3	16	30	Basalt Porfiri	100-250	75.2	0.2-0.6	3-10	0.1-1	Slight.R	>5	Slight.W	Wet	Favour	60	55
	Rating			12	17	10	2	4	3	2	5	7	-2		

Source: Calculation

It was found that this area consists of lithology in the form of residual soil (clay and highly weathered sand), porphyryc lava basalt and pyroclastic volcanic breccia. Generally rock mass has a slightly weathering rate, and RMR value between 54-69 present category of class No.II-III describe as fair to good rock [Table 3].

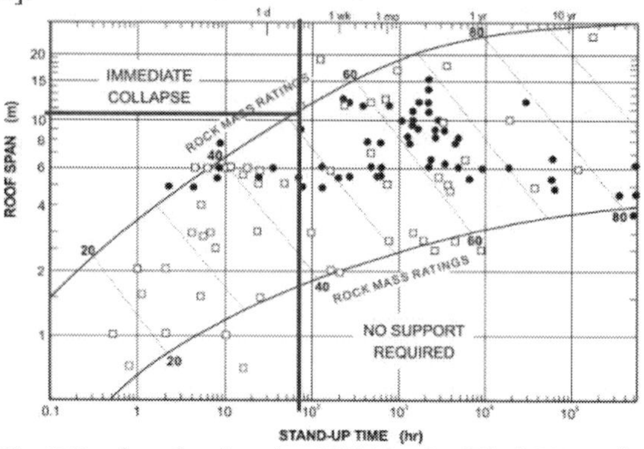

Fig. 9.Stand-up time based on BW.05 point (Plinth&Tunnel)

Based on [Fig.9], the lowest RMR value at BW.05 borehole point is correlated using stand-up time chart by Bieniawski. The roof span required is obtained from the tunnel planning roof span of 10 m, which has a stand-up time of 70 hours without rock support system and immediate collapse for 5 days.

Table 11. Lithology description and rock mass rating on BW-07 (Outlet Tunnel)

No	Depth (m) From	Depth (m) To	Lithology	RMR (Bieniawski, 1989) Classification Parameters: UCS	RQD(%)	Joint Spacing	Cond.of discontinuities: Disc.	Sep.	Roughs	Infilling	Weather	G.Water	AJ	RMR Total	GSI (Hoek Brown, 2002)
1	0	15	Soil Clay												
	Rating														
2	16	24	Breccia Piro.	100-250	59.4	0.2-0.6	3-10	0.1-1	Slight.R	>5	Slight.W	Drip	Favour	53	48
	Rating			12	13	10	2	4	3	2	5	4	-2		
3	25	40	Basalt Porfiri	100-250	82.9	0.3-10	3-10	0.1-1	Slight.R	>5	Slight.W	Drip	Favour	62	57
	Rating			12	17	15	2	4	3	2	5	4	-2		

Source: Calculation

It was found that this area consists of lithology in the form of residual soil (clay and highly weathered sand) and pyroclastic volcanic breccia. Generally rock mass has a slightly weathering rate, and RMR value between 53-62 present category of class No.II-III describe as fair to good rock [Table 3].

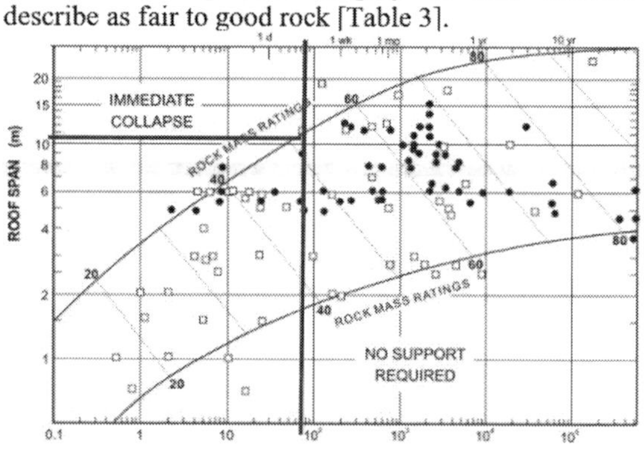

Fig. 10.Stand-up time based on BW.07 point (Outlet Tunnel)

Based on [Fig.10], the lowest RMR value at BW.05 borehole point is correlated using stand-up time chart by Bieniawski. The roof span required is obtained from the tunnel planning roof span of 10 m, which has a stand-up time of 70 hours without rock support system and immediate collapse for 5 days.

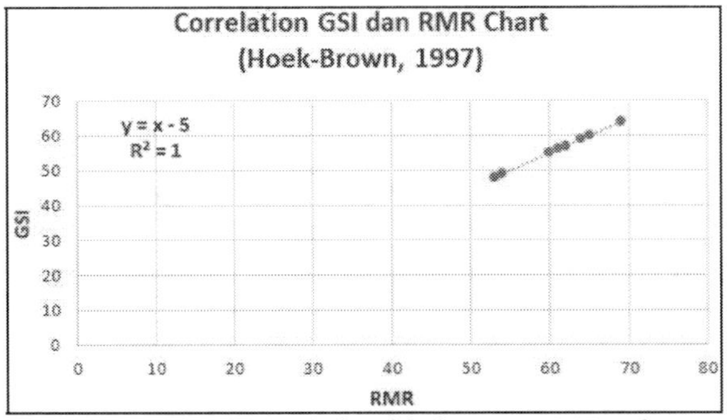

Fig. 11. Graph of the relationship between GSI and RMR

Based on the enrapture of bore data in the 3 points area described above, obtained an average RMR rock mass classification value of 60, categorized as rock mass class no.II with a description of fair rocks[Table.3]

Refers to the tunnel's excavation method and support system guide (Bieniawski, 1989). in [Table.7], the recommended excavation method is:

a) Excavation by drill and blast;
b) Top heading and bench: 1,5-3 meter advance in top heading;
c) Commence support after each blast;
d) Commence support 10 meter from the face.

And then can also be recommended support system as follows below:

a) Rock bolts (20 mm diameter, fullly bonded): systematic bolts 4 meter long, spaced 1,5-2 meter in crown and walls with wiremesh in crown;
b) Shotcrete: 50-100 mm in crown, and 30 mm in sides.

Tunnel excavation by drill and blast methods is conducted by digging from inlet to outlet. Heavy equipment needed to support the work with conditions in the field are Jumbo Drill, Schaeff Tunnel Loader, and Dump Trucks.

Blasting work items in diversion tunnels include:

a) Explosive storage (Handak) retrieval
b) Depth survey and drilling;
c) Charging;
d) Blasting;
e) Smoke and dust cleaning;
f) Post-explosion rock cleaning and material loading.

The process of digging tunnels by drill and blast methods in one day can do blasting 1 time, and the results of excavation obtained 1,5-2 meter, so to dig a tunnel along 370 meter takes approximately 6 months.

5 Conclusion and recommendation

The research area consists of geological lithology in the form of residual soil (clay and weathered sand), porphyry lava basalt and pyroclastic volcanic breccia. Generally rock mass has slightly weathering rate.[Table 9-11].

Intact rock has Uniaxial Compressive Strength (UCS) values ranging from 100-250 Mpa and >250 Mpa and belongs to the category of rocks with very good strength. Geological Strength Index (GSI) rating between 48-64 and Rock Mass Rating (RMR) between 53-69 both presents categories of fair to good rocks class III-II [Table.5].

The average RMR value at every borehole point above is correlated using a stand-up time chart by Bieniawski[Fig.5]. The roof span required is obtained from the tunnel planning roof span of 10 meters, with a stand-up time of 70 hours without support system.

Refers to the tunnel's excavation method and support system guide in [Table.7], the recommended excavation method is excavation by drill and blast on top heading and bench tunnel face.

And then make recommendation tunnel's support system using rock bolts and wiremesh-shotcrete combination without steel ribs support.

The author gratefully acknowledges to the Ministry of Public Works and Housing for the support of research funding through the 2020's Superspecialist Scholarship Program at Geological Engineering of Tunnel and Underground Structure, Master by Research Program of Geological Engineering, Gadjah Mada University.

References

1. Z.T.Bieniawski, InJ.A.Hudson(Ed.) *Classification of rock masses of rock engineering: The RMR system and future trends, comprehensive rock engineering*, **3**:553-574 (1993)
2. D. Deere, R. Miller, *Engineering Classification and Index Properties for Intact Rock*, in New Mexico: Technical Report No. AFWL-TR-65-116, Air Force Weapons Laboratory, Kirkland Air Force Base (1979)
3. E. Hoek, P. Marinos, M. Benissi, Bull Eng Geol Environ **57**, 151–160 (1998)
4. P.G. Marinos, V. Marinos, E. Hoek, *The Geological Strength Index (GSI):A characterization Tool for Assessing Engineering Properties of Rock Masses*, 87-94 (2007).
5. Balai Besar Wilayah Sungai Pompengan-Jeneberang, *Soil and Rock Investigation Diversion Tunnel of Pamukkulu Dam*, Takalar, in Geological Report (2020)
6. R. Sukamto, S. Supriatna, *Peta Geologi Lembar Ujungpandang,Bantaeng dan Sinjai, Sulawesi*, (1982)

The Study of fluviomarine rice field productivity and its effect on farmer's income in Mojowarno Village, Kaliori District, Rembang Regency

Azizatul Maghfiroh*, *Sudrajat* Sudrajat

Department of Environmental Geography, Faculty of Geography, Universitas Gadjah Mada, Indonesia

Abstract. Fluviomarine rice field is a type of agricultural land resulting from past marine activity processes with alluvium deposition material. This study specifically discusses the productivity of the rice field agricultural system by linking the physical components of the land based on the characteristics of geomorphological conditions (physiography, soil type, elevation contour, slope, and land use) with social interactions by local farmers. This study wants to analyze the productivity of the fluviomarine rice field farming system on agricultural income in Mojowarno village. This study uses 72 households data sample with the cluster random sampling method, which is based on the fluviomarine form cluster unit. The data analysis used a descriptive quantitative analysis. The results of this study are the fluviomarin rice fields farming system inMojowarno Village is a rainfed farming system with a planting frequency of only 2 times a year. Meanwhile, the average productivity of the land was only 5,307 tons /hectar, while the average income received by farmers was only Rp 5,879,631.00 /year. The regression statistical test results show that farmers' income is significantly influenced by land productivity. This study is expected to be used as a reference in preparing the development strategy for increasing farmer's income.

1 Introduction

Agriculture is one of the important sectors in an economic development. This occurs becauseagriculture functions as a food supply and job opportunities for residents in rural areas. In general, the agricultural systems developed in Indonesia are wetland systems, moorlands, yards, and plantation systems [5]. The agricultural system varies in Indonesia are greatly, both spatially and temporally. These variations of agricultural systems become very interesting part of the study of agricultural geography. One of the main concerns of agricultural geography on these variations in agricultural systems is study the causal relationship of a growing agricultural system in an area. The bio-physical component and socio-demographic component will affect the agricultural system that is formed in an area. Therefore, understanding a complex agricultural system because it is formed from the interaction between physical and non-physical aspects, a geographic approach will be easy to understand.

One of the agricultural systems that very interesting in the study of geography is the fluviomarine rice fields agricultural system. This occurs because the fluviomarine rice fieldsare a type of agricultural land that located in the fluviomarine formations resulting from past marine activity processes with alluvium sediment material, which the population uses for agricultural activities. One of these types of land is located in the coastal area of the North Java Sea and it is behind the tidal flats with a rainfed lowland farming system. Utilization of fluviomarine lowland is the form of a rainfed agricultural system because it only relies on rainwater (rainfed rice). As the result, the agricultural system on fluviomarine fields is characterized by lower productivity than irrigated fields. In fact, rice plants only can plant once a year during the rainy season (September-December) [4].

The condition of the growing agricultural sector in fluviomarine rice fields will have an influence on the land productivity and farmers income. Productivity is a reflection of the quantity of production produced by agricultural land per unit area of land in a certain period [7]. Based on the Decree of the Minister of Agriculture No. 472 / Kpts / Rc.040 / 6/2018 concerning the Location of National Agricultural Areas which determine the national rice development area that Rembang is not a district that is a priority area for rice development [8]. The main problem in lowland rice cultivation in Kaliori District, Rembang Regency is low productivity. The average productivity of lowland rice has only reached 3.22 tonnes / ha. The physical characteristics of rice fields and the agricultural system implemented by farmerscause the level of productivity between rice fields to vary.

The low productivity of the fluviomarine rice fields cannot be separated from the characteristics of the fluviomarine plain land which can be seen from the type

* Corresponding author: azizatul.m@mail.ugm.ac.id

of land that is developed, the availability of water sources, the ability of farmer's resources, and the availability of capital. Rice productivity in fluviomarine land cannot be separated from the influence of physical conditions; such as soil fertility conditions, types of commodities, the level of technology application, and socio-economic conditions of farmers [1]. The level of productivity of rice fields has an impact on the amount of income received by farmers. Therefore, research on the productivity of fluviomarine rice fields is very interesting. Basedon it, this study aims to analyze the fluviomarine rice fields agricultural system, to analyze theproduction and productivity of the fluviomarine rice fields, to analyze the agricultural income received by farmers, and to analyze the effect of fluviomarine rice fields productivity on agricultural income.

2 Methods

2.1 Location

This research was located in Mojowarno Village, which is one of the northern coastal zones of Java. Administratively, Mojowarno Village is included in the Kaliori District, Rembang Regency. Mojowarno Village consists of 3 hamlets, namely; Dukuh Samben, Dukuh Mojo, and Dukuh Cering. Based on the physical characteristics, there are two types of land resources in the Mojowarno area, namely; tidal flats and fluviomarin plains. The fluviomarine plain covers an area of 164.32 hectares and it used for lowland agricultural activities. Most of the residents of Mojowarno Village who work in agriculture are still of productive age, amounting to 64.41%.

2.2 Data

Data collected in the form of primary data and secondary data. Primary data comes from interviews. Primary data collection is carried out directly through interviews with farmers using a structured questionnaire. In addition, indept interviews were conducted with the head of farmer groups and government officials in Mojowarno Village. Meanwhile, field observation activities are carried out by direct observation of agricultural production activities. Secondary data were collected from the village office, BPS and other agencies. Secondary data were collected in the form of administrative shapefile data for Mojowarno Village, map of terrain in Kaliori District at a scale of 1: 50,000, Land Map of Rembang Regency at a scale of 1: 80,000, google earth images, SRTM images, and geological shapefiles for Mojowarno Village. Secondary data in this study are to support research activities.

2.3 Population and sample

The population in the study were all heads of family (KK) farmers in Mojowarno Village andown rice fields. In 2020, there were 257 heads of farming families in Mojowarno Village, spread over 7 clusters of land units. To determine the size of the sample size using the Slovin formula [9], as follows:

$$n = \frac{N}{N.d^2+1} \quad [9] \qquad (1)$$

Information:
n = the number of research samples
d = tolerance limit for research sampling error
N = the number of the study population
The results of the calculation obtained a sample size of 72 households. Taking the number of samples from each cluster of fluviomarin landform units in a proportionate manner. Mathematically it can be formulated as follows:

$$Psi = \frac{Xi}{Y} X K \qquad (2)$$

Psi = the number of sample proportions of rice land owners in landform units-i
Xi = the number of heads of household for owners of rice fields in landform units-i
Y = the number of head of household of rice fields (257 KK)
K = the number of research samples (72 KK)
The distribution of the number of samples from each cluster of fluviomarine landforms can beseen in **Table 1** below.

Table 1. The number of samples of farmer households based on land unit clusters

Number	Land Unit Code	The number of Samples (Head of Family)
1	A0112.N.w	8
2	A013.F.w	7
3	A013.N.w	7
4	B03.R.w	19
5	B03.U.pw	8
6	B03.U.w	15
7	B03.F.w	8
Grand total		72

Taking respondents from each cluster by random sampling. The random sampling methodprovides an equal chance of the population to be selected as a sample in a particular area [16].

2.4 Data analysis

The processing data of this research use several processing software including Microsoft Excel, ArcGIS, dan SPSS. Then, the method of data analysis using descriptive statistical analysis approaches, spatial analysis, and inferential statistical analysis. The data from interview with respondents about an agricultural systems, production and productivity of rice fields, and agricultural incomes are tabulating to be form of frequency tables, pie charts, or bar charts. Descriptive statistical analysis used to analyze the fluviomarine lowland agricultural system which was presented in the frequency distributiontable, while agricultural income was analyzed using the following equation:

$$NR = TR - TC$$
$$TR = PxY \qquad [12]$$
$$TC = TFC + TVC \qquad (3)$$

Information:
NR = Net Revenue (farmer's income)
TR = Total Revenue (total farmer revenue)

TC = Total Cost (total cost of agricultural production)
P = price per unit of product
Y = total product
TFC = Total Fixed Cost
TVC = Total Variable Cost

Analysis of farmers' income levels, from the results of using fluviomarine rice fields, the following mathematical equations are used:

$$\text{Interval} = \frac{maximum\ value - minimum\ value}{number\ of\ categories}\ [14] \quad (4)$$

Information:
Low category = minimum value+interval
Moderate category = low category value+interval
High category = moderate category value+interval

Meanwhile, to analyze the productivity of fluviomarine rice fields, the following mathematical equation is used:

$$\text{Productivity} = \frac{total\ production\ (ton)}{land\ area\ (ha)}\ [12] \quad (5)$$

To see the spatial distribution of productivity and agricultural income from fluviomarin rice fields, it was analyzed using a cluster approach [11]. Fluviomarine land units are differentiated based on land shape, soil type, lithology, elevation contours, and land use. The differentiated landformdata arranged in ArcMap layers according to administrative boundaries. The spatial distribution presented of the map of productivity and agricultural income in fluviomarine lowland with a scale of 1: 10,000.

The presentation of the productivity and agricultural income map for rice fields are presented with an area symbol that represent a sample of rice fields in fluviomarine landform units. The symbol of area is divided into three different color gradations to distinguish agricultural productivity and income classes. Analysis of the pattern of productivity and agricultural income is compiled from an overlay of landform map with productivity levels and agicultural income of rice fields. Furthermore, an analysis of the pattern of agricultural productivity and income is carried out based on clusters of landform units associated with a sample of rice field area in each of these landform unit clusters.

Analysis of the effect of fluviomarine rice field productivity on agricultural income using simple linear regression analysis using SPSS software. Before using simple linear regression, the data were tested using the classical assumption test with normality test, linearity test, and heteroscedasticity test. Mathematically, the simple linear regression equation model is used:

$$Y = \beta 0 + \beta X + \mathcal{E} \quad (6)$$

Information :
Y = dependent variable (farmer income)
X = independent variable (lowland productivity)
β0 = constant
β = regression constant / independent variable coefficient
Ɛ = error

To determine the strength of the relationship between variables analyzed using the correlation coefficient. The type of correlation used the Pearson product moment (r) [13], as follows:

$$r_{xy} = \frac{n\sum XiYi - (\sum Xi)(\sum Yi)}{[n\sum Xi^2 - (\sum Xi^2)][n\sum Yi^2 - (\sum Yi^2)]}\ [13] \quad (7)$$

Information :
r_{xy} = correlation coefficient
Y = dependent variable (farmer income)
X = independent variable (lowland productivity)

Meanwhile, to determine the magnitude of the influence of the independent variables (rice field productivity) on the dependent variable (farmer income), it is analyzed using the following mathematical problems:

$$Kd = r^2 x\ 100\%\ [13] \quad (8)$$

Information:
Kd = determination coefficient
r = correlation coefficient

If the determinant coefficient value is close to 100%, the independent variable (productivity) is able to explain all the information needed to predict the dependent variable (income). The results of linear regression analysis and correlation were analyzed descriptively quantitatively. A more indepth analysis is supported by information from the farmer groups and the village information management officers. This information obtained during the data collection process, especially secondary data that available at the village office.

3 Result and discussions

3.1 Characteristic of farmer households of fluviomarine rice field agriculture

In this study, the characteristics of farmers were analyzed by demographic and socio- economic characteristics. Demographic characteristics include age, sex, number of household members, and number of working household members, while socio-economic characteristics include factors of education, length of farming, and land area. Respondents in this study are farmers who carry out agricultural activities, especially fluviomarine rice fields. The farmer respondents can act as private landowners, tenant farmers, or sharecroppers with a profit-sharing system. The farmers carry out farming activities in fluviomarine fields from generation to generation in addition to carrying out farming activities in ponds.

According of Table 2 it is known that the average age of the farmers is> 35 years. This phenomenon shows that currently the agricultural sector is less attractive to young workers. Meanwhile, the level of education is mostly still low (87.5% have primary school education/equivalent). The farming experience ranges from 32 years, while the average rice field is 0.29 hectares. The results of this study illustrate that agricultural activities are carried out byfarmers who are old with low education and narrow land ownership. Based on the results of interviews with the village head, the large of old age farmer in this village is due to the lack of regeneration of young farmers. The farmer's childern prefer to continue living outside the city either continue their education or work as factory workers. So that no one continues to work on the rice fields.

Table 2. Demographic and Sosio-Economic Characteristic of Farmer's Fluviomarine Rice Fields in Mojowarno Village at 2020.

Charac-teristic	Description	Information
Demographic	Average age of respondents (years)	55
	Sex	Man
	The average number of household members (people)	3
	The average number of working household members (people)	2
Sosio-Economic	Respondent's education (people):	
	a. No school	1
	b. Elementary school	63
	c. Junior High School	7
	d. Senior High School	0
	e. Bachelor	1
Sosio-Economic	The average time of farming (years)	32
	The average area of rice land (ha)	0.29

3.2 Agricultural system in fluviomarine rice field

The rice field system that developed in Mojowarno village is dry rice fields or called rainfed lowland agriculture. This is because the rice fields in the research location have limited water sources and only depend on the availability of rainwater. The variations in the components of the agricultural system in a region can be grouped into five components, namely; land processing techniques, planting frequency, types of planting commodities, planting season, and cropping patterns [3]. The results showed that the farmers in the location research were using tractors to cultivate rice fields. Farmers think that using tractors to cultivate rice fields will speed up their work, even though they need more capital.The frequency of planting in fluviomarine rice fields averaged 2 times a year with the main rice crop commodity (**Table 3**).

Table 3. Frequency of planting on fluviomarine rice fields in Mojowarno Village.

MT1 (Nov-Feb)	MT2 (Mar-Jun)	MT3 (Jul-Oct)
Rice	Rice	Fallow

In **Table 3**, it can be seen that the planting season of rice (MT) in Mojowarno Village is divided into 3 times a year, namely; MT1 (the beginning of the rainy season in early November), MT 2 (the season before the dry season in mid-March-late June), and MT3 (the dry season at the beginning of July - the end of October). The results of the study found that78% of farmers in Mojowarno Village applied a rice-fallow cropping pattern. This happens because the water source relies on only rain water. As a result, the time for planting rice is highly dependent on the arrival of the rainy season which on average falls in November-Junewhile the dry season starts in July until October. The difference in conditions between theseseasons allows the occurrence of gaps in agricultural production between seasons [10]. An effective, efficient, sustainable, and high-productive agricultural system can be ralized by utilizing land based on the ability and sustainability of rice fields as well as the characteristics of the agricultural system carried out by farmers.

3.3 Fluviomarine rice field based on land cluster

The topography of the fluviomarine plain and alluvial plain is sloping to flat. Alluvial plainsformed from river sediment are generally found in the central part of Mojowarno Village which is associated with the river flow in the middle of the village, while the fluviomarine plains generally dominate all of the rice fields. Geological conditions of rice fields generallyconsist of past fluviomarin deposits in the form of clay and sand deposits. Rice fields like this are generally found in flat areas with slopes of 0-2% and in limited areas are found in 2-6% very sloping areas and 6-13% sloping areas. The total area of rice fields in Mojowarno Village is 138.47 ha. The rice fields are divided into 7 clusters of rice fields based on the characteristics of the terrain (physiography, slope, and land use) (**Table 4** and **Figure 1**).

Table 4. Characteristics of fluviomarin rice fields in Mojowarno Village

Clus-ter	Code	Physiography	Slope	Land Use
1	A0112.N.w	Curved floodplain in meander	little flat (0-2%)	Rice field
2	A013.F.w	Alluvial terraced plain	Flat (0-2%)	Rice field
3	A013.N.w	Alluvial terraced plain	little flat (0-2%)	Rice field
4	B03.R.w	The northern coast of Java fluviomarine plain	Sloping slope (6-13%)	Rice field
5	B03.U.pw	The northern coast of Java fluviomarine plain	very gentle slope (2-6%)	Settlement and rice field
6	B03.U.w	The northern coast of Java	very gentle	Rice field

 https://doi.org/10.1051/e3sconf/202132504003

		fluviomarine plain	slope (2-6%)	
7	B03.F.w	The northern coast of Java fluviomarine plain	Flat (0-2%)	Rice field

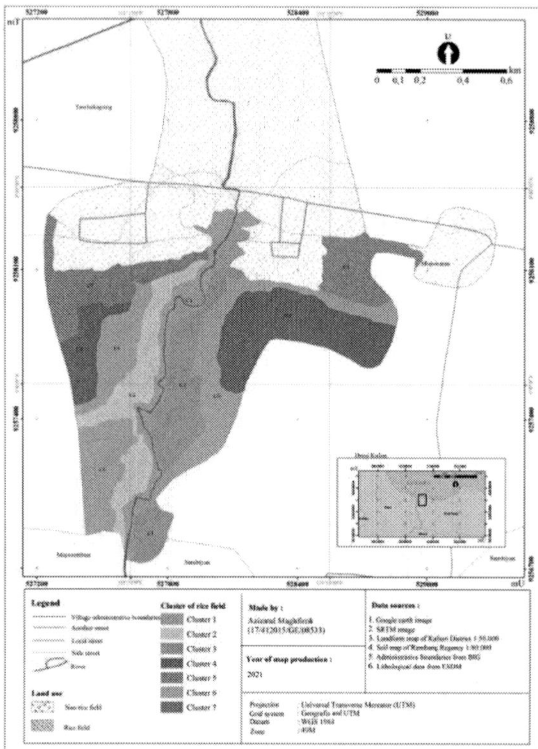

Fig. 1. Map of fluviomarine rice field distribution according to clusters in Mojowarno Village.

All rice fields in cluster 1are often flooded when there is excess water in the rainy season. This is because the location of rice fields is on a meandering curve so that it will experience excess water in the rainy season. The commodity cultivated in this rice fields is dominated by rice plants because of its proximity to the river flow so that the water source is able to meet the planting process. Some of the rice fields in the cluster 2 and cluster 3 have the potential to be flooded when the water supply is abundant in the rainy season, especially in the rice fields around the river meanders. The main commodity cultivated in the first and second planting seasons is rice plants, while in the third planting season is crops like a mung beans and corn are planted. The rice fields in cluster 4 have no potential for flooding due to its sloping topography. In addition, the distance from water source to rice fields is relatively far, which affects water control efforts, especially in the dry season.

Meanwhile, some rice fields in cluster 5 have the potential to be flooded when the water supply is abundant in the rainy season. The ability of farmers to plant until 2 times a year is supported by efforts to control water during the planting process, especially in the dry season. The sloping topography of cluster 6 rice fields causes the appearance of rice fields to have terraced bund with no potential for flooding. In some land, the bunds of the rice fields are used to grow crops such as cassava and corn. The location of rice fields in cluster 7 which is far away from the main river that minimizes the potential for flooding. Meanwhile, the location of this land is too close to residential areas, causes an intensive land management. The main commodity cultivated in the first and second planting season is rice, while in the third planting season is crops with mung beans or uncultivated.

3.4 Productivity and production of fluviomarine rice field

Rice field productivity is a comparison of production yields with land area in a certain time period. The results of the study found that the fluviomarine rice fields in Mojowarno Villagewere able to produce 1.9-5.5 tons of milled dry unhulled rice per year. In this study, the productivity of rice fields is grouped into 3 (three), namely low productivity ranging from 4.36-4.92 tonnes / ha, moderate productivity ranging from 4.93-5.49 tonnes / ha, and high productivity ranging from 5.50 tonnes / ha. -6.03 ton / ha. More clearly the production of planting area and productivity of fluviomarine rice fields can be seen in **Figure 2** below.

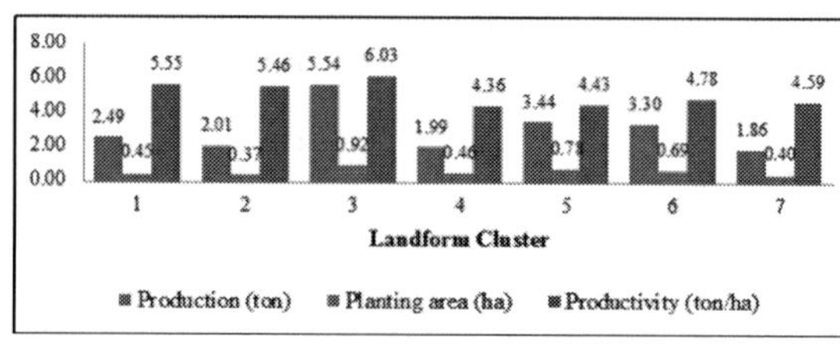

Fig. 2. Production, planted area, and productivity of fluviomarine rice fieldsaccording to land clusters.

In **Figure 3**, it can be seen that the distribution of rice field productivity in the village according to the cluster is dominated by low productivity levels. Low productivity is in 4 (four) clusters, namely cluster 4, cluster 5, cluster 6, and cluster 7 with a total land area of 79.95 ha or about 64.32%. For rice fields with high productivity are found in cluster 3 and cluster 1 with an average productivity value of 5.55-6.03 tonnes / ha, while rice fields with moderate productivity are found in cluster 2 with an average productivity of 5.46 tonnes. /Ha.Meanwhile, it can be seen that the productivity of fluviomarin rice fields in Mojowarno Village is on average 5.03 tonnes / ha. The results of this study are in line with the findings of Jonharnas & Sitindaon (2017) in Deli Serdang Regency, which is 5.44 tonnes / ha.

This study also found that the spatial pattern of productivity distribution of rice fields showed a grouping tendency, based on the similarity of physiographic types. This can be seen in lowproductivity rice fields clustered in cluster 4, cluster 5, cluster 6, and cluster 7 with a slope between 0-2 m and flat topography. The low average productivity ahead of the dry season isdue to the fact that the cluster only has limited water sources, so agricultural land often becomes dry. In addition, it is also caused by the formation of rice fields which is influenced by the formation of the origin of the past marine process which leaves a salt content (salinity). This condition will ultimately affect the ability of plants to grow and develop in producing agricultural production. More clearly, the distribution of rice field productivity can be seen in **Figure 3** below.

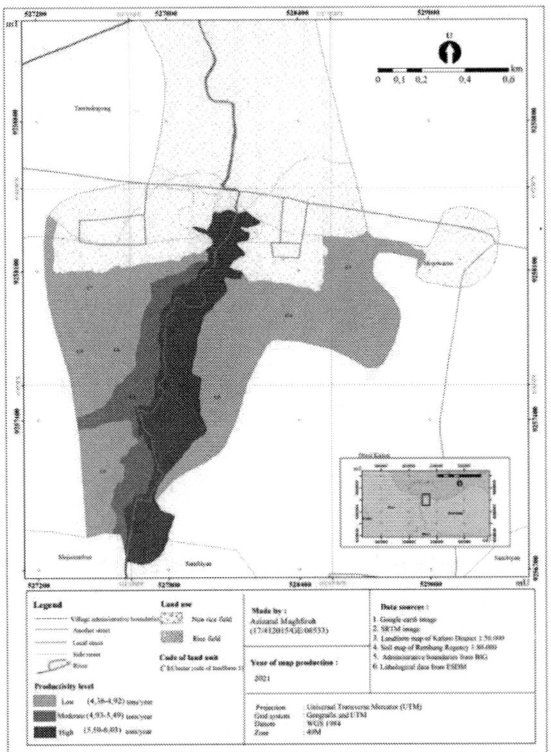

Fig. 3. Distribution map of rice field productivity according to land cluster in MojowarnoVillage, Kaliori District, Rembang Regency

For the medium level of productivity, the cluster 2 rice fields are located west of the main river with a flat slope (0-2%). Meanwhile, high productivity levels are clustered in cluster 1 and cluster 3 rice fields, which are located east of the main river. The rice fields to the west of the river have a lower slope than the rice fields to the east of the river with flat to slightlyflat slopes (0-2%). This condition causes the rice fields to the west of the river when the rainy season has the potential to be flooded. In fact, there are several rice fields that are submerged in water so that the commodity crops are damaged which has the potential to reduce the areaof harvested land and its production. As a result, it has an impact on the low level of productivity of rice fields to the west of the river compared to the east. Even though cluster1,cluster 2, and cluster 3 have the opportunity to always be flooded during the rainy season, during the dry season they benefit because they can use river water to irrigate the land of ricefields easily. Another factor that causes low rice production on the west side of the river is due to the process of washing the soil during floods and high salt deposits that are depositedin the soil. These two phenomena ultimately lead to a decrease in soil fertility. Based on this,it is quite logical that the rice production on the west side of the river is lower than that in theeast side of the river.

3.5 Fluviomarine rice field agricultural income

Analysis of income received by farmers from the use of fluviomarin rice fields in this study was used to analyze the amount of profits received by farmers in one year. Factors that affectthe size of farmers' income include the frequency of farming activities in one year, the levelof production or productivity of rice fields, and the prevailing price of agricultural commodities. The income received by farmers is determined by the amount of production generated from farming and the amount of costs incurred during cultivating agricultural activities. The results of the study found that the average income of the fluviomarin rice field farmers in Mojowarno Village was still low, namely Rp. 5,879,631.00 / year with an average of 0.29 hectares of rice fields. The distribution of farmer income from farming products in Mojowarno Village can be seen in **Figure 4** below.

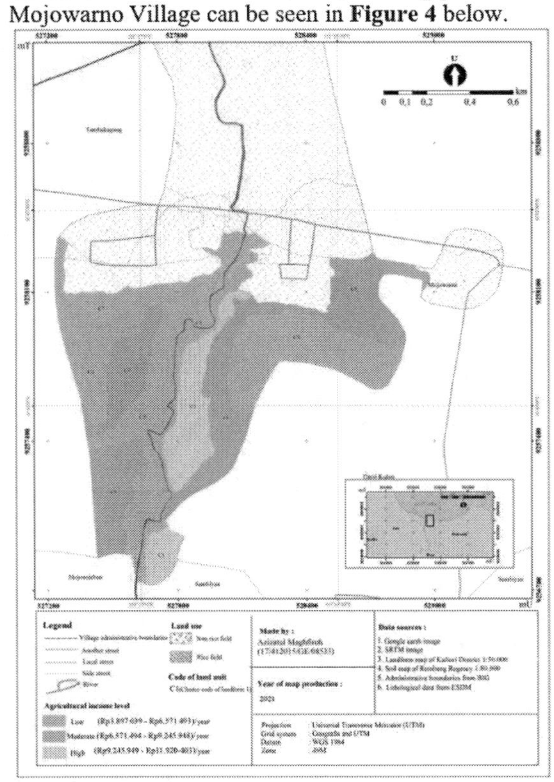

Fig. 4. Distribution map of agricultural revenue according to the land cluster in Mojowarno Village, Kaliori District, Rembang Regency

In **Figure 4**, it can be seen that the distribution of fluviomarine rice field farmer's income levels according to clusters is quite varied. High income farmers are found in cluster 3, farmers classified as medium income are found in cluster 4 and cluster 5, while farmers classified as low income are found in cluster 1, cluster 2, cluster 6, and cluster 7. Variations in the amount of income received by land farmers fluviomarin fields are caused by wide variations and production. In cluster 3,

 https://doi.org/10.1051/e3sconf/202132504003

the farmers cultivate an average rice field of 0.4258 ha with an average production of 5.535 tons. In cluster 4 and cluster 5 the average land areacontrolled is 0.325 with an average production of 2.715 tons. Meanwhile, in cluster 1, cluster2, cluster 6, and cluster 7, the average area of rice fields controlled by farmers was only 0.2444 ha with an average production of 2.415 tonnes. This high income received by farmersin each cluster area is closely related to the area and production of land. The results of the study are in line with the results of research by Kusnadi, et al., (2011) which stated that the more rice fields cultivated for agricultural activities the higher the production and the higherthe farmer's income and vice versa [6].

3.6 Effect of fluviomarine rice field productivity on farmer's income

High land productivity will reflect high yields of land and will ultimately result in high income. Based on this, to determine the magnitude of the influence of rice field productivity on farmers' income, a linear regression statistical test was carried out. Simple linear regression statistical tests can be analyzed by performing the classical assumption test first which includes the assumptions of normality, linearity, and heteroscedosticity. In **Figure 5**, the graph of the relationship between productivity variables and farmer income variables is linear.

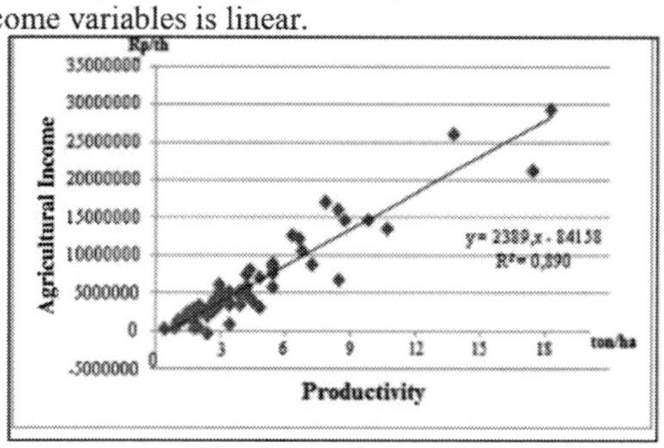

Fig. 5. Linearitas graph

The results of the linear regression test showed that the t-count value was 23.807 more than the t-table 1.9944 so that it can be seen that the productivity variable of rice fields has an effect on the variable of farmer's income. Meanwhile, the significance value of 0.000 less than 0.05 at the 95% confidence level indicates that the regression model can be used to explain significantly the effect of the level of productivity on farmers' income. The constantvalue is -841,580,634 and the regression coefficient value is 2,389,387 so that the regression equation is Y = -841,580,634 + 2,389,387X. This shows that every 1% increase in the productivity of rice fields, the farmer's income will increase by 2,389,387, assuming other factors are constant. The regression coefficient is positive, so it can be said that the direction of the influence of the rice field productivity variable with the farmer income variable is positive. Meanwhile, seen from the value of R Square (coefficient of determination), the value is 0.890. The R square value of 0.890 means that the land productivity variable is able to have an effect on farmers' income by 89%, while the magnitude of other variables not examined in this study has an effect of 11%. The results of this study are in line with the results of research by Baizhurah (2014) in East Seunagan, Nagan Raya Regency, which found that farmers' income was significantly affected by the R Square value of 0.990 [2]

4 Conclusion

The fluviomarin rice field farming system in Mojowarno Village is a rainfed farming systemwith a planting frequency of only 2 (two) times a year so that the dominant cropping pattern is rice-rice-fallow. On average, the farmers have a rice field area of 0.29 ha with productionreaching 1,856-5,535 tons of unhulled rice / year. Meanwhile, the average productivity of the land was only 5,307 tons / ha, while the average income received by farmers was only Rp 5,879,631.00 / year. The regression statistical test results show that farmers' income is significantly influenced by land productivity. The effect of land productivity on farmer income is positive. This means that the increase in the productivity of fluviomarine rice fields will be followed by an increase in the income received by farmers.

On this occasion the authors would like to thank the UGM Publisher and Publication Board (PBB) for providing RTA 2021 Grant funds, so that the authors can attend the ICST 2021 seminar and complete the thesis. And also really appreciate reviewers for suggestions to make this article better. Without all of these supports, this research would not be successful. Hopefully, this article will be useful for the future.

References

1 S. Arsyad, Konservasi Tanah dan Air (2nd ed.) (IPB Press, Bogor, 2018)
2 N. Baizhurah, Analisis Pengaruh Produksi Padi Sawah Terhadap Pendapatan Masyarakat di Kecamatan Seunagan Timur Kabupaten Nagan Raya. Skripsi (Universitas Teuku Umar, Aceh, 2014)
3 L. Fresco, Cassava in Shifting Cultivation: A Systems Approach to Agricultural Technology Development in Africa (Development Oriented Research in Agriculture, Royal Tropical Institute, Netherlands, 1986)
4 J. H. Jonharnas,. S. H. Sitindaon, Jurnal Agroteknologi, **7**(2), 15-20 (2017)
5 E. Kusmiadi, Pengertian dan Sejarah Perkembangan Pertanian. In Pengantar Ilmu Pertanian (pp. 1–28) (2014)
6 N. Kusnadi, N. Tinaprilla, S. H. Susilowati, A. Purwoto, Jurnal Agro Ekonomi, **1**(29), 25–48 (2011)
7 S. Mangkuprawira, Manajemen Mutu Sumberdaya Manusia (Ghalia Indonesia, Bogor, 2007)
8 Badan Penelitian dan Pengembangan Pertanian. Sentra Produksi. Accessed date 22 September 2020 from:

http://inaagrimap.litbang.pertanian.go.id/index.php/sentraproduksi/tanaman- pangan/padi. (2018)
9 Ridwan. Metode dan Teknik Penyusunan Tesis. (CV Alfa Beta, Bandung, 2005)
10 Satoto, Y. Widyastuti, U. Susanto, M. J. Mejaya, IPTEK Tanaman Pangan, **8**(2) (2013)
11 Scott,.Warmerdam. Spatial Statistics for Public Health and Safety (ESRI, San Diego, 2006)
12 E. Silamat, Yuwana, M. Z. Yuliarso, Agrisep, **14**(2), 197–215 (2014)
13 Sugiyono. Metode Penelitian Kuantitatif Kualitatif dan R&D (Alfabeta, Bandung, 2012)
14 J. Supranto, Statistik Teori dan Aplikasi (Erlangga, Jakarta, 2000)
15 S. H. Susilowati, Forum Penelitian Agro Ekonomi, **34**(1), 35–55 (2016)
16 Triyono. *Teknik Sampling Dalam Penelitian Sosial*. Lokakarya Penelitian Sosial Fakultas Adab IAIN Suka Yogyakarta, XI(March), 2–9 (2018)

Utilization of Plastic Waste Polyethylene (PET) for reinforced subgrade

Roberth Evander Meidudga[1], *Fikri* Faris[1*], *Hary* Christady Hardiyatmo[1], and *Afif* Agri Diansyah[1]

[1]Department of Civil Engineering and Environmental, Gadjah Mada University, Indonesia

Abstract. Plastic waste is a serious problem for the survival of all creatures on Earth. Recycling is one solution to reduce the negative impact of environmental pollution caused by plastic waste. A lot of plastic waste that pollutes the Earth comes from plastic bottle waste (PET). There had been many studies that discussed the use of plastic bottle waste (PET); however, this research used the plastic bottle waste (PET) as a geocell for subgrade reinforcement. To this date, the use of plastic bottle waste as geocell is still rare. The main purpose of this study was to determine the effect of adding geocell of plastic bottle waste on the subgrade using a capping layer that had a low bearing capacity. This research was expected to reduce the impact of environmental pollution due to plastic bottle waste, and be an alternative solution for additional reinforcement in the capping layer. Therefore, it could increase the ultimate bearing capacity and reduce the thickness of the cover layer on subgrades that had a CBR value below 5%. The results of the analysis using the Plaxis program showed that the geocell from plastic bottle waste could be used to save the use of capping layer materials by up to 65%.

1 Introduction

Between 1950 and the present, about 6.3 billion tons of plastic waste have been generated worldwide. As the human population grows, the demand for products that use plastic also increases. Recently, plastic waste production reaches 380 million tons per year. The disposal of plastic into the environment has had a negative impact, one of which is on international waters [1].

Indonesia still has problems with waste management, second worst in the world after China. Previous studies on plastic waste that pollute the environment show that Indonesia is responsible for 15% of plastic waste in global waters [2]. One type of plastic waste that is widely circulated and hurts the environment is Polyethylene Terephthalate (PET). This type of plastic is plastic with code 1. Its use as a primary material for packaging bottles reaches 30% of the world use [3].

Research and development of cellular (geocell) confinement system began with the U.S. Army Corps of Engineering in September 1975 to test the feasibility of contructing tactical bridge approach roads over soft ground [4]. Geocell made of geosynthetics such as geotextiles or geogrids, are thermally welded or mechanically bonded interconnected pocket structures in the form of mattresses used with in-filled soil. General reinforcing mechanisms of geocell is confining the in-fill soil from shearing away and derive anchorage resistance through the surrounding soil against the applied load [5]. The geocell mattress consists of a series of interlocking cell constructed from polymer. The more recent advancement of reinforcement of reinforced soil is to provide three-dimensional confinement to the soil using geocells. The geocell foundation mattress consists of a series of the interlocking cell constructed from polymer geogrids, which contains and confines the soil within its pockets. It intercepts the potential failure planes because of its rigidity and forces them deeper into foundation soil, thereby increasing the bearing capacity of the soil [6]. In pavement and road construction, the effect of geocell-reinforced recycled asphalt pavement (RAP) bases over weak subgrade under cyclic plate loading and found that geocell has improved the performance of RAP bases over weak subgrade as compared with the unreinforced base section and geocell significantly increased the percentage of resilient deformation of the RAP base. The geocell reinforcement reduced the vertical stresses transferred to the subgrade by distributing the load over a wider area [7]. The bearing capacity improvement of gravel base layer in road constructions using geocell and concluded that geocell layer placed within the gravel base layer of asphalt paved construction reduced the vertical stresses on subgrade during vehicle crossing about 30 % and increased the layer modulus of the gravel base layers compared to an unreinforced layer. As a result, the measured deflections on the asphalt surface were also reduced [8].

Transforming plastic bottle waste (PET) into geocell is a solution to overcome the existing plastic waste issue. Considering the enormous potential in the use of PET-type plastic (mineral water bottles), this study focused on the use of plastic bottle waste (PET) as a subgrade reinforcement. Plastic bottle waste planted under a road layer that resembles a geocell is expected to increase the ultimate bearing capacity of the subgrade to stabilize the subgrade and minimize the settlement.

*Corresponding author: fikri.faris@ugm.ac.id

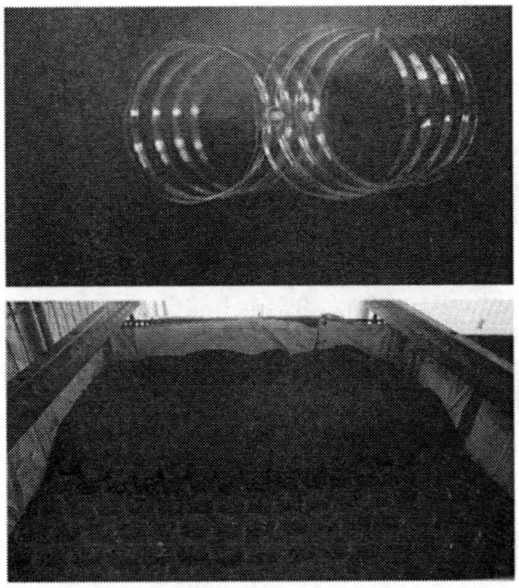

Fig. 1. Geocell plastic bottle waste

The geocell used in this study was a geocell made from PET plastic bottle waste (600 ml mineral water bottle). The bottle was cut into two height variations, 50 mm and 75 mm, then the pieces of the bottle were joined by using a blind rivet measuring 2.4 mm × 6 mm in a square shape. Figure 1 shows the geocell of the assembled PET plastic bottle waste.

2 Experimental methods

2.1 Plate load test

Laboratory scale plate load, test was carried out according to ASTM D-1194, and by adding a plastic bottle waste geocell to the subgrade [9].

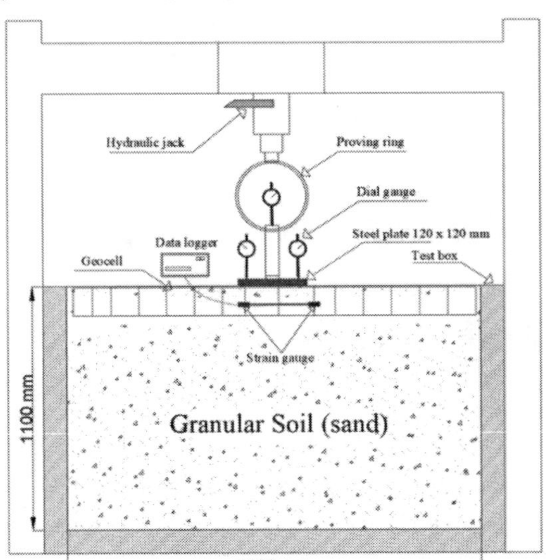

Fig. 2. Schematic view of the test setup.

The test was conducted in the laboratory of the Department of Civil and Environmental Engineering, Gadjah Mada University. The loading test was conducted to determine the ultimate soil bearing capacity and the settlement at the laboratory scale. In this case, maximum soil holding capacity and settlement with geocell reinforcement and without reinforcement from bottle waste. The subgrade used during laboratory testing was sand with a classification (USCS) SP (poorly graded sand) Friction angle ϕ =32° and *Cohesion, c* = 0 kN/m². The loading test results are presented in Table 1.

The ultimate bearing capacity is defined as the maximum load per unit area where the soil can still support the load without collapsing, which is expressed in the following equation [10].

$$q_u = \frac{P_u}{A} \quad (1)$$

in which *qu* is the ultimate bearing capacity (kN/m²), *Pu* is the maximum load (kN), and *A* is the plate area or foundation (m²).

Bearing capacity ratio (BCR) is the ratio of the carrying capacity value in the presence of geosynthetic materials or geocells from PET plastic bottle waste and the carrying capacity values without geosynthetic material or geocell from PET bottle. The equation for the bearing capacity ratio (BCR) is as follows [11].

$$BCR = \frac{q_r}{q_o} \quad (2)$$

in which BCR is the bearing capacity ratio, q_r is the bearing capacity value in the presence of geosynthetic reinforcement (kN/m²), and q_o is the bearing capacity value without geosynthetic reinforcement (kN/m²).

Table 1. The results of the bearing capacity of the soil and its decrease.

Variation of loading test	Ultimate soil bearing capacity (kN/m²)	Settlement (mm)
Unreinforced	416,67	8,93
Reinforced geocell 50 mm	1111,11	17,59
Reinforced geocell 75 mm	1250	15,153

The results of the loading test in the laboratory can be seen in Table 1. It shows that the increase in the ultimate soil bearing capacity increased when geocells were added from plastic bottle waste. The average increase in the carrying capacity of the soil after adding the geocell reinforcement of plastic bottle waste was 183.82%. The bearing capacity ratio (BCR) value on average increased to 2.94.

Table 2. The results of soil bearing capacity and its settlement

Variation of loading test	Soil bearing capacity (kN/m²)	Settlement (mm)
Unreinforced	416,67	8,93
Reinforced geocell 50 mm	416,67	3,01
Reinforced geocell 75 mm	416,67	3,035

It can be seen in Table 2 that the percentage of settlement experienced a decline if given the similar load of about 416.67 kN/m² for each variation of the loading test. The decrease without reinforcement was 8.93 or 100% to 3.0225 mm or 33.85%. This decrease proves that the geocell from plastic bottle waste can affect the characteristics of sandy soil by increasing the bearing capacity and declining the settlement.

The increase in soil-bearing capacity and settlement is affected by the height of the geocell and locking of the sandy soil in the geocell structure, thus, there is no excessive lateral force on the sand. The increase in the ultimate bearing capacity occurred when a 75 mm geocell was used. This is because the area to accommodate and lock the sandy soil in the geocell is slightly wider. This condition reduces the lateral forces of the sandy soil caused by the load on it.

2.2 Tensile strength test

After the loading test was conducted, a tensile test was also carried out for the plastic bottle material using the ASTM E8 reference using universal testing machine (UTM). After testing the tensile material on 9 samples, the average working force was 101.59 N with the maximum strain of 3.79 mm or 37.97% of the original form.

From the results, an approach was conducted by using an equation referred to previous studies [12], which defines geocells as a composite material or plate with the following equation,

$$E_g = 4(\sigma_3)^{0.7}(Ke + 200M^{0,16}) \tag{3}$$

in which E_g is the young's modulus of the geocell-reinforced soil, σ_3 is the average horizontal stress at the mid-height of the geocell layer, Ke is the young's modulus parameter of the unreinforced soil and M is the line of intersection of the geocell material modulus.

The geocell connection test (blind rivet) was carried out simply by giving the load gradually on three plastic bottles that had been cut and hung. The previously mentioned load was by increasing load of 1 kg per 30 sec on the hung geocell until it failed at the connection or broke. From the results of 30 samples tested for connection, the average load that could be held by the connection on this PET bottle waste geocell was 14.56 kg. The geocell connection test is presented in Figure 3.

Fig. 3. Geocell connection test (blind rivet).

After trial and error in the Plaxis program, the values obtained were E_g = 107259 kN/m, E_gA= 5362.96 kN/m, and EI = 1.117 kN/m². These values were used in the material input parameter that defined the geocell material as a plate in the Plaxis 8.6 program.

2.3 Subgrade analysis with Plaxis 8.6 program

The subgrade analysed in this study was a subgrade with a CBR value below 5%. One example is silt (*lanau*) which was given special treatment by increasing the thickness of the capping layer (granular soil) — classified by USCS as sand with poor grade or SP — of 300 mm to stabilize the above pavement [13].

2.3.1 Traffic load input parameters on Plaxis

The traffic load used the "D" load intensity, with uniform load (BTR) q = 9 kN/m² on the entire body of the road, as referred to [14].

2.3.2 Soil input parameters on Plaxis

The soil parameter included in the Plaxis was silt soil. The material model used in this research was Mohr-Coulomb. In accordance with the results of the soil study, there was attempt to input the following silt (*lanau*) parameters in the Plaxis program.

$$\gamma_{unsat} = 15\ (\text{kN/m}^3) \quad \nu = 0{,}30 \quad \psi = 0\ (^\circ)$$
$$\gamma_{sat} = 18\ (\text{kN/m}^3) \quad c = 40\ (\text{kN/m}^2)$$
$$\phi = 10\ (^\circ) \quad E = 3000\ (\text{kN/m}^2) \tag{4}$$

The above parameters were used for the input of soil material on the silt (drained) subgrade. The parameters of the soil on the road layer and the overburden layer above the subgrade refer to [15].

2.3.3 Geocell input parameters on Plaxis (plate)

$$EA = 5362{,}960\ \text{kN/m} \quad w = 0{,}850\ \text{kN/m}$$
$$EI = 1{,}117\ \text{kN/m}^2 \quad \nu = 0{,}250$$
$$d = 0{,}050\ \text{m} \tag{5}$$

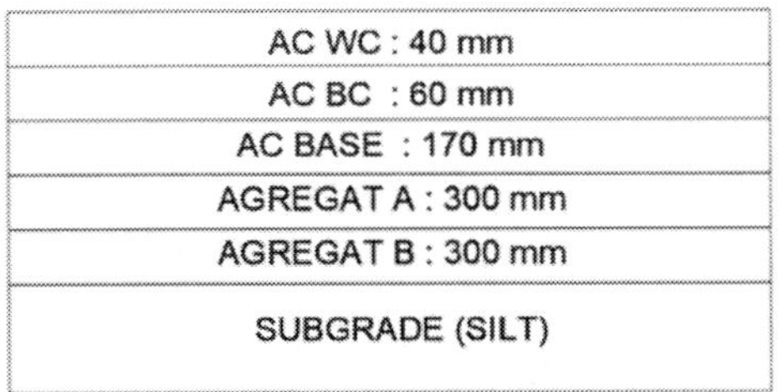

Fig. 4. Road layer.

3 Result and discussion

3.1 Total displacements (Extream U_{tot})

A comparison of the settlement that occurred in the subgrade before and after the installation of geocells from plastic bottle waste can be seen in Figure 4 and Figure 5. It was found that the geocell from the plastic bottle waste can reduce the settlement due to the load above the subgrade up to 15,27 mm or 57,26 % of the total settlement without geocell reinforcement. The largest settlement reduction was occurred in the middle part of the loading area. The reduction effect was confirmed by both plane strain and axisymmetric model by using Plaxis FEM software.

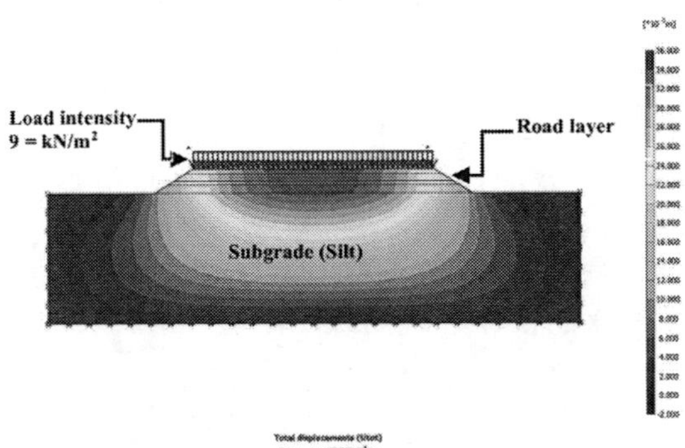

Fig. 5. Extreme U_{tot} unreinforced geocell plane-strain (35,73*10^{-3}m).

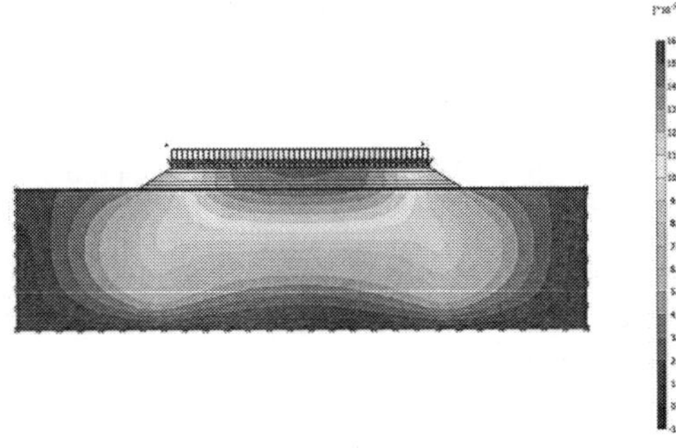

Fig. 6. Extreme U_{tot} reinforced geocell plane-strain (15,27*10^{-3}m).

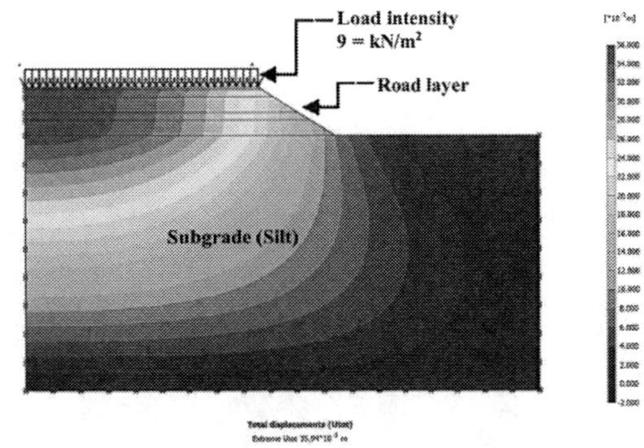

Fig. 7. Extreme U_{tot} unreinforced geocell axisymmetry (35,94*10^{-3}m).

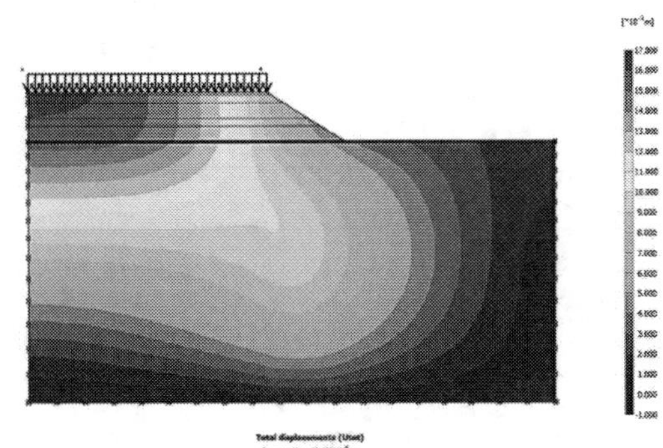

Fig. 8. Extreme U_{tot} reinforced geocell axisymmetry (16,42*10^{-3}m).

3.2 Ultimate bearing capacity

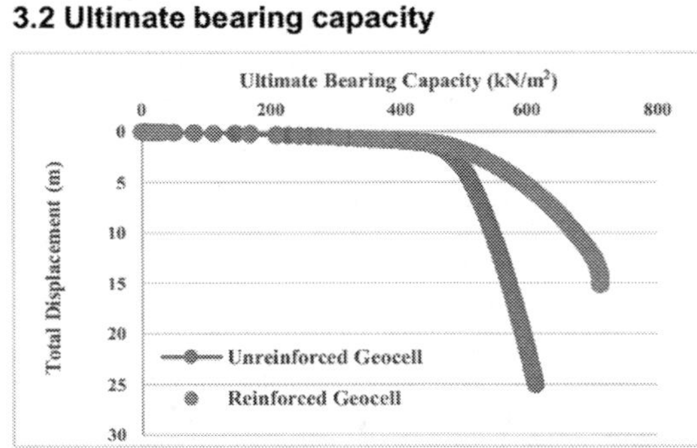

Fig. 9. Curve ultimate bearing capacity Plaxis 300 mm capping layer.

In Figure 9 the total displacements that occurred were more than 25 mm. This is because in order to obtain the ultimate bearing capacity value in the Plaxis program, the subgrade must receive as much load as possible until the subgrade collapses. A trial of uniform loading was carried out on the road body that tested at 1000 kN/m^2. When the subgrade collapses, the comparison of the bearing capacity between the unreinforced geocell and the reinforced geocell can be seen as the result curve Plaxis in Figure 9. More details of the results curve ultimate bearing capacity capping layer of 300 mm are presented in Table 3.

Table 3. Result of curve Plaxis capping layer 300 mm.

Result Plaxis	Ultimate bearing capacity (kN/m^2)	BCR
Unreinforced geocell PET	613,534	1
Reinforced geocell PET	712,911	1,162

The results of the ultimate bearing capacity on Plaxis had increased up to 16.2%. These results prove an increase occurred when geocells were added to the subgrade. In addition, the results validate the results of the experiments in the laboratory (plate load test) where geocells from plastic bottle waste can increase bearing capacity and decline the settlement.

3.3 Analysis of capping layer thickness efficiency

In this analysis, we tried reducing the thickness of the capping layer material up to 50% followed by adding the geocell material from the plastic bottle waste. The purpose of the analysis was to observe the behaviour of the geocell when the capping layer thickness is only 50% (150 mm).

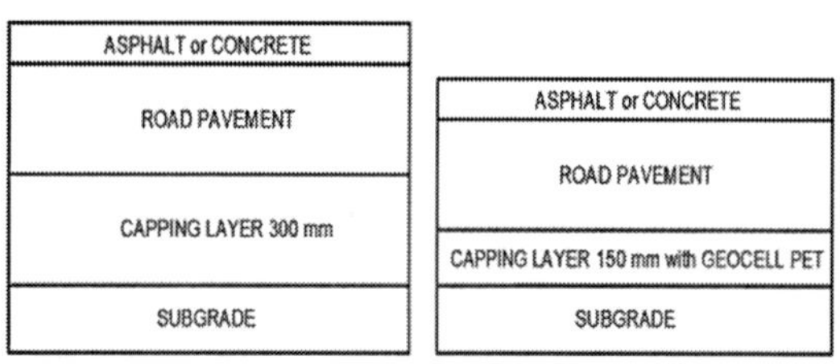

Fig. 10. Trial 50% less capping layer.

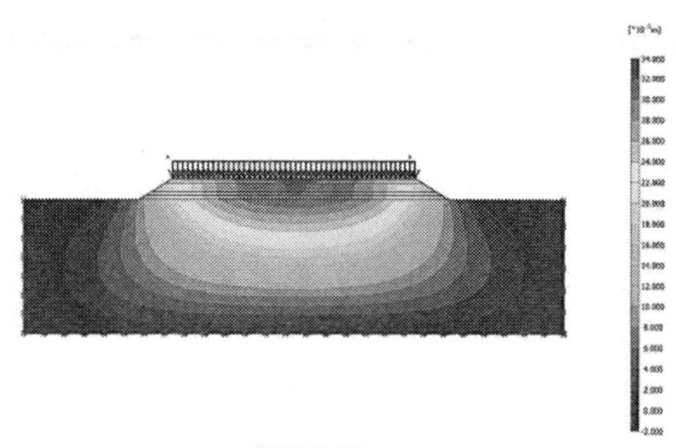

Fig. 11. Extreme U_{tot} unreinforced geocell plane-strain (32,39*10^{-3}m).

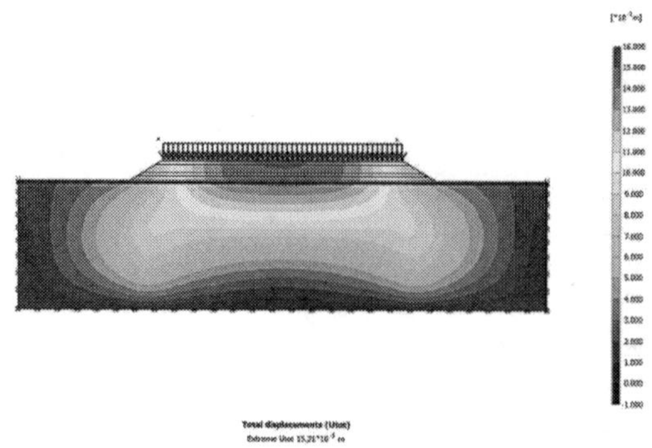

Fig. 12. Extreme U_{tot} reinforced geocell plane-strain (15,21*10^{-3}m).

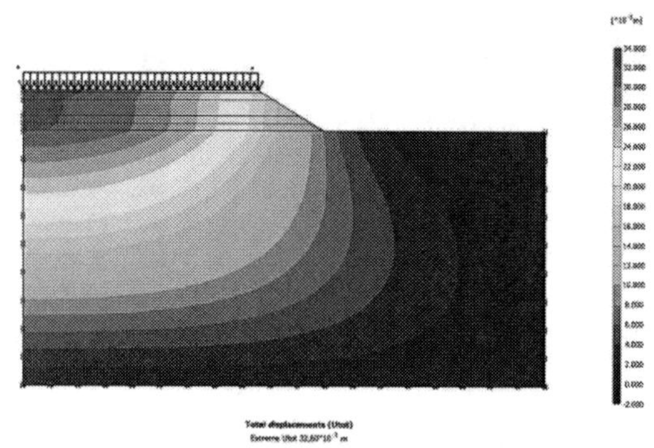

Fig. 13. Extreme U_{tot} unreinforced geocell axisymmetry (32,60*10^{-3}m).

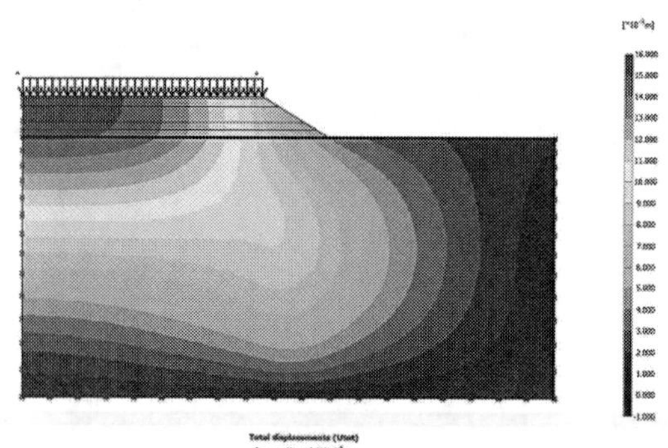

Fig. 14. Extreme U_{tot} reinforced geocell axisymmetry (15,81*10^{-3}m).

After the capping layer was reduced to 50% (150 mm) by applying the same load when loading the capping layer of 300 mm above the road pavement, the geocell can reduce the settlement to 53,04% of the total settlement without geocell reinforcement. The results are shown in Figure 11 and Figure 12.

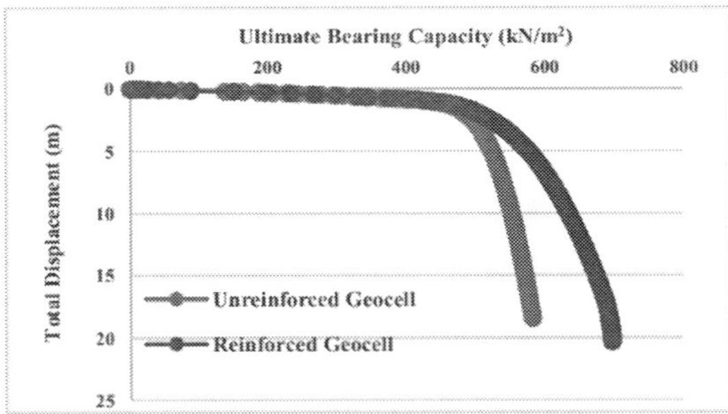

Fig. 15. Curve ultimate bearing capacity Plaxis 150 mm capping layer.

The results of the curve from the Plaxis program showed the ultimate bearing capacity behaviour with and without geocell reinforcement were relatively similar. This is due to there was only one parameter was made for the soil material and loading in the Plaxis program, with only the thickness of the capping layer that was different. The results of the 150 mm Plaxis capping layer curve are shown in Table 4.

Table 4. Result curve Plaxis capping layer 150 mm

Result Plaxis	Ultimate bearing capacity (kN/m^2)	BCR
Unreinforced geocell PET	582,435	1
Reinforced geocell PET	696,733	1,197

The results of the Plaxis program in Table 4 show that the performance of geocell from plastic bottle waste could increase the bearing capacity of the soil up to 19.7% when the capping layer was reduced to 150 mm.

Tried to reduce the capping layer with a thickness of 100 mm by Plaxis analysis

Fig. 16. Trial 65% less capping layer.

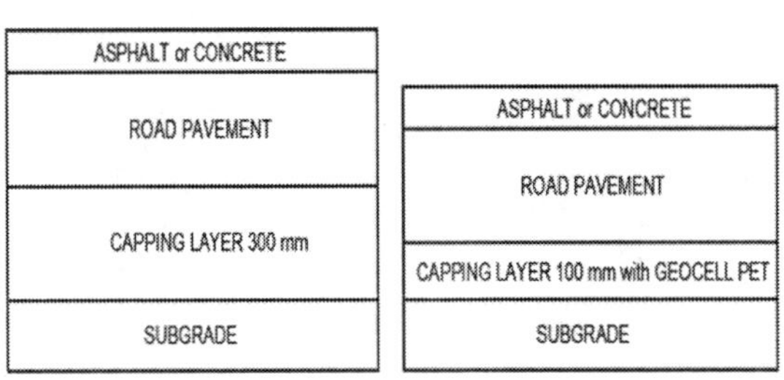

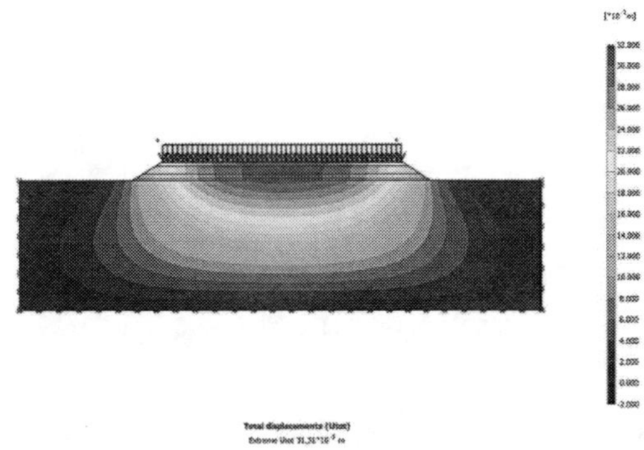

Fig. 17. Extreme U_{tot} unreinforced geocell plane-strain ($31,31*10^{-3}$m).

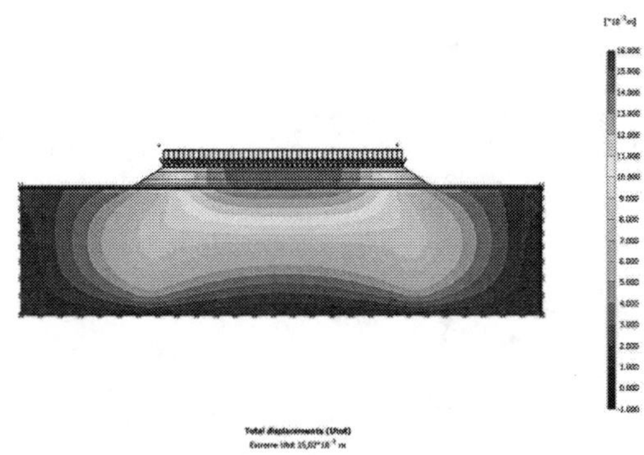

Fig. 18. Extreme U_{tot} reinforced geocell plane-strain ($15,02*10^{-3}$m).

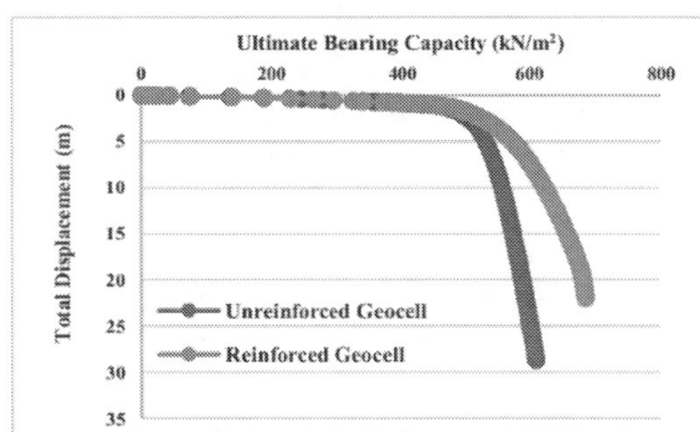

Fig. 19. Curve ultimate bearing capacity Plaxis 100 mm capping layer.

Table 5. Result curve Plaxis capping layer 150 mm.

Result Plaxis	Ultimate bearing capacity (kN/m^2)	BCR
Unreinforced geocell PET	610,818	1
Reinforced geocell PET	686,784	1,197

The results of the Plaxis program in Figure 19 show that the performance of geocell from plastic bottle waste could increase the bearing capacity of the soil up to 11,94 % when the capping layer was reduced to 100 mm.

https://doi.org/10.1051/e3sconf/202132504004

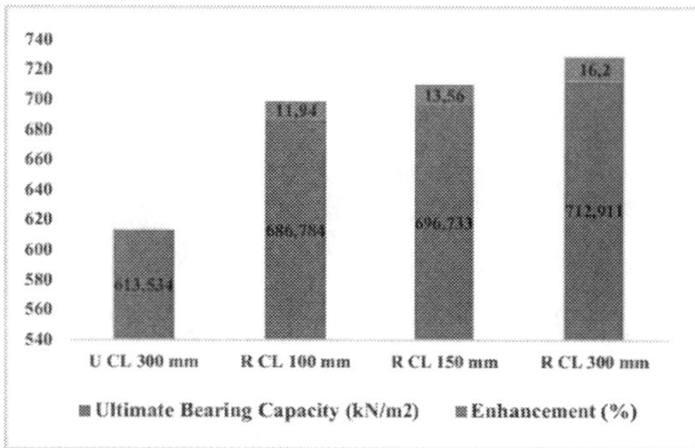

Fig. 20. Percentage on ultimate bearing capacity increase with capping layer of 100 mm, 150 mm, and 300 mm.

The percentage comparison of ultimate bearing capacity when unreinforced capping layer 300 mm (U CL 300 mm), reinforced capping layer 100 mm (R CL 100 mm), reinforced capping layer 150 mm (R CL 150 mm), and reinforced capping layer 300 mm (R CL 300 mm) can be seen in Figure 20. These percentages show that the strengthening of the 100 mm capping layer using plastic bottle waste geocells with a settlement in the thickness of the capping layer can be applied to road pavements that have low bearing capacity.

4 Conclusion

The results of this study were aimed to save the environment from the negative impacts of plastic bottle waste that damaged the environment on land, sea, and air pollution due to the burning of plastic waste. The results of the analysis using Plaxis showed that the geocell of plastic bottle waste could increase the bearing capacity up to 11,94%. This could be an alternative to increase the efficiency up to 65% by reducing capping layer thickness. Therefore, geocells from plastic bottle waste could provide a solution to save granular materials, which consequently reduce the transportation cost of materials from quarries to project site and as an alternative for plastic bottle waste management to prevent the environment pollution.

Acknowledgment is presented to Dr. Eng. Fikri Faris for his trust in providing the opportunity to complete this manuscript, also to Prof. Hary Christady Hardiyatmo who has provided many inputs and the Department of Civil and Environmental Engineering, Gadjah Mada University.

List of symbols

q_u	ultimate bearing capacity (kN/m^2)
P_u	maximum load (kN)
A	plate area or foundation (m^2)
BCR	bearing capacity ratio
q_r	bearing capacity value without geosynthetic reinforcement (kN/m^2)
q_o	bearing capacity value in the presence of geosynthetic reinforcement (kN/m^2)
E_g	young's modulus of the geocell-reinforced soil (kN/m^2)
σ_3	confining pressure (kN/m^2)
K_e	young's modulus parameter of the unreinforced soil (kN/m^2)
M	line of intersection of the geocell material modulus (kN/m)
γ_{unsat}	unsaturated soil (kN/m^3)
γ_{sat}	saturated soil (kN/m^3)
ϕ	friction angle (°)
ν	poisson's ratio of soil
ψ	angle of dilatancy (°)
c	cohesion of soil (kN/m^2)
E	young's modulus (kN/m^2)

References

1. O. Alabi, K. Olagbonjaye, O. Awosolu, O. Alalade, Toxicol Risk Assess. **5**, 1 (2019)
2. P. Purwaningrum. JTL. **8**, 141–147 (2016)
3. Y. Sofiana, Humaniora. **1**, 331–337 (2010)
4. Dhane, D. Kumar, A. Priyadarshe. *Geocell : an emerging technique of soil reinforcement in civil engineering field*, IOSR-JMCE, in Nasional Converence on Advances in Engineering, Technology & Management (2015)
5. Krishna, A. Biswas, *Geocell reinforced fondations, in training course on introduction to geosynthetics and their applications*, December 14 2015 NIT Surat India (2015)
6. S.K. Dash, N.R. Krishnaswamy, K. Rajagopal, Geotextiles and Geomembran. **19**, 235–256 (2001)
7. J. K. Thakur, J. Han, S. K. Pokharel, R. L. Parsons, Geotextiles and Geomembran **35**, 14–24 (2012)
8. A. Emersleben, N. Meyer, *Bearing capacity improvement of gravel base layer in road constructions using geocell*, in International Association for Computer Methods and Advances in Geomechanics IACMAG, 1–6 October 2008, Goa India (2008)
9. A. Hedge, T.G. Sitharam, Geotextiles & Geomembran **43**, 171181 (2015)
10. H.C. Hardiyatmo, Fondasi I, Edisi 4 (Gadjah Mada University Press, Yogyakarta, 2020)
11. H.C. Hardiyatmo, Geosintetik untuk rekayasa jalan raya, edisi 2 (Gadjah Mada University Press, Yogyakarta, 2015)
12. I. Mehdipur, M. Ghazavi, R. Ziaie , Geotextiles & Geomembran, **37**, 23–34 (2013)
13. H. C. Hardiyatmo, Perancangan perkerasan jalan & penyelidikan tanah, edisi 2 (Gadjah Mada University Press, Yogyakarta, 2015)
14. SNI (1725:2016). Pembebanan Untuk Jembatan, 39 (2016)
15. Manual Desain Perkerasan (Revisi 2017). Kementrian PUPR, 7–10 (2017)

Cleat properties of the Tanjung Agung coalfield, South Sumatra Basin: implications to fluid flow

Muhammad Addiansyah[1*], *Donatus* Hendra Amijaya[1], and *Ferian* Anggara[1]

[1]Geological Engineering Department, Gadjah Mada University, Indonesia

Abstract. Observation of the coal cleats of the Muara Enim Formation in Tanjung Agung coalfield aims to characterize the cleat properties and interpret their implications for fluid flow. This study used the scanline method to measure cleat attributes in the form of orientation, aperture, spacing, filler, and cleat connectivity. The face cleat was dominated by NE-SW direction as an indication of the effect of paleostress, while butt cleat with NW-SE direction indicated the effect of stress release. The presence of one major fault and three minor faults trending NW-SE in the mine pit is interpreted as the main controlling factor for cleat development in the study area. These cleats have an aperture ranging from 0.4 - 2.3 mm, good connectivity, and are not filled, so they are considered to be analogous to fluid flow paths with transition and turbulent flow types with relatively NE-SW direction.

1 Introduction

South Sumatra Basin is one of the most productive coal producing basins in Indonesia. The stratigraphic unit that acts as a coal-bearing formation in the South Sumatra Basin is the Muara Enim Formation which was deposited during the regression phase during the Late Miocene - Early Pliocene [1].

Tanjung Agung Coalfield is one of the coal mining areas of the Muara Enim Formation using the open pit method, so that there are many coal outcrops that can be observed for the purposes of various thematic studies (Figure 1). One of the interesting properties of coal to observe is the cleat because it can provide an overview of the characteristics of the cleat attribute and its ability to flow fluids, especially gas.

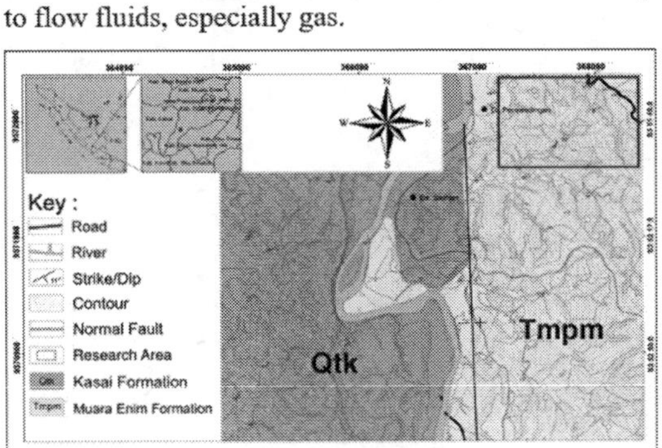

Fig. 1. The research area is in the Muara Enim Formation [2].

A number of previous studies stated that cleats can develop through many mechanisms in various periods of coal formation. These mechanisms include the dehydration, drying or devolatilization process during ripening which indicates a decrease in cleat space along with increasing coal rank [3]. The stress on coal from tectonic activity is also considered to have an effect on cleat formation because face cleats are considered to be extensional (opening-mode cleats), so its direction is often perpendicular to the direction of the minimum horizontal stress (minimum horizontal stress) and parallel to the direction of the maximum horizontal stress (maximum horizontal stress) [3]. Meanwhile butt cleats are interpreted to form during relaxation from the initial stress (Figure 2) [4]. Observation of cleats generally shows that the cleat is perpendicular to the direction of the thrust fault and fold hinge axes, so this phenomenon supports the hypothesis that cleats can be formed in response to tectonic activity.

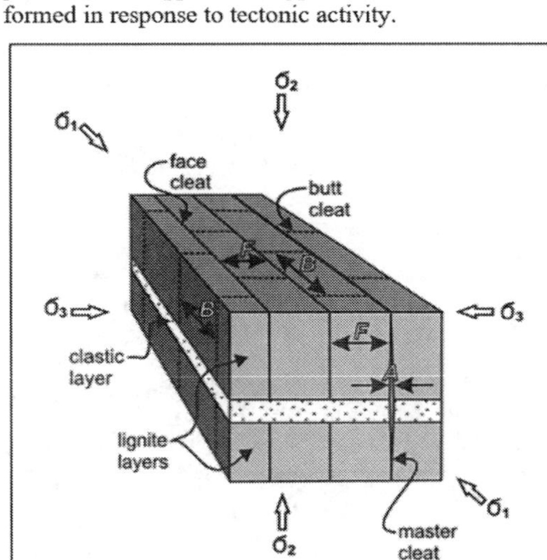

* Corresponding author: muhammadaddiansyah@mail.ugm.ac.id

https://doi.org/10.1051/e3sconf/202132504005

Fig. 2. Illustration of cleat attributes and its relationship to stress [5].

Based on their formation, cleats are classified into endogenic and exogenic cleats. Endogenic cleats are cleats that are formed during the coal process due to dehydration, devolatilization, and matrix shrinkage. Meanwhile, exogenic cleats are generated from external forces such as tectonic activity. Apart from the endogenic processes mentioned above, cleat orientation shows the influence of tectonic stress [4-5]. Another classification is based on the relative cleat distance proposed dividing the cleats into three groups; macrocleat (measuring decimeter to centimeter), mesocleat (measuring millimeter), and microcleat (measuring micrometer) [5].

Observation of cleats on coal will be very useful in understanding coal permeability which can contribute to various engineering activities that utilize coal cleats. The coal permeability would be higher along the face cleat than along the butt cleat. Coal permeability through the cleat network is influenced by the cleat opening, space, orientation, height, and length. Coal permeability will generally increase with increasing cleat openings and decreasing cleat spacing which will have an impact on connectivity between cleats. Based on this interpretation, the cleat attributes that need to be investigated in the context of understanding fluid flow in coal are space, length, height, connectivity, aperture, level and type of mineral filling, and orientation.

This study uses low rank coal in the Muara Enim Formation of the South Sumatra Basin as the object of research. The coal is exposed in the mining area, so fresh coal can be found along the outcrop. The coal is unformally divided into four parts (B1, B2, B3, and B4) which are exposed in the mine openings of a company in Tanjung Agung District, Muara Enim Regency, South Sumatra (Figure 3). The selection was based on the potential for CBM resources in the South Sumatra Basin which reached 183 TCF (equivalent to 41% of the total CBM resources in Indonesia), so it is necessary to carry out detailed observations of the coal cleats of the Muara Enim Formation in various areas for activities [6]. CBM exploration-production as well as other geological extraction activities in the Muara Enim Formation coal which utilizes coal cleats.

Fig. 3. Coal outcrop of the Muara Enim Formation in Tanjung Agung coalfield.

2 Methods and data

This study used a descriptive-observative approach with data sources obtained from field observations. Cleat data acquisition uses the scanline method along the coal seam. The scanline method is used to anticipate duplication of cleat data measurements and validate face and butt cleat data collected from the field prior to further analysis [7].

The object of this research is the measurement results of cleat attributes which include orientation, aperture, spacing, filler, and cleat connectivity using a measuring tape, calipers, and a geological compass. Meanwhile, cleat length and height are not measured because it cannot be measured completely in the coal outcrop. Cleat measurement using the scanline method in one of the mining areas in Tanjung Agung produces 400 cleat data (Table 1 and Table 2).

Table 1. Measurements results of face cleat attributes.

Seam	Face Cleat			
	Orientation (N...°E)	Aperture (mm)	Spacing (cm)	Freq
B1	213-226	0.6-1.15	1.72-2.74	50
B2	197-223	0.6-0.9	1.42-2.83	50
B3	194-219	0.5-1.7	0.56-2.93	50
B4	193-206	0.45-2.3	0.65-2.74	50

Table 2. Measurements results of butt cleat attributes.

Seam	Butt Cleat			
	Orientation (N...°E)	Aperture (mm)	Spacing (cm)	Freq
B1	294-314	0.4-0.65	1.66-2.63	50
B2	278-313	0.45-0.85	1.52-2.86	50
B3	278-216	0.4-0.85	1.27-2.86	50
B4	283-304	0.4-0.75	1.295-1.9	50

The measurement results will be variables in the data analysis to determine the cleat property and its implications for gas flow. The cleat property is correlated with the classification of flow types based on their geometry and connectivity. The analysis is also accompanied by stereographic analysis to match the direction of regional forces acting around the study area, so that the genesis of the cleat formation can be interpreted.

3 Result and discussion

The coal in the study area has a total thickness of 12.8 meters with the characteristic of being dominated by banded dull coal lithotype, tuff parting, and the position of the layer is N114°E/14°. The developed face cleat is generally in a NE-SW direction, while the butt cleat is in a NW-SE direction. The general orientation of the face cleat has the same relative direction with the orientation of the fault structure around the mine pit (Figure 4).

https://doi.org/10.1051/e3sconf/202132504005

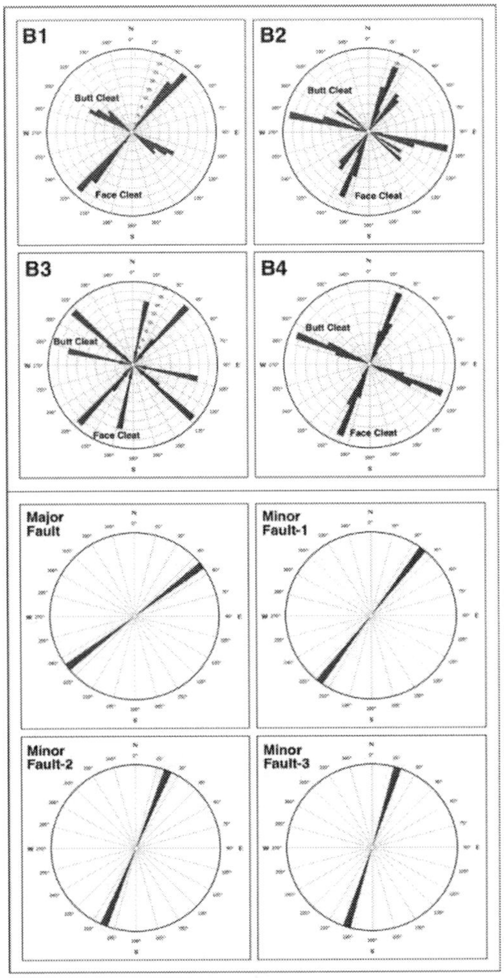

Fig. 4. The orientation similarities between face cleats and fault structures in the mine pit.

The formation of cleats in coal has started to occur by many mechanisms in various periods of coal, but the presence of one mayor fault and three minor faults trending NE-SW in the mine pit are interpreted as the main controlling factor for cleat development in the study area (Figure 5). This interpretation occurs with about cases found in cleats in Tanjung Enim Coalfield which have cleat face orientation conditions and patterns that show relative directions in Quadran III [8].

Fig. 5. Minor faults in the mine pit (blue bag as parameter).

Cleat aperture ranges from 0.4 - 2.3 mm, good connectivity, and not filled. The aperture range of the cleats is included as a macroscopic fracture in the gas flow pattern model in the fracture-pore system with the transitional flow and turbulent flow types [8]. The cleat condition is considered to be analogous to a good fluid flow path with an estimated NE-SW relative flow direction (Figure 6).

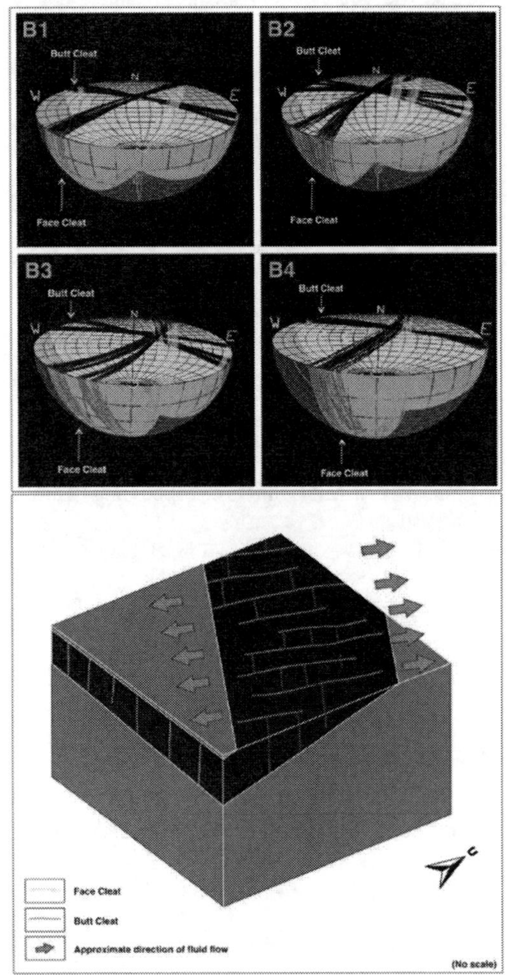

Fig. 6. 3D cleat stereographic analysis and illustration of the approximate direction of fluid flow.

4 Conclusion

The coal in the study area has cleat properties that are influenced by tectonic activity so that it has an orientation that is in the same direction NE-SW as the faults in the pit and a qualified aperture as a fluid flow path with good connectivity between cleats. The cleat property is very important as an analogue of the direction of exploration and production of various fluids in the coal and as a reference for the type of treatment of coal in engineering activities related to the coal cleats of the Muara Enim Formation in Tanjung Agung District, Muara Enim Regency, South Sumatra.

https://doi.org/10.1051/e3sconf/202132504005

We thank the Ministry of Education, Culture, Research and Technology for supporting this research program and funding. We also thank PT Sriwijaya Bara Priharum in Tanjung Agung for providing the opportunity to conduct research in its working area.

References

1. I.B. Sosrowidjojo, Indonesian Mining J. **16**, 2 (2013)
2. F. Fadliansyah, E. Dimas, F. Fachrudin, S. Nalendra, J. Teknik Patra Akademika **8**, 1 (2017)
3. M.S. Brook, B.W. Hebblewhite, R. Mitra, Intl. J. of Coal Geology **168** (2016)
4. G.K.W. Dawson, J.S. Esterle, Intl. J. of Coal Geology **82** (2010)
5. M. Widera, Geologos **20**, 1 (2014)
6. Kementerian Energi dan Sumber Daya Mineral, *Peta Jalan Kebijakan Gas Bumi Nasional 2014-2015*. Jakarta: Kementerian ESDM (2014)
7. S.N. Jati, Idarwati, I.M. Putra, F.A. Nursiwan, J. of Phys: Conference Series **1363** (2019)
8. S.N. Jati, E. Sutriyono, E.W.D. Hastuti, *Coal properties and cleat attributes at Tanjung Enim coalfield in South Palembang Sub-basin, South Sumatra,* in AIP Conference Proceedings **2245** (2020)
9. S. Liu, S. Sang, Q. Zhu, H. Liu, H. Gao, J. Cent. South Univ. **21** (2014)

Suspended solids and bacteria removal mechanisms in ceramic filter and pervious concrete filter: a review

Ekha Yogafanny Boulven[1,2*], *Radianta* Triatmadja[1], *Budi* Kamulyan[1], *Fatchan* Nurrochmad[1], and *Intan* Supraba[1]

[1]Civil and Environmental Engineering Department, Faculty of Engineering, Gadjah Mada University, Yogyakarta, Indonesia
[2]Environmental Engineering Department, Faculty of Mineral Technology, University of Pembangunan Nasional "Veteran" Yogyakarta, Indonesia

Abstract. The need for drinking water with affordable access is increasing nowadays. The poor water quality problems can be solved by several water treatment methods, i.e. ceramic filter (CF) and pervious concrete filter (PCF). Those two mentioned technologies work based on the pores that exist within the filter. This study aimed to review (1) the influence of the material composition of the CF and PCF on hydraulic characteristics (porosity, permeability, pore size, pore structure, pore connectivity/ distribution) and its removal effectiveness, and (2) the removal mechanisms of suspended solids and bacteria based on its hydraulic characteristics. The study reviewed 45 literature, including books, reports, and published articles. The type, mineral, and sources of clay and the type, shape, proportion, and size of combustible materials on CF will affect its hydraulic characteristic. The shape, size, and type of aggregate, the ratio of water to cement, and the ratio of aggregate to cement on PCF will affect its hydraulic characteristics. The removal mechanisms of suspended solids are straining on the surface and trapping on the deadlocked pores. On the other hand, the bacteria removal mechanisms strain the bacteria in the dirt layer and trap the bacteria in the pores.

1 Introduction

The need for drinking water with affordable access is increasing nowadays, especially in developing countries and rural-remote areas where piped drinking water is not accessible. The current Covid-19 pandemic forces us to improve our resilience and immune system by increasing access to health services and vaccinations and basic hygiene and sanitation [1]. In developing countries, there are some ways to get water. Often, once the water sources are found, it is contaminated, especially by suspended solid and pathogens; hence it is not suitable for drinking water.

People applied several water treatment methods to solve a poor water quality, especially for decentralized water treatment systems in rural areas. A decentralized water treatment system is a water treatment that does not require a piping network so that it is considered to be able to overcome the problems of topographic and remote area barriers in the supply of clean water in developing countries [2,3]. This system consists of point-of-use and point-of-entry. A point-of-use system is a water treatment system that is carried out right before the water is used. Meanwhile, point-of-entry is a water treatment system carried out before water enters the household water system [4–6]. Some of the technologies used in point-of-use systems are chlorination, coagulation, solar disinfection, ceramic water filter, bio-sand filter [7], membrane filtration, fibre/fabrics filter [3,8,9], and intermittent slow sand filtration [10–13]. Moreover, a new-promising technology called pervious concrete filter (PCF) can be used to reduce turbidity from the water reaching 95% efficiency [14].

Ceramic filter (CF) is a point-of-use technology that is widely used in developing countries. CF is made from a combination of clay and combustible materials such as rice husks, sawdust, coffee husks, and peanut husks [7]. The pores in the CF come from combustible organic matter that burns during the combustion process. The type and size of combustible material and the mixing process will affect the pore size, pore distribution, porosity, and permeability of the filter. These pores will drain the water and trap the contaminant from the contaminated water. This technology has been proven successful in reducing bacteria reaching >99% or >2 LRV [7,15,16], viruses [17], protozoa [18], organic and inorganic compounds [19]. Unfortunately, high energy is needed reaching 900 °C – 1000 °C in the ceramic firing process.

Pervious concrete filter (PCF) is one of the concrete applications in water purification. This technology is made from a mixture of sand and cement in a specific ratio, with no heat energy needed. The pores within the PCF are derived from the proper mix design composition of sand, cement, and water. This composition will affect the porosity, permeability, pore size, and pore distribution formed in the filter. This filter utilizes the pores in the concrete to drain water while holding the impurities that will remove from the

* Corresponding author: ekha.yogafanny@mail.ugm.ac.id

contaminated water. PCF can reduce turbidity in water up to 95% and E.Coli bacteria up to 98.71% or Log Removal Value (LRC) = 2 [14,20–22], especially with a finer sand composition.

Those two mentioned technologies work based on the pores that exist within the filter. The hydraulic characteristics, i.e. porosity and permeability, play an essential role in increasing the efficiency of those filters in reducing the contaminants. Besides, the pore size, pore distribution, and the open connection between pores also affect the effectiveness of those two technologies. The different hydraulic characteristics formed on these two technologies cause the removal effectiveness to vary due to other particle removal mechanisms. The pore size, the permeability, the material used in the filter will affect the particle removal mechanisms. Understanding this relationship will improve the development of these two technologies as filters, especially for the PCF – new technology in a water filter.

This paper reviews (1) the influence of material composed of the CF and PCF on hydraulic characteristics (porosity, permeability, pore size, etc.) and its removal effectiveness, and (2) the removal mechanisms of suspended solids and bacteria based on its hydraulic characteristics.

2 Research methods

The method applied in this study is a literature review based on 45 works of literature, including books, reports, and published articles. The review consists of identifying the hydraulic characteristics based on the material composition of CF and PCF and its removal effectiveness. Moreover, the identification of suspended solids and bacteria removal mechanisms is also conducted according to its hydraulic characteristics.

3 Result and discussion

3.1 Material composition and production

3.1.1 Ceramic filter

The primary material used to make ceramic water filters (CF) is clay. Clay has its characteristics depending on the location of the quarry. Yakub et.al. (2013) made the CF from a mixture of clay (Cedar Heights Redart Airfloated Clay, Pittsburgh, PA) and sawdust (80% of oak and 20% of cedar) [23]. The clay is then mill and sieved through a 30 mesh or about 0.55 mm opening. This step is taken to ensure no non-clay particles such as sand or organic matter [24]. The composition of clay is observed by X-ray fluorescence (XRF) analysis.

The combustible material must be added to clay to produce the pores so that the water can flow through and the contaminant can be trapped in it. The combustible material consists of sawdust, rice husks, agriculture by-products [7,25], and other organic materials such as coffee and peanut husks [24]. This organic material will burn during the combustion process and will leave pores in the ceramic. The size and proportion of the combustible material plays a significant role in the formation of pores and thus impacts the flow rate and removal efficiency.

Making ceramics filters begins with the preparation of the materials (clay and combustible material). The proportions of these two ingredients are arranged in such a way, for example, 50:50, 65:35, to 75:25. These two materials are milled and sieved according to the expected sizes, for example, a 30-mesh (0.55mm) size for both clay and combustible material or sieved the sawdust as combustible material by 35 – 1.000 mesh of size [23]. After the fixed composition is ready, the mixing process is done manually or using a machine by pouring the water into it [26].

After the dough is finished, the next step is moulding into the expected shape; usually, the flowerpot shape presents in Figure 1 by pressing it according to the desired shape and thickness. The filter that has been formed is then air-dried. The next step is firing the filter in the kiln. The firing process of the filter consists of seven stages; they are (1) air drying at 19°C – 25°C, 59% of humidity, 15 days, (2) water smoking at 20°C – 120°C, (3) decomposition at 120°C – 350°C, (4) burn-out combusts at 350°C – 450°C, (5) ceramic change at 350°C – 700°C, (6) carbon burns out at 700°C – 900°C, (7) vitrification at <1000°C [7,27]. The summary of material composition and production of two filters is given in Table 1.

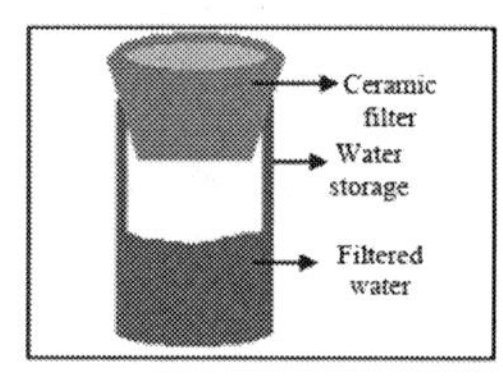

Fig. 1. Frustum shape or flower pot shape ceramic filter.

3.1.2 Pervious concrete filter

Pervious concrete is porous concrete that has high permeability and low compressive strength. This pervious concrete can be used as a filter to purify the water, so it is then called a pervious concrete filter (PCF). PCF is made from a mixture of sand (aggregate), cement, and water with a specific composition that causes pores in the PCF. PCF applications in terms of water treatment have been studied, resulting in filter effectiveness in reducing turbidity by 95% and bacteria by 98.7% [14,20–22].

The success of the use of PCF for water treatment purposes is largely determined by the hydraulic characteristics of the PCF, which is formed by a mixture of sand, cement, and water. The smaller the size of the sand used, the smaller the pore size that will form in it. The greater the ratio of sand and cement, the greater the porosity of the PCF will be. The size of the sand used varies from 1 – 2 mm [14], 2.36 – 13.2 mm [28], 5 – 10 mm [29], 10 – 12.4 mm [30], 0.15 – 1 mm [21], 0.15 – 4.45 mm [31], and 9.5 – 12.5 mm [32].

The type of sand commonly used is type 1 Portland cement. The water to cement ratio used varies from 0.27 to 0.4. The water to cement ratio of 0.33 performed an ideal permeability and compressive strength for the 10 and 12.5 mm of sand size [30]. The sand to cement ratio (M) used varies from M 3.7 to M 12.

To get a good PCF, sand, cement, and water that have been weighed according to a specific composition are mixed using the following procedure [33]:

a) 5 – 10% of cement mixed with aggregate for one minute.

b) The remaining cement and water are added to the mixture.

c) The mixture is stirred for three minutes, then rested for two minutes, and stirred again for another two minutes.

d) The finished dough is put into the cast by rodding on each layer of dough inserted. After all the dough is put into the cast, vibration is carried out for 5 – 7 seconds.

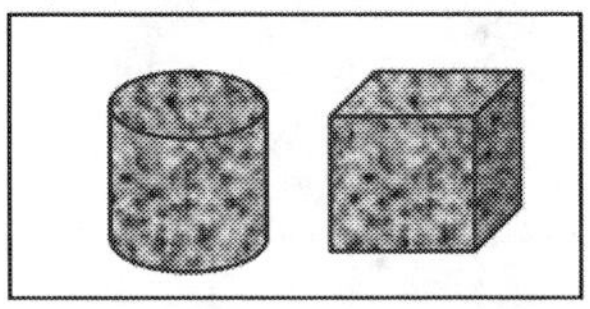

Fig. 2. Cylinders and cubic shape of PCF.

Table 1. Material composition and production of ceramic filter and pervious concrete filter.

Criteria	Ceramic Filter	Pervious Concrete Filter
Material	Main material = Clay Combustible material = sawdust, rice husks, recycled paper, coffee husks, peanut husks	Aggregate (sand), Cement, Water
Size of material	Clay (± 0.05 mm or 30 mesh) Combustible material (35 – 1000 mesh of sizes) (23)	Aggregate = from 1 – 2 mm (14), 2.36 – 13.2 mm (28), 5 – 10 mm (29), 10 – 12.4 mm (30), 0.15 – 1 mm (21), 0.15 – 4.45 mm (31), and 9.5 – 12.5 mm (32).
Composition	Clay : Combustible material 50 : 50, 65 : 35, 75 : 25 (23)	A/C = 3.7 – 12 W/C = 0.27 – 0.4 (30)
Production process	Sieving – Mixing – Moulding – Air drying 15 days – Firing (7, 26, 27)	Sieving – Mixing – Casting – Air drying

3.2 Hydraulic characteristics

The hydraulic characteristics of PCF studied in this study are porosity and permeability. Porosity is one of the essential factors in determining the performance of CF and PCF. In PCF, porosity is determined by many factors, including aggregate size, type of aggregate, aggregate gradation, type of cement, aggregate-cement ratio, water-cement ratio, type and dose of admixture, and method of mixing and formation [28]. The main characteristics of pores in pervious concrete include porosity, number of pores, pore size, relationship, pore distribution, and tortuosity [28]. In CF, porosity is determined by the type and size of the combustible material, composition of materials, firing temperature, etc. [23,34].

The porosity of CF, ε, was measured via water absorption [35] and calculated by Eq. 1.

$$\varepsilon = \frac{m_w m_d}{m_w m_h} \quad (1)$$

where m_w is the wet mass of CF, m_d is the dry mass, and m_h is the hydrostatic mass.

The porosity of PCF, φ, is determined by Eq. 4 [29].

$$\varphi = 1 - \frac{(w_2 - w_1)}{\rho V} \quad (2)$$

where w_2 (kg) is the weight of the oven-dry sample, w_1 (kg) is the weight of the underwater sample, ρ (kg/m^3)is water density at room temperature, and V (m^3) is the volume of sample.

Permeability is an essential factor that plays important roles in both CF and PCF. The permeability value describes the filter's ability to drain the water. The permeability is strongly influenced by the pore structure, pore size, and porosity in a filter. The pore structure, size, and distribution of pores in PCF depend on the composition of the material, the aggregate size, the mixing method, and the formation of the concrete filter [28]. Whereas in CF, the pore structure, pore size, and distribution depend on the composition of the material, the type and size of combustible material, the heating temperature, the mixing method, etc [23]

The permeability coefficient of the CF, k, was calculated from Eq. 1 [35].

$$k = \frac{Q\mu L}{A\rho g h} \quad (3)$$

Where Q is the flow rate of water in CF, μ is the viscosity of the water, L is the length or thickness of filter base, A is the cross-sectional area of the base, ρ is the density of water, g is gravitational constant, and h is the height of water above the base of CF.

The permeability coefficient, k, of the PCF can be calculated using Eq. 4.

$$k = \frac{aL}{At} \ln \frac{h_1}{h_2} \quad (4)$$

where k is the coefficient of permeability (cm/second), a is the surface area of the column/chamber (cm^2), L is the length/height of the porous concrete sample (cm), A is the area of the porous concrete sample (cm^2), h_1 is the initial height of the water (cm), h_2 is the final height of the water (cm), and t is the duration of time it takes for the water to descend from the water level h_1 to h_2 (seconds). The equation (2) can be used if only the permeability measurement uses the falling head method. The PCF sample is tested in the form of cubes or cylinders.

Through the pores in the CF and PCF filters, water can flow through the filter so that pollutants larger than the pores will be strained on the surface. With a certain pore structure and connectivity in the filter, the pollutants carried by the feed water can be trapped in the pores.

https://doi.org/10.1051/e3sconf/202132504006

3.3 Suspended solid and bacteria

Total Suspended Solids (TSS) and bacteria are the dominant types of pollutants in the water. Total suspended solids consist of inorganic materials such as sand, silt, and clay. However, bacteria and algae are also often counted as TSS. According to the Soil Science Society of America, and has a size of 2 – 0.05 mm, silt is 0.05 – 0.002 mm, and clay is < 0.002 mm. Bacteria and algae have a length between 0.0005 and 0.004 mm (0.5 – 4 μm) [36]. Inorganic materials and bacteria, algae, and other decomposed organic materials are often found in water. It is necessary to know the concentration of bacteria and TSS in water. Turbidity is one indicator in determining the level of water clarity. The more particles contained in the water, the greater the turbidity value. These particles are not only suspended solids but can also be dissolved solids. Although TSS and turbidity are closely related, turbidity does not directly indicate the TSS value. However, many researchers use the turbidity parameter to determine the level of water clarity as one of the performance indicators of a filter in removing pollutants in water, especially TSS.

3.4 Effect of material composition on hydraulic characteristics and removal effectiveness

3.4.1 Ceramic filter

Many studies have proven that ceramic filter is user-friendly technology that can effectively reduce contaminants from suspensions to viruses. The high effectiveness of ceramic filters in reducing pollutants is caused by many factors, including clay material, combustible material (composition, type, and size), firing temperature, a surface coating method, and raw water quality.

In CF fabrication, several things need to be considered according to the clay material used, even though it will indirectly affect the removal effectiveness. Some of these things are clay physical properties (shrinkage, plasticity, workability), availability of clay sources, the leachability of heavy metals, and inorganic material. Although clay has no direct effect on the pore structure of CF, many studies have found that different clay sources have various effects on porosity, permeability, and bacterial removal effectiveness.

The combustible material is an essential parameter in determining the removal effectiveness of CF. The types of organic materials used in the manufacture of CF are usually sawdust [23,35], rice husks, and recycled paper fibres in various forms [7]. The type of combustible material will affect the pore size distribution because of the different shapes of the kind of material.

Moreover, the proportion of clay and the combustible material also affect the porosity and permeability. The use of combustible material, i.e. 20% and 15%, yields different flow rates and removal effectiveness. When the amount of combustible material is high, namely 20%, the CF flow rate is fast, i.e. 13.9 L/h, with > 2 LRV (> 99%) for bacterial removal. Conversely, when the amount of combustible material is reduced to 15%, the flow rates decrease to 5.9 L/h with high removal effectiveness reaching > 4 LRV (> 99.99%) for bacteria removal [7]. The proportion of clay and combustible material (sawdust) on CF and its effect on hydraulic characteristics was studied by Yakub et al. (2013) [23]. That study analyzed three different proportions (percentage) of clay and sawdust, i.e. 50:50, 65:35, and 75:25. The results showed that the CF with 50% of sawdust produced a higher porosity reaching 46%, and achieved 99.99% of E.Coli bacteria removal or 3.16 – 8.17 of log removal value (LRV). On the contrary, the CF with 35% of sawdust yields 39% porosity and 99.99% of E.Coli bacteria removal or 5.82 – 6.9 of LRV. Using the sawdust's size of 35 – 1,000 mesh in that experiment, Yakub et al. (2013) found that most of the pore size in that three different compositions were in the range of 0.05 – 1 μm [23]. Nigay et al. (2018) studied the mixture of clay and the three different types of additive i.e. activated carbon as organic material and hydroxyapatite as inorganic material and a mix of both materials. The CF with 20 % of activated carbon produced 52% porosity and 43 mD of permeability higher than the CF without additive. The CF with 20% of hydroxyapatite yields 40% of porosity and 30 mD of permeability. The increased removal of nitrate (organic) and Lead (inorganic) was reached by the CF with 20 % of activated carbon and hydroxyapatite, i.e. 5.7 and 39.5 mg/g, respectively [19]. The higher the percentage of combustible material, the higher the CF's porosity and permeability [19,35]. The porosity, permeability, and pore size distribution will affect the difference in flow rates and removal effectiveness.

Apart from the types and proportion of the combustible material, the size of the material also determines the porosity, pore size, and flow rates. The smaller the material size, the smaller the pore size and flow rate, but the higher is the removal effectiveness. Conversely, the larger the material used, the larger the pore size and leads to an increase in flow rates, which usually reduces the removal effectiveness.

3.4.2 Pervious concrete filter

The aggregate size affect the size of the pores in PCF; thus, it also affects the permeability. The larger the aggregate size, the larger the pore size will be. In addition to the aggregate size, the water to cement ratio and the aggregate to cement ratio also affect the permeability and compressive strength of PCF. The greater the water to cement ratio and the smaller the aggregate to cement ratio leads to increased compressive strength and decreased permeability.

Based on the results of a study by Neamitha & Supraja (2017) and Ali & Kareem (2014), the best values of the water to cement ratio for pervious concrete are almost the same, i.e. 0.33 and 0.32, with the aggregate size of 10 mm – 12.5 mm and 9.5 mm – 12.5 mm respectively. At low water to cement ratio values, the concrete mixture will be very dry, causing an increase in porosity and permeability. If the value of the water to cement ratio is high, a "drawdown" effect will be formed where water fills the pores of the concrete, especially at the bottom part, which results in a decrease in the permeability of the porous concrete [30,32].

The compressive strength is generally inversely proportional to its permeability. This condition is due to the large cement filling the pores, covering each aggregate grain. This condition will strengthen the bonding of each grain, thus increase the compressive strength. On the other hand, it will block pores in pervious concrete to automatically reduce the permeability. Therefore, the amount of cement content and the aggregate to cement ratio are significant in PCF manufacture. The smaller the value of the aggregate-cement ratio, the stronger the concrete will be so that the compressive strength will be greater, but the permeability will be more negligible [14]. Unfortunately, the higher the aggregate-cement ratio, the easier the concrete cracks due to the lower compressive strength of PCF.

The composition of materials in the manufacture of PCF is determined based on the purpose of PCF use. Several previous studies have examined pervious concrete as a water filter to remove several pollutants, including turbidity and coliform bacteria. Small aggregate sizes are used for the experiment of pervious concrete as a water filter for suspended solids such as (Triatmadja, 2008) which used an aggregate size of 1 – 2 mm [14], Kamulyan et al. (2011) used an aggregate size of 0.425 – 0.850 mm, 0.850 – 1 mm, and 1 – 2 mm [37], and even smaller if it is used to remove coliform bacteria such as using an aggregate size of 0.15 – 1 mm [21]. The size of the aggregate is not the only key to the successful performance of PCF. The aggregate to cement ratio and the water to cement ratio also need to be adjusted and tested to produce pervious concrete that could work appropriately according to its function.

The different ratio of water-cement (0.24 – 0.32) in the aggregate size of 2.36 – 12.5 mm was analyzed by Costa et al. (2021) [38]. With admixture, the lower water to cement ratio (0.24) resulted in a higher porosity, i.e. 28.72% and permeability, i.e. 2.01 cm/s. On the other hand, the comparison of with and without admixture on pervious concrete with water to cement ratio of 0.3 found that the pervious concrete without admixture produced a higher porosity, i.e. 26.67%, than that with admixture, i.e. 24.29%. Moreover, the permeability in pervious concrete without admixture was also higher, i.e. 1.7 cm/s, than that with admixture, i.e. 0.99 cm/s. The influence of the water to cement ratio on the porosity was also proven quantitatively by the binary image of the pervious concrete surface. The low water to cement ratio presented more interconnected pores than the high one [38].

The influence of aggregate to cement ratio (M) on porosity and permeability was investigated by Zhang et al. (2020) [28]. With the uniform aggregate size, the higher the aggregate to cement ratio, the higher the porosity and permeability of pervious concrete.

The image processing of sliced pervious concrete to investigate the pore structure was conducted by Ni et al. (2021) [39]. The study analyzed the 2-dimension pore structure of pervious concrete using the software, i.e. Image-Pro Plus (IPP) and Industrial CT scan (scanner machine model XTH320. This study using three different aggregate sizes, i.e. 1.18 – 2.36 mm, 2.36 – 4.75 mm, and 4.75 – 9.5 mm. The M ratios used in this study were M 2.9 to M 4.1. The results showed that the pore sizes ranged 1 – 3 mm, 3 – 5 mm, and 6 – 8 mm for the aggregate sizes of 1.18 – 2.36 mm, 2.36 – 4.75 mm, and 4.75 – 9.5 mm, respectively [39]. The image processing shows that the larger the aggregate size, the more significant the large pores formed on the pervious concrete. The higher the M ratio, the higher the number of pores formed in the pervious concrete.

Pervious concrete can be used as a water and wastewater filter according to its porosity, permeability, pore size, pore connectivity, and material composition. Kim et al. (2017) reported that pervious concrete reduced the concentration of phosphorus and suspended nitrogen by trapping on the pores and the dissolved phosphorus by adsorption [40]. The study conducted by Shabalala et al. (2017) found that the concrete filter was successful in removing the heavy metals (Al, Fe, Mn, Co, and Ni) present in the acid mine drainage (AMD) due to the precipitation mechanism and reduce the acidity of AMD due to substitution mechanism [41]. The decrease in Lead (Pb) by pervious concrete was analyzed by Muthu et al. (2018) and found that pervious concrete can remove about 97.2% to 99.1% concentration of Pb in solution [42]. A study conducted by Maadji et al. (2016) found that PCF with composition M = 4, the diameter of 10 cm and height of 20 cm was able to reduce turbidity to less than 5 NTU and E. Coli bacteria reached 98.71% or 2 LRV[20,22]. Kamulyan et al. (2012) found that the PCF with 0.425 – 0.850 mm aggregate size removed turbidity from the feed water with the effectiveness ranging from 87.8% – 93.9% in 0.5 m/h of velocity [43]. Velocity affected the removal effectiveness that the slower the velocity of water in PCF, the higher is the turbidity removal effectiveness [44].

3.5 Removal mechanisms of suspended solids and Bacteria

According to the hydraulic characteristics (porosity and permeability) and pore size of CF and PCF, the removal mechanism could vary depending on the size of contaminants in the water. This study investigates the contaminants that probably exist in the water source, i.e. suspended solids and bacteria. The total suspended solids (TSS) consists of particles size larger than two µm. Suspended solids usually consist of inorganic materials such as clay, silt, or sand, but organic materials such as plankton and decomposed algae. Bacteria have various sizes, i.e. ~1 – 3 µm.

Commonly, the ceramic filter has low permeability. This condition is due to the small size of the pores formed on it. The CF has a pore size ranging from 0.005 - 0.1 µm [19] and 0.05 - 1 µm [23]. Based on the porosity, pore size, the additive/ composition, and clay structure of the CF, the contaminants, including bacteria, chemicals, and viral, can be removed by several capture mechanisms, i.e. trapping, substitution, and adsorption [35].

In terms of suspended solids, the removal mechanism that may occur in the filter is straining.

Straining happens when the particle size is larger than the pore size in the filter; for example, the suspended solids (particle size > 2 μm) are retained in the surface of CF (pore size ~0.005 μm – 1 μm) [45]. A smaller particle such as bacteria (~1 μm – 3 μm) may pass the pores on the surface of CF. However, the pores may trap the bacteria due to wide ranges of pore sizes from nano-scale pores to micro-scale pores and pore structure on the CF.

The removal mechanisms of suspended solids and bacteria in PCF were also controlled by the hydraulic characteristics especially pore size and pore connectivity. The previous studies found that, generally, the pore size of PCF was larger than that of CF. These pores characteristics affect the permeability value of the PCF, where the PCF permeability value is greater than the CF permeability.

Suspended solids can be removed from the PCF by a straining mechanism based on the pore size and pore connectivity. Suspended solids will be trapped and stuck in the deadlocked pores. Bacteria may be removed from the feed water if the dirty layer formed on the PCF. The dirty layer will then capture the bacteria, thereby reducing the number of bacteria in the outlet. Bacteria will also trap in the tiny pores formed in the PCF. However, to improve the effectiveness in bacteria removal, the PCF needs to be made using a small aggregate size, i.e. less than ~1 mm with appropriate proportions of cement, aggregate, and water.

4 Conclusion

Ceramic filter mostly consists of clay and combustible material such as sawdust. In some cases, the addition of organic (such as activated carbon) and inorganic (such as hydroxyapatite) to the ceramic was objected to improve the removal effectiveness. The type, mineral, and sources of clay and the type, shape, proportion, and size of combustible materials will affect its hydraulic characteristic (porosity, permeability, pore size, pore structure, pore connectivity/ distribution). The larger the size of combustible material, the larger the proportion of large pores formed in CF, thus increasing the porosity and permeability. The greater proportion of combustible material, the greater is the porosity and the permeability of the ceramic filter. The higher porosity and permeability, the smaller is the effectiveness of CF in reducing the contaminant.

The pervious concrete filter consists of aggregate, cement, and water. The addition of additive and admixture was sometimes added to improve the mechanical strength and performance of PCF in removing the contaminants. The shape, size, and type of aggregate, the ratio of water to cement, and the ratio of aggregate to cement will affect its hydraulic characteristics (porosity, permeability, pore size, pore structure, pore connectivity/ distribution). The larger the aggregate size, the larger the proportion of large pores formed in PCF, increasing the porosity and permeability of PCF and decreasing the ability to remove the small-sized contaminants. The smaller the aggregate to cement ratio, the smaller the pore size formed in the PCF. By forming connected, semi-connected, and closed pores in the PCF, the contaminants can be removed according to the contaminant size.

The removal mechanisms of suspended solids and bacteria for CF and PCF are similar to the same "cast type" filter. The pores are the main feature in those two filters that retain the contaminants, especially those larger than 1 μm. The removal mechanisms of suspended solids (>2 μm of size) are straining on the surface and trapping on the deadlocked pores. On the other hand, the bacteria (~1 – 3 μm of size) removal is straining in the dirty layer and trapping the bacteria in the pores.

So far there is no information regarding the performance of PCF constructed with combined value of composition (M) in stages. In real aplication such stages of PCF element of different value of M could probably be used to enhace the capacity of trapping or straining the contaminant depending on their sizes. This may prolonge the use of the PCF before it has to be backwashed.

The authors are thankful for the financial support given by The Indonesia Endowment Fund for Education (LPDP) through the scheme of Beasiswa Pendidik 2020.

References

1. World Health Organization, World Health Statistics 2020: Monitoring Health for The SDGs (2020)
2. A. Battilani, M. Steiner, M. Andersen, S. N. Back, J. Lorenzen, A. Schweitzer, A. Dalsgaard, A. Forslund, S. Gola, W. Klopmann, F. Plauborg, and M. N. Andersen, Agric. Water Manag. **98**, 385 (2010)
3. C. K. Pooi and H. Y. Ng, Npj Clean Water **1**, (2018)
4. United Stades of America, US Environ. Prot. Agency EPA 815 (2006)
5. F. Klingel and S. Diener, Guidebook for the Implementation of Decentralised Water Supply Systems in Moldova (Skat Consulting Ltd, Moldova, 2014)
6. Sustainable Infrastructure Society, Guide Book : Planning and Implementation of " Point of Entry " and " Point of Use " Water Treatment Systems in British Columbia (2007)
7. H. Yang, S. Xu, D. E. Chitwood, and Y. Wang, Front. Environ. Sci. Eng. **14**, 1 (2020)
8. Joseph Mark Brown, Effectiveness of Ceramic Filtration for Drinking Water Treatment in Cambodia (2007)
9. D. Xu, L. Bai, X. Tang, D. Niu, X. Luo, X. Zhu, G. Li, and H. Liang, Sci. Total Environ. **691**, 322 (2019)
10. A. Silva and S. Fuchs, (2015)
11. G. S. Logsdon, R. Kohne, S. Abel, and S. LaBonde, J. Environ. Eng. Sci. **1**, 339 (2002)
12. E. Yogafanny, S. Fuchs, and U. Obst, J. Sains &Teknologi Lingkung. **6**, 107 (2014)
13. J. F. Schijven, H. H. J. L. Van den Berg, M. Colin, Y. Dullemont, W. a. M. Hijnen, A. Magic-

Knezev, W. a. Oorthuizen, and G. Wubbels, Water Res. **47**, 2592 (2013)
14. R. Triatmadja, 2008, **1** (2008)
15. H. Zhang and V. Oyanedel-Craver, J. Hazard. Mater. **260**, 272 (2013)
16. A. R. Bielefeldt, K. Kowalski, and R. S. Summers, Water Res. **43**, 3559 (2009)
17. M. Y. M. Soliman, D. van Halem, and G. Medema, Int. J. Hyg. Environ. Health 224, (2020)
18. A. R. Bielefeldt, K. Kowalski, C. Schilling, S. Schreier, A. Kohler, and R. S. Summers, Water Res. **44,** 1482 (2010)
19. P. Nigay, A. Nzihou, C. E. White, and W. O. Soboyejo, **144**, 1 (2018)
20. R. Maadji, R. Triatmadja, and F. Nurrochmad, in Proc. 37th IAHR World Congr. (2017), pp. 2705–2713
21. R. Maadji, R. Triatmadja, F. Nurrochmad, and Sunjoto, in Proc. Second IAHS Panta Rhei Int. Conf. Water Syst. Knowl. Innov. Its Pract. Dev. Ctries. (2017), p. 44
22. R. Maadji, R. Triatmadja, F. Nurrochmad, and Sunjoto, in Proc. Int. Semin. Water Resil. a Chang. World (2016), pp. 711–721
23. I. Yakub, D. Ph, A. Plappally, D. Ph, M. Leftwich, D. Ph, K. Malatesta, D. Ph, K. C. Friedman, S. Obwoya, D. Ph, F. Nyongesa, D. Ph, A. H. Maiga, D. Ph, M. Asce, A. B. O. Soboyejo, D. Ph, S. Logothetis, W. Soboyejo, and D. Ph, 08544, 986 (2013)
24. J. Rayner, Current Practices in Manufacturing of Ceramic Pot Filters for Water Treatment, 2009
25. U. E. Ekpunobi, S. U. Agbo, and V. I. E. Ajiwe, J. Environ. Chem. Eng. **7**, 102791 (2019)
26. H. Yang, Z. Yan, X. Du, L. Bai, H. Yu, A. Ding, G. Li, H. Liang, and T. M. Aminabhavi, Chem. Eng. J. **382**, 123033 (2020)
27. E. A. Zereffa and T. B. Bekalo, Mater. Sci. Appl. Chem. **34**, 69 (2017)
28. Y. Zhang, H. Li, A. Abdelhady, and J. Yang, Constr. Build. Mater. 263, 120614 (2020)
29. Y. Qin, H. Yang, Z. Deng, and J. He, Adv. Mater. Sci. Eng. 2015, **1** (2015)
30. M. Neamitha and T. M. Supraja, Int. Ref. J. Eng. Sci. ISSN **6**, 2319 (2017)
31. M. M. Taghizadeh, a. Torabian, M. Borghei, and a. H. Hassani, Int. J. Environ. Sci. Technol. **4**, 505 (2007)
32. M. K. Ali and Q. M. S. Kareem, Int. Jourbal Enhanc. Res. Sci. Technol. Eng. **3**, Issue 1, 107 (2014)
33. K. Wang, V. R. Schaefer, J. T. Kevern, and M. T. M. M. T. Suleiman, NRMCA Technol. Forum Focus Pervious Concr. **1** (2006)
34. P. M. Nigay, A. A. Salifu, J. D. Obayemi, C. E. White, A. Nzihou, and W. O. Soboyejo, **146**, 1 (2020)
35. P. Nigay, A. A. Salifu, J. D. Obayemi, C. E. White, A. Nzihou, and W. O. Soboyejo, **145**, 1 (2019)
36. W. J. Rogan, M. T. Brady, H. J. Binns, J. a. Forman, C. J. Karr, K. Osterhoudt, J. a. Paulson, J. R. Roberts, M. T. Sandel, J. M. Seltzer, R. O. Wright, E. Blackburn, M. Anderson, S. Savage, N. B. Ragan, P. Spire, J. a. Bocchini, H. H. Bernstein, J. S. Bradley, C. L. Byington, P. H. Dennehy, M. C. Fisher, R. W. Frenck, M. P. Glode, H. L. Keyserling, D. W. Kimberlin, W. a. Orenstein, L. G. Rubin, R. S. Baltimore, J. a. McMillan, B. Bell, R. Bortolussi, R. D. Clover, M. a. Fischer, R. L. Gorman, R. D. Pratt, J. S. Read, B. Gellin, J. R. Starke, J. Swanson, C. J. Baker, S. S. Long, L. K. Pickering, E. O. Ledbetter, H. C. Meissner, and J. Frantz, Drinking Water from Private Wells and Risks to Children (2009)
37. B. Kamulyan, F. Nurrochmad, R. Triatmadja, and Sunjoto, in 4th ASEAN Environ. Eng. Conf. (2011)
38. F. B. Pereira da Costa, L. M. Haselbach, and L. C. P. da Silva Filho, Constr. Build. Mater. **268**, 121084 (2021)
39. T. Ni, W. Ma, Y. Yang, J. Yu, J. Liu, C. Jiang, and C. Gu, Constr. Build. Mater. **267**, 121052 (2021)
40. G. M. Kim, J. G. Jang, H. R. Khalid, and H. K. Lee, Constr. Build. Mater. **136**, 1 (2017)
41. A. N. Shabalala, S. O. Ekolu, S. Diop, and F. Solomon, J. Hazard. Mater. **323**, 641 (2017)
42. M. Muthu, M. Santhanam, and M. Kumar, Constr. Build. Mater. **174**, 224 (2018)
43. B. Kamulyan, F. Nurrochmad, and R. Triatmadja, in Annu. Eng. Semin. "Towards Sustain. Eng. (2012)
44. B. Kamulyan, F. Nurrochmad, R. Triatmadja, and S. Sunjoto, in Int. Conf. Sustain. Dev. Water Wastewater Treat. (Yogyakarta, Indonesia, 2009)
45. S. a. Bradford, J. Simunek, M. Bettahar, M. T. Van Genuchten, and S. R. Yates, Environ. Sci. Technol. **37**, 2242 (2003)

 https://doi.org/10.1051/e3sconf/202132504007

IoT-based Estimation System for Microcystis aeruginosa Cyanobacteria in Laguna de Bay using an Arduino-controlled Spectrophotometric Device

Lawrence D. Alejandrino [1], *Jessica Joy D. Jocson* [1], *Micah Romina R. Mirarza* [1], *Ericson D. Dimaunahan* [2], *Ramon G Garcia* [1], *Flordeliza L Valiente* [1]
[1] School of Electrical, Electronics and Computer Engineering, Mapua University Manila, Philippines
[2] Department of Physics, Mapua University Manila, Philippines

Abstract. Laguna de Bay, the largest freshwater lake in the Philippines, provides livelihood to the fishermen and serves as a source of potable water to the locals. However, freshwater quality has degraded, whereas one of the main contributors are Cyanobacteria that produce cyanotoxins. Existing studies that uses a similar device are either too expensive or too bulky. The purpose of this study is to estimate the cyanobacteria concentration by using a low-cost 16-channel spectrophotometric device to determine the level of severity efficiently. Using Linear Regression, the dataset is modelled by the algorithm to estimate the number of cyanobacteria present on the water sample, while Support Vector Machine (SVM) algorithm for severity level classifier. This study achieved high accuracy in estimating the cyanobacteria using linear regression and classifying the level of severity by support vector machine.

1 Introduction

Laguna de Bay has been experiencing environmental problems, resulting in the presence of Microcystis aeruginosa, also called the blue-green algae. Cyanobacteria is a photosynthetic bacterium that has a blue-green colour. Microcystis aeruginosa from the phylum of Cyanobacteria found in Laguna de Bay is a vitally necessary factor to freshwater and marine ecosystems [1]. This phylum produces neurotoxins and hepatotoxins, which are fatal to creatures such as wild animals and birds, although, for humans, it can only cause skin irritations, fever-like symptoms, gastroenteritis, and a variety of allergic reactions [2]. Albeit the lake can still sustain fisheries, it can be exposed to contamination from pollution.

Fishers and locals are aware of the algal blooms that occur every dry season due to uncontrollable heat. Thus, they prefer to harvest early to avoid the occurrence of a fish kill. Yet, the cyanobacteria are invisible to the human eye, making it a problem for early harvest since algal blooms vary depending on the year [3]. Chlorophyll-a is a light-absorbing pigment of the cyanobacteria species, so many researchers and scientists are using spectrophotometers to detect cyanobacteria [4]. However, typical commercial spectrophotometers are bulky, power-consuming, and expensive. The portable spectrophotometer price in the market costs up to 500 USD, which is expensive. Also, manual recording, cell count, and analysis of water samples are time and energy-consuming.

A study developed a low-cost visible light spectrophotometer capable of measuring absorbance and transmittance of various materials liquid solution within the visible range of electromagnetic spectrum [5]. Materials used are the following: Arduino Uno, Stepper Motor and Driver, breadboard, op-amp, photodiode, 9V battery, push button, LED, resistors, battery cap, switch, DVD, and test tube with a total of 21.9 USD. Another study built an LED spectrophotometer with a photodiode circuit to determine iron, manganese, and fluoride in drinking water [6]. They concluded that there is no practical difference between the LED spectrophotometers and in comparison, to the four commercial spectrophotometers they used as a basis. Multichannel spectrophotometry allows better sensitivity, corrects interferences, and measures multiple components in a solution. [7] A 16-channel spectrophotometer will produce a more sensitive output, making it easier to identify disparities that interferences could cause to the test. A study showed how to calculate the concentration of cyanobacteria by using the parameter absorbance [4]. This study also showed the correlation of concentration to cell density (cells/mL). Support vector machine is for remote sensing data classification. According to Vapnik, Support Vector Machine is a machine learning method used for small datasets and classification and regression during data analysis. It contains features such as margin maximization and kernel technology that adopted in high-dimensional feature space. High fitting accuracy and a small quantity of tuneable parameters are also some of the valuable properties of SVM [8].

The primary purpose of this study is to develop a device that can estimate the number of cyanobacteria in Laguna de Bay by spectrophotometry and access the data remotely and determine its level of severity through machine learning. Specifically, this research aims to: (1) design a low-cost 16 channel spectrophotometer that measures the intensity of light as a function of

 https://doi.org/10.1051/e3sconf/202132504007

wavelength; (2) implement internet of things (IoT) by interfacing and accessing the ThingSpeak website or mobile application which displays the graphed data; (3) estimate cyanobacteria concentration using linear regression approach; and (4) determine the severity level by training the support vector machine.

This study proposes a fast approach in cyanobacteria estimation since the dataset is viewed and analysed remotely with the aid of the internet. Since not all cyanobacteria contaminated water is visible to the human eye, conducting a study about early detection of M. aeruginosa can prohibit them from inhabiting the waters, subsequently damaging the ecosystem and livelihoods dependent on the lake. The local government may establish strict environmental policies on certain parts near the lake detected by the spectrophotometer to be harmful.

This study covers the estimation of cyanobacteria species, M. aeruginosa, by spectrophotometry. The three severity levels observed are the minor, moderate, and massive cyanobacteria concentrations. The parameters correlated to these are absorbance and cell density. A stable internet connection is required to transmit the data to the cloud. Testing is best done within the day to preserve the fresh amount of cyanobacteria cells.

2 Methodology

2.1 Conceptual Framework

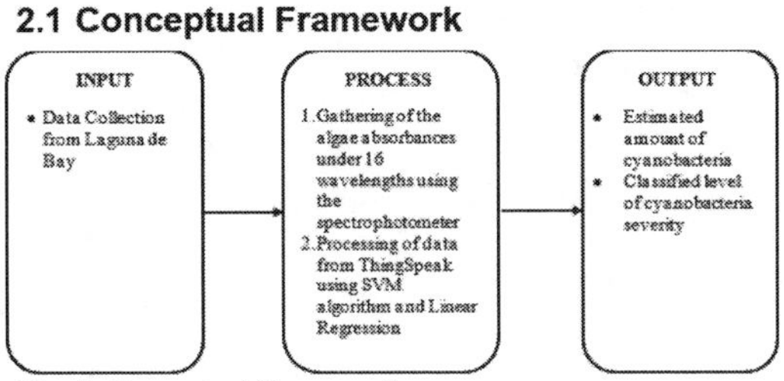

Fig. 1. Conceptual Framework

2.2 Prototype Design and Set-up

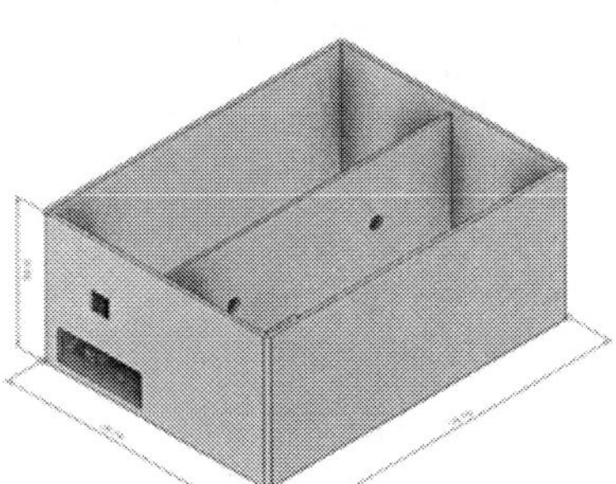

Fig. 2. Prototype Dimensions and Set-up

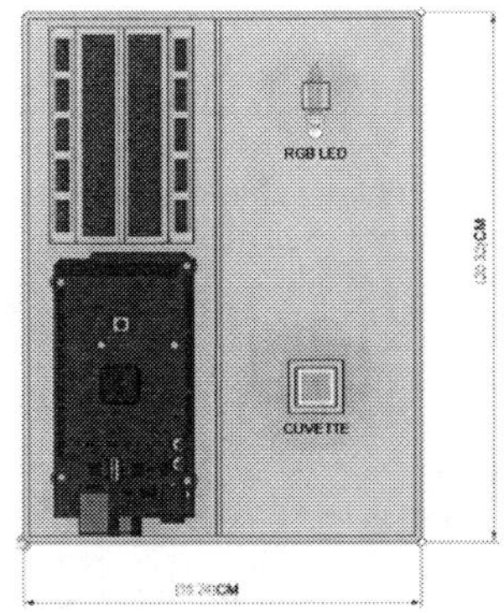

Fig. 3. Prototype Isometric View

2.2.1Prototype Enclosure

The enclosure for the prototype is made of plastic which is 3D printed to ensure that the components inside are properly positioned. The enclosure is divided into two to isolate the light from the Arduino and the RGB LED. The plastic is black in colour to minimize the reflection of light inside the enclosure that may affect the reading of the sensor.

2.2.2Microcontroller

An Arduino Mega 2560 is a microcontroller board based on the ATmega2560 (datasheet). This microcontroller will be used to collect the data from the photodiode which is connected to pin A0 of the Arduino.

2.2.3Light Source

The light source that the researchers will be using is an RGB LED. This is because a single RGB LED can produce a wide array of colors or wavelengths. This can be done by programming the RGB LED using the Arduino. Commercially available RGB LEDs which can have a wavelength range of 380 to 780 nm. Although, wavelengths at the upper end of the UV range (10-400nm) produce results that are accurate and cost-efficient since the absorption of these wavelengths is lower. One piece of RGB LED will be used for the prototype.

2.2.4Sensor

The sensor to be used is a photodiode. A photodiode reacts to visible and near infrared radiation and it acts like a miniature solar panel that converts the light energy coming from the RGB LED into voltage. In this study, it is used to measure the absorbance of the M. aeruginosa obtained from the Laguna de Bay. The sensor will be connected to the analog input of Arduino Mega 2560 and will be placed inside the enclosure together with the light source, cuvette, and other components. BPW34 is the model of the photodiode that will be used in this study.

2.2.5 Cuvette

A cuvette is a small container that is similar to test tube, but it has a square cross-sectional area that is sealed on one end. It is important to use a cuvette instead of a test tube since the shape of a test tube is round, causing refraction leading to losses in the light transmitted through the sample. The sample of water from Laguna de Bay that will be tested will be held by a cuvette when being tested in the spectrophotometer.

2.2.6 ESP8266 Wi-Fi Module

This Wi-Fi module is attached to the Arduino as it allows the Arduino to transmit data to the cloud using ThingSpeak.

2.2.7ThingSpeak

ThingSpeak is a website which stores and analyses data from IoT devices. It is configured through a code that enables the communication between the cloud and the prototype.

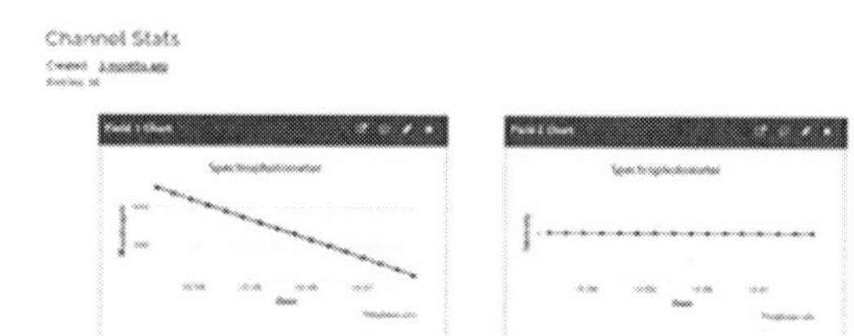

Fig. 4. Sample Output from ThingSpeak

2.3 Data Gathering

The samples to be used for testing will be taken from Laguna de Bay at multiple sites during the afternoon, the sample collector must wear proper PPE and to avoid exposure to the cyanotoxins. The water samples collected will be transported in containers placed in a dark cooler at +4°C in order to preserve the quality of cyanobacteria [9]. In case the storage time is longer than 12 hours, quick-freezing of samples to -20°C is suggested [10].

Each sample will be placed in 3.5ml quartz cuvettes and tested in the spectrophotometer. The data obtained from the spectrophotometer is transmitted to ThingSpeak, made available for remote access and for machine learning.

Beer's Law states that the absorbance of light-absorbing material is proportional to its concentration in solution. The intensity of the light passing through the sample cell is also measured for that wavelength - given the symbol, I. If the value of I is less than that of Io, then the sample has absorbed some of the light, neglecting reflection of light off the cuvette surface.

Absorbance cannot be measured directly, thus the transmittance of a solution is measured instead. In the diagram below, the relationship of percent transmittance and absorbance is shown. This shows that if all the light transmitted is absorbed, the percent transmittance is zero and the absorbance is infinite, and when the percent transmittance is 100, the absorbance is infinite.

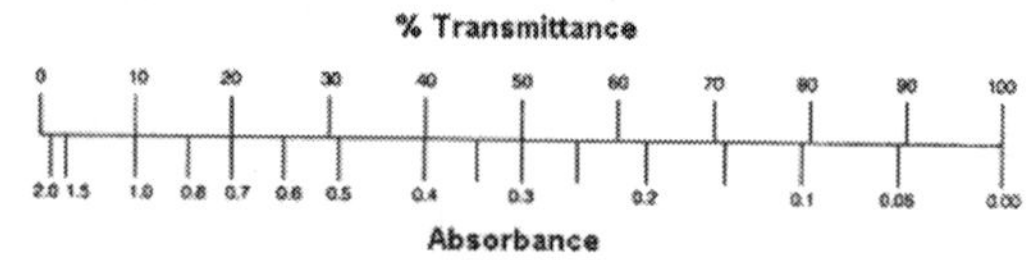

Fig 5. Relationship between % transmittance and absorbance

Transmittance is measured by the fraction of the transmitted light by the incident light. Io is obtained by recording the amount of light passing through an empty cuvette while I is obtained by recording the amount of light transmitted through a cuvette with the sample.

$$T = I/I_o \tag{1}$$

The absorbance is then calculated by the logarithm of the fraction of the incident light by the transmitted light.

$$A = \log(I_o/I) \tag{2}$$

The researchers were able to obtain 108 data observations from the spectrophotometer through the Beer-Lambert Law. These will serve as the dataset for the SVM algorithm.

2.4 Spectrophotometer Testing

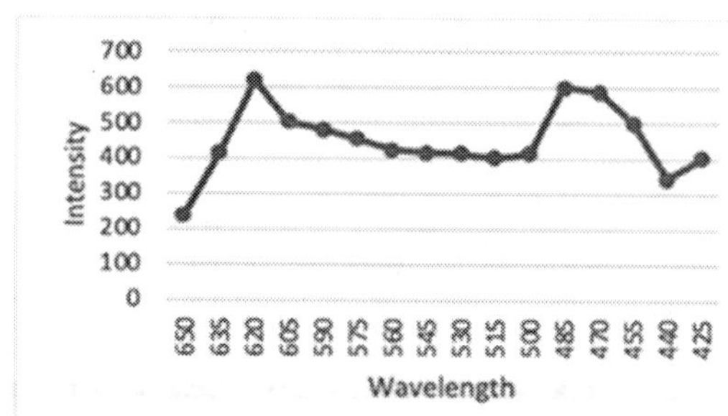

Fig. 6. Average Values of Intensity

In testing the samples that have been gathered from Laguna de Bay, the researchers need to determine the peak value of the intensity and its equivalent wavelength. After conducting some test, the average value for the intensity peaked at wavelength 620 nm. The 620 nm wavelength will be used in determining the absorbances of the cyanobacteria.

2.5 Calibration

Calibration is used to compare the results from a much reliable method or device to the results from the prototype. It will help the researchers determine if the prototype needs an adjustment to lessen the error. The dataset will be divided into three classifications which are Minor, Moderate, and Massive. For Minor, the

https://doi.org/10.1051/e3sconf/202132504007

concentration should be less than 100,000 cells/ml, Moderate ranges from 100,000 to 500,000 cells/ml and Massive should be above 500,000 cells/ml. 3-point calibration will be used in the dataset in which the researchers will get a sample from each of the three categories. Only two of the samples coming from the Moderate and Massive classification will be tested in a laboratory because the sample from Minor will be zero in value.

TABLE I. SAMPLES TESTED IN PROTOTYPE

Absorbance	Concentration	Classification
0	0	Minor
0.0158	191,811 cells/ml	Moderate
0.0484	593,224 cells/ml	Massive

TABLE II. SAMPLES TESTED IN LABORATORY

Absorbance	Concentration	Classification
0	0	Minor
0.0158	191,811 cells/ml	Moderate
0.0484	593,224 cells/ml	Massive

Computing the percentage error will be the difference between the concentration of the samples tested in spectrophotometer (Actual Value) and the samples tested in laboratory divided by the samples tested in laboratory (Expected Value).

$$\%error = \frac{|Actual\ Value - Expected\ Value|}{Expected\ Value} x100 \quad (3)$$

For the minor classification, the percentage error would be 0%, moderate classification would be 0.95%, and massive classification would be 0.55%.

2.6 Estimation of Cyanobacteria Concentration

There are many ways to determine the concentration of cyanobacteria but one of them using a spectrophotometer. Using a spectrophotometer is not an ideal and an indirect method in determining the concentration of cyanobacteria but it gives you an estimate on what its exact value should be [x]. In spectrophotometry, the relationship between absorbance and concentration is linear. When the value of the absorbance increases, the value of the concentration also increases and vice versa. The formula below is the result of a linear regression that will be used to correlate the absorbance to concentration at 620 nm wavelength for all the samples [4].

$$y = (8.13x10^{-8})x + (2x10^{-4}) \quad (4)$$

2.7 Data Analysis

The support vector machine algorithm or SVM is a type of supervised machine learning algorithm. This algorithm is capable of analyzing and processing data for classification and regression analysis. SVM is the algorithm that will be used for the classification of the level of severity of M. aeruginosa. In this research the samples will be classified into three classes, which are minor, moderate, and massive. The data points in the 3D plane are the frequencies emitted by the M. aeruginosa and it is separated by the hyperplane.

From the data obtained using the spectrophotometer, the dataset is divided into three. 60% of this dataset is allotted for training, 20% for validation, and 20% for testing. SVM will be administered using MATLAB.

2.7.1 Validation Process

Figure 5 and 7 shows the scatter plot for the validation process and testing process. Each of these scatter plots are comprised of a random 20% from the dataset, so there are 22 observations plotted for each process. The scatter plot presents the distribution of the data observations' absorbances and their SVM classifiers. These classifiers are shown here in the colors blue, orange and yellow, whereas blue represents massive, orange represents minor, and yellow represents moderate.

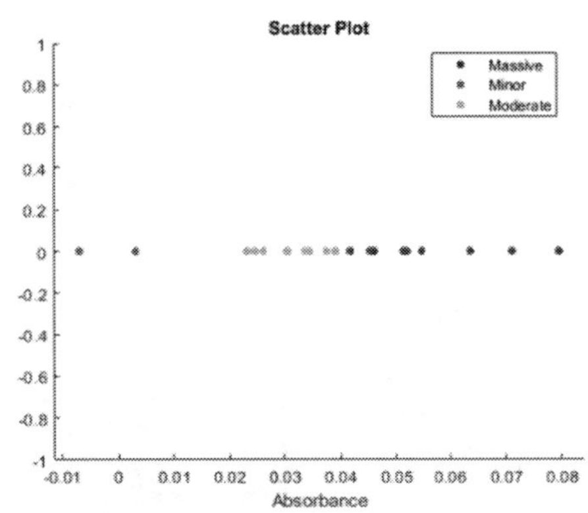

Fig. 7. Scatter Plot for Validation

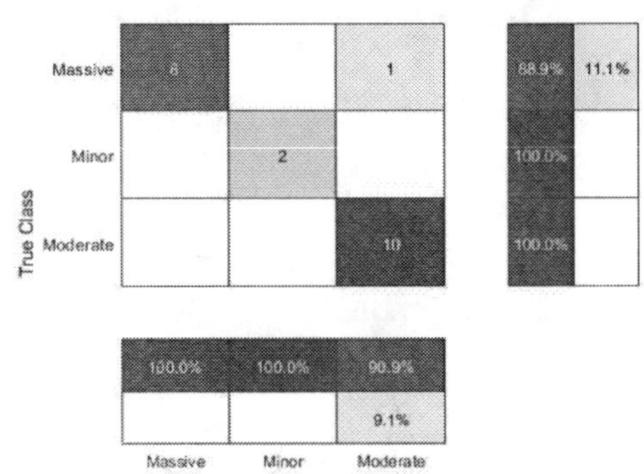

Fig. 8. Confusion Matrix for Validation

 https://doi.org/10.1051/e3sconf/202132504007

Figure 8 displays the classification and misclassification of each level using the trained support vector machine. There is a total of 22 observations for the validation test. 22 (20%) of it are randomly selected for validation test. For the massive concentration, there were a total of 9 data observations wherein 8 (88.9%) are classified correctly by the support vector machine. The remaining 1 (11.1%) is misclassified as either minor or moderate. 2 data observations are minor and all (100%) of it are correctly identified by the support vector machine. The last data observation is a total of 11 moderate concentration. In this observation, 10 (90.9%) is correctly classified and 1 (11.1%) is misclassified by the SVM. In total, the SVM validation accuracy is at 95.24%.

2.7.2 Testing Process

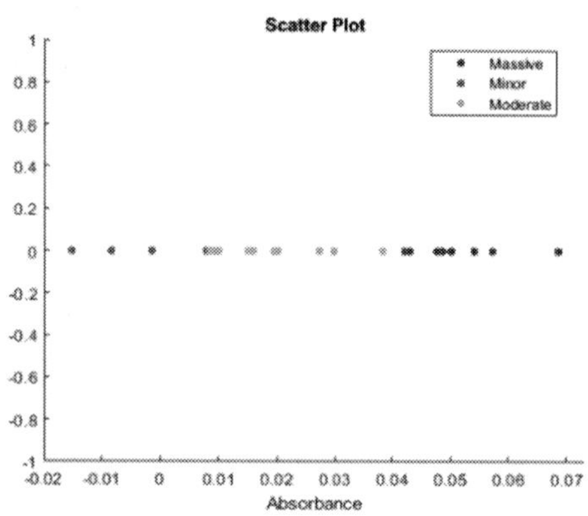

Fig. 9. Scatter Plot for Testing

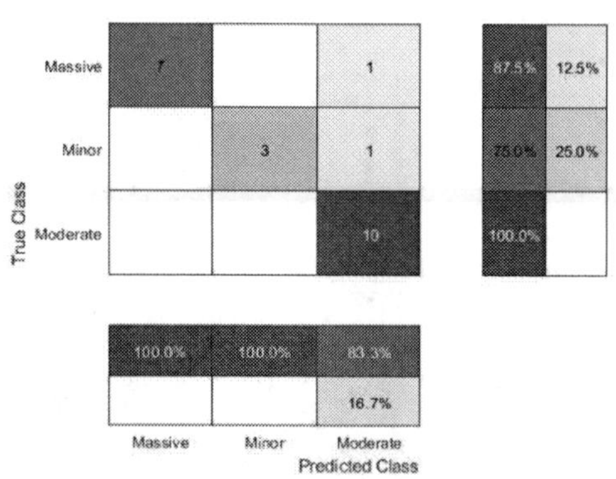

Fig. 10. Confusion Matrix for Testing

To protect against overfitting, validation and testing process are implemented and compared. 20% validation test and 20% actual test is observed to protect against overfitting. Figure Y displays the classification and misclassification by the trained support vector machine. There is a total of 108 observations for the actual test. For the massive concentration, there were a total of 8 data observations and 7 (87.5%) are correctly classified by the support vector machine. This level had the highest prediction accuracy among the 3 levels. The remaining 1 (12.5%) is misclassified as either minor or moderate.

For the minor concentration, there were 4 data observations and 3 (75%) are classified correctly by the support vector machine. The remaining 1 (25%) of the minor data observation were misclassified as moderate or massive. In the last 12 data observations, 12 are moderate and 10 (83.3%) of it are classified correctly. The other 2 (16.7%) left were misclassified as either massive or minor.

Out of the 22 data observations, 20 were classified as correct. Thus, the SVM test accuracy is 90.91%. Table A shows that the prediction model used is accurate since the validation and test process has a high accuracy.

TABLE III. VALIDATION AND TESTING ACCURACY

Absorbance	Concentration	Classification
Validation	95.24%	4.76%
Testing	90.91%	9.09%

3 Conclusion

The researchers were able to develop a low-cost spectrophotometer that can measure the absorbance of cyanobacteria. Wireless communication between the prototype and the cloud was achieved. The linearity between absorbance and concentration was also proven based from the results using Linear Regression and SVM. The researchers also accomplished the classification of the samples into three categories labelled as Minor, Moderate, and Massive using Support-Vector Machine algorithm through MATLAB. Lastly, the estimation of the cyanobacteria concentration using the gathered data was also achieved.

4 Recommendation

For further development of the low-cost spectrophotometer, it is recommended that the case design is compatible with an in-situ set-up for real-time testing of scientists and researchers. For this kind of set-up, the suggested power source is a solar-powered battery. Last, it is recommended to look for a more stable sensor than the photodiode.

References

1. S.H. Kim, Y. He, E.H. Lee, J.H. Kim, *Portable Fluorometer for Cyanobacteria Detection,* in IEEE Sensors Journal **17(8)**, 2377-2384 (2017)
2. L. Backer, W. Carmichael, B. Kirkpatrick, C. Williams, M. Irvin, Y. Zhou, Y.S. Cheng, Recreational exposure to low concentrations of

microcystins during an algal bloom in a small lake. Marine drugs **6(2)**, 389-406 (2008)

3. C. Rosales, A. Rollon, Factors Affecting the Occurrence and Growth of Toxic Cyanobacteria Microcystis Aeruginosa in Laguna De Bay, Philippines. Journal of Environmental Science and Engineering, 1109-1116 (2011)
4. A. Agberien, B. Ormeci, Monitoring of Cyanobacteria in Water Using Spectrophotometry and First Derivative of Absorbance, 12 (2020)
5. E. Ashfaque, M. Alam, M. R. Islam, I.J. Faria, *Development and validation of a low-cost visible light spectrophotometer,* in 2017 4th International Conference on Advances in Electrical Engineering (2017)
6. M.W. Prairiem S.H. Frisbie, K.K. Rao, A.H. Saksri, S. Parbat, *An accurate, precise, and affordable light emitting diode spectrophotometer for drinking water and other testing with limited resources* in in PLOS ONE **15(1)** (2020)
7. D.L. Witte, *Introduction to Multiple-Wavelength Spectrophotometric Measurements in Analytical and Clinical Chemistry,* in Contemporary Topics in Analytical and Clinical Chemistry. Springer, Boston, MA (1982)
8. S.X. Lu, J. Meng, G.E. Cao, *Support vector machine based on a new reduced samples method*, in 2010 International Conference on Machine Learning and Cybernetics (2010)
9. J.L. Graham, K.A. Loftin, A.C. Ziegler, M.T. Meyer, Guidelines for design and sampling for cyanobacterial toxin and taste-and-odor studies in lakes and reservoirs: U.S. Geological Survey Scientific Investigations Report 2008, 5038 (2008)
10. World Health Organization, US Environmental Protection Agency, Commission of the European Communities. *Chapter 10: Cyanobacteria and Algae*, in Monitoring Bathing Waters - A Practical Guide to the Design and Implementation of Assessments and Monitoring Programmes, 19 (2000)
11. W. Hu, Dry Weight and Cell Density of Individual Algal and Cyanobacterial Cells for Algae Research and Development, 7 (2014)

Mineralogy of hydrothermal breccia cement of Humpa Leu East porphyry copper-gold prospect, Sumbawa Island, Indonesia

Okki Verdiansyah[1,2], *Arifudin* Idrus[1*], *Lucas Donny* Setijadji[1], *Bronto* Sutopo[3], *I Gde* Sukadana[4]

[1]Geological Engineering Department, Universitas Gadjah Mada, Indonesia
[2]Geological Engineering Department, Institut Teknologi Nasional Yogyakarta, Indonesia
[3]Geomin unit of PT. Antam Tbk, Indonesia
[4]Center for Nuclear Minerals Technology, National Nuclear Energy Agency, Indonesia

Abstract. Indonesia is a country that has several world-class copper-gold deposits, particularly in eastern Sunda arc. The Hu'u complex has several prospects in the surface as lithocap of extensive epithermal style alteration, but some were detected associated with porphyry beneath the surface. The study focuses on hydrothermal breccia cement as a factor influencing the porphyry system in the Hu'u district. The methods used is mineralogical analysis with petrography and μ-XRF elemental mapping on hydrothermal breccia samples. The Hu'u district is interpreted as a paleo-volcano; a member of the Old Volcanics Rocks Formation. The Humpa Leu East lithology consists of pre-volcanics unit (lava and pyroclastics), diorite, andesite-μ-diorite, and tonalitic intrusion at the depth. Hydrothermal alteration evolved from tonalite body to outward, consist of potassic, inner propyllitic and overprinted by phyllic and advanced argillic. Several phases of hydrothermal activities occur in this system, including the hydrothermal breccia phase associated with complex fluids. The hydrothermal cement of Humpa Leu East porphyry at least have three phases of mineralogical assemblages and possibly influencing the mineralization. The mineralogical assemblage of hydrothermal cement in HLE consists of quartz-feldspar-plagioclase-biotite as a high-temperature phase; then followed by epidote-sericite-chlorites-anhydrite-carbonates in medium temperature; there are aluminum-rich clay minerals interpreted as gibbsite. Mineralization occurs in three phases including chalcopyrite-magnetite, bornite-chalcopyrite and chalcopyrite-sphalerite phases. The occurrences of chalcopyrite at all phases indicate the stability of intermediate sulfidation proses in Humpa Leu and as a possible factor to answer the abundant copper in the Hu'u hydrothermal fluid system
Keywords: porphyry, gold, copper, mineralogy, μ-XRF, Indonesia

1 Introduction

Indonesia is a country that has several world-class copper-gold deposits such as Batu Hijau, Martabe, Grasberg - Ertsberg, and the discovery of Onto Copper-Gold Deposits in Hu'u Complex in 2013 [1]. The mineralization emplaced on the Tertiary or Oligocene to Pleistocene magmatic – volcanic belt of Sunda-Banda arc, especially at the eastern side [2,3]. The deposits classify as world-class with more than > 5 million ounces gold equivalent, which is dominantly associated with porphyry-epithermal type.

Hu'u complex has several prospects in the surface as low to medium metal's grade lithocap of large epithermal style alteration, but some were detected associated with porphyry beneath the surface. In 2013, the exploration team succeeded in obtaining a large deposit in a very large hydrothermal alteration complex. The Humpa Leu East (HLE) is located in Sumbawa Island, administratively lies on Hu'u sub-district, Dompu district, Nusa Tenggara Barat province, Indonesia (Fig. 1). The Humpa Leu (East) prospect has 257 Mt resources potential with a grade at 0.2% Cu and 0.2 g/t Au [4].

The regional geology of Hu'u consists of a Tertiary volcanic unit [5,6] in the outer volcanic complex, with magmatism still active. The latest research in Hu'u complex was done by Burrows et al. (2020) and reporting that the U-Pb zircon age for the andesite that caps the volcano-sedimentary host rocks provides a maximum age of 0.838 ± 0.039 Ma, with a slightly younger porphyry zircon crystallization age of 0.688 ± 0.053 Ma. The mineralization in this 'young' ages to be interested in exploration, and the character of deposits in Hu'u still open up opportunities for new discovery in this belt or Pleistocene ages volcanism or magmatism.

The mineralization of precious metals is mainly affected by the fluids as the metals carrier agent in this system. Therefore, the mineralogical study of hydrothermal breccia in the porphyry system can be used to interpret the hydrothermal fluids system [7]. μ-XRF is used in many applications working with multiple material types, including examining minerals to determine the element distribution in geological samples [8,9,10]. This study is focused on mineralogical analysis

* Corresponding author: arifidrus@ugm.ac.id

 https://doi.org/10.1051/e3sconf/202132504008

in hydrothermal breccia cement using μ-XRF, which aims to determine the character and paragenesis of hydrothermal system in HLE.

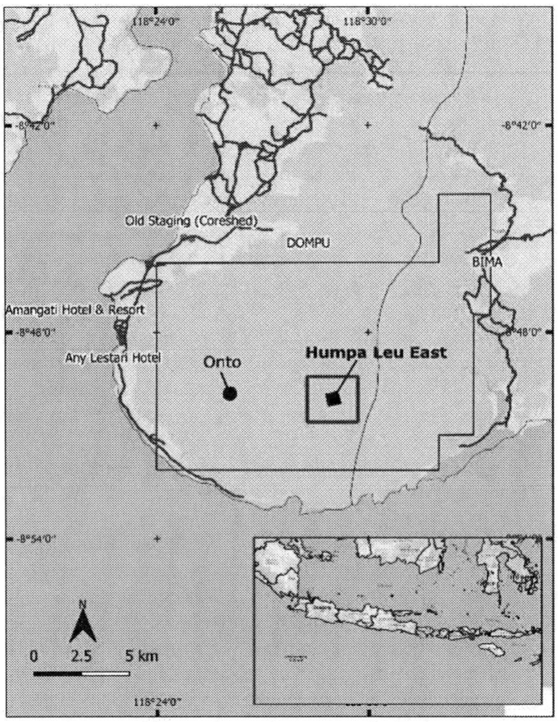

Fig. 1. Location of the study area.

2 Methods

The research uses several steps to interpret geological framework from surface alteration and lithological map and then subsurface from drill core analyses and interpretation. Then, the research materials are focused on hydrothermal breccia samples, then analysed by the petrographic method by polarizing microscope and elemental analysis by μ-XRF. Total samples analysed is two sample from two zone of breccia, then analysed of elemental distribution, mineral chemistry from pint analysis, and quantitative analysis with AMICS software. The μ-XRF analysis uses M4-Tornado Bruker unit at Centre for Nuclear Minerals Technology National, Nuclear Energy Agency (BATAN) laboratory, Jakarta, Indonesia. The spectrometer was equipped with an X-ray tube and a vacuum pump and was operated at 50-kV voltage and 600-μA current. Finally, we use AMICS software for quantitative mineralogical analysis from μ-XRF data to increase confidence about mineral distribution and petrogenesis.

3 Results

3.1 Geology of Humpa Leu East

Humpa leu East is part of the Hu'u volcanic system with typical lithology as volcanic product, lava, pyroclastic and diatremes, and spotted subvolcanic intrusion. The lithology of Humpa Leu East consists of andesitic domes, pre-volcanic unit (lava and pyroclastics), diorite, μ-diorite, andesite subvolcanic intrusion, and tonalitic intrusion at the depth (Fig 2). The lithological unit in surface is seen as interfingering of lava and pyroclastics; in lower elevation such as rivers or valleys, the dioritic rock can be seen. Tonalitic intrusion or quartz diorite was founded by deep drilling at >100 m from the surface. The top of tonalitic intrusive rock only occurs in elevation 200 m above sea level (m.asl) to the depth that's characterized pencil-like intrusive caused by the width of the intrusive body about 100-150 m narrow to depth. Otherwise, It's can be similar types Onto deposits but different in size that's as a big intrusive body width of about 1-1.3 km and still bigger to depth.

The Humpa Leu prospect is the clear porphyry type prospect in this district with more deep drilling. The Humpa Leu (East) prospect has 257 Mt of resources potential with a grade at 0.2% Cu and 0.2 g/t Au [4]. Hydrothermal alteration evolved from tonalite body to outward, consist of biotite-feldspar-sericite-magnetite-chalcopyrite-bornite±digenite (potassic) and epidote-actinolite-chlorite-magnetite-chalcopyrite (inner propyllitic) and overprinted by sericite-illite-kaolinite-chalcopyrite-pyrite±sphalerite (phyllic) and alunite-dickite-pyrite-chalcopyrite (advanced argillic). The mineralization is controlled by disseminating chalcopyrites and other copper-sulfides, quartz–anhydrite veins, and hydrothermal breccia. The texture of mineralization is associated with quartz veins stockwork, parallel pseudobands, and cement in hydrothermal breccia. In this research, we try to characterize the cement of hydrothermal as the mineralization trigger in this system

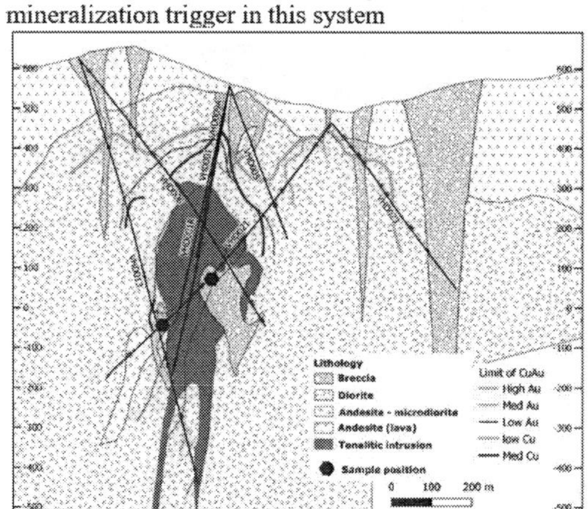

Fig 2. Interpretation of east-west section geological model of Humpa Leu East and the sample described in this study.

https://doi.org/10.1051/e3sconf/202132504008

3.2 Hydrothermal Breccia

Several phases of hydrothermal occur in this system, including the hydrothermal breccia phase associated with complex fluids. The samples of hydrothermal breccia were taken from about 500 m from the surface or at -100 to 100 m above sea level. The hydrothermal breccia mineralogy is associated with potassic type mineralogy, commonly consisting of feldspar, quartz, biotite, and magnetite. In addition, HLE hydrothermal breccia characters are encountering high-temperature hydrous silicates, retrograde silicates, and sulfate minerals.

The hydrothermal cement of Humpa Leu East porphyry at least has three phases of mineralogical assemblages and possibly influencing the mineralization. A mineralogical assemblage of hydrothermal cement in HLE consists of quartz-feldspar-plagioclase-biotite-magnetite as high-temperature phase; then followed by epidote-sericite-chlorites-anhydrite-carbonates in medium temperature, and there are aluminum-rich clay minerals interpreted as gibbsite. All three phases are associated with copper sulfide that is chalcopyrite-bornite, chalcopyrite-magnetite, and chalcopyrite-sphalerites. Those minerals can describe clearly in petrographic analysis and reinforcement by elemental mapping analysis and quantitative mineralogical analysis from µ--XRF scanned data (Fig 3).

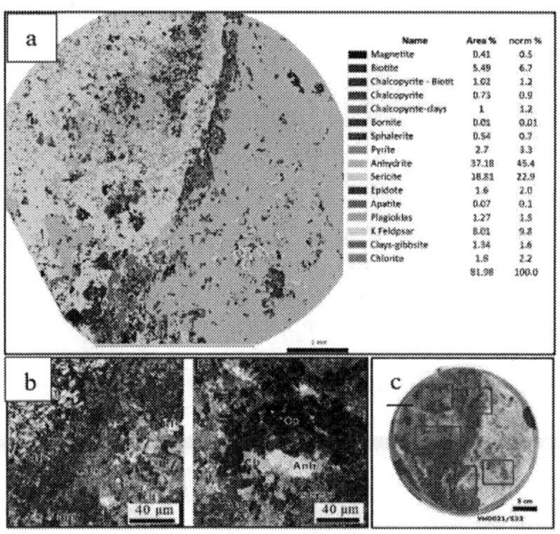

Fig 3. Mineralogical analysis of cement hydrothermal of sample VHD021/523. (a) A quantitative mineralogy from µ--XRF data shows two dominant variation consist of calcium sulfate group in right side, silicate groups in left side, and also clays in bottom (b) petrographic analysis shows cement mineral, consist of quartz (qz), K Feldspar (kfs), plagioclase (pl), sericite (ser), anhydrite (anh), epidote (ep), carbonate (cb), chlorite (chl), clays (cl), an opaque (op), (d) polish section for µ--XRF analysis and box position of detailed µ--XRF analysis.

3.3 Mineralogical phase

The observed hydrothermal breccias showed the presence of two main components in the hydrothermal solution or fluids. The initial component is siliceous fluids that produce silicate mineral groups such as quartz-feldspar-biotite and the dissemination of copper minerals such as chalcopyrite and bornite. The second component is calcium-rich sulfate and less carbonate, which produces the minerals, anhydrite and a little carbonate, and gypsum. The mineralization formed due to contact of the solution phases shows the oriented distribution of Fe-Mg-rich silicates minerals such as biotite-phlogopite-chlorite and the formation of chalcopyrite-magnetite and sphalerite mineralization. This contact phase shows the existence of an intermediate sulfidation state mechanism in the porphyry system.

3.3.1 Chalcopyrite-magnetite

Chalcopyrite is a mineral present from the late-magmatic phase, early phase, to late or epithermal phase in the Humpa Leu porphyry system. Chalcopyrite shows chemical formula $Cu_{1-1.2}Fe_{1-3.6}S_{1-1.6}$, and in the some case associated with bornite with formula $Cu_{3.6-6}Fe_{1-8.8}S_{1-3.6}$ (some detected as mix formula with chalcopyrite). The iron oxide minerals occur as hematite-magnetite-pyrrhotite, commonly occurring in the early phase as a massive veinlet and disseminates associated with quartz vein, followed by chalcopyrite. In hydrothermal breccia, magnetite occurs as disseminated and accumulate mineral with chalcopyrite and some pyrite. The magnetite extensively grows in the silicates phase with other sulfides, less abundant in the anhydrite phase. The silicates gathered as cement include quartz, sericite-feldspar, biotite, epidote, chlorite, and clays (Fig 4).

Meanwhile, copper (Cu) disseminates strongly in the silicates phase due to chalcopyrites and possibly of bornite minerals. The analysis of ferrous oxide mineral shows a variation of Fe element with formula $Fe_{0.8-1.6}O_{0.8-3}$ as a description of presence iron-hematite-magnetite, but still dominantly by magnetite type. The magnetite contains several anomalies of the element, such as 0.06 – 0.25% Cu, 0.03-0.57 % Zn that occurs if sulfur presents.

The silicate cement in the siliceous phase occurs to envelop the metal minerals in irregular to pseudo bands pattern, in sharp contact with the anhydrite phase. From elemental mapping analysis, the copper (Cu) occurs in associated with feldspar-sericite minerals; also the magnetite but more closely associated with biotite-quartz. Moreover, it all showed high aluminum and potassium anomaly in spot analysis due to copper target.

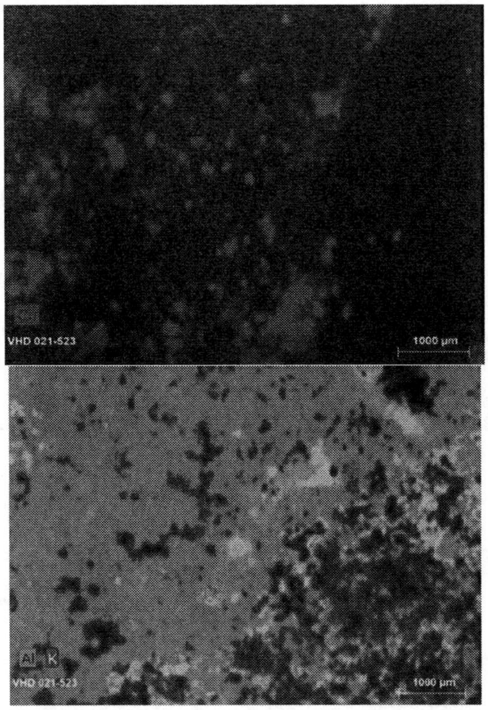

Fig. 4. The siliceous phase of cement breccia from sample VHD021/523 *(at position 3 of Fig 3c),* shows the distribution of copper minerals and related to Aluminum-potassium minerals cement. (picture analysis at position 4 of Fig 3c),

3.3.2 Bornite-chalcopyrite

As has been explained in the previous paragraph, bornite may be present simultaneously as chalcopyrite. The presence of bornite in the HLE porphyry system is difficult to find because it has a very fine size grain (<20 µ-ns). In hydrothermal cement, bornites disseminated associated with chalcopyrite and surrounded by silicate minerals, namely biotite-chlorite and sericite (Fig 5). As shown in Figure 3, it can be seen that the bornite is only present 0.01% in the sample; it shows that the hydrothermal breccia is part of the transition from high to intermediate sulfidation stage. It is strengthened by occurrences of retrograde minerals caused by a decrease in temperature, such as chlorite-sericite-clays.

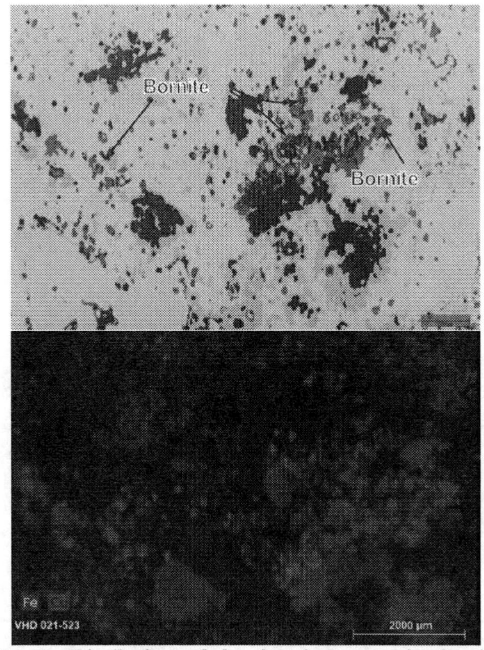

Fig 5. Distribution of bornite that's associated with chacopyrite and feldspar. Table of point analysis shows object 1 as bornite with empirical formula as $Cu_{3.6}FeS_2$, and the other was mix with feldspar minerals. (picture analysis at position 5 of Fig 3c)

3.3.3 Chalcopyrite-sphalerite

One of the characteristics of porphyry mineralization in HLE is the occurrence of sphalerite, together with chalcopyrite in the porphyry phase and especially in the epithermal phase. In the hydrothermal breccias, sphalerite is associated with chalcopyrite, mainly at the contact margin between the siliceous and calcium sulfate stage. Sphalerite occurs associates with the distribution pattern of chalcopyrite-biotite-chlorite (fig 6) and chalcopyrite-magnetite (Fig 7). The occurrence of sphalerites indicates intermediate sulfidation state condition during the hydrothermal process, influence the grade of gold (Au) in several phases.

https://doi.org/10.1051/e3sconf/202132504008

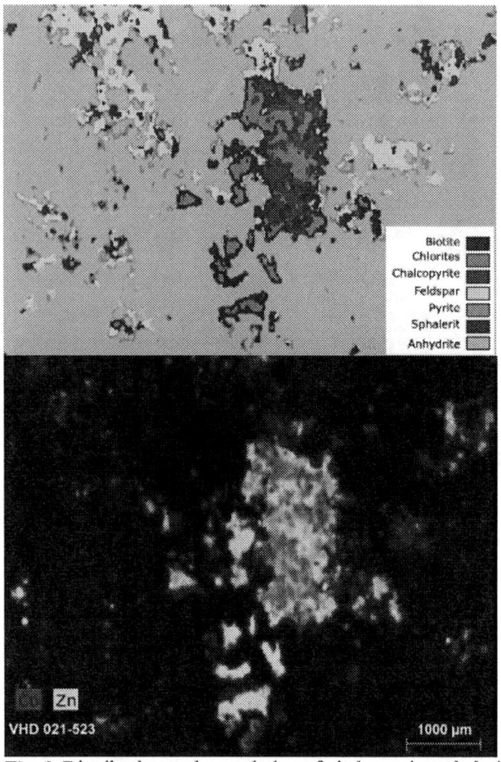

Fig 6. Distribution and association of chalcopyrite-sphalerite that's associated with biotite-chlorite-feldspar (picture analysis at position 2 of Fig 3c)

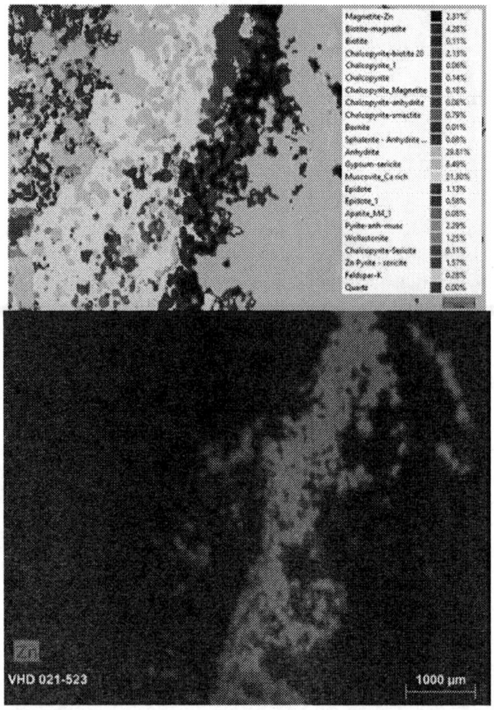

Fig 7. Distribution and association of chalcopyrite-magnetite-sphalerite that's associated with biotite-chlorite-feldspar-sericite-epidote-anhydrite in contact of siliceous phase and sulfate phase (picture analysis at position 1 of Fig 3c)

4 Interpretation & Discussion

The condition of Hydrothermal breccia shows the complexity of the temperature formed in the porphyry system in HLE. In this one event, the change conditions prove an immediate process of mixing meteoric and magmatic water and causing mineralization by the redox system process. The initial process is represented by the presence of chalcopyrite-magnetite or hematite, followed by chalcopyrite-bornite and then chalcopyrite-sphalerite. The occurrences of chalcopyrites at all phase indicate the stability of intermediate sulfidation proses in Humpa Leu and as a possible factor to answer the abundant copper in the Hu'u hydrothermal fluid system.

This study is a part of PhD study results of the first author at the Geological Engineering Department at Universitas Gadjah Mada. We thank to all geologists of PT. Sumbawa Timur Mining, Mr. Yulindra, Mr. Kristison, Mr. Nurtyas, Mr. Nurhadi wibowo and Mr. Doughlas. Special thanks to the management of Institut Teknologi Nasional Yogyakarta for the powerful support to us.

References

1. D. R. Burrows, M. Rennison, D. Burt, and R. Davies, "The Onto Cu-Au Discovery, Eastern Sumbawa, Indonesia: A Large, Middle Pleistocene Lithocap-Hosted High-Sulfidation Covellite-Pyrite Porphyry Deposit," Econ. Geol., vol. 115, no. 7, pp. 1385–1412, (2020)
2. J. C. Carlile and A. H. G. Mitchell, "Magmatic arcs and associated gold and copper mineralization in Indonesia," J. Geochemical Explor., vol. 50, no. 1–3, pp. 91–142, (1994).
3. A. Maryono, R. L. Harrison, D. R. Cooke, I. Rompo, and T. G. Hoschke, "Tectonics and Geology of Porphyry Cu-Au Deposits along the Eastern Sunda Magmatic Arc, Indonesia," Econ. Geol., vol. 113, no. 1, pp. 7–38, Jan. (2018),
4. PT. Sumbawa Timur Mining, "Paparan geologi, panas bumi dan geohidrologi - Presentasi Project Hu'u," (unpublished report).
5. A. Sudradjat, A. Mangga, and N. Suwarna, "Peta Geologi Regional Lembar Sumbawa, Nusa Tenggara skala 1:250.000." Pusat Penelitian dan Pengembangan Geologi, Bandung, (1998).
6. H. Sundhoro et al., "Survei Panas Bumi Terpadu (Geologi, Geokimia dan Geofisika) Daerah Hu'u, Kabupaten Dompu, Provinsi Nusa Tenggara Barat," in Proceedings Kolukium Hasil Lapangan (2005)
7. E. A. Agorhom, Z. Swierczek, W. Skinner, and M. Zanin, "Combined QXRD-QEMSCAN mineralogical analysis of a porphyry copper-gold ore for the optimization of the flotation strategy," XXVI Int. Miner. Process. Congress IMPC, 2012)

8. B. Xu, G. Kou, B. Etschmann, D. Liu, and J. Brugger, "Spectroscopic, raman, empa, micro-xrf and micro-xanes analyses of sulphur concentration and oxidation state of natural apatite crystals," Crystals, vol. 10, no. 11 (2020)
9. L. Germinario, R. Cossio, L. Maritan, A. Borghi, and C. Mazzoli, "Textural and Mineralogical Analysis of Volcanic Rocks by μ-XRF Mapping," Microsc. Microanal., vol. 22 (2016).
10. M. Haschke, U. Rossek, R. Tagle, and U. Waldschläger, "Fast Elemental Mapping with Micro-xrf," (2012).

Analysis of engineering geology conditions based on the quality of rock mass in the Tiga Dihaji Dam's diversion tunnel, South Sumatera, Indonesia

Zekrinaldi[1], *Ferian* Anggara[1*], and *Hendy* Setiawan[1]

[1]Gadjah Mada University, Sleman, Yogyakarta, Indonesia

Abstract. This research has examined the rock mass quality case study in the Tiga Dihaji Dam's diversion tunnel. Observations of geological conditions were carried out on the surface and subsurface of the study site and show that the study area consists of tuffaceous sandstone and carbonate interbeds. The method of this study is based on the classification of the Geological Strength Index (GSI), Rock Mass Rating (RMR), and the Q-system. The results indicate that tuffaceous sandstone has a GSI value of 15 - 87.5 (very poor - very good), RMR 48 - 82 (fair - very good), and Q-system 0.01 – 60.0 (exceptionally poor - very good). Meanwhile, carbonate interbeds have a low value, with a GSI value of 10.5 - 77.5 (very poor to very good), RMR 17.0 – 56.0 (very poor - fair), and Q-system 0 - 35.2 (exceptionally poor - good). Moreover, a correlation was made between rock mass quality for conditions in the study area. The correlation between GSI and RMR was obtained by the equation GSI = 2.2885RMR - 82.567 (R^2 = 0.6653), RMR and Q-system RMR = 2.0175ln(Q) + 63.061 (R^2 = 0.4987), and GSI and Q-system GSI = 7.2119ln(Q) - 54.578 (R^2 = 0.8095).

1 Introduction

The condition of the Selabung river during the rainy season has a large discharge. It even often becomes a problem both along the river channel and the surrounding areas. Meanwhile, in the dry season, the river flow has a slight discharge. The surrounding area becomes dry and causes water shortages in the agricultural area. For this reason, efforts were made to overcome this condition by constructing the Tiga Dihaji dam, which can be used as a reservoir for water during the rainy season.

The Tiga Dihaji Dam is located in the Selabung River, South Ogan Komering Ulu, South Sumatera, Indonesia (Figure 1). In constructing the Tiga Dihaji Dam, a diversion tunnel is needed to divert the flow of river water during the construction process. Based on the previous study [1], the diversion tunnel is on the right side of the river with approximately 595 m and a circular shape with a 7 m diameter. The elevation of the inlet tunnel is at +221.0 m above sea level, while the outlet tunnel is located at +218.41 m above sea level.

Appropriate techniques and technologies are required in the design and construction of a tunnel [2]. Geological detailed and engineering geology investigations on the surface and subsurface were carried out in the construction of this tunnel to determine the rock mass classification. The rock mass classification is an empirical method to identify the parameters that influence the behavior of these rocks and divides them into groups based on their characteristics [3]. The rock mass classifications analyzed in this study are Geological Strength Index [4], Rock Mass Rating [3], and Q-system [5].

Many previous researchers have published the correlation between the quality of rock mass (Table 1). However, none of the correlations can be used in general at specific locations due to the difference in geological conditions. So, it is necessary to correlate the quality of the rock mass, specifically at the study area.

Table 1. Correlation of rock mass quality by previous researchers.

Authors	Empirical Correlation	Eq
Hoek and Brown (1997) [6]	$GSI = RMR - 5$	(1)
Ali et al. (2014) [7]	$GSI = 0.9932RMR - 4.913$	(2)
Zhang et al. (2019) [8]	$RMR = 0.827GSI + 15.394$	(3)
Bieniawski (1976) [9]	$RMR = 9\ln(Q) + 44$	(4)
Cameron-Clarke and Budavari (1981) [10]	$RMR = 5\ln(Q) + 60.8$	(5)
Vásárhelyi and Deák (2010) [11]	$RMR = 7.10\,(Q) + 68$	(6)
Laderian and Abaspoor (2012) [12]	$RMR = 8.15\ln(Q) + 44.88$	(7)

* Corresponding author: ferian@ugm.ac.id

 https://doi.org/10.1051/e3sconf/202132505001

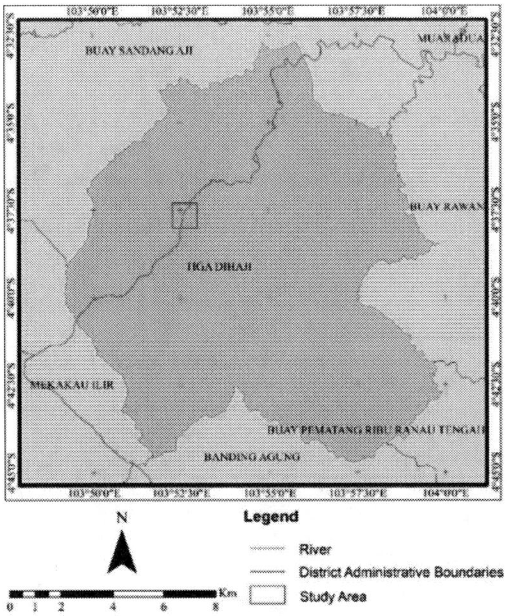

Fig. 1. Location map of the study area.

2 Methodology

2.1 Geological conditions

Observations of geological conditions were carried out on the surface and subsurface of the study site. Observations of surface geological conditions covering lithological conditions and geological structures and the results can be used as a reference for subsurface conditions. Meanwhile, observations under subsurface geological conditions are carried out on the drilling results in lithology and parameters in calculating rock mass quality.

In the study area, the rock mass quality analysis used subsurface data from the drilling results as many as three sample points, namely DIV-1 (tunnel inlet) with a depth of 65 m, DIV-2 (center of the tunnel) with a depth of 60 m, and DIV-3 (tunnel outlet) a depth of 60 m.

2.2 RQD

Deere initiated Rock Quality Designation (RQD) in 1963 [13]. RQD is a simple method for predicting rock mass quality obtained from drilling results and is considered rock mass. The RQD value is calculated by adding up the total length of the core pieces, which is more or equal to 10 cm to the total length of the core drilled (Figure 2). A good RQD value indicates the intact rock mass. The RQD value is used in determining the classification of other rock masses.

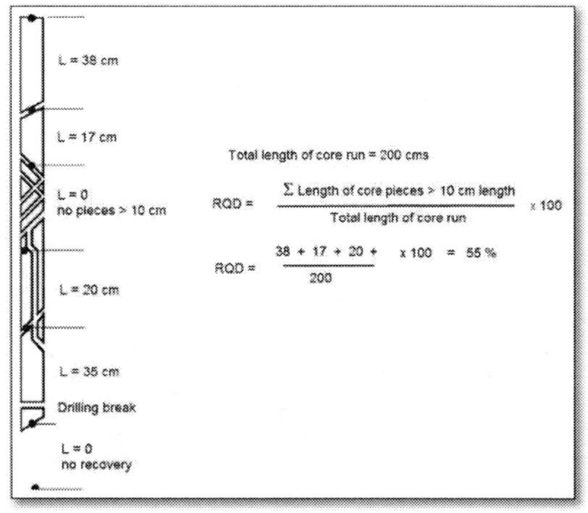

Fig. 2. Calculation of RQD value.

2.3 GSI

The Geological Strength Index (GSI) is obtained from the joint condition value and the RQD value. The value of the joint condition refers to the classification of the joint condition [3]. The GSI calculation uses the following Equation 8 [4].

$$GSI = 1.5Jcond + {}^{RQD}/_{2} \tag{8}$$

Where:

GSI = Geological Strength Index
Jcond = Joint Condition
RQD = Rock Quality Designation

2.4 RMR

Rock Mass Rating (RMR) is a classification of rock mass using an empirical method in determining the assessment, which is used to evaluate the resistance of rock mass. Bieniawski classified rock mass from this method by summing up six parameters [3]. These parameters are rock strength, drill core quality, spacing of joints, conditions of joints, groundwater conditions, and discontinuity orientation. These parameters are obtained from subsurface observation data and laboratory results.

2.5 Q-system

The Rock Mass Quality (Q) System, also known as The Tunneling Quality Index, was first proposed Norwegian Geotechnical Institute (NGI). The Q-system is one of the classifications of rock masses based on a case study of more than 200 tunnels and caverns, mainly in Norway and Finland [5].

The Q-system is a function of the six parameters, which are stated in Equation 9.

$$Q = \frac{RQD}{Jn} x \frac{Jr}{Ja} x \frac{Jw}{SRF} \tag{9}$$

Where:

RQD = Rock Quality Designation

Jn = Joint set number
Jr = Joint roughness number
Ja = Joint alteration number
Jw = Joint water reduction factor
SRF = Stress Reduction Factor

2.6 Correlation between GSI, RMR, dan Q-system

The results of the calculated rock mass quality are correlated through statistical analysis and regression equations. For the correlation representative, the coefficient of determination (R^2) is used to analyze the results obtained. The value of the coefficient of determination is in the range of 0 (zero) and 1 (one). If the coefficient of determination is close to 0 (zero), it means that the model's ability to explain the dependent variable is minimal. Meanwhile, suppose the value of the coefficient of determination of the variable is close to 1 (one). In that case, it means that the model's ability to explain the dependent variable is getting stronger.

3 Results and discussion

3.1 Geological conditions

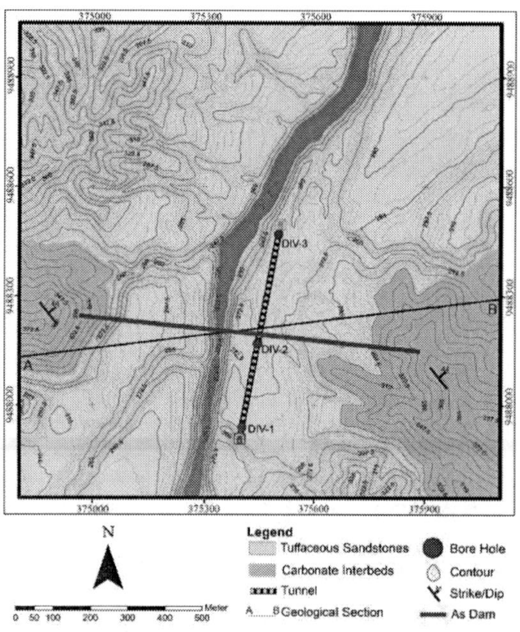

Fig. 3. Geological map of the study area.

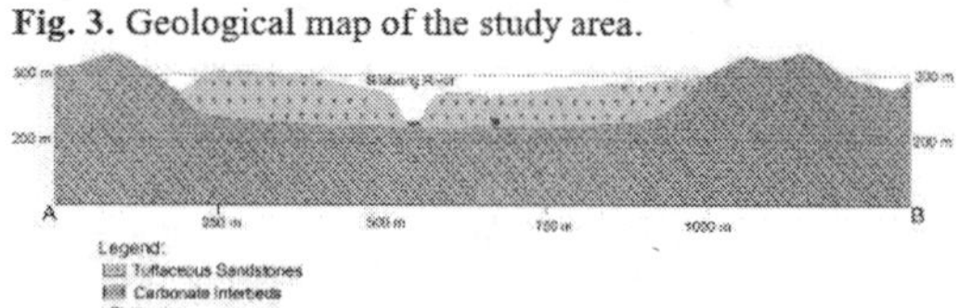

Fig. 4. Geological sections A – B.

According to the regional geology map of Indonesia on the Baturaja sheet [14], the regional geology of the study area is in the Ranau formation and the Gumai formation. The geological investigations show the tunnel is in the tuffaceous sandstone and carbonate interbeds, as shown in Figure 3 and Figure 4. In general, the tuffaceous sandstone is light grey to dark grey in color grain size of fine sand - medium sand, massive sedimentary structure, moderate sorting, packed closed, grain shape subrounded - subangular, low-level weathering, the minerals that make up the rock are still visible such as quartz, plagioclase, biotite, hornblende, and ash-sized materials, and weight of $\gamma = 1.4 - 2.2$ gr/m^3, triaxial c = 11.8 – 68.8 kg/cm^2 ϕ = 34.8^0 – 58.2^0, uniaxial compressive strength $\sigma_c = 94.3 - 306.3$ kg/cm^2.

Meanwhile, carbonate interbeds consist of carbonate sandstone, carbonate siltstone, and carbonate claystone with 3 - 30 cm of thickness and slope to the northeast (N308°E/44°). This thin layer causes this rock to become brittle and also weak between the layering areas.

The core data analysis results show that the tunnel is between tuffaceous sandstone and carbonate interbeds (Figure 5).

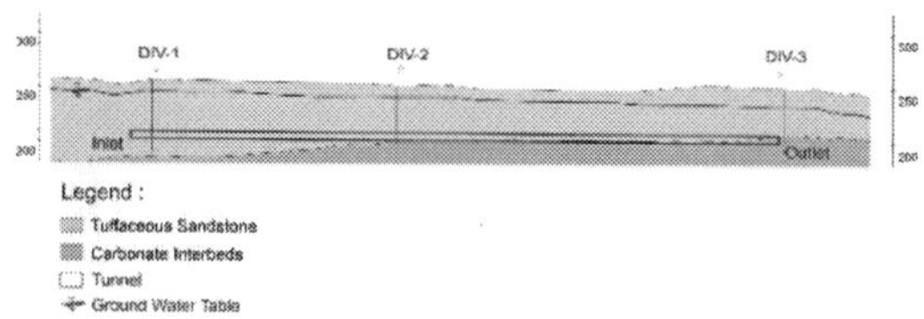

Fig. 5. Geological sections in the tunnel.

3.2 RQD

The RQD value of tuffaceous sandstone (Figure 6) obtained from this analysis ranges from 0 - 100 or 76.35 average value, while the RQD of carbonate interbeds is between 0 - 80 or with an average value of 15.37 (very poor quality). The average RQD value of the carbonate interbeds (Figure 7) is very bad due to its thin layers, which causes many cores to be lost during drilling activities.

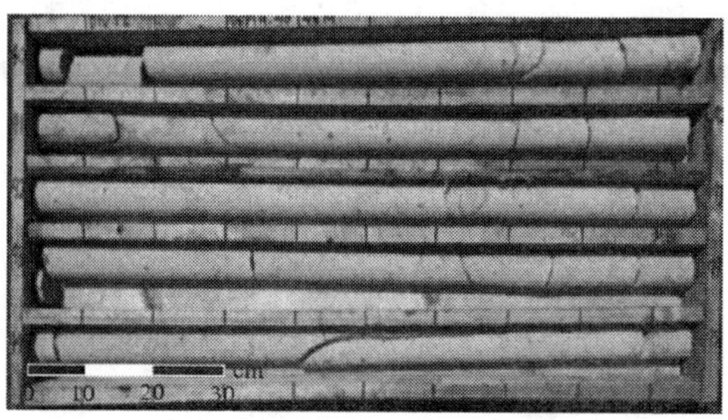

Fig. 6. Core example at DIV-1 (depth 40 – 45 m).

3.3 GSI

In the study area, the value of joint condition for tuffaceous sandstone has an average value of 25. The joint surface is quite rough with cracks <1 mm, and the rocks are high to slightly weathered. The GSI value for tuffaceous sandstone is between 15 - 87.5 or with an average value of 74.50 (good quality). Meanwhile, the joint condition of the carbonate interbeds has an average value of 7, interpreted from surface geology due to the cores being destroyed (lost). The GSI values ranging from 10.5 to 77.5, or the average is 26.16 (poor quality).

Fig. 7. Core example at DIV-2 (depth 55 – 60 m).

3.4 RMR

The RMR value for tuffaceous sandstone is 48.0 – 82.0, or an average of 68.63 (good quality), with UCS value rating is 2, RQD rating is 17, spacing of discontinuities rating is 20, condition of discontinuities rating is 25, and groundwater condition is 4. Meanwhile, the RMR value of carbonate interbeds is 17.0 – 56.0 or an average of 26.22 (poor quality), with UCS value rating is 2, RQD rating is 10, spacing of discontinuities rating is 5, condition of discontinuities rating is 10, and groundwater condition is 4.

3.5 Q-system

The analysis indicates that tuffaceous sandstone has a Q value between 0.01 – 60.0 or an average value of 32.04 (good quality), while carbonate interbeds have a Q value between 0.01 - 35.2 an average of 6.77 (medium quality).

As a comparison, the rock mass quality of each classification is attached in Table 2.

Table 2. Summary of rock mass quality.

Drill point	Lithology Unit	Average Value / Quality		
		GSI	RMR	Q-system
DIV-1	Tuffaceous sandstone	78.98 Very Good	69.46 Good	34.263 Good
DIV-2	Tuffaceous sandstone	75.59 Good	68.81 Good	32.484 Good
	Carbonate interbeds	10.50 Very Poor	17.00 Very Poor	0.01 Exceptionally Poor
DIV-3	Tuffaceous sandstone	66.74 Good	67.69 Good	28.306 Good
	Carbonate interbeds	36.97 Poor	32.56 Poor	11.413 Good

3.6 Correlation between GSI, RMR, dan Q-system

The quality of the rock mass that is correlated is 153 points on the tuffaceous sandstone. There is no correlation for carbonate interbeds due to insufficient data and poor quality.

The correlation between GSI and RMR is shown in Figure 8 by comparing the results between the study areas with the Hoek and Brown (1997) [6], Ali et al. (2014) [7], and Zhang et al. (2019) [8]. Correlation is linear, where the correlation in the study area has a more vertical line. The relationship between GSI and RMR in the study area is stated in Equation 10 with a coefficient of determination of 0.6653.

$$GSI = 2.2885RMR - 82.567 \ (R^2 = 0.6653) \quad (10)$$

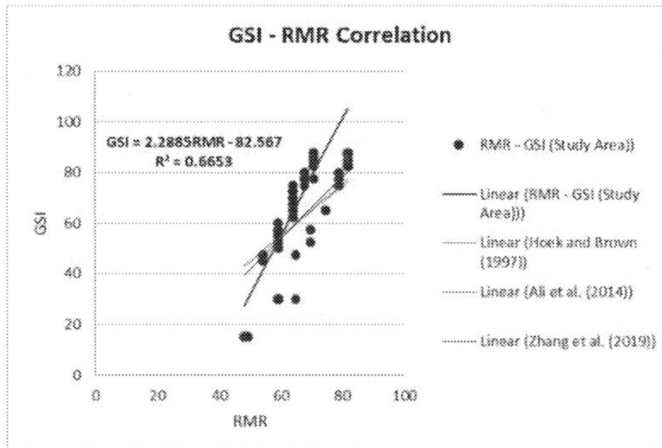

Fig. 8. GSI - RMR correlation graph.

The relationship between RMR and Q-system (Figure 9) shows a line of correlation the results in the study area with Bieniawski (1976) [9], Cameron-Clarke and Budavari (1981) [10], Vásárhelyi and Deák (2010) [11], and Laderian and Abaspoor (2012) [12]. The correlation line is in the form of the logarithm, with a coefficient of determination of 0.4987. The correlation between RMR and Q-system in the study location is stated in Equation 11.

$$RMR = 2.0175ln(Q) + 63.061 \ (R^2 = 0.4987) \quad (11)$$

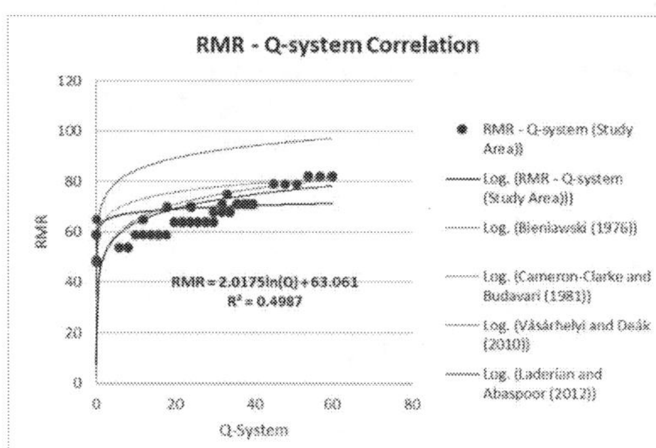

Fig. 9. RMR - Q-system correlation graph.

The relationship between GSI and Q-system is shown in the graph in Figure 10. Correlation is logarithmic, with a coefficient of determination of

0.8095. The relationship between GSI and Q-system in the study area is stated in Equation 12 below.

$$GSI = 7.2119 ln(Q) - 54.578 \ (R^2 = 0.8095) \quad (12)$$

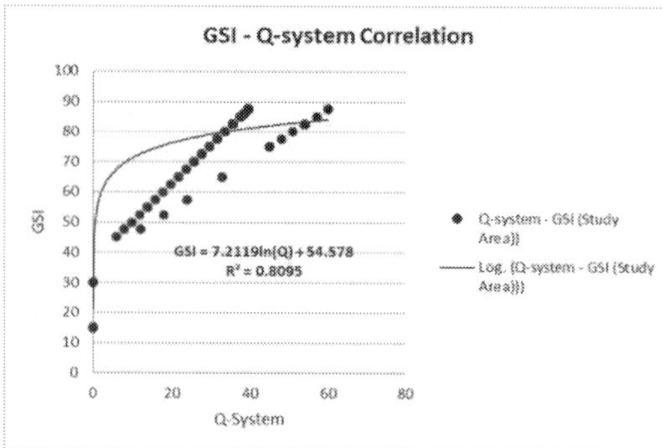

Fig. 10. GSI – Q-system correlation graph.

4 Conclusion and recommendation

The research area is in the Tiga Dihaji Dam's diversion tunnel, with a tunnel length of 595 m. The geological conditions in the research area consist of tuffaceous sandstone and carbonate interbeds. The results of the rock mass quality analysis show that the value of tuffaceous sandstone had a GSI value of 15 - 87.5 (very poor - very good), RMR 48 - 82 (fair - very good), and Q-system 0.01 – 60.0 (exceptionally poor - very good). Meanwhile, carbonate interbeds have a low value, with a GSI value of 10.5 - 77.5 (very poor to very good), RMR 17.0 – 56.0 (very poor - fair), and Q-system 0.01 - 35.2 (exceptionally poor - good). From these results, the values of rock mass strength, especially tuffaceous sandstone, were correlated in this study, namely GSI - RMR, RMR - Q-system, and GSI - Q-system. The correlation between GSI and RMR is obtained by the equation GSI = 2.2885RMR - 82.567 (R^2 = 0.6653), RMR and Q-system RMR = 2.0175ln(Q) + 63.061 (R^2 = 0.4987), and GSI and Q-system GSI = 7.2119ln(Q) - 54.578 (R^2 = 0.8095). This equation will be more accurate if more sample points are analyzed.

It is recommended that further studies be undertaken in the following area: the stability of the tunnel by using a numerical method with the input of reinforcement parameters based on the rock mass that has been obtained.

Thanks to the River Basin Organization (BBWS in Indonesian) in Sumatera VIII for allowing the study at the Tiga Haji Dam, and Mr. I Gde Budi Indrawan for guidance during data collection in the study area.

References

1. PT Virama Karya, *Detail engineering design Bendungan Tiga Dihaji*, (PT Virama Karya, Palembang, 2016)
2. M. Sharifzadeh, F. Kolivand, M. Ghorbani, S. Yasrobi, Tunneling & Underground Space Tech. **35**, 178–188 (2013)
3. Z. T. Bieniawski, Engineering rock mass classifications: a complete manual for engineers and geologists in mining, civil and petroleum engineering, (A Wiley-Interscience Publication, New York, 1989)
4. E. Hoek, T. G. Carter, M. S. Diederichs, *Quantification of the geological strength index chart*, in Proceedings of the 47th US Rock Mechanics/Geomechanics, Symposium San Francisco, 23-26 June 2013, USA (2013)
5. N. Barton, R. Lien, J. Lunde, Springer-Verlag **6**, 4:189–236 (1974)
6. E. Hoek, E. T. Brown, Inter. J. of Rock Mechanics & Mining Sci. **34**, 8:1165–1186 (1997)
7. W. Ali, N. Muhammad, M. Tahir, Inter. J. of Sci. Eng. & Tech. **3**, 1292–1296 (2014)
8. Q. Zhang, X. Huang, H. Zhu, J. Lia, Tunnelling and Underground Space Tech. **83**, 73–81 (2019)
9. Z. T. Bieniawski, *Rock mass classification in rock engineering*, in Proceedings Symposium Exploration Rock Engineering, Johannesburg, 97-106 (1976)
10. I. S. Cameron-Clarke, S. Budavari, Engineering Geology **17**, 19–53 (1981)
11. B. Vásárhelyi, F. Deák, *Comparing the different Rock Mass Classifications on the Access Tunnels of Bátaapáti, Hungary*, in Conference: Geologically Active, 3773–3780 (2010)
12. A. Laderian, M. A. Abaspoor, Tunneling and underground space tech. **27**, 149–158 (2012)
13. D. U. Deere, Rock Mechanics and Eng. Geology **1**,16–22 (1963)
14. S. Gafoer, T. C. Amin, R. Pardede, Peta geologi lembar Baturaja, Sumatera (Pusat Pengembangan dan Penelitian Geologi, Bandung, 1993)

Identification of precipitation diurnal process using PERSIANN over Central Java Area

Muhammad Fianggoro[1], *Emilya* Nurjani[1], and *Andung Bayu* Sekaranom[1,2*]

[1]Department of Environmental Geography, Faculty of Geography, Universitas Gadjah Mada, Yogyakarta-Indonesia
[2]Research Center for Disaster, Universitas Gadjah Mada, Yogyakarta-Indonesia

Abstract. The diurnal cycle is a process that repeats daily due to earth rotation. This cycle could significantly affect the hydrological conditions and lithosphere conditions, including humans. This diurnal cycle is also likely to influence seasonal patterns in the region. The diurnal cycle has a close relationship with the time-series pattern and distribution of rainfall. It often affects hydrometeorological disasters in the form of floods and droughts. This paper provides an overview of early identification related to the diurnal cycle of rain in the Central Java region - Indonesia. Rainfall estimation data from The Precipitation Estimation from Remotely Sensed Information using Artificial Neural Networks (PERSIANN) is used as the primary input in the analysis. The research results show that there are differences in rainfall patterns between the northern, central, and southern parts of the study area

1 Introduction

The diurnal cycle is a pattern that repeats itself every day for twenty-four hours [1]. This pattern is caused by the rotation of the earth on its rotating axis. The diurnal cycle is one of the most basic forms of meteorological patterns. The diurnal cycle is also considered a cycle on a local scale or a regional scale with a daily period [2]. The diurnal cycle is a daily cycle closely related to local climatic conditions. It has a mutual influence on climate phenomena that occur in an area, especially in the form of an archipelago [3].Temperature, humidity, and wind are the main drivers of the diurnal cycle, affecting rainfall. One of the parameters that are often influenced the diurnal cycles is wind, which includes land and sea breezes. These two types of wind have different characteristics and directions. In addition, characteristics of the nearby ocean potentially affect the amount of cloud formation and rainfall that propagate overland [4]. Although the rainfall pattern that occurs in the long term might be similar, it is possible to have different diurnal patterns. The distance between land and sea will also have implications for the distribution of rain and the characteristics of the diurnal cycle repetition.

Another parameter that is considered to affect the diurnal cycle is topography, where the shape or relief of the earth's surface also plays a role in regulating rainfall distribution [5]. The other variables that are considered significant are elevation and land use. The height of the surface of heterogeneous elevation for the central part of Java is allegedly contributing to the distribution of rainfall and influencing the diurnal cycle that occurs in the most central region of Java. The elevation parameter also has a direct relationship with temperature. Further, the direction to the slope has a relationship with meteorological parameters because it is related to humidity transport as carried by the wind. Apart from elevation, land use can directly or indirectly affect the diurnal cycle, mainly due to the urban heat island. The degree of fluctuation in land-use change dynamics can affect hydrological variables [6], which indirectly affects the repetition pattern of the diurnal cycle.

Central Java Region consists of Central Java and Yogyakarta Province, is flanked by two types of the sea with different characteristics, and provides a different tendency to study. The condition of the area, which has a dense population, encourages the need for precipitation-related analysis, particularly related to the disaster. The precipitation is closely related to the hydrological cycle, which is the main foundation for determining the potential of water resources and disasters. precipitation can affect the hydrological process over the land surface [7]. In this case, the diurnal cycle is an essential factor representing general meteorological conditions that recur in the two provinces.The diurnal cycle potentially influences the rainfall that occurs in Java, especially in the Central Java Region. At present, Central Java is mostly affected by hidrometeorological disasters, particularly floods. Diurnal cycle therefore is related to precipitation system that significantly affect the hydrological conditions and lithosphere conditions, including the local population. However, the understanding of the effects of this diurnal cycle is still not fully understood. There are a gap in the diurnal rainfall related research because most of the rainfall measurement is based on daily data.

* Corresponding author: andungbayu@geo.ugm.ac.id

 https://doi.org/10.1051/e3sconf/202132505002

This paper contributes to the gap by using satellite estimation data to understand the process in detail. This diurnal cycle also has a significant possibility and influences the existing seasonal patterns in the region. Therefore, this paper aims to determine the diurnal cycle of rainfall in the Central Java region that particularly have implication to flooding.

2 Data and method

The diurnal pattern of rainfall can be measured by processing precipitation estimation data obtained from satellites orbiting above the earth's atmosphere [8,9]. This measurement provides rainfall estimation products in the area covered by the satellite's sensors. The satellite data used in this paper is data from the Center for Hydrometeorology and Remote Sensing (CHRS) at the University of California, Irvine (UCI). The product is Precipitation Estimation from Remotely Sensed Information Using Neural Networks (PERSIANN). In more detail, the product used is the PERSIANN-Cloud Classification System (PERSIAN-CCS) which has a spatial resolution of 0.04 ° x 0.04 ° which, when converted in kilometers, are about 4 km x 4 km over the tropics.

PERSIANN was developed with an artificial neural network to estimate the rain rate from Infrared (Infrared/IR) image data [10], with adjustment of network parameters based on the derived microwaves and ground measurement data. This product is able to estimate the level of precipitation around the world with coverage of 60°S to 60°N. PERSIANN-CCS is the rainfall data estimate satellite output from CHRS with the highest resolution currently available and has been supported by an artificial neural network system that can process information more quickly and accurately [11,12]. The data available on the data access portal states that the PERSIAN-CCS rain data is available from January 2003 until the present. Rainfall data is available in the formation of different time intervals according to the needs of the analysis from hourly to daily.

The PERSIANN data was chosen because it has a reasonably detailed spatial resolution and has a temporal resolution of up to hours. Researchers and experts in Indonesia quite often use PERSIANN data as an instrument to estimate the value of rainfall in an area. Several studies using PERSIAN-CCS rainfall data include research conducted by [11] regarding the trend of extreme daily rainfall in the Bangka Belitung Islands. Another study conducted by Pratiwi et al. [13] used PERSIAN-CSS data to predict observational rain data using Cross-Correlation. Comparison of the PERSIANN-CCS data with surface measuring instrument data is known to produce minor differences. Thus, it is sufficient to represent the rainfall on the surface (14).

The diurnal cycle analysis of rainfall in this paper mainly uses the PERSIAN-CCS product. The data is downloaded on a predetermined temporal scale, namely in 2015 to 2019. The five years from 2015 to 2019 reflect the many global climate phenomena that affect the study area. Some examples of phenomena that influence the study area are El-Nino and La-Nina. The El-Nino phenomenon significantly affects land conditions on the island of Java. This climatic phenomenon often causes drought in Java and several other areas directly adjacent to the Indian Ocean and the Pacific Ocean. Another interesting phenomenon in the research data period is tropical cyclones, which occurred several times in the 2016-2019 period.

The analysis was carried out by making a spatial slice (cross-section) of rainfall statistics carried out on time composite data. This analysis aims to obtain the transverse rainfall profile of an area that was sliced. Another output that can be identified is the value of rainfall distribution according to the area being sliced. Determination of the cross-section is based on the variation in rain distribution and topographic conditions. In this case, it consists of one east-west section and one north-south section (Fig. 1). Because the data used consist of temporal dimension, this step will be repeated for each time variation resulting in the composite time information. The output is then analyzed descriptively by emphasizing other aspects that can support the argument regarding the rain distribution output.

The temporal resolution used is 1 hour which is then averaged every hour. Graphical representation is processed to identify the changes of each temporal variation of the rain data at each hour. The average rainfall data was carried out by overlaying the composite raster of the rain distribution obtained from the rainfall estimation product in the three-year temporal period at the same hour. The hourly mean data were then compared with the other hours in the same area. This output is then analyzed descriptively based on the pattern of extracted data.

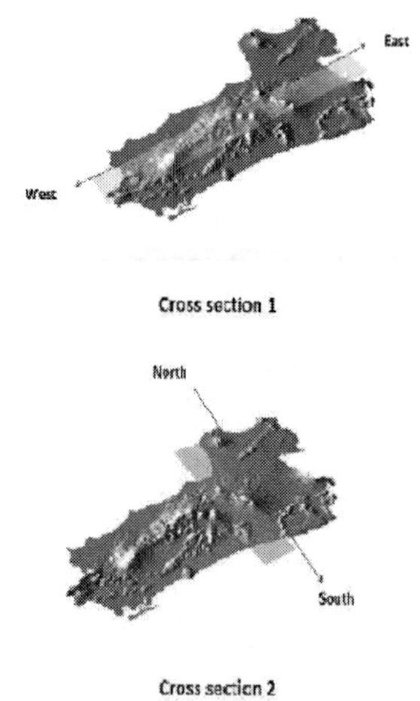

Fig. 1. Cross sections for diurnal cycle analysis in the study area, including a) East-West, and b) North-South. (Source: Processing of elevation data from ETOPO: https://www.ngdc.noaa.gov/mgg/global/

3 Result and discussion

3.1 Spatial distribution of precipitation

As a maritime country, Indonesia is one of the regions in the world that has a high level of rainfall [15-17]. Java Island, especially the central part of Java, has variations in rainfall that are distinctly different each season. The central part of the island of Java is identified as having a monsoon type of rainfall. The data was taken from the central part of the island of Java on a five-year temporal scale starting from January 1, 2015, to December 31, 2019. The data obtained were then processed to obtain hourly rain data at the location. The hourly rainfall raster data is then combined based on the hour for five years to get the average hourly rainfall during the 2015-2019 period.

The results of the PERSIANN-CCS rainfall raster composite data were then shown every three hours to facilitate the identification of the diurnal cycle (Fig. 2). The results of the composite rainfall data for PERSIAN-CCS for five years showed several variations with quite clear differences. Judging from the spatial variation, periods with evenly distributed rainfall with a high enough intensity often occur at 12.00 and 15.00; for other hours, it is alleged that the distribution of rainfall is lower so that there is no rain at all.

A slicing method was used for the study area to compile a cross-profile for the East-West and North-South directions to facilitate the identification process of the diurnal cycle. The transverse profile is made as an initial analysis step that categorizes the temporal scale of rain before six-hour intervals, namely 06.00, 12.00, 18.00, and 24.00 local times. This categorization is done to see the differences from the morning to midnight. The axis variables used are the length of the slicing area on the horizontal axis and the hourly rainfall value in millimeters per hour on the vertical axis.

3.2 Cross Section Profile of Precipitation Rate

Fig. 3 shows that the transverse profile of rainfall in the East-West direction shows the highest rainfall at 12.00 and 18.00 local time. The distribution of rain is relatively even in each region. At this location, the most frequent light rain occurs at noon. The low precipitation phases for this section are in the night at 24.00 p.m. until morning at 06.00 a.m. During the last five years, rainfall in this area is indicated to have occurred frequently from noon to late in the evening. This shows the role of sea breezes that blow towards the land, which can bring rain clouds from the ocean and fall on the land. Meanwhile, land winds can bring back clouds and evaporation that occurs on land back to the oceans. It indicates that the diurnal pattern of land winds and sea breezes can regulate the cycle of rain occurring in the region.

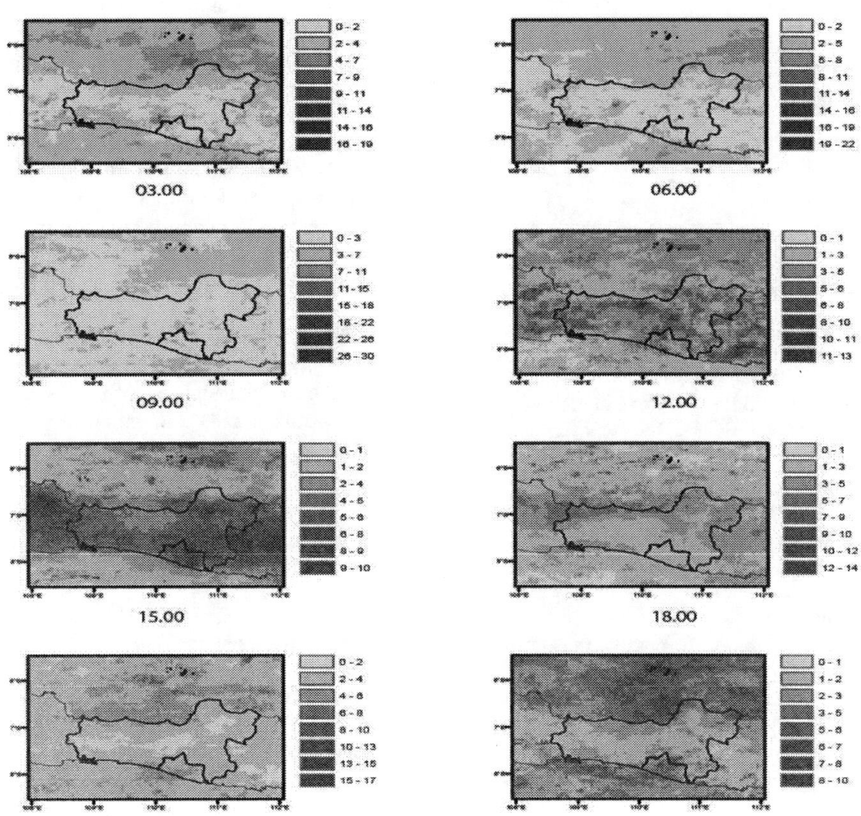

Fig. 2. Differences in the spatial pattern of average hourly rainfall (in mm/hr) at 3.00 a.m. to 24.00 p.m. local time during 2015-2019 (source: PERSIANN data processing)

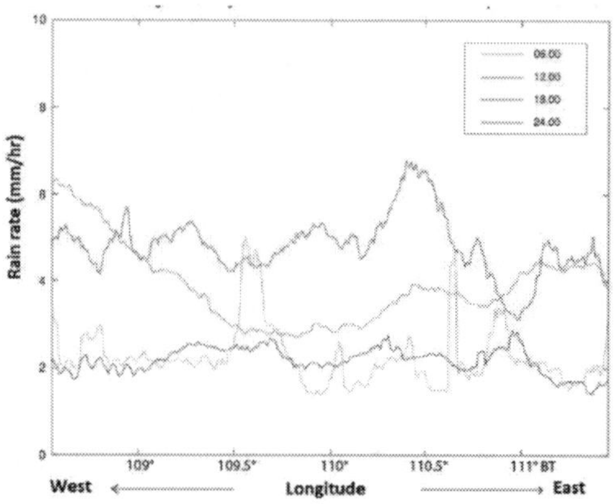

Fig. 3. Graph of the cross profile of average rainfall in east-west direction (source: PERSIANN data processing)

The cross-sectional rainfall profile for the North-South direction appears to have more fluctuations in the mean annual rainfall per hour (Fig. 4). This location intersects right between Mount Merbabu and Mount Merapi. The high elevation across the mountain can affect rainfall in intensity and peak rain time [4]. In terms of the average annual rainfall during 2015-2019, this location has moderate rain intensity with a value exceeding 6 mm/hour at 12.00 local time for areas around latitude 7° to 7.5°S. Based on the transverse profile, the lowest rain intensity occur at 18.00 local time for the 7° to 8° S. At this location, rain occurs more frequently in the afternoon to the night.

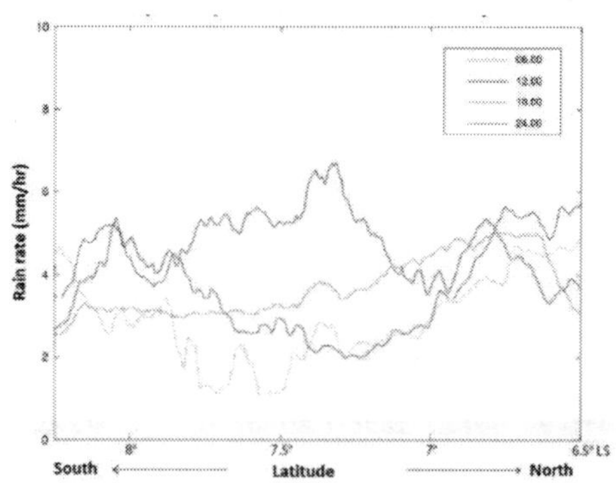

Fig. 4. Graph of the cross profile of average rainfall in north-south direction (source: PERSIANN data processing)

The cross-sections of the east-west and north-south directions above show that the average rainfall conditions for the Central Java region have a diurnal pattern where the rain intensity moves up in the morning and has a moderate intensity in the afternoon. The rain intensity began to fall again from the afternoon and reached the lowest at night until the early hours. This phenomenon follows the diurnal pattern of land winds and sea breezes, which have the most significant role in bringing the humidity from the oceans in the north and south of the island of Java to the mainland in the morning to noon. In contrast, at night, the land winds bring back land circulation to the sea. At night, the land cools faster due to the loss of longwave radiation. At the same time, water remains warmer at night, so that the daily pressure pattern reverses, and land breeze forms as air moves to areas of lower pressure above the sea.

4 Conclusion

The above results show that over the study area, the spatial pattern of repetition of rain with an intensity of up to 7 mm / hour occurs between 10.00 a.m. to 3.00 p.m. Further, it is widely distributed in the central region of Java Island, especially in an area dominated by highlands and mountains. In the central part, the rain tends to be higher in the afternoon compared to the area close to the ocean. On the other hand, (in the southern and northern regions) rain occurs at a lower intensity than in the central part.

The diurnal cycle of rainfall shows that the intensity of the rain moves up in the morning and reaches moderate intensity in the afternoon. The intensity of the rain falls back down at night until early morning. This phenomenon is probably influenced by land and sea breezes, which have the most significant role in bringing the accumulation of clouds from the oceans in the north and south of Java to the mainland in the morning to noon. In contrast, at night, the land breeze brings back the circulation of the land to the sea. In regional content, the similar result are found in Sumatra and Borneo [1,15]. Therefore, high amount of late afternoon precipitation could potentially floods that occur among the country.

The next step suggested for related research is identifying factors that influence the cycle of repetition of rainfall in the region. Topography, which includes elevation, direction towards slopes, the distance of land to the coastline, and land use, has influenced rainfall patterns. The four topographic parameters mentioned have a role in changing the diurnal trend of rainfall on land, especially in the study area. Atmospheric factors, including meteorological parameters, predominantly temperature, humidity, wind direction, and speed, are also predicted to affect [18]. The analyzed meteorological parameters have a relationship with one another which significantly affects the distribution of rain both spatially and temporally

This publication is supported by grants of "Hibah Dosen Mandiri" by Faculty of Geography Universitas Gadjah Mada in 2021. The grant entitled "Modeling the Diurnal Process of Precipitation over Central Part of Java Island" is investigated by Dr. Andung Bayu Sekaranom and teams.

References

1. H. Ichikawa, T. Yasunari, J. Clim., **19**, 1238–1260 (2006)
2. R. Ariani, *The influence of the madden-julian oscillation on diurnal cycle of rainfall over sumatera*. in Thesis, 1–54 (2014)
3. N.Suaydhi, Jurnal Sains Dirgantara, **14** , 57 (2017)
4. A. Nanda, P. Septy, *Characteristics of Diurnal Rainfall in Sangkapura-Bawean and Citeko-Bogor*

Meteorological Station based on regional and local influence. in Proceeding of Seminar Nasional Fisika Dan Aplikasinya, November 2015, Indonesia (2015)

5. D. Subarna, Seminar Sains Atmosfer, **1**, 73–79 (2013)
6. M. Wibowo, Teknologi Lingkungan, **6**, 283–290 (2005)
7. M. A.Solera-Garcia, N. A. Chappell, W. Tych, IAHS-AISH Publication, **308**, 356–361 (2006)
8. A. B. Sekaranom, H. Masunaga, J. Appl. Meteorol. Climatol. **56**, 1867–1881 (2017)
9. A. B. Sekaranom, H. Masunaga, J. Appl. Meteorol. Climatol., **58**, 37–54 (2019)
10. R. R. E. Vernimmen, A. Hooijer, Mamenun, E. Aldrian, A. I. J. M.Van Dijk, Hydrology and Earth System Sciences, **16**, 133–146 (2012)
11. A. Fadholi, Jurnal Ilmu dan Inovasi Fisika, **4**, 12–22 (2020)
12. A. B. Sekaranom, , E. Nurjani, M. P. Hadi, M. A. Marfai, Quaest. Geogr., **37**, 97–114 (2018)
13. D. W. Pratiwi, J. Sujono, A. P. Rahardjo, *Evaluation of satellite rainfall data for rain-rate estimation using cross correlation method*, in Proceeding of Seminar Nasional Sains Dan Teknologi, 1-2 November 2017, Indonesia (2017)
14. Y. Hong, D. Gochis, J. T. Cheng, K. L. Hsu, S. Sorooshian, Journal of Hydrometeorology, **8** (3) (2007).
15. J.-H. Qian, J. Atmos. Sci., **65**, 1428–1441 (2008)
16. A. B. Sekaranom, E. Nurjani, R. Harini, A. S. Muttaqin, Indonesian Journal of Geography, **52**, 143–153 (2020)
17. A. B. Sekaranom, E. Nurjani, *The development of Articulated Weather Generator model and its application in simulating future climate variability*, in proceeding of IOP Conference Series: Earth and Environmental Science, **256** (2019)
18. A. B. Sekaranom, E. Nurjani, I. Pujiastuti, *Cloud structure evolution of heavy rain events from the East-West Pacific Ocean: A combined global observation analysis*, in proceeding of IOP Conference Series: Earth and Environmental Science, **148** (2018)

https://doi.org/10.1051/e3sconf/202132505003

Stable cross-sectional dimensions for navigation channel in the estuary of Sangatta River, Indonesia

Tommy Sutarto*

Politeknik Negeri Samarinda, Civil Engineering Department, 75131 Samarinda, Kalimantan Timur, Indonesia

Abstract. The objectives of this study are: 1) to capture the existing condition of the river channel in Sangatta River estuary, including the bathymetry of the river bed, and the cross-sectional dimensions of the river, 2) to design a stable navigation channel in Sangatta River estuary that allows less maintenance dredging and improves navigability of the channel during a critical condition when the water level is at the lowest, and 3) to test the effectiveness of the stable cross section in term of the availability of navigable depth. This research is divided into five stages, namely: 1) preparatory stage and preliminary survey, 2) main surveys, 3) hydrological and hydraulic analysis, 4) waterway design, and 5) dredging plan. The hydraulic analysis also includes determining the water level and depth profiles before and after dredging. HEC-RAS software is used to simulate, for each month, the water level and depth profiles in the channel at a critical condition when the sea water level is at low water spring (LWS) coincides with the river flow of 50% probability of occurrence, Q50. A stable channel for Sangatta River estuary is characterized with a bed slope, S = 0.00015, a flow depth, D = 3.6 m, a bed width, b = 76.3 m and a sidewall slope, m = 1.5. The simulation results show that the stable cross sections of the river provide sufficient water depth for navigation, even during the critical condition.

1 Introduction

The depth of navigation channel is generally designed only based on vessel size without considering river morphodynamic processes, such as bank erosion [1, 2, 3], scouring, and siltation, which dictate the cross-sectional dimensions of the river. More often, contribution of bank erosion to total sediment load and deposition in the channel are underestimated due to the difficulties of observing and measuring this process in the nature [4, 5]. As a result, navigation channels often become shallow quickly and require more frequent dredging at a high cost [2].

Alluvial channel cross sections show minimum change in shape and size over time if their dimensions correspond to the dimensions of stable channel. An alluvial channel is in stable, regime or dynamic equilibrium state when the rate of sediment transport is approximately equivalent to the existing sediment load [6, 7], this regime is characterized by less bank erosion, i.e., mass failure and fluvial erosion [8], minimum scouring, siltation, and lateral migration.

Stable slope and cross-sectional dimensions can be predicted using various methods, based on dominant discharge and bed sediment size [6, 9, 10]. Dominant discharge can be estimated by analyzing historical flow and sediment transport data. Unfortunately, most of the time, those type of data are not available or lacking [11]. This is one of the reasons why the stable cross-sectional dimensions are rarely taken into consideration in determining the depth of navigation channels.

In this manuscript, a navigation channel is designed following stable channel dimensions. The effectiveness of that channel in term of the availability of navigable depth is tested. The dominant discharge is determined by analyzing the rainfall data which are well recorded. This method is applied at Sangatta River estuary, Indonesia.

Sangatta River is one of the main rivers that cross Kutai Timur Regency, Kalimantan Timur Province, Indonesia. More specifically, the river passes through three districts, namely Rantau Pulung, Sangatta Utara and Sangatta Selatan. The latter two districts are the center of government and economy in Kutai Timur Regency.

Sangatta River used to be an inland waterway for vessels transporting commodities, especially fish, into inland areas, such as Sangatta Utara and Sangatta Selatan, which are now growing into residential, government, and business centers. Unfortunately, severe siltation occurs at several reaches of the river, especially at the estuary. Here, the water depth is significantly reduced, making the river unnavigable more frequently, especially during the spring tide when the water level is at the lowest also called low water spring (LWS). This circumstance has even changed the portrait of the economic activity in Sangatta Utara and Sangatta Selatan.

Siltation has also significantly reduced the drainage capacity of the river. It is not uncommon the river overflow and inundate Sangatta Utara and Sangatta Selatan District.

* Corresponding author: tommysutarto@polnes.ac.id

To solve the problems, it is necessary to dredge the river channel at the estuary of the Sangatta River. This effort will provide two benefits at once, i.e., restoring the required depth for navigation, and reducing the intensity of river floods at the middle stream reach, especially in strategic areas, e.g., Sangatta Utara and Sangatta Selatan.

Currently, local government is planning to dredge the Sangatta River, including the estuary of the river. However, there are no results of a mature study that can be referred to develop a reliable dredging plan.

This manuscript provides important information required for dredging the channel in the estuary of the Sangatta River. The objectives of this study are: 1) to capture the existing condition of the river channel in Sangatta River estuary, including the bathymetry of the river bed, and the cross-sectional dimensions of the river, 2) to design a stable navigation channel in Sangatta River estuary that allows less dredging and improves navigability of the channel during a critical condition when the water level is at the lowest, and 3) to test the effectiveness of the stable cross sections in term of the availability of navigable depth.

2 Study site

The dredging location is at the estuary of Sangatta River. The survey, river hydraulics study, and design is focused at this location (Figure 1). However, hydrological study involving the prediction of flood and average flow hydrographs certainly require collection of data at the Sangatta watershed scale (Figure 2).

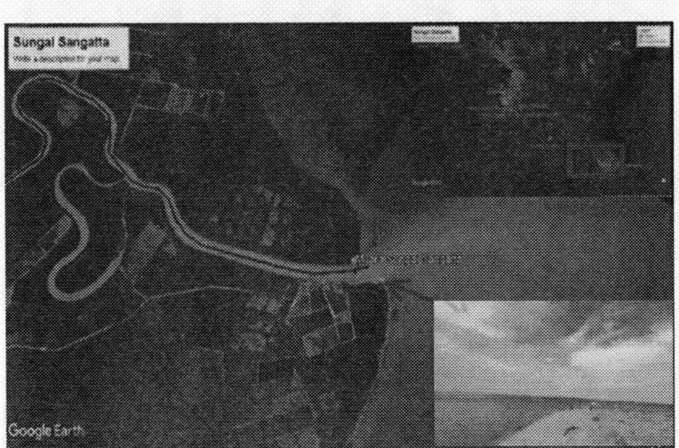

Fig. 1. Estuary of Sangatta River (Google Earth).

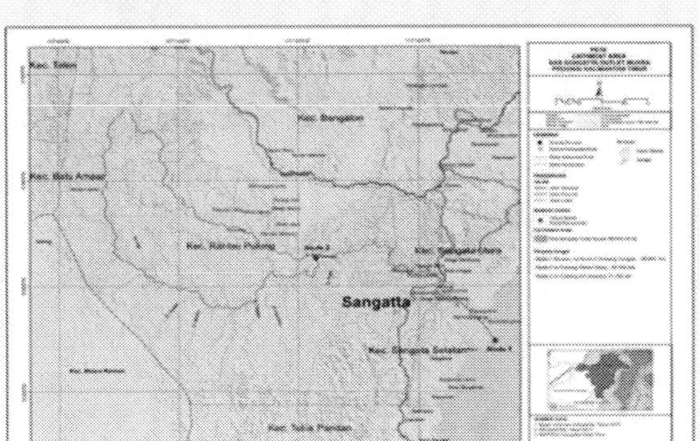

Fig. 2. Sangatta River basin.

The upstream and middle stream reaches of Sangatta River Basin are located in Rantau Pulung District, while the downstream reach is in Sangatta Utara and Sangatta Selatan District (Figure 2).

3 Methodology

This research is divided into five stages, namely: 1) preparatory stage and preliminary survey, 2) main surveys, 3) hydrological and hydraulics analysis, 4) waterway design, and 5) dredging plan.

The preparation stage consists of administrative preparation and technical preparation. Administrative preparations aim at forming a survey team and survey plan, as well as communicating with related agencies. Technical preparations include collecting supporting data such as map, tide table, and known topographic reference points, planning the sounding lines in the river, and determining the location of tidal observation stations, and scheduling for personnel and equipment mobilization.

A preliminary survey is conducted to picture the existing conditions of water body, riverbank, and riparian zone, and to identify any potential problems that could hamper survey activities later.

The next stage is the main surveys, which is focused on the estuary of the river. Here, a topographic survey of the riverbank is performed using a Total Station (Figure 3a). The topography of the riverbank is measured as far as 30 m from the riverbank toward the land, and 1200 m long from STA 0 + 000 (the mouth of the river) upstream ward to STA 1 + 200.

Also, a river bed bathymetric survey is conducted using echo sounding technique (Figure 3b and 3c). Echo sounding was conducted on the 3rd week of December 2020 along 1200 m from STA 0 + 000 upstream ward to STA 1 + 200. The water depth data, obtained from echo sounding technique, are corrected with the change in water level due to tide during measurement. For this reason, the water depth measurement is conducted simultaneously with tide observation (Figure 3d).

Fig. 3. Main surveys. a). Topographical survey. b) Bed bathymetry survey. c) Echo sounder. d) Peil scale for tide observation.

Bed sediment is collected and brought to the laboratory for testing its gradation. The other concern in this phase is determining the safe disposal location for dredged material.

Furthermore, hydrological analysis is performed to determine the design rainfall depth, flood hydrograph, dominant discharge, potential evapotranspiration, and the average flow hydrograph in the estuary of the Sangatta River. Following hydrological analysis, hydraulic analysis is conducted to determine stable cross-sectional dimensions which are used as reference information for navigation channel design and dredging plan. The hydraulic analysis also includes determining the water level profile and channel depth before and after dredging. HEC-RAS software is used to simulate the water level profile in the channel, every month, corresponding to a critical condition which is represented by LWS level at the sea coincides with a flow of 50%-probability of occurrence, Q_{50}, in the channel.

The last two phases of the study, i.e. waterway design, and dredging plan. The dredging plan should address the dredging method, disposal location, volume, cost, and schedule. These are not discussed in this manuscript due to space limitation.

4 Results and discussion

4.1 Channel bathymetry

Figure 4a presents the contours of river bed obtained from bathymetric survey. Also, Figure 4b demonstrates the longitudinal profile of riverbed and bank top obtained from bathymetric and topographic survey. The height of the river bank is between 3 and 5 m. The thickness of siltation at the bed is increasing from the upstream (STA 1 + 200) to the mouth (STA 0 + 000). River sedimentation is seen to reach a thickness of up to 2.0 m at STA 0+000.

4.2 Rainfall depth

Historical river flow data are not available for the studied reach because there is no measurement station near the location. Fortunately, the rainfall data is well recorded. In this study, the flood and average flow hydrograph are estimated from rainfall data with a rainfall-runoff model, e.g., synthetic unit hydrograph model. Descriptions on this method can be referred to [12, 13].

Frequency distribution analysis is performed for the daily rainfall data recorded in last 10 (ten) years, it is found that log-normal distribution can represent the frequency distribution of annual maximum daily rainfall in Sangatta watershed. The relation curve for maximum daily rainfall depth vs. return period is demonstrated in Figure 5. The rainfall depth of 2-year storm is equal to 76.49 mm.

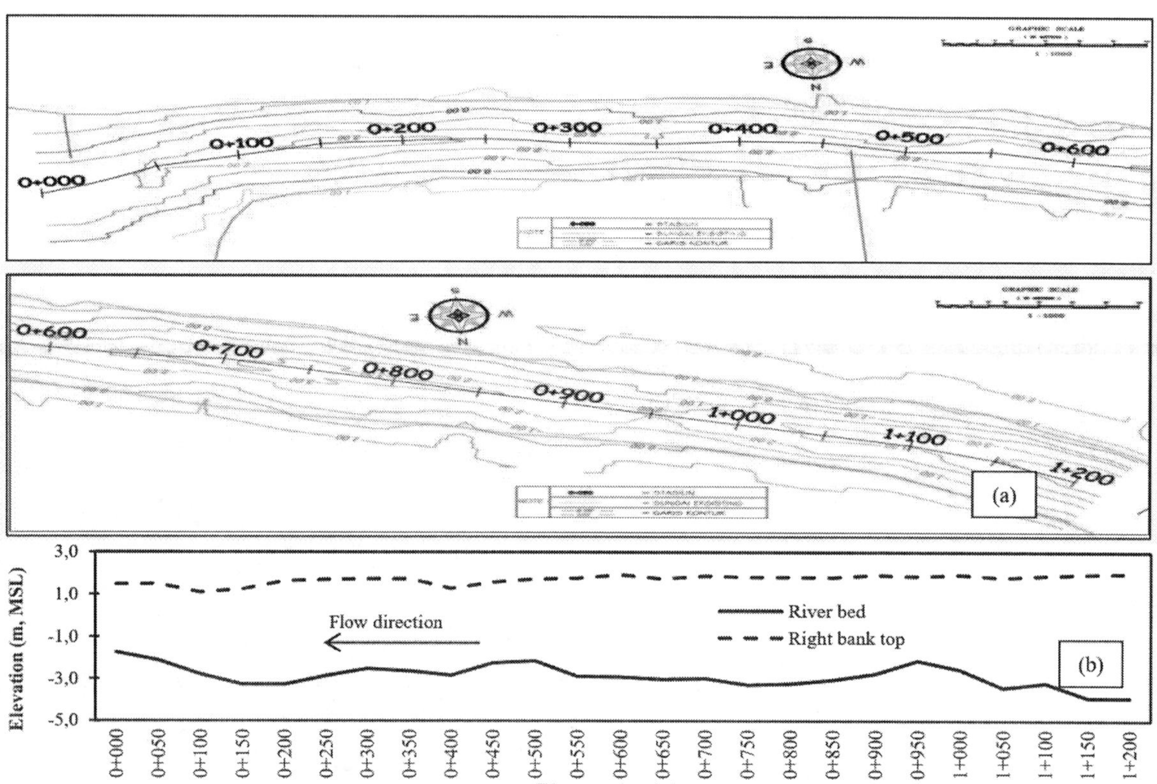

Fig. 4. Results of topographic and bathymetric surveys from STA 0+000 until STA 1+200. a) River bed contour b) Longitudinal profile of river bed and right bank top (December 2020).

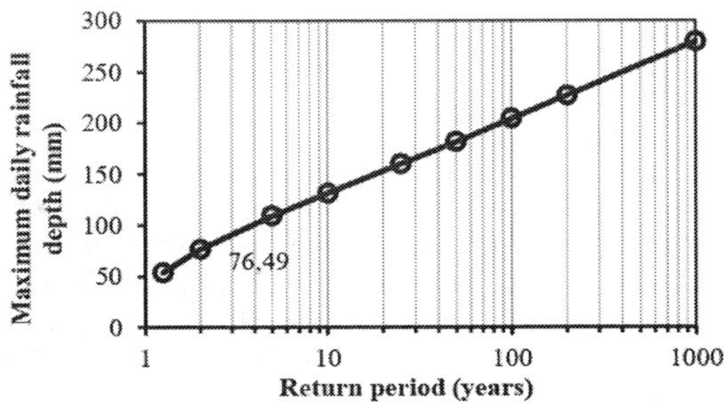

Fig. 5. The relation curve for maximum daily rainfall depth vs. return period in Sangatta watershed.

Hourly rainfall distribution can be estimated using Mononobe equation [14]:

$$I_t = \frac{R}{T}\left(\frac{T}{t}\right)^{2/3} \quad (1)$$

where: I_t = average rainfall intensity from the initiation of rain up to time-t (mm/hour), T = duration of rain event (in this case 4 hours), dan R = rainfall depth (mm).

Next, rainfall depth at time-t can be determined by:

$$R_t = tI_t - (t - \Delta t)I_{(t-\Delta t)} \quad (2)$$

where: R_t = rainfall depth at time-t (mm), Δt= time interval (hour), I_t = average rainfall intensity from the initiation of rain up to time-t (mm/hour), dan $I_{(t-\Delta t)}$= average rainfall intensity from the initiation of rain up to time-$(t - \Delta t)$ (mm/hour).

After hitting the ground surface, a proportion of rain drops turn into runoff and flow into river, this proportion of rain is called effective rainfall, R_e. The other proportion of rain infiltrate underground and make a base flow. The effective rainfall, R_e, can be estimated by equation:

$$R_e = CR \quad (3)$$

where: C = runoff coefficient whose value is largely determined by the type of land cover and cultivation. A table of C values for various land cover is given in [15] and [16].

Using equations (1), (2) and (3) the hourly rainfall and infiltration distribution of 2-year storm is obtained as shown in Figure 6.

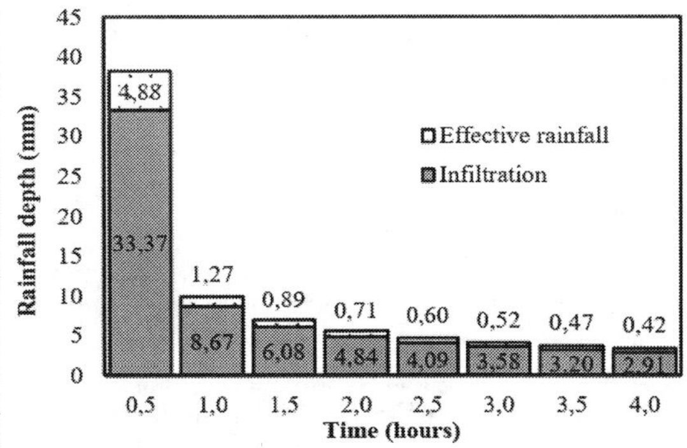

Fig. 6. A 76.49-mm rainfall depth (2-year storm) is distributed within 4-hour rain duration.

4.3 Flood hydrograph

The Nakayasu synthetic unit hydrograph model is used to develop a 2-year flood hydrograph for study site (Figure 7). The input parameters for this model are effective rainfall depth (Figure 6), area of watershed, and river length. A complete description of this procedure can be seen in [12], [13], [17], and [18]. Based on Figure 7, the peak flow for 2-year flood, $Q_{2yr.}$, is 423.78 m^3/s

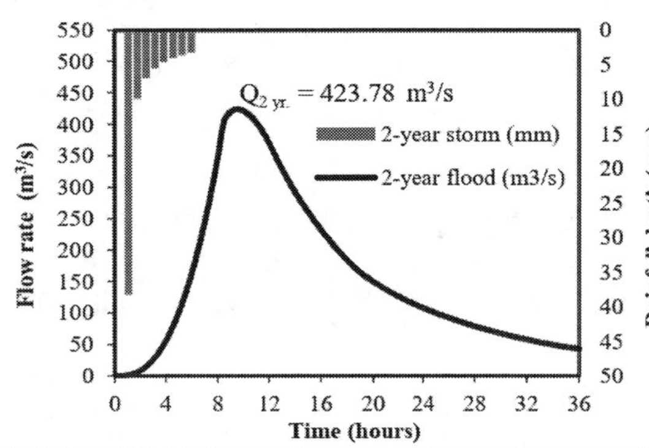

Fig. 7. A 2-year flood hydrograph for study site.

Earlier studies conducted in Indonesia, e.g., [18] and [19] have found that synthetic unit hydrograph model overestimated the peak flood by 30 to 50%. This deviation could be attributed to the assumptions used to construct the unit hydrograph concept and the complexity of governing parameters, e.g. nature of rain, and watershed characteristics. For this reason, the peak flood is corrected to become $Q_{2yr.}$ = 313.91 m^3 /s which is 35% less.

4.4 Dominant discharge

River dredging is intended to restore the stable cross-sectional dimensions of the river. The dimension of the stable section is determined by the channel forming discharge which is equivalent to the dominant discharge, Q_d. Dominant discharge is the discharge that contributes the largest to the average annual sediment

 https://doi.org/10.1051/e3sconf/202132505003

transport [20]. The Q_d is estimated to be nearly equal to bank full discharge, Q_b, which is also equivalent to the 2-year flood, $Q_{2yr.}$, [6, 21-25]. Shortly, it can be written $Q_d \approx Q_b \approx Q_{2yr.} \approx$ 313.91 m³/ s.

4.5 Average flow

The monthly average flow in the river is obtained using the hydrological model introduced by Mock [26]. This approach is implemented due to the fact that historical flow data are not available. Detail explanation on Mock model can be referred to [26-28]. Input parameters for the model are monthly rainfall depth, P, and potential evapotranspiration, ET_o. The monthly average flow hydrograph for the period 2009 to 2019 are shown in Figure 8.

4.6 Flow of 50%-probability of occurrence

A flow with 50% probability of occurrence, Q_{50}, is the flow that has a chance to be equaled or exceeded as much as 50 percent. The probability of occurrence, Pr, can be determined following the Weibull formula:

$$Pr = \left(1 - \frac{m}{n+1}\right) x100 \qquad (4)$$

In this equation, Pr = percent (%) probability that a given flow will be equaled or exceeded, m = rank of the inflow value, with 1 being the smallest possible value, n = total number of events or data points on record.

By ordering the average flow data from smallest to largest, and using equation (4), it is found that Q_{50} correspond to the flow at rank 6 (m = 6) in Table 1.

4.7 Stable cross section

The geometric variables that characterize a stable cross section are flow width, B, flow depth, D, and bed slope, S, [29]. These variables are dependent variables whose values are influenced by independent variables, namely dominant discharge, Q_d, and diameter of bed sediment, d_{50} [6, 7, 10].

The bed slope, S, of a stable channel can be approximated by:

$$S = \frac{f^{5/3}}{1830 {Q_d}^{1/6}} \qquad (5)$$

where: Q_d = dominant discharge (ft³/det), f = silt factor, $f = 1.6{d_{50}}^{1/2}$, and d_{50} = bed sediment diameter (mm).

Laboratory sieve and hydrometer tests for bed sediment samples give d_{50} = 0.55 mm. Substituting d_{50} = 0.55 mm and Q_d = 313.91 m³/s = 11,085.63 ft³/s into equation (6) gives S = 0.00015.

Rational method developed by Chang [9,30] is applied to determine the dimensions of stable channel. River cross section is considered to have a trapezoidal shape with side wall slope V:H = 1:1.5. Three parameters, namely S, d_{50}, and Q_d are known parameters to compute the dependent parameters, i.e., B and D using these equations:

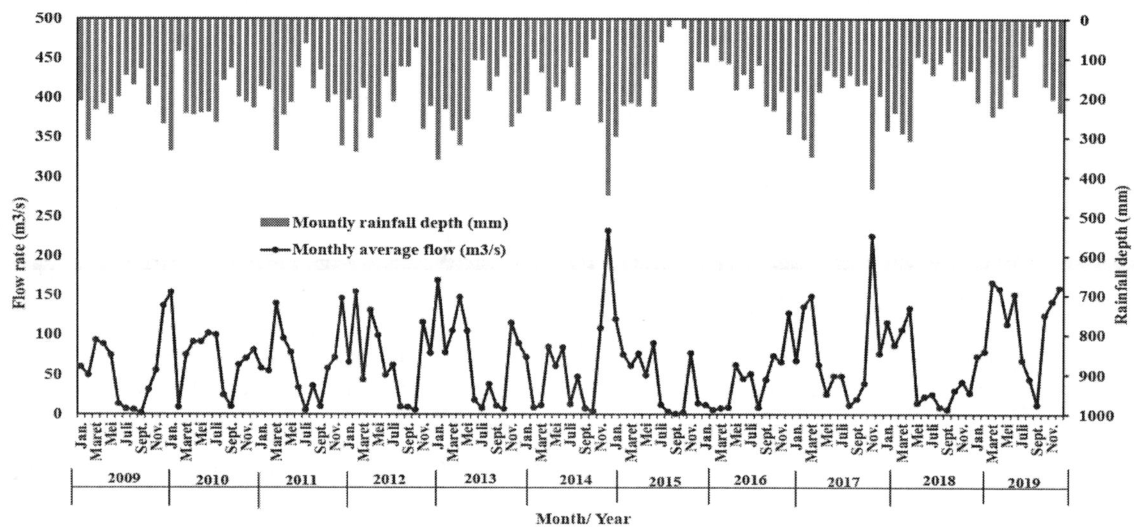

Fig. 8. Monthly rainfall depth and monthly average flow obtained from Mock hydrological model for the period 2009 – 2019

Tabel 1. Determination of flow with 50%- probability of occurrence, Q_{50}.

Ranking	Flow rate (m³/s)											
	Jan	Feb	Mar	Apr	Mei	Jun	Jul	Ags	Sep	Okt	Nov	Des
1	11.4	5.4	7.4	8.1	13.8	12.8	4.7	2.9	0.4	1.7	40.7	13.5
2	57.9	*8.2*	*11.2*	62.0	25.0	17.4	6.7	5.6	1.3	*3.6*	55.2	26.7
3	59.7	8.5	43.4	76.2	48.9	21.8	7.6	8.4	6.0	5.0	65.7	76.3
4	65.6	49.2	60.9	*85.0*	*60.6*	33.3	11.6	8.6	*7.2*	6.7	70.3	76.6
5	67.1	54.3	75.1	88.3	62.1	44.6	*12.4*	9.1	8.8	29.7	71.1	80.7
6	***71.6***	***74.9***	***93.1***	***91.2***	***74.2***	***48.0***	***24.9***	***11.2***	***9.3***	***31.0***	***76.9***	***89.3***
7	72.8	77.5	105.1	95.2	77.8	49.2	48.0	24.1	9.3	38.8	*108.2*	127.3
8	115.7	78.9	106.4	130.9	90.9	*84.0*	50.6	35.7	10.6	57.7	115.0	136.8
9	119.9	86.6	139.1	133.5	99.2	89.4	61.7	37.4	11.5	62.2	115.8	146.1
10	153.6	135.6	148.5	147.4	104.9	101.9	67.7	44.2	18.8	73.4	142.2	159.4
11	169.2	154.5	166.7	158.1	113.8	151.2	99.7	*47.5*	43.6	124.6	225.0	*231.5*
Q_{50} (m³/s)	***71.6***	***74.9***	***93.1***	***91.2***	***74.2***	***48.0***	***24.9***	***11.2***	***9.3***	***31.0***	***76.9***	***89.3***

$$\frac{S_c}{{d_{50}}^{1/2}} = \frac{0.00238}{{Q_d}^{0.51}} \quad (6)$$

$$B = 4.17\left(\frac{S}{{d_{50}}^{1/2}} - \frac{S_c}{{d_{50}}^{1/2}}\right)^{0,05} {Q_d}^{0,5} \quad (7)$$

$$D = 0.055\left(\frac{S}{{d_{50}}^{1/2}} - \frac{S_c}{{d_{50}}^{1/2}}\right)^{-0,3} {Q_d}^{0,3} \quad (8)$$

where: S_c = critical bed slope corresponding to bed load threshold, B and D are in feet.

Substituting d_{50} = 0.55 mm, S = 0.00015 and Q_d = 11,085.63 ft³/s into equation (6), (7), dan (8) give B equals 285.83 ft or 87.12 m, and D equals 11.81 ft or 3.6 m.

4.8 Required navigation depth

It is expected the water depth in Sangatta River estuary is still sufficient for fishing vessels of 30-Gross Tonnage (GT) to navigate even when the water surface is at the lowest level in the channel. A 30-GT vessel typically has a total length (LOA) of 18.5 m, a width of 4.5 m, and a draft of 1.5 m [31].

The water surface, in the channel, is at the lowest level when the sea water level is at low water spring (LWS) coincides with the river flow rate that is equal to Q_{50}.

The required navigation depth, H, can be determined as (Figure 9) [32]:

$$H = d + (\text{the largest among } R \text{ or } G_g) + G_s + S + P + K \quad (9)$$

where: d =draft, R = net free space, G_g = vertical movement due to wave (about 0.3 times wave height), G_s = vertical movement due to squat = $k\,V$, k = coefficient influenced by vessel size, V= navigation speed km/jam, S = siltation depth between two successive dredging, approximate error in dredging, P = 0.05 m, and dredging tolerance, K = 0.10 m.

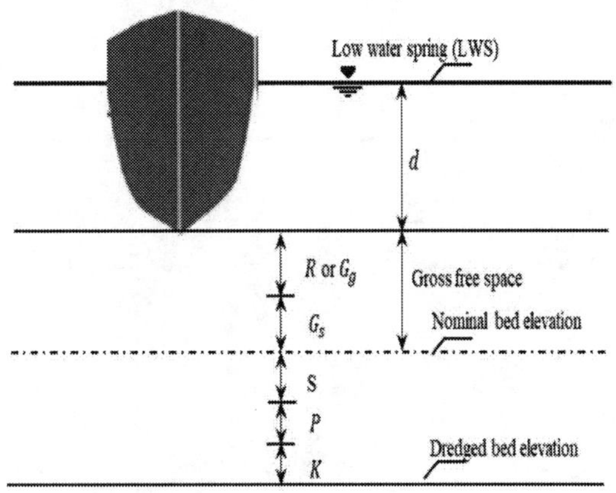

Fig. 9. Water depth for river navigation.

Given LOA < 86 m and river bed is dominated by sands, then the R value is equal to 0.2 m [32]. If the allowable speed, V, for navigating the channel is 6 knot or 11.112 km/hour, then the wave height is 0.286 m and G_g = 0.3 x 0.286 = 0.086 m [32]. For a vessel with LOA < 86 m, the k equals 0.017, and G_s = 0.017 x 11.112 = 0,189 m [32]. Given S = 0.4 m, P = 0.02 m and K = 0.10 m, substitute those values into equation (9) to obtain the required navigation depth H = 1.5 + 0.2 + 0.189 + 0.4 + 0.02 + 0.10 = 2.41 m.

4.9 Available water depth in critical condition

HEC-RAS software is used to model the surface water profile and water depth in the channel, before and after maintenance dredging, at the lowest water level that occurs when the sea water level is at LWS level coincides with the river flow rate that is equal to Q_{50}.

Initially, the geometry of the channel, from STA 0+100 until STA 1+200, is modeled. This includes modeling the channel reach, cross sections, and surface roughness. A georeferenced aerial image is used to facilitate the depiction of channel reach (Figure 10a) and cross sections (Figure 10b). The geometries of pre-dredged cross sections are obtained from the survey, while the dimensions of post-dredged cross sections follow the stable channel dimensions.

Next, 1D steady flow modeling is performed 12 times, each representing the flow conditions per month from January to December. The Q_{50} value (Table 1) is entered as an inflow at the upstream end, namely STA 1+200. The upstream boundary condition is the normal water level for a bed slope of 0.00015, while the downstream boundary condition is the water level at LWS level. The LWS level per month is presented in Figure 11, obtained from TNI-AL Hydro-Oceanography Service (2020). The lowest LWS is -1.5 m (msl) or 1,5 m below mean sea level. This occurs in March and October (Figure 11).

Figure 12 presents a comparison of channel cross-section modeled in HEC-RAS between pre-dredging and post-dredging condition at river station STA 0+100. The pre-dredged cross section is obtained from the survey, while the dimensions of post-dredged cross sections follow the stable channel dimensions.

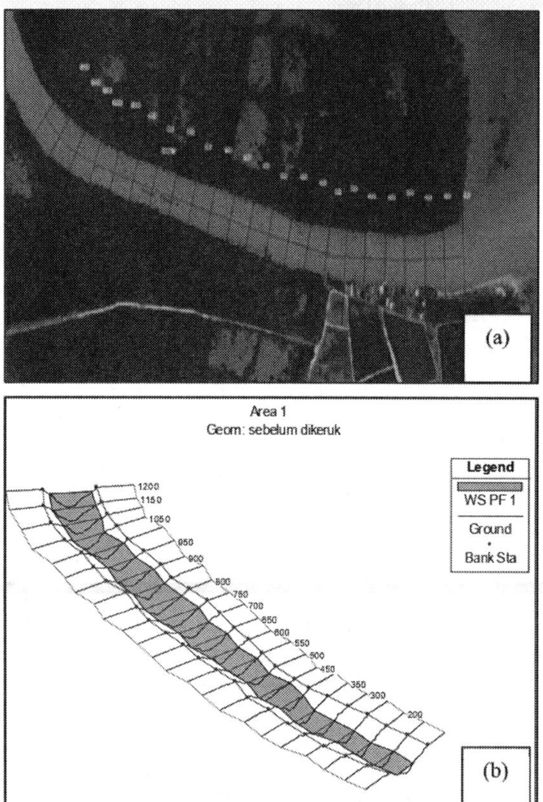

Fig. 10. Channel geometry modeling. a) Georeferenced aerial image is used to draw the channel reach. b) Channel cross sections from STA 0+100 to 1+200.

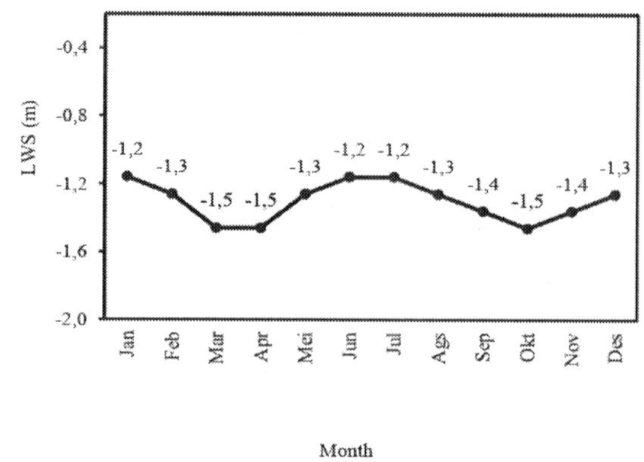

Fig. 11. Low water spring (LWS) elevation per month.

Figure 12 presents a comparison of channel cross-section modeled in HEC-RAS between pre-dredging and post-dredging condition at river station STA 0+100. The pre-dredged cross section is obtained from the survey, while the dimensions of post-dredged cross sections follow the stable channel dimensions.

Also, Figure 12 demonstrates a comparison of water depth, for each month, between pre-dredging and post-dredging condition obtained from HEC-RAS simulations. The figure presents only the simulation results from January until April due to space limitation. It can be observed, the water depth significantly increases after maintenance dredging.

Figure 13 shows a comparative graph of water depth before and after dredging at STA 0+100 from January to December. It shows the lowest depth is 1.29 m before dredging and 2.56 m after dredging. Both occurred in October. Also, the water depth significantly increases, by a factor of 2, after maintenance dredging.

Figure 14 shows longitudinal profile of channel bed and water surface before and after maintenance dredging for the critical condition in October. Note, before dredging, the bed elevation is high, especially at STA 0+000 and 0+050, such that the bed and water level almost coincide, creating a very limited water depth along the channel. This condition indicates siltation has occurred in the channel making it unnavigable.

On the other hand, the post-dredged channel geometry characterized by stable slope and cross sectional dimensions and lowered bed elevation has improved the water depth along the channel. Also note, there is a decrease in water level up to 0.6 m due to bed lowering, this is obviously has a positive impact on flood control at the middle stream.

Month	*Pre-dredging*	*Post-dredging*
Jan	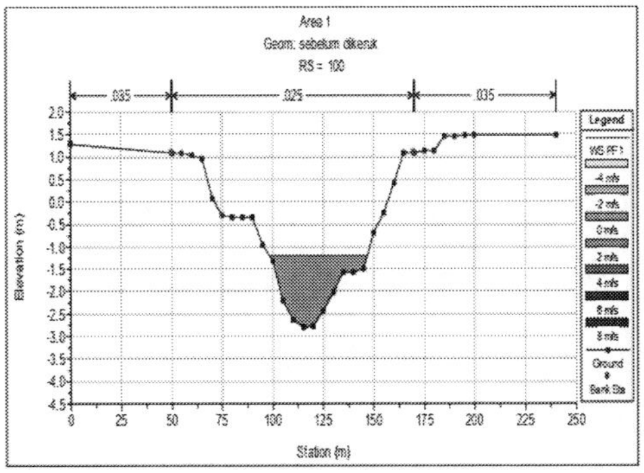	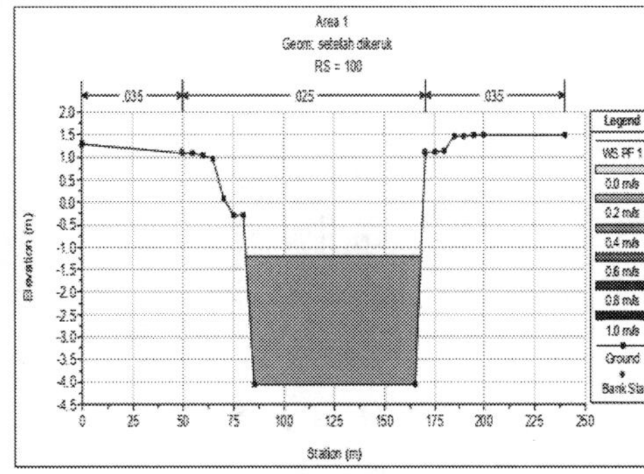
Feb	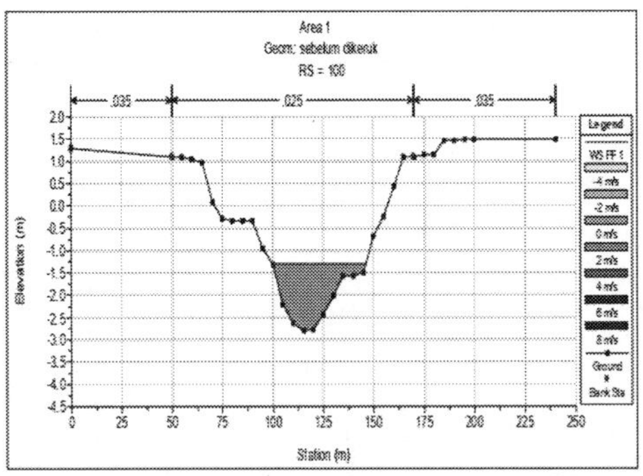	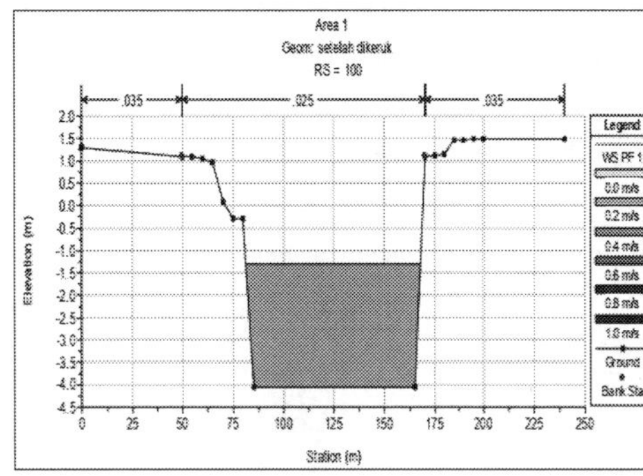
Mar	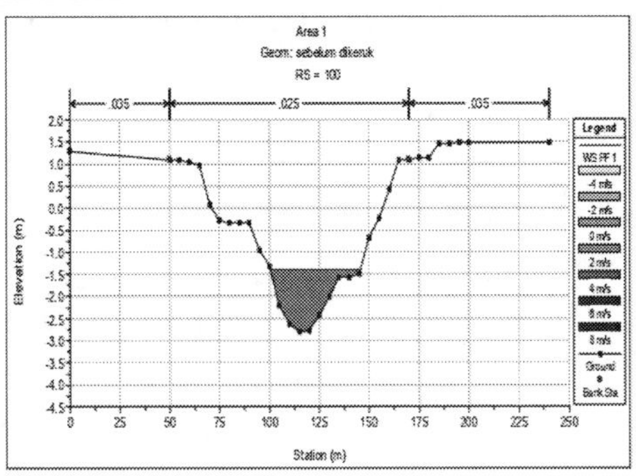	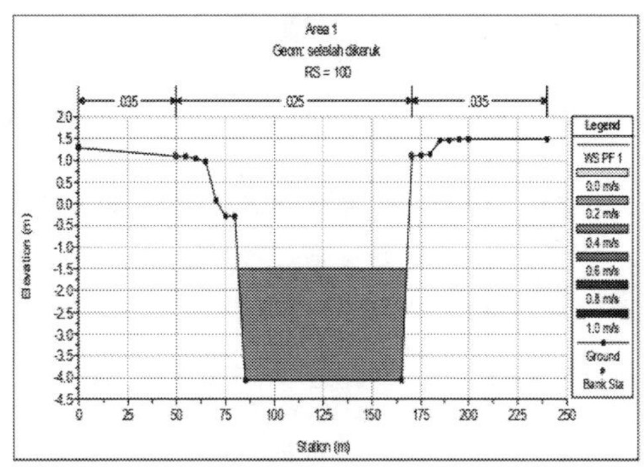
Apr	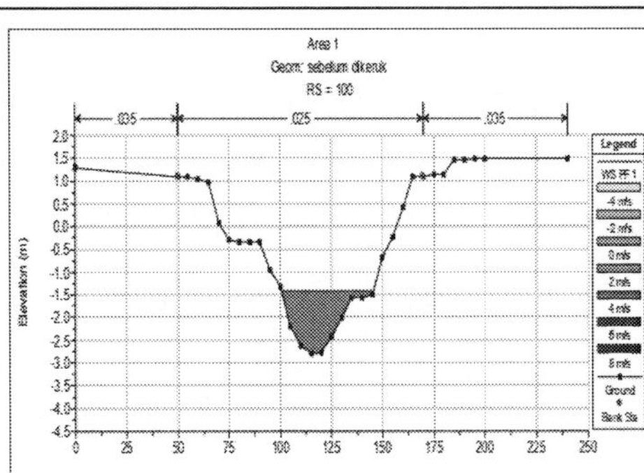	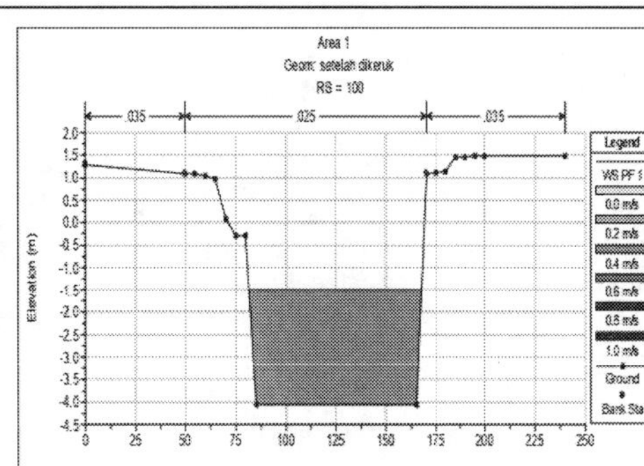

Fig. 12. Cross-section and water-depth comparison between pre-dredging and post-dredging condition at river station STA 0+100 for a critical condition when the water surface, in the channel, is at the lowest level.

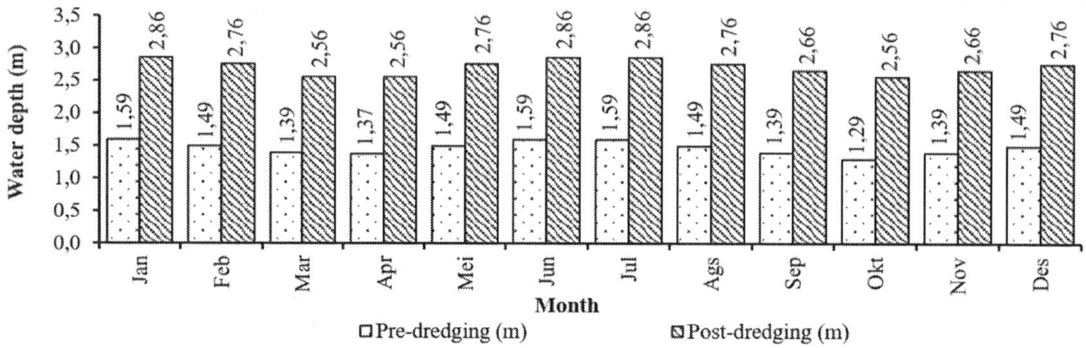

Fig. 13. A water depth comparison between pre-dredging and post-dredging at river station STA 0+100 for the critical condition.

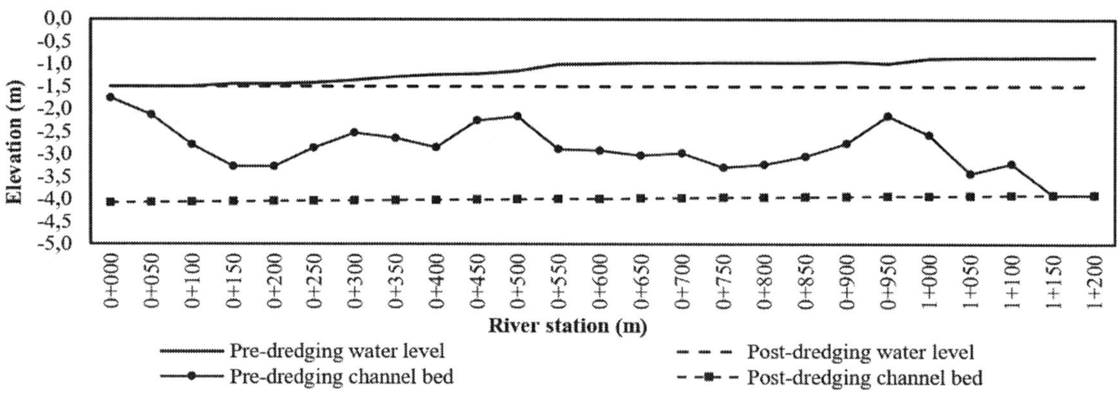

Fig. 14. Longitudinal profile of channel bed and water surface before and after dredging for the critical condition in October.

4.10 The effectiveness of stable cross sections

The effectiveness of stable cross-sections in term of providing required navigation depth (H = 2.41 m) is tested, especially for a critical condition when the water surface is at the lowest level in October.

Figure 15 portrays a comparison among pre-dredging and post-dredging water depth in the channel, as well as required navigation depth. Prior to dredging, the water depth is insufficient for 30-Gross-Tonage fishing vessel to navigate. This circumstance occurs in almost all river stations, except STA 1+050, 1+150 and 1+200. The minimum depth is 0.26 m at STA 0+000.

An opposite situation occurs as the geometry of the channel follows the stable slope and cross-sectional dimensions, herein represented by post-dredged cross sections. The water depths at all river stations are greater than or equal to the required navigation depth (H = 2.41 m, Figure 15). The lowest water depth is 2.41 m at STA 1+150 and 1+200. Thus, the stable slope and stable cross-sectional dimensions are acceptable and they can be referred for developing navigation channel in Sangatta River estuary

It is also interesting to observe that the water depth decreases after dredging at STA 1+050, 1+150, and 1+200.This can be attributed to the lowering of water level as an implication of the decrease in bed elevation. This drop in water level certainly has a positive impact on controlling floods at the middle stream.

5 Conclusions

The siltation in Sangatta River estuary has reached a thickness up to 2 m based on bathymetric survey conducted in December 2020. This situation has caused Sangatta River estuary unnavigable more frequently, especially when the water surface is at the lowest level during the spring tide in October. In this state, the minimum water depth is only 0.26 m much less than the required navigation depth for a 30-Gross-Tonage fishing vessel, (H = 2.41 m).

The stable channel geometry for Sangatta River estuary is characterized with the following dimensions: bed slope, S = 0.00015, water depth, D = 3.6 m, bed width, b = 76.3 m and side wall slope, m = 1.5.

The results of HEC-RAS models show that the water depth in the channel is sufficient for a 30-Gross-Tonage fishing vessel to navigate, even when water surface is at the lowest level, if the channel geometry follows the stable slope and stable cross-sectional dimensions. Thus, the dimensions of stable channel are acceptable and they can be referred for developing navigation channel in Sangatta River estuary.

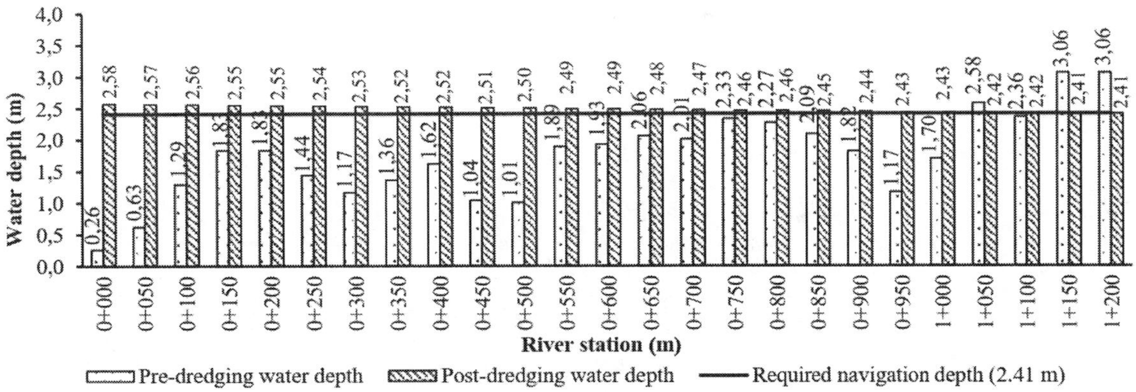

Fig. 15. A comparison among pre-dredging, post dredging, and required water depth for the critical condition in October.

Acknowledgements

This study is part of a project entitled Survey, Investigation, and Design (SID) for Sangatta River Normalization funded by Office of Public Works and Public Housing, Kalimantan Timur, Indonesia under contract No. 602/Bid-SDA/KPA/989.M/XI/2020, 10 November 2020.

References

1. A.N.T. Papanicolaou, T.E. Sutarto, C.G. Wilson, E.J. Langendoen, *Bank stability analysis for fluvial erosion and mass failure*, in Proceedings of the 2014 World Environmental and Water Resources Congress, 2014, pp. 1497–1508 (2014).
2. V.T. Nguyen, J.h. Zheng, J.s., Zhang, Water Science and Engineering 6(2), 178-188 (2013).
3. T.E. Sutarto, A.N.T. Papanicolaou, C.G. Wilson, IOP Conference Series: Earth and Environmental Science 451(1), 012093 (2020). doi:10.1088/1755-1315/451/1/012093.
4. A.N.T. Papanicolaou, C.G. Wilson, A.G. Tsakiris, T.E. Sutarto, F. Bertrand, M.,Rinaldi, S. Dey, E. Langendoen, Earth Surface Processes and Landforms 42(11), 1717–1732 (2017).
5. T.E. Sutarto, *A short review on various techniques for measuring stream bank soil erosion strength*, in Proceedings of 2018 International Conference on Applied Science and Technology, iCAST, 2018, pp.252–258 (2018). doi: 10.1109/iCAST1.2018.8751560.
6. P.Y. Julien, River Mechanics, p. 158 (University Press, Cambridge, UK., 2002).
7. R.J. Garde, K.G.R. Raju, Mechanics of Sediment Transportation and Alluvial Stream Problems (Willy Eastern Limited, New Delhi, 1977)
8. T.E.Sutarto, Procedia Engineering (125), 368–375 (2015).
9. H.H. Chang, Journal of Hydraulic Divison 106(HY5): 873-891 (1980).
10. S. Shaghaghi, H. Bonakdari, A. Gholami, O. Kisi, J. Shiri, A.D. Binns, B. Gharabaghi, Journal of Hydrology (2018), doi: https:// doi.org/10.1016/j.jhydrol.2018.09.057.
11. A.W. Salami, S.O. Bilewu, A.M. Ayanshola, and S.F. Oritola, Journal of American Science 5(4):23-32 (2009).
12. D.A. Chin, Water Resources Engineering, 2nd ed. (Pearson Prentice Hall, USA, 2006).
13. L.W. Mays, Water Resources Engineering. 1st ed. (John Willey & Sons, Inc., USA, 2001).
14. S. Priambodo, Suhardjono, L. Montarcih, and E. Suhartanto, MATEC Web of Conferences 276, 04012 (2019).
15. V. Novotny and H. Olem, Water Quality: Prevention, Identification, and Management of Diffuse Pollution (John Wiley & Sons, Inc., New York, 1994).
16. C. Asdak, Hidrologi Dan Pengelolaan Daerah Aliran Sungai (Gajah Mada University Press, Yogyakarta, 2007), ISBN: 979-420-737-3.
17. C.D. Soemarto, Hidrologi Teknik, 2nd ed. (Penerbit Erlangga, Jakarta, 1993).
18. I.G. Tunas, Pengembangan model hidrograf satuan sintetik berdasarkan kharakteristik fraktal daerah aliran sungai, *Dissertation*, (Institut Teknologi Sepuluh November, Surabaya, 2017).
19. N.F. Margini, D.A.G Nusantara, M.B. Ansori, Jurnal Teknik Hidroteknik, 2(1), 41-45 (2017).
20. M.G. Wolman, J.P. Miller, Journal of Geology 68, 54-74 (1960).
21. E. D. Andrews, J. of Hydrology, 46, 311-330 (1980).
22. L.B. Leopold, A View of the River, p.290 (Harward University Press, 1994).
23. F.D. Shields, R.R. Copeland, P.C. Klingenman, M.W. Doyle, A. Simon, J. of Hydraulic Engineering, 129, 575-584 (2003).

24. Ministry of Public Works Republic of Indonesia, Modul Dasar-Dasar Perencanaan Banjir Pelatihan Pengendalian Banjir, Pusat Pendidikan dan Pelatihaan Sumber Daya Air dan Konstruksi PUPR (2017).
25. V. Ferro, P. Porto, Geomorphology 139-140, 313-321 (2012).
26. F.J. Mock, *Land Capability Appraisal Indonesia: Water Availability Appraisal*. Basic Study Prepared for the FAO/UNDP Land Capability Appraisal Project, AGL: SF/INS/72/011 Basic Study I. Bogor: UNDP-FAO Of The United Nations (1973).
27. S.S. Sachro, Sugianto, H. Budieny, Media Komunikasi Teknik Sipil 19 (1) (2013).
28. I. Komariah, and T. Matsumoto, Journal of Water Environment Technology, 17(4), 203-217 (2019).
29. H. Afzalimehr, M. Abdolhosseini, V.P. Singh, J. Hydrol Eng. 15(10), 859-864 (2010). doi: 10.1061/(ASCE) HE.1943- 5584.0000260.
30. H.H. Chang, Journal of Irrigation and Drainage Engineering, 111(1), 36-43 (1985).
31. B. Triatmodjo, Perencanaan Pelabuhan. p. 143-147 (Beta Offset, Yogyakarta, 2010).
32. Ministry of Transportation Republic of Indonesia, Laporan Akhir studi standarisasi di bidang lalu lintas dan angkutan transportasi ASDP (2010).

Spatial Distribution of Omah Alas, Tanjungsari District, Karst of Gunung Sewu

Hilary Reinhart[1*], *Shabrina* Tamimi[2], and *Nadhine* Salsa Maulita[3]

[1]Department of Development Geography, Faculty of Geography, Universitas Gadjah Mada
[2]Alumnae of Master of Architecture Study Program, Department of Architecture and Planning, Faculty of Engineering, Universitas Gadjah Mada
[3]Urban and Regional Planning Program Study, Department of Architecture and Planning, Faculty of Engineering, Universitas Gadjah Mada

Abstract. Amid the arid characteristics and limited soil and land resources in the karst area to support the agriculture sector, the settlers in it strive and establish Omah Alas settlement as part of the adaptation and livelihood. Nevertheless, the study on the spatial distribution of Omah Alas as a unique type of settlement where every single house is scattered in a distance and located in the field is very few. This study specifically examines the spatial distribution of Omah Alas in Karst of Gunung Sewu and the factors behind it. Specifically, this study takes place at the Kemadang and Banjarejo Village, Tanjungsari District, and targets the farmers for the population and using purposive sampling aims those who spend their daily life in their field. The data are obtained using field survey, interview, and observation for the primary data and Peta Rupa Bumi Indonesia (RBI) for the secondary data. Those data are analyzed using nearest neighborhood analysis and kernel density. The results show how the village setting influences the accumulation of Omah Alas while the Omah Alas spatial distribution is random due to the irregularity of landform in the study area. Land use and land ownership also affect spatial distribution.

1 Introduction

In general, humans have interacted with the karst landscape and extracted its resource from time through time. As a landscape, karst provides water, dwelling places like rock shelter, rocks as a building material, and land to be cultivated. This practice of cultivation leads humans to develop a certain agricultural system in karstic areas worldwide [1-4]. Agriculture is considered one the most fundamental activities for the human to maintain their livelihood and shapes the society they live in since it provides the basic needs of food and energy. It requires several elements from the ecosystem such as water, soil, seed, and land. Every element that is needed can be found in the karst area. Nevertheless, karst itself is a fragile environment [5] with its vulnerability [6]. As an interaction between humans and the karst landscape, agriculture inevitably applies a certain threat to the karst environment and eventually will affect the human as well. Agriculture is a complex system full of uncertainty hence farmers and peasants at Karst of Gunung Sewu are urged to adapt accordingly to the environment.

The unique morphology and morphostructure [7, 8] of sinusoid hills in this area demand optimization of the doline area and the sloppy hills of the karst to be used. Not only facing its morphology setting, farmers and peasant at Karst of Gunung Sewu also must overcome the challenges from the thin soil, scarcity of water, and weather uncertainty while controlling the environmental impact of agriculture to the environment. The agriculture at Karst of Gunung Sewu could impact the karst environment by eroding its soil and leads to rocky desertification [9]. Further, agriculture practice is also one of the causes of groundwater pollution in the karst area [10] due to its poor filtration and hydrological characteristics. All those obstacles force farmers at Karst of Gunung Sewu to conduct a particular livelihood to adapt and sustain their life a daily basis [11]. Sustainable livelihood is the fundamental requirement since the karst area is highly fragile and must be managed with care and precaution. The whole process of adaptation is originally included in karst management and further, produces specific adaptation from economical adaptation [12] to landscape and physical modification [13].

Previous studies have successfully identified an adaptation toward climate change [14, 15]. The hydrological characteristics with conduit and porous bedrock make run-off at the surface is very limited hence it is very difficult to build a proper irrigation system. Agriculture depends mostly on rainfall. Thus, the agriculture sector is prone to climate change and may result to crop failure or even arid. The adaptation toward land scarcity was studied by Uhlig [16] in Karst of Gunung Sewu discovered how agriculture utilized and optimized the area of sloppy hills known as highland

* Corresponding author: hilary.reinhart@ugm.ac.id

agriculture and the doline area or known as lowland agriculture. Each slope part contains different species and has its biodiversity which is a special characteristic in the karst landscape [17]. Another form of human adaptation toward the physical environment is the settlement.

The settlement as the mode of adaptation had been practiced since prehistoric times [18] and agriculture is one of the factors that affect it [19]. Apart from that, settlement is also adjusted by concerning social and cultural function [20]. As a unique landscape, the karst environment also determines the settlement from its typology to its quality. The factors that determine settlement in the karst area are (i) physical, (ii) man, (iii) society, (iv) shell, and (v) network [21] while the factors that have a significant role in the quality of settlement are socio-economic and daily revenue [22]. In short, the intrinsic characteristic of karst is very fundamental to the relation of adaptation – agriculture – settlement.

This paper aims to scrutinize the spatial distribution settlement of Omah Alas in Tanjungsari District, Gunungkidul Regency, and the factors that affect it. The term of Omah Alas was coined by Tamimi [23] and refers to temporary buildings, off the grid in the field (or 'Alas', farmers in Tanjungsari District refer to their field as 'Alas'. 'Alas' means forest in English or 'hutan' in Indonesian) that are mostly dwelled. This phenomenon is unique and very limited studies have ever been conducted toward it. Omah Alas have strategic roles in agricultural activities where they provide shelter and make the work time for the farmers more efficiently since they do not have to commute daily from their permanent house in the village hence cut the production cost while helping the land resource management. Omah Alas is also effective in maintaining farmers's assets such as farming utensils, food crops, and livestock [23]. To examine the Omah Alas, this paper would grasp the notions of settlement and deal with the research question of how Omah Alas spatial distribution is and what are the influences of that spatial distribution.

2 Methods

To answer the research question and achieve the research purpose, this paper will use the cartography and Geographic Information System tool to help visualize and analyze the data. The analysis is done using Nearest Neighborhood Analysis (NNA) and the kernel density to determine the spatial distribution pattern of Omah Alas

2.1 Research Location

2.1.1 Karst of Gunung Sewu

Karst area, consisting of carbonate and evaporate rocks, encompasses roughly 20% of the non-ice earth's surface [24]. This landscape is mainly controlled by climate, lithology, and land cover factors. Thus, every karst region in the earth has its unique properties, strongly depending on the climate and lithology where it is situated as they would affect the dissolution rate and rock weathering to the morphology of the karst. One of the most prominent karst areas in the world is Karst of Gunung Sewu, tropical karst located at Southern Part of Java Island and extends to almost 120 km from west to east covering three regencies of Gunungkidul, Pacitan, and Wonogiri in three different provinces of DI Yogyakarta, Central Java, and East Java. Karst of Gunung Sewu is well known for its significant value of morphology, geology, and archaeology [25]. It has been inhabited since the Paleolithic era [26] and along with the advancement of civilization, agriculture is practiced in Karst of Gunung Sewu. Visualization of Karst of Gunung Sewu landscape can be seen in Fig. 1 below.

Fig. 1. Karst of Gunung Sewu Landscape (source: documentation, 2015)

2.1.2 Kemadang and Banjarejo Village, Tanjungsari District

Kemadang and Banjarejo Village are two of the villages in Tanjungsari District, Gunungkidul Regency with a total area of 19 km^2 for Kemadang and 16 km^2 for Banjarejo Village and occupy 27% and 23% of the total Tanjungsari District respectively making them largest and the second-largest villages in it [27]. Both villages are located on the southern coast of Gunungkidul Regency and have many famous beaches like Baron, Kukup, Sepanjang, and Watukodok Beach. With total of 7025 residents and spread in its 17 sub-villages (*dusun*), Kemadang Village also the most populous village in Tanjungsari Districts. Meanwhile, Banjarejo population is 5439 and spread into 21 sub-villages. Most of the villagers in Kemadang and Banjarejo are small scale-farmers and also work in the informal tourism sector like a kiosk or small restaurant in the beach area. The research location is shown in Figure 2 below.

 https://doi.org/10.1051/e3sconf/202132505004

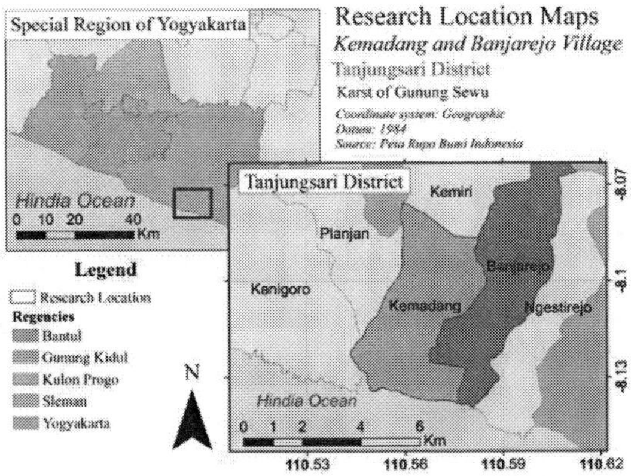

Fig. 2. Research Location Map (source: Analysis, 2021)

2.2 Data Collection

The data required for this research are divided into primary data of the location of 'Omah Alas', demography data, the profile of each 'Omah Alas' owners, and secondary data of Basic Geospatial Information in Rupabumi Maps (*Peta RBI*) including topography, land use, and transportation and monograph of Tanjungsari District. The primary data are obtained from a field survey conducted from January-March 2020 and the secondary data are obtained from an institutional survey or internet. During the field survey, the location of 'Omah Alas' is acquired by the plotting using GPS and the social data of demography and profile are collected by using in-depth interview using interview guidelines designed to conceive the relation between the owner with the 'Omah Alas' and the context of 'Omah Alas' in their environment.

We do purposive sampling or the sampling for certain criteria [28] for the farmers in the research location and select those whose daily life is mostly spent in the field hence satisfy the definition of Omah Alas. We manage to collect 30 samples of the houses located in the southern part of Kemadang and Banjarejo Village.

2.3 Data Analysis

The relation and context of Omah Alas will be analyzed using spatial analysis. Spatial analysis is a fruitful tool in seeing a Spatio-temporal dimension of phenomena and had been widely applied to examine the settlement in many fields of study [29]. In this research, the spatial analysis is only used to apprehend the spatial context related to the Omah Alas. Analysis of spatial elements of Omah Alas purposes to view the point spatial distribution of the settlement and used to help to describe the socio-cultural context of 'Omah Alas'.

For the spatial distribution pattern, we used nearest neighborhood analysis to determine the aggregation of Omah Alas location. The nearest neighborhood analysis consists of three patterns of clustered, random, and scattered patterns and calculated using equation (1)

$$N_n = \frac{D_{obs}}{\frac{1}{2}\sqrt{\frac{a}{n}}} \qquad (1)$$

with N_n is the nearest neighbor value, D_{obs} is the mean of observed nearest neighbor, a is the area of observation and n is the total points. We also employed the Kernel Density to view the aggregate of the Omah Alas at the study area. Kernel Density works by using the radius of random features to determine the density [30]. It goes by equation (2).

$$KD = \frac{1}{nh}\sum_{i=1}^{n} K(\frac{a-a_i}{h}) \qquad (2)$$

Where KD is the Kernel Density, h is the bandwidth, 1 is the point, and n is the number of points.

3 Results and discussions

3.1 Spatial Distribution of Omah Alas

3.1.1 Village Setting

The settlement pattern in the Karstic area differs from other landscapes due to the slow rate of development, low availability of land, density, and geomorphological setting. The map of the settlement pattern in study area displays the linear pattern along the main road. A relatively nucleated pattern is shown in the center of the village where the village administrative office is located. Omah Alas and the settlement at the field are dispersed due to the land setting and the number of settlers. This pattern is also affected by the main occupation of the residence. The nucleated pattern is dominated by farmers whose fields are located distances away.

People who live in the linear pattern of settlement along the Baron Road mostly open a small grocery store providing snacks and beverages for the passerby. This activity is boosted by the tourism activity where most of the visitors use Baron Road as the primary access whenever they visit the beach area. Meanwhile, the pattern for the Banjarejo Village is relatively more scattered than Kemadang Village due to a lack of economical activity. The village pattern of Kemadang and Banjarejo can be seen in Fig. 2 below.

 https://doi.org/10.1051/e3sconf/202132505004

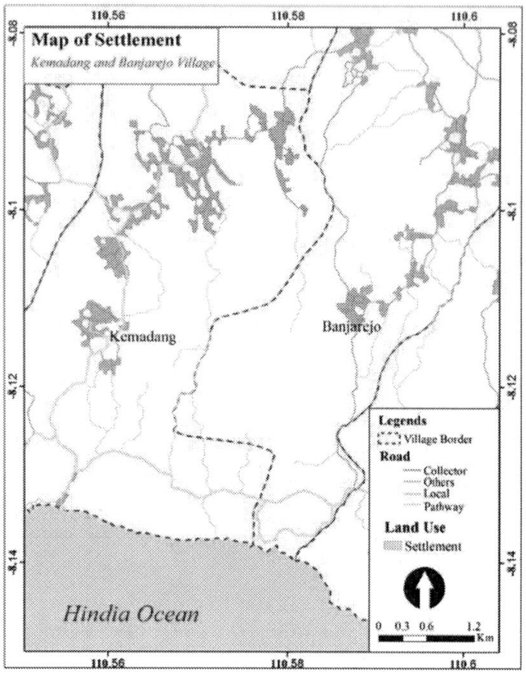

Fig. 3. Settlement Pattern of Kemadang Village (Source: Analysis, 2021)

Even though the administrative location of Omah Alas is on the Kemadang and Banjarejo Village. All of the Omah Alas dwellers have a house in the Kemadang Village. Therefore the factors that affect the Omah Alas location mostly originate from the Kemadang Village where the Omah Alas dwellers have their house and registered. Afterward, the condition of Kemadang Village as one of the primary tourism becomes one of the affecting factors. For this reason, we focus more on the Kemadang Village setting than Banjarejo.

The presence of tourism attraction particularly at Kemadang Village mixed with coastal and farming land creates various job sectors: agricultural, fisheries, and tourism. Each sector is then grouped into a particular sub-village or sector [31]. As the coastal area is located in the southern part of the village, people in the southern mostly work in the maritime or fisheries sector while people in the northern sector work in the agricultural sector. The division of the sector can be seen in Table 1.

Table 1. Sectors of Activity at Kemadang Village

Sector	Sub-Village	Activity
Central Kemadang	Kemadang Lor, Kemadang Kidul, Pucung, Tenggang, Nglaos, Karanglor, Ngasem, Karangasem	Agricultural
Kelor	Kelor Lor, Kelor Kidul	Agricultural
Kayubimo	Kayubimo	Agricultural
Sumuran	Sumuran, Watubelah, Kemadang Kulon	Tourism, Services
Rejosari	Rejosari, Ngepung	Maritime, Tourism, Services

(Source: [31], Analysis, 2021)

3.1.2 Spatial Distribution of Omah Alas

Kemadang Village pattern affects the properties of Omah Alas particularly the land ownership and the location of Omah Alas themselves. The agricultural sub-village is the Kelor and Central Kemadang sector and farmers from those sectors dominate the Omah Alas affecting linear patterns with direction to the south where the fields are mostly located. Farming activity is the main cause that attracts settlements to go souther. With most of the settlement is developed in the northern part, it consumes more space and land in the north and pushes farmers to move to the southern part hence causes Omah Alas to accumulate in the south. The settlement also develops following the morphological characteristic. From the maps shown in Figure 1, the settlement is adjusted according to the contour line. Morphology of Karst Gunung Sewu, especially in the research area is the polygonal karst [7] with lineaments of N-S and influenced by the tectonic movement and local fault. The spatial distribution can be seen in Fig. 3 below.

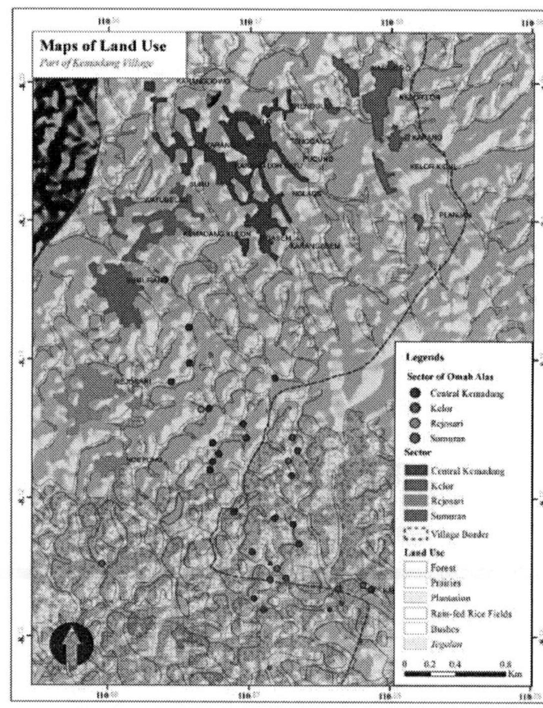

Fig. 4. Distribution of Omah Alas in Southern Part of Kemadang and Banjarejo Village (Source: Analysis, 2021)

We discovered 30 Omah Alas dispersed in the field. From these 30 Omah Alas, 9 of the dwellers stay in the Omah Alas a daily basis although they owe house in the village (*dusun*) while the others stay for only half a day and go back home when dusk and return to the field at dawn. Hence, the owner of Omah Alas can be divided into two categories: the full-time owners who stay overnights and half-time owners who use Omah Alas only as a resting place in the day and go home when dusk [23]. From these categories, the full-time owners mostly come from Central Kemadang and Kelor sectors. Farmers from those two sectors also own most of the field. To see how the aggregate of the Omah Alas, the Kernel Density Map could be seen on Figure 5.

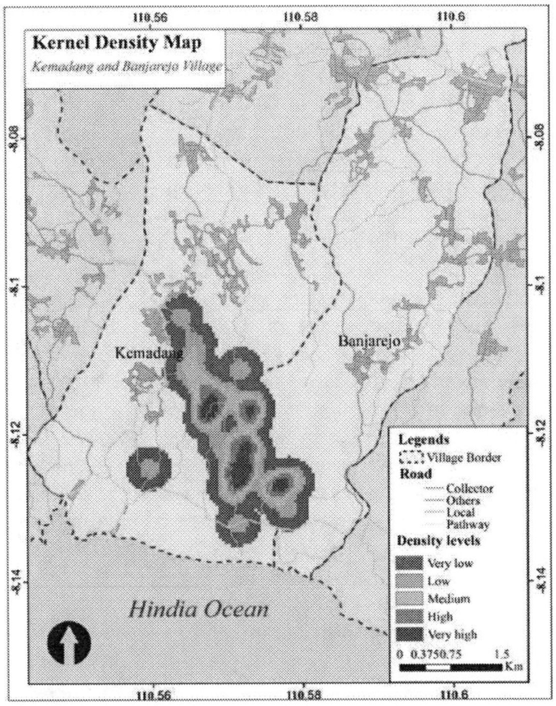

Fig. 5. Kernel Density Distribution of Omah Alas (Source: Analysis, 2021)

The Kernel Density map exhibits Omah Alas is denser in the borderline of the village. It also confirms that the southern the area, the more Omah Alas will be. There are several hotspots where Omah Alas are aggregated to a certain point.

From the settlement pattern, the Omah Alas in the south are the extension of the linear pattern of the Central Kemadang and Kelor sector. Meanwhile, the Rejosari and Sumuran sector farmers occupy a relatively small area in the eastern field of the sector and because of the temporality manners, the pattern of Rejosari and Sumuran are not too related to the presence of Omah Alas. To determine the spatial distribution pattern, we run the nearest neighborhood analysis, and the result is shown in Fig. 6 and Table 2 below.

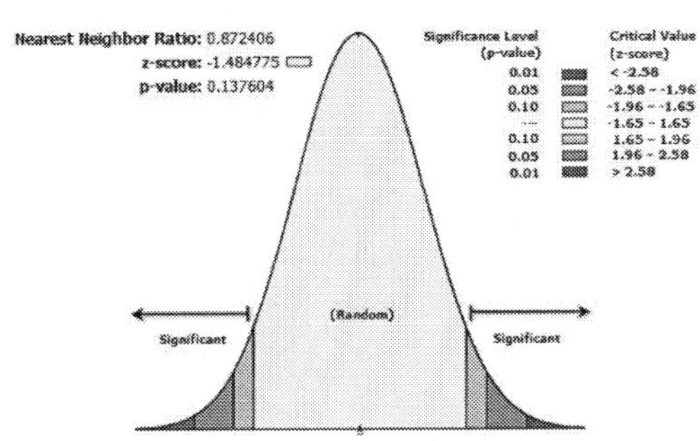

Fig. 6. Nearest Neighborhood Analysis Result (Source: Analysis, 2021)

Table 2. Results of Nearest Neighborhood Analysis

Observed Mean Distance	175.0803 Meters
Expected Mean Distance	200.6867 Meters
Nearest Neighbor Ratio	0.872406
z-score	-1.484775
p-value	0.137604

(Source: Analysis, 2021)

The nearest neighborhood calculation generates the Observed Mean Distance at 175 meters. This is the number of the average distance between one Omah Alas to other Omah Alas and also the average length of the size of a doline in karst Gunung Sewu which shows how the landform influences the distance between Omah Alas. however, the z-score of -1.48 is located on the insignificance area of the normal distribution of significant value which categorizes the spatial distribution into random pattern although the Omah Alas are mostly located in the closed depression. This could happen due to the irregularity of the landform in the karst area. The location of Omah Alas can be seen in Figure 7.

Fig. 7. Aerial View of Omah Alas (source: Google Earth Image and Analysis, 2021)

The random pattern also gives a clue on how the location of Omah Alas is chosen without considering the distance between other Omah Alas. Every Omah Alas dweller is pivoting only to their assets: land and livestock. Sharing workforce in daily work seldomly occurs and happens only when harvest season arrives and most of sharing work are done by involving family relationships like the farmer's children or relatives instead of the Omah Alas neighbor. This spatial pattern and distribution complements the previous study on spatial characteristics of Omah Alas which shows an individual pattern of Omah Alas and limited to the local settings [32].

3.2 *Omah Alas* and Karst

Omah Alas is a specific term and refers to the hut in the middle of the field that uses for semi-permanent settlements. It is a common phenomenon in the southern area of Karst of Gunung Sewu, specifically in

Tanjungsari District. Tamimi, Rahmi, and Ikaputra [33] elaborates on how the Omah Alas as a building functions accordingly with agriculture as the inherent economical activities among the settlers. Omah Alas also serves the two types of daily livelihood: the settlers and the commuters. This pattern of daily livelihood is based on several considerations and yields some consequences. The picture of Omah Alas location setting example can be seen in Fig. 8.

Fig. 8. Omah Alas at The Foothill (source: Documentation, 2021)

The reason for the settlers to stay on Omah Alas is mostly practical and related to the distance from the house in the village [23]. This is the first factor of the village setting. It would take time to go to the field in the dawn whereas they must be hasty to be in the field. If they got too late, the weather would be too hot to work, and it is considerably inconvenient for them. In addition, although they could use a mode of transportation such as a motorcycle or public transportation, they renounce that idea because they need to expend additional cost and not to mention some of them cannot ride the motorcycle. Another reason for the Omah Alas to stay in their field is due to the presence of pests that could demolish their plant. Unlike other fields, a threat for the farmers in Tanjungsari District is big mammals like monkeys or porcupines. Monkeys and their herd have become the biggest problem in recent years. They usually attack in the evening and cannot be stopped using regular equipment. Farmers must use fireworks to produce explosive sounds to repel the monkey. It requires full-time surveillance. For non-practical reasons, some Omah Alas settlers admit they feel serenity in the field while gazing at the plant and crop. They can enjoy themselves without burden or anxiety. Omah Alas contains dual functions for the settlers. Besides the dwelling and serves daily needs such as dishes, break, and night sleep, it also plays a role in supporting agricultural activities. Farmers build temporary storage (*lumbung*) for crop and livestock integrated with the main building since it would be much easier to manage and guard the crop and livestock if their place is adjacent.

The random dispersed pattern of Omah Alas is also affected by the irregularity point at the presence of doline or closed depression uniquely happens in the karst area or the morphological factor. Doline has a critical role for agricultural activity and livelihood since it is the only morphological unit in karst area that deposits soil and retains waters. The soil in karst area is terra rosa and originated from the weathering of limestone hence, this soil is very rich in calcium. Because the deposition of soil occurs in doline, the soil thickness is relatively higher and fits to be cultivated. Water is also considerably abundant in doline area. Rainfall that drops in the karst would flow into this depression and be collected securing water demand for the crop. Furthermore, flat and plain morphology in doline makes it easier to be adjusted and managed. A combination of these factors is the key factor for the agricultural setting in the karst area.

Regardless, sole contingency on doline is not enough and there's still space available at the conical hill slopes. Hillslope in the karst area is not suitable for agriculture plants like paddy or corn as the main commodity. It has very thin soil because the soil material is constantly eroded and washed away to the doline and the slope also tackles cultivation. Farmers in Karst of Gunung Sewu overcome these obstacles by building *terrasering* system [34]. To retain soil and water on the terrace. Nevertheless, this method doesn't work well on paddy and corn because it still can be easily swept away by the water flow which runs fast on slopes thus farmers plant hardwood trees like teaks or acacias. This system had already been shown by Nibbering [34] and he divided this as low-land agriculture and high-land agriculture.

Another factor that influences the spatial pattern is land ownership. The limited land resource makes the land distribution is considerably hard to be done. Physical barriers such as karstic hills create a constraint in dividing farming land into the same amount. Hence, the farmers who want to expand their land must accuse irregular forms to achieve the area they need. Further, omah alas an only be built on the righteous property hence make its distribution is in accordance to the land ownership.

In short, the factors that affected Omah Alas spatial distribution is shown in Table 3.

Table 3. Factors in the Spatial Distribution of Omah Alas

No	Factors	Explanation
1	Village settings	Field and the house in the permanent settlement is distant.
2	Morphology	Omah Alas's location follows flat morphology and is located in doline as the most suitable morphology for cultivation with its thick soil and water retention.
3	Land use	Omah Alas are bounded in the agricultural land use due to the high numbers of agricultural activities
4	Land ownership	Omah Alas can only be built on the righteous properties.

(Source: [32], Analysis, 2021)

The aggregate of Omah Alas forms a social and certain neighborhood network and becomes a unique pattern. A study on settlement patterns by Kaho and Giyarsih [22] evaluates how the house's qualities in karst area but are only limited to the village area and does not take the Omah Alas type into account. Another research in the settlement pattern at karst area of Pati, conducted by Anatasari and Pradoto [21] suggests that settlement in karst area tends to approach a water source. However, because of the difference in hydrological characteristics, Omah Alas at Karst of Gunung Sewu is not attracted to water sources. There are barely any water sources in the surrounding area of Omah Alas and for domestic needs, the settlers rely on the rainwater container (*Penampung Air Hujan* (PAH)).

Moreover, we argue that Omah Alas will be hardly developed into a village or settlement due to the remoteness and agricultural land/space demand; they are not the seed of settlement. This argumentation lays on the fact that the surrounding land of Omah Alas is private ownership and has been used as crop fields. However, with the very rapid infrastructure expansion in the coastal region, houses will be vastly emerge causing land-use change. Rapid development in the village and transition to the higher level (village to a small town) also may cause the development although in the karst area, the development will be driven mainly by the morphological features as shown in the China karst where the rural-urban relationship is heavily configured by the elevation aspect [35]. Nevertheless, with the relatively similar altitude, the development in the Karst of Gunung Sewu and the case of Omah Alas will be more driven by the presence of doline and karstic hills for the path or access of interaction between the two regions. A further and thorough observation is desperately required to trace the evolution of Omah Alas into a further form.

4 Conclusion

We conclude that Omah Alas are distributed in the random pattern and aggregated in the southern part of the research location where several high densities are found with the average distance among the Omah Alas is 175 meters. This pattern is influenced by the physical factor of the Karst Area and the settlement pattern in Kemadang Village as the host-village of Omah Alas. The settlement pattern of Omah Alas is also affected by the land use and the land ownership where the Omah Alas are only located in the agricultural land use and in the righteous properties or ownership of the land.

We thank Kemadang Village residents for their hospitality and willingness to provide the data we need. The data are also part of the Shabrina Tamimi's thesis titled "*Arsitektur Omah Alas Petani Subsisten di Desa Kemadang, Gunungkidul*".

References

1. E. Burri, B. Castiglioni, U. Sauro, Int. J. Speleol. **28**, 33 (1999)
2. L. Song, Int. J. Speleol. **28**, 139 (1999)
3. D. Gillieson & M. Thurgate, Int. J. Speleol. **28**, 149 (1999)
4. I. Bárány-Kevei, Int. J. Speleol. **28**, 89 (1999)
5. G. Veni, H. DuChene, N. C. Crawford, C. G. Groves, G. N. Huppert, E. H. Kastning, R. Olson, B. J. Wheeler, Living with Karst: A Fragile Foundation (American Geological Institute, 2001)
6. P. Forti, *The scientific and socio-economic importance of karst and caves and their vulnerability*, in Brief for GSDR (2015)
7. E. Haryono, M. Day, J. of cave and karst studies **66**, 62 (2004)
8. H. D. Tjia, Indonesian J. on Geoscience **8**, 75 (2013)
9. A. Sunkar, Media Konservasi **13**, 1 (2008)
10. H. Reinhart, A. Rifani, Jurnal Ilmu Lingkungan **17**, 2 (2019)
11. M. Baiquni, *Livelihood Strategy of The Community in A Karst Village, Gunung Sewu, Indonesia* in Proceedings Asian Trans-Disciplinary Karst Conference, Yogyakarta, January 7-10th (2011)
12. J. C. Scott, Moral Ekonomi Petani: Pergolakan dan Subsistensi di Asia Tenggara (LP3ES, 1976)
13. C. Coxon, *Agriculture and Karst* in P. E. van Betnen (ed.), Karst Management, Springer, Dordrecht (2011)
14. E. D. Suryanti, Sudibyakto, M. Baiquni, Jurnal Kebencanaan Indonesia **2**, 3 (2010)
15. W. Widiyanti, A. Dittman, Procedia Environmental Sciences **20**, 693 (2014)
16. H. Uhlig, GeoJ. **4**, 31 (1980)
17. R. A. Suwasono, P. Mateus, Sutedjo, IOP Conf. Series: Earth and Environmental Science **144** (2018)
18. P. Toth, P. Demjan, K. Griacova, Documenta Praehistorica **38**, 307 (2011)
19. P. V. Kirch, A. S. Hartshorn, O. A. Chadwick, P. M. Vitousek, D. R. Sherrod, J. Coil, L. Holm, W. D. Sharp, *Environment, agriculture, and settlement patterns in a marginal Polynesian landscape* in Proceedings of the National Academy of Sciences **101**, 9936 (2004)
20. P. Khadiyanta, S. Dewantari, *Settlement adaptation by reshaping dwellings in the degrading area at Genuk Disctrict of Semarang City, Indonesia* in Proceeding of International Conference, Intelligent Planning Towards Smart Cities, 3-4 November 2015, Surabaya, Indonesia, (2016)
21. A. T. Anatasari, W. Pradoto, Jurnal Pembangunan Wilayah & Kota **12**, 251 (2016)
22. H. E. D. P. R. Kaho, S. R. Giyarsih, Jurnal Bumi Indonesia **7**, 1 (2018)
23. S. Tamimi, Arsitektur Omah Alas Petani Subsisten di Desa Kemadang, Gunungkidul (Thesis, Department of Architecture and Regional Planning

Faculty of Engineering Universitas Gadjah Mada, 2020)

24. D. C. Ford, P. Williams, Karst Hydrogeology and Geomorphology (John Wiley & Sons, 2007)
25. E. Haryono, Suratman, *Significant Features Of Gunung Sewu Karst As Geopark Site* in 4th International UNESCO Conference on Geopark, April 12-15, 2010, Langkawi (2010)
26. T. Simanjuntak, Gunung Sewu in prehistoric times (Gadjah Mada University Press, 2002)
27. BPS, Kecamatan Tanjungsari Dalam Angka 2018, (2019)
28. M. Singarimbun, S. Effendi, Metode Penelitian Survai (LP3ES, 1982)
29. H. Subagio, R. H. Koestoer, Y. M. Zagloel, L. Darmajanti, IOP Conf. Series: Earth and Environmental Science **280** (2019)
30. N. U. Khan, W. Wan, S. Yu, ISPRS Int. J. Geo-Inf. **9**, 70 (2020)
31. I. Fauzi, G. Radjiman, Antara Dua Dunia: Strategi Bertahan Hidup Penduduk Desa Kemadang, Kecamatan Tanjungsari, Kabupaten Gunungkidul, (Thesis, Department of Architecture and Regional Planning Faculty of Engineering Universitas Gadjah Mada, 2014) https://repository.ugm.ac.id/131086/
32. S. Tamimi, D. H. Rahmi, *Omah Alas as an Adaptive Dwelling in Karst Area: Spatial Characteristics*, J. Built Env. Studies **1**, 45 (2020)
33. S. Tamimi, D. H. Rahmi, Ikaputra, Journal of Architectural Discourses **17**, 104 (2020)
34. J. W. Nibbering, Upland cultivation and soil conservation in limestone regions on Java's south coast: Three historical case-studies in P. Boomgaard, F. Colombijn, and D. Henley (ed.) Paper Landscapes explorations in the environmental history of Indonesia (KITLV Press, 1997)
35. J. Li, Z. Cai, L. Duan, J. Earth Sci. Res. J. **24**, 267 (2020)

E3S Web of Conferences **325**, 06001 (2021) https://doi.org/10.1051/e3sconf/202132506001
ICST 2021

Analysis of Economic Inequality and Human Development: A Case Study of West Sumatra Province in 2015-2019

Muhammad Fikram Palembang, *Ismanita* Prahara, *Muhammad* Rizky Shiddiq Nugraha, *Relinda* Dewi Astabella, *Revia* Ardyaning Pramesti, *Dimas* Maula Hayat, and *Ratih* Fitria Putri*

Department of Environmental Geography, Faculty of Geography, Gadjah Mada University, Indonesia

Abstract. West Sumatra Province is one of the provinces with diverse human resources. This situation causes the need for human resource management to encourage economic growth. This study aims to identify economic development, economic inequality and human development, as well as the relationship between economic inequality and human development in West Sumatra Province in 2015-2019. The basic data used in the analysis are GRDP and the Human Development Index (IPM) using quantitative descriptive analysis methods. Economic development was analyzed using Location Quotient (LQ) and Klassen typology, while economic inequality was using Entropy Theil. The research showed that the Klassen typology results were dominated by developed but slow regions which amounted to nine regencies/cities and based on LQ showed that urban areas had a base sector in the secondary and tertiary sectors, while rural areas had a base sector in the tertiary sector. The HDI has increased which is accompanied by an increase in entropy, indicate that high GRDP is not evenly distributed. In addition, there is a strong relationship between economic inequality in terms of GRDP per capita and HDI. This shows an increase in economic growth that triggers improvements in the health and education sectors so that it has a good impact on the quality of human development but needs an even distribution.

1 Introduction

Economic growth in the development process is influenced by the quality of resources in a human region. Simply put, the better the quality of human resources, the better the regional economic growth. The focus of economic development in order to achieve better development is focused on reducing poverty, inequality, increasing capital income, and reducing unemployment [1]. One of the causes of economic development being hampered is because one of these focuses is still difficult to achieve, the problem is economic inequality. This problem cannot be separated from human development, because good human development will lead to good economic growth. In accordance with statement by the United Nations Development Program (UNDP) through the concept trickle down which explains that economic development will lead to human development, on the contrary in the concept of virtuous circle it explains that human development will lead to economic development. The two components are interconnected as described in the concept of a triangle circle, if one of these components goes well then the other components will follow [2].

West Sumatra Province is located along the west coast of the central part of Sumatra Island, bordered by the Bukit Barisan plateau in the east, and a number of offshore islands in the south part. Amount of 5.534.472 people spread over 19 administrative areas consisting of 7 cities and 12 regencies. Most of the population in West Sumatra Province work in agriculture, which is 36,22 percent. Furthermore, other sectors that dominate are the trade sector amount of 19,10 percent [3]. The geographical condition of West Sumatra Province which is not entirely in the form of highlands makes this area dominated by the agricultural sector. The dominance of the agricultural sector in this province triggers economic growth and human development to become increasingly complex and diverse. Based on previous research [4], there was an increase in development inequality in West Sumatra was influenced by economic growth, investment, labor and balancing funds. These conditions encourage the economic growth of West Sumatra Province which needs more attention in order to achieve the goals of regional development to improve the welfare of the community and the quality of regional human resources [5].

An analysis of inequality and human development in West Sumatra Province is needed to evaluate the level of economic and human growth in the area. The results of the analysis can be used as an evaluation material for human resource management and economic development. Therefore, the purpose of this study was to determine the condition of economic development, economic inequality and human development and to determine the relationship between economic inequality and human development in West Sumatra Province from 2015 to 2019.

*Corresponding author: ratihfitria.putri@ugm.ac.id

2 Method

The research was carried out in West Sumatra Province which consisted of 12 regencies involving Kepulauan Mentawai, Pesisir Selatan, Solok, Sijunjung, Tanah Datar, Padang Pariaman, Agam, Lima Puluh Kota, Pasaman, Solok Selatan, Dharmasraya, Pasaman Barat and 7 cities involving Padang, Solok, Sawah Lunto, Padang Panjang, Bukittinggi, Payakumbuh and Pariaman (Fig. 1). The research was conducted using a descriptive-quantitative analysis method by processing secondary data for 2015-2019. The data used in this study consist of secondary data published by Bureau of Statistik (BPS) Indonesia and West Sumatra, involving data of Gross Regional Domestic Product (GRDP) by regencies/city in West Sumatra, Gross Domestic Product (GDP) of Indonesia, number of population, and human development indicators through Human Development Index (HDI). The inequality of West Sumatra measured by Entropy Theil, Location Quotient (LQ), and Klassen Typology. Furthermore, correlation between economic inequality and human development was analyzed with Pearson Correlation.

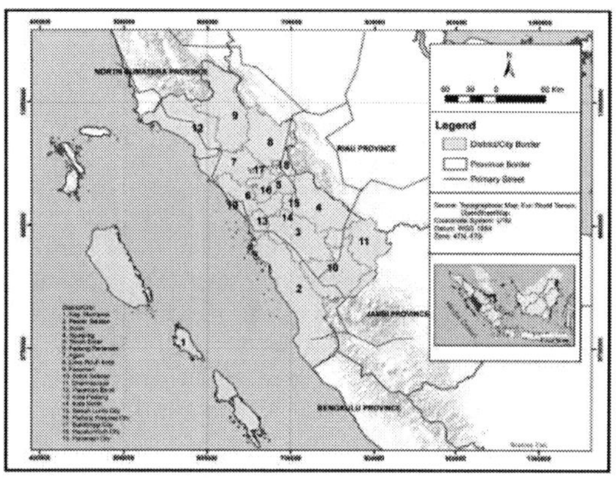

Fig 1. Administration Map of West Sumatra

2.1 Entropy Theil

Theil Entropy Index is used to measure economic inequality between regions [6]. Economic inequality in West Sumatra Province is measured using the Theil Entropy Index formula [7] as follows:

$$I\ Theil\ in = \sum (yi / Y) \times \log [(yi / Y) / (xi / X)] \quad (1)$$

Description:

I Theil in	:	Economic inequality inside the region
yi	:	Gross Regional Domestic Product (GRDP) per capita of *i* regency/city
xi	:	Population of *i* regency/city
Y	:	Average GRDP per capita of the province
X	:	Population of the province

Inequality

$$I\ Theil\ out = \sum (y/Y) \times \log [(y / Y) / (x / X)] \quad (2)$$

Description:

I Theil out	:	Economic inequality outside the region
Y	:	Gross Domestic Product (GDP) per capita of the province
x	:	Population of the province
Y	:	Average GDP per capita of the country
X	:	Population of the country

The greater Theil Entropy indicates the higher economic inequality between regions. On the other hand, the smaller Theil Entropy Index indicates low economic inequality between regions.

2.2 Location Quotient (LQ)

Location Quotient (LQ) is an indirect approach that is used to measure the performance of the base sector which is the growth driver by measuring the relative concentration or degree of specialization of economic activity through a comparative approach to the economy of a region [8]. The formula used to calculate LQ, namely:

$$LQ = (Vik / Vk) / (Vip / Vp) \quad (3)$$

Description:

Vik	: GRDP Agricultural sector Regency
Vk	: GRDP Total all sectors in Regency
Vip	: GRDP Agricultural sector Province
Vp	: GRDP Total all sectors Province

The results of the LQ calculation produce 3 criteria, namely:

- LQ > 1, meaning that sector i is the basis or source of economic growth in region k.
- LQ = 0, meaning that sector i is not the base sector in region k.
- LQ < 1, meaning that sector i is not the base sector or a source of economic growth in region k.

2.3 Klassen Typology

The Klassen typology is carried out by comparing the regional economic growth with the economic growth that is the reference in the area and comparing the regional GRDP per capita with the GRDP per capita which is the reference for the region [9]. The Klassen typology can be shown in the following matrix [10]:

Table 1. Klassen Typology Matrix.

	Y area > Y reference	**Y area < Y reference**
r area > r reference for	Advanced and fast-growing region	Potential region or can still grow

		rapidly
r region < r reference for	Advanced region but depressed	The region is relatively lagging behind

Description:

Y region : GRDP per capita area

Y reference : GRDP per capita which is the reference for the region

r region : Regional economic growth

r reference : Growth economy is the reference area of the

Calculation of Typology Klassen results will produce four categories of areas:

- Fast advancing and growing rapidly.
- Regions that are fast progressing but slow to develop.
- Developing areas, in contrast to potential areas.
- Lagging regions.

2.4 Pearson Correlation

Correlation is used to determine the relationship between variables. Pearson correlation has a value ranging from -1 to 1, with the following formula:

$$r = \frac{xy - \frac{(\sum x \sum y)}{n}}{\sqrt{(\sum x^2 - \frac{(\sum x)^2}{n})(\sum y^2 - \frac{(\sum y^2)}{n})}} \quad (4)$$

The results of the Pearson correlation are described by the notation r with the criteria in Table 2 below [11]:

Table 2. Criteria for the results of the Pearson Correlation.

r value	Criteria
0.00 - 0.19	Very weak correlation
0.20 - 0, 49	Weak correlation
0.50 - 0.69	Fair correlation
0.70 - 0.79	Strong correlation
0.80 - 1.00	Very strong correlation

3. Result and Discussion

3.1 Economic Inequality in West Sumatra Province

3.1.1 Entropy Theil

Theil Entropy Index explains the economic inequality of a region based on per capita income. If the value is closer to 0 (zero), the distribution of income is more equal [12]. Theil Entropy Index measures economic inequality inside the Province of West Sumatra and economic inequality outside the Province of West Sumatra. Based on the calculation of the Theil Entropy Index outside the Province of West Sumatra in 2015-2019 using the variables of GRDP and the population of the Province of West Sumatra then compared to the GRDP and the total population of Indonesia, it is known that the Theil Entropy Index of the Province of West Sumatra is increasing.

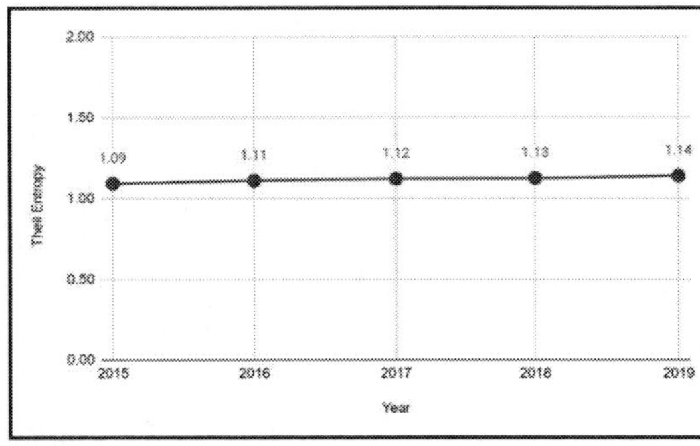

Figure 2. Theil Entropy Index of West Sumatra Province 2015-2019.

Figure 2 shows the increase in the Theil Entropy Index of West Sumatra Province in 2015-2019. Theil Entropy Index of West Sumatra Province increased by 0.05 for 5 years (2015-2019). The increase in Theil's Entropy Index indicates that the economic inequality of West Sumatra Province against other provinces in Indonesia is getting bigger. The value of the Entropy Index is influenced by GRDP and population. The large GRDP and small population cause the low value of Theil's Entropy. A low Theil Entropy Index indicates low economic inequality. GRDP per capita of West Sumatra Province increased in 2015-2019, however, its value is still below the average GDP per capita of Indonesia. Thus, the GRDP per capita of West Sumatra Province is still below the GRDP per capita of other provinces. The value of GRDP per capita of West Sumatra Province and the average GDP per capita of Indonesia can be seen in Figure 3.

Table 3. GRDP per capita of West Sumatra Province and GDP per capita of Indonesia

Year	GDRP of West Sumatra	GDP of Indonesia
2015	27,044.14	38,138.93
2016	28,164.93	39,246.29
2017	29,310.69	40,495.93
2018	30,477.76	41,997.49
2019	31,669.59	43,195.44

Economic inequality between regencies/cities in the Province of West Sumatra is calculated using the Theil-in Entropy Index (I Theil In). Based on the results, it is known that the Theil Entropy Index of regencies/cities in West Sumatra Province is at a low to moderate level. Urban (city) has higher economic inequality than rural (regency). Regencies/cities in West Sumatra Province that have a low Theil Entropy Index are South Pesisir Regency, Solok Regency, Tanah Datar Regency, Agam Regency, Pasaman Regency, and Pasaman Barat Regency, while regencies/cities that have a moderate Theil Entropy Index are Sijunjung Regency, Padang Pariaman Regency, Lima Puluh Kota Regency, South Solok Regency, Dharmasraya Regency, Padang City, Solok City, SawahLunto City, Padang Panjang City, Bukittinggi City, Payakumbuh City dan Pariaman City. The spatial distribution of Theil Entropy Index levels in West Sumatra Province can be seen on the Theil Entropy Index Map of West Sumatra Province (Figure 3).

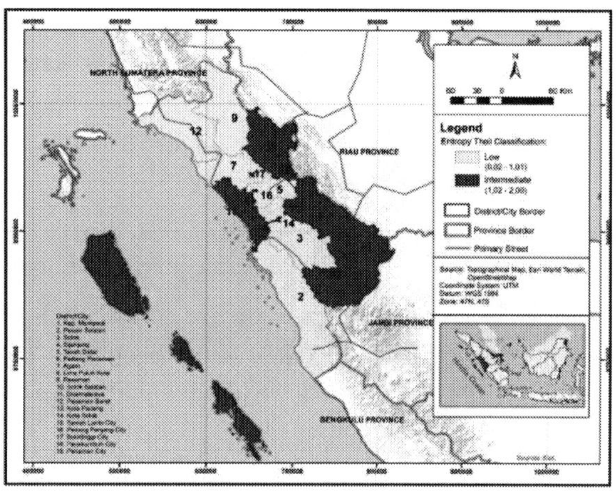

Fig. 3. Theil Entropy Index Classification Map of West Sumatra Province by Regency/City 2015-2019.

3.1.2 *Location Quotient (LQ)*

Economic inequality arises due to the absence of equity in economic development which can be caused by differences in the main pillars of the economy as well as differences in the base sector in each region. The basic sector is a sector that has an important role in economic growth because it is directly related to the demand for goods and services from outside the region. Economic basis theory states that the rate of economic growth of a region is determined by the magnitude of the increase in exports from that region. Thus, basic activities have an important role as the main driver of the economy in a region. Economic inequality based on differences in the base sector in West Sumatra Province can be seen through the Location Quotient value.

Table 4. The results of the LQ calculation by regencies/cities in West Sumatra Province in 2015-2019.

Regency/Municipal	Primary Sector	Secondary Sector	Tertiary Sector
Kepulauan Mentawai	1,73	0,29	0,72
South Pesisir	1,55	0,80	0,80
Solok Selatan	1,59	0,54	0,82
Sijunjung	1,64	0,32	0,84
Tanah Datar	1,25	1,16	0,87
Padang Pariaman	1,03	1,24	0,95
Agam	1,28	1,18	0,85
Limapuluh Kota	1,56	0,75	0,80
Pasaman	1,92	0,46	0,70
South Solok	1,46	0,53	0,88
Dharmasraya	1,41	0,58	0,90
West Pasaman	1,57	0,38	0,57
Padang City	0,29	1,41	1,24
Solok City	0,23	0,55	1,41
Sawah Lunto City	0,57	1,38	1,12
Padang Panjang City	0,20	1,02	1,35
Bukittinggi City	0,05	0,73	1,46
Payakumbuh City	0,27	0,58	1,39
Pariaman City	0,73	0,79	1,15

Based on Table 4 the results of LQ calculations in Sumatra Province West 2015-2019 shows that the primary sector is the economic base sector in the Kepulauan Mentawai, South Pesisir, Solok, Sijunjung, Tanah Datar, Padang Pariaman, Agam, Limapuluh Kota, West Pasaman, South Solok, Dharmasraya, and West Pasaman. It is because the primary sector has an LQ>1 index, meaning that the regency is more specialized or more dominant in the primary sector than at the provincial level, where this sector in the economy has a comparative advantage and is categorized as a leading base sector. The secondary sector is the economic base sector in Tanah Datar Regency, Padang Pariaman Regency, Agam Regency, Padang City, Sawah Lunto City, and Padang Panjang City because the secondary sector has an LQ index > 1, meaning that the primary sector is the basis or source of economic growth, where these regencies/cities are more specialized in the secondary sector than at the provincial level. While the tertiary sector into sectors of the economic base in the city of Padang, Solok, Sawah Lunto, Padang Panjang, Bukittinggi, Payakumbuh and Kota Pariaman for the tertiary sector has LQ> 1, meaning that the tertiary sector into the base or be a source of economic growth, which the regency/city more specialized in the tertiary sector than at the provincial level.

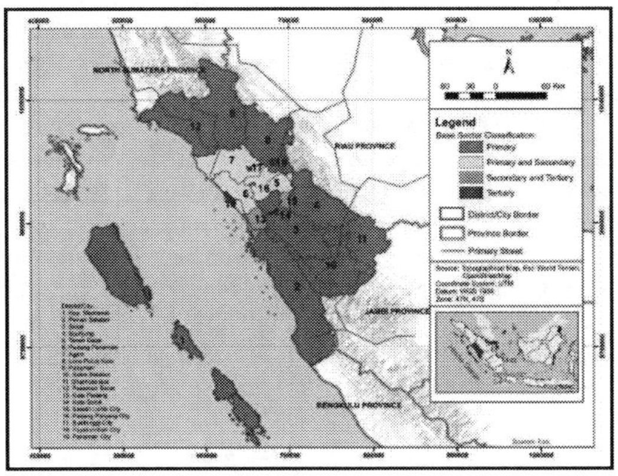

Fig 4. Base sector distribution map in West Sumatra Province.

Based on the results of the analysis of the economic base sector as shown in Figure 4 the distribution map of the base sector, it was found that there is a very significant difference between urban and rural areas, where urban areas have an economic base sector in secondary and tertiary sectors, while rural areas have an economic base sector in the primary sector. This is due to differences in the content of natural resources, geographical conditions, the rate of infrastructure growth as well as differences in demographic conditions as well as differences in urban and rural areas. Rural areas contain natural resources and geographical conditions that are suitable for the development of the primary sector (agriculture, livestock, fisheries and forestry), while urban areas have a lot of infrastructure development and have human resources who have high knowledge, so that the sector that develops a lot are industry, trade and services sectors.

3.1.3 Klassen Typology

An area can be said to be developed and developing, one of which is by using GRDP data, which is the amount of added value produced by all business units in an area or is the total value of goods and services produced by all economic units in a region [13]. Then to see the developments in an area, it can be seen from the development of the GRDP of the previous year with the current year. Then to find out whether the area is included in the category of developed, developing or underdeveloped, the Klassen typology analysis can be used. West Sumatra Province has 19 regencies and cities. The results of the calculation of the klassen typology of each city regency in the province of West Sumatra with the reference value used is the average of the whole, or the value of the GRDP and economic growth of the province of West Sumatra it self from 2015-2019, has a distribution as shown in Figure 5. Graph distribution category the number of regencies/cities in West Sumatra Province from 2014 to 2019.

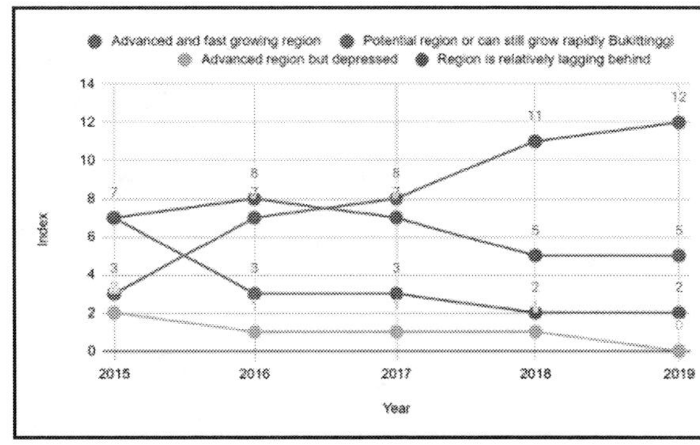

Fig 5. Graph of the distribution of categories for the number of regencies/cities in West Sumatra Province 2014-2019.

Based on the graph, it can be seen that areas that are very prime in terms of progress and development have decreased every year. The decline occurred in seven regencies/cities in West Sumatra Province in 2015 and the remaining two regencies/cities in 2019. This also occurred in potential regions or can still grow rapidly and relatively lagging behind regions, although the decline was not too significant. The decline in the number of regencies/cities classified as advanced and fast-growing regions in underdeveloped areas, in 2015 from eight regencies/cities to only five regencies/cities in 2019. The decline in relatively lagging behind regions which originally had two regencies/cities in 2015 became none at all in 2019. Meanwhile, what has increased rapidly is the number of potential regions, in the sense that the area is developed but has slow economic growth, initially there were only three regencies/cities in 2015 increasing to 12 regencies/cities in 2019.

The decline in the advanced and fast-growing region showed a negative shift in value. This is because the number of very prime areas has decreased, of course, it is an indication of problems in the regency and city. However, in the category of potential region or can still grow rapidly and relatively lagging behind region, the decline was followed by an increase in advanced regions but depressed. This is a shift in value in a positive direction. It means the area has a GRDP that is above the reference value so that it is included in the category of developed regions.

One of the factors that makes it difficult to maintain an area in prime condition is the value of GRDP. Most regions, in fact, experienced a decline to potential conditions with high GRDP values but slowing economic growth. If the value of GRDP is low and there is a significant increase in the following year, it will certainly make a spike in economic growth in an area. So that when the GRDP value is already in the advanced category, or greater than the reference

 https://doi.org/10.1051/e3sconf/202132506001

GRDP, it will certainly be difficult to continue to maintain economic growth from the existing GRDP. When GRDP is high, it will tend to experience stability in which economic growth is calculated as a percentage. This of course causes the existing percentage to decrease even though the magnitude of the value increases. This increase resulted in stable growth and was below the reference value, but the existing GRDP was already above the reference value, so that those in the prime sector with advanced GRDP and fast growth slowly fell to potential sectors with advanced GRDP but slow economic growth.

Table 5. Result of Classification Klassen Typology based on the rate of GRDP and GRDP per capita by Regency / City in West Sumatra Province from 2015 to 2019.

	Y1>y	Y1<y
r1>r	**Quadrant I** ***Advanced and fast growing region***	**Quadrant II** ***Potential region or can still grow rapidly***
	Bukittinggi City and Padang Panjang City	Pasaman Barat
r1<r	**Quadrant III** ***Advanced region but depressed***	**Quadrant IV** ***The region is relatively lagging behind***
	Kepulauan Mentawai, Padang Pariaman, Agam, Dharmasraya, Padang City, Solok City, Sawah Lunto City, Payakumbuh City, Pariaman City	South Pesisir, Solok, Sijunjung, Tanah Datar, Limapuluh Kota, Pasaman, South Solok

Table 5 shows the Klassen typology category in each regency and city as a result of the average category in 2015-2019 in the province of West Sumatra. Most of the regencies and cities in the province of West Sumatra are in the III Quadrant and there are only two cities that fall into the prime category or the I quadrant. This is only natural because both are big cities in West Sumatra Province. While in the second quadrant there is only one regency, the rest are in the last quadrant, namely the fourth quadrant which is the lagging quadrant, and the average for the last five years shows that there are seven regencies that fall into that category.

3.2 The Conditions of Human Development in the Province of West Sumatra

Discussion on the Human Development Index (HDI) in the Province of West Sumatra can be reviewed on the conditions of HDI in each regency. Based on the calculation results, in general, all regencies/cities in 2016-2019 experienced an increase in the index. The increase in HDI during that period (Figure 6) was also accompanied by a higher theil entropy (Figure 2) and increase of GRDP (Figure 7). This phenomenon indicates a high value of GRDP does not always indicate a low Theil Entropy. In other words, a high GRDP indicates that there is no even distribution of income in West Sumatra Province. This seems to be against the basic concept of calculating HDI. The HDI calculation takes into account economic indicators so that the higher the economic status, the higher the HDI obtained. This phenomenon indicates that there are indicators that are more dominant than economic indicators, considering that there are indicators of education and health in the HDI calculation.

The HDI condition of each regency contributes to the HDI in general at the provincial level. Figure 7 explains that in the period 2015-2019 in West Sumatra increased levels of HDI. At first glance, the increase in HDI at both the provincial and regency levels is good news for development in this province, although there is a note for development in this province. The note in question is economic inequality that still occurs and has an increasing trend every year.

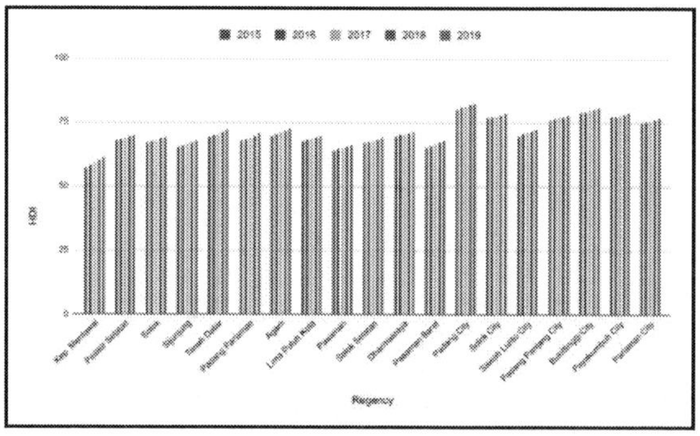

Fig 6. Graph of the Human Development Index (HDI) of West Sumatra Province in 2015-2019.

3.3 Relationship between Economic Development and Human Development in the Province of West Sumatra

Economic growth as described by GRDP per capita has a relationship with human development described by HDI. Economic growth and human development there is a two-way relationship, where economic growth increases human development. However, on the other hand, increasing human development makes it possible to increase economic growth [14] [15]. This happened in West Sumatra Province which can be seen in Figure 6. The figure shows the pattern of per-capita GRDP development and human development in West Sumatra Province from 2015 to 2019. This condition indicates that economic development will affect the increase in human development, or otherwise, the increase in human development will affect the increase in economic development.

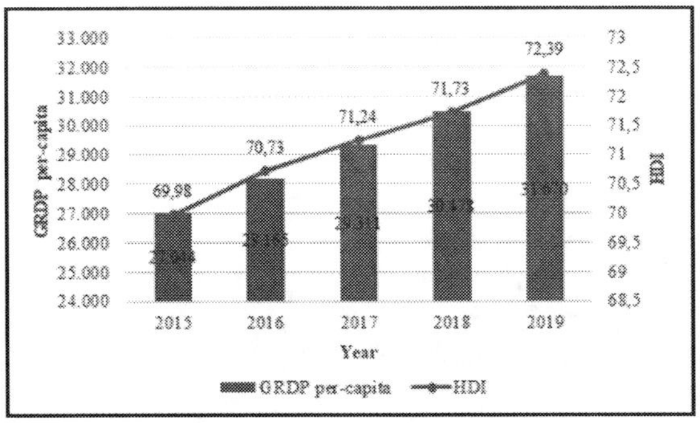

Fig 7. Graphic of GRDP per-capita and HDI in West Sumatra Province 2015-2019.

The relationship between economic development and human development is analyzed based on regencies/cities in West Sumatra Province. The analysis was carried out using Pearson correlation with GRDP as a variable of economic development and HDI as a variable of human development. This analysis was conducted to determine the increase in the economic development of each region will affect the increase in human development in the region. The following are the results of the Pearson GRDP correlation with HDI in West Sumatra Province from 2015 to 2019.

Table 6. Pearson Correlation Results between economic development variables (GRDP) and human development (HDI) by regency/city in West Sumatra Province 2015-2019.

Regency/city	Pearson Coefficient	Significancy
Kepulauan Mentawai	1,000	0,000
South Pesisir	0,986	0,002
Solok	0,989	0,001
Sijunjung	0,995	0,000
Tanah Datar	0,986	0.002
Padang Pariaman	0,962	0,009
Agam	0,999	0,000
Limapuluh Kota	0,992	0,001
Pasaman	0,991	0,001
South Solok	0,995	0,000
Dharmasraya	0,977	0,004
West Pasaman	0,998	0,000
Padang City	0,996	0,000
Solok City	0,994	0,001
Sawahlunto City	0,996	0,000
Padang Panjang City	0,993	0,001
Bukittinggi City	0,994	0,000
Payakumbuh City	0,971	0,006
Pariaman City	0,996	0,000

According to the Pearson correlation analysis presented in Table 6, it shows that the correlation between GRDP per capita and the development index has a strong relationship and is statistically significant because both are at a significance value less than equal to 0,05. It means high economic growth causes human development to also increase. This happens because HDI is one of the indicators that influence economic growth, therefore the Human Development Index (HDI) has a big influence on increasing economic growth. This is reinforced by the results of previous research [16], economic growth and human development have a mutually influencing relationship.

Based on the results of the analysis of the relationship between economic growth and human development, each regency/city in West Sumatra Province in 2015-2019 can be used as a basis for determining the direction of development and policies that can be taken by the government to improve regional development, especially for regencies/cities that are relatively less developed. Therefore, there is a need for a balance between economic growth and human development. The balance of the two can create an even distribution of income and the allocation of public spending in the fields of education and health can improve human development so as to increase regional development.

4 Conclusion

The results of this study indicate that the condition of economic development based on the calculation of Location Quotient (LQ) in urban areas has a base sector in the secondary and tertiary sectors, while rural areas have a leading base sector in the tertiary sector. Economic development from Klassen's typology results, West Sumatra Province there are two cities that are included in developed and developing areas, one district is included in the slow growing developed area, eight districts/cities are classified as developing areas, and seven districts/cities are classified as underdeveloped areas. Economic inequality in terms of the Theil Entropy Index of Sumatra Province against other Provinces in Indonesia is increasing in 2015-2019. This condition is caused by GDP per capita of West Sumatra Province in 2015-2019 is still below Indonesia's GDP per capita although it has increased every year. In addition, this study analyzed the conditions of human development obtained by increasing the index. The increase in HDI during that period was also accompanied by a higher theil entropy which illustrates that high GRDP doesn't indicate an even distribution of income in the Province of West Sumatra. Economic development has a strong relationship with the Human Development Index based on the results of the Pearson correlation test. This relationship shows that an increase in economic growth causes an increase in the Human Development Index in each regency/city in West Sumatra Province. This condition is caused by increased economic growth triggering improvements in health and education aspects so that it has a good impact on the quality of human development.

Acknowledgement

This Research is part of the lecture block of Human and Economic Resource Analysis held by Department of Environmental Geography, Faculty of Geography, Universitas Gadjah Mada. We would like to thank for the guidance to all of lecturer of Human and Economic Resource Analysis.

References

1. M. P. Todaro, S. C. Smith. Economic Growth, 11th Edition (Prentice Hall, New Jersey, 2011)
2. UNDP. Human Development Report 2001, Making New Technologies Work for Human Development (Oxford University Press, New York, 2001)
3. Badan Pusat Statistik. Provinsi Sumatra Barat dalam Angka 2021 (BPS, Yogyakarta, 2021)
4. B. Bakri, S. Syafrizal, H. Aimon. Analisis Ketimpangan Pembangunan antar Kabupaten/Kota di Sumatra Barat dan Kebijakan Penanggulangannya. Jurnal Kajian Ekonomi, **4**(7). (2016)
5. World Bank. World Development Report: The State in Changing World (The World Bank, New York, 1997)
6. H. Theil. Economics and Information Theory. (North-Holland Publishing Company, Amsterdam, The Netherlands, 1967)
7. G. L. Ying. China's Changing Regional Disparities during the Reform Period. Economic Geography, **24**, 59-70 (2000)
8. L. Arsyad. Ekonomi Pembangunan (UPP STIM YKPN, Yogyakarta, 2010)
9. E.S. Rahayu. Aplikasi Tipologi Klassen Pada Strategi Pengembangan Subsektor Pertanian Tanaman Pangan di Kabupaten Boyolali. Journal of Rural and Development, **1** (2010)
10. T. Widodo. Perencanaan Pembangunan: Aplikasi Komputer (Era Otonomi Daerah) (UPP STIM YKPN, Yogyakarta, 2006)
11. Suliyanto. Ekonomika Terapan: Teori dan Aplikasi dengan SPSS. (Andi, Yogyakarta 2014)
12. A. Nalle, H. Wulakada. Analysis of Economic Growth and Inequality Based On Economic Geographic Review in East Nusa Tenggara. DOARJ - Directory of Academy Resource Journal, **3** (9): 62-75 (2019)
13. Badan Pusat Statistik. Provinsi Sumatra Barat dalam Angka 2010 (BPS, Yogyakarta, 2010)
14. G. Ranis, F. Stewart, A. Ramirez. Economic growth and human development. World development, **28**, 197-219 (2000)
15. N. E. A. M. Ţ. U. Daniela-Mihaela, C. Oana-Georgiana. Correlations between human development and economic growth. Annals of the „Constantin Brâncuşi" University of Târgu Jiu, Economy Series. Volume **1**, 118-122 (2015)
16. E. Elistia, & B. A. Syahzuni. The correlation of the human development index (HDI) towards economic growth (GDP per capita) in 10 ASEAN member countries. JHSS (Journal of Humanities and Social Studies), Volume **2** (2), 40-46. (2018)

Characterization and identification of clay-material provenance from Bayat ceramic home industry in Klaten, Central Java, Indonesia

Nugroho Imam Setiawan[1*], *Indra* Perdana[2], *Vidya* Syarifah[1], *Pra Cipta* Buana[2], *Milzam Dzulfiqar* Setyowiyoto[1], *Rizki* Istinanda[2], *Didit Hadi* Barianto[1], and *Chusni* Ansori[1,3]

[1]Geological Engineering Department, Engineering Faculty, Universitas Gadjah Mada, DI Yogyakarta, 55281, Indonesia
[2]Chemical Engineering Department, Engineering Faculty, Universitas Gadjah Mada, DI Yogyakarta, 55281, Indonesia
[3]Research Centre for Geotechnology, Indonesian Institute of Sciences, Jl. Sangkuriang, Kompleks LIPI, Bandung, 40135, Indonesia

Abstract. Bayat Ceramic Home Industry in Klaten, Central Java, Indonesia is produced from three primary locations of Melikan, Paseban, and Ndolon. The clay-sources of the ceramic were mined from Bayat and surrounding areas, which is geologically named Jiwo Hills. Characterization and identification of clay material sources of the ceramic products from these areas were done in to understand the correlation between provenance and production processes. This paper presents the first contribution on Bayat ceramic by investigating the mineralogical composition using petrography and X-Ray Diffraction (XRD) analyses. Understanding the characteristics and compositions of clay material source of ceramic in Bayat can be used to determine other clay source locations or quarry that have the similar characteristics to current clay-materials. The clay minerals of ceramic products from Melikan and Paseban are halloysite, smectite, and kaolinite, while from Ndolon are smectite and kaolinite. Mono- and poly-crystalline quartz with the lithic fragments of phyllites were found in Melikan and Paseban ceramics. Ceramic from Ndolon is rich in plagioclase and lithic andesite. Clay provenance of ceramic products from Melikan and Paseban might be derived from weathered low-grade metamorphic and igneous rocks from West Jiwo Hills. While from Ndolon might be derived from weathered igneous or volcanic rocks from Kebo-Butak Formation. The amorphous silica found in the ceramic from Melikan, Paseban, and Ndolon suggests the firing process between 700–800 °C. Thus, our results provide the database of clay source characteristics to determine the following possible clay-quarry location for the raw materials of ceramic products in this location.

1 Introduction

The Bayat ceramic home industry in Klaten, Central Java, Indonesia, has been known local to international markets for their quality and characteristics [1]. Some of the physical characteristics of the ceramic from Bayat is blackish-shine brown in color and smooth surface. The Bayat ceramics are mainly crafted by the residents of Melikan and Paseban Villages, which administratively belong to Wedi and Bayat Subdistricts, respectively. According to historical stories among the villagers, the ceramic industry in Bayat has existed for 600 years ago with the production techniques were studied for generations [2]. They created their unique traditional techniques of oblique ceramic wheel to shaping their ceramic [3]. The clay-sources of the ceramic also mined from their surrounding areas, which geologically named Jiwo Hills.

The Jiwo Hills in Bayat is one of the areas that crop out the basement rocks in Central Java [4, 5]. This area is an isolated hill, which approximately has 30 km^2 expose metamorphic, igneous, and sedimentary rocks ranging from Cretaceous to Neogene ages [4, 6]. Most of the rocks in this area are highly weathered, which is indicated by the soil layer's thickness and alluvium deposit. Therefore, the clay-materials from the weathering processes are plenty.

The Bayat ceramic home industry in this area is mainly produced from three areas of Melikan, Paseban, and Ndolon. Administratively, Ndolon is a hamlet part of the Paseban Village. The Melikan and Paseban are located on the northern side, while Ndolon is located on the southern side of the Bayat-Cawas Street (Figure 1). Craftsmen in these areas follow their predecessors to determine the location of the raw clay-materials. They do not have detailed information of the ceramic clay-provenance composition. Understanding the composition is important as a guide to determine the future location of clay-quarry if this current quarry location is mined out.

This research aims to characterize and identify clay material sources of the ceramic products from these areas to understand the correlation between provenance and production processes. The results of this contribution are important to be used as suggestions for formulating and determining the possible source of the raw material for ceramic and other policies related to the

* Corresponding author: nugroho.setiawan@ugm.ac.id

development of the ceramic industry in Bayat in the future.

2 Regional geology

Jiwo Hills located in the Bayat area, which approximately 45 km NE from Yogyakarta (Figure 1). The Jiwo Hills is divided by the NE-SW trend of Dengkeng River into two areas of West and East Jiwo (Figure 1). The basement and older rocks exposed in this area are low-grade metamorphic rocks, mainly composed of various phyllites, marble, and quartzite with the age of Early Cretaceous [5, 6].

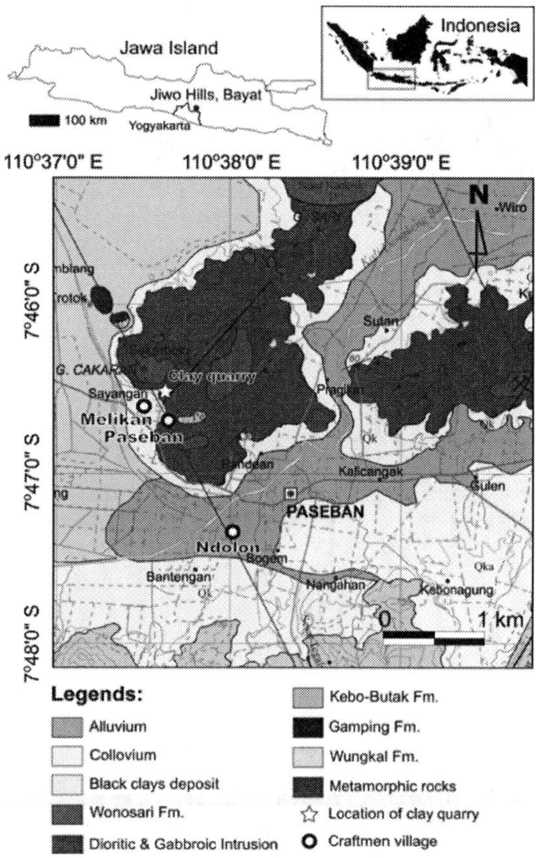

Fig. 1. Research location and geological map of Jiwo Hills. Modified from Barianto et al. [4].

Barianto et al. [5] suggest that Eocene sedimentary rocks of Wungkal and Gamping Formations (quartz sandstone and nummulitic limestone), and Oligocene – Early Miocene volcaniclastic rocks of Kebo-Butak Formation, were unconformably overlain the metamorphic rocks. Moreover, the Oligo-Miocene of dioritic to gabbroic igneous rocks intruded all previously settled formations in this area. During Middle to Late Miocene, carbonate rocks of the Wonosari Formation conformably overlain the volcaniclastics Kebo-Butak Formation. Four general major faults with the orientations of NE–SW, N–S, NW–SE, and E–W were recognized in this area [5].

Being in the tropical country with the history of exhumation and uplifting processes since Cretaceous to Quarternary, this area was significantly experiencing a denudation and weathering processes. It is indicated by difficulty to get fresh outcrop and by characteristics of the thickness of the soil layer and alluvium deposit (Figure 1).

3 Methodology

Fieldwork and interview with the craftsmen in Melikan, Paseban, Ndolon, and surrounding areas were performed to collect the primary data for this research. In-depth interviews were conducted with several craftsmen, including Mr. Rusmantoro from Melikan, Mr. Budi from Paseban, and Mr. Kawit from Ndolon.

Generally, there are three materials related to the ceramic product, which are (1) clay-sized from the quarry, (2) mixed-material between clay and sand, and (3) ceramic products. The clay-sized raw materials of the Melikan and Paseban areas are derived from the same quarry, which derived from the alluvium deposit (Figures 1 & 2a). We collected five samples of clay-sized raw materials from this area.

Ceramic products from Ndolon having different quarry locations from Melikan and Paseban. Based on the information from the craftsmen, the clay-sources materials were collected from a quarry in the Gunung Kidul area, southern part of the Bayat. In this recent paper, we could not get samples directly from the quarry. However, we analyzed 1 sample that already collected by the craftsmen in Ndolon. Furthermore, we also analyzed 1 sample per-locations of mixed clay-sand materials and ceramic products.

Fig. 2. a. Samples of clay-sized soil derived from the clay-quarry in Melikan and Paseban. **b.** The semi-traditional furnace from concrete and red bricks in Melikan and Paseban. **c.** Common ceramic products of Melikan and Paseban

All materials from each location of Melikan, Paseban, and Ndolon were analyzed for their mineralogical composition using XRD in the Central Laboratory of Geological Engineering Department, Universitas Gadjah Mada. The methods used were bulk and separated clay analyses. The samples preparation for XRD analyses was following Wicaksono et al. [7]. The minerals were determined using JADE software. The last step corrected manually match with JCPDS database.

Furthermore, the ceramic product samples were also observed and analyzed their mineralogical composition under the polarization microscope in Optical Geology Laboratory, Geological Engineering Department, Universitas Gadjah Mada. The samples preparation for petrography analyses was following Setiawan et al. [8]. The abundance of the minerals in the XRD and petrography analyzes were determined qualitatively.

4 Result and discussion

4.1 Physical characteristics

The final products of the ceramic depend on the material sources and processing. The main steps of making Bayat ceramics were explained in Safitri et al. [3], which are simplified by preparing the raw materials (clay), shaping the products, drying, and firing. In general, Melikan and Paseban ceramic products are identical. The craftsmen in these areas share their clay-sources from the same quarry (Figures 1 & 2a).

The ceramics in these areas were firing in the semi-traditional kiln, which builds from concrete and red bricks (Figure 2b). Furthermore, the ceramics from these areas have blackish-shine brown color with a smooth surface. The blackish-shine brown color is a unique characteristic of the ceramic products from these areas, are made by a smoke-firing process using mindi or munggur leaves [3, 9]. Without a smoke-firing process, the ceramics from these areas show standard reddish-brown color. The craftsmen from these areas make traditional functional ceramics (i.e., cooking pot, plates, bowls, mug) and decorative ceramics (Figure 2c).

Ceramic products from Ndolon having different clay source location from Melikan and Paseban, which are from Gunung Kidul area. The ceramics in this location were firing in more traditional open system without kiln structure. The firing held on the ground covered by straw and dried leaves (Figure 3a). The result shows light reddish-brown color with irregular dark burnt effect on the surface of ceramics (Figure 3b). The surface of the ceramics is rougher compared to the products from Melikan and Paseban. Furthermore, the craftsmen from Ndolon mostly only make traditional functional ceramics (i.e., stove, cooking pot, bowls).

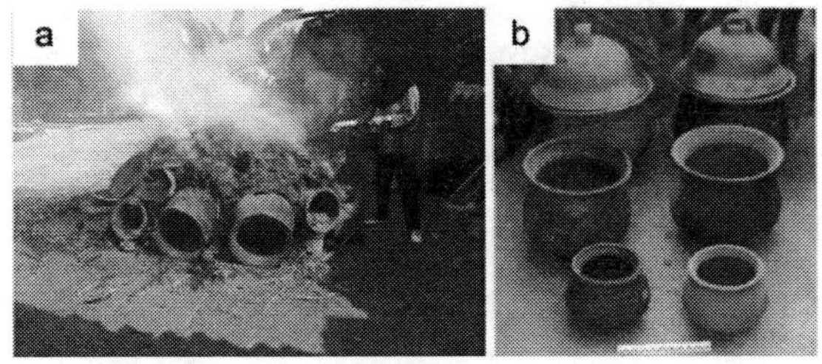

Fig. 3. a. Traditional firing system in Ndolon. **b.** Ceramic products with open firing systems [9].

4.2 Mineral composition

4.2.1 Clay composition

Twelve samples were analyzed using XRD, nine from Melikan and Paseban, and three samples from Ndolon (Table 1). The material types of the samples were clay-sized soil derived from quarry, mixed between clay and sand, and ceramic products. The XRD analyses reveal that the first material of clay-sized soil derived from the quarry in Melikan and Paseban compose of quartz, albite, halloysite, smectite, and kaolinite in all samples. Some samples show plagioclase, amphibole, and pyrite (Table 1). Quartz, albite, and amphibole are commonly found in low-grade metamorphic rock provenance, while plagioclase is commonly found in igneous rock. These results are in agreement with the lithology in West Jiwo Hills, which compose of low-grade metamorphic rocks and igneous intrusions (Figure 1) [4, 5].

Table 1. Resume of XRD analyses bulk and clay methods.

No	Sample No	Location	Type	Composition									
				Qz	Ab	Pl	Amp	Am. Sil	Hal	Lau	Sme	Kal	Pyr
1	NIS 1M*	Melikan	Quarry clay	●		□			□		□	□	
2	NIS 2M*	Melikan	Quarry clay	●	○		□		□		□	□	
3	NIS 3M*	Melikan	Quarry clay	●	□		□		□		□	□	□
4	NIS 5M	Melikan	Clay	●								□	□
5	NIS 3P	Paseban	Clay	●	○		□		□		□	□	□
6	NIS 1N	Ndolon	Clay	●		○					□	□	
7	NIS 6M	Melikan	Mix Clay-Sand	●	○		□		□		□	□	□
8	NIS 1P	Paseban	Mix Clay-Sand	●		○			□	□	□	□	
9	NIS 2N	Ndolon	Mix Clay-Sand	●		●					□	□	
10	NIS A	Melikan	Ceramic	●	○		□	□	□	□			
11	NIS B	Paseban	Ceramic	●	○		□	□	□	□		□	
12	NIS C	Ndolon	Ceramic	○		●	□	□	□				□

Remark
● Abundance ○ Moderate □ Rare
Qz: quartz, Ab: albite, Pl: plagioclase, Amp: amphibole, Am. Sil: amorphous silica, Hal: halloysite, Lau: laumontite, Sme: smectite, Kal: kaolinite, Pyr: pyrite
*) Quarry clay same sources Melikan and Paseban

Furthermore, the appearance of pyrite indicates that the source provenance experienced hydrothermal alteration. Setiawan et al. [5] and Idrus et al. [10] confirmed that the igneous rocks in Jiwo Hills experienced hydrothermal alteration and skarnization. Other minerals of halloysite, smectite, and kaolinite are considered as common low-temperature clay minerals that might be developed during weathering processes [11].

The clay-sized soil from Ndolon has simple compositions of quartz, plagioclase, smectite, and kaolinite (Table 1). Quartz is a common mineral that can present in other provenances, while plagioclase might be derived from igneous rock provenance. As mentioned before, smectite and kaolinite are considered as common low-temperature clay minerals that might be formed by weathering processes

4.2.2 Mixed clay-sand composition

The second material is mixed between clay and sand. The sand was added and mixed with the clay to increase the strength of the ceramics during firing. The ratio between sand and clay in Melikan and Paseban is 1:10 [3]. The XRD analyses show that the second material from Melikan and Paseban has a composition of quartz, albite, halloysite, smectite, kaolinite in all samples. Some samples show plagioclase, amphibole, laumontite, and pyrite (Table 1). Clearly, a different result is shown by the appearance of laumontite mineral in the second material composition. The laumontite was possibly added from the sand. The craftsmen bought the sand from the sand-gathered community, which was obtained from the channel bar of Mundu River in Brakal, Wedi, Klaten [3]. The laumontite commonly formed in the hydrothermal system (under low temperature (<220 °C), neutral pH hydrothermal condition, low fCO_2), and in low-grade regional and thermal metamorphism [11]. Both forms of the laumontite are possibly found in this location, especially in the West Jiwo Hills (Figure 1), where the sand and clay were collected. The laumontite might be derived from the hydrothermal alteration related to dioritic and gabbroic intrusions and low-grade metamorphic rocks of the basement complex [4, 5, 10].

Meanwhile, the XRD analyses of the second material from Ndolon reveal no different composition from the first material of clay-sized soil, which are quartz, plagioclase, smectite, and kaolinite (Table 1). The sand was collected from Sekarbolo River with a mixing ratio between sand and clay is 1:1.

4.2.3 Ceramic composition

The third material of ceramic products from Melikan and Paseban was analyzed with XRD and petrography analyses. The resume of petrography analyses can be seen in Table 2. Based on the XRD analyses, the ceramic from Melikan and Paseban compose of quartz, albite, amphibole, amorphous silica, halloysite, and laumontite. One sample shows kaolinite (Table 1; Figures 4a & 4b). While petrography analyses show a more detailed composition of mono- and poly-crystalline quartz, plagioclase, clay, opaque mineral, pyroxene, hematite, and rock fragment of phyllite. Some samples show amphibole and biotite (Table 2, Figures 5a-5d).

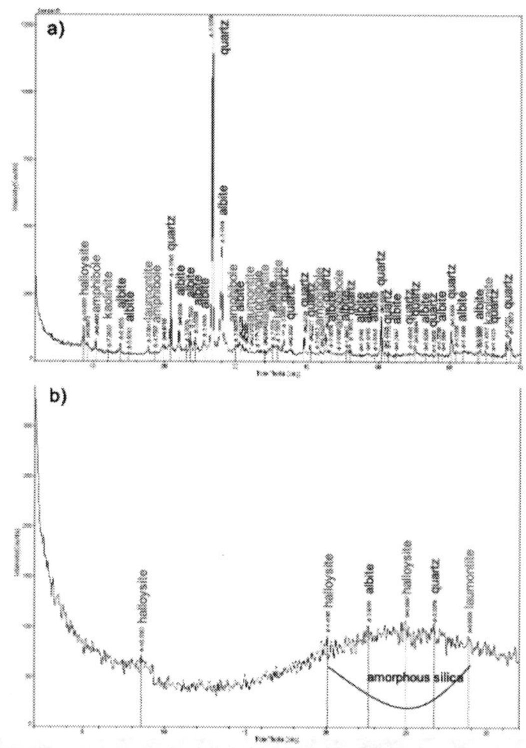

Fig. 4. Peak representative of XRD analysis ceramic product from Melikan or Paseban **a.** bulk and **b.** clay methods.

Table 2. Resume of petrography analyses in ceramic products.

No.	Sample No	Location	Composition											
			mQz	pQz	Bt	Pl	Hbl	Cly	Opq	Px	Hem	Chc	And	Phy
1	NIS A	Melikan	○	●		□		●	□	□	□			○
2	NIS B	Paseban	□	○	□	□		●	□	□	○			□
3	NIS C	Dolon			○	●	○	●	□	○	○	□	○	

Remark
- ● Abundance (> 25%)
- ○ Moderate (5 – 25%)
- □ Rare (< 5%)

Abbreviation
mQz: monocrystalline quartz; pQz: polycrystalline quartz; Bt: biotite; Pl: plagioclase; Hbl: hornblende; Cly: clay minerals; Opq: opaque minerals; Px: pyroxene; Hem: hematite; Chc: chalcedony; And: lithic andesite; Phy: lithic phyllite

The amorphous silica found in XRD analysis might be a new mineral formed by the firing process, which indicates temperature 700–800 °C [12]. The presence of mono- and poly-crystalline quartz, biotite, and phyllite fragments (Figure 5d) from the petrography analyses in the ceramic from Melikan and Paseban clearly indicate that the provenance is from low-grade metamorphic rocks. However, other minerals such as plagioclase and pyroxene might indicate that the provenance is not only from metamorphic rocks but also igneous rocks. Both possible provenances are found as outcrops in West Jiwo Hills [4, 5]. Furthermore, opaque mineral and hematite might be derived from provenance that experienced hydrothermal alteration or oxidation process.

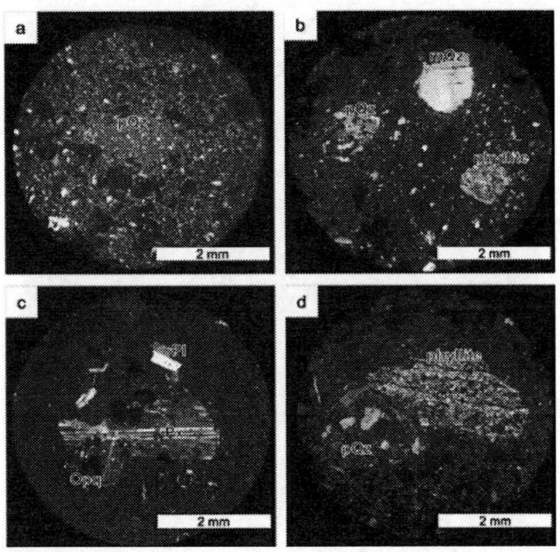

Fig. 5. Representative petrographical analysis of ceramic product from Melikan and Paseban show composition of **a.** and **b.** mono- (mQz) and poly-crystalline quartz (pQz); **c.** clinopyroxene (Px), plagioclase (Pl), opaque mineral (Opq); and **d.** lithic phyllite.

The ceramic product from Ndolon were also analyzed in XRD and petrography. Based on the XRD analyses (Table 1; Figures 6a & 6b), ceramic from Ndolon have a composition of quartz, plagioclase, hornblende, halloysite, pyrite, and amorphous silica. Similar to the ceramic from Melikan and Paseban, the amorphous silica might be a new mineral formed by the firing process, which indicate temperature 700–800 °C [12].

Meanwhile, the petrography analyses reveal that the ceramic composition is biotite, plagioclase, clay, opaque minerals, pyroxene, hematite, chalcedony, and rock fragment of andesite (Table 2; Figures 7a-7d). The presence of andesite rock fragments, plagioclase, and pyroxene suggests a strong indication that the provenance of ceramic in Ndolon is derived from igneous or volcanic rocks (Figures 7a & 7b). This is in agreement with the information from the craftsmen in Ndolon that the clay-quarry is located in the southern part of Ndolon (Gunung Kidul area), which possibly compose of weathered igneous or volcanic rocks from the Kebo-Butak Formation (Figure 1).

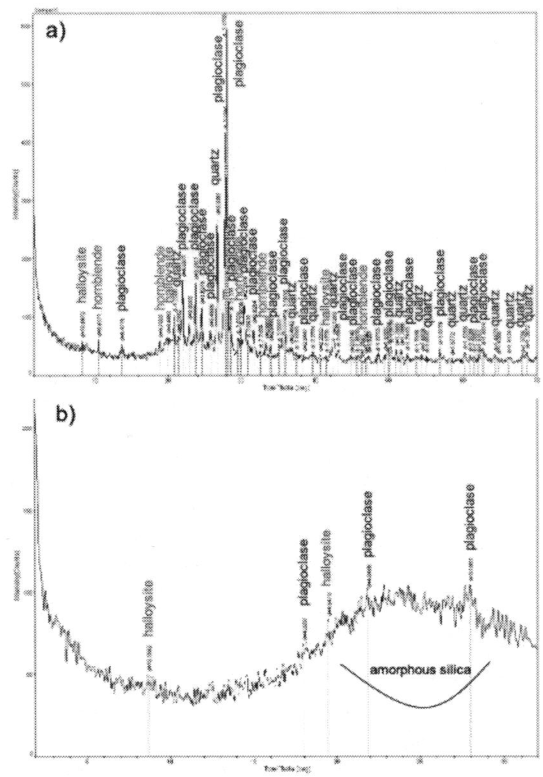

Fig. 6. Peak representative of XRD analysis ceramic product from Ndolon **a.** bulk and **b.** clay methods.

Furthermore, chalcedony (Figure 7d) formed where silica deposition is very rapid, and generally at less than 190 °C, though rarely it can be formed at higher temperature as amorphous silica (700–800 °C) [12]. Therefore, the chalcedony could be formed during the firing processes or simply highly-saturation silica condition from the provenance.

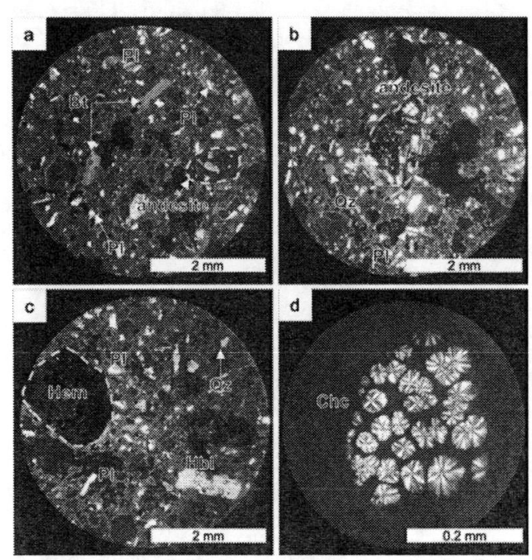

Fig. 7. Representative petrographical analysis of ceramic product from Ndolon compose of **a.** biotite (Bt), plagioclase (Pl), and clinopyroxene (Px); **b.** lithic andesite, quartz (Qz), and plagioclase (Pl); **c.** lithic hematite (Hem), hornblende (Hbl); and **d.** chalcedony (Chc).

5 Suggestion and recommendation

Historically, the ceramic home industry in Bayat developed 600 years ago [2]. Understanding the characteristics and compositions of clay material source of ceramic in Bayat can be used to determine other clay source location or quarry that have similar characteristics to current clay-materials. This further process requires geochemical and geological mapping studies of potential clay-sources in Bayat and the surrounding area. The identification of the composition in the ceramic also provides qualitative information of the formation temperature, for example, the presence of amorphous silica formed at 700–800 °C [12]. The technique and temperature of firing in the kiln have been studied for generations. It is necessary to conduct a more in-depth study related to whether the firing temperature is optimum or not to produce a better quality of ceramic. This process requires further experimental studies of heating clay-materials in the laboratory and directly measuring the firing process in close or open kilns.

6 Conclusion

Based on the XRD and petrography analyses, the clay minerals of ceramic products from Melikan and Paseban are halloysite, smectite, and kaolinite, while Ndolon is smectite and kaolinite. Mono- and poly-crystalline quartz with a lithic fragment of phyllites were found in Melikan and Paseban ceramic. The ceramic from Ndolon is rich in plagioclase and has a composition of lithic andesite.

Clay provenance of ceramic products from Melikan and Paseban might be derived from weathered low-grade metamorphic and igneous rocks from West Jiwo Hills. While from Ndolon might be derived from weathered igneous or volcanic rocks from Kebo-Butak Formation.

The amorphous silica that found in all the ceramics from Melikan, Paseban, and Ndolon might be formed by the firing process of 700–800 °C.

We would like to thank Bayat ceramic home industry communities in Melikan, Paseban, and Ndolon, and Klaten Agency for Regional Development (Bappeda) for their assistance during interviews and fieldwork. This research was financially supported by the Ministry of Research and Higher Education of Indonesia under the RTA program and Collaboration Community Service (PKM) Grant from Engineering Faculty, Universitas Gadjah Mada.

References

1. Y. Hartanto, S.P. Gustami. Sosiohumanika. **3**, 427-438 (2003), in Bahasa
2. P. Yustana. J. Penelitian Seni Budaya. **6**, 14-32 (2014), in Bahasa
3. W. Safitri, T.H. Retnowati, B. Prihadi. Adv. Soc. Sci., Ed. Hum. Res. **444**, 34-40 (2020)
4. D.H. Barianto, U. Margono, S. Husein, M.I. Novian. *Geological map of Wonosari Quadrangels, Jawa (1408-31), Scale 1:50,000* (Geology Research and Development Centre of Indonesia, 2017)
5. N.I. Setiawan, Y. Osanai, C. Prasetyadi, Seminar Nasional Kebumian, **6**. 11-23 (2013)
6. C. Prasetyadi. *Evolusi tektonik Paleogen Jawa bagian Timur*, (Dissertation, unpubls., Institut Teknologi Bandung, 2007), in Bahasa
7. D. Wicaksono, N.I. Setiawan, W. Wilopo, A. Harijoko. Seminar Nasional Kebumian. **10**, 1864-1880 (2017), in Bahasa
8. N.I. Setiawan, Sariyanto, A.A. Saputro. Seminar Nasional Kebumian. **7**, 378-388 (2016), in Bahasa
9. D.C. Pakarti. *Teknik Putaran Miring dan Perkembangan Keramik Bayat Klaten*, (Skripsi, unpubls., Universitas Sebelas Maret, 2012), in Bahasa
10. A. Idrus, I.W. Warmada, B. Junitia. Seminar Nasional Kebumian. **6**, 37-47 (2013)
11. G.J. Corbett. *Epithermal Au-Ag and porphyry Cu-Au exploration*, (Short course manual, unpubls, 2017)
12. M.M. Haslinawati, K.A., Matori, Z.A. Wahab, H.A.A. Sidek, A.T. Zainal. Int. J. Basic Appl. Sci. IJBAS, **9**, 22-25 (2009)

E3S Web of Conferences 325, 06003 (2021) https://doi.org/10.1051/e3sconf/202132506003
ICST 2021

Correlation study between physical disability and covid-19 cases in Sleman, Indonesia

Ihsan Weda Prabaswara[1], *Agustin Putri* Samsiana[2], *Nabilah Luthfatur* Rohmah[1], *Nabiyya* Perennia[1], *Nasrudin Dwi* Jatmiko[1], *Nawang* Setyoningsih[2], *Mya Dwi* Rostika[3], and *Ratih Fitria* Putri[1*]

[1]Environmental Geography Department, Faculty of Geography, Universitas Gadjah Mada, Yogyakarta, 55281, Indonesia
[2]Development Geography Department, Faculty of Geography, Universitas Gadjah Mada, Yogyakarta, 55281, Indonesia
[3]Department of International Relations, Faculty of International Relations, Daito Bunka University, Japan

Abstract. The spread of the highly infectious COVID-19 has severely impacted the global community. Population density and intense physical interaction are the catalysts of the virus spread. Many regulations were issued to halt the spread of the virus, including the limitation of people's mobility and physical interaction. With the limitation of mobility and physical interaction, many people with disabilities are struggling to access their needs since they depend on their caregivers. The lack of inclusive policies in the health system itself also ends up deprioritizing the disabled. This article aims to identify the correlation between physical disability and COVID-19 cases and identify the spatial distribution of the correlation in Sleman regency, Indonesia. This article presents the correlation with the qualitative approach using Pearson product-moment correlation method. The data used in this article is secondary data obtained from official government agencies. The findings of this study showed that Pearson correlation between physical disability and COVID-19 cases is 0,04346 and classified as very low positive.

1 Introduction

The waves of COVID-19 pandemic have severely impacted the community throughout the world, including Indonesia. The pandemic is highly infectious and airborne, making it even more widespread [1]. Intense physical interaction between people performs as the catalysts in the spreading of the virus which causes the disease. The virus rapidly spread in high density area [2].

Sleman Regency which is situated in the Special Region of Yogyakarta, has recorded the rise of cases through 2020 between April and December with the highest record of infected people being 109 cases in a day on December 16th, according to COVID-19 Regency Task Force. To counter the spreading of the virus, the local government has authorized the community to restrict their social activities and to obey the health protocols with various regulations, such as Keputusan Bupati Sleman No. 25/Kep.KDH/A/2020, Peraturan Bupati Sleman No. 37.1/2020, and Instruksi Bupati Sleman No. 443/01032/INSTR/2020 and No 5/INSTR/2021. This regulation was meant to halt the spread of the virus by limiting people's mobility and physical interaction; and ensure that the community is protected from the virus. However, the implementation of the regulation has presented its own challenges.

One of the more common analyses is the effect of economic class on COVID-19 vulnerability, as people with higher economic privilege can afford better medical care and to stay home without losing their means of living. One of the less common studies is how people with disabilities are affected by the pandemic. People living with disabilities (PLWD) are affected in mainly three aspects: accessibility to information, accessibility to services, and data collecting. First, PLWD has less access to information about the pandemic because not all relevant information is presented in a format that is accessible to them. Second, due to restrictions of mobility, PLWD can lose access to services they need such as caregivers or therapy. This is especially concerning since many PLWD have comorbidities that can make them even more vulnerable to the virus. Lastly, data collection on the pandemic does not often consider PLWD and thus also affects the inclusive aspect of planning for policies to deal with the pandemic [3]. With their added vulnerability, PLWD should be a case for concern. Mapping techniques have been a crucial tool during the pandemic. Tracking the spread of the virus is essential to prevent the further expansion. Analyzing the pandemic

* Corresponding author: ratihfitria.putri@ugm.ac.id

spatially could help identify areas that have higher levels of infection, thus helping to identify which areas are in need of the most help or need to be quarantined. Using GIS, we can also identify what types of people are infected and how they are treated. In this study, mapping will be used to present the patterns of COVID-19 infection and disability in order to achieve better understanding spatially within the region of Sleman District. The COVID-19 infection mapping provides a visualization of the pandemic conditions in Sleman. This is useful in knowing the spatial distribution of COVID-19 infection in this area so that we know which sub-district is most affected by the COVID-19 pandemic. This identification is useful for knowing areas that need attention regarding the COVID-19 pandemic.

Despite various pieces of legislation grouping PLWD as a vulnerable group to consider during disaster management, a previous study conducted in the USA has shown that the correlation between COVID-19 and PLWD is negative [4]. This study which is conducted by dr. Chakraborty uses the same methodology albeit with much more data and slightly different variables. In this paper, we will try to apply the same methodology, but specifically to persons with physical disabilities to see if we would be able to generate the same results.

2 Data and Method

2.1 Data

Variables used in this study were data on the number of COVID-19 confirmations which were used as the main variable and became the Y variable, while the X variable is based on the number of PLWD. The secondary data used in this study are mentioned below.

a. COVID- 1 9 Confirmation Data per district in Sleman Regency (per 31 December 2020) from government website (Badan Pusat Statistik, *Yogyakarta Tanggap* COVID-19, and Dashboard COVID-19 Kabupaten Sleman)
b. Physical Disability Data per district in Sleman Regency (for 2nd Semester of 2020) from government website (Governance Bureau of Regional Secretariat of Yogyakarta)

Data collection carried out by collecting secondary data related to the variables to be analyzed through the linked site. This is done to follow health protocols during a pandemic. After that, the data processed using Microsoft Excel, so that the data is obtained in the form of tables and graphs which are then analyzed using descriptive quantitative methods.

2.2 Pearson Correlation

Pearson Product-Moment Correlation Coefficient is a method used to find the strength of the relationship between the variables to be studied with a range from -1 to 1 [5]. This method has been used in the United States of America, Spain, Italy, Germany, France, United Kingdom, Belgium, New Zealand, Turkey [6], Poland, Portugal [7] , and India [8]. The two variables investigated in the Pearson Product-Moment Correlation Coefficient method are usually given the symbols the variable X and variable Y. The correlation is then calculated using formula (1).

$$r = \frac{n\,\Sigma xy - (\Sigma x)(\Sigma y)}{\sqrt{[n\,\Sigma x^2 - (\Sigma x)^2][n\,\Sigma y^2 - (\Sigma y)^2]}} \tag{1}$$

where, (r) represents the Pearson Product-Moment Correlation Coefficient, (n) represents the amount of data pairs between X variable and Y variable, (Σx) represents the total sum of X variable, (Σy) represents the total sum of Y variable, (Σx^2) represents the square of total sum of X variable, (Σy^2) represents the square of total sum of Y variable, and (Σxy) represents the product of the total sum of X Variable and Y Variable. The results of the statistical test can fall under one of three categories as follows.

a. Positive correlation
   Correlation between variables can be considered positive if an increase in value of variable X is followed by an increase in valueof variable Y, and the decrease in value of variable X is followed by a decrease of valuein variable Y.
b. Negative Correlation
   Correlation between variables can be considered negative if an increase in value of variable X is followed by a decrease in value of variable Y, and the decrease in value of variable X is followed by an increase of valuein variable Y.
c. No Correlation
   Variables might have no correlation when the decrease and increase in both values happen at random. This indicates that the variables do not affect one another.

Through the aforementioned approach and analysis, it can be declared that this research uses descriptive quantitative research methods. The descriptive research method in conducting analysis covers the COVID-19 phenomenon in Sleman and its correlation with PLWD. The analysis and approach reflect that this research is using quantitative descriptive methods.

2.3 Spatial Analysis

Spatial analysis data were collected through online procedure. The administrative boundaries data of sub-districts within the area were downloaded from Ina-Geoportal, provided by Geospatial Information Agency (Badan Informasi Geospasial). PLWD distribution and

COVID-19 case maps were composed utilizing QGIS 3.18.3. software. Both PLWD and COVID-19 data were classified into three categories, low, moderate and high using equal interval classification provided by the software. Equal interval classification interprets the input data into three categories with equal amount of range [9].

2.4 Social Vulnerability

The indicators used to evaluate social vulnerability during times of disaster are population density and population of vulnerable groups. These indicators are regulated by the National Body of Disaster Management (BNPB) through Peraturan Kepala BNPB Nomor 2 Tahun 2012. According to this regulation, the indicators for vulnerability are: poverty ratio, gender ratio, disability ratio and age group ratio. Here, disability is categorized into physical disability, vision impared, hearing-speech impaired, and mental disability. However. in the case of COVID-19, physical disability is not taken into account since the lack of mobility equals less exposure. Therefore, people who have physical disabilities would not be relevant when evaluating social vulnerability during the COVID-19 pandemic.

3 Discussion

3.1 Pearson Correlation

Between the variables, PLWD and confirmation cases of COVID-19, the correlation results showed the number of cases is 0.04346. The correlation result of 0.04346 can show a value that is more than zero, which means that the correlation is positive. However, if it is observed directly through Figure 1. that the scatterplots are far away from the line so that the correlation number is approximately zero as shown in Table 1. Because, a value of zero indicates no relationship between the two variables being compared. Thus, the change in one variable is not directly proportional or inversely proportional to the other variable and correlation is classified as very low positive.

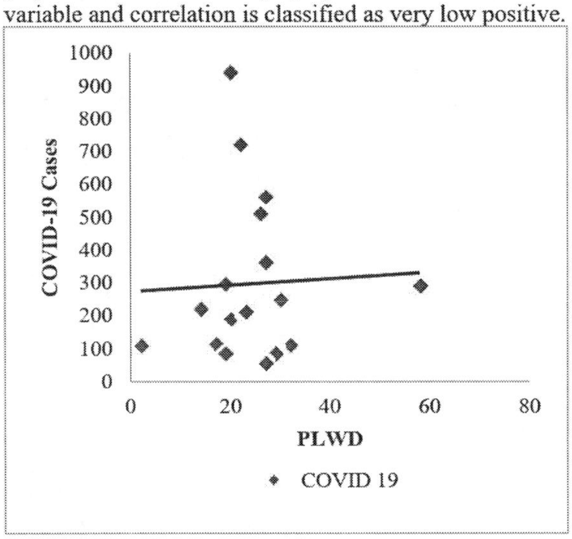

Fig. 1. Pearson Correlation Chart between Variable PLWD and Confirmed Cases of COVID-19

The observation results from Figure 1 which shows an image of the irregular correlation can also be observed from the inconsistency of the correlation, where there is a correlation with a high number of confirmed COVID-19, but the number of PLWD in the area is not many. On the other hand, there is also a low number of confirmed cases of COVID-19, but there are a lot of PLWD. Thus, value of the graph shows an inconsistent up and down correlation. The correlation results for each District in Sleman Regency presented in Table 1. The highest confirmed numbers of COVID- 19 are in several districts, including Depok District with 938 people, Ngaglik District with 718 people, Gamping District with 557 people, and Mlati District with 507 people. Meanwhile, the lowest number of COVID-19 was in Cangkringan District with 51 people, Pakem District with80 people, and Turi District with 82 people. The highest number of disabilities was in Ngemplak District with 58 people, Moyudan District with 32 people, and Tempel District with 30 people. Meanwhile, the lowest number of people with disabilities is in Minggir District with 2 people, Sleman District with 14 people, and Prambanan District with 17 people. From the data in areas with high numbers of confirmed COVID-19, it shown relatively low or moderate disability rate. The highest number of disabilities in each sub-district did not show a high number of confirmed positive COVID-19 aswell.

Table 1. Table of Data and Calculation Results for each District in Sleman Regency

No.	District	PLWD Physical	Covid-19 Cases	Pearson Correlation
1	Moyudan	32	106	0.043461309
2	Minggir	2	105	
3	Seyegan	20	185	
4	Godean	19	291	
5	Gamping	27	557	
6	Mlati	26	507	
7	Depok	20	938	
8	Berbah	23	208	
9	Prambanan	17	109	
10	Kalasan	27	357	
11	Ngemplak	58	287	
12	Ngaglik	22	718	
13	Sleman	14	217	
14	Tempel	30	245	
15	Turi	19	82	
16	Pakem	29	80	
17	Cangkring an	27	51	

3.2 Spatial Distribution

The distribution of PLWD in Sleman District represented Figure 2. Population of PLWD is higher on the east and southwest side of Sleman, with the

https://doi.org/10.1051/e3sconf/202132506003

highest disabled populations located in the Ngemplak sub-district. Areas with a high population of PLWD are not likely to be the center of regional activities. and located near the border with neighboring districts that are not the center of regional activities. Being far from the center of regional activities, the location is expected to be less crowded and more peaceful, so that PLWD do not experience disturbance from the surroundings.

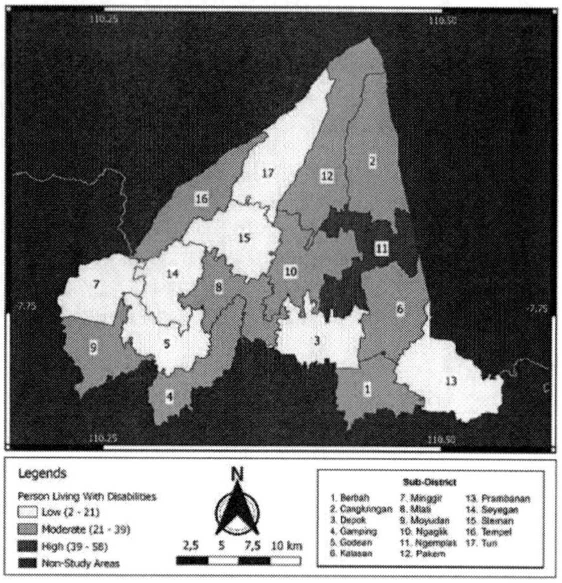

Fig. 2. Spatial Distribution of PLWD in Sleman

Distribution of COVID-19 cases in Sleman is influenced by the activities of the people. As shown in Figure 3., Depok sub-district record the highest COVID-19 cases. The mentioned sub-district is the center of public activities in Sleman with various commercial districts and public spaces. Although we saw restriction of public spaces activities such as reducing the operating hours of public spaces and stores, Sleman doesn't implement strict lockdown regulation to keep local business running. Thus, there is still a possibility of COVID-19 disease could easily spread within the population in the area with high public physical contact. Sub-districts with low COVID-19 cases were mostly rural areas with less physical activity compared to sub-districts with high COVID-19 cases. Activities in the rural sub-districts most dominantly occurred within the population of the respective areas, while activities in the Depok sub-district involves people from the outside of the area as well.

COVID-19 cases in Sleman Regency that are more concentrated on Ngaglik and Depok Sub-districts indicate that COVID-19 cases are easier to spread in public activity centers, which are commercial and densely populated areas. This phenomenon supports the research conducted by Duran-Polanco and Siller [10] which states that public areas are prone to the spread of COVID-19 because there are possibilities of people interacting physically. PLWD tend to live in areas that are not the centers of public activity, such as in Ngemplak Sub-district, which has the highest distribution of PLWD. This is a positive phenomenon because people with disabilities have a lower probability of contracting the COVID-19 virus that is spreading in the community. Nevertheless, attention to cases of COVID-19 for PLWD is still needed, especially in Ngaglik Sub-district but with persons with moderate disabilities, having high COVID-19 cases.

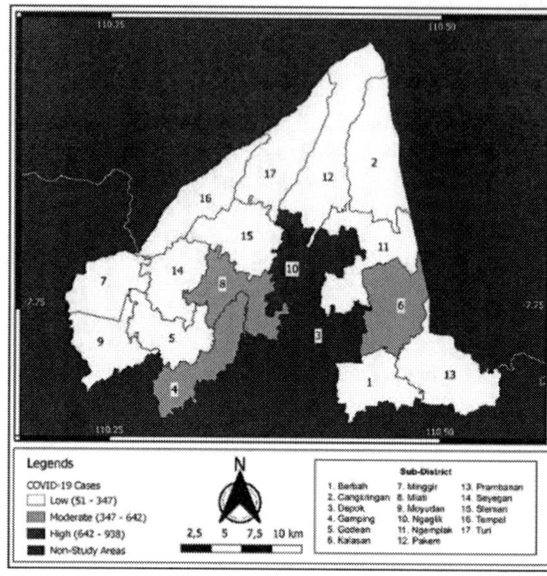

Fig. 3. Spatial Distribution of COVID-19 Cases in Sleman

3.3 Social Vulnerability

The Pearson calculations show that the data does have a positive correlation. The number is too small to be regarded as significant, however we should recognize a few limitations in our study. First, the data analyzed is comparing PLWD populations with the number of infected by COVID-19 people overall, not the number of PLWD infected. If we were to use the latter such as similar studies before us, the correlation might be much stronger. On the contrary, a similar study from the US showed a negative correlation despite using the proper data [4]. Second, this study only analyses physical data disability, instead of disability as a whole. One of the effects of the pandemic is a restriction of mobility, which is an already familiar experience to people with physical disabilities. In fact, this restriction could be an advantage for them since now most services and facilities operate remotely, in comparison to life before the pandemic when not all services would operate this way. Not all persons with physical disabilities suffer from similar afflictions and some might require regular physical therapy or assistance which is often inaccessible in our work from home society.

In accordance to the previous study [4], there has been a decrease in neurorehabilitation rates for stroke, as much as 50% in Italy. When services reopened, the rates of neurorehabilitation saw an increase, due to complications of COVID-19 and postponed medical appointments for those already with reduced mobility due to neurological inflictions. This could lead to a further slowing down of medical services that are already congested due to the pandemic. The publication also concluded that remote services for this kind of treatment aren't effective - in person appointments are still necessary [11]. Although people suffering physical disabilities may have less exposure to the virus, they still feel the side-effects the pandemic has brought upon them.

The challenges faced by PLWD as a vulnerable group, especially during the pandemic, should be given serious attention. Even in normal circumstances, PLWD have higher rates of depression than most of the population [12]. People with physical disabilities have specific needs, therefore depend on medical services and caregivers [13]. The social restrictions and compulsory mask-wearing that have been set by the government limits the interactions between PLWD and their caregivers. The combination of these conditions made PLWD a vulnerable group during the COVID-19 pandemic [14].

4 Conclusion

This study shows that the correlation between PLWD and COVID-19 has no correlation. This phenomenon is caused by the results which show the number of cases is 0.04346. A value of zero indicates that there is no relationship between the two variables being compared. So, that it can be said that the correlation shows or can be classified as very low positive or uncorrelated. The graph shows an irregular correlation of the inconsistency and the correlation, where there is a correlation with a high number of confirmed COVID-19, but the number of PWLD in the area is not large. On the other hand, there are also few confirmed cases of COVID-19, but many PLWD. The graph value shows an inconsistent up and down correlation. PLWD, especially people with physical disabilities, have lower risk to COVID-19 due to lack of mobility and lower exposure. Thus, including people with physical disabilities as one of indicators of social vulnerability is insignificant. In the case of COVID-19, people with chronic illness are the vulnerable group which need more attention in order to be able to reach medical treatment that is often constrained by limited access.

The spatial distribution of PLWD was concentrated on the east side of Sleman, not likely to be the center of public activities. The location is expected to be less crowded and more peaceful, so that PLWD do not experience disturbance from the surroundings. Distribution of COVID-19 in Sleman was centralized in Depok Sub- district, which served as the center of public activities. The low COVID-19 cases were recorded in rural areas because it has lower physical contact activities within the area compared to Depok Sub-district.

Acknowledgment

This study is a part of Program Riset Besar (Grand Research Program) conducted by Geography Study Club, Faculty of Geography, Universitas Gadjah Mada. We would like to thank the institution that provided us with the online data used in this study, as well as the Faculty of Geography, Universitas Gadjah Mada for their support.

References

1. R. K. Mohapatra, L. Pintilie, V. Kandi, A. K. Sarangi, D. Das, R. Sahu, and L. Perekhoda, Chem. Biol. Drug Des. **96**, 1187 (2020)
2. H. Coşkun, N. Yıldırım, and S. Gündüz, Sci.Total Environ. **751**, (2021)
3. H. Kuper, L. M. Banks, T. Bright, C. Davey, and T. Shakespeare, Wellcome Open Res. **5**, 1 (2020)
4. J. Chakraborty, Disabil. Health J. **14**, 101007 (2021)
5. N. Al-Rousan and H. Al-Najjar, Eur. Rev. Med. Pharmacol. Sci. **24**, 4565 (2020)
6. D. Gondauri and M. Batiashvili, Electron. J.Gen. Med. **17**, 17 (2020)
7. A. Strzelecki, A. Azevedo, and A. Albuquerque, Healthc. **8**, (2020)
8. A. K. Upadhyay and S. Shukla, Diabetes Metab. Syndr. Clin. Res. Rev. **15**, 993 (2021)
9. Paul Bolstad, *GIS Fundamentals : A First Text on Geographic Information System5 Th Edition* (2016)
10. L. Durán-Polanco and M. Siller, Annu. Rev. Control (2021)
11. L. Leocani, K. Diserens, M. Moccia, and C. Caltagirone, Eur. J. Neurol. **27**, e50 (2020)
12. M. V. S. Gudlavalleti, N. John, K. Allagh, J. Sagar, S. Kamalakannan, S. S. Ramachandra, B. R. Shamanna, C. Noe, H. B. Pant, F. Soji, and S.Mani, BMC Public Health **14**, 1 (2014)
13. N. Scherer, I. MacTaggart, C. Huggett, P. Pheng, M. U. Rahman, and J. Wilbur, BMJ Open **11**, 1(2021)
14. A. Lebrasseur, N. Fortin-Bédard, J. Lettre, E. L. Bussières, K. Best, N. Boucher, M. Hotton, S. Beaulieu-Bonneau, C. Mercier, M. E. Lamontagne, and F. Routhier, Disabil. Health J. **14**, (2021)

 https://doi.org/10.1051/e3sconf/202132506004

Determinants of low birth weight and its relationship with stunted toddler status in Magelang Regency, Indonesia

Sri Rum Giyarsih*, *Ratih Fitria* Putri, *Agustinur* Saputri, *Esti* Nugraheni, and *Nia* Setiyawati

Universitas Gadjah Mada, Yogyakarta, 55281, Indonesia

Abstract. Stunting contributes to a decrease in the quality of life. One of the potential factors which causes stunting is low birth weight (LBW). The objectives of this study are (1) to obtain representations of the characteristics of mothers and stunted toddlers; (2) to analyze the influence of maternal age at birth, gestational age, maternal hygiene and health behavior, and maternal nutrition knowledge related to LBW; and (3) to identify the relationship between LBW and stunted toddlers in Magelang Regency. This study used a mix-method approach that combined primary data and in-depth interviews. The binary logistic regression results found that maternal age at birth, gestational age, maternal hygiene and health behavior, and maternal nutritional knowledge had a significant effect to LBW on stunted toddlers.

1 Introduction

The condition of the mother before pregnancy and during early pregnancy is a phase that determines the level of development of future generations. The mother's health and the nutrition that she obtains will affect the condition of the baby born. A mother whose nutritional intake is insufficient and who is experiencing chronic energy deficiency has the potential to give birth to a baby with low birth weight. Such infants are at risk of experiencing malnutrition, the impact of which will continue in the next phase of life [1].

Growth in toddlerhood is one measure of the degree of health in adult life. Stunting is one of the growth disorders in childhood and is a chronic nutritional problem that Indonesia continues to face. The cause of stunting is an imbalance between the needs and actual nutritional intake of the body in the long run. Toddlers who are stunted may suffer irreversible severe physical and cognitive impairment [2]. Stunting has also become a global concern, with the inclusion of a stunting prevalence indicator in the second of the Sustainable Development Goals (SDGs) in Target 2.2, namely the elimination of all malnutrition problems by 2030 and of toddler stunting by 2025.

In 2018, all regions in Indonesia had a high (20 - <30 percent) or very high (≥ 30 percent) number of cases of stunting, apart from for DKI Jakarta, whose level was 17.6 percent (moderate). The results of the Nutrition Status Monitoring (NSM) of 2017 showed that 28.5 percent of toddlers faced stunting problems in Central Java. By 2018, the region was included in the very high category for the prevalence of stunted toddlers (31.2 percent), a figure which exceeded the national average of 30.8 percent. From the 2017 NSM data, it can also be seen that the stunting rate of Magelang Regency reached 37.6 percent, second worst among the regencies/cities in Central Java [3]. However, based on the results of National Basic Health Research (Riskesdas) 2018, the stunting rate in Magelang Regency had decreased to 29.69 percent [4].

Stunting in toddlers can be caused by various factors, one of which is LBW, which is the cause of around 20 percent of incidents [1]. According to the World Health Organization, low birth weight refers to babies under 2,500 grams [5]. Prendergast and Humphrey state that birth weight and birth length are associated with subsequent growth in childhood [6]. If this is added to other factors such as insufficient nutritional intake and if the health of toddlers is not well maintained, meaning they experience illness or infection, the risk of experiencing stunting is higher [7].

The causes of the incidence of LBW are multidimensional, making it difficult to take precautions. However, in general it is related to characteristics of the mother's pregnancy, such as prematurity; pregnancy/birth at risk (maternal age < 20 or > 35 years); socioeconomic conditions that lead to lack of nutrition during pregnancy and no pregnancy checks; and poor lifestyle during pregnancy such as smoking, drinking alcohol and taking drugs [8]. Donal et al. state that the incidence of LBW is related to risk factors to maternal health identified by traditional examinations (traditional birth attendants) or antenatal examinations, such as drinking alcohol or smoking [9].

Stunting and LBW can also have an impact on the economy of the country and of households. Stunting can cause potential economic losses due to decreased productivity of around 3-14 trillion rupiah, equivalent to 0.04-0.16 percent of Indonesia's Gross Domestic Product [10]. The results of the study by Sicuri et al. of low-income countries show that reducing the prevalence

* Corresponding author: srirum@ugm.ac.id

of LBW will save both healthcare and household costs [11]. Therefore, this study intends to determine the characteristics of mothers and stunted toddlers; to assess the determinants of LBW in stunted toddlers in terms of maternal characteristics; and to identify the relationship between the incidence of LBW and the status of stunted toddlers in Magelang Regency.

2 Methodology

The research used a combination of quantitative and qualitative methods, with the quantitative being the main approach. The results from this method were analyzed qualitatively, which allowed verification of the findings and ensuring that the resulting analysis was more comprehensive and in-depth.

The quantitative method used primary data from field surveys conducted in 2019, together with a structured interview method for the families of stunted toddlers, and secondary data from the Magelang District Health Office. On the other hand, the qualitative data collection technique used involved in-depth interviews with three informants, namely a village midwife, a coordinating midwife, and a public health center nutrition officer. The sample comprised 242 families of stunted toddlers. Independent variables of maternal characteristics were employed, which were indicated by maternal age at birth, gestational age, maternal hygiene and health behavior, and mother's nutrition knowledge. Maternal hygiene and health behavior were reflected by the habit of washing hands with soap and maternal nutritional knowledge is approached with knowledge about 4 healthy 5 perfect nutrition. The first objective of the study was analyzed descriptively with tables, the second with binary logistic regression, and the third with a contingency coefficient. The results of the data processing were then strengthened by the results of the in-depth interviews.

3 Results and discussion

3.1 Characteristics of mothers and stunted toddlers

Observing the characteristics of mothers of stunted toddlers, the majority of babies were born without a history of LBW; sufficient gestational age; mothers gave birth at a non-risk age (20-35 years); they demonstrated hygienic and healthy behavior by washing their hands with soap; and had knowledge about 4 healthy 5 perfect nutrition. As shown in Table 1, the number of mothers who gave birth to stunted toddlers with a history of LBW at a risky age was 19.23 percent, while mothers who gave birth with no risk numbered 7.37 percent. 36 percent of stunted toddlers who experienced LBW were born prematurely, while 6.91 percent were born at full term. 9.05 percent of mothers of stunted toddlers who displayed hygienic and healthy behavior in their daily life had stunted children with LBW, while the figure for mothers of stunted toddlers with LBW who did not show hygienic and healthy behavior was 19.05 percent. 8.88 percent of mothers who had knowledge of nutrition gave birth to LBW babies, while 17.86 percent of those who had no knowledge of nutrition did so.

As can be seen in Table 2, 33.49 percent of stunted toddlers who did not experience LBW had a very short body posture, while 62.50 percent of those who did experience LBW had such a posture. This indirectly shows that a higher proportion of toddlers with a history of LBW develop with with very short stature compared to those who did not experience LBW.

3.2 Influence of maternal age at birth, gestational age, maternal hygiene and health behavior, and maternal nutritional knowledge on stunted toddlers' LBW

The stages of the binary logistic regression consisted of a model significance test, model accuracy test and partial test. The significance of the model using Omnibus Test showed a p-value of (0.000) < α (0.05), meaning that at least one of the four independent variables significantly influenced the incidence of LBW. The model accuracy test using the Hosmer-Lemeshow test showed a p-value of (0.344) > α (0.05), indicating that the model was fit and it was possible to progress to the partial test (Wald test). Table 3 shows that maternal age at birth, gestational age, and maternal hygiene and health behavior significantly influenced the incidence of LBW, at α = 0.05, while maternal nutritional knowledge was significant at α = 0,10. Broadly speaking, all the independent variables significantly affected the incidence of LBW as their p-values were < α.

Table 1. Number of low birth weight incidents of stunted toddlers according to mothers' characteristics in Magelang Regency, 2019.

Mother's Characteristics		Incidence of LBW		Total
		Yes	No	
Maternal age at birth	At risk (< 20 or > 35 years)	10 (19.23%)	42 (80.77%)	52 (100%)
	No risk (20-35 years)	14 (7.37%)	176 (92.63%)	190 (100%)
Gestational age	Premature (<9 months)	9 (36.00%)	16 (64.00%)	25 (100%)
	Normal period (≥ 9 months)	15 (6.91%)	202 (93.09%)	217 (100%)
Maternal hygiene and health behavior	No	4 (19.05%)	17 (80.95%)	21 (100%)
	Yes	20 (9.05%)	201 (90.95%)	221 (100%)
Maternal nutritional knowledge	No	5 (17.86%)	23 (82.14%)	28 (100%)
	Yes	19 (8.88%)	195 (91.12%)	214 (100%)

Table 2. Tabulation of incidence of LBW and status of stunted toddler in Magelang Regency, 2019.

LBW	Stunting Status	
	Stunted	Severely Stunted
(1)	(2)	(3)
No	145 (66.51%)	73 (33.49%)
Yes	9 (37.50%)	15 (62.50%)
Total	154 (63.64%)	88 (36.36%)

Table 3. Estimation of parameters and odds ratio for the LBW incidence of stunted toddlers.

Independent Variable	Dependent Variable: Incidence of LBW		
	B	Sig.	Exp (B)
(1)	(2)	(3)	(4)
Constant	-3.377	0.000	0.034
Maternal Age at Birth (X_1)			
No risk	-	-	-
At risk	1.087	0.028	2.967
Gestational Age (X_2)			
Normal period (≥ 9 months)	-	-	-
Premature (< 9 months)	2.461	0.000	11.722
Maternal Hygiene and Health Behavior (X_3)			
Yes	-	-	-
No	1.309	0.046	3.704
Maternal Nutritional Knowledge (X_4)			
Yes	-	-	-
No	1.142	0.062	3.134

The tendency of mothers to give birth to LBW babies at a risky age was 2.967 times higher than that of mothers who gave birth at a non-risky age, assuming that the other independent variables remained the same. Mothers who are pregnant in their teens are still at their growth stage [12], so their nutritional intake is divided between the mother and the fetus [13]. This has an impact on the development of fetal weight in the womb. The incidence of LBW babies is mostly in young couples due to the incomplete development of the female reproductive organs and insufficient maternal nutrition during pregnancy [14]. Pregnancy that occurs when the mother is less than 20 years old is a direct factor that increases the risk of LBW [15] and an indirect factor that affects stunting status in children [16]. Women who give birth at > 35 years old are more at risk [17-18] because of decreased organ functions and general health, so they are prone to bleeding, prolonged labor and LBW babies [19]. In addition, pregnant women over 35 years of age are also associated with decreased nutrient absorption, which has an impact on the insufficient intake for both the mother and the fetus [20].

Maternal age affects the physical and mental readiness of the mother in preparing for pregnancy and childbirth. The results of the interviews with the village midwife at one of the public health centers in Magelang Regency showed that maternal age at birth was a determining factor in low birth weight. Pregnancy at non-ideal ages can cause anemia and chronic energy deficiency.

> "In Magelang Regency there are still mothers under 20 and over 35 years old who give birth. Lack of parental supervision of children, free use of the internet, and cultural factors and beliefs, such as the prohibition against refusing marriage proposal and not participating in family planning due to the assumption that 'many children have a lot of fortune' are the cause."

The tendency of mothers giving birth to LBW babies prematurely (< 9 months) is 11.722 times higher than women giving birth at full term (≥ 9 months). Preterm birth is the dominant cause of the incidence of LBW [21-23]. According to Manuaba, when gestational age is less than 9 months, the formation of the fat storage system is disrupted, which causes risk to LBW babies [12]. Through in-depth interviews, the village midwife stated that premature gestation was one of the risk factors for LBW.

> "Gestational age especially premature is one of the factors affecting LBW cases and stunting in this region."

Mothers who were not accustomed to hygienic and healthy behavior tended to give birth to LBW babies 3.704 times more than mothers who displayed such behavior, assuming other independent variables remaining the same. One of the risk factors in increasing the chance of LBW is mothers with a history of disease during pregnancy [24]. Washing hands with soap is one of ten types of household hygiene and health behavior that acts as self-protection and family protection to prevent germs and diseases from entering the body [25]. Mothers who do not demonstrate such behavior in their daily life tend to be more susceptible to infection and disease affecting the mother and baby, thus increasing the risk of LBW.

The tendency of mothers who do not know about 4 healthy 5 perfect nutrition to give birth to LBW babies is 3.134 times higher than for mothers who do know this, assuming other independent variables to be the same.

Mothers who know about such nutrition tend to be more careful in maintaining nutritional intake during pregnancy, which will affect their nutritional status and increase the weight of the fetus in the womb until birth. Nutritional intake during pregnancy is a menu designed to meet the needs for micronutrients and macronutrients that function in the growth of the mother and fetus. Iron, calcium, zinc and vitamins are examples of the ingredients in the 4 healthy 5 perfect intake needed by the mother to reduce the risk of LBW [26–29].

From the results of the interviews with the coordinator of midwife and nutrition officer at one of the public health centers in Magelang Regency, information was obtained showing that maternal nutrition knowledge will determine their nutritional intake during pregnancy, which is an important factor in determining LBW. Mothers' lack of knowledge of nutrition will affect the condition of their pregnancy because of insufficient nutritional intake, leading to anemia and chronic energy deficiency, which increase the risk of LBW. Pregnant women who do not know about 4 healthy 5 perfect nutrition tend to pay less attention to nutritional intake, for example protein, meaning they lack protein, as does consequently the fetus.

> "Maternal factors such as anemia and chronic energy deficiency, gestational age especially premature, twin baby, too low or too high maternal age, nutritional intake that is influenced by maternal knowledge, and maternal mental readiness are factors that influence cases of LBW and stunting in this region. A mother can be anemic because of low understanding of nutrition that affect to her nutritional intake. Premature birth due to twin baby, nutritional factors, and economic factors -not many cases- are more likely to be maternal knowledge."

3.3 Relationship between LBW and the status of stunted toddlers

The relationship between the incidence of LBW and stunting status was statistically tested by the contingency coefficient. The results of the association test showed that the association between LBW and stunting status was statistically significant because of the p-value of $(0.005) < \alpha\ (0.05)$.

LBW disrupts the growth and development process over the long term and leads to a high risk of experiencing growth failure or stunting as a consequence. LBW indicates that the fetus has experienced slow growth in the womb, that continues until the baby is born. The growth of LBW babies is slower than ones with normal birthweight, due to the impaired digestive functions experienced by them, which then causes obstacles to the absorption of nutrients into the body [30]. Slow growth in early life makes it difficult for LBW babies to achieve the same optimal growth as those with normal birthweight if this is not balanced with adequate nutritional intake (such as fat and protein) after birth [7, 19]. LBW also causes toddlers to be more susceptible to infectious diseases, such as diarrhea and ARI, so that their growth is not optimal.

Based on the results of the interviews with the village midwife, it was established that LBW is one of the risk factors for stunting in Magelang Regency.

> "The factors that cause stunting are LBW, nutritional intake when pregnant and when the child is born, economic factors, parenting styles, culture, and provision of complementary feeding too early."

Magelang Regency Health Office's Chief of Family Health and Nutrition Section also added that LBW will affect the length of the baby (usually born short) and inhibit their growth, so that the risk of stunting is higher.

> "LBW has an effect on cases of stunted toddlers, toddlers born with low body weight are usually low in height too. Therefore, it needs prevention before birth."

This result is congruent with Putri and Qomariyah's research findings. They found that one of the factors related to stunting in Purwomartani was birth weight. Purwomartani is a village with the highest incidence of stunting in Kalasan, Sleman, Yogyakarta in the surrounding Magelang areas [31].

4 Conclusion

In general, the representation of the characteristics of the mothers of stunted toddlers in Magelang Regency was good, with more mothers giving birth to stunted toddlers at a non-risky ages, at full-term gestation, with hygienic and healthy behavior, and with knowledge of nutrition, especially 4 healthy 5 perfect. The results show that maternal age at birth, gestational age, maternal hygiene and health behavior, and maternal nutritional knowledge significantly influenced the incidence of LBW. The association test showed a significant relationship between LBW and stunting status, while the results of the in-depth interviews also confirmed the findings obtained from the quantitative method.

The authors would like to thank the Magelang Regency Government, the Magelang District Health Office, the informants and the related stakeholders who contributed to providing data and information, and supporting the smooth running of this research.

References

1. Ministry of Health Republic of Indonesia, *Situasi Balita Pendek (Stunting) di Indonesia* (Data and Information Publication of Ministry of Health Republic of Indonesia, Jakarta, 2018)
2. UNICEF, WHO, & World Bank Group, *Levels and Trends in Child Malnutrition: Key Findings of the 2020 Edition of UNICEF/WHO/World Bank Group Joint Child Malnutrition Estimates* (WHO, Geneva, 2019)
3. Ministry of Health Republic of Indonesia, *Buku Saku Pemantauan Status Gizi (PSG) Tahun 2017* (Ministry of Health Republic of Indonesia, Jakarta, 2018)

4. R. Saputra, Jumlah stunting di Kabupaten Magelang menurun. URL: http://beritamagelang.id (2019)
5. World Health Organization, *International Statistical Classification of Diseases and Related Health Problems, 10th Revision* (World Health Organization, Geneva, 2011)
6. A.J. Prendergast, J.H. Humphrey, Paediatrics and International Child Health **34**, 250-265 (2014), doi:10.1179/2046905514Y.0000000158.
7. Supariasa, *Pendidikan dan Konsultasi Gizi* (EGC, Jakarta, 2012)
8. I. Ebtanasari, Hubungan berat badan lahir rendah (BBLR) dengan kejadian stunting pada anak usia 1-5 tahun di Desa Ketandan Kecamatan Dagangan Kabupaten Madiun, Thesis (STIKES Bhakti Husada Mulia, Madiun, 2018)
9. D. Donal, H. Hartono, M. Hakimi, O. Emilia, Indones. J. Geogr. **50**, 11–24 (2018), doi:10.22146/ijg.15951.
10. B.S. Renyoet, D. Martianto, D. Sukandar, Jurnal Gizi dan Pangan **11**, 247–254 (2016), doi:10.25182/jgp.2016.11.3.%p.
11. E. Sicuri, A. Bardají, B. Sigauque,M. Maixenchs, A. Nhacolo, D. Nhalungo, E. Macete, P.L. Alonso, C. Menéndez, PloS One **6**, e28744 (2011), doi:10.1371/journal.pone.0028744.
12. Manuaba, *Ilmu Kebidanan, Penyakit Kandungan dan Keluarga Berencana untuk Pendidikan Bidan* (EGC, Jakarta, 1998)
13. A. Wemakor, H, Garti, T. Azongo, H. Garti, A. Atosona, BMC Research Notes **11**, 877 (2018), doi:10.1186/s13104-018-3980-7.
14. C.M. McDonald, K.P. Manji, R. Kupka, D.C. Bellinger, D. Spiegelman, R. Kisenge, G. Msamanga, W.W. Fawzi, C.P. Duggan, The Journal of Nutrition **143**, 204–214 (2013), doi:10.3945/jn.112.168682.
15. H. Saloojee, H. Coovadia, Maternal age matters: for a lifetime, or longer, Lancet Glob Health, **3**(7), e342–e343 (2015).
16. N. Indrasari, Jurnal Ilmiah Keperawatan Sai Betik **8**, 114-123 (2012), doi:10.26630/jkep.v8i2.152.
17. R.A. Mahumud, M. Sultana, A.R. Sarker, Distribution and Determinants of Low Birth Weight in Developing Countries, J Prev Med Public Health, 50, 18-28 (2017)
18. M.S. Schimmel, R. Bromiker, C. Hammerman, L. Chertman, A. Ioscovich, S. Granovsky-Grisaru, A. Samueloff, D. Elstein, The effects of maternal age and parity on maternal and neonatal outcome, Arch Gynecol Obstet, **291**(4), 793–798 (2015).
19. B.Y.M. Badjuka, Afiasi: Jurnal Kesehatan Masyarakat **5**, 23-32 (2020), doi:10.31943/afiasi.v5i1.94.
20. Y. Vivatkusol, T. Thavaramara, C. Phaloprakarn, Int. J. Womens Health **9**, 347–352 (2017), doi:10.2147/IJWH.S128941.
21. L.D. Hailu, D.L. Kebede, Determinants of Low Birth Weight among Deliveries at a Referral Hospital in Northern Ethiopia, BioMed Research International, 1–8 (2018)
22. I. Trisnawati, H. Salimo, B. Murti, Biopsychosocial and Economic Determinants of Low Birth Weight in Jambi, South Sumatera: Path Analysis, Journal of Maternal and Child Health, **3**(1), 1-10 (2018)
23. H. Sulistyoningsih, *Gizi untuk Kesehatan Ibu dan Anak* (Graha Ilmu, Yogyakarta, 2011)
24. A. Triana, Jurnal Kesehatan Komunitas **2**, 193–198 (2014), doi:10.25311/jkk.Vol2.Iss5.73.
25. Department of Health Republic of Indonesia, *Rumah Tangga Sehat dengan Perilaku Hidup Sehat dan Bersih* (Department of Health Republic of Indonesia, Jakarta, 2009)
26. B.W. Chaffee, J.C. King, Paediatr. Perinat. Epidemiol. **26**, 118–137 (2012), doi:10.1111/j.1365-3016.2012.01289.
27. L.M. De-Regil, C. Palacios, A. Ansary, R. Kulier, J.P. Peña-Rosas, Cochrane Database Syst. Rev. **2**, CD008873 (2012), doi:10.1002/14651858.CD008873.pub2.
28. A. Imdad, Z.A. Bhutta, Paediatr. Perinat. Epidemiol. **26**, 138–152 (2012), doi:10.1111/j.1365-3016.2012.01274.x.
29. A. Imdad, Z.A. Bhutta, Paediatr. Perinat. Epidemiol. **26**, 168–177 (2012), doi:10.1111/j.1365-3016.2012.01312.x.
30. Soetjiningsih, I.G.N.G. Ranuh, *Tumbuh Kembang Anak* (EGC, Jakarta, 2013)
31. F.A.D. Putri, N. Qomariyah, Factors Associated to Stunting in Children Aged 6-24 Months in Yogyakarta, Jurnal Cakrawala Promkes, **2**(1), 16-23 (2020)

Factors influencing the compliance of workplace safety culture in the government linked company (GLC)

Abdul Aziz Mat Isa[1*], *Warishah* Abdul Wahab[1], *Rohayu* Che Omar[1*], *Mohd Zaid* Mohd Nordin[2], *Hairin* Taha[1], *Rasyikin* Roslan[1]

[1]Institute of Energy Infrastructure, Universiti Tenaga Nasional,43000 Kajang, Malaysia
[2]College of Graduate Studies, Universiti Tenaga Nasional, Kajang, 43000, Malaysia

Abstract. This research focused on five factors; the management commitments towards safety, availability of safety rules and procedures, safety communication and feedback, the effectiveness of safety training, and the acquisition of safety knowledge that contributes to compliance of safety culture in the government-linked companies (GLC), by exploring various dimensions by which converge into achieving organizational safety performance outcomes. The importance of safety performance in the GLC is to ensure the employees have the right safety culture, a safe working environment, and comply with safety rules and procedures. The contribution of the findings of this study may provide knowledge of existing safety management practices in Malaysia's Government Linked Company (GLC) and facilitate the organizations in evaluating the importance of safety management on their performance. This study also emphasizes the relationship between the five factors mentioned towards the compliance of safety culture in the organization. Besides that, it also provides an empirical review on which components of safety management practices have a better relationship with the organizational safety culture. Based on the results obtained, there is a positive correlation between management commitments towards safety, availability of safety rules and procedures, safety communication and feedback, the effectiveness of safety training, and acquisition of safety knowledge with regards to employees' safety compliance towards a better safety culture in the organization.

1 Introduction

All accidents occur due to multiple causes. Therefore, it is crucial to identify the causes of accidents so that it can prevent from reoccurring in the future hence minimize the injury, ill-health and also cost to the business. By identifying the factors of the accident at the workplace, it is essential to investigate those that occurred and considering the appropriate preventative measures in reducing the risk of them happening again.

Multiple studies have reported the impact of safety management on a safe environment in the workplace. A review of early research by [1] reveals that management commitment to safety was the primary factor in successful safety programs as well as other factors. More safety staff, safety committees, and safety training contributed to low accident rate companies [2]. They also confirmed that management commitment to safety is essential. Factors such as management involvement, selected promotional efforts, workforce characteristics, and physical conditions played roles in injury frequencies [3]. Safety behaviours and positive outcomes via workers' safety knowledge and motivation were found to have a positive relationship with management commitment [4].

Successful health and safety programs may benefit the company monetarily in several different ways such as reducing down insurance costs, reducing costly litigation and disability claims, increasing the productivity of employees, leading to more contracts as well as reducing the number of compliance inspections and associated penalties. It has been reported that companies not equipped with safety and health programs recorded 30% more accidents than those with the programs [5]. The problems related to safety are due to inconsistencies in applying operating procedures as well as due to non-existence procedures [6].

Communication is the critical part of promoting safety initiatives among employees- Like most workplace initiatives, effective employee communication and a resulting high engagement with the initiative is key to its success. An earlier study suggests that both management communication and employee feedback are essential for suggesting safety improvements and reporting near misses as well as unsafe conditions and practices [7-9]. It shows that poor communication occurs due to multiple causes for example lack of information or knowledge, goals and priorities are not clearly explaining and failing to ask questions when something is unclear. Preconceived ideas/close-mindedness, jumping to conclusions, not understanding others' needs, losing patience and allowing discussions to become heated, time pressure, failure to explore all options, and poor communication patterns are also part of poor communication. The empathic communication style, which is characterized

* Corresponding author: azan@uniten.edu.my, rohayu@uniten.edu.my

by compassionate, concerned, and considerate verbal behaviour to prevent poor communication in an organization was proposed by [10]. The effectiveness of safety communication profoundly influences safety culture in an organization. The empathic communication style would improvise safety performance, better morale, and enhanced organizational culture.

Safety training and education are vital in teaching safe attitudes and providing feedback on the effectiveness of current safety procedures. This section measures the level and effectiveness of safety training in terms of developing and sustaining a work environment conducive to incident and injury-free. A study by [11] shows that safety training can change workers' safe behaviour and attitude. Prevention can be carried out successfully when employees understand better about the work's potential hazards and risks.

Safety knowledge is the crucial knowledge that must be practiced to ensure a safe working environment. Health knowledge means information and skills that relate to preserving or maintaining people free of illness, injury [12,13]. Failure to utilize this knowledge in day-to-day practice may increase accident probability and may lead to an accident. This knowledge ranges from standard operating procedures, safety barriers, as well as emergency procedures. The knowledge of safety also includes general competencies such as training and understanding everyone's roles and any approved code of practice and guidance.

Practical safety training is a fundamental element in every successful organization. Safety training includes accident prevention programs and any occupational safety and health programs. Training improves employees' knowledge and attitudes [14]. An organization with a favourable climate for safety would provide sufficient training to have productive employees [15]. The knowledge gain from the training may save lives. For instance, in a chemical hazard workplace, a significant accident may occur due to minor errors (human or technical) such as out-of-control chemical reactions. During these situations, good knowledge of the processes, associated dangers, and methods to prevent them are crucial for workers in such plants [14]. Safety culture needs to be a part of organizational culture that is supported by the management and affects workers' attitudes and behaviours to an organization's ongoing safety performance. Based on past research and other relevant items, the independent variable identified are the management commitments towards safety, availability of safety rules and procedures, safety communication and feedback, the effectiveness of safety training, and acquisition of safety knowledge while the dependent variable is the compliance of safety culture in the organization. Thus, the relationships between the five variables are outlined in Figure 1.

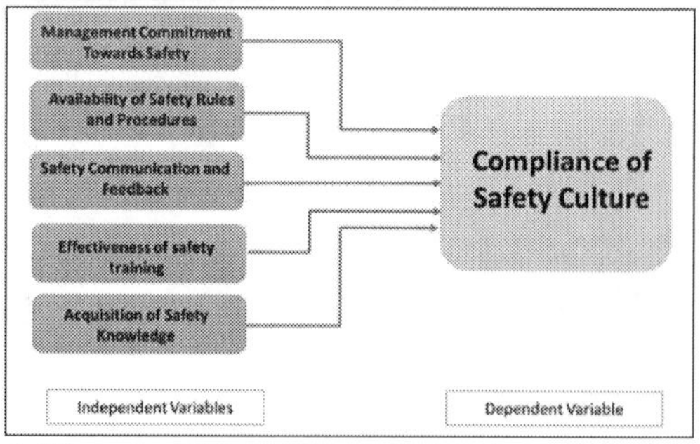

Fig. 1. Theoretical framework.

The specific objectives of the research are to investigate the relationship of the management commitments towards safety, availability of safety rules and procedures, safety communication and feedback, the effectiveness of safety training, and acquisition of safety knowledge towards the compliance of safety culture in Malaysia's government-linked company (GLC). Through literature review gap analysis, there five hypotheses outlined describe as:

i. H1: There is a significant relationship between the management commitment towards safety and compliance of safety culture in the GLC.
ii. H2: There is a significant relationship between the availability of safety rules and procedures and compliance with safety culture in the GLC.
iii. H3: There is a significant relationship between safety communication and feedback and compliance of safety culture in the GLC.
iv. H4: There is a significant relationship between the effectiveness of safety training and compliance of safety culture in the GLC.
v. H5: There is a significant relationship between the acquisition of safety knowledge and compliance with safety culture in the GLC.

2 Materials and methods

The methodology of this study is divided into three main steps Quantitative Research Method, Data Collection, and Data Analysis, as shown in Figure 2 below.

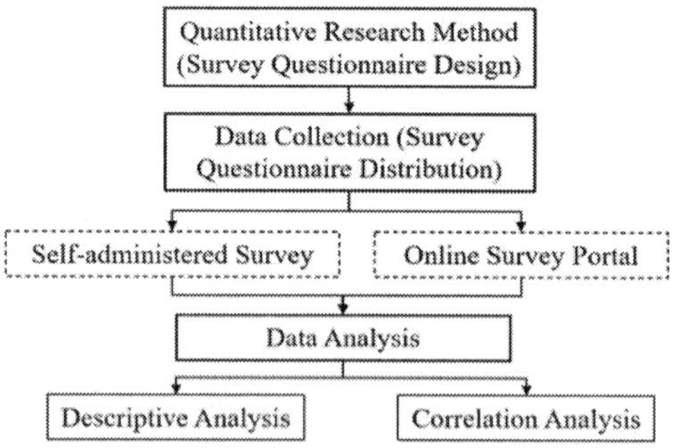

Fig. 2. Research methodology.

2.1 Quantitative research method

A qualitative method was employed in conducting this study. SPSS Version 23.0 was used in this research. A

Likert scale was used to scaling responses in survey research. Through the questionnaires, respondents have been asked to indicate the level of agreement on the 6 level agreement scale, namely "1 = strongly disagree", "2 = disagree", "3 = slightly disagree", "4 = slightly agree", "5 = agree", and "6 = strongly agree". A response indicating the least favorable degree is given the least score (1), and the most favorable is given the highest score (6). By combining both positive and negative items, respondents are forced to consider the question and provide a more accurate response which would reduce potential biases. Acquiescent bias occurs when participants tend to agree with all the questions in the survey. On a 6-point scale, the participants would have answered all with 5's and 6's. Extreme response bias is when the participants provide mostly high or all low ratings. On a 6-point scale, all the answered rate as 6 or 1. By alternating positive and negative items, the extreme serial responder might be reduced [16]. When a reverse-scored item, 1's is turned into 6's, 2's is turned into 5's, 3's is turned into 4's, 4's is turned into 3's 5's is turned into 2's, and 6's is turned into 1's. Below is a simple mathematical formula to calculate for reverse-score items:

$$Reverse\ Score(x) = Max(x) + 1 - x \quad (1)$$

2.2 Data collection (survey questionnaire distribution)

Data collection for this study is done by employing a self-administered survey method and using the online survey portal https://docs.google.com ("Google Docs"). The survey period was carried out in April 2017 for approximately a month. Survey questionnaires were distributed to 200 executives and 200 non-executives in the GLC Distribution Network Department. However, by the end of the survey period, 257 responses were collected, which equals 64.3% out of the total employees who participated.

2.3 Reliability analysis of the questionnaire

Cronbach's Alpha was used in the study to determine the reliability of the questionnaire. The widely accepted social science cut-off for the internal consistency Cronbach α is 0.7 or higher for a set to be considered as a scale [14]. According to [17], Cronbach's Alpha value of more than 0.60 is acceptable, and this method has been widely used in previous research. The guide to interpreting Cronbach's Alpha for dichotomous questions is tabulated in Table 1.

Table 1. Cronbach's Alpha interpretations.

Cronbach's Alpha	Reliability
$\alpha \geq 0.9$	Excellent
$0.9 > \alpha \geq 0.8$	Good
$0.8 > \alpha \geq 0.7$	Acceptable
$0.7 > \alpha \geq 0.6$	Questionable
$0.6 > \alpha \geq 0.5$	Poor
$0.5 > \alpha$	Unacceptable

2.4 Validity of the instrument

Validity is used to check whether the instrument is measuring the right things that are supposed to measure [18]. Content validity analysis includes any validity approaches that emphasize the test content. This evaluation aims to measure the independent variables and the dependent variable. Furthermore, the analysis of how the statements is built was also carried out. In this analysis, it is found that each statement is free from double-barrelled, ambiguous, recall-dependent, leading, loaded, and socially desirable. The questions are established to keep the questions straightforward, compact, and easy to understand. All the statements in the questionnaire are relevant and valid.

3 Data analysis

3.1 Descriptive analysis

Descriptive statistics were used to provide useful information on the research focus and describe the basic features of the data. Mean analysis was one of the descriptive statistical methods employed by using SPSS in this study. The suggested scale model by [19] is used to interpret the mean score, as shown in Table 2 below:

Table 2. Mean score interpretation.

Mean Score	Interpretation
1.00 – 1.80	Strongly Disagree
1.81 – 2.60	Disagree
2.61 – 3.20	Slightly Disagree
3.21 – 4.20	Slightly Agree
4.21 – 5.20	Agree
5.21 – 6.00	Strongly Agree

3.2 Correlation analysis

Correlation analysis was used to determine the relationship between quantifiable variables to determine the relationship between variables. Correlation coefficients range from-1.00 to +1.00, with -1.00 indicating a perfect negative correlation, while +1.00 indicates a perfect positive correlation. A value of 0 indicates no correlation at all. A positive correlation coefficient indicates that there is a positive linear relationship between the variables: as one variable increases the value, so does the other. A negative value indicates a negative linear relationship between variables: as one variable increases the value, the other variable decreases the value [20]. The r values present the strength of a relationship, while the p-value indicates whether there is a significant correlation between variables or not. [21] suggested that an absolute value of r of 0.1 is classified as small, an absolute value of 0.3 is classified as medium, and 0.5 is classified as significant. The summary of the strength of the relationship between the two variables is tabulated in Table 3.

Table 3. Strength of the correlation.

Correlation, r	Strength of Relationship
± .50 - 1.00	Strong relationship
± .30 - ± .49	Moderate relationship
± .10 - ± .29	Weak relationship

4 Results

4.1 Reliability analysis of the questionnaire

In reliability analysis, the fundamental rule is that the higher the value of Cronbach's Alpha, the better is the reliability of the data to analyze. Results of the Cronbach's Alpha value for each variable are recorded as below:

Table 4. Reliability analysis.

Variable Subject	No. of Item	Cronbach's Alpha
Management commitment to safety	5	0.790
Availability of safety rules and procedures	5	0.816
Communication on safety	5	0.850
Effectiveness of safety training	5	0.760
Acquisition of safety knowledge	5	0.877
Compliance with safety culture	9	0.903

Table 4 shows the reliability analysis of the questionnaire. The Cronbach's Alpha value obtained ranged between 0.760 and 0.903. Therefore, the reliability of the questionnaire is accepted and can be used for further analyses throughout the research.

4.2 Descriptive analysis

Based on the interpretation defined in Table 5, it can be interpreted that the respondents mostly answer with "agree" since the range of mean is between 4.4047 to 5.0484. The average value of the mean for all the variables is 4.6239.

Table 5. Mean analysis.

Variables	N	Mean	Std. Deviation	Mean Scale Interpretation
Management Commitment	257	4.4047	0.78141	Agree
Safety Rules & Procedures	257	4.5518	0.71053	Agree
Safety Communication	257	4.5401	0.82632	Agree
Safety Training	257	4.4062	0.74496	Agree
Safety Knowledge	257	4.7922	0.75079	Agree
Safety Compliance	257	5.0484	0.59759	Agree
Overall	257	4.6239	0.56852	Agree

*Note: Based on the Likert scale from 1 (Strongly Disagree) to 6 (Strongly Agree)

The highest mean between all the independent variables towards the compliance of safety culture in the GLC is the acquisition of safety knowledge with 4.7922 while the lowest mean is management commitment towards safety with 4.4047.

A normality test is vital for data analysis. It is employed to determine whether sample data of the questionnaires have been drawn from a normally distributed population. SPSS statistical tests, such as the T-test and the one-way and two-way ANOVA, require a customarily distributed sample population. The distribution is measured by studying its Skewness and Kurtosis. Generally, Skewness determines the relative size of the two tails whether the data are symmetric distribution while Kurtosis is a measure of the heaviness of the tails of a distribution. According to [22], data normality means that the distribution of the data is normally distributed with mean = 0, standard deviation = 1, and asymmetric bell-shaped curve. [22] stated that the Skewness value should be within the range ±1 for normal distribution and the Kurtosis value should be within range ±3 for standard distribution measure while values exceeding five indicates data are non-normally distributed [23]. According to [24] posit that the acceptable range for Skewness and Kurtosis is between -1 and 1. However, values between ±2 are in many cases are acceptable [25]. Table 6 below shows the descriptive statistics data which are used to describe the general overview and basic features of the data collected in this study. Based on the result, it can be concluded that the data distribution is normal.

Table 6. Descriptive statistics.

	N	Min.	Max.	Mean	Std. Deviation	Skewness		Kurtosis	
	Statistic	Statistic	Statistic	Statistic	Statistic	Statistic	Std. Error	Statistic	Std. Error
Management Commitment	257	2.20	6.00	4.4047	.78141	-.245	.152	-.490	.303
Safety Rules & Procedures	257	2.40	6.00	4.5518	.71053	-.114	.152	-.098	.303
Safety Communication	257	1.80	6.00	4.5401	.82632	-.308	.152	-.205	.303
Safety Training	257	2.60	6.00	4.4062	.74496	.019	.152	-.435	.303
Safety Knowledge	257	2.00	6.00	4.7922	.75079	-.737	.152	1.376	.303
Safety Compliance	257	3.33	6.00	5.0484	.59759	-.248	.152	-.438	.303
Valid N (listwise)	257								

4.3 Correlation analysis

Pearson correlation test has been conducted to measure the existence (given by a p-value) and strength (given by the coefficient r between (-1 and +1) of a linear relationship between two variables. If the outcome is significant, we conclude that a correlation exists between the variables. The results of the correlation are tabulated as per Table 7 below.

Table 7. Correlation results.

		Safety Compliance	Management Commitment	Safety Rules & Procedures	Safety Communication	Safety Training	Safety Knowledge
Safety Compliance	Pearson Correlation	1	**.298****	**.329****	**.382****	**.380****	**.454****
	Sig. (2-tailed)		.000	.000	.000	.000	.000
	N	257	257	257	257	257	257
Management Commitment	Pearson Correlation	**.298****	1	.663**	.668**	.524**	.317**
	Sig. (2-tailed)	.000		.000	.000	.000	.000
	N	257	257	257	257	257	257
Safety Rules & Procedures	Pearson Correlation	**.329****	.663**	1	.727**	.688**	.518**
	Sig. (2-tailed)	.000	.000		.000	.000	.000
	N	257	257	257	257	257	257
Safety Communication	Pearson Correlation	**.382****	.668**	.727**	1	.672**	.452**
	Sig. (2-tailed)	.000	.000	.000		.000	.000
	N	257	257	257	257	257	257
Safety Training	Pearson Correlation	**.380****	.524**	.688**	.672**	1	.517**
	Sig. (2-tailed)	.000	.000	.000	.000		.000
	N	257	257	257	257	257	257
Safety Knowledge	Pearson Correlation	**.454****	.317**	.518**	.452**	.517**	1
	Sig. (2-tailed)	.000	.000	.000	.000	.000	
	N	257	257	257	257	257	257

**. Correlation is significant at the 0.01 level (2-tailed).

Table 7 generally, shows that all five independent variables namely; management commitment towards safety, availability of safety rules and procedures, communication on safety, the effectiveness of safety training, and acquisition of safety knowledge are significantly correlated to the dependent variable (compliance of safety culture). For this study, five (5) hypotheses for correlation were tested.

Table 4 shows the reliability analysis of the questionnaire. The Cronbach's Alpha value obtained ranged between 0.760 and 0.903. Therefore, the reliability of the questionnaire is accepted and can be used for further analyses throughout the research.

i. H1: There is a significant relationship between the management commitment towards safety and compliance of safety culture in the organization. Based on Table 7, shows that there is a significant positive relationship between the variable management commitment towards safety and employee's compliance towards safety culture, where $r = 0.298$, $n = 257$, $p < 0.00$. Thus, from the correlation, it can be concluded that hypothesis H1 is accepted.

ii. H2: There is a significant relationship between the availability of safety rules and procedures and compliance with safety culture in the organization. Based on Table 7, shows that there is a significant positive relationship between the variable availability of safety rules and procedures and employee's compliance towards safety culture, where $r = 0.329$, $n = 257$, $p < 0.00$. Thus, from the correlation, it can be concluded that hypothesis H2 is accepted.

iii. H3: There is a significant relationship between safety communication and feedback and compliance of safety culture in the organization. Based on Table 7, shows that there is a significant positive relationship between the variable communication on safety and employee's compliance towards safety culture, where $r = 0.382$, $n = 257$, $p < 0.00$. Thus, from the correlation, it can be concluded that hypothesis H3 is accepted.

iv. There is a significant relationship between the effectiveness of safety training and compliance of safety culture in the organization. Based on Table 7, shows that there is a positive relationship between the variable effectiveness of safety training and employee's compliance towards safety culture, where $r = 0.380$, $n = 257$, $p < 0.00$. Thus, from the correlation, it can be concluded that hypothesis H4 is accepted.

v. There is a significant relationship between the acquisition of safety knowledge and compliance with safety culture in the organization. Based on Table 7, shows that there is a positive relationship between the variable acquisition of safety knowledge and employee's compliance towards safety culture, where $r = 0.454$, $n = 257$, $p < 0.000$. Thus, from the correlation, it can be concluded that hypothesis H5 is accepted.

Table 8. Interpretation of correlation scale.

Variable Subject	Correlation, *r*	Strength of Relationship
Management commitment to safety	0.298	Weak relationship
Availability of safety rules and procedures	0.329	Moderate relationship
Communication on safety	0.382	Moderate relationship
Effectiveness of safety training	0.380	Moderate relationship
Acquisition of safety knowledge	0.454	Moderate relationship

From Table 8. the correlation, r results for independent variable management commitment towards safety value of '0.298' shows a weak relationship between the compliance towards safety culture practices in the GLC. Even though the relationship is weak, management commitment still has a relationship with the dependent variables and was positively correlated with the compliance of safety culture. The correlation results for other independent variables (availability of safety rules and procedures, communication on safety, the effectiveness of safety training, and acquisition of safety knowledge) statistically display a moderate relationship between the compliance of safety culture practices in the GLC. The correlation, r results are between '0.329' to '0.454' shows that the independent variables were positively correlated with dependent variables. This result reveals statistically that any positive change in all five independent variables (management commitment towards safety, availability of safety rules and procedures, communication on safety, the effectiveness of safety training, and acquisition of safety knowledge) lead to increase the employees' compliance towards implementing safety culture in the organizational.

5 Discussion

5.1 Management commitment related to safety

Based on the descriptive studies, it is shown that the average value of the mean for variable management commitment is 4.4047. The majority of the workers were agreed that the management was committed to developing a safety culture in the organization. This result indicates that there is still room for improvement for management commitments practices towards safety in the GLC.

Based on the correlation analysis, management commitment is the only independent variable that resulted in a weak relationship as compared to the other four (4) independent variables. It also did not contribute towards safety culture since the enforcements were not strictly carried out by the top management such as management by walk about (MBWA) and safety quality enforcement (SQE). Apart from that, the roles of the senior management and supervisor at the site were very minimal to assure the safe performance practiced by workers.

A successful safety management commitment requires a delicate balance of safety cultural acceptance, best practices in the work processes, and the right degree of system and technology support. The safety performance improves further if the focus is in the right direction and the implementation of the strategy. Management commitment will encourage both management and employees to create a workplace that is as safe as possible. The involvement of the top management will lead to better safety performance by demonstrating their commitment to safety.

5.2 Availability of safety rules and procedures

Based on the descriptive studies, it is shown that the average value of the mean for variable management commitment is 4.5518. The majority of workers were agreed that the safety rules and procedures are available at the company and very useful in developing safety cultures in the organization. This result indicates that there are still rooms for improvement for safety rules and procedures to influence safety culture in the organization.

The result from the correlation analysis shows that there is a moderate and positive relationship between the availability of safety rules and procedures towards compliance with safety culture. Safety rules and procedures are vital components of the internal risk control system to ensure customers, employees and owners do not get injured during business operations. It can help to prevent accidents from happening in the business environment and allow employees to work quicker and improve the level of productivity.

5.3 Safety communication and feedback

Based on the descriptive studies, it is shown that the average value of the mean for variable management commitment is 4.5401. From the results, it appears that all respondents were agreed that safety communication and feedback are important in creating a safety culture in the organization.

From the correlation analysis, there is a moderate and positive relationship between the availability of safety communication and feedback towards safety with compliance of safety culture. The more active, the organizational communication related to safety, the more improved the compliance of safety culture and safety performance.

5.4 Effectiveness of safety training

Based on the descriptive studies, it is shown that the average value of mean for variable safety training is 4.4062. The workers were agreed that the safety training provided by the company was effectively developing a safety culture among the workers. However, there is still room for improvement in the effectiveness of safety training implementation in the GLC.

The findings from correlation analysis revealed that there is a moderate and positive relationship between the effectiveness of safety training towards safety with compliance of safety culture.

A practical safety training is absolutely a worthy investment considering the loss in staff productivity (lost time injury), property damage, and cost arising from an accident, which could have been prevented. Many injuries and property damage can be prevented with the sharing of a lesson learned for accidents that happened by implementing practical safety training and refresher courses every year. Safety training helps employees to give safety its due importance with a useful job analysis and risk hierarchy control techniques. Safety training also helps the employees to choose the right protective clothing and equipment (PPE) to safeguard themselves against any hazardous job task and activity.

5.5 Acquisition of safety knowledge

Based on the descriptive studies, it is shown that the average value of mean for variable safety training is 4.7922, which is also the highest mean among all the independent variables. From the result, it can be concluded that most workers were agreed that the acquisition of safety knowledge is vital in developing compliance towards a safety culture.

From the correlation analysis, it can be seen that there is a positive relationship between the acquisition of safety knowledge towards safety with compliance of safety culture. Safety knowledge recorded the highest correlation, r compared to the other four (4) independent variables and it is the only factor that has significance and contributes to the dependent variable which is the compliance of safety culture.

Awareness of safety knowledge stimulates productivity and produces better work outcomes. The employee with good safety knowledge increases their work productivity and provide them with a better understanding of manoeuvring through the related safety issues in their daily task and activity. With the management's support and persistence, changes in the organization's safety culture have taken place in self-learning initiative and knowledge communication. The familiarity with this safety knowledge acquisition helps the employee practice and apply for the safety course during the employee's working hours. Due to that, they would need to learn the right processes, that refer with regards to the safe manner in which how the task is supposed to be carried out with a safe manner, to refresh themselves with the safety rules and procedures, and other purposes related to the nature of their job responsibilities. This knowledge has proven to be meaningful; ultimately, the employees acquire knowledge to take necessary precautions when delivering their daily tasks, and they know how to decrease the risk of accidents in the workplace.

6 Conclusion

This research is focused on five factors; the management commitments towards safety, availability of safety rules and procedures, safety communication and feedback, the effectiveness of safety training, and the acquisition of safety knowledge that contributes to compliance of safety culture in the government-linked companies (GLC), by exploring various dimensions by which converge into achieving organizational safety performance outcomes. The importance of safety performance in the Malaysia GLC is to ensure the employees have the right safety culture, a safe working environment, and comply with safety rules and procedures. The findings provide good evidence and support for researchers to explore the effect of the attributes such as safety management, rules and procedures, communication and feedback, and knowledge with regards to employees' safety compliance towards a better safety culture in the organization.

Theoretically, the acquisition of safety knowledge is a very crucial element in the safety management system since it contributes 24.2% towards compliance with safety culture in the GLC. Therefore, it can be concluded that management practices should continue to focus on how to ensure that the employees are equipped with all the required knowledge. The GLC needs to revaluate its existing and future safety management strategies as well. Apart from that, this study also provides an empirical review on which components of safety management practices have a better relationship with the organizational safety culture.

Safety performance is very vital to achieve GLC goals, which are geographical expansion. The GLC aims to excel in all its business areas domestically as well as globally, and it can be achieved through an excellent safety performance which reflects a high competency in managing the energy industry. A good result of safety performance gives a reasonable perception by the public, government, stakeholders, business partners, and competitors. It's the safety culture within the organization that drives towards achieving these goals.

The contribution of the findings of this study may provide the improvement of existing safety management practices in the GLC. It facilitates the organization in evaluating the importance of safety management on their overall productivity performance in the company by creating a good safety culture. Based on the study, there is a positive correlation between management commitments towards safety, availability of safety rules and procedures, safety communication and feedback, the effectiveness of safety training, and acquisition of safety knowledge with regards to employees' safety compliance towards a better safety culture in the organization. To conclude this study, the findings and result analysis indicate that the level of compliance with safety culture is the fundamental approach to enhance overall safety performance in the organization.

The research presented in this paper is funded by Tenaga Nasional Berhad (TNB) through a post-graduate university scholarship at UNITEN and UNITEN R & D Sdn. Bhd. under

URND Seeding Fund (U-TE-RD-19-19). The authors would like to thanks Universiti Tenaga Nasional (UNITEN) and UNITEN R & D Sdn. Bhd. for the opportunity and sponsorship in completing this study.

References

1. A. Cohen, Safety Research **9**, 168–178 (1977)
2. M. Smith, H. Cohen, A. Cohen, R. Cleveland, J. of Safety Research **10**, 5–15 (1978)
3. R. Simonds, Y. Shafai-Sahrai, J. of Safety Research **9**,120–127 (1997)
4. M.S. Christian, J. C. Bradley, J. C. Wallace, M. J. Burke, J. of Applied Psy. **94**, 5:1103–1127 (2009)
5. D. Goetsch, Occupational safety and health for technologists, Engineers, and Managers. Seventh Edition (Prentice-Hall, USA, 2011)
6. S. Hood, Developing operating procedures: 9 steps to success (Accident Prevention, 1994)
7. N.H. Zakaria, N. Mansor, Z. Abdullah, Business & Management Review **2**, 5:75–88 (2012)
8. H. Y. Chong, S. L. Thuan, Inter. J.of Occupational Safety &Ergonomic **20**, **3**:503–513 (2014)
9. S. Brisk, S. Workplace accidents: How to avoid them and what to do when they happen, URL: Business.com (2016)
10. J. Williams, Improving safety communication skills: becoming an empathic communicator (The American Society of Safety Session 716, Blacksburg, Virginia, 2006)
11. M. Hallowell, J. Gambatese, J. Construction Safety Risk Mitigation. J. Const. Eng. Manage **135**, 12:1316–1323 (2009)
12. J. Wintle, P. Moore, N. Henry, S. Smalley, G. Amphlett, Plant ageing, management of equipment containing hazardous fluids or pressure (Health and Safety Executive, research report RR509, Norwich, 2006)
13. Z. Hassan, C. Subramaniam, Md. Lazim Mohd. Zin, F. Mohd. Shamsudin, S. Ramalu, Academy of Entrepreneurship J **25**, 2S (2019)
14. M. Vinodkumar, M. Bhasi, Accident Analysis & Prevention, **42** (2010)
15. S. Zakaria, S. Inter. Education Studies **9**, 5 (2016)
16. J. Sauro, J. Lewis, *When designing usability questionnaires, does it hurt to be positive?*, in CHI, May 7–12, 2011, Vancouver, BC, Canada (2011)
17. D. Cooper, P. Schindler, Business research methods (12 ed.) (Irwin Professional, Boston, Massachusetts, USA, 2013)
18. A. Bryman, D. Cramer, Quantitative data analysis with SPSS 12 and 13. a guide for social scientists (Taylor & Francis e-Library, 2005)
19. K. Moidunny, The effectiveness of the National Professional Qualification for Educational Leaders (NPQEL) (The National University of Malaysia, Malaysia, 2009)
20. J. Cortina, J. of Applied Psy. **78**, 98–104 (1993)
21. J. Cohen, Statistical power analysis (2nd ed.) (Erlbaum, Hillsdale NJ, 1998)
22. B. Byrne, B. Structural equation modeling with AMOS: basic concepts, applications, and programming (Routledge, New York, 2010)
23. P. Bentler, EQS 6 Structural equations program manual (CA: Multivariate Software, Inc., Encino, 2006)
24. C. Dwork, V. Feldman, M. Hardt, T. Pitassi, O. Reingold, A. Roth, *Preserving statistical validity in adaptive data analysis*, in Proceedings of the Forty-Seventh Annual ACM on Symposium on Theory of Computing, Portland, Oregon, USA (2015)
25. D. George, P. Malery, SPSS for Windows step by step. A simple guide and reference fourth edition (11.0 update) (La Sierra University, California, 2005)
26. A. Ek, Safety culture in sea and aviation transport, Doctoral thesis

Gender empowerment approaches: interventions through Qanun number 6 of 2009 in Aceh, Indonesia

Adrianto Shifa Al Aro, *Aini* Saniyah, *Fahima* Ulumia, *Nur* Azkiyatuz Zahro, *Nur* Ayumi, *Milta* Charennina, *Wikan* Amarudin Fahmi, dan *Ratih* Fitria Putri*

Department of Environmental Geography, Faculty of Geography, Universitas Gadjah Mada

Abstract. Aceh is a province in Indonesia that makes Islamic sharia as one of the bases in policy determination, one of which is policies related to gender empowerment. It is interesting to be studied about the condition of gender empowerment in Aceh, especially after the ratification of Qanun Aceh No. 6 of 2009 on Women Empowerment and Protection. The study used descriptive qualitative analysis of secondary data obtained from BPS Aceh Province. The results showed that the implementation of Qanun No. 26 of 2009 related to the empowerment and protection of women is still not done to the maximum, either explicitly or implicitly. This is evidenced by the condition of Aceh Gender Empowerment Index which has a volatile trend from 2010 – 2019 with an index ranging from 52.06 to 67.4. The trend is strongly influenced by the three components of GEI, namely women's involvement in parliament, women's involvement as professionals, and women's income donations. The involvement of women in parliament still does not meet the minimum figure of 30%. The involvement of women as professionals has a volatile trend. Meanwhile, women's income contributions always increase every year.

1 Introduction

The term 'gender' comes from the Latin word "genus" meaning birth, family or idea [1]. In the social field, gender includes a role in the social development of society that is not determined by gender. Therefore, every man and woman must have equal rights in social life. However, physical differences make people think that men have a higher level of social life in society, whereas the rights of men and women in social life are actually the same [2].

At the heart of the various definitions and meanings of gender is the thought that men and women are on the same platform in social life. In general, world-class institutions are attuned to the idea that the role of women is often subjugated by social norms. The imbalance causes the gender gap i.e., both men and women lag their peers on key developmental indicators such as education and income. Although many development practitioners still confuse the approach of gender equality with women's empowerment, the gender approach has gained significant place at least at the time of policy formulation.

Women in some developing countries suffer from gender inequality. Countries such as Yemen, Chad, and Pakistan have been ranked at the bottom of the Global Gender Gap Index according to the World Economic Forum. In Indonesia, for example, there are 96% male literacy, but only 90% women are literate. A total of 86% of men participate in the labor market, but only 53% of women do so. It also occurs in income conditions between men and women. Men on average earn US$6,903, but women only get US$ 2,985. Only one in five legislators, senior officials, and managers are women [3]. The approach of empowering women especially by means of education and extensive training to develop their capacity is essential for other empowerment measures that need to be taken by the government.

Aceh is a province in Indonesia that makes Islamic sharia as one of the basic in policy determination, one of which is the policy that governs the empowerment and protection of women. This policy is published in Qanun Aceh No. 6 of 2009 which is made on the basic of several considerations that refer to the Quran, Al-Hadith, and some state regulations in the form of law and government regulations. One of the provisions written in Qanun Aceh is that the Government of Aceh provides the widest possible access for women to participate in the labor market, economic activities, and public space. This makes the issue of women empowerment in Aceh Province an interesting thing to be raised considering the role of women and Islamic sharia is an inseparable thing [4].

This study aims to find out the condition of gender empowerment in Aceh Province, especially after the publication of Qanun Aceh Number 6 of 2009. This study explains the Gender Empowerment Index (GEI) and

* Corresponding author: ratihfitria.putri@ugm.ac.id

trends over the last ten years along with its three main components in Aceh Province. GEI's main components include the involvement of women in parliament, donations of women's income, and women as professionals. GEI condition in Aceh Province is described in the form of a graph chart based on the year and administrative area of the district/ city.

2 Data and Methods

The research method used is qualitative with a review of previous literature studies. The data and information obtained are analysed using descriptive analysis. The selection of research subjects was conducted using regional case studies at the Aceh Province level.

This study uses the publication data of Aceh Province in figures from 2011 - 2021 collected through secondary data from the Central Statistics Agency of Aceh Province, the publication of the Ministry of Women Empowerment and Child Protection in the 2020 Gender-Based Human Development Report, and the legislation of the Ministry of Law and Human Rights. The data used in this study include:

1. The Gender Empowerment Index data and its components include women's involvement in parliament, involvement of women as professionals, and donations of women's income in society.
2. Regulation of the Special Region of Aceh Province in the form of Qanun Aceh Number 6 of 2009 concerning The Empowerment and Protection of Women.

3 Results and Discussions

3.1 Gender empowerment index

Gender Empowerment Index (GEI) is one of the main indicators in measuring the success of human development in an area. The index is derived from the arithmetic average of three main components, namely the involvement of women in parliament, women as professionals, and the contribution of women's income [5]. Based on this, it can be said that the dynamics of GEI are closely related to the level of female participation in public spaces and economic activities. GEI in each province in Indonesia has a number and trends are quite varied. There are provinces with GEI numbers that are larger than the national GEI, there are also provinces with GEI that are worth less than the national GEI, one of which is the GEI of Aceh Province which is presented in figure 1.

The picture shows that the Gender Empowerment Index (GEI) of Aceh Province is experiencing a patterned volatile trend. The increase in GEI figures continuously occurred in 2010-2016 and 2017-2018. GEI's figure increased by 14 points during 2010-2016, from 53.4 in 2010 to 67.4 in 2016, while in 2017-2018 there was an increase of 0.32 points. The increase in GEI figures indicates that the role of women in Aceh Province in decision making, economic activity, and employment is increasing, and gender disparity is decreasing. The decrease in GEI figures occurred in the period 2016-2017 and 2018-2019. The decrease in 2016-2017 was 0.1 while in 2018-2019 there was a decrease of 0.29 points. The decrease in GEI figures in Aceh Province indicates that the role of women in decision making, economic activities, and employment is decreasing, and the disparity of gender roles is increasing.

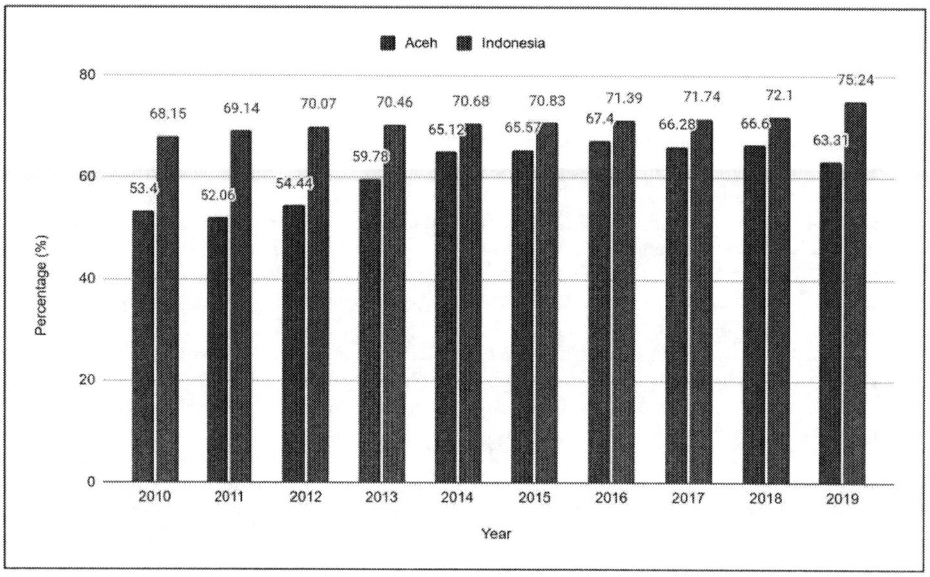

Fig. 1. Gender Empowerment Index (GEI) in Aceh Province and Indonesia 2010-2019
Source: BPS, processed (2021)

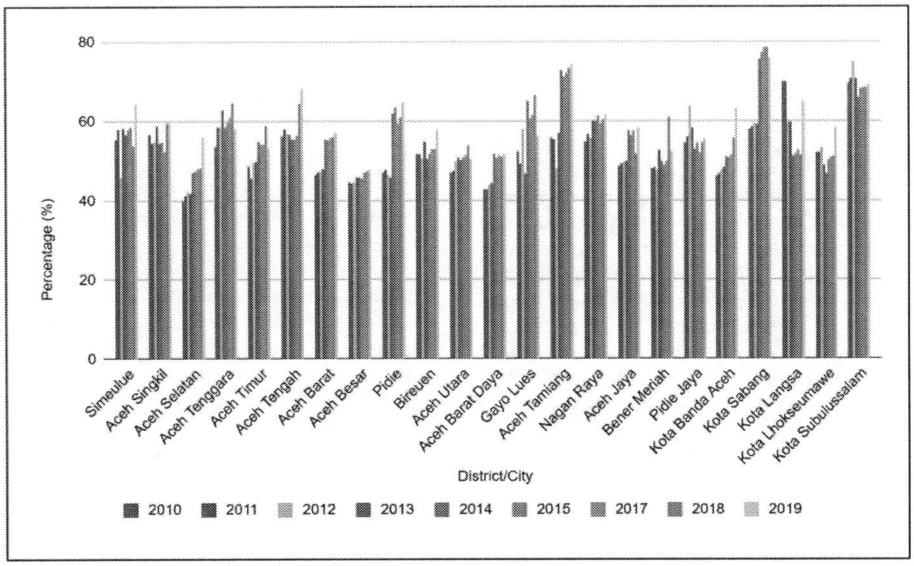

Fig. 2. Gender Empowerment Index (GEI) in Aceh Province by District/City 2010-2019
Source: BPS, processed (2021)

The fluctuating trend of GEI figures in Aceh Province shows that throughout 2010-2019 there was a dynamic level of participation and awareness of women in public spaces and economic activities.

The distribution of Gender Empowerment Index (GEI) in Aceh Province by district and city changed in 2010 and 2019, as shown in figure 2. In 2010, there were two administrative districts that had a number of $60 \leq GEI < 70$, namely Subulussalam and Langsa, while other districts and cities had GEI numbers < 60. This shows that the level of women's participation in public spaces and economic activities in Subulussalam and Langsa city is quite good / middle class compared to other cities and districts. Other cities and districts with low GEI levels (GEI < 60) in Aceh Province showed that in 2010, women's awareness of the importance of contributions in public spaces and economic activities was still likely to be low. The distribution of GEI numbers in 2019 shows that the administrative areas that have the number $60 \leq GEI < 70$ are Banda Aceh, Subulussalam, Central Aceh Regency, Pidie, Nagan Raya, and Simeulue. Administrative areas with GEI numbers ≥ 70 are Sabang city and Aceh Tamiang regency, while other districts and cities have GEI numbers < 60.

Based on the distribution of GEI figures in 2010 and 2019, it is known that Sabang city and Aceh Tamiang district experienced a very significant increase in GEI figures compared to other districts and cities in Aceh Province. Aceh Tamiang district had a GEI number of 55.78 in 2010 and rose to 74.39 in 2019, an increase of 18.61 points over the last nine years. Sabang city has GEI number of 57.92 in 2010 and rose to 76.01 in 2019, an increase of 18.09 points over the last nine years. The large number of GEI increases even exceeded the increase in GEI figures of Aceh province and nationally. The rapid increase in GEI numbers in Sabang and Aceh Tamiang districts is influenced by the high involvement of women in parliament compared to other districts/cities. This shows that the level of women's awareness of the importance of contributing to public spaces is increasing along with wide-open opportunities and opportunities for women to participate in political activities and public spaces.

The efforts of the Aceh Provincial Government related to increasing gender empowerment in Aceh Province can be proven by the existence of Qanun Aceh Number 6 of 2009 on Women Empowerment and Protection. Qanun is a type of local legislation that regulates the administration and life of the people in Aceh Province. Qanun Aceh is a regulation that applies throughout Aceh Province, while Qanun only applies in certain districts/cities. Based on the qanun, women's empowerment and protection aim to strengthen women's participation in the development process, improve the quality of life of women to be able to play a balanced role with men in various aspects of life that are relevant, provide assurance to women to be able to fulfill their rights as human beings in all aspects life and provide a sense of security to women in all aspects of life. The Qanun becomes a regulation that clarifies the previous regulation, namely Law Number 11 of 2006 on the Government of Aceh which provides an opportunity for representation.

Unlike GEI in Aceh Province which still tends to be low and has a fluctuating trend, Central Kalimantan Province has the highest GEI value compared to other provinces in Indonesia in recent

 https://doi.org/10.1051/e3sconf/202132506006

years. The GEI of Central Kalimantan Province in 2010 was 68.62 and has a tendency to increase every year to 83.20 in 2019 [6] which is also the highest GEI value in Indonesia. Not much different from Aceh Province, Central Kalimantan also has a legal foundation that regulates gender empowerment, namely Central Kalimantan Governor Regulation No. 4 of 2014 on Gender Mainstreaming in Regional Development in Central Kalimantan. Although both have regulations governing gender empowerment, the success of GEI Aceh province and central Kalimantan seems to show a considerable gap. Reporting from an article written by [7] from the Office of Women Empowerment, Child Protection, Population Control and Family Planning of Central Kalimantan Province web, the success of GEI in Central Kalimantan Province is due to the strong government's commitment in the implementation of Central Kalimantan Governor Regulation No. 4 of 2014. The government is considered to have a strong commitment in ensuring equal rights between men and women in the economic, political, socio-cultural, and legal fields so as to encourage the realization of gender equality and justice in Central Kalimantan Province.

Similar to Aceh Province, another province that still has a low GEI in Indonesia is Papua Province. In 2010 the GEI of Papua Province was only 55.42 but the numbers showed a steady increase every year, until in 2018 the GEI of Papua Province was 68.71 [6]. The development strategy in Papua Province so far has been directed at empowering the community in general and increasing the role of the private sector in community affairs that have not been optimally carried out by local governments. The role of private institutions in absorbing women's employment in Papua Province has been quite high. Although the increase has not been so significant, it can be seen that women's participation is already quite high. Until 2019, women's participation in the private sector reached 42.35%. The legal basis for gender empowerment in Papua can be seen from the role of the government by issuing Law No. 21 of 2000 on Special Autonomy, and is still in the evaluation process. Under the provisions of this special autonomy law, women get improved certainty over their participation in determining local policies through the Papuan People's Council (MRP). In this assembly representing the Papuan people who are segregated into various ethnic groups, women's groups get a quota of one third (1/3) of the seats in the MRP. This opens up opportunities for women to be able to erode various gender inequalities caused by social and cultural aspects.

3.2 Women's Involvement in Parliament

Parliament in governance in Indonesia is synonymous with legislative institutions. In addition, everything that has something to do with government is to tighten leadership or leadership. Today, the empowerment of women in various dimensions of life gets global attention, including in the field of government, especially for observers of development issues. This is in line with the fact that women are often considered to be discriminated against in terms of taking on strategic roles including in politics.

In a democratic country, the involvement and participation of women and men in public policy drafting positions occupied the same position. This is in line with efforts to produce a decision that impacts the balance of society at all levels, as well as the realization of achieving goals and justice. However, in the implementation in the field, it seems that women are still more objects than subjects. The cause of this persistence of the phenomenon is due to patriarchal thinking that makes women's confidence to take a role in politics very low. According to Law Number 11 of 2006, the representation of women in parliament is at least 30% [8]. The representation of women in parliament is also very important in public decision making. This will bring different women's perspectives in seeing and solving various public problems, because women will think more holistically and responsive to gender.

Figure 3 shows women's involvement in parliaments in Aceh and Indonesia provinces. Based on the chart, the percentage of women's involvement in parliament in Aceh Province is lower than in Indonesia. The trend of the percentage of women's involvement in parliament in Aceh Province is quite volatile from year to year. The percentage of female involvement in parliament in 2010 was 5.8% and then decreased in 2011 to 4.35%. That figure then increased to 5.8% in 2012. The number of women involved in parliament then increased significantly in the following year, namely in 2013 to 2019 with a percentage ranging from 10.14% to 16%. The increase in the percentage of women's involvement in the parliament had a positive impact after the enactment of the Qanun on empowerment and protection for women in 2009, especially in article 8 governing women's rights in politics.

Under the article, women are entitled to political positions, both in the executive and legislative environment proportionally. Women are also entitled to conduct various political activities in accordance with the laws and regulations and are entitled to occupy the position of every level of management of political parties and local political parties. In addition, women are also entitled to be nominated as members of the legislature by national political parties as well as local political parties in accordance with their abilities.

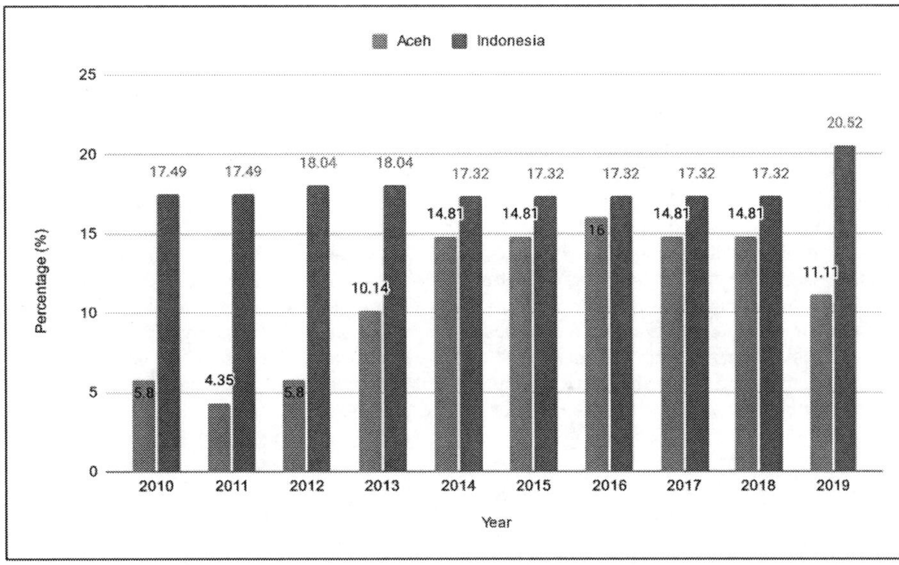

Fig. 3. Women's Involvement in Parliament in Aceh Province and Indonesia 2010-2019
Source: BPS, processed (2021)

The involvement of women in parliament in Aceh Province by district/city is represented in figure 4. Based on the chart, the percentage of female involvement in parliament in each district/city varies greatly. The percentage trend each year is also very varied because there are districts/cities that experience decreases and increases. Aceh Tamiang district has the highest average involvement of women in parliament in Aceh Province at 23.33% because in 2010-2019 it had a percentage that is between 13.33 to 36.67%. Sabang city also has a high average of 20% with the percentage of women's involvement in parliament around 10-30%.

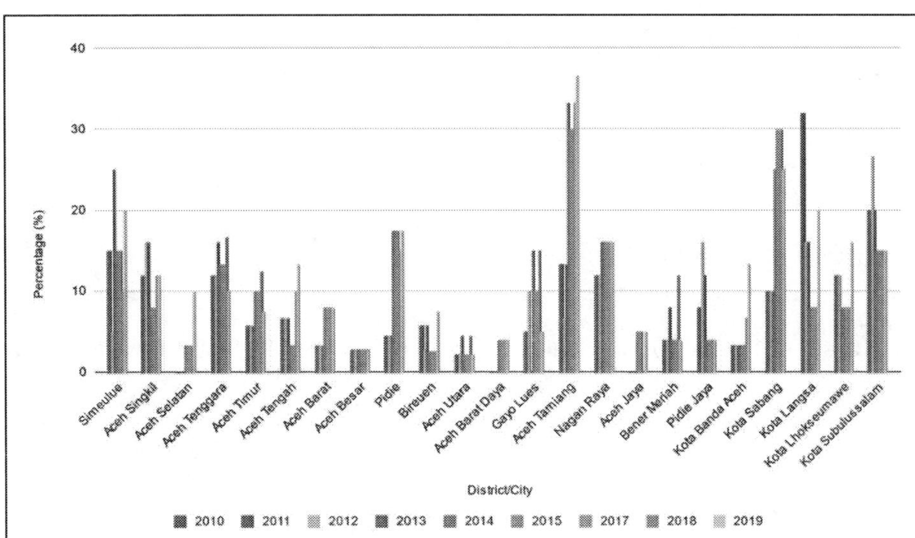

Fig. 4. Women's Involvement in Parliament in Aceh Province by District/City year 2010-2019
Source: BPS, processed (2021)

Meanwhile, the average percentage of female involvement in other district/city parliaments is around 2.6-17.96%, In general, the involvement of women in parliament in Aceh Province in 2010 to 2019 has not qualified at least 30% because the percentage in 2010 to 2019 is only about 4.35% to 16%. When associated with the existence of Qanun Aceh Number 6 Year 2009, the application of Qanun is not spread evenly in each district/city to create a gap. This can be because in article 4 of Qanun Aceh Number 6 of 2009, women empowerment and protection activities are carried out and coordinated by the Aceh Device Work Institute or District/ City. The condition causes a difference in the implementation of women's empowerment and protection in each district/city and the impact is the percentage of women's involvement in parliament is also different.

Rule related to women's empowerment is also carried out in other countries, one of which is in Kenya. Kenya has an important role in the development of women's empowerment in the world because Kenya was once the site of the Third World Conference of Women in 1985 that was organized by the United Nations in Nairobi. The conference provides a great opportunity for women in Kenya, especially to improve gender equality [9]. Meanwhile, Kenya was also one of the countries that ratified article 9 of the Maputo Protocol on October 6, 2010. The article discusses women's equality, which has to do with the representation of women in parliament. Meanwhile, there was a change in the constitutional rules in 2010, namely the two third gender rule. Under the rules, 2/3 of lawmakers cannot have the same gender. This means that there are opportunities for women, one of which is related to the provision of special seats for women. Both have greatly affected the number of representations in Parliament in Kenya which increased from 7.1% in 2005 to 19.7% in 2015 [10]. This condition shows that the role of the government is very important in increasing women's empowerment.

3.3 Women's Income Contribution

The Gender Empowerment Index (GEI) shows whether women can actively participate in economic and political life. GEI focuses on participation, by measuring gender inequality in the economic, political, and decision-making fields. One of GEI measurement indicators is the female income contribution figure. The income of female workers is still far enough away with male workers or there is still inequality. In Indonesia, women's income contribution is always lower compared to men. Indicators of women's income contribution are influenced by many factors, including education level, work experience, number of family dependents, variable age and working hours [11].

The number of women working in Indonesia as well as in some areas such as Aceh Province will continue to increase every year from 2010 to 2019 as shown in figure 5. The percentage of women's income contribution in Aceh province has increased every year although not very significant, from 31.67% in 2010 to 34.59% in 2019. This is because along with the increasing learning opportunities for women, the success of government programs in the field of family planning, the increasing number of day care centres and the increasingly sophisticated technology to support the dual role of women, as housewives and as workers. The increased desire of women to work not only affects the job market, but also affects the economy and well-being of the woman herself and her family. The increasing income of housewives, the more welfare, nutritional quality, and health of the whole family [12].

The increase in the value of women's income contribution in Aceh Province is still smaller or below the national standard which ranges from 33.5% - 37.1%. This is due to several things. First, the slow and unaligned growth of the industry population. Aceh province is still improving in developing its regional industry. Currently, there is practically no reliable regional industry as a source of livelihood for Acehnese people. Business movements were hampered after the tsunami that destroyed many potential areas.

Second, women are still considered to have competence under men, so women are often less prioritized as employees. In addition, employment opportunities for the male population are more widely open than those of female residents. Many primary school-age residents (7-15 years old) who do not attend school for financial reasons, choose to work in various informal sectors [13]. The achievement of education for women in Aceh Province is not in line with their achievements in the world of work. Although more and more women get the opportunity to go to school, they have difficulty getting suitable jobs after graduation.

Job openings are few and women choose to relent to men when competing for certain positions. The lack of job openings in Aceh leads to a low availability of female workers in accordance with their areas of expertise, so women's income contributions tend to be lower and there are gaps.

The gender-based wage gap is the real thing that harms women by suppressing their incomes. Efforts to address the wage gap between genders must be made by analysing the economic use of a region first. It is intended to know whether or not there is a case between the use of economic resources for women and men in terms of education and career choices [14]. Furthermore, women who have their own income will have more choices. Women who have a large contribution to the family economy will tend to play a role in decision making in their household.

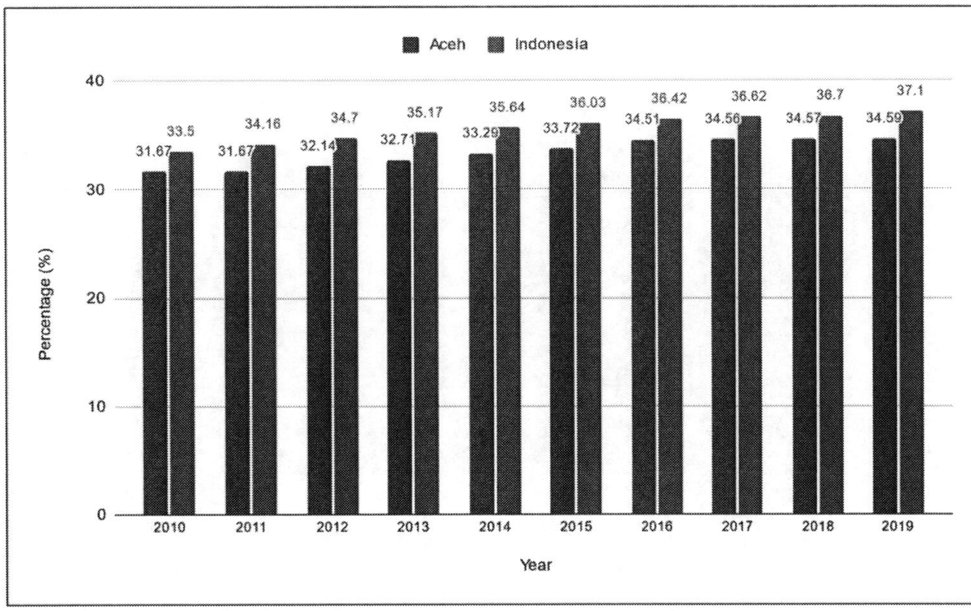

Fig. 5. Women's Income Contribution of Aceh and Indonesia Provinces in 2010-2019
Source: BPS, processed (2021)

The role and participation of women in the world of work, has managed to contribute quite large to the welfare of the family, especially in the field of economics. In Aceh Province, the value of women's income contribution has varying conditions in each district or city and looks volatile every year from 2010 to 2019. Districts that have a high value of women's income contributions, including Central Aceh Regency, Bireuen Regency, Gayo Lues Regency, and Aceh Jaya Regency. The district is largely dominated by rural areas with most of the poor. Based on figure 6, it can be interpreted that the number of female residents entering the job market is greater than the male population and the number of female residents is indicated more in rural areas.

Rural women are more aware and understand their position in addition to being housewives also have a role to help families in fulfilling the needs of life that tend to do any work [15]. Women in rural areas are mostly involved or working in trade and agriculture, namely as rough laborers of coffee plantations and farm laborers. However, the contribution of women's income in this region plays a big role because in general the income of women families belonging to the poor category can increase and can help increase the overall income of the family.

In contrast to rural areas, in urban areas that have the highest percentage of female income contribution is Subulussalam City. Banda Aceh, which is the urban center and the capital of Aceh Province, has a percentage of women's income contributions that tend to be moderate. This is because the work that many women in Banda Aceh are working in the informal sector due to the level of education and skills(skills) that have not been able to enter the formal market, so the contribution of income generated is relatively small [16]. In addition, most of the conditions in the household in urban areas the largest income contribution is still held by the man/husband who is the head of the household.

When it is associated with Qanun Aceh Number 6 of 2009, there is an article that supports the increase of women's income contribution through empowerment in the economic field, namely article 15. Based on this article, the government of Aceh and districts /cities facilitate women's access to economic resources, employment opportunities, markets, and trade. In addition, the government also provides access to employment, adequate working conditions, and control of economic resources. The Government of Aceh also provides business services, training, and access to markets, information, and technology, especially for low-income women. However, the implementation of qanun is still not maximized because although there has been an increase in the percentage of income contributions each year, the increase has not been significant.

 https://doi.org/10.1051/e3sconf/202132506006

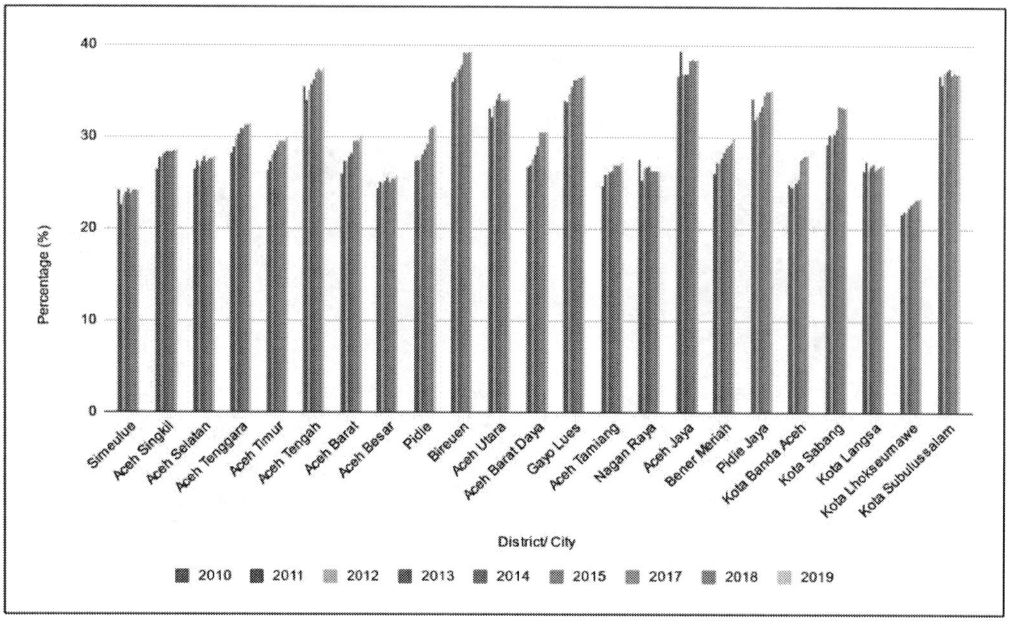

Fig. 6. Aceh Province Women's Income Contribution by District/City 2010-2019
Source: BPS, processed (2021)

3.4 Involvement of Women as professionals

The involvement of women as professionals is driven by birth rate, life expectancy, literacy rate, average female school time, urbanization rate and women's contribution to household income [17]. In 2009, Aceh Province implemented a regulation that regulates the involvement of women as professionals in both the formal and informal sectors. The regulation in question is Qanun Aceh Number 6 of 2009 concerning The Empowerment and Protection of Women. The impact of the implementation of the regulation, in Aceh Province in 2010 to 2019, has a percentage of women's involvement as professionals that always exceeds national trends.

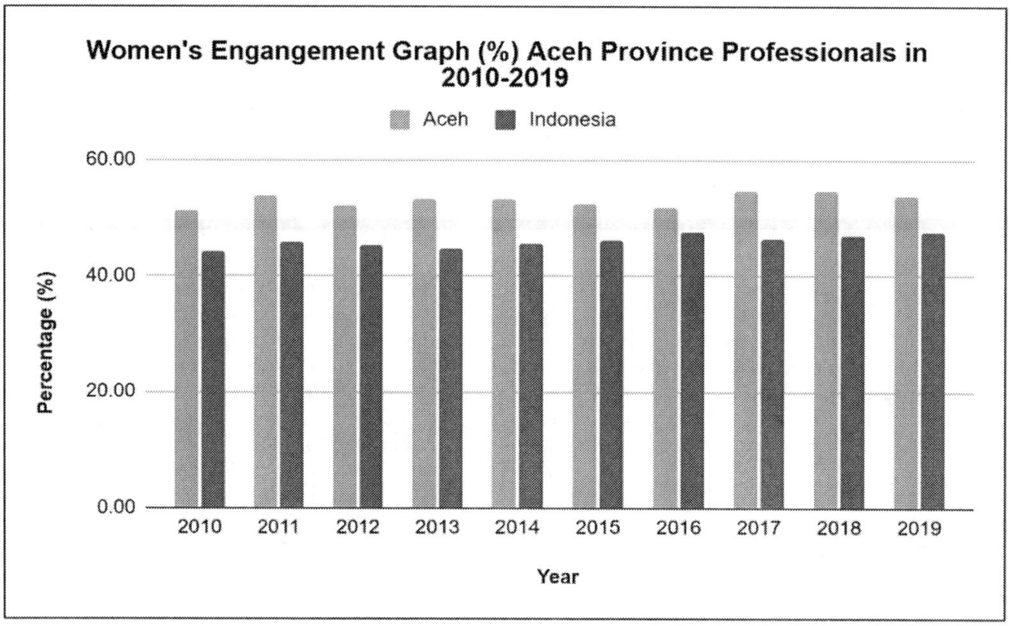

Fig. 7. Women as Professionals (%) in Aceh Province 2010-2019
Source: BPS, processed (2021)

When viewed at the provincial level, based on Figure 7. the percentage of female involvement as professed personnel experienced a patterned fluctuating trend. The percentage increase in the percentage of women's involvement as professionals continued to occur in 2010-2011, 2013-2014, and 2016-2017. During 2010-2011 the percentage of women's involvement as professionals increased by 2.53% from 51.14% in 2010 to 53.67%. In 2013-2014, the percentage of women's involvement as professionals in Aceh Province increased by 0.14% from 53.14% in 2013, in 2014 to 53.28%. Meanwhile, in 2016-2017 the percentage of women's involvement as professionals increased by 3.01% from 51.7% in 2016, in 2017 to 54.71%.

The increasing percentage of women as professionals indicates that women's crucifixion in making decisions and participating in the economy is increasingly aligned with men. This is in accordance with article 14 paragraph 1 in Qanun Aceh No. 6 of 2009 which states that the Government of Aceh and the Government of district/city and other institutions must provide the widest possible opportunity for women to occupy various positions at all levels of work professionally and proportionally.

The decrease in the percentage of female involvement as a professional force occurred in 2011-2012, 2014-2016, and 2017-2019. The percentage of involvement of women as professionals in 2011-2012 decreased by 1.64%. The percentage decrease in the percentage of female involvement as professionals in 2014-2016 was 1.58%. Meanwhile, the percentage of women's involvement as professionals in 2017-2019 decreased by 0.79%.

The percentage graph of women's involvement in professionals by district/city in Aceh Province from 2010-2019 is used to determine its distribution and distribution in each region (figure 8). The percentage of female involvement in professionals in Bireun District in 2010-2019 was the highest in Aceh Province. Even the percentage has reached 63.99%. While the district that has the lowest percentage of female involvement in professionals is in Simeulue District, which is 37.78%. There is a high gap between districts/cities in Aceh Province. This is influenced by several factors including the type of employment, the participation rate of the labour force, and the last level of education that is completed in the area. The increasing percentage of the female workforce should be accompanied by an increase in gender-biased employment to absorb the labour force without gender discrimination.

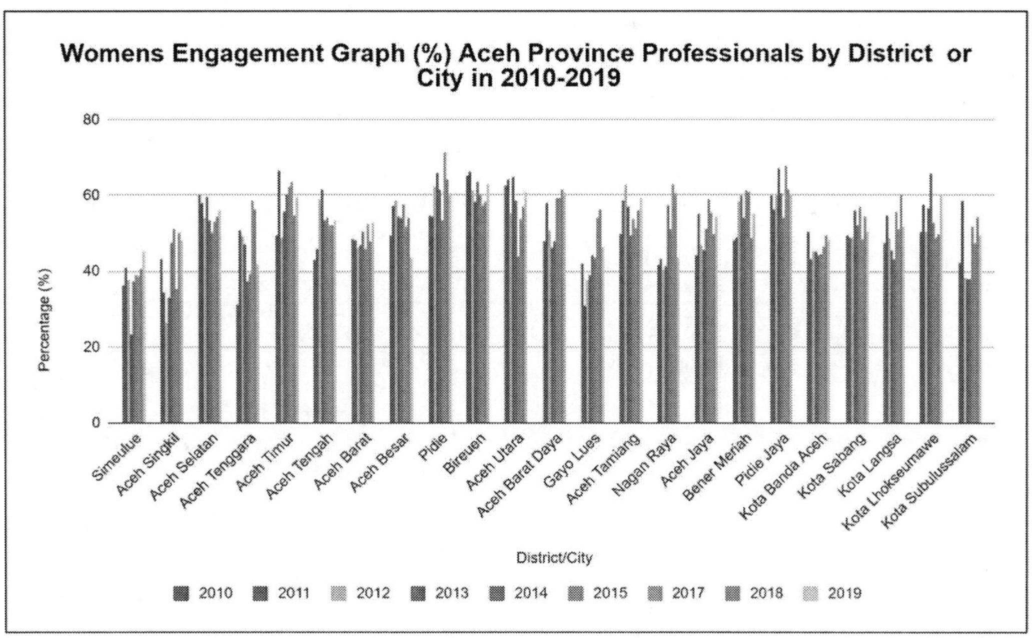

Fig. 8. Women as Professionals (%) in Aceh Province by District or City 2010-2019
Source: BPS, processed (2021)

4 Conclusions

The results of the analysis showed that the implementation of Qanun Aceh Number 6 of 2009 related to gender empowerment in Aceh Province has not been carried out to the maximum in 2010-2019. This can be seen from the condition of the gender empowerment index in Aceh Province which experienced a volatile trend from 2010-2019 with an index ranging from 52.06 to 67.4 which is still below the national figures. The three components of the gender empowerment index show quite different results. The component of women's involvement in parliament in Aceh Province still does not meet the minimum percentage of 30% and there is a percentage gap for some district and municipal administrative areas. The component of women's income contribution tends to increase the percentage figure throughout 2010-2019 which is not so significant and has met a minimum percentage of 30%, but the value is still below the national figure. The component of women's involvement as professionals has a volatile trend and has met a minimum percentage of 30% with a value above the national figure.

References

1. M. K. Singh, Womens. Stud. Int. Forum, 71 (2018)
2. D. Richardson, Conseptualising Gender, in Introducing Gender and Women's Studies, no. June, Basingstoke, Palgrave Macmillan (2015)
3. World Economic Forum, Global gender gap report 2013: Ethiopia, Geneva [Website] (2013) (papers2://publication/uuid/BA7EA38E-351D-450B-A186-7B7C581087DF, accessed 9 June 2021)
4. Pemerintah Provinsi Aceh, Qanun Aceh Nomer 6 Tahun 2009 tentang Perlindungan dan Pemberdayaan Perempuan (2009)
5. I. M. Dini, Fajriyah, Y. Mahdiah, E. Fahmadia, I. Lukitasari, Pembangunan Manusia Berbasis Gender 2020 (2020)
6. BPS, Indeks Pemberdayaan Gender (IDG) 2010-2020 [Website], (2020) (https://www.bps.go.id/indicator/40/468/6/indeks-pemberdayaan-gender-idg-.html , accessed 9 June 2021)
7. Astusti, D. Dileli, Kalimantan Tengah Predikat Indeks Pemberdayaan Gender Tertinggi di Indonesia [Website], (2021) (https://dp3appkb.kalteng.go.id/artikel/kalimantan-tengah-predikat-indeks-pemberdayaan-gender-tertinggi-di-indonesia.html , accessed 9 June 2021)
8. Pemerintah Republik Indonesia, Undang Undang Nomor 11 Tahun 2006 Tentang Pemerintahan Aceh (2006)
9. J. Biegon, Gender Equality and Political Processes in Kenya: Challenges and Prospects (Strathmore University Press, 2016)
10. O. Ojigo, The Impact of Article 9 of the Protocol to the African Charter on Human and Peoples' Rights on the Rights of Women in Africa on women's political participation in Africa. In Gender Equality and Political Processes in Kenya: Challenges and Prospects. Strathmore University Press (2016)
11. D. A. Putrie, A. Rahman, Analisis dan Pemodelan Pendapatan Pekerja Perempuan di Indonesia Menggunakan Data Panel, in Seminar Nasional Official Statistics 2019: Pengembangan Official Statiistics dalam Mendukung Implementasi SDG's (2018)
12. H. Mudzhar, Anto, Wanita dalam Masyarakat Indonesia: Akses, Pemberdayaan, dan Kesempatan (Sunan Kalijaga Press, Yogyakarta, 2001)
13. E. Andina, Pencapaian Millenium Development Goals Bidang Pemberdayaan Perempuan dan Perlindungan Anak (Penerbit Pusat Pengkajian Pengolahan Data dan Informasi, Jakarta, 2012)
14. J. Schieder, E. Gould, Women's Work' and the Gender Pay Gap, Washington DC, 2016. [Online]. Available: epi.org/110304.
15. C. F. Oktari, Sartiyah, J. Ilm. Mhs., **5**, 2, (2020).
16. R. Zahara, Karakteristik Pengangguran Perempuan Berdasarkan Pendidikan di Kota Banda Aceh (Banda Aceh, 2018)
17. W. Rajagukguk, J. Ketenagakerjaan, **10**, 1, (2015)

Impact of employee age and work experience on safety culture at workplace

Abdul Aziz Mat Isa[1*], *Warishah* Abdul Wahab[2], *Rohayu* Che Omar[1*], *Mohd Zaid* Mohd Nordin[2], *Hairin* Taha[1] and *Rasyikin* Roslan[1]

[1]Institute of Energy Infrastructure, Universiti Tenaga Nasional,43000 Kajang, Malaysia EDP Sciences, Editorial Department, 91944 Les Ulis Cedex A, France
[2]College of Graduate Studies, Universiti Tenaga Nasional, Kajang, 43000, Malaysia

Abstract. Occupational accidents can be caused by different factors and can have many consequences, such as minor or fatal injury, equipment damage, permanent disability, or, in some cases, fatality. Therefore, it is essential to identify the causes of accidents to prevent them from recurring in the future and minimize injury, ill-health, and business costs. It is crucial, by determining the causes of the incident in the workplace, to analyse those that happened and take practical preventive steps to minimize the likelihood of them occurring again. A safe and accident-free working environment can allow the company to function efficiently and effectively. This study examined whether the age and experience of the employees, as determined by the demographics, have significant differences in the mean scores of the compliance towards safety culture within the company.

1 Introduction

1.1 Age and work experience factor

Safety practices in world industries are unique due to demographics, legislation, and stakeholder contractual agreements. Implementation of safety, health, and environmental management system is no widespread in developing countries [1], and Malaysia is not exceptional.

There are different generation cohorts in employment in a large organization, and each generation possesses different values, preferences concerning work motivation, and changes with their work environment. The individual in these generation cohorts is motivated by believing that they each share a different set of values and attitudes [2]. The main reason employees' age is considered an essential factor in this study is that technological and competency varies between age groups which affects their inclination on compliance of safety culture in the organization. Concerning generational differences, it is a significant area of concern and entails the management of an organization to analyse to leverage and better understand the behaviour of these generational groups [3].

In terms of age, multiple reasons may affect workers' productivity base on the years of work experience, physical abilities, work demands, academic background, family and care commitments, motivation and energy, loyalty, and personality factors. It has been reported that the declination of job performance is directly correlated to workers' age. However, it is less impactful on professions that do not require abilities by age, such as physical demand jobs [4].

According to [5], safety culture perceptions are influenced by demographic factors such as age, education, and experience. Workers who had been working for almost two decades are better informed and had better views about the importance of safety. On the other hand, employees at an early stage of their career would be interested in safety but gradually decrease due to company policies. They would prioritize safety at the peak of their career. Maturity may play roles later in their service life, equipped with experience that helps them address safety aspects and identify hazardous situations. A study indicated a positive relationship between age and safety perception. The study found that those older workers had the best perceptions on safety, indicated the highest level of job satisfaction, were the most compliant with safety procedures, and recorded the lowest accident involvement rate [6].

Some may be used to routine jobs described in their job scope, deterring empowerment from participating in a safety knowledge-sharing environment. May it be a different age group or work experience; every employee belongs to diverse generational cohorts. It is found that generational diversity existed in the workplace since different employees have clear expectations for their careers, work habits, life goals, and attitudes [7]. This research also emphasized that younger generations are less committed to organizations and more committed to their profession. Consequently, job commitment is one of the human aspects of employees' satisfaction and

* Corresponding author: azan@uniten.edu.my, rohayu@uniten.edu.my

morale, which inevitably affects their safety compliance and safety performance.

From a practical perspective, understanding age-related perceptions of workplace safety would influence management's decisions regarding workers' adaptability, work effectiveness, accident frequency, and implementation of safety management policies. Job position at different levels within an organizational structure is another important consideration of safety practice and performance evaluation. Job position abided by specific job titles and job descriptions enfold employees with a particular set of roles, expectations, duties, and responsibilities, to the extent that it may limit workers to unleash their full potentials.

1.2 Compliance with safety culture

Safety compliance is defined as adhering to safety policies and procedures and engaging in the required safety behaviour [8]. Safety compliance sets out clear guidelines that employees should follow to create a safe working environment. Safety compliance consists of required or expected behaviours to be part of the obligations and responsibilities of the assigned role, i.e. intra-role behaviours related to safety. These behaviours range from complying with standard work procedures, wearing personal protective equipment, and reporting incidents [9]. Compliance with safety and commitment have had a significant impact on work-related injuries [10]. Besides, people, the environment, and the economy may have to face the consequences of industrial or occupational accidents. Therefore, it is essential to solving the safety problems at the managerial level to prevent accidents and reduce economic losses [11].

Management should implement compliance programs consisting of periodic monitoring and auditing, particularly the high-risk areas identified by the risk assessment. These compliance programs focus on risk control measures and means of controlling workers' behaviour to comply with prescribed safe practices. The roles of program managers are to monitor their operations, keep up to date with current legal and regulatory requirements, translate this information into written guidance for their staff, train their staff on how to follow them, and keep in mind that they do so accordingly [12]. In short, there should be frequent external auditor audits to verify any inadequacy in meeting the safety guidelines [13]. The involvement of management and employees' compliance with safety procedures is essential for creating a safer environment in the workplace.

On the contrary, safety literature also suggested that a safe environment does not necessarily depend on compliance on its own. Compliance is complex and dynamic environments may, in certain circumstances, include rules and procedures that appear to be ambiguous, contradictory, complicated, confusing, or inadequate (non-applicable) to existing situational constraints. It has also been reported that certain factors that may cause severe or loss of time injuries are uncontrollable to employees (e.g., the environment or machinery). However, employees have the power to display safety behaviours at work, and these behaviours are critical indicators of accidents and injuries at work [14].

Safety culture is a subset of corporate culture within the organization. It has characteristics that can support safety, quality, productivity and directly affect safety performance. Safety culture is the attitude, beliefs, perceptions, and values of workers on safety issues when carrying out their day-to-day operations to achieve an interdependent safety culture. An example of a safety culture is the "first safety" practice, reflecting the importance of safety in the way people believe, think and act in creating a safe working environment. Safety culture needs to be part of the organizational culture that underpins management and affects workers' attitudes and behaviour to the organization's ongoing safety performance.

The study filled the gaps in the literature by testing the following two (2) hypotheses; (H1) There is a significant difference between the age of employees and the compliance of the safety culture within the organization under study; (H2) There is a substantial difference between the work experience of employees and the compliance of the safety culture within the organization under investigation.

2 Materials and methods

The research methodology of this study is divided into three phases, including the Survey Questionnaire Design, Survey Questionnaire Distribution, and Inferential Statistics, as shown in Figure 1.

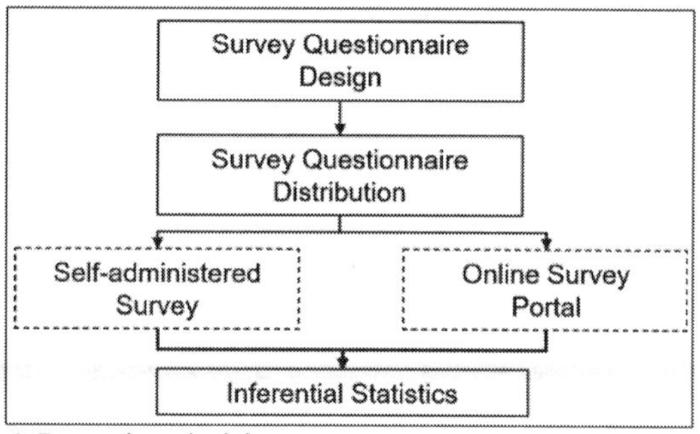

Fig. 1. Research methodology.

2.1 Survey questionnaire design

The questionnaire design process follows various steps, such as developing the conceptual framework, the construction of questionnaire questions, the selection of the proper questionnaire design, and the technical implementation of the softcopy questionnaires.

2.2 Survey questionnaire distribution

Data collection for this study is done using a self-governing survey method and the https:/docs.google.com ('Google Docs') online survey portal. Questionnaire surveys were distributed to the 200 executives and 200 non-executives from one

Government Link Company (GLC) in order to explore the relationship between variables and help to confirm or reject the assumptions derived for this study. However, by the end of the survey period, only 257 responses were collected, which equals 64.3% out of the total employees who participated. The survey period was approximately one month in April 2019.

There are a few advantages to using an online survey. Less time is needed to conduct an online survey compared to traditional manual research methods. It is also not expensive, as it was unnecessary to spend on postage. Besides, online surveys take less time to enter the information in the database as the responses are transferred automatically, and the results are available at any time. The margin of error is significantly reduced with online surveys.

Currently, most people prefer to respond to online surveys using smartphones to complete the survey anytime and anywhere. It also provides convenience to respondents as it is mobile and adopts a green approach with less or no use of papers. The questionnaire in this study was designed to be concise and straightforward, in the format of pre-coded closed-ended questions. The questionnaire was also prepared in bilingual form (English and Malay). A great deal of attention focuses on the wording, structure, and layout of the questionnaire to avoid any misunderstanding among the respondents in answering all the questions.

2.3 Inferential statistics

Inferential statistics help to make judgments and to observe the difference between the mean groups in this study. Most of the significant inferential statistics come from the General Linear Model, which includes the T-Test, Variance Analysis (ANOVA), Covariance Analysis (ANCOVA), and Regression Analysis. In this study, ANOVA was used to provide information on mean group variability levels and to compare mean group variability. The primary purpose of ANOVA is to determine whether there is an interaction between the mean groups, such as independent variables, and demographic information on the dependent variable.

The factors that influence the dependent variable that is hypothesized earlier can be investigated. The F-Test One Way ANOVA test is performed to compare the variability of scores between different groups. For this study, the test will determine whether the age and years of experience of employees have significant differences in the mean safety compliance scores for Government Link Company (GLC) safety culture.

According to Rumsey (2016), in the end, all hypothesis tests use a p-value to weigh the strength of the evidence. The p-value of less than 0.05 indicates strong evidence against the null hypothesis that the null hypothesis must be rejected. A significant p-value higher than 0.05 shows weak evidence against the null hypothesis.

3 Results

This section may be divided into subheadings. It should provide a concise and precise description of the experimental results, their interpretation, and the experimental conclusions that can be drawn.

Survey questionnaires were distributed to 200 executives and 200 non-executives of the Government Link Company (GLC) networks. However, by the end of the survey period, 257 responses were collected, equivalent to 64.3 percent of the total staff involved. Compared to the total population of 10,431 in this Government Link Company (GLC) network, the respondents represent 2,5% of the total population with an error margin of 5% and a confidence level of 95%. In the study, two research questions need to be answered. Research Question 1 questioned whether there is a significant difference between the age of employees and the safety culture of the organization understudy? Whereas the second research question asked whether there was a significant difference between the work experience of employees and the compliance of the safety culture within the organization understudy?

An inferential test is being conducted to determine whether the age and years of experience of employees based on demographics have significant differences in the mean scores of safety culture compliance within the Government Link Company (GLC) network.

Table 1 shows the One-Way ANOVA as a result of compliance of safety compliance within the organization based on the age group of employees. The ANOVA One-Way study was conducted to determine whether population demography has significant differences in mean scores on Government Link Company (GLC) network safety culture compliance.

Table 1. One-way ANOVA result of safety compliance based on the age group of the employees.

Age group	N	Mean	Std. deviation	Std. error	95% confidence interval for mean		Min.	Max.
					Lower bound	Upper bound		
22 years and below	2	5.4444	0.78567	0.55556	-1.6146	12.5034	4.89	6.00
23 – 28 years	13	5.3248	0.44781	0.12420	5.0542	5.5954	4.22	5.78
29– 34 years	85	5.0719	0.59061	0.06406	4.9445	5.1993	3.67	6.00
35 – 40 years	77	4.9957	0.57832	0.06591	4.8644	5.1269	3.89	6.00
41 years and more	80	5.0194	0.63688	0.07121	4.8777	5.1612	3.33	6.00
Total	257	5.0484	0.59759	0.03728	4.9750	5.1218	3.33	6.00

ANOVA

	Sum of Squares	df	Mean Square	F	Sig.
Between Groups	1.635	4	0.409	1.147	.335
Within Groups	89.787	252	0.356		
Total	91.422	256			

The interpretation of the ANOVA test is that the significant value (p-value) of less than the 5-007 value of 0.05 indicates strong evidence against the null hypothesis that there is a significant difference. A sizeable significant value (p-value) greater than the 5-007 value of 0.05 indicates weak evidence against the null hypothesis that there is no significant difference.

From Table 1, one-way ANOVA shows that there is no statistical difference between the average age group of employees. The result shows that the significant value (p-value) is 0.335, which is greater than 0.05 (p-value>0.05) for the employee age group [F (4, 252) = 1.147, p = 0.335]. The above analysis indicates no significant difference between the age group of employees and safety compliance with the organization's safety culture.

Post-hoc comparisons using Tukey HSD in the ANOVA test indicated that the maximum mean score for the employee age group within 22 years and below (M = 5.44, SD = 0.79) was not significantly different from the lowest mean score for the employee age group between 35 and 40 years (M = 5.00, SD = 0.58). The small difference in the mean value of the score between the age group of employees supports the result of p-value > 0.05. It shows that there is no significant difference between the age group of employees and safety compliance with the organization's safety culture. Therefore, the H1 hypothesis: State significant difference comparing the age of the employees and the compliance of safety culture within the organization under study is being rejected.

Table 2 below shows the One-Way ANOVA result of the compliance of safety compliance within the organization based on the employees' work experience. One-way ANOVA analysis of variance showed no statistical difference between the mean numbers of the employees' work experience. The result shows that significant value (p-value) is 0.592 which is greater than α value of 0.05 (p-value > 0.05) for the work experience group of employees [F (6, 250) = 0.773, p = 0.592]. The analysis indicates that there is no significant difference between the work experience of employees and the safety compliance towards the safety culture in the organization.

Table 2. One-way ANOVA Result of Safety Compliance based on the employees' work experience.

Work experience group	N	Mean	Std. Deviation	Std. Error	95% confidence interval for mean		Min.	Max.
					Lower Bound	Upper Bound		
5 years and less	14	5.0952	0.71327	0.19063	4.6834	5.5071	3.89	6.00
6 – 10 years	80	5.0417	0.54559	0.06100	4.9203	5.1631	3.67	6.00
11 – 15 years	59	5.1299	0.55177	0.07183	4.9862	5.2737	4.00	6.00
16 – 20 years	36	4.8981	0.67944	0.11324	4.6683	5.1280	3.89	6.00
21 – 25 years	49	5.0884	0.64867	0.09267	4.9021	5.2748	3.44	6.00
26 – 30 years	5	5.1556	0.50062	0.22388	4.5340	5.7772	4.67	6.00
31 years above	14	4.9048	0.59574	0.15922	4.5608	5.2487	3.33	6.00
Total	257	5.0484	0.59759	0.03728	4.9750	5.1218	3.33	6.00

ANOVA

	Sum of Squares	df	Mean Square	F	Sig.
Between Groups	1.664	6	0.277	.773	.592
Within Groups	89.758	250	0.359		
Total	91.422	256			

Post-hoc comparisons using Tukey HSD in the ANOVA test indicated that the highest mean score was recorded for employees with work experience between 26 years and 30 years of age (M = 5.16, SD = 0.50), which was not significantly different from the lowest mean score for employees with work experience between 16 and 20 years of age (M = 4.90, SD = 0.68). The small difference in the mean value of the score between the employees ' work experience group supports the p-value > 0.05. It shows that there is no significant difference between employees ' work experience and safety compliance with the organization's safety culture. H2 hypothesis rejected: there is a significant difference between the work experience of employees and the compliance of the safety culture within the organization under the study.

4 Discussion

The findings of this study show no significant difference between the age and work experience group of employees and safety compliance with the organization's safety culture. The worker with an increased number of years of working and remaining working at an older age may have implications for safety performance and sustainable work. Older workers are often more reliable than younger workers and usually have a wide range of expertise, knowledge, skills, and a higher level of commitment; however, older workers are reduced to age-old physiological systems such as muscle strength, stature, dexterity, and mobility. However, some older workers may be more robust than their younger colleagues because they are specific to individuals [16].

According to laboratory-based studies, some cognitive abilities will decrease with age, such as memory and response time. However, safety work performance is unlikely to be affected by extensive work experiences, such as improved safety judgment, job-specific safety knowledge, work experience skills, and high self-motivation to learn. In the case of Government Link Company (GLC), workers over 55 years of age have the option of continuing to work until 60 years due to increases in the official retirement age. It will be difficult to predict the effects that changes in age and work experience may have on the performance of safety at work; those workers who continue to work longer may be exposed to occupational hazards.

5 Conclusion

Age and work experience are not the only factors that affect safety performance at work, although health and physical ability may change with age. In certain circumstances, employees with a health problem can adjust their daily work and activities to continue to work. The attitude of employees is the most significant barrier to working with a health problem rather than the health condition itself. Health is influenced by several other external factors, including lifestyle, exercise, and nutrition, in which age is not the only factor determining the level of health. Consideration of the workability of the immediate manager and supervisor is the most appropriate way to determine whether or not any change in the health or capacity of the worker puts them at an increased risk and risk from their work. Each individual's resources shall be considered in deciding the requirements of the employer's job. Older and experienced workers need to be identified as valuable assets to the company. Generally, this organizational group tends to be more resourceful in completing their work and with a high degree of determination to ensure that their tasks are well managed, regardless of age or work experience.

The research presented in this paper was funded by Tenaga Nasional Berhad (TNB) through a post-graduate university scholarship (UNITEN) and UNITEN R & D Sdn. Bhd. under URND Seeding Fund (U-TE-RD-19-19). The authors would like to thanks Universiti Tenaga Nasional (UNITEN) and UNITEN R & D Sdn. Bhd. for the opportunity and sponsorship in completing this study.

References

1. E. Koehn, R. Kothari, C. Pan, J. of Construction Engineering & Management **121**, 3:261–265 (1995)
2. E. Parry, P. Urwin, IJMR **13**, 1:79–96 (2011)
3. M.S. White, C. Burns, and H. A. Conlon Workplace Health & Safety, SAGE Journal **66**, 10: 493–498 (2018)
4. V. Skirbekk, Ageing Horizons **8**, 4–8 (2008)
5. R. Masood, R. Choudry (2012). *Investigation of demographic factors relationship with safety climate*, in 48th ASC Annual International Conference Proceedings (2012)
6. C. Pfeifer, J. Wagner, Contemporary Economics **8**, 1:25–46 (2014)
7. A. Singh, B. Gupta, Benchmarking: An International Journal, 2015 **22**, 6:1192–1211 (2015)
8. Y. Feng, E. A. L. Teo, F. Y. Y. Ling, S. P. Low, International Journal of Project Management **32**, 6:932–943 (2014)
9. M. Martinez, F. Gracia, I. Tomas, J. Peiro, Risk Analysis **34**, 7 (2014)
10. J. Nahrang, F. Morgeson, D. Hofmann, A meta-analytic investigation of individual and contextual influences on workplace safety, satisfaction, and well-being (The Society for Industrial and Organizational Psychology, San Francisco, 2008)
11. Z. Lin, Using TOC thinking process tools to improve safety performance (Department of Computer Science, Shangluo University, Shaanxi, China, 2009)
12. R. Kusserow, J. of Health Care (2017)
13. P. Mitropoulos, G. Cupido, M. Namboodiri, J. Constr. Eng. Manage. **135**, 9:881–889 (2009)
14. A. Ghahramani, H. R. Khalkhali, Safety & Health at Work **6**, 2:97–103 (2015)
15. D. Rumsey, Statistics for dummies 2nd edition. (John Wiley & Sons. Inc, United States, 2006)
16. J. O. Crawford, A. Davis, H. Cowie, K. & Dixon, K. The ageing workforce: implications for occupational safety and health (European Agency for Safety and Health at Work, 2016)

Predicting the impact of COVID-19 on fertility in the Special Region of Yogyakarta, Indonesia

Muhammad Arif Fahrudin Alfana [1*], *Hidayati Nur* Rohmah[1], *Salma Nurulhuda*[1] *and Mohammad Isnaini* Sadali[2]

[1]Department of Environmental Geography, Faculty of Geography, Universitas Gadjah Mada, Indonesia
[2]Department of Development Geography, Faculty of Geography, Universitas Gadjah Mada, Indonesia

Abstract. The aim of this research is to explain the predicted impact of Covid-19 on fertility in the Special Region of Yogyakarta (DIY). The indicators used are the number of pregnancies and the number of births. To get pregnancy and birth predictions, various basic data were processed using the Spectrum program using the DemProj and Famplan modules, then analyzed by descriptive analysis. The basic data is collected from the Central Statistics Agency (BPS) and other relevant agencies. The result shows that birth and pregnancies in DIY will increase. This prediction is based on the decreasing use of short -term contraceptive methods during the pandemic. In the worst-case scenario, the rise in pregnancies between 2019-2020 would be 67,862 cases, meanwhile births are expected to increase by 37.56 thousand cases during the same period.

1 Introduction

The impact of Covid-19 has affected many aspects of life, including demography [1-4]. One aspect of demography that is affected is fertility [5]. It is predicted that the effect on fertility would increase [4, 5]. The rise in fertility is measured by the number of births and pregnancies.

Prediction on the rise of births is based on the decreasing use of contraceptive methods during the Covid-19 pandemic [7, 8]. During the pandemic, the use of contraceptive tools in Indonesia is expected to decline up to 10 percent [7, 9]. Meanwhile, Maurizio predicted that there will be a reduction of 20 to 40 percent in the use of short-term contraceptive tools (condom, pill, and injection) during the Covid-19 pandemic [10]. Additionally, there is a decrease in family planning service visitation during the Covid-19 pandemic which also contributes to the rise in fertility [11].

The Special Region of Yogyakarta is one of Indonesia's provinces that is affected by Covid-19. On March 21st, 2020 there are only 5 newly confirmed cases, but the trend keeps on increasing until there is a surge to 27.251 cases on February 25, 2021 [12]. The high spikes in cases will potentially affect the fertility aspect. Herawati stated that the Covid-19 pandemic will lead to a baby boom in the Special Region of Yogyakarta [13].

The concern surrounding the increase of fertility in the Special Region of Yogyakarta is an interesting issue to be discussed post 2000. This is because the fertility in the Special Region of Yogyakarta has been fluctuating since 2000 which is reflected in the Total Fertility Rate (TFR). In the period of 2000-2017, fertility in the Special Region of Yogyakarta experience an increase [14-16], before declining to 1.9 in 2019 [17]. Backed by evidence that Covid-19 has the potential to increase fertility, it raises the concern of the government of the Special Region of Yogyakarta. Therefore, predictions with several scenarios need to be conducted to determine the right policies so that the Special Region of Yogyakarta could achieve its TFR target of 2.1 by 2022.

This study aims to elaborates the prediction of Covid-19 effects on fertility in the Special Region of Yogyakarta using birth and pregnancy indicators. Birth and pregnancy data is processed with indirect methods to yield TFR value. The predicted TFR could then be used as an input for policy determination. Hence, the information regarding the predicted effect on birth and pregnancy becomes essential as input for stakeholders in constructing development scenarios and population planning, especially for the National Population and Family Planning Board (BKKBN) of the Special Region of Yogyakarta.

2 Data and Method

This research is conducted quantitatively and analyzed by descriptive analysis. The Special Region of Yogyakarta is chosen as a study area that is analyzed at the provincial level. The research discussed the prediction of the number of births and pregnancies in 2019-2022 due to the effect of Covid-19 pandemic. The prediction is the result of data processing using Spectrum software. Spectrum is an open source software which was also used by UNFPA to predict the Covid-19 impacts on Maternal Health and Family Planning in Indonesia.

* Corresponding author: arif.fahrudin@ugm.ac.id

The data used in this study are obtained from the publications by the Central Bureau of Statistics of the DIY and other secondary sources. These data are then inputted and processed in Spectrum Program Module DemProj and FamPlan. As a limitation of the study, predictions of pregnancy and birth in this study were only built on the assumptions of Maurizio's model, without taking into account other factors such as economic, social and other factors. The model is adopted from the scenario of Maurizio in predicting the Covid-19 impacts on Maternal Health and Family Planning [10]. Maurizio modeled 2 scenarios of Covid-19 impacts in Indonesia in general as follows [10]:

1. Best case scenario which uses the assumption that there is a 20 percent decrease in the usage of short-term contraceptive methods (condom, pill, and injection) in 2020 from the 2019 baseline. The condition would return to normal in the following years of 2021 and 2022 with the estimated numbers similar to those of 2019.
2. Worst case scenario sees a 40 percent drop in short-term contraceptive use (condom, pill, and injection) in 2020 from baseline. The condition would return to normal in the following years of 2021 and 2022 with the estimated numbers similar to those of 2019.

The two scenarios of the Covid-19 impacts then compared to the full coverage scenario, where all main healthcare services are at the same condition as pre-pandemic levels (in this scenario, healthcare services are assumed to be the same for 2019-2022). The result of the data processing using this scenario are:

1) Projection of pregnancies in the Special Region of Yogyakarta in 2019-2022 using the best case scenario, worst case scenario, and the full coverage scenario.
2) Projection of birth cases in the Special Region of Yogyakarta in 2019-2022 using the best case scenario, worst case scenario, and the full coverage scenario.

3 Result and Discussion

3.1 Projection of the Number of Pregnancies in the Special Region of Yogyakarta 2019-2022

The projected number of pregnancies in the Special Region of Yogyakarta continues to rise based on the full coverage scenario, from 109,142 pregnancy cases in 2019 to 112,284 in 2020. With this scenario, there is an increase of 3,142 pregnancies during 2019-2020. This number continues to increase in 2021-2022 (as seen on Figure 1).

The result shows a stark difference in 2019-2020 if compared to the two scenarios conducted by Maurizio. The best case scenario result indicates an increment of 35,297 pregnancies in 2019-2020. Meanwhile the worst case scenario shows an increase of 67,862 cases in 2019-2020. Based on these results, the two scenarios proposed by Maurizio [10], which is reduction in short-term contraceptive use (condom, pill, and injection), has a significant impact on the surge of pregnancy cases in the Special Region of Yogyakarta.

The predicted pregnancy numbers in the Special Region of Yogyakarta consist of intended and unintended pregnancies. The result of data processing in Spectrum software signifies that unintended pregnancies in the Special Region of Yogyakarta is abundant. The number of unintended pregnancies in the three scenarios is around 60 percent of the total pregnancy cases (as shown in Table 1). This is in accordance with the general conditions that around more than 50 percent of pregnancies are considered unintended pregnancy [18].

Table 1. Intended and Unintended Pregnancies in Special Region of Yogyakarta 2019-2022, dengan Tiga Skenario

	Pregnancies		
	Intended	Unintended	Total
DIY-Full Scenario			
2019	41,230	67,911	109,142
2020	40,292	71,992	112,284
2021	45,599	76,076	121,674
2022	50,532	80,182	130,714
DIY-Best Scenario			
2019	41,230	67,911	109,142
2020	55,934	88,504	144,439
2021	48,579	76,076	124,655
2022	51,631	80,182	131,813
DIY-Worst Scenario			
2019	41,230	67,911	109,142
2020	71,401	105,603	177,004
2021	50,727	76,076	126,803
2022	52,387	80,182	132,569

Source : secondary data processing

Stover and Winfrey stated that the high cases of unintended pregnancies are mainly caused by an increase in unmet need and failure in contraceptive methods [19]. The findings of Hapsari et al. and Spagnoletti et al. is evidence that the failure in contraceptive methods caused the occurrence of unintended pregnancies in the Special Region of Yogyakarta [20, 21]. Moreover, the decline in Contraceptive Prevalence Rate (CPR) also caused the high cases of unintended pregnancies [22].

The CPR scenario in the Special Region of Yogyakarta (as seen on Figure 2) is the result of data processing using a Spectrum program with three scenarios. Comparison between the result from CPR scenario and pregnancy scenario shows that the reduction in CPR causes an increase in unintended pregnancies in the Special Region of Yogyakarta, which is in accordance to Lule et al. [22]. Based on Figure 1 and Figure 2, it can be concluded that as CPR decreases in 2020, both in the best case and worst case scenario, will be followed by an increase in pregnancy cases.

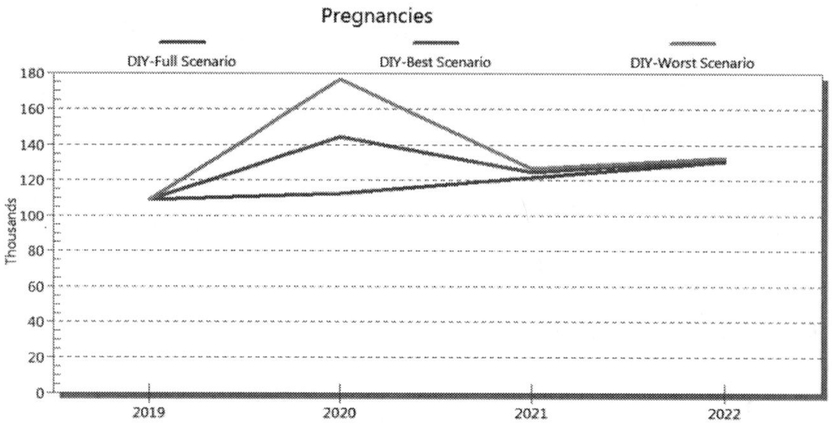

	2019	2020	2021	2022
DIY-Full Scenario	109.14	112.28	121.67	130.71
DIY-Best Scenario	109.14	144.44	124.65	131.81
DIY-Worst Scenario	109.14	177.00	126.80	132.57

Fig. 1. Projection of Pregnancy Cases in the Special Region of Yogyakarta, with Three Scenarios

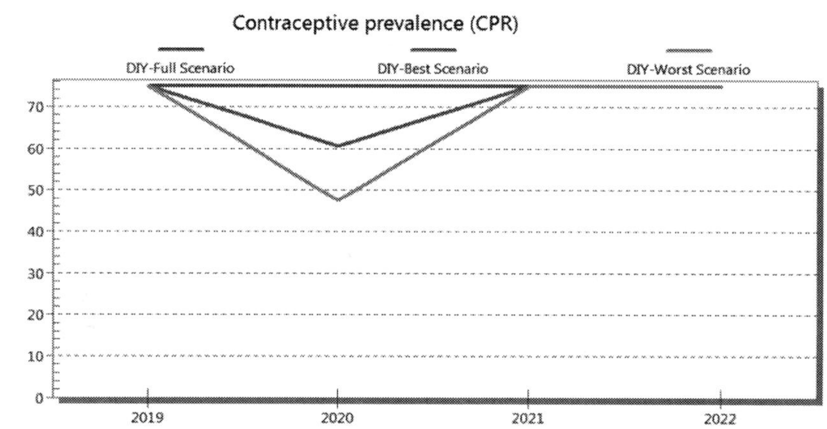

	2019	2020	2021	2022
DIY-Full Scenario	75.01	75.01	75.01	75.01
DIY-Best Scenario	75.01	60.48	75.01	75.01
DIY-Worst Scenario	75.01	47.51	75.01	75.01

Fig. 2. Contraceptive Prevalence Rate (CPR) in Special Region of Yogyakarta 2019-2022, with Three Scenarios

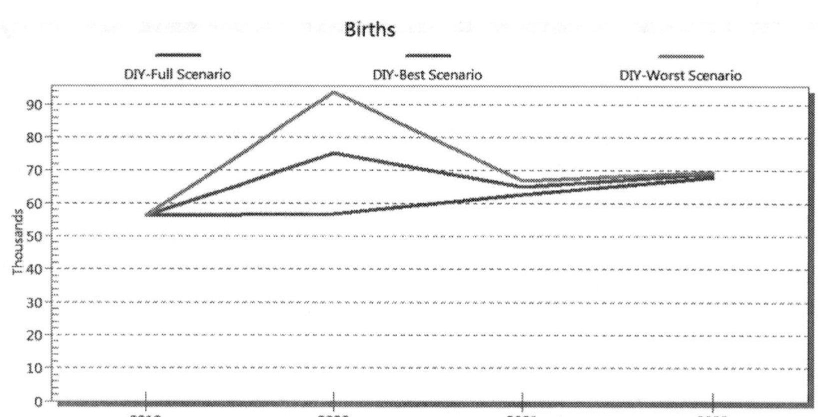

	2019	2020	2021	2022
DIY-Full Scenario	56.24	56.65	62.49	68.02
DIY-Best Scenario	56.24	75.21	65.09	68.97
DIY-Worst Scenario	56.24	93.80	66.96	69.63

Fig. 3. Projection of Birth Cases in the Special Region of Yogyakarta 2019-2022, with Three Scenario

3.2 The Number of Births in the Special Region of Yogyakarta 2019-2022

The projection of birth cases in the Special Region of Yogyakarta with the full coverage scenario conveys that there's a rise in 2019-2022. The predicted birth cases range between 56-68 thousand cases (as seen on Figure 3). This value is close to the result of a projection conducted by Indonesia Central Bureau of Statistics, Indonesian Ministry of Development and Planning (Bappenas) and UNFPA, where birth cases are predicted to be around 56-58 thousand cases in 2019-2022 [23].

In the best case scenario and worst case scenario, there's a significant rise of birth cases in 2019-2020. The best case scenario sees a predicted increase of 18.97 thousand birth cases in 2019-2020. Meanwhile the worst case scenario sees a predicted increase of 37.56 thousand birth cases. Compared nationally, the birth cases in Indonesia are predicted by BKKBN to rise by 300-450 thousand cases [24]. Based on this, the Special Region of Yogyakarta contributes around 6-10 percent of total predicted birth cases nationwide.

3.3 The Challenges in fertility aspect due to Covid-19 in Yogyakarta Special Region

Adoption of Maurizio [10] scenario in the case of the Special Region of Yogyakarta gives a good overview of the challenges about fertility due to Covid-19 impacts. As in predicted pregnancies, predicted birth caused by the lowering use of short-term contraceptive methods results in a spike of cases in the Special Region of Yogyakarta. Therefore, the provision of contraception during the pandemics needs to be maintained. This is due to the high potential of rise in pregnancies and births during the Covid-19 pandemic.

That condition corresponds with the findings of Caruso, Rapisarda & Minona and Coombe et al. where several users of short-term contraceptives such as pills decide to discontinue the method when the Covid-19 pandemic occurred [25, 26]. Although sexual activities continue during the pandemics. As a result, some of these users experience unintended pregnancies [25]. Even on a broader scope, the findings of Riley et al. predicted that Covid-19 already caused a reduction in contraceptive use in low and middle income countries and caused pregnancies and births to increase [28].

Predictions of increasing pregnancies and births in the Special Region of Yogyakarta illustrate that the challenges ahead surrounding fertility become an interesting topic to discuss. The Special Region of Yogyakarta's performance as one of the provinces with the best fertility indicator would be disrupted if this condition is not properly controlled [14, 28]. Hence, it's imperative to take steps in minimizing the negative impacts.

Witono and Parwodiwiyono stated that the Special Region of Yogyakarta data indicates a decrease in short-term contraceptive use in the beginning of the Covid-19 pandemic [28]. The lowering use is mostly for pill and injection methods. Based on this evidence, the scenario proposed by Maurizio is very likely to occur in the Special Region of Yogyakarta. Therefore, the provision of contraceptive use during the pandemic needs to be maintained, especially for pills and injections.

Nanda et al. elaborates that the Covid-19 pandemic highlights several challenges ahead, including in the Special Region of Yogyakarta [18]. One example is ensuring the continuity of access and contraceptive services, including counseling and joint decision-making for couples regarding fertility. Nanda et al. also stated that there needs to be several adaptations towards the existing system [18]. This is to ensure that the population's reproductive rights and the state's interest can go hand-in-hand while awaiting for the healthcare system to recover to pre-Covid-19 levels in a short time [29].

Nanda et al. particularly mentioned that adaptive measures can be done through maximizing the "no-touch" approach [18]. The details include using Telehealth for counseling and screening using remote communication methods such as SMS, WhatsApp, video call, or phone call. In addition, access to family planning information already provided such as in the "Klik KB application" or optimization of e-Government tools already provided by the government of the Special Region of Yogyakarta also become important. If this can be implemented by all parties, then the negative impacts of fertility challenges such as surges in pregnancy and birth cases could be minimized.

4 Conclusion

The Special Region of Yogyakarta is one of the provinces in Indonesia with the best fertility achievement. Unfortunately, this performance is expected to be disrupted due to the Covid-19 pandemic. Based on the prediction done by adapting Maurizio approach, the number of pregnancies and births in the Special Region of Yogyakarta is expected to rise. For the pregnancy indicator, the best case scenario prediction shows an increase of 35,297 cases in 2019-2020. Meanwhile the worst case scenario shows an increase of 67,862 pregnancy cases in 2019-2020. For the birth indicator, it is predicted that in 2019-2020, the cases would increase by 18.97 thousand in the best case scenario. Meanwhile in the worst case scenario, the increase is predicted to be 37.56 thousand birth cases in 2019-2020. This indicates that the Special Region of Yogyakarta contributes around 6-10 percent from the total births predicted nationwide.

This finding conveys that fertility in the Special Region of Yogyakarta is expected to increase due to the Covid-19 pandemic. Therefore, several policies need to be immediately implemented to minimize the negative outcomes. One proposed policy is to take adaptive measures towards the current system through maximizing the "no-touch" approach. Optimizing the existing e-Government tools provided by the government of the Special Region of Yogyakarta is also important so that this fertility issue could be controlled properly in the future.

Acknowledgement

This paper is part of "Hibah Dosen Mandiri" Faculty of Geography, UGM entitled "Proyeksi Dampak Covid-19 Terhadap Perubahan Indikator Fertilitas Di Daerah Istimewa Yogyakarta" funded in 2021. The author would like to thank several parties including Dr. Sukamdi and Dr. Agus Joko Pitoyo, for their guidance and input during the writing.

References

1. Khalili, M. A., Leisegang, K., Majzoub, A., Finelli, R., Panner Selvam, M. K., Henkel, R., Mojgan, M., & Agarwal, *A. Male Fertility and the COVID-19 Pandemic: Systematic Review of the Literature*. The world journal of men's health **38**, 4 506–520. (2020).
2. Yang, X., Yu, Y., Xu, J., Shu, H., Liu, H., Wu, Y., & Wang, Y. *Clinical course and outcomes of critically ill patients with SARS-CoV-2 pneumonia in Wuhan, China: a single-centered, retrospective, observational study*. The Lancet Respiratory Medicine (2020)
3. Kikkawa, A., and Suan, E. B. *Trends and patterns in intra-ASEAN migration. In Skilled Labor Mobility and Migration*. Edward Elgar Publishing. (2019).
4. Alfana, M.A.F., Pitoyo, A.J., Rohmah, H.N., dan Laudiansyah, R. *Permasalahan Manajemen Bencana di Indonesia: Urgensi Pendataan Para Perantau, Belajar Dari Kondisi Selama Pandemi Covid-19*. Disajikan dalam Webinar Nasional "Perspektif Problematik Manajemen Bencana di Indonesia". Surakarta: Pusat Studi Bencana Universitas Negeri Surakarta. (2020).
5. Aassve, A., Cavalli, N., Mencarini, L., Plach, S. and Livi Bacci, M. *The COVID-19 Pandemic and Human Fertility*. Science **369,** 6502 370-371 (2020).
6. Anser, M.K., Yousaf, Z., Khan, M.A., Voo, K.H., Nassani, A.A., Alotaibi, S. M., Abro, M. M. Q. and Zaman, K. *The impacts of COVID-19 measures on global environment and fertility rate: double coincidence*. Air Qual Atmos Health 13, 1083–1092. https://doi.org/10.1007/s11869-020-00865-z. (2020).
7. Azanella, L. A. *BKKBN Memprediksi Angka Kehamilan Melonjak Selama Pandemi Corona, Ini Alasannya*. Kompas.com. https://www.kompas.com/tren/read/2020/05/09/150000965/bkkbn-memprediksi-angka-kehamilan-melonjak-selama-pandemi-corona-ini? page=all. (2020).
8. Aly, J., Haeger, K.O., Christy, A.Y. and Johnson A.M. Contraception access during the COVID-19 pandemic. *Contracept Reprod Med*. **8,** 5 17. PMID: 33042573; PMCID: PMC7541094. (2020)
9. Yusita, I., Noprianty, R., Kurniawati, R., Rofiasari, L., dan Anriani, A. *Women Assistance in Fertile Age Couples in the use of Contraception as an Effort to Suppress Baby Booms During the Covid-19 Pandemic*. Jurnal Peduli Masyarakat: **2,** 4 181-186. (2020)
10. Maurizio, F. *The Impact of Covid-19 on Maternal Health and Family Planning in Indonesia*. Bangkok: UNFPA Asia Pacific Regional Office. (2020).
11. Purwanti, S. *Dampak Penurunan Jumlah Kunjungan Kb Terhadap Ancaman Baby Boom Di Era Covid-19*. Jurnal Bina Cipta Husada **16,** 2 105-118. (2020).
12. https://corona.jogjaprov.go.id/data-statistik
13. Herawati, D., Rosyada, D. F., Pratiwi, R. P., & Wigati, E. N. *Family Planning Services by Midwifery of Private Midwifery Practice in Yogyakarta During the Pandemic Period Of Covid-19*. Jurnal Ilmu Kesehatan Masyarakat, **11,** 2 123-135 (2020).
14. Alfana, M. A. F., Giyarsih, S. R., Aryekti, K. dan Rahmaningtias, A. *Fertilitas Dan Migrasi: Kebijakan Kependudukan Untuk Migran Di Kabupaten Sleman*. Jurnal Natapraja. **3**, 1 17-24 (2015)
15. Alfana, M. A. F., Laudiansyah, R., Fauziyanti, N. U., Sadali, M. I., Ismayuni, N., and Maulana, R. Demographic transition and population ageing in Yogyakarta. in *E3S Web of Conferences* **200**. doi:10.1051/e3sconf/202020004002 (2020)
16. Bappeda (Development Planning Agency at Sub-National Level) Daerah Istimewa Yogyakarta. *Laporan Keterangan Pertanggungjawaban Gubernur 2018* Yogyakarta: Bappeda Daerah Istimewa Yogyakarta (2019)
17. Bappeda (Development Planning Agency at Sub-National Level) Daerah Istimewa Yogyakarta. *Laporan Keterangan Pertanggungjawaban Gubernur 2019*. Yogyakarta: Bappeda Daerah Istimewa Yogyakarta (2020)
18. Nanda, K., Lebetkin, E., Steiner, M. J. Yacobson, I. & Dorflinger, L. J. *Contraception in the Era of COVID-19*. Global Health: Science and Practice. **8**, 2 166-168 (2020)
19. Stover, J., Winfrey, W. *The effects of family planning and other factors on fertility, abortion, miscarriage, and stillbirths in the Spectrum model*. BMC Public Health **17,** 775 (2017)
20. Hapsari E.D., Widyawati, Nisman W.A., Lusmilasari L., Siswishanto R., Matsuo H. *Change in contraceptive methods following the Yogyakarta earthquake and its association with the prevalence of unplanned pregnancy* Contraception **79**, 4 316-322 (2009)
21. Spagnoletti, B.R.M., Bennett, L.R., Kermode, M. and Wilopo, A.S. *I wanted to enjoy our marriage first... but I got pregnant right away': a qualitative study of family planning understandings and decisions of women in urban Yogyakarta, Indonesia*. BMC Pregnancy Childbirth **18,** 353 (2018)

22. Lule, E., Singh, S. and Chowdhury, S. A. *Fertility Regulation Behaviors and Their Costs: Contraception and Unintended Pregnancies in Africa and Eastern Europe & Central Asia*. HNP discussion paper series. World Bank: Washington D.C. (2007)
23. Indonesia Central Bureau of Statistics, Indonesian Ministry of Development and Planning (Bappenas) and UNFPA. *Proyeksi Penduduk Indonesia 2015-2045 Hasil SUPAS 2015* Jakarta : Indonesia Central Bureau of Statistics, Indonesian Ministry of Development and Planning (Bappenas) and UNFPA (2018)
24. Iswara, A. J.. Prediksi 450.000 *Bayi Indonesia Lahir Usai Pandemi, Begini Kata Media Asing*. https://www.kompas.com/global/read/2020/05/22/191329570/prediksi-450000-bayi-indonesia-lahir-usai-pandemi-begini-kata-media-asing?page=all (2020)
25. Caruso, S., Rapisarda, A. M. C. & Minona, P. *Sexual activity and contraceptive use during social distancing and self-isolation in the COVID-19 pandemic* The European Journal of Contraception & Reproductive Health Care **25**, 6 445-448 (2020)
26. Coombe, J., Kong, F., Bittleston, H., Williams, H., Tomnay, J., Vaisey, A., Malta, S., Goller, J., Temple-Smith, M., Bourchier, L., Lau, A., & Hocking, J. S. *Contraceptive use and pregnancy plans among women of reproductive age during the first Australian COVID-19 lockdown: findings from an online survey*. The European Journal Of Contraception & Reproductive Health Care : The Official Journal Of The European Society Of Contraception 1–14 (2021)
27. Riley, T., Sully, E., Ahmed, Z., & Biddlecom, A. *Estimates of the Potential Impact of the COVID-19 Pandemic on Sexual and Reproductive Health In Low- and Middle-Income Countries*. International Perspectives on Sexual and Reproductive Health **46,** 73-76 (2020).
28. Dasgupta, A., Kantorová, V., & Ueffing, P. *The impact of the COVID-19 crisis on meeting needs for family planning: a global scenario by contraceptive methods used.* Gates open research **4,** 102. (2020)
29. Witono, W., & Parwodiwiyono, S. *Kepesertaan Keluarga Berencana Pada Masa Awal Pandemi Covid-19 Di Daerah Istimewa Yogyakarta. PANCANAKA Jurnal Kependudukan, Keluarga, Dan Sumber Daya Manusia* **1,** 2 77-88 (2020)

Spatial distribution and correlation of dating and premarital sexual experience among male adolescents in Java Islands, Indonesia

Seri Aryati*, *Muhammad* Rafif, and *Zaima* Hilmi
Department of Environmental Geography, Universitas Gadjah Mada, 55281 Bulaksumur, Yogyakarta

Abstract. The high percentage of male adolescents who had premarital sex and dating in Indonesia on the Java Islands makes it a serious problem. This is exacerbated by the results of research which show that the experience of dating from an early age can increase the likelihood of adolescents engaging in sexual intercourse. This study aims to analyze the correlation between dating experiences and premarital sex as measured by age at first initiation, as well as to determine the spatial distribution of prevalence and correlation. This research is a quantitative statistical study using the Spearman Rank correlation test and also using a map to describe the spatial distribution. The results showed that there are 84 percent of male adolescents in Java who started dating at the age of less than 19 years, also doing sexual intercourse in that age range. This shows male adolescents have experienced a process of greater sexual maturity, so that they have a higher urge to engage in sexual behavior. This condition is in line with the results of the correlation test, where there are 4 provinces in Java which have a positive and significant correlation between the two variables, including Banten, Jakarta, Central Java and West Java. It means the younger the age at first dating adolescents, the younger they will have premarital sex.

1 Introduction

Adolescence is a transitional period in the human life span that connects childhood and adulthood [1]. Biologically, adolescence is marked by the growth and development of primary and secondary reproductive organs. Meanwhile, psychologically it is characterized by the formation of erratic attitudes and feelings, desires and emotions [2]. This condition encourages adolescents to be in a crisis position in their search for their identity, so they are filled with high curiosity. One of the most highlighted in the process of finding identity is the negative impact it causes, namely the problem of shifting sexual behavior of adolescents. A shift in adolescent sexual behavior can be shown from premarital sexual behavior. As the most populous island in Indonesia, Java Islands has population of adolescents aged 15-24 years more than 24,3 million and this number is 54 percent of total adolescents in Indonesia based on Indonesian Population Census 2020. Data from the 2017 Indonesian Demographic and Health Survey (IDHS) shows a percentage of adolescents who had premarital sex at 7,6 percent, especially for male adolescent who tended to have more premarital sexual intercourse. From that number, 42,4 percent male adolescent who had premarital sexual intercourse lives in Java Islands.

Many studies have shown that sex in adolescents has a very broad impact, one of them is increasing the risk of sexually transmitted diseases. Sex before the age of 17 has a four to fivefold risk of developing disease [3]. Premarital sexual behavior in adolescents is a serious problem. Premarital sexual behavior related to behavior in all forms of behavior from sexual relations with the opposite sex by adolescents before marriage. This behavior can occur by adolescent dating behavior, starting from kissing to touching sensitive areas, which in turn can lead to sexual intercourse [4]. Dating behavior in the current era has become common among adolescents. The 2017 IDHS data shows that 82 percent of adolescents in Indonesia claim to have dated, of which 48 percent are adolescents who live in Java Islands. From this figure, it is known that in dating, 85 percent of adolescents hold hands, 50 percent have embraced, 31 percent have kissed lips and 17 percent have touched or aroused sensitive body parts of his/her couple. As previously mentioned, The study shows that there is a positive and significant correlation between age at first dating and adolescent sexual behavior in Indonesia through the Chi-Square test wirh a p-value of $0.004 < 0.005$ [5].

*Corresponding author: seri.aryati@ugm.ac.id

Another research conducted by Oktriyanto and Alfiasari (2019), they used data from the Survey of Performance Indicators of the National Medium Term Development Plan Program of the National Population and Family Planning 2015. The research aims to analyze dating and sexual behavior among adolescents in Indonesia. The resulth through the Chi-Square test with a p value of 0.000, show that dating adolescents have a higher likelihood of having premarital sex when compared to non-dating adolescents. This is confirmed through the results where adolescents who are in a relationship holding hands, hugging, kissing lips and touching sensitive areas of the body increase the likelihood of having sex before marriage. The OR value of the correlation of each dating behavior with premarital sex behavior has the lowest value of 4,564 and the highest value of 44,966 for variables touching sensitive areas. This shows that teenagers who touch sensitive areas of his/her couple are more likely to have premarital sex by 44,966 times [6].

This is supported by research data from Bhatta, et al. (2013) conducted in Jhapa district, the eastern part of Nepal on secondary school adolescents. His research shows that 32,6 percent of male adolescents had premarital sex, where more than half, 57,1 percent had sex with his girlfriend [7]. In addition, a study conducted in Pune District, Maharashtra, India by Alexander, et al. (2007) showed that 37 - 49 percent of adolescents who have relationships in dating admitted had sexual intercourse, this number is much higher when compared to adolescents in general, which is only around 15 - 17 percent [8].

The impact due to the positive correlation between dating behavior and premarital sex in adolescents will get worse if the adolescents already know the dating behavior and live it from a younger age. The 2017 IDHS report shows that, on average 45 percent of young women and 44 percent of young men are dating between the ages of 15-17. An even more surprising number is that, there are 74 percent of male adolescents who had sex for the first time between the ages of 15-19 [9]. Therefore, this study will analyze the correlation between dating and premarital sexual behavior as measured by the age at first dating and the age at first sexual intercourse. The gap of this study from previous studies sited in the use of data. This study used secondary data so that it may not be able to explain real-time conditions. Several previous studies in Indonesia, including research conducted by Sari and Rokhanawati (2018), Oktriyanto and Alfiasari (2019) also used secondary data. However, there are differences in the methods used in the research of Oktriyanto and Alfiasari (2019) where in describing the relationship between the two variables dating and premarital sexual experience, they use the Chi Square test, which means that the data is nominal or ordinal data, so in their research only to show relationship between the two variables. Another study conducted by Bhatta, et al. (2013), they also used the Chi Square test to see the relationship between the two variables. In addition, considering that there is no research that examines these two variables in terms of spatial distribution, this study will also examine the spatial distribution of correlation and also its prevalence.

Spatial analysis is currently widely used in various fields to see the distribution of phenomena in an area and even to see the influence of the region on these phenomena. In the research of Tesema, et al. (2020) where spatial distribution aims to identify the distribution and factors of abortion in women of childbearing age in Ethiopia [17]. There is also a study by Sharma, et al. (2021) which uses one of the spatial analysis in describing the spatial disparity of the prevalence of unmet need in India [18]. Therefore, this paper is expected to provide an illustration that there is an interesting relationship between the age of first dating and having premarital sexual intercourse among male adolescents in Java Islands. In addition, through the spatial depiction of the correlation value between the two variables, it can be an interesting study, especially for socio-human research and even in the field of reproductive health.

2 Method

This study uses secondary data as an analysis tool. The data used are the results of the 2017 Indonesian Demographic and Health Survey (IDHS). The IDHS is conducted by the Central Statistics Agency (BPS) in collaboration with the National Population and Family Planning Agency (BKKBN) and the Ministry of Health (Kemenkes). The IDHS is also part of the international Demographic and Health Surveys (DHS) Program whose funding is provided by the United States Agency for International Development (USAID).

The IDHS data consists of various components. In order to study the dating behavior of adolescents, the 2017 IDHS component of Adolescent Reproductive Health (KRR) is used. In it, there is a discussion about unmarried women and men aged 15-24 years. The population of the study was 7246 unmarried males aged 15-24 years. The total sample of respondents in this study were 400 adolescent respondents who had a history of dating and had a history of premarital sex. This is in accordance with the variables used in this study were age at first dating and age at first sexual intercourse. The first age variable to have sexual intercourse is male adolescents who admit to having premarital sex according to age. All of these variables were then processed using descriptive statistics and analyzed descriptively quantitatively.

This research uses the Spearman rank correlation test. The spearman rank correlation test is one of the non-parametric statistical tests used to determine the relationship between two variables, namely the independent variable and the dependent variable [10]. This correlation test is used for small amounts of data and uses a minimal ordinal variable scale. In the use of the Spearman correlation test, the data used do not need to be normally distributed.

Based on this, the use of the Spearman correlation test uses the following equation:

$$r_s = 1 - \frac{6\,\Sigma\, d_i^2}{n\,(n^2 - 1)}$$

Explanation:

r_s = Rank Spearman Coefficient Correlatiom
d = the difference between the pairs of levels
n = number of pairs [10]

The results of the Spearman rank correlation test were analyzed using 2 hypotheses [11], namely:
H0 = there is no relationship between X and Y variables
H1 = there is a relationship between the X and Y variables.
The hypothesis test uses a confidence level of 95% or an error rate (α) of 5% (0.05) or 99% with an error rate of 1% (0.01) with the following conditions:

1. Sig. > 0.05 / 0.01 = H0 is accepted, there is no relationship between X and Y variables
2. Sig. <0.05 / 0.01 = H1 is accepted, variable X has a significant relationship with variable Y [12]

The correlation coefficient value can be used to measure the strength of the relationship between variables [13]. The correlation coefficient can be interpreted through **table 1** as follows:

Table 1. Correlation Coefficient Index

Coefficient Interval	Correlation Level
0,80 - 1,00	Very Strong
0,60 - 0,79	Strong
0,40 - 0,59	Fairly Strong/Strong Enough
0,20 - 0,39	Weak/Low
0,00 - 0,19	Very Weak/Low

Source : [13]

3 Result and Discussion

The results and discussion in this research consist of 3 parts, namely the results and analysis of the spatial distribution of the prevalence of adolescents who have dated and adolescents who had premarital sex, cross-tabulation analysis between the age at first dating and the age at first having premarital sex, and the last one is a discussion of the correlation result based on Spearman Rank correlation test.

3.1 Spatial Distribution of Dating and Premarital Sex Among Male Adolescent in Java Islands

As explained in the method, the variables used are male adolescents who have been dating and have had premarital sex at the age of 15-24 years. Through the **figure 1** it is known that the spatial distribution of the percentage of male adolescents who have dated in Java. The results show that the highest percentage of male adolescents who have dated is in West Java Province with a percentage value of 83 percent. Then followed by the provinces of Jakarta, Central Java, Yogyakarta and Banten with a percentage in the range of 70-80 percent. The lowest percentage is occupied by East Java Province with a value of 67.8 percent.

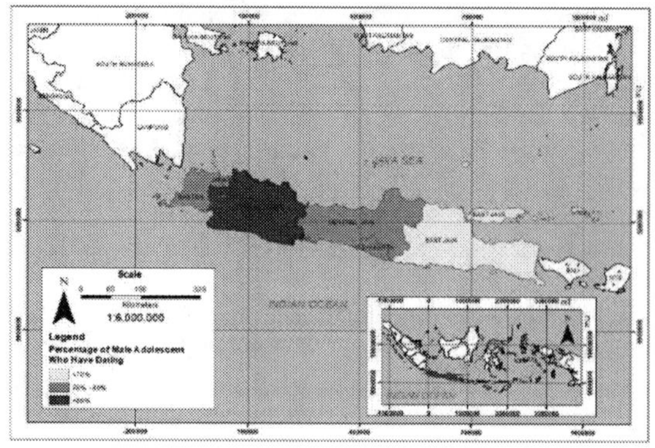

Fig. 1. Spatial Distribution Map of the Percentage of Male Adolescent Who Have Dating

In contrast to the spatial distribution of the percentage of male adolescents who have dated, the spatial distribution of the highest percentage of male adolescents who have had premarital sex is Yogyakarta Province as seen in the **figure 2** Yogyakarta Province is the highest with a percentage value of 8.5 percent of the number of successful respondents interviewed. The next position is occupied by the Provinces of Jakarta, West Java, Central Java and East Java with a percentage value of 4-8 percent . Then, the lowest percentage is occupied by Banten Province with a value of 1.9 percent.

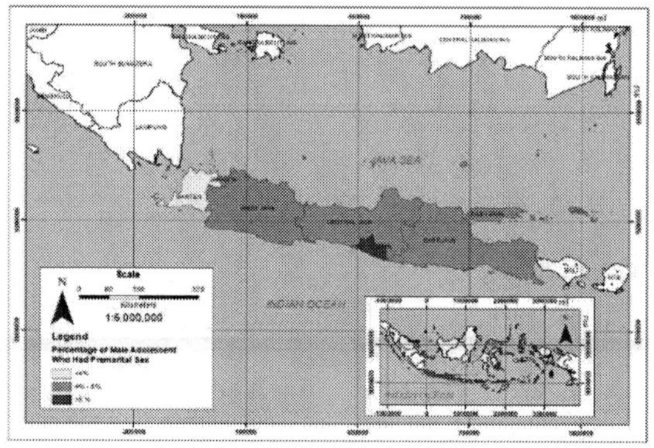

Fig. 2. Spatial Distribution of the Percentage of Male Adolescent Who Had Premarital Sex

3.2 Cross-Tabulation of Dating and Premarital Sex Among Male Adolescent in Java Islands

As already mentioned in the Method section, the indicator of dating in this study is measured based on age at first girlfriend and the indicator for premarital sex is age at first sexual intercourse. Thus, the cross-tabulation analysis shows the percentage of adolescents according to age who had dating and premarital sex. This percentage is calculated based on the total number of sample respondents.

Table 2 shows that there are 7,5 percent of adolescents who had a history of dating at first time under the age of 15, it turns out that they had premarital sex at the age of less than 15 too. There are 32 percent of adolescents who had dating for the first time at the age of less than 15, they also had premarital sex intercourse at that age range. The largest percentage is 44,5 percent adolescent who had a history of dating for the first time at the age of 15-19 years, were adolescents who had premarital sex with in the same age range.

The table can also show that overall there are 84 percent of adolescents who have had the first premarital sexual intercourse at the age of under 19 and all are adolescents with a history of dating for the first time at the age of under 19 as well. In simple terms, it can indicate that male adolescents in Java with a history of dating for the first time at a relatively young age, will tend to have premarital sex. However, this statement will be described based on the results of the correlation test to determine the strength of the correlation, its direction and significance.

Tabel 2. Cross-tabulation between the percentage of adolescents in dating and premarital sex history based on age

		Age at first sexual intercourse			
		<15	15-19	>19	TOTAL
Age at first girlfriend	<15	7,5	32	2,75	42,25
	15-19	0	44,5	12,5	57
	>19	0	0	0,75	0,75
TOTAL		7,5	76,5	16	100

Source: 2017 IDHS Data Processed

3.3 Correlation Result Based On Spearman Rank Correlation Test.

Data testing was carried out based on the results of interviews with 400 unmarried male adolescent respondents with variables in the form of dating behavior as measured from the first age of dating and premarital sex behavior which was measured based on the first age at premarital sexual intercourse. The results of calculations using the Rank Spearman correlation analysis show that there is a significant relationship between the two variables in the Provinces of Banten, Jakarta, West Java and Central Java, while in the Provinces of Yogyakarta and East Java there is no significant correlation between the two variables as seen in **table 3.**

Table 3. Correlation Coefficient and Significant Value of Spearman's Correlation Test between Age First Dating and Age First Sexual Intercourse Among Male Adolescent.

No.	Province	Correlation Coefficient	Significant Value
1	Jakarta	0,719**	0,000
2	West Java	0,362**	0,000
3	Central Java	0,602**	0,000
4	Yogyakarta	0,072	0,777
5	East Java	0,168	0,122
6	Banten	0,538*	0,044

* Correlation is significant at the 0.05 level (1-tailed).
** Correlation is significant at the 0.01 level (2-tailed).
Source: Research Result, 2021

Through this table, it is also known that the correlation coefficient value shows the strength of the relationship between the two variables used. The strong relationship is shown in the Province of Jakarta with a correlation coefficient value of 0.719 and the Province of Central Java with a correlation coefficient value of 0.602. A fairly strong relationship is shown in Banten Province with a correlation coefficient value of 0.538. Meanwhile, Yogyakarta and East Java Provinces did not find a significant correlation between the two variables.

The correlation value in the category is quite strong, indicating that there is indeed a relationship between the age at first dating and the age at first sexual intercourse. This is indicated by the correlation value that is close to the value of 1, which indicates that the correlation is getting stronger, as happened in Central Java and Jakarta Provinces. Meanwhile, the absence of a significant correlation between the two variables in the Provinces of East Java and Yogyakarta indicates that the age at first dating did not affect the age at first sexual intercourse. In terms of spatial distribution in Java, this correlation is shown in the **figure 3.**

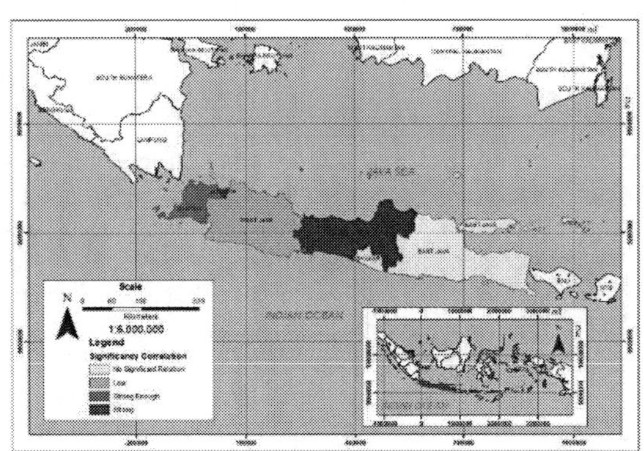

Fig. 3. Spatial Distribution Map of Correlation Age at First Dating and Age at First Sexual Intercourse Among Male Adolescent in Java

The existence of a positive and significant correlation between age at first dating and age at first sexual intercourse has been demonstrated in some previous studies. With a correlation coefficient of 0,023 and a significance of 0,004, it shows that the younger the age at first dating adolescents, the more likely they are to have premarital sex [5]. Other studies have also shown that premarital sex performed at less than 15 years of age is associated with sexual experiences at an early age, one of them is experience in dating [14].

The younger adolescent in dating encourages adolescents to engage in premarital sexual behavior at a relatively young age. This study shows that adolescents tend to have the first sexual intercourse between the ages of 15-19 and the age when they first started dating. Although other studies have shown that there is a relationship between age and sexual behavior, which is adolescents aged 20-24 are 2,3 times more likely to have sexual intercourse than adolescents aged 15-19 years. This is because adolescents aged 20-24 years experience the process of sexual maturity earlier, so they have a greater sexual drive [15]. Therefore, based on the 2017 IDHS, male adolescents in Java Island tend to experience sexual maturity earlier at the age of 15-19 years, considering that teenagers have started dating for the first time and had sex for the first time in that age range.

The risk of sexually transmitted diseases is one of the threats with the increasing prevalence of adolescents engaging in premarital sex in Indonesia, especially Java. Moreover, this risk will increase if in sexual intercourse adolescents do not use contraceptives. Previous studies have found that many adolescents do not practice safe sex [16]. This is also proven from the 2017 IDHS, where 49 percent of male adolescents in Java do not use protection to prevent pregnancy when they first practice sexual intercourse. Furthermore, the results of this study which show that there are four provinces in Java Islands that have a significant correlation, including Banten, West Java, Jakarta, and Central Java, are expected to be a reference for the government, especially in these four provinces, that further studies are needed. It is intended that appropriate programs and policies can be adopted to overcome this phenomenon.

4 Conclusion

Based on IDHS 2017, there are 84 percent of male adolescents in Java who started dating at the age of less than 19 years, also doing sexual intercourse in that age range. This shows that at that age, male adolescents have experienced a process of greater sexual maturity, so that they have a higher urge to engage in sexual behavior. This condition is in line with the results of the correlation test between the age of first dating and the age of first having premarital sexual intercourse.

Based on the results of the correlation test between the variables of age at first dating and age at first sexual intercourse in Java Island, it is known that there are 4 provinces that have a significant correlation, namely Banten, West Java, Jakarta, and Central Java, while there are 2 provinces that have no significant correlation that are provinces of East Java and Yogyakarta. The existence of a significant relationship in these 4 provinces indicates that the age at first dating increased the prevalence of the age at first premarital sexual intercourse. This is in line with the correlation coefficient which shows a positive value, which means that the younger the age at first dating adolescents, the younger they will have premarital sex.

This condition is exacerbated by the discovery of the fact that 49 percent of male adolescents in Java who have premarital sex do not use protective equipment for preventing pregnancy when they first have sex. The increased risk of transmission of sexually transmitted diseases (STDs) and early pregnancy can be a serious threat if the right policies or measures are not determined. Therefore, this condition needs the government's attention to be able to review this problem by formulating related policies, one of which is the socialization of early sex education so that adolescents understand better about the dangers of STDs and pregnancy outside of marriage.

Acknowledgment
I would like to thank Faculty of Geography Universitas Gadjah Mada for the support of conducting this research.

References

1. J.W. Santrock, Adolescence (2003)
2. K.B. Hidayati, M. Farid, Persona.J.Psi.Indo. **5**, 2 (2016)
3. F. Kasim, J.Studi Pemuda **3**, 1 (2014)
4. L. Suwarni, D. Ismail, Y.S.Prabandari, MG. Adiyanti, IJPHS **4**, 3 (2015)
5. D.E.Sari, D. Rokhanawati, JHTAM **1**, 1 (2018)
6. Oktriyanto, Alfiasari, KEMAS **15**, 1 (2019)
7. D.N. Bhatta, A.K.Koirala, N. Jha, Health Reinassance **11**, 2 (2013)
8. M. Alexander, L. Garda, S. Kanade, S. Jejeebhoy, B. Ganatra, Int. Fam. Plan. Perspect. **33**, 4 (2007)
9. National Population and Family Planning Board, Center Bureau of Statistics, Ministry of Health, Indonesia Demographic and Heatlh Survey 2017, (2018)
10. G. E. Dirgantara, M. F. Akbar, R. D. Bekti, JSIK, **3**, 2 (2018)
11. D, Apriyanto, R. Harini, JBI, **1**, 3 (2012)
12. R. Ilmiyah, E. Darminto, JB, **4**, 1 (2020)
13. F. N. Sihombing, JPP, **5**, 1 (2017)
14. B.M. Magnusson, J.A. Nield, K.L. Lapane, BMC Public Health **15**, 98 (2015)
15. D. Rusmiati, S.P.Hastono, J. National Public Health **10**, 1 (2015)
16. L.P.Wong, PLoS One **7**, 12 (2012)
17. Tesema, G. A., Mekonnen, T. H., & Teshale, A. B. PLos One **15**, 6 (2020)
18. Sharma H, Singh SK, Srivastava S. SAGE Open (2021)

Spatio-temporal assessment of rice self-sufficiency in West Sulawesi Province, Indonesia

Heein Yang[1], *Amanda* Rastiana Putri[2], *Ali* Muta'aly Muhammad[2], *Alim* Rohman[2], *Hanifah* Nur Fadhilah[2], *Ihsan* Weda Prabaswara[2], *Thalia* Mutiara Fikri[2], and *Ratih* Fitria Putri[2*]

[1]Departement of Electrical and Computer Engineering, Ajou University, Korea

[2]Environmental Geography, Department of Environmental Geography, Faculty of Geography, Universitas Gadjah Mada, Indonesia

Abstract. Agriculture is an important economic sector in human life. One of the analyses of agricultural resources is food adequacy which is closely related to food availability and food needs. The research location of this study is West Sulawesi Province which has the potential for economic development to improve the welfare of its people and be able to implement sustainable development. The purpose of this study is to determine rice sufficiency from 2016 to 2020 for each district and to projected it for several years. The research was conducted by quantitative methods and analyzed descriptively and spatially. The data collection method is documentation method by literature studies to collect secondary data from Central Statistics Bureau (BPS) of West Sulawesi Province. In general, rice sufficiency in West Sulawesi Province from 2016 to 2020 is included in the sufficient category. However, for several years there were two districts with inadequate amounts of rice, which were Majene and North Mamuju. The projection results in 2025 and 2030 also show the same thing, Majene and North Mamuju are still in the insufficient category and it will get worse in 2030. So in general, West Sulawesi Province is predicted to be unable to meet food needs.

1 Introduction

Food is one of the basic needs and includes human rights that play an important role in maintaining the life and livelihood of human beings [1]. Based on Law Number 18 of 2012 [2] concerning food, it is explained that food is a mandatory responsibility for the state in realizing the availability, affordability, and fulfillment of sufficient, safe, quality, and nutritionally balanced food consumption, both at the national and individual levels. evenly throughout the region. Food has a role in realizing quality human resources in order to carry out daily life. Food consumption patterns have changed from time to time caused by changes in household income, public awareness of food and nutrition and community behavior [3]. Therefore, it is important to meet food needs through its availability. Availability of food that is lower than the need, can create national food instability [4]. In addition, food also occupies a high priority in economic development, because the failure of food production can cause price spikes and scarcity in the market which has an impact on socio-political turmoil [5].

West Sulawesi Province is the last province to be established on the island of Sulawesi. The province was established in 2004 as part of decentralization in Indonesia. West Sulawesi Province consists of six regencies, namely Majene Regency, Polewali Mandar Regency, Mamasa Regency, Mamuju Regency, North Mamuju Regency (Pasangkayu), and Central Mamuju Regency which are the result of division in 2014. Food consumption among districts in West Sulawesi Province can vary depending on the environment including local resources and culture, tastes, and income of the community [6]. Sufficient food in an area is necessary to meet the needs and assist regional development. Therefore, as a newly formed region, West Sulawesi must maintain its sustainability, including its food resources, to further develop the region.

A person's survival is related to food sufficiency since food is an important source of nutrition for the human body to achieve prosperity and improve individual health. Fulfillment of food sufficiency is essential to improve human resource quality by providing a sufficient amount of nutritious, safe, and affordable food for everyone in the respective region [6]. Sufficient food in an area can be used as an indicator of the fulfillment of public consumption. A region with enough food sufficiency will provide better public prosperity compared to a region without sufficient amount of food [7]. Thus, adequate amounts of food are essential for the survival and development of a community.

Food sufficiency is one of the determinants of economic stability. Thus, food self-sufficiency becomes a development framework that can encourage the development of other sectors in the region [8]. According to BPS [9] (2015), the agricultural sector is still the dominant sector in the economic structure in West Sulawesi. The agricultural sector has a contribution of 39.71% to the Gross Domestic Product (GDP) in West

* Corresponding author : ratihfitria.putri@ugm.ac.id

Sulawesi. Absorption of labor in the agricultural sector in West Sulawesi in 2020 was around 337,770 people, an increase compared to the previous year which reached 297,366 people [10]. In general, food sufficiency aims to achieve food availability, price stability, as well as affordability for the community [11]. One of the government's efforts in realizing food sufficiency is carried out through Government Regulation (PP) number 68 year 2002 [12] concerning food sufficiency which states that the provision of food is carried out to meet the needs of household food consumption which continues to grow from time to time.

Rice as a strategic food commodity needs to be guaranteed its availability to meet the food needs of the community. Decrease in agricultural land area in agricultural activities will have an impact on food sufficiency. This is because the availability of rice can affect economic conditions, especially as rice is still the main food commodity [13].

The second goal of the SDGs is to eliminate, achieve food sufficiency and good nutrition, and promote sustainable agriculture. Sustainable agriculture is the management of agricultural resources carried out to meet the needs of present and future generations, by caring for and improving the quality of the environment and preserving natural resources [14]. One of the indicators in achieving sustainable agriculture is the proportion of productive and sustainable agricultural areas [15] [16].

The main problem in fulfilling food self-sufficiency in an area is the change in land use and a decrease in agricultural productivity [17]. The area of agricultural land is one of the important factors that affect food availability, where currently the area is fixed and even tends to decrease [18]. A decrease in the area of agricultural land is always associated with an increase in population due to the economic development of a region [19]. West Sulawesi Province experienced a population growth of 8.63% in 2016 to 2020 or an increase of around 112,751 people over the last 5 years. The increase in population demands the conversion of agricultural land to other land uses. From 2015 to 2019, agricultural land, especially rice fields, has decreased by around 2353 hectares. As a result, rice production in West Sulawesi Province from 2016 to 2020 tends to decrease [10]. The increase in population should be balanced with increasing rice production, but in fact the ability to supply rice is increasingly limited. With these conditions, it is feared that food self-sufficiency, especially rice in the future, will not be achieved if the amount of availability continues to decline and is not higher than the need for rice [20].

The study was conducted to determine rice self-sufficiency in West Sulawesi in 2016-2020 and the results of the calculation can be used to represent the condition of agricultural resources and determine appropriate agricultural policies. Furthermore, this study also aims to estimate the rice self-sufficiency of West Sulawesi Province in 2025 and 2030.

2 Methods

2.1 Study area

West Sulawesi Province has an area of 16.937,16 square kilometers. The northern part is bordered by Central Sulawesi Province, the southern part is bordered by South Sulawesi Province and Mandar Bay, bordered by Central Sulawesi Province and South Sulawesi Province in the east and Makassar Strait in the west. The map of West Sulawesi Province can be seen in **Figure 1.**

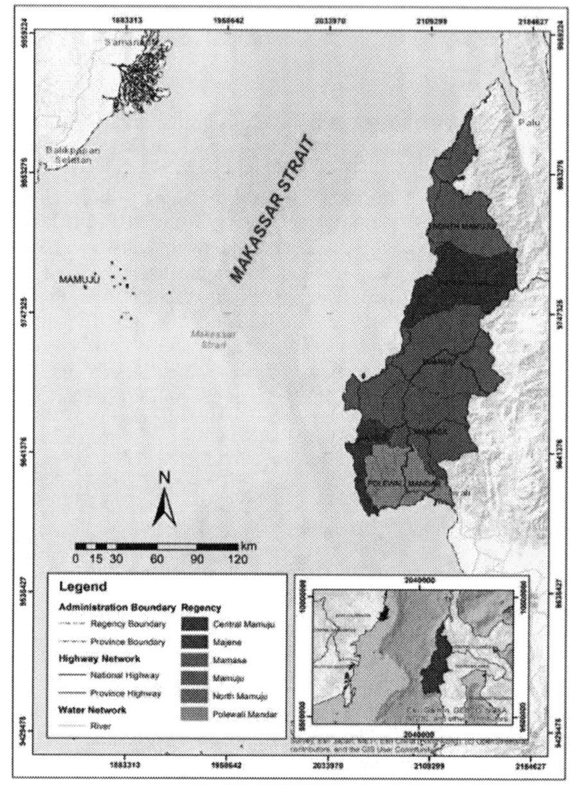

Fig. 1. West Sulawesi map

West Sulawesi Province consists of six regencies, namely Majene Regency, Polewali Mandar Regency, Mamasa Regency, Mamuju Regency, North Mamuju Regency (Pasangkayu), and Central Mamuju Regency which are the result of regional expansion in 2014. West Sulawesi was chosen as the study location because food crops are one of the dominant agricultural sub-sectors with 46.35% of the population working in the agricultural sector [21]. One of the priorities of the West Sulawesi government program is to increase food security to support the availability of staple food for the population [22]. Potential agricultural land area of 1,429.29 square kilometers, community cultural factors and supportive agro-climatic conditions make West Sulawesi potential to become one of the national food barns in Eastern Indonesia.

2.2 Population and sample

The research was conducted by quantitative methods and analyzed descriptively and spatially. Quantitative method is used to calculate the rice sufficiency value in West Sulawesi 2016-2020. The population in this study is the Province of West Sulawesi. The sample in this study was 6 regencies in West Sulawesi, namely Majene, Polewali Mandar, Mamasa, Mamuju, North Mamuju, and Central Mamuju.

The data collection method was a documentation method by literature studies to collect secondary data from BPS of West Sulawesi Province including the publication of West Sulawesi Province in Figures for 2017-2021 and the publication of the Study of Consumption of Staples in 2017 by *BPS* which was accessed online. Administrative boundaries data of West Sulawesi Province collected from Badan Informasi Geospasial (*BIG*), downloaded through the Ina-geoportal website.

2.3 Rice sufficiency value

The concept of rice sufficiency is a comparison between availability and needs. The rice sufficiency value is calculated using the variables annual rice production, population per year, and annual rice consumption per capita. Rice sufficiency value is calculated using **Equation (1).**

$$S = \frac{P}{(n \times C)/1000} \quad (1)$$

where, S is rice sufficiency value, P is annual rice production (kg/ha/year), n is population of West Sulawesi Province (People/year), and C is annual rice consumption per capita (kg/year). The calculated S with value above 1 means the rice supply is sufficient, while the value below 1 means rice supply is insufficient.

2.4 Projection of rice sufficiency value

The projection of rice sufficiency in 2025 and 2030 takes into account the value of availability and needs of rice. This research uses the assumption that every population has the same value of rice consumption needs. Projection of rice sufficiency value is calculated using **Equation (2).**

$$Pn = Po\,(1 + r)^n \quad (2)$$

where, Pn is projection of rice sufficiency value, Po is annual rice production at year 0 (kg/ha/year), n is the number of years between 0 to n, and r is average availability of rice in n.

Data analysis is based on the results of calculations and spatial data on rice sufficiency of West Sulawesi Province from 2016 to 2020. Tabular data including rice production data, population data, and rice consumption per capita were processed utilizing Microsoft Office Excel software. The presentation of data calculation results of rice sufficiency using tables and diagrams while rice sufficiency maps were made using ArcMap 10.5.

3 Results and discussion

3.1 Rice sufficiency value of West Sulawesi Province

Rice production from 2016 to 2020 tends to decrease. Rice production in 2016 reached 548,536 tons, increased in 2017 to 667,097 tons. There was a significant decrease in rice production in 2018 and 2019 as much as 326,196 tons and 300,142 tons, respectively. Meanwhile, in 2020 there was an increase in rice production of 319,166 tons. The number of residents in West Sulawesi Province from 2016 to 2020 continues to increase as the region develops. There were 1,306,480 people in 2016, 1,330,960 people in 2017, 1,355,550 people in 2018, 1,380,260 people in 2019, and 1,419,230 people in 2020. The growth of population from 2016 to 2020 affects the food sufficiency in West Sulawesi as the growth of food demand forms a relatively similar pattern as population growth. Annual estimation of rice consumption per capita in West Sulawesi Province from 2016 to 2020 is 126.4 kg.

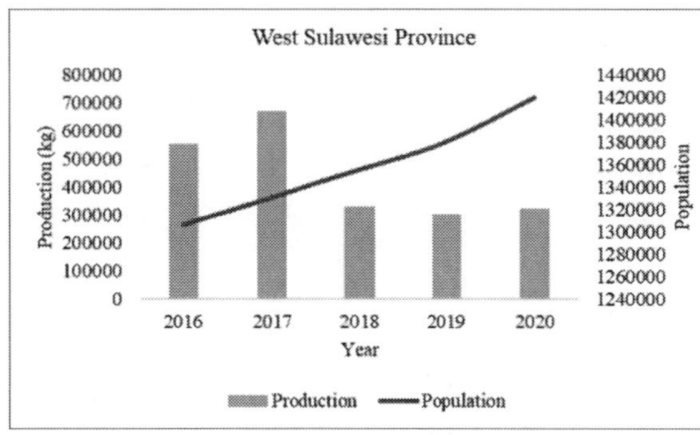

Fig. 2. Comparison of rice production and population in West Sulawesi Province 2016-2020

Based on the amount of rice production, population, and consumption per capita, it can be determined the value of rice sufficiency from 2016 to 2020. Based on the calculation results, it can be seen that the rice commodity from 2016 to 2020 is classified as sufficient with a value greater than 1. Although it is said to be sufficient, the value of the sufficiency of rice commodities in West Sulawesi tends to decline, especially in 2018. This happens because rice production has decreased while the population has increased so that the sufficiency of rice in 2018 has decreased. The land conversion of paddy fields to non paddy fields caused the reduction of rice production. The area of paddy fields and harvested land can affect the availability of rice, the larger the area of harvested land, the greater the production of rice produced [16]. Based on data in 2010-2019, the land use with the narrowest area is paddy fields. Changes in the area of paddy fields tend to fluctuate or have experienced increases and decreases in the area in certain years. In 2010-2014 there was a decrease in the area of around 1200 ha, while in 2015-2019 it was around 2000 ha.

Table 1. Rice sufficiency value in each district

District	Rice sufficiency value				
	2016	2017	2018	2019	2020
Majene	0.74	1.00	0.19	0.19	0.17
Polewali Mandar	4.09	4.57	3.27	3.00	2.89
Mamasa	4.46	5.64	3.55	2.74	2.73
Mamuju	4.14	4.22	1.22	1.34	1.74
North Mamuju	1.38	1.82	0.40	0.10	0.12
Central Mamuju	3.43	6.03	1.04	1.19	1.20
West Sulawesi Province	3.32	3.97	1.90	1.72	1.78

Source: Data processing (2021)

The sufficiency of rice commodities by district from 2016 to 2020 tends to be sufficient, although several districts tend to be insufficient, namely Majene Regency and North Mamuju Regency. Food sufficiency for rice commodities in Majene Regency in 2016, 2018, 2019, and 2020 was not sufficient, but in 2017 it was sufficient. The rice commodity in Majene Regency which is not fulfilled is dominated by a significant decrease in rice production while the population continues to increase.

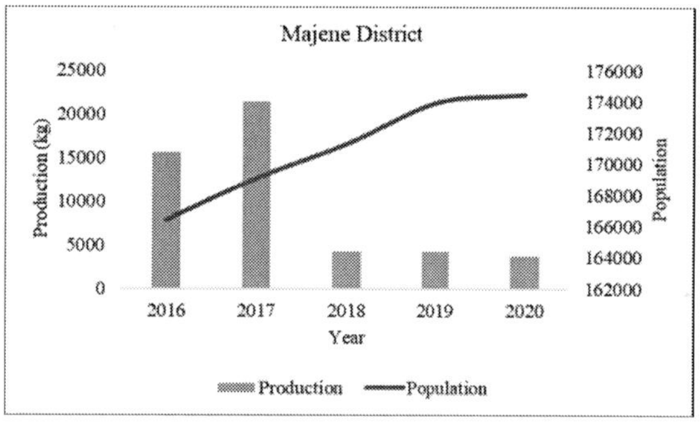

Fig. 3. Comparison of rice production and population in Majene District 2016-2020

The magnitude of the decline in rice production from 2017 to 2018 was 17,178 tons, where the decline was quite large. Food sufficiency in 2017 was considered sufficient because rice production in 2017 increased by 5,765 tons even though the population also increased. North Mamuju Regency tends not to have its food needs fulfilled, only in North Mamuju Regency in 2016 and 2017 the food needs of rice commodities are fulfilled while in 2018, 2019, and 2020 it is not fulfilled. This is because there was a significant decrease in production in 2018 which was 29,484 tons of rice, in 2019 there was a decrease in the production of 6,333 tons of rice, while in 2020 there was an increase in production of 594 tons of rice. Changes in the amount of rice production are accompanied by an increase in population so that when rice production decreases while the population increases, the sufficiency of food commodities can not be fulfilled.

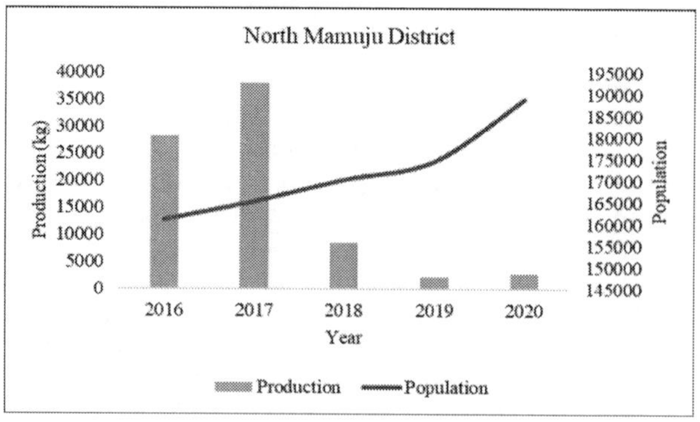

Fig. 4. Comparison of rice production and population in North Mamuju District 2016-2020

3.2 Trend of rice sufficiency value

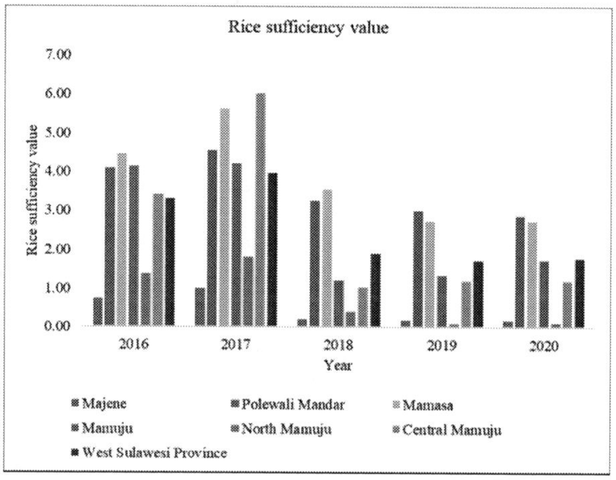

Fig. 5. Trend of rice sufficiency value in West Sulawesi Province 2016-2020

Rice commodities in Polewali Mandar Regency, Mamasa Regency, Mamuju Regency, and Central Mamuju Regency from 2016 to 2020 are sufficient. Although it is said to be sufficient, the value of the sufficiency of rice commodities in West Sulawesi tends to decrease, especially in 2018. This is because rice production has decreased significantly while the population has increased so that rice sufficiency in 2018 has decreased. Polewali Mandar Regency rice production tends to decrease although in 2017 it increased, in 2018 and 2019 it decreased, while it increased again in 2020. The amount of rice production in Polewali Mandar Regency is still quite sufficient to meet the food needs of the population. Likewise in Mamasa Regency and Mamuju Regency, which experienced a decrease in production in 2018 and 2019 and then increased in 2020 with a population that continues to increase, but can still meet the food sufficiency of the population. Central Mamuju Regency rice production increased from 2016 to 2017 and in 2018 decreased, but rice production increased again in 2019 and 2020 so that food availability was still sufficient for the population.

Based on the food sufficiency map of West Sulawesi Province in 2016 and 2020, it can be seen that there is a change in food sufficiency from 2016 to 2020. However, this change leads to inequality in food sufficiency in West Sulawesi Province. This is indicated by food sufficiency which is almost evenly distributed in all districts in 2016 where only Majene District has insufficient food sufficiency. Meanwhile, in 2020 there was a change in food sufficiency where there was one district that initially had sufficient food sufficiency to become insufficient, namely North Mamuju Regency. This inequality in food sufficiency is due to the decline in rice production while the population is increasing so that food sufficiency tends to decrease.

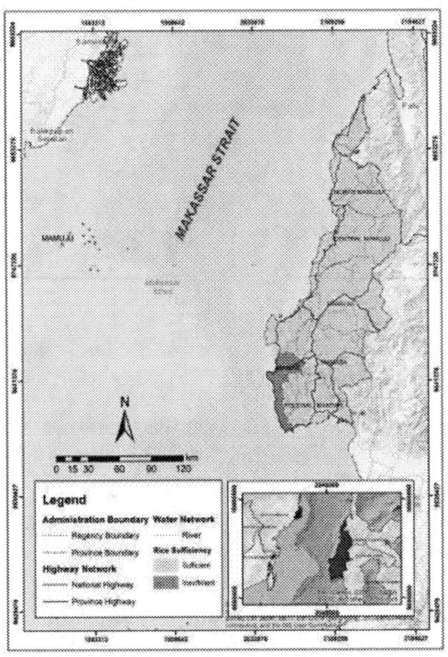

Fig. 6. Spatial distribution of rice sufficiency value in West Sulawesi Province in 2016

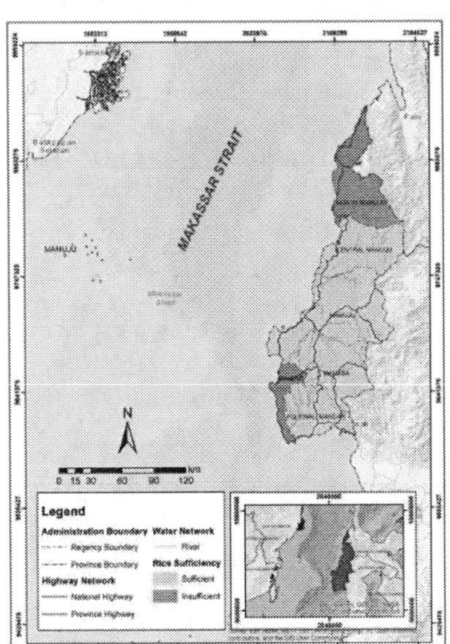

Fig. 7. Spatial Distribution of rice sufficiency value in West Sulawesi Province in 2020

One of the goals of the SDGs is to achieve sustainable agriculture. However, based on the results of the calculation of food sufficiency between 2016 and 2020, it shows that there is a shortage of main food sources. If in 2016 there was only 1 district experiencing shortages, in 2020 there will be 2 districts. This shows that the Province of West Sulawesi does not yet have a sustainable agricultural system.

3.3 Projection of rice sufficiency value in 2025 and 2030

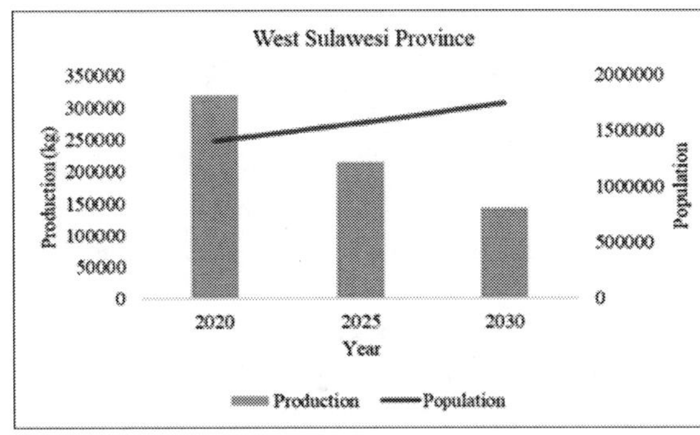

Fig 8. Comparison of rice production and population in West Sulawesi Province 2020, 2025, and 2030

The results of the projection of rice production in West Sulawesi in 2020, 2025, 2030 show that the amount of rice production is decreasing but the population is increasing. The increase in population resulted in the conversion of agricultural land into non-agricultural land. Population increase has a positive effect on land conversion [23]. Conversion of land that continues to occur will then have an impact on rice production. If there are no policies implemented, the results of rice production cannot meet the needs of the population. This can be a challenge for the future [13].

Table 2. Rice sufficiency value in each district

District	Rice Sufficiency Value	
	2025	2030
Majene	0.07	0.03
Polewali Mandar	1.99	1.38
Mamasa	1.75	1.13
Mamuju	1.14	0.74
North Mamuju	0.03	0.01
Central Mamuju	1.33	1.48
West Sulawesi Province	1.07	0.64

Source: Data processing (2021)

https://doi.org/10.1051/e3sconf/202132506010

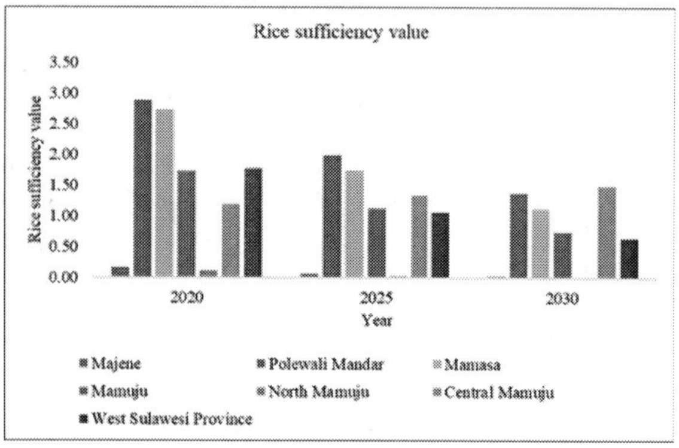

Fig. 9. Trend of rice sufficiency value in West Sulawesi Province 2020-2030

The results of the calculation of food adequacy in West Sulawesi in 2025 and 2030 for each district/city have varying values (Table 2). There are regencies/cities that can still have enough food and cannot have enough food. Districts projected in 2025 to be able to meet their food requirements then in 2030 cannot be sufficient, namely Mamuju Regency, while districts projected in 2025 and 2030 cannot be sufficient for food, namely Majene Regency and North Mamuju Regency. What is quite interesting from the results of the calculation of the projection of each district is that almost all districts have insufficient values for 2030, in contrast to Central Mamuju Regency which actually experienced an increase in the value of food sufficiency in the next five years projection period.

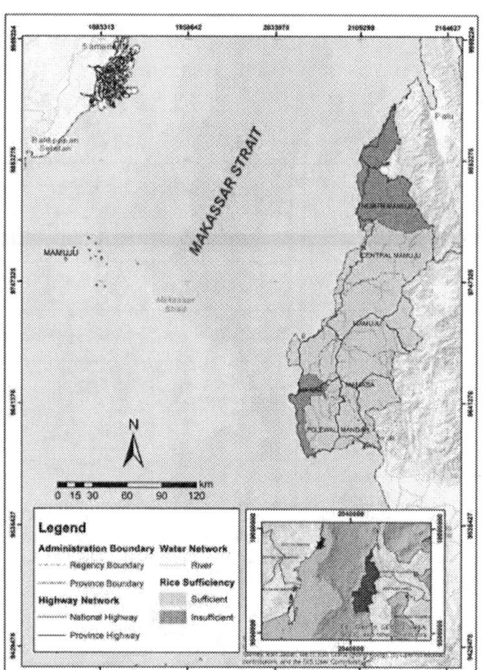

Fig. 10. Spatial distribution of rice sufficiency value in West Sulawesi Province in 2025

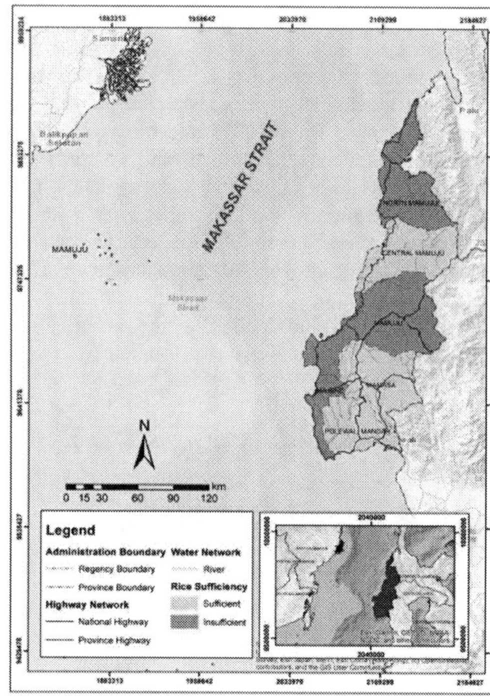

Fig. 11. Spatial distribution of rice sufficiency value in West Sulawesi Province in 2030

In general, it can be seen on the map over the next five years, namely in 2030 there will be an addition to three districts that fall into the insufficient category as shown in (Fig. 11). This will continue to happen over time if no preventive steps are taken from the government or related agencies to prevent it. One solution to deal with insufficient food needs is through the use of technology. The existence of rapid technological developments can offer a good solution to address the problem of sustainable food needs in the future [24]. Those who ignore global food projections and scenarios in the future tend to view this problem as a supply problem and do not focus on increasing food production through technology [25]. Based on this statement, it would be better if the parties responsible for handling the problem of food needs in West Sulawesi Province should not ignore the prediction of the condition of food needs in the future.

4 Conclusion

The necessity of food sufficiency must be considered by every stakeholder including government, public institutions like non-governmental organizations (NGO), and the community itself. The increase in food production is related to the physical, social, economic, and environmental aspects, while food demand is in line with the increasing population, economic development, and environmental dynamics.

Food sufficiency for rice commodities by district from 2016 to 2020 tends to be sufficient, although there are several districts that tend to be insufficient, namely Majene Regency and North Mamuju Regency. There is a change in food sufficiency from 2016 to 2020, but this change leads to inequality. The projection results in 2025 and 2030 also show the same thing, namely Majene Regency and North Mamuju Regency are still in the insufficient category. Even in 2030 it is predicted that it will get worse with the addition of one other district that is included in the insufficient category, so that in general the province of West Sulawesi is predicted to be unable to meet food needs.

The implementation of regional policies about food management is necessary to improve food sufficiency, especially in North Mamuju and Majene Regency. The policies should attract investment in the agricultural sector, facilities and infrastructure development, technology utilization, capacity building of human resources in the agricultural sector, as well as the expansion of partnership networks, especially between farmers, government, and private sector. The development of sustainable agriculture and local wisdom is also expected to support the increase in food production and productivity so as to improve the quality and sufficiency of food.

References

1. Y. Y. Kurniawan, A. Daerobi, B. Sarosa, and Y. P. Pratama, J. Ilmu Ekon. Terap. **03**, 1 (2018)
2. Pemerintah Republik Indonesia, (2012)
3. S. Apriani and Y. F. Baliwati, J. Gizi Dan Pangan **6**, 200 (2011)
4. A. S. Somantri, P. Luna, I. Widi, and B. Waryanto, Inform. Pertan. **29**, 95 (2020)
5. D. H. Azahari, Anal. Kebijak. Pertan. **6**, 174 (2016)
6. R. Febriamansyah and S. Tarumun, J. Gizi Dan Pangan **10**, 233 (2015)
7. B. Arifin, Risiko Dan Ketahanan Pangan Di Daerah Sentra Padi Kabupaten Pinrang, Universitas Gadjah Mada, 2012
8. Dinas Kesehatan Provinsi Sulawesi Barat, *Rencana Aksi Daerah Pangan Dan Gizi Provinsi Sulawesi Barat Tahun 2015-2019* (2016)
9. Badan Pusat Statistik, *Sulawesi Barat Dalam Angka 2015* (2015)
10. Badan Pusat Statistik, *Provinsi Sulawesi Barat Dalam Angka 2021* (2021)
11. L. Abdullah, L. Wado, and K. X. Cenerawasih, J. Ketahanan Nas. **25**, 178 (2019)
12. Presiden Republik Indonesia, (2002)
13. F. R. Ramdani and R. Harini, J. Bumi Indones. **14**, 1 (2019)
14. J. M. Mawara, Pros. Semin. Nas. 2017 Fak. Pertan. UMJ 77 (2017)
15. Bappenas, *Metadata Indikator Pilar Pembangunan Sosial Edisi II* (2020)
16. N. U. Fauziyanti, M. A. F. Alfana, and R. F. Putri, IOP Conf. Ser. Earth Environ. Sci. **451**, (2020)
17. F. T. Ahsani and D. Ardian, Semin. Nas. Off. Stat. 196 (2019)
18. M. Rachmat, Forum Penelit. Agro Ekon. **33**, 1 (2015)
19. P. Alexander, M. D. A. Rounsevell, C. Dislich, J. R. Dodson, K. Engström, and D. Moran, Glob. Environ. Chang. **35**, 138 (2015)
20. L. Muta and F. I. Rahmatullah, J. Bumi Indones. **6**, 1 (2017)
21. P. Antonius, 2 (2021)
22. S. Mulia and D. Pasambe, *Prospek Pengembangan Pertanian Di Propinsi Sulawesi Barat : Menyoroti Dinamika Pembangunan Pertanian Kawasan Timur Indonesia* (Badan Penelitian dan Pengembangan Pertanian (Balitbangtan), Jakarta, 2008)
23. R. Harini, Indones. J. Geogr. **44**, (2012)
24. Food and Agriculture Organization, *The Future of Food and Agriculture: Trends and Challenges* (Rome, 2014)
25. M. van Dijk, T. Morley, M. L. Rau, and Y. Saghai, Nat. Food **2**, 494 (2021)

The Poverty Index Mapping in the Barlingmascakeb Area in 2019

Fitroh Amaniah[1], *Fani* Setiati[1], *Mya Dwi* Rostika[2] and *Ratih Fitria* Putri[1*]

[1] Department of Environmental Geography, Faculty of Geography, Universitas Gadjah Mada, Jl. Sekip Utara, Bulak Sumur, Yogyakarta 55281, Indonesia
[2] Department of International Relations, Faculty of International Relations, Daito Bunka University, Japan

Abstract. Poverty is one of the serious problems in the national development process in Indonesia. Development basically aims to lead to a better situation and create a just and prosperous society. In the concept of Population-Centered Development, populations are not only treated as "objects" but also "subjects" of development. However, the incessant development carried out in various regions has not fully brought the welfare of the population. One of them is Central Java province which is a poverty pocket in Indonesia with a poverty percentage of 10.80 in 2019. Especially in the Barlingmascakeb area, namely Banjarnegara, Purbalingga, Banyumas, Cilacap, Kebumen, in the same year the percentage of poverty was relatively higher, respectively 14.76 percent, 15.03 percent, 12.53 percent, 10, 73 percent, and 16.82 percent. This study aims to examine the poverty index spatially in the Barlingmascakeb area in 2019 and find out the dominant indicators that determine the poverty index as one of the evaluation materials for the development process in the region. The data used to measure the poverty index is to use indicators on several dimensions of the Population-Centered Development Index (PCDI). The participation dimension is represented by the EPR (Employment to Population Ratio) indicator and the level of labor force participation rate; the dimension of partiality represented by the percentage of RREB (Regional Revenue and Expenditure Budget) for poverty reduction and other indicators related to poverty, namely income per capita. The spatial distribution of the poverty index is known through Geographic Information System (GIS) software.

1 Introduction

Poverty is a population problem that is familiar to developing countries, one of which is Indonesia. According to [1] poverty is one of the serious problems in the national development process in Indonesia. Poverty is considered as a form of development problem due to the negative impact of unbalanced economic growth so that it widens the income gap between communities and between regions [2]. The poverty of an area is an indicator of the success of the development of an area because the lower the poverty level of an area indicates an increase in the welfare of the population.

Barlingmascakeb is one of the regionalization areas in Central Java which includes Cilacap Regency, Banyumas Regency, Purbalingga Regency, Banjarnegara Regency, and Kebumen Regency. This is regulated in regional regulation number 6 of 2010 concerning the Spatial and Regional Planning of Central Java Province. The goal is to establish communication, coordination, and cooperation between one region and another so that equitable development can be achieved [3]. According to [4] states that the Barlingmascakeb area has a fairly high gini ratio index level. Rapid economic growth that is not accompanied by economic equality has the potential to lead to higher inequality and poverty.

Based on data from the Central Bureau of Statistics [5] from 2016 to 2019, the percentage of poverty in the Barlingmascakeb area tends to decrease from year to year (Fig. 1). In 2019, the poverty rate of each district in Barlingmascakeb was 14.76 percent respectively; 15.03 percent; 12.53 percent; 12.53 percent; 10.73 percent, and 16.82 percent. Kebumen Regency is the district with the highest poverty rate among other districts in Barlingmascakeb, while the lowest poverty is in Cilacap Regency. According to [4], Cilacap Regency is included in the main industrial area in Central Java after Semarang and Surakarta. Although the percentage of poverty decreases every year, there are still many people who live below the poverty line [6].

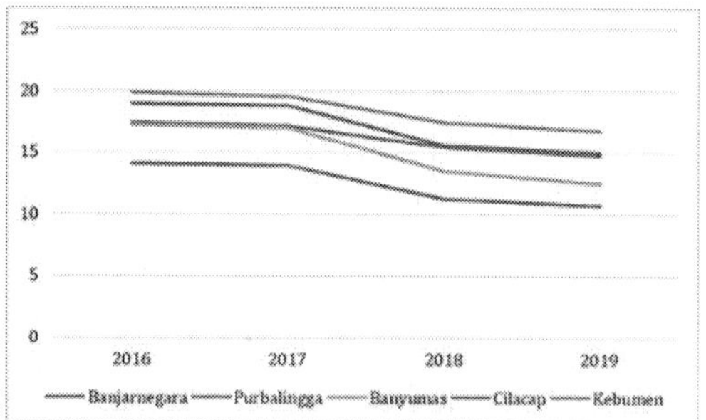

Fig.1. Percentage of Poverty by District in the Barlingmascakeb Area 2016-2019
Source: Central Bureau of Statistics, (2021)

* Corresponding author: ratihfitria.putri@ugm.ac.id

There are many factors that can affect poverty in an area. Gross Regional Domestic Product (GRDP) is one of the factors that reflects the productivity and income of the community in an area. If GRDP increases, productivity and income will increase which will then be able to reduce the number of poor people [7]. In addition, the participation of the working labor force and the availability of employment opportunities have an effect on poverty. Narrow job opportunities causing inequality in the distribution of income and will further create a high level of poverty [8]. On the other hand, the Regional Revenue and Expenditure Budget (RREB) poverty is one step to alleviate poverty [9].

GRDP per capita, employment to population ratio, labor force participation rate, and the percentage of RREB in poverty reduction are some of the variables that make up a population-oriented development index that can be used to determine the magnitude of the poverty index in an area, including poverty in Barlingmascakeb. According to [10], population-centered development is a development concept that places the population as the center of development, namely as subject and object. The poverty index which refers to population-centered development is oriented to population participation in accordance with potential development so as to improve the quality of the population which is the goal of development [11].

Based on the explanation above, the purpose of this research is to examine the poverty index spatially in the Barlingmascakeb area in 2019 and find out the dominant indicators that determine the poverty index as one of the evaluation materials for the development process in the region. The existence of accurate poverty information in the Barlingmascakeb area is expected to be able to succeed in poverty reduction programs and direct development that is more focused and on target.

2 Method

2.1 Location

Location of this research is in the Barlingmascakeb area (Banjarnegara, Purbalingga, Banyumas, Cilacap, and Kebumen). Barlingmascakeb is one of eight strategic areas in Central Java with fast economic growth. Barlingmascakeb's economic growth rate is the second largest after the Bregas area (Brebes, Tegal, and Slawi). However, the level of welfare among districts in this region is not evenly distributed [12]. In addition, the poverty rate in the Barlingmascakeb area in 2019 was still higher than the Central Java province poverty rate of 9.20 percent. Based on these things, the Barlingmascakeb area was chosen as the research location to determine the poverty index in 2019.

2.2 Data

This study uses secondary data in the form of employment data, GRDP data, population data, and RREB data for poverty reduction in each district of the Barlingmascakeb area. 2019. These data are sourced from the publications of the Central Bureau of Statistics (CBS) at the district/city level of Barlingmascakeb and CBS at the National level. Data processing is carried out using Microsoft Excel to calculate the employment to population ratio index, labor force participation rate (LFPR), and GRDP per capita by district/city in the Barlingmascakeb area. From these indices, a poverty index between districts in Barlingmascakeb was obtained in 2019. The map of the distribution of poverty in the Barlingmascakeb area was processed using ArcGIS software. The results of data processing were analyzed descriptively qualitatively by explaining the poverty index in the Barlingmascakeb area.

2.3 Research Design

This research begins with data collection from statistics data from CBS and shapefile data from the geospatial information agency. The next stage is to process the data to calculate the poverty index between districts in the Barlingmascakeb area. The last is the data analysis stage which is carried out descriptively to explain the phenomenon of poverty that occurs in the Barlingmascakeb area. The output of this research is in the form of tables, graphs, and maps of the distribution of the poverty index in Barlingmascakeb district. The design of this research is shown in the following flow chart (Fig. 2)

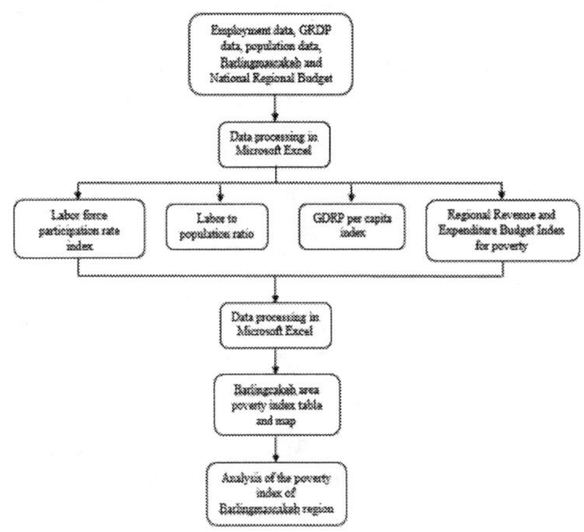

Fig. 2. Research flow chart.

3 Result and Discussion

Poverty is a classic problem in developing countries which to overcome it must be investigated from the smallest unit (based on data availability), namely the district. The trend of poverty that continues to decline in Indonesia is the result of the government's efforts to alleviate the problem of poverty which must be appreciated, including in Central Java Province, which is one of the pockets of poverty in Indonesia. The reduction in the poverty rate at the national and provincial levels that is not matched by a decrease at the district level which is the smallest unit will make the poverty problem vulnerable to increase and of course not investigate the problem from its root. This can be seen in the Barlingmascakeb area, which is an area that will be the center of growth in Central Java, which has a poverty rate that tends to decrease in each district but the figure is still above the province, this shows that development policies and poverty alleviation programs have not been successful and a development evaluation is needed. in each district in the Barlingmascakeb area.

One of the indicators that can be used as evaluation material for improving the quality of development policies and programs as well as poverty alleviation is the poverty index. The poverty index consists of several indicators, namely employment to population ratio (EPR), labor force participation rate (LFPR), percentage of RREB for poverty reduction, and GRDP per capita. Based on Table 1, the highest and lowest employment to population ratios in the Barlingmascakeb area are Banjarnegara Regency and Cilacap Regency, namely 68.91 and 57.70. The interpretation of this figure is that in 2019 in Banjarnegara Regency of 100 people aged 15 years and over there were around 69 people who worked and in 2019 in Cilacap Regency of 100 people aged 15 years and over there were around 58 people who worked. While the employment to population ratio of Central Java Province in 2019 was 65.47. This means that of the five districts in the Barlingmascakeb area, there are only two districts that are above the provincial figure, namely Banjarnegara Regency at 68.91 and Purbalingga Regency at 66.07.

The same is true for indicators of the level of labor force participation. Only two districts in the Barlingmascakeb area have LFPR figures above the province, namely Banjarnegara Regency and Purbalingga Regency. The two indicators, namely EPR and LFPR are indicators related to labor. This means that these districts must formulate policies that can encourage participation and employment indicators such as the preparation of quality labor resources, expansion of job opportunities, easy access to capital and markets, as well as improving the quality of infrastructure, licensing, and investment. An increase in the ratio of the workforce and the level of labor force participation will affect the level of individual income and the welfare of society in general through the multiplayer effect arising from the expansion of employment opportunities.

Table 1. Employment to Population Ratio, Labor Force Participation Rate, Percentage of RREB for Poverty Reduction, and GRDP per Capita by District of Barlingmascakeb Area in 2019.

Regency/City	*Employment to Population Ratio*	i	Labor force participation rate	i	Regional revenue and expenditure budget for poverty (%)	i	*Gross Regional Domestic Product per capita* (million)	i	Poverty Index
Banjarnegara	68,91	0,42	71,78	0,42	45,63	0,73	23,40	0,02	0,40
Purbalingga	66,07	0,36	70,33	0,38	49,47	0,79	26,68	0,03	0,39
Banyumas	64,67	0,33	67,50	0,32	53,27	0,86	31,87	0,03	0,38
Cilacap	57,70	0,17	62,37	0,19	51,62	0,83	66,06	0,08	0,32
Kebumen	61,91	0,27	65,53	0,27	52,29	0,84	23,35	0,02	0,35

Percentage of Regional Revenue and Expenditure Budget (RREB) for poverty reduction programs in each district in the Barlingmascakeb area has exceeded the Central Java provincial government's budget allocation for poverty reduction, which is 45.23%. This is good for accelerating poverty alleviation in each district of the Barlingmascakeb area. The district with the highest RREB allocation for poverty reduction is Banyumas Regency, which is 53.27%, while the district with the lowest RREB allocation for poverty reduction is Banjarnegara Regency, which is 45.63%. The higher the RREB allocation for poverty alleviation in the district, the more programs that can be run to reduce poverty levels.

The highest GRDP per capita in the Barlingmascakeb area, namely Cilacap Regency, far exceeds other regencies at 66.06 million and the lowest is Kebumen Regency at 23.35 million. This shows the need to increase efforts to increase regional income which is still low, especially in Kebumen Regency, Banjarnegara Regency, and Purbalingga Regency by utilizing and maximizing leading sectors and developing potential sectors.

Judging from the large value of the indicators that make up the poverty index in the Barlingmascakeb area, each index has a fairly high gap, especially the RREB indicator with GRDP per capita (Fig. 3). The Barlingmascakeb area has a relatively high RREB for poverty alleviation, but on the other hand the level of per capita income is still very low. Thus, in terms of poverty alleviation in the Barlingmascakeb area, it is necessary to prioritize indicators that are still low in order to accelerate poverty alleviation.

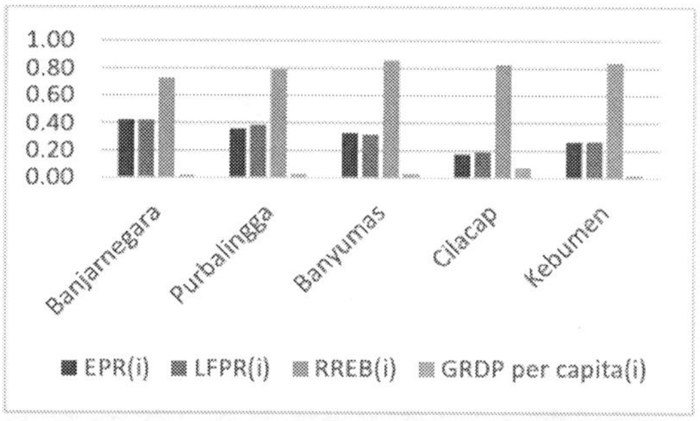

Fig. 3. Employment to Population Ratio Index, Labor Force Participation Rate (LFPR) Index, Percentage Index of RREB for Poverty Reduction, and GRDP Index per Capita by District of Barlingmascakeb Area in 2019.

Based on the calculation results show that the highest poverty index in the Barlingmascakeb area is in Banjarnegara Regency at 0.40 and the lowest poverty index is in Cilacap Regency at 0.32 (Table 2). This is influenced by the high employment to population ratio index and the labor force participation rate of Banjarnegara Regency and indirectly also shows the effectiveness of employment policies for poverty reduction in this district even though the percentage of RREB for poverty reduction is the smallest compared to other districts in the Barlingmascakeb area. The opposite occurs in Cilacap Regency which has a relatively high percentage of the APBD for poverty reduction and the highest per capita income, but this is not matched by an increase in the quality of employment which is the main element driving the regional economy as well as the strongest element for poverty alleviation which can be seen from the EPR and LFPR indicators. The lowest LFPR in the Barlingmascakeb area makes Cilacap Regency the district with the lowest poverty index in the Barlingmascakeb area.

Table 2. Inter-Regency Poverty Index in the Barlingmascakeb Area in 2019

Regency/City	Poverty Index	Class
Banjarnegara	0,40	Moderate
Purbalingga	0,39	Moderate
Banyumas	0,38	Moderate
Cilacap	0,32	Low
Kebumen	0,35	Moderate

The distribution of the poverty index in the Barlingmascakeb area is relatively uniform with a moderate poverty index in four districts, namely Banjarnegara, Purbalingga, Banyumas and Kebumen Regencies. Meanwhile, Cilacap Regency has a low poverty index (Fig. 4). The higher the poverty index, the easier it is to get out of the poverty trap.

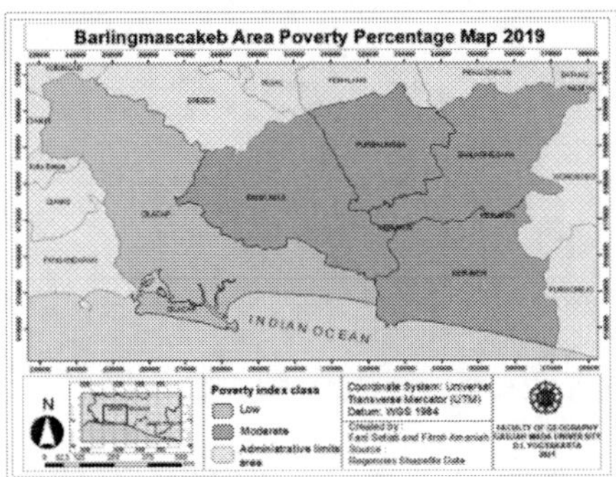

Fig. 4. Map of the Distribution of Poverty in the Barlingmascakeb Area in 2019

The current poverty index shows that there is still homework for the government to continue to improve poverty alleviation programs in the Barlingmascakeb area. Each district has different characteristics, so in poverty alleviation it is necessary to consider the factors that affect poverty based on the indicator values that make up the poverty index in the district concerned. Thus, the poverty alleviation program in the Barlingmascakeb area run by the government will be more targeted.

4 Conclusion

Based on the explanation above, it can be concluded that:

1. Spatially, the poverty index in the Barlingmascakeb area is in the moderate range, except for Cilacap Regency which has a low poverty index. The poverty index of each district in the Barlingmascakeb area is 0.40, respectively; 0.39; 0.38; 0.32; and 0.35.
2. The most dominant factor influencing the formation of the poverty index in the Barlingmascakeb area is income per capita (GRDP per capita) which is still low in each district. The high percentage index of the RREB indicates that the government is aware that it is quite good in terms of alleviating poverty problems in the Barlingmascakeb area.

Suggestions

From this study, the authors have inputs that need to be considered by the government of Barlingmascakeb area, namely:

In terms of formulating poverty alleviation policies in the Barlingmascakeb area, especially in Cilacap and Kebumen Regencies, it is necessary to prioritize programs for expanding employment opportunities and improving the quality of human resources in order to absorb the workforce. which is much more. Some things that can be done include opening skills training, courses, improving educational facilities, empowering local communities, and others. Improvements in the labor market system will directly increase people's real incomes which can gradually reduce poverty.

References

1. Solikatun, Supono, Masruroh, Y., & Zuber, A., *Kemiskinan Dalam Pembangunan*. Jurnal Analisa Sosiologi, **3**(1), 70–90 (2014)
2. Rahman, P.A., et al., *Kemiskinan Dalam Perspektif Ilmu Sosiologi*. Jurnal Pendidikan Tambusai, **3**(6), 1542–1548 (2019)
3. Bachtiar, P. A. *Analisis Pengaruh Tenaga Kerja Terdidik,Tenaga Kerja Tidak Terdidik, Dan Realisasi Belanja Modal Pemerintah Terhadap Pertumbuhan Ekonomi (Studi Kasus: Barlingmascakeb, Subosukawonosraten, dan Kedungsepur)*. Universitas Diponegoro (2013)
4. Hapsari, S. H., *Analisis Ketimpangan Pembangunan dan Pertumbuhan Ekonomi Antar Kabupaten di Kawasan Barlingmascakeb Tahun 2010-2016*, Universitas Muhammadiyah Yogyakarta (2018)
5. Badan Pusat Statistik (BPS), *Persentase Penduduk Miskin Menurut Kabupaten / Kota (Persen), 2019-2020,* Diakses pada tanggal 17 Juni 2021 di https://www.bps.go.id/indicator/23/621/1/persentase-penduduk-miskin-menurut-kabupaten-kota.html (2021)
6. Badan Pusat Statistik (BPS), *Profil Kemiskinan di Indonesia Maret 2019*, BPS (2019)
7. Wijaya, H., Istiqomah, I., & Arintoko, A., *Analisis Faktor - Faktor yang Mempengaruhi Kemiskinan (Studi Kasus di Kabupaten Banjarnegara, Cilacap, Purbalingga, Kebumen, dan Banyumas),*

Jurnal Ilmiah Universitas Batanghari Jambi, **20** (2), 451 (2020)

8. Tambunan, T., Perekonomian Indonesia: Teori dan Temuan Empiris, Ghalia Indonesia (2001)
9. Koestedjo, E. H., *Strategi Penanggulangan Kemiskinan Daerah Kabupaten Gresik Tahun 2017*, Journal.Uwks.Ac.Id, **18** (1) (2018)
10. Tjiptoherijanto, P., *Menuju Pembangunan Berwawasan Kependudukan.* Jurnal Populasi, **11** (1) (2000)
11. Saputra, W., Kebijakan Prespektif Dalam Pembangunan Nasional Yang Berwawasan Kependudukan, (2012)
12. Sutrisno, A., *Analisis Ketimpangan Pendapatan Dan Pengembangan Sektor Unggulan Di Kabupaten Dalam Kawasan Barlingmascakeb Tahun 2007-2010.* Economics Development Analysis Journal, **1**(1) (2012)

Livelihood of street vendors in Yogyakarta amidst the COVID-19 pandemic

Agus Joko Pitoyo[1,2*], *Akbar Abdul* Rokhim[2], *Ikhwan* Amri[2], and *Bagas* Aditya[2]

[1]Center for Population and Policy Studies, Universitas Gadjah Mada, Yogyakarta, 55281, Indonesia
[2]Department of Environmental Geography, Universitas Gadjah Mada, Yogyakarta, 55281, Indonesia

Abstract. This study aims to find out the characteristics of street vendors in Yogyakarta City before and during the COVID-19 pandemic, the livelihood strategies performed by the street vendors during the COVID-19 pandemic, and the relationship between street vendors characteristics and their livelihood strategies during the COVID-19 pandemic. The authors used a quantitative method with a total of 83 respondents. The results indicated age characteristics ranging from 19-68 years, the last education completed by the street vendors was dominated by senior high school or equivalent, the status in the household was dominated by the head of the household, and the number of household members was 1 to 9 people. The COVID-19 pandemic induced significant impacts on the employment situation in the household, working hours, and income. The livelihood strategies performed by the street vendors amidst the COVID-19 pandemic were survival strategy (73.49%) and consolidation strategy (26.51%). This study found that the relationship between the characteristics of street vendors and livelihood strategies varied.

1 Introduction

The COVID-19 pandemic that has hit Indonesia since March 2020 causes a major impact on the Indonesian economy. The government policies related to social distancing and working from home (WFH) have weakened Indonesian economic activities and had an impact on all sectors, one of them is street vendors. As one of the crucial sectors, street vendors play a key role as an economic actor who rely on their activities based on human activities to meet their daily needs through buying and selling activities and simple equipment or technology, and thus when the activities stop, it becomes a blow to street vendors because the facilities they use to earn profits are reduced to the point where they are lost.

A study conducted by the Indonesian Institute of Sciences showed that 94.69% of macro, small and medium enterprises experienced a decline in income [1]. Micro-enterprises, which include street vendors, experienced a decrease in income of more than 75% by 43.3%. The number is relatively high since the possibility of losing their jobs and becoming poor is getting higher.

The Statistics Indonesia conducted a study related to socio-demographic conditions during the COVID-19 pandemic, showing that the sector experiencing the greatest impact was retail, including street vendors. The COVID-19 pandemic caused workers' conditions to fall into the worst condition, 2.52% of workers experienced layoffs and 22.74% did not work, and 70.53% of the population had low incomes of less than IDR 1,800,000 who had become the most affected ones by the COVID-19 pandemic [2]. A preliminary study conducted in Indonesia also proved that the pandemic crisis disrupted the incomes of most informal workers [3, 4].

The locations of street vendors in Yogyakarta City are generally agglomerated in the area of tourism sites, monuments, palaces, station, and stadium. Besides, street vendors are also often found around formal sector areas, such as malls, large shops, schools, and offices [5]. The street vendors in Yogyakarta are often associated with tourist attractions since it is easy to find buyers, so the income is higher as well [6, 7]. Therefore, when the COVID-19 pandemic occurs which reduces community activities outside the home, especially in tourist locations, the impact felt by the street vendors is very big.

The impact of the COVID-19 pandemic has made street vendors need several strategies to meet their daily needs. The strategies must be carried out in order to meet and improve the life needs of street vendors both in the short and long term by regulating, selecting, and utilizing their capabilities [8]. The absence of social and health protection for street vendors and low-income groups makes them very vulnerable during a crisis.

The present study aims to (1) find out the characteristics of street vendors in Yogyakarta amidst the COVID-19 pandemic; (2) find out the livelihood strategies of street vendors in Yogyakarta amidst the COVID-19 pandemic; and (3) find out the relationship between the characteristics of street vendors in Yogyakarta and their livelihood strategies amidst the COVID-19 pandemic.

* Corresponding author: aguspit@ugm.ac.id

2 Method

The data were collected using a survey by conducting online (via google form) and offline structured interviews. The survey results obtained 83 street vendors in Yogyakarta City, Indonesia. The street vendors are those with the status of business owners and not only workers (laborers), sellers of goods and/or services in temporary or non-permanent buildings [9, 10]. The sampling technique was non-probability sampling in the form of convenience sampling. Convenience sampling is part of non-probability sampling which is commonly used by selecting samples based on the ease of accessing samples that meet the basic criteria for the sample to be used [11, 12].

The authors used a quantitative descriptive method by describing the characteristics of street vendors before and during the COVID-19 pandemic, the livelihood strategies performed by street vendors, and the relationship between the characteristics of street vendors and their strategies amidst the COVID-19 pandemic. The characteristics of street vendors are age, last education level, status in the household, number of household members, number of household members who worked before and during the COVID-19 pandemic, number of working hours before and during the COVID-19 pandemic, and their income before and during the COVID-19 pandemic.

This study adopts White's theory to classify livelihood strategies, i.e., survival, consolidation, and accumulation. The survival livelihood strategies performed by the street vendors are those in the form of prioritizing food needs, looking for side jobs, and utilizing social assistance. The variables in the consolidation strategy are in the form of double earning patterns and priority health needs, while the accumulation livelihood strategies are mostly carried out by wealthy entrepreneurs who have a lot of resources so that they can increase their assets significantly.

3 Results and discussion

3.1 Characteristics of street vendors

Table 1 presents the characteristics of street vendors in Yogyakarta before and during the COVID-19 pandemic. The age of street vendors varies from 19 years old to 68 years old. Street vendors included in the age group of 53 to 58 years or older tend to do their work permanently in one place, while the age group of 19 to 35 years or younger tend to move their activities from one place to another. The age group of 53 to 68 years has a fairly large proportion compared to that of the age group of 19 to 35 years. This indicates that being a street vendor is still in demand by the elderly because of the ease of doing the work and does not require a big amount of workforce [13].

The last education level of senior high school or equivalent had the largest proportion which was then followed by junior high school or equivalent, elementary school or equivalent, academy/diploma/university, and those who did not attend a school or did not complete their education at elementary level. The education level of street vendors in Yogyakarta tends to be low. This shows that the street vendors in Yogyakarta are workers with low education and have limited abilities and skills.

The status of street vendors in the household consists of the head of household, wife or husband, parent/parent-in-law, and son/daughter, which was dominated by the head of the household consisting of 59 street vendors. The number of household members of the street vendors ranged from 1 to 9 members with the most number of households group was 1-3, i.e., 49 street vendors.

The number of working household members shows the more income that the household gets, so it has an impact on increasing the fulfillment of household needs. The number of household members who worked with only one person is the largest proportion both before the COVID-19 pandemic and during the COVID-19 pandemic, i.e., 39.76% and 46.99%. It shows that street vendors are a vulnerable sector in terms of income and economic independence [14].

The number of working hours shows the level of productivity of street vendors which is closely related to their income. The street vendors who work for less than 11 hours per day have the largest proportion both before and during COVID-19, i.e., 72.29% and 83.13%. It indicates that the street vendors have the same working hours as formal workers.

The street vendors who chose to reduce their working hours due to the implementation of the WFH policy and social distancing have caused a decrease in the number of buyers. Meanwhile, the street vendors who chose to increase their working hours are generally those who work only at certain times, e.g., at night or in the morning, thus increasing their working hours by selling their products from morning to noon or afternoon and starting selling from noon or evening for the street vendors who usually sell at night. Besides, the street vendors who increased their working hours had a work location that is not far from their homes.

The monthly income earned by street vendors in Yogyakarta before and during COVID-19 was still dominated by low income under the Regional Minimum Wage of Yogyakarta, i.e., IDR 2,004,000 [15]. The largest proportion before the COVID-19 pandemic ranged from IDR 1,000,001 to IDR 2,000,000, while the monthly income during the COVID-19 pandemic was under IDR 1,000,000. It confirms that street vendors are very vulnerable economic actors even before the COVID-19 pandemic. During the COVID-19 pandemic, there have been no street vendors who have an income above IDR 4,000,000.

Table 1. Characteristics of street vendors before and during the COVID-19 pandemic

Characteristics	Number	Proportion	Characteristics	Number	Proportion
Age			**Number of Household Members**		
• 19-35	19	22.89%	• 1-3	49	59.04%
• 36-52	41	49.40%	• 4-6	29	34.94%
• 53-68	23	27.71%	• 7-9	5	6.02%
Last Education			**Status in the Household**		
• No schooling completed	5	6.02%	• Head of household	59	71.08%
• Elementary school/equivalent	12	14.46%	• Wife/husband	18	21.69%
• Junior high school/equivalent	15	18.07%	• Parent/parent-in-law	1	1.20%
• Senior high school/equivalent	44	53.01%	• Son/daughter	5	6.02%
• Academy/diploma/ university	7	8.43%			
Number of Household Members who Worked Before the COVID-19 Pandemic			**Number of Household Members who Worked during the COVID-19 Pandemic**		
• 1	39	46.99%	• 1	33	39.76%
• 2	27	32.53%	• 2	31	37.35%
• 3	11	13.25%	• 3	14	16.87%
• 4	6	7.23%	• 4	5	6.02%
Daily Working Hours Before the COVID-19 Pandemic			**Daily Working Hours During the COVID-19 Pandemic**		
• <11	60	72.29%	• <11	69	83.13%
• 11-15	21	25.30%	• 11-15	12	14.46%
• >15	2	2.41%	• >15	2	2.41%
Monthly Income (IDR) Before the COVID-19 Pandemic			**Monthly Income (IDR) During the COVID-19 Pandemic**		
• ≤1,000,000	19	22.89%	• ≤1,000,000	52	62.65%
• 1,000,001-2,000,000	31	37.35%	• 1,000,001-2,000,000	20	24.10%
• 2,000,001-3,000,000	17	20.48%	• 2,000,001-3,000,000	9	10.84%
• 3,000,001-4,000,000	8	9.64%	• 3,000,001-4,000,000	2	2.41%
• >4,000,000	8	9.64%	• >4,000,000	0	0.00%

3.2 Livelihood strategies performed by street vendors during the COVID-19 pandemic

The livelihood strategies performed by street vendors in Yogyakarta are presented in Table 2. It shows that 61 respondents were classified as street vendors who used survival strategies and 22 respondents are classified as street vendors who used consolidation strategies. They were classified as the street vendors who used survival strategy due to limited access to selling, whether to get raw materials or for marketing. The government policies during the COVID-19 pandemic have made people's activities outside the home very limited, resulting in a decrease in income [16]. Also, the absence of street vendors who were included in the accumulation strategy shows that the COVID-19 pandemic has greatly impacted the lives of street vendors. The results were in line with a study conducted by Armansyah et al. which confirms that street vendors in normal conditions are dominated by groups with survival strategies and it is difficult to improve their livelihood strategies, i.e., to enter into a consolidation strategy [17].

Table 2. Livelihood strategies performed by street vendors

Livelihood Strategies	Number	Proportion
Survival	61	73.49%
Consolidation	22	26.51%
Total	**83**	**100.00%**

Table 3 shows that 61 street vendors who prioritized food needs, four street vendors who did side jobs, and 39 street vendors who used social assistance. This is in line with a study conducted by Wildayana et al. which shows that the group of survival strategy prioritizes the income to fulfill their primary needs to survive [18]. The side jobs included massage therapists, parking officers, boarding house guards, and craftsmen. The side jobs are still included in the informal sector.

https://doi.org/10.1051/e3sconf/202132506012

Table 3. Survival strategy

Strategy	Activity	Number	Proportion
Survival (n = 61)	Fulfilling primary needs as a priority	61	100%
	Doing side job	4	6.6%
	Using social assistance	39	63.93%

Table 4 presents the main strategies performed in the consolidation strategy, i.e., prioritizing health needs carried out by 19 street vendors and double earning pattern carried out by 15 street vendors by prioritizing the fulfillment of health needs since the fulfillment of basic needs, especially food, has been met. Besides, it increasingly raised people's awareness to maintain and prioritize their health over work. The income of the consolidation strategy tended to be stable due to the availability of sufficient money deposits and investments made in trading.

Table 4. Consolidation strategy

Strategy	Activity	Number	Proportion
Consolidation (n = 22)	Prioritizing health needs	19	86.36%
	Double earning pattern	15	68.18%

The double earning pattern was also the reason the income of street vendors in the consolidation strategy group was relatively more stable and secure. This is in line with a study conducted by Wildayana et al. which states that increasing livelihood strategies to a higher level is always followed by various sources of income [18]. Besides, a study conducted by Widodo explains that the double earning pattern in the consolidation strategy aims to develop the household economy [19].

3.3 The relationship between characteristics of street vendors and their livelihood strategies

Table 5 shows that the age group of 53-68 years was dominated by 18 street vendors using survival strategy, while the consolidation strategy was only used by five street vendors. In line with a study conducted by Wildayana et al., older people tend to use survival strategy because of their low level of survival and their low productivity due to relatively weak physical conditions and tends not to live with family members [18].

This study found that the 36-52 years old group who adopt a survival strategy has a fairly high proportion because the COVID-19 pandemic has a direct impact on the economy. However, in relative terms, the proportion of survival strategy in the older group is higher than that of the other groups. The present study is similar to the study conducted by Alfana which shows that the older age group dominates the survival strategy due to the decrease in physical abilities [20]. Besides, the age group of 19-35 years and 36-52 years tend to have higher resilience in trading during the crisis due to their physical strength and other strategic activities that can be carried [21].

Table 5. The relationship between age and livelihood strategies of street vendors

Age		Livelihood Strategies	
		Survival	**Consolidation**
19-35	Number	13	6
	% of Total	15.66	7.23
36-52	Number	30	11
	% of Total	36.14	13.25
53-68	Number	18	5
	% of Total	21.69	6.02
Total		**83**	
%		**100.00**	

Table 6 shows that the street vendors with the last education level of not attending school/not completing elementary school were included in the group with a survival strategy. It shows that the last education completed by the street vendors affected the livelihood strategies they used. The higher the education, the more the strategy will increase because it has more information and more developed capabilities so as to increase competition in trading [22, 23].

Table 6. The relationship between education and livelihood strategies of street vendors

Last Education		Livelihood Strategies	
		Survival	**Consolidation**
No schooling completed	Number	5	0
	% of Total	6.02	0.00
Elementary school/ equivalent	Number	9	3
	% of Total	10.84	3.61
Junior high school/ equivalent	Number	11	4
	% of Total	13.25	4.82
Senior high school/ equivalent	Number	31	13
	% of Total	37.35	15.66
Academy/ diploma/ university	Number	5	2
	% of Total	6.02	2.41
Total		**83**	
%		**100.00**	

The existence of street vendors with an academic/diploma/university as their last education level which was included in the group with a survival strategy shows that education does not always influence livelihood strategies. This is in line with a study conducted by Armansyah et al. that the final education level of street vendors did not necessarily improve their livelihood strategies [17].

Table 7 shows that more than 70% of the status in street vendors' households was the head of the household. The status of the head of the household affects the choice of livelihood strategies since it bears the burden of meeting the household life needs, and the larger the household members, the greater the needs needed, so the livelihood strategies become an option [24]. The status of the household as a son or daughter was included in a consolidation strategy even during the COVID-19 pandemic because it is not too burdensome to meet the needs of the household life.

Table 7. The relationship between status in the household and livelihood strategies of street vendors

Status in Household		**Livelihood Strategies**	
		Survival	**Consolidation**
Head of household	Number	43	16
	% of Total	51.81	19.28
Wife/ husband	Number	16	2
	% of Total	19.28	2.41
Parent/ parent-in-law	Number	1	0
	% of Total	1.20	0.00
Son/ daughter	Number	1	4
	% of Total	1.20	4.82
Total		**83**	
%		**100.00**	

Table 8 shows that the livelihood strategies performed by household members consisting of 1-3 people were dominated by survival strategy during the COVID-19 pandemic since it only consisted of a small family with the head of the household working alone, so meeting the necessities of life is quite difficult.

In contrast, the number of household members consisting of 4-6 people and 7-9 people that used the consolidation strategy was higher than the number of household members consisting of 1-3 people because there were more household members who worked, so the fulfillment of household needs had been shared. A high number of household members can lead to a double earning pattern and domestic workers so as to improve the livelihood strategies [25, 26].

Table 9 shows that there were three household members who worked that mostly used survival strategy. It indicates that the number of household members who worked during the COVID-19 pandemic might not necessarily improve their livelihood strategies. This is in line with a study conducted by Khatiwada et al. that there are other factors in the form of the high number of household members which will further increase the burden to meet the needs of the household [22].

Table 8. The relationship between the number of household members and livelihood strategies of street vendors

Number of Household Members		**Livelihood Strategies**	
		Survival	**Consolidation**
1-3	Number	39	10
	% of Total	46.99	12.05
4-6	Number	19	10
	% of Total	22.89	12.05
7-9	Number	3	2
	% of Total	3.61	2.41
Total		**83**	
%		**100.00**	

Table 9. The relationship between the number of household members who worked and livelihood strategies of street vendors

Number of Household Members		**Livelihood Strategies**	
		Survival	**Consolidation**
1	Number	30	9
	% of Total	36.14	10.84
2	Number	20	7
	% of Total	24.10	8.43
3	Number	9	2
	% of Total	10.84	2.41
4	Number	2	4
	% of Total	2.41	4.82
Total		**83**	
%		**100.00**	

Table 10 shows that an increase in the number of working hours during the COVID-19 pandemic did not improve livelihood strategies because the COVID-19 pandemic causes very limited activities that are extremely different from normal conditions. This is in contrast to a study conducted by Widjajanti et al. [5] and Samosir [27] that the longer working hours, the better the livelihood strategies. The street vendors who increased their working hours were classified as the group of 9-14 hours and the group of 15-20 hours, so they could improve their livelihood strategies by being sellers who opened their shop near their homes that had been dominated by food vendors.

Table 11 shows that the income of the street vendors was dominated by 52 street vendors with a proportion of 62.64% and it was then categorized as a survival strategy, i.e., 49 street vendors (59.04%) and a consolidation strategy, i.e., three street vendors (3.61%).

The street vendors with income under IDR 1,000,000 were classified as a group with consolidation strategy since their status in the household is son or daughter. Therefore, the income they got was used for personal needs only. Table 11 also shows that income affected the livelihood strategies they used. This is in line with a study conducted Alfana that income affects the level of livelihood strategies that are performed for their own needs [20].

Table 10. The relationship between the number of working hours and livelihood strategies of street vendors

Daily Working Hours		Livelihood Strategies	
		Survival	Consolidation
2-8	Number	36	13
	% of Total	43.37	15.66
9-14	Number	24	7
	% of Total	28.92	8.43
15-20	Number	1	2
	% of Total	1.20	2.41
Total		**83**	
%		**100.00**	

Table 11. The relationship between daily incomes and livelihood strategies of street vendors

Incomes		Livelihood Strategies	
		Survival	Consolidation
≤1,000,000	Number	49	3
	% of Total	59.04	3.61
1,000,001-2,000,000	Number	10	10
	% of Total	12.05	12.05
2,000,001-3,000,000	Number	2	7
	% of Total	2.41	8.43
3,000,001-4,000,000	Number	0	2
	% of Total	0.00	2.41
Total		**83**	
%		**100.00**	

4 Conclusion

This study captured an overview of the impact of the COVID-19 crisis on 83 street vendors in Yogyakarta. In general, the characteristics of age ranged from 19-68 years old, the last education had been dominated by senior high school or equivalent, the status in the household had been dominated as the head of the household, and the number of household members was 1 to 9 people. The pandemic has caused real socio-economic disturbances to the livelihoods of street vendors, as evidenced by the dynamics of the numbers of workers in a household, daily working hours, and monthly incomes.

The livelihood strategies of street vendors in Yogyakarta consisted of the survival strategy performed by 61 street vendors (73.49%) which had been dominated by the fulfillment of food needs, the use of social assistance, and side jobs. Meanwhile, the consolidation strategy was performed by 22 street vendors (26.51%) which had been dominated by the fulfillment of health needs as a priority and double earning pattern.

The relationship between livelihood strategies and characteristics of street vendors in Yogyakarta including age, education, status in the household, and the number of household members varied. Meanwhile, the number of household members who worked during the COVID-19 pandemic had decreased and they had used a survival strategy. Also, the number of working hours and income had decreased. The fewer the number of working hours and income, the more likely it is to be classified as a survival strategy.

This research was funded by RTA (*Rekognisi Tugas Akhir*) Program 2020 from Universitas Gadjah Mada, grant number 2607/UN1/DITLIT/DIT-LIT/PT/2020.

References

1. LIPI, Survei Kinerja UMKM di Masa Pandemi COVID 19, URL: http://lipi.go.id/berita/survei-kinerja-umkm-di-masa-pandemi-covid19/22071. (2020)
2. BPS, *Hasil Survei Sosial Demografi Dampak COVID-19* (BPS, Jakarta, 2020)
3. A.J. Pitoyo, B. Aditya, I. Amri, E3S Web of Conferences **200**, 03014 (2020), doi: 10.1051/e3sconf/202020003014
4. A.J. Pitoyo, B. Aditya, I. Amri, A.A. Rokhim, Ind. J. Labour Econ. (2021), doi:10.1007/s41027-021-00333-x
5. R. Widjajanti, Teknik **30**, 162-170 (2012), doi:10.14710/teknik.v30i3.1892
6. I.G.M.Y. Antara, L., Muta'ali, W.P. Mulyani, Majalah Geografi Indonesia **29**, 149-153 (2015)
7. E.W. Ariyadi, L.T. Adishakti, D. Kristiadi, Manusia Dan Lingkungan **12**, 140–149 (2005)
8. M. Susanto, S. Sudrajat, Jurnal Bumi Indonesia **6** (2017)
9. Government of Yogyakarta City, Regulation of Yogyakarta Mayor No. 37/2007 concerning Arrangement of Street Vendors in Special Area of Malioboro – A. Yani, 1 (2010)
10. A.J. Pitoyo, Populasi **10**, 73-97 (1999), doi:10.22146/jp.12485

11. B. Albert, T. Tullis, D. Tedesco, *Planning the Study* (Morgan Kaufmann, Burlington, 2010), doi:10.1016/B978-0-12-374892-8.00002-8
12. C. Wilson, *General Interviewing Issues* (Morgan Kaufmann, Burlington, 2014), doi:10.1016/B978-0-12-410393-1.00006
13. A.D. Nasrulloh, Strategi penghidupan pedagang kaki lima di Kawasan Stasiun Lempuyangan Kota Yogyakarta, Undergraduate Thesis (Universitas Gadjah Mada, Yogyakarta, 2019)
14. CORE, Waspada lonjakan pengangguran dampak pandemi COVID-19: Lima rekomendasi CORE, URL: https://www.coreindonesia.org/view/468/waspada-lonjakan-pengangguran-dampak-pandemi-COVID-19-lima-rekomendasi-core. (2020)
15. Government of Yogyakarta Special Region, Keputusan Gubernur Daerah Istimewa Yogyakarta Nomor 257/KEP/2019 tentang Penetapan Upah Minimum Kabupaten/Kota (2020)
16. A. Majithia, Impact of COVID-19, URL: https://www.wiego.org/impact-COVID-19. (2020)
17. Armansyah, Sukamdi, A.J. Pitoyo, Russ. J. Agric. Soc.-Econ. Sci. **95**, 104-110 (2019), doi:10.18551/rjoas.2019-11.13
18. E. Wildayana, M.E. Armanto, Z. Idrus, I.A. Radiatmoko, SRICOENV **2001**, 1-9 (2018)
19. S. Widodo, Sosial Humaniora **15**, 10-20 (2011), doi:10.7454/mssh.v15i1.890
20. M.A.F. Alfana, Strategi penghidupan rumahtangga miskin perdesaan (kasus di tiga desa Kabupaten Kulon Progo Daerah Istimewa Yogyakarta), Thesis (Universitas Gadjah Mada, Yogyakarta, 2014)
21. Ajifah, Struktur, strategi, dan resiliensi nafkah pedagang kaki lima di Pasar Anyar Bogor, Undergraduate Thesis (Institut Pertanian Bogor, Bogor, 2018)
22. S.P. Khatiwada, W. Deng, B. Paudel, J.R. Khatiwada, J. Zhang, Y. Su, Sustainability **9** (2017), doi:10.3390/su9040612
23. D. Xu, J. Zhang, G. Rasul, S. Liu, F. Xie, M. Cao, E. Liu, Sustainability **7**, 4850-4869 (2015), doi:10.3390/su7054850
24. E. Handoyo, A.B. Setiawan, JEJAK: Jurnal Ekonomi dan Kebijakan **11**, 173-188 (2018)
25. E. Alviawati, Majalah Geografi Indonesia **27**, 104-117 (2016), doi:10.22146/mgi.15919
26. R. Amalia, Struktur dan strategi nafkah pedagang makanan di sektor informal daerah suburban Kabupaten Bogor, Undergraduate Thesis (Institut Pertanian Bogor, Bogor, 2013)
27. R.A. Samosir, Analisis pendapatan pedagang kaki lima sektor informal di Kecamatan Semarang Tengah Kota Semarang, Undergraduate Thesis (Universitas Diponegoro, Semarang, 2015)

Travel characteristics of suburban residents during the COVID-19 pandemic (case study of Sonosewu, Bantul Regency)

Dewina Inggar Santi[1], and *Sri* Rum Giyarsih[1*]

[1]Department of Environmental Geography, Faculty of Geography, Gadjah Mada University, Yogyakarta, 55281, Indonesia

Abstract. The existence of COVID-19 pandemic had an impact on residents' travel, including in Sonosewu Hamlet, Bantul Regency, Indonesia because the government has implemented recommendation to reduce residents' mobility in order to accelerate the handling of COVID-19. This research is interesting because there was no research that specifically examines the characteristics of residents' travel during COVID-19 pandemic in the suburbs of Yogyakarta. The purpose of this study was to identify the travel characteristics of the residents of Sonosewu during the COVID-19 pandemic. The data collection was carried out through questionnaire surveys. The data processing and analysis were carried out by using descriptive statistical methods. Based on the results obtained from 328 respondents, it is known that the travel frequency from November 30 to December 6, 2020, tended to be high with 6 to 7 trips a week. The purpose of travel was mostly to work and shopping, and the main travel destinations are Bantul Regency and Yogyakarta City. The result of this research will be useful for the development of knowledge, especially in the field of urban geography and can be taken into consideration by the government for the development of suburban areas.

1 Introduction

Suburban area has several characteristics, such as its location directly adjacent to the city area, the population is oriented towards the city, the rural area where residents work in the city, and the residents need life both in the city and in the village [1]. Based on the characteristics, residents will do travel. Population travel is defined as a one-way movement from the point of origin to the destination using certain modes of transportation [2]. According to Meyer and Miller, it can be seen from some attributes, such as purpose and frequency [3].

Sonosewu Hamlet, Ngestiharjo Village, Kasihan Subdistrict, Bantul Regency is one of the suburban areas because it is directly adjacent to the city of Yogyakarta (Figure 1). This is beneficial for the residents of Sonosewu because there are options for socio-economic facilities to meet the needs of life both in the surrounding area and in the city of Yogyakarta so that the opportunity to travel is even higher. However, this does not always apply in all conditions, especially during the COVID-19 pandemic.

The COVID-19 pandemic is caused by the Severe Acute Respiratory Syndrome Coronavirus-2 (SARS-CoV-2) virus which attacks the respiratory tract and is a dangerous infectious disease where transmission comes from human-to-human transmission through droplets, either directly or indirectly [4]. Romdiati revealed that population travel as the cause of the spread of the virus was evidenced at the beginning of this virus spreading to residents who traveled internationally, then spreading to local transmission [5]. This shows that travel and direct social interactions will exacerbate the spread of COVID-19. Therefore, the government issued several policies and appeals in order to prevent the wider spread of the virus and accelerate the handling of the COVID-19 pandemic.

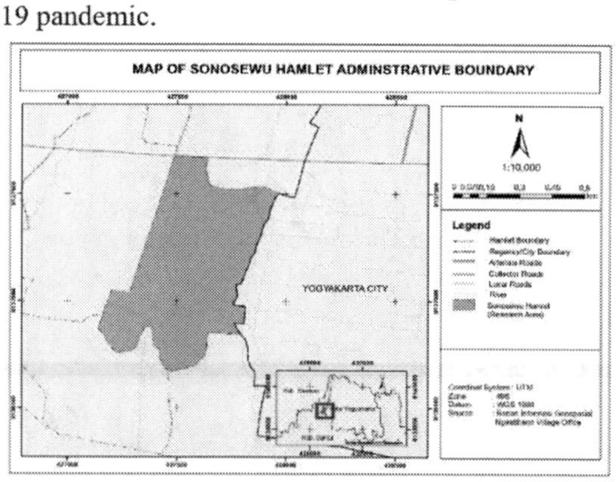

Fig. 1. Map of Sonosewu Hamlet administrative boundary.

Kasihan Subdistrict during the COVID-19 pandemic by the Bantul Regency Health Office was classified as a moderate risk zone with a value of 2.0 or level 3 with the criteria that the area had a high risk of spreading COVID-19, and the spread and potential of the virus were out of control [6]. Based on this condition, the Bantul Regency Government has implemented several sector implementations in this zone, including advising people to stay at home, temporarily closing educational facilities, carrying out social and physical distancing, and so on [7]. With these policies, it is hoped that

* Corresponding author: srirum@ugm.ac.id

residents can reduce social activities and change their travel behavior.

Changes in population activity during the pandemic, one of which is indicated by the results of the identification of movements by Google COVID-19 Community Mobility Reports [8]. As of June 27, 2020, there was still a decrease in the movement to several places despite the stipulation of a new adaptation period in the Special Region of Yogyakarta (Figure 2). The movement to several places is still decreasing compared to the baseline of measurement, namely the median value from January 3 to February 6, 2020. Also, data from the Office for the Coordination of Humanitarian Affairs shows that there was a decrease in the average daily movement in Bantul Regency from March 1 2020 to July 5 at -23.1% [9].

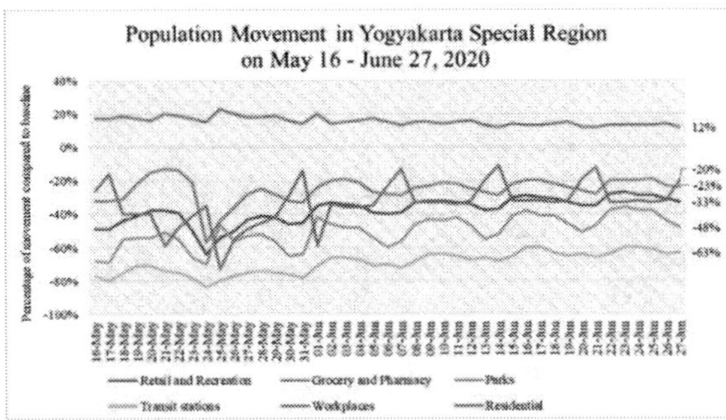

Fig. 2. Population movement in Yogyakarta Special Region to various places compared to the basic measurement. Source: Google COVID-19 Community Mobility Report.

Based on research problems, this study aims to identify the travel characteristics of the Sonosewu population during the COVID-19 pandemic in terms of frequency and purpose. This research can be used as a source of information regarding the frequency of trips made by residents for one week during the pandemic, and the purpose of these trips can be known as a matter of consideration for the government in determining policies during the COVID-19 pandemic. In addition, the results of this research will also be useful for the development of knowledge, especially in the field of urban geography.

2 Research method

The population in this study were Sonosewu residents with an age range of 18 to 56 years old. The consideration of this age group is related to the range of productive age. Residents aged 18 years old are assumed to be able to travel independently supported by possession of a driver's license, while 56 years old is the limit before retirement age and in this group, most of them constitute the workforce.

The sampling technique used probability sampling, that is, each research unit from the population has the same opportunity to be used as a sample [10]. The process of determining the sample used the simple random sampling method, where samples are taken from the population randomly without paying attention to the strata [10]. This is because all populations are homogeneous in terms of being affected by the COVID-19 pandemic and subject to policies and appeals related to restrictions on social activities, travel restrictions, and physical distancing. The determination of the number of samples was based on calculations in the Issac and Michael formula with an error rate of 5% [10]. Based on the calculation of 2,244 population members, the number of respondents was 328 people.

The overall data used in this study are primary data obtained by using a systematic and written questionnaire to the respondents. The use of a questionnaire is more efficient if the researcher knows the variables to be measured and is more suitable if the number of respondents is large enough [10]. This is in line with this study because the number of respondents is quite large. The data collection process was carried out by distributing questionnaires to respondents to obtain data on the travel characteristics for one week, namely from November 30 to December 6, 2020. Data processing and analysis were carried out using descriptive statistical techniques. Data analysis was carried out to describe the travel characteristics of the population in terms of frequency, destination, and distribution of destination locations during the COVID-19 pandemic.

3 Results and discussion

Coronavirus disease or COVID-19 is an infectious disease caused by the Severe Acute Respiratory Syndrome Coronavirus-2 (SARS-CoV-2) virus which first emerged in December 2019 in Wuhan, China [4]. Until now, COVID-19 has become a pandemic that has hit many countries around the world, including Indonesia. The first recorded case of COVID-19 in Indonesia was reported on March 2, 2020, with two positive cases [11]. The spread and transmission of COVID-19 are fast and widespread between humans, increasing the number of confirmed cases.

The COVID-19 pandemic has an impact on almost all sectors of life, including education, economy, and even has an impact on travel behavior. The object identified in this study was a home-based trip, where the home trip is the beginning of the activity and will eventually return home [2]. This study was represented by a sample of 328 respondents with details of 163 males and 165 females who were identified during one week during the COVID-19 pandemic from November 30 to December 6, 2020. It was recorded that there were 1,335 trips.

3.1 Travel frequency

Travel frequency is the number of trips in time units. The time used in this study was one week during the COVID-19 pandemic. This parameter is most highlighted because the government has appealed to limit activities outside the home or to reduce the number of trips in the context of efforts to accelerate the handling of COVID-19. Population mobility is correlated with the growth rate of COVID-19 cases, where a decrease in population travel has a significant positive relationship with reduce case growth of COVID-19 [12]. De Vos revealed that the existence of several policies during the COVID-19 pandemic allowed the frequency of travel for the population to be less because many had worked from home, and some outside activities decreased in frequency [13].

The travel frequency of the Sonosewu residents in one week is presented in Table 1. From November 30 to December 6, 2020, Sonosewu residents tended to travel with high frequency (43%), which is between 6 to 7 times a week. As many as 35% of the population traveled in low frequency, namely not traveling or 0 times to 2 trips a week, with details of 113 people.

Table 1. Travel frequency in Sonosewu.

Travel Frequency in a Week	Classification	Population
0 – 2	Low	113
3 – 5	Moderate	74
6 – 7	High	141

The population of Sonosewu mostly traveled on weekdays, and on weekends tended to be less (Table 2).

Table 2. The number of daily trips during COVID-19 pandemic.

Day and Date	Number of trips
Monday, 2020-11-30	205
Tuesday, 2020-12-01	209
Wednesday, 2020-12-02	190
Thursday, 2020-12-03	205
Friday, 2020-12-04	204
Saturday, 2020-12-05	176
Sunday, 2020-12-06	146

3.2 Travel purpose

The travel of the population was mostly carried out for work, as shown in Figure 3. Accumulatively, the largest proportion of residence-based travel destinations for people of Sonosewu was travel to work (63%), followed by destinations to shopping (19%), social interests (16%), school or campus (3%), and the lowest was for recreation (2%).

The findings in this study are different from Tamin's study, which stated that one of the main purposes of travel is to educational places [14]. The findings of this study prove that the purpose of traveling to those places was not the main activity. This was because educational activities were carried out online during the COVID-19 pandemic. Trips to education (i.e., to school or campus) made by the residents of Sonosewu amounted to 35 trips or 3% of the total trips made mostly for practical activities. These activities can be carried out by the results of adjustments to the Joint Decree of the 4 Ministers regarding the Guidelines for Implementation of Learning [15], where the implementation continues to apply the health protocol.

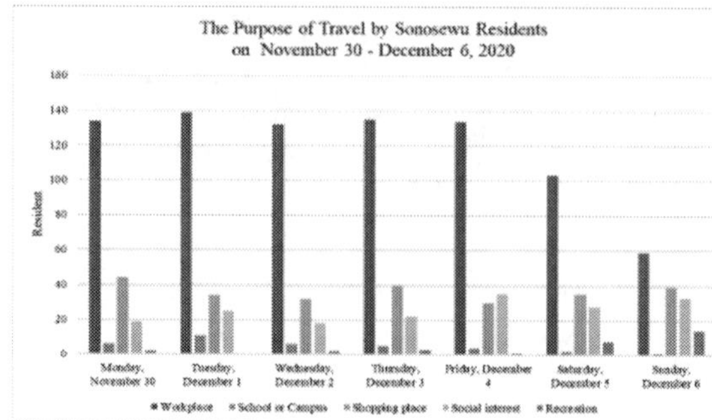

Fig. 3. Purpose of daily population travel.

The purpose of work travel was difficult to control in times of pandemic. The policies set by the government for work from home could not be applied entirely to all sectors, especially informal workers [16]. The characteristics of the residents of Sonosewu, where most of the population work as laborers, traders, and private employees, caused the application of 'work from home' not implemented. This condition occurred due to the pressure to fulfill basic needs during the pandemic.

3.3 Location of travel destination

The distribution of destinations for residents to travel varies widely and can be determined by several factors. Based on Figure 4, the destination locations for one week of travel during the COVID-19 pandemic to Bantul Regency and Yogyakarta City were the same, namely 520 trips each, then 255 trips to Sleman Regency, and the rest to other regions, such as Kulon Progo, Magelang, Solo, Kebumen, Lombok, Madiun, Denpasar, Gunungkidul, and Tasikmalaya [17].

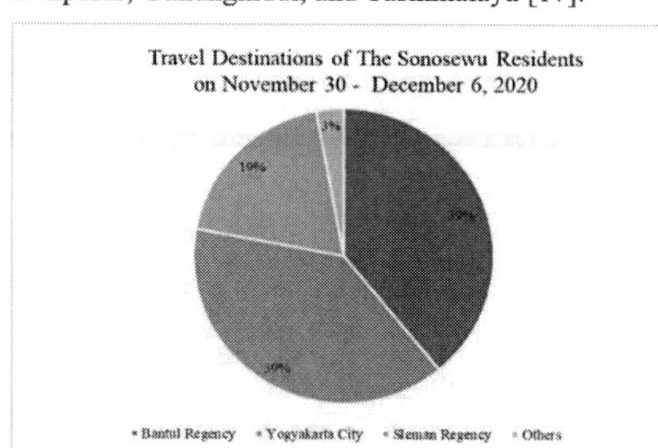

Fig. 4. Travel destinations during COVID-19 pandemic.

Table 3. Number of daily trips.

Destination Places	Location				
	Sonosewu	**Bantul Regency (other than Sonosewu)**	**Yogyakarta City**	**Sleman Regency**	**Others**
Workplace	135	167	309	207	18
School or campus	1	19	10	5	0
Shopping place	69	32	126	25	2
Social interest	54	31	65	16	14
Recreation	1	11	10	2	6
Total	**260**	**260**	**520**	**255**	**40**

Given the high number of trips, it can be said that Yogyakarta City and Bantul Regency have been able to meet the living needs of the people of Sonosewu. The high number of people traveling to Yogyakarta City is due to high accessibility. The population mobility will tend to increase along with the short distance [18]. However, in the aspect of potential characteristics between Bantul Regency and Yogyakarta City, it is certainly different, including the availability of socio-economic facilities that encourage travel.

Based on Table 3, the purpose of trips to work and shopping to Yogyakarta City was higher than in Sonosewu itself or in other hamlets in Bantul Regency. The city of Yogyakarta is known as a center of economic activity, trade and services, and government services so that it provides high job opportunities. Also, one of the destinations for residents to travel to shopping in Yogyakarta City is the Pasar Legi. Christaller in the central place theory argues that residents will look for the closest, inexpensive and accessible place to fulfill their needs to meet their needs [19]. This is following the characteristics of Pasar Legi, which is relatively close to Sonosewu and has a cheaper price. On the other hand, the purpose of the trip to the Bantul Regency is higher than Yogyakarta City, namely for recreational purposes and social interests. There are many recreational purposes in Bantul Regency apart from Sonosewu Hamlet, including for tourism purposes on the South Coast.

The aspect of population travel is one of the limited aspects during the COVID-19 pandemic. Overall, it is known that the residents of Sonosewu during the COVID-19 pandemic had a relatively high frequency of travel. This phenomenon does not mean that people are ignoring instructions from the government to stay at home. Despite the high frequency, people traveled for important purposes, such as workplaces and shopping places. Aspects of social activities that have also become the government's spotlight during the COVID-19 pandemic can be described through travel with the aim of social interests. Based on the results, it is known that there were quite a lot of trips for social purposes but the distance to the destination is relatively close, namely in Sonosewu Hamlet and other areas in Bantul Regency.

4 Conclusion

During the COVID-19 pandemic, the people of Sonosewu tended to have a high frequency of travel, namely 6 to 7 times a week from November 30 to December 6, 2020. Most of the travel destinations during the COVID-19 pandemic were to go to work (63%) and shopping places (19%). The location of the destination of the Sonosewu residents' travel to Yogyakarta City was as many as those made to Bantul Regency, which is 520 times.

This paper is part of an undergraduate thesis written by the first author under the guidance of the second author. The author would like to thank the respondents in Sonosewu, Bantul Regency who were willing to fill out the questionnaire in this study.

References

1. Muhlisin, Dinamika Permukiman **2**, 73–82 (2009)
2. J. de D. Ortuzar, L.G. Willumsen, Modelling Transport (4th Edition) (John Wiley & Sons, United Kingdom, 2011)
3. A. Nugroho, Hubungan antara karakteristik sosio ekonomi demografi dan karakteristik perjalanan dengan pemilihan moda untuk maksud bekerja (Studi kasus: Kelurahan Cijaura, Kelurahan Margasari dan Kelurahan Manjahlega), Thesis (Universitas Komputer Indonesia, Bandung, 2019)
4. WHO, Pertanyaan dan jawaban terkait Coronavirus, URL: https://www.who.int/indonesia/news/novel-coronavirus/qa/qa-for-public. (2021)
5. H. Romdiati, COVID-19 Mengubah mobilitas penduduk, URL: http://kependudukan.lipi.go.id/en/berita/53-mencatatcovid19/893-covid-19-mengubah-pola-mobilitas-penduduk. (2020)
6. Dinas Kesehatan Kabupaten Bantul, Zonasi risiko COVID-19 menurut kecamatan berdasarkan data

https://doi.org/10.1051/e3sconf/202132506013

dari 1-14 September 2020 ke Kabupaten Bantul, URL: https://corona.bantulkab.go.id/. (2020)

7. Pemerintah Kabupaten Bantul, Implementasi sektor pada zonasi risiko COVID-19, URL: https://corona.bantulkab.go.id/. (2020)
8. Google, COVID-19 community mobillity reports, URL: https://www.google.com/covid19/mobility/. (2020)
9. OCHA, Average daily reduction in people's movement range from 1 Mar 2020 - 5 July 2020, URL: https://data.humdata.org/dataset/movement-range-maps. (2020)
10. Sugiyono, Metode penelitian pendidikan (pendekatan kuantitatif, kualitatif, dan R&D) (Penerbit Alfabeta, Bandung, 2017)
11. R. Nuraini, Kasus Covid-19 pertama, masyarakat jangan panik, URL: https://indonesia.go.id/narasi/indonesia-dalam-angka/ekonomi/kasus-covid-19-pertama-masyarakat-jangan-panik. (2020)
12. H. S. Badr, H. Du, M. Marshall, E. Dong, M. M. Squire, L. M. Gardner, Lancet Infect. Dis. **20**, 1247–1254 (2020)
13. J. De Vos, Transportation Research Interdisciplinary Perspectives **5**, 1–3 (2020)
14. O.Z. Tamin, Perencanaan dan Pemodelan Transportasi (2th ed.) (Penerbit ITB, Bandung, 2000)
15. Kementerian Pendidikan dan Kebudayaan Republik Indonesia, Penyesuaian keputusan bersama empat menteri tentang panduan pembelajaran di masa pandemi COVID-19, URL: https://www.kemdikbud.go.id/main/blog/2020/08/penyesuaian-keputusan-bersama-empat-menteri-tentang-panduan-pembelajaran-di-masa-pandemi-covid19. (2020)
16. A. J. Pitoyo, B. Aditya, I. Amri, E3S Web of Conferences **200**, 03014 (2020)
17. D. I. Santi, Hubungan persepsi penduduk tentang bahaya pandemi COVID-19 dengan frekuensi perjalanan penduduk Dusun Sonosewu selama satu minggu, Thesis draft (unpublished)
18. I. B. Mantra, Mobilitas penduduk sirkuler dari desa ke kota di Indonesia (Pusat Penelitian Kependudukan Universitas Gadjah Mada, Yogyakarta, 1999)
19. R. A. Medina, *Pola spasial pemilihan lokasi belanja kebutuhan primer penduduk Kecamatan Bogor Tengah*, in 8th Industrial Research Workshop and National Seminar, 527–535 (2017)

COVID-19 and Gender Role in Unemployment: How Women in Special Capital Region of Jakarta is Less Likely to be Unemployed During Economic Disturbance

Jamal Habibur Rahman[1], *Faradisa Hilya* Mumtaza[1], *Rahmatullah Tri Nanda* Putra[1], *Fachrurizal Sai* Kintoro[1], *Rezis* Rahayuli[1], *Anisa Octa* Nur'aini[1], *Nurul* Ramadhani[1], *Mya* Dwi Rostika[2], Abraham *Cardenas* Tristan[3] and *Ratih Fitria* Putri[1*]

[1]Department of Environmental Geography, Faculty of Geography, Universitas Gadjah Mada, Indonesia
[2]Department of International Relations, Faculty of International Relations, Daita Bunka University, Japan
[3]Geomatics Engineering, Faculty of Engineering, Autonomous University of San Luis Potosí, México

Abstract. The gender gap that occurs in Special Capital Region of Jakarta, between men and women has an impact on aspects of employment related to labor absorption. Increasing the number of optimal employment opportunities, both in terms of productivity and efficiency, can determine the success of economic development in a region. This study was conducted to see how likely it is that women are unemployed, in relation to the employment aspect in Special Capital Region of Jakarta which is based on secondary data from BPS and analyzed with a descriptive quantitative approach. The results showed that the open unemployment rate (TPT) in Special Capital Region of Jakarta in 2019-2020 experienced a significant increase in 2020, namely 4.86% for men and 4.5% for women. Different results are shown in the variable labor force participation rate (TPAK), which tends to experience a fluctuating trend with the participation rate of men being higher than women. In general, the COVID-19 pandemic has an impact on employment, which in turn has an impact on unemployment. During the COVID-19 pandemic, the unemployment rate was dominated by men, due to the high wages of men compared to women. In addition, women's labor force participation rates are relatively low compared to men's, causing women to have little chance of being unemployed. Female workers have higher resilience than male workers, this is due to the higher saturation of women in informal economic activities with entrepreneurship. Entrepreneurial activities tend to have more resilience than formal workers because there is no risk of termination of employment from the employer.

1 Introduction

The global crisis due to the COVID-19 pandemic has caused countries in the world to take policies to limit people's activities. The restrictions were made as an effort to contain the wider spread of COVID-19. This condition has an impact on various sectors. In addition to the health sector, the economic sector is one of the sectors most seriously affected by COVID-19. Based on a scenario [1], Gross Domestic Product (GDP) in 2020 in several countries experienced a decline. The COVID-19 disruption to the global economy has had an impact on the employment situation, such as reduced income, reduced working hours, work from home regulations and increased unemployment.

Several studies that have raised the issue of gender equality in the impact of COVID-19 on employment have shown that women are the most affected group compared to men [2]. Gender gap research on income, expenditure, savings, and job loss by [3] in 6 countries consisting of China, Japan, South Korea, Italy, United Kingdom, and United States because of COVID-19 shows that women are 24% more likely than men to lose their jobs, and 50% more than men to experience a decline in income. In Germany, Singapore, and the United States, transitions to unemployment, reduced working hours, and transitions to working from home are more common among women than men [4].

In Indonesia, COVID-19 was first reported on March 3, 2020. Since then, COVID-19 cases have spread to 34 provinces in Indonesia. Economic activities were significantly affected by the policies taken by the government in tackling COVID-19. The policy of large-scale restrictions during the COVID-19 pandemic caused severe economic losses [5]. Special Capital Region of Jakarta as the economic center in Indonesia has a high number of workers. The number of workers in Special Capital Region of Jakarta in 2019 based on the National Labor Force Survey (commonly known in Indonesia as SAKERNAS) in August 2019 reached

*Corresponding author: ratihfitria.putri@ugm.ac.id

 https://doi.org/10.1051/e3sconf/202132506014

4,836,980 people with a higher male Employment to Population Ratio (EPR) of 76.72 percent compared to 45.01 percent for women [6]. However, with the COVID-19 pandemic, it is indicated that employment stability in DKI Jakarta in 2020 will be disrupted.

Disruption of employment stability in Special Capital Region of Jakarta due to COVID-19 has an impact on employment. The supply of labor will have a greater amount than the amount of labor needed. Employment from a gender perspective is often a consideration. This study aims to see how likely it is for women to be unemployed, related to the employment aspect. It is analyzed with a descriptive quantitative approach based on the variables of the Labor Force Participation Rate (commonly known in Indonesia as TPAK or Tingkat Partisipasi Angkatan Kerja) and the Open Unemployment Rate (commonly known in Indonesia as TPT or Tingkat Pengangguran Terbuka) in 2019 - 2020 in Special Capital Region of Jakarta.

2 Methodology

2.1 Research Location

This research is located in Special Capital Region of Jakarta with coordinates between 06° 08' latitude and 106° 48' east longitude and the area is 661.52 km2. Special Capital Region of Jakarta is bordered by Banten Province in the west, east and south by West Java Province, and borders the Java Sea in the north. Administratively, Special Capital Region of Jakarta is divided into 6 cities/regencies, namely the Thousand Islands, Central Jakarta, East Jakarta, West Jakarta, North Jakarta, and South Jakarta (Figure 1). Special Capital Region of Jakarta is crossed by 13 rivers which empties into Jakarta Bay with the Ciliwung River being the main river that divides the city into two parts.

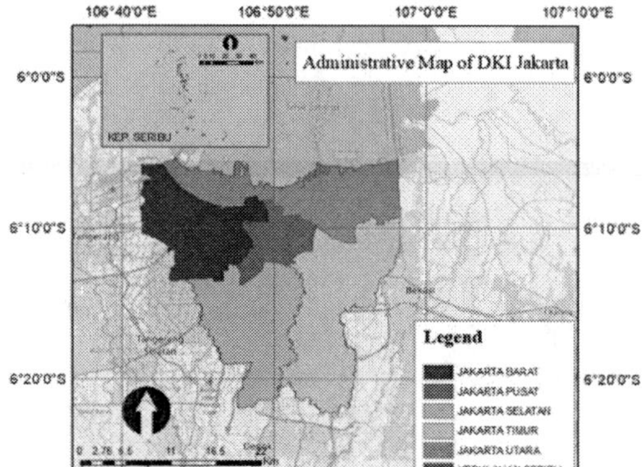

Fig. 1. Administrative of Special Capital Region of Jakarta.

Jakarta city is an area that is prone to land subsidence as a result of the excessive use of groundwater in the area [7]. One of the factors causing land subsidence is the high population density in Jakarta city. Population growth and increasing economic activity over time require an increase in the need for space. The highest population density in Jakarta is reached 15.964 people/km^2. Economic activity is one of the pull factors for the urbanization phenomenon in the region. Population growth and high activity development in Jakarta city can not be followed by the speed of development of infrastructure and other urban facilities. The need for residential land in Jakarta is increasing in line with the increasing population density. The density of settlements is a phenomenon that can be seen as one of the impacts of the high population density in Jakarta city.

The Special Capital Region of Jakarta, which is the center of the Indonesian economy, has a major contribution to national economic growth. Regional economic growth is closely related to the number of workers that can be absorbed by business fields, both in the formal and informal sectors. The COVID-19 pandemic has caused a decrease in job opportunities in Special Capital Region of Jakarta. The decrease in the number of people working is influenced by an increase in the number of workers in the informal sector while the number of workers in the formal sector has decreased. Nevertheless, the formal sector still dominates the absorption of labor in Special Capital Region of Jakarta.

2.2 Data and Resource

This study uses secondary data in the form of data on the number of the workforce, the population aged 15-64 years, the number of working people, and the minimum wage in Special Capital Region of Jakarta in 2019-2020. The data is sourced from the Central Statistics Agency (commonly known in Indonesia as BPS or Badan Pusat Statistik) Special Capital Region of Jakarta. Data collection techniques were carried out through literature studies from several publications, scientific articles, and journals related to labor conditions in Special Capital Region of Jakarta. The results of the employment data processing are presented in the form of tables, graphs, and maps related to the spatial distribution pattern.

2.3 Method

Calculations related to employment aspects in Special Capital Region of Jakarta in 2011-2019 are based on several variables that can describe employment conditions in Special Capital Region of Jakarta, such as the Labor Force Participation Rate (TPAK) and Open Unemployment Rate (TPT). TPAK is the ratio between the labor force and the total working age population. The TPAK variable is used to determine the percentage of the economically active working age population in Special Capital Region of Jakarta.

TPT is the ratio between the number of unemployed and the number of the labor force. The TPT variable is used to determine the percentage of the workforce that is included in unemployment. The results of the calculation of the four variables were then analyzed using quantitative-descriptive analysis methods. This method is carried out by describing data related to employment in Special Capital Region of Jakarta which has previously been calculated using several equations.

3 Result and Discussions

An important aspect in human life is employment where employment does not only affect the economic side, but also involves the social side. Viewed from the economic side, employment emphasizes human needs and work as an effort to fulfill the needs of daily life, while the social side emphasizes community recognition of individual abilities. The increasing number of population due to birth, death and migration factors can affect the working age population, labor force and labor market conditions in an area. The increase in the number of working age population and the labor force must be in line with the increase in the labor market and job opportunities so that it is expected to reduce unemployment and poverty levels.

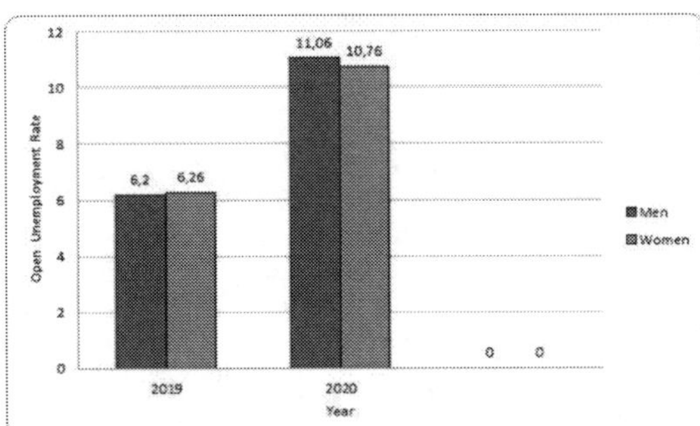

Fig. 2. Special Capital Region of Jakarta Province Open Unemployment Rate 2019 - 2020

This Open unemployment is a workforce who really does not have a job. Open unemployment can also be defined as working age population who are not working, are looking for work, preparing for a business, are looking for work and already have a job but have not started working. Unemployment can occur due to the high growth of the workforce that is not matched by the existence of jobs so that employment tends to be small. The open unemployment rate is the percentage of the number of unemployed by the total labor force.

The absorption of both male and female workers in the Special Capital Region of Jakarta Province continues to experience a fluctuating trend. The open unemployment rate (TPT) in 2020 for both men and women experienced a sharp increase. In 2020, male unemployment reached 11.06% and female unemployment reached 10.76%. Both have increased compared to the previous year, where in 2019 the open unemployment rate (TPT) for men reached 6.2% and women reached 6.26%. Based on this description, there has been an increase in the open unemployment rate (TPT) respectively by 4.86% for men and 4.5% for women. This illustrates that the open unemployment rate (TPT) for men has increased faster than women. The increase in open unemployment is a result of the COVID-19 pandemic. So, it can be said that in Special Capital Region of Jakarta Province, men are classified as more vulnerable to impacts, especially related to employment during the COVID-19 pandemic compared to women. The implementation of *Pembatasan Sosial Berskala Besar* or PSBB (literally Large-Scale Social Restrictions) in various regions has caused economic activity to be paralyzed. In addition, the COVID-19 pandemic has also caused many employees/workers to be laid off in the formal and informal sectors. Layoffs in the informal sector reached 56.7% of the total workforce.

The vulnerability of men to the open unemployment rate during the pandemic can be caused by the higher wages of men compared to women. The average wage before the pandemic was recorded that in February 2020 the average wage for Special Capital Region of Jakarta was Rp. 4,560,084 and decreased in August 2020 to IDR 4,224.72. The average wage has decreased because the percentage of people working full-time has decreased due to COVID-19. The average wage for both men and women is above the national average. The average wage for women, which is below the average for men, shows that men's wages are higher than women's, making them more vulnerable to open unemployment during the COVID-19 pandemic. This is in line with [8] who revealed that the minimum wage imposed in districts/cities has a significant positive effect on the unemployment rate.

A significant increase in the number of open unemployment in Special Capital Region of Jakarta occurred in August 2020 due to the COVID-19 pandemic. Many workers have stopped working from the company due to the impact of the COVID-19 pandemic or the implementation of large-scale social restrictions (PSBB). The COVID-19 pandemic has greatly impacted employment in several business fields, where business fields contribute 11.01% of employment in Special Capital Region of Jakarta [9].

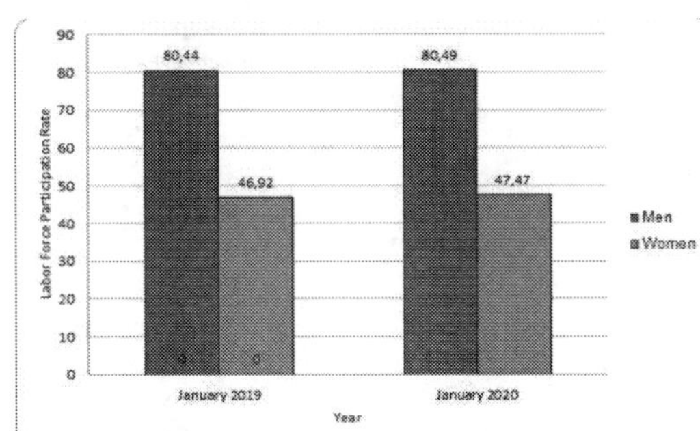

Fig. 3. Special Capital Region of Jakarta Province Labor Force Participation Rate 2019 - 2020

The labor force participation rate (TPAK) for both men and women also experienced a fluctuating trend. Based on the figure, the labor force participation rate (TPAK) of men is always higher than that of women. This can be seen in August 2019, where male TPAK reached 80.44% and female TPAK reached 46.9%. In addition, it can also be seen in August 2020 where the TPAK for both men and women experienced a not too significant increase. Male TPAK in August 2020 reached 80.49% while female TPAK reached 47.47%. Both experienced an increase compared to August 2019 of 0.05% for men and 0.55% for women.

The conditions in Jakarta which show that female workers are not more affected by the pandemic than male workers are supported by the findings of [10]. According to their findings, the COVID-19 pandemic has a negative impact on male workers and young workers, which is due to the tendency of employers to terminate employment relationships with young male workers and not yet married, assuming the low socio-economic side effects that may arise. In addition, the fact that women are more likely to work in the informal sector can also be taken into consideration, in which the sector is more resilient to economic disturbance than the formal sector. This is in accordance with the research of [11] in India, which showed that female workers in the informal sector were not negatively affected by the pandemic compared to male workers. Interestingly, they also found the fact that female workers in the formal sector were greatly affected by the pandemic.

Contrary to these findings, [12] found that on average in some Asia Pacific countries such as Bangladesh, Maldives, Nepal, and Thailand, there was a tendency that female workers were more likely to experience a negative impact on COVID-19 than male workers, both from the formal and informal sectors. This condition occurs due to the lockdown regulations and the appeal to stay at home set by the local government which is enough to force informal workers not to work amid uncertainty in living income, so it is feared that it will have a bad effect on mental health. The difference in conditions found in previous studies is supported by the statement of [13], which is based on the 2008 economic crisis. The statement [13] is that during the economic crisis, male workers in industrialized countries experienced a worse impact than women in terms of job losses. As for developing countries, the impact will depend on which economic sector is more affected. This shows that the hypothesis made in Jakarta does not necessarily apply in all regions, even though these regions have more or less the same criteria for employment conditions, so further research needs to be done.

4 Conclusion

Jakarta is facing a rather unique employment issue during the COVID-19 pandemic, especially in gender-related unemployment among the workers. Women in Jakarta are experiencing less severe unemployment than men, an unseen phenomenon in other countries and regions. Most prominently, this distinctive inclination is pushed by higher resilience among female workers, due to higher saturation of women in informal economic activities with self-employment, such as small hawkers and petty service providers. This particular type of activity tends to have more resilience than formal workers due to no risk of termination from the employer. Yet, this explanation is still unfit to most cases with the similar informal dominance. Women in Bangladesh, Maldives, Nepal, and Thailand are still exposed to higher chances of being unemployed during an economic crisis, while being a state with a notoriously informal-saturated economy. Therefore, an acknowledgment is presented that this specific issue in Spesial Japital Region of Jakarta still needs further research and explanation.

References

1. W.J. McKibbin & R. Fernando. The Global Macroeconomic Impacts of COVID-19: Seven Scenarios. SSRN Electronic Journal **1** (19) (2020)
2. N. Kabeer, S. Razavi, & Y.V.D. Rodgers. Feminist Economic Perspectives on the COVID-19 Pandemic. Feminist Economics **27** (1–2) 1–29 (2021)
3. H.A.H. Dang, & C.V. Nguyen. Gender inequality during the COVID-19 pandemic: Income, expenditure, savings, and job loss. World Development **140** (1) (2021)
4. M. Reichelt, K. Makovi, & A. Sargsyan. The impact of COVID-19 on gender inequality in the labor market and gender-role attitudes. European Societies **23** (1) 228–245 (2021)
5. R.E. Caraka, Y. Lee, R. Kurniawan, R. Herliansyah, P.A. Kaban, B.I. Nasution, P.U. Gio, R.C. Chen, T. Toharudin, & B. Pardamean. Impact of COVID-19 large scale restriction on environment and economy in Indonesia. Global Journal of Environmental Science and Management **6** (1) 65–84 (2020)
6. Special Capital Region of Jakarta Central Statistics Agency, Employment Executive Summary of Special Capital Region of Jakarta Province August 2019, Special Capital Region of Jakarta : Central Statistics Agency (2020)
7. R F Putri, S Wibirama, Sukamdi and S R Giyarsih. Population condition analysis of Jakarta land deformation area. IOP Conf. Series: Earth and Environmental Science, **148**, 012007 (2018)
8. R. Pitartono, & H. Banatul. Analysis of the Unemployment Rate in Central Java 1997 - 2010. Dipenegoro Journal of Economics **1** (1) 1-10 (2012)
9. BPS, Impact of COVID-19, Special Capital Region of Jakarta Unemployment Reaches Double Digits, *Official News Statistics*, BPS Special Capital Region of Jakarta Province (2020)
10. Z. Yu, Y. Xiao, Y. Li, Emerg. Mark. Financ. Trade, **56**, 2390 (2020)
11. S. Desai, N. Deshmukh, S. Pramanik, Fem. Com. **27**, 152 (2021)
12. P.A. Seck, J.O. Encarnacion, C. Tinonin, S. Duerto-Valero, Fem. Com. **27**, 123 (2021)
13. United Nations Women, The Global Economic Crisis and Gender Equality, New York: UN Women (2014)

 https://doi.org/10.1051/e3sconf/202132507001

Land-Use Change Analysis of National Southern Cross Road Route Development in Bantul Regency

Utia Kafafa[1*], *Rika* Harini[1]

[1]Department of Environmental Geography, Faculty of Geography, Universitas Gadjah Mada, Indonesia

Abstract. The National Southern Cross Road Route well known as Jalan Jalur Lintas Selatan (JJLS) in Indonesia. The segment which crosses Bantul Regency has 16.65 km long. The development of JJLS will have various kinds of impacts on the community around the road infrastructure. This study wants to know how the impact of the development that saw from the spatial dynamics of land-use change. The study area is located in Poncosari, Gadingsari, Srigading, Tirtohargo, and Parangtritis Village. We process the data using the map overlay technique then analyze it with quantitative descriptive. The results of this study are the development of JJLS in Bantul Regency has not significantly affected the land-use changes. It is evident from the location of the land-use change which is far from the road and tends to be disheveled. The Land-use changes that occur do not form a specific pattern such as the pattern of conversion of agricultural land into non-agricultural land. The less significant changes in land use are caused by the condition of the JJLS which is not yet fully connected. Moreover, the status of land ownership around JJLS is Sultan Ground so a special permit is needed if the community wants to use the land.

1 Introduction

Infrastructure development aims to realize balanced inter-regional and equitable distribution of development results [1]. The National Southern Cross Road Route well known as Jalan Jalur Lintas Selatan (JJLS) in Indonesia. It is a national road construction program that connects five provinces, namely East Java, DIY, Central Java, West Java, and Banten. The Southern Cross Road (JJLS) The development of JJLS passes through the coastal areas of the Daerah Istimewa Yogyakarta (DIY), namely in the Kulonprogo, Bantul, and Gunungkidul Regency.

The development of JJLS is in line with the DIY government's agenda to develop coastal areas with equal distribution of accessibility by the paradigm of *Among Tani Dagang Layar* [2]. This new paradigm will improve people's welfare with development from land to maritime [3]. Jalan Jalur Lintas Selatan (JJLS) crosses Bantul Regency for 16.65 km. This area is strategic because it connects Kulonprogo and Gunungkidul Regency. The area around JJLS is designated for cultivation areas such as trade and services, food crop agriculture, and settlements.

The JJLS development process will certainly have various kinds of impacts on the lives of the people around the road infrastructure. Infrastructure such as roads can play a positive role in development, especially economic growth and can create new jobs for the surrounding community [4]. Road infrastructure plays an important role by providing mobility for the efficient movement of people, goods, and services [5].

In addition to the various positive impacts that can be felt, negative impacts will also appear as side effects of development [6]. Road construction can provide considerable economic and social benefits. However, if poorly planned or executed, such projects can lead to various problems such as serious cost overruns, corruption, environmental damage, and social and political conflicts [7].

The negative impacts of infrastructure development include reduced productive agricultural land, reduction in green open land, and damage to the environment around the construction of road infrastructure [8]. The construction of roads in the research location has displaced several productive agricultural areas. Changes in land use will trigger socio-economic changes in the community along with changes in adaptation patterns in meeting the needs of life.

This research wants to know how the impact of this development is seen from the spatial dynamics of land-use change. The spatial approach in this study is expected to be able to identify and analyze areas around development that are experiencing rapid growth or vice versa. The practical benefits of this research are expected to provide a spatial description of the impact of JJLS development in Bantul Regency and can be used as evaluation material to formulate development policies for the government.

2 Methods

* Corresponding author: utiakafafa@mail.ugm.ac.id

https://doi.org/10.1051/e3sconf/202132507001

2.1 Location

This research is located along with the JJLS development in Bantul Regency. The locations include Poncosari Village which is part of the administrative area of Srandakan District, Gadingsari and Srigading Village which is part of the Sanden District administrative area, and Tirtohargo and Parangtritis Village which are part of the Kretek District administrative area. The reason for choosing the location was because the area experienced a direct impact from the JJLS development. Fig. 1. shows the location of the research.

Fig. 1. Map of Research Location.

2.2 Data

Satellite images are recorded images of the earth's surface obtained from satellite media [9]. Interpretation of satellite images is an activity of analyzing and identifying objects recorded in the image. This study uses the interpretation of Google Earth imagery in 2013 and 2020 to determine the spatial dynamics of land-use changes that occur around the JJLS development. The year was chosen to represent the year before and after road construction.

2.3 Data Processing

Processing of land-use change is carried out using the ArcMap software application. Land-use change is identified through an overlay of the initial year's land use map (2013) and the final year's land use map (2020). Map overlay process can be seen in Fig.2. The next step is calculating the percentage of land-use change area. The processing result data then presented with a table of changes in the land-use area and a map of changes in land use around JJLS. Field observations were also carried out to validate the land use map.

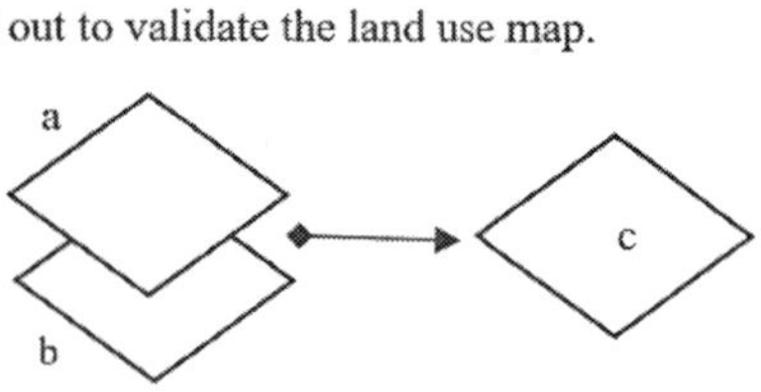

Fig. 2. (a) Land-use map in 2013, (b) land-use map in 2020, (c) Map of land-use change.

2.4 Data Analysis

Overall, this research uses a descriptive analysis. Descriptive analysis is an analysis that describes the general state of the data. This analysis includes several things such as frequency distribution, measurement of central tendency, and measurement of variability [10]. The analysis process in this research is divided into map overlay analysis and quantitative descriptive analysis. The map overlay analysis describes changes in the types and percentage of land use that occurred in 2013 and 2020 for each area passed by JJLS.

3 Result and Discussions

3.1 Land-use change in Poncosari Village

Based on the calculation results, the land-use in Poncosari Village has an area of 1286 Ha. From Table 1., we know that several types of land use have increased and some have decreased during 2013-2020. Mixed garden area shrank by 2% and open land area shrank by 7%. Agricultural land has increased by 2%, settlements also increased by 2%, and bodies of water increased by 4%.

Table 1. Percentage of the land-use area in Poncosari Village.

Land-Use	2013		2020	
	Ha	%	Ha	%
Forest	21.53	2	21.45	2
Mixed Garden	225.4	18	202.6	16
Agricultural Land	561.6	44	597.4	46
Built Up Area	3.77	0.3	3.77	0.3
Open Land	142.7	11	56.78	4
Open Land (Beach)	13.37	1	17.12	1
Settlements	189.1	15	213.1	17
Livestock	5.5	0.4	5.5	0.4
Shrubs	29.41	2	29.41	2
Pond	26.82	2	24.59	2
Water Body	66.78	5	114.2	9
Grand Total	1286	100	1286	100

The distribution of land-use change can be observed in Fig. 3. The points where land-use changes occur are quite spread out throughout the village and not only around JJLS. It indicates that the development of the JJLS does not significantly affect changes in land use around the road foundation. It also can be observed that changes occur in the vicinity of the road, namely changes in open land into settlements, open land into ponds, ponds into open land, and forests into open land.

Land-use changes were mostly found in the west or along the Progo River. It is because the river morphology has dynamic properties according to time and space. This process of change has occurred since the river was formed and still ongoing. The driving factors include erosion and deposition, land use, run-off, the presence of tides (backwater), the type of river cliff-forming material, and transportation [11].

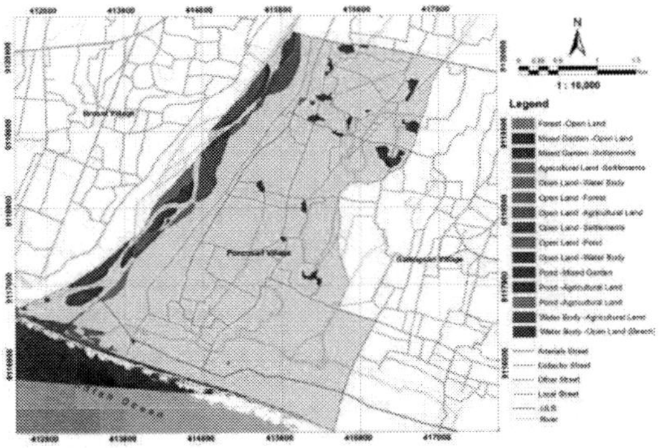

Fig. 3. Map of land-use change in Poncosari Village 2013-2020.

3.2 Land-use change in Gadingsari Village

In general, land-use in Gadingsari Village is classified into 10 types. It consists of forest, mixed gardens, agricultural land, built-up land, open land, open land (beach), settlements, livestock, shrubs, and ponds. The total area is 779.64 Ha. Table 2. shows that the area of mixed garden land use has shrunk by 6%. Settlements increased by 2%, while the area for agricultural land was fixed during 2013-2020.

Table 2. Percentage of the land-use area in Gadingsari Village.

Land-Use	2013		2020	
	Ha	%	Ha	%
Forest	37.35	5	41.57	5
Mixed Garden	200.09	26	159.34	20
Agricultural Land	312.14	40	310.69	40
Built Up Area	0	0	1.43	0.2
Open Land	37.58	5	30.9	4
Open Land (Beach)	6.43	1	6.43	1
Settlements	152.98	20	183.7	24
Livestock	3.95	1	2.55	0,3
Shrubs	24.41	3	38.09	5
Pond	4.71	1	4.94	1
Grand Total	779.64	100	779.64	100

The distribution of land-use change in Gadingsari Village can be observed in Fig. 4. Land-use changes are widespread and vary at several points. There is a pattern where the change from mixed garden to settlement occurs mostly in the western part of Gadingsari Village. This is because it is close to the city center so the demand for land is getting higher. These changes may be driven by the increasing population so that the higher demand for land, especially for residential land. Based on the results of data processing, it proves that the existence of JJLS development has a slight impact on land-use change but the change has less effect on the surrounding environment. This is because the changes that occur are not changes from land to settlements/ built-in land which can become the center of growth.

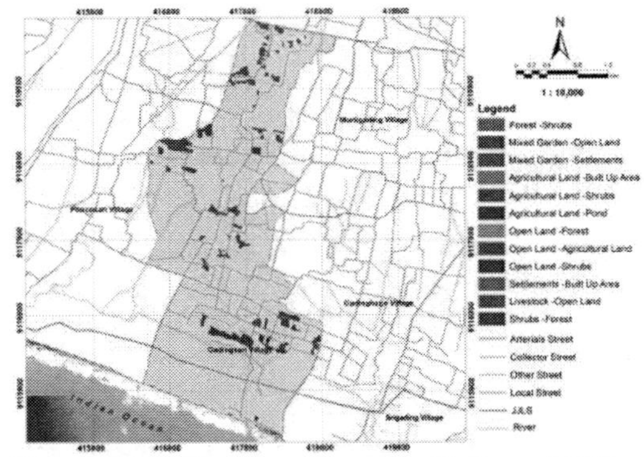

Fig. 4. Map of land-use change in Gadingsari Village 2013-2020.

3.3 Land-use change in Srigading Village

The land use in Srigading Village consists of mixed garden land use, agricultural land, built-up land, open land, open land (beach), settlements, shrubs, ponds, and bodies of water. Land uses that experienced an increase in percentage area from 2013 to 2020 were open land (beach) which increased by 1% and settlements by 2%. Otherwise, the land use that experienced a shrinkage percentage was mixed gardens by 1% and open land by 1%. Table 3. shows the land-use change dynamically during 2013-2020.

Table 3. Percentage of the land-use area in Srigading Village

Land-Use	2013		2020	
	Ha	%	Ha	%
Mixed Garden	84.72	11	77.38	10
Agricultural Land	452.48	58	449.02	58
Open Land	15.69	2	11.01	1
Open Land (Beach)	19.36	2	20.84	3
Settlements	179.01	23	191.86	25
Livestock	4.4	1	4.4	1
Shrubs	3.24	0.4	3.24	0.4
Pond	0.41	0.1	3.04	0.4
Water Body	18.85	2	17.37	2
Grand Total	778.16	100	778.16	100

The location of land-use change that occurred in Srigading Village can be observed in Fig. 5. Changes from mixed gardens to settlements occurred in the northern part. The northern part is getting closer to the center of government and economic activity. It makes the community easier to access public facilities such as government offices, markets, and hospitals so that the conversion of agricultural land to non-agricultural land is prevalent in the northern part.

As for the entire JJLS itself, there is no significant change in land use. It is caused by the demand for land around JJLS in Bantul Regency is not too high due to roads that are not fully connected. Furthermore, the location around JJLS is quite far from the center of government activities and the village economy.

Another factor that causes the insignificant change in land use is related to the status of land ownership. The status of land ownership around JJLS is divided into the status of the land with ownership rights and the status of the sultan's ground. Sultan Ground island owned by the Ngayogyakarta Hadiningrat Sultanate and its management must obtain official permission from that party. Srigading Village is also one of the villages where shrimp pond relocation destinations. The shrimp ponds are scattered around the Samas Lagoon.

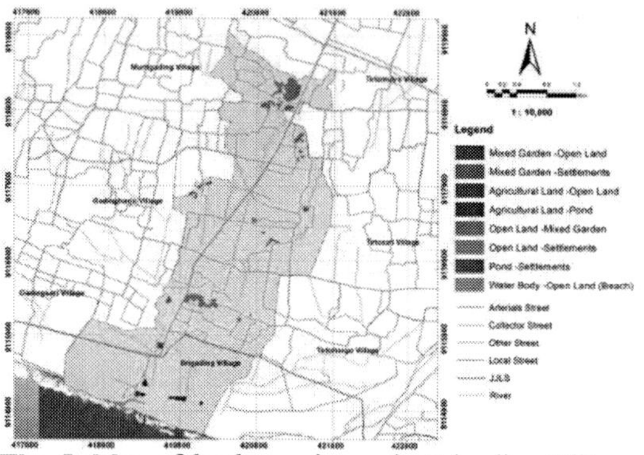

Fig. 5. Map of land-use change in Srigading Village 2013-2020.

3.4 Land-use change in Tirtohargo Village

Tirtohargo Village has 388.56 Ha total area of land use. Land use experienced a percentage increase from 2013 to 2020, namely agricultural land which increased by 2%. Otherwise, land use experienced a percentage decrease, namely open land and water bodies each by 1%. Table 4. shows the land-use change dynamically during 2013-2020.

Table 4. Percentage of the land-use area in Tirtohargo Village

Land-Use	2013		2020	
	Ha	%	Ha	%
Mixed Garden	39.52	10	37.33	10
Agricultural Land	218.83	56	224.28	58
Open Land	17.94	5	16.11	4
Open Land (Gisik)	5.78	1	5.78	1
Settlements	56.66	15	58.85	15
Livestock	1.67	0	1.67	0
Water Body	48.16	12	44.54	11
Grand Total	388.56	100	388.56	100

The distribution of land-use change locations in Tirtohargo Village can be observed in Fig. 6. The points where land-use change occurs are in the upper eastern and western parts and not only around JJLS. Changes in land use that occur in Tirtohargo Village include mixed gardens into settlements, open land into agricultural land, bodies of water into agricultural land, bodies of water into open land. Land-use change from mixed gardens occurs along the local road in the west.

The conversion of open land to agricultural land occurred in the southern part near the Depok Lagoon. It can be observed that the changes that occur around the JJLS body include water bodies becoming open land. This change of water bodies into open land is likely due to the construction of the Kretek II Bridge which will connect JJLS in the Samas-Sanden area with Depok-Parangtritis. The plan is that this bridge will be built along 800 meters to facilitate access for residents and is a national strategic project for the development of JJLS. The Kretek II Bridge will later become the entrance to the DIY region from the southern part.

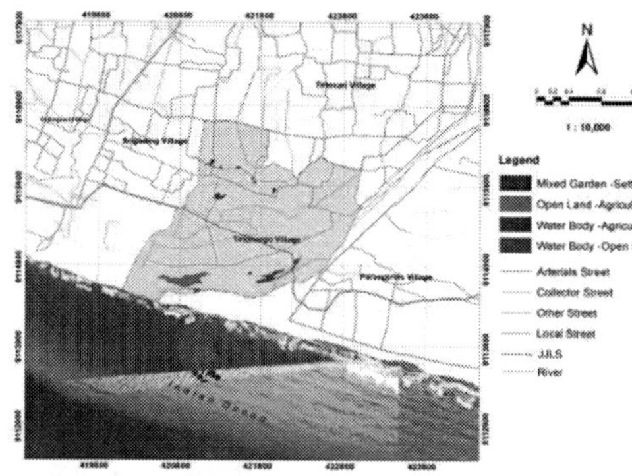

Fig. 6. Map of land-use change in Tirtohargo Village 2013-2020.

3.5 Land-use change in Parangtritis Village

Parangtritis Village has extensive land uses in the form of the sand dune. The process of forming a sand dune is influenced by the interaction of wave and wind forces [12]. The sand dunes in Parangtritis Village are of the barchan type [13]. A barchan type is a rare form found in wet tropical climates. Apart from being used as a conservation area, the sand dune is also used as a place of tourism.

In 2020, the type of land use was still the same as in 2013. However, there were depreciation and an increase in the area of land use. Land use that has decreased include mixed gardens, which shrank by 4%, and land use that experienced an increase in area, namely land use in the form of settlements which increased by 1%. Table 5. shows the percentage of land-use change in Parangtritis Village during 2013-2020.

Table 5. Percentage of the land-use area in Parangtritis Village

Land-Use	2013		2020	
	Ha	%	Ha	%
Forest	7.48	1	7.48	1
Mixed Garden	312.88	28	269.36	24
Agricultural Land	190.77	17	190.77	17

Continued from Table 5.

Land-Use	2013		2020	
	Ha	%	Ha	%
Open Land	0,64	0	32.9	3
Open Land (Sand dunes)	354.2	32	354.2	32
Settlements	150.09	13	160.07	14
Livestock	62.95	6	62.95	6
Pond	7.59	1	7.59	1
Water Body	31.13	3	31.13	3
Grand Total	1117.09	100	1117.09	100

The distribution of land-use change locations in Parangtritis Village can be observed in Fig. 7. The location points for land-use change occur mostly in the eastern part of the village. This is because the eastern part has a local road, namely Parangtritis Road. Many growth centers will appear around the road so that the demand for land will increase.

Through Fig. 7. it can be observed that the changes that occur around the JJLS agency include mixed gardens into open land and mixed gardens into settlements. Changes from mixed plantations to open land often occur in road segments that are still under construction or not yet connected. The road segment is located in Grogol VIII Hamlet, part of Grogol IX Hamlet, and Grogol X Hamlet. JJLS in Parangtritis Village itself passes through 4 hamlets, were in one of the hamlets, namely in Depok Hamlet, the road conditions are already operating and used by both local people and tourists who traveled in Parangtritis.

The road segment which is still in development planning will later connect Bantul Regency with Gunungkidul Regency and will be called the Parangtritis-Girijati section. The road passes through the hill on the east side of Parangtritis Village and the road design follows the contours of the existing hill. The plan is to build 18 curves so that they will be known as 18 turns [14]. Kelok 18 will become a tourism potential in DIY because it has an attraction in the form of beautiful scenery in the form of beach views. Road users and tourists can stop around the winding section to enjoy the view.

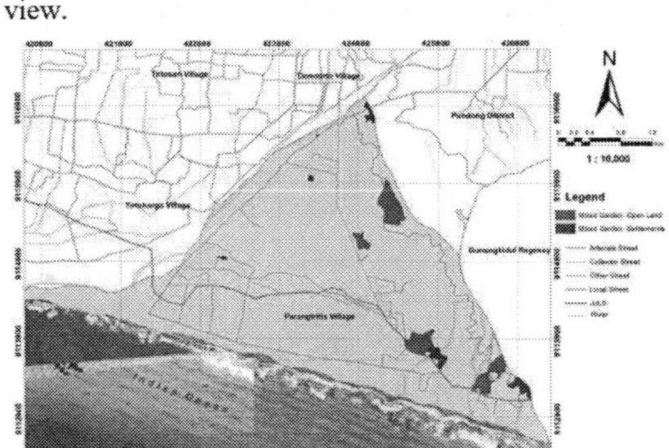

Fig. 7. Map of land-use change in Parangtritis Village 2013-2020.

4 Conclusion

In general, the existence of JJLS development in Bantul Regency does not significantly affect the spatial dynamics of land-use changes in each of the villages it passes. The less significant changes in land use are caused by the condition of the JJLS which is not yet fully connected and the status of land ownership which is the Sultan Ground. Most of the changes along the JJLS body occurred in Parangtritis Village and the least occurred in Poncosari Village.

The author expresses her deepest gratitude to RTA-UGM 2021 which has funded this research. And also really appreciate reviewers for suggestions to make this article better. Without all of these supports, this research would not be successful. Hopefully, this article will be useful for the future.

References

1. W. D. Sasongko, War. Penelit. Perhub. **28**, 92 (2018)
2. O. D. Sakarov and G. Kaharap, in *Pros. Semin. Nas. XII "Rekayasa Teknol. Ind. Dan Inf. 2017" Sekol. Tinggi Teknol. Nas. Yogyakarta* (2017), pp. 453–459
3. I. Chairunnisa, R. Rijanta, and S. Hasanati, J. Bumi Indones. **6**, (2017)
4. M. Ja'far, *Infrastruktur pro Rakyat: Strategi Investasi Infrastruktur Indonesia Abad 21* (Pustaka Tokoh Bangsa, Yogyakarta, 2007)
5. C. P. Ng, T. H. Law, F. M. Jakarni, and S. Kulanthayan, IOP Conf. Ser. Mater. Sci. Eng. **512**, (2019)
6. T. D. Mena, W. P. Tyas, and R. E. Budiati, J. Kesehat. Masy. **7**, 156 (2019)
7. M. Alamgir, M. J. Campbell, S. Sloan, M. Goosem, G. R. Clements, M. I. Mahmoud, and W. F. Laurance, Curr. Biol. **27**, R1130 (2017)
8. L. Prapti, E. Suryawardana, and D. Triyani, J. Din. Sos. Budaya **17**, 82 (2015)
9. H. S. Yunus, *Metodologi Penelitian Wilayah Kontemporer* (Pustaka Pelajar, Yogyakarta, 2010)
10. B. B. Wiyono, *Statistik Pendidikan: Buku Bahan Ajar Mata Kuliah Statistik* (FIP UM, Malang, 2001)
11. R. '. Kurniawan, S. '. Sutikno, and B. '. Sujatmoko, J. Online Mhs. Fak. Tek. Univ. Riau **4**, 1 (2017)
12. H. D. . Wijayanti and F. Abubakar, Pros. Semin. Nas. XI Rekayasa Teknol. Ind. Dan Inf. 2016 50 (2016)
13. Sunarto, *Geomorfologi Dan Kontribusinya Dalam Pelestarian Pesisir Bergumuk Pasir Aeolian Dari Ancaman Bencana Agrogenik Dan Urbanogenik* (Fakultas Geografi. Universitas Gadjah Mada., Yogyakarta, 2014)
14. A. Yulianto, J. Media Wisata **16**, 1689 (2018)

https://doi.org/10.1051/e3sconf/202132507002

Settlement Quality Mapping Analysis Using Google Earth Imagery and GIS in Sorosutan Subdistrict, Yogyakarta City

Mega Rakhmatika[1*], *Chafda* Larasati[1], *Abraham Cardenas* Tristan[2], *Ratih Fitria* Putri[1]

[1]Department of Environmental Geography, Faculty of Geography, Universitas Gadjah Mada, Indonesia
[2]Faculty of Engineering, Autonomous University of San Luis Potosi, San Luis Potosi 78290, Mexico

Abstract. Sorosutan Subdistrict is located in Umbulharjo Subdistrict, Yogyakarta City. The development of Sorosutan Subdistrict is so rapid as indicated by the density of settlements and the complexity of the problems in it. One of these problems is related to settlement conditions. The condition of Sorosutan Subdistrict which has a slum area needs to be investigated on the quality of its settlements through assessment, weighting, and interpretation of Google Earth Imagery. The stage of determining the quality of settlements in this research is interpreting the imagery and determine the scoring of each parameter of settlement quality using GIS software. Based on the analysis of the results of data processing, it was found that the quality of settlements in Sorosutan Subdistrict is partly of medium quality of 54.27%, while good quality is 23.05% and bad quality is 22.68%. The distribution of good and bad quality settlements dominates the western and southern parts of Sorosutan Subdistrict. The distribution of settlement quality dominates the eastern part of Sorosutan Subdistrict. Hamlets which are decided as slum settlements in these qualities of settlements assessment have bad qualities in hamlet numbers 004 and 015.

1 Introduction

The growth of population and development in urban areas is quite rapid. The population growth has resulted in an increasing need for land as a place to live. The development of the urban area is in line with the increase in population so that it raises problems in housing and settlements [1].

The problem of settlements may hinder the existence of good environmental quality. The settlement development that is not organized and not in accordance with the land use triggers the emergence of slum settlements. Based on Law No. 1 of 2011 [2], slum settlements are settlements that are not suitable for habitation due to irregularity of buildings, high building density, and quality of buildings, facilities, infrastructure that do not meet the requirements. Slums that are not handled will further worsen the quality of settlements.

Sorosutan Subdistrict is located in Umbulharjo Subdistrict, Yogyakarta City. Sorosutan Subdistrict has a population of 15,594 people in 2020 with an area of 1.68 km^2. The population density of Sorosutan Subdistrict is 9283 people/km^2.

The development of Sorosutan Subdistrict is so rapid as indicated by the density of settlements and the complexity of the problems in it. One of these problems is related to settlement conditions. Based on Yogyakarta Mayor's Decree Number 158 of 2021 [3], several hamlets in Sorosutan Subdistrict are slum settlements. These are located in the neighborhood/hamlet respectively, namely 012/004, 036/015, 040/010, 043/011, 061/016, 062/016, and 065/016.

The condition of Sorosutan Subdistrict which has a slum area needs to be investigated on the quality of its settlements through assessment, weighting, and interpretation of Google Earth Imagery. The assessment, weighting, and interpretation are carried out on settlement areas which include settlement density, settlement layout patterns, road shelter trees, the width of settlement entrances, road surface conditions, and the location of settlements in the surrounding environment. This research can be used as urban spatial planning, especially in Sorosutan Subdistrict.

2 Methods

The stage of determining the quality of settlements in this research is interpreting the imagery and determine the scoring of each parameter of settlement quality using GIS software. The first stage is interpreting blocks settlement and non-settlement. Settlement blocks' boundaries are delineated based on hamlets' boundaries and differences in settlement characteristics. The scoring of each parameter in the settlement block is used as a determinant of settlement quality. The data used in this research is Sorosutan Subdistrict Boundary Map, building data, and Google Earth Imagery 2021. The data analysis techniques used in the research are:

* Corresponding author: mega.rakhmatika@mail.ugm.ac.id

2.1 Assessment of The Settlements Quality

The assessment of the settlements quality uses parameters sourced from the Directorate General of Human Settlements, Department of Public Works (2006). Here are the parameters which can be seen from Google Earth Imagery:

2.1.1 Settlement Density

The density of settlements assessed in the research is based on the density of buildings in a settlement block. The density of settlements is measured qualitatively by looking at the uniformity of settlement blocks. The settlement density can be calculated by the formula (1) and the parameters of the settlement density are in Table 1.

$$\text{Settlement Density} = \frac{\sum \text{The entire roof area}}{\sum \text{The entire settlement area}} x100\% \quad (1)$$

Table 1. Classification and Score of Settlement Density

Criteria	Classification	Score
The average settlement density in a settlement block is rare (density <40%)	Good	3
The average settlement density in a settlement block is medium (density 40%-60%)	Medium	2
The average settlement density in a settlement block is dense (density >60%)	Bad	1

Source: Directorate General of Human Settlements, Department of Public Works (2006).

2.1.2 Building Layout Pattern

The building layout pattern is determined by the layout regularity and the building size. Buildings that have the same relative size and follow a certain pattern are grouped in the same settlement block unit. The formula and table for the classification of building regularity patterns can be seen below.

$$\text{Building Layout Pattern} = \frac{\sum \text{Regular buildings}}{\sum \text{Buildings}} x100\% \quad (2)$$

Table 2. Classification and Score of Building Layout Patterns

Criteria	Classification	Score
>50% of the existing buildings in a settlement block are orderly arranged	Good	3
25%-50% of the existing buildings in a settlement block are orderly arranged	Medium	2
<25% of the existing buildings in a settlement block are orderly arranged	Bad	1

Source: Directorate General of Human Settlements, Department of Public Works (2006).

2.1.3 Trees of Settlements' Shelter

Trees around settlements are used to shelter the entrance to settlement block and reduce pollution caused by vehicles. The classification of shelter trees in each settlement block is determined as shown in the Table 3.

Table 3. Classification and Score of Settlements' Shelter Trees

Criteria	Classification	Score
If the settlement block has a shelter tree	Good	2
If the settlement block does not have a shelter tree	Bad	1

Source: Directorate General of Human Settlements, Department of Public Works (2006).

2.1.4 The Width of Settlements' Road

The width of the road referred to in this research is the average width of the road which connects the local road with the main road in a settlement block. The following table of criteria to determine the quality of the road width of a settlement block.

Table 4. Classification and Score of Settlements' Road Width

Criteria	Classification	Score
Average road width >6m	Good	3
Average road width between 4-6m	Medium	2
Average road width <4m	Bad	1

Source: Directorate General of Human Settlements, Department of Public Works (2006).

2.1.5 Road Surface Condition

The condition of the road surface can be determined from the type of paving material, which is based on the percentage of the asphalt hardened road condition for the entire road. Assessment of road surface condition through Google Earth Imagery is carried out by paying attention to the hue of the object.

Calculation of road surface condition in each settlement block is calculated using the formula (3) The classification used for assessing the road surface condition is listed in the Table 5.

$$\text{Road Surface Condition} = \frac{\sum \text{Paved road length}}{\sum \text{Entire road length}} \text{x}100\% \quad (3)$$

Table 5. Classification and Score of Road Surface Condition

Criteria	Classification	Score
>50% of the length of the existing road in the settlement block is hardened with asphalt.	Good	3
25% - 50% of the length of the existing road in the settlement block is hardened with asphalt.	Medium	2
<25% of the length of the existing road in the settlement block is hardened with asphalt.	Bad	1

Source: Directorate General of Human Settlements, Department of Public Works (2006).

2.1.6 The Location of Settlement

The assessment of the settlement location parameter is based on the proximity of the settlement to the source of pollution. The following table of criteria to determine the quality of location of a settlement:

Table 6. Classification and Score of Settlement Location

Criteria	Classification	Score
If the location of the settlement is far from the source of pollution activities (industry, factory, and market)	Good	3
If the location of the settlement is not directly affected by the pollution source activities (industry, factory, and market)	Medium	2
If the location of the settlement is close to sources of air and noise pollution (industry, factory, and market)	Bad	1

Source: Directorate General of Human Settlements, Department of Public Works (2006).

2.2 Distribution of Settlement Quality Classes in Sorosutan Subdistrict

The assessment of the settlement quality class is based on the total score. The total score is obtained from the sum and multiplication of the score of each determining parameter with a weighing factor. Calculation of the total score in each settlement block is calculated using the formula (4).

$$\text{Total Score} = (A\times3)+(B\times1)+(C\times2)+(D\times3)+(E\times2)+(F\times2) \quad (4)$$

Information:
A: Score of Settlement Density
B: Score of Settlement Layout
C: Score of The Settlements' Shelter Tree
D: Score of The Road Width
E: Score of The Road Surface Condition
F: Score of Settlement Location
Source: Directorate General of Human Settlements, Department of Public Works (2006).

After determining the highest and lowest of the total score, then determine the class interval (range) with the formula (5).

$$\text{Class Interval} = \frac{(\text{Highest score} - \text{Lowest score})}{\text{Number of classes}} \quad (5)$$

Source: Directorate General of Human Settlements, Department of Public Works (2006).

3 Results and Discussions

The quality of settlements of an area is determined by several parameters. These parameters can be interpreted using Google Earth Imagery, namely population density parameters, settlement layout parameters, road width parameters, road conditions parameters, shelter tree parameters, and settlement location parameters.

3.1 Settlements Quality

3.1.1 Settlement Density Parameters

Determination of settlement density can be seen from the imagery by looking at the presence of buildings that are close to each other. Density assessment is done by finding the roof area which is then divided by the area of the settlement block. According to the diagram (Fig. 1), it is found that Sorosutan Subdistrict is dominated by settlement blocks with medium density with an area of 94,8 ha or 78.93% of the total area of settlements in Sorosutan Subdistrict. The density of settlements in the good category is still a small part of the area, which means that buildings with low density that make space to move freely and wide road access are still rare in this Subdistrict. The density of these settlements resulted in poor air circulation due to the lack of trees around the settlements. The density level is either only an area of 11,1 ha or about 9.27% of the total administrative area of Sorosutan Subdistrict.

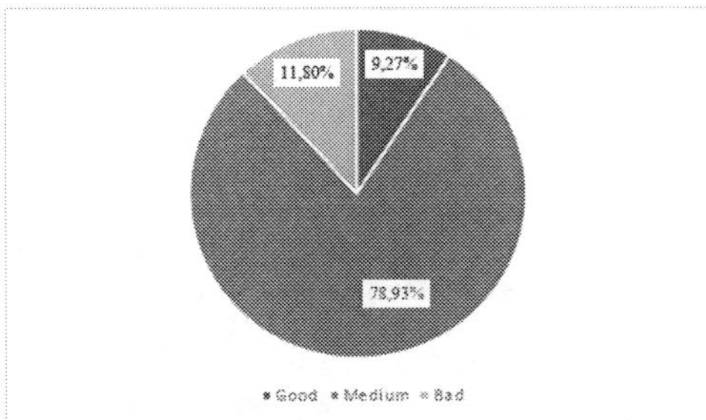

Fig. 1. Percentage of Block Quality Area on Settlement Density Parameters

3.1.2 Settlement Layout Pattern

The settlement layout pattern is assessed based on the regularity between buildings in the settlement block. The regularity of the building is interpreted based on the pattern and shape of the building. The settlement layout pattern is assessed as good, medium, and bad.

Sorosutan Subdistrict mostly has a bad settlement layout pattern. The total settlements area of Sorosutan Subdistrict is dominated by bad settlement layout patterns covering an area of 64.84% (Figure 3.1.1). Meanwhile, the layout pattern of good settlements has a score of 18.45% (Fig. 2) and medium has a score of 16.68% (Fig. 2). Neat and well-organized buildings are influenced by the factor of vacant [4]. The settlement pattern of Sorosutan Subdistrict which is dominated by bad conditions can be caused by the decrease of vacant land.

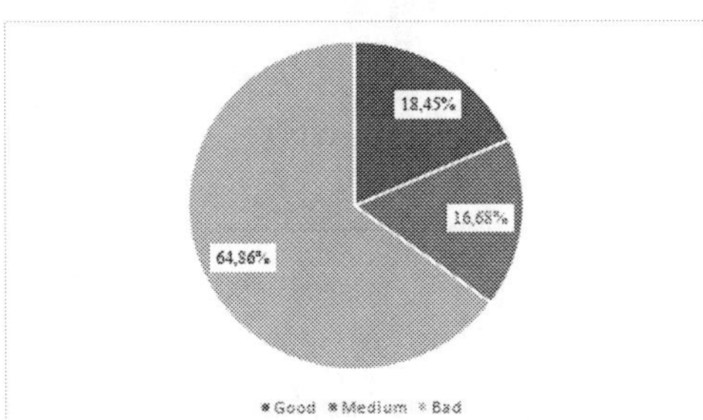

Fig. 2. Percentage of Block Quality Area on Settlement Layout Pattern Parameters

3.1.3 Shelter Tree Parameters

Shelter trees have an influence on the comfort level, settlement blocks with lots of trees have less heat. In addition, shelter trees are also used to reduce air pollution from vehicles. With Google Earth Imagery, the shelter tree measurement is seen from a green hue which looks clustered due to the thick leaves. Shelter tree parameters are only divided into two, namely good and bad.

Based on the diagram (Fig. 3.) shelter trees with good class dominate the settlements in Sorosutan Subdistrict, which are 106,6 ha with a percentage of 88.78%. However, some areas are still classified as bad with an area of 13,4 ha and a percentage of 11.22% of the total area of settlements in Sorosutan Subdistrict.

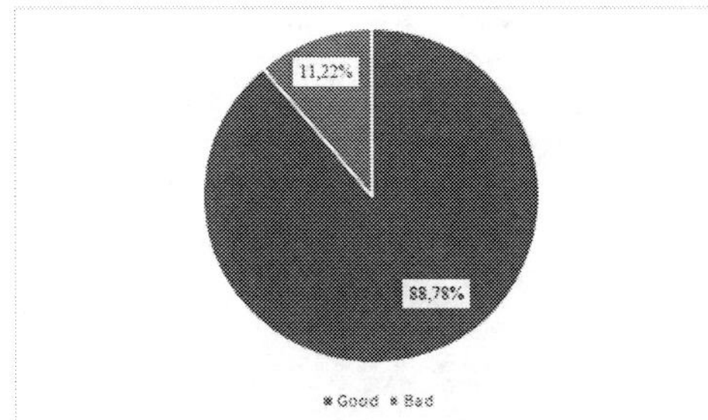

Fig. 3. Percentage of Block Quality Area on Shelter Tree Parameters

3.1.4 The Width of Settlement Entrance

The width of the entrance to the settlement indicates the ease of mobility and accessibility, especially for transportation. The width of this road is calculated based on the width of the entrance to the settlement block from the main road. The width of the entrance to settlements with a score of more than 6 meters is categorized as good, while the bad category has a width of fewer than 4 meters. The width of this road is known based on imagery interpretation and measurements.

The total block area of 38.87% (Fig. 4) Sorosutan Subdistrict has a good quality road width. This shows that many roads leading to settlements can be passed by 2-3 cars freely, this means that it is easy for transportation to pass. Settlement blocks with a medium quality road width is an area of 36.91% (Fig. 4) and bad quality is an area of 24.21% (Fig. 4).

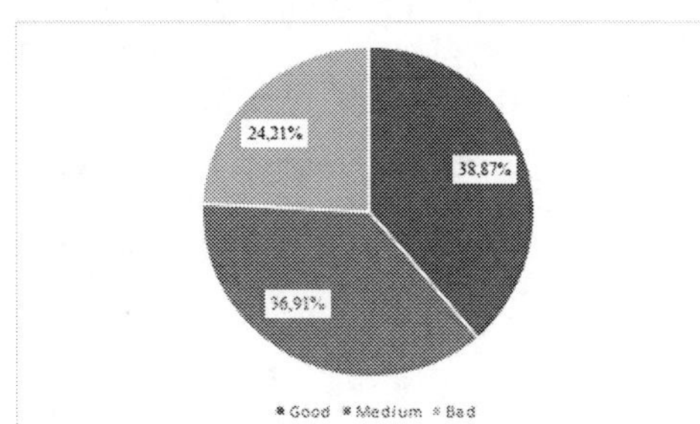

Fig. 4. Percentage of Block Quality Area on The Width of Settlement Entrance Parameters

3.1.5 Road Surface Condition Parameters

The condition of the road surface is categorized as good if the road is made of asphalt. Roads that are classified as asphalt can make it easier for road users to drive and feel comfortable. This asphalt road is often found in urban areas. Parameters of road surface conditions are related to the width of the road. A wide road can be assumed to have been paved with asphalt.

https://doi.org/10.1051/e3sconf/202132507002

Sorosutan Subdistrict has good road surface conditions in all settlements. This shows that there are no settlement blocks less than 50% whose roads are still in the form of soil. The condition of the road in Sorosutan Subdistrict has been 100% completely paved with asphalt (Fig. 5). The condition of the road which is paved with asphalt shows that the quality of the settlement is good because it facilitates accessibility.

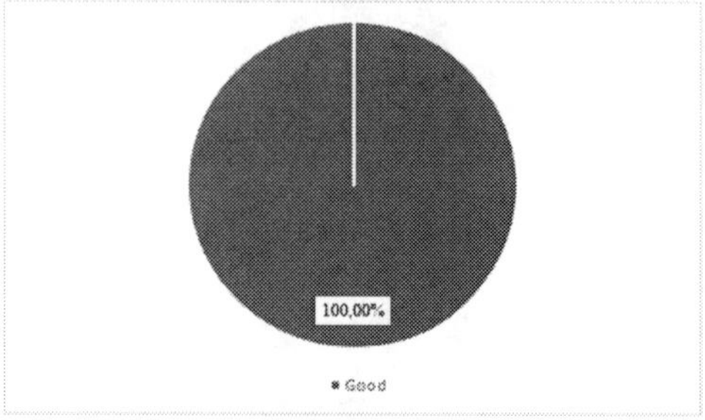

Fig. 5. Percentage of Block Quality Area on Road Surface Condition Parameters

3.1.6 Location

Settlement location parameters are used in determining the quality of settlements on the basis of pollution. Pollution in settlements comes from factories, industries, and markets. The location of settlements against pollution sources is classified into good, medium, and bad classes. The location of settlements that are closer to the source of pollution, the quality of the location is getting worse.

Sorosutan Subdistrict is a part of the center of Yogyakarta. The city that acts as the center of these activities certainly has a variety of industries, factories, and markets as sources of settlement pollution. This condition can be seen from the area of the Sorosutan Subdistrict which has poor quality in location parameters. This poor quality dominates the total settlement blocks area in the Subdistrict by 71.54% (Fig. 6). Medium quality has a percentage of 16.02% (Fig. 6) and good quality has a percentage of 12.41% (Fig. 6) of the total settlement block area of Sorosutan Subdistrict.

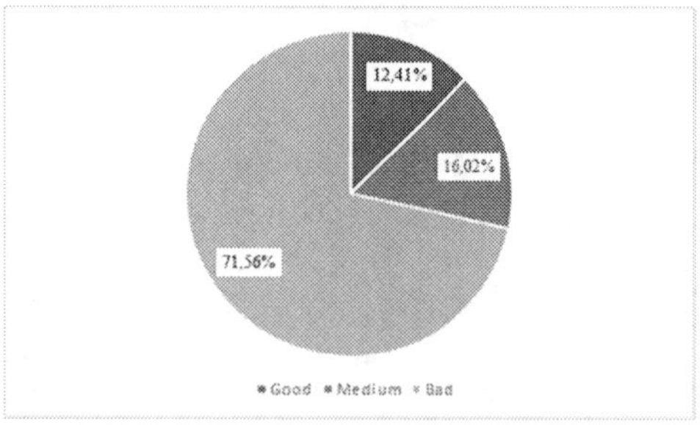

Fig. 6. Percentage of Block Quality Area on Location Parameters

3.2 Distribution of Settlement Quality in Sorosutan Subdistrict, Umbulharjo District, Yogyakarta City

Assessment of settlement quality is obtained from six parameters determining the quality of settlements through imagery interpretation. These parameters are settlement density, building layout, road width, road surface conditions, settlement locations, and shelter trees. The parameter is then calculated the total score according to the weight of each parameter to obtain an assessment of the quality of settlements in an area. After knowing the total score in each settlement block, a classification of settlement quality classes is produced. The results of the classification of each class can be seen in Fig. 7.

The result of the area of each settlement class calculation shows that Sorosutan Subdistrict is dominated by the quality of medium settlements, then good and the last is bad. According to the calculation, the medium settlement quality class has an area of 65,19 ha or 54.72% of the total area of settlements in Sorosutan Subdistrict (Fig. 7). Settlement blocks with medium quality spread in Sorosutan Subdistrict. As shown in the (Fig. 8) the green hue is dominating Sorosutan Subdistrict.

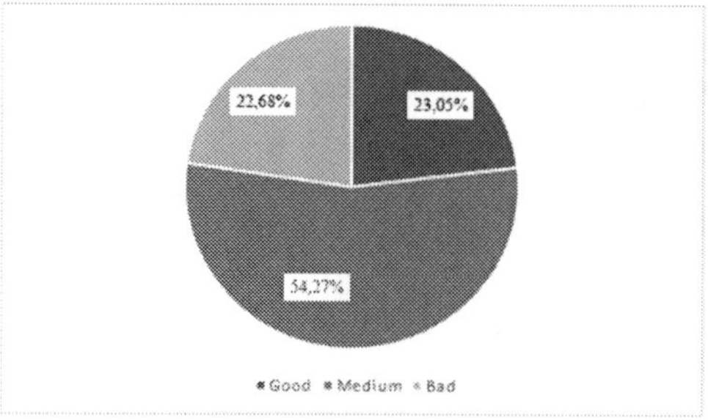

Fig. 7. Percentage of Block Quality Area

Based on the Fig. 8, Sorosutan Subdistrict has 15 settlement blocks with good grades. The settlement block covers an area of 27,6 ha or 23.05% of the total area of settlements in Sorosutan Subdistrict. Settlement blocks that are classified as good are blocks 4, 10, 12, 13, 15, 16, 18, 20, 22, 23, 24, 26, 29, 30, 34. The criteria for good settlement quality are low settlement density, there are protective trees, road width > 6 m, regular settlement pattern, the location of the settlement is far from pollution sources, has a road surface that has been paved with asphalt, and has good road access. Medium-class settlement blocks in Sorosutan Subdistrict are spread throughout the region and are dominated by the eastern part of Sorosutan Subdistrict.

The blocks are blocks 5, 6, 7, 8, 9, 14, 19, 28, 32, 35, 37, 40, 41, 42. The medium class has criteria for the quality of settlements being not too dense, there are shade trees, road width 4 – 6 m, the settlement pattern is quite regular, the location of the settlement is far from pollution sources, has an asphalt surface, and good road access. The low-class settlement block has an area of

https://doi.org/10.1051/e3sconf/202132507002

27,2 ha or 22.68% of the total area of settlements in Sorosutan Subdistrict. The settlement blocks are 1, 2, 3, 11, 17, 21, 25, 27, 31, 33, 36, 38, 39. The distribution of bad settlement quality dominates the western and southern parts of Sorosutan Subdistrict. These bad quality settlement blocks are located close to factories, industries, and markets. Other criteria for settlement blocks include high density, irregular settlement patterns, no protected trees, and difficult road access. Indiscriminate development can cause some parameters that determine the quality of settlements to be bad, such as the width of the entrance being small due to too tight construction. Hamlets designated as slums in the assessment of the quality of these settlements have low quality in hamlet number 004 and 015.

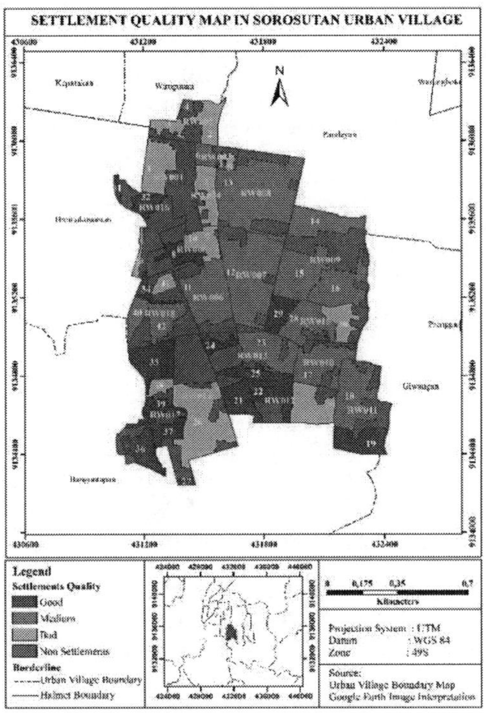

Fig. 8. Settlements Quality Map in Sorosutan Hamlet Subdistrict in each Hamlet

4 Conclusions

The quality of settlements in Sorosutan Subdistrict is determined by the density of settlements, settlement layout patterns, road shelter trees, the width of the settlement entrance, the condition of the settlement roads, and the location interpreted through Google Earth Imagery. The quality of settlements in Sorosutan Subdistrict is partly of medium quality of 54.27%, while good quality is 23.05% and bad quality is 22.68%. The distribution of good and bad quality settlements dominates the western and southern parts of Sorosutan Subdistrict. The distribution of settlement quality dominates the eastern part of Sorosutan Subdistrict. Hamlets which are decided as slum settlements in these qualities of settlements assessment have bad qualities in hamlet numbers 004 and 015.

References

1. W. T. Prasetyo and S. Rahayu, Tek. Perenc. Wil. Kota **2**, 293 (2013)
2. Undang-undang Nomor 1 Tahun 2011, *Perumahan Dan Kawasan Permukiman* (Indonesia, 2011), pp. 1–136
3. Keputusan Walikota Yogyakarta Nomor 158 Tahun 2021, *Penetapan Lokasi Perumahan Kumuh Dan Permukiman Kumuh* (Yogyakarta, 2021), pp. 1–19
4. M. Farizki and W. Anurogo, Maj. Geogr. Indones. **31**, 39 (2017)

Spatial pattern analysis of land cover changes using Fragstat in Kendal Delta, Kendal Regency, Central Java

Nur Rizki Fitri Hastari[1], *Esya* Rachma Ningrum[1], *Puspasari* Setyowati[1], *Hayyun* Nadia[1], *Qooi* Insanu Putra[1], *Sofia* Lutfiana[1], *Muhammad* Rois Akbar[1], *Heni* Dwi Lestari[1], *Muhammad* Fauzan Ramadhan[2], *Lucia* Sandra Budiman[2], *and Eko* Haryono[1*]

[1] Departement of Environmental Geography, Faculty of Geography, Universitas Gadjah Mada, 55281, Sleman, Yogyakarta, Indonesia

[2] Master Program in Geography, Faculty of Geography, Universitas Gadjah Mada, 55281, Sleman, Yogyakarta, Indonesia

Abstract. The northern coastal plains of Kendal Regency are deltaic landform whose processes are strongly influenced by fluvial (fluvial dominated deltas). The area of land in the Kendal Delta area continues to experience changes due to regional developments that occur as Kendal Regency is designated as one of the national activity center in terms of strategic and industrial economies. Various developments can trigger various land conversions so that studies are needed to identify changes occurring. The purpose of this study is to identify land cover patterns based on geomorphological aspects and analyzed land cover changes based on landscape structures in Delta Kendal in 1989 and 2021. Changes in the Kendal Delta landscape structure were analyzed using landscape metrics approach. The results showed that the land cover pattern in the Kendal Delta from 1989 to 2021 has a change. Based on the calculation from landscape metrics, land cover in Delta Kendal that has a relatively high patch density tends to have a high level of interference, and vice versa. In general, there is a relationship if there is an increase in the value of PD and ED, there will be a decrease in the value of MNN and IJI.

1 Introduction

The increase in population and the pressure of urban development causes the need for space to be greater and the use of land to increase [1]. Central Java Province has designated Kendal as one of the center of national activities that functions as a strategic economic and industrial area [2]. The development of urban areas causes the need for land to increase and land use changes occur [3]. Kendal Regency in 2012-2018 experienced physical development of the region causing land conversion in the form of residential and industrial land [4].

The northern coastal plain of Kendal Regency is a delta landform. The presence of the Quaternary fault causes the formation of a basin that has decreased, this basin is filled with deltaic and fluvial deposits (fluvio-deltaic basin). The delta formed on the north coast of Kendal is a fluvial-dominated delta, which is a delta that is strongly influenced by fluvial landforms [5]. The delta ecosystem has an important role in maintaining environmental balance because it has great benefits from its very fertile physical properties due to the alluvial sedimentation process that carries high nutrients [6]. The flat morphology of the delta with a slope of 0-8% makes it easier to use land [7]. This flat morphology is used as settlements, swamps, shrubs, gardens, rainfed rice fields, irrigated rice fields, rivers, fields, and ponds [7]. In addition, Delta Kendal has a strategic position and has the potential to be developed for various uses such as industrial areas, tourism, aquaculture, and settlements.

Landscape structure is a pattern of landscape that is determined from the type of use and structure in the form of size, shape, arrangement and distribution of elements of the landscape [8]. Landscape elements can be defined as separate entities or referred to as 'patches', whereas landscape patterns are described as the use of metrics to measure patches (e.g. size, shape and isolation), mosaic (patch richness and connectivity diversity) and characteristics [9]. Patch can also be defined as the smallest unit of analysis that has homogeneous properties, such as rice fields, industrial areas [10]. The scale of analysis can be in the form of land cover, land use, soil and geomorphology.

Changes in use to built-up land are one of the characteristics of urbanization [11]. Uncontrolled land use changes will cause environmental, social, and economic problems [12]. Changes in land use that are not controlled and are not in accordance with environmental ethics will cause damage. One of the variables to determine changes in land use is to use land cover. Land cover can provide very important information for modelling and understanding natural phenomena that occur on the earth's surface [13]. Land cover information is obtained through data from remote sensing which can record land cover dynamics spatially and temporally. Analysis of the development of land

* Corresponding author: e.haryono@ugm.ac.id

cover changes in an area can be done through the use of remote sensing data in the form of multitemporal satellite images [14]. Multitemporal land cover data can show patterns of land conservation over time.

Landscape metrics is a method used to analyze the growth of an area using remote sensing (PJ) and Geographic Information Systems (GIS) [15]. Landscape metrics can be used in landscape ecological assessments to support planning and decision making in landscape management [16] and can be used to measure landscape structure and structural complexity. Landscape metrics is a quantitative index that describes the compositional and spatial aspects of the landscape based on map data, remote sensing, and Geographic Information System analysis [9].

The landscape metrics method can be used to analyze many things, one of which is changes in land cover patterns [17]. According to [16], landscape metrics can help in understanding landscape changes from various perspectives (visual, ecological, and cultural). Metric analysis can be based on Area/Density/Edge metrics, Shape metrics, Core area metrics, Isolation/Proximity metrics, Diversity metrics, Evenness metrics, Connectivity metrics, Dominance metrics [18]. The landscaping metrics method used is able to easily describe the dynamics of changes in land cover patterns in more detail [17]. This study aims to identify land cover patterns based on geomorphological aspects and analyze land cover changes based on landscape structures in Delta Kendal.

2 Methods

2.1 Research Location

Kendal Regency is one of the regencies in Central Java Province which has 20 sub-districts. Kendal Regency is located between 60° 32' - 70° 24' South Latitude and 109° 40' - 110° 18' East Longitude. Administratively, it is bordered by the Java Sea in the north, Semarang City in the east, Semarang Regency in the southeast, Temanggung Regency in the south and Wonosobo Regency in the southwest and Batang Regency in the west (Fig 1.). Kendal Regency has an area of 100.233 Ha with the largest area being Singorojo District.

Topographic conditions in Kendal Regency are divided into three categories, namely mountainous areas, hilly areas and beaches. The mountainous area is located in the southern part of the district with an altitude between 0-2579 masl. The hilly area covers the central part of the district. Lowland areas and beaches are located in the northern part of the district with an altitude between 0-10 meters above sea level [19]. Kendal Regency consists of various landforms such as landforms from volcanic, fluvial and marine processes.

The location of this research was conducted in the coastal area of Kendal Regency which is a coastal delta. The Delta Coast is supported by four main watersheds, from East to West, namely the Blorong watershed, the Bodri watershed, the Damar watershed, and the Kuto watershed. Of the four watersheds, the largest and most dominant is the Bodri watershed. Therefore, the coast of Kendal Regency is better known as the coast of the Bodri Delta. The research was conducted at this location because the development of the coastal morphology of the Bodri Delta is quite dynamic until now. There are several locations that experience abrasion (such as on the coast of Kartika Jaya) and there are those that experience accretion. Overall, the coast of the Bodri Delta, Kendal Regency tends to experience growth.

Changes in the shoreline of the coastal area of Kendal Regency during the period 1864 -1973 (109 years) experienced varied growth [20]. During this period, the Kendal Regency coast experienced developments that tended to move towards the East. The mouth of the Bodri River has also changed its position. Initially in 1864 the estuary of the Bodri River passed through the Margorejo area (west of the current Bodri River estuary), then in 1910 the estuary of the river had shifted to the East and experienced growth and its estuary branched.

This coastal development continued until 1946 where the direction of the Bodri River estuary had shifted to the North and Northeast. Furthermore, the mouth of the Bodri River continues to experience changes, namely in 1973 the river channel had two branches, one empties into the North and Northeast, the other empties into the Northwest until now. In some places it expands towards the sea (accretion) as a result of a fairly large sedimentation process, but in other places there are coastal areas that experience the opposite problem, namely abrasion, for example in the coastal area of Kartika Jaya.

2.2 Research Method

The research was conducted using quantitative methods that emphasize geographic information system techniques in its implementation. This research is descriptive quantitative where the analysis is described from the results of processing various numerical data to identify the characteristics and patterns of change of the phenomenon under study. Research is classified as deductive research, which is carried out by observing a phenomenon and then comparing it with existing theories. The final result of this research is to quantify the ecological structure of the coastal landscape in Kendal Regency. To realize these results, the stages of research that will be carried out are: (1) Inventory and Data Collection (2) Data Processing and (3) Data analysis.

2.3 Research Data

The data used are secondary data in the form of Landsat images and various spatial data which are described in detail in Table 1. Data collection was carried out by studying literature and downloading geospatial data openly (open source) on the United States Geological Survey (USGS) website, Center for Development and Geological Research, and the Geospatial Information Agency. Data processing is carried out using a number of software including ArcGIS, Fragstat, and Google Earth.

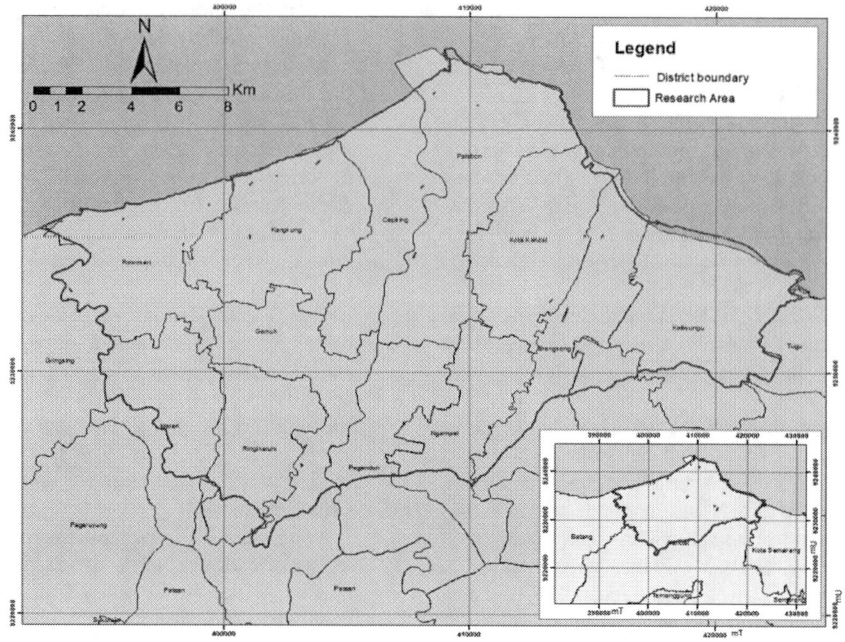

Fig. 1. Administration map of Delta Kendal, Kendal Regency

Table 1. Research Data

Data	Format	Source
Landsat Imagery 5 TM 30 m resolution in 1989	Raster	*United States Geological Survey*
Landsat Imagery 8 OLI 30 m resolution in 2019	Raster	*United States Geological Survey*
Google Earth Imagery	Raster	*Google Earth*
Alos Palsar Imagery	Raster	*United States Geological Survey*
Geological Map Scale 1:100.000	Vektor	Geological Research and Development Center
Soil Type Map Scale 1:100.000	Vektor	Agricultural Research and Development Agency of the Ministry of Agriculture
Indonesia Topographic Map scale 1:25.000	Vektor	Geospatial Information Agency

2.4 Data Processing

First steps of data processing are to make land cover maps in 1989 and 2021. The 1989 land cover map is the result of interpretation from the Landsat 5 TM imagery with a resolution of 30 meters in 1989, while the 2021 land cover map is sourced from the 2019 Landsat imagery. Geometric and atmospheric correction for Landsat imagery was built using ArcGIS. Classification of land cover made by interpretation of composite band 754, where band 7 is SWIR-2 spectrum, band 5 is NIR spectrum, and band 4 is red spectrum. Satellite image visual interpretation based on spatial object recognition. Object characteristic recognized by elements of interpretation such as hue, color, shape, size, texture, shadow, pattern, and object appearance association.

Interpretation of landsat imagery used for validating land cover. The results of the interpretation of the Landsat imagery are then validated. Land cover map of 1989 validated by google earth imagery in 1989. Meanwhile, land cover map of 2019 validated by field validation in May 2021. The validation land cover map of 2019 is based on sample points that have been made with stratified random sampling. Each land cover class has 10 sample points so that the total sample points are 330 points (Fig 2). The post-field stage is to make a land cover map for 2021 according to the validation results. Processing the data from the validation of the land cover using the confusion matrix. Processing of land cover change data in 1989 and 2021 was carried out using landscape metrics in the Fragstat software.

2.5 Research Analysis

The changes in the landscape structure of the Delta Kendal were analyzed using Fragstat software. The input data for the analysis of changes in the landscape structure of the Delta Kendal in Kendal Regency is a land cover map on the Delta Kendal landform in 1989 and 2021. The analysis of changes in landscape structure is carried out by analyzing patches on the Fragstat software using a number of landscape metric parameters including Edge Density (ED) , Patch Density (PD), Mean Shape Index (MSI), Mean Nearest Neighbor (MNN), and Interspersion and Juxtaposition Index (IJI).

Edge Density (ED) is the sum of the lengths (m) of all edge segments involving the **appropriate patch** type, divided by the total landscape area (m2), multiplied by 10,000 (to convert to hectares) [21], with the formula as in equation 1.

$$\mathrm{ED} = \frac{\sum_{k=1}^{m'} e_{ik}}{A}(10,000) \qquad (1)$$

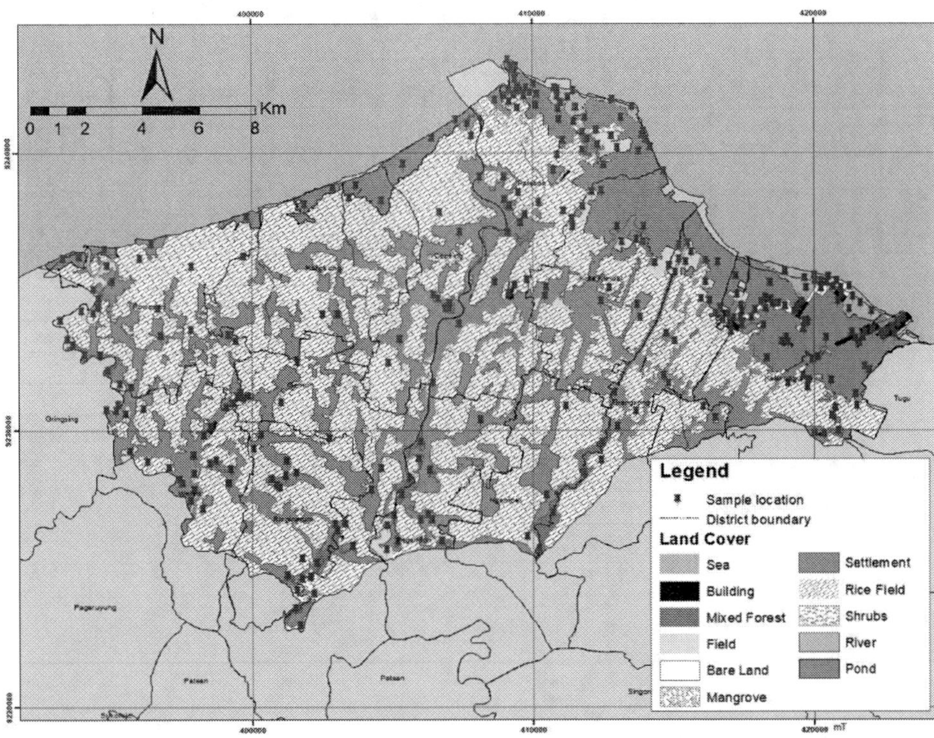

Fig. 2. Sample points map

Patch Density (PD) is the number of patches of the appropriate patch type (NP) divided by the total landscape area, multiplied by 10,000 and 100 (to convert to 100 hectares). The PD value can also be used as an index to the heterogeneity of a landscape formation [21], with the formula as in equation 2.

$$PD = \frac{n_i}{A}(10{,}000)\,(100) \qquad (2)$$

Mean Shape Index (MSI) is the sum of each patch circumference divided by the square root of the patch area (hectare) for each patch class (class level) and adjusted to the standard polygon or square. Mean Nearest Neighbor (MNN) is the nearest neighbor distance from individual patches to similar patches, namely the class in each landform in meters. Meanwhile, the Interspersion and Juxtaposition Index (IJI) is a measure of the relative interspersion of each landform class.

3 Result and Discussion

3.1 Accuracy Test of the Classification

The accuracy test of the classification results was carried out to test the accuracy of the land cover map generated from the digital classification process with sample points from the field results. Based on the results of the confusion matrix calculation on the land cover map in 2021, the kappa value is 0.853 and the overall accuracy value is 86.74%, proving that the land cover map is reliable and feasible. From 330 sample points, not all can be checked due to limited access. In general, land cover change in the field changed into settlement and industrial building growth in Kendal (Tabel 2).

3.2 Land Cover Pattern Analysis

Land cover from time to time continues to change due to land conversion (Fig. 3. and Fig. 4.). Figure 3 shows the land cover at Delta Kendal in 1989 was dominated by vegetation cover in the form of rice fields, fields, mixed forest, and shrubs. Figure 4 shows that in 2021 land cover in the Delta Kendal has experienced a change from the dominance of vegetation cover to built-up area which consists of construction and buildings. Table 3 explains the changes in the land cover area in Delta Kendal. In 1989, the largest land cover was rice fields with an area of 182.92 km2, followed by settlements, mixed forest, ponds, and fields which each had an area of 33.35, 29.43, 25.52, and 23.02.

Land cover decreases significantly are mixed forest, field, and shrubs. There are 50.63 km2 decreased. Meanwhile, Settlement is land cover that increased significantly, it is about 45,1 km2. The highest settlement increase happened in the District of Brangsong, Pegandon, Ngampel, Patebon, part of Kaliwungu, and Kendal City. Research results [4] shows that between 2007 - 2018 development of settlement areas in Brangsong, Patebon, Kaliwungu, and Kendal City are high enough. Industrial area growth is a sign of change from agricultural land to non-agricultural land [22] which is proved by Kendal Industrial Area (Kawasan Industri Kendal - KIK) growth in northern Kaliwungu District [23].

Tabel 2. Confusion matrix Calculation of Kendal Land Cover Map in 2021

Land cover		Observation										Classification overall	User accuracy
		A	B	C	D	E	F	G	H	I	J		
Interpretation	A	31	0	0	0	0	0	0	0	0	0	31	100%
	B	0	40	0	0	0	0	0	2	0	0	42	95,24%
	C	0	0	28	1	0	0	0	1	0	0	30	93,33%
	D	0	0	0	42	0	4	1	2	1	0	52	80,77%
	E	0	6	0	0	28	0	1	1	0	0	36	77,78%
	F	0	0	0	0	0	30	0	0	0	0	30	100%
	G	0	6	0	0	0	1	71	3	1	0	83	85,54%
	H	0	1	1	0	0	0	0	63	0	0	65	96,92%
	I	0	2	0	0	0	2	0	2	34	0	52	65,39%
	J	0	1	0	0	3	0	3	2	0	45	54	83,33%
Truth overall		31	57	29	43	31	39	76	78	39	52	475	
Recall		100%	70,18%	96,55%	97,67%	90,32%	76,92%	93,42%	80,77%	87,18%	87%		
OA		86,74%											
Kappa		0,851											

Information :

Information :

A : River D : Field G : Settlement J : Shrubs
B : Building E : Bare Land H : Rice Field
C : Mixed Forest F : Mangrove I : Pond

KIK development based on spatial plans of Kendal Regency of 2011 - 2031 as an effort to improve the regional economy in the eastern coastal area of Kendal Regency. High land cover change from production land to built-up land triggered by industrial and city growth in Kendal Regency.

In addition, land cover that experienced another significant increase was the increase in ponds by 12.61 km2. This increase was accompanied by a decrease in mangrove and rice fields. The largest increase in pond land is in the Patebon, Brangsong, Kaliwungu, and Kendal sub-districts caused by land suitability for pond activities. For example, in Brangsong District, the land suitability for shrimp and milkfish ponds in the S1 (Very Suitable) class is 39.68% and S2 (Sufficiently Appropriate), 60.32% of the total area [24]. Kendal District based on the results of research by [25] shows the suitability for a name shrimp aquaculture ponds due to the condition of the waters. Thus, the land cover pattern formed in the lower delta plain of Delta Kendal is formed following the suitability of the land owned.

Table 3. Land Cover Area of Delta Kendal in 1989 and 2021.

No	Land Cover	Area (km2)	
		Year 1989	Year 2021
1	Rice field	182.92	179.00
2	Building	1.53	2.19
3	Field	23.02	6.00
4	Shrub	11.77	1.64
5	Pond	25.52	38.13
6	Mixed Forest	29.43	5.95
7	Mangrove	4.76	1.94
8	Settlement	35.35	80.45
9	Bare land	0.87	1.46

3.3 Landscape Metrics Analysis

Land cover change analysis can also be done using landscape metrics analysis, which is to find out changes in fragments that dominate a landscape [26]. The matrix analysis used includes Patch Density (PD), Edge Density (ED), Mean Shape Index (MSI), Mean Nearest Neighbor (MNN), and Interspersion and Juxtaposition Index (IJI) as presented in Table 4. Temporal analysis to determine the dynamics of land cover change in the Delta Kendal in each landform.

Patch Density (PD) is an indicator of fragment area. A fragment is a homogeneous area that can be distinguished from the surrounding area [26]. The land cover in 1989 which had the highest PD value was a field of 0.127 but in 2021 it experienced the most significant decrease in PD value to 0.065. In addition to the field that experienced a decrease in the value of PD is the mixed forest. This decrease indicates a decrease in the area of fragments triggered by a decrease in land area. The land cover that experienced a significant increase in the PD value was a settlement from 0.037 to 0.254, which means that settlement experienced an increase in the area of the fragments. This is due to the sporadic growth of settlements. A fragmented matrix can cause an edge or edge area with a large area [26].

The land cover that experienced a significant decrease in the ED value was bare land, about 17,235. In 1989 the bare land had the highest ED value, but in 2021 it experienced the most significant decrease in the ED value, this is because the bare land experienced an uneven increase in area. These changes can affect changes in habitat biodiversity which can later affect changes in resources [26].

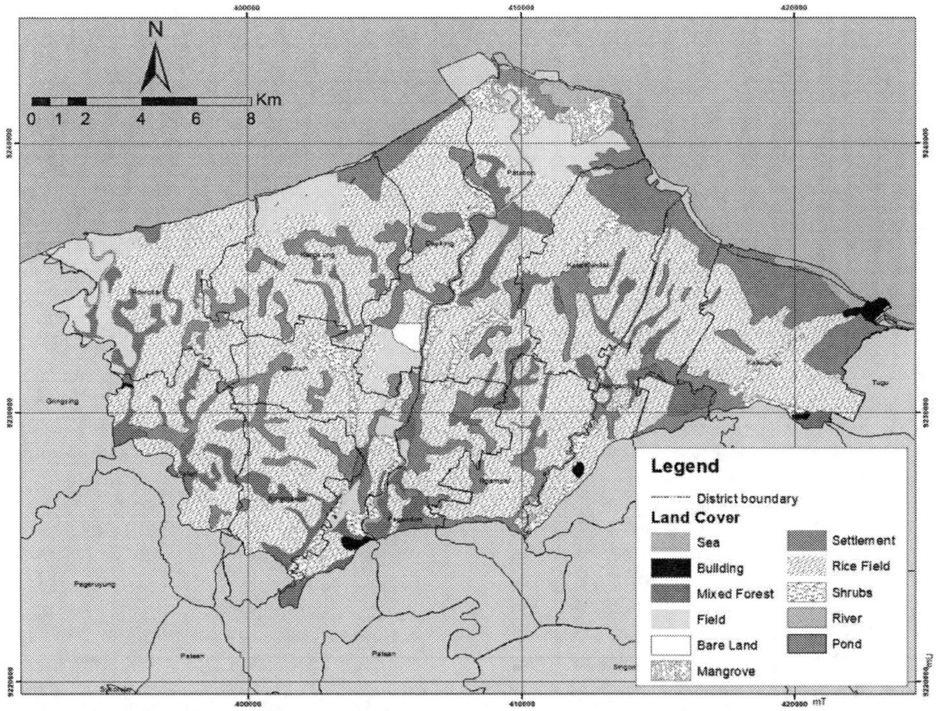

Fig. 3. Land Cover Map of Delta Kendal in 1989

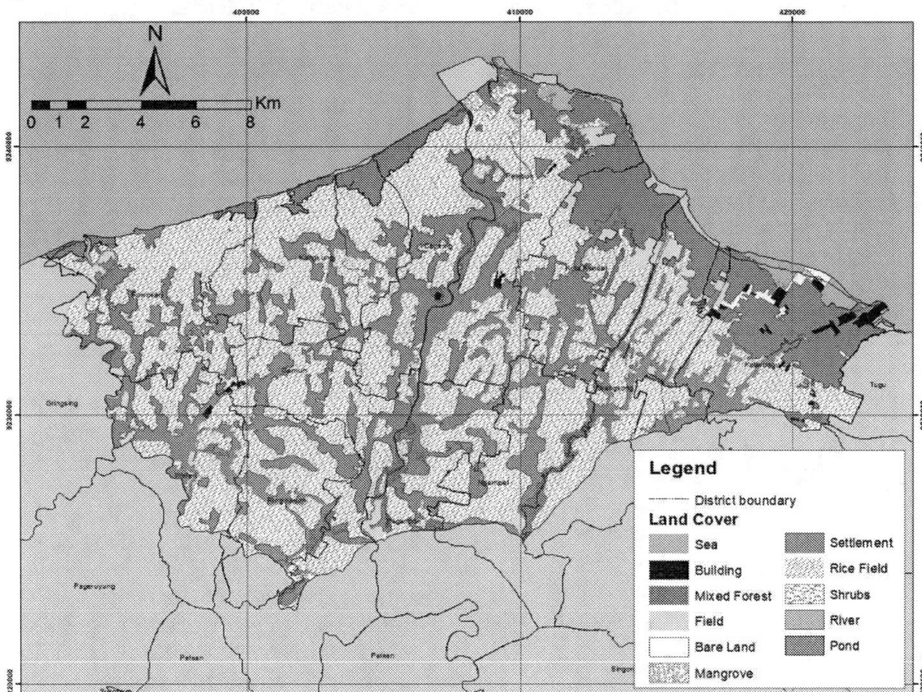

Fig. 4. Land Cover Map of Delta Kendal in 2021

The land cover that experienced a significant increase in the ED value was the rice field and settlement, this shows that the land cover has a fairly high edge density when viewed from the edge aspect [27]. A high ED value indicates that the edge area is very wide, which indicates that there is fragmentation or a change in structure from homogeneous to heterogeneous [26]. Fragmentation in rice fields occurs due to changes in rice field land into settlements, while fragmentation in settlements occurs due to the sporadic development of settlements.

The Mean Shape Index (MSI) is an indicator of the shape of the fragment. The shape of the patch becomes more regular if the MS value is close to 1 with a square or circle shape [27]. In 1989 the MSI values for all land cover tended to be high, even reaching 3, indicating an

Table 4. Fragstat Result in Delta Kendal in 1989 and 2021

No	Land Cover	PD (/100 ha)		ED (m/ha)		MSI		MNN (m)		IJI (%)	
		1989	2020	1989	2020	1989	2020	1989	2020	1989	2020
1	Rice field	0,050	0,099	3,713	25,677	2,286	0,115	655,73	199,07	63,331	32,025
2	Building	0,034	0,065	3,481	1,146	3,267	0,316	1206,76	684,89	80,895	65,202
3	Field	0,127	0,065	9,177	2,082	1,936	0,231	402,47	1061,33	36,161	72,315
4	Shrubs	0,016	0,040	0,353	0,617	1,372	0,276	6488,45	1726,63	64,560	40,575
5	Pond	0,006	0,028	0,739	4,439	2,088	0,084	421,02	271,71	55,518	85,759
6	Forest Mix	0,124	0,105	8,692	2,737	2,082	0,302	520,14	838,98	50,970	55,604
7	Mangrove	0,037	0,046	3,706	0,835	1,856	0,265	1585,05	403,85	84,416	68,548
8	Settlement	0,037	0,254	1,876	25,342	1,443	0,210	293,35	323,94	60,372	30,985
9	Barren	0,028	0,040	18,053	0,817	2,969	0,263	225,02	786,66	68,078	67,711

irregular land cover structure. Land cover is still dominated by productive land and settlements and there is no urbanization phenomenon. In 2021 the MSI value, in general, decreased due to changes in land structure to become more regular due to urban planning and urbanization.

The MNN value indicates the level of land cover spatial isolation and is an indicator of landscape connectivity. The higher the MNN value, the higher the level of spatial isolation between patches. Shrub's land cover in 1989 had the highest MNN value, which was 6588.4 meters, then it decreased significantly to 1726,635 in 2021. This was due to a significant reduction in land area due to the conversion of land to built-up land. The land cover that experienced an increase in the MNN value was the field from 402.46 in 1989 to 1061.326 in 2021. This increase indicates an increase in spatial isolation due to a significant reduction in land area and land fragmentation.

IJI is an indicator of land fragmentation stating the level of land cover interspersion. In general, the IJI value in 2021 is lower than in 1989. This indicates that the degree of spatial intermixing between patches has decreased. The highest level of interspersion was found in the mangrove forest land cover in 1989, which was 84.4%. Meanwhile, the lowest interspersion rate was found in dryland cover/fields in 1989 at 36.1%. Based on the calculation results, it can also be seen that land cover that has a relatively high patch density tends to have a high level of interference, and vice versa. In general, there is a relationship if there is an increase in the value of PD and ED, there will be a decrease in the value of MNN and IJI and vice versa.

Overall, it shows that every land cover in the Delta Kendal undergoes structural changes. The largest land cover change occurred in the residential land cover, from 30.67 km2 in 1989 to 64.5 km2 in 2021. The rapid growth of settlements was influenced by the development of KIK and the construction of the Semarang-Batang Toll Road. The development of industry triggers an increase in population density due to the increase in migrant workers entering Kendal Regency due to factors that attract employment opportunities. In addition, the presence of the Semarang-Batang Toll Road can increase the potential for land-use change [28]. The use of agricultural and non-agricultural land will turn into toll roads and trigger changes in agricultural land to non-agricultural areas around toll roads due to the increase in population. Access development tends to support economic activities and open transportation links between regions so that it becomes a factor that attracts residents to settle in Kendal Regency. Residents who settle down definitely need a place to live so that the development of settlements in the Kendal Regency becomes more rapid.

4 Conclusion

The land cover pattern of the Kendal Delta in 1989 showed that the land cover was dominated by vegetation cover in the form of rice fields, fields, forest mix and shrubs, while in 2020 the land cover in the Kendal Delta experienced a change, namely from the dominance of vegetation cover developing into built-up land such as settlements and buildings.

The quantification of the Kendal delta landscape structure was carried out using patch density (PD), edge density (ED), mean shape index (MSI), mean nearest neighbor (MNN), and interspersion and juxtaposition index (IJI). Based on the overall calculation results indicate that every land cover in the Kendal delta undergoes structural changes. Rice field and settlement land cover increased in PD and ED values and decreased MNN values. This is due to the increase in land settlement sporadically and the fragmentation of paddy fields. Field land cover, forest mix decreased PD and ED values and increased MNN values. This is due to a significant reduction in land area.

Based on the calculation results, it can also be seen that land cover that has a relatively high patch density tends to have a high level of interference, and vice versa. In general, there is a relationship if there is an increase in the value of PD and ED, there will be a decrease in the value of MNN and IJI.

References

1. R. H. Dauhari, Prinsip-Prinsip Penataan Ruang Laut Pesisir dan Pulau-Pulau Kecil, Makalah

Seminar Nasional Kelautan Ikatan Mahasiswa Geodesi, Institut Teknologi Bandung (2001)

2. Bappeda Kabupaten Kendal, Rencana Terpadu dan Program Investasi Infrastruktur Jangka Menengah (RPI2JM) Bidang Cipta Karya Kabupaten Kendal 2016-2020 Bab V, Badan Pembangunan Daerah Kabupaten Kendal (2015)
3. K. Kusrini, S. Suharyadi, S. R. Hardoyo, Perubahan Penggunaan Lahan dan Faktor yang Mempengaruhinya di Kecamatan Gunungpati Kota Semarang, Majalah Geografi Indonesia, 25(1), 25-40 (2011)
4. K. I. Aulia, S. Subiyanto, B. Sudarsono, Analisis Arah Perkembangan Fisik Wilayah Kabupaten Kendal Menggunakan Sistem Informasi Geografis, Jurnal Geodesi Undip, 8(1), 486-495 (2019)
5. U. M. Lumbanbatu, Perkembangan Dataran Pantai (Coastal Plain) Daerah Kendal Provinsi Jawa Tengah, Jurnal Geologi dan Sumberdaya Mineral, 19(4), 225-237 (2009)
6. T. B. Sanjoto, Perubahan Spasial Delta Sungai Bodri Sebagai Basis Zonasi Tata Ruang Pesisir Kabupaten Kendal, Doctoral dissertation, Program Pascasarjana Undip (2012)
7. Sriyanto, S. Sanjoto, Tjaturahono, Arahan Komoditas Unggulan Perikanan Tambak di Pesisir Kabupaten Kendal (Evaluasi Kesesuaian Lahan Budidaya Perikanan Tambak), Jurnal Geografi (Media Pengembangan Ilmu dan Profesi Kegeografian, 14(2), 66-74 (2018)
8. U. Walz, Landscape structure, landscape metrics and biodiversity, Living Reviews in Landscape Research, 5(3), (2011)
9. J. A. Kupfer, Landscape ecology and biogeography: Rethinking landscape metrics in a Post-Fragstats landscape, Progress in Physical Geography, 36(3), 400–420 (2012)
10. A. Wijaya, Analisis Dinamika Pola Spasial Penggunaan Lahan Pada Wilayah Terdampak Kenaikan Muka Air Laut Di Kota Pekalongan, Institut Teknologi Sepuluh Nopember Surabaya (2017)
11. H. Han, C. Yang, J. Song, Scenario simulation and the prediction of land use and land cover change in Beijing, China, Sustainability, 7(4), 4260-4279 (2015)
12. D. W. Wardani, P. Danoedoro, B. Susilo, Kajian perubahan penggunaan lahan berbasis citra penginderaan jauh resolusi menengah dengan Metode Multi Layer Perceptron dan Markov Chain, Majalah Geografi Indonesia, 30(1), 9-18 (2016)
13. Y. C. Liang, Y. Zeng, E. C. Peh, A. T. Hoang, Sensing-throughput tradeoff for cognitive radio networks. IEEE transactions on Wireless Communications, 7(4), 1326-1337 (2008)
14. A. R. Darmawan, N. Puspaningsih, M. B. Saleh, Kajian Perubahan Tutupan Lahan dengan Menggunakan Metode Multi Layer Perceptron dan Logistic Regression Di Taman Nasional Gunung Ciremai (2018)
15. B. Bhatta, Analysis of Urban Growth and Sprawl from Remote Sensing Data, Canada: Springer (2010)
16. E. Gökyer, Understanding Landscape Structure Using Landscape Metrics. Advances in Landscape Architecture (2013)
17. E. Uuemaa, M. Antrop, J. Roosaare, R. Marja, Ü. Mander, Landscape metrics and indices: an overview of their use in landscape research, Living reviews in landscape research, 3(1), 1-28 (2009)
18. K. McGarigal, B. J. Marks, FRAGSTATS: Spatial Pattern Analysis Program for Quantifying Landscape Structure, General Technical Report (GTR) (1995)
19. Dinas Pekerjaan Umum, Penyusunan Rencana Terpadu dan Program Investasi Infrastruktur Jangka Menengah (RPI2JM), Bidang Cipta Karya Kabupaten Kendal 2016-2020 (2016)
20. E. C. F. Bird, O. S. Ongkosongo, Environmental changes on the coasts of Indonesia, UNU (1980)
21. M. Ramdhan, Patch Analysis Menggunakan Fragstat (Studi Kasus Area Jambi), Kementerian Kelautan dan Perikanan (2015)
22. H. Han, C. Yang, J. Song, Scenario simulation and the prediction of land use and land cover change in Beijing, China, Sustainability, 7(4), 4260-4279 (2015)
23. M. N. Sadewo, I. Buchori, Simulasi perubahan penggunaan lahan akibat pembangunan kawasan industri kendal (KIK) berbasis cellular automata, Majalah Geografi Indonesia, 32(2), 142-154 (2018)
24. D. R. Setianingrum, A. Suprayogi, Analisis Kesesuaian Lahan Tambak Menggunakan Sistem Informasi Geografis (Studi Kasus: Kecamatan Brangsong, Kabupaten Kendal, Provinsi Jawa Tengah), Jurnal Geodesi Undip, 3(2), 69-80 (2014)
25. A. A. Awanis, S. B. Prayitno, V. E. Herawati, Kajian Kesesuaian Lahan Tambak Udang Vaname Dengan Menggunakan Sistem Informasi Geografis Di Desa Wonorejo, Kecamatan Kaliwungu, Kendal, Jawa Tengah, Buletin Oseanografi Marina, 6(2), 102-109 (2017)
26. L. B. Prasetyo, Pendekatan Ekologi Lanskap Untuk Konservasi Biodiversitas, Bogor: Fakultas Kehutanan Institut Pertanian Bogor (2017)
27. A. W. Wijaya, A. P. Wijaya, A. I. Rahmawati, E. Paryani, H. D. Lestari, I. Amri, L. Ardianti, S. R. Putri, E. Haryono, Analisis Ekologi Bentanglahan di Taman Nasional Baluran dan Sekitarnya, Majalah Geografi Indonesia, 34(1), 34-42 (2020)
28. A. Z. A. Tyo, B. Sudarsono, F. J. Amarrohman, Analisis Kesesuaian Perubahan Penggunaan Lahan Terhadap Rencana Tata Ruang Wilayah (RTRW) di Kecamatan Ngaliyan Pasca Pembangunan Jalan Tol Semarang-Batang. Jurnal Geodesi Undip, 10(1), 11-20 (2020)

Best Practices of Capital City Relocation in Various Countries: Literature Review

Rini Rachmawati[1]*, *Eko* Haryono[2], *Rizki* Adriadi Ghiffari[1], *Hilary* Reinhart[1], *Farah* Dewi Permatasari[3], and *Amandita* Ainur Rohmah[4]

[1]Department of Development Geography, Faculty of Geography, Universitas Gadjah Mada, Yogyakarta, Indonesia
[2]Department of Environmental Geography, Faculty of Geography, Universitas Gadjah Mada , Yogyakarta, Indonesia
[3]Graduate School on Environmental Science, Universitas Gadjah Mada, Yogyakarta, Indonesia
[4]Graduate Program of Regional Development, Faculty of Geography, Universitas Gadjah Mada, Yogyakarta, Indonesia

Abstract. Capital city relocation issue is currently emerging due to the condition and problem Indonesia encounters. Study toward some cases of capital city relocation could be crucial input for analysis and comparison. This paper aims to review capital city relocation examples. Literature review shows there are variances in background and driving factors internally or externally, and infrastructure needs for city development. The driving factors consist of environmental pressure, pollution, urban heat island, housing, urban sprawl, and disaster. The infrastructures required for the development of a new capital city are information and communication technology (ICT), physical infrastructure, transportation, and open green space. From the analysis, Myanmar's Yangon to Naypyidaw relocation and Malaysia's Kuala Lumpur to Putrajaya relocation are the best practices in capital city development that boosts new activities and urban extension. Burma shows the infrastructure's role in development of a new capital city and Kuala Lumpur reveals how ICT implementation helps to establish a new capital city. This paper is expected to fill the research gap from previous research in terms of reviews related to the need for ICT in capital city preparation. This indicates that this paper has followed the trend of ICT advancement. The next research and publication will strengthen the use of ICT as one of the variables.

1 Introduction

The capital city relocation issue is currently becoming crucial due to several urges namely people density, environmental degradation, and urban discomfort as living space. These problems are highly important to be identified through literature review.

Capital city relocation study has already been conducted from various perspectives such as Geographical Science, Development Geography, Environmental Geography, and Urban Planning. Capital city relocation driving factors are needed to be studied from the literature review. Analysis on infrastructure development also must be considered in supporting new capital city development. Examples of required infrastructures are public facilities which consist of hospitals, art, and sports facilities, library, market, urban transportation, and housing.

The new capital city requires a novel urban development strategy as described by the National Urban Development Project (NUDP). NUDP elaborates urban development that integrates infrastructure through three schemes: encouraging harmonious policy and institution in city development agenda, increasing integrated planning for city investment based on spatial planning framework, and formulating criteria and mechanism that prioritize investment, identification of funding gap and enhancing local government's financial management capacity.

Information and Communication Technology (ICT) on urbanization is not merely a people or residents' movement but also a social adaptation in associating traditional behavior with urban life characters [1]. It also relates to technological changes which are caused by the human behavior model in the city [2]. The city undergoes rapid development in many aspects including ICT. People in the city are quickly engaged with the lifestyle changes, particularly with high technology-based and at the end, the information technology brings lifestyle changes [3]. Hence, ICT usage triggers urbanization and it recently has encompassed all the globe especially in urban areas. Researchers try to match the ICT application with urban structure. Therefore, ICT applications must be included in this research.

Review toward capital city relocation literature is important to fathom initial problems and conditions from several countries. It also can be a scientific base and reference before field survey. A literature review is also expected to identify background, strength, and weakness as supporting factors and constraints, also the plan that has been implemented in the capital city relocation.

* Corresponding author: rinirachma@ugm.ac.id

2 Background of Capital City Relocation

Capital city is very dense, and it is a place where history was written amid high political and economic functionality. For centuries, several governments have decided to move their capital city across the globe. Some capital cities were planned to be located in less-developed regions to distribute the development [4].

Authorities also relocate their capital city to decrease the big city control on them. Thus, cooperation or power-sharing would be created and facilitate competition and interest-sharing which is geographically concentrated. It is expected to minimize civilian conflicts [5]. State also moves their city for reasons like geographical aspects which are related to whether the security and government center is strategic or not. Another reason is to distribute the development, security reason, and historical background [6].

After World War 2, 16 countries have relocated their capital city [7, 8]. During the 20th century, Montenegro was the first country to relocate the capital city from Cetinja to Podgorica in 1946 and Malaysia was the latest to move its capital city from Kuala Lumpur to Putrajaya in 2000. In the 21st century, there are two countries that move their capital city: Tanzania and Myanmar. Tanzania even moved its capital city twice -a unique case, from Dar Es Salaam to Dodoma in 1973 and back to Dar Es Salaam again in 2016. A list of countries that relocate to the capital city can be seen in Table 1.

Table 1. List of Countries that Relocate the Capital City

Year	Countries	Previous Capital City	New Capital City
1946	Montenegro	Cetinje	Podgorica
1956	Brazil	Rio de Janeiro	Brasilia
1957	Mauritania	Saint Louis	Nouakchott
1959	Pakistan	Karachi	Islamabad
1961	Botswana	Mafeking	Gaberone
1963	Libya	Benghazi	Tripoli
1965	Malawi	Zomba	Liliongwe
1970	Belize	Belize City	Belmopan
1973	Tanzania	Dar es Salaam	Dodoma
1974	Guinea Bissau	Boe	Bissau
1975	Nigeria	Lagos	Abuja
1982	Srilanka	Colombo	Sri Jayawardenapura Kotte
1983	Cote d'Ivoire	Abidjan	Yamoussoukro
1990	Germany	Bonn	Berlin
1997	Kazakhstan	Almaty	Astana
2000	Malaysia	Kuala Lumpur	Putrajaya
2005	Myanmar	Rangoon	Naypyidaw
2016	Tanzania	Dodoma	Dar Es Salaam

Sources: [7, 8]

In the South-East Asia Region, Myanmar and Malaysia are examples of countries that relocate their capital city. In the case of Myanmar, capital city relocation was done under geographical motive and warfare history, representing Myanmar's military response toward invasion probability [9]. Moreover, Naypyitaw as the new capital city was selected because the location was more centered than Yangon. The Naypyidaw capital city structure is composed of a city center surrounded by eight small cities. This city is managed by Naypyitaw Development Center (NPTDC) and this area is called Naypyidaw Union Territory (NUT). Naypyitaw's location is also strategic because Naypyitaw International Airport is only 16 km from town and is supported by Yangon-Naypyidaw Highway and those two facilities provide intercity transportation [10].

Malaysia, instead of totally relocating its capital city, only transferred some of its administrative function to Putrajaya. Administrative center relocation creates a better chance of having a more well-planned town center complemented by modern technology and adequate infrastructure to increase the government's efficiency and productivity [11]. Kuala Lumpur to Putrajaya relocation was also driven by the motivation to enforce national identity and gaining global recognition. Putrajaya is a good example of Malaysia's urban design and exhibits various contemporary urban development philosophies, especially the effort to manifest sustainable and smart cities [11]. To summarize, the background of Myanmar and Malaysia's capital city relocation can be seen in Table 2.

Table 2. Summary of Myanmar and Malaysia Capital City Relocation

Aspects	Myanmar	Malaysia
Main pattern	Physical Infrastructure	ICT
Part of Relocation	Total relocation	Partial relocation
Reasons	• Warfare (hard element) • Centrality	• Urban expansion (soft element)

- National identity
- International recognition

Source: Analysis (2021)

3 Driving Factors of Capital City Relocation

3.1 Environmental Pressures

Population growth in urban regions steadily increases and in 2030, it is expected that 60.4% of the world population would live in cities, especially capital cities [12]. The capital city is basically different from the common city because it has three exclusive main elements of political power, reverence, and urban environment [13]. The capital city attracts people with its employment chances, economic factors, and entertainment [14, 15] and the increasing population inevitably gives heavy pressure on the environment along with the growth of demand and needs for every person. However, the environment's capability to provide the resources needed is naturally limited [16] and depends on the biophysical characteristics of the area. This concept of limitation is well-known as urban carrying capacity [17, 18].

Maintaining population growth to be still below the carrying capacity is imperative to ensure the environment can still bestow people's livelihood. If this limit is surpassed, there will be environmental deterioration entailed by disaster and eventually cut the population number through mortality or migration [19]. Environmental deterioration is indicated by pollution, land-use change, and health problems [20]. All these factors can be categorized as internal factors because it comes from the inside of a city. On the contrary, external factors arise from outside of city boundaries such as climate change and sea-level rise as can be seen in Figure 1.

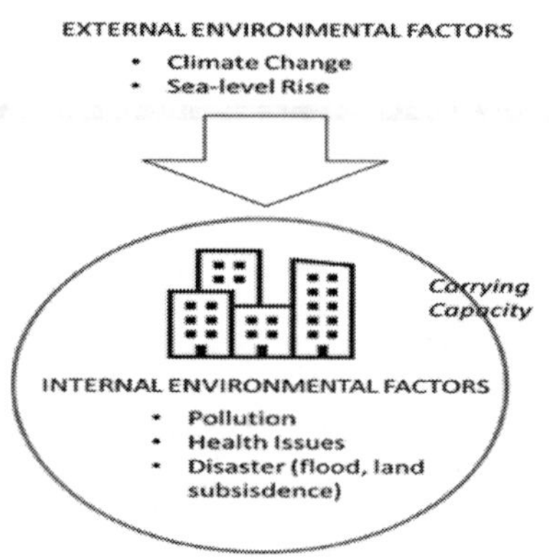

Fig. 1. External and internal environmental pressures (source: analysis, 2021)

Cases of Kuala Lumpur and Yangon relocation cannot be unattached from environmental factors such as urban carrying capacity although its main considerations are most economical and geopolitical. Urbanization in each capital city is highly rapid, it initiates environmental problems like pollution, urban heat island, housing, and urban sprawl, and natural disaster [21, 22]. Those problems would lead to more severe impacts including air, water, and land resources depletion.

3.2 Air Pollution

Air pollution in Kuala Lumpur and Yangon is heavily influenced by urban expansion [23, 24] which directly increases pollutant emission from transportation [25-27]. For the anticipation, a better integrated and sustainable transportation system must be employed with cleaner fuel and a cutback of vehicle numbers. Furthermore, green space also could help lessen the pollutants. These strategies could be merged into green city design. Capital city expansion such as Kuala Lumpur-Putrajaya is likely to help prevent pollution in a region and combined with green space, eliminate the pollutant from the air. Capital city expansion also relieves congestion as one of the main sources of emission.

3.3 Urban Heat Island

UHI is formed when a significant temperature difference between a high-density built environment and a high-vegetation area occurs. The temperature difference can reach up to 4°C higher and in some area at Great Kuala Lumpur like Kampung Baru, it could reach 5°C [28, 29]. Building's height difference and their dispersion trap hot air and prompt the accumulation. UHI leads to thermal discomfort, affecting public health, economical disadvantages, and pollution [30] and in Kuala Lumpur, it has been a big concern. In Yangon, Land Surface Temperature (LST) happens to keep rising as the impact of infrastructure and vegetation loss. Capital city relocation should be designed to avoid the UHI spot by regulating building heights since it is extremely difficult to adjust existing building height.

3.4 Housing and Urban Sprawl

Rapid urbanization in Kuala Lumpur and Yangon demands vast land for living and activity space. It directs to the decreasing land availability and on the other side tightens the competition for the remaining land, ascending its price [31, 32]. Land scarcity also forces Kuala Lumpur and Yangon residents to seek for the land outward, causing urban sprawl. Urban sprawl degrades environmental and resources quality as the impact of higher transportation usages and vegetation loss [33].

At Yangon, urban growth follows a multicentral pattern and expanding without a firm policy causing land competition [32, 34, 35]. As the impact, gentrification and marginalization happen which causes slums and declining residents' quality of life [36]. Capital city relocation could add land supply and spare urbanization pressure toward the environment.

https://doi.org/10.1051/e3sconf/202132507004

3.5 Disaster

Throughout history, natural disasters like volcanic eruptions and floods are the main driving factors in capital city relocation [13]. Ancient capital cities are usually built adjacent to the shore or river delta [37] due to the accessibility but it is prone to a natural disaster.

Both Kuala Lumpur and Yangon are located near the coast where Yangon lies in the Irrawaddy River Delta. This location puts Yangon in an open exposure to flood, cyclone, and sea-level rise and the risk is getting higher along with the urban expansion [38, 39]. Yangon is also threatened by land subsidence [40] caused by heavy burden and groundwater exploitation. Kuala Lumpur, although its location is relatively far 'deeper' than Yangon, is frequently hit by flash floods caused by poor drainage, land-use change, and naturally because Kuala Lumpur sits in the Klang Flood Plain [41, 42]. Relocation of the capital city could minimize the exposure by selecting a safer location and the new capital city could be designed to be more adaptive to the disaster.

3.6 Geographic Centrality

Centrality is related to the equal distance affordability amongst regions in a country [43]. Based on that consideration, the capital city is ideally located at the centroid point of a country hence it can reach the whole country and people more efficiently [44]. The example of centrality can be seen in Malawi, Brazil, and Burma cases where the centrality of Naypyidaw connects the highland of Mandalay with the lowland part of Yangon [45]. The summary of environmental factors can be seen in Table 3.

Table 3. Environmental Factor that Affects Capital City Relocation

Factor	Intensity	Resource	Explanation
Pollution	Low	Air Resources, Water Resources	Air and water pollution decrease public health
UHI	Low	Air resources	Has negative impact towards thermal comfort Increases emission along with the usage of air conditioner and cooling system
Housing	Low	Land resources	Higher land price Marginalization Slums
Urban Sprawl	Medium-high	Land resources, air resources	Vegetation loss Increasing emission from transportation
Disaster	High	Water resource, air resource, land resource	Infrastructure damage Economical loss Casualties

Source: Analysis (2021)

4 Infrastructure for capital city development

A complete and modern infrastructure construction has a noteworthy impact on government functionality. Technological development in this digital era in the term of digital governance becomes a notable factor to organize a state. Nowadays, the government is obliged to have easy and quick information access to strengthen central government control. Dynamic urban development yields innovation opportunities in managing governmentality. Therefore, innovation is fundamentally beneficial for the public and institution.

Governmentality must adapt to the latest technology by adopting the concept of Internet of Things (IoT), cloud computing, to Artificial Intelligence. Latest technology alleviates administration loads thus the function of planning, organizing, and monitoring-evaluation will run more effectively. ICT-based also could support public services accessibility and smart city implementation [46].

Capital city is supposed to be livable and smart for its people [47]. Government institutions in the new capital city must employ ICT that could provide competitiveness for the governments where they must perform effective, efficient, and transparent services for the people [46, 48, 49]. ICT could also contribute to education, health, commerce, and the government sector. Smart cities also elevate people's quality of life by accelerating economic development [47, 50].

From the regional development and environment point of view, infrastructure of the capital city should be planned under demand and good space capability consideration. One of the government visions in relocating the capital city according to Chin [51] and Moser [11] is to build a city that will be able to reflect the natural and cultural heritage of the country. It is also important to optimize technological capacity combined with state-of-the-art facilities to anticipate future challenges. Currently ICT has been implemented in all aspects of urban development and plays an important role in the future of cities. Among them are related to the provision of government services and social services, such as health, economy, and education in the form of ICT-based applications that support smart city implementation [52].

In the Kuala Lumpur case, the relocation of an administration that is close to Cyberjaya results in a chance to amalgamate the latest technology with smart city [51].

Infrastructure must address sustainability and people's interests. Putrajaya is designated as Garden City that harmonizes city development and management with surrounding nature. Putrajaya development considers urban life quality with open-space recreation hence the people could experience various entertainments, sports, and recreational activities. Many buildings in Putrajaya are designed to be attractive as icons and create comfort for the people. Capital city relocation that puts identity and character which roots on locality and tradition forward and center is essentially compatible with Smart City. Putrajaya is highly regarded as a remarkable work as a new administrative center.

Another aspect that determines the accomplishment of development of a new capital city is the accessibility. Putrajaya as Malaysia's new capital city is located within 20 kilometers from Kuala Lumpur International Airport (KLIA) and 25 kilometers from Kuala Lumpur. The distance helps the employees to reach their office. Putrajaya is also supported by integrated transportation modes including bus, train, and monorails. The infrastructure requirements can be seen at Figure 2.

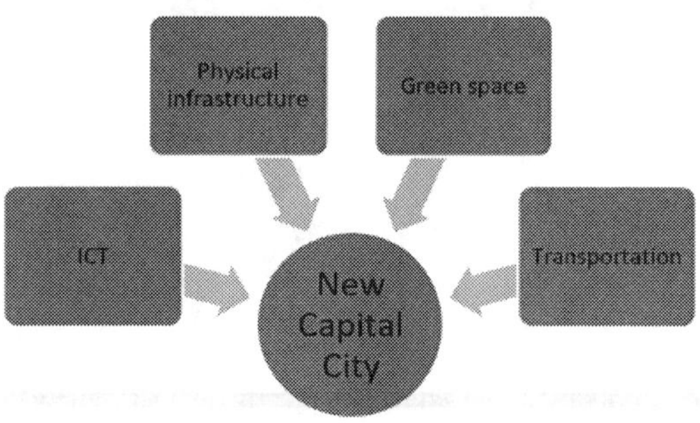

Fig. 2. Infrastructure for New Capital City (source: analysis, 2021)

5 Conclusion

Literature review shows there are various capital city relocation backgrounds, encouraging and discouraging factors internally or externally namely environmental pressure, pollution, UHI, housing and urban sprawl, natural disaster, and location. The infrastructure required for the urban development to be the capital city comprises ICT, social and regional development, and physical and environmental aspects. Capital city relocation of Myanmar from Yangon to Naypyidaw could be referred to as best practices during capital city development that becomes a new activity in regional development. Kuala Lumpur to Putrajaya relocation is the best practice in implementing ICT.

Acknowledgments

The authors would like to thank Universitas Gadjah Mada for supporting the research through Scheme of Penelitian Dasar (PD) in 2021, NUMBER 2169/UN1/DITLIT/DIT-LIT/PT/2021. Research Funds from the Budget Implementation Entry List (DIPA), Deputy for Strengthening Research and Development, Ministry of Research and Technology/National Research and Innovation Agency (Deputi Bidang Penguatan Riset dan Pengembangan, Kementerian Riset dan Teknologi/Badan Riset dan Inovasi Nasional) for the 2021 Budget Year.

References

1. R. Paddison, Handbook of Urban Studies, London: Sage Publications, (2001)
2. P.L. Knox and L.M. Carthy, Urbanization: An Introduction to Urban Geography, 2nd ed, (2005)P.L.
3. Knox and S.A. Marston, Place and Regions in Global Context Human Geography, 3rd ed, New Jersey: Pearson Education Inc, (2004)
4. Richard, Katherine Schulz. Capital City Relocation. accessed from https://www.thoughtco.com/capital-city-relocation-1435389 on August 18th 2019.
5. A. Potter, Locating the government: Capital cities and civil conflict, Research and Politics, **4**, 4 (2017)
6. Ilman, 2015
7. M. Mubaroq and A. Solikin, Review on the Financing Scheme of Indonesia's Capital City Relocation Plan: Lessons Learned from Brazil, Malaysia, and Tanzania," in *ICOFEB 2018*, Lhokseumawe: EAI, (2019)
8. D. K. Ishenda, and S. Guoqing, Determinants in Relocation of Capital Cities J. of Public Administration and Governance 9 **4**, 200 (2019)
9. D. Preecharushh, Naypyidaw: New Captial of Burma, Bangkok: White Lotus (2009)
10. A. M. bin A. Manan and H. Suprayitno, Preliminary Overview of Several Capital Relocations in Relationship with a Plan of Indonesian Capital Relocation J. Infrastruct. Facil. Asset Manag, **2**, 1 (2020)
11. S. Moser, Putrajaya: Malaysia's new federal administrative capital. Cities **27**, 4 (2010)
12. United Nation, World Population Report 2018, (2019)
13. V. Rossman, Capital Cities: Varieties and Patterns of Development and Relocation, Routledge: Amsterdam, (2017)
14. Ischak, Urbanisasi Dan Dampaknya Terhadap Lingkungan. Humaniora **13**, 2 (2001)
15. I. Suntajaya, Piramida, Faktor-Fator Yang Mempengaruhi Terjadinya Urbanisasi Di Provinsi Bali. **10**, 2 (2014)

16. H. Wahyuningsih, Perhitungan Ruang Terbuka Hijau Perkotaan Jenis Publik (Studi Kasus: Kota Surakarta). J. Arsit. dan Perenc, **1**, 2 (2018)
17. Y. Su, H. Xue, H. Liang, An Evaluation Model for Urban Comprehensive Carrying Capacity: An Empirical Case from Harbin City. Int. J. Environ. Res. Public Health **16**, 367 (2019)
18. Y. Wei, C. Huang, J. Li, L. Xie, An evaluation model for urban carrying capacity: A case study of China's mega-cities. Habitat Int., **53** (2016)
19. A. R. Khavarian-Garmsir, A. Pourahmad, H. Hataminejad, and R. Farhoodi, Climate change and environmental degradation and the drivers of migration in the context of shrinking cities: A case study of Khuzestan province, Iran. Cities Soc., **47** (2019)
20. C. S. S. Ferreira, R. P. D. Walsh and A. J. D. Ferreira, Degradation in urban areas. Environmental Science & Health 5, pp. 19–25, (2018)
21. H. C. Siong, Urban governance and rapid urbanization issues in Malaysia. Jurnal Alam Bina **13**, 4 (2008)
22. C. Wang, S. W. Myint, P. Fan, M. Stuhlmacher, J. Yang, The impact of urban expansion on the regional environment in Myanmar: a case study of two capital cities. Landscape Ecol., **33** (2018)
23. O. H. L. Ling, K. H. Ting, A. Shaharuddin, A. Kadaruddin and M. J. Yaakob, Urban growth and air quality in Kuala Lumpur city, Malaysia. EnvironmentAsia **3**, 2 (2010)
24. M. A. B. A. Tajudin, M. F. Khan, W. R. W. Mahiyuddin, R. Hod, M. T. Latif, A. H. Hamid, S. A. Rahman, M. Sahani, Risk of concentrations of major air pollutants on the prevalence of cardiovascular and respiratory diseases in urbanized area of Kuala Lumpur, Malaysia. J. Ecotoxicology and Environmental Safety, **171** (2019)
25. M. F. Khan, A.H. Hamid, M. A. Bari, A. B. A. Tajudin, M. T. Latif, M. S. M. Nadzir, M. Sahani, M. I. A. Wahab, Y. Yusup, K. N. A. Maulud, M. F. Yusoff, N. Amin, MdAkhtaruzzaman, W. Kindzierski, P. Kumar, Airborne particles in the city center of Kuala Lumpur: Origin, potential driving factors, and deposition flux in human respiratory airways. J. Sci. of the Total Env., **650** (2019)
26. L. N. Huy, N. T. K. Oanh, T. T. Htut, O. M. T. Hlaing, Emission Inventory for On-road Traffic Fleets in Greater Yangon, Myanmar. Atmospheric Pollution Research, **11** (2020)
27. W. Y. Aung, M. Noguchi, E. P. Yi, Z. Thant, S. Uchiyama, T. Win-Shwe, N. Kunugita, O. Mar, Preliminary assessment of outdoor and indoor air quality in Yangon city, Myanmar. Atmospheric Pollution Research, **10** (2019)
28. Lin Yola, 2018. Impact of Urban Configurations on Microclimate and Thermal Comfort in Residential Area of Kuala Lumpur. Dissertation. Faculty of Built Environment Universiti Teknologi Malaysia
29. S. A. Zaki, N. S. Azid, M. F. Shahidan, M. Z. Hassan, M. Y. M. Daud, N. A. A. Bakar, M. S. A. Ali, F. Yakub, Analysis of Urban Morphological Effect on the Microclimate of the Urban Residential Area of Kampung Baru in Kuala Lumpur Using a Geospatial Approach. Sustainability, **12** (2020)
30. I. S. M. Elsayed, Mitigation of the urban heat island of the city of Kuala Lumpur, Malaysia. Middle-East J. of Scientific Research **11**, 11 (2012)
31. N. Wahi, R. M. Zin, V. Munikanan, I. Mohamad, S. Junaini, Problems and Issues of High Rise Low Cost Housing in Malaysia. IOP Conf. Series: Materials Science and Engineering **341** (2018)
32. M. Sabrié, Yangon "Emerging Metropolis": Challenges for the Authorities and Resilience of the Yangonites. Moussons: Recherche en sciences humaines sur l'Asie du Sud-Est, **33** (2019)
33. M. A. Naeema, Malaysia Sustainable Cities Program, Working Paper Series (2016)
34. T. Sritarapipat, W. Takeuchi, Modeling urban expansion in Yangon, Myanmar using Landsat time-series and stereo GeoEye Images. IOP Conf. Series: Earth and Environmental Science **37** (2016)
35. S. S. Moe, Current Trends of Urban Development in Yangon City and Its Implications on the Environment. GMSARN Int. J., **3** (2009)
36. M. Matelski, M. Sabrié, Challenges and Resilience in Myanmar's Urbanization: A Special Issue on Yangon. Moussons: Recherche en sciences humaines sur l'Asie du Sud-Est, **33** (2019)
37. B. Karsono, Historical analysis of urbanity on the waterfront. ARSITEKNO, **2** (2013)
38. G. Fernandez, A. M. Tun, K. Okazaki, S. H. Zaw, K. Kyaw. Factors influencing fire, earthquake, and cyclone risk perception in Yangon, Myanmar. Int. J. of Disaster Risk Reduction, **28** (2018)
39. L. Tierolf, H. de Moel, J. van Vliet, Modeling urban development and its exposure to river flood risk in Southeast Asia. Computers, Environment and Urban Systems, **87** (2021)
40. T. v. d. Horst, M. M. Rutten, N. C. v. d. Giesen, R. F. Hanssen, Monitoring land subsidence in Yangon, Myanmar using Sentinel-1 persistent scatterer interferometry and assessment of driving mechanisms. Remote Sensing of Environment, **217** (2018)
41. N. Samsuri, R. A. Bakar, T. Unjah, Flash Flood Impact in Kuala Lumpur – Approach Review and Way Forward, Int. J. of the Malay World and Civilisation (Special Issue 1), **6** (2018)
42. T.R. Bhuiyan, M. I. H. Reza, E. A. Choy, J. J., Pereira, Facts and trends of urban exposure to flash flood: A case of kuala lumpur city, Community, Environment and Disaster Risk Management, **20** (2018)
43. F. Homburg, Capital Cities, Journal of Geography **19**, 1, pp. 8–15 (1920)
44. D. Potts, Capital Relocation in Africa: The Case of Lilongwe in Malawi, The Geographical J. **151**, 2 (1985)
45. D. M. Seekins, 'Runaway chickens' and Myanmar identity, City **13**, 1 (2009)
46. R. Rachmawati, P. Rachmadani, V. N. Anifa, and F. Lutfiana, Various ICT-based applications and their uses to support smart city implementation in the Regency of Blora, in E3S Web Conf, 200, (2020)

 https://doi.org/10.1051/e3sconf/202132507004

47. Kementerian Luar Negeri, Pustaka Aditya: Perpindahan Ibu Kota Negara di Mata Diaspora Jepang, (2020)
48. E D Canedo, A P M do Vale, R L Patrão, L C de Souza, R M Gravina, V E dos Reis, F L L Mendonça and R T de Sousa, Jr., Information and Communication Technology (ICT) Governance Proceses: A Case Study, Information **11**, 1, (2020)
49. R. Rachmawati, E. R. Ramadhan, and A. 'Ainur Rohmah, Aplikasi Smart Province 'Jogja Istimewa': Penyediaan Informasi Terintegrasi dan Pemanfaatannya, Maj. Geogr. Indones **32**, 1 (2018)
50. R. Rachmawati, ICT-Based Innovation in the Smart City Masterplan and Its Relation to Regional Planning, in IOP Conf. Ser. Earth Environ. Sci, 328, 1, (2019)
51. H. S. Chin, Putrajaya Administrative Centre of Malaysia Planning Concept and Implementation, Sus. urban devt Gov. Conf. Seoul, pp. 1–20, (2006)
52. R. Rachmawati, A.D. Sari, H.A.R. Sukawan, I.M.A. Widhyastana, and R.A. Ghiffari. The Use of ICT-Based Applications to Support the Implementation of Smart Cities during the COVID-19 Pandemic in Indonesia. *Infrastructure* **6**, 119 (2021)

Peri-Urbanization in DIY and Its Relationship to Sustainable Agricultural Lands Protection Program

Tita Thalia Nurcahyani[1], *and Djaka* Marwasta[1*]
[1]Faculty of Geography, Universitas Gadjah Mada, 55281, Sleman, Yogyakarta, Indonesia

Abstract. One of the main issues for cities in the world today is peri-urbanization. The rate of urbanization that occurred as an impact of industrialization resulted in a rapid population increase of urban areas. The increase of urban population causes the need of space for housing and other urban facilities also increase simultaneously, while administratively, urban area is relatively constant. The consequences is, the annexation of productive agricultural lands (especially rice fields) in the suburbs by urban functions. This study attempts to examine peri-urbanization and its relationship with sustainable agricultural land protection programs. This study aims to (1) assess the level of peri-urbanization in DIY; and (2) Elaborate the relationship between peri-urbanization and sustainable agricultural land protection programs. The spatial approach is used in this study considering that space is an integral part of a city which is a container for various urban activities and functions. This research is also in line with the ideas of a "sustainable city" that has been proclaimed by the United Nations (UN) and all countries in the world. The results of the study show that the implementation of regional regulations on sustainable agricultural land is quite effective in holding back the pace of peri-urbanization.

1. Introduction

One of the main issues for cities in the world today is the expansion of cities towards the outskirts [1], [2]. Urban physical development that goes beyond city administrative boundaries has an impact on the conversion of productive agricultural land into built-up land, decreased agricultural production, energy waste, increased pollution, and social problems in suburban areas [3]. Conversion of productive agricultural land in the suburbs is one of the impacts of urbanization. As a result, there has been a decrease in agricultural production because the area of agricultural land has decreased and the quality of human resources in the agricultural sector has tended to decline [4], [5]. This phenomenon is even more prevalent in developing countries that still rely on the agricultural sector in the national economy.

The high population growth in cities of developing countries has resulted in higher competition for land tenure in cities. The consequence of this phenomenon is the occurrence of peri-urbanization, on the grounds that land is still readily available and cheap to obtain [6]. This peri-urbanization is supported by the improvement of transportation facilities and infrastructure that connect the center and the suburbs of the city. The consequences of peri-urbanization include an increase in the level of energy use due to the increasing use of private modes of transportation. The increase in the number of motorized vehicles has the impact of increasing levels of pollution, especially air pollution [7], [4], [8], [9], [10]. The high level of air pollution is increasingly felt in the city center and the inner membrane of the city, due to the increasing tendency of traffic jams.

Some of the empirical facts that have been researched are; according to the Natural Resource Defense Council report in 2002, in the United States between 1960 and 1990, more than twice as much land has emerged as a result of the encroachment of the metropolitan city, while the population has only increased by less than half [11], [3]. It is noted that the average addition of newly built land is 148 hectares per hour. Kushner in 2006 presents the fact that in Atlanta between 1975 and 1992 there was a reduction of 1,540 square kilometers of vegetated land to built up land as a result of urban expansion. Atlanta is the least populated city among other cities in America (only a quarter of the population density of Los Angeles).

This periurbanization phenomenon also occurs in Indonesia, especially in Java. One of the areas experiencing periurbanization symptoms is DIY. The following is a graph of population numbers and population growth rates in Sleman, one of the districts in DIY that experienced the highest periurbanization.

* Corresponding author: jakamar@ugm.ac.id

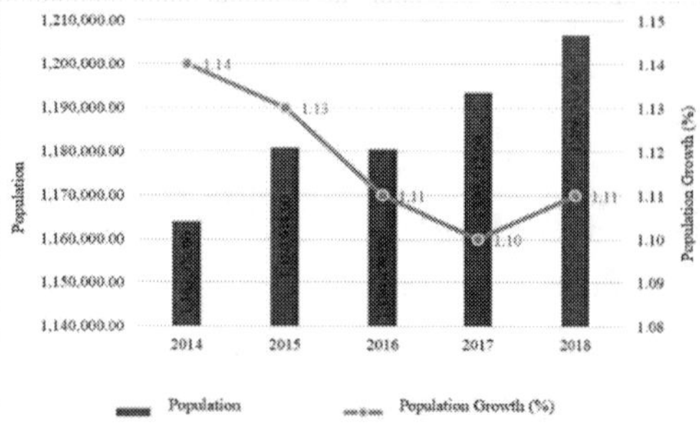

Fig. 1. Population and Its Growth Graphs of Sleman.

This research attempts to examine peri-urbanization and its relationship with sustainable agricultural land protection programs. The spatial approach will be used in this study considering that space is an integral part of a city which becomes a container for various urban activities and functions. This research is also related to the ideals of a "sustainable city" that has been proclaimed by the United Nations (UN) and all countries in the world [12]. The study conducted in this research emphasizes the spatial aspects, especially the analysis of spatial processes, and their impact on physical and social space.

2. Research Problem

Urban Sprawling is a dilemma problem faced by many metropolitan cities in developing countries. Meet the two primary human needs (food and housing) at the same time as well as same space, become a big dilemma. Fulfilling food needs requires productive agricultural lands, while providing the needs for housing forces city residents to seize fertile agricultural lands.

This phenomenon is very worrying for the future of farmers and the agricultural sector. Agricultural production is degraded because productive agricultural lands are converted into built-up land. Limitation of productive agricultural lands conversion into built up areas is required firmly, but it must be based on appropriate and accountable reasons.

DIY is an example of an area experiencing this massive land conversion occurence. To minimize the conversion of agricultural land to non-agricultural land, the Yogyakarta Regional Government have established regional spatial plans listed in the DIY Regional Regulation No.5 of 2019. Simultaneously, Sleman Regency as a part of DIY also launch the Regional Regulation of Sleman Regency No.12 of 2012. Spatial planning is very important, because Sleman Regency has high agricultural potential and is the main source of food production to support regional and national food supply. DIY government has also established a sustainable food agriculture area as part of the policy to protect sustainable food agriculture land, and is listed in the DIY Regional Regulation No. 10 of 2011 concerning the Protection of Sustainable Food Agricultural Land.

From the description above, there are two problems faced in the research area: (1) What is the peri-urbanization rate in DIY ?; and (2) How does the PERDA-regulated sustainable agricultural land protection program affect peri-urbanization process? Based on that research problems, this study aims to: (1) assess the level of peri-urbanization in DIY; and (2) Elaborate the relationship between peri-urbanization and sustainable agricultural land protection programs.

3. Research Methods

Based on the research objectives, this research is an exploratory-descriptive study, it aims to determine the process of periurbanization and its effects on environmental conditions. The analytical approach used is a deductive approach. In the deductive approach, it starts with a general study of phenomena first, then is used as a basis for understanding more specific phenomena.

In relation to the research object, this research used survey method, which is to observe and measure various variables used to assess the process and effect of periurbanization. The data collection technique uses a sampling methods, which is in the form of an area sample. The analysis used is mainly spatial analysis. The household sampling technique used was purposive sampling. The reason for determining the selected sample is its level of importance.

Based on the analytical method, this research is a quantitative research. The location of this research is DIY. Specifically, the research location is focused more in the Sleman Regency. This location was chosen based on the consideration of the phenomenon of peri-urbanization and the existence of a local regulation on sustainable agricultural land protection. The condition of the research area is in the form of fertile agricultural lands around the city of Yogyakarta which have a high level of productivity, but have become the locus for the expansion of the urban area of Yogyakarta. As a result, many agricultural lands have been invaded and converted into developed land. This phenomenon is quite in accordance with the objectives of the research which is to examine the process and impact of periurbanization.

To be able to study the peri-urbanization process carefully, administrative boundaries and morphological boundaries are used together. The boundaries used to determine the morphological boundaries of urban refer to Pryor (1968), which is areas that have more than 75% urban land use. In addition, this study also uses the types of urban physical growth as suggested by [13], [14], [15] as a reference for determining the urban development process in the study area.

This research is including in the field of geography with the main focus on the spatial approach. The main object of study study is the phenomenon of urban encroachment towards the outskirts (peri-urbanization) with an emphasis on assessing urban function invasion, tracing the variables and factors involved in the process of encroachment, and their effects on agricultural land.

The identification of spatial process is carried out by tracing the increasing of urban built land in the last 10 years (2011-2020). Multi-time satellite imagery is used as a data source as the basis for spatial analysis by overlapping maps between two time basis. This identification process is intended to obtain a picture of the peri-urbanization that has occurred. The activities carried out in this research include: interpretation of digital images in the laboratory, observation and physical measurements in the field, and data analysis. Activities in the laboratory include: maps making, maps overlying, compiling a field observation checklist and tracing secondary data, compiling a questionair for informants, and determining area samples.

Field activities carried out included interviews of several informants, especially from the bureaucracy and farmers. Secondary data documentation and field photographs were also carried out during field activities. Research activities were resumed in the laboratory for the process of field data input, data editing, statistical analysis, and presentation of some research information. To support all research activities in accordance with these procedures, several sets of materials and tools, variable measurements, procedures for determining the population, samples and units of analysis are needed, as well as data analysis procedures.

The data used in this study are primary and secondary data. Primary data is data obtained from direct observations and/or measurements in the field or identified from research materials directly in the laboratory. Some of the primary data collected in this study are maps of satellite imagery interpretation, field observation data, and data taken from interviews with selected respondents. Secondary data is data obtained from another sources, either from existing publications and reliable data validity and accuracy, or data obtained from previous research results. Some of the secondary data collected in this study include thematic maps, tables and or graphs (especially population data), and publications from various agencies as well.

The stages carried out in this research include (1) the pre-field stage; (2) stage of field activities; and (3) the post-field stage. The pre-field stage is a series of activities related to preparation to support the implementation of field activities, covering a variety of activities ranging from literature studies, tracing secondary data, preparation of research materials to be used in the field (list of fields and field maps), and sample determination. area and informants.

Some indicators and how to measure a sustainable agricultural land protection program is obtained by comparing the area of agricultural land stipulated in the PERDA as sustainable agricultural land with changes in actual land use. This comparison is taken by using of the Geographical Information System. The sustainable indicator is determined by the percentage of the area of sustainable agricultural land that still remains as existing agricultural land.

The data analysis framework in this study consists of two types, namely spatial analysis and correlation analysis. The spatial analysis is intended to answer the first objective of the study, which is the identification of peri-urbanization process. Correlation analysis is used to elaborate on the second research objective, which is to examine the relationship between peri-urbanization and the implementation of sustainable agricultural land protection programs. Figure 2 shows the flow of research activities stages.

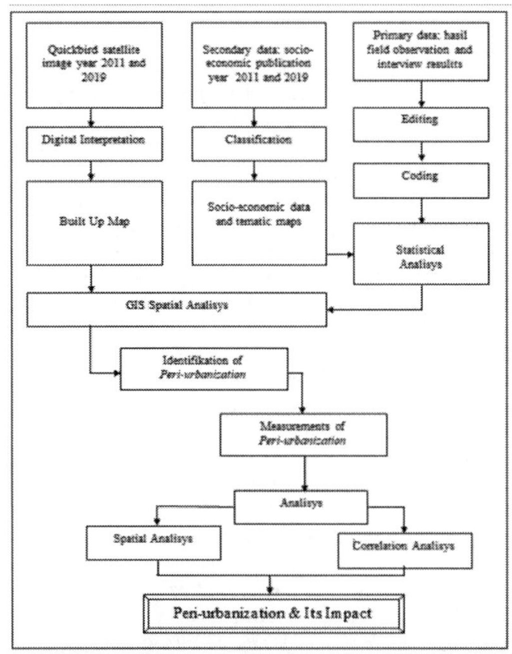

Fig. 2. Research Methods Flow Chart

To determine the agricultural land shortening process, it is carried out by using a spatial approach. The data that has been obtained are processed using ArcGIS software and processed using the overlaying method of zoning maps for sustainable food agriculture and land use maps in Sleman Regency to determine the area of agricultural land. Then the data obtained is calculated using the Conversion Rate formula (V) as follows:

$$V = \frac{L_t - L_{t-1}}{L_{t-1}} \times 100 \quad (1)$$

V = agricultural land shortening (%)
Lt = agricultural area in year t
Lt-1 = agricultural area before year t

A negative V value indicates that existing agricultural land are experiencing land conversion or depreciation.

4. Literature Reviews

The world's population continues to increase from year to year. According to the United Nations, the world's population has doubled in a period of 50 years between 1950, from 3 billion to 6 billion in 2000 [16]. The high total population growth of the countries in the world is mostly contributed by the population growth in developing countries.

In the context of developing countries, the high population growth in urban areas is more dominant due to the uncontrolled rate of urbanization. Cities as centers of socio-economic and cultural activities have experienced growth in line with population growth, both non-physical (social, economic, political and cultural) and physical growth. However, many cities that are experiencing growth actually have a negative impact on the physical and sociocultural environment in and around them [17]. Expectations regarding a city that can provide a safe, comfortable, and prosperous place for its residents have not been realized because the growth of the city, especially the growth in population and building density, is not as expected.

In a global context, a new form of urban development is characterized by the appearance of a new concentration of labor and urban activities in areas outside the city center, as a result of the flexibility of production systems and technology [18], [19]. In the era of 1980 to 1990, from several dictionaries appeared several terms related to urbanization such as edge city, megalopolis, technoburbs, flexspace, peperoni-pizza cities, a city of realms, superburbia, disurb, perimeter cities, outer cities, technopolis, heteropolis, exopolis and perimetropolitan bow waves. In France, the term periurbanisation is used to refer to any place outside the city and its surroundings with a function as a place to live. The term rurbanisation was introduced around 1970 which is used to be the same as periurbanisation.

Urban area is a term used to identify an area that has a large population, is based on a non-agricultural economy, and its land use is dominated by developed land [20], [21], [22]. The definition of a city or urban area can be viewed from various perspectives, including juridical-administrative, morphological (physical), demographic (population), population density, functional in organic areas, and socio-economic [23]. Although urban areas can be viewed from various perspectives, in general the boundaries of urban areas are always related to: (1) population (number and or population density); (2) spatial function (economic basis); and (3) morphology.

An urban area is an area that has an atmosphere of modern life and life which is characterized by, among others: (1) Facilities and infrastructure for urban life (buildings, housing, transportation, communication, industry, recreation, etc.); and (2) specific social, economic, cultural, political, demographic [24] [25] [26]. To be categorized as an urban area, an area must meet certain requirements, which is called the urban threshold. Each country has its own urban threshold standard. For example, it can be stated that the urban threshold standard according to the SMSA (Standard of Metropolitan Statistic Area) in effect in the United States includes: (1) a minimum population of 50,000 people; (2) non-agricultural land use is more than 75%, (3) workers in the city center are more than 15% or workers who live in the city center are more than 25%, (4) there are public transportation facilities; and (5) more than 50% of high density settlements (according to a certain threshold), in this case the United States uses a minimum population density of 1000 people per square mile [30] in [27].

In Indonesia, Law Number 18 of 1965 has been stipulated which regulates the boundaries of an area as a city or an urban area, namely an area with a population of more than 50,000 people [28]. Based on the hierarchy, there are three city hierarchies according to this Law, namely Kota Praja with a population of between 50,000 and 75,000, Municipal City with a population of between 75,000 and 100,000, and Kota Raya with a population of more than 100,000. In line with the high population growth, especially in urban areas, this limit changes. Based on Law Number 5 of 1974, for regions in Java, a minimum limit of 150,000 people is defined as cities, while outside Java, the minimum limit is 50,000 people [23].

There are several terms related to the definition of urban, including city, urban areas, and urbanity. City is a term to indicate a certain area associated with the authority of regulating territory by the city government [1]. An urban area is an area bounded by juridical administrative boundaries and is under the authority of the city government. Urbanity is areas that have the characteristics of a city, both in physical, social, economic and cultural terms. In Law No. 22/1999, it is explained that what is meant by urban areas is an area that has non-agricultural main activities, with the arrangement of the functions of the area as a place for urban settlements, centralization and distribution of government services, social services, and economic activities.

Conceptually, urban areas are easy to define, but in practice the delineation of urban boundaries is not easy to do. To limit urban areas, especially from a morphological (physical) point of view, a clear definition of urban land is required. Urban land can generally be divided into functional and physical definitions. Urban functionally related to industrial activities, settlements, trade, services, and so on. Physically associated with high population density and land cover (land use), which is characterized by built-up lands.

Some experts have described quantitatively to limit urban land use, one of which is Pryor. The concept put forward is based on the percentage of urban land use, the percentage of rural land use and the percentage of distance from the main urban land to the main rural land which are members of the rural-urban land use triangle. Urban areas are characterized by residential land, commercial land, and industrial land with a higher density than the average density for rural areas. Rural areas are characterized by a lower population density, a higher proportion of vacant land and agricultural land, as well as changes in land use and non-dynamic daily population mobility to urban areas [2].

Pryor made a model based on a gradual transformation from city to village or vice versa village to city, where the farther away from the real urban area, the more blurred the urban appearance and the clearer the distant decay principle. Morphologically, the closer to the urban area, the dominance of urban land use forms is getting bigger. There are 4 gradual zones from urban to rural areas, namely:

(1) Urban areas
(2) Suburban areas
(3) Rural areas
(4) Rural areas

Based on the Pryor model, suburban areas are areas located between urban built-up land and areas with a maximum urban land use area of 50%, and vice versa are rural areas.

One of the interesting phenomena of peri urban urban areas in Asia is the existence of commuters. Commuters in developing countries are low-income people who are trying to find additional income in cities. Over time, it turns out that in this peri-urban area pockets of very good quality and luxurious settlements appear, so that the activities of moving from the suburbs are also carried out by the upper class, although there are not many of them.

5. Results and Discussion

Two important findings in this study are (1) periurbanization process that occurred in DIY; and (2) the relationship between the implementation of regional regulations for the protection of perpetual agricultural land and the periurbanization process. The discussion about the findings of this study focuses on that two issues.

5.1 Periurbanization Process

This research was conducted in Sleman Regency, DIY. Sleman Regency is an area that has the largest fertile agricultural land and at the same time has experienced intensive land conversion. Figure 3 shows an administrative boundary of Sleman Regency.

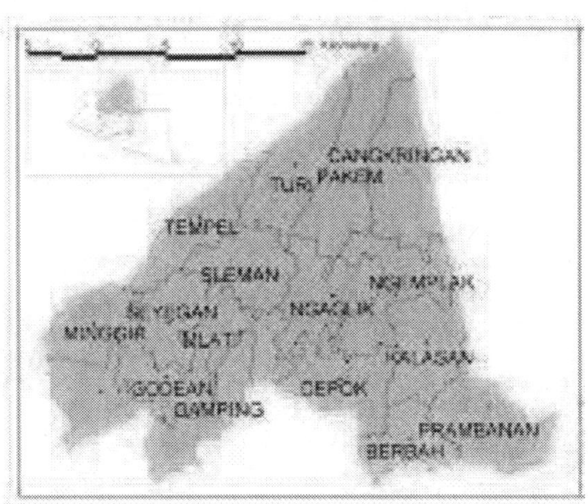

Fig. 3 Administrative Map of Sleman Regency

The conversion of fertile agricultural land to built up land mainly occurs in areas adjacent to the city of Yogyakarta. This process is known as periurbanization. The positive impact of the periurbanization process is an increase in economic growth, and this has happened in Sleman Regency. Figure 4 is a graph of the economic growth rate for 5 years.

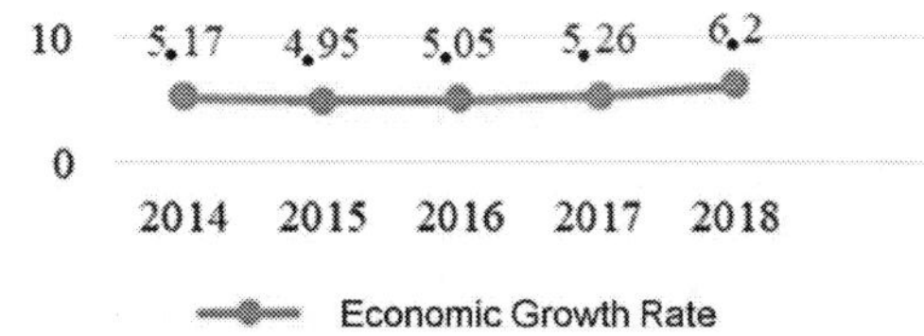

Fig. 4 Economic Growth Rate of Sleman Regency

Based on the graph, it can be seen that the rate of economic growth in Sleman Regency is always increasing, and this is due to the conversion of agricultural land to developed land.

Referring to BPS data for 2020, the area of agricultural land always tends to decline, on the other hand, developed land always tends to increase. Table 1 shows the changes in land area for each land use.

Table 1. Landuse Change in Sleman Regency (Ha)

No	Landuse	2014	2015	2016	2017	2018
1	Housing	18,591	18,627	18,756	18,786	18,823
2	Ricefield	24,719	24,628	24,577	24,550	24,517
3	Dry Field	3,924	3,922	3,922	3,922	3,918
4	Forest	53	53	53	53	53
5	Bare land	1,264	1,264	1,264	1,264	1,264
6	Others	8,932	8,989	8,911	8,908	8,908
	Total	57,483				

Source: BPS Sleman, 2020 [29]

Most of the conversion of agricultural land to developed land occurs on fertile land. Thus, it can be stated that periurbanization has a positive impact on the urban economy, but has a negative impact on agricultural production. To anticipate the worsening impact of periurbanization on agricultural production, the government has enacted a regional regulation on perpetual agricultural land. The hope is that food security will not be threatened by the uncontrolled periurbanization process.

The regional regulation has stipulated that protected lands do not change their function. The procedure is that the government will not issue title certificates if the agricultural land which is designated as perpetual rice field is converted into developed land. Figure 5 shows the distribution of paddy fields designated as perennial agricultural land in Sleman Regency.

 https://doi.org/10.1051/e3sconf/202132507005

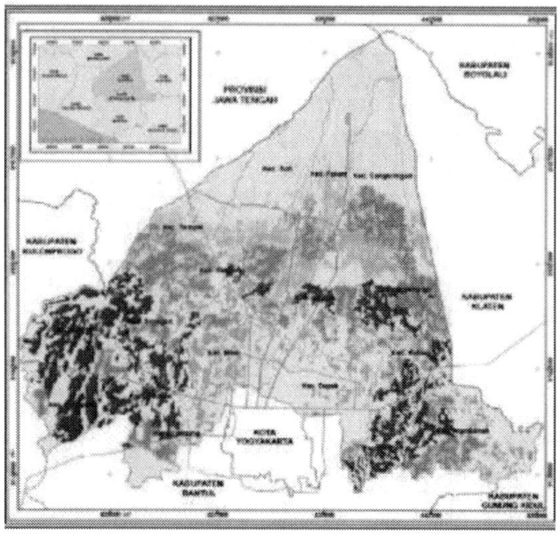

Fig. 5 Sustainable Agricultural Land Distribution

5.2 Relationship between periurbanization and agriculture land protection

Since the Yogyakarta Special Region and Sleman regencies enacted regional regulations on perpetual agricultural land, the rate of periurbanization has decreased. In Figure 5, it can be seen that since 2013, the rate of conversion of agricultural land has decreased. This is the impact of the implementation of the Perpetual Paddy Field Regional Regulation.

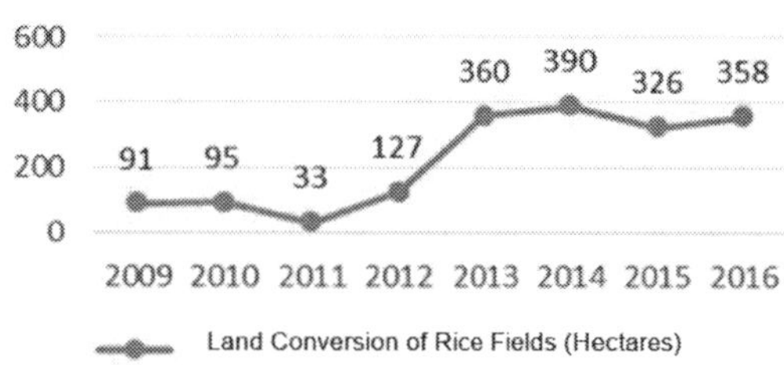

Fig. 6. Trends of Land Conversion in Sleman

Along with the times and population growth, land conversion is mostly done to meet human needs. Change in land use or land use change or commonly referred to as land conversion is a change in the function of part or all of the land area from its original function to another function one time to the next

Land conversion is the process of changing the use of part or all of the land from its original function to other functions. The agricultural land that decreased causes a decreasing of agricultural production. The conversion of paddy fields is the most intensive land use change for conversion to settlement as a result of population growth. Agricultural land conversion generally occurs in two stages, namely the land owner relinquishing his rights to non-farmers, then followed by the new owner using the agricultural land for non-agricultural activities.

Land conversion tends to be in line with population growth. The rate of population growth will indirectly affect the conversion of agricultural land to developed land and settlements. The high rate of population growth will be directly proportional to the growth rate of settlements. The factors that directly affect land conversion are the growth of infrastructure development including transportation facilities, land growth for industry, and growth of residential facilities, while indirectly influenced by population, economic structure, urbanization, and consistency in the implementation of the spatial plan.

The increase in population results in increased population pressure on land and an impact on land use changes. Changes in land use in general can occur because of the need to meet the needs of an increasing population and increasing demands for a decent quality of life. There are three important factors that cause agricultural land conversion, namely: 1) external factors which are the main factors caused by the dynamics of urban development, demography and economy; 2) internal factors, namely factors arising from farmers and agricultural land owners due to the crisis in their socio-economic conditions; 3) the policy factor, namely the regulatory aspect that regulates everything concerning agricultural land that has been determined by the central and regional governments.

Land conversion from agriculture to non-agriculture can have a negative impact on agriculture, either directly or indirectly. The direct impact is a decrease in agricultural production capacity and an increase in food problems, while the indirect impacts include damage to the irrigation system in agricultural production areas due to built-up land, loss of investment that has been invested in building reservoirs, irrigation networks, and printing rice fields. In addition, land conversion also has a negative impact on social activities, such as loss of livelihoods as farmers, reduced job opportunities and income opportunities for farming and all economic activities that intersect with farming. The impact of land conversion will be permanent because most of the converted land has never been returned to agricultural land.

Sustainable agriculture is the management of resources to meet basic human needs for clothing, food and signage while maintaining the quality of the natural environment itself and preserving it. To create sustainability, an appropriate policy is needed. Policy is an effort or action that is strategic in the long term and comprehensive to achieve the desired goals. The central and local governments have made several policies for land control and protection. However, at the present time we still encounter a lot of land that is still being converted.

6. Conclusions

1. Periurbanization rate in DIY, especially in Sleman Regency, tent to be very high before year 2013, and declining since year 2013.

2. Regulation of sustainable agricultural land protection program (PERDA) affected to reducing peri-urbanization process.

Very high appreciation to Faculty of Geography Universitas Gadjah Mada, which provide an opportunity to conduct this research, and to all of the informants involved in this research.

References

1. Yunus, H.S., Megapolitan: Konsep, Problematika, dan Prospek, Yogyakarta: Pustaka Pelajar (2006)
2. Yunus, H.S., Dinamika Wilayah Peri Urban: Determinan Masa Depan Kota, Yogyakarta: Pustaka Pelajar (2008)
3. Paterson, R., Handy, S., Kockelman, K., Bhat, C., Rajamani, J., Jung, J., Banta, K, Desai, U., and Waleski, J., Techniques for Mitigating Urban Sprawl, Center for Transportation Research, The University of Texas at Austin (2003)
4. Kahn, M.E., The Environmental Impact of Suburbanization. Journal of Policy Analysis and Management; 19(4):569-586. (2000)
5. Hess, G., Just What is Sprawl, Anyway? Journal of City and Regional Planning; 26 (2):pp 11-26. (2001)
6. Yunus, H.S., Perubahan Penggunaan Lahan di Daerah Pinggiran Kota: Kasus di Pinggiran Kota Yogyakarta; Disertasi, Yogyakarta: Universitas Gadjah Mada, Yogyakarta (2000)
7. Freeman, L., The Effects of Sprawl on Neighborhood Social Ties. Journal of the American Planning Association; 67(1):69-77(2001)
8. Duncan, J.E., The Search for Efficient Urban Growth Patterns. Tallahassee, FL: Florida Department of Community Affairs (1989)
9. Burchell, R.W. and Shad, N.A., The Evolution of the Sprawl Debate in the United States, West Northwest Winter; 5(2):137-160 (1999)
10. Frank, J.E., The Costs of Alternative Development Patterns: A Review of the Literature, Washington, DC: Urban Land Institute (1989)
11. Benfield, F.K, Raimi, M.D., and Chen, D.D.T., Once There Were Greenfields:How Urban Sprawl is Undermining America's Environment, Economy and Social Fabric, Washington DC: Natural Resources Defense Council (1999)
12. Yunus, H.S., Pembangunan Kota Berkelanjutan: Permasalahan dan Strategi Pencapaiannya, Yogyakarta: Universitas Gadjah Mada, Yogyakarta (2004)
13. Ellman, T., Infill: the cure for sprawl? Arizona Issue Analysis 146. [online] URL:http://www.urbanfutures.org/p82897.html (1997)
14. Antrop. M., Changing Patterns In The Urbanized Countryside Of Western Europe, Landscape Ecology, Vol 15. Pp: 257-270 (2000)
15. Wilson, E.H., Development Of A Model To Quantify and Map Urban Growth, University of Connecticut (2002)
16. United Nations, World Urbanization Prospects The 2001 revision Data Tables and Highlights. ESA/P/WP.173.http://www.un.org/esa/population/publications/ wup2001/wup2001dh.pdf (accessed 20 Jan. 2006)
17. Parker, S., Urban Theory and the Urban Experience: Encountering the City, London: Routledge (2004)
18. Fujita, M., Krugman, P., and Venables, A.J., The Spatial Economy:Cities, Regions, and International Trade, Cambridge: MIT Press (1999)
19. Pendall, R., Fulton, W. and Harrison, A., Losing Ground to Sprawl? Density Trends in Metropolitan America. Fair Growth: Connecting Sprawl, Smart Growth, and Social Equity; Georgia World Congress Center, Atlanta Georgia. Washington, DC: Fannie Mae Foundation (2000)
20. Pacione, M., Urban Geography: a Global Perspective (2nd Edition), London: Routledge (2005)
21. Herbert, D.T., and Thomas, C.J., Urban geography: A First Approach, Chichester: Wiley (1982)
22. Witherick, M., Ross, S., and Small, J., A Modern Dictionary of Geography (4th edition), New York: Oxford University Press (2001)
23. Yunus, H.S., Manajemen Kota: Perspektif Spasial, Yogyakarta: Pustaka Pelajar (2005)
24. Knox. P.L., Urban Geography: An Introduction to Urban Geography, New Jersey: Prentice Hall (1994)
25. Bintarto, Interaksi Desa-Kota dan Permasalahannya, Jakarta: Ghalia Indonesia (1989)
26. Bintarto, Urbanisasi dan Permasalahannya, Jakarta: Ghalia Indonesia (1986)
27. US Census Bureau, [Web Page]; accesed 15 Oktober 2008; URL:http://www.census.gov (2000)
28. Yunus, H.S., Klasifikasi Kota, Yogyakarta: Pustaka Pelajar (2005)
29. BPS, Statistik Propinsi DIY 2020, BPS Propinsi DIY (2020)
30. Pryor, R.J.. Defining The Rural-Urban Fringe. Social Force. Vol. 47. Issue 2. Pp: 202-215. (1968)

Study of the service functions of health facilities in Yogyakarta Special Province

Mohammad Isnaini Sadali[1], *Raden* Rijanta[2], *Lutfi* Mutaali[2], *and Andri* Kurniawan[2]

[1]Doctoral Program, Faculty of Geography, Universitas Gadjah Mada, Yogyakarta, Indonesia
[2]Department of Development Geography, Faculty of Geography, Universitas Gadjah Mada, Yogyakarta, Indonesia

Abstract. The provision of public service facilities in settlement and regional environments is the right of the community to support a decent life and livelihood also to improve the quality of life. Therefore, the availability of health facilities must be considered so that every Indonesian can obtain their right to health services. This study aims to analyze the availability of health service facilities and the service function of health facilities in Yogyakarta Special Province. As a differentiator with previous research, the researcher conducted a analyze comprehensively by combining analysis of the centrality index and the service function to determine the hierarchy of health service centers. The location of this research is in the Yogyakarta Special Province with the smallest area analysis unit is the sub-district. The results showed that the availability of low order goods health service facilities was generally fulfilled in all sub-districts of Yogyakarta Special Province. Cities are still areas of high concentration of health services, this is reinforced by the highest centrality index and hierarchy of health facilities in Yogyakarta City.

1 Introduction

In accordance with the mandate of The 1945 Constitution (UUD) and Article 28 H of the Amendment of 1945 Constitution (UUD), health services are one of the basic rights of every Indonesian, so every citizen has the right to get good and decent health services. The existence of health service facilities affects the public health status of a country. Law Number 36 of 2009 concerning Health explains that a health service facility is a tool and/or place used to carry out health service efforts, whether promotive, preventive, curative, or rehabilitative carried out by the government, local government, and/or the community [9].

Indonesia's total population in 2019 reaches 268,074,565 people [4] and requires facilities and health personnel as many as possible so that they can be evenly distributed throughout the region [12]. Based on The Minister of Health Regulation Number 75 of 2014 concerning Puskesmas (Community Health Centre), every Puskesmas in 2019 can provide services according to standards [10]. Therefore, it is necessary to increase health services in terms of the number and types of health facilities, health human resources, health financing, and disease control. The distribution of medical personnel and health facilities such as Puskesmas and hospitals is currently still piling up in the city. The central area of activity is the central of regional activity including service facilities (including health facilities) tend to be more complete in cities [1]. The high number of Puskesmas in each region cannot guarantee whether these facilities are adequate to support public health in the regions.

The commitment of the Indonesian Government in providing more adequate health services for the community in the next 5 years is demonstrated by the implementation of the National Health Insurance by BPJS Kesehatan (Health Social Security Agency). This policy needs to be balanced with monitoring and evaluation to ascertain whether the National Health Insurance policy by BPJS Kesehatan can improve the both quality of health services and access to reach them for all Indonesian citizens with the principle of justice [2][11]. The government has 3 (three) basic plans, namely: development of health personnel from 2011 to 2025, improvement of the BPJS system which is currently still under development, and expansion of health infrastructure. The availability of health insurance (insurance) must be balanced with good health quality [15].

In order to strengthen primary health services, there are three indicators related to the implementation of Puskesmas in the 2015-2019 RPJMN and the Ministry of Health's Strategic Plan 2015-2019, namely: 1) Sub-Districts that have at least 1 accredited Puskesmas; 2) the number of non-inpatient Puskesmas and in-patient Puskesmas that provide services according to standards; and 3) the number of Puskesmas in cooperation with the Blood Transfusion Unit and hospitals in providing blood services to reduce the Maternal Mortality Rate (MMR). Through the strengthening of primary health services, it is hoped that it will be able to improve the level of public health in each region. Of course, human resource support and health infrastructure are needed to achieve these targets.

One of the health efforts made by the Provincial Health Office of the Yogyakarta Special Province

https://doi.org/10.1051/e3sconf/202132507006

(DIY) in carrying out its duties and functions is basic health services and referrals. Basic health service facilities are carried out by the Puskesmas and equal health facilities, while referral health services are managed by the Hospital. There are 121 Puskesmas in DIY, consisting of 45 nursing and 76 non-nursing public health centres [5]. In 2019, the achievement of improving the quality of health facilities in DIY was shown by the accreditation of all Puskesmas. The highest number of Puskesmas that provides nursing is in Bantul Regency (17 Puskesmas), while the smallest is in Yogyakarta City (2 Puskesmas). Yogyakarta City has a very close distance between Puskesmas and there are several hospitals that provide nursing facilities so that the Puskesmas that provide nursing in Yogyakarta City is deemed unnecessary. Meanwhile, the access and distance of several areas in Bantul Regency are quite far from the hospital, so that a Puskesmas with provide nursing is needed.

This research was started by identifying the number and mapping the existence of health service facilities in Yogyakarta Special Province so that the distribution could be seen for the sake of further analysis. The existence of health service facilities in Yogyakarta Special Province needs to be analyzed the level of availability and the function of its services to ensure public access to health services. This study seeks to provide an explanation regarding the distribution of health service facilities in Yogyakarta Special Province and to compare the service functions of health facilities in each region.

2 Method

This research uses quantitative research methods, with secondary data as the main data and the analysis is carried out in an explanative manner. The research location is administratively located in the Yogyakarta Special Province. Some of the main secondary data collected to support this research are:

- Total population.
- Number and types of health service facilities in 2015 and 2019.
- Regency and City in Figures in 2015 and 2019.
- Yogyakarta Special Province (DIY) Health Profile.
- Planning documents.

The determination of the 5-year period (2014-2019) in this analysis refers to the issuance of The Minister of Health Regulation Number 75 of 2014 which has a target that in 2019 health services can meet the standards and the numbers meet the needs of the community. Analysis of service functions can be seen from the comparison between the availability of health service facilities and comparison variables. This can be seen from the size of actual users, potential users, total population, and by standard comparisons. The number of units and types of health service facilities in DIY is considered as the principle of serviceability (threshold). This principle is conceptually referred to from the theory of the central place presented by Christaller, 1933 [6]. The standard of serviceability of health facilities in Indonesia refers to the Minister of Public Works Regulation No. 41/PRT/M/2007 concerning Guidelines for Technical Criteria for Cultivated Areas and SNI 03-1733-2004 on Procedures of Planning for Housing Environment in Urban Areas.

The basis for providing health facilities is the number of people served by health facilities. The basis for this provision will also consider the spatial approach of the existing environmental units or groups. Of course, this can be related to the formation of building groups/blocks which will be formed according to the context of the environment. Whereas the placement of the provision of health facilities [7] will consider the reach of the service area radius related to the basic needs of health facilities that must be met to serve in certain areas. Table 1 shows the minimum population to support the provision of a health facility in the region The greater the serviceability of the facility, the greater the minimum population requirement (threshold) that must be met.

Table 1. Total Population Supporting Health Facilities

Type of Health Facility	Number of Supporting Population
Hospital	240.000
Maternity Hospital	30.000
Polyclinic	30.000
Puskesmas (Community Health Centre)	120.000
Supporting Puskesmas	30.000
Pharmacy	30.000
Posyandu (Integrated Health Centre)	1.250

Note: Reference is taken from SNI 03-1733-1989, Planning Procedures for Urban Housing Areas.
Source: SNI 03-1733-2004, Procedures of Planning for Housing Environment in Urban Areas.

3 Result

3.1 Availability of Health Service Facilities in Yogyakarta Special Province

Regional development as a central activity will be oriented towards providing infrastructure or supporting facilities in it capable of supporting the activities of the area [3]. The availability of health facilities is needed to provide health services to the community and has a very strategic role in accelerating the improvement of public health status as well as controlling population growth. The government needs to encourage the improvement of the quality of health service facilities, both public and private health facilities in order to meet the expectations and needs of the community [13]

The availability of health facilities in DIY can be measured by looking at the availability of hospital facilities, maternity hospitals, polyclinics, health centres, auxiliary health centres, pharmacies, and Posyandu. The suitability of the availability of health facilities is measured by considering the number of residents and existing health facilities. Measurement of the availability of health facilities in DIY is also carried out between 2015 and 2019 so that progress could be seen. There are three categories of suitability for the availability of health facilities, namely: highly fulfilled, fulfilled, and unfulfilled. The highly fulfilled category is obtained if the number of existing health facilities exceeds the ideal number of health facilities. Fulfilled categories are obtained if the number of existing health facilities is the same as the ideal number of health facilities. The unfulfilled category is obtained if the number of existing health facilities is less than the ideal number of health facilities.

The sub-districts in regency/city which are predominantly rural-urban have a better level of availability of health facilities. For example, the availability of maternity hospitals in sub-districts in Gunungkidul has a "not fulfilled" category. On the other hand, the availability of maternity hospitals in the sub-districts in the Yogyakarta City has been "fulfilled" and even "highly fulfilled". Apart from the rural-urban category, the difference in the level of availability of health facilities can also be seen from the types of facilities. Health facilities with a higher level of service (high order goods services), such as hospitals and maternity hospitals, are generally not fulfilled according to regional needs. Meanwhile, the availability of health facilities with a lower level of service (low order goods services) such as Puskesmas, pharmacies, and Posyandu, tends to exceed the ideal number required by the regions.

The development of the availability of health facilities in DIY from 2015 to 2019 was volatile. Generally, there has been a decrease in the number of low-order health facilities such as Posyandu, but they still meet their ideal needs. On the other hand, there has been an increase in the number of higher order health facilities such as hospitals and maternity hospitals. The amount of increase is not significant but is able to meet the ideal need for health facilities in the area. Fig. 1 shows a number of sub-districts which meet the ideal number of health facilities in each regency/city in DIY. In general, a number of sub-districts that do not meet the ideal number of health facilities are always smaller than the number of sub-districts that do or are highly fulfilling.

Based on the type of health facilities, regency/city that meet the ideal number of hospital facilities are Yogyakarta City, Bantul Regency, Kulon Progo Regency, and Gunungkidul Regency. Meanwhile, in Sleman Regency, there are still 2 sub-districts that do not meet the ideal number of hospitals. Furthermore, districts/cities that meet the ideal number of maternity hospitals are Yogyakarta City, Sleman Regency, Bantul Regency, and Kulon Progo Regency, while Gunungkidul Regency does not meet the ideal number of maternity hospitals needed at all. The availability of many polyclinics does not meet the ideal number of needs per sub-district, including 2 sub-districts in Yogyakarta City, 6 sub-districts in Sleman Regency, 4 sub-districts in Bantul Regency, 5 sub-districts in Kulon Progo Regency, and 2 sub-districts in Gunungkidul Regency. Details of the classification of the suitability of the availability of health facilities and their distribution in the districts/cities of Yogyakarta Special Province according to the type of facilities can be seen in Table 2.

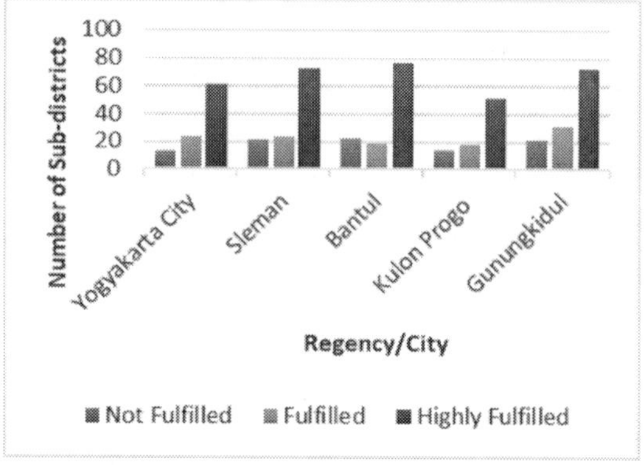

Fig. 1. Graph of the Number of Sub-districts that Meet the Ideal Number of Health Facilities in DIY in 2019

Source: Secondary Data Analysis, 2020

Table 2. Sub-districts that Meet the Ideal Number of Health Facilities in Yogyakarta Special Province in 2019

Type Facilities	Yogyakarta City			Sleman			Bantul		
	NF	F	HF	NF	F	HF	NF	F	HF
Hospitals	0	6	8	2	13	2	0	8	9
Maternity Hospitals	4	7	3	12	3	2	16	0	1
Polyclinic	2	5	7	6	3	8	4	5	8
Puskesmas	0	1	13	0	2	15	0	0	17
Supporting Puskesmas	7	5	2	1	3	13	1	3	13
Pharmacies	0	0	14	0	0	17	0	3	14
Posyandu	0	0	14	1	0	16	2	0	15
Total	13	24	61	22	24	73	23	19	77

Type Facilities	Kulon Progo			Gunungkidul		
	NF	F	HF	NF	F	HF
Hospitals	0	6	6	0	14	4
Maternity Hospitals	6	5	1	18	0	0
Polyclinic	5	5	2	2	6	10
Puskesmas	0	0	12	0	0	18
Supporting Puskesmas	0	0	12	0	0	18
Pharmacies	3	2	7	2	11	5
Posyandu	0	0	12	0	0	18
Total	14	18	52	22	31	73

Information: Not Fulfilled (NF); Fulfilled (F); Highly Fulfilled (HF)

Source: Secondary Data Analysis, 2020

The need for health facilities at a lower level/order of service (low order goods services), such as Puskesmas, has been met in every sub-district in DIY. However, some of The Supporting Puskesmas are still not fulfilled, namely 7 sub-districts in Yogyakarta City,

1 sub-district in Sleman Regency, and 1 sub-district in Bantul Regency. Pharmacy facilities are still not fulfilled in several sub-districts, namely 3 sub-districts in Kulon Progo Regency and 2 sub-districts in Gunungkidul Regency. Then, the availability of Posyandu facilities tends to exceed the ideal number needed in each sub-district. The availability of health facilities greatly determines the coverage of health services. Variations in the level of health facilities (primary, secondary and tertiary) and the range/coverage of health services can ensure good quality health services [17].

3.2 Service Functions of Health Facilities in DIY

The analysis used to explain the service functions of health facility is the centrality index and the hierarchy of service functions. The Centrality Index is used to see the service capability of a centre in terms of a number of health facility units in the service centre. The centrality index analysis takes into account a number of health facility units, so the assumption used is that the region with the most health facility units is the highest order and is designated as a health service centre. The Centrality Index not only looks at a number of functions or facilities, but also based on the frequency of the existence of health functions or facilities.

The service function of health facilities provides an indication of the quality and adequacy of services, so that the better the serviceability, the better the quality of the facilities. In addition, the service function is also able to show the allocation of space or areas in need of health facilities based on a number of inhabitants. The hierarchy of the health facilities service centre is measured using the cumulative value of the centrality index and service function. There are three hierarchies that describe the structure or hierarchy of health service centres in an area, how many functions there are, how many types of functions are served and how many people are served, and how much is the frequency of the existence of a function in a residential area. Hierarchy I is the centre of health area, while hierarchy II and III act as a hinterland [8]. Below (Table 3 and Table 4) are the results of the analysis of the hierarchy of health service centres in DIY.

Table 3. Hierarchy of Health Facilities Service Centres by Regency/City in DIY Province in 2015

Regency/ City	Centrality Index	Service Function	Cumulative (IS+SF)	Hierar-chy
Bantul	6.29	2.64	8.92	II
Kulon Progo	5.43	1.76	7.19	II
Gunungkidul	2.14	1.13	3.27	III
Sleman	5.54	3.36	8.90	II
Yogyakarta	98.02	22.68	120.70	I

Source: Secondary Data Analysis, 2020

Table 4. Hierarchy of Health Service Centre Facilities by Regency/City in DIY Province in 2019

Regency/ City	Centrality Index	Service Function	Cumulative (IS+SF)	Hierar-chy
Bantul	6.29	2.67	8.96	II
Kulon Progo	5.43	1.85	7.28	II
Gunungkidul	2.14	1.11	3.26	III
Sleman	5.54	3.41	8.96	II
Yogyakarta	98.02	22.62	120.64	I

Source: Secondary Data Analysis, 2020

At the provincial level, Yogyakarta City is a health service centre (hierarchy I) with a cumulative value of 120.64 in 2019. Sleman Regency, Bantul Regency, and Kulon Progo Regency are included in hierarchy II hinterlands with each cumulative value of the centrality index and service function in in 2019, namely 8.96; 8.96; and 7.28. Last, Gunungkidul Regency is included in the hinterland hierarchy III, with a cumulative value that is much lower than the hierarchy I service centre, which is only 3.26 in 2019. Observed from a number of types and health facility units, it can be said that each regency/city has a complete range of 7 types of health facilities including hospitals, maternity hospitals, polyclinics, Puskesmas, Supporting Puskesmas, pharmacies, and Posyandu. However, when viewed from a number of health facility units, the higher the hierarchy of health service centres, the higher the number of health facility units available. The difference in the cumulative value of the centrality index and the service functions of health facilities is very significant between the service centre and its hinterland, indicating inequality in development, especially related to the availability of health facilities. There are a few things that cause inequality in health facilities in the regional are the availability, geographic conditions, population distribution, as well as regional policy [14]. The cumulative value of the centrality index and service function from 2015 to 2019 also does not show any significant development for the hinterland area to approach the order of health service centres.

A review of the results of the analysis of the service functions of health facilites at the regency/city level is able to show the sub-districts that play a role as a health service centre and its hinterland. Yogyakarta City as a provincial health service centre has a cumulative value for the Centrality Index and Service Function of each sub-district in the range 17.43 to 68.05 in 2019. Sub-districts with hierarchy I health service status are Danurejan (68.05); Ngampilan (66.90); Pakualaman (65.99); and Gedongtengen (58.93). Sub-districts with hierarchy II health service status are Kraton hierarchy (47.55); Gondoamanan (43.47); Wirobrajan (43.47); Mergangsan (41.50); Jetis (38.36); and Mantrijeron (35.58). Then, Sub-districts with hierarchy III health service status are Tegalrejo (30.76); Gondokusuman (27.53); Kotagede (25.48); and Umbulharjo (17.43). The sub-district that experienced an increase hierarchy was only Gedongtengen. It was in hierarchy II in 2015 then became in hierarchy I in 2019 and there were no sub-districts that experienced decrease hierarchy (see Fig. 2).

https://doi.org/10.1051/e3sconf/202132507006

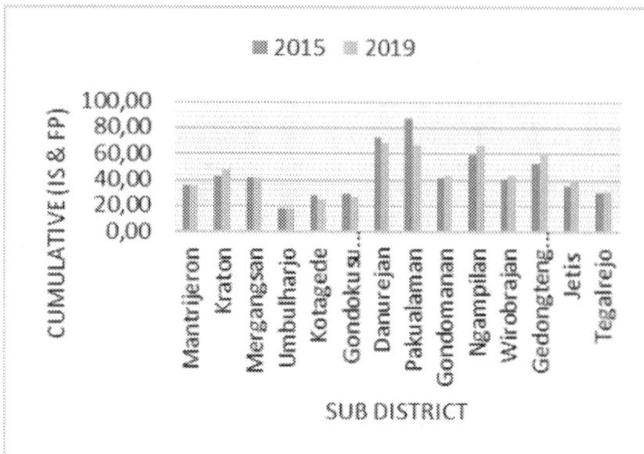

Fig. 2. Graph of Cumulative Value (Centrality Index and Service Function) of Health in Yogyakarta City

Sleman Regency has the cumulative value of the Centrality Index and Service Function of each sub-district in the range of 10.37 to 25.63 in 2019 (see Fig. **3**). Sub-districts with hierarchy I health service status are Gamping (25.63); Berbah (24.91); Mlati (24.82); and Sleman (19.72). In the hinterland hierarchy II, there are Godean (19.02); Seyegan (18.05); Ngemplak (17.75); Moyudan (17.73); Ngaglik (17.07); Depok (16.64); and Tempel (16.58). Then, sub-districts with the hinterland hierarchy III are Minggir (14.91); Kalasan (14.86); Prambanan (14.24); Turi (11.85); Pakem (11.69); and Cangkringan (10.37). In 2015-2019, the sub-district that experienced an increase hierarchy from hierarchy III to hierarchy II was Ngemplak. There are also sub-districts that experienced the increase hierarchy from hierarchy II to hierarchy I, namely Berbah and Sleman. Sub-districts that experienced decrease hierarchy from hierarchy II to hierarchy III were Minggir and Kalasan, then Sub-districts that experienced decrease hierarchy from hierarchy I to II were Moyudan, Seyegan, and Godean.

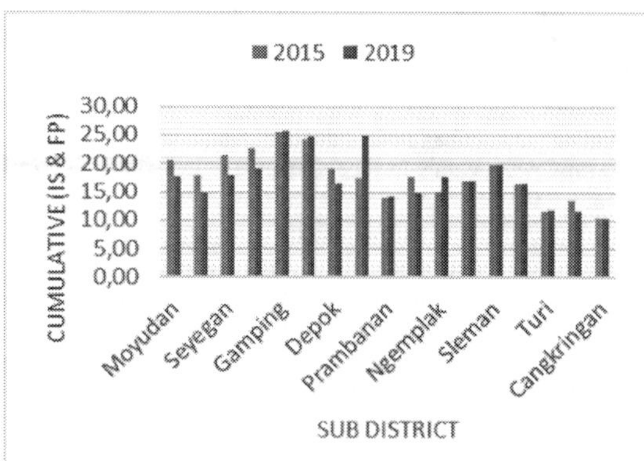

Fig. 3. Graph of Cumulative Value (Centrality Index and Service Function) of Health Facilities in Sleman Regency

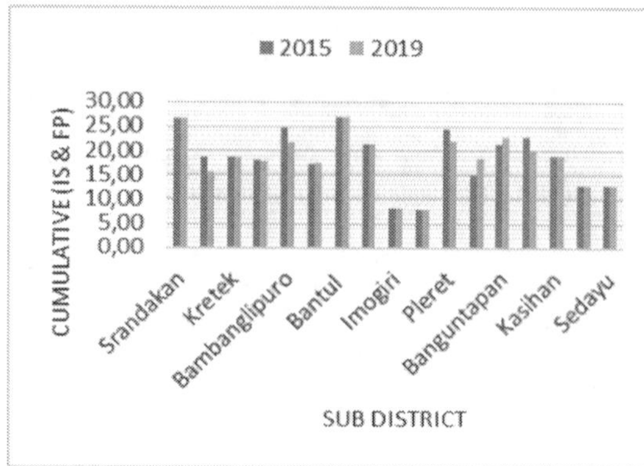

Fig. 4. Graph of Cumulative Value (Centrality Index and Service Function) of Health in Bantul Regency

Fig. 4 above shows that Bantul Regency has the cumulative value of the Centrality Index and Service Function of each sub-district in the range of 7.88 to 26.83 in 2019. Sub-districts with hierarchy I health service status are Bantul (26.83); Srandakan (26.55); Banguntapan (22.71); Pleret (21.92); Bambanglipuro (21.61); and Jetis (21.35). In hinterland hierarchy II, there are Sewon (19.86); Kasihan (18.83); Kretek (18.66); Piyungan (18.22); Pundong (17.86); Pandak (17.45); and Sanden (15.52). Then, sub-district in hierarcy III are Pajangan (12.96); Sedayu (12.81); Imogiri (8.30); and Dlingo (7.88). In 2015-2019 the sub-district that experienced an increase hierarchy from hierarchy III to hierarchy II was Piyungan. There were also sub-districts that had increased hierarchy from hierarchy II to hierarchy I, namely Jetis and Banguntapan. The sub-district that experienced a decreased hierarchy from hierarchy I to hierarchy II was Sewon.

Kulon Progo Regency has the cumulative value of the Centrality Index and Service Function of each sub-district in the range of 4.45 to 20.95 in 2019 (Fig. **5**). Sub-districts with hierarchy I health service status are Wates (20.95); Temon (19.24); Lendah (18.58); Galur (17.85); and Nanggulan (16.54). In the hinterland hierarchy II, there are Panjatan (13.87) and Sentolo (11.52). Then, sub-districts with the hinterland hierarchy III are Kalibawang (9.61); Pengasih (8.33); Samigaluh (7.75); Girimulyo (5.72); and Kokap (4.45). In 2015-2019 the sub-district that experienced an increase hierarchy from hierarchy II to hierarchy II was Sentolo. The sub-district that experienced a decreased hierarchy from hierarchy II to hierarchy III was Kalibawang.

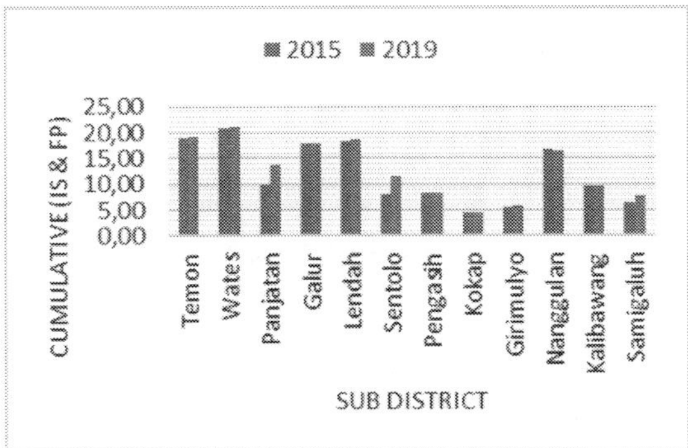

Fig. 5. Graph of Cumulative Value (Centrality and Service Function Index) of Health Facilities in Kulon Progo Regency

The range of cumulative values for the Centrality and Service Function Index for each sub-district in Gunungkidul Regency is 9.32 to 24.43 in 2019 (see Fig. **6**). Sub-districts with hierarchy I health service status are Ngawen (24.43); Wonosari (21.53); Paliyan (19.10); Semin (17.92); and Karangmojo (17.45). Sub-districts with hierarchy II are Patuk (15.88); Nglipar (15.24); Rongkop (14.01); Gedangsari (13.56); Playen (13.24); Tanjungsari (12.92); and Saptosari (12.84). Then, sub-districts with hierarchy III are Purwosari (12.36); Girisubo (12.21); Bake (11.19); Tepus (11.00); Semanu (10.89); and Ponjong (9.32). In 2015-2019 the sub-districts that experienced the increase hierarchy from hierarchy III to hierarchy II were Saptosari and Rongkop. There was also sub-district that experienced an increase hierarchy from hierarchy II to hierarchy I, namely Semin. Then, the sub-district that experienced a decrease hierarchy from hierarchy II to hierarchy III was Semanu.

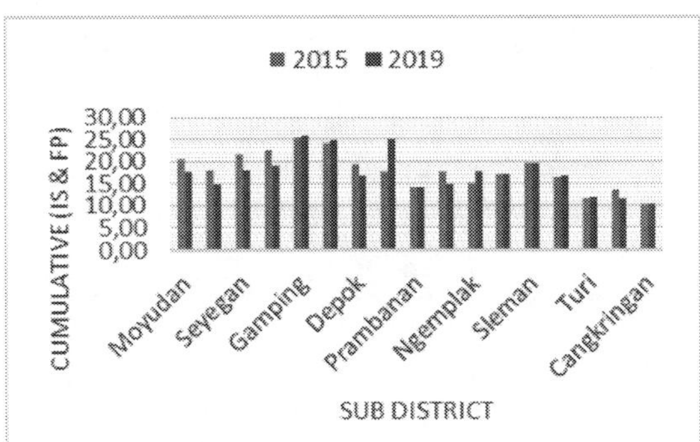

Fig. 6. Graph of Cumulative Value (Centrality Index and Service Function) of Health Facilities in Gunungkidul Regency

4 Closing

The provision of health service facilities in residential area and regional is the right of the community to support a decent life and livelihood for the sake of improving the quality of life. The suitability between the availability and regional basic needs of service facilities can be one of the benchmarks for the success of regional development. The provision of health facilities in Indonesia currently uses population size as a comparison variable (ratio of health facilities and total population). Thus the minimum need for health facilities can be determined by considering serviceability.

The results showed that there were health facilities with a higher level of service (high order goods services) that were not fulfilled according to regional needs. Meanwhile, health facilities with a lower level of service (low order goods services) such as Puskesmas, pharmacies, and Posyandu tend to exceed the ideal number required by the region. The city centre is still the centre for regional health services. This is indicated by the high index of centrality and service function in the Yogyakarta City, so that it is included in hierarchy I (the highest). Sleman Regency, Bantul Regency, and Kulon Progo Regency are hinterland areas with hierarchy II, while Gunungkidul Regency is the area with the lowest hierarchy at the Provincial level.

This research is part of the Doctoral Dissertation of Geography Study Program, Universitas Gadjah Mada. Thanks to the promoter team who have provided input for this research, also to Rifan and Zara who have supported this research.

References

1. H. S. Yunus, *Struktur Tata Ruang Kota* (Pustaka Pelajar, Yogyakarta, 2002)
2. Yandrizal, D. Suryani, B. Anita, and H. Febriawati, JKKI, **3**, 2 (2014)
3. M. I. Sadali, F. Intizhar and Aisyah, MKG, **18**, 2 (2017)
4. Secretariat General of the Indonesian Ministry of Health, *Profil Kesehatan Indonesia Tahun 2019*, (Ministry of Health's, Jakarta, 2020)
5. Yogyakarta Special Region Provincial Health Office, *Profil Kesehatan Daerah Istimewa Yogyakarta Tahun 2018* (Yogyakarta Special Region Provincial Health Office, Yogyakarta, 2019)
6. W. Christaller, *Central Place in Southern Germany* (Prentice-Hall, New Jersey, 1933)
7. K. M. Curtin and R. L. Church, JGS, **9** (2007)
8. D. Pumain, in *Hierarchy in Natural and Social Sciences*, edited by D. Pumain (Springer, Dordrecht, 2006), pp. 169-222
9. Republic of Indonesia, *Undang-Undang Nomor 36 Tahun 2009 tentang Kesehatan* (State Secretariat, Jakarta, 2009)
10. Republic of Indonesia, *Peraturan Menteri Kesehatan Nomor 75 Tahun 2014 tentang Puskesmas* (State Secretariat, Jakarta, 2014)
11. G. Rushton, *Optimal Location of Facilities*, (N.H.: COMPress, Wentworth, 1979)
12. Susilowati, Ketidakmerataan Akses Pelayanan Kesehatan Rawat Jalan di Indonesia, Doctoral

https://doi.org/10.1051/e3sconf/202132507006

dissertation, Universitas Gadjah Mada, Yogyakarta, 2004

13. Selma A.S. Siahaan, JPPPK, **2**, 2 (2018)
14. Suman, S. Bhutani, IJARD, **2**, 6 (2017)
15. S. S. Endartiwi, and P. D. Setianingrum, JIIK, **6**, 1 (2019)
16. R. McCollum, M. Taegtmeyer, L. Otiso, M. Mireku, N. Muturi, T. Martineau, and S. Theobald, Int J Equity Health, **18**, 65 (2019)
17. M. C. Sheff, A. A. Bawah,,P. O. Asuming, P. Kyei, M. Kushitor, J. F. Phillips, and S. P. Kachura, Glob Health Action, **13**, 1 (2020)

The bionomic of road users: a new perspective of individual mobility approach

Pranoto Dirhan Putra [1], *Sri Rum Giyarsih*[2], *and Djaka Marwasta*[2*]
[1] Faculty of Engineering, Atma Jaya Yogyakarta University, 55281. Sleman, Yogyakarta, Indonesia
[2] Faculty of Geography, Universitas Gadjah Mada, 55281, Sleman, Yogyakarta, Indonesia

Abstract. Mostly, transport studies are centered on the factors of transportation rather than human movements. In a similar vein, communication technology revolution influence to people mobility. The performance of traffic pattern does not tend to deliver utmost needs of people mobility, because less attention of the substance of inhabitant movements. The importance of bionomic patterns of road users is based on the fact that behavior approach and adaptation will create pattern of people mobility. The aim of this research is to determine whether they adapt to nearby environment or not. Taking this, variables surveyed are individual attribute, time, duration, location, vehicle, communication to transport and purpose of trips. Investigation is conducted by providing questionnaire. The main result of this study are that the bionomic pattern of road users of selected population shows that they prefer to use simple mode vehicle to transport such as motor bike and walking. Moreover, these choices are preferred to be taken as a response to higher dense of traffic population and worsen road network system. It would be wise for planning if it considers potential effect not only to traffic but also regular road users need including goods supply, proper access and facilities for unmotorized vehicle.

1. Introduction

Mobility has been facing new challenges even when most of cities are still growing and technology is surely rounding into rapid-continuous advancement [1], [2], [3]. Indeed, individual movements are being more random in time and more countless destinations [4]. Also, albeit the slogan of reducing energy consumption and pollution has declared in many countries by generating sustainable transport, the quality of mobility of persons or groups in road networks of towns is obviously far from environmentally sound [5].

Then again, most of transport studies are focused on the elements of transportation rather than "something" substance that create human movements such as traffic [6], city planning [7] and computer modelling [8], [9], [10]. In line with this, global development in communication technology obviously gives a new role to people mobility [11], [12], [13], [14] Since then, this transformation has produced mobility which is more casual and faster [15].

Nevertheless, road, which was intended to connect various hierarchy of needs, has become a kind of disaster instead of supporting it, [16], [17], [18], [19], [20].

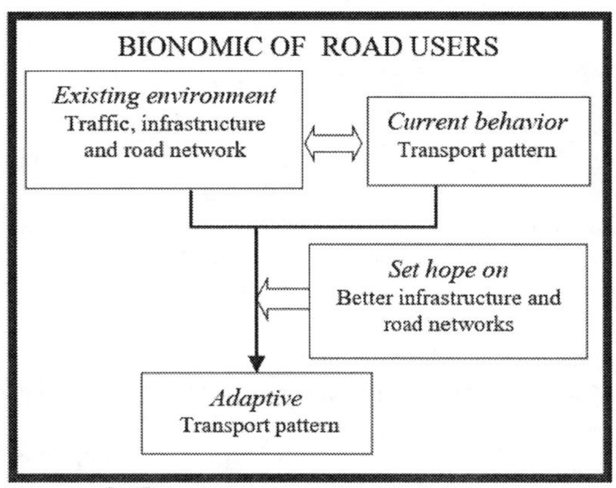

Source: [40]
Fig. 1. The concept of bionomic due to this study

Back an argument to bionomic, this idea responds to reality that today's human mobility is more random and technology-oriented as depicted in Figure 2 [21]. It illustrates something like transformation of networks planning from setting distance to computer modelling. In addition, higher growth vehicle population in most of global cities has given impact on more complex traffic disruption, increasing pollution dan rising delay [22], [14], [12]. Thus, this occurrence should be anticipated using appropriate planning approach for better future environment [23]. Subjects linked to transportation are more individuals and more various [24]. So, behavior of

* Corresponding author: jakamar@ugm.ac.id

road users, which is defined as a result of adaptation to environment, is obviously in line to be investigated deeper.

2. Research Problem

Nowadays, the performance of traffic pattern does not tend to deliver utmost needs of people mobility [25], [26]. This perhaps occur because less attention of the substance of inhabitant movements such as the bionomic of road users. In addition, this phenomenon may also happen as the presence of the gap between transportation required by society and the existence of infrastructure [27] (Irawati, et al., 2012).

Therefore, the importance of bionomic patterns of road users is based on the fact that behavior approach and adaptation will create something like core pattern of people mobility [28], [29]. Figure 1 displays the idea of bionomic approach for road users which is purposed in this study. This is initiated by the actual circumstance that most of road users hope for better infrastructure and road networks despite there is a gap between existing infrastructure and various individual behavior [28]. The interaction among the three elements of these will produce adaptation as a respond to them.

The main aim of this survey is delivering questions to determine whether they adapt to nearby environment or not. Taking this, variables surveyed are individual attribute, time, duration, location, vehicle, communication to transport and purpose of trips. Investigation is conducted by providing questionnaire.

3. Research Methods

Selected residents to be studied is population who stay in Condongcatur village, Depok Subdistrict, Sleman Regency, Special Region of Yogyakarta, Indonesia. Sample's age is chosen in between 17 to 60 years old. This is assumed that they tend to take destination and route more independent rather than out of selected age range. Also, this productive oldness tends to transport intended for fulfill daily needs. It might be because better physical condition and career opportunity. This age range is classified into four groups which are 17-25years old, 26-35 years old, 36-45 years old and 45-60 years old. Number of samples taken is 80 inhabitants

Data, then, plotting into graphs so as to determine the features of road users in relation to their adaptation to infrastructure environment. At this point, the bionomic of them can be interpreted by comparing among charts.

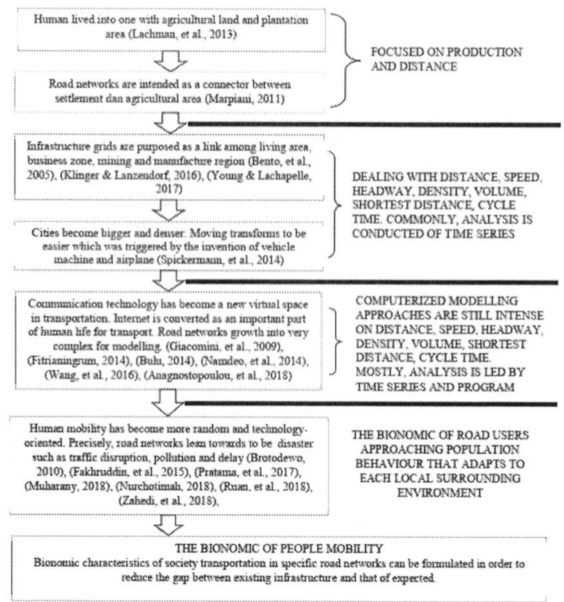

Source: [40]
Fig. 2. The argument for the standing of bionomic of road users related to this study

4. Literature Reviews

The outline of human bionomic is clearly associated with social effort to fulfill essential needs of living and the process to achieve it [31]. Along with the worldwide development of cities, transportation has taken major influence on quality of life [32], [14] (Zahedi, et al., 2018; Lai, 2018). Presently, the crossing point among both of them is becoming more crucial as described in Figure 3 that shows the layout between individuals, transportation and destination purposes [33], [34].

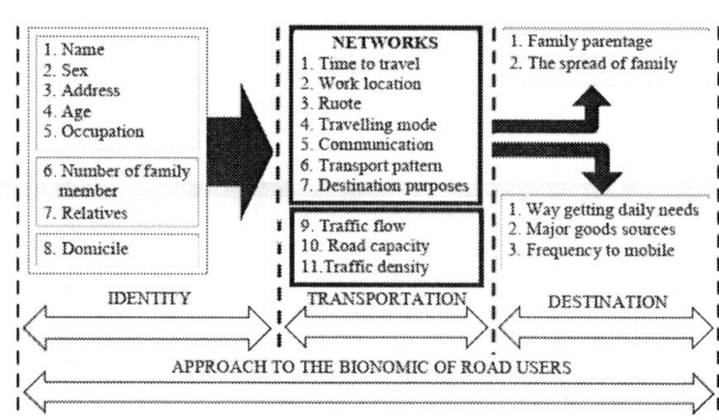

Source: [40]
Fig. 3 modification of the bionomic approach of road users

This figure can be construed that wider such problems of networks and its components, bigger gap between individual demand and journey's end [35], [36], [37]. Also, it explains that nowadays networks and transportation has taken over a key role for society

and towns with the intention most of living rudiments such as travelling cost, time, energy and environment impact [38], [39], [41] (Conway, et al., 2019; Mahauta &

Andrieub, 2019; Gusnita, 2010). Concerning this, investigation slightly starts on the portray of selected family characteristics.

5. Results and Discussion

Families inside this particular study area have between zero to six children which is 4.5%, 14.8%, 38.6%, 18.2%, 15.9%, 4.5% and 3.4% respectively as shown in Figure 4. Whereas, Figure 5 represents the percentage of family member in range of 2 members (9.1%), 3 members (11.4%), 4 members (44.3%), 5 members (18.2%), 6 members (10.2%), 7 members (4.5%) and 8 members (2.3%). While, most of respondent ages are 46-65 years (77.3%) and the rest is at the age of 36-45 years (21.6%), 26-35 years (1.1%) as displayed in Figure 6. But there is no respondent with age interval of 17-25 years (0.0%).

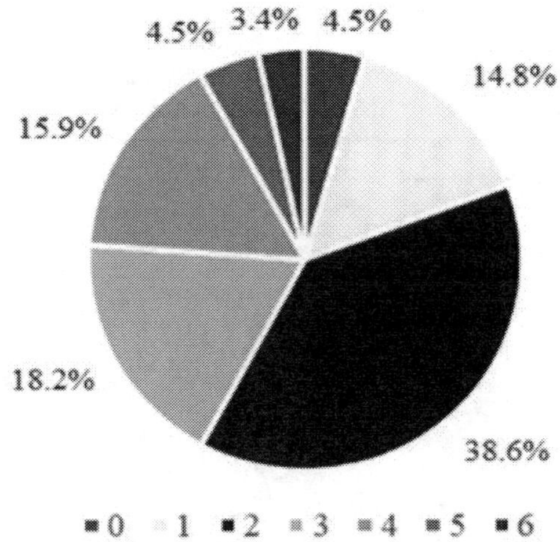

Source: [40]
Fig. 4 Percentage of number of children

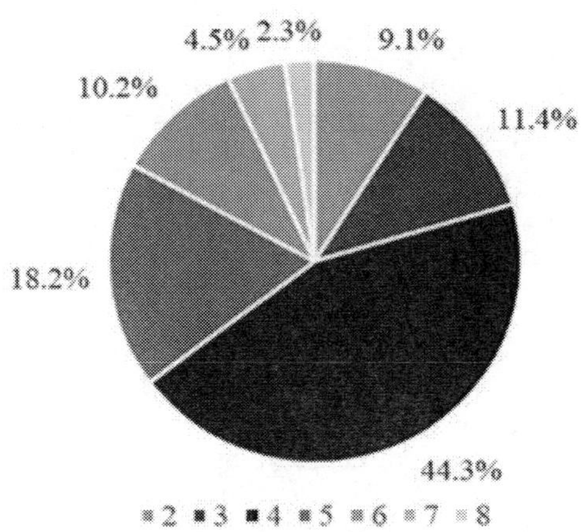

Source: [40]
Fig. 5 Percentage of family member

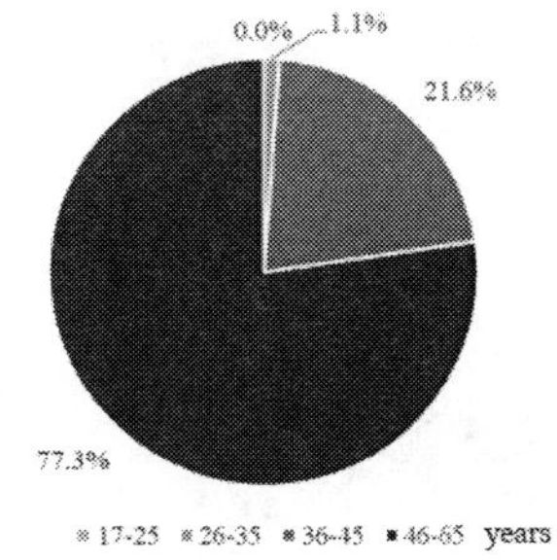

Source: [40]
Fig. 6 Percentage of respondent age

Resident occupation is categorized into labor (19.3%), seller (2.3%), farmer (1.1%), private business (50.0%), house wife (4.5%), retired (8.0%), public officer (6.8%) and unemployed (8.0%) as shown in Figure 7.
Whereas, most of respondents (89,8%) are coming from selected study area or, in other words, they are not moving to other cities for living. Home town of inhabitants is displayed in Figure 8. Likewise, these majority samples have relatives who are living in the same area (75.0%) as shown in Figure 9.

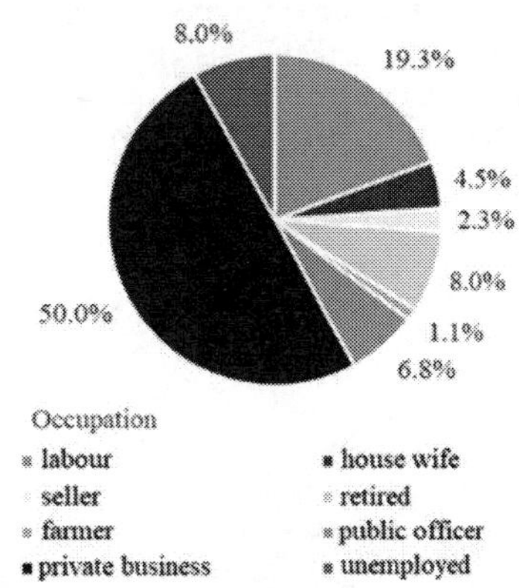

Source: [40]
Fig. 7 Percentage of respondent occupation

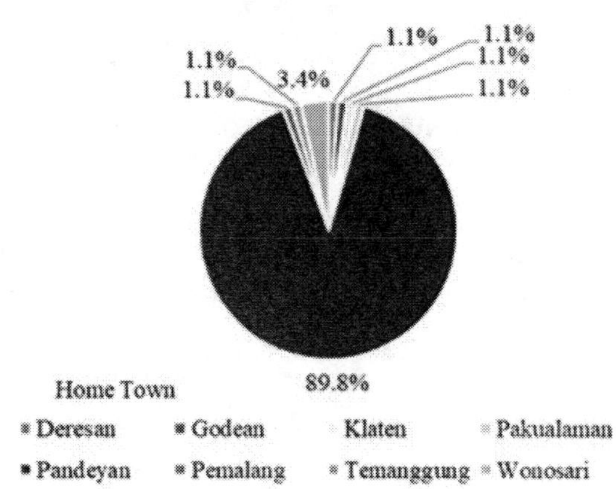

Source: [40]
Fig. 8 Percentage of respondent;s home town

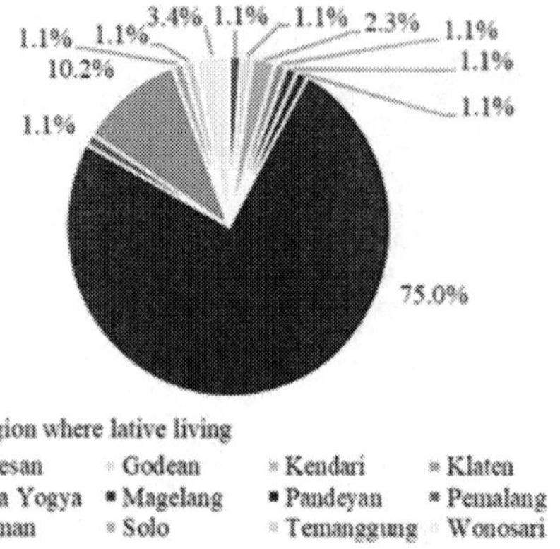

Source: [40]

Fig. 9 Percentage of respondent native living

Distribution type of sources for primary commodity that respondents usually visit are traditional market (17.8%), modern market (20.1%), home market (21.1%), self-service market (20.4%), supermarket (20.1%) and others (0.5%) as revealed in Figure 10. This distribution, which is almost comparable for all market, may occur because they are located close to the village although respondents confront with higher dense traffic. Yet, regarding to Figure 11, home market is the most place visited by locals (18.9%) for every day shopping following by traditional market (2.8%), self-service market (2.3%), super market (0.3%) and other (0.3%). But, in unplanned schedule scale, modern market is the most visited by residents (19.9%) continued by supermarket (18.6%), self-service market (15.9%), traditional market (14.6%) and home market (2.0%).

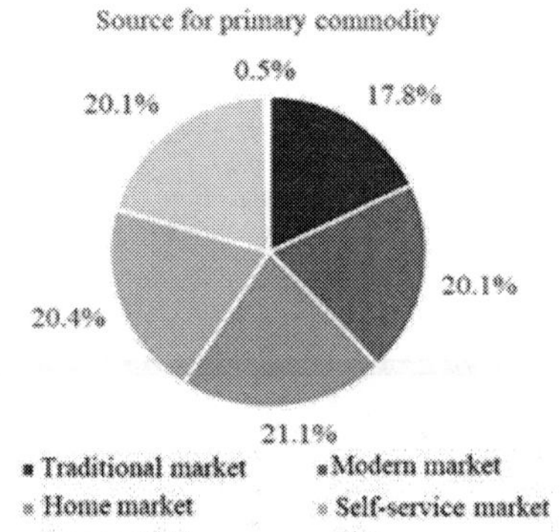

Source: [40]

Fig. 10 Source for primary commodity

Opposing phenomena between home market which is most visited daily and modern market which is visited without day (Figure 10 and Figure 11) shows that there is a kind of obstacle that probably tends to reduce people's ability to reach them. This is in line with the presumptively stated in Figure 3 that road networks have a tendency to be such as barrier for users.

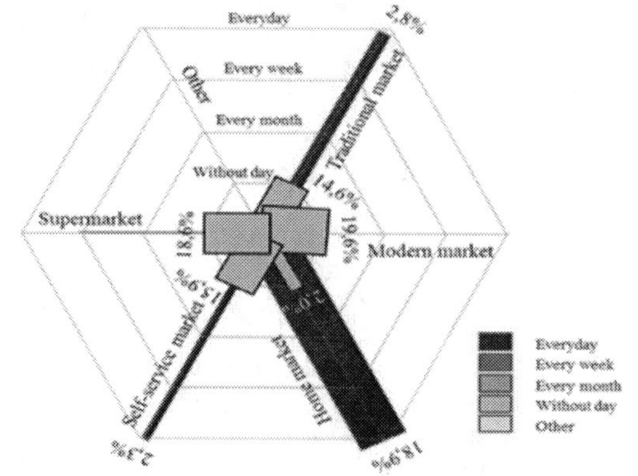

Source: [40]

Fig. 11 Distribution frequency of visiting source for primary commodity

Next, motor cycle is the most favorite vehicle for riding (40.8%) following by walking (31.2%), bicycle (23.6%) and car (4.5%) as portrayed in Figure 12. Surprisingly, none of the respondent takes public transport for travelling despite of proper mass transport passing by surrounding their living area. Again, disreputable road networks could be charged with as casual factor in this occurrence.

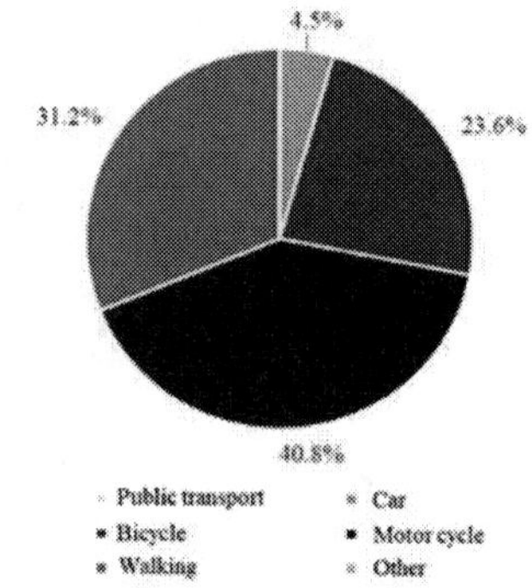

Source: [40]

Fig. 12 Proportion of transport mode using by inhabitants

Ensuing, travelling purpose is dominated finding food (21.5%) at around 12.00 p.m. that is followed by work reason (9.8%) at around 06.00 a.m. and recreation (0.8%) at about 06.00 a.m. as revealed in Figure 13. This diagram clearly shows distribution pattern of resident movement versus time to do such things. Shortly, there is three critical periodic time which is 06.00-07.00 a.m., 12.00-13.00 p.m. and 18.00-19.00 a.m. local time.

Similar way, regarding to Figure 14, there is evidence that walking at the period of 12.00-13.00 p.m. is the highest proportion (17.5%) following by the same mode of transport at 18.00-19.00 p.m. (15.3%). The second biggest favorite mode is motor cycle (8.5%) at 07.00-08.00 a.m. It can be questioned why they prefer to choose these two modes of transport in their routine activities even so better means of carriage is basically available. It perhaps occurs in respond to the condition of worsen traffic situation and degraded performance of infrastructure. Although it might more than one motive, then again, there is mark that local road users obviously adapt their environment and adjust their behavior.

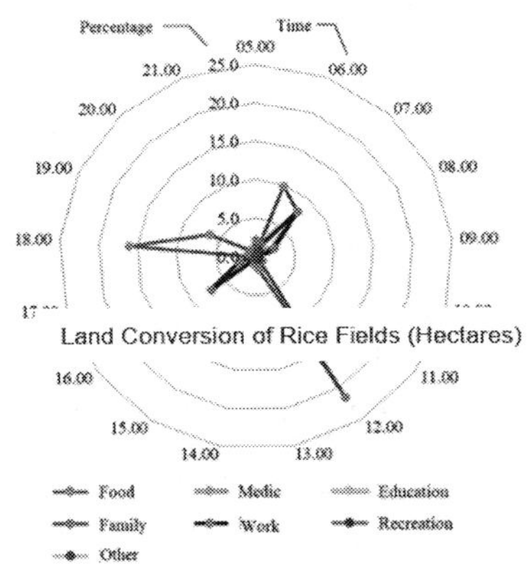

Source: [40]

Fig. 13. distribution of travelling purposes from Monday to Sunday

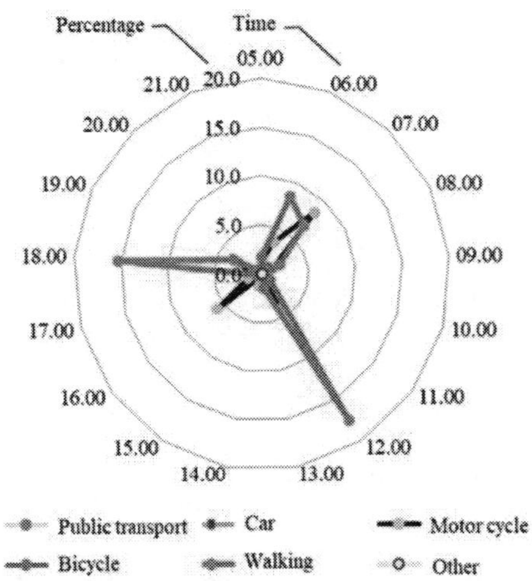

Source: [40]

Fig. 14. Proportion of transport mode from Monday to Sunday

It can be clearly construed form Figure 13 and Figure 14 that respondents tend to have both similar pattern of travelling purposes for activities and similar pattern mode for moving if we look it deeper related to time which is 06.00-07.00, 12.00-13.00 and 18.00-19.00 local time. Combining data collected above, it can be summarized that road users have bionomic characteristic as an outcome of adaptation to their environment that form unique behavior.

6. Conclusion

Finally, the bionomic pattern of road users of selected population in this study shows that they prefer to use simple mode vehicle to transport such as motor bike and walking. Also, this is taken in favorite of time at 06.00-07.00, 12.00-13.00 and 18.00-19.00 local time. Moreover, these choices are preferred to be taken as a response to higher dense of traffic population and worsen road network system. Therefore, most of them favor home market to fulfill their need for primary goods rather than other market.

As a suggestion, it would be wise for planning if it considers potential effect not only to traffic but also regular road users need including goods supply, proper access and facilities for unmotorized vehicle (Putra, 2021).

Acknowledgment

This paper is a tad part of dissertation which is entitled The Bionomic Approach of Transport Users in Road Network System. This research is funded with RTA (Rekognisi Tugas Akhir) grant scheme of UGM year 2021. It is written by the first author under a good hand supervision of second author and marvelous guidance of third author. It is innumerable thanks for all sage reviewers who has delivered prudently suggestions in order to be the pink of perfection of this paper.

References

1. Mustafa, A., Analisis Permasalahan Kinerja Jaringan Jalan Kota Yogyakarta Pada Saat Liburan Dan Usulan Pemecahannya. Yogyakarta: Perpustakaan Pusat UGM (2016)
2. Deng, Y. & Srinivasan, S., Urban Land Use Change and Regional Access: A Case Study in Beijing, China. Habitat International, Volume 51, February 2016, pp. 103-113 (2016)
3. McCan, B., Completing Our Streets, The Transition to Safe and Inclusive Transportation Networks. The Center for Resource Economics (2013)
4. Brownleea, A. E. et al., A Fuzzy Approach To Addressing Uncertainty In Airport Ground Movement Optimisation. Transportation Research Part C, Volume 92 (2018), p. 150–175(2018)
5. Azizalrahman, H. & Hasyimi, V., A Model for Urban Sector Drivers of Carbon Emissions. Sustainable Cities and Society, Volume 44, January 2019, pp. 46-55 (2019)
6. Guo, F., Polak, J. W. & Krishnan, R., Predictor Fusion for Short-Term Traffic Forecasting. Transportation Research, Volume Part C 92 (2018), p. 90–100 (2018)
7. Dejeammes, M., Urban Mobility Plans and Accessibility. Journal of Transport and Land Use, Volume 2, No. 2 (Spring 2009),, pp. 67-78 (12 pages) (2009)
8. Felician, C., H. M. & Nishinari, K., Correction: A Universal Function for Ccapacity of Bidirectional Pedestrian Streams: Filling The Gaps in The Literature. PLoS ONE, Volume 4(4):e0216314 • April 2019 (2019)
9. Yu, H., Tseng, H. E. & Langaria, R., A Human-Like Game Theory-Based Controller For Automatic Lane Changing. Transportation Research, Volume Part C 88 (2018), p. 140–158 (2018)

10. Febrianda, Y., Mahmudah, N. & Muchlisin, Pemodelan Lalu Lintas Mengunakan PKJI 2014 Dan SoftwareVissim 9 pada Simpang APILL Madukismo, Ring Road Selatan, Yogyakarta. Yogyakarta: Program Studi Teknik Sipil, Fakultas Teknik, Universitas Muhammadiyah Yogyakarta (2014)
11. Peprah, C., Amponsah, O. & Oduro, C., A System View of Smart Mobility and Its Implications for Ghanaian Cities. Sustainable Cities and Society, Volume 44 (2019), p. 739–747 (2019)
12. Frankhauser, P., Tannier, C., Vuidel, G. & Houot, H.,. An Integrated Multifractal Modelling to Urban and Regional Planning. Computers, Environment and Urban Systems, Volume 67, January 2018, pp. 132-146 (2018)
13. Long, Y.,. Big/Open Data for Urban Management. Journal of Urban Management, Volume 4 (2015), p. 73 (2015)
14. Lai, S.-K., Toward A General Theory of Cities. Journal of Urban Management, Volume 7 (2018), pp. 43-45 (2018)
15. Longo, A., M. Z. & Navathe, S. B., The Unified Chart of Mobility Services: Towards A Systemic Approach to Analyze Service Quality in Smart Mobility Ecosystem. Journal of Parallel and Distributed Computing, Volume 127, May 2019, pp. 118-133 (2019)
16. Berdica, K., An Introduction to Road Vulnerability: What Has Been Done, is Done and Should Be Done. Journal Transport Policy, Volume 9 (2002), pp. 117-127 (2002)
17. Parker, S. A., Guide to Planning Resources on Transportation and Hazards. Transportation Research Board of The National Academies, Transit Cooperative Research Program (2009)
18. Jenelius, E., User Inequity Implications of Road Network Vulnerability. Journal of Transport and Land Use, Volume 2 (3/4) [Winter 2010], p. 57–73 (2010)
19. Morellia, V. G. et al., Towards Intelligently-Sustainable Cities?. Journal of Land Use, Mobility and Environment, Volume 1 (2013), pp. 73-86 (2013)
20. Mudjanarko, S. W., Perencanaan Transportasi Kota. Surabaya: Program Studi Teknik Sipil, Fakultas Teknik Universitas Narotama (2016)
21. M. Lemos, C., Agent-Based Modeling of Social Conflict, From Mechanisms to Complex Behavior. s.l.:Springer Witherick, M., Ross, S., and Small, J., 2001, A Modern Dictionary of Geography (4th edition), New York: Oxford University Press (2018)
22. Chen, J. et al., What Makes The Difference in Construction Carbon Emissions between China and USA?. Sustainable Cities and Society, Volume 44, January 2019, pp. 604-613 (2019)
23. Marzanoa, G., Lizutb, J. & Siguencia, L. O., Crowdsourcing Solutions for Supporting Urban Mobility. Procedia Computer Science, Volume 149 (2019), p. 542–547 (2019)
24. Maerivoeta, S. et al., A Field Trial on Smart Mobility. Procedia - Social and Behavioral Sciences, Volume 54 (2012), p. 926 – 935 (2012)
25. Adirinekso, G. P., Karakteristik Penggunaan Pelaku Perjalanan Dalam Pemilihan Moda Transportasi Pekerja di Kota Jakarta Barat. Jurnal Riset Manajemen dan Bisnis, Volume 11, No. 1, Juni 2016, pp. 73-89 (2016)
26. Batty, M. & Marshall, S., Thinking Organic, Acting Civic: The Paradox of Planning for Cities in Evolution. Landscape and Urban Planning, Volume 166 (2017), p. 4 – 14 (2017)
27. lrawati, I., Herman, Wigunawan, D. & Rahardjoa, T. D., Solusi Pemecahan Masalah Transportasi Di Kota Bandung Berdasarkan Pendapat Pakar. Bandung, Lembaga Penelitian dan Pengabdian Kepada Masyarakat (LPPM). (2012)
28. Hu, X., J. W. & Wang, L., Understanding the Travel Behavior of Elderly People in the Developing Country: A Case Study of Changchun, China. Social and Behavioral Sciences, Volume 96 (2013), p. 873 – 880 (2013)
29. Dacko, S. G. & Spalteholz, C., Upgrading the City: Enabling Intermodal Travel Behaviour. Technological Forecasting & Social Change, Volume 89 (2014), p. 222–235 (2014)
30. Ding, H., Yang, M., Wang, W. & C. X., Simulating and Analyzing the Effect on Travel Behavior of Residential Relocation and Corresponding Traffic Demand Management Strategies. Korean Society of Civil Engineers, Journal of Civil Engineering, Volume (2018) 22(2), pp. 837-849 (2017)
31. Campos, W. G. & Schoereder, J. H., Ecology, Behavior and Bionomics, Comparison of the Efficiency of Flight-Interception Trap Models for Sampling Hymenoptera and Other Insects. Anais da Sociedade Entomológica do Brasil, Volume 29(3), pp. 381-389 (2000)
32. Zahedi, S., Batista-Foguet, J. M. & Wunnik, L. v., Exploring the Public's Willingness to Reduce Air Pollution and Greenhousegas Emissions from Private Road Transport in Catalonia. Science of the Total Environment, Volume 646 (2019), p. 850–861 (2018)
33. Perraa, V.-M., Sdoukopoulosa, A. & Pitsiava-Latinopoulou, M., Evaluation of Sustainable Urban Mobility in The City of Thessaloniki. Transportation Research Procedia, Volume 24 (2017), p. 329–336 (2016)
34. Carlow, V. M., Limits—Urban Density and Mobility Networks in West Berlin during the Period of Containment. Sustainability, Volume 6, pp. 7452-7465 (2014)
35. Zwierzchowska, I. et al., Introducing Nature-Based Solutions Into Urban Policy – Facts and Gaps. Case

Study of Poznań. Land Use Policy, Volume 85 (2019), pp. 161-175 (2019)

36. Zhou, H. & Gao, H., The Impact of Urban Morphology on Urban Transportation Mode: A Case Study of Tokyo. Case Studies on Transport Policy (2019)
37. Zhen, F. et al., Analyzing Urban Development Patterns Based on The Flow Analysis Method. Cities, Volume 86, March 2019, pp. 178-197 (2019)
38. Conway, T. M., Almas, A. D. & Coore, D., Ecosystem Services, Ecological Integrity, and Native Species Planting: How to Balance tThese Ideas in Urban Forest Management?. Urban Forestry & Urban Greening, Volume 41, May 2019, pp. 1-5 (2019)
39. Mahauta, V. & Andrieub, H., Relative Influence of Urban-Development Strategies and Water Management on Mixed (Separated and Combined) Sewer Overflows in The Context of Climate Change and Population Growth: A Case Study in Nantes. Sustainable Cities and Society, Volume 44, January 2019, pp. 171-182 (2019)
40. Putra, P. D., Pendekatan Bionomic Pelaku Transportasi Pada Sistem Jaringan Jalan. Yogyakarta: Universitas Gadjah Mada (Unpublished Doctoral Thesis Proposal) (2021)
41. Gusnita, D., Green Transport: Transportasi Ramah Lingkungan Dan Kontribusinya Dalam Mengurangi Polusi Udara. Berita Dirgantara, Volume 11 No. 2 Juni 2010:, pp. 66-71 (2010)

Study of agricultural economic potential in West Kalimantan using Regional Analysis Techniques

Salma Nurulhuda[1], *Mery* Askarina[1], *Si'ta* Romadhoniastri[1], *Afanin Fatkha* Azahra[1], *Deatry Kharisma* Karim[1], *Muhammad Nadafa* Isnain[1], and *Ratih Fitria* Putri[1*]

[1]Environmental Geography Department, Faculty of Geography, Universitas Gadjah Mada, Yogyakarta, 55281, Indonesia

Abstract. The agricultural resource is one of the most critical sectors in the economy, which contributes a high portion of the growth and development of a particular region. Differences in characteristics and potential between regions give rise to the basic and non-basic sectors, which then affect income contribution. This research aims to analyze the economic potential of the agricultural sector of the regencies and cities in West Kalimantan Province as basic or non-basic sector. Both basic and non-basic sector analysis can be carried out using quantitative-descriptive methods using Shift-Share and Location Quotient (LQ) analysis. The analysis was carried out using secondary data from the Central Statistics Agency (BPS) publication from 2015 to 2020. The results of the LQ analysis show the agricultural sector in Sambas, Bengkayang, Landak, Mempayah, Sanggau, Ketapang, Sintang, Kapuas Hulu, Sekadau, and Kayong Utara are the basic sector. Meanwhile, the agricultural sector in Kubu Raya, Pontianak, and Singkawang are non-basic sector. Agriculture in Melawi Regency has changed from a basic sector in 2015, to a non-basic sector in 2020. The results of the Shift-Share analysis show the agricultural sector in Sambas, Bengkayang, Landak, Mempayah, Sanggau, Ketapang, Kapuas Hulu, Sekadau, Kubu Raya, Pontianak and Singkawang have progressive or advanced growth, while Sintang, Melawi and Kayong Utara have slowed growth.

1 Introduction

Each region has different characteristics and potential. This difference certainly explains the differences in economic activity and regional income. One of the leading economic sectors in the agricultural sector is its significant contribution to the growth and development of a region. In addition, agriculture is also a sector that always exists because it provides the primary human need, namely food needs. It is shown based on the Central Statistics Agency data, which shows Indonesia's Gross Domestic Product (GDP) in the national agricultural sector at a constant base price in 2020 of 1,378.1 trillion rupiahs, contributing 12.9 percent of the national GDP 10,722.4 trillion rupiahs. However, this sector is not a definite leading sector in every region.

Leading sectors from various regions need to be studied to determine which economic sectors have the most influence on economic growth that will generate the most income. [1] explained that this leading sector could be identified based on economic theory.

This theory distinguishes the economic sector into two: the leading/basic sector and the non-basic sector. [1] explains that the more basic sectors an area has, the demand for goods and services from that area will also increase so that the flow of the economy is maintained. West Kalimantan is one of the provinces in Indonesia that has suitable land for agricultural cultivation and has rainfall > 2000 mm/year [2]. Therefore, the agricultural sector has an essential role in the economy in West Kalimantan Province. It can be seen from the most considerable contribution of Gross Regional Domestic Product (GRDP) on Constant Prices and GRDP on Current Prices in West Kalimantan in 2020, which was obtained from the agricultural sector, reached 32,340,499.90 million rupiahs or covering 24% of the total GRDP of constant price. Meanwhile, GRDP at prevailing prices in the agricultural sector also produced the most considerable contribution in 2020, which was 44,765,052.30 million rupiahs or 20.9% of the total GRDP at current prices [3].

* Corresponding author: ratihfitria.putri@ugm.ac.id

This research aims to analyze the economic potential of the agricultural sector of the regencies and cities in West Kalimantan Province as basic or non-basic sector. Basic and non- basic sectors can be calculated using several methods, including the Location Quotient (LQ) and Shift-Share methods. The LQ method is carried out by comparing the roles between sectors in regional economic activities [4]. In contrast, the Shift-Share method is used to determine the development of the economic sector and its comparison with other agricultural sectors.

2 Data and Method

This research uses the quantitative-descriptive method. The descriptive method helps to describe in an authentic, systematic, and accurate way of the empirical evidence the relationship between the investigated issues. The quantitative method, on the other hand, helps to quantify in a numerical way of data collection, data analysis, and data visualization [5]. This study collects secondary data of the Gross Regional Domestic Product (GRDP) in West Kalimantan from the year 2015 to 2020. The data compares the total and agricultural sector of the GRDP in regencies/cities within West Kalimantan Province to the total and agricultural GRDP in a provincial scale at a constant price. The data is obtained from the Central Statistics Agency (BPS).

2.1 Location Quotient Analysis

Location Quotient (LQ) analysis is a tool to determine the leading economic sector in a region. The leading economic sector is synthesized by comparing the proportion of certain economic sector in regencies/cities scale to the proportion of certain economic sector in the provincial scale [6]. Here is the following formula.

$$LQ = [(Yik/Yk) / (Yip/Yp)] \quad (1)$$

LQ= Location Quotient
Y_{ik}= GRDP of certain sector in regencies/cities scale
Y_k = total GRDP in regencies/cities scale
Y_{ip}= GRDP of certain sector in provincial scale
Y_p = total GRDP in provincial scale

The LQ value is then determined by this condition: LQ > 1, means that the sector/sub sector is a leading sector.

LQ = 1, means that the sector/sub sector has an equal role with other sectors
LQ < 1, means that the sector/sub-sector is a non-leading sector.

2.2 Shift-Share Analysis

Shift-share analysis is used to examine the economic development in a certain sector of a certain region. This method is able to determine the relative comparison between one economic sector to another [7]. This method comprises of share component (National share or N) and shift component (Proportional shift or P and differential shift or D). Here are the formulae of each component.

$$Nj = Yj0\,(Yt/Y0) - Yj0 \quad (2)$$

$$Pj = [(Yit/Yi0) - (Yt/Y0)].Yij0 \quad (3)$$

$$Dj = [Yijt - (Yit/Yi0)].Yij0 \quad (4)$$

N_j= National share or N
P_j= Proportional Shift or P
D_j= Differential Shift or D
Y_{j0}= initial total GRDP in regencies/cities
Y_0= initial total GRDP in province
Y_t= final GRDP in province
Y_{i0}= initial agriculture based GRDP in province
Y_{it} = final agriculture based GRDP in province
Y_{ij0}= initial agriculture based GRDP in regencies/cities
Y_{ijt} = final agriculture based GRDP in regencies/cities

https://doi.org/10.1051/e3sconf/202132507008

3 Results and Discussion

3.1 Location Quotient Analysis

Table 1. Location Quotient of West Kalimantan Province

Regency/City/Province	Year						Avg.
	2015	2016	2017	2018	2019	2020	
Sambas	1.48	1.49	1.52	1.52	1.52	1.51	1.51
Bengkayang	1.46	1.49	1.28	1.54	1.56	1.53	1.48
Landak	1.58	1.59	1.62	1.62	1.62	1.59	1.60
Mempayah	1.20	1.21	1.23	1.21	1.19	1.16	1.20
Sanggau	1.52	1.53	1.52	1.52	1.53	1.55	1.53
Ketapang	1.33	1.27	1.28	1.28	1.26	1.20	1.27
Sintang	1.17	1.15	1.13	1.11	1.12	1.13	1.14
Kapuas Hulu	1.08	1.09	1.10	1.08	1.08	1.08	1.09
Sekadau	1.65	1.67	1.69	1.68	1.67	1.68	1.67
Melawi	1.01	1.01	1.02	0.99	0.98	0.98	1.00
Kayong Utara	1.38	1.38	1.37	1.33	1.31	1.31	1.35
Kubu Raya	0.67	0.66	0.67	0.66	0.66	0.67	0.67
Pontianak	0.07	0.07	0.07	0.07	0.07	0.07	0.07
Singkawang	0.55	0.58	0.59	0.57	0.57	0.56	0.57
West Kalimantan Province	1.77	1.77	1.76	1.81	1.84	1.85	1.80

Source: Central Statistics Agency/Badan Pusat Statistik, 2015–2020 (modified)

The Location Quotient Analysis is used to figure out whether the agricultural sector in each regency/city of West Kalimantan Province is considered the basic or non-basic sector. The result of the Location Quotient calculation for the period of 2015–2020 shows that most of the regencies in West Kalimantan Province has a basic sector in agriculture, which is indicated by the LQ value above 1 as seen on Table 1. There is one regency and two cities that has an LQ value less than 1. The regency that has an LQ value below 1 is Kubu Raya. The agricultural sector in Kubu Raya only contributes around 16% of the province GRDP or around 2,844,645.16 million rupiahs in 2020.

Aside from Kubu Raya regency, the cities in West Kalimantan, Pontianak and Singkawang both also have LQ value less than 1, which means that agriculture is not a basic sector in the two cities. The proportion of agricultural contribution to the province GRDP for Pontianak and Singkawang cities are 2% and 13% respectively, or 405,875.37 and 890,507.24 million rupiah according to data retrieved from Central Statistics Agency.

Assessing the trend throughout the years, the LQ tends to fluctuate. The highest value can be found in Sekadau Regency, with the figure of 1.69 in 2017. The lowest value of 0.07 belongs to Pontianak City, which shows stagnation throughout 2015–2020. Melawi Regency shows an interesting trend, in which it was a basic sector in the year 2015–2017 with the LQ value of 1.01 and 1.02 but then the figure drops to 0.99 and 0.98 in the year 2018–2020. The average LQ of Melawi Regency for the whole period of 2015–2020 is at 1.00 so it can still be considered a basic sector in agriculture. The map seen on Figure 1 shows that Melawi Regency is the only area that experience a shift from basic sector in 2015 to a non-basic sector in 2020 for the agricultural sector. [8] stated that the agricultural transformation process is tied to labor participation and how it is affected by rural-urban migration.

 https://doi.org/10.1051/e3sconf/202132507008

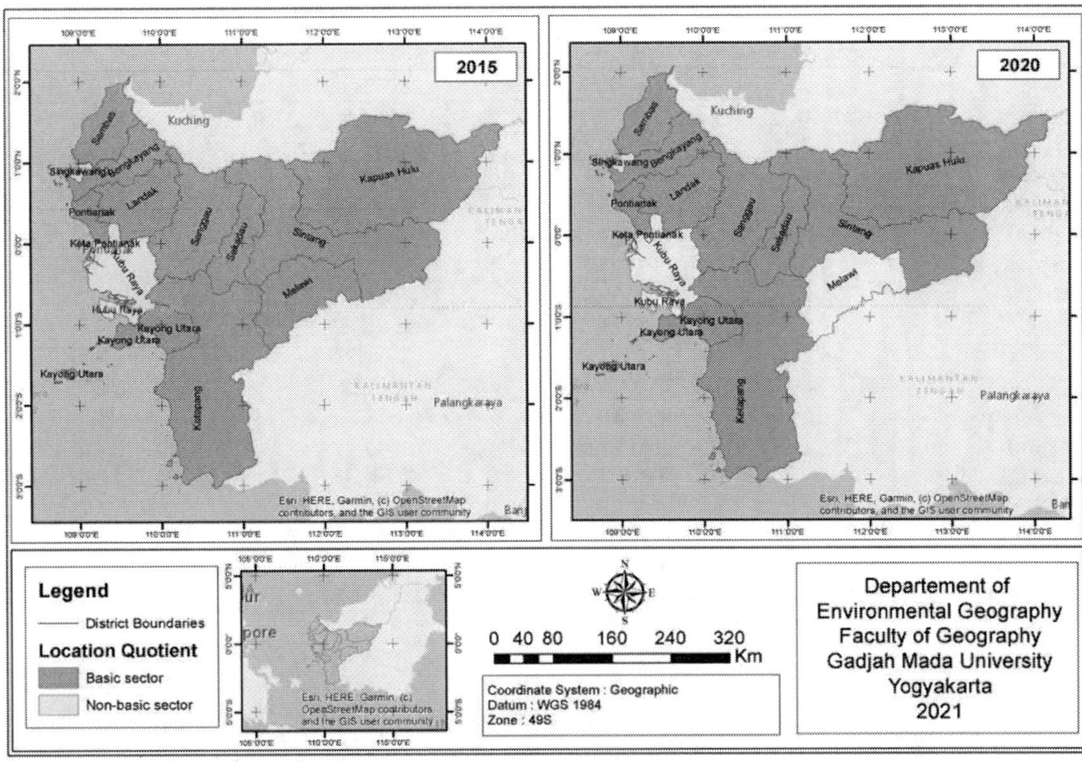

Fig. 1. Location Quotient map of agricultural sector in West Kalimantan 2015–2020

3.2 Shift-Share Analysis

Shift-Share Analysis can be used to show the development of economic sectors in West Kalimantan. Moreover, this analysis also show how it is related to other sectors. Shift-Share Analysis of regencies and cities in West Kalimantan is carried out by calculating the components of National Share (N), Proportional Shift (P), and Differential Shift (D) in 2015–2020. The N value of all regencies and cities in West Kalimantan shows a positive value which indicates a growth in each region. The P value in all regencies and cities in West Kalimantan Province shows a positive value, which indicates that the agricultural sector in the region has a fast growth rate. Positive D values are found in Bengkayang, Landak, Sanggau, Sekadau, and Kubu Raya Regencies which means those regencies have a competitive advantage in the agricultural sector compared to other regencies. Meanwhile, Sambas, Mempayah, Ketapang, Sintang, Kapuas Hulu, Melawi, Kayong Utara regencies, Pontianak City, and Singkawang City have negative D value which means these regencies and cities have lower competitiveness than other regencies and cities in West Kalimantan. The Shift-Share value (N, D, and P components) for each regency/city is shown on Fig.2.

Based on the results from the shift-share analysis, it shows a positive correlation between the N value and P value. The regencies and cities in West Kalimantan Province which have fast agricultural growth have also fast-growing agricultural sector. However, the value of D can distinguish the rate of growth of each regency and city from the existence of competitiveness in the agricultural sector. Agricultural competitiveness can be affected by the level of production and consumption of agricultural products. In this case, there appears a different demand and supply of agricultural products in each regency and city, thus making the competitiveness of agricultural products different.

https://doi.org/10.1051/e3sconf/202132507008

Table 2. Shift-Share Analysis Result of West Kalimantan Province

Regency/City/Province	Average		
	Nj (2015-2020)	Pj (2015-2020)	Dj (2015-2020)
Sambas	469,727.98	30,514.26	-1,583.51
Bengkayang	204,204.92	13,454.53	14,692.82
Landak	241,304.94	16,826.71	4,581.23
Mempayah	176,171.79	9,165.91	-167.45
Sanggau	461,466.28	28,877.46	16,819.31
Ketapang	582,761.58	33,616.65	-9,864.44
Sintang	328,208.73	15,051.59	-19,895.01
Kapuas Hulu	222,459.97	10,052.22	-9,232.35
Sekadau	144,823.46	10,587.30	11,523.21
Melawi	117,039.72	4,747.71	-6,564.06
Kayong Utara	86,810.50	4,701.98	-6,090.47
Kubu Raya	614,457.96	17,545.85	7,770.91
Pontianak	866,767.43	2,398.10	-869.48
Singkawang	236,136.27	5,729.05	-1,120.72
West Kalimantan Province	4,752,341.54	203,269.33	0.00

Source: Central Statistics Agency/Badan Pusat Statistik, 2015–2020 (modified)

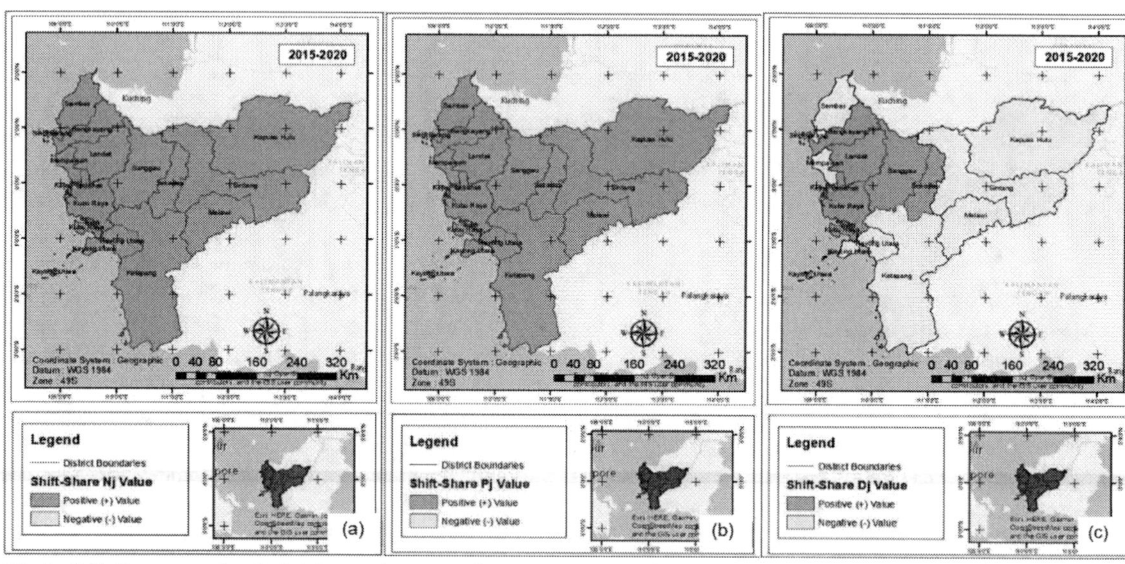

Fig. 2. Shift-Share map of agricultural sector in West Kalimantan 2015–2020 for each components (a) N value (b) P value (c) D value

Agricultural sector competitiveness in West Kalimantan Province is determined by comparing agricultural GRDP in each year. Agricultural GRDP is compared at a constant price according to regencies and cities in West Kalimantan Province. The constant price difference in agricultural GRDP in every year gives impact to the competitiveness of agricultural products. On other word, constant price difference is a tool for measuring whether the competitiveness rises or falls.

The lower of the constant price difference of agricultural products each year, the lower the competitiveness. Agricultural GRDP shows that regencies with the lowest constant prices for agricultural products have a negative D value. In this case, the growth of the agricultural sector is slow, as happened in the regencies of Sambas, Mempayah, Ketapang, Sintang, Kapuas Hulu, Melawi, Kayong Utara, Pontianak City and Singkawang City.

The regency with the highest competitiveness is Bengkayang Regency, with a D value of 14,692.82. Meanwhile, the regency with the lowest competitiveness was Sintang Regency, with a value of D -19,895.01. From these differences, it can indicate that the effect of differences in the growth of the agricultural sector is due to the fairly large imbalance of constant prices for agricultural product.

4 Conclusion

The result of the Location Quotient calculation shows that most of the regencies in West Kalimantan Province has a basic sector in agriculture, which is indicated by the LQ value above 1. Aside from the regencies, the cities in West Kalimantan, Pontianak and Singkawang both also have LQ value less than 1, which means that agriculture is not a basic sector in the two cities. The highest value can be found in Sekadau Regency, with the figure of 1.69 in 2017. The lowest value of 0.07 belongs to Pontianak City, which shows stagnation throughout 2015–2020.

Shift-share analysis show how it is related to other sectors. So, the regency has a competitive advantage in the agricultural sector compared to other regencies. From the results of the shift-share analysis, it shows a positive result between the N value and P value. This shows that all regencies and cities in West Kalimantan Province have fast agricultural growth in line with the agricultural sector which is also growing fast. To determine the competitiveness of the agricultural sector in the 2015–2020 in West Kalimantan Province, a comparison of agricultural GRDP in that year is needed. The lower the constant price difference of agricultural products each year, the lower the competitiveness.

From the discussion on LQ and Shift-Share analysis, it shows that agricultural growth in the district is faster than in the city. This is evidenced by the fact that Pontianak City and Singkawang City aren't the - basic sectors in agriculture, because they have the lowest LQ value. Likewise, the results of the Shift-Share analysis show a negative competitive level. Meanwhile, the Sanggau and Sekadau districts have fast agricultural growth, because these areas have a basic sector in agriculture with high competitive level according to LQ and Shift-Share analysis.

Acknowledgement

This Research is part of lecture block of Human and Economic Resource Analysis held by Department of Environmental Geography, Faculty of Geography, Universitas Gadjah Mada. We would like to thank for the guidance to all of lecturer of Human and Economic Resource Analysis.

References

1. L. Muta'ali, Teknik Analisis Regional Untuk Perencanaan Wilayah Tata Ruang dan Lingkungan, Yogyakarta: BPFG (2015)
2. R. Rizieq, Agrosains **5**, 77-91 (2008)
3. BPS Kalimantan Barat, Kalimantan Barat dalam Angka Tahun 2021. Pontianak: Badan Pusat Statistik, 2020.
4. R. S. Septianingrum *et. al.,* IOP Conf. Ser.: Earth Environ. Sci., **451** (2020)
5. S. Siyoto, A. Sodik, Dasar Metodologi Penelitian, Yogyakarta: Literasi Media Publishing (2015)
6. Y. Yurliana, M. R. R., S. Rachmadi, Jurnal Perspektif Pembiayaan dan Pembangunan Daerah **3**, 115-128 (2015)
7. E. P. Yuendini, I. N. Rachmi, N. N. A. Puspitasari, R. Harini. M. A. F. Alfana, Jurnal Geografi **16**, 128-136 (2019)
8. D. Ge, H. Long, W. Qiao, Z. Wang, D. Sun, R. Yang, Journal of Rural Studies **76,** 85-95 (2020)

Aquifer system and groundwater potency in coastal area of Kretek, Bantul Regency, Indonesia

La Ode Saleh Isa[1] & *Ignasius* Loyola Setyawan Purnama[1]

[1]Faculty of Geography, Universitas Gadjah Mada, Yogyakarta, Indonesia

Abstract. The existence of groundwater in coastal areas really needs attention because of its vulnerability to sea water intrusion. Utilization and good management are needed to maintain its sustainability. The purpose of this research is to determine the aquifer system in the study area and to calculate the groundwater potency. The aquifer system is determined by processing geoelectric sounding data using Rockworks 16 software. Groundwater potency is calculated using a static approach. The safe yield of groundwater is calculated by multiplying of groundwater fluctuation, village area and the specific yield. Water requirement is calculated base on domestic, livestock and tourism water requirement. The results showed that the Kretek District had two aquifer units, namely the fluvio marin plains and sand dunes. In the fluvio marin plains, most of the constituent material is sand with clay as inserts, while in the aquifer unit the sand dune has constituent material in the form of unconsolidated sand originating from the Merapi Volcano. The groundwater potency of the study area is 234,448,0000 m^3/year and the safe yield is 11,772,400 m^3/year. With the water requirement for domestic, livestock and tourism are 555,044 m^3/year, the groundwater potency in the study area is still sufficient. However, it is recommended that the well drilling does not exceed a depth of 40 meters, because at several locations there has been detected an interface. In addition, in several locations there were also detected connate water originating from ancient marines during the deposition process of the fluvio marin plains in the past.

Keywords: aquifer, groundwater, coastal area, Kretek District, Bantul Regency

1 Introduction

In accordance with the concept of the hydrological cycle, the amount of water on earth is constant [1-2]. Lack and excess of water in a place is actually just a matter of uneven distribution, either according to space or time, or because of its unbalanced form (solid, liquid, gas).

According to [3-4], along with the advancement of civilization, human needs in various aspects are increasing, including the requirement of water sources. Currently, there are more and more problems related to the need for and supply of water resources [5-6]. The thing expected is the existence of water resources that are sufficient in number, of good quality and evenly distributed in space and time. Therefore, in water resource management, what must always be strived for is efforts to make water available where it is needed with good and sustainable quality [7].

Groundwater is one part of natural resources whose existence is very much needed by humans, because until now the majority of the population's daily water sources are still sourced from groundwater [8-9]. However, the availability of groundwater is not always the same from one place to another, likewise its utilization is not the same.

Groundwater is found in various geological formations, especially in aquifers [10-11]. Aquifers are rock formations that can store and transmit water. In general, aquifers cover a large area and can be seen as underground water reservoirs. The water that enters this reservoir is either natural or artificial, and flows out by gravity or through artificial wells. The annual volume of water entering and leaving is only a small part of the total storage capacity.

The usage of groundwater resources for various purposes must pay attention to the total water contained in a groundwater basin. By considering a basin as a natural underground reservoir, the use of groundwater in one place will certainly also affect the overall groundwater contained in the basin. Therefore, the concept of equilibrium between the amount of water entering and leaving a basin must be considered in the utilization of groundwater.

The safe yield of a groundwater basin is the amount of water that can be taken from the groundwater basin without causing unwanted effects. Therefore, to preserve groundwater, groundwater extraction must be limited to safe yield [12-13].

Kretek District is one of the districts in Bantul Regency which has a coastal area. Geomorphologically, the area is located in the fluvio-marine plains of Merapi Volcano and sand dunes. Due to the existence of Parangtritis tourism object, this area is growing rapidly with a relatively dense population than the surrounding area. As a result, the water requirement for daily needs also increases, most of which comes from groundwater. This research focuses on Tirtohargo and Parangtritis Village which are the coastal areas of this district (Figure 1).

https://doi.org/10.1051/e3sconf/202132508001

Figure 1. Research location [14]

2 Method of research

2.1 Calculation of water availability

The results of geoelectric data processing on field sounding resulted in several subsurface lithology layers. To make it easier to delineate the groundwater aquifer layer at the study site, further data processing was carried out by using 3D modeling and 2D cross section of the prediction point using Rockworks 16 software.

2.2 Calculation of groundwater potency and its safe yield

Groundwater potency is calculated using a static approach :

$$Vat = Sy \times Vak \quad (1)$$

where Vat is groundwater volume that can be extracted from the aquifer, Sy is specific yield and Vak is aquifer volume, obtained by multiplying between village area and aquifer thickness. The safe yield of groundwater is calculated by multiplication of groundwater fluctuation, village area and the specific yield.

$$\text{Safe yield} = F.\ A.\ Sy \quad (2)$$

where F is the groundwater fluctuation and A is village area.

2.3 Calculation of water requirement

Water requirement is calculated base on domestic, livestock and tourism water requirements. The water requirement for domestic is determined according to the number of the population and the amount of water demand of each person/day. Due to the research area is a village with a fairly dense population, the water requirement of 100 liters/person/day is determined.

Water requirement for livestock is calculated based on the number of livestock and water consumption of each livestock/day. The standard of water requirement for big livestock (cow, buffalo, horse) is 40 liters/livestock/day; small livestock (goat, sheep) is 5 liters/livestock/day; and poultry is 0.6 liter/livestock/day [15].

3 Result and discussion

3.1 Aquifer system of the research area

Kretek district has two aquifer units, i.e. the fluvio marin plains and sand dunes. In the fluvio marin plains, most of the constituent material is sand with clay as inserts. The bedrock that underlies this aquifer system is limestone and marl of the Sentolo Formation.

According to MacDonald and Partners (1984), the Fluvio Marin Plains Aquifer unit is a major aquifer, which is a highly capable aquifer to provide adequate water for domestic and industrial purposes. Observing the characteristics of the aquifer, the aquifer permeability is classified in the fast criteria reaching 59.9 m/day, while the transmissibility of the aquifer and pescific discharge reaches 921.9 m^2/day and 4.00 m^3/sec/m.

Sand dune aquifer units are located along the coast of the study area. The width of this aquifer ranges from 3 to 4 km, with the constituent material in the form of unconsolidated sand originating from Merapi Volcano. The material is carried out to sea by the Opak River and thrown back on land by the wave and wind.

This aquifer has a thickness between 20 - 45 meters. The further north the thickness decreases to form a wedge structure, with the base of the aquifer in the form of clay from marine deposits. Above the surface, this aquifer system is characterized by alternating between beach ridge and swales. The material in the swale is relatively smoother than the beach ridge.

Referring to the results of the electrical sounding carried out by [17-18], in more detail the research area can be made a 3D model as shown in Figure 2.

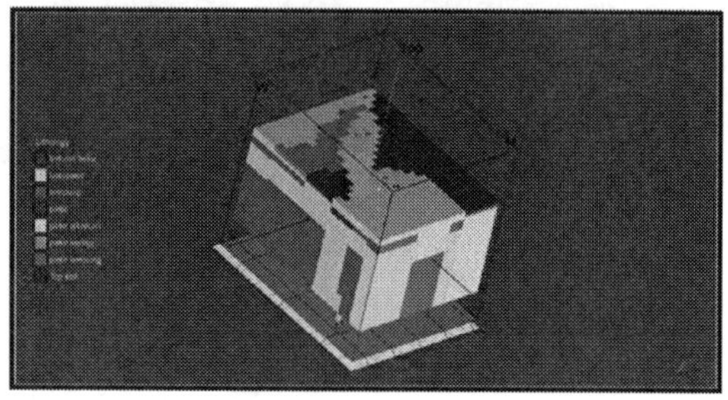

Figure 2. 3D lithology profile of the research location

In Figure 2, it can be seen that there are variations in subsurface lithology at each geoelectric sounding point. Clay and alluvium sand materials dominate the research area. Clay sand material becomes aquitard in this area, while alluvium sand material becomes the main aquifer constituent rock. In general, the aquifer system in the study area is divided into two formations, namely the fluvio marin plains and the sand dune. The fluvio marin plains is grouped into two facies, namely coastal sediment and river sediment. Beach sediment is composed of clay, sand, and gravel material, while river sediment is composed of clay, silt and fine sand, so that clay and alluvium sand materials are almost scattered in all research locations. Sand dune is composed of fine sand and coarse sand which is a manifestation of the marin and aeolin processes, so that alluvium sand material is also present in these landforms. Further explanation will be discussed in each cross section.

Cross section A-A' has a direction from north to south and is composed of variations in the subsurface lithology layer (Figure 3). Alluvium sand material dominates this cross section with different layer thicknesses. This material becomes the groundwater aquifer at the research location. The best groundwater potential is found at points G9, G5, and G8 because they have a fairly thick layer. However, at point G8 there is connate water at a depth of 20-30 m and below that depth the alluvium sand layer is saturated with groundwater again. Connate water at point G8 is salt water originating from marines during the past deposition of fluvio marin plains. Clay sand saturated with groundwater is found at points G12, G14, and G15 with varying thicknesses. Furthermore, igneous rock was inserted at point G14. This material is thought to be a lava flow from Mount Merapi.

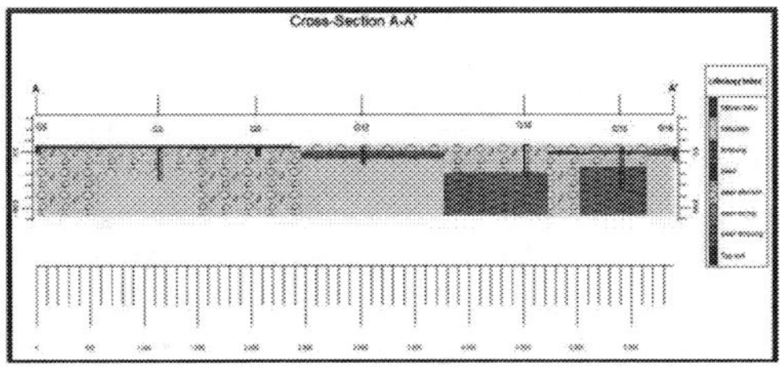

Figure 3. Cross section A – A'

In cross section B-B', there are several layers of subsurface lithology (Figure 4). Alluvium sand beds are present at all measurement points and become groundwater aquifers at the study sites with varying layer thicknesses. However, at point G8, there is connate water in the alluvium sand material, which originated from the marines during the deposition process of the fluviomarin plains in the past. Below this point, the alluvium sand layer is again saturated with groundwater. Clay sand layers are also present at almost all measurement points with varying layers. At point G2, there is an interface between the groundwater and saltwater zones at a depth of 40.4 - 64.7 m. In addition, at depths below 64.7 m, there is a layer of clay sand in salty conditions. This phenomenon is caused by the sounding point which is in the beach ridge. For points G3 and G4, clay sand is saturated with groundwater. When viewed from the lithological layer, the research locations at these points have high groundwater potential.

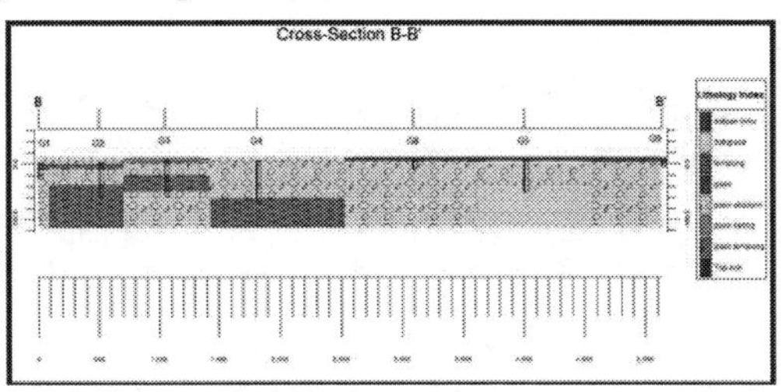

Figure 4. Cross section B – B'

In the C-C' cross section, alluvium sand material which is saturated with groundwater is found at all sounding points with different layer thicknesses (Figure 5). This material is groundwater aquifer at the research location. At the sounding point G8 with depth of 2.66 - 11.6 m, alluvium sand is saturated with groundwater containing connate water originating from marines during the past deposition of the fluvio marin plain. However, below 11.6 m, the alluvium sand is saturated with groundwater again. Clay sand material at points G6 and G7 also contains saturated groundwater, but at point G6 there is brackish water at a depth of more than 100 m. The brackish water at this depth is caused by the interface between groundwater and sea water because the position of the G6 measurement point is in the beach ridge. When viewed from the lithological layer, the research locations at these points have high groundwater potential.

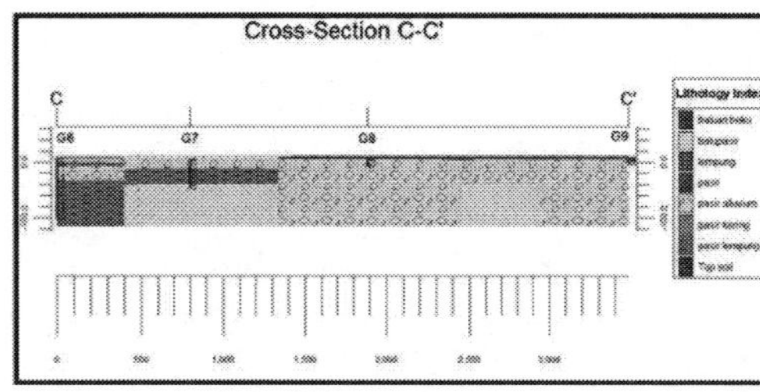

Figure 5. Cross section C – C'

3.2 Groundwater potency

As described in the research method, groundwater potency in the study area is determined based on the groundwater potency in each district, by multiplying of the thickness of the aquifer, the specific yield, and the area of each district. The safe yied of groundwater extraction is the result of the multiplication of the area, the specific yield and the fluctuation of groundwater. The results of the calculation of these two parameters are shown in Table 1.

Table 1. Groundwater potency in each district

No.	Village	Area (m^2)	Sy (%)	Aquifer Thickness (m)	Groundwater Fluctuation (m)	Groundwater Potency (m^3/year)	Safe Yield (m^3/year)
1.	Tirtohargo	3,620,000	38	40	2	55,024,000	2,751,200
2.	Parangtritis	11,870.000	38	40	2	180,424,000	9,021,200

Source : Kretek District in Number 2020 and calculation result

Based on the results of the calculations in Table 1, it can be seen that the research area has the groundwater potency 235,448,000 m^3/year, with the safe yield of the groundwater extraction 11,772,400 m^3/year. Parangtritis Village has higher groundwater potency and safe yields than Tirtohargo Village.

3.3 Water requirement

The amount of water needed for domestic requirement in the study area is 1,091,000 liters/day. Tirtohargo Village has a domestic water requirement of 303,300 liters/day and Parangtritis District has a domestic water requirement of 787,700 liters/day (Table 2).

The requirement for water for livestock in the study area is 63,062 liters/day. In Tirtohargo Village the water requirement is 24,825 liters/day, while in Parangtritis Villlage is 38,237 liters/day.

Table 2. The water requirement of domestic and livestock in the study area in 2019

No.	Village	Domestic		Livestock			
		Population (person)	Water Requirement (liters/day)	Big	Small	Poultry	Water Requirement (liters/day)
1.	Tirtohargo	3033	303,300	554	0	4441	24,825
2.	Parangtritis	7877	787,700	860	0	6396	38,237

Source : Kretek District in Figure 2020 dan calculation result

3.4 Evaluation of groundwater potency and safe yield

The results of the calculation show that the amount of water requirement in the study area is 421,233 m^3/year, while the safe yield of groundwater extraction are 11,772,400 m^3/year. Based on this data, it can be said that the groundwater extraction that has been carried out by residents has not exceeded the safe yield. However, it should be noted that both village are tourist areas. Based on the data, tourist visits in this area are around 1,338,112 people per year, so if the number of tourists is also taken into account, the additional water requirement added by 133,811 m^3/year to 555,044 m^3/year.

However, it is recommended that the well drilling does not exceed a depth of 40 meters, because at point G2, there is an interface in a depth of 40.4 - 64.7 m. Beside that, in G6 there is brackish water at a depth of more than 100 m. In several locations there were also detected connate water originating from ancient marines during the deposition process of the fluvio marin plains in the past. For example at point G8 there is connate water in a depth of 20-30 m

4 Conclusion

1. Kretek district has two aquifer units, i.e. the fluvio marin plains and sand dunes. In the fluvio marin plains, most of the constituent material is sand with clay as inserts. The bedrock that underlies this aquifer system is limestone and marl of the Sentolo Formation. Sand dune aquifer units are located along the coast of the study area. The width of this aquifer ranges from 3 to 4 km, with the constituent material in the form of unconsolidated sand originating from Merapi Volcano.
2. The groundwater potency in the study area is 234,448,0000 m^3/year and the safe yield is 11,772,400 m^3/year. With the water requirement for livestock and tourism of 555,044 m^3/year, the groundwater potency in the study area is still sufficient. However, it is recommended that the well drilling does not exceed a depth of 40 meters, because at several locations has been detected an interface. In addition, in several locations there were also detected connate water originating from ancient marines during the deposition process of the fluvio marin plains in the past.

Acknowledgement

This article is result of a study financed by the Final Assignment Recognition (Rekognisi Tugas Akhir/RTA) Programme Universitas Gadjah Mada Fiscal Year 2021.

Reference

1. Davie, T. Routledge, Taylor & Francis Group (2008).
2. Purnama I L S. Indonesian Journal of Geography 40 (2) : 153-166 (2008).
3. Senanayake I P., Dissanayake D M D O K., Mayadunna B B & Weerasekera W L. Geoscience Frontiers 7 (2016) : 115-124 (2016).
4. Belkhiri L., Tiri A & Mouni L. Groundwater for Sustainable Development 11 (2020) 100473 (2020).
5. Yangxiao Zhou & Wenpeng Li. Geoscience Frontiers 2(2) (2011) 205e214 (2011).
6. Rahmawati, N., Vullaume, J.F., Purnama, S. Journal of Hydrology 494 (2013) : 146-159 (2013).
7. Purnama, I L S. Indonesian Journal of Geography, 51(2), 206–216 (2019).
8. Messene G M. Journal of Environment and Earth Science Vol.7, No.2, 2017 : 32-39 (2017).
9. Purnama, I L S and Cahyadi A. Taiwan Water Conservancy 67 (4) : 57-67 (2019).
10. Todd, D.K., Mays, L.W. John Wiley & Sons, New York (2005).
11. Purnama I L S. 2010. Hidrologi Air Tanah. Penerbit Kanisius.
12. Sophocleous M. Natural Resources Research, Vol. 9, No. 2, 2000 : 99-110. (2000).
13. Purnama I L S., Marfai, M.A. Journal of Natural Resources and Development. 2012 (02) : 25-32 (2012).
14. Statistik Center Bureau of Bantul Regency. Kretek Distrik in Figures 2020.
15. Departemen Pekerjaan Umum. Proyek Pembinaan Pengelolaan Sumberdaya Air, Direktorat Pendayagunaan Sumberdaya Air, Direktorat Jenderal Pengairan, Departemen Pekerjaan Umum, Jakarta (1997).

https://doi.org/10.1051/e3sconf/202132508001

16. MacDonald & Partners. Directorate General of Water Development. Groundwater Development Project (P2AT) Indonesia (1984).
17. Putri F W. Thesis. Fakultas Geografi UGM (2008).
18. Kurniadhini F. 2016. Skripsi. Fakultas Geografi UGM (2016).

https://doi.org/10.1051/e3sconf/202132508002

Determination of the Water Quality Index (NFS WQI) of water bodies in the Huasteca Potosina, Mexico

Hilda Guadalupe Cisneros Ontiveros[1], *Nahum* Andres Medellín Castillo[1,2], *Abraham* Cárdenas Tristán[1], *Marisol* Castro Cárdenas[1], *César* Arturo Ilizaliturri Hernández[2], *Ratih* Fitria Putri[3*]

[1] Faculty of Engineering, Autonomous University of San Luis Potosi, San Luis Potosi 78290, Mexico
[2] Multidisciplinary Graduate Program in Environmental Sciences, Autonomous University of San Luis Potosi, San Luis Potosi 78210, Mexico.
[3]Department of Environmental Geography, Faculty of Geography, Universitas Gadjah Mada, Indonesia

Abstract. The water quality of water bodies in Huasteca Potosina was assessed by the National Sanitation Foundation Water Quality Index (NFS WQI) depending on nine water quality parameters include physical, chemical, and biological properties. The water quality obtained at the sampling site 50-90 in the NSF-WQI range shows the medium quality. Results revealed that the Moctezuma River in Tanquian de Escobedo has low water quality with a score of 50, the Tancuilin and Axtla Rivers have good water quality, and the Moctezuma River (Tampamolón Corona), Amajac River, Valles River, Panuco River, Patitos Lagoon, Plan de Iguala Lagoon were all located in medium water quality. This indicates that the water quality of the Huasteca Potosina is irregular, due to the presence of contaminating agents coming from fertilizers, pesticides, domestic water, and residual water from the communities adjacent to the rivers and lagoons, causing an environmental impact to the aquatic ecosystems of the place. However, this research shows the importance of making use of the water quality index in this region, this to relate the environmental impacts responsible for the contamination of surface water and interpret the experimental data of this water quality monitoring to facilitate decision making and consider frequent monitoring of the quality of water bodies in the Huasteca Potosina.

1 Introduction

The state of San Luis Potosi, Mexico, is divided into four zones: Altiplano, Centre, Media, and Huasteca. The study zone for this work focused on the Huasteca, which is east of the state. Figure 1 shows the zones, where the green shaded area represents the study zone. The Huasteca Potosina, which includes 19 of the 58 municipalities in San Luis Potosi state, has an area of approximately 12,500 km2 and represents 17.3% of the total state surface [1].

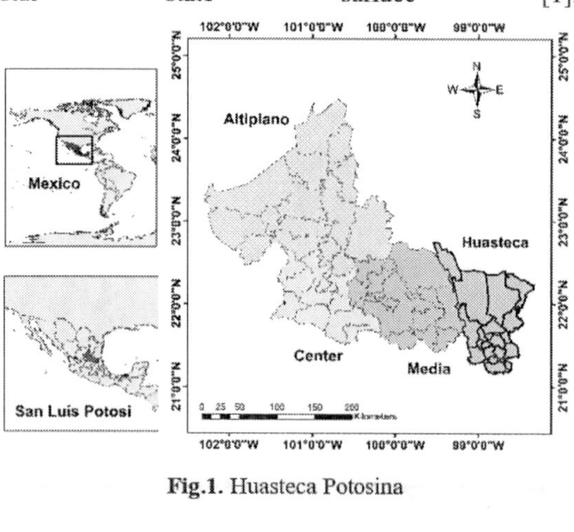

Fig.1. Huasteca Potosina

The Huasteca Potosina is in the hydrological region Panuco No. 26, according to the classification of CONAGUA. This area is divided into two portions: Upper and Lower Panuco. The Huasteca is in the lower portion, where a good number of runoffs are generated, tributaries that, in some way, are of great importance for the Panuco River. This river originates in the State of Mexico, enters the Sierra Madre Oriental on a rugged topography, which is more notable as the current descends, until the confluence of the Tempoal and Tampaón rivers. It receives the name of the river Panuco and continues with that designation until its mouth in the Gulf of Mexico, downstream from Tampico, Tamaulipas [2]. Its relief is slightly undulating with altitudes that oscillate approximately between 50 and 3000 meters above sea level. The hydrological region No. 26 is subdivided by four basins: Panuco River, R. Tamuin, R. Tamesi, and R. Moctezuma. The sub-basins that comprise the Huasteca Potosina are R. Panuco, R. Tamesi, Río Tampaon, R. Valles, R. Puerco, R. Mesillas, R. de Los Naranjos, R. Gallinas, R. Moctezuma, R. Axtla, R. Amajac, and R. San Pedro.

The Water Quality Index (WQI) became evolved in 1970 through the National Sanitation Foundation (NSF) of the United States, using Rand Corporation's Delphi research technique [3]. The WQI was formulated by the NFS as a quantitative aggregation of diverse selected

* Corresponding author : ratihfitria.putri@ugm.ac.id

and weighted water exceptional parameters to symbolize the excellent expert judgment of 142 professional respondents into one index [4]. Using a weight factor to distinguish the importance (weight-inferred and determined by experts) of every parameter for the future result is one of quantitative method with mathematical equations.

The INSF has the characteristic of being a multi-parameter index and is based on three studies. In the first one, 35 contamination variables included in the index were tested; the experts gave their opinion about them and classified them in three categories according to the parameter should be: "not included," "undecided," or "included." Among those included, they had to give a score from 1 to 5, according to their greater or lesser importance, with one being the most significant. They also had the opportunity to include more variables. In a second study, all the experts' answers were compared, so that the answers could be modified if deemed appropriate. As a result of this second study, nine identified variables of significant importance were identified: Fecal Coliforms, Dissolved Oxygen, pH, Temperature, BOD_5, Phosphates, Nitrates, Deviation, Total Solids, and Turbidity.

Finally, in the third study, participants were asked about developing an assessment curve for each variable. Water quality levels ranged from 0 to 100, located on the ordinates and the different levels of the variables on the abscissae. Each participant made the curve they thought represented the variation in water quality caused by the variables' level of contamination. These curves were known as "Functional Relations" or "Function Curves" [6-7].This study aims to evaluate the water quality status of water bodies of Huasteca Potosina based on NSF-WQI.

2 Method

2.1. The study area

Table 1 shows the water bodies for which the INSF was determined.

Table 1. Geographic location of the sites sampled in the Huasteca.

	Municipality	Water body	Geographical coordinates	
			Latitude	Longitude
S1	Tamazunchale	Amajac River	21°14'46"	98°46'33"
S2	Matlapa	Tancuilin River	21°21'25.3"	98°51'51"
S3	Axtla de Terrazas	Axtla River	21°26'02"	98°52'39"
S4	Axtla de Terrazas	Moctezuma River	21°25'55"	98°49'46"
S5	Tampamolon Corona	Moctezuma and Claro Rivers	21°29'28"	98°47'51"
S6	Tanquian de Escobedo	Moctezuma River	21°35'32"	98°39'22"
S7	Tamuin	Patitos Lagoon	22°2'30.7"	98°46'17"
S8	Tamuin	Tampaon River	22°00'04"	98°46'25.27"
S9	Ebano	Plan de Iguala Lagoon	22°02'16"	98°27'28"
S10	Veracruz	Panuco River	22°02'12"	98°24'0"
S11	Cd. Valles	Valles River	21°59'12"	99°01'15"
S12	Tamasopo	Puente de Dios River	21°55'48"	99°24'59"

2.2. Water samples collection and analysis

Water samples were taken at the sites corresponding to the map in Figure 2 in the period of August 2019, 1.5 L of the sample were taken with sterile polyethylene containers, stored at 4°C for later transport until analysis. At the sampling time, the bottles were rinsed three times with the river or lake water before the final sample was taken (see Figure 3).

The physicochemical and microbiological parameters evaluated for each water body were fecal coliform (FC), dissolved oxygen (DO), pH, temperature (T), biochemical oxygen demand (BOD), phosphates (PO_4), nitrates (NO_3), total solids (TS), and total turbidity (Tur.). The parameters measured in situ were carried out with the help of a multiparametric probe.
The parameters analyzed ex-situ were carried out in the Environmental Chemistry Laboratory of the Faculty of Chemical Sciences UASLP.

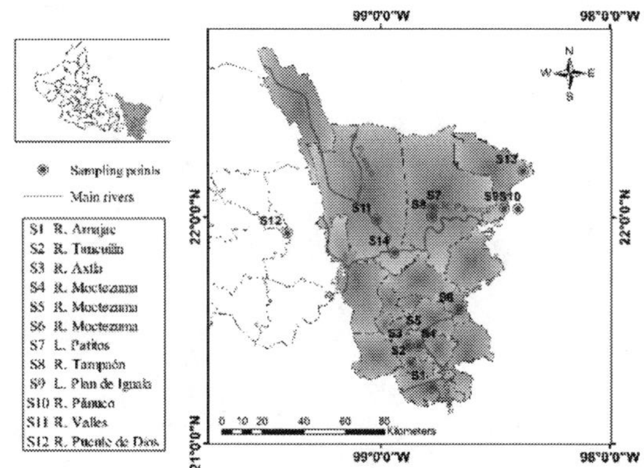

Fig.2. Geographic location of sampling points in Huasteca Potosina

Fig.3. Water sampling in the Moctezuma River

2.3. The national sanitation foundation's water quality index (NSF-WQI)

The index developed by [7-9] is on a 100-point scale representing the results of nine variables, such as BOD, DO, NO3, PO4, T, Tur., TS, pH, and FC. The index calculator is to be had online: http://www.water-research.net/index.php/water-treatment/water monitoring/monitoring-the-quality-of-surfacewaters [10].

https://doi.org/10.1051/e3sconf/202132508002

This index represents water quality in general and does not look at water use capacities. Several index parameters have exclusive importance, so a weighted average is beneficial for calculating the index, Table 2. The temperature extrade refers back to the temperature di between upstream and downstream control sites [8, 11].

Table 2. Weight scores of the nine NSF-WQI parameters

Parameters	Weighted mean
DO, mg/L	0.17
FC, CFU/100mL	0.16
pH	0.11
BOD, mg/L	0.11
T, °C	0.1
NO_3, mg/L	0.1
PO4, mg/L	0.1
Tur., NTU	0.08
TS, mg/L	0.07

The weighted score (Wi) should be expanded with the aid of using the subindex value (SI) of the parameter acquired with the aid of using the NSFWQI; then it is going to be summed with the index equation [8].

$$NSF\ WQI = \sum_{i=1}^{n} SI_i W_i \qquad (1)$$

where NSF-WQI is the score of the water quality index; SI is the sub-index value, and Wi is the weighted score (all calculated through the index calculator to be had online).

The wide variety received from making use of the index is classed in 5 scale classes as in Table 3 [7, 12].

Table 3. NSF-WQI categories

Range	Quality
0-25	Very bad
26-50	Bad
51-70	Medium
71-90	Good
91-100	Excellent

3 Results and discussions

3.1. Water quality parameters

The World Health Organization (WHO) (2011) water consumption standards [13] to compare with the quality data of the Huasteca Potosina water bodies, which presented in Table 4.

The pH scale used quantifies the alkalinity and acidity of the water. The pH range in this study was 7.6-8.8. The Huasteca water bodies' pH levels tend to be alkaline because there are naturally occurring carbonates and bicarbonates in solution, typical of karst systems in the geomorphology of the study area.

DO is important to aquatic existence for respiration, and most organisms have a really perfect scope of DO. Results for DO concentrations in water bodies were 4.7 to 10.5 mg/L, and saturation percentages ranged from 61.6 to 142.2%. Water bodies with concentrations greater than 5 mg/L indicate sufficient oxygen for most aquatic organisms. High levels of DO are conducive to improved water quality.

Temperature can determine the charge of biochemical responses withinside the aquatic ecosystem [14]. The water temperature range is 25-33.4 °C; the variation with the ambient temperature was about 9°C.

Turbidity is a degree of the clarity of water. Turbid water reduce light infiltration and impacts photosynthesis and aquatic life. High total solids can enhance the temperature of the water temperature because stable substances get warmness from daylight [15]. The minimum value of turbidity was in the R. Puente de Dios with a value of 0.2 NTU; on the contrary, the maximum value of turbidity was in Patitos lagoon with a 190 NTU value. TS levels varied from 154 to 2006 mg/L.

BOD is the amount of oxygen used by bacteria and fungi to decompose organic compounds in water for five days [16]. In the Huasteca water bodies, the BOD levels were from 0 to 8.0 mg/L; this indicates that the higher the BOD concentration, the lower the DO content, causing damage to aquatic organisms. Water bodies with BOD concentrations lower than 3mg/L favor water quality [17].

For plants and aquatic creatures, nitrate is an important supplement to be able to use nitrogen. Soil erosion, agricultural waste to domestic waste, is a source of nitrate [18]. Nitrate concentrations in Huasteca water bodies are relatively low according to WHO criteria [13], because the ranges varied from 0.3 to 19.6 mg/L.

Another important supplement for plants and creatures in their growth is phosphate. However, a high concentration of phosphate in water is also not good because it can increase eutrophication. Phosphate can come from detergents, fertilizers and waste [19]. The concentration of phosphates in the Huasteca water bodies 0.2 to 8.9 mg/L.

The presence of fecal coliform bacteria in water is an indicator that the water has been contaminated with sewage and other possible pathogenic organisms. The results of this study indicate that river water contains fecal coliform bacteria exceeding the water quality criteria.

Table 4. Mean water quality parameters values of rivers and lakes of Huasteca Potosina.

	WHO [13]	S1	S2	S3	S4	S5	S6
T		29.8	29.5	29.5	28.9	31.5	30.4
pH	6.5-8.5	8.2	7.9	7.9	8.4	8.8	8.7
TS	1000	396	154	208	838	624	631
DO		6.0	7.7	6.0	4.8	10.5	4.7
Sa%		78.4	101.8	78.2	63.3	142.2	63.8
Tur.	<5	16.0	2.5	4.7	5.0	2.6	5.8
PO_4		0.2	0.3	0.2	8.9	5.9	8.3
NO_3	50	5.8	2.7	0.7	21.7	13.7	19.6
BOD		4.2	0.0	5.8	4.9	5.0	8.0
FC		240.0	0.0	4.0	43.0	4.0	93.0

	WHO [13]	S7	S8	S9	S10	S11	S12
T		34.9	29.0	33.4	32.8	31.6	25.6
pH	6.5-8.5	8.5	7.9	8.2	8.7	7.9	7.6
TS	1000	552	914	2006	838	764	1600
DO		6.8	4.8	4.8	8.6	4.9	5.0
Sa%		95.9	62.4	66.8	119.7	67.1	61.6
Tur.	<5	190.0	4.9	62.1	4.3	2.8	0.2
PO_4		1.0	1.5	0.7	3.7	0.6	0.5
NO_3	50	11.0	1.0	10.3	10.3	0.3	2.3
BOD		5.8	5.0	6.4	6.5	2.8	5.8
FC		460	93	240	150	75	15

3.2 Application of NSF-WQI

Considering the application of NSF-WQI, the water quality in the river is at a medium level based on the categories in Table 3, as presented in Table 5 below. The Tancuilin and Axtla rivers located in the municipalities of Matlapa and Axtla de Terrazas, respectively, obtained good quality with an NSF-WQI score of 90 and 82, respectively. In contrast, the Moctezuma River in the municipality of Tanquian de Escobedo obtained an NFS score of 50, placing the river with low water quality. Water bodies such as the Amajac River, Montezuma river, Patitos lagoon, Tampaón river, Plan de Iguala lagoon, Panuco river, Valles river, and Puente de Dios river were placed in the medium category, with the NFS-WQI value range of 55-68.

Table 5. NSF-WQI values of rivers and lakes of Huasteca Potosina

Place	NSF-WQI value	Class
S1	67	Medium
S2	90	Good
S3	82	Good
S4	55	Medium
S5	57	Medium
S6	50	Bad
S7	53	Medium
S8	63	Medium
S9	54	Medium
S10	56	Medium
S11	67	Medium
S12	68	Medium

4 Conclusion

This study demonstrates the advantages of using WQI in assessing overall water quality and summarizing large amounts of data into a single value. The water from the rivers and lagoons of the Huasteca Potosina can be consumed by the community after traditional treatment; however, the results show that anthropogenic activities contaminate the rivers and lagoons, and the NFS WQI values were 50-90, which indicates a medium and good water quality in the case of the Tancuilin and Axtla rivers. The use of WQI is more systematic and provides a comparative assessment of water quality from the sampling stations. It is likewise beneficial for community to recognize water quality and be a beneficial device in water quality management.

References

1. Instituto Nacional de Estadística y Geografía (INEGI). *Anuario estadístico y geográfico de San Luis Potosí 2017*. Available in: www.inegi.org.mx (2017)
2. Instituto Nacional de Estadística y Geografía (INEGI), & Cadena, A. *Hidrología*. In *Síntesis Geográfica del Estado de San Luis Potosí* (p. 4). Instituto Nacional de Estadística y Geografía. (1985)
3. Ball, R. Church, R. *Water Quality Indexing And Scoring.Journal Of The Environmental Engineering Division*, American Society Of Civil Engineers, 106, EE4, 757-771. (1980)
4. Hussein Ewaid, S. *Water quality evaluation of Al-Gharraf river by two water quality indices.* Applied Water Science, 7(7), 3759–3765. https://doi.org/10.1007/s13201-016-0523-z (2017)
5. Fernandez, N. J., & Solano, F. *Capítulo III: indices de calidad (ICAs) y de contaminación (ICOs) del agua de importancia mundial*. In Índices De Calidad Y De Contaminación Del Agua (pp. 43–113). Universidad de Pamplona (2007)
6. Ott, W. *Environmental Indices, Theory And Practice, AA Science*, Ann Arbor, Michigan. (1978)
7. Brown R., Mcclelland N., Deininger R., Tozer R. *"A Water Quality Index- Do We Dare?" Water And Sewage Works*. October. P. 339-343 (1970)
8. Abbasi T, Abbasi SA *Water quality indices*. Elsevier, Amsterdam, p 384 (2012)
9. American Public Health Association (APHA) *Standard methods for the examination of water and wastewater*, 27th edn. Washington, DC (2012)
10. [NSF] National Sanitation Foundation International. The Public Health and Safety Organization. Avaiable in : https://www.nsf.org/ (Accessed on september of 2020).
11. Rai RK, Upadhyay A, Ojha CSP, Singh VP. *The Yamuna River basin.* Water Science and Technology Library 66:307–356 (2012)
12. USGS *National field manual for the collection of water quality data, techniques of water resources, book 9*, handbooks for water resources investigations, pp 1539 (2015)
13. World Health Organization (WHO) *Guidelines for drinking water quality*, 4th edn. Available in: http://www.who.int/water. Accessed August 2020 (2011)
14. Simon FX, Penru Y, Guastalli AR, Lorens J, Baig S. *Improvement of the analysis of the biochemical oxygen demand (BOD) of Mediterranean seawater by seeding control*. Talanta 85(1):527–532 (2011)
15. Fulazzaky MA.*Water quality evaluation system to assess the status and the suitability of the Citarum*

river water to different uses. Env Monit Assess 168(1–4):669–684 (2010)

16. Naubi I, Zardari NH, Shirazi SM, Ibrahim NFB, Baloo L. *Effectiveness of water quality index for monitoring Malaysian river water quality*. Pol J Environ Stud 25(1):231–239 (2016)
17. CONAGUA, *Statistics of Water in Mexico*. Edition 2013. México. 2014.
18. Ell MJ. *Total suspended solids (TSS).* In: NDDO Health (ed), North Dakota, USA (2008)
19. Viessman W, Hammer MJ. *Water supply and pollution control*, 7th edn. The University of Michigan, Pearson Prentice Hall, USA (2005)

Engineering Geological Characteristics Based on Rock Mass Rating (RMR) and Geological Strength Index (GSI) in Jlantah Dam Intake Tunnel, Karanganyar District, Central Java Province

Doni Apriadi Putera[1], *Heru* Hendrayana[1] and *I Gde Budi* Indrawan[1*]

[1]Geological Engineering Departement, Engineering Faculty, Gadjah Mada University, Sleman, Yogyakarta, Indonesia

Abstract. This paper presents the results of a geological engineering investigation in the form of rock mass characterization at the Jlantah Dam Intake Tunnel. The study was carried out through technical geological mapping, core drill evaluation and supported by laboratory test data. The determination of rock mass classification at the research site has been carried out using the Rock Mass Rating (RMR) method, but it is necessary to use another method that is more suitable based on rock mass for weak rocks, namely using the Geological Strength Index (GSI) method.The rock mass quality will be used as a parameter in determining the excavation method and tunnel support system that will be used in the Jlantah Dam intake tunnel. The results showed that the research area consisted of lithology in volcanic breccias and tuff lapilli. GSI rock mass value at the research location ranged from 15 - 65, while the RMR value ranged from 24 - 70. The correlation between RMR and GSI in the study area is different when compared to Hoek and Brown (1997) but has similarities with Zhang et al (2019).

1 Introduction

The research location is at the Jlantah Dam construction site. Administratively, this dam is located in Tlobo Village, Jatiyoso District, Karanganyar Regency, Central Java Province (Figure 1).

Several site investigations have been carried out in the design stage of the Jlantah Dam construction [8]. Still, the study of rock mass characterization in the intake tunnel was not carried out in detail. A detailed description of rock mass needs to be carried out for the tunnel's safety because the characteristics of the rock mass influence the stability of the portal slope, the determination of the excavation method, and the tunnel support system.

The determination of rock mass classification at the research location has been carried out using the Rock Mass Rating (RMR) method, while the rocks at the study site are lapilli tuff rocks in moderately weathered conditions. Determination of rock mass classification should be made based on rock mass classification for weak rocks, for example by applying the Geological Strength Index (GSI) [7], in order to obtain an appropriate rock support system.

2 Regional Geology

Based on the Geological Map of the Ponorogo Sheet compiled by Sampurno and Samudro [10] , the rock formations that contain the tunnel location and its surroundings consist of Lava Lawu (Ella) Jobolarang Lava (Qvjl), and Lava Sidoramping (Qvsl) formations (Figure 2). Mapping in the field resulted in the four rock and soil units that make up the research area. The four rock and soil units from old to young are: Volcanic Breccia Unit, Slightly and Moderately weathered Lapilli Tuff Unit and Silt Sand Soil Unit. The rock units in the research area are included in the Lawu Volcano sediment with a Holocene age.

From the results of the surface geological mapping that has been carried out along the Jlantah River and the Puru River, starting from the plan of the Jlantah Dam to the inundation plan area, it is known that the regional geological structure pattern covering the investigation area has a west-east straight line and is not a fault. The local geological structures found are the stocky structure of the breccias and poor bedding on the breccias sandstone insert.

3 Methodology

The main reason for using RMR is its ease and flexibility in various practical purposes in engineering [2]. According to Bieniawski [2], the rock mass classification of the RMR method uses 6 (six) parameters with

 https://doi.org/10.1051/e3sconf/202132508003

measurements that can be carried out in the field and interpretation of drill hole data. These parameters include 1) Uniaxial Compressive Strength (UCS) of intact rock material, 2) Rock Quality Designation (RQD), 3) discontinuity spacing, 4) discontinuity conditions, 5) groundwater conditions, and 6) discontinuity orientation. The rock mass quality parameters are used to determine the RMR value at the research site based on Bieniawski [2].

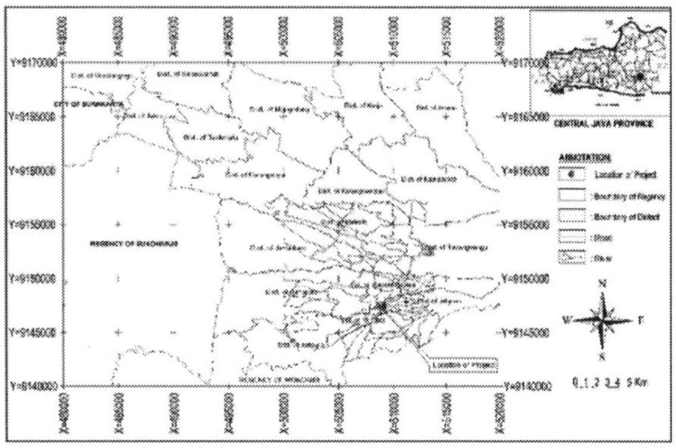

Fig. 1. Location of Jlantah Dam intake tunnel

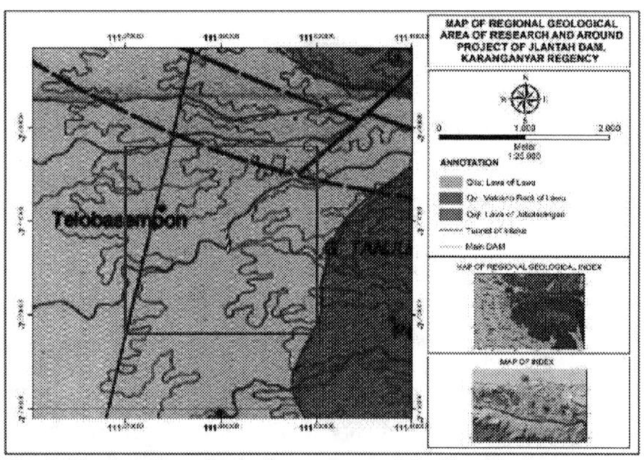

Fig. 2. Regional geological map around the jlantah dam

The GSI introduced by Hoek and Marinos [7] provides figures which, when combined with intact rock properties, can be used to estimate the reduction in rock mass strength for different geological conditions. In subsequent developments, GSI is incorporated into two main parameters, namely the structure of the block properties and surface conditions in the form of roughness, weathering, or alteration and fill as described by Hoek at al [5]. According to Hoek et al. [5], to calculate the subsurface GSI value using the formula:

$$GSI=1.5\ JCond_{89} + RQD/2 \quad (1)$$

The Jcon (joint condition) value in equation (1) refers to the classification of the common disease according to Bieniawski [2].

The rock mass classification using the GSI system is an extension of the rock mass classification of the RMR system. The parameters used for GSI are generally based on RMR. Research on the correlation between RMR and GSI has been carried out by several researchers, namely Hoek and Brown [4], Ceballos and Olalla [3] and Zhang et al [11]. The RMR_{89} value is the basic RMR value [2] by providing a groundwater value with a weight of 15 (dry conditions) and a common orientation value with a weight of 0 (zero). The minimum value for the rock mass classification of RMR 1989 is 23. Hoek and Brown [4] explain that rock mass with GSI> 25 and RMR> 23, the GSI value can be calculated by the following Equation (2):

$$GSI = RMR_{89} - 5 \quad (2)$$

Zhang et al [11] investigated the correlation of RMR with GSI at the surface of the tunnel excavation and produced equation (3)

$$RMR_{89} = 0.827\ GSI + 15.394 \quad (3)$$

Ceballos and Olalla [3] examined the correlation between RMR and GSI based on the rock types, namely igneous, metamorphic and sedimentary rocks and based on outcrops, namely natural, mining and excavated for road project work. The resulting equation in general is as follows:

$$GSI = 1.13\ RMR - 11.63 \quad (4)$$

4 Results and Discussion

Observations of rock outcrops indicate that the research area consists of 4 rock units from old to young, namely: Volcanic Breccia Unit, Slightly and Moderately weathered Lapilli Tuff Unit, and Silt Sand Soil Unit. The map of the distribution of rock units and the geological cross-section of the research location can be seen in Figures 3 and 4. Examples of outcrops of lapilli tuff and volcanic breccias are shown in Figures 5 and 6. The path of the intake tunnel is located in the Tuff Lapilli Unit.

Based on subsurface data (core drill, Fig. 7), the rock mass classification in the research area has rock discontinuity with high to slightly weathering rates. The GSI value of rock mass in the research area ranges from 15 - 65, while the RMR value ranges from 24 - 70. The relationship between the GSI value and the rock mass RMR in the research area can be expressed by the following equation (Figure 8):

$$RMR = 0.827\ GSI + 14.493 \quad (5)$$

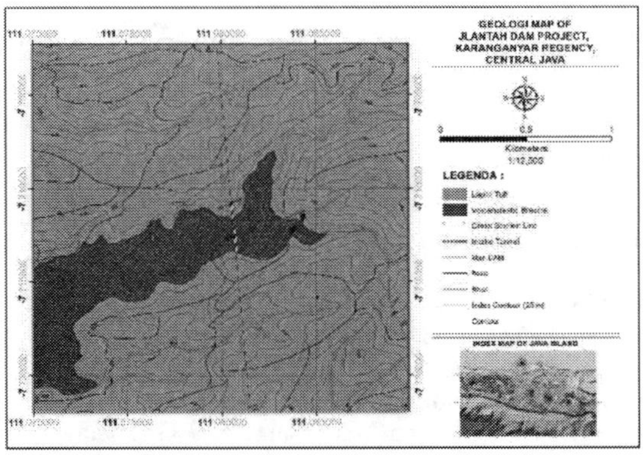

Fig. 3. Geological map of the research area

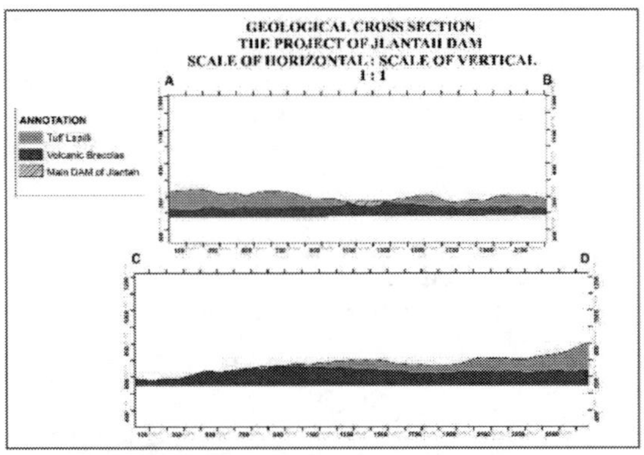

Fig. 4. Geological cross section of Jlantah dam

Fig. 5. Outcrop of Lapilli tuf slightly weathered

Fig. 6. Outcrop of volcanic breccia slightly weathered

Equation (5) obtained in this research is different from the empirical equation (2) proposed by Hoek and Brown [2] however, it has similarities with Zhang et al [11] empirical equation and when seen in Figure 9, the linear line of equation 5 is close to the linear line of the Ceballos and Olalla [3] equation

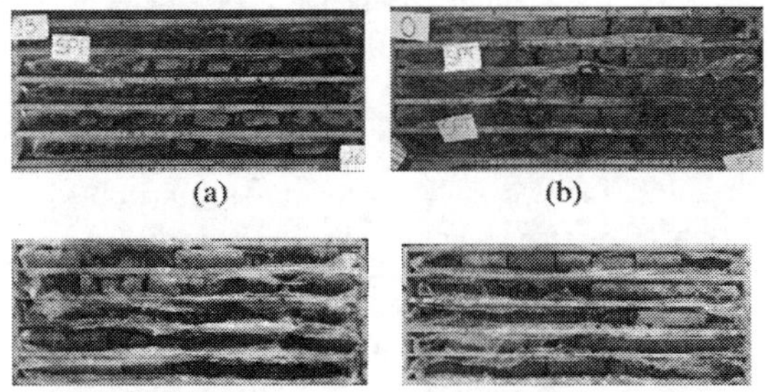

Fig. 7. Typical drill cores at the research location, material : (a) residual soil ; (b) lapilli tuf poor quality ; (c) lapilli tuf with moderate quality ; (d) volcanic breccia with moderate quality

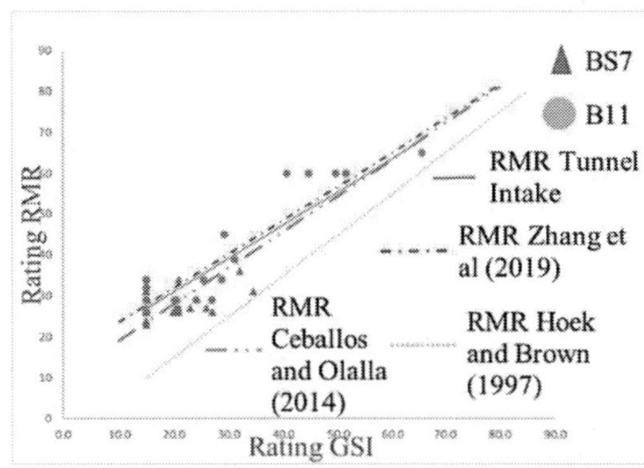

Fig. 8. The correlation equation between RMR_{89} and GSI at the research location was compared with the correlation between RMR_{89} and GSI Hoek and Brown [4], RMR and GSI Zhang et al [11] and RMR and GSI Ceballos and Ollala [3]

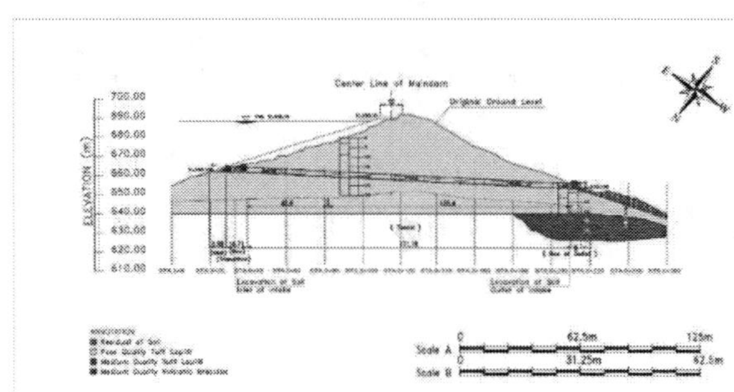

Fig. 9. Longitudinal section of engineering geological rock mass subsurface of the intake tunnel

Table. 1. Engineering geological characteristics of the rock mass at the research area

Lithology	Characteristics
Lapilli tuff with poor quality	Brownish (moderately weathered) - whitish (slightly weathered - fresh). Medium grained - coarse. Good - medium sorting, containing fragments of rock fragments (litihik 15%) with a size of 2 ~ 64 mm, pumice, volcanic glass, 10% quartz mineral, 75% dominant glass, tapered mineral form. This lithology has a moderate level of weathering, GSI values 15-27 RMR_{89} 24-31, UCS ranges from 1 - 25 Mpa. Water content 16.01% dry density 0.82 gr / cm^3
Lapilli tuff with moderate quality	Brownish (moderately weathered) - whitish (slightly weathered - fresh). Medium grained - coarse. Good - medium sorting, containing fragments of rock fragments (lithik 15%) with a size of 2 ~ 64 mm, pumice, volcanic glass, 10% quartz mineral, 75% dominant glass, tapered mineral form. This lithology has a slightly weathering level, GSI values 28 - 43 RMR_{89} 32 - 39, UCS ranges from 5 - 25 MPa. Water content 13.3%, dry density 1.17 gr / cm^3
Volcanic breccia with moderate quality	Brown to gray, with fragments consisting of various volcanic materials (andesite, andesite porphyry), gravel-sized fragments - rooted, poorly sorted, angled to an angle of responsibility. Abundant base period (Matric Supported) in medium - very coarse volcanic sand, silica cement tuff. This lithology has fresh to slightly weathered conditions, GSI values 44-66, RMR_{89} 40-70, UCS ranges from 25 - 50 MPa. Water content 10.6% dry density 1.57 g / cm^3.

The quality of rock mass in the research area can be divided into rock mass classes based on lithology and weathering. The quality of rock mass at the research location is divided into three, namely poor quality lapilli tuff (RMR_{89} 24 - 31), medium quality lapilli tuff (RMR_{89} 32 - 39), and medium quality volcanic breccias (RMR_{89} 40 - 70). The description of the characteristics of each rock mass engineering geological unit is shown in Table 1. In general, the rock mass in the research area has a mild to a high level of weathering. Whole-rock has a UCS value ranging from 1 to 50 MPa and is extremely weak to medium-strong strength category. Figure 9 shows that the intake tunnel alignment is in poor quality tuff lapilli.

5 Conclusions and Recommendations

The research area consisted of lithology in the form of volcanic breccias and tuff lapilli. Rock masses have a mild to a high degree of weathering. Whole-rock has a UCS value ranging from 1 to 50 MPa and is extremely weak to medium-strong strength category. The GSI value of the rock mass in the study area ranges from 15 - 65, while the RMR value ranges from 24 - 70. The intake tunnel is located in tuff lapilli rock with moderate weathering conditions.

The relationship between the GSI value and the RMR of rock mass in the research location can be expressed by equation 3 have results that are quite close to the research of Zhang et al [11].

Further research is needed, including determining the stability of the portal slope, the determination of the excavation method, and the tunnel support system based on the rock mass characterization results of this study.

Thank you very much to BBWS Bengawan Solo for research and supervising lecturers of Masters in Geological Engineering, Gadjah Mada University, so that this journal can be completed. We acknowledge PT. Waskita Karya (Persero) for the financial support for education and research through the super-specialist program held by PUPR in the year 2020.

References

1. PUSGEN, *Indonesia earthquake source and hazard map* (*Peta sumber dan bahaya gempabumi Indonesia*). Indonesia: Minister for Public Works and Housing (2017)
2. Z.T. Bieniawski, *Engineering Rock Mass Classification*, NewYork (1989)
3. F. Ceballos and C. Olalla, *Relationship between RMRb and GSI based on in situ data*. Technical University of Madrid, Spanyol (2014)
4. E. Hoek and E. T. Brown, *Practical estimates of rock mass strength*. International Journal of Rock Mechanics and Mining Sciences, **34**,8: 1165-1186 (1997)
5. E. Hoek, T. G. Carter, M. S. Diederichs, *Quantification of the geological strength index chart*. US Rock Mechanics/Geomechanics Symposium, San Francisco (2013)
6. ISRM, *Suggested methods for the quantitative description of discontinuities in rock masses: international cociety for rock mechanics*. International Journal of Rock Mechanics and Mining

Science and Geomechanics Abstracts, **15**: 319-368 (1978)
7. P. Marinos and E. Hoek, *GSI: a geologically friendly tool for rock mass strength estimation.* International Society for Rock Mechanics and Rock Engineering (2000)
8. PT. Aditya, *Studi Investigasi Tambahan Untuk DD Waduk Jlantah*, Kartasura (2017)

Evaluation of groundwater usage in relationship to groundwater vulnerability to sea water intrusion in Cilacap Coastal

Salsabiila Bayu Nugrahaeni[1], *Ignasius* Loyola Setyawan Purnama[1] and *Vincentia* Anindha Primacintya[2]

[1]Faculty of Geography, Universitas Gadjah Mada, Indonesia
[2]Accounting Programme, Master of Science and Doctoral, Faculty of Economics and Business, Universitas Gadjah Mada, Indonesia

Abstract. Besides being able to cause land subsidence, excessive groundwater use in coastal areas can also cause to sea water intrusion. The purpose of this study is to evaluate the use of groundwater in the study area in relation to its vulnerability to sea water intrusion. Because groundwater in the study area is used for domestic, industry and livestock purposes, the water use that is taken into account is the use of water for the three sectors. The amount of water used for domestic purposes is calculated based on the population and the amount of water needed of each person per day. The amount of water use for industry is calculated based on the number of industrial employees and water usage of each employee per day. Water use for livestock is calculated based on the number of livestock and water use of each livestock per day. The results of this water usage calculation are then linked to the criteria for groundwater vulnerability to sea water intrusion and the depth of the interface. Observing the relationship between groundwater usage and the vulnerability of groundwater to sea water intrusion and the depth of its interface, Tegal Kamulyan, Cilacap and Sidakaya villages, all of which are located in South Cilacap District, need attention. The three village are classified as moderate vulnerability to sea water intrusion and shallow interface depth, but their water usage is quite high. For this reason, it is necessary to make efforts to find other water sources for domestic, industry and livestock requirement other than groundwater.

Keywords: groundwater, sea water intrusion, Cilacap

1 Introduction

The larger the population requires more and more facilities and infrastructure, one of which is the availability of water sources. Because groundwater is still the main source of water for domestic and industrial needs, groundwater extraction will also be even greater. Without being balanced with an equivalent recharge, the increased usage of groundwater often causes a decrease in the groundwater level. Meanwhile, on the other hand, a lot of open land as infiltration areas has been converted into residential areas [1-2].

Coastal aquifers have an important role in meeting the needs of fresh water for the community [3]. Even in some countries, aquifers in coastal areas supply fresh water for industrial, agricultural, and drinking water for the community [4-5]. Likewise, in several Mediterranean regions with low rainfall, groundwater in coastal areas is the main supply for household needs. Therefore, the existence of aquifers in coastal areas has an important role for both humans and the environment.

Population growth in this area encourages increased fulfillment of fresh water needs [6]. However, if groundwater extraction exceeds the safe yield, the volume and surface of the groundwater will decrease, so that the hydrostatic pressure will weaken. In this situation, sea water can push and flow to the mainland and there is sea water intrusion. The boundary of freshwater and sea water on land is called an interface [7-10].

The interface area is a zone of water transition and diffusion between fresh and salt water, and salt water will be under fresh water [11-14] . If the pressure of fresh groundwater from the land against sea water is getting stronger, the interface will move towards the sea. However, if the groundwater's hydrostatic pressure weakens, sea water will push ashore and seawater intrusion occurs [15-17].

According to [18], it can be said that seawater intrusion has become a problem in cities located in coastal areas, because it causes changes in the type of groundwater chemistry or even the quality of its water. As a result, groundwater as the main source of drinking water for the majority of the population cannot be utilized because it tastes salty.

Among the other southern coastal areas of the Java Island, Cilacap is the most developed area [19]. Urban development is often identified with increasing population and settlements. In relation to water use, population growth results in increased use of water for daily life. Because groundwater is still the main source of water for domestic and industrial needs, groundwater extraction through dug wells is also increasing. Without being balanced with an equivalent additive, the increased use of groundwater often causes the groundwater level to fall and even in coastal areas it can cause sea water intrusion.

2 Methodology

In this research, the calculated water usage is the usage of water for domestic, industrial and animal husbandry purposes, because the three sectors make use of groundwater. The use of water for domestic needs is determined according to the size of the population and the amount of water needed per person per day. According to Mangku Sitepoe [in 20] , water demand in big cities is generally more than 150 liters/person/day, in medium cities 80-150 liters/person/day, districts cities 60-80 liters/person/days and villages ranging from 30-60 liters/person/day. Based on these criteria, because North Cilacap, Central Cilacap and South Cilacap are considered medium cities, the water usage is determined to be between 120 liters/person/day.

Industrial water use is calculated based on the number of industrial employees and water usage of each employee per day [21]. The standard of water requirement for industrial employees is 500 liters/employee/day.

Water use for livestock is calculated based on the number of livestock and water usage of livestock in each day, where the types of livestock that are calculated using water are cattle-buffalo-horse (big livestock), sheep (small livestock) and poultry. Standard water requirements for big livestock are 40 liters/livestock/day, small livestock 5 liters/livestock/day and poultry 0.6 liter/livestock/day [22].

The results of this water usage calculation are then linked to the criteria for groundwater vulnerability to sea water intrusion and the depth of the interface from Purnama's research results [23] (Figure 1). Quantitative descriptive analysis is used to give an explanation of the relationship between these three aspects.

3 Result and discussion

3.1 Location, boundary and area

This research was carried out in a coastal area that was formerly known as the Administrative City of Cilacap, which is astronomically located at 108°04'30 "- 109°30'15" East Longitude and 07°00'20 "- 07°45'12" South Latitude. Administratively it covers North Cilacap District, Central Cilacap District and South Cilacap District, which are now part of the Cilacap Regency. The administrative boundaries of the research area are Jeruklegi District (north), Kesugihan District (east), Indian Ocean (south) and Donan River (west).

The research area has an area of 50.18 km^2, with details of North Cilacap covering 18.84 km^2, Central Cilacap covering an area of 22.15 km^2 and South Cilacap covering an area of 9.19 km^2. Judging from the topography, the research area is a sloping lowland.

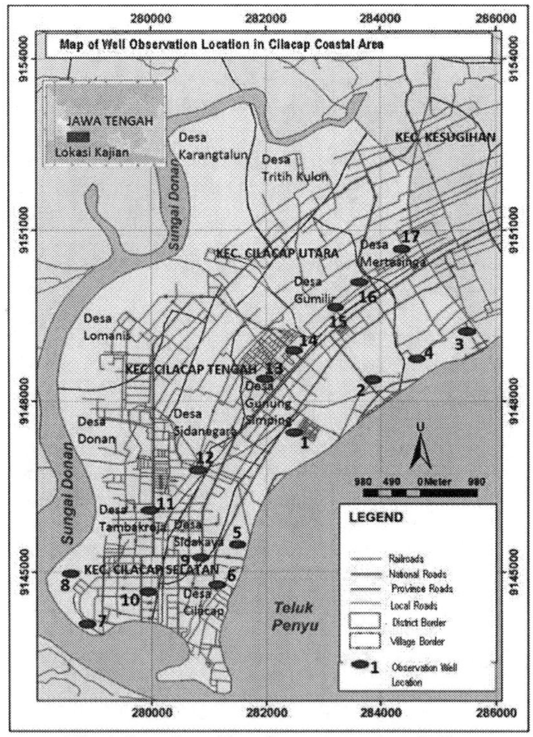

Fig. 1. Location of interface measurement [23]

3.2 Climate

To see climate patterns, climate classification is used. Among climate variables, rainfall is the most variable, so that climate classifications are often based on patterns and distribution of rainfall. Climate classification based on the distribution of monthly rainfall has been carried out in various ways, including the Schmidt-Ferguson System which is based on the calculation of the average wet and dry months year after year.

Based on the climatic division of Schmidt-Ferguson, the research area has a climate of B, which is a wet type with an average annual rainfall of 3,379 mm. The lowest average monthly rainfall of 136 mm occurs in August and the highest average monthly rainfall of 385 mm occurs in January. The other types are divided by Koppen. In the Koppen classification, beside from using rainfall data, the division of climate types is also based on air temperature. According to Koppen, climate is divided into five main climate types, i.e. climate type A or tropical rain climate, climate type B or dry climate, climate type C or temperate climate, climate type D or snowfall climate and climate type E or polar climate. Furthermore, climate type A is further divided into Af type or humid tropical, Am type or tropical rain and type Aw or tropical savanna. Based on the division of climate types from Koppen, the climate type on the coast of Cilacap is Af, which is a humid tropical climate, with the lowest average monthly temperature of 25.4°C occurring in August and the highest average monthly temperature of 27.4°C occurring in April.

3.3 Geology

Pay attention to its geological conditions, the research area is located in the coastal alluvial plain, river alluvial plain and swamp alluvial plain which are quaternary in age. The coastal alluvial plain is in the form of off-shore material with a material thickness of between 2 to 8 meters. River alluvial plains in the form of river alluvial deposits (silt, sand and gravel) are found in valleys, while alluvial plains of swamps in the form of clay, silt and sand are found in the Tritih and North Mertasinga areas.

Observing the stratigraphy, in the study area and its surroundings there are 3 rock units, namely coastal alluvium rock units, rivers and valleys, tuff marl rock units and tuff sandstone units. (Wongsosentono, 1974). The coastal alluvium units, rivers and valleys are exposed in the foothills of the Serayu Delta. The coastal alluvium forms an arc-shaped coastal embankment. Between one embankment and another embankment, it is separated by plains with clay, silt and silt material.

The alluvium of rivers and valleys in the form of fine-grained deposits is found in areas far from the river, while those in the form of coarse-grained deposits are found near rivers. The tuff marl rock units are scattered in the Jeruklegi and Tritih areas with material in the form of marl with an insertion of silt and sand, while the tuff sandstone units are composed of alternating sandstones with tuffs.

3.4 Geohydrology

Based on 3 cross sections of lithology drilled by the Directorate of Geology in 1970, groundwater in the study area is found in unconfined aquifers with sand material found up to a depth of 14-18 meters from the ground level. Under this layer of sand is a layer of clay.

Taking into account the material that makes up the aquifer, in terms of the porosity value, it is classified as large, namely between 39% to 42%, while the specific yield of aquifer is between 3 to 28%. Looking to the permeability value, it is classified as fast with a value of 2.5 to 45 m/day.

3.5 Population, industry and water usage

Based on data from North Cilacap in Number of 2019, Central Cilacap in Number of 2019 and South Cilacap in Number of 2019, the total population in the study area is 198,765 people. South Cilacap has the highest population of 83,329 people, followed by North Cilacap with 81,524 people and the lowest is Central Cilacap with 33,912 people.

Looking to Table 1, the total water demand for domestic usage in the study area is 23,851,800 liters/day. South Cilacap has the largest domestic usage namely 9,999,480 liters/day, followed by North Cilacap with 9,782,880 liters/day and Central Cilacap with 4,069,440 liters/day

Table 1. Total population, number of industries and water usage

District/Village	Population (people)	Number of Industry Employees				Water Usage (liters/day)
		Large	Medium	Small	Home Industry	
North Cilacap						
Kebon Manis	9,864	0	0	90	81	1,633,680
Gumilir	17,389	0	0	161	352	2,343,180
Mertasinga	19,073	0	62	83	430	2,545,260
Tritih Kulon	22,325	0	0	63	575	2,997,500
Karang Talun	12,873	672	0	27	253	2,020,760
Central Cilacap						
Kutawaru	10,341	0	0	218	266	1,482,920
Lomanis	1,505	1,855	0	32	33	1,140,600
Donan	6,144	0	0	186	363	1,011,780
Sidanegara	9,433	0	0	257	134	1,327,460
Gunung Simping	6,489	0	0	102	44	851,680
South Cilacap						
Tambakreja	22,343	781	0	65	62	3,150,160
Tegalreja	12.301	0	0	295	200	1,723,620
Sidakaya	11,015	0	0	32	124	1,399,800
Cilacap	18,903	0	105	41	161	2,341,360
Tegal Kamulyan	18,767	1,094	32	125	145	2,877,540

Source : North Cilacap, Central Cilacap and South Cilacap in Number of 2019

Based on the scale of its activities, industry can be divided into four types, namely large industry, medium industry, small industry and home industry. The largest number of large industries is in South Cilacap, namely 4 industries with 1,875 employees, while in Central Cilacap and Sputh Cilacap, they have 2 and 1 large industries, respectively. For medium industry, South Cilacap also has the highest number of medium industries, namely 4 industries with 137 employees, while for small industries and home industries it is evenly distributed in the research area.

Based on the number of employees, the largest usage of water for industry is in Central Cilacap, amounting to 1,745,000 liters/day. The usage of water for industry in

South Cilacap is 1,631,000 liters/day and in North Cilacap is 1,424,500 liters/day.

3.6 Number of livestock and its water usage

In the research area, it turns out that there are also many livestock businesses, both large livestock, small livestock and poultry. The types of big and small livestock that are cultivated by the population include dairy cows, ordinary cows, buffalo, horses, goats and sheep, while the types of poultry that are cultivated are village chickens, purebred chickens, ducks and manila ducks.

The usage of water for livestock in the study area is 48,164 liters/day. The highest water usage was found in North Cilacap with 28,920 liters/day, followed by North Cilacap with 10,929 liters/day and Central Cilacap with 8,315 liters/day. Details of the number of livestock and water use for livestock are shown in Table 2.

Table 2. Number of livestock and water usage

District/Village	Number of Livestock			Water Usage (liters/day)
	Big Livestock	Small Livestock	Poultry	
North Cilacap				
Kebon Manis	45	44	4368	4,701
Gumilir	52	61	4199	4,904
Mertasinga	62	72	8908	8,185
Tritih Kulon	73	61	5457	6,499
Karang Talun	41	48	4985	4,631
Central Cilacap				
Kutawaru	65	253	780	4,333
Lomanis	7	58	283	730
Donan	34	84	307	1,964
Sidanegara	12	31	223	769
Gunung Simping	4	45	223	519
South Cilacap				
Tambakreja	33	361	1798	4,203
Tegalreja	25	0	717	1,430
Sidakaya	0	81	816	895
Cilacap	0	154	1750	1,820
Tegal Kamulyan	0	184	2769	2,581

Source : [24] and calculation result

3.7 Evaluation of groundwater usage to sea water intrusion

Based on the accumulation of the three sectors, the highest water usage was in the North Cilacap of 11,569,300 liters/day, followed by South Cilacap at 11,492,480 liters/day and the least was in the Central Cilacap of 5,14,440 liters/day (Table 3). Observing the relationship between water usage and the vulnerability of groundwater to seawater intrusion and the depth of its interface, Tegal Kamulyan, Cilacap and Sidakaya villages, which are located in South Cilacap, need attention. Tegal Kamulyan Village is classified as moderate vulnerability with an interface depth of 39.20 to 63.58 meters, but its water usage is quite high, reaching 2,880,121 liters/day. Likewise in the Cilacap and Sidakaya Villages. Cilacap Village is classified as moderate vulnerability with an interface depth of 26.68 meters from the ground level, but the usage of groundwater reaches 2,343,180 liters/day. For Sidakaya Village, it is classified as moderate vulnerability with an interface depth of 57.12 meters from the ground level and the amount of water usage 1,400,695 liters/day. In these three village it is necessary to reduce the use of groundwater.

Table 3. Water Usage, groundwater vulnerability to sea water intrusion and interface depth

District/Village	Category of Groundwater Vulnerability to Sea Water Intrusion	The Depth of Interface (m)	Water Usage (liters/day)
North Cilacap			
Kebon Manis	Low	87.78	1,638,381
Gumilir	Low	129.74	2,348,084
Mertasinga	Far from shore is low	86.54 to 110.23	2,553,445
	Close from shore is moderate	64.12 to 72.28	

Tritih Kulon	Not a coastal area	Not calculated	3,003,999
Karang Talun	Not a coastal area		2,025,391
Central Cilacap			
Kutawaru	Not a coastal area	Not calculated	1,487,253
Lomanis	Not a coastal area	Not calculated	1,140,600
Donan	Not a coastal area	Not calculated	1,013,744
Sidanegara	Low	95.11	1,328,229
Gunung Simping	Moderate	86.12	852,199
South Cilacap			
Tambakreja	Low	78.62 to 101.19	3,154,363
Tegalreja	Low	44.52	1,725,050
Sidakaya	Low	57.12	1,400,695
Cilacap	Moderate	26.68	2,343,180
Tegal Kamulyan	Moderate	39.20 to 63.58	2,880,121

Source : [23] and calculation result

4 Conclusion

Observing the relationship between water use and the vulnerability of groundwater to sea water intrusion and the depth of its interface, Tegal Kamulyan, Cilacap and Sidakaya villages, which are located in South Cilacap District, need attention. Because the three urban villages are classified moderate vulnerability to sea water intrusion, the interface depth is quite shallow but the water usage is quite high. For this reason, it is necessary to make efforts to find other water sources for domestic, industrial and livestock demand other than those from groundwater.

Acknowledgement

This manuscript is the result of research funded by Final Assignment Recognition (Rekognisi Tugas Akhir/RTA) Programme Universitas Gadjah Mada Fiscal Year 2021.

Reference

1. Gwalema, S.R. The Indonesian Journal of Geography 43 (2) : 123-140 (2011).
2. Purnama I L S., Rahardjo N & Sulaswono B. Jurnal Rekayasa Lingkungan 4 (3) : 131-140 (2008).
3. Pousa, J., Tosi., L., Kruse, E., Guaraglia, D, Bonardi, M., Mazzoldi, A, Rizzetto, F, Schnack, E. Environ. Geol. 51 : 1307–1316 (2007).
4. Sappa, G. Journal of African Earth Sciences (2015). 105. 64–84 (2015).
5. Comte, J C. Journal of Hydrology: Regional Studies (2016) 5 : 179–199 (2016).
6. Tillman, F.D., Leake, S.A. Hydrogeol. J. 18, 1515–1524. (2010).
7. Celico, Fulvio, Capuano, Paola, De Felice, Vincenzo, Naclerio, Gino. Hydrogeol. J. 16 :1441–1451 (2008).
8. Basack, S., Bhattacharya, A.K., Sahana, C., Maity, P. Issue 3 (5) : 80-90 (201).
9. Purnama I L S., Marfai, M.A. Journal of Natural Resources and Development. 2012 (02) : 25-32 (2012).
10. Rahmawati, N., Vullaume, J.F., Purnama, S. Journal of Hydrology 494 (2013) : 146-159 (2013).
11. Barlow, Paul M. Geological Survey, Virginia (Chapter 1) (2003).
12. Vandenbohede, A., Lebbe, L. Hydrogeol. J. 14 : 462–472 (2006).
13. Young Kim, Kue, Suk Park, Yoon, Pyo Kim, Gee. Hydrogeol. J.17 : 617–629 (2009).
14. Purnama I L S., Febriarta., Cahyadi, A., Khakhim, N., Ismangil, L. & Prihatno, H. Jurnal Geografi, 11(22) : 155 – 165 (2013).
15. Rotzoll, Kolja, Oki, Delwyn S., El-Kadi, Aly I. Hydrogeol. J. 18 : 1425–1436 (2010).
16. Dayal, B., Chauhan, R.S. Biological Forum-An International Journal 2 (2) : 36-37 (2010).
17. Mukherjee. G., Debbarma, C., Paul, S., Dey, S. Indonesian Journal of Geography 43 (2) : 171-181 (2011).
18. Marandi, A., Vallner, L. Estonian Journal of Earth Sciences 59 (4) : 277-287 (2010).
19. Purnama I L S and Cahyadi A. Taiwan Water Conservancy 67 (4) : 57-67 (2019).
20. Priyana Y dan D. Safriningsih. Forum Geografi 19 (1) : 81-87 (2005).
21. Nippon Koei, Co., Ltd. Departemen PU dan BAPPENAS, Jakarta (1993).
22. Departemen Pekerjaan Umum. Proyek Pembinaan Pengelolaan Sumberdaya Air, Direktorat Pendayagunaan Sumberdaya Air, Direktorat Jenderal Pengairan, Departemen Pekerjaan Umum, Jakarta (1997).
23. Purnama I L S. Indonesian Journal of Geography, 51(2), 206–216 (2019).
24. North Cilacap, Central Cilacap and South Cilacap in Number of 2019. Badan Pusat Statistik Kabupaten Cilacap (2019).

Geostatistical analysis of static level evolution between 1995-2005 and 2007-2017 in San Luis Potosí Aquifer, México

Maria Ines Navarro-Hernández[1], *Abraham Cardenas* Tristán[1], *Ratih Fitria* Putri[2*]

[1]Center for Research and Graduate Studies, Faculty of Engineering, San Luis Potosí Autonomous University, México
[2]Department of Environmental Geography, Faculty of Geography, Universitas Gadjah Mada, Indonesia

Abstract. Knowing the piezometric levels evolution of an aquifer is essential for planning the use of water resources in the cities of central México, such as San Luis Potosí, whose population depends almost entirely on groundwater. Applying geostatistical methods from georeferenced data can reliably provide information on the spatial variability of valley piezometry. The objective of this study is to evaluate the evolution of the static levels of the aquifer in San Luis Potosí Valley, through an interpolation analysis by the kriging method. For both periods, the experimental variogram was adjusted to the Gaussian model and the method with the best results was ordinary kriging. From this analysis it was obtained that the decay of the static levels is more critical in the central zone of the valley, while the zones of the northwest and towards the south, the aquifer levels have shown an important recovery; the important values for the decrease in static levels coincide with the areas where the population density is greater.

1 Introduction

Since the 1970s, geostatistics has emerged as a science that is responsible for the study of phenomena with spatial correlation, to be able to predict the values of the variables in places where sampling was not carried out or places for which there are no data. Statistical methods use stochastic theories for interpolation and the distribution of uncertainty. An important aspect of geostatistics is that it uses spatial autocovariance structures, represented by semivariograms, which indicates the degree of similarity of values that belong to a regionalized variable, taking into account the distance that this attribute is concerning others that influence about him. To be able to predict the interpolation of the values of points that were not sampled, geostatistics relies on several methods, such as the Kriging method [1].

Throughout the geostatistics history, geostatistical methods have been widely used to analyze spatial processes and more recently space-time processes in almost all applied sciences, such as hydrogeology, environmental engineering in studies of pollutants concentrations (air, water, soil) and meteorology for climate predictions, etc. [2], [3].

2 Study area

The SLP valley is located in the physiological province between the Sierra Madre Oriental and the Sierra Madre Occidental. It has an approximate altitude of 1850 masl and occupies an extension of 1980 km2; It is surrounded to the west by the Sierra de San Miguelito and to the east by the Sierra de Álvarez [4]. This basin is located in the southwestern part of the state of San Luis Potosi (Figura 1), and within the area covered by this valley are totally or partially the municipalities of San Luis Potosí, Soledad de Graciano Sánchez, Mexquitic de Carmona and Cerro de San Pedro [5], with a current population of 1,221,556.

The average annual temperature is 17.5 ° C, with precipitation of approximately 402 mm per year and average annual evaporation of 2038.7 mm, this confirms a type of dry or semi-arid climate because evaporation far exceeds the precipitation [5].

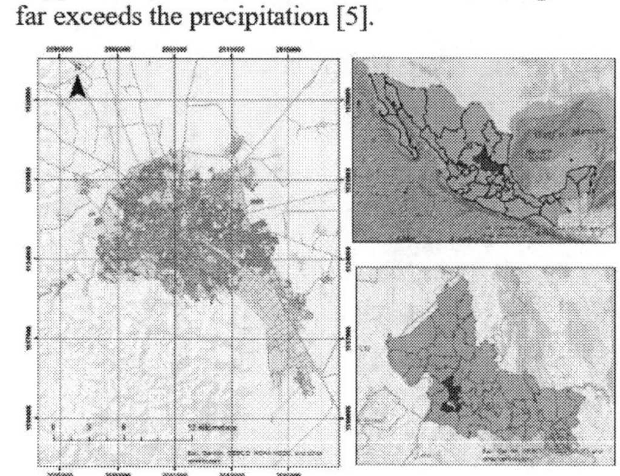

Fig. 1. Location of the study area.

Since 1995, there is a dropped cone in the city of San Luis Potosí, between 1971 and 1995 the drop in static levels was 95 m. In the period 1995-2001, this cone descended 25 m more towards the northern part mainly,

* Corresponding author: ratihfitria.putri@ugm.ac.id

even though there were areas in which the levels were kept constant [6].

by 1960 59% of the water for domestic use came from surface water and the rest was extracted from the aquifer, in 2005, this amount had increased to 92% and only 8% of the water destined for domestic use came from superficial water bodies [6]. The increase in the extraction of groundwater has been accelerated; and this went from being 0.97 m3/s in 1970 to 4.1 m3/s for 2007 [7].

Currently, several factors exacerbate this decrease in piezometric levels, one of them being the accelerated population growth and the expansion of agricultural and industrial activities. However, the low permeability of the lithological materials that border the valley, the low precipitation and high evaporation of the area that characterizes semi-arid climates, erosion, and deforestation, as well as the paving of recharge areas, affect the impediment of retention and infiltration of runoff water into the aquifer. This has caused the deficit in the hydrological balance between the period from 1995 to 2002 to have increased from 36 million m3/year to 42.5 million m3/year [8].

The volume of water withdrawn for 2006 was 92.2 million m3/year. This was obtained from the exploitation of 125 supply sources so that the depths of these wells vary between 270 and 700 meters. [9].

3 Methodology

Kriging is a geostatistical interpolation technique that returns the best linear unbiased prediction of intermediate data and avoids clustering effects. This method assumes that the data is generated from the Gaussian distribution with the correlation structure defined by the estimated semivariogram [10]. The semivariogram is a function that allows analyzing the spatial behavior of a variable within an area. The experimental semivariogram is defined by the following mathematical expression [11]:

$$(h) = \frac{1}{2N(h)} \sum_{\gamma=1}^{N(h)} [Z(X_i + h) - Z(X_i)]^2 \tag{1}$$

Where, Z(X_i) is the value of the variable at site x; Z(X_i+h) is another sample value separated from the previous one by a distance h, and N is the number of pairs that are separated by the distance h. The semivariance function is calculated for various distances h. Thus, the experimental semivariogram corresponds to an average distance between pairs of sites within each interval and not to a specific distance h. To interpret the semivariogram, the criterion is that the smaller the distance between the sites, the greater the spatial correlation between the sample values. [12].

Ordinary kriging does not require information beyond the measurement and its geographical coordinates, it is based on the assumption that the variation is random and spatially dependent, and that the random process is intrinsically stationary, with a constant measurement and a variance that depends only on the separation of distance and direction between places, but not in the absolute position. Kriging prediction is a linear sum of data that can have one, two, or three dimensions [13].

Assuming that the following sampling points, X1, X2, …, XN have been recorded for the values of a random variable Z to give N data, Z(Xi),i=1,2,…, N. For spot kriging, Z is predicted at any new point X0, using the equation:

$$Z(X_0) = \sum_{i=1}^{N} \lambda_i Z(X_i), \tag{2}$$

Where λ_i are the weights, to guarantee that the estimate is impartial, the weights must add 1.

$$\sum_{i=1}^{N} \lambda_{y0} = 1 \tag{3}$$

The expected difference is $E[Z(x_0) - z(x_0)] = 0$ and the variance of the prediction are calculated using the equation:

$$Var\,[Z(x_0)] = E[\{Z(x_0) - z(x_0)\}^2]$$
$$= 2\sum_{i=1}^{N} \lambda_i \gamma(x_i - x_0) - \sum_{i=1}^{N}\sum_{j=1}^{N} \lambda_i \lambda_j \gamma(x_i - x_j) \tag{4}$$

Where $\gamma(x_i - x_0)$ is the semivariance of Z between the sampling point x_i and the target point x_0, and the quantity $\gamma(x_i - x_j)$ is the semivariance between the i-nth and j-nth sampling points. These semi-variances are derived from the variogram model, to ensure that the variances are not negative. If a target point is also a sampling point, kriging returns the observed value and the estimated variance will be zero [11].

As is known, Kriging is a geostatistical method that uses a set of geostatistical tools to predict the values in places where there was no sampling. It is important to highlight that to model with these methods, it is first necessary to verify that the data complies with the conditions, analyzing elements such as normal distribution without trend, histogram, and normal QQ graph, to determine the best way to model the spatial structure of the data, that is, the type of semivariogram (spherical, exponential, etc.) and from this, to be able to select the best kriging method [14].

Two geostatistical analyzes were performed in this study, the first is an analysis of the piezometric evolution from 1995 to 2005; the second corresponds to the analysis of the piezometric evolution for the period from 2007 to 2017. The field information that supports these statistical analyzes was provided by CONAGUA, however, the piezometric measurements for the year 2017 are found on the official website of the INEGI, in a hydrogeological study of the San Luis Potosí aquifer.

In the second instance, two databases were prepared, which include the measurements of the piezometric changes for both periods; this database contains information on the coordinates in UTM, and the static levels for each well in both years, so that in the 1995-2005 analysis 22 wells were used, while for the 2007-2017 period 24 wells were used. The third phase of this analysis involves making an exploratory analysis of the variable evolution of the piezometric level, for example, observing the distribution of the data, since, in case of not having a normal trend, some transformations will be necessary to achieve this behavior.

The fourth phase is the application of geostatistical methods, for this, it is necessary to make a structural analysis, that is, study the semivariogram of the primary variable, choose the semivariogram that best fits, and subsequently, the type of Kriging that best fits to the behavior of the data. In the final stage, once the best method to simulate the data has been determined, the spatial prediction map is constructed through the ArcGIS geostatistics tool, and finally, interpretations about the phenomenon are made.

4 Results and Discussion

4.1 Analysis of the evolution of the piezometric level for the period 1995-2005

In this first analysis that includes ten years, 22 wells distributed in the San Luis Potosí Valley were used. Thus, it was pertinent to carry out an exploratory study of the data, in this case, of the variable that contains the changes in the piezometric level between 1995 and 2005. Therefore, a histogram was generated, as shown in Fig. 2 where, in the upper right corner appear the values for the measures of central tendency, asymmetry, and kurtosis measures.

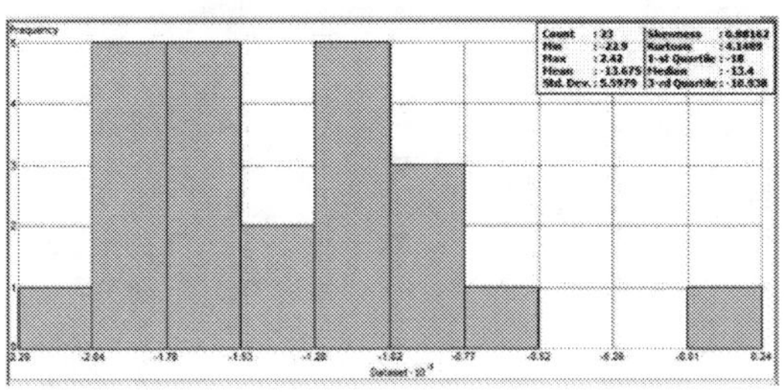

Fig. 2. Histogram of change in static level 1995-2005.

As observed in the histogram, if a curve could be constructed, this would be of the leptokurtic type, that is, a little pointed, which translates into more homogeneous data, this can be validated with the value of kurtosis that is 4.14, what which means that it is very close to 3, which is the homogeneity value. On the other hand, the skewness is 0.881, and when this value is close to zero, it can be said that it is a low skewness; however, being positive indicates that the skew is to the right of the curve or histogram.

Next, a trend analysis is performed, which consists of a three-dimensional graph of the data of the variable studied. In Fig. 3, it is observed how these values are distributed, which are also projected on each of the axes, with this, the projections of the trends (blue line and green line) can be observed.

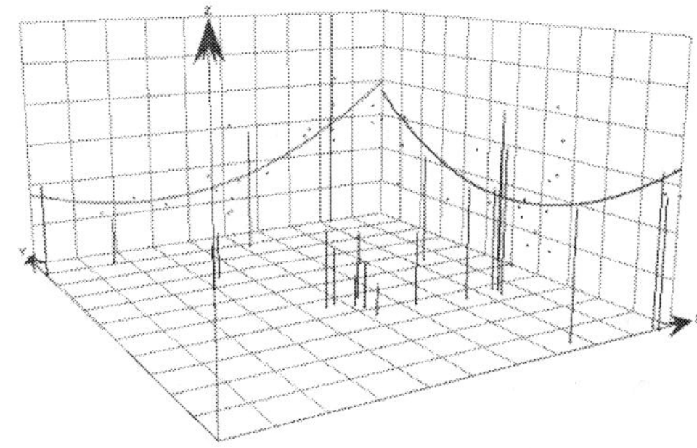

Fig. 3. Trend analysis for static levels 1995-2005.

For the projection of the XZ plane (green line) it is noted that the trend increases from west to east, that is, from this information we can begin to deduce that the piezometric decays are towards the east and central part of the area of study. On the other hand, concerning the projection of the YZ plane (blue line), a significant drift and greater dispersion are observed, indicating that the main decays are in the central part, while to the north and south the level remains stable or increases.

Then, simple kriging is carried out, for which no transformation was done to normalize the data, nor was it necessary to eliminate any trend. Simple kriging is a method in which the value of the mean must be known, in this case, that value is -13.7; It also requires knowing the covariance, however, the disadvantage of this type of kriging is that it produces very generalized results. For this simple kriging analysis, the semivariogram model that best fit the data was the Gaussian (this model is commonly used in data with normal distribution), as shown in Fig. 4, anisotropy was also incorporated. After performing simple kriging, ordinary kriging was constructed, for which the same Gaussian-type semivariogram model was maintained; ordinary kriging, unlike simple kriging, does not require knowing the mean, but it must be constant; Another difference is that it uses more complex equations for its construction and generates more detailed information, despite exaggerating the outliers and consequently presenting "portholes" effects in the interpolation results. Finally, universal kriging was also performed taking into account the first and second-order drift, as observed in the trend diagram, this type of kriging is similar to ordinary kriging, with the difference that it takes into account the trend of the data, so this is subtracted from the results. The disadvantage of this type of kriging is that it is stricter than the previous ones, therefore, it does not interpolate areas where there is no data, and this poses a problem for the study area since there is very little information to the west of the VSLP, therefore this surface lacks interpolation.

https://doi.org/10.1051/e3sconf/202132508005

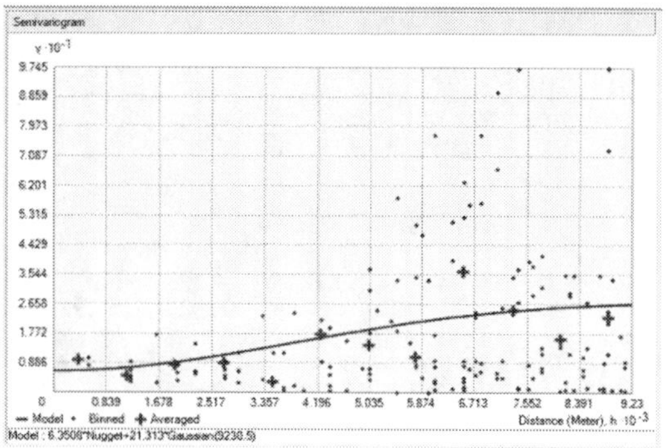

Fig. 4. Semivariogram, static level data 1995-2005.

When completing the construction of the three types of kriging, it is necessary to make a comparison between the values of the mean, the root means, the standardized mean, the root mean square error, and the standard mean error, to discriminate which method is which provides better results, or in other words, which method offers better interpolation. When comparing simple kriging with ordinary kriging, it was found that ordinary kriging shows better results in its predictions, even though the differences are minimal, because when comparing the mean, the one that is closest to zero is that of ordinary kriging, that is, it is less scattered, they both have a positive quadratic error, which means that they are undervaluing the prediction data, however, ordinary kriging is closer to unity. Fig. 5 shows the prediction map of the piezometric evolution level in San Luis Potosí valley, with ordinary kriging.

As can be seen in the prediction map built from ordinary kriging, there is a concentration of wells points with piezometric measurements for both years (1995 and 2005) in the central area, as a consequence, the prediction error spreads to the west, where there is less information available, therefore, the kriging prediction is sharper towards that extreme. On the other hand, from these results, we can confirm what had been inferred from the trend analysis in the data exploration stage, and that is that the area where there is more piezometric decay is in the central area and the recovery of the level Static occurs more to the northeast of the study area. This can be attributed to the fact that during this decade many agricultural wells that were located towards the northeast were closed due to the decrease in the need for irrigation water as a consequence of the growth of the municipal seat. Thus, by decreasing agricultural activity in that area, it could be a favorable scenario in this area for recovery of the aquifer's static levels.

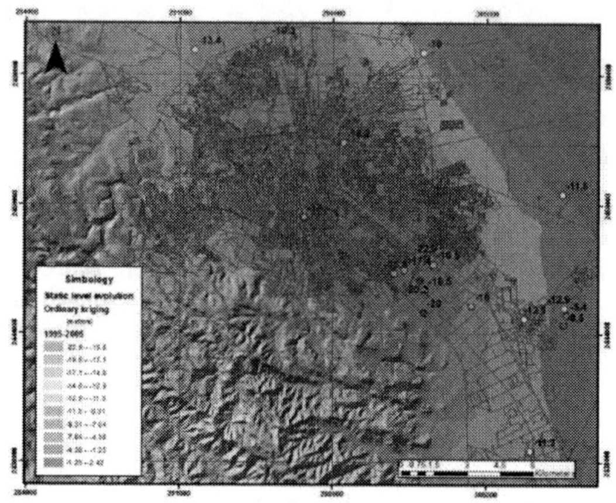

Fig. 5. Prediction map of the change in static level 1995-2005.

4.2 Analysis of the piezometric level evolution for the period 2007-2017.

For the second analysis for the 2007-2017 period, 24 wells distributed in the San Luis Potosí Valley were used. Fig. 6 in the upper right corner shows the values for the measures of central tendency, asymmetry, and kurtosis measurements. When observing the histogram, it is noted that if a curve could be constructed, it would be of the leptokurtic type, that is, a little pointed, which translates into more homogeneous data, so this can be validated with the value of kurtosis that is 3.36, which means that it is very close to 3, which is the homogeneity value. On the other hand, the skewness is -0.246, and this value, being close to zero, can be said to be low skewness; however, and being negative, indicates that the skew is to the left of the curve or histogram.

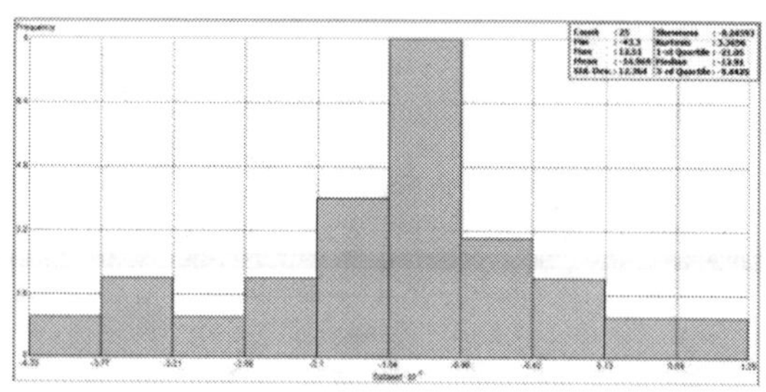

Fig. 6. Histogram of change in static level 2007-2017.

When performing a trend analysis, which consists of a three-dimensional graph of the data from studied variable (Fig. 7), it can be seen how these values are distributed, which are also projected on each of the axes.

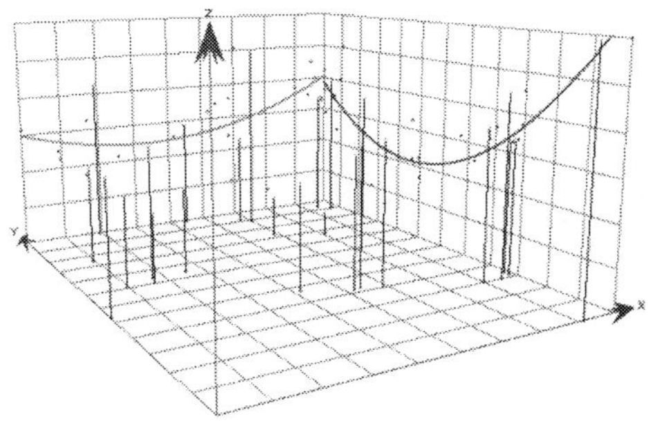

Fig. 7. Trend analysis for static levels 2007-2017.

For the projection of the XZ plane (green line), we note that the trend increases from west to east, that is, that from this information we can begin to deduce that the piezometric decays are towards the east and central part of the study area. On the other hand, for the projection of the YZ plane (blue line), a significant drift and greater dispersion are observed, which indicates that the main decays are in the central part, while towards the south the static levels are considerably recovered.

Next, we proceed to build simple kriging, without the need to do any kind of transformation to normalize the data, nor was it necessary to eliminate any trend. For this simple kriging analysis, the semivariogram model that best fit the data was the Gaussian, as shown in Fig. 8. Anisotropy was also taken into account.

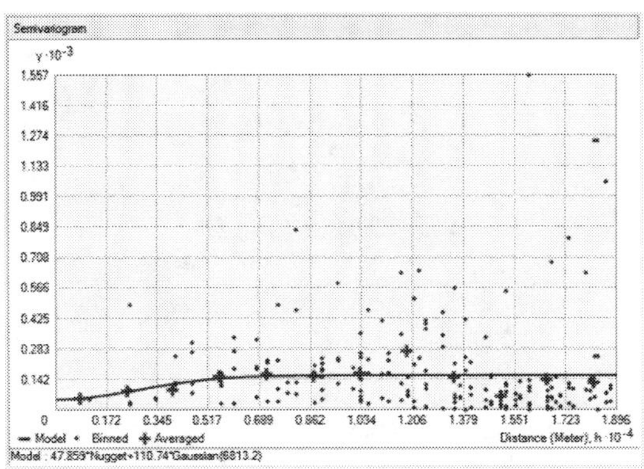

Fig. 8. Semivariogram, static level data 2007-2017.

After simple kriging, ordinary kriging was performed, this was adjusted to the same Gaussian-type semivariogram model; ordinary kriging, unlike simple kriging, does not require knowing the mean, but it must be constant. Finally, universal kriging was also carried out, taking into account the first and second-order drift as observed in the trend diagram. When completing the construction of the three types of kriging, a comparison was made between the mean values, the root means, the standardized mean, the mean root error, and the standard mean error, to be able to discriminate the method with the one with the best results, or in other words, which method offers the best interpolation. When comparing simple kriging with ordinary kriging, it was obtained that the ordinary one shows better results in its predictions because when comparing the mean, the mean that is closest to zero is that of ordinary kriging, that is, it is less dispersed, for, on the other hand, they both have a positive squared error, which means that they are undervaluing the prediction data, however, the ordinary is closer to unity.

Fig. 9 shows the prediction map of the evolution of the piezometric level in San Luis Potosí valley, with ordinary kriging. From this prediction map made with ordinary kriging, we note that there is a concentration of wells points with piezometric measurements for the years 2007 and 2017 in the central-eastern area, as a consequence, the prediction error propagates to the West and Southwest, where less information is available, therefore, the kriging prediction is sharper towards that extreme. On the other hand, from these results, we can confirm what had been inferred from the trend analysis in the data exploration stage, and that is that the area where there is more piezometric decay is in the central-eastern area (mainly the urban spot) and the recovery of the static level in these years occurs more towards the northeast and southeast of the study area.

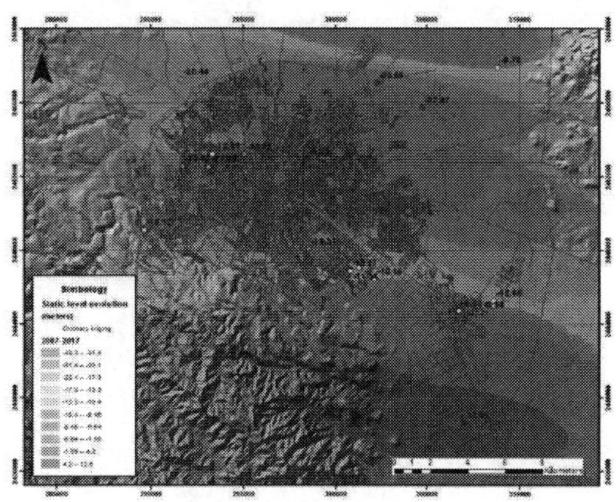

Fig. 9. Prediction map of the change in the static level 2007-2017.

5 Conclusion

When comparing these results of the changes in the piezometric levels for the period of 1995-2005 and 2007-2017, we observe that the zone of the decay of the static levels is maintained towards the center of the metropolitan zone of the San Luis Potosí Valley. And this phenomenon also affects the municipality of Soledad de Graciano Sánchez; With the difference that this area of influence is currently migrating to the east, the reason for this may be the accelerated increase of the population during the last decade towards the newly affected areas.

Also, it is important to note that the area with the greatest decays in the piezometric levels coincides with the area where there is the greatest thickness of deformable materials (quaternary deposits of sand, silt, and clay, poorly or not consolidated) and the greatest population accumulation. Thus, monitoring changes in the static levels of an aquifer is essential for planning and managing water resources. Furthermore, such pronounced changes in the static levels of the aquifer combined with geological factors can generate physical phenomena of deformation such as the phenomenon of subsidence, which has affected the civilian population in the San Luis potosí valley and this has caused significant economic losses related to damage to buildings.

This research was supported by the National Council of Science and Technology (CONACYT) and the Center for Research and Postgraduate Studies of the Autonomous University of San Luis Potosí. Piezometric level data was provided by the National Water Commission (CONAGUA) and the National Institute of Statistics and Geography (INEGI).

References

1. Burrough P. A., Environ. Ecol. Stat. **8**, 361 (2001)
2. S. De Iaco, M. Palma, and D. Posa, Spat. Stat. **14**, 546 (2015)
3. E. A. Varouchakis, *Background of Spatiotemporal Geostatistical Analysis: Application to Aquifer Level Mapping* (Elsevier Inc., 2018)
4. E. L. Flores-márquez, I. K. Ledesma, and C. Arango-galván, 425 (2011)
5. J. Martínez-Hernández, Identificación De La Recarga Natural E Inducida En La Cuenca De San Luis Potosí, Universidad Autónoma de San Luis Potosí, 2008
6. V. Trueba, A. Jiménez, and Q. Nguyen, *Estudio Para Determinar El Modelo de Explotación de Agua Disponible En El Valle de San Luis Potosí, Considerando La Demanda Actual y Futura, Bajo Un Marco de Sustentabilidad* (San Luis Potosí, 2008)
7. T. Hergt, J. Castro-Larragoitia, A. Cardona-Benavides, and J. J. Carrillo-Rivera, Ing. Hidraul. En Mex. **24**, 37 (2009)
8. M. C. Noyola-Medrano, J. A. Ramos-Leal, E. Domínguez-Mariani, L. F. Pineda-Martínez, H. López-Loera, and N. Carbajal, Rev. Mex. Ciencias Geol. **26**, 395 (2009)
9. CONAGUA, *Actualización de La Disponibilidad Media Anual de Agua En El Acuífero San Luis Potosí (2411) Estado de San Luis Potosí* (San Luis Potosí, 2015)
10. L. Zou, J. Kent, N. S. Lam, H. Cai, Y. Qiang, and K. Li, 1 (2015)
11. A. M. Oliver and R. Webster, *Basic Steps in Geostatistics:The Variogram and Kriging* (2015)
12. R. Giraldo, Univ. Nac. Colomb. 94 (2011)
13. F. E. Ikuemonisan, V. C. Ozebo, and O. B. Olatinsu, Geod. Geodyn. 1 (2020)
14. M. J. Abedini, M. Nasseri, and D. H. Burn, Comput. Geosci. **41**, 136 (2012)

Hydrogeological Characterization of the Gunungsewu Karst Area in the Upper Reach of Gremeng Karst Drainage Basin, Indonesia

Ahmad Cahyadi[1,2*], *Tjahyo Nugroho Adji*[1,2], *Eko Haryono*[1,2], and *Margaretha Widyastuti*[1,2]

[1]Karst Research Group, Faculty of Geography, Universitas Gadjah Mada, Yogyakarta, Indonesia
[2]Department of Environmental Geography, Faculty of Geography, Universitas Gadjah Mada, Yogyakarta, Indonesia

Abstract. The definition of a karst drainage basin in the aquifer of the karst area is an absolute thing to do as a water resource management unit. This research aimed to characterize the hydrogeology of the Gunungsewu Karst Area in the upper reach of the Gremeng Karst Drainage Basin. For this purpose, it was divided into three stages, namely geological survey, speleological survey, and artificial tracer test. The results indicate that the area observed lies in two or more geological formations: Semilir Formation, composed of sandstone and tuff from the ancient volcano Wonodadi eruptions, and Wonosari Formation, in which carbonate rocks consisting of massive coral limestone and bedded chalky limestone predominate. In volcanic rocks, the surface rivers have developed into allogenic streams flowing from underlying beds exposed updip. The contact between the two different formations results in the formation of ponors and springs. The artificial tracer test revealed two underground river systems with single conduits controlled by the allogenic rivers and large slope gradients in the study area.

1 Introduction

Following the groundwater basin's conceptual development in hydrogeology in the 1960s, karst hydrologists/hydrogeologists began to draw attention to the conceptualization and investigation of a karst drainage basin [1]. However, scholars have termed karst drainage basins as, among others, karst catchment areas, karst watersheds, springsheds, hydrogeological systems, karst basins, and underground river systems to date. Each refers to a definition of the basin as surface and subsurface areas that recharge an underground river, conduit, or spring in a karst landscape [2].

In-depth knowledge of a karst drainage basin is pivotal in understanding local water resource potentials and characteristics. Karst drainage basins with known extent and boundaries make all relevant hydrological analyses possible, especially related to the water balance of a karstic system [3,4]. Karst drainage basins with known extent and boundaries make all relevant hydrological analyses possible, especially related to the water balance of a karstic system [5,6]. However, determining the basins is more complicated and takes a longer time than delineating surface watersheds [7]. Some of the reasons include topographical and hydrological features that often have mismatched boundaries, variable groundwater tables during the rainy or dry season, dissolution tunnels with varying developments over time, many tunnels in an underground river system (e.g., maze cave), and the ease with which subsurface stream piracy occurs in underground rivers or conduits because of cave collapse, earthquakes, or sediment deposition [8,9].

Gunungsewu Karst Area on Java Island is one of the Indonesian karst landscapes with a drainage basin definition that has existed for a long time. Further, the Ponjong Hydrogeological Subsystem makes up the basin's northern and central parts. MacDonalds and Partners started a survey in 1979 and completed their research report in 1984 [10], marking the beginning of karst area definition in the subsystem. However, the results are considered limited—potentially raising conceptual and analytical issues—because the research incorporated the entire Ponjong Hydrogeological Subsystem into the Bribin Karst Drainage Basin despite the absence of any tracer tests in the region's allogenic rivers. Subsequent studies by the Faculty of Forestry, Universitas Gadjah Mada (1993)[11] Adji (2009)[12], Gunawan (2013)[13], and Widyastuti (2014)[14] still include the former as part of the latter.

Cahyadi et al. (2019; 2020)[15,16] divide the northern segment of the Ponjong Hydrogeological Subsystem into two karst drainage basins: Beton and Gremeng, but have not discussed it in detail. Therefore, the current research set out to complement previous studies that merely separated the subsystem's northern part from the Bribin Karst Drainage Basin. It has a specific objective to describe one of the primary stages of defining a karst drainage basin, that is, hydrogeological characterization. For this purpose, it focuses on the upper reach of the Gremeng Karst Drainage Basin, which is located in the northernmost

* Corresponding author: ahmadcahyadi@geo.ugm.ac.id

part of the Ponjong Hydrogeological Subsystem in the Gunungsewu Karst Area.

2 Methods

This research was conducted in the upper reach of the Gremeng Karst Drainage Basin, the northern edge of the Ponjong Hydrogeological Subsystem in the Gunungsewu Karst Area. Administratively, it belongs to Sawahan Village, Ponjong District, Gunungkidul Regency. This location has four allogenic watersheds with streams flowing from outside the karst area: Grembel River, Seropan River, Garuda River, and Kalimati River (Figure 1).

The research method had three stages: geological survey, speleological survey or cave mapping, and artificial tracer test. In the geological survey, the distributions of limestone, allogenic springs, allogenic rivers, catchment areas, and geological structures were mapped. Subsequently, the speleological survey mapped the locations of caves and cave passages (if feasible). The third stage was a tracer test using tracing dyes in solution to determine the connectivity between ponors, sinkholes, or caves with springs or underground rivers in the study area. The artificial tracer test produced breakthrough curves (BTCs) based on the fluorometer readings, the shape of which helps to analyze the characteristics of developed tunnels [17].

3 Results and Discussion

The rocks found in the study area can be broadly divided into two groups: volcanic and carbonate (Figure 1). The volcanic rocks consist of sandstone and tuff, which are part of the Semilir Formation deposited 20–16 million years ago in marine and terrestrial environments [18,19]. The constituent rocks are likely sourced from the ancient volcano Wonodadi [20]. In locations composed of volcanic rocks, the surface rivers have developed into allogenic streams in the observed karst drainage basin.

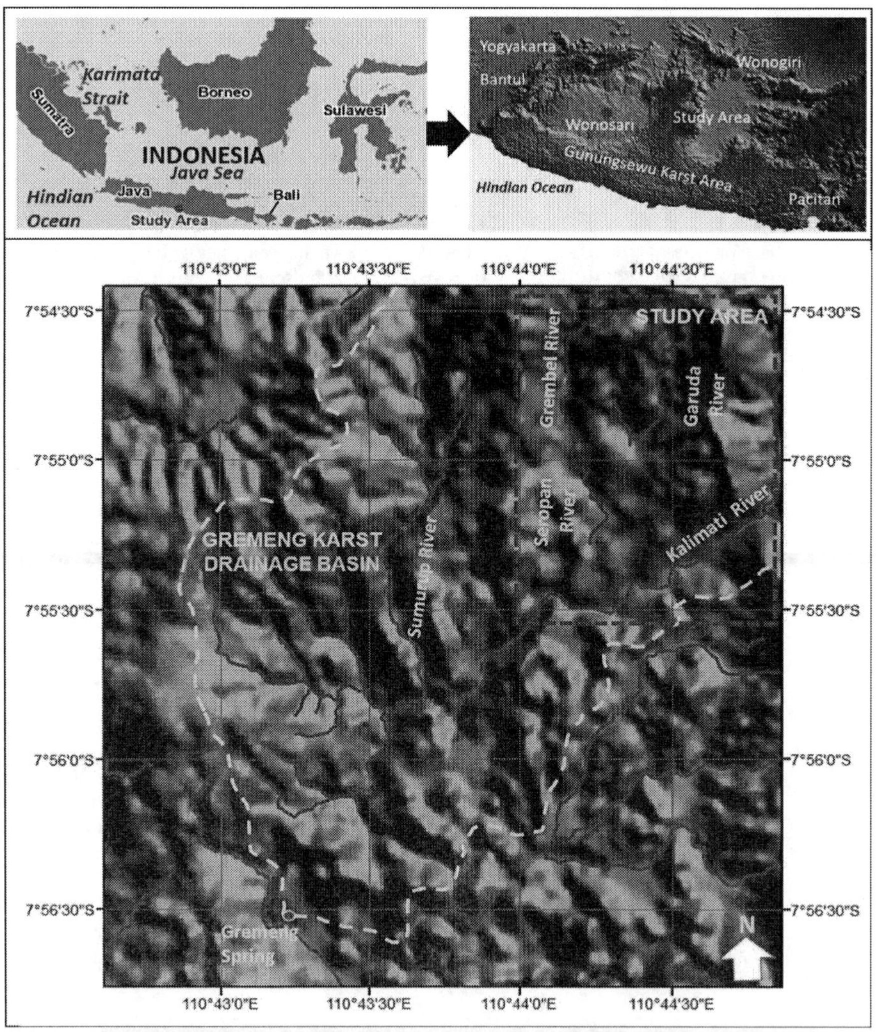

Fig. 1. The Map of Study Area.

The carbonate rocks making up the study area consist of massive coral limestone and bedded chalky limestone, members of the Wonosari Formation—which was deposited in the post-volcanism period (i.e., the Middle to Late Miocene) or younger than the Semilir Formation. The contact between Wonosari and Semilir Formations results in two hydrogeological formations: ponors through which allogenic rivers enter the underground river system and springs as the output of underground rivers (Figure 2). The allogenic rivers in question are allogenic streams flowing from underlying beds exposed updip [15,21].

The artificial tracer tests were conducted for three locations: Garuda Ponor to Tempuran Spring, Seropan Ponor to Tempuran Spring, and Grembel Ponor to Sutorenggo Spring. The tracer dyes were introduced at the same time at Garuda Ponor, Seropan Ponor, and Grembel Ponor, with the fluorometers installed at Sutorenggo Spring and Tempuran Spring. Because of the simultaneous injection, three tracer dyes were used to provide distinguishable readings after excitation.

The tracer tests revealed that the karst drainage basin observed had two main underground river systems: the system connecting Grembel River - Grembel Ponor - Sutorenggo Spring - Seropan River - Seropan Ponor - Tempuran Spring (Figure 3) and another system connecting Garuda River - Garuda Ponor - Tempuran Spring (Figure 4). However, without any accessible routes to Tempuran Spring, the speleological survey at this site was not feasible, so that the exact location of the meeting point between the two underground river systems is unknown.

Another result of the artificial tracer test is BTCs, as shown in Figure 5. BTCs of the three locations formed three separate single curves with steep rising and falling limbs, indicating that the conduits developing along the underground rivers have large slope gradients. Similar results were found in the conduit of the Beton Karst Drainage Basin, which shares the same geological characteristics with this research, located at the transition between Semilir and Wonosari Formations and in the same hydrogeological subsystem (i.e., Ponjong Hydrogeological Subsystem)[16].

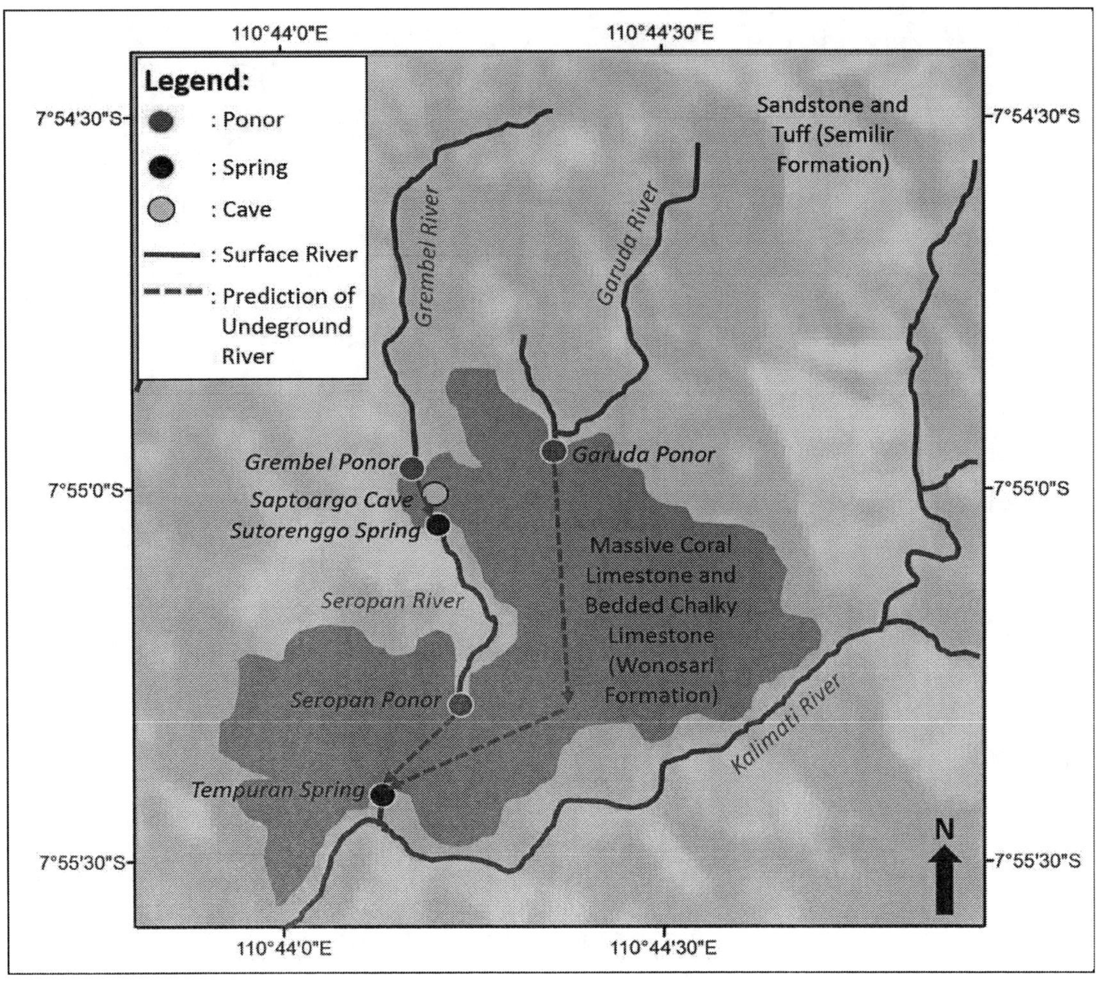

Fig. 2. The Rock Formation Distribution Map and Predition of Underground River Flow Direction in Study Area.

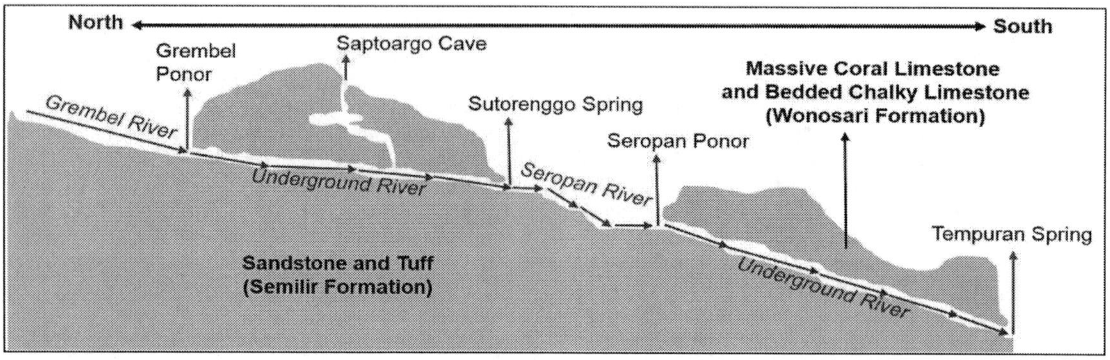

Fig. 3. Scheme of the Underground River System Connecting Grembel Ponor to Tempuran Spring.

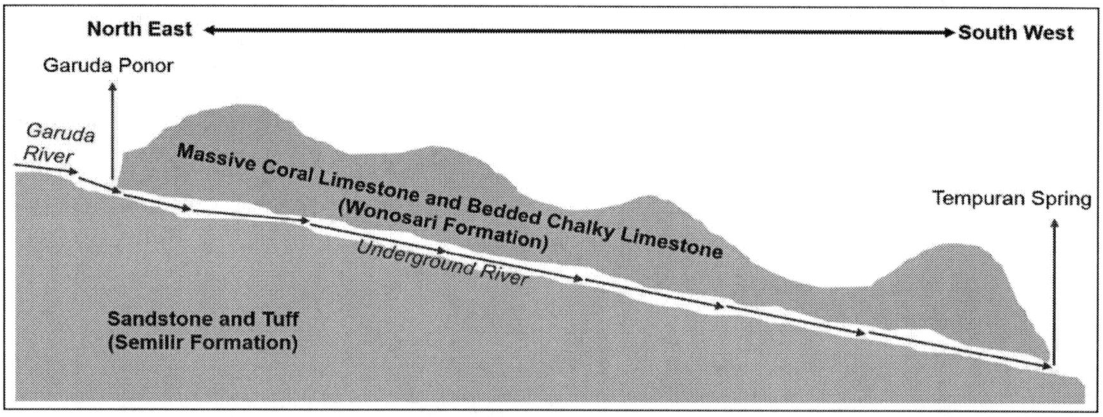

Fig. 4. Scheme of the Underground River System Connecting Garuda Ponor to Tempuran Spring.

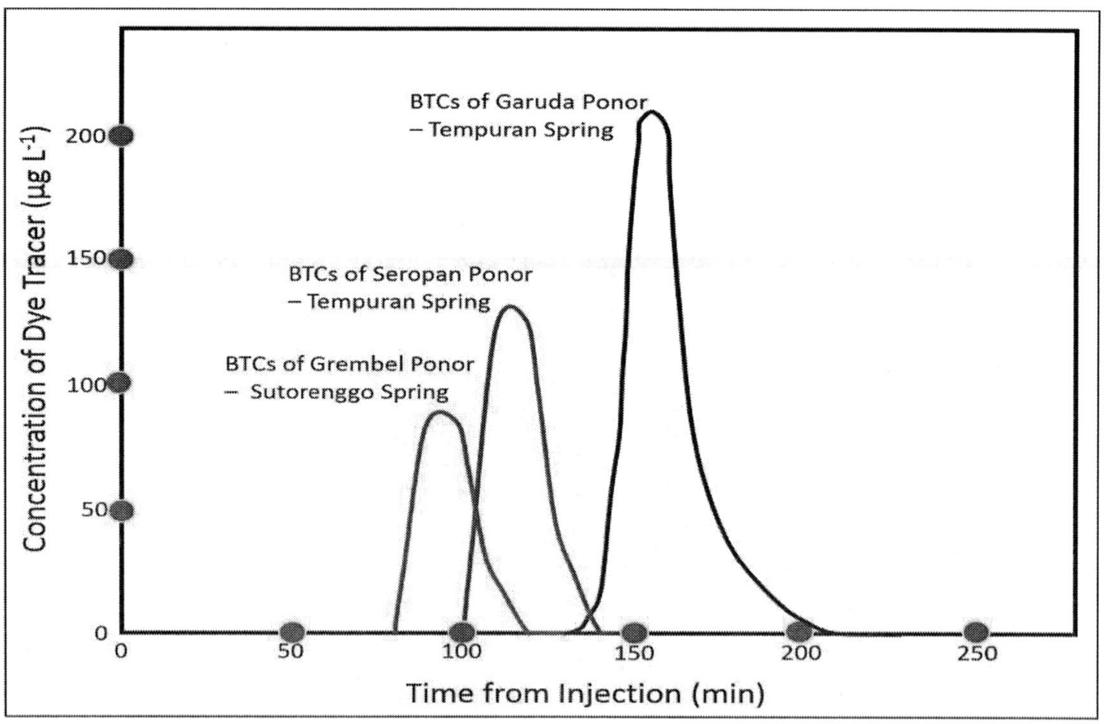

Fig. 5. Breakthrough Curves of the Artificial Tracer Test Results in the Study Area

Single BTCs also indicate that the three locations have single conduits. As confirmed by Palmer (2001, 2007)[22,23] and Cahyadi et al. (2020)[15], underground river systems affected by allogenic recharge often produce single conduits, especially in areas with a significant difference in elevation.

In this study, no underground river system was identified other than the one that empties into the Tempuran Spring. This is inferred from the absence of other springs as a system output from an underground river. In addition, from all existing ponor and caves in the study location, it has been detected that they have connections with the Tempuran Springs as the main output of the underground river system at the study site.

4 Conclusions

The research location consists of two rock formations, namely Semilir Formation (mainly comprising volcanic rocks: sandstone and tuff) and Wonosari Formation (with massive coral limestone and bedded chalky limestone as predominant constituent rocks). The contact zone between the two formations creates allogenic rivers flowing from underlying beds exposed updip. Broadly speaking, the two identified underground river systems merge to form a single system that exits Tempuran Spring. This results in a single conduit, which is a characteristic of a cave formed due to the influence of an allogenic river and a relatively large slope gradient.

Acknowledgments

This research uses some of the data analyzed and presented in the first author's dissertation in the Geography Study Program, Faculty of Geography, Universitas Gadjah Mada, Indonesia. It received funding from Universitas Gadjah Mada through the Grant Scheme for Final Assignment Recognition Program (*Hibah Program Rekognisi Tugas Akhir*) in 2021.

References

1. White, W.B. Fifty Years of Karst Hydrology and Hydrogeology: 1953-2003. In Hormon, R.S. Wicks, C. (eds). Perspectives on Karst Geomorphology, Hydrology and Geochemistry- A Tribute Volume to Derek C. Ford and William B. White. Geological Society of America (2006)
2. Quinlan, J.F. Ewers, R.O. Subsurface Drainage in the Mammoth Cave area. In White, W.B. White, E.L. (eds.) Karst Hydrology Concepts from the Mammoth Cave Area. New York : Van Nostrand Reinhold (1989)
3. Bonacci, O. Jukić, D. Ljubenkov, I. Definition of Catchment Area in Karst: Case of the Rivers Krčić and Krka, Croatia. Hydrological Sciences Journal, **51**(4), 682-699 (2006)
4. Widyastuti, M. Riyanto, I.A. Naufal, M. Ramadhan, F. Rahmawati, N. Catchment Area Analysis of Guntur Karst Spring Gunung Kidul Regency, Java, Indonesia. IOP Conrence Series: Earth and Environmental Science, **256** (2019)
5. Chen, X. Sun, Y. Huang, R. Role of Hydro-geochemical Functions on Karst Critical Zone Hydrology for Sustainability of Water Resources and Ecology in Shoutwest China. Acta Geochim, **36**(3), 494-497 (2017)
6. Zhang, Z. Chen, X. Cheng, Q. Li, S. Yue, F. Peng, T. Waldron, S. Oliver, D.M. Soulsby, C. Coupled Hydrological and Biogeochemical Modelling of Nitrigen Transport in the Karst Critical Zone. Science of the Total Environment, **732,** (2020)
7. Cahyadi, A. Fatchurohman, H. Pratiwi, E.S. Metode-metode Identifikasi Karakteristik Daerah Tangkapan Air Sungai Bawah Tanah dan Mataair Kawasan Karst. In Marfai, M.A. Widyastuti, M. (eds.). Pengelolaan Lingkungan Zamrud Khatulistiwa. Yogyakarta : Penerbit Pintal (2013)
8. Thrailkill, J. The Inner Bluegrass Karst Region. In Dougherty, P.H. (ed). Caves and karst of Kentucky. Kentucky : Geological Survey (1985)
9. Taylor, C.J. Greene, E.A. Hydrogeologic Characterization and Methods Used in the Investigation of Karst Hydrology. In Rosenberry, D.O. LaBaugh, J.W. (eds). Field Techniques for Estimating Water Fluxes Between Surface Water and Ground Water. Virginia : U.S. Geological Survey (2008)
10. MacDonald and Partners. Greater Yogyakarta Groundwater Study. Yogyakarta: Directorate General of Water Resources Development of the Republic of Indonesia (1984)
11. Faculty of Forestry, Universitas Gadjah Mada. Penyusunan Arahan Konservasi Tanah dan Air di Daerah Tangkapan Air Gua Bribin Kabupaten Gunung Kidul, D.I. Yogyakarta. Research Report. Yogyakarta : Forestry Service of Yogyakarta Special Province (1993)
12. Adji, T.N. Variasi Spasial Temporal Hidrogeokimia dan Sifat Aliran untuk Karakterisasi Sistem Karst Dinamis di Sungai Bawah Tanah Bribin Kabupaten Gunungkidul, Daerah Istimewa Yogyakarta. Dissertation. Faculty of Geograpy, Universitas Gadjah Mada (2009)
13. Gunawan, T. Kontribusi Citra Penginderaan Jauh dalam Kajian Batas Daerah Aliran Sungai (DAS) dalam Rangka Pengelolaan Sumber Daya Air (Kasus di Gunungkidul Daerah Istimewa Yogyakarta). Research Report. Faculty of Geograpy, Universitas Gadjah Mada (2013)
14. Widyastuti, M. Kajian Kerentanan Airtanah terhadap Pencemaran di Daerah Karst Gunungsewu (Studi di Daerah Aliran Sungai Bawah Tanah Bribin, Kabupaten Gunungkidul dan Wonogiri0. Dissertation. Faculty of Geograpy, Universitas Gadjah Mada (2014)

https://doi.org/10.1051/e3sconf/202132508006

15. Cahyadi, A. Haryono, E. Adji, T.N. Widyastuti, M. Riyanto, I.A. Naufal, M. Ramadhan, F. Allogenic River in the Hydrogeological System of Gremeng Cave, Gunungsewu Karst Area, Java Island, Indonesia. IOP Conference Series: Earth and Environmental Science, **448** (2020)
16. Cahyadi, A. Haryono, E. Adji, T.N. Widyastuti, M. Naufal, M. Ramadhan, F. Agniy, R.F. Riyanto, I.A. Analisis Konektivitas dan Karakteristik Lorong pada Sistem Hidrogeologi Mataair Beton, Kawasan Karst Gunungsewu, Kabupaten Gunungkidul dengan Uji Perunutan. Jurnal Geografi, **12**(2), 105 – 114. (2020)
17. Cahyadi, A. Agniy, R.F. Analisis *Breakthrough Curve* untuk Karakteristerisasi Pelorongan di Sistem Sungai Bawah Tanah Pindul Kabupaten Gunungkidul. Proceeding of Pertemuan Ilmiah Tahunan Ke-1 Perhimpunan Ahli Airtanah Indonesia (PIT-PAAI). Bandung, November 16-17, 2016, 375 – 385 (2016)
18. Akmaluddin, Setijadji, D.L. Watanabe, K. Itaya, T. New Interpretation on Magmatic Belts Evolution during the Neogene-Quaternary Periods as Revealed from Newly Collected K-Ar Ages from Central-East Java – Indonesia. Proceeding of Joint Convention IAGI-HAGI-PERHAPI, Surabaya (2005)
19. Surono. Sedimentasi Formasi Semilir di Desa Sendang, Wuryantoro, Wonogiri, Jawa Tengah. Jurnal Sumber Daya Geologi, **28**(1), 29-41 (2008)
20. Bronto, S. Mulyaningsih, S. Hartono, G. Astuti, B. Waduk Parangjoho dan Songputri: Alternatif Sumber Erupsi Formasi Semilir di Daerah Eromoko Kabupaten Wonogiri, Jawa Tengah. Jurnal Geologi Indonesia, **4**(2), 79 – 92 (2009)
21. Ford D. Williams P.W. Karst Hydrogeology and Geomorphology. Chicester: John Willey and Sons (2007)
22. Palmer, A.N. Dynamics of Cave Development by Allogenic Water. Acta Carsologica, **30**(2*),* 13-32. (2001)
23. Palmer, A.N. Cave Geology. Dayton, Ohio: Caves Books. (2007)

Identification of Karst Underground River Catchment Areas with Artificial Tracer Tests and Water Balance in Banteng Cave Springs (Karst Gombong Selatan, Central Java)

Muhammad Qodri Al-Ghozali[1,2*], *Tjahyo Nugroho* Adji[1,2], *Eko* Haryono[1,2], *Ahmad* Cahyadi[1,2], *Romza Fauzan* Agniy[1,2], *Gangsar Edi* Laksono[1], *Alpine Prima* Priambada[1,2], *Aulia Ika* Rahmawati[1,2], *Danung Shodikh* Mahrizkhal[1,2], *Andy* Setiawan[1,2], *Danang Riza* Fauzi[1,2], *Elisabeth Supi* Astuti[1,2], *Rahmat Dwi* Putra[1,3], *Muslih* Biladi[3]

[1]Karst Research Group, Faculty of Geography, Universitas Gadjah Mada, Yogyakarta, Indonesia

[2]Department of Environmental Geography, Faculty of Geography, Universitas Gadjah Mada, Yogyakarta, Indonesia

[3]Department of Science and Geographic Information, Faculty of Geography, Universitas Gadjah Mada, Yogyakarta, Indonesia

Abstract. The karst hills of Gombong Selatan have abundant potential water resources, especially in locations that have underground springs and rivers. The connectivity between the subsurface passageways that is difficult to know can threaten the potential of water resources, one of which is due to pollution caused by uncontrolled human activities. Therefore, identification of catchment systems and boundaries of water catchment areas in karst aquifers is needed that can contribute to sustainable water resources management policies. This catchment identification needs to be conducted because previous studies have never explored the eastern side of this karst area. This study aims to (1) define the underground river flow connectivity (upstream-downstream) of Banteng Cave; and (2) limiting the water catchment area of Banteng Cave. The method used to determine the subsurface connectivity system was carried out through an artificial tracer test, while the catchment area was delineated using a water balance approach. The results showed that the underground river of Banteng Cave has connectivity with Lake Blembeng, as evidenced by a change in watercolour after the tracer test and breakthrough curve (BTC) analysis of the tracing test results. BTC analysis shows that the Banteng Cave passageway has one main passage and does not have a tunnel branch. The estimated area of the Banteng Cave catchment used a water balance approach, which is 141.73 hectares. The Banteng Cave karst catchment conditions are dominated by the formation of valleys and karst cones accompanied by the appearance of valleys and karst hills that are quite evenly distributed, indicating that the Banteng Cave karst catchment is included in the advanced karst development phase. Furthermore, this research contributes significantly to increase knowledge regarding the characteristics of void karst development in aquifers which in the future are very important for determining water resources management policies.

1 Introduction

Karst is a landscape that undergoes a more dominant process of dissolving rocks than other landforms. The complex interaction process between soluble rock, carbon dioxide in the atmosphere, and water flow makes karst landscapes distinctive [1]. The uniqueness of the karst area is characterized by the presence of eksokarst and endokarst phenomena. Eksokarst is a type of surface karst formation, whereas endokarst is a karst formation that develops below the surface [2].

Gombong Selatan has various landscape conditions as a form of the influence of the geology of its constituents and its hydrological system. One of the landscapes in Gombong Selatan is the karst landscape. The area occupies mainly in the north and a small part in the south. The Karst Gombong Selatan area has the potential for water resources in underground rivers, in watery caves and often appear as springs. It is recorded that there are 63 caves and springs scattered throughout this karst area [3].

The potential of water resources in the Gombong Selatan karst area can be threatened due to uncontrolled human activities. One of the problems that have occurred in the karst area is the presence of potassium contamination in the springs of Banteng Cave, Pakuran Village, Buayan District, which resulted in the death of many fish and livestock of residents around the cave. This incident occurred because of fishing in Blembeng Lake using potassium. It can be assumed that there is a flow relationship between Lake Blembeng and the Banteng Cave spring [4]. Therefore, studies related to water resources in these locations are critical.

One of the efforts to protect karst water resources is by knowing the characteristics of the karst aquifer. The connectivity characteristics between voids in the karst aquifer can be identified using an artificial tracer test. The data obtained can also be helpful in knowing the extent of the karst catchment area as a basis for water resources management in the karst area. Based on this background, the objectives of this study are (1) to define the connectivity of the Banteng Cave underground river flow

* Corresponding Author: m.qodri.alghozali@mail.ugm.ac.id

(upstream-downstream); and (2) delineating the underground river catchment of the Banteng Cave karst.

2 Site Description

The karst landscape in the Gombong Selatan Hills is a tropical karst type [5], which is affected by high levels of precipitation and evaporation to form a relatively rounded closed basin [6]. The limestone formation in the area is the Kalipucang Formation. Based on the development of the existing karst morphology, this location has entered into an adult phase, which is characterized by the development of karst aquifer characteristics such as underground rivers, sinkholes and ponors, conduit passages, and sinking streams [5].

The Banteng Cave is administratively located in Pakuran Village, Buayan District, Kebumen Regency, Central Java. Banteng Cave is located at an altitude of 331 masl with absolute positions of 49M 330360 (easting) and 9148422 (northing). Banteng Cave is a horizontal cave with an underground river system. The underground river flows throughout the year (perennial) and always overflows during the rainy season. The passages that develop in the cave are relatively narrow and short, with the direction of the passage tending to the west, and can only be explored as far as 25 meters, which then has a low-roofed passage and ends at a sump [6]. This sump is the location where the water appears in the form of a narrow pool in a cave that cannot be explored anymore.

3 Methods

3.1. Artificial Tracer Tests

A tracer test is a helpful method to obtain information about subsurface flow connectivity and groundwater flow characteristics using a tracer agent [7]. The type of tracer used is a substance that is not harmful to humans and the environment, one of which is uranine.

Tracing tests in this study were initiated by installing a fluorometer type GGUN FL30 in Banteng Cave as a monitoring location, which was then poured into the Surupan Cave, which is the outlet of Blembeng Lake. The location of tracer pouring is shown in Figure 1. The distance between Surupan Caves and Banteng Caves is about 876 meters. The Banteng Cave system tracing monitoring was carried out from 23 August 2020 to 27 August 2020.

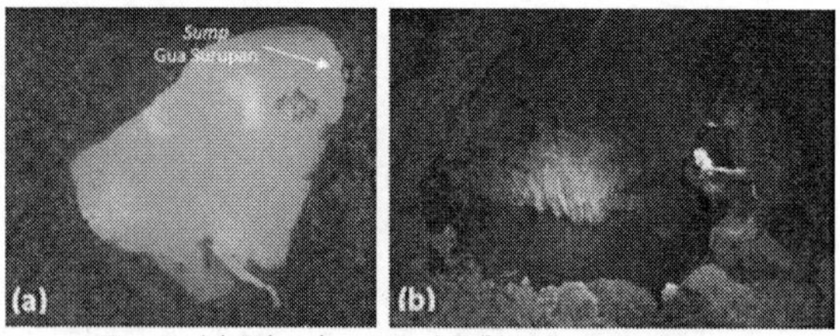

Fig. 1. (a) Blembeng Lake (b) Surupan Cave

Discharge data, target peak concentration, and horizontal distance between the injection point and the point of observation are used in calculating the quantity of tracer poured, according to the formula [8]:

$$M = 1.9 \times 10^{-5}(LQC)^{0,95} \qquad (1)$$

where:
M : quantity of tracer substance required (kg);
L : the distance from the injection point to the observation point (km);
Q : discharge (L/s);
C : peak concentration target (μg/L)

The analysis of the results of the fluorometer recording was then processed using the help of QTRACER2 software [10]. The software is able to calculate and estimate the flow properties of underground rivers quantitatively [10], which includes the average flow velocity, Reynold number and recovery value.

3.2. Catchment Area Estimation

Determining the catchment area's boundary using a water balance calculation is considered quite relevant and logical with the resulting catchment area [11]. The water balance concept illustrates that a hydrological system can be determined by the amount of water entering and leaving the system in a certain period [12]. Determining the catchment area using a water balance approach requires several parameters: discharge, rainfall, temperature, crop coefficient, and evapotranspiration. Discharge data is based on previous research [13], while rainfall and temperature data are obtained from *en.climate-data.org* between 1999-2019. The temperature data were then processed using the Blaney-Criddle method [15] to produce evapotranspiration data. The Blaney-Criddle evapotranspiration calculation is formulated as follows [12,16]:

$$PET = \Sigma Kt \times Kc \times t \times p/100 \qquad (2)$$

where:
PET : Potential evapotranspiration;
Kt : 0.0173t – 0.314;
t : average monthly temperature (°F);
Kc : crop coefficient;
p/100 : the monthly percentage of daylight hours in a year (%);

The value of the crop coefficient (Kc) is adjusted to land use in the catchment area. The classification of Kc values is presented in Table 1.

Table 1. Value of Kc based on land use

Information	Kc
Mixed garden	0.80
Moor	0.90
Settlement	0.00
Irrigated rice fields	1.15

Shrubs	0.80
Rainfed rice fields	0.80
Grass	0.80

Source: [16]

Water balance assumes that rainwater (P) is the main input, then evapotranspiration (E) as a factor of water loss due to evaporation and transpiration by sunlight, and discharge (Q) of springs as the main output from the karst catchment area. This study assumes that the rainwater that enters the system directly goes to the springs to ignore the storage change factor (Δs). Thus, the mathematical calculation of the water balance is presented in formula [2]:

$$Q = P - E \tag{3}$$

where:
Q : spring discharge (m^3/sec);
P : rainfall (mm/year);
E : evapotranspiration (mm/year)

Determination of the catchment area using discharge data during the measurement, which is converted into volume data using the equation:

$$V = Q \times time\ unit \tag{4}$$

$$V = \frac{(Q(n)+Q(n+1)}{2} + \frac{Q(n+1)+Q(n+2)}{2} + \dots + \frac{Q(m-1)+Q(m)}{2} \tag{5}$$

Thus, the water balance mathematical calculation changes to:

$$V = (P - E) \times A \tag{6}$$

where:
V : discharge volume (m^3/year);
P : rainfall (m/year);
E : evapotranspiration (m/year);
A : catchment area (m^2)

4 Result and Discussion

4.1. Artificial Tracer Tests

Tracing results show that there is connectivity between Surupan Cave and Banteng Cave. This connectivity is evidenced by the change in colour to green in the underground river and the results of fluorometer recording at the Banteng Cave monitoring location. The condition of the Banteng Cave during the tracer test is shown in Figure 2. The tracer test results show that the Banteng Cave system was recharged by the mainstream originating from Blembeng Lake. Blembeng Lake is a doline with the morphology of a closed basin filled with water from underground river inlets and rainwater. The lake also has an outlet in the form of a sinking stream called Surupan Cave, the entrance to the underground river system. The direction of the sinking stream stretching tends to be northeast.

Fig. 2. The appearance of uranine in the Banteng Cave

The breakthrough curve (BTC) results of the tracing test are presented in Figure 3. The formed BTC shows one single peak, which means that there is only one single passageway system and no flow branching. The highest peak concentration was 329.57 µg/l with travel time to the peak concentration was 23 hours.

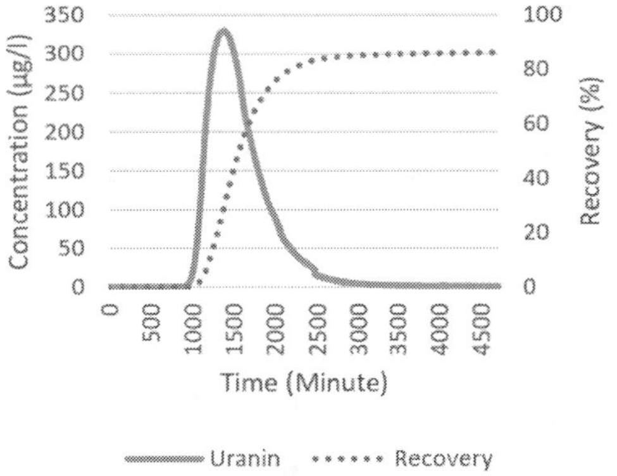

Fig. 3. BTC of the Surupan Cave-Banteng Cave

The BTC analysis in the Banteng Cave system was carried out during the dry season so that it has a slow flow rate, which is supported by a discharge calculation that shows a value of 0.06 l/sec. Previous measurements by [14] stated that the discharge in Banteng Cave was 30 l/sec. The analysis of some of the transport parameters using QTRACER2 in the Banteng Cave system is presented in detail in Table 2.

Table 2. BTC analysis of Banteng Cave

Observation point	Banteng cave
Injection point	Surupan cave
Tracer substance	Uranine
quantity (gram)	100.00
Peak concentration (µg/l)	329.57
First time of detection (min)	815.00
distance (m)	876.00
Reynold number estimation	8978.00
Average velocity (km/day)	1.18
recovery (%)	86.08

The average velocity is a measure by which most of the mass of the substance moves in the channel [10], in this case it is also known as advection. Based on the calculation, the value of advection during the tracer test is 1.18 km/day. Research [17] in Linjiang, China has an advection value range of 0.8-2.3 km/day. When compared, the advection value in Banteng Cave shows relatively fast velocity and turbulent flow properties.

Turbulent flow conditions can also be detected through the Reynold number value. The turbulent flow is indicated by the Reynold number value> 4000 [18]. Based on the results of processing the Reynold number value in the Banteng Cave, it shows a value of 8,978. These results reinforce the notion that the flow in the sub-surface system is turbulent. Turbulent flow can also be influenced by the geometry of the cave passages that are not uniform, such as the condition of the Banteng Cave passageway, which has narrow, elongated and widened passages, and there is a sump in the cave [7].

Furthermore, the recovery value shows the concentration of tracer substances that are still recorded at the monitoring location. Based on data processing, the recovery value in Banteng Cave was 86.076%. In comparison, a study by [11] at Beyyayla Spring, Turkey, showed a recovery of 63.1% when the flow conditions were low. The recovery value in Banteng Cave is high; thus, it strengthens the notion that the passage system in the cave only has one main passage, and there is no flow branching in the underground river flow system [11]. The high recovery value also reflects that the karst aquifer in the system has developed [18].

4.2. Catchment area delineation

Temperature and rainfall data representing the research location were obtained from *en.climate-data.org*, recorded in 1999-2019 (Table 3). The calculation of the amount of temperature and average rainfall in 20 years is 2,484 mm and 25.5 °C. The data is then used to calculate potential evapotranspiration using the Blaney-Criddle method [14].

Table 3. Average temperature and rainfall at the study site, 1999-2019

Month	temperature (°C)	rainfall (mm)
January	25.6	359
February	25.7	298
March	25.9	294
April	26.0	225
May	25.8	128
June	25.2	99
July	24.5	68
August	24.5	46
September	25.2	83
October	25.8	163
November	25.8	324
December	25.6	397

Source: *en.climate-data.org* (2020)

The potential evapotranspiration value used is the accumulation of evapotranspiration every month. The results of the calculation of the potential evapotranspiration value in Banteng Cave are presented in Table 4. Based on these calculations, the Banteng Cave catchment has an evapotranspiration value of 1,212.16 mm/year. The evapotranspiration value is then used to calculate the karst catchment area with the water balance concept. Based on calculations using water balance, the estimated area of the Banteng Cave catchment is 141.73 ha. The results of the analysis of the water balance of the Banteng Cave system are presented in Table 5. The estimated catchment area that has been determined is then delineated by on-screen digitizing to estimate the spatial catchment boundary. The digitization process takes into account topographic conditions and tracing test results. The estimation results of the Banteng Cave catchment are presented in Figure 4.

Table 4. Calculation of the evapotranspiration value

Month	T (°C)	Kc	Kt	P (%)	U (mm)	E (mm/y)
January	25.6	0.8	1.04	8.85	144.98	1212.16
February	25.7	0.8	1.04	8.12	133.74	
March	25.9	0.8	1.05	8.68	144.48	
April	26.0	0.8	1.05	8.10	135.54	
May	25.8	0.8	1.04	8.20	135.79	
June	25.2	0.8	1.02	7.94	127.28	
July	24.5	0.8	1.00	8.20	126.63	
August	24.5	0.8	1.00	8.37	129.24	
September	25.2	0.8	1.02	8.10	129.90	
October	25.8	0.8	1.04	8.68	143.72	
November	25.8	0.8	1.04	8.40	139.09	
December	25.6	0.8	1.04	8.85	144.98	

Table 5. Banteng cave catchment based on water balance

Sub-surface system	V (m^3/y)	E (mm/y)	P (mm/y)	Area (ha)
Banteng Cave	1802623.28	1212.16	2484	141.73

Several previous studies related to determining the catchment area estimate using water balance have also been carried out in the Gunung Sewu karst area, namely, the Guntur Spring, which found an area of 30.7 ha [12] and [19] in Pindul Cave, resulting in a catchment area of around 10,000 - 20,000 ha. Knowledge of watershed boundaries is essential to determine the type of karst groundwater management that is susceptible to pollution [15].

The estimated catchment area of Banteng Cave tends to be smaller than that of the Gunung Sewu karst system. This is influenced by water limitations, leading to a narrow underground river system, which is reflected in the relatively small discharge. The area of the Banteng Cave catchment includes several villages, including Pakuran Village, Wonodadi Village, and Watukelir Village.

The condition of the Banteng Cave catchment is dominated by karst topography with the characteristics of a conical hill and a closed depression (cockpit). The karst cone pattern is elongated with many peaks which are relatively uniform and have steep slopes. The valley pattern forms a closed basin that is elongated and relatively narrow. Based on this appearance, and also supported by the development of the karst aquifer below the surface as evidenced by the artificial tracer test in Banteng Cave, it strengthens research [7] that the Gombong Selatan karst area is in a mature karst development phase.

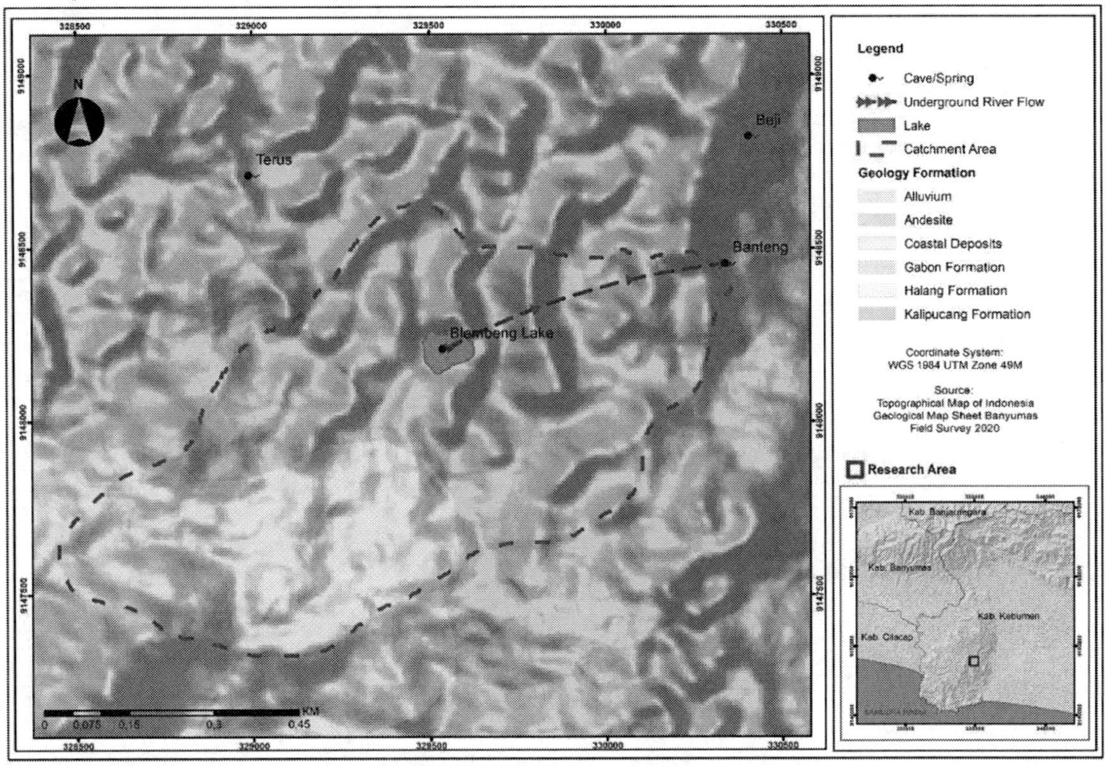

Fig. 4. Estimated catchment boundary in the Banteng Cave system

5 Conclusion

The connectivity between the underground river flows of Blembeng Lake-Banteng Cave has been proven through tracing tests. BTC records show that the Banteng Cave system has one main passageway and does not branch off the passageway to other springs. The quantitative analysis of the curve also indicates that the passage development has developed and is highly classified.

Determination of karst catchment estimation using water balance shows that the area of the Banteng Cave catchment is 141.73 ha. Catchment conditions dominated by karst cones and valleys, accompanied by subsurface aquifers, indicate that the Banteng Cave catchment is included in the mature karst development phase.

References

1. Sudarmadji, E. Haryono, T. N. Adji, M. Widyastuti, R. Harini, A. Cahyadi, and H. Nugraha, *Seri Bunga Rampai Ekologi Lingkungan Kawasan Karst Indonesia Jilid 1* (Deepublish, Yogyakarta, 2012)
2. D. Ford and P. Williams, Karst Hydrogeology and Geomorphology (Chichester: John Wiley & Sons, Ltd, 2007)
3. G. E. Laksono, Kajian Kerusakan Lingkungan Berbasis Indeks Ketergangguan Karst Di Kawasan Karst Karangbolong Kabupaten Kebumen, Universitas Gadjah Mada, 2019
4. Acyntyacunyata Speleogical Club, *Laporan Gombong Speleology Expedition,* (2014)
5. E. Haryono, Acta Carsologica., **46**, 63 (2017)
6. E. Haryono and T. N. Adji, *Geomorfologi Dan Hidrologi Karst* (Karst Student Forum, Fakulty of Geography, Universitas Gadjah Mada, Yogyakarta, 2004)
7. N. Goldscheider, Int. J. Speleol, **37**, 27, (2008)
8. S. Worthington and C. Smart, Eng. Environ. Impact Karst (2003)
9. M. Field, U.S. Environ. Prot. Agency, 1–179, (2002).
10. H. Aydin, M. Ekmekçi, and M. E. Soylu, Environ. Earth Sci. **71**, 3921 (2014)
11. M. Widyastuti, I. A. Riyanto, M. Naufal, F.

Ramadhan,, N. Rahmawati, Int. Conf. Environ. Resour. Manag. Glob. Reg., **256,** (2019)

12. B. Triatmodjo, *Hidrologi Terapan*, (Beta Offset, Yogyakarta, 2010)
13. Ruswanto, I. Badri, and A. Anwar, Inventarisasi Geologi Lingkungan Kawasan Karst Gombong, Kabupaten Kebumen, Jawa Tengah (Dep. ESDM, Bandung, 2002)
14. R.G. Allen, W.O. Pruitt, J. Irrig. Drain. Eng, 112, 139-155, (1986)
15. R. F. Agniy, Pengelolaan Sumberdaya Air Berdasarkan Sifat Akuifer Karst di Kawasan Karst Jonggrangan, 2019
16. J. Doorenbos, W. O. Pruitt,. FAO Irrig. Drain, **24,** (1977)
17. J. Qi, M. Xu, , X. Cen, L, Wang, Q. Zhang, Water, **10** (2018)
18. E. Shashi Menon. *Fluid Flow in Pipes. Transmission Pipeline Calculations and Simulations Manual,* Gulf Professional Publishing (2015)
19. E. P. Lestari, & M. Widyastuti, J. BUMI Indo, **6,** 4 (2017).

Regional Potential Analysis Based on Agricultural Commodities of Food Crops and Their Contribution to the Economy in the Karangsambung-Karangbolong Geopark Area, Indonesia: A Location Quotient Approach

Rika Harini[*], *Rina* Dwi Ariani, *Bowo* Susilo and *Evita* Hanie Pangaribowo

Faculty of Geography, Universitas Gadjah Mada, Indonesia

Abstract. The Karangsambung-Karangbolong Geopark Area is located in Kebumen, one of the districts with the highest poverty rate in the Province of Jawa Tengah, Indonesia. A regional economy can be improved by optimizing existing potentials in the agricultural sector, especially food-crop commodities. This research set out to determine 1) the basic or leading food crops per sub-district, (2) the relatively potential food crops to maintain, and 3) the rating of the carrying capacity of facilities in maintaining relatively potential commodities. This research used secondary data from the BPS-Statistics Indonesia, such as food-crop commodity production and infrastructure conditions. It employed the Location Quotient (LQ) method, the LQ's strength criteria analysis and scalogram. The analysis results showed that each sub-district has distinct basic commodities, but rice, cassava and sweet potatoes were the best potential commodities to develop. The Karangsambung-Karangbolong Geopark Area is equipped with facilities and infrastructure with fairly good to good carrying capacities supporting the development of the three commodities.

1. Introduction

Kebumen is the district with the highest poverty rate in the Province of Jawa Tengah, Indonesia (16.82%), far above the average poverty rate of the province (10.8%) [1]. Poverty has a negative and significant effect on economic growth [2]. According to the updated data in 2020 by BPS-Statistics Indonesia, the district's economic growth is -1.46% [3]. Poverty reportedly results in a lack of investment and physical capital, leading to lower economic growth. The inadequate investment is attributed to sub-standard human and natural resource management [4], [5].

Regional economic conditions are dependent on several factors, including the potential of natural resources owned. Regions rich in natural resources are believed to have the capacity to achieve greater economic growth than ones with less natural resources [6], [7], [8], [9], [10]. However, the failure to appropriately manage such abundance will bring about the resource curse hindering economic growth [11], [12] or even causing negative economic growth. Therefore, the resource curse must be overcome with regional economic development through joint activities between the government and the local community by optimally utilising and managing resources to stimulate economic growth [13].

Many factors play pivotal roles in determining regional economic growth, including natural resources, human resources, industrialisation and technological innovation. Technological innovation can strengthen the positive impact of several variables that affect economic growth and help to reduce negative impacts, especially the environmental impact [12]. Agriculture, forestry and fishery are sectors with the largest share contributing to the regional economy of Kebumen District, i.e., 21.79% [14]. Based on these data, it can be concluded that the three sectors are the basis of regional economic development in Kebumen.

The development of regional potentials through a commodity approach, especially food-crop agricultural commodities, is not perceived merely in terms of economy but also facility supporting such commodity development. According to Nikijuluw (2013), the economic potential in a region becomes insignificant for the regional economic development if there is an absence of optimal development and utilisation. The economic base theory addresses how to overcome various problems concerning agricultural commodity-based potentials in regional development by focusing on the demand for goods and services from other regions as the main determinants of economic growth. This theory also identifies a region's basic or non-basic development sectors [18], [19].

* Corresponding author: rikaharini@ugm.ac.id

This research was designed to determine 1) the basic or leading commodities, 2) the potential commodities to maintain and 3) the carrying capacity of facilities supporting the leading commodities. For this purpose it employed Location Quotient (LQ), the analysis of the LQ's strength criteria and scalogram.

2. Methodology

This research was conducted in the Karangsambung-Karangbolon Geopark Area, Kebumen District, the Province of Jawa Tengah, Indonesia. This location was selected purposively based on two considerations: its natural resources potential and its socioeconomic condition as the district with the highest poverty rate in the province. The research used secondary data—i.e., details and information not directly obtained from the sources [18]—collected from several publications issued by the BPS-Statistics Indonesia for Kebumen District.

The first research objective was analysed using the Location Quotient method to determine the potential of economic activity of particular agricultural commodities to indicate basic and non-basic sectors by comparing the capabilities of the same commodities in a wider area (see Equation 1) [19]:

$$LQ = \frac{(^{vi}/_{vt})}{(^{Vi}/_{Vt})} \qquad (1)$$

Where:
LQ = LQ (ratio value) of food-crop commodities in each sub-district
vi = Average production of each food-crop commodity in Sub-district i
vt = Total average production of overall food-crop commodities in Sub-district i
Vi = Average production of each food-crop commodity in Kebumen District
Vt = Total average production of overall food-crop commodities in Kebumen District

The proviso of this method is that if LQ > 1, it means that the commodity becomes the basic or leading commodity; the results can be used to meet the region's needs and those of other regions. If LQ < 1, it indicates that the commodity is considered non-basic or does not have advantages. The production of such commodity cannot be used to meet the region's needs; thus, it requires supplies from other regions. Meanwhile, if LQ = 1, it means that the commodity is considered non-basic and does not have advantages in which it is still capable of meeting the needs of the region but is unable to provide supplies to other regions in need [20], [21].

The second research objective was analysed based on the LQ's strength criteria. The analysis criterion is that food-crop commodities that occupy the most base areas are relatively potential commodities to maintain. The basis of the analysis is sub-district; the more sub-districts categorised as the basis for a certain commodity, the more reliable it is categorised as relatively potential sectors to develop.

The third research objective was analysed using a scalogram. The advantage of the analysis is that it is able to score the facilities owned by each sub-district [22]. Further, according to Budiharsono (2001), the scalogram analysis categorises the carrying capacity of facilities into three, namely:

1. Category 1 means that the carrying capacity of facilities in the region is good
2. Category 2 means that the carrying capacity of facilities in the region is relatively good
3. Category 3 means that the carrying capacity of facilities in the region is poor.

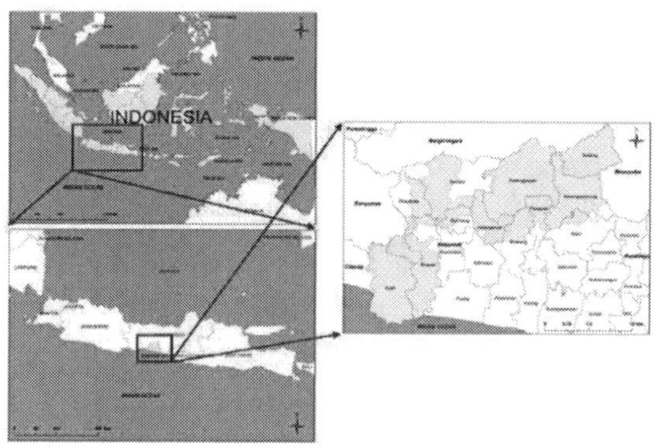

Fig. 1. Research Site Map

3. Results and discussion

3.1 Basic Food Crops in the Karangsambung-Karangbolong Geopark Area

The main policy that needs implementation in regional economic development was to optimally arrange regional development priorities based on the region's potential. These potentials vary across regions; hence, the dominant leading potential needs to be determined [24].

The Location Quotient approach was used to identify basic or leading food-crop commodities in the Karangsambung-Karangbolong Geopark Area by calculating their LQs. LQ > 1 is a criterion for basic commodities, while LQ < 1 is for non-basic commodities. Table 1 summarises the basic food-crop commodities in the Geopark area.

Table 1. Basic Food Crops in the Karangsambung-Karangbolong Geopark Area in 2009–2018

NO	FOOD-CROP COMMODITIES	SUB-DISTRICTS
1.	Rice	Ayah, Buayan, Rowokele, Gombong, Pejagoan, Alian
2.	Maize	-
3.	Soybeans	Ayah, Buayan, Rowokele, Gombong, Pejagoan
4.	Cassava	Sempor, Karanganyar, Karanggayam, Sruweng, Karangsambung, Sadang
5.	Sweet potatoes	Buayan, Sempor, Karanganyar, Sruweng, Pejagoan, Karangsambung
6.	Groundnuts	-
7.	Green beans	Buayan, Sempor, Gombong, Karanganyar, Sruweng

Source: Secondary data analysis (2021)

The analysis results in Table 1 indicate that the sub-districts in the Karangsambung-Karangbolong Geopark Area had different basic food crops owing to their respective characteristics, e.g., land fertility, geographical location, human resources and existing facilities and infrastructure. Rice, cassava and sweet potatoes were the basic commodities in most sub-districts. Basic agricultural commodities are commodities that have great potential to be marketed outside the economic boundaries of the production area because of the excess yield amounts even after being used to meet the region's needs. Therefore, they potentially provide incomes for the region and are more feasible to develop than non-basic commodities. Meanwhile, maize and groundnuts were non-basic commodities in the entire Geopark area. In conclusion, not all commodities in a region can be used as leading commodities. It requires distinct management of agricultural resources in each region.

The bar chart in Figure 1 shows the LQs of each observed commodity in the Karangsambung-Karangbolong Geopark Area. Soybeans, cassava and sweet potatoes were commodities with LQ > 1, while rice, maize and groundnuts had LQ < 1. The remainder, green beans, was the only commodity showing LQ = 1.

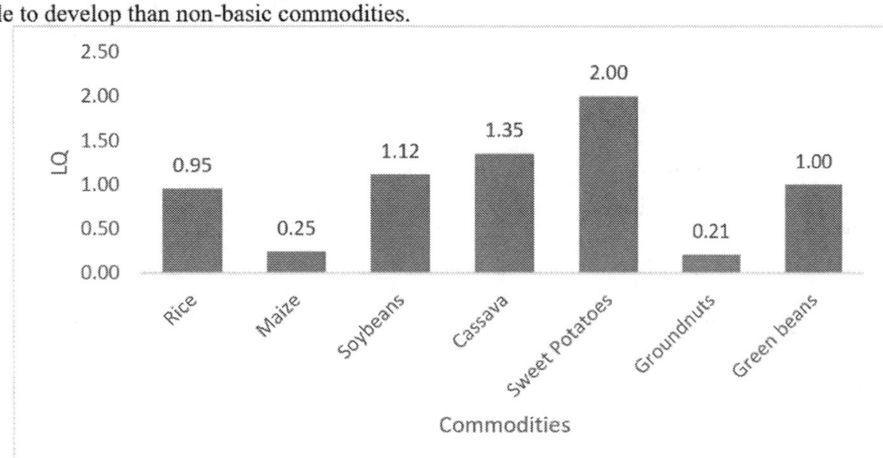

Fig. 2 Average Location Quotients of Food-Crop Commodities in the Karangsambung-Karangbolong Geopark Area

The commodity with the highest LQ (2.00) was sweet potatoes, indicating that this food crop has been able to meet the needs in the Geopark area while also providing supplies to other regions. Other food crops categorised as basic or leading commodities were cassava (1.35) and soybeans (1.12). Leading commodities are commodities with added values and large production that can increase regional economic growth. They also have substantial multiplier effects on other economic activities, the development of the surrounding areas and other commodities with high market demand [25]. The food crops that were not categorised as basic commodities were rice, maize, groundnuts and green beans (LQ ≤ 1). These results indicate that the first three commodities have not been able to meet the region's needs and thereby require supplies from other regions (LQ < 1), while green beans have been able to meet the region's needs despite but unable to provide supplies to other regions (LQ = 1).

The selection of the basic commodities can change at any time, depending on the production level during the analysis. Suppose during the analysis a specific commodity is selected or determined as a basic or non-basic commodity and the government adopts this selection and makes efforts to develop and increase its production by adopting technology and expanding land, it is highly likely that the commodity changes its status (from non-basic to basic and vice versa) in the following year. Agricultural production is generally influenced by several factors, including nature, labour, capital and management [26]. The development of basic or leading commodities is enforced with the issuance of Law Number 18 of 2012 on Food, stating that the local government has responsibility for the sustainable development of local food production in the region [27].

3.2 Relatively Potential Food Crops to Maintain in the Karangsambung-Karangbolong Geopark Area

The Location Quotient's strength criteria were analysed to determine the relatively potential food-crop commodities to maintain in the Karangsambung-Karangbolong Geopark Area. The idea is food-crop commodities that occupy the most base areas are relatively potential commodities to maintain at the research location. Figure 3 shows the relatively potential food-crop commodities to maintain in the Karangsambung-Karangbolong Geopark Area.

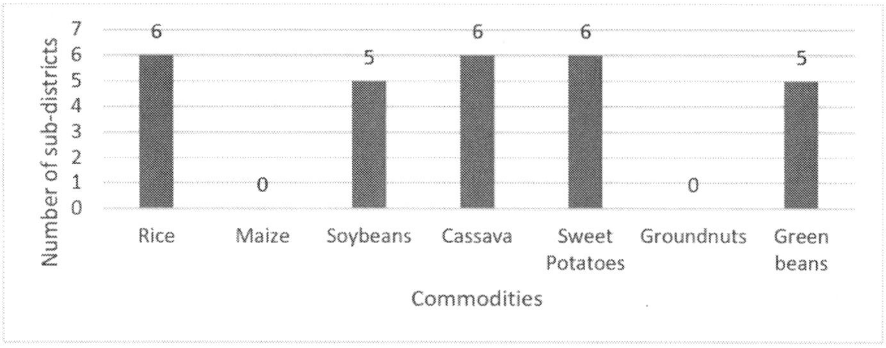

Fig. 3 Relatively Potential Food-Crop Commodities to Maintain in the Karangsambung-Karangbolong Geopark Area

Based on Figure 3, it can be seen that the relatively potential food-crop commodities to develop in the Karangsambung-Karangbolong Geopark Area were rice, cassava and sweet potatoes. These three commodities occurred as the basic food crops in six of the twelve sub-districts.

Rice was categorised as the relatively potential commodity to develop in the Geopark area. Despite its general LQ (0.95), rice became the basic commodity in six sub-districts. Some of the efforts that can be carried out to support this food crop as the leading commodity are 1) using excellent seeds, 2) reducing the use of pesticides to eradicate pests and 3) improving facilities and infrastructure.

Cassava has the potential to develop as the leading commodity because it produces higher yields than the average regional production. Some of the efforts that can be carried out to maintain it as the leading commodity are 1) disseminating information about the use of excellent seeds and 2) using cassava as raw materials for industry and animal feed.

Likewise, sweet potatoes also show the potential to develop as the leading commodity since the crop's average regional production is lower than that of the Geopark area. Sweet potatoes can continue to have said potential by 1) disseminating information about the use of excellent seeds and 2) using sweet potatoes as raw materials for snacks.

https://doi.org/10.1051/e3sconf/202132508008

Fig. 4 Agricultural Land in the Karangsambung-Karangbolong Geopark Area

The existence of leading agricultural commodities in the Karangsambung-Karangbolong Geopark Area will certainly improve the regional economy since the agricultural sector is the mainstay of livelihood in the rural areas. It can absorb a greater workforce than other sectors, provide foods, raw materials and capital and survive in times of crisis [28], [29]. With leading commodities, the agricultural sector also plays a crucial role in the regional economy as it 1) has abundant natural resources, 2) shores up regional income and 3) becomes the livelihood that many people depend on [30].

3.3 Rating of the Carrying Capacity of the Facilities to Maintain Relatively Potential Commodities in the Karangsambung-Karangbolong Geopark Area

According to Novrilasari (2008), the carrying capacity of facilities in a region can affect regional development [31]. Areas with good facilities will be more developed than those with poor facilities. The relatively potential food-crop commodities to develop in the Geopark area are undoubtedly supported by sufficiently available facilities to sustain regional economic activities.

The centre of growth and service in a region will provide benefits to it. According to Sidik (2016), the scalogram is a region's concentration analysis that identifies key facilities with a hierarchy and characters typical of a developing region [32]. The scalogram analysis was used to rate the regional development in terms of facilities that supported food-crop commodities with relative potential for being developed in the Geopark area. The facilities include the length of the road and the distance to the city centre, markets and irrigation. Table 2 shows the distribution of the carrying capacity of facilities to maintain the relatively potential commodities in the Karangsambung-Karangbolong Geopark Area.

Table 2. Distribution of the Carrying Capacity of Facilities to Maintain the Relatively Potential Commodities in the Karangsambung-Karangbolong Geopark Area

Sub-districts	a	b	c	d	Total Scores	Categories
Ayah	3	1	3	1	8	2
Buayan	2	1	3	2	8	2
Rowokele	2	1	2	1	6	2
Sempor	3	2	1	3	9	1
Gombong	1	2	2	3	8	2
Karanganyar	1	2	2	1	6	2
Karanggayam	3	1	1	1	6	2
Sruweng	1	3	2	3	9	1
Pejagoan	1	3	1	3	8	2

Sub-districts	a	b	c	d	Total Scores	Categories
Alian	1	3	2	2	8	2
Karangsambung	2	2	1	1	6	2
Sadang	2	1	1	2	6	2

Source: Secondary data analysis, 2021
Notes: a: road length (km)
b: distance to the city centre (km)
c: market (unit)
d: irrigation

Based on Table 2, it can be inferred that the highest total score of facilities from all sub-districts in the Geopark area was nine (9), and the lowest was six (6). The score marks the carrying capacity of facilities to maintain relatively potential food-crop commodities. Sub-districts with a total score of 3–5 were included in Category 3 (poor), while ones with a total score of 6–8 were in Category 2 (relatively good). Lastly, those with a total score of 9–12 fell into Category1 (good).

Sub-districts in Category 1 had facilities with a good carrying capacity to support the relatively potential commodities in the Karangsambung-Karangbolong Geopark Area. There were two sub-districts included in this category: Sempor and Sruweng. Sub-districts in Category 2 had a relatively good carrying capacity of facilities to maintain the relatively potential commodities. There were ten sub-districts included in this category: Ayah, Buayan, Rowokele, Gombong, Karanganyar, Karanggayam, Pejagoan, Alian, Karangsambung and Sadang. Sub-districts that fell into Category 3 had facilities with a poor carrying capacity to maintain the relatively potential commodities. However, none of the twelve sub-districts in the Geopark area was included in this category.

Rice, cassava and sweet potatoes are potential commodities to maintain in the Geopark area since these three food crops were identified as the basic commodities in most sub-districts. Also, the sub-districts that became the basis for these commodities are located in the Geopark area and have a good carrying capacity of facilities to support farming activities for these commodities.

4 Conclusion

The sub-districts in the Karangsambung-Karangbolong Geopark Area have different basic food-crop commodities based on their respective geographical conditions. Rice, cassava and sweet potatoes are concluded as the basic and relatively potential commodities in most sub-districts. Ten sub-districts located in the Geopark area have a relatively good carrying capacity of facilities, and two of which have a good carrying capacity to support the relatively potential commodities.

References

1. Databoks (2020, 28 Maret). Wilayah dengan Tingkat Kemiskinan Tertinggi di Jawa Tengah". Diakses 31 Mei 2021, dari https://databoks.katadata.co.id/datapublish/2020/03/28/wilayah-dengan-tingkat-kemiskinan-tertinggi-di-jawa-tengah/
2. H. Siregar dan D. Wahyuniarti, Dampak Pertumbuhan Ekonomi Terhadap Penurunan Jumlah Penduduk Miskin, Jurnal Ekonomi Politik dan Keuangan. (2007) 54-56
3. DPRD Kabupaten Kebumen (2021, 29 Maret). LKPJ Bupati Tahun 2020: Pertumbuhan Ekonomi Minus 1,4 Persen, Pendapatan Daerah Tercapai 103 Persen. Diakses 31 Mei 2021, dari https://dprd-kebumenkab.go.id/2021/03/29/lkpj-bupati-tahun-2020-pertumbuhan-ekonomi-minus-14-persen-pendapatan-daerah-tercapai-103-persen/
4. R. Breuning and O. Majeed, Inequality, poverty and economic growth, International Economics, 161 (2020) 83-99
5. M. Ravallion, Why Don't We See Poverty Convergence?, American Economic Review, 102, 1 (2012) 504-524
6. M. Asrif, K. B. Khan, M. K. Anser, A. A. Nassani, M. M. Q. Abro and K. Zaman, Dynamic interaction between financial development and natural resources: Evaluating the 'Resource curse' hypothesis, Resources Policy, 65 (2020), 101566
7. Z. Khan, M, Ali, L. Jinyu, M. Shahbaz and Y. Siqun, Consumption-based carbon emissions and trade nexus: Evidence from nine oil exporting countries, Energy Economics, 89 (2020), 104806
8. Z. Khan, M. Hussain, M. Shahbaz, S.Yang and Z.

https://doi.org/10.1051/e3sconf/202132508008

Jiao, Natural resource abundance, technological innovation, and human capital nexus with financial development: A case study of China, 65 (2020), 101585
9. R. A. Dwumfour and M. Ntow-gyamfi, Natural resources, financial development and institutional quality in Africa: Is there a resource curse?, Resources Policy, 59 (2018) 411-426
10. S. Rahim, M. Murshed, S. Umarbeyli, D. Kirikkaleli, M. Ahmad, M. Tufai and S. Wahab, Do natural resources abundance and human capital development promote economic growth? A study on the resource curse hypothesis in Next Eleven countries, Resources, Environment and Sustainability, 4 (2021) 100018
11. K. Ahmed, M. K. Mahalik and M. Shahbaz, Dynamics between economic growth, labor, capital and natural resource abundance in Iran: An application of the combined cointegration approach, Resources Policy, 49 (2016) 213-221
12. Q. Shao, Nonlinear effects of marine economic growth and technological innovation on marine pollution: Panel threshold analysis for China's 11 coastal regions, Marine Policy, 121 (2020) 104110
13. Subandi, Ekonomi Pembangunan, Cetakan Kesatu, Alfabeta, Bandung, 2011.
14. Badan Pusat Statistik, Kabupaten Kebumen dalam Angka 2021, BPS, Kebumen, 2021.
15. M. L. Jhingan, Ekonomi Pembangunan dan Perencanaan, Raja Grafindo Persada, Jakarta. 2007.
16. L. Arsyad, Ekonomi Pembangunan, Edisi Kelima, UPP STIM YKPN, Yogyakarta, 2015.
17. J.B. Nikijuluw, Analisis Sektor Ekonomi Unggulan Kabupate/Kota di Provinsi Maluku, Cita Ekonomika, 7, 2 (2013) 196-303.
18. L. Muta'ali, Teknik Analisis Regional untuk Perencanaan Wilayah Tata Ruang dan Lingkungan, BPFG, Yogyakarta, 2015
19. E. P. Yuendini, I. N. Rachmi, N. N. A. Puspitasari, R. Harini dan M. A. F. Alfana, Analisis Potensi Ekonomi Sektor Pertanian dan Sektor Pariwisata di Provinsi Bali Menggunakan Teknik Analisis Regional, Jurnal Geografi, 16, 2 (2019) 128-136
20. Sugiyono, Metode Penelitian Kuantitatif dan R & D, Alfabeta, Bandung, 2016.
21. Bendavid, Regional and Local Economic Analysis for Practionersm, Praeger Publisher Inc, New York, 1991.
22. B. Oksatriandhi, E. B. Santoso, Identifikasi Komoditas Unggulan di Kawasan Agropolitan Kabupaten Pasaman, Jurnal Teknik POMITS, 3, 1 (2014) 8-11
23. N. Baladina, R. Anindita, R. Iskandar, Sukardi, Identifikasi Potensi Komoditas Pertanian Unggulan dalam Penerapan Konsep Agropolitan di Kecamatan Poncokusumo, Jurnal AGRISE, 13, 1 (2013) 30-41
24. Suning, Identifikasi Sektor Ekonomi Basis (Unggulan) dan Hierarki Pusat Pelayanan Berdasarkan Tingkat Kemampuan Fasilitas Dalam Rangka Pengembangan Wilayah (Studi Kecamatan Kasiman dan Kecamatan Padangan Kabupaten Bojonegoro), Teknik Waktu, 9, 2 (2011) 25-33.
25. S. Budiharsono, Teknik Analisis Pembangunan Wilayah Pesisir dan Lautan, Pradnya Pramita, Jakarta, 2001.
26. Sjafrijal, Perencanaan Pembangunan Daerah Dalam Era Otonomi Daerah, Rajawali Pers, Jakarta, 2014.
27. Soekartawi, Agribisnis Teori dan Aplikasinya, Raja Grafindo Persada, 2005.
28. R Harini dan B. Susilo, Kajian Spasial Perubahan Iklim Terhadap Produksi Pertanian, Jurnal Agribisnis, 1, 1 (2017) 14-20
29. D. Novrilasari, Analisis Sektor Unggulan dalam Meningkatkan Perekonomian dan Pembangunan Wilayah Kabupaten Kuantan Singingi, Institut Pertanian Bogor, Bogor, 2008.
30. R. Rosyanti, Analisis Pendapatan Usaha Tani, Pemasaran dan Nilai Tambah Ubi Kayu, Agribisnis, IPB Bogor, 2011
31. Todaro, Pembangunan Ekonomi Dunia Ketiga, Erlangga, Jakarta, 2011
32. R. Harini, C. Ramadhan and F. S. Rudi, Agricultural resource potential for improving food security and community welfare in Penimbun Village, Karanggayam, Kebumen, Central Java, IOP Conference Series: Earth and Environmental Science, 451 (2020) 012041
33. Undang-Undang Republik Indonesia Nomor 18 Tahun 2012 tentang Pangan
34. J. Sidik, Keragaan Relatif dan Karakteristik Perkembangan Kecamatan di Kabupaten Bangkalan, Media Trend, 11, 1 (2016) 20-34.

The analysis of spring potency for domestic uses (Case study: Wanadadi Sub-district, Banjarnegara-Indonesia)

Gita Oktaviani Fadilah[1] dan *Margaretha* Widyastuti[1]*
[1]Department of Environmental Geography, Faculty of Geography, Universitas Gadjah Mada, 55281 Yogyakarta, Indonesia

Abstract. Water is a fundamental human need, and one of the sources constantly available in rainy and dry seasons is spring. The residents of Wanadadi Sub-district (Banjarnegara, Indonesia) rely on spring water as an alternative source, specifically to deal with water scarcity in dry seasons. This research was intended to investigate the potential of the springs based on their availability and quality to meet domestic water requirements. There has never been a study focusing on the springs in the sub-district; therefore, the results can give an overview of their potential and suitable conservation strategies. This research also enriches scientific knowledge of spring characteristics in structural depressions and hill slopes. Spring availability was determined by measuring the discharge of 34 springs and comparing the results with the total water needs. Spring water quality was identified by testing the water samples of eleven springs, each representing a village in the sub-district, and comparing the parameter values with the standards for sanitation and hygiene purposes issued in the Decree of the Minister of Health No. 32/2017. The results showed that based on water availability the springs in five villages have good potential. Meanwhile, in terms of quality the springs in all villages have very low potential.

Keywords: spring discharge, spring water quality, the potential of springs, Wanadadi Sub-district

1 Introduction

Water is one of the essential substances on earth used to meet diverse fundamental human needs in, among others, the domestic sector, fisheries, agricultural practices, and transportation. Its quantity and quality determine to what extent a source can fulfill the population's demands. Water volume and discharge indicate the amount of water available for this purpose. Meanwhile, quality restricts the intended use of a water source and is generally estimated by evaluating the physical, chemical and biological properties as manifestations of natural conditions and anthropogenic interferences that can affect human and aquatic system health [1].

Water can be extracted from various sources, such as groundwater, rivers, lakes, and precipitation. Springs are also among the water sources available and thus widely used in rainy and dry seasons [2]. Springs are the emergence of subterranean water that is concentrated at the surface of the ground in a perceptible current and are distinguished from seepage [3]. The emergence can be caused by gravitational and non-gravitational forces. Gravity springs include depression springs, contact springs, artesian springs, springs in impervious rocks and solution tubular/cavern springs [4]. Springs usually appear on volcanic slopes (particularly at slope breaks), in specific geological structures like fractures and faults and in karst areas.

The annual increase in population size inevitably raises total water requirements and risks of water scarcity issues. There are frequent reports of shortages in the domestic sector of Wanadadi Sub-district, especially during the dry season when water levels in household and communal wells are lower. To deal with this, the residents extract water from springs as an alternative to the dried wells. For this reason, the research sought to identify the potential of springs in the sub-district to meet local demands in terms of quantity (availability) and quality. The spring water availability was determined by comparing the measured discharges with the total water needs calculated according to the provisions set by the National Standardization Body of Indonesia in SNI 19-6728.1-2002. As for quality, it was examined by comparing the parameter values measured in the field and the laboratory with the water quality standards for sanitation and hygiene purposes issued in the Decree of the Minister of Health No. 32 of 2017.

Research on the potential availability and quality of water from these springs is new and has never been conducted in Wanadadi Sub-district. The current

* Corresponding author: mwiwik@ugm.ac.id

research can provide an overview to the community and government authorities regarding the potential of springs and a reference to regulate spring water utilisation and formulate spring conservation strategies. It also enriches scientific information about spring characteristics (i.e., water availability and quality) in structural hill slopes and depressions.

2 Research Location

Wanadadi is a sub-district in the Banjarnegara District area. Astronomically, it is located at 343722–353804 mE and 9181607–9189774 mN. It administratively consists of eleven villages: Tapen, Kasilib, Karangjambe, Wanadadi, Wanakarsa, Karangkemiri, Gumingsir, Linggasari, Medayu, Kandangwangi and Lemahjaya. Geologically, the sub-district lies on three geological formations: alluvium (Qa), terrace deposits (Qt) and the breccia of the Ligung Formation (QTbi) (Fig. 1).

Based on the hydrological condition of the groundwater itself, the sub-district is part of the discharge area of the Purwokerto-Purbalingga Groundwater Basin. It is composed of three hydrogeological units: alluvial deposits, terrace deposits and old volcanic complexes in the form of volcanic breccias and agglomerates [5]. Meanwhile, according to the Schmidt-Fergusson classification the sub-district has a C climate or wet climate.

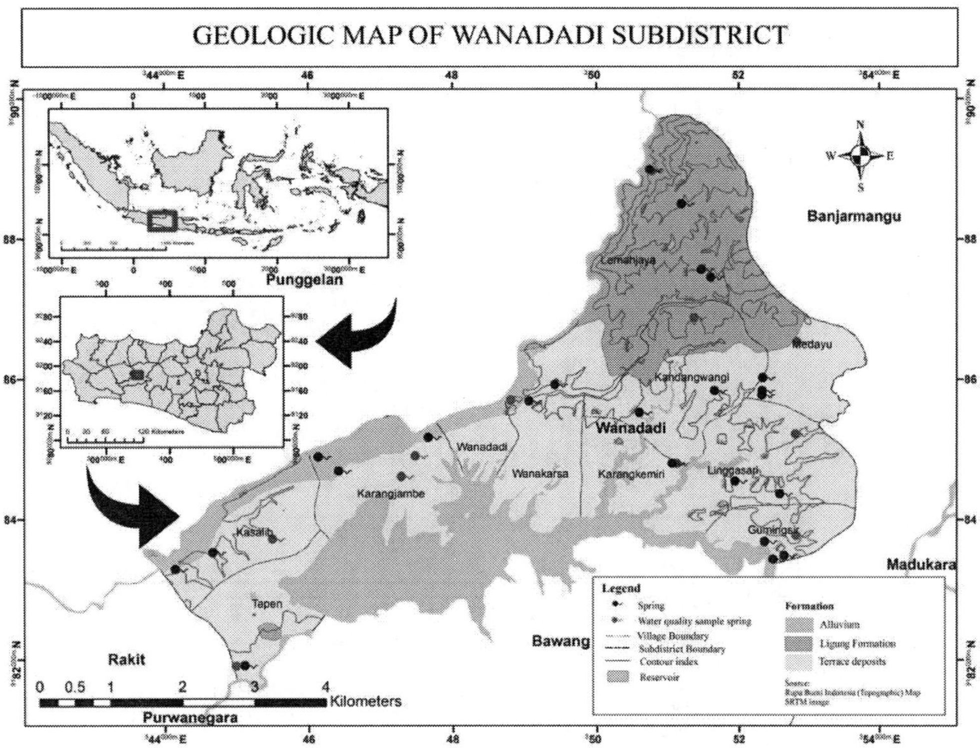

Fig. 1 Geologic map of Wanadadi Sub-district

3 Method

The spatial distribution and discharge of the 34 springs in the Wanadadi Sub-district were observed directly in the field. Spring discharge was measured in the dry season to determine the lowest amount of water available to meet the local domestic requirements. Because the springs appear as concentrated outflows (like water coming out of a faucet), volumetric gauging is suitable for measuring their discharges. In this method flow rate or discharge, Q (m^3/s), is the volume of water filling a measuring glass, V (m^3), per unit of time, t (s), or as presented in Eq. 1 below [6].

$$Q = V/t \qquad (1)$$

In this context flow rates indicate the quantity of water released by a spring. These data were further categorised into eight classes of potential according to Meinzer [4], from I (largest) to VIII (smallest). The total discharge (availability) of all springs in one unit of analysis, i.e., village, was calculated to obtain water availability in one year. This figure was then compared with the domestic water consumption of the population of each village, which is different between urban and rural areas. Because the sub-district shows the characteristics of a rural region, the consumption was assumed to be 60 L/day/person. The annual domestic water requirement, Q DMI, is the number of days in one year (365 days) times population size and 60 L, or as shown in Eq. 2 below [7]. Here, population size is the number of people living in one unit of analysis.

$$Q\ (DMI) = 365\ days \times \Sigma\ population \times 60\ L \qquad (2)$$

The springs are assumed to release a fixed amount of water (availability) from 2020 until 2030, while the water requirement is assumed to increase with population growth. The population data in the next ten years were obtained using the geometric population projection. This method is widely applied because it is easy to use, and the projected data are considered close to the actual condition [8]. It assumes that the population growth rate in each year does not change [9]. The geometric growth was calculated using Eq. 3 below [9]:

$$Pn = Po\,(1 + r)^n \qquad (3)$$

where Pn is the population after n years, Po is the population at time zero, r is the rate of growth and n is time in years.

Here, population projection was used to calculate the amount of domestic water required in the projected year. This figure was compared with the fixed amount of water released from springs (availability) throughout the years of observation to determine the potential of springs as a domestic water supply, assuming they are the only water source used in each village. This procedure enabled the calculation of the annual water balance between water availability and requirements for domestic use in 2020 (at the time of the discharge measurement) and 2030 (projected water availability).

For the water quality analysis, samples were collected from eleven springs, each representing the village of Wanadadi. Some data were tested directly in the field, namely taste, odour, temperature, TDS and pH. Meanwhile, some others were analysed in the laboratory: chemical properties other than pH and biological properties, i.e., total coliforms. In the study area the springs emerge through one or several openings on the ground surface and are accumulated and stored in a tank or container. Therefore, samples were collected from the tap into the sample bottle and preserved for further analysis in the laboratory.

All water quality data were analysed and presented in maps to illustrate the distribution of springs along with their physical, chemical and biological properties. To identify the springs' potential, each parameter value of these properties was also compared with their respective allowable levels in water used for sanitation and hygiene purposes, as described in the water quality standards in the Decree of the Minister of Health No. 32 of 2017.

4 Results and discussion

The results of the discharge measurements at 34 springs and the water quality tests at 11 selected springs demonstrate the potential of these water sources to meet the basic household needs in Wanadadi Sub-district.

4.1 Spring discharge

The springs in Wanadadi have a vital part in fulfiling the local domestic water needs. Their discharge indicates to what extent they can play this role. To meet the increasing water demand, spring water availability must be identified to create a basis for planning and conserving water sources in the future.

Table 1. The spring discharges of Wanadadi Sub-district and their Meinzer's classification

Villages	Springs	Discharge (L/s)	Meinzer's Classification	Villages	Springs	Discharge (L/s)	Meinzer's Classification
Tapen	Tapen 1	Not measurable	-	Kandangwangi	Kandangwangi 1	0.01	VII
	Tapen 2	0.18	VI		Kandangwangi 2	0.02	VII
Kasilib	Kasilib	0.10	VI		Wanatawang	0.11	VI
	Kasilib 2	Not measurable	-	Karangkemiri	Karangkemiri 1	0.18	VI
	Karangpucung	0.05	VII		Karangkemiri 2	0.22	VI
Karangjambe	Karangjambe 1	0.14	VI	Medayu	Medayu 1	Not measurable	-
	Karangjambe 2	0.69	VI		Medayu 2	0.08	VII
	Bangkong	0.58	VI		Medayu 3	0.14	VI
Wanadadi	Punthukrandu	0.64	VI		Medayu 4	0.04	VII
	Gandu	Not measurable	-	Linggasari	Linggasari 1	0.21	VI
Wanakarsa	Wanakarsa 1	Not measurable	-		Linggasari 2	0.36	VI
	Wanakarsa 2	0.18	VI		Linggasari 3	0.01	VII
Lemahjaya	Lemahjaya 1	Not measurable	-	Gumingsir	Gumingsir 1	Not measurable	-
	Tampingan 1	0.20	VI		Gumingsir 2	0.10	VII
	Tampingan 2	0.13	VI		Gumingsir 3	0.03	VII
	Karabaok 1	Not measurable	-		Gumingsir 4	0.04	VII
	Karabaok 2	Not measurable	-				
	Karabaok 3	0.73	VI				

According to the discharge classification by Meinzer [4], the 34 springs observed fell into different classes. Nearly half of them (16) were categorised into class VI (0.1–1 L/s), nine were class VII (0.01–0.1 L/s), and the remaining nine springs were not gaugeable because they were covered with cement (Table 1). Based on classes VI and VII criteria, the springs in the sub-district were concluded as low-yielding. Several factors like geological structures, surface relief and climatic conditions are responsible for the small discharge [10]. The springs in Wanadadi are formed when the topography intersects relatively shallow water tables; thus, the amount of water released from the springs is not significant. Also, the slope where the springs emerge is fairly steep, and their discharges were measured in dry months (August-September) where there is normally a decline in groundwater recharge (seasonal factor). Therefore, despite the small discharge, all springs have been classified as perennial flows.Specific geological structures like faults and fractures are known factors of spring discharge [11]. However, because these geological structures are not found in the study area, their influence is negligible. Also, different rock constituents of aquifers (i.e., the Ligung Formation, terrace deposits and alluvium formation) do not lead to substantial variation in spring discharges. This is evident from the discharge classification results, which vary between mere two classes: VI and VII.

4.2 Spring water quality

Quality determines the suitability of spring water for domestic use. Here, the parameters observed were of physical properties: taste, odour, temperature and TDS and chemical properties: nitrate (NO_3-N), nitrite (NO_2-N), sulfate (SO_4) and pH. The water quality of each spring is presented in the map in Fig. 2.

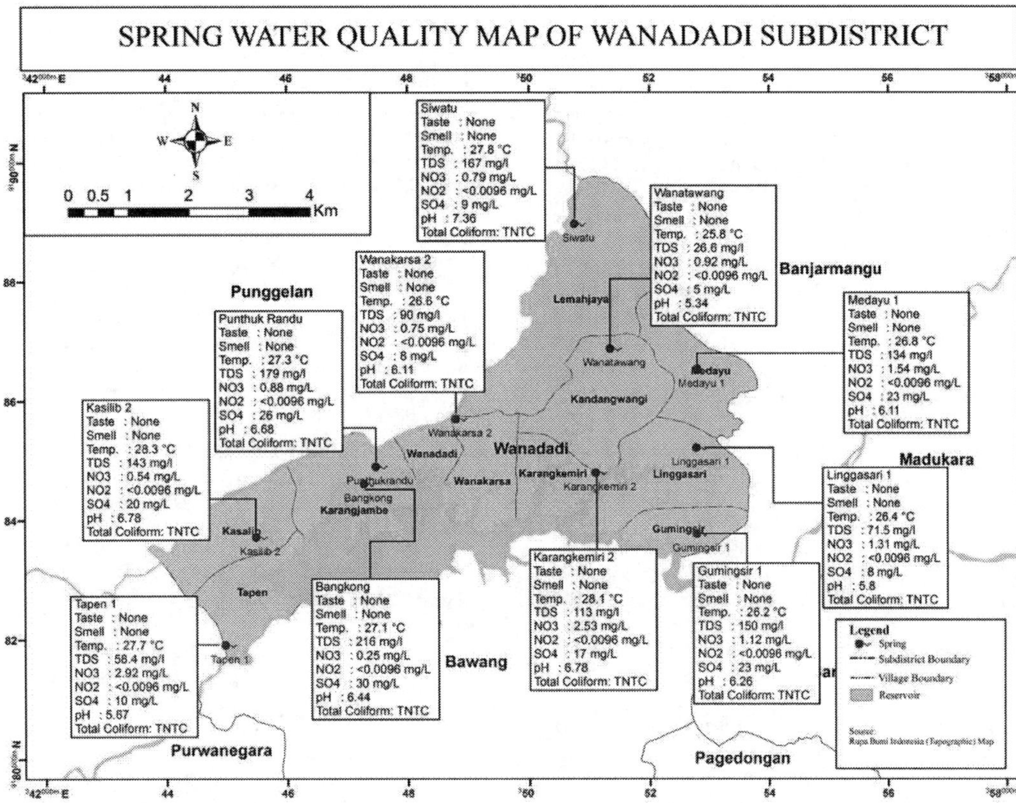

Fig. 2 The water quality map of the eleven sampled springs in Wanadadi Sub-district

Direct measurements in the field revealed that the water at eleven sampled springs was physically tasteless and odourless, with no specific indication of any harmful substances. The measured spring temperature was in the range of 25.8–28.3°C. The air temperature average of the Banjarnegara District in 2020 was 24.33°C [12]. At some selected springs, the temperatures deviated less than 3°C, or within the acceptable range for sanitation and hygiene purposes. Any temperature exceeding the upper limit indicates the presence of dissolved chemicals in large quantities or organic matter decomposition, thus making a body of water unsuitable for consumption [13].

TDS (total dissolved solids) intercorrelates with other water quality parameters: turbidity, hardness, alkalinity and conductivity [14]. The lowest TDS, 26.6 mg/L, was identified at Wanatawang Spring (Kandangwangi Village), whereas the highest, 216 mg/L, was at Bangkong Spring (Karangjambe Village). With this range of TDS, all samples met the water quality standard for sanitation and hygiene, i.e., below 1000 mg/L. A particularly high TDS can make water taste bitter and increase hardness and corrosivity; both are signs of decreased water quality [14].

Nitrates found in water sources are usually attributable to chemical contamination from residues of urea in fertiliser [15] and are sensitive to the entry of waste generated by livestock and crop farming [16]. Contamination of a source of water by nitrates (NO_3^-N) and nitrites (NO_2^-N) is of great concern due to their detrimental impacts on health [16]. Even in small doses, excess nitrate can cause mild illnesses, such as goiter. If present in lethal doses, both compounds can cause methemoglobinemia, i.e., the conversion of more than 10% hemoglobin to methemoglobin, resulting in dysrhythmias, heart attack, coma and death [17]. The nitrate contents (NO_3^-N) of the spring water samples did not exceed their maximum allowable presence in water for sanitation and hygiene purposes, <0.0096 mg/L.

Sulphate (SO_4) in groundwater is mainly derived from sulphate mineral deposits in the rock constituents of an aquifer, although in some cases industrial waste has been identified as a source of excess sulphates [18]. Drinking water containing this compound tastes bitter with a metallic aftertaste and has laxative effects [19]. The laboratory analysis results showed sulphate contents varying from 5 mg/L to 30 mg/L, which are far below its upper limit value, 400 mg/L.

pH measures the intensity of the acidity or alkalinity of a dilute liquid and represents its hydrogen ion concentration [20]. The pH values of the 11 springs ranged from 5.34 to 7.36. Most of the samples had acidic pH of lower than 6.5, which can increase metal object corrosivity, give water an unpleasant taste, induce the toxicity of several chemicals and disrupt health [20].

Total coliform is a measure of biological water quality. The test used the standard membrane filter/plate count procedure number 9222 by the American Public Health Association (AHPA), and all samples showed TNTC results (too numerous to count), meaning that the total bacterial colonies per membrane filter at the time of testing were above 200 colony-forming units (CFU) per 100 ml [21]. It also indicates that the number of colonies has exceeded the upper limit allowed for each plate to produce a reliable calculation and does not match the actual count. Coliform bacteria in a water source can cause various diseases, such as diarrhea and nausea [20].

4.3 Spring potential

4.3.1 Water balance

Water balance demonstrates the potential of springs to fulfil domestic water needs in terms of availability or quantity. Table 2 summarises the results of the water balance calculation for each village. The total annual discharge was measured directly at most of the sampled springs. Only a few were not gaugeable because the spring water collection chambers were closed; therefore, their discharge was estimated from pumping capacity. For instance, Tapen 1 and Kasilib 2 Springs have been widely used by the communities in Tapen and Kailib Villages and installed with a jet pump with a capacity of 75 L/minute. Other examples include Medayu 1 and Gumingsir 1 Springs, whose water can be extracted with the available shallow groundwater pumps at a capacity of 33 L/minute.

Table 2. The water balance of springs in Wanadadi Sub-district in 2020 and 2030

Years	2020				2030			
Villages	Spring discharge (m^3/year)	Water requirements (m^3/year)	Water balance (m^3/year)	Notes	Spring discharge (m^3/year)	Water requirements (m^3/year)	Water balance (m^3/year)	Notes
Tapen	84564.58	49400.91	35163.67	Surplus	84564.58	50442.57	34122.01	Surplus
Kasilib	83527.16	53795.54	29731.62	Surplus	83527.16	54929.86	28597.31	Surplus
Karangjambe	44578.68	40832.49	3746.18	Surplus	44578.68	41693.48	2885.20	Surplus
Wanadadi	20358.41	67796.51	-47438.10	Deficit	20358.41	69226.05	-48867.64	Deficit
Wanakarsa	5765.04	60376.44	-54611.40	Deficit	5765.04	61649.52	-55884.48	Deficit
Lemahjaya	33519.97	108717.34	-75197.37	Deficit	33519.97	111009.72	-77489.75	Deficit
Kandangwangi	4529.78	67487.35	-62957.56	Deficit	4529.78	68910.36	-64380.58	Deficit
Karangkemiri	12557.53	52890.11	-40332.58	Deficit	12557.53	54005.34	-41447.81	Deficit
Medayu	164493.95	58035.58	106458.37	Surplus	164493.95	59259.30	105234.65	Surplus
Linggasari	18146.90	52470.53	-34323.63	Deficit	18146.90	53576.91	-35430.01	Deficit
Gumingsir	91877.97	33125.33	58752.63	Surplus	91877.97	33823.80	58054.16	Surplus
Total	563919.98	644928.15	-81008.16	Deficit	563919.98	658526.92	-94606.93	Deficit

Based on the calculated water balance in 2020, five villages experienced a water surplus: Tapen, Kasilib, Karangjambe, Medayu and Gumingsir. Meanwhile, the springs with discharges lower than the domestic water requirements created a water deficit in six other villages, assuming that the residents only used water sourced from the local springs for their daily activities. Even with the same spring discharge in the first and last years of observation, the projected water balance for 2030 showed an identical pattern: a water surplus in five villages. It can be concluded that the springs in these five villages have a good potential in terms of quantity and will be able to meet the domestic water needs of their residents in the future, at least for the next ten years

4.3.2 Suitability with the water quality standards

To assess the suitability of the spring water for sanitation and hygiene purposes, the research referred to the water quality standards issued in the Decree of the Minister of Health No. 32 of 2017. Samples of spring water were tested for taste, odour, temperature, TDS, nitrite, nitrate, pH, sulphate and total coliform. Upon comparing the test results with the said standards, it was found that the eleven springs selected to represent each village had poor potential in terms of quality (Fig. 3). The taste, odour, TDS, nitrite, nitrate and sulphate were within the standards, but the temperature, pH and total coliform of several samples were either higher or lower than their allowable presence.

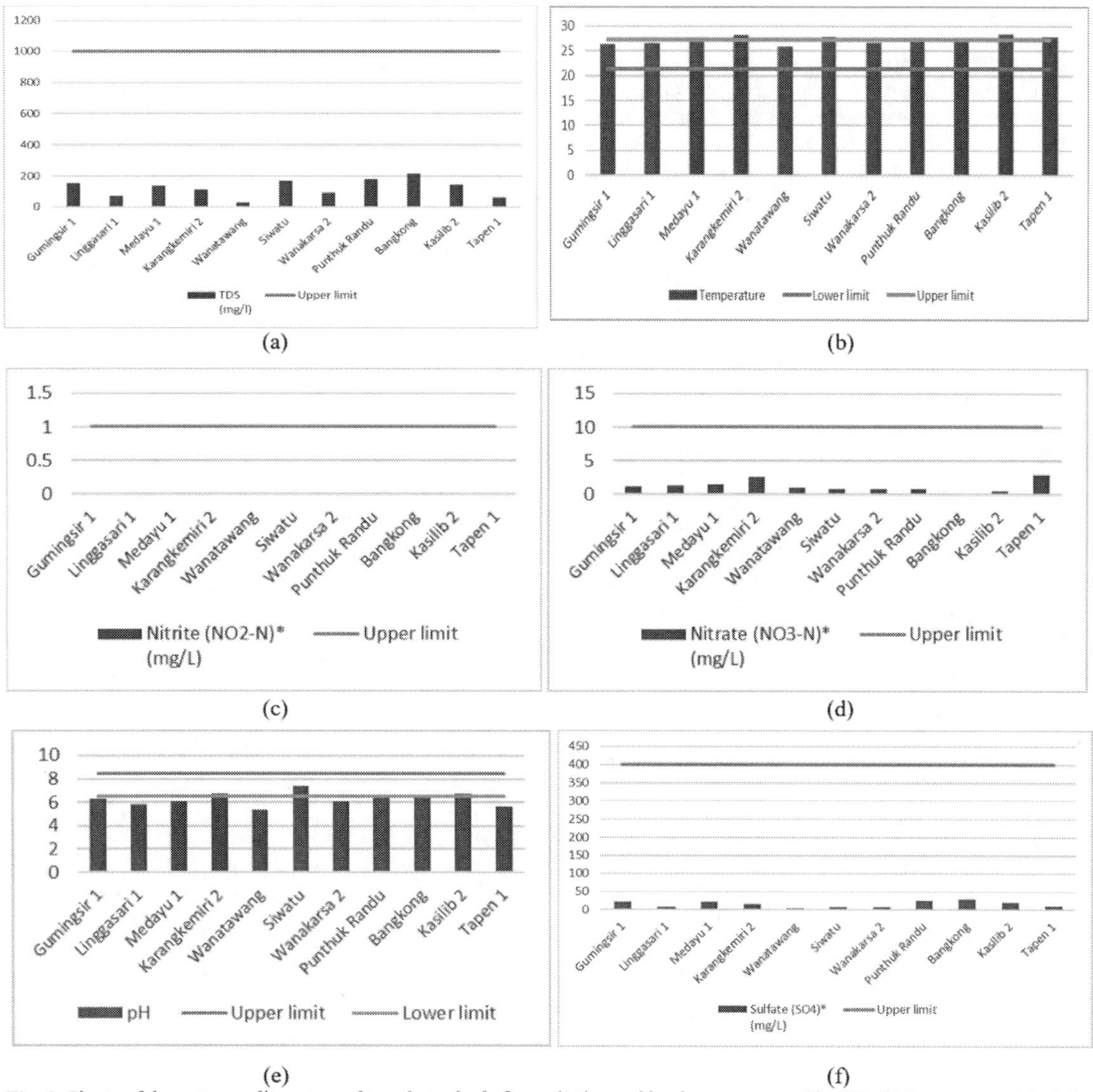

Fig. 3. Charts of the water quality test results and standards for sanitation and hygiene purposes: (a) TDS, (b) temperature, (c) nitrite content, (d) nitrate content, (e) pH and (f) sulphate

Total coliforms were found to be particularly high in all samples, as evident from the TNTC results in the microbial enumeration test. TNTC, or too numerous to count, means that the number of colonies observed is above 200 CFU/100 ml, exceeding the upper limit for water used in sanitation and hygiene activities, 50 CFU/100 ml. Most of the springs sampled for water quality testing are left uncovered and exposed to the surrounding elements, except for Medayu 1, Tapen 1 and Kasilib 2 Springs. Without cover structures and proper management, a water source is at high risk of contamination by the waste by-products of domestic activities, livestock farming and crop farming. Research conducted in Kenya proves that livestock and wild animals can increase total coliform counts in unprotected drinking water sources [22]. Another influencing factor is soil; for example, high *E. Coli* counts in Babadan (Umbulharjo, Yogyakarta) are associated with alluvial soils composed of clay layers. Alluvial soils generally have high porosity but low percolation and permeability, which barely filter out *E. coli* bacteria that tend to move vertically in groundwater [23]. In the Wanadadi Sub-district, Wanakarsa 2 Spring is located in an alluvial deposit complex, while most of the other springs are in the terrace deposit formation and the breccia of the Ligung Formation.

Similarly, pH was also the parameter that did not meet the water quality standards for sanitation and hygiene at six springs. Gumingsir 1, Linggasari 1, Medayu 1, Wanatawang, Wanakarsa 2, Bangkong and Tapen 1 had slightly acidic water below the lower limit value, 6.5. Many scholars have found several water bodies with low pH; for example, 85% of the volcanic springs in São Miguel, Azores, have pH 5–7 [24]. Acid spring water can result from geological conditions at and around the location; for example, groundwater in Massachusetts is naturally acidic because of soil and rock factors [25]. The springs sampled and analysed in the water quality testing were located in the Ligung Formation, terrace deposits and alluvial deposits composed of volcanic materials with intermediate acidity levels [26]. Also, low pH can indicate the influence of surrounding human activities on water bodies, such as residential areas, markets, plantations and waste disposal [27]. For instance, Cirebon coastal waters Indonesia have pH values from 6.06 to 6.36 due to high detergent levels (0.01–0.02 mg/L) in the rivers that feed into these waters [28]. In Wanadadi, some springs are close to the settlements: Gumingsir 1, Medayu 1, Wanakarsa 2 and Tapen 1. Linggasari 1 Spring is located near fish ponds, Bangkong is near rice fields and settlements, while Wanatawang is in the middle of a plantation. Anthropogenic activities in the vicinity can generate and introduce waste into the springs and affect their pH.

Like pH, the temperatures did not meet the water quality standards at several springs. The water temperatures at Karangkemiri 2, Kasilib 2, Siwatu and Tapen 1 deviated about 3–4°C from the air temperature, or higher than the upper limit value. However, this does not classify the springs in Wanadadi Sub-district as hot springs, which otherwise have substantially higher temperatures, e.g., 45–80°C at some hot springs in southern Thailand [29] and 41–99°C at some hot springs in western Malaysia [30]. The high temperatures measured at the springs in Wanadadi can be influenced by solar radiation at the measurement time.

5 Conclusion

The research in Wanadadi Sub-district has found that the springs in Medayu Village have the highest water availability (a total discharge of 164493.95 m^3), while those in Kandangwangi Village have the lowest (4529.78 m^3). In 2020 the residents in Lemahjaya Village had the highest water requirement (108717.34 m^3), but those in Gumingsir Village had the lowest (33125.33 m^3). Based on the water balance, the springs in five villages (Tapen, Kasilib, Karangjambe, Medayu and Gumingsir) have great potentials in that the amount of water released could fulfil the domestic water needs in 2020. The water balance projection for 2030 also shows the same results even though the water needs of each village have increased. In terms of quality, all springs have one to three parameters that do not meet the standards for sanitation and hygiene purposes set in the Decree of the Minister of Health No. 32 of 2017. The parameters in questions are pH (as detected at Gumingsir 1, Linggasari 1, Medayu 1, Wanatawang, Wanakarsa 2, Bangkong and Tapen 1) and temperature (Karangkemiri 2, Kasilib 2, Siwatu and Tapen 1). As evident from the TNTC results (too numerous to count) in the microbial enumeration test, all samples contain excessively high total coliforms (above 200 CFU/100ml). Based on the availability and quality aspects, it can be concluded that not all springs in each village have good potential to sufficiently provide water for domestic activities. However, villages with surplus water balance can improve the local springs' potentials with proper water quality treatments.

Acknowledgement

The authors would like to thank Universitas Gadjah Mada (UGM) for funding this research through the Final Project Recognition Grant scheme (*Hibah Rekognisi Tugas Akhir*, RTA) in 2021. Gratitude is also extended to the Faculty of Geography UGM for providing the opportunity to complete this research and disseminate the results through this article. These supports are invaluable to the preparation and completion of the research.

References

1. D. Chapman, *Water Quality Assessments - A Guide to Use of Biota, Sediments and Water in Environmental Monitoring*, Second Edi (F & FN Spon, London, 1996)
2. Sudarmadji, D. Darmanto, M. Widyastuti, and S. Lestari, J. Mns. Dan Lingkung. **23**, 102 (2016)
3. M. F. N. Said and Sudarmadji, J. Bumi Indones. **3**, 1 (2014)
4. D. K. Todd and L. W. Mays, (2005)
5. A. M. . Meyerink, Serayu Val. Proj. **4**, 25 (1980)
6. Badan Standardisasi Nasional, *Standar Nasional Indonesia: Perencanaan Sistem Penyediaan Air Minum* (Badan Standardisasi Nasional, 2012)
7. Badan Standardisasi Nasional, *Standar Nasional Indonesia: Penyusunan Neraca Sumber Daya – Bagian 1: Sumber Daya Air Spasial* (Indonesia, 2002)
8. Ditjen Cipta Karya DPU, *Buku Panduan Pengembangan Air Minum* (Jakarta Selatan,

2007)
9. Badan Pusat Statistik, *Pedoman Penghitungan Proyeksi Penduduk Dan Angkatan Kerja* (Jakarta, 2010)
10. Z. Michalczyk, S. Chmiel, S. Głowacki, J. Sposób, and B. Zielińska, Ecohydrol. Hydrobiol. **20**, 599 (2020)
11. L. W. Santosa, Forum Geogr. **20**, 68 (2006)
12. Badan Pusat Statistik Kabupaten Banjarnegara, *Banjarnegara Dalam Angka* (Badan Pusat Statistik Kabupaten Banjarnegara, Banjarnegara, 2021)
13. F. Majid, H. Santjoko, and Narto, *Pasir, Zeolit Dan Arang Aktif Sebagai Media Filtrasi Untuk Menurunkan Kekeruhan, TDS Dan e-Coli Air Sungai Selokan Mataram Yogyakarta* (Yogyakarta, 2019)
14. V. Kothari, S. Vij, S. K. Sharma, and N. Gupta, Environ. Sustain. Indic. **9**, 100093 (2021)
15. Abdurrivai and S. N, J. Sulolipu Media Komun. Sivitas Akad. Dan Masy. **17**, 1 (2017)
16. Y. Kawagoshi, Y. Suenaga, N. L. Chi, T. Hama, H. Ito, and L. Van Duc, Sci. Total Environ. **657**, 146 (2019)
17. A. E. Manampiring, Karya Ilm. 1 (2009)
18. P. Das, K. Maya, and D. Padmalal, Geothermics **90**, 101997 (2021)
19. G. Prasad, A. S. Reshma, and M. Vinodini Ramesh, Mater. Today Proc. (2021)
20. I. G. A. P. R. Yoga, N. P. W. Astuti, and N. N. A. Sanjaya, Higiene **6**, 53 (2017)
21. APHA, *APHA Method 9222: Standard Methods for the Examination of Water and Wastewater* (1992)
22. D. Trajano Gomes da Silva, J. Ebdon, J. Okotto-Okotto, F. Ade, O. Mito, P. Wanza, E. Kwoba, T. Mwangi, W. Yu, and J. A. Wright, Int. J. Hyg. Environ. Health **230**, 113602 (2020)
23. E. Winata and E. Hartantyo, J. Fis. Indones. **17**, 8 (2013)
24. P. Freire, C. Andrade, R. Coutinho, and J. V. Cruz, Sci. Total Environ. **466–467**, 475 (2014)
25. United States Environmental Protection Agency New England, *Healthy Drinking Waters for Massachusetts* (Massachussets, 2007)
26. J. A. E. Tjahjono, in *Kolok. Direktorat Inventar. Sumber Daya Miner. DIM* (Jakarta, 2002)
27. Sudarto, S. F. Retnowaty, Y. Fitri, and A. Suroso, J. Phot. **5**, 63 (2015)
28. T. Susana, Indones. J. Urban Environ. Technol. **5**, 33 (2009)
29. W. Ngansom, K. Pirarai, and H. Dürrast, Geothermics **85**, 101746 (2020)
30. H. Baioumy, M. Nawawi, K. Wagner, and M. H. Arifin, J. Volcanol. Geotherm. Res. **290**, 12 (2015)

The Impact of Global Climate Change to Climate Condition of Bengkulu Watershed, Indonesia

Gita Ivana Suci Lestari Faski[1,2] and *Ignasius Loyola Setyawan* Purnama[1]

[1] Faculty of Geography, Universitas Gadjah Mada, Yogyakarta, Indonesia
[2] Indonesian Agency for Meteorology, Climatology and Geophysics, Bengkulu Climatology Station, Indonesia
e-mail : gita.ugm2019@gmail.com

Abstract. Global climate change that occurred in this century can affect the pattern of rain and increase in temperature on earth. This study aims to determine and analyze the increase in rainfall, air temperature, potential evapotranspiration and actual evapotranspiration in the Bengkulu watershed. For this reason, the regional rainfall is calculated using the Thiessen Polygon, the mean air temperature of the watershed based on the median elevation, potential evapotranspiration using the Thornthwaite Method and actual evapotranspiration using the basis of the difference in rainfall to potential evapotranspiration. The results showed that every year there was an increase in rainfall, air temperature, potential evapotranspiration and actual evapotranspiration in the Bengkulu Watershed. In the 2009-2013 period, the average annual rainfall of 3,581 mm increased to 3,641 mm in the 2014-2018 period. For air temperature, the average monthly air temperature in the Bengkulu Watershed for the 2009-2013 period was 25.8°C, while the air temperature in the 2014-2018 period was 26.1°C. This means that in a period of 5 years there is an increase in temperature of 0.3°C. Furthermore, due to the increase in air temperature, there was an increase in the average monthly potential evapotranspiration from the 2009-2013 period to the 2014-2018 period, namely from 1,493 mm to 1,537 mm, while for actual evapotranspiration there was an increase from 1,486 mm to 1,518 mm.

1 Introduction

According to United Nation Environmental Program (UNEP) experts, climate change is a major environmental problem of this century. Some indications of the impact of climate change are the increased air temperature and changing rainfall patterns [1]. As a result of changes in climate elements, the condition of water resources can change [2].

The amount of water on earth is constant, as is the amount of water that can be accommodated in a watershed. The existence of a phenomenon of shortage and excess of water in an area is only a matter of its unequal distribution, or because the phases of its form in the hydrological cycle are not balanced. Therefore, the problem in water management is that water can be sufficiently available where it is needed with good and sustainable quality [3-4].

In water management in a watershed, the concept of the hydrological cycle in its entirety must be understood to study rain as an input that falls on the surface of the watershed and returns to the sea again through various processes, stages and changes in form [5-6]. In this phenomenon the most important processes are rainfall, air temperature and evapotranspiration, which then affect surface runoff in the watershed.

Rainfall is an important aspect as a cause of runoff including floods, because rain is the main source of surface runoff [7-9]. The size of the rainfall and runoff and the distribution of runoff over time can also be used as the basis for the operation of water structures. Rain is water that falls to the earth's surface in the form of droplets due to the condensation of water vapor in the atmosphere. The factors that influence the distribution of rain are latitude, elevation, distance from water sources, location from mountains, wind direction and physical properties of air masses.

Air temperature is a measure of the heat of the earth's surface and the atmosphere due to solar radiation [10]. The factors that influence the difference in temperature from one place to another are the angle of incidence of sunlight, long exposure time, elevation and geographical conditions of the area. Air temperature greatly affects evapotranspiration, namely the process of water loss due to evaporation and transpiration.

Regarding evapotranspiration, there are two dimensions of evapotranspiration, namely potential evapotranspiration (Ep) and actual evapotranspiration (Ea). Potential evapotranspiration is the amount of water lost due to evapotranspiration if there is sufficient water available or in ideal conditions, while actual evapotranspiration is the actual evapotranspiration that occurs.

Astronomically, the Bengkulu watershed is located at coordinates 102°14'39 "- 102°35'00" East Longitude and 3°37'6 "- 3°50'33" South Latitude [11]. According to [12], the average annual rainfall in the Bengkulu watershed is around 3,118 mm/year, so that the

Bengkulu River, which is the main river of the Bengkulu Watershed, experiences two floods during the rainy season [13]. Because in the process of rainfall to flow in a watershed is influenced by the characteristics of the watershed climate, flood events can be analyzed through aspects of rainfall, air temperature and evapotranspiration in the watershed [14].

2 Methodology

This study is based on secondary data, namely monthly rainfall and air temperature data in the Bengkulu watershed in 2009-2018 which were obtained from the Bengkulu Climatology Station BMKG. For rain, the rainfall data used is monthly rainfall data at six rain stations around the Bengkulu watershed area which is then processed into regional average monthly rainfall data using the Thiessen Polygon method. The rain gauge used is the Observatory (Obs) gauge, while the location of rainfall station used in this study is shown in Figure 1 and Table 1 and the percentage of the Thiessen Polygon area to the sub-watershed is shown in Table 2.

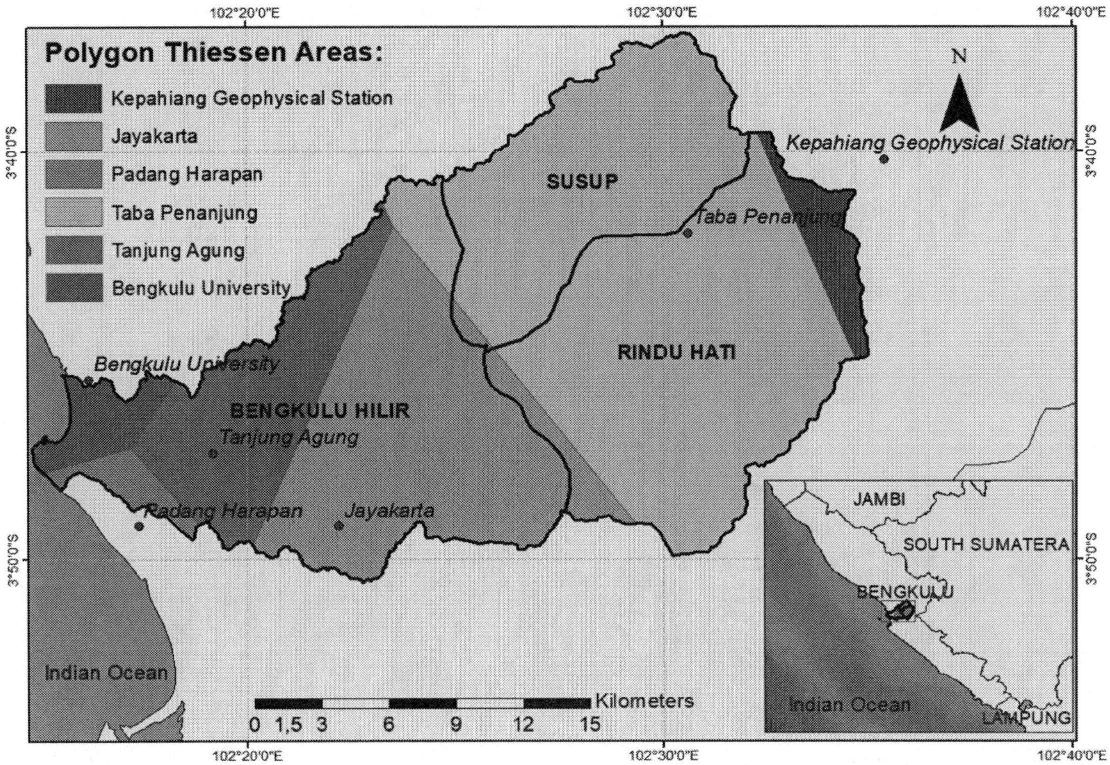

Fig. 1. Location of rainfall station [15].

Tabel 1. Location and elevation of rainfall station.

Rainfall Station	South Latitude	East Longitude	Elevation (m)
Bengkulu University	3°45'36"	102°16'12"	29
Tanjung Agung	3°47'32"	102°19'13"	20
Padang Harapan	3°49'18"	102°17'11"	30
Jayakarta	3°49'12"	102°22'12"	39
Taba Penanjung	3°42'00"	102°30'36"	149
Kepahiang Geophysic Station	3°40'11"	102°35'20"	517

Source : [15]

Table 2. Percentage of Thiessen Polygon area to area of sub-watersheds in Bengkulu Watershed.

Rainfall Station	Sub Watershed (%)		
	Rindu Hati	Susup	Bengkulu Hilir
Bengkulu University	0	0	6.64
Tanjung Agung	0	0	29.40
Padang Harapan	0	0	4.16
Jayakarta	5.46	1.47	54.90
Taba Penanjung	86.06	98.53	4.90
Kepahiang Geophysic Station	8.47	0	

Source : [15]

Monthly air temperature determination is based on air temperature data from the nearest station as a reference and is calculated using the Mock method. Determination of the air temperature in the area using the median elevation

Potential evapotranspiration is calculated using the Thornthwaite equation, while actual evapotranspiration is determined based on the amount of rain and potential evapotranspiration. In the wet month (rainfall > potential evapotranspiration), the actual evapotranspiration is assumed to be equivalent to the potential evapotranspiration, while in the dry month (rainfall < potential evapotranspiration), the actual evapotranspiration is determined by increasing the amount of rainfall that occurs with changes in soil moisture storage which is deficit.

3 Results and discussion

3.1 Rainfall

The pattern of annual rainfall during 2009 to 2018 in the three sub-watersheds in the Bengkulu watershed can be shown in Table 3 and Figure 2.Based on Table 3 and Figure 2, it can be seen that every year the average annual rainfall in the Bengkulu watershed has increased.

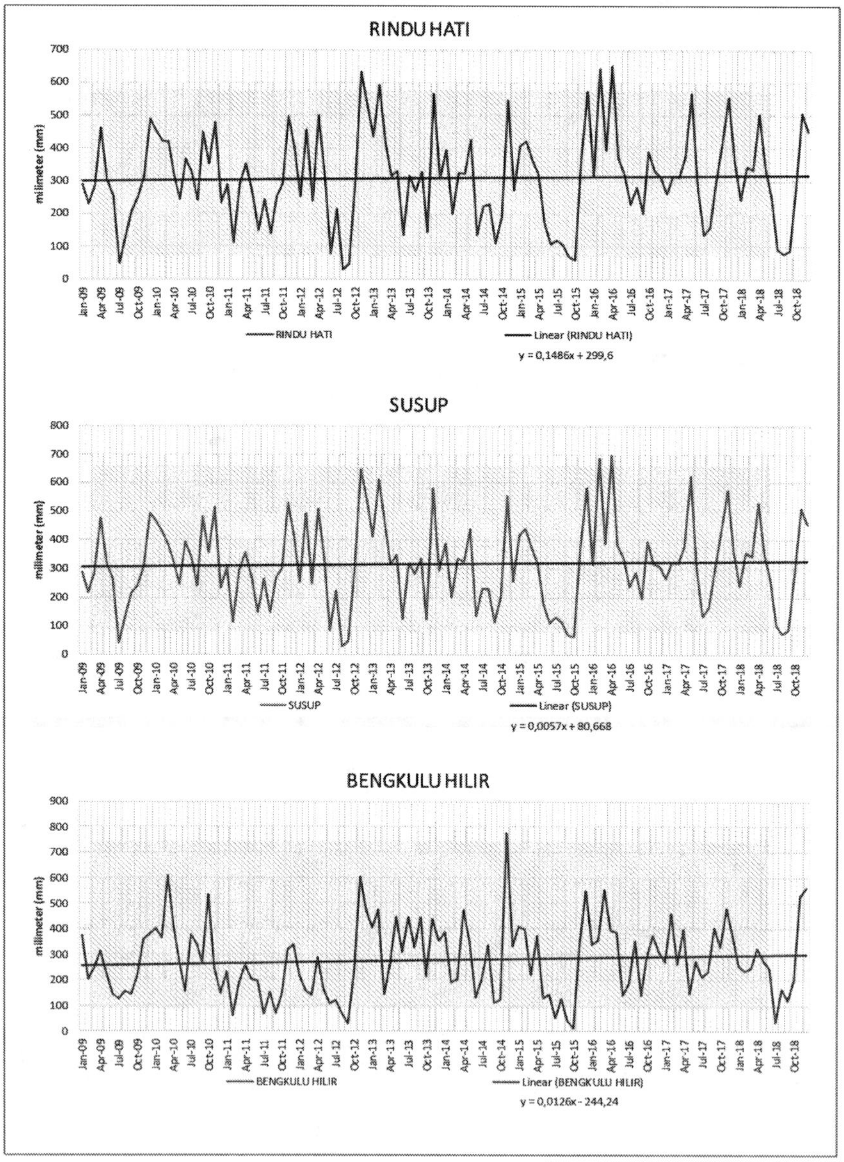

Fig. 2. Monthly rainfall 2009-2018 [15].

In the 2009-2013 period, Susup Sub-watershed had the highest average annual rainfall of 3,801 mm, while Bengkulu Hilir Sub-watershed had the lowest average annual rainfall of 3,225 mm. In the 2014-2018 period, the Susup Sub-watershed also had the highest average annual rainfall of 3,780 mm, while the Bengkulu Hilir Sub-watershed also had the lowest average annual rainfall of 3,455 mm. Observing the average rainfall per period, in the 2009-2013 period the average annual rainfall of 3,581 mm increased to 3,641 mm in the 2014-2018 period.

Table 3. Annual rainfall for the 2009-2013 and 2014-2018 periods.

Periods	Annual Rainfall in Sub-watershed (mm)		
	Rindu Hati	Susup	Bengkulu Hilir
Period 1			
2009	3,256	3,277	2,891
2010	4,304	4,409	4,123
2011	3,282	3,416	2,240
2012	3,582	3,706	2,598
2013	4,160	4,195	4,272
Average	3,717	3,801	3,225
Period 2			
2014	3,370	3,406	3,622
2015	3,088	3,164	2,798
2016	4,441	4,548	3,823
2017	4,037	4,181	3,845
2018	3,510	3,602	3,189
Average	3,689	3,780	3,455

Source : [15]

3.2 Air temperature

Looking at Figure 3, every year the average monthly air temperature in the Bengkulu Watershed has increased. Because it has the highest median elevation value, the average monthly air temperature in the Rindu Hati Sub-watershed is the lowest, while the highest average air temperature is found in the Bengkulu Hilir Sub-watershed (Table 4). Based on Table 4, it can also be seen that the average monthly air temperature in the Bengkulu Watershed for the 2009-2013 period was 25.8°C, while the air temperature in the 2014-2018 period was 26.1°C. This means that in a period of 5 years in the Bengkulu Watershed there is an increase in temperature of 0.3°C.

Table 4. Average monthly air temperatures for the 2009-2013 and 2014-2018 periods.

Periods	Average Monthly Air Temperature (°C)		
	Rindu Hati	Susup	Bengkulu Hilir
Period 1			
2009	25.1	25.7	26.4
2010	25.3	26.0	26.6
2011	25.1	25.7	26.4
2012	25.3	25.9	26.6
2013	25.3	25.9	26.5
Average	25.2	25.8	26.5
Period 2			
2014	25.3	25.9	26.6
2015	25.5	26.2	26.8
2016	25.7	26.3	26.9
2017	25.4	26.0	26.7
2018	25.4	26.0	26.7
Average	25.5	26.1	26.7

Source : [15]

https://doi.org/10.1051/e3sconf/202132508010

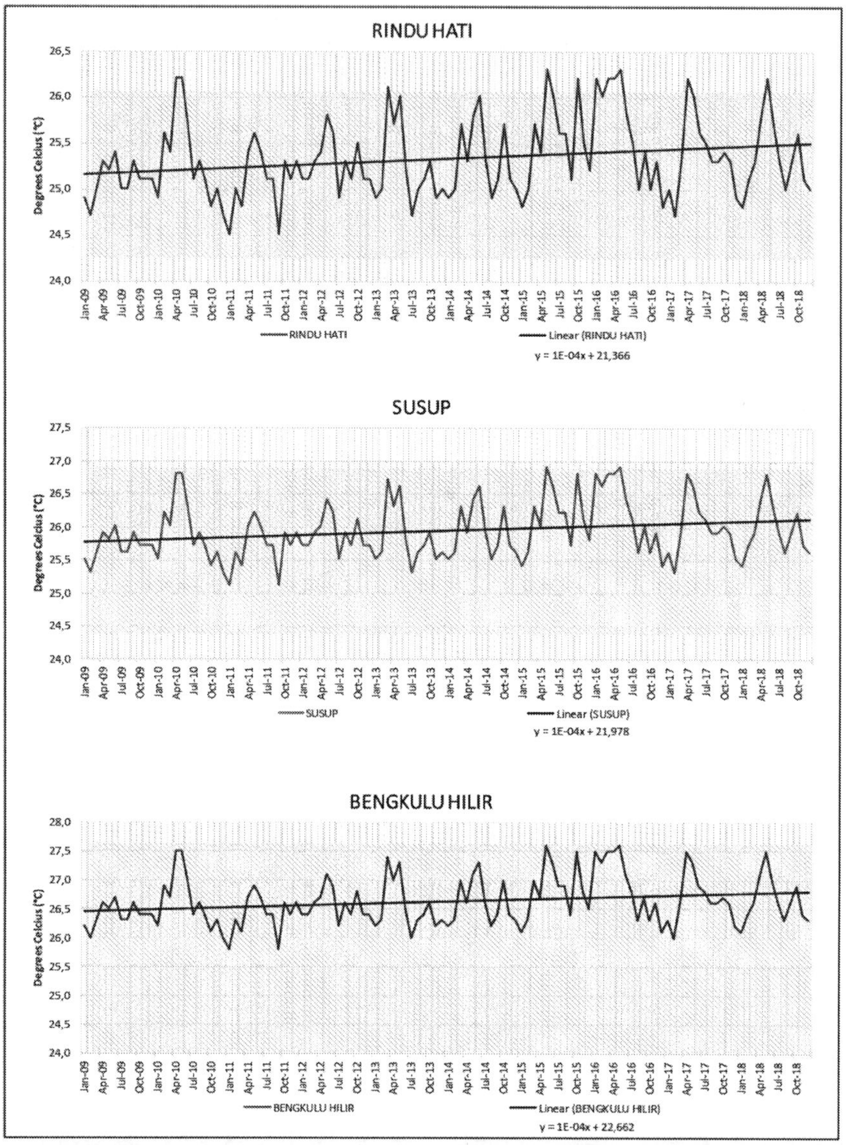

Fig. 3. Monthly air temperature in 2009-2018 [15].

3.3 Evapotranspiration

In general, each year the potential and actual monthly evapotranspiration in the Bengkulu Watershed also increases (Figure 4). However, at some point in time the opposite may also have happened. For example, the actual evapotranspiration in September 2012 decreased, but in October 2015 only the Bengkulu Hilir Sub-watershed experienced a decline. This phenomenon can occur due to the effect of low rainfall of less than 50 mm, whereas for October 2015 only actual evapotranspiration in Bengkulu Hilir Sub-watershed decreased because the rainfall in October 2015 in Bengkulu Hilir sub-watershed was only 12 mm.

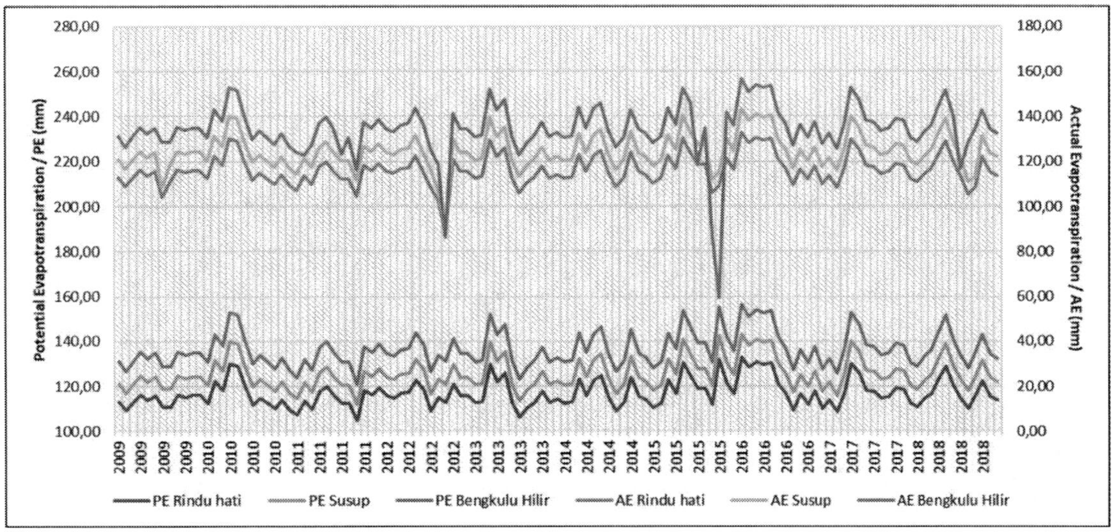

Fig. 4. Monthly potential and actual evapotranspiration in 2009-2018 [15].

The average potential and actual evapotranspiration in each sub-watershed are different due to the difference between the average potential and actual evapotranspiration in each of these sub-watersheds. The difference that occurs shows that the rainfall in that month is smaller than the potential evapotranspiration in the sub-watershed.

In the period 2009-2013, Rindu Hati Sub-watershed and Susup Sub-watershed experienced differences in the average potential evapotranspiration and actual evapotranspiration from June to September, while for Bengkulu Hilir Sub-watershed it occurred from February to March and June to September (Figure 5). In the 2014-2018 period, the difference between potential evapotranspiration and actual evapotranspiration in Rindu Hati and Susup Sub-watersheds occurred from June to October, while in Bengkulu Hilir Sub-watershed it occurred from May to October (Figure 6).

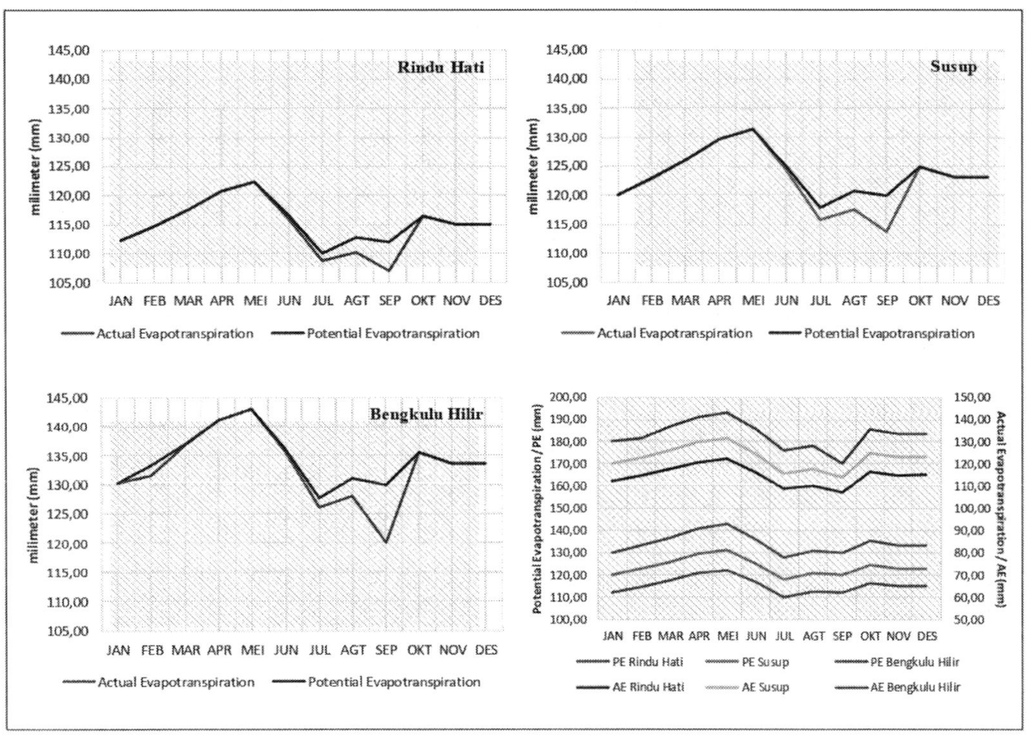

Fig. 5. Monthly potential and actual evapotranspiration in 2009-2013 [15].

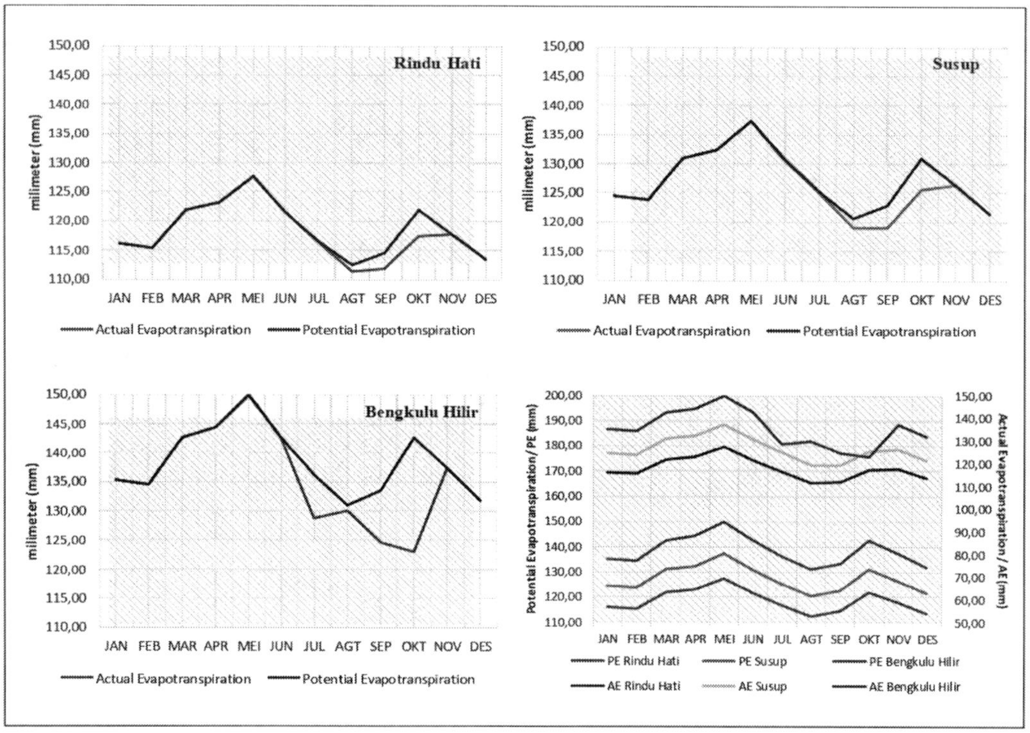

Fig. 6. Monthly potential and actual evapotranspiration in 2014-2018 [15].

On an annual basis, the average potential evapotranspiration in Rindu Hati Sub-watershed in the 2009-2013 period as well as the 2014-2018 period is the lowest. The highest average potential evapotranspiration occurs in Bengkulu Hilir Sub-watershed (Table 4). For actual evapotranspiration, the lowest average actual evapotranspiration also occurred in the Rindu Hati sub-watershed, while the highest actual evapotranspiration also occurred in Bengkulu Hilir Sub-watershed (Table 5). When viewed by time, there is a tendency for an increase in the average annual potential evapotranspiration from the 2009-2013 period to the 2014-2018 period throughout the watershed, from 1,493 mm to 1,537 mm. For actual evapotranspiration there is an increase from 1,486 mm to 1,518 mm.

Table 5. Annual potential pvapotranspiration in the Bengkulu Watershed.

Periods	Annual Potential Evapotranspiration in Sub-watershed (mm)		
	Rindu Hati	Susup	Bengkulu Hilir
Period 1			
2009	1,367	1,464	1,588
2010	1,405	1,507	1,639
2011	1,368	1,464	1,589
2012	1,395	1,496	1,625
2013	1,391	1,491	1,620
Average	1,385	1,481	1,612
Period 2			
2014	1,402	1,503	1,634
2015	1,437	1,543	1,681
2016	1,456	1,565	1,706
2017	1,411	1,514	1,647
2018	1,407	1,509	1,641
Average	1,422	1,527	1,662

Source : [15]

Table 5. Annual actual evapotranspiration in the Bengkulu Watershed.

Periods	Annual Actual Evapotranspiration in Sub-watershed (mm)		
	Rindu Hati	Susup	Bengkulu Hilir
Periods 1			
2009	1,360	1,454	1,588
2010	1,405	1,507	1,639
2011	1,368	1,464	1,569
2012	1,355	1,446	1,562
2013	1,391	1,491	1,620
Average	1,391	1,472	1,595
Period 2			
2014	1,402	1,501	1,631
2015	1,407	1,507	1,515
2016	1,456	1,565	1,706
2017	1,411	1,514	1,647
2018	1,394	1,491	1,624
Average	1,414	1,516	1,624

Source : [15]

4 Conclusion

1) Every year there is an increase in the average monthly rainfall in the Bengkulu Watershed. In the 2009-2013 period, the average annual rainfall of 3,581 mm increased to 3,641 mm in the 2014-2018 period.
2) Every year the average monthly air temperature in the Bengkulu Watershed has also increased. The average monthly air temperature in the Bengkulu watershed for the 2009-2013 period was 25.8°C, while the air temperature in the 2014-2018 period was 26.1°C. This means that in a period of 5 years in the Bengkulu Watershed there is an increase in temperature of 0.3°C.
3) Due to the increase in air temperature, there was an increase in the average annual potential evapotranspiration from the 2009-2013 period to the 2014-2018 period, namely from 1,493 mm to 1,537 mm, while for actual evapotranspiration there was an increase from 1,486 mm to 1,518 mm.

Acknowledgement

This manuscript is part of a study for thesis entitled " Water Balance Analysis of Bengkulu Watershed in Relation to the Watershed Physical Characteristics", financed by Final Assignment Recognition (Rekognisi Tugas Akhir/RTA) Programme Universitas Gadjah Mada year 2021.

References

1. Purnama I L S & Cahyadi A. Foum Geografi 33 (2) : 140-152 (2019)
2. Purnama I L S., Salamah S D., Wandari K A & Primacintya V A. E3S Web of Conferences 200, 02017 (2020)
3. Saptarshi, P.G and Raghavendra, R.K. J. Indian Soc. Remote Sens. 37 : 693-704 (2009)
4. Murtiono U.H. Forum Geografi 23 (1) : 11-24 (2009)
5. Hamilton, L.S dan King, P.N. Gadjah Mada University Press, Yogyakarta (1986)
6. Suprayogi, S., Purnama, I L S & Darmanto, D. Gadjah Mada University Press (2014)
7. Wannielista, M., R. Kersten and R. Eaglen. John Wiley and Sons Inc., New York (1997)
8. Purnama I L S. 2008. Indonesian Journal of Geography 40 (2) : 153-166 (2008)
9. Purnama I L S. Advances in Social Science, Education and Humanities Research, volume 79 (2017)
10. Davie T. Routledge, Taylor & Francis Group, London (2008)
11. Andriansyah O. & Mustikasari R. In Telapak (2011)
12. Harmiati, Apriyanti, H., Supriyono, Triyanto, D., & Alexsander. Jurnal Ilmu Pemerintahan : Kajian Ilmu Pemerintahan Dan Politik Daerah, 3 (2), 136–148 (2018)
13. Gunawan, G. 2017. Analisis data hidrologi Sungai Air Bengkulu menggunakan metode statistik. Inersia, 9 (1), 47–58
14. Tamba, C., Fauzi, M., & Suprayogi, I. 2016. Kajian potensi ketersediaan air menggunakan model neraca air bulanan Thornthwaite-Mather (studi kasus : Sub DAS Subayang Kampar Kiri Hulu). Jom FTEKNIK, 3(2), 1–8
15. Faski G I S L. Thesis. Fakultas Geografi Universitas Gadjah Mada (2021)

The influence of aquifer material to the groundwater potency in Ngawi Regency

Ignasius Loyola Setyawan Purnama[1*], *Laelina* Rahmawati[1], *Vincentia* Anindha Primacintya[2] and *Erick* Febriarta[3]

[1]Faculty of Geography, Universitas Gadjah Mada, Indonesia
[2]Accounting Programme, Master of Science and Doctoral, Faculty of Economics and Business, Universitas Gadjah Mada, Indonesia
[3]Faculty of Enginering, Janabadra University, Indonesia

Abstract. Until now, groundwater is still the main source of water for domestic needs. Unfortunately not all places can contain groundwater. Groundwater potential in an area is largely determined by the rock layers that make up the aquifer in the area. This study aims to analyze the hydrogeological conditions of the study area and to analyze the potential for groundwater in each geological unit. To achieve this goal, this study is based on the analysis of existing secondary data, such as rock type data obtained from drill data and geological maps. Apart from these data, the main secondary data from this research is geosonar measurement data that has been done previously. In this geosonar estimation, VLF (Very Low Frequency) electromagnetic technology is used, which is a geophysical method that uses very low radio communication frequency signals (15-30 kHz) to determine the anomalies of the electrical properties of rocks. The results showed that hydrogeologically, Ngawi Regency can be divided into two, namely areas with high groundwater potency in the central and southern parts and areas with rare groundwater in the northern part. In addition, it is also known that there is a relationship between the aquifer material and the groundwater potency. Alluvial deposits which are dominated by sand and gravel have high groundwater potency, while the Kalibeng Formation which is composed of solid marl and Anggota Banyak Kalibeng Formation composed of andesite breccias is a scarce groundwater area.

Keywords: groundwater, aquifer, geosonar

1 Introduction

Water is a chemical compound that is only composed of two types of atoms, namely H and O atoms. The formula for water molecules is also simple, namely H_2O. Even so, water is one of the most important resources for human life and other living things on earth [1]. Without water, there is no life on this earth.

The existence of water on earth is determined by a process called the hydrological cycle, which is the process of circulating water on earth, from the sea to the air, rivers, lakes and back to the sea. The phenomenon of the hydrological cycle is quite complicated because it involves various processes, various stages and various changes in form. In this phenomenon the most important processes are evaporation, transpiration, rain and surface runoff, while the changes in form that occur can be in the form of liquids, solids or gases.

In accordance with the concept of the hydrological cycle, the amount of water on earth is constant. The lack and excess of water in an area is actually just a matter of its unequal distribution both spatially and in terms of time. For example, in one year the amount of water in an area is actually sufficient, but because the water becomes flooded in the rainy season, then during the dry season there is a drought. Lack and excess of water can also occur due to the unbalanced phases of its form in the hydrological cycle.

It is true that water is a natural resource that can be renewed, however the existence of water resources is not limitless. The existence of water resources is limited according to space and time, both in quantity and quality. The thing that is desired is the availability of water in sufficient quantities, available at all times and of good quality.

Generally, there are three types of water, namely rainwater, surface water (river water, lake water) and groundwater. Because of its better quality, humans prefer to use groundwater to support their daily needs than other types of water [2]. Therefore, according to [3] groundwater is the most important part of water resources. In addition, in Indonesia the use of groundwater for household needs is also preferred because of its more secure availability, which is less affected by seasonal differences.

Currently, integrated and sustainable water resource management (including groundwater) has become a major issue for the world community [4]. In water management efforts, the assessment of groundwater potential is important [5].

Unfortunately not all places have adequate groundwater content. The size of the groundwater content in a place depends on the presence of rock layers that can store water [6]. The properties of rock that greatly affect its ability to store and transmit groundwater are its porosity and permeability. It can be seen that groundwater potency in an area is largely determined by the aquifer material in the area [7].

Sand and loose gravel are two types of rock that have the property of being able to store water and drain the excess. Therefore, an area that has these two rock types will have the opportunity to store large amounts of water [8]. In fact, it can be said that 90% of groundwater occupies the rock layers, especially those found in alluvial deposits [9].

Apart from alluvial deposits, although most of them have lower potential, some other rock formations can also function as water storage such as limestone, volcanic rock, sandstone and igneous and metamorphic rocks. Limestone is a type of rock that has unique

*Corresponding author : Setyapurna@geo.ugm.ac.id

https://doi.org/10.1051/e3sconf/202132508011

properties. If in other rocks, the older the grain is, the older the limestone is, the more pores it has due to the dissolution of water. Therefore, the characteristics of limestone (density, porosity and permeability) as water storage also vary widely, especially depending on their age. At different ages the degree of compression and fracture development is usually different. In young limestone, the holes that are formed are usually microscopic in size, while in old limestone caves and rivers will be formed underground, so that old limestone has the possibility to have high groundwater potential [10].

There are various types of volcanic rock, both with affanitic and paneric grains, both those containing low or high ferromagnesian elements. Some examples of volcanic rocks are granite, syenite, diorite, trachyte, andesite and basalt. From several examples of these volcanic rocks, only basalt rock can function as a good aquifer. This is because the basalt can have many pores, leaving gas holes. In addition, basalt rock also often experiences many cracks at the time of its formation. The former gas holes and cracks will be occupied by groundwater from rainwater that falls over the basalt rocky area.

Sandstones are rocks formed by cemented sand. Because of the cement, the sandstones are in a compressed state. Likewise with conglomerate rocks. Conglomerate rocks are formed as a result of cemented rounded gravel and crust. However, if the rock has experienced a lot of cracks, it will also be able to store water even in a limited amount.

Ngawi Regency is one of the regency in East Java Province whose territory borders the Province of Central Java. Astronomically this district is located at 110 ° 10 '- 111 ° 40' East Longitude and 7 ° 21 '- 7 ° 31' South Latitude, while administratively divided into 19 districts. The area of Ngawi Regency is 1,295.98 km².

Based on the topography, in general it varies considerably from flat, wavy, hilly to high mountains, with an altitude between 40 m to 3,031 m above sea level. In the middle is a plain area which is fertile agricultural land, in the south it is a hilly and mountainous area, while in the north it is a row of the Kendeng Mountains with limestone rocks which are less fertile.

Ngawi Regency is one of the districts in East Java Province whose territory borders the Province of Central Java. Astronomically this district is located at 110 ° 10 '- 111 ° 40' East Longitude and 7 ° 21 '- 7 ° 31' LS, while administratively divided into 19 districts. The area of Ngawi Regency is 1,295.98 km².

Based on the topography, in general it varies considerably from flat, wavy, hilly to high mountains, with an altitude between 40 m to 3,031 m above sea level. In the middle is a plain area which is fertile agricultural land, in the south it is a hilly and mountainous area, while in the north it is a row of the Kendeng Mountains with limestone rocks which are less fertile.

Judging from its physiographic condition, Ngawi Regency is located in the central depression zone of Java Island [11]. The designation as a depression zone refers to the basin area between the mountainous geography of the north and south of this district. In more detail, regionally, the physiography of Ngawi Regency can be divided into 3 parts. The southern part is an area that is still influenced by the physiographic formation of quaternary volcanoes, the middle part is influenced by fluvial processes, while the northern part is in the form of the Kendeng anticlinorium.

Judging from the lithology, the rocks that make up Ngawi Regency are several groups of Quaternary and Tertiary Rock Formations from the Late Miocene to the Holocene Period. Late Miocene lithology is dominated by rock formation groups that are in the physiography of the northern side of this district, while on the south side it tends to be a group of rocks with a younger age, namely Pleistocene to Holocene age. It is possible that different types of constituent rocks will affect groundwater conditions in this regency, because the presence and size of groundwater content in a place is highly dependent on the presence of rock layers that are capable of storing water. The purpose of this study is to analyze the hydrogeological conditions of the study area and to analyze the potential for groundwater in each geological unit

2 Methodology

This research is based on the analysis of existing secondary data, such as rock type data obtained from drill data and geological maps. Apart from these data, the main secondary data from this research is geosonar sounding data that has been done previously. In this geosonic estimation, VLF (Very Low Frequency) electromagnetic technology is used, which is one of the geophysical methods that determines the anomalies of the electrical properties of the ground/bedrock using very low radio communication frequency signals (15-30 kHz).

The principle of VLF is to measure the disturbance in the magnetic field from radio waves, which is caused by subsurface geological features, so that by analyzing the nature of the magnetic field interference, anomalous areas below the surface can be identified. The magnetic field emitted by the sonar transmitter causes an electric current which will induce a conductive geological unit, so that Eddy currents will arise in the secondary magnetic field. This secondary magnetic field will then interact with the main magnetic field originating from the transmitter which will then create a disturbed magnetic field due to shifting of phrases and changing directional orientation of the main field. The disturbed magnetic field is measured by the VLF receiver and because this disturbance is caused by subsurface conditions, it is possible to interpret the subsurface conditions including groundwater conditions from the nature of the disturbance.

To determine the hydrogeological condition of the research area, a qualitative descriptive analysis was carried out based on geological and hydrogeological data, while to determine the influence of aquifer material to groundwater potency, a quantitative descriptive analysis was carried out with the help of tables. Furthermore, from the two analyzes, an integrated analysis can be carried out related to the groundwater potency in the research area.

3 Result and discussion

3.1 The geology of Ngawi Regency

Rock is a collection of similar or non-similar minerals that are bonded loosely or solidly. Based on their genesis, rocks can be grouped into igneous, sedimentary and metamorphic rocks. Igneous rock is rock that occurs from freezing liquid silica solution and incandescent known as magma [12]. Sedimentary rocks are rocks formed from rocks that have existed as a result of the denudation process, the result of chemical reactions or the activity of organisms [13]. Metamorphic rocks are rocks that come from other rocks that experience changes in texture and mineral composition in the solid phase as a result of changes in physical conditions in the form of pressure, temperature or both.

Based on its texture, igneous rocks can be divided into faneric igneous rocks, porphyritic igneous rocks and affanitic igneous rocks. Examples of igneous rock from acidic to ultra-alkaline are granite, syenite, diorite, gabbro as well as peridodite and serpentine. Examples of porphyritic igneous rocks ranging from acidic to alkaline are micro granite, micro syenite, micro diorite and dolerite, while examples of affanitic igneous rocks ranging from acidic to alkaline are obsidian rhyolite, trachyte, andesite and basalt. Apart from these three types of igneous rock, there are also known pyroclastic igneous rocks resulting from volcanic eruptions in the form of ash, lapilli and bomb.

Sedimentary rocks can be divided into 2 types, namely clastic sedimentary rocks and non-clastic sedimentary rocks. Clastic sedimentary rocks are formed from re-deposition of detritus/original rock fragments through a process of compaction, cementation, recrystallization, autogenesis and metasomatism. Examples of clastic sedimentary rocks are shale, argillite, sandstone, orthokuarsite, graywacke, breccia and conglomerates.

Non-clastic sedimentary rocks are formed as a result of chemical reactions or from the activities of organisms such as evaporite sediments, precipitation sediments and coal sediments. Examples of evaporite sediments are gypsum and salt rock (halite). Examples of precipitation sediments are limestone, travertine and tuff, while examples of coal sediments are coal, black shale and bitumen.

Based on their structure, metamorphic rocks can be divided into metamorphic rocks and non-foliated metamorphic rocks. Examples of metamorphic foliated rocks are marble, hornstone, quartzite and granulite, while examples of non-foliated metamorphic rocks are slate, philit, milonite, skis, gneis and migmatics.

Physiography of Ngawi Regency according to [11] is the central depression zone of Java Island. This depression zone refers to the basin area between the north and south positive hysiography of this regency. Regionally Ngawi Regency has a physiography that is divided into 3 parts from the south to the north. The south side is the part that is still influenced by the physiographic formation of quaternary volcanoes. While the northern part is filled with physiography in the form of the Kendeng Anticlinorium.

Regionally, the structural condition of Ngawi Regency has a geological structure that is clustered on the south and north sides. The south side is indicated in the form of a fault structure that is in an area that is still affected by the fault group around Lawu Volcano. The northern side of Ngawi Regency is dominated by a regional structure in the form of folds and there are shear and upward faults. This condition cannot be separated from the physiographic conditions of the northern side of this district which are a row of hills which are part of the Kendeng Anticlinorium.

According to Geological Map, the lithology of Ngawi Regency consists of several groups of quaternary and tertiary rock formations that are scattered in the study area. The age of the lithology that is scattered in this regency ranges from the age of the rocks during the late Miocene to the Holocene. Late Miocene lithology is dominated by rock formation groups located in the physiography of the north side of this regency. Meanwhile, the south side tends to be a group of rocks with formations ranging from plistocene to holocene. The rock formations and rock groups that have a wide distribution in Ngawi Regency are (1) Alluvial Deposits which are composed of clay, silt, sand and gravel. This rock formation is deposited along the floodplains of the Lusi, Wulung and Bengawan Solo rivers. (2) Lawu Deposits are composed of volcanic sandstones, volcanic siltstone, volcanic breccia and lava. This rock formation overlaps the older formation in the southern Kendeng Lane, (3) the Tambakromo Formation which is composed of claystone, marl and limestone. Claystone is dark gray, soft, non-layered, in some places a sandy texture. Marl is light gray in color, as a thin insert (2-8 cm). Limestone is light gray, as a thin insert (2-8 cm) containing fossils of Cibicides, Robulus, Rotalia beccarii, Globorotalia tosaensis, Globorotalia truncatulinoides, Globigerinoides fistulosus and Pulleniatina obliquiloculata. This formation is in the Late Pliocene upper-Plistocene age with a sedimentary environment in neritic, 350 meters thick and overlaps in line with the Selorejo Formation, (4) the Kalibeng Formation composed of solid marl and local sandstone insertions (20-50 cm), which are tuff and limestone. In several places at the bottom and middle there are breccias which are multi-member and at the top of the limestone clitic members. This formation contains bentonic foraminifera including Cassidulina, Cibicides, Nodosaria and Planulina. Planktonic foraminifera fossils are Globorotalia crassaformis, Globorotalia plesiotumida, Globorotalia tosaensis and Pulleniatina obliquiloculata which show the age of the Late Miocene-Early Pliocene with a depositional environment in the upper deep-bathial neritic. It is up to 5000 meters thick and overlaps in harmony with the Kerek Formation and (5) Anggota Banyak of Kalibeng Formations consisting of andesite breccias, light gray in color, andesite fragments with a little tuff, fragment size 0.3-12 cm, hard, open-packed, solid. Tuff breccia gray-white, tufa fragments and a little andesite, fragment size 0.2-10 cm, open box, solid. Underwater channel depositional environment with a thickness of between 8-25 meters.

3.2 Aquifer discharge from geosonar measurement results

To determine the groundwater potency in the study area, 66 geosonar soundings points were carried out. Determination of measurement points is carried out based on the extent of rock formations. The measurement results are shown in Figure 1 and Table 1.

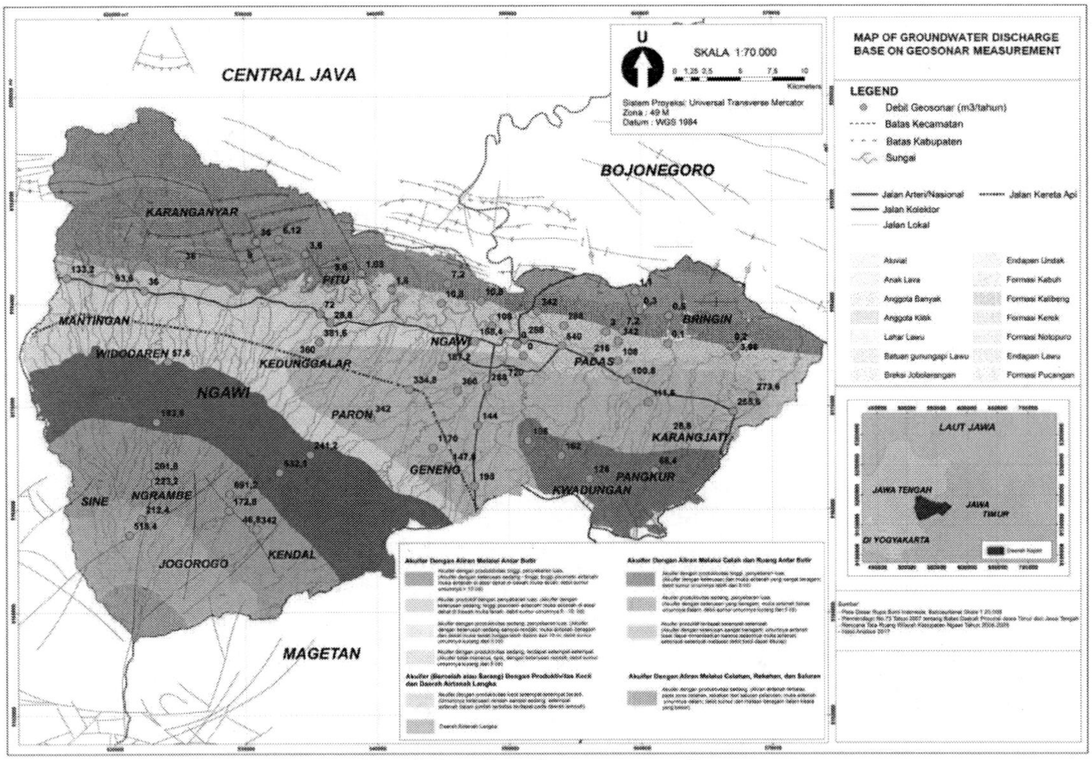

Fig. 1. Location of geosonar soundings

Table 1. The relationship between aquifer material and groundwater potency in Ngawi Regency

Rock Formation	Material	Data from Hydrogeological Map		Groundwater Discharge (m^3/year)
		Aquifer Characteristic	Aquifer Productivity	
Lawu Deposit	Volcanic sandstones, volcanic siltstone, volcanic breccia and lava.	Aquifer with flow through fracture and spaces between grains	High productivity with wide distribution	57.6 183.6 532.8 341.2
			Medium productivity with a wide distribution	223.2 201.6 46.8 342.0 172.8 691.2
			Productive but localized	518.4 212.4
Alluvial Deposit	Clay, silt, sand and gravel	Aquifer with flow between grains	High productivity with wide distribution	108.0 162.0 126.0 68.4

			Medium productivity with wide distribution	133.2 93.6 36.0 360.0 381.6 28.8 72.0 158.4 108.0
			Productive with wide distribution	342.0 198.0 147.6 1170.0 144.0 334.8 360.0 288.0 720.0 187.2 100.6 111.6 28.8 255.6 272.6
Tambakromo Formation	Claystone, marl and limestone	Aquifer with joint and diaklas	Little local productivity matters	288.0 342.0 288.0 540.0 216.0 108.0 342.0 3.9 3.0 0.1
Kalibeng Formation	Marl, solid and local sandstone inserts are tuff and limestone	Aquifer with joint and diaklas	Area with scarce groundwater	36.0 9.0 36.0 6.12 3.6 3.6 1.1 1.8 7.2 10.2 10.8
Anggota Banyak Kalibeng Formation	Andesite breccias, andesite fragments with a little tuff, tuff breccias, tuff fragments and a little andesite	Aquifer with joint and diaklas	Area with scarce groundwater	7.2 1.1 0.3 0.5 0.2

Base on Figure 1 and Table 1, it can be seen that the Lawu Sediment with volcanic sandstone aquifer, volcanic siltstone, volcanic breccia and lava is the aquifer with the highest groundwater potency with an average discharge of 307.9 m^3/year. Next is the Tambakromo Formation with aquifer material are claystone aquifer, marl and limestone which has an average discharge of 213.1 m^3/year. Followed by Alluvial Deposits with aquifer material are clay aquifer, silt, sand and gravel having an average discharge of 193.1 m3 / year.

After these three rock formations, other rock formations in Ngawi Regency are the Kalibeng Formation and the Anggota Banyak of the Kalibeng Formation. The Kalibeng Formation is composed of solid marl and local sandstone inserts that are tuff and limestone, while the Anggota Banyak of the Kalibeng Formation are composed of andesite breccias, andesite fragments with little tuff, tufa breccias, tuff fragments and a little andesite. Both of these rock formations are scarce groundwater areas indicated by the average aquifer discharge of 11.4 m^3/year and 1.86 m^3/year.

Based on this description, it can be said that there is a relationship between the aquifer material with the groundwater potency in the study area. Alluvial Deposits which are composed of sand and gravel have high groundwater potency. However, Lawu Sediment which is an aquifer with flow through fracture and spaces between grains and the Tambakromo Formation, which is a joint and diaklas aquifer, can also be an aquifer with high groundwater potency.

4 Conclusion

1) Hydrogeologically, Ngawi Regency can be divided into two area, i.e. areas with high groundwater potency that located in the central and southern parts and areas with scarce groundwater in the northern part.
2) There is a relationship between aquifer material and the potency of groundwater. Alluvial Deposits which are dominated by sand and gravel have high groundwater potential, while the Kalibeng Formation which is composed of solid marl and Anggota Banyak of the Kalibeng Formations composed of andesite breccias are scarce groundwater areas.

Acknowledgement

This manuscript is the result of research funded by the Community Fund of the Faculty of Geography, Universitas Gadjah Mada Fiscal Year 2021.

References

1. A. A. Fenta, A. Kifle, Hydrogeology Journal **23**, 195-206 (2014)
2. A. M. Hoque, M. M. Hoque, K. M. Ahmed. Hydrogeology Journal **15**, 1523-1534 (2007)
3. S. A. Naghibi, H. A. Pourghasemi. Environ Monit Asses **188**, 44-54 (2015)
4. I. S. Al-Salamah, Y. M. Ghazaw, A. R. Ghumman. Environ Monit Assess., **173**, : 851-860 (2011)
5. R. J. Lubis, Y. Sakura, R. Delinom. Hydrogeology Journal **16** (2008)
6. M. Wanielista, R. Kersten, R. Eaglin. *Hydrology : Water Quantity and Quality Control.* New York : John Wiley and Sons Inc. (1997)
7. D. K. Todd, L. W. Mays. *Groundwater Hydrology*. New York : John Wiley & Sons (2005)
8. T. Davie. *Fundamentals of Hydrology*, London : Routledge, Taylor & Francis Group, (2008).
9. C. W. Fetter, *Applied Hydrogeology,* New York : Macmillan Publishing Company (1988).
10. K. R. Rushton, *Groundwater Hydrology : Conceptual and Computational Models,* Chichester, West Sussex : John Wiley & Sons Ltd, (2003).
11. R. W. V. Bemmelen, *The Geology of Indonesia : General Geology of Indonesia and Adjacent Archipelagoes*, The Hague : Martinus Nijhoff (1970)
12. D. S. Graha, *Batuan dan Mineral*. Bandung : Penerbit Nova (1987)
13. J. Katili, *Pengantar Geologi Umum*. Bandung : Balai Pendidikan Guru (1959)

The Influence of Watershed Morphometry to River Discharge Variation (Case Study of Bengkulu Watershed, Indonesia)

Gita Ivana Suci Lestari Faski[1,2] and *Ignasius Loyola Setyawan* Purnama[1]

[1] Faculty of Geography, Universitas Gadjah Mada, Yogyakarta, Indonesia
[2] Indonesian Agency for Meteorology, Climatology and Geophysics, Bengkulu Climatology Station, Indonesia
e-mail : gita.ugm2019@gmail.com

Abstract. Morphometric analysis in a watershed is a requirement in analyzing hydrological conditions in a watershed, because the properties of river networks and their relationship to each other will affect the condition of water resources in the watershed. This study aims to analyze the variations in the discharge of two rivers in the upper Bengkulu watershed and the morphometric aspects that influence it. For this reason, several morphometric aspects are calculated, such as the area of the watershed, circulation ratio, median elevation and slope. The results showed that the Rindu Hati River had a higher monthly discharge than the Susup River. The sizeable difference in discharge between the Rindu Hati and Susup Rivers is due more to differences in the size of the area. As is known, the total area of the Rindu Hati Sub-watershed is 184.2 km^2, while the Susup Sub-watershed is 103.6 km^2. With a large area, the Rindu Hati Sub-watershed will be able to accommodate a large amount of rainwater as well.

1 Introduction

A watershed is a unitary area of an ecosystem that is bounded by a topographical water separator and functions as a collector, storage and channel for water, nutrient sediments in the river system, all of which come out through one outlet [1-4]. This definition implies that a watershed is a landscape area with a topographical boundary in the form of a line connecting the highest points in the area. In addition, referring to this definition implies that in addition to being a hydrological unit, watershed is also a unitary ecosystem area. As an ecosystem unit, in the watershed there will also be dynamic interactions between elements of living things, the biophysical environment and chemical elements. Likewise, there is a balance phenomenon between the input and output of matter and energy [5-6].

In natural resource management efforts, watershed is often used as a unit of analysis because of the three components of the ecosystem, namely abiotic (soil, water), biotic (vegetation) and culture (humans and their social relationships), water is the most obvious aspect of its movement (following the hydrological cycle). and its territory. However, problems arise related to the increase in watershed damage during the 21st century [7-8]. According to [9], the impact assessment of various land management will help in identifying watershed management efforts, whereas according to [10], data related to soil loss and sedimentation are very useful for identifying the location of soil erosion in conservation efforts. In the end, integrated watershed management is needed that requires holistic thinking and involves the participation of stakeholders [11-14]

As previously mentioned, morphometric analysis in a watershed is a requirement in assessing hydrology, because the behavior of river networks and their relationship with each other affects the water resources in the watershed [15-16]. Watershed morphometry, which includes the area, shape and density of river flows, affects the infiltration system, water storage and surface runoff, including sediment transported in the runoff [17]. Watershed with a large area will tend to have a greater amount of rainfall, so that the amount of flow that is discharged from the watershed is also greater. A watershed that has a round shape will tend to have a higher flood discharge, because the rainfall that falls into the watershed will quickly reach the outlet. Likewise, watersheds that have many river branches will have a higher peak discharge because water will enter the river more quickly and easily [18].

The Bengkulu Watershed is located in two regency, namely Central Bengkulu Regency and Bengkulu City. which consists of several districts, namely Taba Penanjung, Karang Tinggi, Pondok Kelapa and Talang Empat Districts in Central Bengkulu Regency, as well as Gading Cempaka, Muara Bangka Hulu, and Segara Bay Districts in Bengkulu City. The Bengkulu watershed is divided into 3 sub-watersheds, namely the Rindu Hati sub-watershed and the Susup sub-watershed in the upstream part and the Bengkulu Hilir sub-watershed at the downstream part. The upstream part of the watershed is in the Bukit Barisan mountains and drainage into the Indian Ocean. With the relatively short distance between the upstream and downstream and the narrow river width, the Bengkulu Watershed has a relatively steep to very steep slope with high river flow velocity [19].

With the physical characteristics of the watershed, as well as the damage in the upstream part due to coal mining activities and changes in land use from protected forests to production forests, the time of concentration of river flow in the watershed is getting smaller. As a result, the runoff coefficient increases causing an increase in river discharge which can cause flooding [20]. Therefore, the management of water resources in the Bengkulu watershed is very important because of these problems. Watershed morphometric analysis forms the basis for watershed management policies.

2 Methodology

The area of the watershed is determined based on its watershed boundaries, which are delineated on the topographic map. Furthermore, the actual watershed area is determined by multiplying the area of the watershed in the topographic map with the map scale. As is known, the wider the watershed area, the greater the river discharge, because the rainfall captured in the watershed will also be greater.

The shape of the watershed can be represented through a circulation ratio (Rc) value which is determined by the following equation:

$$\mathrm{Rc} = \frac{\mathrm{A}}{\mathrm{Ac}} \quad (1)$$

where Rc is the circulation ratio, A is the area of the watershed, Ac is the area of a circle with a circumference equal to the circumference of the watershed.

Median elevation is the point of elevation of the watershed which will divide the watershed area by an equal area. The method of determination is done by making a hypsometric curve, namely the curve of the relationship between the height and the watershed area above that height. The median elevation was determined to be 50% or ½ the area of the watershed. In a watershed study, the median elevation needs to be known, among other things, to determine the average temperature in the watershed. In addition, the median elevation is also often used as a parameter that refers to the height of the watershed.

The slope of the watershed is determined by analyzing the contour lines. The area with the total length of the large contour lines will have a steeper slope of the watershed than the area with the shorter total length of the contour lines.

3 Results and discussion

3.1 Watershed morphometry

The Bengkulu Watershed has an area of 500.5 km^2, which is divided into the Rindu Hati Sub-watershed covering an area of 184.2 km^2, the Susup Sub-watershed covering an area of 103.6 km^2 and the Bengkulu Hilir Sub-watershed covering an area of 212.7 km^2. Because the wider the area, the greater the river discharge, the Bengkulu Hilir River will likely have the largest discharge.

The shape of the watershed can be identified from a circulation ratio (Rc) value. A Rc value greater than 0.5 indicates that the area has high relief and a permeable surface, so that the water flow lag time is longer, while a low Rc value indicates low relief and a watertight surface, resulting in a faster water flow lag time (Mokarram and Sathyamoorthy, 2015).

According to Vienastra (2018), if the Rc value gets closer to 1, the watershed will tend to be round, but if the Rc value approaches 0, the watershed will be elongated. A round watershed will have a high runoff discharge because it will drain water faster to the outlet, while an elongated watershed will take longer to drain water to the outlet. Rindu Hati and Susup Sub-watersheds have Rc values of 0.59 and 0.55, so they tend to be round in shape, while Bengkulu Hilir Sub-watersheds have Rc 0.35 so that they are elongated.

Drainage density (Dd) is a measure that shows the length of the river in the watershed area per one square kilometer. Therefore, the greater the drainage density value, the better the drainage system in the area. According to Linsley (1959), if the drainage density value is less than 1, it indicates that the drainage system in the watershed is not good and there is inundation. If the drainage density value is between 1 and 5, it indicates a good drainage system and no inundation. If the drainage density value is more than 5, it indicates that the drainage system is not good and there is drought.

The value of drainage density in Rindu Hati Sub-watershed is 0.004 m/m^2, Susup Sub-watershed 0.006 m/m^2 and Bengkulu Hilir Sub-watershed 0.005 m/m^2, so that the flow density in all sub-watersheds is included in the low flow density category. According to Antoko and Sukmana (2007), watersheds with a drainage density value < 0.25 m/m^2 have a low capacity to store water and have the potential to experience inundation in the downstream side of the river. As a result, the runoff volume will be high and the discharge will increase.

The Bengkulu Watershed has an elevation of 0-1,025 m above sea level. Rindu Hati Sub-watershed has a height between 16-1,025 m, Susup Sub-watershed 17-975 m and Bengkulu Hilir Sub-watershed 0-427 m. Judging from its median elevation, Rindu Hati Sub-watershed has a median elevation of 257 m, Susup Sub-watershed 155 m and Bengkulu Hilir Sub-watershed 41 m. The altitude of the place will affect the amount of rainfall due to orographic effects, so based on the altitude of the location, the Rindu Hati Sub-watershed and Susup Sub-watershed will have a greater amount of rain and a larger runoff volume.

Judging from its slope, most of the Rindu Hati Sub-watershed has an 8-40% slope or is in the form of a wavy, hilly and steep area, likewise most of the Susup Sub-watershed also has an 8-40% slope. However, Bengkulu Hilir Sub-watershed has a slope of 0-15% or is gentle and hilly, so that the water flow in the Bengkulu Hilir River tends to be slower and the water infiltration into the ground is more.

3.2 River discharge

In general, the Rindu Hati River has a higher monthly discharge than the other two rivers. In the 2009-2013 period, the average discharge of the Rindu Hati River was 163.48 m^3/second, then decreased to 161.29 m^3/second in the 2014-2018 period. The Susup River has a discharge of 91.69 m^3/second in the 2009-2013 period and decreased to 90.24 m^3/second in the 2014-2018 period. However, for the Bengkulu Hilir River, there was an increase in discharge, which amounted to 131.89 m^3/second in the 2009-2013 period to 149.49 m^3/second in the 2014-2019 period. This increase in river discharge is likely due to changes in land use.

It can be said that by considering the pattern (Figure 1), the variations in the monthly discharge of the Rindu Hati and Susup Rivers in 2009-2018 are similar, while the Bengkulu Hilir River has a slightly different pattern of discharge variations.

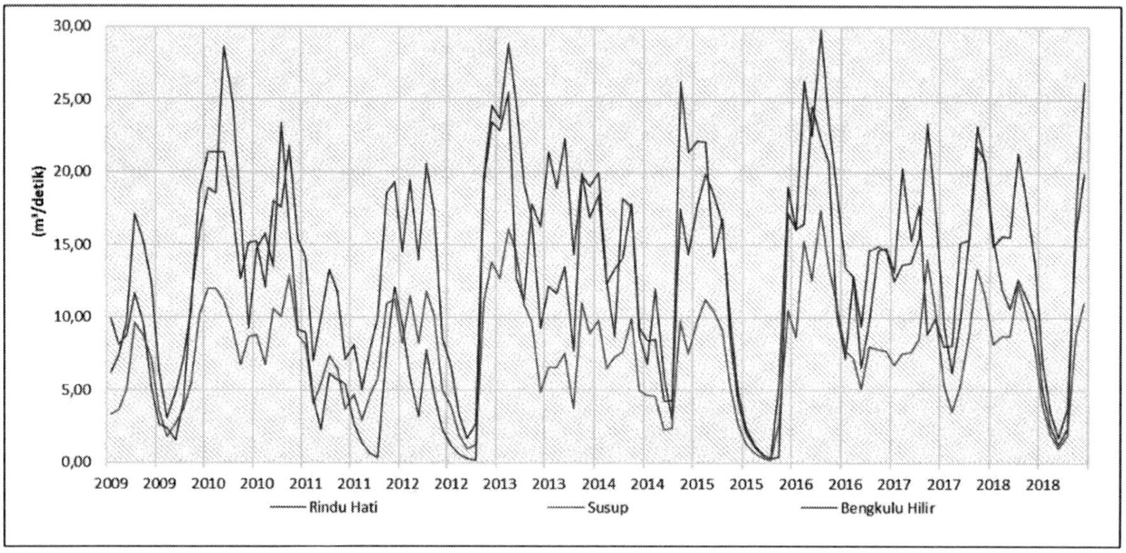

Fig. 1. Variations of river discharge in the Bengkulu Watershed 2009-2018 [21].

3.3 Watershed morphometry and river discharge

Beside watershed morphometry, other aspects that can affect river discharge are land use and rainfall volume. The dominant land use in Rindu Hati and Susup Sub-watersheds is relatively the same, namely dry land and bush farming, while in Bengkulu Hilir Sub-watershed, apart from dry land and bush farming, it also includes settlements, rice fields and open land. (Figure 2).

For annual rainfall, Rindu Hati and Susup sub-watersheds also have almost the same amount, namely 3,703 mm and 3,790 mm, while the amount of rainfall in Bengkulu Hilir sub-watershed is lower, namely 3340 mm. Based on the similarity of the data for these two aspects, the analysis of the Rindu Hati and Susup River discharge can be carried out only based on their morphometrics (Table 1).

Table 1. The relationship between watershed morphometry and river discharge.

Sub-Watershed	Watershed Morphometry					River Discharge (m^3/second)
	Area (km^2)	Circulation Ratio	Drainage Density (m/m^2)	Median Elevation (m)	Slope (%)	
Rindu Hati	184.2	0.59	0.004	257	8-40	162.38
Susup	103,6	0.55	0.006	155	8-40	90.96
Bengkulu Hilir	212,7	0.35	0.005	41	0-15	140.69

Source : [21]

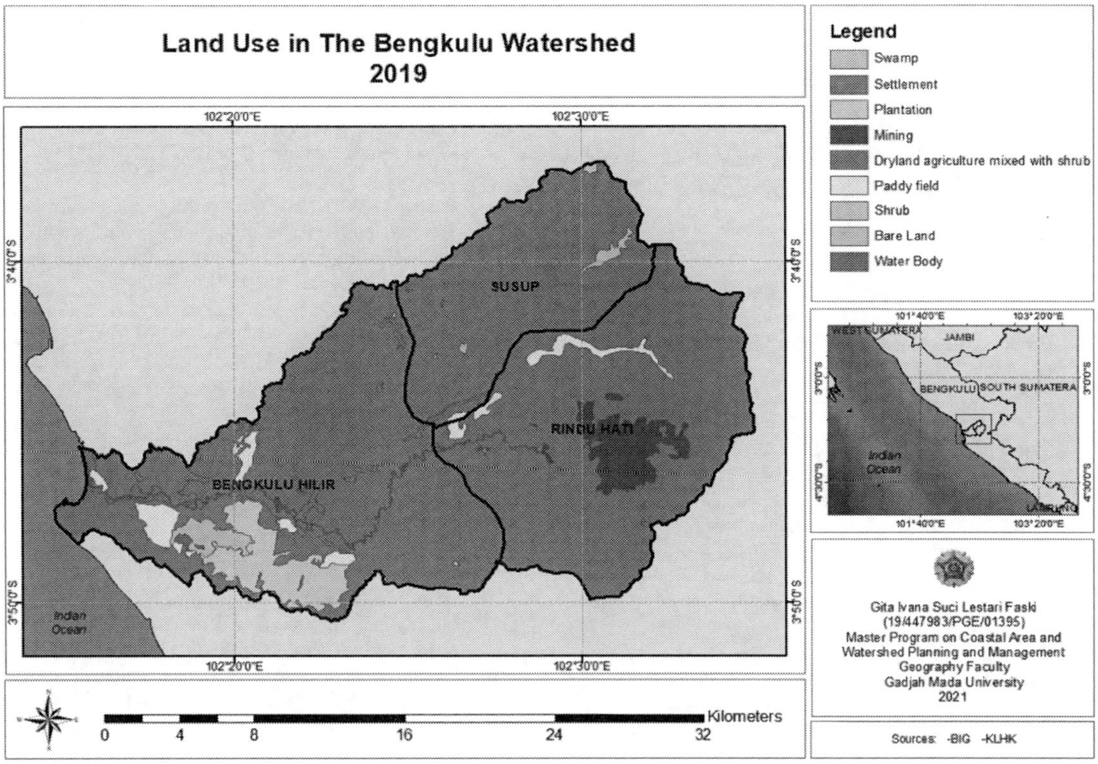

Fig. 2. Land use in the Bengkulu Watershed [21].

Looking at Table 1, it can be seen that the Rindu Hati River has an average monthly river discharge that is much higher than that of the Susup River. The Rindu Hati River has a monthly discharge of 162.38 m^3/second, while the Susup River has a monthly discharge of 90.96 m^3/second. The slope of the two sub-watersheds is the same, namely between 8-40%, so that the slope aspect can be neglected. The circulation ratio values of the two sub-watersheds are also almost the same, namely 0.59 and 0.55, so they tend to be round in shape and will have high discharge because they will flow water more quickly to the outlet. Because the values are similar, the circulation ratio aspect can also be ignored.

The drainage density value in the Rindu Hati Sub-watershed is 0.004 m/m^2, while the Susup Sub-watershed is 0.006 m/m^2, so that the drainage density of the two sub-watersheds is also in the same drainage density category, which is low, so it has the ability to store water which is low and has the potential experiencing inundation in the downstream side of the river.

Regarding the median elevation aspect, Rindu Hati Sub-watershed has a median elevation of 257 m and Susup Sub-watershed of 155 m. Theoretically, altitude can affect the amount of rainfall due to orographic effects, so that the Rindu Hati Sub-watershed should have more rain than the Susup Sub-watershed. In fact, Susup Sub-watershed has a higher amount of rain than Rindu Hati sub-watershed although the difference is not much, namely 3,703 mm and 3,790 mm. This is probably due to the influence of the differences in the area of the two sub-watersheds, where the Rindu Hati Sub-watershed has a larger area than the Susup Sub-watershed.

Based on the description of the four morphometric aspects, it can be seen that the sizeable difference in discharge between the Rindu Hati River and the Susup River is more due to differences in area size. As is known, the total area of the Rindu Hati Sub-watershed is 184.2 km^2, while the Susup Sub-watershed is 103.6 km^2. With a large area, the Rindu Hati Sub-watershed will be able to accommodate a large amount of rainwater as well.

4 Conclusion

1) Rindu Hati River has a higher average monthly discharge than Susup River.
2) The sizeable difference in discharge between the Rindu Hati and Susup Rivers is more due to differences in the size of the area. As is known, the total area of the Rindu Hati Sub-watershed is 184.2 km^2, while the Susup Sub-watershed is 103.6 km^2. With a large area, the Rindu Hati Sub-watershed will be able to accommodate a large amount of rainwater as well.

Acknowledgement

This paper is part of a study entitled " Water Balance Analysis of Bengkulu Watershed in Relation to the Watershed Physical Characteristics", financed by Final Assignment Recognition (Rekognisi Tugas Akhir/RTA) Programme Universitas Gadjah Mada year 2021.

References

1. Wannielista, M., R. Kersten and R. Eaglen. John Wiley and Sons Inc., New York (1997)
2. Davie T. Routledge, Taylor & Francis Group, London (2008)
3. Purnama I L S. Jurnal Rekayasa Lingkungan 6 (3) : 291-301 (2010)
4. Suprayogi, S., Purnama, I. L. S., & Darmanto, D. 2014. Gadjah Mada University Press (2014)
5. Griend, A.A.V. Institute of Earth Sciences Free University, Amsterdam (1979)
6. Asdak, C. Gadjah Mada University Press, Yogyakarta (2005)
7. Lee. R. Gadjah Mada University Press, Yogyakarta (1990)
8. Romulo C L., Posner S.,Cousins S., Fair J H., Bennett D E., Stearns H H., Richards R C., McDonald R I. Nature Communication (2018) 9:4375 (2018)
9. Gashaw T., Worqlul A.W., Dile Y., Addisu S., Bantider A., Zeleke. G. 2020. Heliyon 6 (2020) e04777 (2020)
10. Aneseyee A B., Elias E., Soromessa T., Feyisa G L. Science of the Total Environment 728 (2020) 138776 (2020)
11. Holen, B.V. Environment and Urbanization 13 : 209-222. (2001)
12. Athari Z., Rad G P., Abbasi E., Alibaygia A., Westholm E. Water Policy 19 (2017)1143–1159 (2017)
13. Purnama I L S. Proceeding of International Seminar : Utization of Geospatial Information to Raise Environmental Awareness in Realizing The Nation Character (2012)
14. Purnama I L S. IOP Conference Series: Earth and Environmental Science 451 (2020) 012085 (2020)
15. Rai, P. K., Mohan, K., Mishra, S., Ahmad, A., & Mishra, V. N. Applied Water Science 7 : 217–232 (2017)
16. Purnama I L S., Trijuni S., Hanafi F., Aulia T., Razali R. Magister Perencanaan dan Pengelolaan Pesisir dan DAS Fakultas Geografi UGM, Yogyakarta (2012)
17. Antoko, B. S., & Sukmana, A. Jurnal Penelitian Hutan Dan Konservasi Alam 4 (5) : 485–497. (2007)
18. Suprayogi, S., Purnama, I. L. S., & Darmanto, D. 2014. Gadjah Mada University Press (2014)
19. Satmaidi, E., Muthia, A., Wulandari A. Bina Hukum Lingkungan, 2 (2), 198–214. (2018)
20. Karima A. Thesis. Institut Pertanian Bogor (2017).
21. Faski G I S L. Thesis. Magister Perencanaan dan Pengelolaan Pesisir dan DAS Fakultas Geografi Universitas Gadjah Mada (2021)

 https://doi.org/10.1051/e3sconf/202132508013

The Lithofacies of Pandua Formation Shale in the Andowia Area, North Konawe, Southeast Sulawesi and Its potential as Petroleum source rock

Mawar Towan Lestari Ramli[1], *Hendra* Amijaya[1*], and Akmaluddin[1]

[1]Department Geological Engineering, Faculty of Engineering, Universitas Gadjah Mada, Jalan Grafika 2, Yogyakarta 55281, Indonesia

Abstract. Research on the Late Miocene of Pandua Formation shale in Andowia area, Southeast Sulawesi is fundamental because it is considered to have the potential as a source rock in Manui Basin. This study aimed to determine the lithofacies and its potential as petroleum source rock using megascopic, petrographic, and total organic carbon analyses in Pandua Formation shale. Based on the megascopic and petrographic analysis of outcrops, the shale can be subdivided into 11 lithofacies consists of clayey shale, massive claystone, clastic detritus-rich claystone, massive mudstone, mica-rich mudstone, iron oxide-rich mudstone, low-angle laminated mudstone, massive siltstone, carbon-rich massive siltstone, laminated siltstone, and carbon-rich laminated siltstone. The results of the analysis of 19 samples of shale showed that the total organic carbon (TOC) content was classified as poor to excellent (<0.5% - >4%). The lithofacies with a high concentration of TOC are carbon-rich massive siltstone and carbon-rich laminated siltstone. Both lithofacies were categorized as potentially excellent source rock which the TOC value content is 5.78% and 5.74%.The result implies the better understanding of the depositional environment and hydrocarbon accumulation potential of the Manui basin for future exploration.

1 Introduction

Oil and gas are an energy resource that plays a significantrole in supporting the country's economy. Since in the last ten years, crude oil production has decreased [1], further exploration is required to discover new oil and gas fields. One of them is by developing regional investigations in Eastern Indonesia, particularly in the offshore area of Southeast Sulawesi.

The oil and gas potential of the Southeast Sulawesi area, especially in the Manui Basin, is interesting to be further studied. Manui Basin has an active petroleum system and the possibility of hydrocarbon accumulation[2]. This is supported by the discovery of oil and gas seepage near the coast of this area. The discovery of oilseepage in the Lasolo area and gas seepage in the Wawolesea area, North Konawe Regency (Figure 1) proves that this area has an effective source rock. This shows that it can form and expel hydrocarbons (oil andnatural gas). The correlation of seismic data and well data shows that the Middle-Late Miocene sequence haspotential as a reservoir and source rock. This is also supported by sample geochemical data from the Abuki- 1 well, which shows a high TOC value at intervals equivalent to the Middle-Late Miocene age of 1.64%- 3.74% [2].

One of the challenges in oil and gas exploration in this area is the absence of the same outcrops in the onshore area [3]. Thus, there is still a lack of studies related to Miocene units in Southeast Sulawesi, especially regarding the existence of source rock and its potential. The limited understanding of geology, especially regarding the stratigraphy of the Miocene units in Southeast Sulawesi, is very important reason why it is a very important for further studies.

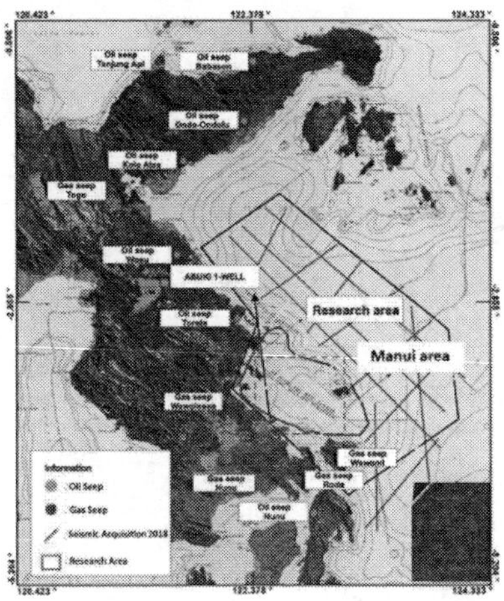

Fig.1. Oil and gas seep in southeast Sulawesi (Badan Geologi, 2019).

* Corresponding author:hamijaya@ugm.ac.id

The Pandua Formation is one of the Late Miocene Sulawesi Molasses Formations, which is potential as a source rock [2]. In the Pandua Formation, several shales are found [4][5]. However, studies related to the shale characteristics of this formation are still very limited. Studies concerning the shale characteristics become the basis for determining the potential of rock as a source of hydrocarbons. This paper aims to determine the shale lithofacies and their potential as petroleum source rock. The shale to be discussed is a general term used to denote fine-grained sediments such as claystone, mudstone, siltstone, and shale [6].

This study was based on field observation, microscopic analysis, and analysis of total organic carbon in several shale samples of the Pandua Formation. The study area is in the Andowia area, North Konawe Regency, Southeast Sulawesi Province, and the location is shown in Figure 2.

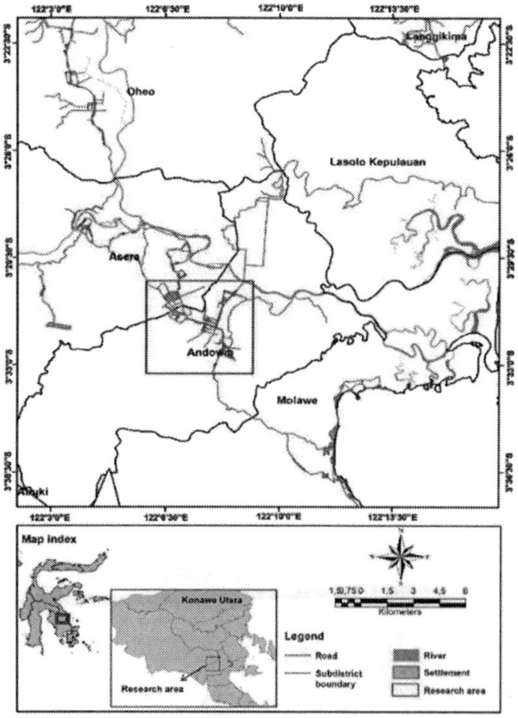

Fig.2. Location of research area.

2 Methods

The method used to determine the lithofacies and organic material content of shale in this study was field observation and laboratory analysis. The field analysis was conducted by observing the outcrops of the Pandua Formation. Measured stratigraphic sections were made to determine shale lithofacies based on geometric, lithological, and sedimentary structure parameters. The measured stratigraphic sections have a thickness of 671 m. The limited outcrops found in the field caused manyblank zones of stratigraphic record (Figure 3). The sample from outcrop was taken for laboratory analysis, which are microscopic and total organic carbon (TOC) content analysis.

Microscopic observation by means of petrographic analysis was conducted to see the types of minerals androck microstructure. Analysis of thin section used a Euromax polarization microscope with a magnificationof 4x to 20x to determine the mineral composition of the shale. Meanwhile, total organic carbon analysis aimed to analyse the content of organic carbon in rock samples. TOC analysis was conducted on 19 selected samples by the combustion method using the Soli TOC cube. For TOC analysis, the sample was crushed into a powder and finer than 100 mesh size. Then the sample was dried inan oven for testing and analysis with a temperature of upto 800°C.

3 Results

3.1 Lithofacies

The shale lithofacies (claystone, mudstone, siltstone, and shale) of the Pandua Formation were determined based on megascopic and microscopic analysis of samples. Megascopically, direct field observations were made of the outcrops to identify the texture, structure and geometric of the sediment. Microscopic observation used petrographic analysis to see the types of minerals and rock microstructure. Based on the lithofacies classification by Potter et al. (1980) [6], there are 11 samples of the Pandua Formation lithofacies which are determined based on the clay content of the rock. This classification is combined with variations of the sedimentary structure and mineral composition of the rock.

3.1.1 Clayey shale

Clayey shale is rare facies available in the Pandua Formation. Facies emerges in the sandstone as a lens forming a flaser structure and partly as a disconnected lens. The flaser structure formed by these facies has a thickness of up to 2 cm. Some facies are non-continuouslenses in sandstones with a thickness of up to 8 cm. Petrologically, these facies have dark reddish-grey colour, nearly composed of all clay-sized material, fissile structure, and very low hardness (Figure 4a). In a thin section, the clay minerals are present with a percentageof 98%. These facies show a very clear fissile structurein thin section analyses (Figure 4b).

3.1.2 Massive claystone

Massive claystone is characterized by dark grey to lightgrey colour, clay grain size, a high hardness, and massive structure in the megascopic sample (Figure 4c). The characteristics of these facies forms a massivehorizontal layer with a thickness of up to 6.4 m and interbedded with varying thickness sandstone. Petrographically, the characteristics of these facies are composed of quartz grains with 100 μm grain sizes with a percentage of 15% (Figure 4d). There are seven samples (sample code ANDH6-8, ANDH6-16, ANDH6-19, ANDH6-21, ANDH6-23, ANDH1-9, and ANDH3-3) from these facies has been analysed for its TOC. The analysis results show that the total organic material content ranges from 0.09% to 0.22%, with the average value of all samples is 0.13% (Table 1).

3.1.3 Clastic detritus-rich claystone

Clastic detritus-rich claystone is characterized by a dark grey colour, a low hardness, and a massive structure in the megascopic sample (Figure 4e). The thickness of these facies is 50 cm. It contains abundant medium- coarse grain size. The characteristics of these facies arecomposed of quartz grains with 220 μm with a percentage of 20%. The sediment lithic presents with 80 μm. The clay matrix is formed with a percentage of 68%. Several other minerals such as chlorite, feldspar, zirconand organic material are present in limited quantities (Figure 4f). A sample (sample code ANDH2-3) from these facies has been analysed for its TOC. The analysis results show that the total organic material content reaches 0.13% (Table 1).

3.1.4 Massive mudstone

Massive mudstone is characterized by a greenish-grey colour, silt-clay grain size, a high hardness, and a massive structure in the megascopic sample (Figure 5a). Petrographically, the characteristics of these facies are composed of quartz grains with 220 μm of grain sizes with a percentage of 25%. These facies are different from other facies because it contains a considerable amount of tabular muscovite minerals about 15% with an abundance of sedimentary clastic detritus (lithic).

3.1.5 Mica-rich mudstone

Mica-rich mudstone is characterized by a light grey colour, silt-clay grain size, a medium hardness, and a massive structure in the megascopic sample (Figure 5c). Muscovite minerals with interference colour present with tabular to bladed shape, 100 μm grain sizesand a percentage of 15%. The clay matrix has a percentage of 40%. Several other minerals, such as feldspar and opaque minerals are present in limited quantities (Figure 5d). TOC analysis was not performed on these facies. However, it is interpreted based on the appearance of colour as having a very low organic material content.

3.1.6 Iron oxide-rich mudstone

Iron oxide-rich mudstone is characterized by a reddish-grey colour, silt-clay grain size, a medium hardness, massive structure, and closed fabric in outcrops samples. The red colour of these is estimated to contain high content of iron oxide (Figure 5e). The thickness ofthis facies is 1,2 m. In thin section, this facies contains abundant medium-coarse grain size. The characteristicsof these facies are composed of rounded quartz grains (up to 420 μm) with a percentage of 35%. The clay matrix has a percentage of 45%. Several other minerals, such as chlorite, feldspar, and zircon, are present in limited quantities (Figure 5f). A sample (sample code ANDH1-12) has been analysed for its TOC. The analysis results show that the total organic material content reaches 0.17%.

3.1.7 Low-angle laminated mudstone

Low-angle laminated mudstone is characterized by brownish-grey colour with a low-angle laminated structure (Figure 5g). The characteristics of these faciesare composed of quartz grains (up to 420 μm) with a percentage of 40%. The presence of this quartz mineral together with muscovite minerals forms a parallel mineral alignment. Muscovite minerals with tabular shape (up to 400 μm) are present with percentage of 12%. The dispersion of muscovite minerals is always associated with clay minerals. The clay matrix has a percentage of 36%. Several other minerals, such asfeldspar and lithic, are present in limited quantities (Figure 5h). A sample (sample code ANDH6-2) has been analysed for its TOC. The analysis results show that the total organic material content reaches 0.12% (Table 1).

3.1.8 Massive siltstone

Massive siltstone is characterized by purplish grey colour with massive structure in outcrops samples (Figure 6a). The thickness of these facies is up to 82 cm.It is characterized by subrounded coarse-medium quartzgrains (up to 300 μm) with a percentage of 47%. The clay matrix is formed with a percentage of 30%. Severalother minerals such as feldspar, lithic, chlorite and zircon grains are present in limited quantities (Figure 6band 6c). A sample (sample codes ANDH2-4) from these facies has been analysed its TOC. TOC analysis shows that the total organic material content reaches 0.14%(Table 1).

3.1.9 Carbon-rich massive siltstone

Carbon-rich massive siltstone is characterized by the appearance of a field that forms a massive layer of reddish-grey colour with a thickness of up to 2.8 m (Figure 6d). The characteristics of these facies have a very fine sand grain size, with a high content of organic material. Organic material is present as carbon fleck in rocks at high concentrations. In addition, organic material is present in several other parts as black crystals, the size of coarse sand grains to very coarse filling the rock cracks to form veins. The crystal granules are defined as the results of silicification that occurs simultaneously with the deposition of organic material. This characteristic is also obtained in several other facies, namely carbon laminated siltstone andlaminated siltstone.

Based on the petrographic analysis, organic material is present in the form of black streaks and partly in the form of brown crystals. The mineral content of these facies is composed of subangular-subrounded quartz with a grain size of 250 µm. The quartz is present quiteabundantly, up to 40%, followed by the presence of large amounts of hematite and opaque minerals, up to 10% and 8%. The matrix is present with a percentage of less than 20%. Several other minerals such as feldspar, chlorite, and zircon are present in a small percentage (Figure 6e and 6f). A sample (sample code ANDH4-14)has been analysed for its TOC. The analysis results show that the total organic material content reaches 5.78% (Table 1).

3.1.10 Laminated siltstone

Laminated siltstone is characterized by a reddish grey to blackish-grey colour with a laminated structure (Figure7a). The thickness of the laminated sediment is 1-3 mm.The presence of a layer on the rock outcrop has the thickness size of 7 cm and the thickest is up to 80 cm. Megascopically, these facies contain organic material that emerges as carbon flecks and black crystal grains that fill rock fractures. Carbon flecks are disseminated on each laminated rock boundary with various concentrations.

The results of the thin section analysis show that thelaminated structure can be seen indicated by the presence of quartz, muscovite, feldspar, and clay minerals which form a parallel alignment (Figure 7b). Meanwhile, the clay matrix is formed with a percentage of less than 20%. Several other minerals such as chlorite, feldspar, muscovite, hematite, lithic, and zircon arepresent in limited quantities (Figure 7c). There are foursamples (sample code ANDH4-5, ANDH4-7, ANDH4-11 and ANDH4-25) have been analysed for its TOC. The analysis results show that the total organic materialcontent ranges from 0.18% to 0.45% (Table 1), with the average value of all samples is 0.34%.

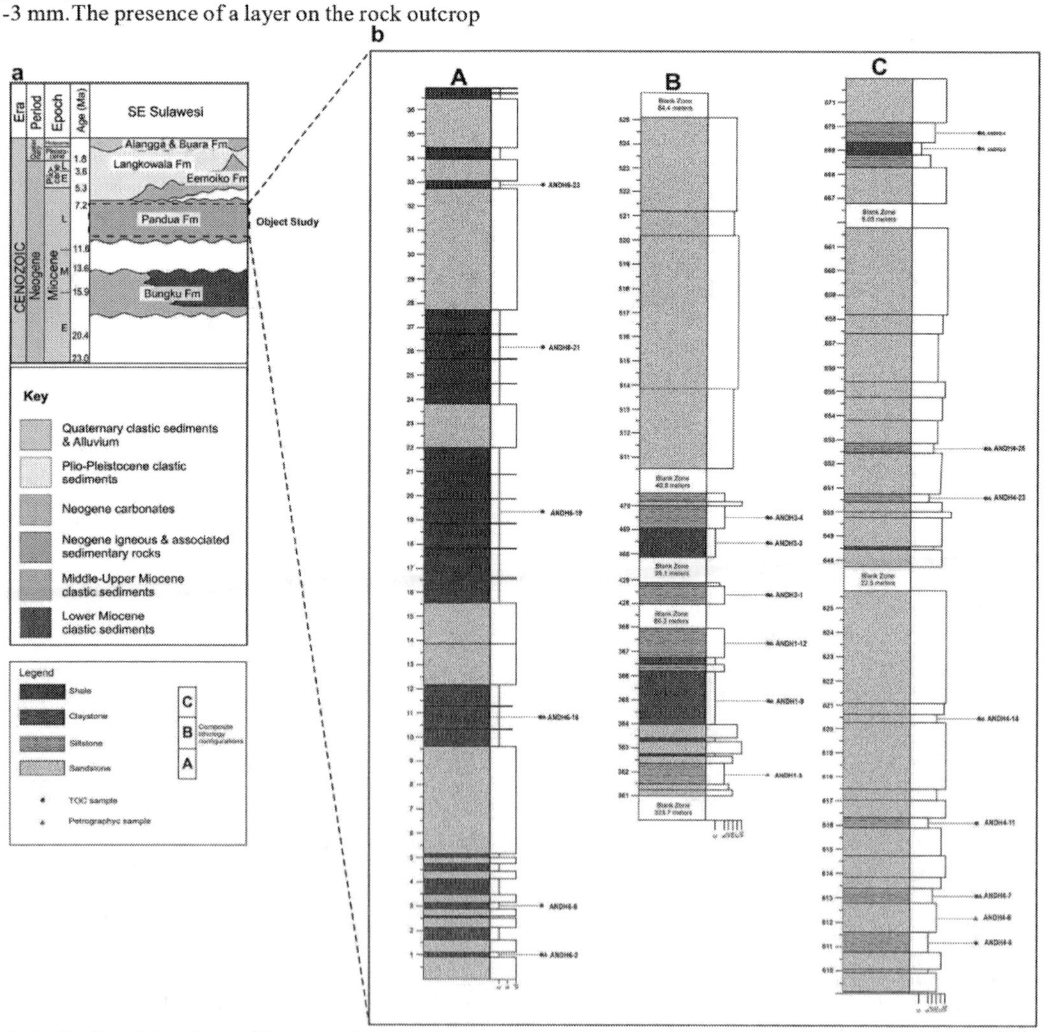

Fig. 3. (a) Stratigraphic column of Neogen sediment in Southeast Sulawesi (modified by Nugraha and Hall, 2018). (b) Stratigraphic measured section column of Pandua Formation in the Andowia area and distribution of the samples.

 https://doi.org/10.1051/e3sconf/202132508013

3.1.11 Carbon-rich laminated siltstone

Carbon-rich laminated siltstone is characterized by a blackish grey to black colour with laminated structure (Figure 7d). The thickness of the laminated sediment is 1-2 mm. Megascopically, these facies contains abundant organic material which spreads evenly in sediment. It is characterized by abundant quartz mineral and organic material with occasional pyrite grains in the outcrop sample. The mineral content of these facies is composed of subangular-subrounded quartz with a grain size up to 500 μm. The quartz is present abundantly, up to 48%, followed by the clay matrix with a percentage of 22%. The organic material is present abundant in the form of black streaks and partly in brown crystals. Several other minerals such as feldspar, chlorite, and zircon are present in a small percentage (Figure 7e and 7f). A sample (sample code NDH4-23) from these facies has been analysed for its TOC. The analysis results show that the total organic material content reaches 5.74%(Table 1).

Fig. 4. Macroscopic and microscopic characteristics of some shale lithofacies (a) Macroscopic appearance of the clayey shale. (b) Clayey shale in a thin section shows the fissile structure (c) Macroscopic appearance of the massive claystone. (d) Massive claystone in a thin section. (e) Macroscopic appearance of the clastic detritus-rich claystone. (f) Clastic detritus-rich claystone in a thin section shows the presence of quartz, chlorite, feldspar, lithic and zircon minerals.

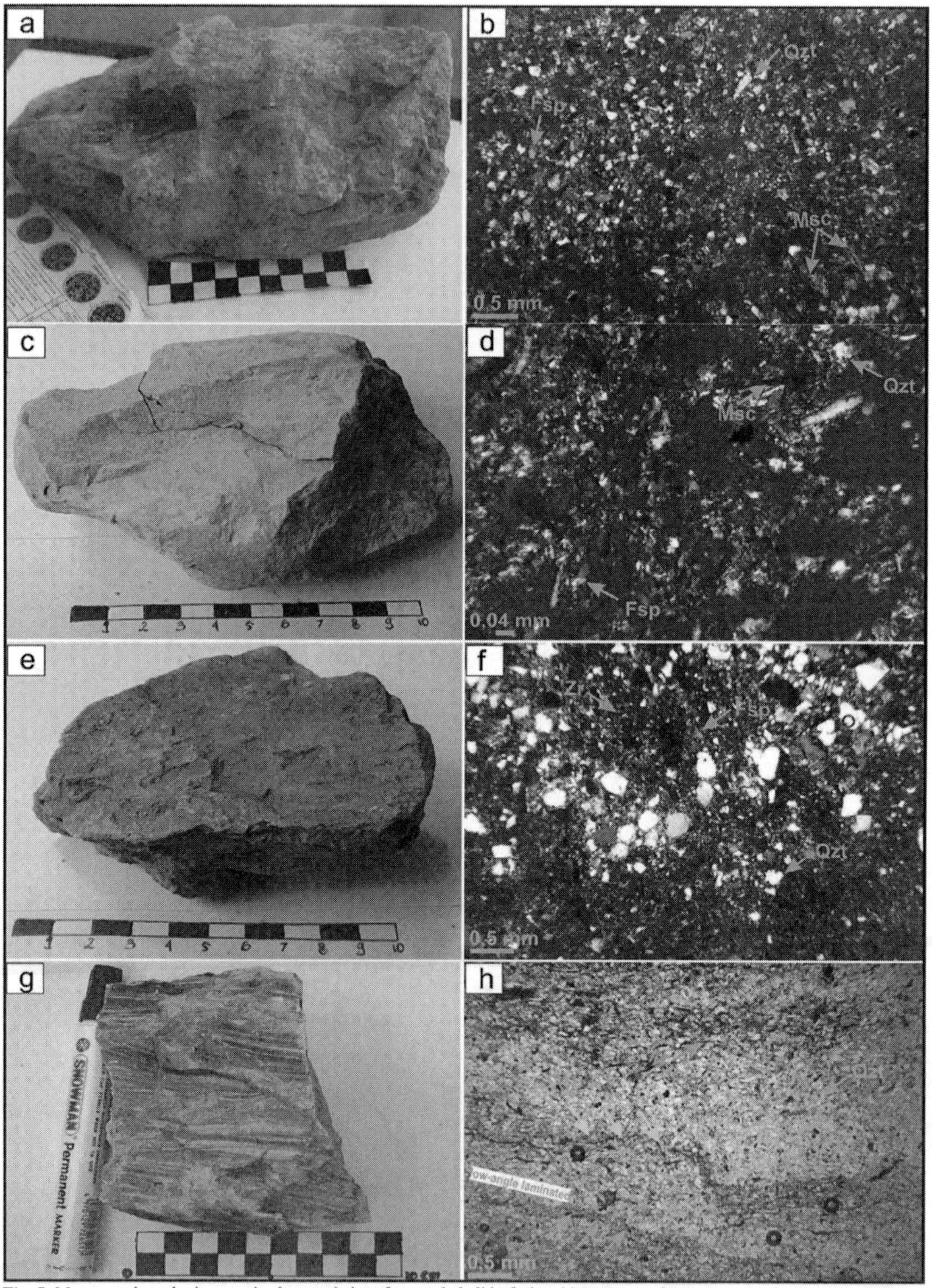

Fig. 5. Macroscopic and microscopic characteristics of some shale lithofacies. (a) Macroscopic appearance of massive mudstone. (b) Massive mudstone in a thin section that shows the presence of quartz, feldspar and muscovite minerals. (c) Macroscopic appearance of mica rich mudstone. (d) Mica rich mudstone in a thin section that shows the presence of quartz, feldspar and muscovite minerals (e) Macroscopic appearance of iron oxide rich mudstone (f) Iron oxide rich mudstone in a thin section shows the presence of rounded quartz, feldspar and zircon minerals. (g) Macroscopic appearance of low-angle laminated mudstone (h) low-angle laminated mudstonea thin section shows the presence of quartz and muscovite minerals with the low angle-lamination structure.

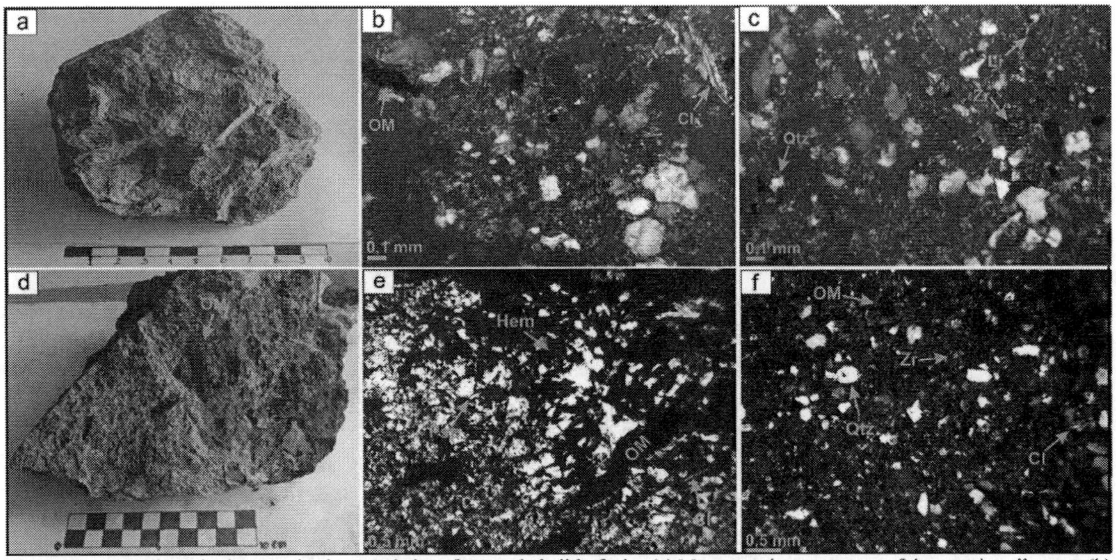

Fig. 6. Macroscopic and microscopic characteristics of some shale lithofacies (a) Macroscopic appearance of the massive siltstone. (b) Massive siltstone in thin section shows the presence of quartz, chlorite and organic materials. (c) Massive siltstone in thin section shows the presence of quartz, lithic, and zircon minerals (d) Macroscopic appearance of the carbon-rich massive siltstone. (e) Carbon-rich massive siltstone in thin section shows the presence of hematite and organic material (f) Carbon-rich massive siltstone in thin section shows the presence of quartz, chlorite, zircon and organic materials.

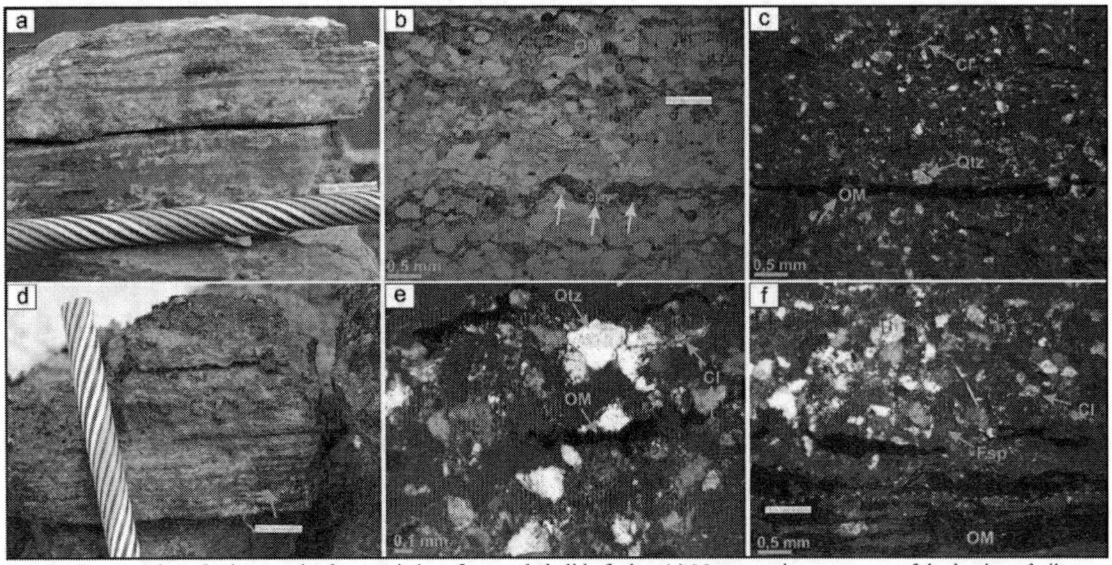

Fig. 7. Macroscopic and microscopic characteristics of some shale lithofacies. (a) Macroscopic appearance of the laminated siltstone facies. (b) Laminated siltstone facies in thin section showing parallel alignments of quartz, clay, and muscovite minerals (c) Laminated siltstone facies in thin section showing the existence of chlorite and organic material formed in parallel (d) Macroscopic appearance of carbon-rich laminated siltstone facies (e) Facies of carbon-rich laminated siltstone in a thin section that shows the existence of mineral organic material at the laminate boundary (f) Facies of carbon-rich laminated siltstone in a thin section that shows the presence of chlorite minerals and alignment of the organic material to form a thin lamination.

3.2 TOC

TOC (Total Organic Carbon) analysis of 19 shales samples of the Pandua Formation was conducted. The result of TOC analysis of the shale in this research is range from 0.08% to 5.78% (Table 1). The facies which contain high value of organic material is carbon-rich massive siltstone facies and carbon-rich laminated siltstone facies. The carbon-rich massive siltstone faciescontain the high content of TOC up to 5.78% and carbon-rich laminated siltstone facies contain TOC up to 5.74%, followed by laminated siltstone facies with the TOC value range from 0.18% to 0.45%, followed by massive claystone facies, iron oxide-rich mudstone facies, massive siltstone facies, low-angle laminatedmudstone facies, clastic detritus-rich claystone facies and massive mudstone facies. The TOC value of the other facies is

Table 1. The distinguished lithofacies with their TOC content of Pandua Formation in Andowia, North Konawe, Southeast Sulawesi

No	Kode sample	Lithofacies	TOC (%)	Quality (Peters and Cassa, 1994)
1	ANDH6-2	Low angle Laminated Mudstone	0.12	Poor
2	ANDH6-8	Massive Claystone	0.09	Poor
3	ANDH6-16	Massive Claystone	0.22	Poor
4	ANDH6-19	Massive Claystone	0.16	Poor
5	ANDH6-21	Massive Claystone	0.13	Poor
6	ANDH6-23	Massive Claystone	0.15	Poor
7	ANDH1-9	Massive Claystone	0.09	Poor
8	ANDH1-12	Iron Oxide-rich Mudstone	0.16	Poor
9	ANDH3-1	Massive Mudstone	0.08	Poor
10	ANDH3-3	Massive Claystone	0.21	Poor
11	ANDH3-4	Massive Siltstone	0.15	Poor
12	ANDH4-5	Laminated Siltstone	0.18	Poor
13	ANDH4-7	Laminated Siltstone	0.45	Poor
14	ANDH4-11	Laminated Siltstone	0.38	Poor
15	ANDH4-14	Carbon-rich Massive Siltstone	5.78	Excellent
16	ANDH4-23	Carbon-rich Laminated Siltstone	5.74	Excellent
17	ANDH4-25	Laminated Siltstone	0.36	Poor
18	ANDH2-3	Clastic Detritus-rich Claystone	0.13	Poor
19	ANDH2-4	Massive Siltstone	0.14	Poor

range from 0.08% to 0.22%. Massive mudstone facies have the lowest TOC value with a percentage of 0.08% (Table 1). Clayey shale facies are interpretedto have similar TOC as laminated siltstone facies because they have almost the same characteristic.

4 Discussion

The lithofacies with high organic material content are the facies carbon-rich massive siltstone and carbon-richlaminated siltstone. In general, these two facies have thecharacteristics of larger grain size (silt-fine sand) with alower content of clay and matrix than the other facies. The other studied stating that the highest organic material commonly exhibits the highest matrix porosity,higher gas in place and lower clay content [7]. Laminated silt-sized which contains detrital grains suchas quartz and feldspar with large particle sized and large densities are interpreted as deposits by gravity flow under the influence of gravitational and hydrodynamic processes [8].

The characteristics of the lithofacies and organic material content are shown by the differences in the sediment structure in Pandua Formation shale. Shales with a microstructure, such as a thin parallel laminated structure (<3 mm), have a higher organic material content than shales with minimal or lacks structure. These can be seen in laminated siltstone facies and carbon laminated siltstone facies. Several researchers also mentioned that the microstructure in the sediments is produced by the preservation of organic material [8].This indicates that the shale with a microstructure has apositive correlation with the abundance of organic material. This relates to the properties and mechanisms of sediment deposition. Other studies are stating that shale with a well-laminated structure and black- coloured domination without bioturbation formed in an anoxic water column and persistent stratified [7].

The characteristics of lithofacies Pandua Formationshale indicating ability to preserve organic material can be recognized from their mineral associations. Lithofacies with pyrite, hematite, zircon and chlorite minerals are associated with high organic material content. Carbon-rich massive siltstone and carbon-rich laminated siltstone contain pyrite, zircon, chlorite minerals and high organic material as much as 5.74% and 5.78%.

Meanwhile, several other facies that do not have similar mineral associations only contain small organicmaterial which are from 0.08% up to 0.22%. This showsthat shale with the presence of pyrite, zircon, and chlorite mineral association positively correlates to the abundance of organic material. The presence of thepyrite mineral commonly used as an indicator to indicate the reducing conditions [9]. These show that the high preservation of organic material in this research was deposited in anoxic/ euxinic conditions [9]. The result indicates that environmental conditions of deposition strongly influence the preservation of organic material. The sedimentation and deposition process affect the organic matter abundance and characteristics of shale mineral types [10].

The lithofacies with a high concentration of TOC are carbon-rich massive siltstone and carbon-rich laminated siltstone. Thus, it has an excellent potential asa source rock to produce hydrocarbons. The TOC analysis shows that Pandua Formation shale exposed inthe Andowia area, North Konawe mostly does not havethe potential as a source rock. Nevertheless, there are some shale shows the potential as excellent source of hydrocarbons.

5 Conclusions

Megascopic and thin section analysis of the Pandua Formation resulted in 11 different lithofacies. The determination of lithofacies was identified based on thetexture, structure, and mineral composition. The lithofacies consist of clayey shale, massive claystone, clastic detritus-rich claystone, massive mudstone, mica-rich mudstone, iron oxide-rich mudstone, low-anglelaminated mudstone, massive siltstone, carbon-rich massive siltstone, laminated siltstone, and carbon-rich laminated siltstone.

The TOC analysis on 19 samples of the Pandua Formation shows that the potential for shale as a sourcerock can be categorized into two groups. The first groupis categorized as "poor". Therefore, it has less potentialto produce hydrocarbons. Meanwhile, the second groupis categorized as "excellent". Thus, it has an excellent to produce hydrocarbons. The results of the TOC analysis show that only some lithofacies of Pandua Formation shale shows the potential as an excellent source rock which are carbon-rich massive siltstone and carbon-rich laminated siltstone. Further studies are needed to identify the distribution of the lithofacies.

The Authors thanks Department of Geological Engineering Universitas Gadjah Mada for the research grant and the laboratory facilities especially the Get in Cicero Laboratory and Sedimentology Laboratory.

References

1. Dewan Energi Nasional, *Outlook Energy Indonesia 2019*, Sekretariat Jenderal Dewan Energi Nasional, Jakarta (2019).
2. Badan Geologi, *Petroleum prospectivity of Salabangka and the Adjacent area, Preliminary Study*, Kementerian Energi dan Sumber Daya Mineral, Bandung (2019).
3. A. Fahruddin, R. Fakhruddin, M. Firdaus, and M. H. Saleh, Indonesia Bulletin of the Geological Society of Malaysia, **70** (2020).
4. A.K. Mubarok, *Geologi dan Analisis Provenan Batupasir Daerah Andowia Bagian Barat dan Sekitarnya, Kabupaten Konawe Utara, Sulawesi Tenggara*, Skripsi, Institut Teknologi Bandung (unpublished), (2016).
5. M.R. Graha, *Geologi dan Provenan Batupasir Daerah Awila Puncak dan Sekitarnya Kabupaten Konawe Utara, Sulawesi Tenggara*, Skripsi, Institut Teknologi Bandung(unpublished), (2017).
6. P. Potter, J.Y. Maynard, W. Pryor, *Sedimentology of Shale*, Springer-Verlag, New York (1980).
7. J. Zaixing, G. Ling, L. Chao, W. Yuan, L. Min, Journal of Paleogeography, **2** (2013).
8. X. Zeng, J. Cai, Z. Dong, L. Bian, Y. Li, Minerals, **8** (2018).
9. G. Xie, S. Liu, G. Xie, W. Hao, Geological Journal, **55** (2019).
10. S. Han, Y. Zhang, J. Huang, Y. Rui, Z. Tang, Minerals, **10** (2020).

AUTHOR INDEX

AUTHOR INDEX

AUTHOR INDEX

EDP Sciences – Web of Conferences
17, Avenue du Hoggar
Parc d'Activité de Courtabouf
BP 112
F-91944 Les Ulis Cedex A
France

ISBN 978-1-7138-3957-6

2nd International Conference on Sustainable Environment and Disaster Management (2nd SUSTAIN 2025)

E3S Web of Conferences Volume 652 (2025)

Malang, East Java, Indonesia
28 August 2025

edp sciences